THE
# WRITER'S
## HANDBOOK
### 2011

CARDIFF
CAERDYDD

Barry Turner has worked on both sides of publishing, as a journalist and author, editor and marketing director. He started his career as a teacher before joining *The Observer* and then moving on to radio and television. His first book, a study of British politics in the early twentieth century, was published in 1970. He has written over twenty books including *A Place in the Country*, which inspired a television series, and a best-selling biography of the actor, Richard Burton. For many years he wrote on travel for *The Times* and now reviews and serializes books for the paper. Among his recent books are *Countdown to Victory* on the last months of World War II and *Suez 1956: The First Oil War*. His latest, *Outpost of Occupation*, the story of the German occupation of the Channel Islands, was published in April. As founding editor of *The Writer's Handbook* he has taken this annual reference title through to its twenty-fourth edition. For eleven years, Barry has edited the annual *Statesman's Yearbook*. He is a founder of the National Academy of Writing at Birmingham City University, where he is also a visiting professor.

# THE
# WRITER'S
# HANDBOOK
# 2011

EDITOR

## BARRY TURNER

MACMILLAN

© Macmillan Publishers Ltd 2010

First published 2010 by
MACMILLAN PUBLISHERS LTD
Houndmills, Basingstoke, Hampshire RG21 6XS and
175 Fifth Avenue, New York, N.Y. 10010
Companies and representatives throughout the world

ISBN: 978-0230-20729-5

This book is printed on paper suitable for recycling and made from fully
managed and sustained forest sources. Logging, pulping and manufacturing
processes are expected to conform to the environmental regulations of the
country of origin.

A catalogue record for this book is available from the British Library.

A catalog record for this book is available from the Library of Congress.

10  9  8  7  6  5  4  3  2  1
19  18  17  16  15  14  13  12  11  10

Credits
Editor *Barry Turner*
Assistant Editor *Jill Fenner*
Editorial Assistant *Kenneth Hadley*
Poetry Editor *Chris Hamilton-Emery*
Contributors   *Pauline Chapman*
*Sara Lloyd*
*George Mann*
*Michelle Paver*
*Peggy Vance*
Tax and Finance Adviser *Ian Spring*

Printed and bound in Great Britain by
Thomson Litho, East Kilbride, Scotland
Data management and typesetting by
HWA Text and Data Management, London

**www.thewritershandbook.com**

# Contents

# So you want to be an author?

*These are challenging times for publishers but there is always a demand for talented writers. Much depends on understanding and knowing how to work the system. Barry Turner offers some advice.*

It was once said of Hollywood, the city of ultimate media risk, that nobody knows anything. Publishers are beginning to feel the same way. Caught between the credit crunch and a technological revolution, they are struggling to come to terms with a market, once noted for long-term predictability, that is now hard to fathom and near impossible to anticipate.

The core of uncertainty is the very future of the printed book, soon to be made obsolete, some argue, by the hand-held or desktop flat screen. The as yet modest sale of ebook readers is almost certain to take off as prices fall and the choice of reading material opens up. The received wisdom among professionals is that digital sales, regardless of format, will surpass ink on paper within a decade.

Meanwhile, the conventional book trade is sending out conflicting images. True, there were more than 120,000 books published last year, six times the total of twenty years ago. But the latest figure is down on the peak year of 2003 when close on 130,000 titles entered the lists.

Volume and revenue growth look healthy enough at first glance. A closer examination reveals that the business is being driven by a few mega-sellers while, overall, costly shop promotions and three-for-two type deals are eating into publishers' profits and authors' royalties.

So where does this leave the first-time author with a mission to succeed? The first thing to say is that however much publishing changes and whatever the pace of change, there will always be a demand for writers. Without them, there is no publishing – print or digital. Moreover, to be able to write lucidly and engagingly is a rare talent with a market value. If you've got what it takes, there will be a publisher ready to take what you've got.

Where aspiring authors invariably go wrong is in not giving sufficient attention to reader choice. Long-winded lectures dished up as novels are not welcome. Those who have spent their college years dipping into Proust or Tolstoy may believe that they have within them a great novel to convulse the literary establishment but the chances are they are wasting their time. Notwithstanding the pretensions of certain cultural pundits, the typical book buyer no longer looks to fiction for great insights into the human condition. Novels can be thought provoking, and all the better for it, but it is entertainment that punches up sales.

What of non-fiction? Here again, it is the skill of holding a reader's attention – in other words, the skill of the storyteller – that makes for success. Serious minded educators and proselytizers find this hard to accept. Even when they have accumulated enough unpublished manuscripts to feed a recycling plant, they remain convinced that one day the world will recant. Sad, really.

This then is the starting point for an author in the making. Study the market. Read what has made other authors successful. What do they offer that you may be missing? I have come across writers who flatly refuse to read across their genre because they don't want to be influenced by a particular style and end up as imitators. This is just too precious. Nobody ever suffered by studying the opposition.

Next, don't get stuck on the quirky, one-off idea. True, the bestsellers that get most publicity are those that come out of the blue – Stephen Hawking's *A Brief History of Time* is still going strong, as is Lynne Truss's *Eats, Shoots and Leaves*. But there is a huge amount of luck in all this. It makes more practical sense to ignore the oddballs and focus on the books that fit some sort of pattern.

Take the airport novel. The outlets in the airport shopping malls are among the most profitable in the country. Sample titles suggest that close on half of the airport sales are in the category of blockbuster fiction with authors such as Robert Ludlum, John Grisham, Jilly Cooper, Stephen King, Patricia Cornwell, Wilbur Smith and Danielle Steel leading the race to the cash register. Hardly surprising when the customers in most need of diversion are mostly long-haul travellers or tourists heading for the beach. The novels with the strongest appeal offer a rattling good story populated by larger than life characters in exotic locations where they engage in plenty of action laced with sex, intrigue, corruption, treachery, betrayal and more sex. An eye-catching cover and a memorable title are standard features.

This may not be the sort of book you want to write. Indeed you may feel that many front-of-shop titles are intended for customers with the intelligence of chewing gum. But at least you know what you are up against, and it will help you think more constructively about the sort of book you *do* want to write.

Start with the idea. What are you trying to say in this book of yours and, more to the point, are you confident enough that buyers will want to pay £15 or £20 to read what you have to say? In other words, does your book have marketing strength? That is the key word – marketing. Books have to be sold and sold hard. Long gone are the days, if they ever really existed, when publishers put literary excellence above all other factors. Now, the first line a publisher looks at is the bottom line. There is nothing inherently wrong in this, incidentally. Publishing is not and never has been a charitable enterprise, although a few years ago there were still innocent souls who thought so. You can find them now, propping up a bar somewhere, wondering what went wrong.

Back to the writer. What must a writer do to get noticed by a publisher? First, check out the publishers who focus on your areas of interest. Aspiring authors waste time and money submitting proposals that are at odds with a publisher's list. Romantic novels do not sit well with computer manuals.

Resist the urge to pitch only to the leading publishers. For one thing, they rarely give much attention to unsolicited material. Their money and sales muscle are largely reserved for the established names on their lists, writers who enjoy an easily recognizable profile when their books appear on the front counter of

Waterstone's. Smaller publishers are more receptive to newcomers. Since they, like the authors they represent, have reputations to make, talent spotting is an essential part of their game plan. Association with a mainstream house can wait for later when there is clear evidence of market appeal. Some of the biggest sellers of our time, from *Watership Down* to *A Year in the Merde*, came to us by this route.

Wherever you decide to start your bid for recognition be business-like. When corresponding with a publisher, you are dealing with someone who has not much time to spare. He has neither the energy nor the optimism to wade through a weighty manuscript which just may turn out to be the blockbuster of the century but more probably will not.

All that is needed at the approach stage is a synopsis, a sample chapter, and a letter of introduction saying who you are and what there is in your life that makes you peculiarly qualified to write this particular book. Previous publications should be mentioned but not, please not, compliments on your literary skills from friends (however influential) and family. Equally irritating to the recipient are those anticipations of sharp practice such as the bold © at the end of a submission specifying 'First British rights only'.

The synopsis should begin with a line or two of what marketing people call the 'unique selling point' (USP). Who is likely to buy this book and why? It is not enough to claim that the author will reach out to the general reader. We all like to think that a mass audience is waiting for us but the reality is that each book has a core appeal on which the sales potential will be judged.

Having settled on a snappy justification, the synopsis can be used to describe the book in some detail. It is impossible to specify length – where a single page may suffice for a beginner's guide to beekeeping, a closely argued case for energy conservation might require several thousand words. An idea for a novel has sparked interest on the strength of one paragraph. What is essential is for the synopsis to be a clear and logical description of the book.

It should end with a few pertinent details. What is the intended size of the book? This has an important bearing on production costs and thus on the sales forecast. The best estimate of size is the number of words, with the average book falling within the 80,000–100,000 bracket.

Will there be illustrations? If so, what sort? Library pictures can be expensive. Commissioned drawings raise the question of the role and standing of the illustrator in the origination of the book.

What about an index? All too often the non-fiction writer brushes aside such petty-fogging questions. Anyone can throw together an index. Wrong. Indexing is a highly skilled task and the quality of an index can make a significant impact on the quality of the book.

When will the manuscript be delivered? Publishers are rightly suspicious of authors who are vague on deadlines. The trouble is that after a few years, a book in the making can move out of range of any feasible sales forecast. Fashion changes, often with bewildering speed. A hot seller next year (linking in to an anniversary, for example, or a current political interest) could easily become a candidate for the remainder shelves if publication is delayed even by a few months. When specifying a delivery date keep in mind the time it takes to print and promote a book – a minimum of six months.

Submissions should be typewritten (with double spacing for the manuscript) and free of messy corrections. Publishers, being human, are liable to be put off by a grubby file of foolscap patched together with sellotape. How many others have seen and rejected this sad little offering? It could well be that email submissions are acceptable. This is worth checking. But on no account turn up in person in the hope of talking directly to an editor. For all the obvious reasons, chutzpah of this order is nearly always counterproductive.

Once a book proposal has been sent off, allow at least a month even for an acknowledgement and up to three months for a considered reply.

Don't be impatient. Writers are often sinned against but they can be unreasonable in their assumption of a quick decision on what, after all, is a risky investment. If, after a decent interval, nothing is heard, a telephone call is justified. But a polite enquiry is more likely to get results than a demand to know 'what the hell is going on'. It helps to have sent material to a named editor. That way you avoid the risk of being sucked into the whirlpool of internal company communications.

There is no harm in canvassing several publishers at the same time. And there is no need to make a secret of so doing. A little friendly competition may help to stimulate interest.

And so to the next leap forward. Assume that a book proposal has sparked a response from a publisher – say, a middle-range publisher with a respectable list of satisfied authors and a good name in the trade. A small celebration may be called for but this is no time to relax.

Every writer likes to think that somewhere out there is a sympathetic counsellor and friend. In theory, the publisher is ideally suited to the role – he knows the problems, understands the pressures and, after all, he is the one who has to make the book work. But publishing is like any other business. The purpose is to make money. There are some well-publicized practitioners of the art who claim that their minds are on higher things, but those who shout loudest about the glories of literature and the evils of materialism invariably end up with the biggest houses and the smartest cars. This is not to say they are dishonest. It is simply that in abiding by the first rule of elementary capitalism, they are out to maximize their profits.

Proceed then, with caution.

For all but well-established authors the best that can be hoped for at this stage is an agreement in principle. The publisher may call for amendments to the proposal – more or fewer words or illustrations, a change of emphasis to help sharpen an argument or clarify an aspect of a plot. The timetable is bound to come up for discussion. Is it realistic? Can the author really turn out 100,000 words in two years? What other work does he have in hand?

If the outcome of the first meeting is encouraging, the next discussion should turn to money. Stories of six-figure advances for unknown authors rarely accord with the facts. Celebs can make substantial money up front but in hard times even they have had the lid of the cashbox slammed down on their fingers. It can happen that a first novel excites such interest as to start an offer war between publishers. But for most of us the sums involved are modest and bordering on derisory. There are non-fiction writers who settle for a few hundred pounds while a tenderfoot novelist is lucky to get more than £1,000. The standard formula is for the advance to equate with the 60 per cent of the estimated royalties payable on the first edition. Unsurprisingly, publishers favour pessimistic calculations.

Whatever its size, the advance is important as a notification by the publisher of serious intent. It should be non-returnable, except when the author fails to deliver a manuscript. Usually, it is split three ways, part on signature of contract, part on delivery of the manuscript and part on publication.

Occasionally, there are good reasons for surrendering the advance in return for other benefits. Macmillan New Writing, a low cost imprint aimed at attracting first-time authors, is a case in point (see p. 96) The forty or so authors so far published under this banner are happy. Though no advances are paid, they get a 20 per cent royalty on sales, twice the standard rate.

It is often assumed that the cheques are more impressive when a literary agent represents the author. This may be so for bestselling authors who put their work up for auction but down the scale there may not be too much room for manoeuvre. Where an agent really proves his worth is in knowing which publishers are most likely to respond to a particular idea and in having an entrée to otherwise unreachable senior editors. Indeed, some publishers, identified in our lists, will only consider material submitted by an agent.

The good agent understands the small print and, to greater advantage, spots the omissions – such as the failure to allow for higher royalties beyond a certain minimum sale. The agented author has a bigger say on promotion budgets, cover design, the timing of publication, print number and on subsidiary rights – the latter capable of attracting earnings long after the book is out of print. The sheer range of potential subsidiary rights is mind-boggling – overseas publication (the publisher will try for world rights but, when an agent is acting, US and translation rights may be reserved to the ultimate benefit of the author), film and television adaptations, audio books, digital retrieval – to mention only the most obvious. Above all, the good agent keeps a watching brief long after the contract has been signed, always ready to challenge the publisher to do better on behalf of his author.

But where is the efficient and sympathetic agent to be found? There is no sure way of matching a writer and agent merely by glancing through the list of names and addresses. The most powerful agencies are not necessarily suitable for a beginner, who may feel the need for the close personal contact offered by a smaller agency. On the other hand, the smaller agency may already have taken on its full quota of newcomers, by definition low earners who must, for a time, be subsidized by the more profitable sector of a client list. The agent who gets the balance wrong is heading for insolvency.

Advice frequently given by the agented to the agentless is to seek out the opinion of authors who have been through the mill and learn from their experiences. Writers' circles and seminars organized by the Society of Authors and the Writers' Guild are fruitful sources of gossip. The initial approach to an agent is precisely the same as that to a publisher. Resist the urge to overload the submission.

It is useful to know from the start what agents charge for their services. A customary figure is 10 per cent but an increasing number go for 15 per cent and a few pitch as high as 17½ or 20 per cent – plus VAT. A VATable author can reclaim the tax. Others must add 17½ per cent to the commission to calculate the agent's deductions from earnings. Reading fees are condemned by writers' associations and spurned by leading agents. It has been argued, by writers as well as agents, that a reading fee is a guarantee of serious intent; that if an agent is paid to assess the value of a manuscript, he is bound to give it professional attention. Sadly, this is not necessarily the case. While there are respectable agents who deserve a reading fee, the regular charging of fees can too easily end up as a means of exploiting the naïve. But some agents do invoice certain administrative costs such as photocopying.

Do not be disappointed if one or even several agents give the thumbs down. They may be overloaded with clients. But even if this is not so, remember that all writing is in the realm of value judgement. Where one agent fails to see talent, another may be more perceptive.

When a writer does strike lucky, the first priority is to arrive at a clear understanding as to the scope of mutual commitment. Will the agent handle all freelance work – including, for example, journalism, personal appearances on radio and television, lecturing – or just books? Will the agent take a percentage of all earnings including those he does not negotiate? This is a touchy subject. Some writers think of their agency as an employment exchange. Any work they find themselves should not be subject to commission. But this is to assume a clear dividing line between what the agent does and what the writer achieves on his own account. In reality the distinction is not always apparent.

Understanding the market: what's needed, by whom, in what form, and in which media, is all part of an agent's job. Once he knows what his client can do, he is able to promote his talents to the people most likely to want to buy. Eventually, offers come out of the blue – an invitation to write for a newspaper, say, an editing job or a chance to present a television programme. It is at this point that the writer is tempted to bypass his agent. 'Why should I pay him, he didn't get me the work?' But the chances are that he did, by making the author a saleable property in the first place.

There are authors who, preferring to remain unagented, show a creditable talent for wheeler-dealing. Others – the misguided or the hopelessly optimistic – enter complex publishing agreements without so much as a glance at the small print.

How can an author be sure that a contract is fair and above board? The writer who handles his own affairs is not entirely alone. A guide to good practice is available to members of the Society of Authors and the Writers' Guild (see Professional Associations and Societies listing). When it comes to signing, you may feel you have had to give way on a few points, but if the general principles set out by the writers' unions are followed, the chances of securing a reasonable deal are much enhanced.

Then there is the question of royalties. The conventional hardback royalty is 10 per cent on the first 2,500 copies, 12½ per cent on the next 2,500 copies and 15 per cent thereafter. On home (mass-market) paperback sales, the minimum royalty should be 7½ per cent, rising to 10 per cent after 30,000 copies.

Greater flexibility in the paying of royalties has followed the heavy discounting common to all the book chains. If an author's work is discounted in the shop, is he entitled to a royalty on the original recommended price or on the marked down price determined by the retailer? Almost certainly the latter but there is no set answer. Much depends on the author's negotiating muscle.

Similarly, generalizations on overseas royalties can be misleading – there are so many different ways in which publishers handle export sales. But as a rule of thumb, if the royalty is calculated on net receipts (when the publisher has sold in bulk at a special price), the percentages should not be less than the home royalty percentage. If the royalty is calculated on the published price, it should be no less than half the home royalty.

As a spot check, confirm four essential points before signing a contract.

First, there should be an unconditional commitment to publish the book within a specified time – say, twelve months from delivery of the typescript or, if the typescript is already with the publisher, from signature of the agreement.

The obligation to publish should not be subject to approval or acceptance of the manuscript. Otherwise what looks like a firm contract may be little more than an unenforceable declaration of intent to publish. Watch that advance payments are not dependent on 'approval' or 'acceptance'.

The commitment to publish takes on added significance given that editors are changing jobs with increasing frequency. An author who has started a book with enthusiastic support from his editor may, when he delivers it, find he is in the hands of someone with quite different tastes and ideas. Provided the book, when delivered, follows the length and outline agreed, the publisher should hold to his contractual obligation.

Second, there should be a termination clause to come into play if the publisher breaks the contract or if the book goes out of print or falls below an agreed level of sales. Some publishers try it on but there is no way that a termination clause should be made dependent on the author refunding any unearned advance.

Third, watch out for an option clause, giving the publisher first refusal on future work. Submitting to what may read like a flattering request could tie you to a deal you may come to regret. If an option clause is unavoidable it should be limited to one book on terms to be mutually agreed (not 'on the same terms') and enforceable only within a specified time limit – say, six weeks after delivery of a novel or of a non-fiction synopsis.

Finally, the author should not be expected to contribute towards the cost of publication. Every writers' organization warns against subsidized or vanity publishing. It is expensive, the quality of production is often inferior to that offered by conventional publishers, and the promises of vigorous marketing and impressive sales are rarely borne out by experience. Self-publishing is another matter. But the author who takes this route will not only have costs to bear but must also be prepared to take on the marketing and promotional roles of the publisher. Many have tried; few have prospered.

At this point, many would-be authors are tempted to give up. It all seems so hopelessly involved. But despite the inevitable disappointments (it is a rare author, even among the top names, who has not had to bear the pain of rejection) the lure of writing as a career or as a part-time occupation remains strong. And it is not just the prospect of fame and fortune that attracts. Just think how many people write, not for money, but simply because they find satisfaction and fulfilment in writing.

One of the excitements of online publishing is that it will give many writers an opportunity for disseminating their work that would be denied them in hard print.

There are also hopeful signs for newcomers in the latest figures for online sales of books. One of the problems for conventional bookshops is finding space to display any but the latest top titles. But with browsing onscreen there is no limit to the number of titles that can be viewed and sampled. It is a hugely significant fact that over 11 per cent of Amazon's sales come from *outside* its leading 100,000 titles. In other words, more authors than ever before are reaching their public. A bright future beckons.

*Data supplied by Nielsen Book © Nielsen Book*

# Writing children's books

*Michelle Paver*

If you're looking at this book, you're probably prepared to do a little reading about the craft of novel-writing – and by doing that, you've already put yourself ahead of the game. I'm not trying to flatter you. But there are still legions of would-be writers who think they can write a novel without any help at all. And maybe a very few of them can, but most of us could do with a few hints when we're starting out.

I know I did. It took me years to get published, writing in the evenings after work, and at weekends; and during that time *The Writer's Handbook* became a trusted friend. It shone a light into the publishing world and gave me the information I needed to send off my submissions – and thereby add to my pile of rejection slips.

What I'm going to do in the next few pages is share some of the things that helped me along the way. If you already know about story-telling, much of this may strike you as blindingly obvious. But it's amazing how easy it is to lose sight of, and it does no harm at all to be reminded of the basics, from time to time.

## Why are you doing this?

Before you start, it's worth pausing to ask yourself: why? Why do you want to write for children? If you think it's easier than writing for adults, think again. Children are ferociously demanding readers, perhaps the most demanding you'll ever have. On the other hand, if there's a story you really want to write, and it's one that you would have wanted to read as a child, that's probably reason enough. And don't worry if you haven't got children of your own, that isn't necessary. What you do need is to be able to remember what it felt like to be a child. What did you care about? What books did you love? What stories gripped you, made you laugh, cry, lie awake in a cold sweat? If you can remember that, you're on your way.

## What to write about?

It's not all that easy to come up with an idea, and as far as I know there are no shortcuts. One thing I have learned, though, is that it helps to let the subconscious do its work, rather than over-analysing early on. By this I mean that instead of thinking in terms of themes or issues, try putting all that to one side and just see what characters float into your mind. For instance, you might feel passionately about the plight of children in care, but thinking in terms of the issues may not give you a very lively story. On the other hand, if a character walks into your head – say, a girl of ten who's had a rough deal – then stick with her. What's her problem? Maybe it turns out to have nothing to do with what you thought you wanted to write about; but since she comes from your subconscious, it's worth trusting her, rather than trying to impose a theme on her.

## Getting to know your main character

I've heard some writers say that they're not interested in plot, only character, but I find this a slightly artificial distinction. It's not as if 'plot' is independent of character. It arises *from* character. Macbeth kills the king because he's ambitious and because ... Well, because he's Macbeth, we could go on, but the point is the murder of the king is part of the plot *and* it arises from the character of Macbeth.

Of course, some books are primarily based on character, while others are mostly plot-led, but for a really involving story, particularly one for children, I think you need both. Your story can have the most fiendishly clever plot and utterly convincing setting, but if your readers don't care about the characters, they're not going to read it.

All of which is a long-winded way of saying that it pays to get to know your main character(s). What do they like and dislike? Do they have a happy family or none at all? Above all, what do they *want*? To me, this is the basis of the story. What does your main character want? It might be to escape an abusive stepfather or a war zone; it might be to lift a curse. But they need to want something, or it'll be hard to care about them.

Once you've got an idea about your main character and what he wants, you can make things difficult for him by creating obstacles to stop him getting it. These can be external, such as wars, bears, plague and the like; and/or they can be internal, such as divided loyalties and deep-rooted fears. How the character tackles these – whether he surmounts them, whether he achieves his aim – that's what gives you your plot.

I know I've made that sound simple, and of course it's anything but. However, that basic idea – a character's aims and how he tackles the obstacles in his way – is at the root of most stories, from *The Odyssey* to *The Hundred and One Dalmatians*.

## To plan or not to plan?

The next step is to ask yourself whether or not you want to do any planning before you start writing. Writers differ hugely on this. Many do little or none, while others (and I'm one of them) do quite a lot. I'm not going to advocate one or the other, as it really does depend on what you're most comfortable with. For myself, I like to have a rough idea of where the story is going, not least because I like to know that I'm working up to a good, satisfying, emotional climax. The caveat is of course that plans can always, always be changed, and often are, depending on how the characters develop. A plan is a guide, not a blueprint.

## Research

Whatever story you write, whether it's set in present-day Manchester or a Stone Age forest, it's got to feel real. I think this is incredibly important, especially for children. They want to feel as if they're inside the story, experiencing it with the characters. They're intolerant of generalities, and quick to sniff out inaccuracy and implausibility. They like specific, convincing, arresting detail. Not too much, of course; just enough to fix the scene vividly in their minds.

This means research. How much you do is up to you, as is how you do it, but I think you need to do enough to feel really confident in what you're writing. I do a lot in libraries, but I also like going on research trips, because my books are set in faraway places, and I can only really get a sense of them – for example, the Arctic tundra or a primeval forest – when I've been there, when I know how the place looks, smells, feels, sounds, tastes. (It's also a great way of escaping interruptions and concentrating purely on the story.)

But by 'research' I don't necessarily mean a long trip to somewhere exotic. If your protagonist's got a Tube journey to make, you might be tempted to rely on the Tube map and imagine the whole thing at your desk. But if you actually did the journey yourself, you might notice all kinds of details that feed into the story in ways you could never have anticipated.

Finally on this, a truism that bears repeating: never overload your story with research! You'll inevitably unearth mountains of amazing stuff that's just crying out to 'go in'. Be ruthless. *Only* allow in what's essential to move your story along, or to establish character, or create atmosphere. I'm sorry if I sound bossy, but this is so important, particularly when writing for children. We've all read prize-winning adult novels which are beautifully written but contain yawn-inducing passages that are just asking to be skipped. A child wouldn't skip, they'd just put the book down and never pick it up again.

## The writing

OK, you've got the bones of a plot and you've done enough research to make your world real, so you're ready to start. Well, poor you, is all I can say. Because I hate, I absolutely hate, those first few pages. And with children's books even more than adult novels, the beginning is crucial. Not the first chapter, not the first page, but the first paragraph, the first line. Somehow you've got to get their attention – not in a gimmicky, over-eager way, but in a way that gets them caring enough to read on, without clunky exposition slowing things down.

Unless you're an amazing writer, you're probably not going to get it right first time. So don't agonize too much over your first few pages. Just put something down. It'll serve its function to get *you* into the story, and you can always go back to it, once you've finished your first draft, and come up with something completely different and much better. I've done that with every book I've ever written.

As for other aspects of style, I don't find that writing for children is very different from writing for adults, and I certainly don't use a simplified vocabulary. I find that it's the subject of the story which dictates the style rather than the age-group.

For instance, I use different language if I'm writing from the point of view of a Stone Age chieftain, compared with that of a wolf.

However, if (in my view) there are no major stylistic differences in writing for children, there are a few things which it's helpful to bear in mind. Surprise them if you can; they love it if the story takes a turn that they didn't see coming. Include plenty of dialogue, a few jokes if you can manage them, and lots of variation in pace and tone. Descriptions should be vivid, but brief. Don't talk down to them, ever. Always give them a reason to turn the page. And don't shrink from the truth. Children are just as interested as adults in the big questions of life and death, loyalty and love.

## Rewrite

Of course, all this is easy to say, and extremely hard to do. So if you've just re-read your first draft and think it's utter rubbish, don't despair! Losing faith in your story is something that happens to us all from time to time. It's horrible and it's just got to be worked through. But as you wallow in the Slough of Despond, it's worth reminding yourself that nobody, but nobody, gets it right first time. And generally, the more rewriting you do, the better it gets. Cut out the boring bits. Tighten up dialogue or exposition (and by tightening up, I mean more cuts). Again, I'm sorry if I sound bossy, but in my view it's an unusual writer whose work doesn't benefit from loads of rewriting.

## Market research and all that

So far, I haven't said anything about markets or age-groups, and that's because I don't find it terribly helpful. There are lots of excellent books out there on how to write novels (and incidentally, I do urge you to take a look at a few of these) – and in some of them you'll find exhortations to study the market. By all means, take a look at other authors' works if you want to, but in my view it isn't essential, and I wouldn't suggest that you make a study of it. If I'd done any market research before writing *Wolf Brother*, I'd have found that there weren't a whole lot of adventure stories for children set in the Stone Age, and what would that have told me? That there was a gap in the market just waiting to be filled? Or that there was no market for that kind of story? Either way, it would have influenced me, and I don't think for the better.

So if you've got an idea for a story and you're keen on it, why not just go ahead and write it, and let agents and publishers work out how to sell it? That way, you'll come up with a fresher, more original story. And you'll have more fun, too.

*Michelle Paver was a City solicitor for thirteen years before she gave up her job to write full-time. She has published five adult historical novels and is the author of the bestselling children's series the* Chronicles of Ancient Darkness, *which has been published in thirty-six languages and sold over three and a half million copies. Her websites are: www.michellepaver.com and www.torak.info*

# Great book. Pity about the reviews

*Barry Turner*

Authors hate being ignored. Who can blame them? Anyone who has worked for a year or more, stringing together 80,000 words with enough pulling power to persuade persons unknown to pay good money to share the enterprise, expects at least a nod of acknowledgement for a job well done. Congratulatory emails from publisher and agent go some way to fortifying the ego but what is really needed is a review – preferably favourable – in a national newspaper or journal.

If only. There are many reasons why reviews are hard to come by, starting with the fact that every year there is an increase in the number of titles competing for attention. A literary editor's book bag could stock a fair sized shop. Indeed, in the great days of Fleet Street there was such an establishment trading almost exclusively in discarded review copies.

In being spoilt for choice, the literary editor is like a publisher surveying the slush pile. What is plucked out for closer inspection has much to do with instant recognition of what is topical or fashionable. Books for small niche markets – a guide to spare parts for Volkswagen Beetles or a history of Fenland drainage – are quickly discarded. A one time literary editor of *The Guardian* has handed down a helpful roll call of subjects to be avoided: 'unreadable academic texts, histories of buses (especially in Croydon), war memoirs, SAS memoirs, romantic novels with Laura Ashley covers, pictorial histories of the Beatles, sequels to Jane Austen books, books about UFOs, death and God'. No list of proscriptions is good for all time. God is back in fashion, paradoxically as a consequence of the attention of militant atheists, but the other no-goes still hold good.

On the plus side is any book which has a peg on which a critic can hang a polemic. A biography of a leading historical figure, for example, grabs the attention of the casual reader along with the opportunity to take issue with the author. Was X or Y or Z all that he was cracked up to be or was he fallible in ways that the critic might adopt as the start for a lively feature?

Anything that links across to the news, such as a current war or an establishment exposé, moves up the line, as do books that tackle the fundamentals of human existence like climate warming. Where does this leave the novel? In a weak position is the answer. The trouble with the novel, from the journalistic point of view, is that it rarely offers anything to write about beyond a digest of the characters and plot. Which is why crime fiction, a hugely popular genre with many of the best contemporary writers to its credit, is invariably relegated to quickie reviews of a paragraph or two. The literary giants whose reputation rests on intellectual reflection command notice, as do authors with a television profile or other claims to fame beyond the world of books. This doesn't leave much space for the rest of us.

Knowing someone with muscle helps a lot. An editor who has dined at your table can be worth a thousand words. It is no coincidence that journalists who write books nearly always secure reviews, though not necessarily flattering ones. Some of the worst notices I have ever read were attached to a novel by a well known columnist who, as far as I know, has never again been tempted to branch out from the day job. But on the whole a place on the inside track can only be an advantage.

It must also be said that there are book lovers who digest reviews with a hefty dose of cynicism. Often, critics are themselves authors who are wary of saying anything that might invite hostility when they next publish a book. As chairman of the Man Booker Prize, Sir Howard Davies took the opportunity to assail critics who shy away from real criticism. 'The only way you can detect that a reviewer doesn't like a book is when they spend the whole time simply describing the plot. They're not brave enough to say "It doesn't work." They are tolerant of untidy novels. They don't care whether they're readable or not.'

Davies stopped short of accusing authors of back scratching but he might, with justification, have given a rough ride to those ubiquitous lists for holiday or Christmas reading where contributors are blatant in delivering recommendations that scream out for a return favour.

Beyond a certain disappointment that literary genius is not immediately recognized, does it matter if our books fail to attract press coverage? The usual moan is of a lost opportunity to attract sales. But it is not altogether certain that reviews do much to influence buying habits. Television has the biggest impact. At the height of their popularity, a nod of approval from Richard and Judy could propel a title to stratospheric sales. Radio is another favourite with marketing managers. When it comes to hard print, an extract in prime position in a paper can do more to shift sales than a raft of reviews, possibly because an extract is more likely to be read.

And here we get to the crux of the issue. Literary editors do their best to spice up their pages but in achieving this they tend to demote the product on which they depend. The review displaces the book by attracting those who enjoy reviews as reviews. There is no question of editors acting as sales agents for publishers except in so far as titles are sometimes sold off the page. 'Book pages belong to journalism, not to marketing,' says Suzi Fay, former literary editor for the *Independent on Sunday*. It might be different if publishers put more money into advertising but none save the best selling of bestsellers qualify for that level of budget.

Whatever we think of the review pages, their future is problematic. The days when a newspaper could subsidize unprofitable if prestigious sections with overall healthy advertising revenue have long gone. The recession has hit the press harder than most while competition from the internet has cut circulation which, in turn, has further encouraged advertisers to look elsewhere to promote their wares.

Reader reviews on Amazon and other retail websites, the equivalent of word of mouth recommendations, always one of the most powerful selling tools, are hugely popular, not least because they empower genuine readers as opposed to supposedly objective critics who invariably turn out to have their own axes to grind. It is here that newcomers and midlist writers can attract notice. And this, after all, is the point of the exercise.

 *The Writer's Handbook recommends:*
*Ed Reardon's Week*

Look on the black side. Just for a moment. (I promise this has a happy ending.) Every writer suffers severe jolts to his self-confidence. A brutally dismissed synopsis, a rejected manuscript, a publisher's call to rewrite whole slabs of painfully constructed composition – these and other humiliations disturb the long night as do thoughts of taking up a less stressful and potentially more profitable occupation such as running a whelk stall. The only comfort is knowing that there is always someone who is worse off. Even a fictional character, suitably disaster-prone, can console the sufferer so that, whatever the tribulations, the pits of despair are still some way off. One such is Ed Reardon, whose edge of reality diary of a writer's week has run up three series in Radio 4's comedy slot, with more to come.

Ed's early career, we are led to believe, was touched by great promise – a published novel, the sale of film rights, scripts for television – but that was all in the eighties. Since then Ed has been on the Cresta Run of decline, kept going by standing in as a writer for celeb memoirs, Christmas joke books and £20 a time evening classes. With an agent for whom lunch is the answer to everything, Ed charts his professional regression, keeping us joyously entertained along the way. Will it come out all right in the end? Will Ed, the distant outsider, surprise us all (and himself) by winning the Man Booker Prize? He deserves every success, as do his creators, Christopher Douglas and Andrew Nickolds, who should win every comedy award going.

Ed Reardon's Week (series 1) *is written by Christopher Douglas and Andrew Nickolds and published by BBC Audiobooks (ISBN 9781408401194)*

# Characterization in popular fiction

*George Mann*

Fiction is about people.

That may sound like a rather obvious statement, but it's amazing how many writers and would-be writers seem to forget this simple truism. Whatever else it is that you think you're writing about – be it killer kitchen appliances, robots, superheroes, sentient spaceships, talking animals – what you're actually doing is writing about people, their relationships to one another and the environment around them.

At its most fundamental level, fiction writing enables us to document the human condition through the medium of stories, to discuss our differing experiences of the world and to say something relevant whilst, hopefully, also giving entertainment. Of course, much of this happens on an unconscious level. Nevertheless, the fact remains: as fiction writers, we write about people. Fundamentally, as people ourselves, we can't do anything else.

So if fiction is all about people, why is it that many writers and would-be writers spend so much time worrying about plot and not the people who populate their stories?

Well, plot is important too – of course it is – but what we must remember at all times is that to be relevant, to be truly successful, the piece you are writing must be *somebody's* story. This somebody – most likely the protagonist of your tale – is probably going to be put through the wringer as your plot unravels around him. But what readers are *really* interested in (even if they haven't yet worked it out for themselves) is how your character responds to that plot, how that fictional person deals with the emotional highs and lows of the story, how it impacts on the character's choices, how it changes him.

It's easy to fall into the trap of valuing plot over character. I know: I've done it myself. It's a particular pitfall of genre fiction. A good science fiction story, for example, often hinges on a single, central conceit. It's a story of an idea (e.g. what happens when all the communication satellites suddenly stop working). The problem with this is that you can be tempted to rely solely on that idea to carry the piece, forgetting that need to root it in character. The idea is so key to what you are writing that it consumes the story and the characters are left watching from the shadows. What you do with the people in your story, how you show the impact of what has happened (e.g. a parent tries desperately to get in touch with his missing child while the aforementioned satellites are down) can make the difference between a good story and a *great* story. Characters give your story relevance; they make it matter. It's the same for crime and mystery writers. It's easy to become so enamoured with the details of a mystery plot – who did what, when – that you forget to ask *why*, or you fail to look at the impact the crimes are having on the other characters, be they subsidiary (e.g. the family members of one of your murder victims) or more central (e.g. your detective).

Cause and effect are two important words that spring to mind here. Think about the motivations that drive a person to behave in a particular way, and then consider the impact that behaviour will have on the other characters that populate your story. Ideally, what you should be striving for is balance, a perfect marriage of character and plot, so that you offer your reader both the thrill of an exciting story and an emotional resonance with the characters that are experiencing your tale.

So what does make a successful character?

That's a difficult question to answer. Is it realism? Not necessarily. Your characters don't have to act and feel in the same way as you or the people around you. Your characters may face extraordinary circumstances. Characters in a fairytale may well have different motivations and mannerisms than a squad of New York cops in a period crime drama. Everyone is different, and the people who populate a fantasy world may have a very different take on life than you, or perhaps more importantly than the person reading your work. Some readers will be comforted by cliché, and cliché can occasionally be a useful tool, a form of shorthand when working within particular genres. All of that is okay. But people generally behave consistently with their environment and culture, and that's one of the key things a writer has to get right. People are a product of their time and place, not of your plot.

By that I mean you should avoid having characters that simply serve the functions of your plot. People are complex, multifaceted. They have personalities and depth, needs and desires, strengths and weaknesses. They make mistakes and bad judgements, as well as heroic gestures or villainous remarks. In life, people are not defined by their status as heroes or villains, and nor should your characters be. They are not ciphers. They require depth and personality.

Your characters must exist outside of your story. Well, okay, they don't, but you should give your reader the perception that they do; that they have had a life that has shaped them to become the people they are now, as the reader will see them in your story. What happened to that pastry chef to cause her to develop such a bizarre phobia of eggs? Why does that shopkeeper have a twitch in his left eye? What led the lonely goat herder to become terrified of rabbits? You might not necessarily give the reader the answer to all of these questions, but it helps if you know it, or at least if you can give the reader the impression that you do. It adds texture to your story, layers of complexity that help to sustain the reader's interest and suspension of disbelief. Your readers may not even notice what you're doing in building up your character's personality, but they will be left with a curiosity to find out what happens to him next. They'll want to keep turning the pages.

It is also important to recognize that there is no such thing as a 'good' or 'bad' person, only 'good' or 'bad' actions, and even these have to be contextualized by culture and environment. The world is not black and white. Villains rarely think of themselves as 'baddies': they believe what they are doing is right. A villain can still be chivalrous towards women, hold doors open for elderly people, be charming and touching and emotional. Likewise, a hero can be obstinate, insufferable and belligerent. Consider the character of Doctor Gregory House from the television series, *House*. Here's a character who is perhaps one of the most odious wretches to ever grace the screen, yet he is charming, intelligent and occasionally kind. He shows weakness as well as strength. He is often wrong, and he is sometimes right. He treats the other characters with disrespect – most of the time. But he also shows signs of being vulnerable and redeemable, and that is why we, as viewers, love to watch him on screen. Part of us wants to see what terrible things he will do next (and how the other characters will react to it), and part of us wants to see him 'saved', to see him reveal that tiny chink in his armour that proves to us he's really a good guy at heart. Essentially, though, as we watch the story unfold on screen, it's the character who engages us, and not necessarily the plot.

Perhaps my favourite character in all of English literature is Zenith the Albino, the 'gentleman crook' from the sadly forgotten Sexton Blake saga, which had its heyday in the 1920s as a popular story paper. Zenith is the perfect example of a multifaceted character. He's the arch nemesis of Sexton Blake (our heroic detective), but he is complex and charming. He always appears in immaculate evening dress, regardless of the climate or the time of day. He is well educated and plays the violin like a master. He lives by his own moral code, which is at odds with the moral code of the wider society within which he operates. Yet this code is not without its merits. Zenith will merrily murder another criminal, but he will work hard to protect the innocent. He will steal from the rich, but he will go out of his way to help an imperilled member of the opposite sex. He rails against Sexton Blake's constant interference in his plans, but has a begrudging respect for the detective and will not kill him in cold blood, even though the opportunity presents itself on numerous occasions. Crucially, Zenith can always be relied upon to act within the limits of this self-imposed code of honour, and it is this consistency and depth that makes him such a successful fictional character; he *feels* real, even though he is, in truth, one of the most outlandish characters one could ever imagine.

So as writers, what can we do to help us develop such multifaceted characters? First of all, reading good characters, seeing how the masters of our art breathe life into their own creations. Take a moment to think about how your favourite writers present their characters to the reader. What distinguishes those characters from all the other fictional characters you've met? What tricks did those writers employ in sketching in the background of the characters you've learned to care so much about?

Observing people, too, is a good exercise, capturing people's mannerisms and body language, their speech patterns and quirks. All of this makes for variety in our own fiction and imbues our characters with a life – with a sense of realism – that enables the reader to recognize them as 'real' people. Think about your character's motivations, too, about the reasons they act as they do, about their interactions with the plot. Why are they part of the story? What is it about these characters that puts them centre stage? Why should the reader care what happens to them?

You could try preparing a small biography of each of your major characters before you start writing, just a couple of paragraphs detailing their likes and dislikes, any distinguishing marks or behaviours, anything from their history that might have a bearing on how they might present themselves during the course of your story. Much of this will likely never see print, but the process itself means you're spending time in the company of your characters, learning what makes them tick, before you ever truly commit them to the page. This way, they emerge fully formed for the reader.

Then, when you have your characters, you need to think about the journey you are sending them on. How is the plot going to change them? What impact is it going to have on their lives? What is it about what's happening that threatens their current routine? To go back to one of my original points: the piece you are writing is somebody's story, and he has to interact with that story on a number of different levels. Your character should not be the same by the end of your piece. Change is necessary – even if things are to be restored to equilibrium, your character should have been on a journey, experienced something new. Otherwise, why are you telling their story? This does not mean you have to torture your protagonists, or throw them out of aeroplanes over the Andes (although, of course, you can). The point is simply that your tale is to take your characters from one place to another, either physically, emotionally or mentally. *This* is the point of your plot – to challenge your characters by somehow altering their circumstances.

And so I return to my original point: fiction is about people. There's no escaping it. Invest in your characters and the rest will follow.

*George Mann is the author of the Newbury and Hobbes fantasy crime series which includes* The Affinity Bridge, The Osiris Ritual *and* The Immorality Engine. *He wrote* Ghosts of Manhattan, *as well as numerous short stories, novellas and an original* Doctor Who *audiobook. His first novel* The Affinity Bridge *was a finalist for the 2008 Sidewise Awards for Long-Form Alternate History.* His website is www.georgemann.wordpress.com/

# Poetic visibility and relevance at the end of the noughties

*Chris Hamilton-Emery*

## Going back to the end of the start

When I think back to the tail end of the nineties, poetry was a rather enclosed, binary world: you could spot the busily warring clans lobbing sonnet grenades at each other as half of our serious writers began the long retreat into the academy and the rest were to follow in barely half a decade. Now almost no serious writer lies outside of a university relationship of one form or another.

It all seemed rather stuffy and locked in battles begun in the seventies, in the wake of the counter culture revolution. *Seemed* stuffy, because no one had much of a view of the art beyond what one encountered on discreet mailing lists or in local workshops in the neighbourhood library. Of course we still had libraries back then and weren't busy shutting them or renaming them as information centres. Poetry felt more isolated and mysterious and private: you didn't exactly fall over poets and a great deal of interaction took place via the postal service in the form of things called letters. Remember them? However, many of us were busy lurking on listservs and getting a taste for what would be unleashed in the middle of the noughties with the birth of social networking. Privacy, solitude, are now seen as quaint.

At the beginning of 1999 the debate had surged about Oxford University Press closing its poetry list, the animosity surrounding this even drawing in the then Arts Minister, Alan Howarth (*who?* you may ask; he is now a life peer). It's hard to remember the shock and exasperation caused by that closure. Would Parliament be as concerned about the closure of a poetry list today?

Later that August we had the last full solar eclipse the UK has enjoyed: it felt as if old things were indeed coming to an end and we were all being swept away with fears of millennium bugs, abhorrence of the dullard's Dome and high hopes for that still young Labour government, even as it carried Britain along with it in a miasma of newly discovered high public emotion, stemming as it did from the death of the Princess of Wales in the election year of 1997. As the moon tipped in front of the sun and millions of us stepped out into the shade to stare, the world of poetry was about to change forever. So what has this decade given us? Let's take a whistle stop tour of online mags, blogging and social networking. In the noughties, poetry went online. Big time.

It was at this point I decided I was going to start a new poetry press with my friend John Kinsella. But far more importantly, the world of technology and especially the web was about to reveal that the world of poetry was far bigger and far more various than anyone had considered. Changes in printing and distribution meant that more people could enter the world as publishers and more people could publish themselves too, as vanity presses changed their clothes, donning white coats as author services businesses and over the decade unleashed a tidal wave of unread books. At the end of that decade we're about to compound this with a tsunami of unread ebooks, too.

We live in an age of lists; that in itself might have formed one item on my list below. Lists masquerading as culture, lists masquerading as consumption. Lists are perhaps an extreme form of consumer guilt, faced as we are with an ever increasing culture industry but no time to partake of it. They can also be nice provocations, though. A new year, a new decade, presents the maximum opportunity for listaholics to regurgitate and reflect, to genuflect before their top fifty, or bow before their meagre top ten. I am not immune. So let's have some reflective nostalgia for poetry in the noughties. Where have we been heading? Did we get there on time?

## Top ten poetic events of the noughties

1.  The Poetry Archive and the work of Andrew Motion in opening up poetry to an entirely new audience, and especially the young. Possibly the most important thing to have happened in the past decade.
2.  Rachel Alexander's reign as Director of Publicity, Faber & Faber, helped propel that business forward, and her work on the arts list has led the way for other poetry publishers to follow. More than anyone else, she has shown how publicity drives poetry's audience, and she's the best illustration of how publicity now drives serious publishing.
3.  The Warwick Writing Programme (founded in 1996) came to define what a British writing school might look like and is now one of the most vital creative writing programmes in the country. David Morley's continuing involvement provides the UK with an exemplary model.
4.  The Poetry School (founded in 1997) has now become an institution helping many writers to develop their skills and understanding, now busy expanding nationally, its continuing impact has been widely felt.
5.  The Forward Poetry Prizes, notorious or judiciously wonderful, depending on whom you talk to, the prizes have grown from strength to strength in the noughties and for many are now the most significant poetry awards in the UK.
6.  The death of Michael Donaghy, which may have led to hagiography in some parts of the contemporary scene, has come to symbolize a form of commitment and vitality — almost all of his workshop students are now in print with substantial careers of their own. Quite a legacy.
7.  *Staying Alive* and *Being Alive*, while sounding like Bee Gees revival acts, these momentous door stops of books changed the landscape. They were accused of introducing a pernicious age of anthologies, yet they cleverly devised a model that worked for booksellers and publishers alike, and these

generous books have become an important cultural event, offering hundreds of thousands of people a new way into poetry. Surprising and magisterial.

8. Salt (founded in 1999) and Shearsman (founded in 1981): the noughties was the decade of new(ish) or revitalized independents, especially those exploiting new technology to capitalize their operations and grow exponentially.

9. Keston Sutherland, the one-time Cambridge undergraduate: maverick, excessive, maddening and wonderful, he rose to prominence in the noughties through a heady mix of serious academic rigour, progressive chutzpah, non-stop transnational networking and a commitment to revitalizing the debate about the art from the deep, deep underground up to the earth's mantle. If academe does not consume him, this next decade may well be even more accustomed to his vigour.

10. Next Generation Poets (2004), divisive for some, derided by others, the promotion, a rerun of the New Generation Poets promotion in the nineties, can be seen as an example of finding formats to promote new talent to the world. Its impulses are no different to Simon Cowell's 'format television' beneath which lies the A&R impulse to draw on new talent and at the same time define a new audience. It continues to show the power of lists.

## The here and now: a story on the brink

You may be reading this invaluable book with a desire to get your poetry out there, into magazines and ultimately, as a debut collection, into print. In many respects it has never been easier for this to happen, in fact you can do it yourself with almost no pain at all and a great deal of success. For some authors this is going to be the preferred choice of doing business, and represents a further collapse of the traditional matrix of writing, publishing, distributing and retailing works of literature. What I'm commenting on is the revolution that began a decade ago as digital manufacturing and the emergence of ebooks on the World Wide Web.

If you want your poetry in print, it can happen within the week ahead, it needn't cost much money. It can be for sale on Amazon or the Book Depository in a few days more. Anyone can buy it, anywhere in the world. It can be shipped to them for free. No one prevents anyone from being in print. This is good news for poets. At least in this one respect, there are no barriers between your keyboard and your readers. Nothing can stop you.

What's more you can do this without ever needing a publisher. The past decade has been the decade of self-publishing and more websites are emerging each month with new models of how to put your work out there, from mulching sites like authonomy.com to self-published ebooks at scribd.com. Amazon is opening its doors to all 'independent authors' (a term I'll coin here) for its Kindle platform, and Apple has now entered into the world of ebooks. Sony and Plastic Logic are soon to follow. Nothing will be the same again.

At the same time, the world of traditional publishing is in a mess. A wonderful mess, full of digital opportunities, but a mess all the same. The biggest challenge is how anyone can make any money out of all this. The collapse of the net book agreement, the rush to deep discounts and huge advances, the emergence of celebrity authors and of course the focus upon bestsellers have created a world where publishers have reduced their lists, booksellers have reduced their range, and everyone has, it seems, turned away from serious literature and serious books to a frivolous and commodified and industrialized world driven by numbers, templates and subgenres. Many would see this as an advancement of the trade, its further modernization. Elsewhere, literature has been busy finding new ways to connect writers and readers off the High Street, away from retailers altogether.

Here in 2010, our poetic frontiers are now too various and too disputed to mean anything vaguely monolithic, our wars and politics have taught us not to trust any claims for pre-eminence: our borders are no longer geographic, but perhaps religious and economic; our societies are virtual; our real neighbours may be 5,000 miles away. Consider it: one of the world's largest countries is now Facebook, and the greater part of some people's daily interactions are with people in the virtual world and our networked society. The idea of the 'local' is now a virtual mythology.

In this new world poetry does not progress, and its variety and outreach are now so wildly disparate, complex and synchronous that no one can claim to harvest its value or pass judgement, at least not with any credibility. We've come to the end of a critical age, critics are on the way out, literary editors are being made redundant and we are in an age of polling, commenting and user-dominance.

Poetry now is, in essence, *beyond* value, and much of it is untradable, given away in a vast international gift culture where readership is fragmentary and communal and bookshops are no longer the place to discover new writing: the world of literature has moved online and in this move the force of literature is primarily a force driven by word of mouth. It's interesting to reflect that if we were to believe the chief poetry prizes of today, we might imagine that there are no more than a dozen poets writing anything of value, perhaps fewer. Some would celebrate that remark. But the truth is that the poetic landscape is too extensive and explosive for anyone to now attend to it with any sense of completeness or satisfaction with their engagement. There's too much to take in, and it's getting bigger each day. No current system can keep up. No literary editor, no team of reviewers, no pundit. The age of judgement has been replaced with an age of publicity: of course you can only choose from what gets put your way. You may not even know what else lurks out there in the vast oceans of literary output. There might even be seething oceans, at that.

As readers, as writers, we must of course select. Power, literary power, now rests in how such selections are managed, and the new war is not one of literary practices and their primacy and largesse, but of constricting readerly consumption: it is about controlling what one sees or discovers. Yet the new age and its merchants are against this censorship, and the new age will win out. It will win out because it will side step human involvement – and what will come to pass will be a systemic force, an augmented range of tools for getting what we want. And what do we want? Very many of us want to be told what to do. One illustration of this is the power of the television book club where almost anyone's selections are better than making our own. Some scoff at this presentiment, but publishers are more than aware of the power of such publicity and are actively supporting this way of controlling visibility and constructing choices. In an age of abundance, the role of publishers is rapidly becoming the role of reception management.

## Poetry's new fragmented readership

I want to consider poetry's readership for a moment. Bear with me.

The past is filled with the destinies of poets. They crowd around us now like paper ghosts: reams of them, packing the night air, packing their truths for safe keeping on our travels into the warm future. The world's shelving cannot hold the bound thin volumes of them all. The millions of dead poets present a special kind of pleading for which we may yet have no ears, no eyes. But we now have archives. Digital repositories. Search engines. Robots.

In 2010 the condition of poetic death is rapidly changing. In the first decade of our new century, our new millennium, we have been busy reclaiming all that was ever committed to paper, and we are in an age of maximum resurrections, if not maximum consumption. What does it mean to be read now? Certainly robots read and index the dead. Is our human reading immersive or passively injudicious? Most poets today continue to write for the future, piling their destinies up for the unborn. How are the unborn different from the dead? I think, if you are considering writing poetry and are using this book to prepare your journey, you should consider two things about readership: *visibility* and *relevance*. You need to be discoverable and you need to matter to someone in some particular way: successful writing is social. The interactions between writer and reader may be seen as participative, interrogative, communal.

The degree of your visibility and the degree of your relevance will dictate the size of your readership, and this fact can be separated out from the idea of products, of books — it's about the literary stream you, as a writer, produce; it's your writing life as a continual singular publication: works are simply stoppages in time.

We may be heading to an age where the fragment is more important than the work. Where users disassemble the work and draw from it the pieces they value and wish to read. Perhaps we will see a new age of serialization, a new age of poems and not collections. We will almost certainly see a world where readers will subscribe to writers and not to intermittent works. The writer will provide a continuous stream of literature from which readers will select their idea of the work, their idea of the relevant material, and writers may occupy this stream, may recognize its demands and certainties. The exclusive or reclusive author, the independent author, may well become a form of literary currency around which social forces move, around which communities of interest may form.

The assembly of literature may well pass to the consumer. Indeed its physical architecture may be unnecessary, an encumbrance, to reading — we can't wait for a slow medium. We're addicted to the impulsive momentary uptake. We may simply choose the right medium at the point of consumption, standing by the espresso machine for the book to fall out, or choosing our own anthology to be delivered in part or as a whole to our handheld device: wireless, global, immediate. This is no future: this is today.

At the same time we are very anxious about who owns all the stuff that we're busy digging up. Never mind the living, who owns the minds of the dead? Google are of course the great undertakers of our age; for dead poets, they are the best hope of a readership yet to come. The future, we have discovered, is an archive. Yet, it may be a perpetual living tomb. Despite our best endeavours towards a 'total literature', our lives are too short to engage with it. We can reproduce it all, but we can't have it all.

We are now moving into an age of total availability. Nothing will be lost, everything is archived, even our spit. We are in an age where privacy and personality are simply a codex and an avatar. All things are kept and cross-referenced and linked and laced with the fabric of our lives. As more of our poetic lives move online, even the detritus of our experience, the bad poems, the edits, the spillage, is saved.

Yes, things have changed a great deal for the dead, and soon the living will have a lot to cope with: so many more conversations to chew over, even more books to consider over toast and marmalade. Will we ever catch up with all our reading? No, for literary production can only escalate exponentially. Our current industry cannot cope with what is coming, and the idea of constraints, of selection, of judgement and of distribution, are rapidly disappearing. With the emergence of ebooks, self-publishing and the networked society, the role of poetry in print, of editorial validation, of lists, is about to change forever. The curatorial value of literature is passing from the few to the indifference of the many. We currently have no systems to manage the vast landscape of creative production. One atavism of this condition is the emergence of talent shows and the democratization of the arts. Anyone can have a go. This is a very good thing.

I'm writing this in 2010, looking back on the enormous shift of that first decade of our century. Think of poetry in 1910, as Edwardian propriety is exploded by modernism, social upheaval and industrial warfare. The population of the earth has increased from 2 billion at the start of 1910 to 7 billion in 2010: we have had similar increases in literary production. Can anyone have imagined it differently? Perhaps some have imagined literary ratios are fixed and immutable. Perhaps those interested in the canon still think that the great writers of each age are few and that their work lasts because of its intrinsic value and not the prejudices of a trade and its economy and the relative size of the literate population.

## The indefinite future

Some people say dogs can't look up because they have no souls — I rather doubt this in my Labrador's case — mind you, she does look vacantly down at her empty dinner bowl and rarely, usefully, up when I throw her slobber-covered ball. But this sacral canine assertion has made me reflect that poets, so often the soulless and the damned of literary history (Dante put several on the first circle of his

*Inferno*), are very often good at looking wistfully back, but are more often intent on looking hopefully forward (to those elusive future readerships and literary salvation). That might sound like a thin excuse for poets to be away with the fairies, as the young Yeats might indeed have wished, out among the daffodils and adrift with the clouds, but actually it's a rather weak segue into a bit of crystal ball gazing, a bit of table knocking, a bit of Hughesian astrology. I'll end this article with my vision of the challenges ahead.

## Top ten changes for poets in the next decade

1.  Print magazines are finished. Spending your hard-earned public funding and lots and lots of pub time to reach as few as twenty people (or even as many as 2,000) is very much a thing of the literary past. The world of poetry magazines will continue to move online. There will be more of them. They will increasingly network with each other.

2.  Critical poetry reviews in newspapers and journals will all but disappear: online features and online comment will take over. The world of reviews will move online and into a model based around community values and communal senses of identity. Online critical sites will unite readerships with journalists and combine these with participative commentary.

3.  Infernal desire machines. We will all become more adept at describing what we want from poetry, what things excite us, and building reader preferences, mapping our desires, to allow publishers and distributors to match the vast literary output to our needs. Get ready for a wave of new topologies of desire. We only want what is relevant to us. This may lead to the end of surprises and the development of sophisticated choice engines.

4.  Fragments over wholes. The idea of books as imaginative works of a particular length with a particular trajectory, pawed over by editors and writers all within the production constraints of a Heidelberg printing press or its successors will be discarded. The work will be disassembled by readers and reading experiences will be fragmentary and creatively selective.

5.  Everyone will become a publisher. We will see more poets becoming publishers as their own work comes to define an area of practice and an audience – communities of poetic practice will emerge, each with the resources to publish like-minded poets to like-minded readers. Not quite cooperatives, more brand-led collectives, we could well see Don Paterson Press Inc. and Simon Armitage Publishing Ltd emerge within a decade. Independent publishers will exponentially increase.

6.  An end to retail poetry. Much bemoaned now, the idea of discovering poetry in bricks and mortar bookshops will disappear. It's doubtful there will be any bookshops left by the end of 2020. The world of literature is inextricably linked to our networked society. Almost all poetry will be read electronically, much of it downloaded for free.

7.  The micromedia business. Publishers focused on staying in business will increasingly turn to the multimedia delivery of poetry. Video, audio and ebooks, print on demand, user-selected content, will become the staple of

most independent presses. Everyone will have on-demand video channels. Most of our poetry experience may well be online and performance based. Poetry will increasingly be what we *see*, as much as what we read or hear.

8. Poetry will return to its roots, as the shared imagination of a specific online community. Poets will be contextualized within new social structures, defining and reflecting a virtual imaginative space to which members adhere. The mind of the poet will increasingly become a public space.

9. Creative writing will create a vast participative infrastructure. We can see the power of this already in the USA, but in the UK the world of writers' programmes is in its infancy. Almost all poets will work within this industry within a decade. It will be almost impossible to sustain an effective writing life outside of it.

10. The biggest challenge and the source of much literary conflict will not lie around getting published but on being read and finding traffic. This will centre upon 'boundaries of ignorance' and how one community can invade the knowledge space of another, taking their writers to broader readerships and stealing followers. This may lead to conversion wars among poets.

*Chris Hamilton-Emery's poetry has appeared widely in magazines including* Magma, Poetry London, Poetry Review, Poetry Wales, PN Review *and* The Rialto. *He has just been anthologized in* Identity Parade: New British and Irish Poets *(Bloodaxe 2010). A first full-length poetry collection,* Dr Mephisto, *was published by Arc in 2002 and a collection of poetry,* Radio Nostalgia, *was published by Arc in 2006. He has edited* Poets in View: A Visual Anthology of 50 Classic Poems, *as well as selections of Emily Brontë, John Keats and Christina Rossetti. Emery is a director of Salt Publishing, an independent literary press based in London and Cambridge. He was awarded an American Book Award in 2006 and Salt Publishing won the Nielsen Innovation of the Year prize in the UK's Independent Publishing Awards 2008.*

# A writer's place in a networked world

*Sara Lloyd*

At the Consumer Electronics Show in Las Vegas in January 2010, the forum for watching emerging technology trends, ebook reading devices came just third in popularity after two hot innovations in television technology: Internet-enabled TV and 3D TV. As one reporter put it, 'ebook devices were everywhere.' In the USA in 2009, publishers experienced exponential growth in ebook sales, some reporting that ebook revenues had grown from a zero base to represent 4–5 per cent of their overall business within the space of two years. Digital pundits predict further explosive growth in this market, to the tune of hundreds of percentage points in forecast year on year growth. In my own local community, neighbours and friends began to ask my advice about which ebook reader they should buy, rather than asking me why anyone would ever want to read on a screen, which had been the question so often put up a year or so ago.

So, whilst ebooks are very, very far from being a mass market, something in the air does seem to be changing – and fast. What is it that has changed and what does all this mean for writers?

Books, especially paperbacks, have provided us with a perfectly formed, low priced and incredibly functional piece of technology which requires no additional hardware to access and read. Most people carry around no more than one or two books at the same time and many enjoy the look, feel and even the smell of a book. There doesn't seem to be a 'problem' with books for technology to solve, in the same way as there was perhaps with music, which, to make it 'portable', meant carrying multiple CDs as well as a device before the MP3 file and the iPod solved all that. Yet despite this, there is an inevitable sense now that we are moving towards a book market in which digital sits firmly at the centre.

What has changed is that over the last ten to fifteen years more or less every aspect of our daily lives has moved significantly online, from shopping to travel bookings to networking, both socially and professionally and everything in-between. Mobile technology and the introduction of broadband and wireless have dramatically increased our expectation that we can access and view every resource we need, any place, any time. In the last five years the music industry and other media industries, especially newspaper publishing and the film industry, have suffered body blows from the unexpected tidal wave of change crashing over their established formats, routes to market, and business models. At the same time a new generation of 'digital natives' is beginning to grow up, one to which this wireless, 'always on' networked world, in which connected devices are the route to all information, entertainment, friends and services, is entirely natural. Against this backdrop it is perhaps less surprising that companies from hardware manufacturers to online retailers and software developers have turned their attention to books and their digital potential and are investing in this area fast.

As part of this sea change in our culture, businesses and society, driven by new technology, a new industrial power base is also emerging, made up of the computing and internet companies whose brands pervade our senses as we navigate through our increasingly digitized daily lives. A great majority of music lovers now have a deep relationship with Apple, as a result of the huge success of iTunes and the iPod; a growing proportion of the population will also have an iPhone or one of the other emerging generation of smartphones, with touch screen and wireless technology. And in 2010 Apple firmly entered the market for ebooks with the iPad and its ebook retail platform, iBooks. The default search page for a huge proportion of internet users is Google, so much so that the accepted verb for 'to search' on the internet is now 'to Google'. And this internet giant is fast becoming the gatekeeper to online books, with its Google Book Search and Google Editions programs. To an increasing number of book buyers, it is to the online retailer Amazon that they first turn when searching for a book, and it is Amazon's ebook reader the Kindle that is so far dominating the consumer market for downloadable ebooks. These three companies in particular, as well as the software developers, technology providers and related supply chain partners who sit around them, will be of increasing significance to readers, authors and publishers, and the impact of their influence on the changing book market is already being felt.

What this means for writers in the short term is that the business landscape into which you are pitching yourselves and your writing is dramatically shifting. The entire economics of the publishing industry is undergoing challenging transformations and the commercial shape of the future is uncertain and continuously changing. On one hand, the evidence to date suggests that ebooks are creating new, supplemental revenue streams, as consumers who have invested in ebook devices indulge in spontaneous digital purchases even as they continue to buy print books; on the other hand, it is unlikely that this pattern will continue indefinitely, and there will be increasing downward pressure on the pricing of ebooks, where the consumer expectation is for a much lower price point and where there is intense competition between ebook retailers for market share. Much will depend on the related market shares and approaches of the Big Three: Apple, Google and Amazon. There may, of course, also be new challenges from the surprise market entrants that can come from the left field in today's fast moving online environment.

As the retail landscape for books shifts away increasingly from the bricks and mortar, high street bookselling of physical books to online and digital bookselling, the online profile and marketing of authors will become more critical. For authors at the outset of your careers, developing a web presence and following is likely to be a selling point for publishers looking to take you on. This might involve setting up your own website or blog, but developing an 'online life' and getting plugged into other, existing blogs and online communities of writers and readers will also be important. The greater your online network the better the digital marketing opportunity to establish you as a writer. In addition, developing ideas for online and digital content and experiences that might supplement and interact with your core writing is also likely to make you an attractive prospect for publishers looking to find writers with an increasing eye to digital opportunities.

At least in the short term, the core of what you do as a writer – writing – is likely to stay very much the same. In the general consumer market, at least, it appears that long form narrative fiction, especially mystery, crime, thriller and science fiction but also literary fiction is the most popular genre for reading on ebook devices. Whilst publishers have experimented with 'enhanced' ebooks, featuring additional multimedia and interactive content, especially in the mobile market, the demand for this has remained small. As the mobile market develops, shorter form fiction and non-fiction, as well as 'bite-sized' reference content, may well increase in popularity, especially at correspondingly lower prices per download, but this trend has not as yet emerged strongly.

Without doubt, in everything you do you will need to consider the increasingly digital and networked environment into which you are 'publishing', the audience which you are aiming to engage and how the experience you are selling will fit into their increasingly digital, wireless and mobile lifestyle. As digital writing and reading develops further into the future you might need to work in partnerships – with technology and other media partners – to deliver engaging 'experiences' that go beyond books for today's digital natives. Whatever happens, you will need to expect your audience to 'talk back' in very direct ways through online and digital networks, and these direct relationships with your readers will become more and more vital in terms of connecting people with your writing.

*Sara Lloyd, Digital Media Director for Pan Macmillan, is responsible for developing the company's digital strategy and programme including ebooks, audio and web development. Sara's career over the last 16 years has spanned newspaper, academic, reference, STM and trade publishing and she has played a key role in transforming many publishing businesses from print to digital.*

# Non-fiction: writing what you know

*Peggy Vance*

In the back-of-beyond world that is illustrated non-fiction, such as guides on sewing, interior design and parenting, being an author or a writer means something quite different from what it means in the world of fiction.

I'll explain. What I do is a bit strange in that I call myself a publisher, but I've also written a lot of books – so sometimes I say I'm an author, depending on who I'm talking to. When I first went into publishing (the day after I'd left university, because in those days you *could actually get a job*) my boss-to-be said, 'Since you went to Cambridge you can be our editor. We need you to produce about thirty books a year. Just start now – there's your desk.' In those days we didn't have computers, just dictating machines, so what I did was just blindly panic and then phone people who weren't authors, they weren't writers; they were experts in their field. That's the big difference: people who write non-fiction actually do something else. So I grabbed the phone and called gardeners and people who were crazy about cats and mystics and astronauts and museum curators (who had clearly never spoken to anyone before) – anyone who might just be able to write what we needed.

## The style

Being an editor is often a route to writing non-fiction because when people don't deliver, you find yourself saying, 'I'll do it, I'll finish the book!' – and you're away. Or else the book gets delivered but the text is so dire you end up rewriting it almost in its entirety. You just have to; that's your job. Writer's block simply isn't an option.

You don't have to be a genius to write non-fiction. What you need – and I didn't know this at the time, which is why 'Is English Peggy's first language?' was a reader's report comment on my first book – is to be able to write as clearly as safety instructions on a plane. If you start writing in a flowery, complex, very inflected way, inevitably the publisher is going to say 'What the HELL. This is UNUSABLE NONSENSE.'

Of course you can use figures of speech, and certainly you can demonstrate your prowess as a writer, but first and foremost you need to communicate information rather than your stylistic brio. That said, any non-fiction writer has to be memorable and illuminating so that the reader will think, 'Oh, *that's* an interesting take on it,' and feels rewarded for the extra effort involved in reading information rather than stories.

## The money

So my first job was really about looking for people who knew enough about the subject and who could write clearly and cheaply. Back in the early eighties the flat-fee offered was generally £100 per 1,000 words – a rate that is (cheekily) sometimes still offered today. The sad truth is that a lot of people know enough – or can find out enough – about a subject to write a little general book for peanuts, and certainly for no royalty. So there's not a lot to be made from writing to commission, fun as it is to churn out a little book and have your name on it. I should know!

In my next editorial job I continued my exciting approach to making a living, one in which I worked all day and wrote all night, banging out a book on mosaics. I didn't *want* to write a book on mosaics; I *needed* to write a book on mosaics. So I wrote it for a derisory flat fee, and, guess what, it was a bestseller. Every other person in the whole world apart from me wanted to do mosaics and we sold a gazillion copies for which I earned zilch. Watch out for that prat trap!

## The commission

There are two key ways in which non-fiction writers get commissioned. One, you're the only person who knows about hanging baskets or whatever and you go to a publishing company and say, 'I've got this great idea and I want to sell it to you. But I *could* take it to your competitor …' In which case obviously you need a really good synopsis with a proposal that is entirely credible, as you do. The other way is that you say, 'I am a writer for hire and can produce text on this, that and the other', so that you become known for delivering very reliable text in certain areas or fields.

To have better prospects as a non-fiction writer you really have to have some sort of specialism, ideally one not shared by loads of other people, so that you can say, 'Well, actually, I'm the only person who really knows about hanging baskets' (or whatever your personal obsession is). You need something special about you that gives you the leverage to say, 'I want more money' or even, 'I want more money – as an advance and royalty.'

However, you can also be the lowest of the low, like me, and write on almost everything. I've written on Buddhism for sheep, Feng Shui for dogs, art, design, interiors and parenting (to my kids' disgust). I know it sounds cynical to take any old commission on for money, but most people who do this really love the process of researching and writing and usually end up knowing a fair bit.

And then there are celebrities. Star authors are little different from any other stars: expensive and money-spinning. They get the best agents, the best publishing houses, the best royalties and the best sales. Sexy chefs, TV personalities, world-famous experts and the like are generally where the real bucks are to be made for everyone in the non-fiction food chain. As a non-fiction prole you'll probably never see anyone actually reading your book, whereas if you're Jamie Oliver or Richard Hamilton, Katie Price or Stephen Hawking, you just might.

## The business

But whoever you are and whatever you write, it pays to understand something of the non-fiction publishing process. Even huge celebrities can't just write whatever they like. Illustrated non-fiction generally requires the writer to work closely with the editors and designers to plan the contents of the book, agree a word count and create a picture list – at the very least. Understanding the constraints of a non-fiction book is critical for success. If your book is to be, say, 192 pages and you've written loads more text than was commissioned, the publisher will think you've wasted her time and not understood the brief. The book might even be flat-planned (thumbnailed) before you get the commission, in which case you have to write to a predetermined structure, usually based on sales requirements. And for illustrated non-fiction, it pays to think visually about how the pictures supplement the text, so that they are vital to the import of the book and are not just ornamental window-dressing. These days non-fiction writers are expected to submit dead-letter perfect text and and – ideally – to be able to receive designed spreads digitally. Being computer literate and properly set up is a distinct advantage.

Almost any non-fiction writer might be asked to be a contributor to a book, rather than a sole author. Some can be a bit uppity about it: 'Who are the other contributors and are they any good and what is their background, etc.?' Publishers are very wary of having a whole ensemble of prima ballerinas. The best contributors are those who love the project for its own sake, aren't freaked out by the idea that their text may sit alongside that of others (sometimes indistinguishable and unattributed) and who are happy to work collegiately with their peers, without obsessing about copyrighting and crediting every scrap of text.

The copyright symbol – it's the bane of my life. Every rubbish proposal ever submitted to me has had 'copyright ME!' stamped all over it. Copyright is automatic; it's innate; it's like breathing. You don't need to stamp copyright all over something, and if you do you just look an amateur. The publisher is invited to think, 'Who'd bother to steal this!'

## The way forward

Networking is vital for non-fiction writers because they're not in a position to say at a party 'I'm a novelist', a statement that can result in a publishing marriage. But they can say, 'I'm really, really interested in the history of the toilet' and, although it may not result in marriage, it could just result in a commission. You have to network that much harder as a non-fiction writer because you may not be thought of as an author, but just somebody who knows something about something. So you've really got to be very pushy and make sure that the very few publishing houses that are likely to be interested in your specialism of Victorian toilets or whatever it is are actually aware of you.

Novelists stick with one publisher at a time. Non-fiction writers don't have to; in fact, it's often better to have a range of publisher clients so that if, for whatever reason, commissions dry up with one, others can pick up the slack. Fiction editors might take their prestige clients with them when they move houses, but non-fiction editors rarely do because the writers, though valuable, are rarely hot properties in the same way.

Lastly, my golden tip, do not hang out with other authors unless it's just for a good time. Spend any spare schmooze money on agents and publishers because they're the ones who can make or break you.

*Peggy Vance is a publisher and the author of a wide range of non-fiction titles, including* The Mosaic Book, Loft Living *and* Gauguin: The Masterworks. *She has made numerous media appearances in the UK and was a guest speaker at* The Writer's Handbook *Event in 2009.*

---

 *The Writer's Handbook recommends:*
*Short Circuit. A Guide to the Art of the Short Story*

After years of neglect, the short story is back in fashion. Online exposure has a lot to do with this. If it is uneconomic to put a standalone short story between hard covers, and a mixed batch is hard to promote when everything in the bookshops is categorized by subject, the Net is ideal for showcasing literary endeavour of whatever length. So it is that the short story has been rediscovered. The question now is whether the impetus is strong enough to sustain a creative new wave. The contributors to this book – all noted practitioners of the art – clearly believe so. That many of them are also teachers helps when it comes to offering practical advice as opposed to the usual philosophical waffle about the unknowable mystery of readable composition.

Each chapter tackles a particular aspect of the short story – characterization, narrative, setting, style, and so on – with ideas for making real what is bouncing about in the imagination. If there is one lesson that springs out from the pages it is that the essence of the short story is brevity. Obvious, you might say. And so it should be. But it remains true that many short stories fail not for the lack of a compelling theme or characters but simply because there is too much verbal padding. The successful author knows, above all, when to shut up.

Short Circuit. A Guide to the Art of the Short Story *is edited by Vanessa Gebbie and published by Salt Publishing (ISBN 9781844717248)*

---

# Confessions of a friendly ghost

*Barry Turner*

To be a successful literary ghost you need to be something of an actor. The big name on the title page may not be up to putting together a life story, or any story for that matter, that can survive public scrutiny – but the ghost will also fail unless the words that purport to come from the big name are convincingly authentic. In other words, the ghost has to assume a false identity for as long as it takes to produce a sellable book.

This can be quite a challenge, believe me. In a long life of earning a meal ticket by literary endeavour I have ghosted half a dozen biographies. They have ranged from Richard Burton as seen by his younger brother, Graham Jenkins, to Elaine Blond, the last daughter of the Marks of Marks and Spencer. Getting inside the personality can take some time. Graham, for example, was understandably over-awed by his famous sibling. When he first spoke for the tape recorder, what had been billed to me by the publisher as hilarious anecdotes of life in the fast lane came over as soft centred reminiscences, barely worthy of reproduction. It was only when I had heard the same stories a dozen times that they began to take on an essential edge that had been missing in earlier versions. By then I was able to give substance to a contradictory character of humble beginnings that inspired his career while at the same time leaving him with overwhelming feelings of self-doubt. For whatever reason, the book was a smash hit with four successive weeks' serialization in the *Sunday Times*.

Elaine Blond could not have been a greater contrast. Hugely wealthy from an early age, she had devoted her life to good works, starting with the Kindertransporte, the 30,000 Jewish children saved from their Nazi oppressors by adoption into British families. Along the way she was a leading backer of advanced plastic surgery for wartime fighter and bomber crews who had suffered debilitating burns. In the 1960s she and her husband Neville Blond were the business brains of the theatrical revolution at the Royal Court.

Well into her eighties when I first met her, Elaine was one of the liveliest, gutsiest characters I have ever encountered. She was famous for her waspish repartee. John Osborne described her as the rudest person he knew outside the staff of British Rail. But in his autobiography where he has other harsh things to say about the Blonds, he neglects to mention that it was Elaine who kept him going with a weekly handout while he was writing *Look Back in Anger*. I was glad to put the record straight.

Unsurprisingly, Elaine Blond sold modestly compared to Burton but the two books had one thing in common – they were hammered by the critics, not because they were badly written or without interest but because a hidden hand was at work in their creation. Calling for a Campaign for Real Books, the biographer and novelist Tim Heald held that 'a real book should be written by the person whose name appears on the cover. It should not be "told to" someone else. It should not be written by the editor, much less by some shadowy team of professional rewrite men'.

Then, as now, I failed to understand the problem. In other pursuits, the name on the invoice rarely acts alone. The artist has his apprentices, the surgeon his nurses, the scientist his research assistant, the lawyer his clerks. What is so particular about a book that its value should be enhanced by the knowledge that its author works in isolation? This is perilously close to arguing that literature is all the better for having been written in a garret.

Then there is this curious assumption that ghosting is new. True, the latest generation of publishing entrepreneurs have exploited ghosting but they did not invent it. Ghosting goes back at least as far as the Bible (how many unnamed scribes were engaged on that masterpiece?). Dr Johnson had 'six amanuenses' to help him write his dictionary, the forerunner of thousands of ghost-created reference works.

It was in the late nineteenth century that ghosting really took off with the boom in self-justifying political and military memoirs. Whether any of those now long-forgotten tomes would have ever surfaced if co-writers had not been involved is open question. The likelihood is that both the nominal authors and their ghosts were destined by the conventions to produce uninspiring literary monuments. Today, most top historians use researchers-cum-writers who rarely get the credit they deserve. Among novelists, where would Jeffrey Archer be without his rewrite team? At what point does an editor become a ghost?

It is arguable that the ghosted biography of a living character can get as close to the truth as autobiography or straight biography. The autobiographer has to live with his prejudices and illusions which means that they are recycled as convincing rationalities. Given adequate sources, the biographer can see through these stratagems but he is invariably too far removed from his subject to be objective. The ghost, on the other hand, can play the interviewer, persuading the subject to say what he might otherwise have left unsaid.

Of course, this is not an all-embracing claim. A ghosted biography can be embarrassingly awful. One turns to the life stories of those titans of the sports field who need assistance to articulate monosyllabically. But this is not to say that a ghosted book is necessarily inferior to any other literary enterprise. Indeed, it may be a sight better.

It is fair to say that the life of a biographical ghost is not always undiluted joy. A wife or husband of a big name can be a problem. Having spent their lives in the shadows when it comes to the point of putting it all down in writing, they have a natural desire to claim their share of the credit. A ghost with another ghost leaning over his shoulder do not make for a good writing team.

For those contemplating an ethereal writing career, there are other warning signals. Sharing the credit may cause heartache. At the very least, the name of the ghost should appear on the copyright line. Thereafter it is all open to negotiation. Frequently nowadays the phrase 'written with' allows the ghost to appear on the title page, though this can turn out to be a double-edged compliment if the ghost, who almost certainly has written every word, is seen as a mere collaborator.

Watch the money. Even though it is the subject, not the ghost, who fronts up on the marketing, anything less than a fifty–fifty split on the proceeds is rarely acceptable. The exceptions are those infrequent books with multi-million sales in prospect when, inevitably, the ghost is in a weak bargaining position.

Checking out the small print for the disposal of subsidiary rights is axiomatic for any contract but in the ghosting context extra vigilance is needed. It would be a foolish ghost who, say, passed over a share of the revenue from newspaper serialization or from audio adaptation. But the top priority is to be clear from the start how far words written by the ghost can be censored. Seeing a life story in print can be an unnerving experience even for the most willing victim, not to mention their partners in love or business. It may be asking too much for the contractual terms of engagement to preclude all amendments but, at the very least, a high degree of mutual trust must be established before work commences.

This can be tested by checking out attitudes to the more obviously sensitive issues. I was once approached by a publisher to ghost a biography of a leading and colourful businessman. In the preliminary talks it emerged that my subject's father who had, to put it mildly, led a controversial life, had died in a five-storey fall from his office window. Did he jump or was he pushed? Neither, said my entrepreneur. It was simply a terrible accident. I checked out the window in question. The sill was at least three feet above the floor. I turned down the job. It did not surprise me that the book was never written.

My last appearance as a literary apparition was in the company of Denholm Elliott or rather, of his widow, Susan. I had long been a fan of Denholm. He was known chiefly as a highly bankable character actor on screen and stage. But the more I talked to Susan, the more depressed I became. It seemed to me that so much of the fun was superficial, a thin disguise for tragedy. Denholm's life had been a mess (it turned out that he had died of Aids), his children were a mess (his daughter took her own life), Susan was a mess (she eventually died in a house fire). When any one of them had to make a decision you could bet your life it was the wrong one. For the first time, I found myself backing away from the real story. Being Susan Elliott was just too painful.

It did not take me long after that to decide that my ghosting days were over. I had lost the enthusiasm for acting a role. Henceforward, as a writer I would be myself. Not a great part, you might say. But it is the one I know – and like – best.

# What makes a bestseller?

*Pauline Chapman*

What makes a bestseller? That is, often literally, the million dollar question and is one which has long taxed the minds of writers, agents and publishers. There are, of course, many who claim to have the answer and who are willing, either for financial gain or out of the goodness of their hearts, to impart it to us. Numerous articles have been written on the subject and there are books and software packages galore. It would appear that the answer is out there for the taking. Fantastic! Now we can all write a bestseller. No more rejection letters, hurray!

If only that were true! Call me an old sceptic but, try as I might, I just can't seem to find the names of the enlightened ones, who want me to read their articles or buy their books or software, on any of the bestseller lists. If they know what makes a bestseller then why are they not using that knowledge to write one, sell it and watch the royalties roll in? Could it be that they don't have the answer after all; that there is no easy answer?

It is true that some phenomenally successful authors write to a formula but, try as you might, and many do, using the same formula will not guarantee a place on the bestseller lists. Let's face it: there is no magic formula, no secret of success. What there is, however, for all aspiring authors, is ream upon ream of good advice. Some people take it and some don't. That's life!

Rather than ask what makes a bestseller we should, perhaps, ask what makes a bestseller different. The clue is in the question. A bestseller by definition is a book which sells more copies than other books. More people want to buy it than any of the other books around. That difference, although an obvious one, is the single, most important point to bear in mind if you want your book to sell. People must want to buy it; not just a few people but lots of people. Your writing must have broad appeal if it's going to appear on any bestseller list. A book about the mating habits of the black-backed gull may be well-written and authoritative but will only appeal to a limited market.

How, then, can we tell whether or not a book has what it takes; the X factor, the *sine qua non* of a bestseller? What we can do is look for those qualities which are common to most bestselling novels. They are qualities which, I would guess, we all look for in a book and are the qualities which make us buy a book, read it and, thereafter, spread the word.

## Plot

Plot or, in other words, a cracking good story is key. Reading a book is like taking a journey from the beginning to the end. Going straight there would be dull. The reader wants the journey to be interesting, possibly exciting, horrifying or amusing and, above all, pleasurable. There should be unexpected twists and turns with secrets being disclosed teasingly en route or saved for a sudden revelation in the final pages. There should be unanswered questions throughout the text which keep the reader turning the pages to find the answers.

## Characterization

Regardless of how many characters appear in works of popular fiction they have one thing in common – believability. One-dimensional characters who do not come to life on the page do not a bestseller make. To engage fully with a piece of fiction it must be peopled with characters the reader can come to know and either love, hate, laugh at or empathize with and, finally, be sorry to leave at the end of the story. This is the quality which makes readers say 'that was *really* good' and which, crucially, makes them recommend the book to their friends.

## Accessibility

A book which is to be read and enjoyed by a large number of people must be written in language which they can readily understand. The use of obscure vocabulary or an overly formal style will alienate a proportion of your prospective market. In other words don't try to be too clever.

## Ease of reading

It is essential that the book be pleasurable to read. Work in a style which comes easily to you and avoid trying to copy from another source. Play to your own strengths; attempting to adopt someone else's style is a recipe for disaster. The finished work will not have a natural feel and is likely to be as much of a chore to read as it was to write. Oh, and *don't* try to be funny if you're not.

## Description and subtlety

Describe events and emotions rather than stating bald facts. If Mary is nervous don't state the fact, describe how her hands are trembling and her breathing becomes shallow. *Show* readers what is happening rather than telling them, and avoid stating the obvious.

## Cliché

Don't use it. 'Evil stepmothers', 'boy wizards' ... if you've heard it before then so has everybody else.

## Title

In the early stages an eye-catching, imaginative title can assist in attracting the initial interest of an agent or publisher. On publication the title is an important factor in making that all-important browser pick up your book rather than another.

Even with all of the above in place your book needs to have that *je ne sais quoi*, that special little spark, which makes people rush out (or sit at their computers) to buy it. Some call it luck, some call it fate, some call it having the right book on the market at the right time, some put it down to talent or plain hard graft. Me, I think that it's a little bit of all of those together with the four Ps: practice, perseverance, passion and patience. There are no secret formulae and no guarantees; we just have to keep writing in the hope that one day we'll discover what makes our very own bestseller!

*This article is the winning entry in* The Writer's Handbook *essay competition.*

*Although a lawyer by profession, currently Pauline Chapman's time is given up to the chaos of life with a husband and three teenage children, to revising her manuscript, entitled* Kaleidoscope Eyes, *a book for children and young teenagers, and to reading contemporary fiction. Having felt compelled to write since childhood, she hopes, one day to find that she too has written that elusive bestseller.*

---

 *The Writer's Handbook recommends:*
*Talking About Detective Fiction*

Among the many who aspire to be novelists, a high proportion are drawn to crime fiction. The combination of mystery, detection (the literary equivalent of the crossword puzzle) and social realism, which is currently all the rage, gives the writer a scope and an undeniable reader appeal that is denied to most other mainstream authors.

As the doyenne of the contemporary detective story, P.D. James is eminently qualified to conduct a masterclass on her craft. Inevitably, given her own predilections, there is much here on the Golden Age of detective fiction with the likes of Dorothy L. Sayers, Ngaio Marsh, Agatha Christie and Marjory Allingham well to the fore, but the creator of poet policeman Adam Dalgleish lines up with the modernists in suggesting that setting and character are more important than plot, however cleverly contrived.

This is emphatically *not* a book on how to knock out a successful crime novel. But those who are ambitious to emulate P.D. James will find in her motivation and methods a potent stimulant to getting on with the job.

Talking About Detective Fiction *by P.D. James is published by Bodleian Library (ISBN 9781851243099)*

---

# Settling accounts

*Ian Spring takes an expert look at the latest Budget and explains how writers can be tax-wise*

When he was Chancellor of the Exchequer, Norman Lamont said tax should be:

• Simple and certain
• Fair and reasonable
• Easy to collect

Unfortunately, over the years, tax law has become ever more complicated and more and more difficult for the 'ordinary' person to understand. Until recently it was inconceivable that tax changes would be passed which had retrospective effect, but this is no longer the case and is happening. Her Majesty's Revenue and Customs (HMRC) now have a new range of powers to enforce compliance with the tax legislation. Penalties, which are a percentage of any additional tax due, are imposed if errors or mistakes are found and are in addition to the extra tax and interest for late payment. The penalty percentage takes into account the severity of the error, ranging from 'careless' to 'deliberate' and the amount of help and cooperation the taxpayer has given in rectifying matters. There is no doubt that HMRC, in their desperation to try to increase the tax take, have become far more aggressive in collecting tax and imposing penalties.

A taxpayer cannot be certain that a tax planning action legally taken now will still be legal in years to come. Tax and its ramifications, such as tax credits, are increasingly viewed by the general public as being unfair and unreasonable and, whilst tax collection is regarded by the Revenue as being easier, this is only because taxpayers are now responsible for paying rather than the Revenue being responsible for collecting.

It is against this background that authors have to comply with the law and complete and submit a tax return each year.

Please note that some of the details included in this article which relate to the 2010/11 tax year, come from the 2010 Budget announced on 24 March 2010. At the time of writing, not all the provisions of that Budget had been enacted. Because of the General Election on 6 May 2010 there may well be a second 2010 Budget and resulting Finance Act which may lead to some of the information being superseded.

## Income tax

### What is a professional writer for tax purposes?

Writers are professionals while they are writing regularly with the intention of making a profit; or while they are gathering material, researching or otherwise preparing a publication.

A professional freelance writer is taxed under section 5 Income Tax (Trading and Other Income) Act 2005. The taxable income is the amount receivable, either directly or by an agent on his behalf, less expenses wholly and exclusively laid out for the purpose of the profession. If expenses exceed income, the loss can either be set against other income of the same or the preceding year or carried forward and set against future income from writing. If tax has been paid on that other income, a repayment can be obtained, or the sum can be offset against other tax liabilities. Special loss relief can apply in the opening years of the profession. Losses made in the first four years can be set against income of up to three earlier years.

Where a writer receives very occasional payments for isolated articles, it may not be possible to establish that these are profits arising from carrying on a continuing profession. In such circumstances these 'isolated transactions' may be assessed under section 687 Income Tax (Trading and Other Income) Act 2005. Again, expenses may be deducted in arriving at the taxable income but, if expenses exceed income, the loss can only be set against the profits from future isolated transactions or other income assessable under section 687.

Authors are taxed under a system called 'Self Assessment', where the onus is on the individual to declare income and expenses correctly. Each writer therefore has to decide whether profits arise from a professional or occasional activity. As already mentioned, the consequences of getting it wrong can be expensive by way of interest, penalties and surcharges on additional tax subsequently found to be due. If in any doubt the writer should seek professional advice.

## Income

A writer's income includes fees, advances, royalties, commissions, sale of copyrights, reimbursed expenses, etc. from any source anywhere in the world, whether or not brought to the UK (non-UK resident or domiciled writers should seek professional advice).

## Agents

It should be borne in mind that the agent stands in the shoes of the principal. It is not always realized that when the agent receives royalties, fees, advances, etc. on behalf of the author those receipts become the property of the author on the date of their receipt by the agent. This applies for income tax and value added tax purposes.

## Expenses

A writer can normally claim the following expenses:

(a) Secretarial, typing, proofreading, research. Where payment for these is made to the author's wife or husband they should be recorded and entered in the spouse's tax return as earned income which is subject to the usual personal allowances. If payments reach relevant levels, PAYE should be operated.

(b) Telephone, faxes, internet costs, computer software, postage, stationery, printing, equipment maintenance, insurance, dictation tapes, batteries, any equipment or office requisites used for the profession.

(c) Periodicals, books (including presentation copies and reference books) and other publications necessary for the profession; however, amounts received from the sale of books should be deducted.

(d) Hotels, fares, car running expenses (including repairs, petrol, oil, garaging, parking, cleaning, insurance, road fund tax, depreciation), hire of cars or taxis in connection with:

    (i) Business discussions with agents, publishers, co-authors, collaborators, researchers, illustrators, etc.

    (ii) Travel at home and abroad to collect background material.

    As an alternative to keeping details of full car running costs, a mileage rate can be claimed for business use. This rate depends on the engine size and varies from year to year. This is known as the Fixed Profit Car Scheme and is available to writers whose turnover does not exceed the VAT registration limit.

(e) Publishing and advertising expenses, including costs of proof corrections, indexing, photographs, etc.

(f) Subscriptions to societies and associations, press cutting agencies, libraries, etc. incurred wholly for the purpose of the profession.

(g) Rent, council tax, water rates, etc., the proportion being determined by the ratio of the number of rooms, used exclusively for the profession, to the total number of rooms in the residence. But see note on capital gains tax below.

(h) Lighting, heating, cleaning. A carefully calculated figure of the business use of these costs can be claimed as a proportion of the total.

(i) Agent's commission, accountancy charges and legal charges incurred wholly in the course of the profession including the cost of defending libel actions, damages insofar as they are not covered by insurance, and libel insurance premiums. However, where in a libel case damages are awarded to punish the author for having acted maliciously, the action becomes quasi-criminal and costs and damages may not be allowed.

(j) TV and video rental (which may be apportioned for private use), and cinema or theatre tickets, if wholly for the purpose of the profession.

(k) Capital allowances are a means of giving businesses tax relief for money spent on equipment, furniture, machinery, etc. There is one set of rules for the purchase of cars and another for all other items. Since 6 April 2009, capital allowances on cars has depended on the car's emissions level.

    (i) Cars with very low carbon dioxide emissions (up to 110g/km) qualify for a 100 per cent first year allowance, i.e. the cost is written off in the year of purchase.

    (ii) Cars with carbon dioxide emissions between 110g/km and 160g/km are treated like all other equipment and attract a 20 per cent allowance, i.e. 20 per cent of the original cost is written off in the first year and then 20 per cent of the written down value in subsequent years.

    (iii) Cars with emission levels over 160g/km have only a 10 per cent allowance.

    (iv) For all other business equipment, e.g. computers, televisions, radios, hi-fi sets, tape and video recorders, dictaphones, office furniture, photographic equipment, etc., all businesses, including authors, are able to claim the Annual Investment Allowance (AIA). The AIA allows relief at 100 per cent on the first £100,000 spent on business equipment each year. This should be more than enough to meet the needs of most authors. If expenditure in a year does exceed £100,000, an allowance of 20 per cent is given against the balance.

All allowances for both cars and equipment will be reduced to exclude personal (non-professional) use where necessary.

(l)   (Lease rent. The cost of lease rent of equipment is allowable; as it also is on cars, subject to restrictions for private use, and for cars with high emission levels.

(m)  Other expenses incurred wholly and exclusively for professional purposes. (Entertaining expenses are not allowable in any circumstances.)

Note: It is essential to keep detailed records. Diary entries of appointments, notes of fares and receipted bills are much more convincing to HM Revenue & Customs who are very reluctant to accept estimates.

**The Self Assessment regime makes it a legal requirement for proper accounting records to be kept. These records must be sufficient to support the figures declared in the tax return.**

In addition to the above, tax relief is available on:

(a)   Premiums to a pension scheme. On 6 April 2006 a new pension-scheme tax regime came into effect which replaced all previous rules for occupational, personal pension and retirement annuity schemes. An individual can make, and tax relief is available on, contributions up to the higher of the full amount of relevant income or £3,600, subject to a maximum annual limit. That limit for 2009/10 was £245,000 and for 2010/11 is £255,000.

(b)   Gift Aid payments to charities – for any amount.

## Capital gains tax

The exemption from capital gains tax which applies to an individual's main residence does not apply to any part of that residence which is used exclusively for business purposes. The appropriate proportion of any increase in value of the residence, since 31 March 1982, can be taxed when the residence is sold, at the capital gains tax rate of 18 per cent.

Writers who own their houses should bear this in mind before claiming expenses for the use of a room for writing purposes. Arguments in favour of making such claims are that they afford some relief now, while capital gains tax in its present form may not stay for ever. Also, where a new house is bought in place of an old one, the gain made on the sale of the first study may be set off against the cost of the study in the new house, thus postponing the tax payment until the final sale. For this relief to apply, each house must have a study and the author must continue his profession throughout. On death there is an exemption of the total capital gains of the estate.

Alternatively, writers can claim that their use is non-exclusive and restrict their claim to the cost of extra lighting, heating and cleaning to avoid any capital gains tax liability.

## Can a writer average out his or her income over a number of years for tax purposes?

Writers are able to average their profits (made wholly or mainly from creative works) over two or more consecutive years. If the profits of the lower year are less than 70 per cent of the profits of the higher year, or the profits of one year (but not both) are nil, then the author will be able to claim to have the profits averaged. Where the profits of the lower year are more than 70 per cent but less than 75 per cent of the profits of the higher year, a pro rata adjustment is made to both years to reduce the difference between them.

It is also possible to average out income within the terms of publishers' contracts, but professional advice should be taken before signature. Where a husband and wife collaborate as writers, advice should be taken as to whether a formal partnership agreement should be made or whether the publishing agreement should be in joint names.

## Is a lump sum paid for an outright sale of the copyright, or part of it, exempt from tax?

No. All the money received from the marketing of literary work, by whatever means, is taxable. Some writers, in spite of clear judicial decisions to the contrary, still seem to think that an outright sale of, for instance, the film rights in a book is not subject to tax. The averaging relief described above should be considered.

### Remaindering

To avoid remaindering, authors can usually purchase copies of their own books from the publishers. Monies received from sales are subject to income tax but the cost of books sold should be deducted because tax is only payable on the profit made.

### Is there any relief where old copyrights are sold?

No. There was relief available until April 2001 but it was then withdrawn. The averaging relief described above should be considered.

## Are royalties payable on the publication of a book abroad subject to both foreign tax as well as UK tax?

Where there is a Double Taxation Agreement between the country concerned and the UK, then on the completion of certain formalities no tax is deductible at source by the foreign payer, but such income is taxable in the UK in the ordinary way. Where there is no Double Taxation Agreement, credit will be given against UK tax for overseas tax paid. A complete list of countries with which the UK has conventions for the avoidance of double taxation may be obtained from the Centre for Non-Residents, HM Revenue & Customs, St John's House, Merton Road, Bootle, Merseyside L69 9BB, or a local tax office.

### Residence abroad

Writers residing abroad will, of course, be subject to the tax laws ruling in their country of residence, and as a general rule royalty income paid from the UK can be exempted from deduction of UK tax at source, providing the author is carrying on his profession abroad. A writer who is intending to go and live abroad should make early application for future royalties to be paid without deduction of tax to the Centre for Non-Residents (address as above). In certain circumstances writers resident in the Irish Republic are exempt from Irish income tax on their authorship earnings.

### Are grants or prizes taxable?

The law is uncertain. Some Arts Council grants are now deemed to be taxable, whereas most prizes and awards are not, though it depends on the conditions in each case. When submitting the Self Assessment annual returns, such items should be excluded but reference made to them in the 'Additional Information' box on the self-employment (or partnership) pages.

### What is the item 'Class 4 N.I.C.' which appears on my Self Assessment return?

All taxpayers who are self-employed pay an additional national insurance contribution if their earned income exceeds a figure which varies each year. This contribution is described as Class 4 and is calculated when preparing the return. It is additional to the self-employed Class 2 contribution but confers no additional benefits and is a form of levy. It applies to men and women below the state retirement age.

### Should an author use a limited company?

The tax regime has made the incorporation of businesses attractive and there can be advantages in so doing even for businesses with relatively low levels of profit, although these advantages have been reduced in recent Budgets. However, there are disadvantages in all cases and there are particular considerations for authors. The advice of an accountant, knowledgeable about the affairs of authors, should be sought if incorporation is contemplated.

## Value added tax

Value added tax (VAT) is a tax levied on:

(a) The total value of taxable goods and services supplied to consumers
(b) The importation of goods into the UK
(c) Certain services or goods from abroad if a taxable person receives them in the UK for business purposes.

The normal rate of tax is 17.5 per cent but for a period from 1 December 2008 to 31 December 2009 this was reduced to a rate of 15 per cent. The 17.5 per cent applied again from 1 January 2010.

### Who is taxable?

A writer resident in the UK whose turnover from writing and any other business, craft or art on a self-employed basis in the year 2010/11 is greater than

£70,000 annually, before deducting an agent's commission, must register with HM Revenue & Customs as a taxable person. Turnover includes fees, royalties, advances, commissions, sale of copyright, reimbursed expenses, etc.

A business is required to register:

- At the end of any month if the value of taxable supplies in the past 12 months has exceeded the annual threshold; or
- If there are reasonable grounds for believing that the value of taxable supplies in the next 12 months will exceed the annual threshold.

Penalties will be claimed in the case of late registration. A writer whose turnover is below these limits is exempt from the requirements to register for VAT but may apply for voluntary registration, which will be allowed at the discretion of HM Revenue & Customs.

A taxable person collects VAT on outputs (turnover) and deducts VAT paid on inputs (taxable expenses) and, where VAT collected exceeds VAT paid, must remit the difference to HM Revenue & Customs. In the event that input exceeds output, the difference will be refunded by HM Revenue & Customs.

## Inputs (expenses)

| *Taxable at the standard rate if supplier is registered* | *Taxable at the zero or special rate* | *Not liable to VAT* |
|---|---|---|
| Rent of certain commercial premises | Books (zero) | Rent of non-commercial premises |
| Advertisements in newspapers, magazines, journals and periodicals | Coach, rail and air travel (zero) | Postage |
| | Agent's commission (on monies from overseas) | Services supplied by unregistered persons |
| Agent's commission (unless it relates to monies from overseas) | Domestic gas and electricity (5%) | Subscriptions to the Society of Authors, PEN, NUJ, etc. |
| Accountant's and solicitor's fees for business matters | Insurance | |
| Agency services (typing, copying, etc.) | | *Outside the scope of VAT* |
| Word processors, typewriters and stationery | | PLR (Public Lending Right) |
| Artists' materials | | Profit shares |
| Photographic equipment | | Investment income |
| Tape recorders and tapes | | |
| Hotel accommodation | | |
| Taxi fares | | |
| Motorcar expenses | | |
| Telephone | | |
| Theatres and concerts | | |

Note: This table is not exhaustive.

## Outputs (turnover)

A writer's outputs are taxable services supplied to publishers, broadcasting organizations, theatre managements, film companies, educational institutions, etc. A taxable writer must invoice all the persons (either individuals or organizations) in the UK to whom supplies have been made, for fees, royalties or other considerations, plus VAT. An unregistered writer cannot and must not invoice for VAT. A taxable writer is not obliged to collect VAT on royalties or other fees paid by publishers or others overseas. In practice, agents usually collect VAT for the registered author, but these authors must remember to do so if they receive income that does not go through their agent, e.g. payments for public readings.

## Remit to Customs

The taxable writer adds up the VAT which has been paid on taxable inputs, deducts it from the VAT received and remits the balance to Customs. Business with HM Customs is conducted through the local VAT offices of HM Revenue & Customs which are listed in local telephone directories, except for VAT returns which are sent direct to the HM Revenue & Customs VAT Central Unit, Alexander House, 21 Victoria Avenue, Southend on Sea, Essex SS99 1AA.

## Accounting

A taxable writer is obliged to account to HM Revenue & Customs at quarterly intervals. Returns must be completed and sent to VAT Central Unit by the dates shown on the return. Penalties can be charged if the returns are late.

It is possible to account for the VAT liability under the Cash Accounting Scheme (leaflet 731), whereby the author accounts for the output tax when the invoice is paid or royalties etc. are received. The same applies to the input tax, but as most purchases are probably on a 'cash basis', this will not make a considerable difference to the author's input tax. This scheme is only applicable to those with a taxable turnover of less than £1,350,000 and therefore is available to the majority of authors. The advantage of this scheme is that the author does not have to account for VAT before receiving payments, thereby relieving the author of a cash flow problem.

If turnover is less than or expected to be less than £1,350,000 it is possible to pay VAT by monthly instalments, with a final balance at the end of the year (see leaflet 732). This annual accounting method also means that only one VAT return is submitted.

## Flat rate scheme

Small businesses can elect to pay VAT under a flat rate scheme (FRS). This is open to businesses with a taxable turnover of up to £150,000 a year. Under the normal VAT accounting rules, each item of turnover and every claimed expense must be recorded and supported by evidence, e.g. invoices, receipts, etc. Under the FRS, detailed records of sales and purchases do not have to be kept. A record of gross income (including zero rated and excepted income) is maintained and a flat rate percentage is applied to the total. This percentage is then paid over to HM Revenue & Customs. The percentage varies from one profession or business to another, but for authors it is 11 per cent when the general VAT rate is 17.5 per cent and 9.5 per cent when the general VAT rate is 15 per cent. In the first year of registration this percent is reduced by 1.

The aim of the scheme is to reduce the amount of time and money spent in complying with VAT regulations, and this is to be welcomed. However, there are disadvantages:

• The detailed records of income and expenses are still going to be required for taxation purposes
• Invoices on sales are issued in the normal way
• The percentage is applied to all business income. So 11 per cent VAT will effectively be paid on income from abroad, zero rated under the normal basis, and PLR, otherwise exempt.

For many authors, normal VAT accounting has imposed a good, timely discipline for dealing with accounting and taxation matters.

## Registration

A writer will be given a VAT registration number which must be quoted on all VAT correspondence. It is the responsibility of those registered to inform those to whom they make supplies of their registration number. The taxable turnover limit, which determines whether a person who is registered for VAT may apply for cancellation of registration, is £68,000 in 2010/11..

## Voluntary registration

A writer whose turnover is below the limits may apply to register. If the writer is paying a relatively large amount of VAT on taxable inputs – agent's commission, accountant's fees, equipment, materials or agency services, etc. – it may make a significant improvement in the net income to be able to offset the VAT on these inputs. A writer who pays relatively little VAT may find it easier, and no more expensive, to remain unregistered.

## Fees and royalties

A taxable writer must notify those to whom he or she makes supplies of the VAT registration number at the first opportunity. One method of accounting for and paying VAT on fees and royalties is the use of multiple stationery for 'self-billing', one copy of the royalty statement being used by the author as the VAT invoice. A second method is for the recipient of taxable outputs to pay fees, including authors' royalties, without VAT. The taxable writer then renders a tax invoice for the VAT element, for which a second payment of this element will be made. This scheme is cumbersome but does involve only taxable authors. Fees and royalties from abroad will count as payments for exported services and will accordingly be zero-rated.

## Agents and accountants

A writer is responsible to HM Revenue & Customs for making VAT returns and payments. Neither an agent nor an accountant nor a solicitor can remove the responsibility, although they can be helpful in preparing and keeping VAT returns and accounts. Their professional fees or commission will, except in rare cases where the adviser or agent is unregistered, be taxable at the standard rate and will represent some of a writer's taxable inputs.

### Income tax

An unregistered writer can claim some of the VAT paid on taxable inputs as a business expense allowable against income tax. However, certain taxable inputs fall into categories which cannot be claimed under the income tax regulations. A taxable writer, who has already claimed VAT on inputs, cannot charge it as a business expense for the purposes of income tax.

### Certain services from abroad

A taxable author who resides in the UK and who receives certain services from abroad must account for VAT on those services at the appropriate tax rate on the sum paid for them. Examples of the type of services concerned include the services of lawyers, accountants, consultants and the provision of information and copyright permissions.

## Inheritance tax

Inheritance tax was introduced in 1984 to replace capital transfer tax, which had in turn replaced estate duty, the first of the death taxes of recent times. Paradoxically, inheritance tax has reintroduced a number of principles present under the old estate duty.

The general principle now is that all assets owned at death are chargeable to tax (currently 40 per cent) except for the first £325,000 (the nil rate band) of the estate and any assets passed to a surviving spouse, civil partner or a charity. The proportion of any unused nil rate band on a person's death can be transferred to the estate of their surviving spouse or civil partner, if he or she died since 9 October 2007. This means that wills, which contain provisions for the nil rate band to go to a discretionary trust (a common tax planning procedure in recent years), should be reviewed.

Gifts made more than seven years before death are exempt, but those made within this period may be taxed on a sliding scale. No tax is payable at the time of making the gift.

In addition, each individual may currently make gifts of up to £3,000 in any year, and these will be considered to be exempt. A further exemption covers any number of annual gifts not exceeding £250 to any one person.

If the £3,000 is not fully utilized in one year, any unused balance can be carried forward to the following year (but no later). Gifts out of income, which do not reduce one's living standards, are also exempt if they are part of normal expenditure.

At death, all assets are valued. They will include any property, investments, life policies, furniture and personal possessions, bank balances and, in the case of authors, the value of copyrights. All, with the sole exception of copyrights, are capable (as assets) of accurate valuation and, if necessary, can be turned into cash. The valuation of copyright is, of course, complicated and frequently gives rise to difficulty. Except where they are bequeathed to the owner's husband or wife, very real problems can be left behind by the author.

Experience has shown that a figure based on two to three years' past royalties may be proposed by HM Revenue & Customs in their valuation of copyright. However, this may not be reasonable and may require negotiation. If a book is running out of print or if, as in the case of educational books, it may need revision at the next reprint, these factors must be taken into account. In many cases the fact that the author is no longer alive and able to make personal appearances, or provide publicity, or write further works, will result in lower or slower sales. Obviously, this is an area in which help can be given by the publishers, and in particular one needs to know what their future intentions are, what stocks of the books remain, and what likelihood there will be of reprinting.

There is a further relief available to authors who have established that they have been carrying on a business for at least two years prior to death. It has been possible to establish that copyrights are treated as business property and in these circumstances inheritance tax 'business property relief ' is available. This relief at present is 100 per cent so that the tax saving can be quite substantial. HM Revenue & Customs may wish to be assured that the business is continuing and consideration should therefore be given to (i) the appointment, in the author's will, of a literary executor, who should be a qualified business person; or (ii), in certain circumstances, the formation of a partnership between the author and spouse, or other relative, to ensure that it is established that the business is continuing after the author's death.

If the author has sufficient income, consideration should be given to building up a fund to cover future inheritance tax liabilities. One of a number of ways to do this would be to take out a whole life assurance policy which is assigned to the children, or other beneficiaries, the premiums for which are within the annual exemption of £3,000. The capital sum payable on the death of the assured is exempt from inheritance tax.

## Tax credits

Working authors may also be eligible to claim Child Tax Credit and Working Tax Credit. Child tax credit is the main way that families receive money for their younger children and for 16–19-year-olds in education. A claim is based on your income and you can claim whether or not you are working. All families (with children) with an income up to £58,175 a year (or up to £66,350 a year if your child is under one) can claim. To qualify, you must be aged 16 or over, usually live in the UK and, if single (or separated), claim according to your individual circumstances. If you are married or a man and woman living together (or a same sex couple from December 2005) you must claim together based on joint circumstances and income.

Working Tax Credit is a payment to top up earnings of the lower paid (whether employed or self-employed, including those who do not have children). To qualify, you must be aged 16 or over, you or your partner must be responsible for a child, and you must work at least 16 hours a week. If you are part of a couple with children, greater amounts may be claimed if you jointly work at least 30 hours a week, provided one of you works for at least 16 hours. Childless couples cannot add their hours together to qualify for the 30-hour element. If you do not have children and you do not have a disability, you must either be aged 25 or over and work at least 30 hours a week or aged over 60 and work at least 16 hours a week. It is also possible that your working tax credit will be increased to assist with the cost of registered or approved childcare.

\*     \*     \*

Anyone wondering how best to order his affairs for tax purposes should consult an accountant with specialized knowledge in this field. Experience shows that a good accountant is well worth the fee which, incidentally, so far as it relates to professional matters, is an allowable expense.

*The information contained in this section has been prepared by Ian Spring of Ian Spring & Co, Chartered Accountants, who will be pleased to answer questions on tax problems. Please write to Ian Spring, c/o* The Writer's Handbook, *34 Ufton Road, London N1 5BX.*

# Glossary

**advance** money paid to author up front for a work.

**AN** advance notice: a document that is used to pass details of a book to bibliographic databases and booksellers in advance of publication. Also known as Title Information Sheet/Advance Title Information/Advance Information.

**backlist** titles that are more than three to six months old.

**bibliographic information** details of a title such as ISBN, price, size, etc.

**blurb** descriptive writing about a book to inform the reader of its content, such as 'jacket blurb'.

**copy** written text, particularly in the context of marketing/advertising.

**copy-editing** reading through text for style as well as grammar, spelling and sense to prepare a manuscript or typescript for typesetting.

**copyright** the exclusive right, belonging to the creator of an original literary, musical, dramatic or artistic work that has involved skill and labour to produce. It also covers sound recordings, films, broadcasts and the typographical arrangement of published works.

**copywriter** a person employed to write copy, such as advertising.

**cover** outside of a book; generally used in reference to the front, unless otherwise specified.

**dust jacket** protective cover for a hardback book.

**ebook** an electronic version of a book, such as can be viewed on a PC, ebook reader or smartphone.

**extent** the number of pages in a book.

**format** for example, hardback, paperback and increasingly ebook.

**frontlist** titles that have recently been published (within approximately the previous three months), or are soon due to be published.

**galleys** bound proof(s) used for publicity purposes, also known as dummies or proofs.

**handover** the point/process of passing a book from editorial to production.

**ISBN** International Standard Book Number: 13-digit (previously 10-digit) number unique to a book, recognized throughout publishing and bookselling.

**list** selection of titles looked after by a particular editor or publisher, usually arranged by subject, genre or audience.

**manuscript** a book in its early stages, when submitted to the editor.

**POD** Print on Demand: when a title is printed as orders come in, rather than holding stock.

**production** the stage prior to publication when the practical details of layout and jackets are finalized and proofreading is done, not just the physical printing of a work.

**proof(s)** later copies of a book (once it is in production) that have not yet been checked for errors. Sometimes bound for publicity use.

**proofreading** the process of checking the typeset pages to ensure that the style has been correctly applied and to check for typographic errors.

**publisher** used to refer to a particular person within publishing, usually senior editorial staff, as well as the company itself.

**recto** right-hand side of a page.

**remainder** a number of books left unsold when demand slows or ceases, sold by a publisher at a discount to remainder bookshops.

**rep** a person employed by a publisher to present new titles to bookshop buyers.

**royalties** agreed percentage of each sale of a book which goes to the author, accounted for in an annual or other agreed frequency royalty statement.

**spec** short for 'specification': technical details of a title, for example page size, length.

**sub** process of presenting title(s) to bookshops prior to publication, that is if a title has been 'subbed' it has been shown to booksellers.

**typescript** typed copy of a manuscript, now used to describe a printout of electronic files.

**typesetting** the process of taking a copy-edited typescript and applying specific type styles and design to produce a proof.

**verso** left-hand side of a page.

**work** a way of referring to a book, usually the concept of the book rather than the physical entity.

| Marginal mark | Instruction | Textual mark |
|---|---|---|
| ✓ | Leave unchanged | ............ under characters to remain |
| Follow matter by ⟨ or ⟨(x2) | Insert in text the matter indicated in the margin (adding an encircled number indicates the number of times the same insert is repeated in the same line without interruption) | ⟨ |
| ⁊ or ⁊(x2) | Delete (adding an encircled number indicates the number of times the same deletion is repeated in the same line without interruption) | / through single character or ⊢——⊣ through all characters |
| ⊙ | Substitute or insert a full stop | / through character or ⟨ where required |
| ⊢—⊣ | Substitute or insert rule (give size of rule as part of marginal mark) | / through character or ⟨ where required |
| ⌐ | Start new paragraph | ⌐ |
| ∽ | Run on (no new paragraph) | ∽ |
| ⌐‿⌐ | Transpose characters or words | ⌐‿⌐ between characters or words |
| ⌐ | Indent or move beginning of line(s) to the right | Textual mark matches marginal mark; vertical lines show position to which text should be moved |
| ⌐ | Cancel indent or move beginning of line(s) to the left | Textual mark matches marginal mark; vertical lines show position to which text should be moved |
| ◡ | Close up. Delete space between characters or words | Linking◡characters |
| ⊥ | Insert or substitute space between characters or words | / or ⟨ through characters where required |
| ⊤ | Reduce space between characters or words | \| between words or characters affected |
| ⊗ | Wrong font (replace by characters of correct font) | Circle character(s) to be changed |
| ≡ | Change to upper case | ≡ under character(s) to be changed |
| ≢ | Change to lower case | ≢ under character(s) to be changed |

# UK Publishers

## AA Publishing
The Automobile Association, Fanum House,
Basingstoke RG21 4EA
☎ 01256 491524   ℻ 01256 491974
AAPublish@TheAA.com
www.theAA.com
Publisher *David Watchus*

Publishes maps, atlases and guidebooks,
motoring, travel and leisure; also illustrated
reference. About 100 titles a year.

## A.&C. Black Publishers Ltd
36 Soho Square, London W1D 3QY
☎ 020 7758 0200   ℻ 020 7758 0222
enquiries@acblack.com
www.acblack.com
Managing Director *Jill Coleman*
Deputy Managing Director *Jonathan Glasspool*

Publishes children's and educational books,
including music, for 0–16 year-olds, visual
natural history, ornithology, nautical, reference,
business, dictionaries, sport, performing arts
and writing books. Acquisitions brought the
Herbert Press' art, design and general books,
Adlard Coles' and Thomas Reed's sailing lists,
plus *Reeds Nautical Almanac*, Christopher Helm,
Pica Press and T&AD Poyser's natural history and
ornithology lists, children's educational publisher
Andrew Brodie and Featherstone Education, and
Whitakers Almanack into A.&C. Black's stable.
Bought by **Bloomsbury Publishing** in May
2000. Acquired **Methuen**'s drama list in 2006.
IMPRINTS **Adlard Coles Nautical**; **Thomas Reed**;
**Christopher Helm**; **T&AD Poyser**; **The Herbert
Press**; **Andrew Brodie**; **Methuen Drama**; **The
Arden Shakespeare**; **Wisden**; **Featherstone
Education**. TITLES *Who's Who*; *Writers' & Artists'
Yearbook*; *Children's Writers' & Artists' Yearbook*;
*Whitaker's Almanack*; *New Mermaids*; *White
Wolves*; *Methuen Drama Modern Plays*; *Arden
Shakespeare*. Around 500 titles in 2009. Initial
enquiry appreciated before submission of mss.
£ Royalties vary according to contract.

## Abacus
▷ Little, Brown Book Group UK

## ABC–CLIO
▷ entry under US Publishers

## Absolute Press
Scarborough House, 29 James Street West, Bath
BA1 2BT
☎ 01225 316013   ℻ 01225 445836
office@absolutepress.co.uk
www.absolutepress.co.uk
Managing Director *Jon Croft*

Founded 1979. Publishes food and wine-related
subjects. About 15 titles a year. No unsolicited
mss. Synopses and ideas for books welcome.
£ Royalties twice-yearly.

## Abson Books London
5 Sidney Square, London E1 2EY
☎ 020 7790 4737   ℻ 020 7790 7346
absonbooks@aol.com
www.absonbooks.co.uk
Publisher *M.J. Ellison*
Manager *Sharon Wright*

Founded 1971 in Bristol. Publishes language and
dialect glossaries. 2 titles in 2008. No unsolicited
mss; synopses and ideas for books welcome.
£ Royalties twice-yearly.

## Academic Press
▷ Elsevier Ltd

## Acair Ltd
7 James Street, Stornoway, Isle of Lewis HS1 2QN
☎ 01851 703020   ℻ 01851 703294
info@acairbooks.com
www.acairbooks.com

Specializing in matters pertaining to the
Gaidhealtachd, Acair publishes books in Gaelic
and English on Scottish history, culture and the
Gaelic language. 75% of their children's books
are targeted at primary school usage and are
published exclusively in Gaelic.
£ Royalties twice-yearly.

## ACC Publishing Group
Sandy Lane, Old Martlesham, Woodbridge
IP12 4SD
☎ 01394 389950   ℻ 01394 389999

sales@antique-acc.com
www.antiquecollectorsclub.com
Managing Director *Diana Steel*

Publishes specialist books on art, antiques decorative arts, architecture, gardening and fashion. The Price Guide series was introduced in 1968 with the first edition of *Price Guide to Antique Furniture*. ACC Publishing Group incorporates three IMPRINTS: **Antique Collectors' Club**; **Garden Art Press** and **ACC Editions** and an international distribution company, ACC Distribution. TITLES *20th Century British Glass* Charles Hajdamach; *Early Georgian Furniture* Adam Bowett; *The Rose* David Austin; *My Favourite Dress* Gity Monsef, Samantha Safer and Robert de Niet; *Photographing Fashion* Richard Lester; *Foale and Tuffin* Iain R. Webb. Also publishes subscription-only magazine, *Antique Collecting*, published 10 times per year. 30–40 titles a year. Unsolicited synopses and ideas for books and magazine articles welcome; no mss.
Ⓔ Royalties twice-yearly, as a rule, but can vary.

## Acumen Publishing Limited

4 Saddler Street, Durham DH1 3NP
Ⓣ 0191 383 1889    Ⓕ 0191 386 3542
steven.gerrard@acumenpublishing.co.uk
www.acumenpublishing.co.uk
Managing Director *Steven Gerrard*

Founded in 1998 as an independent publisher for the higher education market. Publishes academic books on philosophy, classics and history of ideas. No unsolicited mss. Synopses and ideas for books welcome; send written proposal as per Acumen guidelines.
Ⓔ Royalties annually.

## Adlard Coles Nautical
▷ A.&C. Black Publishers Ltd

## African Books Collective

PO Box 721, Oxford OX1 9EN
www.africanbookscollective.com

Founded 1990. Collectively owned by its 17 founder publishers. Exclusive distribution in the UK, Europe and Commonwealth countries outside Africa and in North America through Michigan State University Press for 120 African participating publishers. Concentration is on scholarly/academic, literature and children's books. Mainly concerned with the promotion and dissemination of African-published material outside Africa. Supplies African-published books to African libraries and organizations, and publishes resource books on the African publishing industry. TITLES (published by ABC) *The African Writers' Handbook*; *African Publishers Networking Directory*; *The Electronic African Bookworm: A Web Navigator*; *Courage and Consequence: Women Publishing in Africa*; *African Scholarly Publishing*.

## Age Concern and Help the Aged Books

1268 London Road, London SW16 4ER
Ⓣ 020 8765 7200    Ⓕ 020 8765 7211
orders@acil.org.uk
www.ageconcern.org.uk/bookshop
Approx. Annual Turnover £400,000+

Publishing arm of Age UK. Publishes non-fiction only (no fiction or biographies), including the *Your Rights to Money Benefits*; *We've Made it Easy* series of money management guides, computing titles for the over fifties and a new series of Health Care handbooks. 6 titles a year. Synopses and ideas welcome.

## Airlife Publishing
▷ The Crowood Press Ltd

## Alcemi
▷ Y Lolfa Cyf

## Allen Lane
▷ Penguin Group (UK)

## Allison & Busby

13 Charlotte Mews, London W1T 4EJ
Ⓣ 020 7580 1080    Ⓕ 020 7580 1180
susie@allisonandbusby.com
www.allisonandbusby.com
Publishing Director *Susie Dunlop*
Approx. Annual Turnover £1 million

Founded 1967. Publishes fiction, historical fiction, crime fiction, true crime, pop culture, romance and fantasy. TITLES *Republic* trilogy by Jack Ludlow; *The Railway Detective* series Edward Marston; *Boomsday* Christopher Buckley; *The Morganville Vampire* series Rachel Caine. About 90 titles a year. No unsolicited mss.

✱*Authors' update* **A once small publisher that has now advanced to the middle ranks by dint of a clever selection of titles for the general read.**

## Alma Books Ltd

London House, 243–253 Lower Mortlake Road, Richmond TW9 2LL
Ⓣ 020 8948 9550    Ⓕ 020 8948 5599
info@almabooks.com
www.almabooks.com
Chairman *Elisabetta Minervini*
Managing Director *Alessandro Gallenzi*
Approx. Annual Turnover £450,000

Established in 2005 by founders of **Hesperus Press**. Publishes general fiction, biography, history. 20 titles in 2009. Accepts submissions by post.
Ⓔ Royalties annually.

❋*Authors' update* Alma is Spanish for 'soul', an apt name say the company founders for a publisher that 'regards a book as an aesthetic artefact rather than a mass produced commodity'.

## Alton Douglas Books
▷ Brewin Books Ltd

## Amber Lane Press Ltd
Cheorl House, Church Street, Charlbury OX7 3PR
☎ 01608 810024    🖷 01608 810024
info@amberlanepress.co.uk
www.amberlanepress.co.uk
Chairman *Brian Clark*
Managing Director/Editorial Head *Judith Scott*

Founded 1979 to publish modern play texts. Publishes plays and books on the theatre. TITLES *Whose Life is it Anyway?* Brian Clark; *The Dresser* Ronald Harwood; *Once a Catholic* Mary O'Malley (play texts); *Strindberg and Love* Eivor Martinus. No longer actively publishing. Submissions no longer accepted.
🖷 Royalties twice-yearly.

## Amberley Publishing Plc
The Hampton Centre, Cirencester Road, Chalford, Stroud GL6 8PE
☎ 01285 760030
www.amberley-books.com
Chairman *Charles Fairbairn*
Managing Director *Alan Sutton*
Approx. Annual Turnover £2 million

Founded in 2008 by Alan Sutton, original founder of Alan Sutton Publishing and Tempus Publishing Limited. Publishes local interest and specialist history. Archaeology, military, transport, maritime and aviation. No fiction. DIVISIONS **General History** *Jonathan Reeve*; **Archaeology** *Peter Kemmis Betty*; **Transport & Maritime** *Campbell McCutcheon*; **Local & Regional Interest** *Sarah Flight*. 350 titles in 2009. Unsolicited mss, synopses and ideas welcome by post and email (see website for information).
🖷 Royalties annually.

❋*Authors' update* With an output that is impressive by any standards, Amberley is proving that local history is one area where online does not intrude.

## Amsco
▷ Omnibus Press

## Andersen Press Ltd
20 Vauxhall Bridge Road, London SW1V 2SA
☎ 020 7840 8701    🖷 020 7233 6263
andersoneditorial@randomhouse.co.uk
andersenrights@randomhouse.co.uk
www.andersenpress.co.uk
Managing Director/Publisher *Klaus Flugge*

Editorial Director *Rona Selby*
Editorial Director (Fiction) *Charlie Sheppard*

Founded 1976 by Klaus Flugge and named after Hans Christian Andersen. Publishes children's high-quality picture books and fiction sold in association with Random House Children's Books. Seventy per cent of the books are sold as co-productions abroad. TITLES *Elmer* David McKee; *I Want My Potty* Tony Ross; *little.com* Ralph Steadman; *Cat in the Manger* Michael Foreman; *Preston Pig Books* Colin McNaughton; *Junk* Melvin Burgess; *You* Sandra Glover; *I Love You, Blue Kangaroo* Emma Chichester Clark. Unsolicited mss welcome for picture books; synopsis and first three chapters in the first instance for books for young readers up to age 12. No poetry or short stories. All submissions to be sent by post to the Submissions Department.
🖷 Royalties twice-yearly.

❋*Authors' update* The secret of Andersen's success is publishing children's books that are loved by adults. That and a wicked sense of fun that has been known to upset the education establishment – and hooray for that.

## Andrew Brodie
▷ A.&C. Black Publishers Ltd

## André Deutsch
▷ Carlton Publishing Group

## The Angels' Share
▷ Neil Wilson Publishing Ltd

## Angry Robot
▷ HarperCollins Publishers Ltd

## Anness Publishing Ltd
Hermes House, 88–89 Blackfriars Road, London SE1 8HA
☎ 020 7401 2077    🖷 020 7633 9499
info@anness.com
www.aquamarinebooks.com
www.lorenzbooks.com
www.southwaterbooks.com
Chairman/Managing Director *Paul Anness*
Publisher/Partner *Joanna Lorenz*
Approx. Annual Turnover £15.7 million

Founded 1989. Publishes highly illustrated co-edition titles: general non-fiction – cookery, crafts, interior design, gardening, photography, decorating, lifestyle, health, reference, military, transport and children's. IMPRINTS **Lorenz Books**; **Aquamarine**; **Hermes House**; **Southwater**. About 200 titles a year.

## Anova Books
10 Southcombe Street, London W14 0RA
☎ 020 7605 1400    🖷 020 7605 1401

info@anovabooks.com
firstinitialsurname@anovabooks.com
www.anovabooks.com
www.conwaypublishing.com
Chairman *Robin Wood*
CEO *Polly Powell*
Approx. Annual Turnover £12 million

Anova Books, formerly Chrysalis Books Group, specializes in non-fiction books, publishing under the IMPRINTS **B.T. Batsford**; **Collins & Brown**; **Conway**; **National Trust Books**; **Pavilion**; **Portico**; **Robson Books**; **Salamander**.

**B.T. Batsford** Publisher *Tina Persaud* Founded in 1843 as a bookseller, and began publishing in 1874. 'A world leader in books on chess, arts and craft.' Publishes non-fiction, architecture, heritage, bridge and chess, film and entertainment, fashion, crafts and hobbies.

**Collins & Brown** Publisher *Katie Cowan* Publishes a range of lifestyle categories especially in the areas of craft, health and cookery (Good Housekeeping books).

**Conway** Publisher *John Lee* Publishes naval history, maritime culture, ship modelling and military history, modelling and hobbies.

**National Trust Books** Publisher *Tina Persaud* Imprint of Anova Books since 2005. Publishes heritage, gardening, architecture, gift and cookery-based titles.

**Pavilion** Publisher *Anna Cheifetz* Publishes illustrated books in biography, cookery, gardening, erotica, art, interiors, sport and travel.

**Pavilion Children's Books** Associate Publisher *Ben Cameron* Publishes innovative, informative and fun books for children of all ages.

**Portico** and **Robson Books** Commissioning Editor *Malcolm Croft* Publishes general non-fiction including biography, humour and quirky reference.

**Salamander** Publisher *Frank Hopkinson* Founded 1973. Publishes colour illustrated books on collecting, cookery, interiors, gardening, music, crafts, military, aviation, sport and transport. Also a wide range of books on American interest subjects.

Anova will evaluate proposals but takes no responsibility for unsolicited mss.

## Anthem Press

75–76 Blackfriars Road, London SE1 8HA
☎ 020 7401 4200    🖷 020 7401 4201
info@wpcpress.com
www.anthempress.com
Managing Director *Tej Sood*

Founded in 1992 as an independent publisher of academic research, educational material and innovative works in established & emerging fields. Welcomes mss, synopses and ideas for books. 💷 Royalties annually.

## Antique Collectors' Club
▷ ACC Publishing Group

## Anvil Press Poetry Ltd

Neptune House, 70 Royal Hill, London SE10 8RF
☎ 020 8469 3033    🖷 020 8469 3363
anvil@anvilpresspoetry.com
www.anvilpresspoetry.com
Managing Director *Peter Jay*

Founded 1968. England's oldest independent publishing house dedicated to poetry. Publishes new poets in English and in translation as well as translated classics. AUTHORS include Paul Celan, Carol Ann Duffy, Martina Evans, Michael Hamburger, Dennis O'Driscoll, A.B. Jackson, Stanley Moss and Greta Stoddart. Classic translations include Apollinaire, Baudelaire, Dante, Hikmet, Neruda, Tagore and Verlaine. Preliminary enquiry required for translations. Unsolicited book-length collections of poems are welcome from writers whose work has appeared in poetry magazines or literary journals. Enquire before submitting material. Please enclose adequate return postage.

✱*Authors' update* **With a little help from the Arts Council, Anvil has become one of the foremost publishers of living poets.**

## Apex Publishing Ltd

PO Box 7086, Clacton on Sea CO15 5WN
☎ 01255 428500    🖷 0871 918 4756
mail@apexpublishing.co.uk
www.apexpublishing.co.uk
Managing Director *Chris Cowlin*

Founded 2002. Independent publishing company for unknown and established authors, specializing in autobiography, sport, television and true crime. 170 titles in print. Welcomes unsolicited mss, synopses and ideas for books. Approach in writing to the address above with a copy of the full ms. 💷 Royalties twice-yearly.

## Apollos
▷ Inter-Varsity Press

## Apple Press Ltd

7 Greenland Street, London NW1 0ND
☎ 020 7284 7168    🖷 020 7485 4902
www.apple-press.com
Publisher *Liane Stark*

Founded 1984. Part of **The Quarto Group**. Publishes highly illustrated lifestyle books with particular emphasis on food and drink, home decoration and craft, beauty and fashion, sports and leisure, and gardening.

## Appletree Press Ltd

The Old Potato Station, 14 Howard Street South, Belfast BT7 1AP

☎ 028 9024 3074   📠 028 9024 6756
reception@appletree.ie
www.appletree.ie
Managing Director *John Murphy*

Founded 1974. Publishes gift books, plus general non-fiction of Irish and Scottish interest. TITLES *In St Patrick's Footsteps*; *Ireland's Ancient Stones*; *Scottish Lighthouses*; *Northern Ireland – International Football Facts*. No unsolicited mss; send initial letter, email or synopsis.
⒠ Royalties twice-yearly in the first year, annually thereafter.

## Aquamarine
▷ Anness Publishing Ltd

## Arabia Books
▷ Haus Publishing

## Arc Publications Ltd

Nanholme Mill, Shaw Wood Road, Todmorden OL14 6DA

☎ 01706 812338   📠 01706 818948
arc.publications@btconnect.com
www.arcpublications.co.uk
Editorial Director *Tony Ward*
Publishing Director *Angela Jarman*
Associate Editors *John Kinsella (international), John W. Clarke (UK and Ireland), Jean Boase-Beier (translations), Angela Jarman (music)*

Founded in 1969 to specialize in the publication of contemporary poetry from new and established writers both in the UK and worldwide. AUTHORS Richard Gwyn, Linda France, D.M. Black, Michael Hulse (UK), Tony Curtis (Ireland), bilingual anthologies from Latvia and Macedonia, Kristyna Milobedzka (Poland), Ivana Milankova (Serbia), Behin Keskin (Turkey), Christina Ehin (Estonia), Lev Losseff (Russia), Linda Gregerson (USA), Katherine Gallagher (Australia), Namdeo Dhasal (India). IMPRINT **Arc Music** specializes in profiles of contemporary composers (particularly where none have existed hitherto) and symposia which take a 'new approach' to well-visited territory. Commissioned work only. Up to 20 titles a year. 'Please consult our submissions policy on our website before submitting any work.'
⒠ Royalties as per contract.

## Arcadia Books

15–16 Nassau Street, London W1W 7AB
☎ 020 7436 9898
info@arcadiabooks.co.uk
www.arcadiabooks.co.uk
Managing Director *Gary Pulsifer*
Associate Publisher *Daniela de Groote*
Managing Editor *Angeline Rothermundt*

Approx. Annual Turnover £500,000

Independent publishing house established in 1996. Specializes in quality translated fiction from around the world. IMPRINTS **Arcadia**; **BlackAmber**; **Bliss**; **EuroCrime**; **Maia Press**. TITLES *L'Oréal Took My Home: The Secrets of a Theft* Monica Waitzfelder; *The Hite Report on Women Loving Women* Shere Hite; *The Autobiography of the Queen* Emma Tennant; *No Trace* Barry Maitland. Does not welcome unsolicited material.
⒠ Royalties twice-yearly.

✴*Authors' update* **BlackAmber is dedicated to getting more black writers into the marketplace. Arcadia translations are among the best.**

## Arcane
▷ Omnibus Press

## Architectural Press
▷ Elsevier Ltd

## The Arden Shakespeare
▷ A.&C. Black Publishers Ltd

## Ardis
▷ Gerald Duckworth & Co. Ltd

## Arena Books

6 Southgate Green, Bury St Edmunds IP33 2BL
☎ 01284 754123   📠 01284 754123
arenabooks@tiscali.co.uk
www.arenabooks.co.uk
Chairman *James Farrell*
Managing Director *Robert Corfe*
Approx. Annual Turnover £94,000

Founded 2000. Publishes political science, specialized academic, history, travel and fiction. Planning to publish books examining religious themes. DIVISIONS **Political Science**, **History**, **Specialized Academic** *Robert Corfe* TITLE *The Right to National Self Defence*; **Fiction**, **Travel** *James Farrell* TITLE *The Girl From East Berlin*. 12 titles in 2009. No unsolicited mss. Synopses and ideas welcome; send by post or email. No children's, natural science, cookery or gardening.
⒠ Royalties annually.

## Argentum
▷ Aurum Press Ltd

## Aris & Phillips
▷ Oxbow Books

## Armchair Traveller
▷ Haus Publishing

## Arnefold Publishing

PO Box 22, Maidstone ME14 1AH
☎ 01622 759591   📠 01622 209193
Chairman/Managing Director *I.G. Mann*

Publishes original non-fiction and selected fiction, and non-fiction reprints. No new fiction. IMPRINTS **Arnefold**; **George Mann Books**; **Recollections**. 'Will consider non-fiction titles (which authors find hard to place) on a joint venture/shared profit basis if they are worthy of publication or can be made so. Unsolicited material is welcomed provided it is understood that if not accompanied by return postage it will neither be read nor returned; that no new fiction is wanted and that joint-venture publications may be suggested.'
£ Royalties annually.

## Arris Publishing Ltd
12 Main Street, Adlestrop, Moreton in Marsh GL56 0YN
☎ 01608 658758    📠 01608 659345
gcs@arrisbooks.com
www.arrisbooks.com
Joint Managing Directors *Geoffrey Smith, Michel Moushabeck*
Publishing Director *Victoria Huxley*

Founded in 2001. Publishes politics, history, biography, travel, art and culture. IMPRINTS **Arris Books** *Geoffrey Smith* TITLES *9/11: The New Pearl Harbor*; *Disfigured*; **Chastleton Travel** TITLES *Traveller's Wildlife Guides*; *A Traveller's History of Egypt*; *100 Best Paintings in London*; *World Heritage Sites: Great Britain & Ireland*. About 20 titles a year. No unsolicited mss; synopses and ideas for books welcome. Send introductory letter, synopsis and specimen chapter by post; no email submissions or CDs. No poetry or fiction.
£ Royalties annually.

## Arrow
▷ **The Random House Group Ltd**

## Artech House
16 Sussex Street, London SW1V 4RW
☎ 020 7596 8750    📠 020 7630 0166
p.comans@artechhouse.co.uk
www.artechhouse.com
CEO (USA) *William M. Bazzy*
Commissioning Editor *Penny Comans*

Founded 1969. European office of Artech House Inc., Boston. Publishes books for practising professionals in biomedical engineering, bioinformatics, built environment, computer security, MEMS, nanotechnology, microwaves, radar, antennas and propagation, electro-magnetic analysis, wireless communications, telecommunications, space applications, remote sensing, solid state technology and devices, technology management. 60–70 titles a year. Unsolicited mss and synopses in the specialized areas listed are considered.
£ Royalties twice-yearly

## Ashgate Publishing Ltd
Wey Court East, Union Road, Farnham GU9 7PT
☎ 01252 736600    📠 01252 736736 (Ashgate and Gower)/020 7440 7530 (Lund Humphries)
info@ashgatepublishing.com
www.ashgate.com
www.gowerpub.com
www.lundhumphries.com
Chairman *Nigel Farrow*
Managing Director *Rachel Lynch*

Founded 1967. Publishes business and professional titles under the Gower imprint, humanities, social sciences, law and legal studies under the Ashgate imprint, and art and art history under the Lund Humphries imprint. Acquired Lund Humphries, publisher of art books and exhibition catalogues in 1999. DIVISIONS **Ashgate** *Val Rose* Geography; *Neil Jordan* Social Policy; *Claire Jarvis* Sociology; *Kirstin Howgate, Natalja Mortensen* Politics; *Dymphna Evans* Reference/LIM; *Eric Levy* Law; *Alison Kirk* Law and legal studies; *Meredith Norwich* Visual studies; *Adrian Shanks* Social sciences; *John Smedley* History/Variorum collected studies; *Heidi Bishop* Music; *Erika Gaffney, Ann Donahue* Literary studies; *Tom Gray* History; *Sarah Lloyd* Theology and religious studies; *Guy Loft* Aviation studies. **Gower** *Jonathan Norman* Business, management and training. **Lund Humphries** *Lucy Myers*. Over 700 titles annually. Access the websites for information on submission of material.

## Ashgrove Publishing
27 John Street, London WC1N 2BX
☎ 020 7242 4820
ashgrovepublishing@googlemail.com
www.ashgrovepublishing.com
Chairman/Managing Director *Brad Thompson*

Acquired by Hollydata Publishers in 1999, Ashgrove has been publishing for over 30 years. Publishes mind, body, spirit, health, cookery, sports. 15 titles in 2010. No unsolicited mss; approach with letter and outline in the first instance.
£ Royalties twice-yearly.

## Ashmolean Museum Publications
Beaumont Street, Oxford OX1 2PH
☎ 01865 288070    📠 01865 278106
publications@ashmus.ox.ac.uk
www.ashmolean.org
Chairman *Dr Christopher Brown*
Managing Director *Declan McCarthy*
Approx. Annual Turnover £150,000

Owned by the University of Oxford, Ashmolean Museum Publications was founded in 1972. Publishes popular and scholarly works with a direct relationship to the Ashmolean's rich and diverse collection (art and archaeology, fine art

and exhibition catalogues). 5 titles in 2008. No unsolicited mss but synopses and ideas for books welcome; approach by email in the first instance. No fiction or crime.
£ Royalties annually.

## Ashwell Publishing Limited
▷ Olympia Publishers

## Atlantic Books

26–27 Boswell Street, London WC1N 3JZ
☎ 020 7269 1610    📠 020 7430 0916
enquiries@groveatlantic.co.uk
www.groveatlantic.co.uk
Chairman & Publisher *Toby Mundy*
Managing Director *Daniel Scott*
Editor-in-Chief *Ravi Mirchandani*

Founded 2000. A subsidiary of **Grove/Atlantic Inc.**, New York (see entry under *US Publishers*). Publishes literary and crime fiction, history, current affairs, biography and memoir, politics, popular science and reference books. IMPRINT **Corvus** Fiction. Distributor for the American imprint McSweeney's and co-publisher of Tusker Rock Press. 90 titles a year. Strictly no unsolicited material accepted. Submissions from literary agents only.
£ Royalties twice-yearly.

✱*Authors' update* No better proof of a thriving market for quality books. Veteran publisher Anthony Cheetham is a recent editorial recruit. His imprint, Corvus, will expand the transatlantic business.

## Atlantic Europe Publishing Co. Ltd

Greys Court Farm, Greys Court, Nr Henley on Thames RG9 4PG
☎ 01491 628188    📠 01491 628189
writers@atlanticeurope.com
www.AtlanticEurope.com
www.curriculumVisions.com
Director *Dr B.J. Knapp*

Publishes full-colour, highly illustrated primary school level text books. Not interested in any other material. Main focus is on National Curriculum titles, especially in the fields of religion, science, technology, social history and geography. About 50 titles a year. No unsolicited synopses or ideas.
£ Fees paid.

## Atom
▷ Little, Brown Book Group UK

## Aurora Metro

67 Grove Avenue, Twickenham TW1 4HX
☎ 020 3261 0000    📠 020 8898 0735
info@aurorametro.com
www.aurorametro.com
Managing Director *Cheryl Robson*

Submissions *Neil Gregory*

Founded in 1989 to publish new writing. Over 100 authors published including Germaine Greer, Meera Syal, Benjamin Zephaniah, Carole Hayman. International work in translation and fiction, children's, drama, biography, cookery, non-fiction. About 15 titles a year. No unsolicited mss; send synopses and ideas by email first.
£ Royalties annually.

## Aurum Press Ltd

7 Greenland Street, London NW1 0ND
☎ 020 7284 7160    📠 020 7485 4902
www.aurumpress.co.uk
Managing Director *Bill McCreadie*
Publisher *Graham Coster*
Approx. Annual Turnover £4.5 million

Founded 1976. Acquired by the **The Quarto Group** in 2005. High-quality, illustrated/non-illustrated adult non-fiction in the areas of biography, sport, current affairs, military history, travel guides, music and film. IMPRINTS **Argentum** Practical photography books; **Jacqui Small** High-quality lifestyle books. About 70 titles a year. No unsolicited mss or emailed submissions.
£ Royalties twice-yearly.

✱*Authors' update* Non-fiction for the middle market. If there are still serious readers out there who appreciate quality, Aurum will find them.

## Austin & Macauley Publishers Limited

CGC-33-01, 25 Canada Square, Canary Wharf, London E14 5LQ
☎ 020 7038 8212    📠 020 7038 8100
editors@austinmacauley.com
www.austinmacauley.com
Chief Editor *Annette Longman*
Executive Editor *David Calvert*

Publishes fiction and non-fiction, autobiography, biography, cookery, environmental issues, gardening, history, humour, memoirs, post-1945 fiction, true life experiences, children's, crime, science fiction, fantasy, fitness, health, romance, erotica, sport, self-help, women's and marital issues (including divorce). Expanding educational list: reference books, business studies, cultural matters, current affairs, English, mathematics, mythology, politics, human interest, war, music, theatre, social sciences, cinema and TV. About 70 titles a year. Unsolicited synopses and ideas considered. Accepts email enquiries. Submissions must be by post and include return postage.
£ Royalties annually.

✱*Authors' update* Authors may be asked to cover their own productions costs.

## Authentic Media
9 Holdom Avenue, Bletchley, Milton Keynes
MK1 1QR
☎ 01908 364200
www.authenticmedia.co.uk
Publishing Director *Mark Finnie*
Approx. Annual Turnover £2 million

A division of Biblica. Publishes Christian
books on evangelism, discipleship, biography
and mission. IMPRINT **Paternoster** Founded
1935. Editorial Director *Robin Parry* Publishes
academic, religion and learned/church/life-
related books and journals. Over 50 titles a year.
No fiction, children's/youth books or poetry.
'All manuscript submissions must be made by
email, using the details and form available on the
website.'
Ⓔ Royalties quarterly.

## AuthorHouse UK
500 Avebury Boulevard, Milton Keynes MK9 2BE
☎ 0800 197 4150   ☎ 0800 197 4151
www.authorhouse.co.uk
President & CEO *Kevin Weiss*
Vice President *Terry Dwyer*
UK Managing Director *Tim Davies*

Founded 1997. Acquired by Betram Capital
in January 2007. Publishes fiction, biography
and memoirs, current affairs, history, politics,
humour, travel, health, spirituality, sport and
academic. Provides a comprehensive range of
publishing and promotional services including
editorial, design, production, distribution and
marketing paid for by the author who retains
editorial and creative control.

## Authors OnLine
19 The Cinques, Gamlingay, Sandy SG19 3NU
☎ 01767 652005   ☎ 01767 652005
theeditor@authorsonline.co.uk
www.authorsonline.co.uk
*Submissions:* Freephone 0800 107 2423
Managing Director/Editor *Richard Fitt*
Submissions Editor *Mrs Gaynor Johnson*
Approx. Annual Turnover £250,000

Founded 1998. A service for authors wishing to
self-publish. Publishes new and reverted rights
work in both electronic format via their website
and traditional hard-copy mainly using digital
print-on-demand technology. 120 titles in 2009.
All genres welcome. Submit mss by email or post
(flashdrive or CD-ROM) to the Submissions
Editor.
Ⓔ Payment 60% net, paid quarterly.

## Autumn Publishing, a division of Bonnier Media Ltd
Appledram Barns, Birdham Road, Chichester
PO20 7EQ
☎ 01243 531660   ☎ 01243 538160
autumn@autumnpublishing.co.uk
www.autumnchildrensbooks.co.uk
Managing Director *Michael Herridge*
Editorial Director *Lyn Coutts*

Founded 1976. Division of Bonnier Media Ltd.
Publishes baby and toddler books, children's
activity, sticker and early learning books. About
200 titles a year. Unable to accept unsolicited mss
at this time.

## Avon
▷ HarperCollins Publishers Ltd

## Award Publications Limited
The Old Riding School, The Welbeck Estate,
Worksop S80 3LR
☎ 01909 478170   ☎ 01909 484632
info@awardpublications.co.uk
www.awardpublications.co.uk

Founded 1958. Publishes children's books, both
fiction and reference. IMPRINT **Horus Editions**.
No unsolicited mss, synopses or ideas.

## Azure
▷ Society for Promoting Christian Knowledge

## B.T. Batsford
▷ Anova Books

## Badger Publishing Ltd
Suite G08, Business & Technology Centre,
Bessemer Drive, Stevenage SG1 2DX
☎ 01438 791037   ☎ 01438 791036
orders@badger-publishing.co.uk
www.badger-publishing.co.uk
Publisher *David Jamieson*

Founded 1989. Educational publisher. 45 titles in
2009. No unsolicited mss; synopses and ideas for
books considered. Approach by email.
Ⓔ Royalties twice-yearly.

## Baillière Tindall
▷ Elsevier Ltd

## Bantam/Bantam Press
▷ Transworld Publishers

## Barefoot Books Ltd
124 Walcot Street, Bath BA1 5BG
☎ 01225 322400   ☎ 01225 322401
info@barefootbooks.co.uk
www.barefootbooks.com
Publisher *Tessa Strickland*

Founded 1993. Publishes children's picture books,
particularly new and traditional stories from a

wide range of cultures. No unsolicited mss. See website for submission guidelines.
£ Royalties twice-yearly.

✳*Authors' update* Writers of children's books would do well to keep track of Barefoot which, from small beginnings, is building a quality list that must be the envy of bigger publishers.

## Barrington Stoke

18 Walker Street, Edinburgh EH3 7LP
☎ 0131 225 4113    📠 0131 225 4140
barrington@barringtonstoke.co.uk
www.barringtonstoke.co.uk
Chairman *Ben Thomson*
Managing Director *Sonia Raphael*
Editorial Manager *Kate Paice*

Founded in 1998 to publish books for dyslexic, struggling and reluctant readers. DIVISIONS **Fiction** for 8–13-year olds, reading age 8; **Teenage Fiction** for age 12+, reading age 8; **4u2read.OK** for children aged 8–13, reading age 7; **gr8reads** for age 12+, reading age 7; **fyi** fiction with stacks of facts for ages 10–14, reading age 8; **Reality Check** Non-fiction for ages 10–14, reading age 8; **Reloaded** Myths retold for ages 10–14, reading age 8; **Most Wanted** Fiction for adults, with a reading age of 8; **Solo** for ages 10–14, reading age 6½; **Go!** for ages 10–14, reading age 6. Also books for teachers and resources to accompany fiction books. 60–70 titles a year. Commissioned via literary agents only. No unsolicited material.
£ Royalties twice-yearly.

## BBC Books
▷ The Random House Group Ltd

## BBC Children's Books
▷ Penguin Group (UK)

## BBC Worldwide

Media Centre, 201 Wood Lane, London W12 7TQ
☎ 020 8433 2000
www.bbcworldwide.com
CEO *John Smith*
COO *Sarah Cooper*

The main commercial arm and wholly owned subsidiary of the BBC. Consists of seven core businesses: BBC Worldwide Channels, BBC Worldwide Sales & Distribution, BBC Magazines, Children's & Licensing, BBC Worldwide Content & Production, BBC Worldwide Digital Media, BBC Home Entertainment and BBC Worldwide Global Brands. Leading consumer magazine publisher in the UK with sales of nearly 100 million copies a year. TITLES include *BBC Good Food*, *BBC Gardeners' World* and *Top Gear*. Acquired a 75% stake in **Lonely Planet Publications Ltd** in October 2007.

## Beautiful Books Ltd

36–38 Glasshouse Street, London W1B 5DL
☎ 020 7734 4448    📠 020 3070 0764
office@beautiful-books.co.uk
www.beautiful-books.co.uk
Managing Director *Simon Petherick*
Approx. Annual Turnover £500,000

Founded 2004. Publishes adult fiction and non-fiction. IMPRINTS **Bloody Books** Horror fiction; **Burning House** Contemporary fiction; **Beautiful Books** All other categories. 24–30 titles a year. Unsolicited mss welcome. Approach by post in first instance, with covering letter, brief biography, 1–2-page synopsis and 30–50-page sample of the ms. Include s.a.s.e. for reply (or return of material if required).
£ Royalties twice-yearly.

## Belair
▷ Folens Limited

## Berg Publishers

1st Floor, Angel Court, 81 St Clements Street, Oxford OX4 1AW
☎ 01865 245104    📠 01865 791165
enquiry@bergpublishers.com
www.bergpublishers.com
Managing Director *Kathryn Earle*
Editorial Director *Geraldine Billingham*

Acquired by **Bloomsbury Publishing Plc** in 2008. Publishes scholarly books in the fields of fashion, cultural studies, social sciences, history and humanities. TITLES *Tartan* Jonathan Faiers; *Jews and Shoes* Edna Nahshon; *Making Short Films* Clifford Thurlow; *Bite Me: Food in Popular Culture* Fabio Parasecoli. About 60 titles a year plus 18 journals. No unsolicited mss. Synopses and ideas for books welcome.
£ Royalties annually.

## Berghahn Books

3 Newtec Place, Magdalen Road, Oxford OX4 1RE
☎ 01865 250011    📠 01865 250056
publisher@berghahnbooks.com
www.berghahnbooks.com
Chairman/Managing Director *Marion Berghahn*
Approx. Annual Turnover £700,000

Founded 1994. Academic publisher of books and journals. TITLES *Imperial Germany, 1871–1918*; *Children of Palestine*; *Critical Interventions*; *A Different Kind of War*; *Escape From Hell*. About 90 titles a year. No unsolicited mss; will consider synopses and ideas for books. Approach by email in the first instance. No fiction or trade books. OVERSEAS ASSOCIATE Berghahn Books Inc., New York.
£ Royalties annually.

## Berlitz Publishing
58 Borough High Street, London SE1 1XF
☎ 020 7403 0284　🖷 020 7403 0290
berlitz@apaguide.co.uk
www.berlitzpublishing.com
Managing Director *Katharine Leck*

Founded 1970. Acquired by the Langenscheidt
Publishing Group in February 2002. Publishes
travel and language-learning products only: visual
travel guides, phrasebooks and language courses.
SERIES *Pocket Guides*; *Berlitz Complete Guide to
Cruising and Cruise Ships*; *Phrase Books*; *Pocket
Dictionaries*; *Business Phrase Books*; *Self-teach:
Rush Hour Commuter Cassettes*; *Think & Talk*;
*Berlitz Kids*. No unsolicited mss.

## BFI Publishing
Palgrave Macmillan, Porters North, 4 Crinan
Street, London N1 9XW
☎ 020 7833 4000　🖷 020 7843 4650
bfipublishing@palgrave.com
www.palgrave.com/bfi

Founded 1980. The publishing imprint of the
**British Film Institute**. Publishes academic,
schools and general film/television-related books
and resources. TITLE *The Cinema Book* Pam
Cook. SERIES *BFI Film Classics*; *BFI TV Classics*;
*BFI Screen Guides*; *World Directors*. About 30
titles a year. Email submissions preferred.
£ Royalties annually.

## BFP Books
Focus House, 497 Green Lanes, London N13 4BP
☎ 020 8882 3315
mail@thebfp.com
Chief Executive *John Tracy*
Commissioning Editor *Stewart Gibson*

Founded 1982. The publishing arm of the
**Bureau of Freelance Photographers**. Publishes
illustrated books on photography, mainly aspects
of freelancing and marketing pictures. No
unsolicited mss but ideas welcome.

## BIOS Scientific Publishers
▷ Taylor & Francis Group

## Birlinn Ltd
West Newington House, 10 Newington Road,
Edinburgh EH9 1QS
☎ 0131 668 4371　🖷 0131 668 4466
info@birlinn.co.uk
www.birlinn.co.uk

Founded 1992. Publishes history, military history,
adventure, Gaelic, humour, Scottish history,
Scottish reference, guidebooks and folklore.
IMPRINT **John Donald Publishers** (see entry).
Acquired **Polygon** in 2002 (see entry) and
Tuckwell Press in 2005 and Mercat Press in 2007.

About 170 titles a year across all imprints. No
unsolicited mss; synopses and ideas welcome.
£ Royalties paid.

✱*Authors' update* A thriving regional publisher
with titles that hit the wider market.

## Bitter Lemon Press
37 Arundel Gardens, London W11 2LW
☎ 020 7727 7927　🖷 020 7460 2164
books@bitterlemonpress.com
www.bitterlemonpress.com
Directors *Laurence Colchester,
François von Hurter*

Founded 2003. Specializes in thrillers, *romans
noir* and other contemporary fiction from
abroad. 9 titles in 2009. No unsolicited material;
submissions from agents only.
£ Royalties annually.

## Black & White Publishing Ltd
29 Ocean Drive, Edinburgh EH6 6JL
☎ 0131 625 4500　🖷 0131 625 4501
mail@blackandwhitepublishing.com
www.blackandwhitepublishing.com
Director *Campbell Brown*

Founded 1999. Publishes general non-fiction,
biography, sport and humour as well as selected
fiction and children's books. IMPRINT **Itchy Coo**
Scots language books for children. Send brief
synopsis along with 30 sample pages. 'Do *not*
submit full mss. Submission by email preferred.'
See the website for full submission guidelines.
£ Royalties twice-yearly.

## Black Ace Books
PO Box 7547, Perth PH2 1AU
☎ 01821 642822　🖷 01821 642101
www.blackacebooks.com
Managing Director *Hunter Steele*

Founded 1991. Publishes new fiction, Scottish and
general; some non-fiction including biography,
history, philosophy and psychology. IMPRINTS
**Black Ace Books**; **Black Ace Paperbacks** TITLES
*Succeeding at Sex and Scotland, Or the Case of
Louis Morel* Hunter Steele; *La Tendresse* Ken
Strauss MD; *Count Dracula (The Authorized
Version)* Hagen Slawkberg; *Caryddwen's Cauldron*
Paul Hilton; *The Sinister Cabaret* John Herdman.
No children's, poetry, cookery, DIY, religion.
36 titles in print. 'No submissions at all, please,
without first checking our website for details of
current requirements and submission guidelines.'
£ Royalties twice-yearly.

## Black Lace
▷ The Random House Group Ltd

## Black Spring Press Ltd

Curtain House, 134–146 Curtain Road, London
EC2A 3AR
☎ 020 7613 3066    📠 020 7613 0028
general@blackspringpress.co.uk
Director *Robert Hastings*

Founded 1986. Publishes fiction and non-fiction,
literary criticism, biography. TITLES *The
Lost Weekend* Charles Jackson; *The Big Brass
Ring* Orson Welles; *The Gorse Trilogy* Patrick
Hamilton; *Julian Maclaren-Ross Collected
Memoirs*. About 5 titles a year. No unsolicited
mss.
💷 Royalties annually.

## Black Swan
▷ Transworld Publishers

## BlackAmber
▷ Arcadia Books

## Blackstaff Press Ltd

4C Heron Wharf, Sydenham Business Park,
Belfast BT3 9LE
☎ 028 9045 5006    📠 028 9046 6237
info@blackstaffpress.com
www.blackstaffpress.com
Managing Editor *Patsy Horton*

Founded 1971. Publishes mainly, but not
exclusively, Irish interest books, fiction, poetry,
history, sport, cookery, politics, illustrated
editions, natural history and humour. About 20
titles a year. Unsolicited mss considered, but
preliminary submission of synopsis plus short
sample of writing is essential. Return postage
*must* be enclosed. No submissions by email.
💷 Royalties twice-yearly.

❋*Authors' update* Regional and proud of it, this
Belfast publisher is noted for bringing on young
talent and for 'wonderfully well-presented
catalogues and promotional material'.

## Blackstone Publishers

18 Soho Square, London W1D 3QL
☎ 020 7025 8034    📠 020 7025 8100
mail@blackstonepublishers.com
mail@redarrowbooks.com
www.blackstonepublishers.com
www.redarrowbooks.com
Editor *Nina Rogers*
Publisher *John Kurtz*

Founded 2007. Publishes fiction, including
romance, women's fiction, mysteries, suspense,
science fiction, horror, fantasy, humour, gay,
lesbian, erotica; non-fiction, including memoirs,
biography, history, war, reference; children's,
picture books, young reader and teen fiction.
TITLES *Solitude* Paul Beet; *Confronting Demons*
Ricky Abbott. IMPRINT **Red Arrow Books**.

Unsolicited mss, synopses and ideas for books
(with s.a.s.e.) welcome.
💷 Royalties twice-yearly.

## Blackwell Publishing
▷ Wiley-Blackwell

## Blandford Press
▷ Octopus Publishing Group

## Bliss
▷ Arcadia Books

## Bloodaxe Books Ltd

Highgreen, Tarset NE48 1RP
☎ 01434 240500    📠 01434 240505
editor@bloodaxebooks.com
www.bloodaxebooks.com
Managing/Editorial Director *Neil Astley*

Publishes poetry, literature and criticism,
and related titles by British, Irish, European,
Commonwealth and American writers.
Ninety-five per cent of the list is poetry. TITLES
include seven major anthologies: *Staying Alive*,
*Being Alive*, *Earth Shattering* and *In Person* all
edited by Neil Astley; *The Bloodaxe Book of 20th
Century Poetry* ed. Edna Longley; *Modern Women
Poets* ed. Deryn Rees-Jones; editions of Fleur
Adcock, David Constantine, Helen Dunmore,
Roy Fisher, Jackie Kay, Brendan Kennelly, Mary
Oliver, J.H. Prynne, Peter Reading, Ken Smith,
Anne Stevenson, C.K. Williams and Benjamin
Zephaniah. About 30–40 titles a year. Unsolicited
poetry mss welcome; send a sample of no more
than 10 poems with s.a.s.e., 'but if you don't read
contemporary poetry, don't bother'. No email
submissions of any kind.
💷 Royalties annually.

❋*Authors' update* Assisted by regional Arts
Council funding, Bloodaxe is one of the liveliest
and most innovative of poetry publishers with a
list that takes in some of the best of the younger
poets.

## Bloody Books
▷ Beautiful Books Ltd

## Bloomsbury Publishing Plc

36 Soho Square, London W1D 3QY
☎ 020 7494 2111    📠 020 7434 0151
www.bloomsbury.com
Founder and Chief Executive *Nigel Newton*
Executive Director *Richard Charkin*
Publishing Directors *Alexandra Pringle
(Editor-in-Chief), Kathy Rooney, Sarah Odedina,
Michael Fishwick*
Approx. Annual Turnover £150 million

Founded 1986. Many of its authors have gone
on to win prestigious literary prizes, including
J.K. Rowling's *Harry Potter and the Philosopher's
Stone, Harry Potter and the Chamber of Secrets*

and *Harry Potter and the Prisoner of Azkaban* which won the Nestlé Smarties Book Prize in 1997, 1998 and 1999 respectively. Margaret Atwood's *The Blind Assassin* won the Booker Prize in 2000.

Started Bloomsbury USA in 1998. Acquired **A.&C. Black Publishers Ltd** in May 2000 (see entry), Peter Collin Publishing Ltd in September 2002, Berlin Verlag in 2003, Walker Publishing Inc in New York in 2004, Featherstone Education (A.&C. Black imprint), Berg Publishers and John Wisden & Co. in 2008. Publishes literary fiction and non-fiction, including general reference; also audiobooks. AUTHORS include Margaret Atwood, T.C. Boyle, Sophie Dahl, Jeffrey Eugenides, Neil Gaiman, Daniel Goleman, David Guterson, Sheila Hancock, John Irving, Jay McInerney, Tim Pears, Celia Rees, J.K. Rowling, Ben Schott, Will Self, Donna Tartt, Rupert Thomson, Barbara Trapido, Benjamin Zephaniah. No unsolicited submissions.
£ Royalties twice-yearly.

✱*Authors' update* **Cash rich from Harry Potter, Bloomsbury is investing imaginatively in a range of publishing initiatives including an 'on demand' academic imprint which is to offer free content for non-commercial use and a digital subscription service for libraries. Turnover has been boosted by the acquisition of Featherstone Education and Berg Publishers. The hunt is on for strong selling titles.**

### Blue Door
▷ HarperCollins Publishers Ltd

### BMM
▷ SportsBooks Limited

### Bobcat
▷ Omnibus Press

### Bodleian Library
Osney One Building, Osney Mead, Oxford OX2 0EW
☎ 01865 283850
publishing@bodley.ox.ac.uk
www.bodleianbookshop.co.uk
Head of Publishing *Samuel Fanous*

Founded 1602. Publishes trade and scholarly titles relating to the Bodleian Library collections of over eight million books and manuscripts. TITLES *The Original Frankenstein*; *Book of Beasts: A Facsimile of MS. Bodley 764*; *Postcards from the Trenches*; *The Original Rules of Rugby*; *German Invasion Plans for the British Isles 1940*. 8 titles in 2009. No unsolicited mss. Synopses and ideas for books related to the collections will be considered; approach should be made by post, addressed to the Commissioning Editor.

### The Bodley Head/Bodley Head Children's Books
▷ The Random House Group Ltd

### Boltneck Publications Limited
*Head Office:* Westpoint, 78 Queens Road, Clifton, Bristol BS8 1QX
☎ 0117 985 8709
clive.birch@medavia.co.uk
www.boltneck.co.uk
Managing Director *David Thomas*
Publishing Director *Clive Birch*

Founded 2004. Linked to Medavia Media Agency. Publishes fiction and popular non-fiction on topical subjects or people of interest. DIVISION **Medavia Publishing** *Clive Birch* TITLES *Married to Albert*; *No Big Deal*; *A Decent Man*; *Kirsty – Angel of Courage*. No unsolicited mss.
£ Royalties paid annually.

### Book Guild Publishing
Pavilion View, 19 New Road, Brighton BN1 1UF
☎ 01273 720900   📠 01273 723122
info@bookguild.co.uk
www.bookguild.co.uk
Chairman *George M. Nissen, CBE*
Managing Director *Carol Biss*

Founded 1982. Publishes fiction, children's and non-fiction, including human interest and biography/memoirs. Expanding mainstream list. Subject areas: **Human Interest** TITLES *Mud and the City* Jessica Fellowes; *Not for Wimps* John Parry; **Biography** TITLES *Cupboard Love* Laura Lockington; *The Touch of Durrell* Jeremy Mallinson; *A PG Wodehouse Treasury* Mark Hichens; **General Non-Fiction** TITLE *Rugby Union – the men who make the game* Ian Smith; **Fiction** TITLES *The Tail of Augustus Moon* Melanie Whitehouse; *No 1 Chesterfield Square* Nick Jones; **Children's** TITLE *The Monies* John Mariani. About 90 titles a year. Unsolicited mss, ideas and synopses welcome.
£ Royalties twice-yearly.

✱*Authors' update* **There are two sides to Book Guild Publishing, not always coexisting in happy harmony. The favoured image is of a conventional publisher with a strong list of general titles. But also on offer is a service where authors cover the costs of publication. Unlike many vanity publishers, Book Guild Publishing clearly sets out its terms of agreement. Still, the outlay can be high relative to sales and first-time authors should not be misled into believing that they have found an easy route to fame and fortune.**

### Border Lines Biographies
▷ Seren

## Boulevard Books & The Babel Guides

71 Lytton Road, Oxford OX4 3NY
℡ 01865 712931
info@babelguides.co.uk
www.babelguides.co.uk
Managing Director *Ray Keenoy*

Specializes in contemporary world fiction by young writers in English translation. Existing or forthcoming series of fiction from Brazil, Italy, Latin America, Low Countries, Greece, and elsewhere. The Babel Guides series of popular guides to fiction in translation started in 1995. SERIES *Babel Guides to Fiction in Translation* Series Editor *Ray Keenoy* TITLES *Eminent Hungarians*; *Babel Guide to Italian Fiction in Translation*; *Babel Guide to the Fiction of Portugal, Brazil & Africa in Translation*; *Babel Guide to French Fiction in English Translation*; *Babel Guide to Jewish Fiction*; *Babel Guide to Brazilian Fiction*; *Babel Guide to Welsh Fiction*. Suggestions and proposals for translations of contemporary fiction welcome. Also seeking contributors to forthcoming Babel Guides (all literatures).
Ⓔ Royalties annually.

## Bound Biographies Limited

Heyford Park House, Heyford Park, Bicester OX25 5HD
℡ 01869 232911    ℻ 01869 232698
office@boundbiographies.com
www.boundbiographies.com
Managing Director *Michael Oke*
Editorial Head *Dr A.J. Gray*
Approx. Annual Turnover £250,000

Founded in 1992 to assist in the writing and production of private life stories and corporate histories. Print-on-demand facilities to provide short runs of paper or hardback books to complement the Bound Biographies leather-bound range. Publishes autobiographies predominantly although novels, poetry and special interest books are considered. TITLES *Red Tails in the Sunset* Bryn Williams; *Think of a Card* Don Lewin; *Looking for the Silver Lining* Martin Barraclough. 40 titles in 2009. Unsolicited material welcome; approach by post, telephone or email.
Ⓔ Royalties annually.

## Bounty
▷ **Octopus Publishing Group**

## Bowker (UK) Ltd

St Andrew's House, 18–20 St Andrew Street, London EC4A 3AG
℡ 020 7832 1700    ℻ 020 7832 1710
sales@bowker.co.uk
www.bowker.com
Managing Director *Doug McMillan*

Part of the Cambridge Information Group (CIG). Publishes bibliographic references used by publishers, libraries and retailers throughout the world to source new book information. TITLES *Books in Print* and *Global Books in Print*. 17 titles in 2009. Unsolicited material will not be read.

## Boxtree
▷ **Macmillan Publishers Ltd**

## Bradt Travel Guides

23 High Street, Chalfont St Peter SL9 9QE
℡ 01753 893444    ℻ 01753 892333
info@bradtguides.com
www.bradtguides.com
Managing Director *Donald Greig*
Editorial Director *Adrian Phillips*
Approx. Annual Turnover £1 million

Founded in 1974 by Hilary Bradt. Specializes in travel guides to off-beat places and quirky travel and wildlife-related titles (see website for full list). Worldwide distribution. About 35 titles a year. No unsolicited mss; synopses and relevant ideas welcome.
Ⓔ Royalties twice-yearly.

✳*Authors' update* Judged to be the best of guide publishers by *Wanderlust* magazine, Bradt may well have hit on the antidote to online travel guides. A list of unlikely tourist destinations such as Burkina Faso and Kabul has found its niche market, but it's been a tough year.

## Brady
▷ **Penguin Group (UK)**

## Breedon Books
▷ **Derby Books Publishing**

## Brewin Books Ltd

Doric House, 56 Alcester Road, Studley B80 7LG
℡ 01527 854228    ℻ 01527 852746
admin@brewinbooks.com
www.brewinbooks.com
Chairman/Managing Director *Alan Brewin*
Company Secretary *Julie Brewin*
Director *Alistair Brewin*

Founded 1976. Publishes books on all aspects of Midland life and history including social, hospital, police, military, transport and family histories as well as biographies. TITLES *Warwickshire's Wild Flowers* Steven Falk (ed.); *Matthew Boulton: Revolutionary Player* Malcolm Dick; *Bombs and Betty Grable* John Wilcox. IMPRINTS **Brewin Books; Alton Douglas Books; History Into Print**. About 30 titles a year. Not interested in children's, poetry, short stories or novels. Approach by letter, but do not send full mss. Unsolicited synopses and ideas welcome with s.a.s.e.
Ⓔ Royalties twice-yearly.

## Bristol Classical Press
▷ Gerald Duckworth & Co. Ltd

## Bristol Phoenix Press
▷ University of Exeter Press

## The British Academy

10 Carlton House Terrace, London SW1Y 5AH
☎ 020 7969 5200 📠 020 7969 5300
pubs@britac.ac.uk
www.britac.ac.uk
Head of Publications *James Rivington*
Events and Publications Assistant *Emily Ray*
Publications Officer *Amrit Bangard*

Founded 1902. The primary body for promoting scholarship in the humanities and social sciences, the Academy publishes many series stemming from its own long-standing research projects, or series of lectures and conference proceedings. Main subjects include history, philosophy and archaeology. SERIES *Auctores Britannici Medii Aevi*; *Early English Church Music*; *Fontes Historiae Africanae*; *Records of Social and Economic History*. About 20 titles a year. Proposals for these series are welcome and are forwarded to the relevant project committees. The British Academy is a registered charity and does not publish for profit.
💷 Royalties only when titles have covered their costs.

## The British Computer Society

First Floor, Block D, North Star House, North Star Avenue, Swindon SN2 1FA
☎ 0845 300 4417 📠 01793 480270
publishing@hq.bcs.org.uk
www.bcs.org/books
Chief Executive *David Clarke*
Head of Publishing & Information Products *Elaine Boyes*
Book Publisher *Matthew Flynn*
Approx. Annual Turnover £20 million (Society)

Founded 1957. BCS is the leading professional and learned society in the field of computers and information systems. Publishes books which support the professional, academic and practical needs of the IT community. 20 titles in 2010. Unsolicited material welcome; submissions form and guide on website page: www.bcs.org/books/writer
💷 Royalties annually.

## The British Library

96 Euston Road, London NW1 2DB
☎ 020 7412 7535 📠 020 7412 7768
blpublications@bl.uk
www.bl.uk
Publisher *David Way*
Senior Editor *Catherine Britton*

Founded in 1979, the publishing arm of The British Library publishes academic, general and illustrated books based on the Library's collections and the history of the book. TITLES *Medieval Dress and Fashion*; *Books as History*; *Codex Sinaiticus: The Story of the World's Oldest Bible*; *John Keats: A Poet and His Manuscripts*. About 50 titles a year. Unsolicited mss, synopses and ideas welcome if related to the history of the book, book arts or bibliography. No fiction or general non-fiction.
💷 Royalties annually.

## The British Museum Press

38 Russell Square, London WC1B 3QQ
☎ 020 7323 1234 📠 020 7436 7315
www.britishmuseum.org
Managing Director *Brian Oldman*
Director of Publishing *Rosemary Bradley*

The book publishing division of The British Museum Company Ltd. Founded in 1973 as British Museum Publications Ltd; relaunched 1991 as British Museum Press. Publishes books based on the British Museum's collection, including ancient history, archaeology, ethnography, art history, exhibition catalogues, guides, children's books and all official publications of the British Museum. TITLES *The First Emperor: China's Terracotta Army*; *The Parthenon Sculptures in the British Museum*; *500 Things to Know About the Ancient World*; *Indian Art in Detail*. About 40 titles a year. Synopses and ideas for books welcome.
💷 Royalties twice-yearly.

## Brown Skin Books

PO Box 57421, London E5 0ZD
☎ 020 8986 1115
info@brownskinbooks.co.uk
www.brownskinbooks.co.uk
Chairman *Dr John Lake*
Managing Director *Vastiana Belfon*

Founded 2002. Publishes 'quality, intelligent erotic fiction by black women around the world' and a series of erotic crime thrillers. TITLES *Body and Soul* Jade Williams; *Personal Business* Isabel Baptiste; *Scandalous* and *A Darker Shade of Blue* Angela Campion; *Playthings* Faith Graham; *Online Wildfire* Crystal Humphries; *Strip Poker* and *Beg Me* Lisa Lawrence; *The Singer* Aisha DuQuesne; *Sorcerer* Tamzin Hall. 2 titles in 2009. No unsolicited mss; synopses and sample chapters welcome. Send by email or post. No poetry.
💷 Royalties twice-yearly.

## Brown, Son & Ferguson, Ltd

4–10 Darnley Street, Glasgow G41 2SD
☎ 0141 429 1234 📠 0141 420 1694

info@skipper.co.uk
www.skipper.co.uk
Chairman/Joint Managing Director *T. Nigel Brown*

Founded 1850. Specializes in nautical textbooks, both technical and non-technical. Also Scottish one-act/three-act plays. Unsolicited mss, synopses and ideas for books welcome.
£ Royalties annually.

## Bryntirion Press

Bryntirion, Bridgend CF31 4DX
☎ 01656 655886     ☏ 01656 665919
office@emw.org.uk
www.emw.org.uk
Publications Officer *Gethin Rhys*
Approx. Annual Turnover £40,000

Owned by the Evangelical Movement of Wales. Publishes Christian books in English and Welsh. 50 titles in 2009. No unsolicited mss.
£ Royalties annually.

## Burning House
▷ Beautiful Books Ltd

## Burns & Oates
▷ The Continuum International Publishing Group Limited

## Business Education Publishers Ltd

Evolve Business Centre, Cygnet Way, Rainton Bridge Business Park, Houghton-le-Springs DH4 5QY
☎ 0191 305 5160
info@bepl.com
www.bepl.com
Managing Director *Mrs A. Murphy*
Approx. Annual Turnover £400,000

Founded 1981. Publishes business education, economics and law for BTEC and GNVQ reading. Currently expanding into further and higher education, computing, IT, business, travel and tourism, occasional papers for institutions and local government administration. Unsolicited mss and synopses welcome.
£ Royalties annually.

## Business Plus
▷ Hachette UK

## Buster Books
▷ Michael O'Mara Books Ltd

## Butterworth Heinemann
▷ Elsevier Ltd

## Cadogan Guides
▷ New Holland Publishers (UK) Ltd

## Calder Publications Ltd

51 The Cut, London SE1 8LF
☎ 020 8948 9550     ☏ 020 8948 5599
info@calderpublications.com
www.calderpublications.com
Managing Director/Editorial Head *Alessandro Gallenzi*

Acquired in April 2007 by independent publishers Oneworld Classics. A publishing company which has grown around the tastes and contacts of John Calder, the iconoclast of the literary establishment. The list has a reputation for controversial and opinion-forming publications. Publishes autobiography, biography, drama, literary fiction, literary criticism, music, opera, poetry, politics, sociology, ENO opera guides. AUTHORS Antonin Artaud, Marguerite Duras, Martin Esslin, Erich Fried, P.J. Kavanagh, Robert Menasse, Robert Pinget, Luigi Pirandello, Alain Robbe-Grillet, Nathalie Sarraute, L.F. Céline, Eva Figes, Claude Simon, Raymond Queneau . No new material accepted.
£ Royalties annually.

## Cambridge University Press

The Edinburgh Building, Shaftesbury Road, Cambridge CB2 8RU
☎ 01223 312393     ☏ 01223 315052
www.cambridge.org
Chief Executive *Stephen R.R. Bourne*
Managing Director, Publishing, Academic and Professional (books and journals in H&SS and STM) *A.M.C. Brown*
Managing Director, Publishing, Cambridge Learning (ELT and Education) *J.M. Pieterse*
Managing Director, Publishing, New Directions *J.G. Tuttle*

The oldest printer and publisher in the world, now a global organization with a regional structure operating in the UK, Europe, Middle East & Africa, in the Americas and in Asia-Pacific. Has warehousing centres in Cambridge, New York, Melbourne, Madrid, Cape Town, São Paulo, Singapore, New Delhi and Tokyo, with offices and agents in many other countries. Publishing includes major ELT courses; tertiary textbooks, monographs and journals; scientific and medical reference; professional lists in law, management and engineering; educational coursebooks; and e-learning materials for schools. Recent ventures in e-pedagogy include the companies Global Grid for Learning and English 360, and recent digital initiatives include large-scale academic print-on-demand and online publishing platforms. Publishes at all levels from primary school to postgraduate research and professional. Also Bibles, prayer books and over 230 academic journals. Around 25,000 authors in 116 countries and between 1,500 and 2,000 new titles a year. Synopses and ideas for books are welcome (and preferable to the submission of unsolicited mss). No fiction or poetry.

£ Royalties twice-yearly.

❋*Authors' update* With rapid growth overseas, profits have been ploughed back into educational software for the digital age.

## Camden Press Ltd

43 Camden Passage, London N1 8EA
☎ 020 7226 4673
Chairman *Bob Borzello*

Founded 1985. Publishes social issues; all books are launched in connection with major national conferences. DIVISION **Publishing for Change** *Bob Borzello* TITLE *Living with the Legacy of Abuse.* IMPRINT **Mindfield** TITLES *Hate Thy Neighbour: The Race Issue*; *Therapy on the Couch.* No unsolicited material. Approach by telephone in the first instance.
£ Royalties annually.

## Campbell Books
▷ Macmillan Publishers Ltd

## Candle
▷ Lion Hudson plc

## Canongate Books Ltd

14 High Street, Edinburgh EH1 1TE
☎ 0131 557 5111    ☎ 0131 557 5211
info@canongate.co.uk
www.meetatthegate.com
*Also at:* Lower Basement, 151 Chesterton Road, London W10 6ET
Publisher/Managing Director *Jamie Byng*
Publishing Director *Anya Serota*
Editorial Director *Nick Davies*
Approx. Annual Turnover £14.7 million

Founded 1973. In 2002, won both 'Publisher of the Year' at the British Book Awards and the Booker Prize with *Life of Pi* by Yann Martel and again, in 2009, was 'Publisher of the Year'. Publishes a wide range of literary fiction and non-fiction. Historically, there is a strong Scottish slant to the list but its output is increasingly international, especially now with its joint venture with **Grove Atlantic** in the US and co-publishing agreement with Text Publishing in Australia. Canongate also has a growing reputation for originating unusual projects (typified by *The Pocket Canons* and more recently by *The Myths* series, launched in 2005). Key AUTHORS include Michel Faber, Yann Martel, Alasdair Gray, John Fante, James Meek and Louise Welsh.
£ Royalties twice-yearly.

❋*Authors' update* Having more than doubled its turnover in less than three years, Canongate is gaining strength in the ebook market.

## Canterbury Press
▷ SCM – Canterbury Press

## Capall Bann Publishing

Auton Farm, Milverton TA4 1NE
☎ 01823 401528    ☎ 01823 401529
enquiries@capallbann.co.uk
www.capallbann.co.uk
Chairman *Julia Day*
Editorial Head *Jon Day*

Founded in 1993, Capall Bann now has well over 300 titles in print. Family-owned and run company, 'operated by people with real experience in the topics we publish' which include British traditions, folklore, animals, alternative healing, environmental, Celtic lore, mind, body and spirit. TITLES *The Magical and Mystical World of Faerie*; *The Seer's Guide to Crystal and Gem Divination*; *Nature's Children – Celebrating the Seasons in a Pagan Family*; *Reaching For the Divine – How to Communicate Effectively With Your Spirit Guides & Loved Ones On the Other Side*; *Aphrodisiacs – Aphrodite's Secrets.* About 40 titles a year. Unsolicited proposals for books welcome. 'Fiction accepted on cooperative terms.'
£ Royalties twice-yearly.

## Capstone Publishing (A Wiley Company)

John Wiley & Sons Ltd, The Atrium, Southern Gate, Chichester PO19 8SQ
☎ 01243 779777
info@wiley-capstone.co.uk
www.wiley.com

Founded 1997. Part of **John Wiley & Sons Ltd**. Publishes entrepreneurship, business skills, sales and marketing, leadership and management, personal finance and economics, self-help, health and lifestyle, humour.

## Capuchin Classics

128 Kensington Church Street, London W8 4BH
☎ 020 7221 7166    ☎ 020 7792 9288
marketing@capuchin-classics.co.uk
www.capuchin-classics.co.uk
Chairman *Tom Stacey*
Managing Director *Max Scott*

Founded 2008. Part of Stacey Arts. Publishes fiction including reprints of classic fiction. 19 titles in 2009. No unsolicited mss; will consider synopses and ideas for books sent by email.
£ Royalties annually.

## Carcanet Press Ltd

4th Floor, Alliance House, 30 Cross Street, Manchester M2 7AQ
☎ 0161 834 8730    ☎ 0161 832 0084
info@carcanet.co.uk
www.carcanet.co.uk
Owner *Folio Holdings*
Chairman *Kate Gavron*
Managing Director/Editorial Director *Michael Schmidt, OBE, FRSL*

Carcanet Press was founded by Michael Schmidt in 1969 and is now in its fifth decade. In 2000 it was named the *Sunday Times* millennium Small Publisher of the Year. Five of its authors have received Nobel Prizes, nine have received the Queen's Gold Medal for Poetry and seven have received Pulitzer Prizes, among many other honours. Carcanet publishes a comprehensive and diverse list of modern and classic poetry in English and in translation. IMPRINTS include **FyfieldBooks**, popularly-priced selections from the great European and American classics and **OxfordPoets**, formerly the poetry list of Oxford University Press. Carcanet also publishes a range of inventive fiction, **Lives and Letters** and literary criticism and issues the controversial and illuminating literary magazine *PN Review*, which appears six times a year. AUTHORS John Ashbery, Eavan Boland, Donald Davie, Natalia Ginzburg, Robert Graves, Elizabeth Jennings, Hugh MacDiarmid, Edwin Morgan, Sinead Morrissey, Les Murray, Frank O'Hara, Richard Price, Frederic Raphael, C.H. Sisson, Charles Tomlinson, Jane Yeh. 40 titles a year. Poetry submissions (hard copy only): 6–10 poems with covering letter and return postage. Prospective writers should familiarize themselves with the Carcanet list before submitting work.
£ Royalties annually.

✳*Authors' update* Now one of the UK's leading literary publishers with a diverse list of modern and classic poetry, Carcanet stays close to the source of literary creativity with its own postgraduate Writing School at Manchester Metropolitan University (see entry under *UK and Irish Writers' Courses*).

## Cardiff Academic Press

St Fagans Road, Fairwater, Cardiff CF5 3AE
℡ 029 2056 0333     ℻ 029 2055 4909
info@drakeed.com
www.drakeed.com
Managing Director *R.G. Drake*

Academic publishers.

## Carlton Publishing Group

20 Mortimer Street, London W1T 3JW
℡ 020 7612 0400     ℻ 020 7612 0401
enquiries@carltonbooks.co.uk
www.carltonbooks.co.uk
Publisher *Jonathan Goodman*
Managing Director *Belinda Rasmussen*
Editorial Director *Piers Murray Hill*
Approx. Annual Turnover £17 million

Founded 1992. Carlton Publishing Group is an independent publishing house. It has three main divisions: **Carlton Books** Illustrated leisure and entertainment books aimed at the mass market. Subjects include history, sport, puzzles, health,

popular science, children's non-fiction, popular culture, music, fashion and design; **André Deutsch** Autobiography, biography, history, current affairs, narrative non-fiction and the arts; **Prion Books** Humour, nostalgia, drink and classic literature. 120 titles per year. No unsolicited mss; synopses and ideas welcome. No novels, science fiction, poetry or children's fiction.
£ Royalties twice-yearly.

✳*Authors' update* **Imaginative illustrated books on popular subjects sell across all language barriers.**

## Carroll & Brown Publishers Limited

20 Lonsdale Road, London NW6 6RD
℡ 020 7372 0900     ℻ 020 7372 0460
mail@carrollandbrown.co.uk
www.carrollandbrown.co.uk
Publisher *Amy Carroll*

Publishes practical parenting, health, fitness, recreations, mind, body and spirit. TITLES *Your Babycare Bible*; *Strong Bones for Life*; *Your Pregnancy Companion*. Synopses and ideas for illustrated books welcome; approach in writing in the first instance, enclosing s.a.s.e. for reply. No fiction
£ Fees or royalties paid.

## Cassell
▷ The Orion Publishing Group Limited

## Cassell Illustrated
▷ Octopus Publishing Group

## Catholic Truth Society (CTS)

40–46 Harleyford Road, London SE11 5AY
℡ 020 7640 0042     ℻ 020 7640 0046
f.martin@cts-online.org.uk
www.cts-online.org.uk
Chairman *Most Rev. Paul Hendricks*
General Secretary *Fergal Martin*

Founded originally in 1868 and re-founded in 1884. Publishes religious books – Roman Catholic; a variety of doctrinal, moral, biographical, devotional and liturgical publications, including a large body of Vatican documents and sources. Unsolicited mss, synopses and ideas welcome if appropriate to their list.
£ Royalties annually.

## Catnip Publishing Ltd

14 Greville Street, London EC1N 8SB
℡ 020 7138 3650     ℻ 020 7138 3658
www.catnippublishing.co.uk
Managing Director *Robert Snuggs*
Commissioning Editor *Non Pratt*
Approx. Annual Turnover £400,000

Independent children's books publisher, founded in 2005. Concentrating on developing fiction for the 8–10 and 9–11 age groups. IMPRINTS

**Catnip** New books from Joan Lingard, Dominic Barker, Sarah Matthias; **Happy Cat** TITLE *Scaredy Squirrel at the Beach* Melanie Watt; SERIES *Happy Cat First Readers* (5–7 years); *Talking it Through.* 24 titles a year. Unsolicited submissions for fiction considered; send outline synopsis, including proposed age range and length, by email. No non-fiction and, currently, only limited interest in picture books.
Ⓔ Royalties twice-yearly.

## CBA Publishing
St Mary's House, 66 Bootham, York YO30 7BZ
Ⓣ 01904 671417    Ⓕ 01904 671384
catrinaappleby@britarch.ac.uk
www.britarch.ac.uk
Publications Officer *Catrina Appleby*
British Archaeology Editor *Mike Pitts*
Young Archaeologists' Club Communication Officer *Nicky Milsted*
Approx. Annual Turnover £115,000

Publishing arm of the **Council for British Archaeology**. Publishes archaeology reports, practical handbooks, *Internet Archaeology*; *British Archaeology* (bi-monthly magazine), *Young Archaeologist* (Young Archaeologists' Club magazine), monographs, archaeology and education. TITLES *Human Remains in Archaeology – A Handbook*; *Europe's Lost World, the Rediscovery of Doggerland*; *Where Rivers Meet.* 8–10 titles a year. Please contact by telephone before submitting mss and proposals.

## CBD Research Ltd
Chancery House, 15 Wickham Road, Beckenham BR3 5JS
Ⓣ 020 8650 7745    Ⓕ 020 8650 0768
cbd@cbdresearch.com
www.cbdresearch.com
Managing Director *S.P.A. Henderson*
Approx. Annual Turnover £500,000

Founded 1961. Publishes directories and other reference guides to sources of information. No fiction. IMPRINT **Chancery House Press** Non-fiction of an esoteric/specialist nature for 'serious researchers and the dedicated hobbyist'. About 6 titles a year.
Ⓔ Royalties quarterly.

## CCV
▷ The Random House Group Ltd

## Cengage Learning Europe, Middle East and Africa (EMEA)
Cheriton House, North Way, Andover SP10 5BE
Ⓣ 01264 332424    Ⓕ 01264 342763
emea.enquiries@cengage.com
www.cengage.co.uk
CEO (Worldwide) *Ron Dunn*
CEO (Cengage Learning EMEA) *Jill Jones*

Publishing Director *Linden Harris*

Cengage Learning EMEA develops and publishes products for the educational, professional and trade library markets. DIVISIONS **Higher and Further Education** Established US imprints and a rapidly expanding indigenous programme; **Library Reference** E-reference publishing for libraries, schools and businesses. TITLES include *Hilgard & Atkinson Introduction to Psychology*; *Drury Management & Cost Accounting*; *Mankiw & Taylor Economics*; *State Papers Online I-IV*. Unsolicited material aimed at students is welcome but email or telephone in the first instance to check out the idea.

## Centaur Press
▷ Open Gate Press

## Century
▷ The Random House Group Ltd

## CF4K
▷ Christian Focus Publications

## Chambers Harrap
▷ Hachette UK

## Chancery House Press
▷ CBD Research Ltd

## Channel 4 Books
▷ Transworld Publishers

## Chapman Publishing
4 Broughton Place, Edinburgh EH1 3RX
Ⓣ 0131 557 2207
chapman-pub@blueyonder.co.uk
www.chapman-pub.co.uk
Managing Editor *Dr Joy Hendry*

A venture devoted to publishing works by 'the best of Scottish writers, both up-and-coming and established, already published in *Chapman*, Scotland's leading literary magazine'. Only publishing one or two titles a year, mainly poetry. Not intending to publish any more titles, at least for 2010, given the current climate but will resume in due course. TITLES *Winter Barley* George Gunn; *Lure* Dilys Rose; *Wild Women Series – Wild Women of a Certain Age* Magi Gibson; *Ye Cannae Win* Janet Paisley. No unsolicited mss. Please note that the magazine, *Chapman*, remains open to all contributions, from anywhere in the world (see entry under *Magazines*).
Ⓔ Royalties annually.

## Charnwood
▷ F.A. Thorpe Publishing

## Chartered Institute of Personnel and Development (CIPD)

151 The Broadway, London SW19 1JQ

☎ 020 8612 6200    🖷 020 8612 6201

publish@cipd.co.uk

www.cipd.co.uk/bookstore

Publishing Manager *Ruth Lake*

Approx. Annual Turnover £3 million

Part of CIPD Enterprises Limited. Publishes on personnel, training and management. A list of around 160 titles including looseleaf and online subscription products, student textbooks and professional information resources. New product proposals welcome.

🄴 Royalties annually or flat fees paid.

## Chastleton Travel
▷ Arris Publishing Ltd

## Chatto & Windus
▷ The Random House Group Ltd

## Cherrytree Books
▷ Evans Brothers Ltd

## Chicken House Publishing

2 Palmer Street, Frome BA11 1DS

☎ 01373 454488    🖷 01373 454499

chickenhouse@doublecluck.com

Chairman/Managing Director *Barry Cunningham*

Children's publishing house founded in 2000. Acquired by Scholastic Inc. in May 2005. Publishes books 'that are aimed at real children' – fiction, original picture books. TITLES Cornelia Funke's *Inkheart* Trilogy. Aiming to publish about 25 titles a year. Due to the large number of submissions received, Chicken House is not accepting any unsolicited picture books or fiction for under 9s; fiction for 9-16 year-olds is only accepted in the form of submissions to the annual The Times/Chicken House Children's Fiction Competition (see entry under *Prizes*).

🄴 Royalties twice-yearly.

✱*Authors' update* Set up as a 'small, creative company', Chicken House is now part of Scholastic. But editorial freedom is preserved 'with no rules about what to publish as long as it's good'.

## Child's Play (International) Ltd

Ashworth Road, Bridgemead, Swindon SN5 7YD

☎ 01793 616286    🖷 01793 512795

office@childs-play.com

www.childs-play.com

Chief Executive *Neil Burden*

Founded in 1972, Child's Play is an independent publisher specializing in learning through play, whole child development, life-skills and values. Publishes books, games, A-V materials and plush toys. TITLES *Big Hungry Bear; There Was an Old Lady; Our Cat Cuddles; Royston Knapper; Sign and Sing-Along; Just Like Me; The Ding Dong Bag; It's Raining, It's Pouring; The Flower; Max and the Doglins; The Ant and the Big Bad Bully Goat; The Wim Wom from the Mustard Mill.* Unsolicited mss welcome but no novels. Send s.a.s.e. for return or response. Expect to wait two months for a reply.

🄴 Outright or royalty payments are subject to negotiation.

## Chimera
▷ Pegasus Elliot Mackenzie Publishers Ltd

## Chris Andrews Publications Ltd (trading as Oxford Picture Library)

15 Curtis Yard, North Hinksey Lane, Oxford OX2 0LX

☎ 01865 723404    🖷 01865 725294

chris.andrews1@btclick.com

www.cap-ox.co.uk

Directors *Chris Andrews, Virginia Andrews*

Founded 1982. Publishes coffee-table, scenic books, calendars, diaries, cards and posters of the *central* area of the UK and the Channel Islands. TITLES *Romance of Oxford; Romance of the Cotswolds; Romance of the Thames & Chilterns.* Also owns the **Oxford Picture Library** (see entry under *Picture Libraries*).

## Christian Focus Publications

Geanies House, Fearn, Tain IV20 1TW

☎ 01862 871011    🖷 01862 871699

info@christianfocus.com

www.christianfocus.com

Owner *Balintore Holdings plc*

Chairman *R.W.M. Mackenzie*

Managing Director *William Mackenzie*

Editorial Manager *Willie Mackenzie*

Children's Editor *Catherine Mackenzie*

Approx. Annual Turnover £1.5 million

Founded 1979 to produce children's books for the co-edition market. Now a major producer of Christian books. Publishes adult and children's books, including some fiction for children but not adults. No poetry. Publishes for all English-speaking markets, as well as the UK. IMPRINTS **CF4K** Children's books; **Christian Focus** General books; **Mentor** Study books; **Christian Heritage** Classic reprints. About 90 titles a year. Unsolicited mss, synopses and ideas welcome from Christian writers. See website for submission criteria.

🄴 Royalties annually.

## Christian Heritage
▷ Christian Focus Publications

## Christopher Helm
▷ A.&C. Black Publishers Ltd

## Chrome Dreams
▷ entry under Audio Books

## Chrysalis Books Group
▷ Anova Books

## Church House Publishing
▷ Hymns Ancient & Modern Ltd

## Churchill Livingstone
▷ Elsevier Ltd

## Churchwarden Publications Ltd

PO Box 420, Warminster BA12 9XB
☎ 01985 840189    📠 01985 840243
enquiries@churchwardenbooks.co.uk
Managing Director *John Stidolph*

Founded 1994. Publishes books and stationery for churchwardens and church administrators including *The Churchwarden's Yearbook* (annual).

## Cicerone Press

2 Police Square, Milnthorpe LA7 7PY
☎ 01539 562069    📠 01539 563417
info@cicerone.co.uk
www.cicerone.co.uk
Managing/Editorial Director *Jonathan Williams*

Founded 1969. Guidebook publisher for outdoor enthusiasts. No fiction or poetry. TITLES *Hillwalker's Guide to Mountaineering*; *Tour of Mont Blanc*; *Coast to Coast Trail*. SERIES include *Alpine Walking*; *International Walking*; *British Long-Distance Trails*, *Cycling* and *Winter Activities*. About 30 titles a year. No unsolicited mss; synopses and ideas considered.
Ⓔ Royalties twice-yearly.

❋*Authors' update* **Defying the dive in travel book sales, Cicerone has taken off with its popular walking guides.**

## Cico Books

20–21 Jockey's Fields, London WC1R 4BW
☎ 020 7025 2280    📠 020 7025 2281
mail@cicobooks.co.uk
Publisher *Cindy Richards*

Founded in 1999 and acquired by **Ryland Peters & Small Limited** in 2006. Publishes highly illustrated lifestyle books covering interiors, crafts, mind, body and spirit and gift. About 30 titles a year. No unsolicited mss but synopses and ideas welcome.
Ⓔ Royalties twice-yearly.

## Cinebook Ltd

56 Beech Avenue, Canterbury CT4 7TA
☎ 01227 731368
info@cinebook.co.uk
www.cinebook.com

Managing Director *Oliver Cadic*

Founded 2005. Publisher of Franco-Belgian 'Ninth Art' (graphic novels) translated into English. Characters include Lucky Luke, Iznogoud, Blaker & Mortimer, Thorgal, Spirou & Fantasio. 36 titles in 2009. No submissions.

## Clairview Books Ltd

Hillside House, The Square, Forest Row RH18 5ES
☎ 0870 486 3526
office@clairviewbooks.com
www.clairviewbooks.com
Managing Director *S. Gulbekian*
Approx. Annual Turnover £100,000

Founded 2000. Publishes non-fiction. General books challenging conventional thinking: mind, body and spirit, current affairs, the arts, science, health and therapy, etc. TITLES *Imperial America* Gore Vidal; *The Biodynamic Food and Cookbook* Wendy Cook; *My Descent Into Death* Howard Storm. 5 titles in 2009. No unsolicited material; send initial letter of enquiry. No poetry or fiction.
Ⓔ Royalties annually.

## Classic
▷ Ian Allan Publishing Ltd

## Clive Bingley Books
▷ Facet Publishing

## Colin Smythe Ltd

38 Mill Lane, Gerrards Cross SL9 8BA
☎ 01753 886000    📠 01753 886469
cs@colinsmythe.co.uk
www.colinsmythe.co.uk
Managing Director *Colin Smythe*
Approx. Annual Turnover £2.4 million

Founded 1966. Publishes Anglo-Irish literature, drama; criticism and Irish history. Acts as agent for Terry Pratchett. About 3 titles a year. No unsolicited mss.
Ⓔ Royalties annually/twice-yearly.

## Collector's Library
▷ CRW Publishing Ltd

## Collins
▷ HarperCollins Publishers Ltd

## Collins & Brown
▷ Anova Books

## Colourpoint Books

Colourpoint House, Jubilee Business Park, Jubilee Road, Newtownards BT23 4YH
☎ 028 9182 6339    📠 028 9182 1900
sales@colourpoint.co.uk
www.colourpoint.co.uk
Partners *Wesley Johnston, Malcolm Johnston*

Founded 1993. Publishes school textbooks and transport (covering the whole of the British

Isles), plus books of Irish, general interest and fiction. About 25 titles a year. Unsolicited material accepted but approach in writing in the first instance; include return postage, please.
£ Royalties twice-yearly.

## Columbia University Press
▷ University Presses of California, Columbia & Princeton Ltd

## Communication and Social Justice
▷ Troubador Publishing Ltd

## Communication Ethics
▷ Troubador Publishing Ltd

## Compendium Publishing Ltd
43 Frith Street, London W1D 4SA
☎ 020 7287 4570    📠 08451 303173
alan.greene@compendiumpublishing.com
www.compendiumpublishing.com
Managing Director *Alan Greene*
Editorial Director *Simon Forty*
Approx. Annual Turnover £4 million

Founded 2000. Publishes and packages for international publishing companies, general illustrated non-fiction: history, reference, hobbies, children's and educational transport and militaria. 70 titles in 2009 No unsolicited mss; synopses and ideas preferred.
£ Royalties twice-yearly or fees per project.

## Condor
▷ Souvenir Press Ltd

## Conran Octopus
▷ Octopus Publishing Group

## Constable & Robinson Ltd
3 The Lanchesters, 162 Fulham Palace Road, London W6 9ER
☎ 020 8741 3663    📠 020 8748 7562
enquiries@constablerobinson.com
www.constablerobinson.com
Chairman *Nick Robinson*
Managing Director *Pete Duncan*
Paperback Publisher *James Garbutt*

Archibald Constable, Walter Scott's publisher and the originator of the '3 decker novel', published his first titles in Edinburgh in 1795. Constable & Robinson Ltd are now in the third century of independent publishing under the Constable name. IMPRINTS **Constable** *Andreas Campomar, Becky Hardie* Current affairs, biography, history, psychology, travel, photography. **Corsair** *James Garbutt* Commercial fiction. **Robinson** SERIES Crime fiction (*Krystyna Green*); the 'Mammoth Book Of' (*Duncan Proudfoot*); 'Brief History' (*Leo Hollis*); the Overcoming psychology SERIES (*Fritha Saunders*); **Right Way** *Judith Mitchell* Food and drink, gardening, self-help, pastimes.

150 titles. Unsolicited sample chapters, synopses and ideas for books welcome. No mss; *no email submissions*. Enclose return postage.
£ Royalties twice-yearly.

✸*Authors' update* A strong list of popular interest titles has proved successful in connecting with the elusive general reader. Bucking the trend, sales increased by a healthy 23% in 2009.

## The Continuum International Publishing Group Limited
The Tower Building, 11 York Road, London SE1 7NX
☎ 020 7922 0880    📠 020 7922 0881
www.continuumbooks.com
CEO *Oliver Gadsby*
Approx. Annual Turnover £10 million

Founded in 1999 by a buy-out of the academic and religious publishing of Cassell and the acquisition of Continuum New York. Publishes academic, religious and general books.

DIVISIONS **Academic Humanities** Publisher *Anna Fleming* Education, social sciences, literature, film and music, philosophy (including the **Thoemmes** imprint), linguistics, biblical studies and theology printed under the **T&T Clark** imprint. TITLES *Getting the Buggers to Behave; Teaching in Further Education; What Philosophy Is; The Guerrilla Film Makers Handbook.* **General Trade & Continuum Religion** Publishing Director *Robin Baird-Smith* Publishes under **Continuum Burns & Oates** (RC books) TITLES *Seven Basic Plots; The Home We Build Together.* IMPRINTS **Continuum; T&T Clark Intl; Thoemmes Continuum; Burns & Oates.** 600 titles in 2009. Unsolicited synopses and ideas within the subject areas listed above are welcome; approach in writing in the first instance. OVERSEAS SUBSIDIARIES The Continuum International Publishing Group Inc., New York and Harrisburg.
£ Royalties twice-yearly.

✸*Authors' update* With this publisher what might be thought to be academic books are often written in an accessible style that appeals to the non-specialist. Strong on controversial religious and social themes.

## Conway
▷ Anova Books

## Corgi
▷ Transworld Publishers

## Corgi Children's Books
▷ The Random House Group Ltd

## Corinthian Books
▷ Icon Books Ltd

## Cornerstone
▷ The Random House Group Ltd

## Coronet
▷ Hachette UK

## Corsair
▷ Constable & Robinson Ltd

## Corvus
▷ Atlantic Books

## Country Publications Ltd

The Water Mill, Broughton Hall, Skipton
BD23 3AG
☎ 01756 701381    🖷 01756 701326
editorial@dalesman.co.uk
www.dalesman.co.uk
Book Editor *Mark Whitley*

Publishers of *Countryman, Dalesman, Cumbria* and *Down Your Way* magazines, and regional books covering the North of England. Subjects include walking, guidebooks, history, humour and folklore. About 10 titles a year. Will consider mss on subjects listed above, relating to the North.
Ⓔ Royalties annually.

## Countryside Books

2 Highfield Avenue, Newbury RG14 5DS
☎ 01635 43816    🖷 01635 551004
info@countrysidebooks.co.uk
www.countrysidebooks.co.uk
Publisher *Nicholas Battle*

Founded in 1976. Publishes local interest books on regional subjects, mostly by English county. These include walking guides, local history and nostalgia. General interest subjects include family history, architecture, military and practical books about the English countryside. Around 500 titles available. About 50 titles a year. Unsolicited mss and synopses welcome but no fiction, poetry, natural history or personal memories. Check the website for suitability before any submission.
Ⓔ Royalties twice-yearly.

## CRC Press
▷ Taylor & Francis Group

## Crème de la Crime Ltd

PO Box 523, Chesterfield S40 9AT
☎ 01246 520835    🖷 01246 520835
info@cremedelacrime.com
www.cremedelacrime.com
Managing Editor *Lynne Patrick*
Approx. Annual Turnover £60,000

Founded in 2003 to discover and publish crime fiction by new authors. Up to 6 titles annually. Welcomes enquiries and unsolicited submissions, 'especially from talented but unpublished authors

who are urged to read the detailed guidelines on the website'.
Ⓔ Royalties annually.

## Cressrelles Publishing Co. Ltd

10 Station Road Industrial Estate, Colwall, Malvern WR13 6RN
☎ 01684 540154    🖷 01684 540154
simon@cressrelles.co.uk
www.cressrelles.co.uk
Managing Director *Leslie Smith*

Publishes a range of local interest books and drama titles. IMPRINTS **J. Garnet Miller** Plays and theatre texts; **Kenyon-Deane** Plays and drama textbooks; **New Playwrights' Network** Plays. About 6–12 new play titles a year. Submissions welcome.

## Crimson Publishing

Westminster House, Kew Road, Richmond TW9 2ND
☎ 020 8334 1600    🖷 020 8334 1601
info@crimsonpublishing.co.uk
www.crimsonpublishing.co.uk
Chairman *David Lester*

Founded 2006. Part of Crimson Business Ltd. Non-fiction publisher of books 'to improve the way you live'. Subjects covered: business, travel, working abroad, gap year, careers and education, parenting.

DIVISIONS **Crimson Publishing**; **Trotman** (see entry). IMPRINTS **Crimson**; **Vacation Work**; **Trotman**; **White Ladder**; **Pathfinder Walking Guides**. 30 titles in 2009. Welcomes carefully considered unsolicited mss, synopses and ideas for books in the subject areas listed above. No fiction or books on topics not listed. Approach by email.
Ⓔ Royalties twice-yearly.

## Crombie Jardine Publishing Limited

Office 2, 3 Edgar Buildings, George Street, Bath BA1 2FJ
☎ 01225 464445
catriona@crombiejardine.com
www.crombiejardine.com
Publishing Director *Catriona Jardine*

Specialists in fun and quirky humour and gift books. TITLES include *That's Life*; *101 Things to Do at University*; *Shag Yourself Slim*; *Unpleasant Words*; *What Shat That?*; *Weird Websites*; *Twitter*; *The Arse in Art*. 20 titles a year. No postal submissions. Send synopsis by email in the first instance.
Ⓔ Flat fee paid; no royalties.

✱*Authors' update* **The ultimate gift books for those with a quirky sense of humour.**

## Crossway
> Inter-Varsity Press

## Crown House Publishing

Crown Buildings, Bancyfelin, Carmarthen
SA33 5ND
☎ 01267 211345    ℻ 01267 211882
books@crownhouse.co.uk
www.crownhouse.co.uk
Managing Director *David Bowman*

Founded 1998. Publisher of Neuro Linguistic Programming (NLP), hypnosis, counselling and psychotherapy, personal, business and life coaching, education, accelerated learning, thinking skills and personalized learning. 25 titles in 2010. 'We are happy to receive enquiries from authors wishing us to consider their work for publication. Initially, please email us your thoughts or ideas in no more than 300 words to brandell@crownhouse.co.uk'.
£ Royalties twice-yearly.

## The Crowood Press Ltd

The Stable Block, Crowood Lane, Ramsbury, Marlborough SN8 2HR
☎ 01672 520320    ℻ 01672 520280
enquiries@crowood.com
www.crowood.com
Chairman *John Dennis*
Managing Director *Ken Hathaway*

Publishes sport and leisure titles, including animal and land husbandry, climbing and walking, maritime, country sports, equestrian, fishing and shooting; also crafts, dogs, gardening, DIY, theatre, natural history, aviation, military history and motoring. IMPRINT **Airlife Publishing** Specialist aviation titles for pilots, historians and enthusiasts. About 70 titles a year. Preliminary letter preferred in all cases.
£ Royalties annually.

## CRW Publishing Ltd

69 Gloucester Crescent, London NW1 7EG
☎ 020 7485 5764    ℻ 0870 751 7254
marcus.clapham@crw-publishing.co.uk
www.collectors-library.com
*Also at:* 6 Turville Barns, Eastleach, Cirencester GL7 3QB
Chairman *Cameron Brown*
Editorial Director *Marcus Clapham*
Approx. Annual Turnover £2 million

Founded 2003. Publishes literary classics, gift, children's, boxed sets. IMPRINTS **Collector's Library; Collector's Library Editions in Colour; Collector's Library Omnibus Editions**. 21 titles in 2009. No unsolicited mss.
£ Royalties annually.

## Custom Publishing
> The Orion Publishing Group Limited

## D&B Publishing

80 Walsingham Road, Hove BN3 4FF
☎ 01273 711443
info@dandbpublishing.com
www.dandbpublishing.com
Joint Managing Directors *Dan Addelman, Byron Jacobs*

Founded 2002. Publishes games books, specializing in gambling, primarily. Unsolicited mss, synopses and ideas welcome; approach by email in the first instance.
£ Royalties annually.

## Darton, Longman & Todd Ltd

1 Spencer Court, 140–142 Wandsworth High Street, London SW18 4JJ
☎ 020 8875 0155    ℻ 020 8875 0133
mail@darton-longman-todd.co.uk
www.dltbooks.com
Editorial Director *Brendan Walsh*
Editorial Assistant/Rights Officer *Will Parkes*
Approx. Annual Turnover £1 million

A leading independent publisher of books on spirituality and religion. Unique among UK religious publishers in being jointly owned and managed by all its staff members. While predominantly Christian, DLT publishes books from different backgrounds and traditions. TITLES *Jerusalem Bible*; *New Jerusalem Bible*; *NRSV Bible: Catholic Edition*; *Return of the Prodigal Son*; *God of Surprises*; *The Enduring Melody*; *Hostage in Iraq*. About 36 titles a year. Information on submissions available on the website.
£ Royalties twice-yearly.

## David & Charles Publishers

Brunel House, Forde Close, Newton Abbot TQ12 4PU
☎ 01626 323200    ℻ 01626 323319
postmaster@davidandcharles.co.uk
www.davidandcharles.co.uk
Managing Director *James Woollam*
Director of Editorial and Design *Alison Myer*

Founded 1960, owned by F+W Media. Publishes illustrated non-fiction for international markets, specializing in needlecraft, crafts, art techniques, practical photography, military history, nostalgia, transport and equestrian. No fiction, poetry or memoirs. TITLES *Ghosts Caught on Film*; *Sew Pretty Homestyle*; *John Howe Fantasy Art Workshop*; *Keeping Bees and Making Honey*; *Memories of Steam*. About 100 titles a year. Unsolicited mss will be considered if return postage is included; synopses and ideas welcome

for the subjects listed above. 'Please allow at least eight weeks for a response.'
Ⓔ Royalties twice-yearly or flat fees.

## David Fickling Books
▷ The Random House Group Ltd

## David Fulton Publishers Ltd
2 Park Square, Milton, Abingdon OX14 4RN
☏ 020 7017 6000  Ⓕ 020 8996 3622
info@routledge.co.uk
www.routledge.co.uk
Owner *Taylor & Francis Group*
Senior Publishers *Bruce Roberts (primary, textbooks & classroom), Anna Marie Kino (early years, secondary & education studies), Alison Foyle (special educational needs)*
Approx. Annual Turnover £2.4 million

Founded 1987. Part of Routledge Education. Publishes textbooks and practical books for students, trainee and working teachers at all levels of the curriculum (foundation stage to post compulsory). Special educational needs books also published in collaboration with the National Association for Special Educational Needs (NASEN); some overlap into other areas of health and social care, therapy, educational psychology, and speech and language therapy. A growing list of Early Years books for pre-school and nursery staff. About 100 titles a year. No unsolicited mss; synopses and ideas for books welcome.
Ⓔ Royalties annually.

## Davies-Black
▷ Nicholas Brealey Publishing

## Debrett's Ltd
18–20 Hill Rise, Richmond TW10 6UA
☏ 020 8939 2250  Ⓕ 020 8939 2251
enquiries@debretts.com
www.debretts.com
Head of Publishing *Liz Wyse*
Chairman *Conrad Free*

Biographical reference plus etiquette, correct form and other social guides; also diaries. Book TITLES include triennial *Debrett's Peerage & Baronetage* and annual *People of Today* (as book, CD-ROM and online); *Debrett's A-Z of Modern Manners*; *Debrett's Guide for the Modern Gentleman*; *Etiquette for Girls*; *Correct Form*; *Debrett's Wedding Guide*. Book and article proposals welcome.

## Dedalus Ltd
Langford Lodge, St Judith's Lane, Sawtry PE28 5XE
☏ 01487 832232
info@dedalusbooks.com
www.dedalusbooks.com
Chairman *Juri Gabriel*
Managing Director *Eric Lane*

Approx. Annual Turnover £175,000

Founded 1983. Publishes contemporary European fiction and classics and original literary fiction. TITLES *The Dedalus Book of Absinthe* Phil Baker; *The Arabian Nightmare* Robert Irwin; *Memoirs of a Gnostic Dwarf* David Madsen; *Music in a Foreign Language* Andrew Crumey (Saltire Best First Book Award winner); *The Father of Locks* Andrew Killeen; *Made in Yaroslavl* Jeremy Weingard; *Mr Dick or the Tenth Book* Jean-Pierre Ohl.

DIVISIONS/IMPRINTS **Original Fiction in Paperback**; **Dedalus Europe 1992–2012**; **Literary Concept Books**. Welcomes submissions for original fiction and books suitable for its list but 'most people sending work in have no idea what kind of books Dedalus publishes and merely waste their efforts'. Author guidelines on website. Particularly interested in intellectually clever and unusual fiction. A letter about the author should always accompany any submission. No replies without s.a.s.e.
Ⓔ Royalties annually.

✱*Authors' update* Put at risk by the withdrawal of its Arts Council grant, Dedalus was rescued by a cash injection from Informa publishing group. Thus, virtue and originality are rewarded.

## Derby Books Publishing
3 The Parker Centre, Mansfield Road, Derby DE21 4SZ
☏ 01332 384235
www.breedonbooks.co.uk

Derby Books Publishing was established in 2009 following Breedon Books going into administration. Publishes football and sport, local history, old photographs, heritage and ghost stories with a local slant. Unsolicited mss, synopses and ideas welcome if accompanied by s.a.s.e. No poetry, children's stories or fiction.

## Despatches
▷ Reportage Press

## Dewi Lewis Publishing
8 Broomfield Road, Heaton Moor, Stockport SK4 4ND
☏ 0161 442 9450  Ⓕ 0161 442 9450
mail@dewilewispublishing.com
www.dewilewispublishing.com
Owners *Dewi Lewis, Caroline Warhurst*
Approx. Annual Turnover £280,000

Founded 1994. Publishes photography and visual arts. TITLES *Common Sense* Martin Parr; *New York 1954–5* William Klein. IMPRINT **Dewi Lewis Media** Non-fiction, biography, sports and celebrity books. TITLES *David Beckham: Made in Manchester* Eamonn and James Clarke; *500*

*Flowers* Roger Camp. About 17 titles annually. Not currently accepting new fiction submissions. For all submissions it is essential to check the website first.
Ⓔ Royalties annually.

### Digital Press
▷ Elsevier Ltd

### DK Eyewitness Travel
▷ Penguin Group (UK)

### Donhead Publishing Ltd
Lower Coombe, Donhead St Mary, Shaftesbury SP7 9LY
☎ 01747 828422    🖷 01747 828522
jillpearce@donhead.com
www.donhead.com
Contact *Jill Pearce*

Founded in 1990 to specialize in publishing architectural conservation books for building practitioners. Publishes building and architecture only. TITLES *Windows*; *Stone Conservation, Plastering*; *Preserving Post-War Heritage*; *McKay's Building Construction*; *Survey and Repair of Traditional Buildings*; *Encyclopaedia of Architectural Terms*; *Cleaning Historic Buildings*; *Brickwork*; *Practical Stone Masonry*; *Conservation of Timber Buildings*; *Surveying Historic Buildings*; *Journal of Architectural Conservation* (3 issues yearly). 6 titles a year. Synopses and ideas welcome.

### Dorling Kindersley Ltd
▷ Penguin Group (UK)

### Doubleday
▷ Transworld Publishers

### Doubleday Children's Books
▷ The Random House Group Ltd

### Drake Educational Associates
St Fagans Road, Fairwater, Cardiff CF5 3AE
☎ 029 2056 0333    🖷 029 2956 0313
info@drakeed.com
www.drakeed.com
Managing Director *R.G. Drake*

Literacy, phonics and language development games and activities. Ideas and scripts in these fields welcome. Also resources for subject areas in the primary school and modern languages at KS2.

### Dref Wen
28 Church Road, Whitchurch, Cardiff CF14 2EA
☎ 029 2061 7860    🖷 029 2061 0507
gwilym@drefwen.com
Chairman *R. Boore*
Managing Director *G. Boore*

Founded 1970. Publishes Welsh language and bilingual children's books, Welsh and English educational books for Welsh learners.
Ⓔ Royalties annually.

### Duncan Baird Publishers
Castle House, 75–76 Wells Street, London W1T 3QH
☎ 020 7323 2229    🖷 020 7580 5692
info@dbairdpub.co.uk
www.dbp.co.uk
Managing Director *Duncan Baird*
Editorial Director *Bob Saxton*
Approx. Annual Turnover £8 million

Founded in 1992 to publish and package co-editions overseas and went on to launch its own publishing operation in 1998. Publishes illustrated cultural reference, world religions, health, mind, body and spirit, lifestyle, graphic design. 70 titles in 2009. No unsolicited mss. Synopses and ideas welcome; approach in writing in the first instance with s.a.s.e. No fiction or UK-only subjects.
Ⓔ Royalties twice-yearly.

### Dunedin Academic Press Ltd
Hudson House, 8 Albany Street, Edinburgh EH1 3QB
☎ 0131 473 2397    🖷 01250 870920
mail@dunedinacademicpress.co.uk
www.dunedinacademicpress.co.uk
Director *Anthony Kinahan*

Founded 2001. Publishes academic and serious general non-fiction. Considerable experience in academic and professional publishing. 18 titles in 2009. Not interested in science (other than earth sciences), poetry, fiction or children's. Check website for active subject areas. No unsolicited mss. Synopses and ideas welcome. Approach first in writing outlining proposal and identifying market. Proposal guidelines are available via the website.
Ⓔ Royalties annually.

### Ebury Publishing/Ebury Press
▷ The Random House Group Ltd

### The Economist Books
▷ Profile Books

### Eden
▷ Transworld Publishers

### Edexcel
▷ Pearson

### Edinburgh University Press
22 George Square, Edinburgh EH8 9LF
☎ 0131 650 4218    🖷 0131 662 0053
www.euppublishing.com
www.eupjournals.com
Chairman *Ivon Asquith*

Chief Executive *Timothy Wright*
Deputy Chief Executive/Head of Book Publishing *Jackie Jones*
Head of Journals *Sarah Edwards*
Senior Commissioning Editor *Nicola Carr*
Commissioning Editors *Carol Macdonald, Esmé Watson*

Publishes academic and scholarly books (and journals): African studies, ancient history and classics, film and media studies, Islamic studies, linguistics, literary criticism, philosophy, politics, Scottish studies and religious studies. About 120 titles a year. Email submissions accepted; consult website for guidelines.
Ⓔ Royalties paid annually.

## Edward Elgar Publishing Ltd

The Lypiatts, 15 Lansdown Road, Cheltenham GL50 2JA
Ⓣ 01242 226934    Ⓕ 01242 262111
info@e-elgar.co.uk
www.e-elgar.com
Managing Director *Edward Elgar*

Founded 1986. International publisher in economics, the environment, public policy, business and management and law. TITLES *Who's Who in Economics* (3rd ed.); *Handbook of Environmental and Resource Economics*; *Who's Who in the Management Sciences*. 280 titles in 2009. No unsolicited mss; synopses and ideas in the subject areas listed above welcome. Approach by letter or email; no telephone inquiries.

## Egmont Press

3rd Floor, Beaumont House, Avonmore Road, London W14 8TS
Ⓣ 020 7605 6600    Ⓕ 020 7605 6601
childrensreader@euk.egmont.com
www.egmont.co.uk
UK Managing Director *Robert McMenemy*
Director of Press *Cally Poplak*

Founded 1878. Publishes children's books. The list publishes authors such as Michael Morpurgo, Jenny Nimmo, William Nicholson, Enid Blyton, Andy Stanton, Julia Golding; illustrators like Helen Oxenbury, Jan Fearnley, Lydia Monks; and classic children's characters including Winnie the Pooh and Tintin. 'We accept unsolicited mss but by email only (address above); anything received in hard copy will be recycled.'

## Egmont Publishing

239 Kensington High Street, London W8 6SA
Ⓣ 020 7761 3500    Ⓕ 020 7761 3510
initialsurname@euk.egmont.com
www.egmont.co.uk
UK Managing Director *Rob McMenemy*
Managing Director, Egmont Publishing Group *David Riley*

Publishes children's books: annuals, activity books, novelty books, film and TV tie-ins. Also, children' magazines. Licensed character list includes Thomas the Tank Engine, Barbie, Mr Men, Postman Pat, Ben 10, Rupert Bear, Winnie the Pooh.

## Eland Publishing Ltd

Third Floor, 61 Exmouth Market, London EC1R 4QL
Ⓣ 020 7833 0762    Ⓕ 020 7833 4434
info@travelbooks.co.uk
www.travelbooks.co.uk
Directors *Stephanie Allen, Rose Baring, John Hatt, Barnaby Rogerson*
Approx. Annual Turnover £250,000

Eland reprints classics of travel literature with a backlist of 60 titles, including *Naples '44* Norman Lewis; *Travels with Myself and Another* Martha Gelhorn; *Portrait of a Turkish Family* Irfan Orga; *Jigsaw* Sybille Bedford; *92 Acharnon Street* John Lucas (winner of the 2008 Dolman Best Travel Book Award). IMPRINTS **Eland** Classic travel literature; **Through Writers' Eyes** Travel literature collections; **Poetry of Place** Pocket anthologies. About 6–12 titles a year. No unsolicited mss. Postcards and emails welcome.
Ⓔ Royalties annually.

## Element
▷ **HarperCollins Publishers Ltd**

## Elliott & Thompson

27 John Street, London WC1N 2BX
Ⓣ 020 7831 5013    Ⓕ 020 7831 5011
www.eandtbooks.com
Publishers *Mark Searle, Lorne Forsyth*
Approx. Annual Turnover £100,000

Founded 2001. Non-fiction publisher of sport, music, history, finance and wine. Non-fiction proposals welcome; see website for guidelines. S.a.s.e. essential.

## Elm Publications/Training (wholly owned subsidiary of Elm Consulting Ltd)

Seaton House, Kings Ripton, Huntingdon PE28 2NJ
Ⓣ 01487 773359
elm@elm-training.co.uk
www.elm-training.co.uk
Managing Director *Sheila Ritchie*

Founded 1977. Publishes textbooks, teaching aids, educational resources and educational software in the fields of business and management for adult learners. Books and teaching/training resources are generally commissioned to meet specific business, management and other syllabuses. About 30 titles a year. Not seeking any new books in the near future.

## Elsevier Ltd

The Boulevard, Langford Lane, Kidlington,
Oxford OX5 1GB
☎ 01865 843000    🖷 01865 843010
www.elsevier.com
CEO, Science & Technology (books & journals)
*Herman van Campenhout*
CEO, Health & Sciences *Michael Hansen*

Parent company Reed Elsevier, Amsterdam.
Incorporates Pergamon Press. Academic and
professional reference books, scientific, technical
and medical books, journals, CD-ROMs
and magazines. IMPRINTS **Academic Press;
Architectural Press; Butterworth Heinemann;
Digital Press; Elsevier; Elsevier Advanced
Technology; Focal Press; Gulf Professional
Press; Made Simple Books; Morgan Kauffman;
Newnes; North-Holland**.

DIVISION **Elsevier (Health Sciences)** 32
Jamestown Road, London NW1 7BY ☎ 020 7424
4200 🖷 020 7483 2293 www.elsevier-health.com
Publishes scientific, technical, medical books and
journals. IMPRINTS **Baillière Tindall; Churchill
Livingstone; Mosby; Pergamon; Saunders**.
No unsolicited mss, but synopses and project
proposals welcome.
💷 Royalties annually.

## Emerging Communications Technologies
▷ Troubador Publishing Ltd

## Emissary Publishing

PO Box 33, Bicester OX26 4ZZ
☎ 01869 323447    🖷 01869 322552
www.emissary-publishing.com
Editorial Director *Val Miller, BA(Hons)*

Founded 1992. Publishes mainly humorous
paperback novels, including the complete set of
Peter Pook novels; fiction; non-fiction. No poetry
or children's. AUTHORS include Frederick E.
Smith and Roscoe Howells. No unsolicited mss or
synopses.
💷 Royalties paid according to contract.

## Emma Treehouse Ltd

The Studio, Church Street, Nunney BA11 4LW
☎ 01373 836233    🖷 01373 836299
richard.powell4@virgin.net
sales@emmatreehouse.com
www.emmatreehouse.com
Co-Directors *Richard Powell, David Bailey*
Approx. Annual Turnover £2.2 million

Founded 1992. Publishes children's pre-school
novelty books. No mss. Illustrations, synopses
and ideas for books welcome; write in the first
instance.
💷 Fees paid; no royalties.

## Empiricus Books
▷ Janus Publishing Company Ltd

## English Heritage (Publishing)

Kemble Drive, Swindon SN2 2GZ
☎ 01793 414453    🖷 01793 414769
www.english-heritage.org.uk
Head of Publishing *John Hudson*
Managing Editor *Robin Taylor*

English Heritage's publishing programme reflects
the organization's core interests in the historic
environment of England, covering archaeology,
architecture and general history titles. 20 titles
a year. 'All titles relate directly to the work of the
organization, so we do not accept unsolicited
material.'

## Enitharmon Press

26B Caversham Road, London NW5 2DU
☎ 020 7482 5967    🖷 020 7284 1787
books@enitharmon.co.uk
www.enitharmon.co.uk
Director *Stephen Stuart-Smith*
Contacts *Isabel Brittain, Jacqueline Gabbitas*

Founded 1967. An independent company with
an enterprising editorial policy, Enitharmon
has established itself as one of Britain's leading
literary presses. 'Patron of the new and the
neglected, Enitharmon also prides itself on the
success of its collaborations between writers and
artists, now published by its associate company',
**Enitharmon Editions**. Publishes poetry, literary
criticism, fiction, art and photography. TITLES
*Venice Fantasies* Peter Blake; *Ludbrooke and
Others* Alan Brownjohn; *Graceline* Jane Duran;
*Collected Poems* UA Fanthorpe; *Piccadilly Bongo*
Jeremy Reed and Marc Almond. No unsolicited
mss.
💷 Royalties according to contract.

## Epworth

Methodist Church House, 25 Marylebone Road,
London NW1 5JR
☎ 020 7486 5502
www.mph.org.uk
www.scm-canterburypress.co.uk
Chair *The Revd Michael Townsend*

Publishes Christian books only: philosophy,
theology, biblical studies, pastoralia, social
concern and Methodist studies. No fiction, poetry
or children's. About 12 titles a year. Unsolicited
mss, synopses and ideas welcome; send sample
chapter and contents with covering letter and
s.a.s.e.
💷 Royalties annually.

## Erotic Review Books

31 Sinclair Road, London W14 0NS
☎ 020 7371 1532    🖷 020 7603 8378

eros@eroticprints.org
www.eroticprints.org
Managing Director *J. Maclean*

Publishes books of erotic art and photography
and erotic literature, and the *Erotic Review*
magazine. 10 titles a year. Unsolicited material
accepted but approach in writing or by email in
the first instance with a summary or first chapter.
Ⓕ Royalties twice-yearly.

## EuroCrime
▷ Arcadia Books

## Europa Publications
▷ Taylor & Francis Group

## Evans Brothers Ltd
2A Portman Mansions, Chiltern Street, London
W1U 6NR
Ⓣ 020 7487 0920    Ⓕ 020 7487 0921
sales@evansbrothers.co.uk
www.evansbooks.co.uk
Managing Director *Brian Jones*
UK Publisher *Su Swallow*
Approx. Annual Turnover £4.5 million

Founded in 1908 by Robert and Edward Evans.
Publishes UK children's and educational books,
and educational books for Africa, the Caribbean
and Latin America. IMPRINTS **Cherrytree
Books; Zero to Ten.** OVERSEAS ASSOCIATES
in Kenya, Cameroon, Sierra Leone; Evans Bros
(Nigeria Publishers) Ltd. About 175 titles a year.
'Submissions welcome but cannot be returned.
Unable to respond to individual proposals that are
not accepted.'
Ⓕ Royalties annually.

## Everyman
▷ The Orion Publishing Group Ltd

## Everyman Chess
▷ Gloucester Publishers Plc

## Everyman's Library
Northburgh House, 10 Northburgh Street,
London EC1V 0AT
Ⓣ 020 7566 6350    Ⓕ 020 7490 3708
books@everyman.uk.com
Publisher *David Campbell*
Approx. Annual Turnover £3.5 million

Founded 1906. Publishes hardback classics of
world literature, pocket poetry anthologies,
children's books and travel guides. Publishes no
new titles apart from poetry anthologies; only
classics (no new authors). AUTHORS include
Bulgakov, Bellow, Borges, Heller, Marquez,
Nabokov, Naipaul, Orwell, Rushdie, Updike,
Waugh and Wodehouse. No unsolicited mss.
Ⓕ Royalties annually.

## Exley Publications Ltd
16 Chalk Hill, Watford WD19 4BG
Ⓣ 01923 474480    Ⓕ 01923 800440
editorial@helenexley.com
www.helenexleygiftbooks.com
Editorial Director *Helen Exley*

Founded 1976. Independent family company.
Publishes giftbooks, quotation anthologies and
humour. No submissions, please.

## Expert Books
▷ Transworld Publishers

## F.A. Thorpe Publishing
The Green, Bradgate Road, Anstey LE7 7FU
Ⓣ 0116 236 4325    Ⓕ 0116 234 0205
Group Chief Executive *Robert Thirlby*
Approx. Annual Turnover £6 million

Founded in 1964 to supply large print books
to libraries. Part of the Ulverscroft Group Ltd.
Publishes fiction and non-fiction large print
books. No educational, gardening or books that
would not be suitable for large print. IMPRINTS
**Charnwood; Ulverscroft; Linford Romance;
Linford Mystery; Linford Western.** No
unsolicited material.

## Faber & Faber Ltd
Bloomsbury House, 74–77 Great Russell Street,
London WC1B 3DA
Ⓣ 020 7927 3800    Ⓕ 020 7927 3801
www.faber.co.uk
Chief Executive *Stephen Page*
Approx. Annual Turnover £15.5 million

Geoffrey Faber founded the company in the
1920s, with T.S. Eliot as an early recruit to
the board. The original list was based on
contemporary poetry and plays (the distinguished
backlist includes Eliot, Auden and MacNeice).
Publishes poetry and drama, children's, fiction,
film, music, politics, biography. In June 2008
launched **Faber Finds**, a new imprint bringing
back classic titles via print on demand. DIVISIONS
**Fiction** *Lee Brackstone, Hannah Griffiths, Angus
Cargill, Sarah Savitt* AUTHORS P.D. James, Peter
Carey, Rachel Cusk, Louise Doughty, Giles Foden,
Michael Frayn, Petina Gappah, Jane Harris,
Tobias Hill, Kazuo Ishiguro, Barbara Kingsolver,
Milan Kundera, Hanif Kureishi, John Lanchester,
John McGahern, Rohinton Mistry, Lorrie Moore,
Andrew O'Hagan, DBC Pierre, Jane Smiley,
Emily Woof; **Children's** *Julia Wells* AUTHORS
Philip Ardagh, Ricky Gervais, Harry Hill, Russell
Stannard, G.P. Taylor; **Film** *Walter Donohue* and
**Plays** *Dinah Wood* AUTHORS Samuel Beckett,
Alan Bennett, David Hare, Brian Friel, Rebecca
Lenkiewicz, Patrick Marber, Harold Pinter, Tom
Stoppard, Woody Allen, John Boorman, Joel and
Ethan Coen, John Hodge, Martin Scorsese; **Music**

*Belinda Matthews* AUTHORS Edward Blakeman, John Bridcut, Humphrey Burton, Jonathan Carr, Rupert Christiansen, James Inverne, Nicholas Kenyon, Richard Morrison, Jann Parry, Michael Tanner, Susan Tomes, Richard Wigmore, Elizabeth Wilson; **Poetry** *Paul Keegan, Matthew Hollis* AUTHORS Simon Armitage, Douglas Dunn, Seamus Heaney, Ted Hughes, Paul Muldoon, Daljit Nagra, Tom Paulin; **Non-fiction** *Neil Belton, Julian Loose* AUTHORS John Carey, Roland Chambers, Simon Garfield, John Gray, Jan Morris, Francis Spufford, Frances Stonor Saunders, Jenny Uglow.
£ Royalties twice-yearly.

⊛*Authors' update* **Expanding well ahead of the norm, one of the few generally recognizable publishing brands is venturing beyond books to embrace events, seminars and other writer-led spinoffs. Faber is the prime mover in the successful Independent Alliance of publishers.**

## Facet Publishing

7 Ridgmount Street, London WC1E 7AE
℡ 020 7255 0590    🖷 020 7255 0591
info@facetpublishing.co.uk
www.facetpublishing.co.uk
Publishing Director *Helen Carley*

Publishing arm of **CILIP: The Chartered Institute of Library and Information Professionals** (formerly the Library Association). Publishes library and information science, monographs, reference and bibliography aimed at library and information professionals. IMPRINTS **Library Association Publishing; Clive Bingley Books; Facet Publishing**. Over 200 titles in print, including *The New Walford* and *AACR2*. 25–30 titles a year. Unsolicited mss, synopses and ideas welcome provided material falls firmly within the company's specialist subject areas.
£ Royalties annually.

## Featherstone Education
▷ A.&C. Black Publishers Ltd

## The Feel Good Factory
▷ Infinite Ideas

## Fernhurst Books
▷ Wiley Nautical

## Fig Tree
▷ Penguin Group (UK)

## Findhorn Press Ltd

305A The Park, Findhorn IV36 3TE
℡ 01309 690582    🖷 0131 777 2711
info@findhornpress.com
www.findhornpress.com
Director *Thierry Bogliolo*

Founded 1971. Publishes mind, body and spirit, New Age and healing. About 24 titles a year.

Unsolicited synopses and ideas via email only if they come within Findhorn's subject areas. No children's books, fiction or poetry. 'We receive over 1,000 submissions annually so please check our website and look at the type of books we publish prior to contacting us.'
£ Royalties twice-yearly.

## Fine Wine Editions
▷ Quarto Publishing plc

## First & Best in Education Ltd

Hamilton House, Earlstrees Court, Earlstrees Road, Corby NN17 4HH
℡ 01536 399005    🖷 01536 399012
editorial@firstandbest.co.uk
www.firstandbest.co.uk
www.shop.firstandbest.co.uk
Publisher *Tony Attwood*
Editor *Anne Cockburn*

Publishers of over 1,000 educational books of all types for all ages of children and for parents and teachers; a series of books on marketing and a new series of books about Arsenal F.C. TITLES *Raising Grades Through Study Skills*; *Business Sponsorship of Secondary Schools*; *Policy Documents for Day Nurseries and Nursery Units*; *Making the Arsenal*; *Direct Marketing to Schools*. IMPRINT **School Improvement Reports**. 'Currently looking for books in our areas of publication.' Check the website for submission guidelines.
£ Royalties twice-yearly.

## Fitzgerald Publishing

89 Ermine Road, Ladywell, London SE13 7JJ
℡ 020 8690 0597
fitzgeraldbooks@yahoo.co.uk
Managing Editors *Tim Fitzgerald, Michael Fitzgerald*
General Editor *Andrew Smith*

Founded 1974. Specializes in scientific studies of insects and spiders. TITLES *The Tarantula*; *Keeping Spiders and Insects in Captivity*; *Tarantulas of the USA*; *Scorpions of Medical Importance* (books) and *Earth Tigers – Tarantulas of Borneo*; *Desert Tarantulas* (TV/video documentaries). 1–2 titles a year. Unsolicited mss, synopses and ideas for books welcome. Also considers video scripts for video documentaries.

## Fitzjames Press
▷ Motor Racing Publications

## Floris Books

15 Harrison Gardens, Edinburgh EH11 1SH
℡ 0131 337 2372    🖷 0131 347 9919
floris@florisbooks.co.uk
www.florisbooks.co.uk

Managing Director *Christian Maclean*
Editors *Christopher Moore, Sally Martin*
Approx. Annual Turnover £550,000

Founded 1977. Publishes books related to the Steiner movement, including The Christian Community, as well as arts & crafts, children's (including fiction and picture books with a Scottish theme), history, religious, science, social questions and Celtic studies. Synopses and ideas are welcome for all books.
Ⓔ Royalties annually.

## Focal Press
▷ Elsevier Ltd

## Folens Limited

Waterslade House, Thame Road, Haddenham HP17 8NT
Ⓣ 0844 576 8115     Ⓕ 0844 576 8116
folens@folens.com
www.folens.com
Chairman *Dirk Folens*
Managing Director *John Cadell*

Founded 1987. A leading educational publisher. IMPRINTS **Folens; Belair**. About 150 titles a year. Unsolicited mss, synopses and ideas for educational books welcome.
Ⓔ Royalties annually.

## Fort Publishing Ltd

Old Belmont House, 12 Robsland Avenue, Ayr KA7 2RW
Ⓣ 01292 880693     Ⓕ 01292 270134
fortpublishing@aol.com
www.fortpublishing.co.uk
Chairman *Agnes Jane McCarroll*
Managing Director *James McCarroll*
Approx. Annual Turnover £95,000

Founded 1999. Publishes general non-fiction, history, sport, crime and local interest. TITLES *Glasgow Then and Now; Great Hull Stories; Evil Scotland; Ten Days That Shook Rangers.* 45 titles in 2009. No unsolicited mss; send synopses and ideas for books by post or email.
Ⓔ Royalties twice-yearly.

## Foulsham Publishers

The Oriel, Thames Valley Court, 183–187 Bath Road, Slough SL1 4AA
Ⓣ 01753 526769     Ⓕ 01753 535003
reception@foulsham.com
www.foulsham.com
Chairman/Managing Director *B.A.R. Belasco*
Approx. Annual Turnover £2.5 million

Founded 1819 and now one of the few remaining independent family companies to survive takeover. Publishes non-fiction on most subjects including lifestyle, travel guides, family reference, cookery, diet, health, DIY, business,

self improvement, self development, astrology, dreams, MBS. No fiction. IMPRINT **Quantum** Mind, body and spirit titles. TITLES *Classic 1000 Cocktails; Brit Guide to Orlando and Walt Disney World; Old Moore's Almanack; Raphael's Astrological Ephemeris.* Around 60 titles a year. Unsolicited mss, synopses and ideas welcome by email.
Ⓔ Royalties twice-yearly.

## Fountain Press

Newpro UK Ltd., Old Sawmills Road, Faringdon SN7 7DS
Ⓣ 01367 242411     Ⓕ 01367 241124
sales@newprouk.co.uk
www.newprouk.co.uk
Publisher *C.J. Coleman*
Approx. Annual Turnover £800,000

Founded 1923 when it was part of the Rowntree Trust Social Service. Owned by the British Electric Traction Group until 1982 when it was bought out by H.M. Ricketts. Acquired by Newpro UK Ltd in 2000. Publishes mainly photography and natural history. TITLES *Photography and Digital 'Workshop'; Antique and Collectable Cameras; Camera Manual* (series). About 2 titles a year. Unsolicited mss and synopses welcome.
Ⓔ Royalties twice-yearly.

✱*Authors' update* **Highly regarded for production values, Fountain has the reputation for involving authors in every stage of the publishing process.**

## Fourth Estate
▷ HarperCollins Publishers Ltd

## Frances Lincoln Ltd

4 Torriano Mews, Torriano Avenue, London NW5 2RZ
Ⓣ 020 7284 4009     Ⓕ 020 7485 0490
firstnamesurnameinitial@frances-lincoln.com
www.franceslincoln.com
Managing Director *John Nicoll*
Approx. Annual Turnover £7 million

Founded 1977. Publishes highly illustrated non-fiction: gardening, art and interiors, architecture, parenting, walking and climbing, children's picture, story and information books; also stationery. DIVISIONS **Adult Non-fiction** *Andrew Dunn* TITLES *Chatsworth* Duchess of Devonshire; *Grow Your Own Vegetables* Joy Larkcom; *Pictorial Guides to the Lakeland Fells* A. Wainwright; **Children's General Fiction and Non-fiction** *Maurice Lyon* TITLE *The Wanderings of Odysseus* Rosemary Sutcliffe, illus. Alan Lee; **Stationery** *Anna Sanderson* TITLES *RHS Diary and Address*

*Book*; *British Library Diary*. About 200 titles a year. Synopses and ideas for books considered. £ Royalties twice-yearly.

### Frank Cass
▷ Taylor & Francis Group

### Franklin Watts
▷ Hachette Children's Books

### Free Association Books Ltd
One Angel Cottages, Milespit Hill, London NW7 1RD
☎ 020 8906 0396    🖷 020 8906 0006
info@fabooks.com
www.fabooks.com
Managing Director/Publisher *T.E. Brown*

Publishes psychoanalysis and psychotherapy, psychology, cultural studies, sexuality and gender, women's studies, applied social sciences. TITLES *Drug Use and Cultural Context* eds. Ross Coomber and Nigel South; *Skin Disease* Ann Maguire; *Caring for the Dying at Home* Gill Pharoah; *A History of Group Study and Psychodynamic Organizations* Amy L. Fraher. Always email in the first instance with an outline. OVERSEAS ASSOCIATE ISBS, USA.
£ Royalties twice-yearly.

### The Friday Project
▷ HarperCollins Publishers Ltd

### Frontline Books
5A Accommodation Road, Golders Green, London NW11 9ED
☎ 020 8455 5559    🖷 01226 734438
michael@frontline-books.com
www.frontline-books.com
Managing Director *Charles Hewitt*

Military imprint of **Pen & Sword Books Ltd**. Publishes a wide range of military history topics and periods, from Ancient Greece and Rome to the present day. Send synopsis and sample chapter giving an outline of the book's theme and contents, chapter headings and word extent, if possible. Indication of sources would be helpful together with number of and type of illustrations, if relevant, plus personal details that underline qualifications for writing the book rather than a formal c.v. Enclose s.a.s.e. for return of material. No fiction.
£ Royalties twice-yearly.

### FyfieldBooks
▷ Carcanet Press Ltd

### G.J. Palmer & Sons
▷ Hymns Ancient & Modern Ltd

### Gaia Books
▷ Octopus Publishing Group

### Galore Park Publishing Ltd
19/21 Sayer's Lane, Tenterden TN30 6BW
☎ 01580 764242    🖷 01580 764142
info@galorepark.co.uk
www.galorepark.co.uk
Managing Director *Nicholas Oulton*

Founded 1999. Publishes school textbooks and revision resources. 12 titles in 2009. No unsolicited mss. Synopses and ideas welcome; send by email.
£ Royalties twice-yearly.

### Garden Art Press
▷ ACC Publishing Group

### Garland Science
▷ Taylor & Francis Group

### Garnet Publishing Ltd
8 Southern Court, South Street, Reading RG1 4QS
☎ 0118 959 7847    🖷 0118 959 7356
dan@garnetpublishing.co.uk
www.garnetpublishing.co.uk
Editorial Manager *Dan Nunn*

Founded 1992 and purchased Ithaca Press in the same year. Publishes art, architecture, photography, archive photography, cookery, travel classics, comparative religion, Islamic culture and history, foreign fiction in translation. Core subjects are Middle Eastern but list is rapidly expanding to be more general.

IMPRINTS **Garnet Publishing** TITLES *Simply Lebanese*; *The Art and Architecture of Islamic Cairo*; *What Did We Do To Deserve This?*; **Ithaca Press** Specializes in post-graduate academic works on the Middle East, political science and international relations. TITLES *The Making of the Modern Gulf States*; *Reform in the Middle East Oil Monarchies*; *The United States and Persian Gulf Security*. About 20 titles a year. Unsolicited mss not welcome; write with outline and ideas plus current c.v. in the first instance.
£ Royalties twice-yearly.

### Geddes & Grosset
144 Port Dundas Road, Glasgow G4 0HZ
☎ 0141 567 2830    🖷 0141 567 2831
Publishers *Ron Grosset, R. Michael Miller*
Approx. Annual Turnover £2.5 million

Founded 1989. Publisher of children's and reference books. IMPRINT **Waverley Books** cookery, humour, nostalgia. Unsolicited mss, synopses and ideas welcome. No adult fiction.

### The Geological Society Publishing House
Unit 7, Brassmill Enterprise Centre, Brassmill Lane, Bath BA1 3JN
☎ 01225 445046    🖷 01225 442836

sales@geolsoc.org.uk
enquiries@geolsoc.org.uk
www.geolsoc.org.uk
Commissioning Editor *Angharad Hills*

Publishing arm of the Geological Society of London which was founded in 1807. Publishes postgraduate texts in the earth sciences. 30 titles a year. Unsolicited synopses and ideas welcome.

## George Mann Books
▷ Arnefold Publishing

## Gerald Duckworth & Co. Ltd
First Floor, 90–93 Cowcross Street, London
EC1M 6BF
☏ 020 7490 7300　🖷 020 7490 0080
info@duckworth-publishers.co.uk
www.ducknet.co.uk
Managing Director *Peter Mayer*
Editorial Director (Academic) *Deborah Blake*
Editor (General) *Mary Morris*
Approx. Annual Turnover £2 million

Founded 1898 by Gerald Duckworth. Original publishers of Virginia Woolf. Other early authors include Hilaire Belloc, John Galsworthy, D.H. Lawrence and George Orwell. Duckworth is a general trade publisher whose authors include John Bayley, Mary Warnock, Joan Bakewell and J.J. Connolly. In addition to its trade list, Duckworth has a strong academic division. Acquired by Peter Mayer in 2003.
IMPRINTS/DIVISIONS **Bristol Classical Press** Classical texts and modern languages; **Ardis** Russian literature; **Duckworth Academic**; **Duckworth General**. No unsolicited mss; synopses and sample chapters only. Enclose s.a.s.e. or return postage for response/return.
🖷 Royalties twice-yearly.

## Gibson Square
☏ 020 7096 1100　🖷 020 7993 2214
info@gibsonsquare.com
www.gibsonsquare.com
Chairman *Martin Rynja*

Founded 2001. Publishes exclusively non-fiction: biography, current affairs, politics, cultural criticism, psychology, history, travel, art history, philosophy. Books must contribute to a general debate. IMPRINT **Gibson Square** TITLES *The Rotten State of Britain* Eamonn Butler; *House of Bush House of Saud* Craig Unger; *Blowing up Russia* Alexander Litvinenko; *Two Lipsticks & a Lover* Helena Frith Powell; *If I Did It* O.J. Simpson. Welcomes synopses and ideas. No fiction. Approach by email; reply sent only if interested.
🖷 Royalties annually.

🕮*Authors' update* Dedicated to publishing books 'on the cutting edge', Gibson Square is one of the few publishers to embrace serious political issues.

## Giles de la Mare Publishers Ltd
PO Box 25351, London NW5 1ZT
☏ 020 7485 2533　🖷 020 7485 2534
gilesdelamare@dial.pipex.com
www.gilesdelamare.co.uk
Chairman/Managing Director *Giles de la Mare*
Approx. Annual Turnover £30,000

Founded 1995 and commenced publishing in April 1996. Publishes mainly non-fiction, especially art and architecture, biography, history, music and travel. TITLES *Short Stories, Vols I, II & III* Walter de la Mare; *Handsworth Revolution* David Winkley; *The Life of Henry Moore* Roger Berthoud; *Becoming an Orchestral Musician* Richard Davis; *Blindness and the Visionary* John Coles; *Tricks Journalists Play* Dennis Barker; *Venice: The Anthology Guide* Milton Grundy; *Inherit the Truth 1939–1945* Anita Lasker-Wallfisch; *Calatafimi* Angus Campbell; *Musical Heroes* Robert Ponsonby. Unsolicited mss, synopses and ideas welcome after initial telephone call.
🖷 Royalties twice-yearly.

## GL Assessment
The Chiswick Centre, 414 Chiswick High Road, London W4 5TF
☏ 0845 602 1937　🖷 0845 601 5358
info@gl-assessment.co.uk
www.gl-assessment.co.uk
Publishing Director *Andrew Thraves*

'The UK's largest independent provider of educational assessments', GL Assessment has supplied schools with tests and assessment services (which are used by over 85% of all UK primary and secondary schools) for 27 years. 20 titles a year. No submissions as products are usually commissioned by the company.
🖷 Royalties twice-yearly.

## Global Publishing
▷ Quarto Publishing plc

## Gloucester Publishers Plc
10 Northburgh Street, London EC1V 0AT
☏ 020 7253 7887　🖷 020 7490 3708
markbicknell@everyman.uk.com
www.everymanchess.com
Managing Director *Mark Bicknell*
Approx. Annual Turnover £1 million

Publishes exclusively academic and leisure books relating to chess. IMPRINT **Everyman Chess** TITLES *Garry Kasparov: My Great Predecessors, Vols 1–7*; *Play Winning Chess* Yasser Seirawan; *Art of Attack in Chess* Vladimir Vukovich. Over 300 titles. No unsolicited material.
🖷 Royalties annually.

## GMC Publications Ltd

166 High Street, Lewes BN7 1XU
℡ 01273 477374    🖷 01273 402866
pubs@thegmcgroup.com
www.thegmcgroup.com
Joint Managing Directors *J.A.J. Phillips,
J.A.B. Phillips*
Managing Editor *Gerrie Purcell*

Founded 1974. Publisher and distributor of craft
and leisure books and magazines, covering
topics such as woodworking, DIY, architecture,
photography, gardening, cookery, art, puzzles and
games, reference, humour, TV and film. TITLES
*Miniature Carving and Netsuke*; *The Organic
Herb Gardener*; *Nikon D-SLR System Guide*.
About 50 illustrated books and 11 magazines a
year. Unsolicited mss, synopses and ideas for
books welcome. No fiction.
🄴 Royalties twice-yearly.

## Godsfield Press
▷ Octopus Publishing Group

## Gollancz
▷ The Orion Publishing Group Limited

## Gomer Press/Gwasg Gomer

Llandysul Enterprise Park, Llandysul SA44 4JL
℡ 01559 363090    🖷 01559 363758
gwasg@gomer.co.uk
www.gomer.co.uk
www.pontbooks.co.uk
Chairman/Managing Director *J.E. Lewis*
Publishing Director *Mairwen Prys Jones*
Editor (adult books in Welsh) *Dylan Williams*
Editor (adult books in English) *Ceri Wyn Jones*
Editors (children's books in Welsh) *Sioned
Lleinau, Rhiannon Davies*
**Pont Books** (children's books in English) Editor
*Viv Sayer*

Founded 1892. Publishes adult fiction and non-
fiction, children's fiction and educational material
in English and in Welsh. IMPRINTS **Gomer**; **Pont
Books**. 125 titles a year. Unsolicited mss welcome.
Prior enquiry required. Mss must have a Welsh
dimension.
🄴 Royalties twice-yearly.

## Gordon & Breach
▷ Taylor & Francis Group

## Gower
▷ Ashgate Publishing Ltd

## Graham-Cameron Publishing & Illustration
▷ entry under UK Packagers

## Granta Books

Twelve Addison Avenue, London W11 4QR
℡ 020 7605 1360    🖷 020 7605 1361
www.granta.com

Publisher *Philip Gwyn Jones*
Submissions *Amber Dowell (Editorial Assistant)*

Founded 1979. Acquired by Sigrid Rausing in
2006. Publishes narrative non-fiction and some
literary fiction. About 25 new titles and 25
paperbacks a year. No unsolicited mss; synopses
and sample chapters welcome.
🄴 Royalties twice-yearly.

✱*Authors' update* **Largely associated with
erudition, there are moves to revitalize the
Granta list with the commissioning of more
mainstream books.**

## Green Books

Foxhole, Dartington, Totnes TQ9 6EB
℡ 01803 863260    🖷 01803 863843
edit@greenbooks.co.uk
www.greenbooks.co.uk
Chairman *Satish Kumar*
Publisher *John Elford*
Approx. Annual Turnover £500,000

Founded in 1987 with the support of a number
of Green organizations. Closely associated
with *Resurgence* magazine. Publishes high-
quality books on a wide range of Green issues,
including economics, politics and the practical
application of Green thinking. No fiction or
books for children. TITLES *The Transition
Handbook* Rob Hopkins; *Timeless Simplicity*
John Lane; *Allotment Gardening* Susan Berger.
No unsolicited mss. Synopses and ideas welcome
but check guidelines on the website in the first
instance.
🄴 Royalties twice-yearly.

## Green Print
▷ The Merlin Press Ltd

## Green Profile
▷ Profile Books

## Gresham Books Ltd

19–21 Sayers Lane, Tenterden TN30 6BW
℡ 01580 767596    🖷 01580 764142
info@gresham-books.co.uk
www.gresham-books.co.uk
Managing Director *Nicholas Oulton*
Approx. Annual Turnover £330,000

Part of the Galore Park Group. Publishes hymn
and service books for schools and churches,
school histories and craft-bound choir and
orchestral folders. No unsolicited material but
ideas welcome.

## Grub Street

4 Rainham Close, London SW11 6SS
℡ 020 7924 3966/7738 1008    🖷 020 7738 1009
post@grubstreet.co.uk
www.grubstreet.co.uk
Managing Director *John Davies*

Founded 1982. Publishes cookery, food and wine, military and aviation history books. About 30 titles a year. Unsolicited mss and synopses welcome in the above categories but please enclose return postage.
£ Royalties twice-yearly.

## Guinness World Records Ltd

3rd Floor, 184–192 Drummond Street, London NW1 3HP
☎ 020 7891 4567   🖷 020 7891 4501
marketing@guinnessworldrecords.com
www.guinnessworldrecords.com
Managing Director *Alistair Richards*

First published in 1955, the annual Guinness World Records book is published in more than 100 countries and 28 languages and is the highest-selling copyright book of all time, with more than three million copies sold annually across the globe. Acquired by Jim Pattison Group (JPG) in February 2008. Contact from prospective researchers, editors and designers welcome.

## Gulf Professional Press
▷ **Elsevier Ltd**

## Gullane Children's Books

185 Fleet Street, London EC4A 2HS
info@gullanebooks.com
www.gullanebooks.com
Publisher *Simon Rosenheim*
Editor *Emily Lamm*
Art Director *Paula Burgess*
Rights Director *Katherine Judge*

Owned by D.C. Thomson, Gullane Children's Books was bought by **Meadowside Children's Books** in 2007 (see entry). Publishes picture books and novelty titles. Unsolicited mss accepted. Picture book mss: 800 words max. (for children up to the age of 7). Response will be sent only if s.a.s.e. enclosed. Do not send original artwork samples (copies only). Address to *The Submissions Editor*. 'Strictly no telephone enquiries for submissions; no fax submissions.'

## Gwasg Carreg Gwalch

12 Iard yr Orsaf, Llanrwst LL26 0EH
☎ 01492 642031   🖷 01492 641502
myrddin@carreg-gwalch.com
www.carreg-gwalch.com
Managing Editor *Myrddin ap Dafydd*

Founded in 1980. Publishes Welsh language; English books of Welsh interest – history, folklore, guides and walks. About 90 titles a year. Unsolicited mss, synopses and ideas welcome if on Welsh history and heritage.
£ Royalties paid.

## Hachette Children's Books

338 Euston Road, London NW1 3BH
☎ 020 7873 6000
www.hachettechildrens.co.uk
www.franklinwatts.co.uk
www.orchardbooks.co.uk
www.hodderchildrens.co.uk
www.waylandbooks.co.uk
Owner *Hachette Book Group Inc.*
Managing Director *Marlene Johnson*
Approx. Annual Turnover £36 million

Formed by the combining of Watts Publishing Group with Hodder Children's Books in April 2005. Publishes children's non-fiction, reference, information, gift, fiction, picture and novelty books and audio books. IMPRINTS **Hodder Children's Books** *Anne McNeil* Fiction, picture books, novelty, general non-fiction and audio; **Orchard Books** *Penny Morris* Fiction, picture books and novelties; **Franklin Watts** *Rachel Cooke* Non-fiction and information books; **Wayland** *Joyce Bentley* Non-fiction and information books. About 1,400 titles a year. Unsolicited material is not considered.
£ Royalties twice-yearly.

## Hachette Livre

French publishing conglomerate which embraces **Hachette UK**, **The Orion Publishing Group**, **Octopus Publishing Group**, **Little, Brown Book Group**. For further information, see imprint entries and entry under *European Publishers*.

## Hachette UK

338 Euston Road, London NW1 3BH
☎ 020 7873 6000   🖷 020 7873 6024
www.hachette.co.uk
Group Chief Executive *Tim Hely Hutchinson*

Hodder Headline was formed in June 1993 through the merger of Headline Book Publishing and Hodder & Stoughton. The company was acquired by WHSmith plc in 1999 and then by Hachette Livre S.A. in October 2004. Renamed Hachette Livre UK in August 2007 and subsequently renamed Hachette UK in October 2008. Purchased John Murray (Publishers) Ltd in 2002. About 2,000 titles a year.

DIVISIONS **Headline Publishing Group** CEO *Jamie Hodder-Williams*, Deputy Managing Director *Jane Morpeth* Publishes commercial and literary fiction (hardback and paperback) and popular non-fiction including narrative non-fiction, food and wine, history, humour and popular science. IMPRINTS **Business Plus**; **Headline**; **Headline Review**; **Little Black Dress**. AUTHORS Lyn Andrews, Emily Barr, Louise Bagshawe, Martina Cole, Janet Evanovich, Victoria Hislop, Wendy Holden, Jonathan

Kellerman, Jill Mansell, Maggie O'Farrell, Sheila O'Flanagan, Andrea Levy, Jennifer Johnson, Simon Scarrow, Jed Rubenfeld, Karen Rose, Pamela Stephenson, Penny Vincenzi.

**Hodder & Stoughton** CEO *Jamie Hodder-Williams*, Deputy Managing Director *Lisa Highton*; **Non-fiction** *Rowena Webb*; **Coronet** *Mark Booth*; **Sceptre** *Carole Welch*; **Fiction** *Carolyn Mays*; **Audio** (see entry under *Audio Books*). Publishes commercial and literary fiction; biography, autobiography, history, self-help, humour, travel and other general interest; audio. IMPRINTS **Hodder & Stoughton**; **Hodder Paperbacks**; **Coronet**; **Sceptre**; **Hodder Faith**. AUTHORS Jean M. Auel, Melvyn Bragg, John Connolly, Jeffrey Deaver, Charles Frazier, Elizabeth George, Sophie Hannah, Thomas Keneally, Stephen King, John le Carré, Andrew Miller, David Mitchell, Jodi Picoult, Rosamunde Pilcher, Mary Stewart, Fiona Walker. No unsolicited mss. **Hodder Faith** Director of Publishing *Wendy Grisham* **Bibles** *Ian Metcalfe* Publishes NIV and Today's NIV Bibles, Christian fiction and non-fiction, biographies, autobiographies, Manga. AUTHORS Philip Yancey, Rob Parsons, Joyce Meyer, J. John.

**Hodder Children's Books** (See **Hachette Children's Books**)

**Hodder Education Group** CEO *Thomas Webster* Publishes in the following areas: **Schools** *Lis Tribe*; **Consumer Education** *Katie Roden*; **Further Education/Higher Education** *C.P. Shaw*; **Health Sciences** *C.P. Shaw*. IMPRINTS **Hodder Education**; **Teach Yourself**; **Philip Allan Updates**; **Chambers Harrap**; **Hodder Arnold**; **Hodder Gibson** (see entry).

**Hachette Books Scotland** Publisher *Bob McDevitt* Commercial and literary fiction and non-fiction.

**Hachette Books Ireland** Managing Director Publishing *Breda Purdue* Commercial and literary fiction and non-fiction (see entry under *Irish Publishers*).

**John Murray (Publishers) Ltd** (see entry).

**Octopus Publishing Group** (see entry).

**Little, Brown Book Group** (see entry).

**The Orion Publishing Group Limited** (see entry).

✱*Authors' update* The leading British publisher by turnover, Hachette is strong on mainstream bestsellers, notably Stephanie Meyer's vampire novels (Little, Brown) accounting for 10% of UK turnover. The losers are midlist authors. At the same time the search is on for new talent which can benefit from imaginative and energetic marketing.

## Halban Publishers

22 Golden Square, London W1F 9JW
☎ 020 7437 9300  📠 020 7437 9512
books@halbanpublishers.com
www.halbanpublishers.com
Directors *Peter Halban, Martine Halban*

Founded 1986. Independent publisher. Publishes biography, memoirs, history, fiction, books on the Middle East and of Jewish interest. 8–10 titles a year. No unsolicited material. Approach by letter in the first instance. Unsolicited emails will be deleted.
💷 Royalties twice-yearly for first two years, thereafter annually in December.

## Halsgrove

Halsgrove House, Ryelands Industrial Estate, Bagley Road, Wellington TA21 9PZ
☎ 01823 653777  📠 01823 216796
sales@halsgrove.com
www.halsgrove.com
Publishers *Simon Butler, Steven Pugsley*

Founded in 1990 and now a leading publisher and distributor of regional books under the **Halsgrove**, **Halstar**, **PiXZ Books** and **Ryelands** IMPRINTS. Publishes books of regional interest throughout the UK: local history (including the *Community History* series), biography, photography and art, mainly in hardback. No fiction or poetry. 200 titles in 2009. Unsolicited mss, synopses and ideas for books of regional interest welcome.
💷 Royalties annually.

## Halstar
▷ **Halsgrove**

## Hamish Hamilton
▷ **Penguin Group (UK)**

## Hamlyn
▷ **Octopus Publishing Group**

## Hammersmith Press Ltd

496 Fulham Palace Road, London SW6 6JD
☎ 020 7736 9132  📠 020 7348 7521
gmb@hammersmithpress.co.uk
www.hammersmithpress.co.uk
Managing Director *Georgina Bentliff*

Founded 2004. Publishes health, nutrition, diet, academic medicine, and 'literary medicine', i.e. literary works that relate to the practice of medicine. TITLES *Trick and Treat* Barry Groves; *Smart Guide to Infertility* Professor Robert Harrison. About 6 titles a year. No unsolicited mss; synopses and ideas for books welcome; approach by email. No fiction or children's material.
💷 Royalties twice-yearly.

## Happy Cat
▷ Catnip Publishing Ltd

## Harlequin Mills & Boon Limited

Eton House, 18–24 Paradise Road, Richmond
TW9 1SR
☎ 020 8288 2800   ⊞ 020 8288 2898
www.millsandboon.co.uk
Managing Director *Mandy Ferguson*
Editorial Director *Karin Stoecker*
Approx. Annual Turnover £20 million

Founded 1908. Owned by the Canadian-based
Torstar Group. Publishes a wide range of women's
fiction including romantic novels.

IMPRINTS **Mills & Boon Modern** *Tessa Shapcott*
(50,000 words) Alpha males and attractive
women swept up by intense emotion and passion
set against a backdrop of international locations,
luxury and wealth. **Mills & Boon Modern
Heat** *Bryony Green* (50,000 words) Delivers a
feel-good experience, focusing on the kind of
relationship that women aged 18 to 35 aspire to.
Young characters in urban settings meet, flirt,
share experiences, have great sex and fall in love,
finally making a commitment that will bind them
forever. Based around emotional issues, other
concerns – e.g. jobs and friendships – are also
touched upon and resolved in upbeat way. **Mills
& Boon Romance** *Kimberley Young* (50,000
words) Contemporary upbeat and heroine-
focused romance, driven by strongly emotional
conflicts which are believable and relevant to
today's women. Capturing the depth of emotion
and sheer excitement of falling in love in a variety
of international settings. **Mills & Boon Medical**
*Sheila Hodgson* (50–55,000 words) Modern
medical practice provides a unique background
to highly emotional contemporary romances.
**Mills & Boon Historical Romance** *Linda Fildew*
(70–75,000 words) Richly textured, emotionally
intense historical novels covering a wide range
of time periods from ancient civilizations up to
and including the Second World War. **MIRA**®
*Maddie West* Focuses on the very best voices in
mainstream fiction. 'We publish books we love
and that we're sure will entertain – from crime
and thrillers to commercial women's fiction
and historical fiction. We're always looking for
outstanding new books and writers to introduce
to readers.' About 900 titles a year. Mills &
Boon romance series tip sheets and guidelines
are available from the Mills & Boon website
(see above), www.eharlequin.com or Harlequin
Mills & Boon Editorial Dept. (telephone or
write sending s.a.s.e.). Agented and unagented
submissions accepted. Send query letter, synopsis
and first three chapters in the first instance. For
MIRA, agented submissions only.

✱*Authors' update* Light romance provides a
good income for writers who can adapt to the
clearly defined M&B formula. Just think, M&B
has a network of 12,000 authors worldwide
and every title scores six-figure sales. But
be warned. M&B gets 2,000 unsolicited
manuscripts a year. Latest initiatives include a
series of rugby romances in partnership with
the Rugby Football Union. It seems that one
third of rugby supporters are women. There
are plans to make up to 2,000 titles available in
digital format.

## HarperCollins Publishers Ltd

77–85 Fulham Palace Road, London W6 8JB
☎ 020 8741 7070   ⊞ 020 8307 4440
www.harpercollins.co.uk
*Also at:* Westerhill Road, Bishopbriggs, Glasgow
G64 2QT Tel 0141 772 3200 Fax 0141 306 3119
Owner *News Corporation*
CEO/Publisher *Victoria Barnsley, OBE*
Publisher *Belinda Budge*
Approx. Annual Turnover £173.4 million (UK)

HarperCollins is one of the top book publishers
in the UK, with a wider range of books than
any other publisher; from cutting-edge
contemporary fiction to enduring classics; from
school textbooks to celebrity memoirs; and from
downloadable dictionaries to prize-winning
literature. The wholly-owned division of News
Corporation also publishes a wide selection of
non-fiction including history, celebrity memoirs,
biography, popular science, mind, body and
spirit, dictionaries, maps and reference books.
HarperCollins is also one of the largest education
publishers in the UK. Authors include many
award-winning and international bestsellers
such as Isabel Allende, Paulo Coelho, Josephine
Cox, Michael Crichton, Cathy Kelly, Judith Kerr,
Doris Lessing, Frank McCourt, Tony Parsons,
Nigel Slater and Diana Wynne Jones. Bestselling
licensed properties include Dr Seuss and Noddy.
About 1,200 titles a year.

**Harper Fiction** Publishing Director *Lynne Drew*
IMPRINTS **HarperCollins** Publishing Directors
*Susan Watt, Julia Wisdom*, Publishing Director
*David Brawn* (Agatha Christie, J.R.R. Tolkien,
C.S. Lewis); **Voyager** Fantasy/science fiction
Publishing Director *Jane Johnson* General,
historical fiction, crime and thrillers and women's
fiction; **Blue Door** Publisher *Patrick Janson-Smith*
Fiction.

**Harper NonFiction** Publisher *Carole Tonkinson*,
Editorial Directors Kate Latham (cookery &
lifestyle), *Natalie Jerome* (popular culture)
IMPRINTS **HarperThorsons/Element** *Katy
Carrington* (mind, body & spirit), *Sally Annett*
(media tie-ins); **HarperSport** Publishing Director
*Jonathan Taylor* Sporting biographies, guides

and histories; **HarperTrue** Publisher *Carole Tonkinson* (inspiring real-life stories).

**Angry Robot** Publishing Director *Marc Gascoigne* Science fiction, fantasy and horror.

**Avon** Managing Director *Caroline Ridding*, Editorial Director *Maxine Hitchcock* Popular fiction.

**HarperCollins Children's Books** Managing Director *Mario Santos*, Publisher *Ann-Janine Murtagh*, Brand & Properties Developer *Claire Harding* Quality picture books for under-7s; fiction for age 6 up to young adult; TV and film tie-ins and properties Publishing Directors *Gillie Russell* (fiction), *Sue Buswell* (picture books).

PRESS BOOKS DIVISION

Managing Director *John Bond*, Associate Publisher *Minna Fry*

IMPRINTS **Fourth Estate** Publishing Director/ Publisher *Nick Pearson* Fiction, literary fiction, current affairs, popular science, biography, humour, travel; **HarperPress** General trade non-fiction and fiction Publishing Directors *Arabella Pike* (non-fiction), *Clare Smith* (fiction); **The Friday Project** Publisher *Scott Pack* Fiction, non-fiction and children's.

COLLINS

Publisher *Hannah MacDonald*, Associate Publisher *Myles Archibald* (natural history) Guides and handbooks, phrase books and manuals on popular reference, art instruction, illustrated, cookery and wine, crafts, DIY, gardening, military, natural history, pet care, pastimes IMPRINTS **Collins**; **Collins Gem**; **Jane's**.

COLLINS GEO

Managing Director *Sheena Barclay* Times World Atlases, Collins World Atlases, maps, atlases, street plans and leisure guides. IMPRINTS **Collins**; **Times Books**.

**Collins Language** Managing Director *Robert Scriven* Bilingual and English dictionaries, English dictionaries for foreign learners Editorial Director *Elaine Higgleton* (dictionaries).

**Collins Education** Managing Director *Nigel Ward* Books, CD-ROMs and online material for UK primary and secondary schools and colleges. Accepts submissions via literary agents only. Alternatively, writers can upload their mss to www.authonomy.com
£ Royalties twice-yearly.

❋*Authors' update* The first publisher to announce plans to publish all major titles in print and digital format simultaneously, HarperCollins has a range of online services to appeal to authors. A website that invites sample chapters has unearthed some talent. Authors

with the Friday Project are to be offered profit-share deals – the first from any UK publisher.

## Harriman House Ltd

3A Penns Road, Petersfield GU32 2EW
☎ 01730 233870  📠 01730 233880
contact@harriman-house.com
www.harriman-house.com
Managing Director *Myles Hunt*
Head of Rights *Suzanne Anderson*
PR & Maketing *Louise Hinchen*

Founded 1994; commenced publishing in 2001. Independent publisher of finance, trading, business and economics books, covering a wide range of subjects from personal finance, small business and lifestyle through to stock market investing, trading and professional guides. 40 titles a year. Unsolicited mss, synopses and ideas for books welcome. Initial approach by email.
£ Royalties twice-yearly.

## Hart Publishing Ltd

16c Worcester Place, Oxford OX1 2JW
☎ 01865 517530  📠 01865 510710
mail@hartpub.co.uk
www.hartpub.co.uk
www.hartpublishingusa.com
Managing Director *Richard Hart*
Approx. Annual Turnover £2.5 million

Founded 1996. Publishes academic and professional law books. Winner of the Independent Publishers Guild's 2009 Academic and Professional Publisher of the Year award.
IMPRINTS **Hart**; **Parker Press**. About 100 titles in 2009. Unsolicited mss, synopses and ideas for books welcome; send by email.
£ Royalties twice-yearly.

## Harvard University Press

Fitzroy House, 11 Chenies Street, London WC1E 7EY
☎ 020 7306 0603  📠 020 7306 0604
info@hup-mitpress.co.uk
www.hup.harvard.edu
Director *William Sisler*
General Manager *Ann Sexsmith*

European office of **Harvard University Press**, USA. Publishes academic and scholarly works in history, politics, philosophy, economics, literary criticism, psychology, sociology, anthropology, women's studies, biological sciences, classics, history of science, art, music, film, reference. All mss go to the American office: 79 Garden Street, Cambridge, MA 02138 (see entry under *US Publishers*).

## Harvill Secker
▷ The Random House Group Ltd

## Harwood Academic
[>] Taylor & Francis Group

## Haus Publishing
70 Cadogan Place, London SW1X 9AH
☎ 020 7838 9055  🖷 020 7235 1999
haus@hauspublishing.com
www.hauspublishing.co.uk
Publisher *Barbara Schwepcke*

DIVISIONS/IMPRINTS **Life&Times** Non-academic
biographies of well-known and lesser-known
personalities TITLES *Tito*; *Simone de Beauvoir*;
*Prime Ministers of the 20th Century*; *Patrice
Lumumba*; *Mussolini*; **Armchair Traveller**
Literary travel series TITLES *Mumbai to Mecca*;
*Venice for Lovers*; *The Liquid Continent: A
Mediterranean Trilogy*; **HausBooks** Hardback
biographies TITLES *Ellen Terry*; *Rommel*;
*Diaghilev*; **Haus Fiction** TITLES *The Apple
in the Dark*; *A Minute's Silence*; *A Matter of
Time*; **Arabia Books** Mainly Arab literature in
translation. 26 titles in 2009. Unsolicited mss,
synopses and ideas within the subject areas above
are welcome; approach by email or post only. No
children's or poetry.
🖺 Royalties twice-yearly.

## Hay House Publishers
292B Kensal Road, London W10 5BE
☎ 020 8962 1230  🖷 020 8962 1239
info@hayhouse.co.uk
www.hayhouse.co.uk
Managing Director *Michelle Pilley*
Health, Self-help & Personal Development,
Lifestyle, Soft Business, Parenting *Carolyn Thorne*
MBS, Spirituality & Spiritual Fiction
*Michelle Pilley*
Approx. Annual Turnover £4.5 million

Founded in 2003. Part of Hay House Inc.
Publishes self-help and personal development,
mind, body and spirit, spirituality, lifestyle, soft
business and spiritual fiction. TITLES *Waking from
Sleep* Steve Taylor; *Men, Money and Chocolate*
Menna Van Praag; *Success Intelligence* Robert
Holden; *The Bloke's Guide to Babies* Jon Smith.
121 titles in 2009. Unsolicited material welcome;
extensive submission details available on the
website. No poetry submissions.
🖺 Royalties twice-yearly.

## Haynes Publishing
Books Division, Sparkford, Near Yeovil BA22 7JJ
☎ 01963 440635  🖷 01963 440023
bookseditorial@haynes.co.uk
www.haynes.co.uk
Managing Director *J Haynes*
Approx. Annual Turnover £35.3 million

Founded 1960. Leading publisher of car and
motorcycle manuals, with operations in the UK,
USA, Australia, The Netherlands and Sweden.
Alongside the long-established manuals written
by in-house technical authors, the UK-based
company also publishes a wide range of books by
external authors on all areas of transport – cars,
motorcycles, motorsport, aviation, military,
maritime, railways, cycling and caravans. In
addition, manuals in many new subject areas are
being developed, with strong ranges being built in
computing, family health, home DIY, gardening,
musical instruments and sport. Book proposals
and submissions are welcome in all the described
subject areas, as are any suggestions for manuals
on new topics. Contact *Mark Hughes*, Editorial
Director, Books Division.
🖺 Royalties twice-yearly.

✱*Authors' update* Having sold Sutton
**Publishing, Haynes is back to its core
programme of publishing 'practical and hands-
on' books along with tie-in products such as
T-shirts and toys.**

## Headline/Headline Review
[>] Hachette UK

## Helicon Publishing
RM plc, New Mill House, 183 Milton Park,
Abingdon OX14 4SE
☎ 01235 823816
helicon@rm.com
www.helicon.co.uk
Business Manager *Martin Hall*

A division of RM plc. Helicon is a general
reference database publisher for the UK, US
and Australian markets. Licenses quality
reference content for online, print, and CD-ROM
publications. TITLES include the flagship single-
volume *Hutchinson Encyclopedia*.

## Helm Information
Crowham Manor, Main Road, Westfield, Hastings
TN35 4SR
☎ 01424 882422  🖷 01424 882817
amandahelm@helm-information.co.uk
www.helm-information.co.uk
Proprietor *Amanda Helm*

Founded 1990. Publishes academic books for
students and university libraries. SERIES *The
Dickens Companions* and *Icons*, a series which
exposes the processes by which a figure, historical
or fictional, achieves iconic status, e.g. Faust,
Robin Hood. Will consider ideas and proposals
provided they are relevant to the series listed
above.
🖺 Royalties annually.

## Helter Skelter Publishing
PO Box 50497, London W8 9FA
☎ 07941 206045

Includes FREE online access to **www.thewritershandbook.com**

info@helterskelterpublishing.com
www.helterskelterpublishing.com
Director *Graeme Milton*

Founded 1995. Publishes music and film books
only. 10 titles a year. Unsolicited mss, synopses
and ideas welcome.

### The Herbert Press
▷ A.&C. Black Publishers Ltd

### Hermes House
▷ Anness Publishing Ltd

### Hertfordshire Publications
▷ University of Hertfordshire Press

### Hesperus Press Limited
4 Rickett Street, London SW6 1RU
☎ 020 7610 3331   🖷 020 7610 3217
info@hesperuspress.com
www.hesperuspress.com
Director *Derrick Holman*

Founded 2001. Publishes new translations,
literary biographies, lesser-known classics and
non-fiction. TITLES *Hyde Park Gate News* Virginia
Woolf; *Sarrasine* Honoré de Balzac; *No Man's
Land* Graham Greene; *Hadji Murat* Leo Tolstoy;
*Aller Retour New York* Henry Miller; *The Watsons*
Jane Austen. *Modern Voices* series launched in
2005; *Brief Lives* series of newly commissioned
short biographies of celebrated literary figures
launched in 2008; contemporary fiction in
translation since 2005; and Worldwide list with
particular interest in Indian subcontinental
literature since 2009. 50 titles in 2009. Does not
accept unsolicited mss.
🄔 Royalties annually.

✱*Authors' update* Much praised for matching
book design with quality content.

### History Into Print
▷ Brewin Books Ltd

### The History Press
The Mill, Brimscombe, Stroud GL5 2QG
☎ 01453 883300   🖷 01453 883233
sales@thehistorypress.co.uk
www.thehistorypress.co.uk
Chairman *Andy Nash*
Managing Director *Stuart Biles*
Editorial Head *Laura Perehinec*
Approx. Annual Turnover £12.3 million

Founded in 2007 having acquired the assets of
NPI Media. The UK's largest local and specialist
history publisher. IMPRINTS include **The History
Press; Jarrold; Nonsuch; Phillimore; Pitkin;
Spellmount; Stadia; Sutton; Tempus**. 550 titles
in 2009. Unsolicited submissions welcome;
approach by email in the first instance. No fiction
or children's books. OVERSEAS SUBSIDIARIES The

History Press Inc; Sutton Verlag; Editions Alan
Sutton; Nonsuch Ireland.
🄔 Royalties twice-yearly.

✱*Authors' update* Still sorting itself out after
a few troubled years but with the continuing
popularity of history titles, all should come
right.

### HMSO
▷ TSO

### Hodder & Stoughton
▷ Hachette UK

### Hodder Arnold
▷ Hachette UK

### Hodder Children's Books
▷ Hachette Children's Books

### Hodder Gibson
2a Christie Street, Paisley PA1 1NB
☎ 0141 848 1609   🖷 0141 889 6315
hoddergibson@hodder.co.uk
www.hoddergibson.co.uk
Managing Director *John Mitchell*

Part of the **Hachette UK** group. Publishes
educational textbooks and revision guides
specifically for Scotland and mainly for secondary
education. About 25 titles a year. Synopses/ideas
preferred to unsolicited mss.
🄔 Royalties annually.

### Hodder Headline
▷ Hachette UK

### Honeyglen Publishing Ltd
15 Kensington West, Blythe Road, London
W14 0JG
☎ 020 7602 2876   🖷 020 7602 2876
Directors *N.S. Poderegin, J. Poderegin*

Founded 1983. A small publishing house whose
output is 'extremely limited'. Publishes history,
philosophy of history, biography and selective
fiction. No children's or science fiction. TITLES
*The Soul of China; The Soul of India; Woman and
Power in History; Lost World – Tibet; A Child of
the Century* all by Amaury de Riencourt; *With
Duncan Grant in South Turkey* Paul Roche;
*Vladimir, The Russian Viking* Vladimir Volkoff;
*The Dawning* Milka Bajic-Poderegin; *Quicksand*
Louise Hide. Unsolicited mss welcome.

### Honno Welsh Women's Press
Unit 14, Creative Units, Aberystwyth Arts Centre,
Aberystwyth SY23 3GL
☎ 01970 623150   🖷 01970 623150
post@honno.co.uk
www.honno.co.uk
Editor *Caroline Oakley*

Founded in 1986 by a group of women who wanted to create more opportunities for women in publishing. A cooperative operation which publishes novels, autobiography, biography, popular women's history, general non-fiction, short stories on commission. Must have a Welsh connection.' Contact us or see website for current calls for short story submissions.' 7 titles a year. Welcomes mss and ideas for books from women who are Welsh or have a significant Welsh connection only. Send as hard copy.
© Royalties annually.

## The Horizon Press

The Oakes, Moor Farm Road West, Ashbourne
DE6 1HD
☎ 01335 347349
books@thehorizonpress.co.uk
Managing Director *C.L.M. Porter*
Approx. Annual Turnover £450,000

Formerly Landmark Publishing Ltd, founded in 1996. Publishes itinerary-based travel guides, regional, industrial countryside and local history. No unsolicited mss; telephone in the first instance.
© Royalties annually.

## Horus Editions
▷ Award Publications Limited

## How To Books Ltd

Spring Hill House, Spring Hill Road, Begbroke, Oxford OX5 1RX
☎ 01865 375794     ℻ 01865 379162
info@howtobooks.co.uk
www.howtobooks.co.uk
Managing Director *Giles Lewis*
Editorial Director *Nikki Read*
Rights Director *Ros Loten*

An independent publishing house, founded in 1991. Publishes non-fiction, self-help reference books. How To titles are practical, accessible books that enable their readers to achieve their goals in life and work. How To authors should have first-hand experience of the subject about which they are writing. Subjects covered include small business and self employment, study skills, cooking and gardening, management, jobs and careers, living and working abroad, creative writing, weddings and property investment. IMPRINT **Spring Hill**. 75 new titles and new editions per year. Submit outline followed by sample chapter.

## Human Horizons
▷ Souvenir Press Ltd

## Hurst Publishers, Ltd

41 Great Russell Street, London WC1B 3PL
☎ 020 7255 2201
hurst@atlas.co.uk
www.hurstpub.co.uk
Chairman/Managing Director/Editorial Head *Michael Dwyer*

Founded 1969. An independent company publishing contemporary history, politics, religion (not theology) and anthropology. About 25 titles a year. No unsolicited mss. Synopses and ideas welcome.
© Royalties annually.

## Hutchinson/Hutchinson Children's Books
▷ The Random House Group Ltd

## Hymns Ancient & Modern Ltd

St Mary's Works, St Mary's Plain, Norwich
NR3 3BH
☎ 01603 612914     ℻ 01603 624483
admin@hymnsam.co.uk
www.scm-canterburypress.co.uk
Publishing Director, SCM Press and Canterbury Press *Christine Smith*
Senior Commissioning Editor, SCM *Natalie Watson*
Approx. Annual Turnover £6 million

Publishing division **SCM-Canterbury Press** (see entry) controls **SCM Press**, **Canterbury Press**, **Church House Publishing** and RMEP IMPRINTS which cover hymn books, liturgical material, academic and general religious books, and multi-faith religious and social education resources for pupils and teachers. Media division **G.J. Palmer & Sons** includes *Church Times* (see entry under *Magazines*); *Third Way* (monthly); *The Sign* and *Crucible* (quarterly). About 100 titles a year. Ideas for new titles welcome but no unsolicited mss.
© Royalties annually.

## I.B.Tauris & Co. Ltd

6 Salem Road, London W2 4BU
☎ 020 7243 1225     ℻ 020 7243 1226
mail@ibtauris.com
www.ibtauris.com
Chairman/Publisher *Iradj Bagherzade*
Managing Director *Jonathan McDonnell*

Founded 1984. Independent publisher. Publishes general non-fiction and academic in the fields of international relations, religion, current affairs, history, politics, cultural, media and film studies, Middle East studies. Distributes **Philip Wilson Publishers** and The Federal Trust worldwide. IMPRINTS **Tauris Academic Studies** Academic monographs in the humanities and social sciences; **Tauris Parke Paperbacks** Trade titles, including history, travel and biography; **Radcliffe Press** Colonial history and biography. Unsolicited synopses and book proposals welcome.
© Royalties twice-yearly.

❋*Authors' update* Imprint Radcliffe Press may require authors to contribute towards costs of publication.

## Ian Allan Publishing Ltd

Riverdene Business Park, Molesey Road, Hersham KT12 4RG

☏ 01932 266600    ℻ 01932 266601
info@ianallanpublishing.co.uk
www.ianallanpublishing.com
Chairman *David Allan*
Managing Director *Iain Aitken*

Specialist transport publisher – atlases, maps, railway, aviation, road transport – military, history, maritime, reference, Freemasonry also sport and leisure. Manages distribution and sales for third party publishers. IMPRINTS **Ian Allan Publishing**; **Classic** Aviation titles; **Midland Publishing** Aviation titles; **OPC** Railway titles; **Lewis**; **Lewis Masonic**. About 100 titles a year. Send sample chapter and synopsis with s.a.s.e.

## Icon Books Ltd

Omnibus Business Centre, 39–41 North Road, London N7 9DP

☏ 020 7697 9695
info@iconbooks.co.uk
www.iconbooks.co.uk
Chairman *Peter Pugh*
Managing Director *Simon Flynn*
Editorial Director *Duncan Heath*

Founded 1992. SERIES *Introducing* Graphic introductions to key figures and ideas in the history of science, philosophy, psychology, religion and the arts. TITLE *Introducing Quantum Theory*. Publishes 'provocative and intelligent' non-fiction in science, politics and philosophy. TITLES *Atom*; *The Truth that Sticks*. IMPRINTS **Corinthian Books** Principally commercial sport-led non-fiction; **Wizard Books** Children's non-fiction and game books including SERIES *Fighting Fantasy* (adventure gamebooks). Submit synopsis only. OVERSEAS ASSOCIATE Totem Books, USA, distributed by National Book Network.
£ Royalties twice yearly.

## IHS Jane's

163 Brighton Road, Coulsdon CR5 2YH

☏ 020 8700 3700
info.uk@janes.com
info.uk@IHSjanes.com
www.janes.com
Managing Director *Michael Dell*
Publishers *Sean Howe, James Green*

Founded in 1898 by Fred T. Jane with the publication of *All the World's Fighting Ships*. Acquired by IHS Inc. in 2007. Recent focus has been on growth opportunities in its core business and in enhancing the performance of initiatives like Jane's online and on in-depth intelligence centres. Publishes Web journals, magazines and yearbooks on defence, aerospace, security and transport topics, with details of equipment and systems; plus directories. Jane's Strategic Advisory Services provides bespoke intelligence to clients. Also *Jane's Defence Weekly* (see entry under *Magazines*).

DIVISIONS **Magazines** TITLES *Jane's Defence Weekly*; *Jane's International Defence Review*; *Jane's Airport Review*; *Jane's Navy International*; *Jane's Islamic Affairs Analyst*; *Jane's Missiles and Rockets*; *Jane's Intelligence Weekly*; *Jane's Industry Quarterly*. **Publishing for Defence, Aerospace**. TITLES *Defence, Aerospace Yearbooks*. **Transport** TITLE *Transportation Yearbooks*. **Security** TITLES *Jane's Intelligence Review*; *Jane's Sentinel* (regional security assessment); *Jane's Police Review*. CD-ROM and electronic development and publication. Over 100 titles a year. Unsolicited mss, synopses and ideas for reference/yearbooks welcome.

## Ilex Press
▷ **The Ilex Press Limited under UK Packagers**

## Imagine That!
▷ **Top That! Publishing Plc**

## Imprint Academic

PO Box 200, Exeter EX5 5YX

☏ 01392 851550    ℻ 01392 851178
anthony@imprint.co.uk
www.imprint-academic.com
Publisher *Keith Sutherland*

Founded 1980. Publishes books and journals in politics, philosophy and psychology for both academic and general readers. Book SERIES include *St. Andrews Studies in Philosophy and Public Affairs* and *Societas: Essays in political and cultural criticism*. 25–30 titles a year. Unsolicited mss, synopses and ideas welcome with return postage only.
£ Royalties annually.

## The In Pinn
▷ **Neil Wilson Publishing Ltd**

## Incomes Data Services
▷ **Sweet & Maxwell**

## Independent Music Press

PO Box 69, Church Stretton SY6 6WZ

☏ 01694 720049    ℻ 01694 720049
info@impbooks.com
www.impbooks.com
www.myspace.com/independentmusicpres
Managing Director *Martin Roach*
Marketing & New Media *David Hanley*

Founded in 1992 by author/publisher Martin Roach. Specializes in publishing first-to-the-

market biographies on bands and artists such as Thom Yorke, Slash, Bruce Dickinson, Dave Gahan, Robert Plant, My Chemical Romance, Johnny Marr, The Killers, John Lydon, Foo Fighters, Mick Ronson and Beastie Boys. Also classic titles such as David Nolan's *I Swear I Was There* and definitive subcultural histories on subjects such as scooter boys, bikers and Two Tone. 8 titles a year. Approach by email, enclosing biography and synopsis only.
€ Royalties twice-yearly.

## Independent Voices
▷ Souvenir Press Ltd

## Infinite Ideas
36 St Giles, Oxford OX1 3LD
☎ 01865 514888   🖷 01865 514777
info@infideas.com
www.infideas.com
Joint Managing Directors *David Grant, Richard Burton*
Approx. Annual Turnover £1.5 million

Over 60 titles in bestselling self-help SERIES 52 *Brilliant Ideas* and 20 titles in newly launched **The Feel Good Factory** IMPRINT covering health and relationships; careers, finance and personal development; sports, hobbies and games; leisure and lifestyle. Commencing a self-publishing/partnership offer for new and established authors in all genres. 'We work with companies and consultants, sharing the risk to bring their books to the global market place in order to promote their services.' 60 titles a year. Synopses and ideas welcome by email or letter, but no unsolicited mss without prior communication.
€ 'Generous royalties paid annually on our partnership publishing programme.'
✱*Authors' update* 'Partnership' may mean authors contributing to production costs.

## Inter-Varsity Press
IVP Book Centre, Norton Street, Nottingham NG7 3HR
☎ 0115 978 1054   🖷 0115 942 2694
ivp@ivpbooks.com
www.ivpbooks.com
Chairman *Ralph Evershed*
Chief Executive *Brian Wilson*

Founded mid-1930s as the publishing arm of Universities and Colleges Christian Fellowship, it has expanded to wider Christian markets worldwide. Publishes Christian belief and lifestyle, reference and bible commentaries. No secular material or anything which fails to empathize with orthodox Protestant Christianity. IMPRINTS **IVP**; **Apollos**; **Crossway** TITLES *The Bible Speaks Today* series and New Testament CD-ROM; *Pierced for Our Transgressions* Jeffery, Ovey and

Sach; *The Living Church* John Stott. About 50 titles a year. Synopses and ideas welcome.
€ Royalties twice-yearly.

## Intercultural Press
▷ Nicholas Brealey Publishing

## Iqon Editions
▷ Quarto Publishing plc

## Isis Publishing
7 Centremead, Osney Mead, Oxford OX2 0ES
☎ 01865 250333   🖷 01865 790358
sales@isis-publishing.co.uk
www.isis-publishing.co.uk

Part of the Ulverscroft Group Ltd. Publishes large-print books – fiction and non-fiction – and unabridged audio books. Together with **Soundings** (see entry under *Audio Books*) produces around 4,000 titles on audio tape and CD. AUTHORS Lee Child, Martina Cole, Terry Pratchett, Susan Sallis. No unsolicited mss as Isis undertakes no original publishing.
€ Royalties twice-yearly.

## Itchy Coo
▷ Black & White Publishing Ltd

## Ithaca Press
▷ Garnet Publishing Ltd

## IVP
▷ Inter-Varsity Press

## J. Garnet Miller
▷ Cresrelles Publishing Co. Ltd

## J.A. Allen
An imprint of Robert Hale Ltd, Clerkenwell House, 45–47 Clerkenwell Green, London EC1R 0HT
☎ 020 7251 2661   🖷 020 7490 4958
allen@halebooks.com
www.allenbooks.com
Publisher *Lesley Gowers*
Approx. Annual Turnover £1 million

Founded 1926 as part of J.A. Allen & Co. (The Horseman's Bookshop) Ltd. Bought by **Robert Hale Ltd** in 1999. Publishes equine and equestrian non-fiction. TITLES *British Horse and Pony Breeds* Clive Richardson; *Equine Biomechanics for Riders* Karin Blignault; *Feet First* Nic Barker and Sarah Braithwaite; *Horse-Friendly Riding* Susan McBane. About 20 titles a year. Mostly commissioned but willing to consider unsolicited mss of technical/instructional material related to all aspects of horses and horsemanship.
€ Royalties twice-yearly.

## Jacqui Small
▷ Aurum Press Ltd

## James Clarke & Co.

PO Box 60, Cambridge CB1 2NT
☎ 01223 350865    🖷 01223 366951
publishing@jamesclarke.co.uk
www.jamesclarke.co.uk
Managing Director *Adrian Brink*

James Clarke and Co Ltd was founded in 1859
in Fleet Street, London, mainly as a magazine
publisher. It produced the highly influential
religious magazine, *Christian World*, which
by the outbreak of the First World War was
selling over 100,000 copies a week and was the
leading nonconformist weekly. It soon began
to publish books and by the 1920s it became an
exclusively book publishing company, building
up a list in religion and in alternative medicine
and spirituality. **The Lutterworth Press** (see
entry) became associated with James Clarke &
Co. in 1984. James Clarke & Co's current list
concentrates on academic, scholarly and reference
works, specializing in theology, history, literature
and related subjects. TITLES *Anglo-Catholic
in Religion: T.S. Eliot and Christianity* Barry
Spurr; *Postsecularism: The Hidden Challenge to
Extremism* Mike King; *Anglicanism: The Thought
and Practice of the Church of England* P.E. More &
F.L. Cross. 20 titles in 2009. Book proposals to be
submitted in writing.

✱*Authors' update* **A leading theological
publisher with high intellectual standards.**

## Jane's
▷ HarperCollins Publishers Ltd

## Janus Publishing Company Ltd

105–107 Gloucester Place, London W1U 6BY
☎ 020 7486 6633    🖷 020 7486 6090
publisher@januspublishing.co.uk
www.januspublishing.co.uk
Managing Director *Jeannie Leung*

Publishes fiction, human interest, memoirs,
philosophy, mind, body and spirit, religion and
theology, social questions, popular science,
history, spiritualism and the paranormal, poetry
and young adults. IMPRINTS **Janus Books** Subsidy
publishing; **Empiricus Books** Non-subsidy
publishing. TITLES *The Naked Emperor; It:
The Architecture of Existence; From Necessity
to Infinity: Interpretation in Language and
Translation; The Performer's Anthology; Surgery
Over the Centuries; The Films of Danny Boyle;
Leape Years; Quiddity; The Campbells From
Camden Town.* 'Two of our authors won the
European Literary Award in 2004.' About 450
titles in print. Unsolicited mss welcome. Agents
in the USA, Europe and Asia.
💷 Royalties twice-yearly.

✱*Authors' update* Authors may be asked to
cover their own productions costs but Janus has
moved into conventional publishing with its
Empiricus imprint.

## Jarrold
▷ The History Press

## Jessica Kingsley Publishers Ltd

116 Pentonville Road, London N1 9JB
☎ 020 7833 2307    🖷 020 7837 2917
post@jkp.com
www.jkp.com
www.singing-dragon.com
*Also at:* 400 Market Street, Suite 400,
Philadelphia, PA 19106, USA
Managing Director and Publisher *Jessica Kingsley*
Senior Acquisitions Editor *Stephen Jones*
Director of Electronics *Jemima Kingsley*

Founded 1987. Independent, international
publisher of books for professionals, academics
and the general reader on autism, disability,
special education, arts therapies, child
psychology, mental health, practical theology
and social work. IMPRINT **Singing Dragon**
Chinese martial arts, Qigong, Traditional
Chinese Medicine and complementary medicine.
150 titles a year. 'We are actively publishing
and commissioning in all these areas. We
welcome suggestions for books and proposals
from prospective authors. Proposals should
consist of an outline of the book, a contents list,
assessment of the market and author's c.v., and
should be addressed to *The Editorial Director*.
Complete manuscript should not be sent, and all
submissions should be addressed to the London
office. Email submissions are encouraged. A
proposals form can be downloaded from the JKP
website.' OVERSEAS SUBSIDIARY in Philadelphia,
USA.
💷 Royalties twice-yearly.

## JM Dent
▷ The Orion Publishing Group Ltd

## John Blake Publishing Ltd

3 Bramber Court, 2 Bramber Road, London
W14 9PB
☎ 020 7381 0666    🖷 020 7381 6868
words@blake.co.uk
www.johnblakepublishing.co.uk
Managing Director *John Blake*
Deputy Managing Director *Rosie Ries*

Founded 1991 and expanding rapidly. Bought the
assets of Smith Gryphon Ltd in 1997 and acquired
Metro Publishing in 2001. Publishes mass-market
non-fiction. No fiction, children's, specialist or
non-commercial. About 100 titles a year. No
unsolicited mss; synopses and ideas welcome.
Please enclose s.a.e.

£ Royalties twice-yearly.

✳*Authors' update* With a tabloid journalist's talent for knowing what makes the front page and the front of shop display, John Blake has led the way in creating the celeb market. Can it last? Here is one publisher who clearly believes so.

## John Donald Publishers Ltd
West Newington House, 10 Newington Road, Edinburgh EH9 1QS
☎ 0131 668 4371  📠 0131 668 4466
info@birlinn.co.uk
www.birlinn.co.uk

Bought by **Birlinn Ltd** in 1999. Publishes academic and scholarly, archaeology, architecture, textbooks, guidebooks, local, and social history. New books are published as an imprint of Birlinn Ltd. About 30 titles a year.

## John Murray (Publishers) Ltd
338 Euston Road, London NW1 3BH
☎ 020 7873 6000  📠 020 7873 6446
firstname.lastname@johnmurrays.co.uk
www.hachettelivre.co.uk
Chief Executive *Jamie Hodder-Williams*
Managing Director *Roland Philipps*
Non-fiction *Eleanor Birne*
Fiction *Kate Parkin*

Founded 1768. Part of **Hachette Livre UK**. Publishes general trade books. 50 titles a year. No unsolicited material; send preliminary letter.
£ Royalties twice-yearly.

✳*Authors' update* Though sharing services with Hodder Headline, Murray has kept to its distinctive list of books for the discriminating general reader.

## John Wisden & Co.
▷ Bloomsbury Publishing Plc

## Jonathan Cape/Jonathan Cape Children's Books
▷ The Random House Group Ltd

## JR Books Ltd
10 Greenland Street, London NW1 0ND
☎ 020 7284 7163  📠 020 7485 4902
jeremyr@jrbooks.com
www.jrbooks.com
Managing Director *Jeremy Robson*
Senior Editor *Lesley Wilson*

Founded October 2006. JR Books publishes a wide range of non-fiction subjects, including biography, politics, music, humour, history, sport and self-help. 60 titles, spring 2009. Unsolicited synopses and ideas (with s.a.s.e.) welcome. Approach by mail. No fiction, children's or academic books.
£ Royalties twice-yearly.

## Kahn & Averill
9 Harrington Road, London SW7 3ES
☎ 020 8743 3278  📠 020 8743 3278
kahn@averill23.freeserve.co.uk
Managing Director *Mr M. Kahn*

Founded 1967 to publish children's titles but now specializes in music titles. A small independent publishing house. No unsolicited mss; synopses and ideas for books considered.
£ Royalties twice-yearly.

## Kenilworth Press Ltd (An imprint of Quiller Publishing Ltd)
Wykey House, Wykey, Shrewsbury SY4 1JA
☎ 01939 261616  📠 01939 260991
info@quillerbooks.com
www.kenilworthpress.co.uk
Managing Director *Andrew Johnston*

Founded in 1989 and acquired by Quiller Publishing Ltd in 2005. Publishes equestrian books only. About 10 titles a year. No unsolicited mss; synopses and ideas for books welcome. Send by post or email.
£ Royalties twice-yearly.

## Kenneth Mason Publications Ltd
The Book Barn, Westbourne, Emsworth PO10 8RS
☎ 01243 377977  📠 01243 379136
info@kennethmason.co.uk
www.kennethmason.co.uk
Chairman *Kenneth Mason*
Managing Director *Piers Mason*

Founded 1958. Publishes diet, health, fitness, nutrition and nautical. No fiction. IMPRINT **Research Disclosure**. Initial approach by letter with synopsis only.
£ Royalties twice-yearly in first year, annually thereafter.

## Kenyon-Deane
▷ Cressrelles Publishing Co. Ltd

## Kevin Mayhew Publishers
Buxhall, Stowmarket IP14 3BW
☎ 01449 737978  📠 01449 737834
info@kevinmayhewltd.com
sue@kevinmayhewltd.com
www.kevinmayhew.com
Chairman/Commissioning Editor *Kevin Mayhew*

Established in 1976. One of the leading music and Christian book publishers in the UK. Publishes religious titles – liturgy, sacramental, devotional and school resources and both sacred and secular sheet music for all major instruments. Does not publish children's stories, novels or autobiographies. 250 titles a year. See website for submission guidelines.
£ Royalties annually.

## Kingfisher
> Macmillan Publishers Ltd

## Kluwer Law International
250 Waterloo Road, London SE1 8RD
☎ 020 7981 0656   📠 020 7981 0587
simon.bellamy@kluwerlaw.com
www.kluwerlaw.com
*Also at:* Zuidpoolsingel 2a, 2408 ZE Alphen aan den Rijn, The Netherlands
Publisher *Simon Bellamy*

Founded 1995. Parent company: Wolters Kluwer Group. Publishes international law. About 200 titles a year, including online, CD-ROM, loose-leaf journals and monographs. Unsolicited synopses and ideas for books on law at an international level welcome.
£ Royalties annually.

## Kogan Page Ltd
120 Pentonville Road, London N1 9JN
☎ 020 7278 0433   📠 020 7278 6129
hkogan@koganpage.com
www.koganpage.com
Chairman *Philip Kogan*
Managing Director *Helen Kogan*
Approx. Annual Turnover £5.4 million

Founded 1967 by Philip Kogan to publish *The Industrial Training Yearbook*. Publishes business and management reference books and monographs, careers, marketing, personal finance, personnel, small business, training and industrial relations, transport. Has initiated a number of electronic publishing projects and provision of EP content. About 130 titles a year. OVERSEAS OFFICES in Philadelphia and New Delhi.
£ Royalties twice-yearly.

✴**Authors' update** For a medium sized publisher it is amazing how much of the business book market is covered by Kogan Page. Authors have benefited from strong overseas sales.

## Kyle Cathie Ltd
122 Arlington Road, London NW1 7HP
☎ 020 7692 7215   📠 020 7692 7260
general.enquiries@kyle-cathie.com
www.kylecathie.co.uk
Editorial Assistant *Vicki Murrell*

Founded 1990 to publish and promote 'books we have personal enthusiasm for'. Publishes non-fiction: cookery, food and drink, health and beauty, mind, body and spirit, gardening, homes and interiors, reference and occasional books of classic poetry. TITLES *Forgotten Skills* Darina Allen; *Italian Cookery Course* Katie Caldesi; *Grow Your Own, Eat Your Own* Bob Flowerdew; *The Pilates Bible* Lynne Robinson; *Liz Earle's Skin Secrets* Liz Earle. About 25 titles a year.

No unsolicited mss. 'Synopses and ideas are considered in the fields in which we publish.'
£ Royalties twice-yearly.

## Ladybird
> Penguin Group (UK)

## Landmark Publishing Ltd
> The Horizon Press

## Laurence King Publishing Ltd
4th Floor, 361–373 City Road, London EC1V 1LR
☎ 020 7841 6900   📠 020 7841 6910
enquiries@laurenceking.com
www.laurenceking.com
Chairman *Nick Perren*
Managing Director *Laurence King*
Commissioning Editor, Architecture *Philip Cooper*
Commissioning Editor, Design *Jo Lightfoot*

Publishes illustrated books in the fields of graphic design, fashion and textiles, architecture, interiors, product design and art. TITLES *A World History of Art*; *The Design Encyclopedia*; *How to Be a Graphic Designer Without Losing Your Soul*; *1000 New Designs and Where to Find Them*; *100 Years of Fashion Illustration* and the *Portfolio* series for students of art and design. About 60 titles a year. Unsolicited material welcome; send synopsis by post or email (commissioning@laurenceking.com).

## Lawrence & Wishart Ltd
99A Wallis Road, London E9 5LN
☎ 020 8533 2506   📠 020 8533 7369
lw@lwbooks.co.uk
www.lwbooks.co.uk
Managing Director/Editor *Sally Davison*

Founded 1936. An independent publisher with a substantial backlist. Publishes current affairs, cultural politics, economics, history, politics and education. TITLES *After Blair*; *Labour Legends Russian Gold*; *After Iraq*; *Making Sense of New Labour*. 10 titles a year.
£ Royalties annually, unless by arrangement.

## The Learning Institute
No. 1 Overbrook Business Centre, Blackford, Wedmore BS28 4PA
☎ 01934 713563   📠 01934 713492
courses@inst.org
www.inst.org
Managing Director *Kit Sadgrove*

Founded 1994 to publish home-study courses in vocational subjects such as garden design, writing and computing. Publishes subjects that show the reader how to work from home, gain a new skill or enter a new career. Interests include home working and self employment, especially in twenty-first century jobs. TITLES *Diploma*

*in Interior Design; Become a Garden Designer.*
Author's guidelines sent on receipt of s.a.s.e. No
unsolicited mss; send synopses and ideas only.
£ Royalties quarterly.

## Leckie & Leckie

4 Queen Street, Edinburgh EH2 1JE
☎ 0131 220 6831    📠 0131 225 9987
enquiries@leckieandleckie.co.uk
www.leckieandleckie.co.uk
Publishing Director *Martin Redfern*

Founded 1989. A division of Huveaux Plc.
Publishes Scottish educational materials. TITLES
*Practice Papers for SQA Exams; Success Guides;
Magical Series; Course Note Series.* 120 titles in
2009. No unsolicited mss; synopses and ideas for
books welcome by email.
£ Royalties annually.

## Legend Press Limited

2 London Wall Buildings, London EC2M 5UU
☎ 020 7448 5137
info@legend-paperbooks.co.uk
submissions@legend-paperbooks.co.uk
www.legendpress.co.uk
Managing Director *Tom Chalmers*

Publishes contemporary, diverse and cutting-edge
fiction. Is 'committed to providing writers with
an advantageous platform from which to develop
and succeed. We focus strongly on reaching new
audiences and on working with other publishers
to drive the independent sector and promote new
and different work within the mainstream market.'
Acquired **Paperbooks Publishing Ltd** in April
2008 (see entry). IMPRINT **Legend Business**. 20
titles a year. Send three chapters and synopsis to
the submissions email address above.
£ Royalties twice-yearly.

⊛*Authors' update* Driven by enthusiasm for
'fresh new fiction from authors who are fun and
inspiring to work with', Legend is expanding
into business books.

## Leo Cooper
▷ Pen & Sword Books Ltd

## Lewis/Lewis Masonic
▷ Ian Allan Publishing Ltd

## Library Association Publishing
▷ Facet Publishing

## Library of Wales
▷ Parthian

## Life&Times
▷ Haus Publishing

## Linden Press
▷ Open Gate Press

## Linford Mystery/Linford Romance/Linford Western
▷ F.A. Thorpe Publishing

## Lion Hudson plc

Wilkinson House, Jordan Hill Road, Oxford
OX2 8DR
☎ 01865 302750    📠 01865 302757
info@lionhudson.com
www.lionhudson.com
Managing Director *Paul Clifford*
Approx. Annual Turnover £8.9 million

Founded 1971. A Christian book publisher, strong
on illustrated books for a popular international
readership, with rights sold in over 170 languages
worldwide. Publishes a diverse list with
Christian viewpoint the common denominator.
All ages, from board books for children to
multi-contributor adult reference, educational,
paperbacks and colour co-editions and gift
books. IMPRINTS **Lion** and **Lion Children's** *Paula
O'Gorman*; **Candle** *Carol Jones*; **Monarch** *Paula
O'Gorman* (see **Monarch Books**). About 175 titles
a year. Unsolicited mss accepted provided they
have a positive Christian viewpoint intended for
a Christian or wide general and international
readership.
£ Royalties twice-yearly.

## Little Black Dress
▷ Hachette UK

## Little Books Ltd

73 Campden Hill Towers, 112 Notting Hill Gate,
London W11 3QW
☎ 020 7792 7929
www.littlebooks.net
Managing Director *Max Hamilton-Little*

Founded 2001. Publishes general trade books,
history, biography, health, natural history.
IMPRINT **Max Press** Fiction titles. No unsolicited
material.
£ Royalties annually.

## Little, Brown Book Group UK

100 Victoria Embankment, London EC4Y 0DY
☎ 020 7911 8000    📠 020 7911 8100
uk@littlebrown.co.uk
www.littlebrown.co.uk
Owner *Hachette Book Group*
CEO/Publisher *Ursula Mackenzie*
Approx. Annual Turnover £70 million

Founded 1988 as Little, Brown & Co. (UK) and
became Time Warner Book Group UK in 2002.
Purchased by Hachette in March 2006 and
renamed Little, Brown Book Group UK. Began by
importing its US parent company's titles and in
1990 launched its own illustrated non-fiction list.
Two years later the company took over former

Macdonald & Co. In 2007 the company bought Piatkus Books. Publishes hardback and paperback fiction, literary fiction, crime, science fiction and fantasy; and general non-fiction including true crime, biography and autobiography, cinema, history, humour, popular science, travel, reference, sport. IMPRINTS **Little, Brown** *Ursula Mackenzie, Richard Beswick, Tim Whiting* Hardback fiction and general non-fiction; **Abacus** *Richard Beswick, Tim Whiting, Jenny Parrott* Literary fiction and non-fiction paperbacks; **Atom** *Darren Nash* Young adult/teen paperbacks; **Hachette Digital** *Sarah Shrubb* (see entry under *Audio Books*); **Orbit** *Darren Nash* Science fiction and fantasy; **Sphere** *Antonia Hodgson, David Shelley, Joanne Dickinson, Adam Strange, Rebecca Saunders, Hannah Boursnell* Mass-market fiction and non-fiction hardbacks and paperbacks; **Virago Press** *Lennie Goodings, Ursula Doyle* (see entry); **Piatkus Books** *Tim Whiting, Gill Bailey, Anne Lawrance, Helen Stanton, Emma Beswetherick, Donna Condon* (see entry). 500 titles in 2009. No unsolicited mss. Approach in writing in the first instance.
Ⓔ Royalties twice-yearly.

✱*Authors' update* **Though now part of the Hachette group, Little Brown continues to be run as an independent entity publishing a wide range of commercial titles, mostly popular by virtue of quality though the mass market vampire novels may give pause for thought.**

## Little Tiger Press

An imprint of Magi Publications, 1 The Coda Centre, 189 Munster Road, London SW6 6AW
☏ 020 7385 6333   📠 020 7385 7333
info@littletiger.co.uk
www.littletigerpress.com
Publisher *Monty Bhatia*
Associate Publisher *Jude Evans*
Submissions Editor *Stephanie Stahl*

Little Tiger Press imprint publishes children's picture and novelty books for ages 0–7. No texts over 1,000 words. About 30 titles a year. Unsolicited mss, synopses and new ideas welcome. See website for submission guidelines.
Ⓔ Royalties annually.

✱*Authors' update* **A thriving small publisher with imaginative output.**

## Liverpool University Press

4 Cambridge Street, Liverpool L69 7ZU
☏ 0151 794 2233   📠 0151 794 2235
a.cond@liv.ac.uk
www.liverpool-unipress.co.uk
Managing Director/Editorial Head *Anthony Cond*

LUP's primary activity is the publication of academic and scholarly books and journals but it also has a limited number of trade titles. Its principal focus is on the arts and social sciences, in which it is active in a variety of disciplines. TITLES *Liverpool 800: Culture, Character and History; Irish, Catholic and Scouse: The History of the Liverpool Irish 1800–1939; Writing Liverpool; Liverpool and Transatlantic Slavery; So Spirited a Town; Public Sculptures of South London; sk-interfaces; The Culture of Capital; Cityscape: Ben Johnson's Liverpool; William Roscoe: Commerce and Culture; A Gallery to Play to: The Story of the Mersey Poets; The Cultural Values of Europe.* 30–40 titles a year.
Ⓔ Royalties annually.

## Livewire Books for Teenagers
▷ The Women's Press

## Lonely Planet Publications Ltd

2nd Floor, 186 City Road, London EC1V 2NT
☏ 020 7106 2100   📠 020 7106 2101
go@lonelyplanet.co.uk
www.lonelyplanet.com/about/our-books
Owner *Lonely Planet (Australia)*
Editorial Head *Imogen Hall*

'For over 30 years, Lonely Planet's on-the-ground research and no-holds-barred opinion by our team of expert travel writers has inspired and guided independent travellers to explore the world around them.' With over 500 titles in print, publications include travel guidebooks, downloadable digital guides, phrasebooks, travel literature, pictorial books and How To guides. No unsolicited mss; synopses and ideas welcome.

✱*Authors' update* **Putting up a valiant fight against online competition, Lonely Planet scores on originality with its Regional and Discover series. There is more emphasis on digital content.**

## Lorenz Books
▷ Anness Publishing Ltd

## Luath Press Ltd

543/2 Castlehill, The Royal Mile, Edinburgh EH1 2ND
☏ 0131 225 4326   📠 0131 225 4324
gavin.macdougall@luath.co.uk
www.luath.co.uk
Director *G.H. MacDougall*

Founded 1981. Publishes mainly books with a Scottish connection. Current list includes fiction, poetry, guidebooks, walking and outdoor, history, folklore, politics and global issues, cartoons, biography, food and drink, environment, music and dance, sport. TITLE *The Bower Bird* Ann Kelley (2007 Costa Children's Book Award). SERIES *On the Trail Of; The Quest For.*

About 30–40 titles a year. Over 200 titles in print. Unsolicited mss, synopses and ideas welcome;

'committed to publishing well-written books worth reading'.
£ Royalties paid.

## Lucky Duck Publishing
▷ Sage Publications

## Lulu.com
www.lulu.com

Founded 2002. A self publishing, print-on-demand service for individuals and companies. Produces a variety of digital content including books (hardback and paperback), dissertations, brochures, yearbooks, product manuals, portfolios, music, software, calendars, posters and artwork with no set-up fee and no requirement to buy copies. Authors supply a digital file of their book and a print-ready version of the file is created by the company. Lulu handles sales and pays the royalty specified by the client with a small mark-up as commission. Offers two retail distribution services for a fee. Full details available on the website.

✳*Authors' update* Authors who find it hard to get into print now have recourse to an option that is not vanity publishing, at least not in the strict sense of the word. Lulu is a website designed to carry anything that writers can throw at it, and for free. If browsers show an interest in a particular title, the book can be produced to order and shipped direct. Lulu takes a twenty per cent share of any royalties. Publishing services such as editing, proofing and marketing must be contracted separately. Anyone joining the game should not expect too much. John Morrison, a Lulu author, warns 'this is not a solution for those who want to sell their books in commercial quantities into Waterstone's'.

## Lund Humphries
▷ Ashgate Publishing Ltd

## The Lutterworth Press
PO Box 60, Cambridge CB1 2NT
☎ 01223 350865   ℻ 01223 366951
publishing@jamesclarke.co.uk
www.lutterworth.com
Managing Director *Adrian Brink*

The Lutterworth Press was founded as the Religious Tract Society in Georgian London, with its headquarters just off Fleet Street, in order to provide improving literature for young people and adults. Since then it has published many tens of thousands of titles, ranging from small pamphlets for children to erudite academic works. It became well known to generations of British children thanks to the publication of the *Boy's Own Paper* and the *Girl's Own Paper*. Since 1984 it has been an imprint of **James Clarke & Co.** (see entry). The

Lutterworth current list concentrates in several areas, and some of the work, although aimed at a general readership, is at quite a high level. TITLES include *Henry Wilson: Practical Idealist* Cyndy Manton; *After War is Faith Possible? An Anthology* G.A. Studdert Kennedy; *The Guide to Suffolk Churches*. 25 titles in 2009. Approach in writing with ideas in the first instance.
£ Royalties annually.

## Macdonald & Co.
▷ Little, Brown Book Group UK

## McGraw-Hill Education
McGraw-Hill House, Shoppenhangers Road, Maidenhead SL6 2QL
☎ 01628 502500   ℻ 01628 770224
www.mcgraw-hill.co.uk
www.openup.co.uk
General Manager *Shona Mullen*

Owned by US parent company, founded in 1888. Began publishing in Maidenhead in 1965, having had an office in the UK since 1899. Publishes business, economics, finance, accounting and social sciences for the academic, student and professional markets. Acquired **Open University Press** in 2002 (see entry). Around 100 titles a year. See website for author guidelines. Unsolicited proposals via website only.

## MacLehose Press
▷ Quercus Publishing

## Macmillan New Writing
▷ Macmillan Publishers Ltd

## Macmillan Publishers Ltd
The Macmillan Building, 4 Crinan Street, London N1 9XW
☎ 020 7833 4000   ℻ 020 7843 4640
www.macmillan.com
Owner *Verlagsgruppe Georg von Holtzbrinck*
Chief Executive *Annette Thomas*
Approx. Annual Turnover £350 million (Book Publishing Group)

Founded 1843. Macmillan is one of the largest publishing houses in Britain, publishing approximately 1,400 titles a year. In 1995, Verlagsgruppe Georg von Holtzbrinck, a major German publisher, acquired a majority stake in the Macmillan Group and in 1999 purchased the remaining shares. In 1996, Macmillan bought Boxtree, the successful media tie-in publisher; in 1997, it purchased the Heinemann English language teaching list from Reed Elsevier, and acquired Kingfisher, the popular children's book publisher, in 2007. The educational publishing division was strengthened by the acquisitions of the Mexican list, Ediciones Castillo and the Argentine company Puerto de Palos. No unsolicited material, except for the Macmillan New Writing list (see details below).

DIVISIONS

**Palgrave Macmillan** Brunel Road, Houndmills, Basingstoke RG21 6XS ☎ 01256 329242 🖷 01256 328339 Managing Director *Dominic Knight*; College *Margaret Hewinson*; Scholarly & Reference *Sam Burridge*; Journals *David Bull* Publishes textbooks, monographs and journals in academic and professional subjects. Publications in both print and electronic format.

**Macmillan Education** 4 Between Towns Road, Oxford OX4 3PP ☎ 01865 405700 🖷 01865 405701 info@macmillan.com www.macmillaneducation. com Chief Executive Officer *Julian Drinkall*, Publishing Directors *Sue Bale* (dictionaries), *Alison Hubert* (Africa, Caribbean, Middle East and East Asia), *Julie Kniveton* (Latin America), *Angela Lilley* (ELT), *Kate Melliss* (Europe) Publishes a wide range of ELT titles and educational materials for the international education market from Oxford and through 30 subsidiaries worldwide.

**Pan Macmillan** 20 New Wharf Road, London N1 9RR ☎ 020 7014 6000 🖷 020 7014 6001 www. panmacmillan.com Managing Director *Anthony Forbes Watson* Publishes under **Macmillan, Mantle, Pan, Picador, Macmillan New Writing, Sidgwick & Jackson, Boxtree, Macmillan Children's Books, Kingfisher, Macmillan Digital Audio, Campbell Books, Young Picador, Rodale; Tor**. No unsolicited mss. Visit the writers' area on www.panmacmillan.com for useful articles and information on books that may be helpful in getting your book published, or visit the Macmillan New Writing website (see below) for details of the MNW programme, which does accept unsolicited mss of first novels.

IMPRINTS

**Macmillan** (founded 1843) Fiction: Publishing Director *Jeremy Trevathan*, Publishing Director, Fiction *Imogen Taylor* Publishes hardback commercial fiction including genre fiction, romantic, crime and thrillers. IMPRINT **Tor** (founded 2003) Senior Commissioning Editor *Julie Crisp* Publishes science fiction, fantasy and thrillers. Non-Fiction: Publisher *Jon Butler*, Editorial Director *Georgina Morley* Publishes serious and general non-fiction: autobiography, biography, economics, history, philosophy, politics and world affairs, psychology, popular science, trade reference titles.

**Mantle** Publishing Director *Maria Rejt* Publishes general, crime, thrillers, literary fiction and some non-fiction.

**Pan** (founded 1947) Paperback imprint for Pan Macmillan.

**Picador** (founded 1972) Publisher *Paul Baggaley* Publishes literary international fiction, non-fiction and poetry.

**Rodale** Editorial Director *Liz Gough* Publishes books on health, fitness, diet, cookery, spirituality.

**Sidgwick & Jackson** (founded 1908) Editorial Director *Ingrid Connell* Publishes popular non-fiction in hardback and trade paperback with strong personality or marketable identity, from celebrity and show business to music and sport. Also military history list.

**Boxtree** (founded 1986) Editorial Director *Jon Butler* Publishes brand and media tie-in titles, including TV, film, music and internet, plus entertainment licences, pop culture, humour and event-related books. TITLES *Dilbert; James Bond; Purple Ronnie; Wallace & Gromit; The Onion.*

**Macmillan Children's Books** (New Wharf Road address) Managing Director *Emma Hopkin*, Publishing Director *Rebecca McNally*; Fiction: Editorial Director *Sarah Dudman*; Non-Fiction and Poetry: Editorial Director *Gaby Morgan*; Picture Books and Gift Books: Editorial Director *Suzanne Carnell*. **Kingfisher** Editorial Director *Martina Challis*; **Campbell Books** Editorial Director *Suzanne Carnell*; **Young Picador** Publishes fiction, non-fiction and poetry in paperback and hardback.

**Macmillan New Writing** (at New Wharf Road address) Publishing Director *Maria Rejt*, Editorial Director, Fiction *Will Atkins* Founded in 2006 as a way of finding talented new writers who might otherwise go undiscovered. MNW publishes full-length novels from authors who have not previously published a novel. All genres considered and all submissions assessed, but mss must be complete. No advance, but the author pays nothing and receives a royalty of 20% on net sales. Submissions must be sent by email only, via the MNW website (www.macmillannewwriting. com).

🖷 Royalties annually or twice-yearly depending on contract.

❋*Authors' update* **With journals leading the way (***Nature* **alone accounts for a quarter of Macmillan's turnover), digital publishing is much to the fore. Among major initiatives** *The New Palgrave Dictionary of Economics* **has a double life, print and online, while Pan Macmillan dominates Waterstone's ebook charts. Pan Mac's new team led by Anthony Forbes Watson has reenergized the fiction list. As part of the expansion Maria Rejt will head Mantle, a new trade imprint. Macmillan New Writing is attracting rising stars. The deal for first-time authors is a fixed contract with no advance but with a twenty per cent royalty.**

A new title appears each month. Crime and children's lists are thriving.

## Made Simple Books
▷ Elsevier Ltd

## Maia Press
▷ Arcadia Books

## Mainstream Publishing Co. (Edinburgh) Ltd

7 Albany Street, Edinburgh EH1 3UG
☎ 0131 557 2959    ℻ 0131 556 8720
enquiries@mainstreampublishing.com
www.mainstreampublishing.com
Directors *Bill Campbell, Peter MacKenzie*
Approx. Annual Turnover £3.7 million

Publishes art, autobiography/biography, current affairs, true crime, health, sport, history, illustrated and fine editions, photography, politics and world affairs, popular paperbacks. TITLES *Soldier Five* Mike Coburn; *The Real Nureyev* Carolyn Soutar; *Woodward's England* Mick Collins. Over 80 titles a year. Ideas for books considered, but they should be preceded by a letter, synopsis and s.a.s.e. or return postage.
Ⓔ Royalties twice-yearly.

## Management Books 2000 Ltd

Forge House, Limes Road, Kemble, Cirencester GL7 6AD
☎ 01285 771441    ℻ 01285 771055
info@mb2000.com
www.mb2000.com
Publisher *Nicholas Dale-Harris*
Approx. Annual Turnover £500,000

Founded 1993 to develop a range of books for executives and managers working in the modern world of business. 'Essentially, the books are working books for working managers, practical and effective.' Publishes business, management, self-development and allied topics as well as sponsored titles. Launched the *In Ninety Minutes* series of compact guide books for managers in 2004, offering advice, ideas and practical help across a range of highly relevant business topics in an hour and a half of study. New ideas for this series are welcome. About 24 titles a year. Unsolicited mss, synopses and ideas for books welcome.

## Manchester University Press

Oxford Road, Manchester M13 9NR
☎ 0161 275 2310    ℻ 0161 274 3346
mup@manchester.ac.uk
www.manchesteruniversitypress.co.uk
Publisher/Chief Executive *David Rodgers*
Head of Editorial *Matthew Frost*
Approx. Annual Turnover £2 million

Founded 1904. MUP is Britain's third largest university press, with a list marketed and sold worldwide. Remit consists mainly of A-level and undergraduate textbooks and research monographs. Publishes in the areas of literature, TV, film, theatre and media, history and history of art, design, politics, economics, sociology and international law. DIVISIONS **Humanities** *Matthew Frost*; **History/Art History** *Emma Brennan*; **Politics and Law** *Tony Mason*. About 140 titles a year. Unsolicited mss welcome.
Ⓔ Royalties annually.

## Manson Publishing Ltd

73 Corringham Road, London NW11 7DL
☎ 020 8905 5150    ℻ 020 8201 9233
manson@mansonpublishing.com
www.mansonpublishing.com
Chairman/Managing Director *Michael Manson*

Founded 1992. Publishes highly illustrated books for study and reference. Subject areas covered include medicine, veterinary medicine, earth science, plant science, agriculture and microbiology. About 10 titles a year. No unsolicited mss; synopses and ideas will be considered.
Ⓔ Royalties twice-yearly.

## Marion Boyars Publishers Ltd

24 Lacy Road, London SW15 1NL
☎ 020 8788 9522
catheryn@marionboyars.com
www.marionboyars.co.uk
Publisher *Catheryn Kilgarriff*

Founded 1975, formerly Calder and Boyars. Publishes fiction, philosophy, psychology, sociology and anthropology, theatre and drama, film and cinema, women's studies. AUTHORS include Georges Bataille, Ingmar Bergman, Heinrich Böll, Jean Cocteau, Carlo Gébler, Julian Green, Ivan Illich, Pauline Kael, Ken Kesey, Kenzaburo Oe, Hubert Selby Jr, Lawrence Potter, Eudora Welty, Judith Williamson, Hong Ying, Elif Shafak and Riverbend. Unsolicited mss not welcome for fiction or poetry; submissions from agents only. Unsolicited synopses and ideas welcome for non-fiction.
Ⓔ Royalties annually.

## Marshall Editions
▷ Quarto Publishing plc

## Martin Books
▷ Simon & Schuster UK Ltd

## Martin Breese International

19 Hanover Crescent, Brighton BN2 9SB
☎ 01273 687555
mbreese999@aol.com
www.abracadabra.co.uk
Chairman/Managing Director *Martin Breese*

Founded in 1975. 12 titles in 2009. No unsolicited submissions.
ⓔ Royalties annually.

## Matador
▷ Troubador Publishing Ltd

## Max Press
▷ Little Books Ltd

## Meadowside Children's Books
185 Fleet Street, London EC4A 2HS
info@meadowsidebooks.com
www.meadowsidebooks.com
Owner *D.C. Thomson*
Publisher *Simon Rosenheim*
Editor *Lucy Cuthew*
Rights Director *Katherine Judge*

Founded 2003. Acquired **Gullane Children's Books** in 2007 (see entry). Publishes picture books, junior fiction and young adult fiction. Unsolicited mss accepted. Picture book mss: 1,000 words max. Fiction mss: short synopsis and first three chapters; address to *The Submissions Editor*. Only successful submissions answered. 'Strictly no telephone enquiries for submissions; no fax submissions.'
ⓔ Royalties annually.

## Medavia Publishing
▷ Boltneck Publications Limited

## Melrose Books
St Thomas Place, Ely CB7 4GG
ⓣ 01353 646608     ⓕ 01353 646602
info@melrosebooks.co.uk
www.melrosebooks.com
Chairman *Richard A. Kay*
Managing Director *Nicholas S. Law*
Approx. Annual Turnover £100,000

Independent subsidy publisher established by **Melrose Press** in 2004. To-date has published general and children's fiction, literature, fantasy, biography, travel, religion and science reference. Welcomes submissions by email or post
ⓔ Royalties 55%.

## Melrose Press Ltd
St Thomas Place, Ely CB7 4GG
ⓣ 01353 646600     ⓕ 01353 646601
tradesales@melrosepress.co.uk
www.melrosepress.co.uk
Chairman *Richard A. Kay*
Managing Director *Nicholas S. Law*
Approx. Annual Turnover £2 million

Founded 1960. Took on its present name in 1969. Publishes biographical who's who reference only (not including *Who's Who*, which is published by **A.&C. Black**). About 10 titles a year.

## Mentor
▷ Christian Focus Publications

## Mercat Press
▷ Birlinn Ltd

## The Merlin Press Ltd
6 Crane Street Chambers, Crane Street, Pontypool NP4 6ND
ⓣ 01495 764100
adrian.howe@merlinpress.co.uk
www.merlinpress.co.uk
Managing Director *Anthony W. Zurbrugg*
Director *Adrian Howe*

Founded 1956. Publishes in the area of history, philosophy and politics. No fiction. IMPRINTS **Merlin Press**; **Green Print**. TITLES *Socialist Register* (annual); the Chartist Studies SERIES; *Positive Education*. About 10 titles a year.
ⓔ Royalties annually.

## Merrell Publishers Ltd
81 Southwark Street, London SE1 0HX
ⓣ 020 7928 8880     ⓕ 020 7928 1199
mail@merrellpublishers.com
www.merrellpublishers.com
Managing Director *Hugh Merrell*
Head of Editorial *Claire Chandler*
Creative Director *Nicola Bailey*
Approx. Annual Turnover £2.8 million

Founded 1993. Publishes art, architecture, design, gardening, interiors, photography, cars and motorcycles. 30 titles in 2009. Unsolicited synopses and ideas for books welcome. Send c.v. and synopsis giving details of the book's target markets and funding of illustrations.
ⓔ Royalties annually.

## Methuen Drama
▷ A.&C. Black Publishers Ltd

## Methuen Publishing Ltd
8 Artillery Row, London SW1P 1RZ
ⓣ 020 7802 0018     ⓕ 020 7828 1244
sales@methuen.co.uk
www.methuen.co.uk
Managing Director *Peter Tummons*
Editorial *Naomi Tummons*

Founded 1889. Methuen was owned by Reed International until it was bought by Random House in 1997. Purchased by a management buy-out team in 1998. Acquired **Politico's Publishing** in 2003 (see entry). Publishes fiction and non-fiction; travel, sport, humour. No unsolicited mss; synopses and ideas welcome. Prefers to be approached via agents or a letter of inquiry. No first novels, cookery books or personal memoirs.

## Metro Publishing
▷ John Blake Publishing Ltd

## Michael Joseph
▷ Penguin Group (UK)

## Michael O'Mara Books Ltd
16 Lion Yard, Tremadoc Road, London SW4 7NQ
☎ 020 7720 8643   📠 020 7627 8953
firstname.lastname@mombooks.com
www.mombooks.com
www.mombooks.com/busterbooks
Chairman *Michael O'Mara*
Managing Director *Lesley O'Mara*
Editorial Directors *Toby Buchan, Louise Dixon, Lindsay Davies*
Publishing Director (Buster Books) *Philippa Wingate*
Approx. Annual Turnover £8 million

Founded 1985. Independent publisher. Publishes general non-fiction, humour, anthologies, miscellanies, royalty and celebrity autobiographies and biographies. TITLES *My Word Is My Bond* Roger Moore; *I Am What I Am* John Barrowman; *A Fart in a Colander* Roy Hudd; *Cheryl Cole: Her Story* Gerard Sanderson; *I Before E* Judy Parkinson; *I Used to Know That* Caroline Taggart; *Aunt Epp's Guide For Life* Christopher Rush. IMPRINT **Buster Books** Children's books TITLES *Off With Their Heads* Martin Oliver; *Fabulous Doodles* Nellie Ryan; *The Girls' Book of Glamour* Sally Jeffrie; board books, colouring books, picture books, novelty books, humour and quirky non-fiction. 74 titles in 2009. Unsolicited mss, synopses and ideas for books welcome. 💷 Royalties twice-yearly.

✳*Authors' update* The joker in the pack, O'Mara Books, now in its twenty-fifth year, has never looked back from the stunningly successful 'true story' of Princess Diana. But the core of the list is broad humour bordering on the scatological.

## Michelin Maps & Guides
Hannay House, 39 Clarendon Road, Watford WD17 1JA
☎ 01923 205240/205250   📠 01923 205241
www.Viamichelin.co.uk
Commercial Director *Ian Murray*

Founded 1900 as a travel publisher. Publishes travel guides, maps and atlases.

## Midland Publishing
▷ Ian Allan Publishing Ltd

## Miller's
▷ Octopus Publishing Group

## Millivres Prowler Limited
Unit M, Spectrum House, 32–34 Gordon House Road, London NW5 1LP
☎ 020 7424 7400   📠 020 7424 7401

info@millivres.co.uk
www.millivres.co.uk

Publishes various magazines including *Gay Times*, *Diva Magazine*. Both *AXM* monthly magazine and *The Pink Paper*, the national free gay newspaper, are published online only.

✳*Authors' update* No longer interested in books; the focus is entirely on journals.

## Mills & Boon
▷ Harlequin Mills & Boon Ltd

## Milo Books Limited
The Old Weighbridge, Station Road, Wrea Green, Preston PR4 2PH
☎ 01772 672900   📠 01772 687727
info@milobooks.com
www.milobooks.com
Managing Director *Peter Walsh*

Founded 1997. Publishes non-fiction: true crime and sport. 12 titles a year. Unsolicited mss, synopses and ideas for books welcome. 'Authors should note our specialist areas. We cannot promise to return all material submitted.' Approach in writing in the first instance. 'If we are interested in pursuing an idea, we will call to talk it through in more depth.' No fiction or poetry. 💷 Royalties twice-yearly.

## Mindfield
▷ Camden Press Ltd

## MIRA®
▷ Harlequin Mills & Boon Limited

## Mitchell Beazley
▷ Octopus Publishing Group

## Monarch Books
Lion Hudson plc, Wilkinson House, Jordan Hill Road, Oxford OX2 8DR
☎ 01865 302750   📠 01865 302757
monarch@lionhudson.com
Editorial Director *Tony Collins*

An imprint of **Lion Hudson plc**. Publishes an independent list of Christian books across a wide range of concerns. IMPRINT **Monarch** Upmarket paperback list with Christian basis and strong social concern agenda including psychology, future studies, politics, mission, theology, leadership, fiction and spirituality. About 35 titles a year. Unsolicited mss, synopses and ideas welcome.

## Morgan Kauffman
▷ Elsevier Ltd

## Mosby
▷ Elsevier Ltd

## Motor Racing Publications

PO Box 1318, Croydon CR9 5YP
☎ 020 8654 2711    🖷 020 8407 0339
john@mrpbooks.co.uk
www.mrpbooks.co.uk
Chairman/Editorial Head *John Blunsden*

Founded soon after the end of World War II
to concentrate on motor-racing titles. Fairly
dormant in the mid 1960s but was reactivated
in 1968 by a new shareholding structure. John
Blunsden later acquired a majority share and
major expansion followed in the 1970s. Publishes
motor-sport history, classic and performance
car collection and restoration, race track driving
and related subjects. IMPRINTS **Fitzjames Press**;
**Motor Racing Publications**. About 2–4 titles a
year. No unsolicited mss. Send synopses and ideas
in specified subject areas in the first instance.
💷 Royalties twice-yearly.

## Murdoch Books UK Ltd

Erico House, 6th Floor North, 93–99 Upper
Richmond Road, London SW15 2TG
☎ 020 8785 5995    🖷 020 8785 5985
info@murdochbooks.co.uk
www.murdochbooks.co.uk
Publisher *Kay Scarlett (Australia)*

Owned by Australian publisher Murdoch Books
Pty Ltd. Publishes full-colour non-fiction: homes
and interiors, gardening, cookery, craft, DIY,
history, travel and narrative non-fiction. About
80 titles a year. Contact inquiry@murdochbooks.
com.au in the first instance.

## Myriad Editions

59 Lansdowne Place, Brighton BN3 1FL
☎ 01273 720000
info@myriadeditions.com
www.myriadeditions.com
Managing Director *Candida Lacey*
Rights Manager *Sadie Mayne*

An independent publishing house, Myriad
Editions was founded as a packager in 1993 and
won international acclaim for its award-winning
'State of the World' atlas series. Remains
committed to mapping the most pressing issues
facing the world today. Started publishing fiction
and graphic non-fiction in 2006 with the aim of
nurturing local writers and showcasing original,
home-grown talent. 2–3 atlases and 6 fiction titles
a year. See website for submission details.
💷 Royalties annually.

## Myrmidon Books

Rotterdam House, 116 Quayside, Newcastle upon
Tyne NE1 3DY
☎ 0191 206 4005    🖷 0191 206 4001
submissions@myrmidonbooks.com
www.myrmidonbooks.com

Chairman/Managing Director *Ed Handyside*
Editor *Anne Westgarth*
Approx. Annual Turnover £200,000

Founded 2006. Publishes adult trade fiction. Full
mss from literary agents only; first three chapters
from unrepresented writers. No short stories,
novellas or works under 65,000 words, children's
books, graphic novels, non-fiction. Submit hard
copy only after checking requirements on the
website.
💷 Royalties quarterly.

## NAG Press Ltd
▷ **Robert Hale Ltd**

## National Trust Books
▷ **Anova Books**

## Nautical Data Ltd

The Book Barn, Westbourne, Emsworth PO10 8RS
☎ 01243 389352    🖷 01243 379136
info@nauticaldata.com
www.nauticaldata.com
Managing Director *Piers Mason*

Founded 1999. Publishes pilots and nautical
reference. No unsolicited mss; synopses and ideas
welcome. No fiction or non-nautical themes.
💷 Royalties twice-yearly.

## NCVO Publications

Regent's Wharf, 8 All Saints Street, London
N1 9RL
☎ 020 7713 6161/HelpDesk: 0800 2798 798
🖷 020 7713 6300
ncvo@ncvo-vol.org.uk
www.ncvo-vol.org.uk/publications
Publications Project Manager *Mike Wright*
Approx. Annual Turnover £250,000

Founded 1928. Publishing imprint of the National
Council for Voluntary Organisations, embracing
former Bedford Square Press titles and NCVO's
many other publications. The list reflects
NCVO's role as the representative body for the
voluntary sector. Publishes directories, good
practice information on management and trustee
development, finance and employment titles and
information of primary interest to the voluntary
sector. TITLES *The Voluntary Agencies Directory*;
*The Good Trustee Guide*; *The Good Campaigns
Guide*; *The Good Financial Management
Guide*; *The Good Employment Guide*; *The Good
Management Guide*; *The Good Membership
Guide*; *The UK Voluntary Sector Almanac*. No
unsolicited mss as all projects are commissioned
in-house.

## Neil Wilson Publishing Ltd
G/2, 19 Netherton Avenue, Glasgow G13 1BQ
☎ 0141 954 8007 📠 0560 150 4806
info@nwp.co.uk
www.nwp.co.uk
Managing Director/Editorial Director *Neil Wilson*
Approx. Annual Turnover £110,000

Founded 1992. Publishes Scottish interest and history, biography, humour and hillwalking, whisky; also Scottish cookery and travel.
IMPRINTS **The In Pinn** Outdoor pursuits; **The Angels' Share** Whisky, drink and food-related subjects; **The Vital Spark** Humour; **NWP** History, biography, reference, true crime. About 6 titles a year. Unsolicited mss, synopses and ideas welcome. No politics, academic or technical.
Ⓔ Royalties twice-yearly.

## Nelson Thornes Limited
Delta Place, 27 Bath Road, Cheltenham GL53 7TH
☎ 01242 267100 📠 01242 221914
info@nelsonthornes.com
www.nelsonthornes.com
Managing Director *Mary O'Connor*

Now part of Infinitas Learning, Nelson Thornes was formed in 2000 by the merger of Thomas Nelson and Stanley Thornes. Educational publisher (AQA exclusively endorsed) of printed and electronic resources, from pre-school to higher education. Unsolicited mss, synopses and ideas for books welcome if appropriate to specialized lists.
Ⓔ Royalties annually.

## New Beacon Books Ltd
76 Stroud Green Road, London N4 3EN
☎ 020 7272 4889 📠 020 7281 4662
newbeaconbooks@btconnect.com
www.newbeaconbooks.co.uk
Managing Director *Sarah White*
Approx. Annual Turnover £90,000

Founded 1966. Publishes fiction, history, politics, poetry and language, all concerning black people. No unsolicited material.
Ⓔ Royalties annually.

## New Holland Publishers (UK) Ltd
Garfield House, 86–88 Edgware Road, London W2 2EA
☎ 020 7724 7773 📠 020 7258 1293 (editorial)
postmaster@nhpub.co.uk
www.newhollandpublishers.com
Managing Director *Steve Connolly*
Publishing Director *Rosemary Wilkinson*
Approx. Annual Turnover £8.5 million

With its headquarters in London, New Holland Publishers (UK) Ltd is the International Publishing Division of Avusa, one of Africa's leading publishing groups, with offices in Australia and New Zealand. Acquired Cadogan Guides in April 2007. Publishes non-fiction, practical and inspirational books across a range of categories, including food and drink, DIY, practical art, crafts, interiors, health and fitness, gardening, humour and gift, reference, history, sports and outdoor pursuits, natural history, general travel books and Globetrotter travel guides. No unsolicited mss; synopses and ideas welcome.

✱*Authors' update* **To meet the challenges of an increasingly competitive market, Cadogan has relaunched its travel guides with the focus more on regions than on countries. The aim is to upstage elementary websites.**

## New Kinglake Publishing
Ashgrove House, Suite S4, Elland, Calderdale HX5 9JB
☎ 0845 508 6032
info@kinglakepublishing.co.uk
www.kinglakepublishing.co.uk
Editorial Director *Harry Taylor*
Approx. Annual Turnover £120,000

Self-publishing service which is paid for by the author. Publishing packages offer promotion and marketing services at varying prices. See website for submission details and pricing. Genres considered include fiction and non-fiction, faith books, biography, memoirs and poetry.

## New Playwrights' Network
▷ Cressrelles Publishing Co. Ltd

## Newnes
▷ Elsevier Ltd

## Nexus
▷ The Random House Group Ltd

## Nicholas Brealey Publishing
3–5 Spafield Street, London EC1R 4QB
☎ 020 7239 0360 📠 020 7239 0370
rights@nicholasbrealey.com
www.nicholasbrealey.com
Managing Director *Nicholas Brealey*

Founded 1992. Independent publishing group focusing on innovative trade/professional books covering business and economics, popular psychology and the increasingly active fields of travel writing and crossing cultures. The group has offices in Boston and includes the IMPRINTS **Davies-Black** and **Intercultural Press**. TITLES *The Trouble With Markets*; *Talent is Overrated*; *Coaching for Performance*; *Your Writing Coach*; *Get to the Top on Google*; *The Cult of the Amateur*; *The Leader's Way* by the Dalai Lama; *Authentic Happiness*; *50 Psychology Classics*; *The 80/20 Principle*; *It's All Greek to Me!*; *Don't Tell Mum I Work on the Rigs She Thinks I'm a Piano Player in* *a Whorehouse.* No fiction, poetry or leisure titles. 30 titles a year. No unsolicited mss; synopses and ideas welcome.

✱*Authors' update* **Looks to be succeeding in breaking away from the usual computer-speak business manuals to publish information and literate texts. Lead titles have a distinct trans-Atlantic feel.**

## Nick Hern Books

The Glasshouse, 49a Goldhawk Road, London W12 8QP

📞 020 8749 4953   📠 020 8735 0250
info@nickhernbooks.demon.co.uk
www.nickhernbooks.co.uk
Chairman/Managing Director *Nick Hern*
Submissions Editor *Matt Applewhite*

Founded 1988. Fully independent since 1992. Publishes books on theatre and film: from practical manuals to plays and screenplays. About 60 titles a year. No unsolicited playscripts. Synopses, ideas and proposals for other theatre material welcome. Not interested in material unrelated to the theatre or cinema.

✱*Authors' update* **An early stop for anyone interested in drama. But the standard barrier is high and much of the output is devoted to scripts.**

## Nielsen Book

3rd Floor, Midas House, 62 Goldsworth Road, Woking GU21 6LQ

📞 01483 712200   📠 01483 712201
info.bookdata@nielsen.com
www.nielsenbookdata.co.uk
Operations Director *Richard Knight*
Senior Manager, Publishing Services *Peter Mathews*

Nielsen Book collects data to maintain a unique database of timely, accurate and content rich records for English-language books and other published media, collected from publishers in over 70 countries. The data includes: price, availability, table of contents, subject classifications, market rights, publisher and distributor details, cover/jacket images, book descriptions, author biographies and reviews. The data is made available as subscription services to the book industry via the **BookData**, **BookNet** and **BookScan** services (see entries under *Miscellany*).

## Nightingale Books

▷ Pegasus Elliot Mackenzie Publishers Ltd

## NMS Enterprises Limited – Publishing

National Museums Scotland, Chambers Street, Edinburgh EH1 1JF

📞 0131 247 4026   📠 0131 247 4012

publishing@nms.ac.uk
www.nms.ac.uk/books
Director *Lesley A. Taylor*
Approx. Annual Turnover £130,000

Publishes non-fiction related to the National Museums of Scotland collections: academic and general; children's; archaeology, history, decorative arts worldwide, history of science, technology, natural history and geology, poetry. TITLES *Hidden Treasures of the Romanovs: Saving the Royal Jewels*; '*Wroughte in gold and silk': Preserving the Art of Historic Tapestries*; *Silver: Made in Scotland*; *Tartan: The Highland Habit*; *Minerals of Northern England*; *Robert Louis Stephenson: The Travelling Mind*; *Bagpipes: A National Collection of a National Instrument*; *Minerals of Scotland*; *Commando Country*; *Weights and Measures in Scotland: a European Perspective* (2005 Saltire Society/National Library of Scotland Research Book of the Year). 12 titles a year. No unsolicited mss; only interested in synopses and ideas for books which are genuinely related to NMS collections and to Scotland in general.
🅵 Royalties twice-yearly.

## Nonsuch
▷ The History Press

## North-Holland
▷ Elsevier Ltd

## Northcote House Publishers Ltd

Horndon House, Horndon, Tavistock PL19 9NQ

📞 01822 810066   📠 01822 810034
northcote.house@virgin.net
www.northcotehouse.co.uk
Managing Director *Brian Hulme*

Founded 1985. Publishes a series of literary critical studies, in association with the British Council, *Writers and their Work*; education management, literary criticism, educational dance and drama. 25 titles in 2009. 'Well-thought-out proposals, including contents and sample chapter(s), with strong marketing arguments welcome.'
🅵 Royalties annually.

## Northumbria University Press

Northumbria University, Trinity Building, Newcastle upon Tyne NE1 8ST

📞 0191 227 4603   📠 0191 227 3387
andrew.peden-smith@northumbria.ac.uk
www.northumbria.ac.uk/sd/central/its/uni_press/
Managing Director *Andrew Peden-Smith*

Northumbria University Press, founded in 2006, publishes non-fiction: cultural studies, language and history. 2009 saw the launch of **Northumbria Press** with its emphasis on popular culture. TITLES *Dictionary of North East Dialect*; *Talk of the North East Coalfield*; *England, My England*;

*Byker Revisited*; *Rock and Roll Tourist*; *The Tin Ring*. 10 titles in 2009. Unsolicited material welcome; approach by email. No poetry.
Ⓔ Royalties twice-yearly.

### Nosy Crow

The Crow's Nest, 704 The Chandlery, 50 Westminster Bridge Road, London SE1 7QY
☎ 020 7953 7677   📠 020 7953 7673
hello@nosycrow.com
www.nosycrow.com
Managing Director *Kate Wilson*

New children's publisher established in 2010 by Kate Wilson, former managing director of Scholastic UK. Publishes fiction and non-fiction books for children aged 0 to 14 as well as apps for IPhones. Welcomes unsolicited material; approach by email.

### Nottingham University Press

Manor Farm, Church Lane, Thrumpton NG11 0AX
☎ 0115 983 1011   📠 0115 983 1003
orders@nup.com
www.nup.com

Nottingham University Press has gained 'international recognition as a publisher of high quality scientific and technical publications'.
TITLES *Recent Advances in Animal Nutrition 2009*; *Perfecting the Pig Environment*; *Advances in Equine Nutrition IV*; *The Alcohol Textbook – 5th Edition*.
Ⓔ Royalties twice-yearly.

### NWP
▷ Neil Wilson Publishing Ltd

### O Books

The Bothy, Deershot Lodge, Park Lane, Ropley SO24 0BE
office1@o-books.net
www.o-books.net
Approx. Annual Turnover £1.5 million

Publishes religion, MBS, philosophy, fiction and other areas. About 100 titles a year. For submissions, go to 'Contact us' section on the website.

### Oak
▷ Omnibus Press

### Oberon Books

521 Caledonian Road, London N7 9RH
☎ 020 7607 3637   📠 020 7607 3629
info@oberonbooks.com
www.oberonbooks.com
Publishing Director *James Hogan*
Managing Director *Charles D. Glanville*
Editor *Kate Longworth*

A leading theatre publisher, Oberon publishes play texts (usually in conjunction with a

production), and books on theatre and dance. Specializes in contemporary plays and translations of European classics. IMPRINTS **Oberon Modern Plays**; **Oberon Classics**. Publishes for the National Theatre, Royal Opera House, Royal Court, English Touring Theatre, LAMDA, Gate Theatre, Bush Theatre, West Yorkshire Playhouse, Chichester Festival Theatre and many other London and regional theatre companies. Publishes over 300 writers and translators including Rodney Ackland, Tariq Ali, Howard Barker, John Barton, Ranjit Bolt, Pam Gems, Tanika Gupta, Sir Peter Hall, Bernard Kops, Adrian Mitchell, Sir John Mortimer, Adriano Shaplin, Laura Wade. An extensive classics list includes performance (edited) versions of Shakespeare including *King Lear*; *A Midsummer Night's Dream*; *Romeo and Juliet* and *Twelfth Night*, as well as new translations of plays by Chekhov, Molière, Strindberg and others. About 80 titles a year. Send mss, with information about forthcoming productions to *Daisy Bowie-Sell* (daisy@oberonbooks.com). 'We do not provide readers' reports or feedback about work submitted.'

✱*Authors' update* With its dedication to live performance, Oberon has confirmed its lead in a niche market by sponsoring the Sheridan Morley Prize for theatrical biography (see entry under *Prizes*).

### Octagon Press Ltd

78 York Street, London W1H 1DP
☎ 020 7193 6456   📠 020 7117 3955
admin@octagonpress.com
www.octagonpress.com
Managing Director *Scott Goodfellow*
Director *Saira Shah*
Approx. Annual Turnover £50,000

Founded 1960. Publishes travel, biography, literature, folklore, psychology, philosophy with the focus on East-West studies.
Ⓔ Royalties annually.

### Octopus Publishing Group

Endeavour House, 189 Shaftesbury Avenue, London WC2H 8JG
☎ 020 7632 5400   📠 020 7632 5405
www.octopus-publishing.co.uk
Chief Executive *Alison Goff*
Group Publishing Director *Denise Bates*

Formed following a management buyout of Reed Consumer Books from Reed Elsevier plc in 1998. Bought by French publishers Hachette Livre in 2001 and acquired Cassell Illustrated and the lists of Ward Lock and Blandford Press; also, Gaia Books acquired in 2004.

**Conran Octopus** 📠 020 7531 8627 info-co@conran-octopus.co.uk www.conran-octopus.co.uk

Publisher *Lorraine Dickey* Quality illustrated lifestyle books, particularly interiors, design, cookery, gardening and crafts.

**Hamlyn** Ⓕ 020 7531 8562 info-ho@hamlyn.co.uk www.hamlyn.co.uk Practical books for home and family, cookery, gardening, craft, sport, health, parenting and pet care.

**Mitchell Beazley/Miller's** Ⓕ 020 7537 0773 info-mb@mitchell-beazley.co.uk www.mitchell-beazley.co.uk Publisher *David Lamb* Quality illustrated reference books, particularly food and wine, gardening, interior design and architecture, antiques, general reference.

**Philip's** Ⓕ 020 7531 8460 george.philip@philips-maps.co.uk www.philips-maps.co.uk World atlases, globes, astronomy, road atlases, encyclopedias, thematic reference.

**Bounty** Ⓕ 020 7531 8607 bountybooksinfo-bp@bountybooks.co.uk Publishing and International Sales Director *Polly Manguel* Bargain and promotional books.

**Cassell Illustrated** Popular reference.

**Godsfield Press** Mind, body and spirit.

**Gaia Books** Alternative health and natural/eco titles.

**Spruce** Gift books.
Ⓔ Royalties twice-yearly/annually, according to contract in all divisions.

✱*Authors' update* **The emphasis is on mainstream trade books at the popular end of the market.**

## Old Street Publishing Ltd

40 Bowling Green Lane, London EC1R ONE
Ⓣ 020 7837 1600    Ⓕ 020 7900 6563
info@oldstreetpublishing.co.uk
www.oldstreetpublishing.co.uk
Managing Director *Ben Yarde-Buller*
Director *David Reynolds*

Founded in 2006 by Ben Yarde-Buller and David Reynolds, co-founder of Bloomsbury Publishing. Publishes eclectic fiction including literary fiction, crime/thriller and translation. General non-fiction and humour 'of the better sort'. No poetry, technical or children's books. About 25 titles a year. No unsolicited mss. Synopses and ideas welcome; approach by email.
Ⓔ Royalties twice-yearly.

## Olympia Publishers

60 Cannon Street, London EC4N 6NP
Ⓣ 020 7618 6424    Ⓕ 020 7002 1100
editors@olympiapublishers.com
www.olympiapublishers.com
Chief Editor *G. Bartlett*
Managing Editor *M.B. Anderson*

Founded 2004. Publishes general fiction, memoirs, human interest, chick lit, military, naval, children's and non-fiction. Unsolicited synopses, poetry and short stories also considered. IMPRINTS **Olympia Publishers** Subsidy publishers; **Ashwell Publishing Limited** Non-subsidy publishers. 40–50 titles a year. Send three sample chapters and detailed synopsis for consideration. Postal submissions must be accompanied by s.a.se.
Ⓔ Royalties twice-yearly.

✱*Authors' update* **Note that this is a two-in-one operation. Submissions turned down by Ashwell for conventional publication are likely to be passed on to Olympia where the author will be asked to contribute to production costs.**

## Omnibus Press

Music Sales Ltd, 14/15 Berners Street, London W1T 3LJ
Ⓣ 020 7612 7400    Ⓕ 020 7612 7545
chris.charlesworth@musicsales.co.uk
www.omnibuspress.com
Owner *Robert Wise*
Editorial Head *Chris Charlesworth*

Founded 1971. Independent publisher of music books, rock and pop biographies, song sheets, educational tutors, cassettes, videos and software. A Division of the Music Sales Group of Companies. IMPRINTS **Amsco; Arcane; Bobcat; Oak; Omnibus; Vision On; Wise Publications**. About 25 titles a year. Unsolicited mss, synopses and ideas for music-related titles welcome.

## Oneworld Publications

185 Banbury Road, Oxford OX2 7AR
Ⓣ 01865 310597    Ⓕ 01865 310598
info@oneworld-publications.com
www.oneworld-publications.com
Editorial Director *Juliet Mabey*

Founded 1986. Independent publisher of intelligent adult non-fiction for both the trade and academic markets, across a range of subjects including current affairs, philosophy, history, politics, popular science, psychology, world religions and fiction. SERIES include *Oneworld Beginners' Guides*; *Coping With* (self-help) and *Makers of the Muslim World*. TITLES *Dance with Chance*; *Oilopoly*; *Can a Robot be Human?*; *Pink Brain, Blue Brain*; *How to Save the Planet*; *Growing up Bin Laden*; *The Book of Night Women*; *The Girl Who Fell from the Sky*. 60 titles a year. No unsolicited mss; synopses and ideas welcome, preferably via email or through the website, but if by post, an s.a.se. is required for return of material and/or notification of receipt. No autobiographies, poetry or children's. From 2009, fiction submissions welcome.
Ⓔ Royalties annually.

## Onlywomen Press Ltd
40 St Lawrence Terrace, London W10 5ST
☏ 020 8354 0796
onlywomenpress@btconnect.com
www.onlywomenpress.com
Editorial Director *Lilian Mohin*

Founded 1974. Publishes feminist lesbian literature: fiction, poetry, literary criticism and children's books. Unsolicited mss and proposals welcome.

## OPC
▷ Ian Allan Publishing Ltd

## Open Gate Press incorporating Centaur Press
51 Achilles Road, London NW6 1DZ
☏ 020 7431 4391     ☏ 020 7431 5129
books@opengatepress.co.uk
www.opengatepress.co.uk
Managing Directors *Jeannie Cohen, Elisabeth Petersdorff*

Founded in 1989 to provide a forum for psychoanalytic social and cultural studies. Publishes psychoanalysis, philosophy, social sciences, politics, literature, religion, environment. SERIES *Psychoanalysis and Society.* IMPRINTS **Open Gate Press**; **Centaur Press**; **Linden Press**. No unsolicited mss.
☏ Royalties annually.

## Open University Press
McGraw-Hill House, Shoppenhangers Road, Maidenhead SL6 2QL
☏ 01628 502500     ☏ 01628 770224
enquiries@openup.co.uk
www.openup.co.uk

Founded 1977 as an imprint independent of the Open University's course materials. Acquired by **McGraw-Hill Education** in 2002. Publishes academic and professional books in the fields of education, sociology, health, psychology, psychotherapy and counselling, higher education and study skills materials. No economics or anthropology. Not interested in anything outside the social sciences. About 100 titles a year. No unsolicited mss; enquiries/proposals only.
☏ Royalties annually.

## Orbit
▷ Little, Brown Book Group UK

## Orchard Books
▷ Hachette Children's Books

## The Orion Publishing Group Limited
Orion House, 5 Upper St Martin's Lane, London WC2H 9EA
☏ 020 7240 3444     ☏ 020 7240 4822
www.orionbooks.co.uk/pub/index.htm
Owner *Hachette Livre*

Chairman *Arnaud Nourry*
Chief Executive *Peter Roche*
Deputy Chief Executive *Malcolm Edwards*
Approx. Annual Turnover £70 million

Founded 1992. Incorporates Weidenfeld & Nicolson, JM Dent, Gollancz and Cassell.

TRADE DIVISION

**Trade** Managing Director *Lisa Milton* IMPRINTS **Orion Fiction** Publishing Director *Jon Wood* Hardcover fiction; **Orion Non-Fiction** Publishing Director *Alan Samson* Hardcover non-fiction; **Orion Children's** Publisher *Fiona Kennedy* Children's fiction/non-fiction; **Gollancz** Editorial Directors *Simon Spanton, Jo Fletcher* Science fiction and fantasy. **Weidenfeld Non-Fiction** Publisher *Alan Samson* General non-fiction, history and military; **Weidenfeld Fiction** Publisher *Kirsty Dunseath* History; **Weidenfeld Illustrated** Editor-in-Chief *Michael Dover*; **Custom Publishing** Director *Elizabeth Bond*.

PAPERBACK DIVISION Managing Director *Susan Lamb* IMPRINTS **Orion**; **Phoenix**; **Everyman**. No unsolicited material.
☏ Royalties twice-yearly.

✴*Authors' update* A policy of concentrating resources on a smaller number of titles has 'resulted in our most successful year ever.'

## Osprey Publishing Ltd
Midland House, West Way, Botley, Oxford OX2 0PH
☏ 01865 727022     ☏ 01865 727017/242009
info@ospreypublishing.com
www.ospreypublishing.com
Managing Director *Rebecca Smart*
Publisher *Kate Moore*

Publishes illustrated history of war and warfare and military aviation, also general military history. Founded 1969, Osprey became independent from **Reed Elsevier** in February 1998 and launched Osprey Publishing Inc. as a wholly owned subsidiary in North America in 2004. Acquired **Shire Publications Ltd** in 2007 (see entry). SERIES include *Men-at-Arms*; *Fortress*; *Campaign*; *Essential Histories*; *Warrior*; *Elite*; *New Vanguard*; *Osprey Modelling*; *Duel*; *Aircraft of the Aces*; *Aviation Elite Units*; *Combat Aircraft*; *Raid*; *Command*. About 130 titles a year including 20 non-series titles a year. No unsolicited mss; ideas and synopses welcome.
☏ Royalties twice-yearly.

✴*Authors' update* A self publishing service with AuthorHouse can offer specialist writers an outlet.

## Oxbow Books

10 Hythe Bridge Street, Oxford OX1 2EW
☎ 01865 241249   📠 01865 794449
editorial@oxbowbooks.com
www.oxbowbooks.com
Editor *Clare Litt*

Founded 1983. Publishes academic archaeology, Egyptology, ancient and medieval history. IMPRINTS **Aris & Phillips** Greek, Latin and Hispanic texts and translations; **Windgather Press** Landscape archaeology and history. About 60 titles a year.

## Oxford University Press

Great Clarendon Street, Oxford OX2 6DP
☎ 01865 556767   📠 01865 556646
enquiry@oup.com
www.oup.com
Chief Executive *Nigel Portwood*
Approx. Annual Turnover £578.7 million

A department of Oxford University, OUP started as the university's printing business and developed into a major publishing operation in the 19th century. Publishes academic works in all formats (print and online): dictionaries, lexical and non-lexical reference, scholarly journals, student texts, schoolbooks, ELT materials, music, bibles, paperbacks, and children's books. Around 6,000 titles a year.

DIVISIONS **Academic and Journals** *M. Richardson* Academic and higher education titles in major disciplines, dictionaries, non-lexical reference and trade books. OUP welcomes first-class academic material in the form of proposals or accepted theses. Academic and research journals. **Education** *K. Harris* National Curriculum courses and support materials as well as children's literature. **ELT** *P.R.C. Marshall* ELT courses and dictionaries for all levels.

OVERSEAS BRANCHES/SUBSIDIARIES Sister company in USA with branches or subsidiaries in Argentina, Australia, Brazil, Canada, China, East Africa, India, Japan, Malaysia, Mexico, Pakistan, Southern Africa, Spain and Turkey. Offices in France, Greece, Italy, Poland, Taiwan, Thailand. 💷 Royalties twice-yearly.

🔲*Authors' update* **With more than eighty per cent of its sales outside the UK, OUP has benefited from market growth in the US. English and bilingual dictionaries are the core activity with OUP capturing over half the English and a third of the bilingual market. Sales of online reference are set to overtake print.**

## OxfordPoets
▷ Carcanet Press Ltd

## Oxygen Books Ltd

19 Cedar Road, Hutton, Brentwood CM13 1NB
☎ 01277 213849 (editorial/rights)
heather@oxygenbooks.co.uk (editorial)
www.oxygenbooks.co.uk
Joint Directors *Malcolm Burgess, Heather Reyes*
Approx. Annual Turnover £60,000

Founded 2008. Publishers of city-based literary anthologies. The *city-pick* SERIES (collections of writing, mainly modern and contemporary) on major world cities includes Berlin, Amsterdam, Dublin, London, Paris, Mumbai. 'Looking to expand our list in directions related to this core project.' 3 titles in 2009. No unsolicited mss; synopses or ideas relating to city writing or unusual kinds of journeys welcome. Send brief email with single page outline and single page c.v. to the editorial email address. No children's, illustrated, genre, chick-lit, self-help or poetry.

## Palgrave Macmillan
▷ Macmillan Publishers Ltd

## Pan
▷ Macmillan Publishers Ltd

## Paperbooks Publishing Ltd

2 London Wall Buildings, London EC2M 5UU
☎ 020 7448 5137
info@legend-paperbooks.co.uk
submissions@legend-paperbooks.co.uk
www.paperbooks.co.uk
Managing Director *Tom Chalmers*

Founded in 2006 by Keirsten Clark. Publishes contemporary, diverse and cutting-edge fiction. 'Focuses strongly on reaching new audiences and on working with other publishers to drive the independent sector and promote new and different work within the mainstream market.' Paperbooks was acquired by fellow independent publisher **Legend Press** in 2008. Send three chapters and synopsis to the submissions email address above. 💷 Royalties twice-yearly.

## Parker Press
▷ Hart Publishing Ltd

## Parthian

The Old Surgery, Napier Street, Aberteifi (Cardigan) SA43 1ED
☎ 01239 612059
parthianbooks@yahoo.co.uk
www.parthianbooks.co.uk
www.parthianbooks.com
*Also:* **Library of Wales** Trinity College, Carmarthen SA31 3EP ☎ 01267 676633
📠 01239 612059 www.libraryofwales.org
Publisher *Richard Davies*
Fiction/Art Editor *Lucy Llewellyn*

Editor, Library of Wales *Professor Dai Smith*

Founded 1993. Publishes many of the younger generation of Welsh writers such as Cynan Jones, Rachel Trezise and Caryl Lewis but 'while what we publish reflects a diverse contemporary Wales it is also based around stories and perspectives from writers who are looking out or write about a wider world. We are always interested in a good story'. Parthian's list is an innovative range of new fiction, poetry and drama. Recent books have won awards such as The Dylan Thomas prize and, in 2009, The Welsh Book of the Year. Also publishes the **Library of Wales** series edited by Dai Smith which includes books such as *Border Country* by Raymond Williams and *Ash on a Young Man's Sleeve* by Dannie Abse. Their list also includes a growing number of titles in translation from Spanish, Catalan, German, Basque and Welsh. 'We have developed good translation links throughout Europe and beyond and our books have appeared in 15 foreign language editions including French, Italian, Spanish, Arabic, Turkish, Portuguese, Chinese and Russian.' A developing art and photography list includes *Coalfaces* by Tina Carr and Anne-Marie Schone. 12–15 titles a year. Synopsis with first three sample chapters welcome. 'An idea of what we do is always an advantage.'
Ⓔ Royalties paid annually.

## Particular Books
▷ Penguin Group (UK)

## Paternoster
▷ Authentic Media

## Pathfinder Walking Guides
▷ Crimson Publishing

## Paul Chapman Publishing Ltd
▷ Sage Publications

## Pavilion/Pavilion Children's Books
▷ Anova Books

## Pearson
Halley Court, Jordan Hill, Oxford OX2 8EJ
Ⓣ 01865 888000 (customer services)/311366
Ⓕ 01865 314641
enquiries@pearson.com
www.pearsonschoolsandfecolleges.co.uk
www.pearsoned.com
www.edexcel.com
*Also at:* Edinbugh Gate, Harlow CM20 2JE
Ⓣ 01279 623623
**Pearson & Edexcel**
190 High Holborn, London WC1V 7BH

The world's largest educational publisher. Publishes across a wide range of curriculum subjects from primary students to professional practitioners. Offices in over 30 countries.

## Pegasus Elliot Mackenzie Publishers Ltd
Sheraton House, Castle Park, Cambridge CB3 0AX
Ⓣ 01223 370012   Ⓕ 01223 370040
editors@pegasuspublishers.com
www.pegasuspublishers.com
Senior Editor *D.W. Stern*
Editor *R. Sabir*

Publishes fiction and non-fiction, general interest, biography, autobiography, children's, history, humour, fantasy, self-help, educational, science fiction, poetry, travel, war, memoirs, crime and erotica. IMPRINTS **Vanguard Press**; **Nightingale Books**; **Chimera** TITLES *The Mengele Journals* Simon Mitchell; *Icarus Over Hong Kong* Gustav Preller; *Six Musicians and a Banjo Player* Paddy Lightfoot; *The Chinaman's Bastard* Amanda Taylor; *Don't Forget the Kettle* Karen Campbell; *The Lincoln Imp* Sue Hampton. Unsolicited mss, synopses and ideas considered if accompanied by return postage.
Ⓔ Royalties annually.

✱*Authors' update* **Liable to ask authors to contribute to production costs.**

## Pen & Sword Books Ltd
47 Church Street, Barnsley S70 2AS
Ⓣ 01226 734222   Ⓕ 01226 734438
editorial@pen-and-sword.co.uk
www.pen-and-sword.co.uk
Commissioning Editors *Peter Coles (aviation), Henry Wilson (military & maritime), Rupert Harding (history, local history & family history), Philip Sidnell (ancient history)*

One of the leading military history publishers in the UK. Publishes non-fiction only, specializing in naval and aviation history, WWI, WWII, Napoleonic, autobiography and biography. Also publishes *Battleground* series for battlefield tourists. IMPRINTS **Leo Cooper**; **Frontline Books** (see entry); **Pen & Sword Aviation**; **Pen & Sword Family History**; **Pen & Sword Maritime**; **Pen & Sword Military**; **Remember When** (see entry); **Seaforth Publishing** (see entry); **Wharncliffe Books** (see entry). About 250 titles a year. Unsolicited synopses and ideas welcome; no unsolicited mss.
Ⓔ Royalties twice-yearly.

## Penguin Group (UK)
A Pearson Company, 80 Strand, London WC2R 0RL
Ⓣ 020 7010 3000   Ⓕ 020 7010 6060
www.penguin.co.uk
Penguin Chairman & Global CEO *John Makinson*
CEO Penguin UK & Dorling Kindersley *Peter Field*
Deputy CEO  *Tom Weldon*
Approx. Annual Turnover £121 million

Owned by Pearson plc. Books for adult and children. Adult subjects include biography, fiction, current affairs, general leisure, health, history, humour, literature, politics, spirituality and relationships, sports, travel and TV/film tie-ins. Adult imprints include Allen Lane, Dorling Kindersley, Fig Tree, Hamish Hamilton, Michael Joseph, Penguin, Penguin Classics, Rough Guides, Viking.

**Penguin General Books** LITERARY DIVISION Managing Director *Joanna Prior* IMPRINTS **Viking** Publishing Director *Venetia Butterfield* Editorial Directors *Tony Lacey, Eleo Gordon, Mary Mount, Joel Rickett, Will Hammond* Founded 1925. Fiction and general non-fiction for adults; **Hamish Hamilton** Publishing Director *Simon Prosser* Editors *Juliette Mitchell, Anna Kelly* Fiction, belles-lettres, biography and memoirs, current affairs, history, literature, politics, travel. No unsolicited mss or synopses; **Fig Tree** Publishing Director *Juliet Annan* Editor *Jenny Lord* Fiction and general non-fiction. No unsolicited mss or synopses; **Penguin Ireland** Managing Director *Michael McLoughlin* (see entry under *Irish Publishers*); **Portfolio Penguin** Editorial Director *Joel Rickett* Business books imprint.

**Michael Joseph** COMMERCIAL DIVISION Managing Director *Louise Moore* Editors *Louise Moore* (commercial women's fiction, celebrity non-fiction), *Stefanie Bierwerth* (crime fiction), *Alex Clarke* (commercial male fiction), *Mari Evans* (commercial historical fiction and women's fiction), *Kate Burke* (commercial women's fiction and crime), *Lydia Newhouse* (commercial women's fiction), *Daniel Bunyard* (commercial non-fiction, popular culture and military), *Lindsey Evans* (cookery), *Katy Follain* (commercial non-fiction, popular culture).

**Penguin Press** DIVISION Managing Director *Stefan McGrath*, Publishing Directors *Stuart Proffitt* (Allen Lane), *Simon Winder* (Allen Lane), *Adam Freudenheim* (Penguin Classics), *Georgina Laycock, Alexis Kirschbaum, Will Goodlad* IMPRINT **Particular Books** *Helen Conford* Serious adult non-fiction, reference, specialist and classics. IMPRINT **Allen Lane**; SERIES *Penguin Classics; Penguin Modern Classics*. No unsolicited mss or synopses.

**Penguin Digital** Managing Director *Anna Rafferty*, Digital Publisher *Jeremy Ettinghausen* Ebooks list.

CHILDREN'S DIVISION Managing Director *Stephanie Barton* Children's paperback and hardback books: wide range of picture books, board books and novelties; fiction, non-fiction, poetry and popular culture. No unsolicited mss or synopses. SERIES *Puffin*

*Classics* and *Puffin Modern Classics*. IMPRINTS **Puffin** Managing Director *Francesca Dow*, Publishing Director (fiction) *Sarah Hughes*, Editorial Director (picture books) *Louise Bologaro*, Editorial (characters) *Kate Hayler*, Editorial Director (Puffin Classics) *Elv Moody*; **Razorbill** Publisher *Amanda Punter* Commercial teen fiction; **BBC Children's Books** Editorial Director *Juliet Matthews*, **Ladybird** and **Warne** Editorial *Nicole Pearson* Pre-school illustrated developmental books for ages 0–6, non-fiction 0–8; licensed brands; children's classic publishing and merchandising properties. No unsolicited mss.

**Dorling Kindersley Ltd** DIVISION www.dk.com Deputy CEOs *Andrew Philips, John Duhigg*, Global Publisher *Miriam Farbey* Illustrated non-fiction for adults and children: gardening, health, medical, travel, food and drink, history, natural history, photography, reference, pregnancy and childcare, film and TV. Age groups: preschool, 5–8, 8+, family.

**Brady** DIVISION Publisher *David Waybright*, Business Development Manager *Brian Saliba* Computer games strategy guides and collectors' editions.

**Travel** DIVISION www.traveldk.com www.roughguides.com Managing Director *John Duhigg*, Publishing Directors *Douglas Amrine* (Dorling Kindersley), Publishing Director *Clare Currie*, Rough Guide Reference *Andrew Lockett*, Digital Publisher *Peter Buckley* Travel guides, illustrated travel books, phrasebooks and digital products; popular culture and lifestyle guides. Includes **Rough Guides** and **DK Eyewitness Travel**. £ Royalties twice-yearly.

✱*Authors' update* One of the few publishers to boast a strong brand image, Penguin is celebrating its 75th anniversary with a new logo and a range of book related products. Of the core business, Dorling Kindersley and Rough Guides both fell back, a reflection of online competition. But Michael Joseph and Hamish Hamilton continue to perform strongly. An ambitious digitization programme is under way. An imaginative promotion by Puffin (70 this year) will see the free distribution of a parent's guide to books for children.

## Pennant Books Ltd

PO Box 5675, London W1A 3FB
☎ 020 7387 6400
info@pennantbooks.com
www.pennantbooks.com
Managing Director *Cass Pennant*
Approx. Annual Turnover £250,000

Independent publisher, founded in 2005. Publishes non-fiction, biography, sport, true crime and fan culture. No fiction or children's books. 24 titles in 2009. No unsolicited mss; send synopses and ideas for books by post. £ Royalties twice-yearly.

## Pergamon
▷ Elsevier Ltd

## Persephone Books

59 Lamb's Conduit Street, London WC1N 3NB
☎ 020 7242 9292  📠 020 7242 9272
sales@persephonebooks.co.uk
www.persephonebooks.co.uk
Managing Director *Nicola Beauman*

Founded 1999. Publishes reprint fiction and non-fiction, mostly 'by women, for women and about women'. TITLES *Miss Pettigrew Lives for a Day* Winifred Watson; *Kitchen Essays* Agnes Jekyll; *The Far Cry* Emma Smith; *The Priory* Dorothy Whipple. 6 titles a year. No unsolicited material. £ Royalties twice-yearly.

## Peter Collin Publishing Ltd
▷ Bloomsbury Publishing Plc

## Peter Owen Publishers

73 Kenway Road, London SW5 0RE
☎ 020 7373 5628/7370 6093  📠 020 7373 6760
aowen@peterowen.com
www.peterowen.com
Chairman *Peter Owen*
Editorial Director *Antonia Owen*

Founded 1951. Publishes biography, general non-fiction, English-language literary fiction and translations, history, literary criticism, the arts and entertainment. 'No genre or children's fiction; the company is taking on no first novels or short stories at present.' AUTHORS Yuri Druzhnikov, Shusaku Endo, Philip Freund, Anna Kavan, Jean Giono, Anaïs Nin, Mervyn Peake, Joseph Roth, Ken Russell, Peter Vansittart. About 25 titles a year including the Peter Owen Modern Classics reprint paperbacks. Unsolicited synopses welcome for non-fiction material. Established authors should contact preferably by email in advance to submit fiction. Mss should be preceded by a descriptive letter and synopsis with s.a.s.e. if posted rather than emailed. OVERSEAS ASSOCIATES worldwide. £ Royalties twice-yearly.

✳*Authors' update* Peter Owen has been described as 'a publisher of the old and idiosyncratic school'. He has seven Nobel prizewinners on his list to prove it.

## Phaidon Press Limited

Regent's Wharf, All Saints Street, London N1 9PA
☎ 020 7843 1000  📠 020 7843 1010
initialandname@phaidon.com
www.phaidon.com
Group Chairman & Publisher *Richard Schlagman*
Chairman *Andrew Price*
Editorial Directors *Amanda Renshaw, Emilia Terragni*

Phaidon Press is one of the world's leading publishers of books on the visual arts, culture and creativity. DIVISIONS (with Editorial Heads) **Architecture and Design** *Emilia Terragni*; **Contemporary Art** *Craig Garrett*; **Photography** *Amanda Renshaw*; **Art & Academic** *David Anfam*; **Cooking/Food** *Emilia Terragni*; **Children's** *Amanda Renshaw*. About 100 titles a year. Unsolicited mss welcome but 'only a small amount of unsolicited material gets published'. £ Royalties twice-yearly.

✳*Authors' update* Phaidon's European scope gives it added appeal in the US for titles that might not otherwise enter the English language market. Commended for tight management and strong marketing.

## Philip Allan Updates
▷ Hachette UK

## Philip Wilson Publishers Ltd

109 Drysdale Street, The Timber Yard, London N1 6ND
☎ 020 7033 9900  📠 020 7033 9922
pwilson@philip-wilson.co.uk
www.philip-wilson.co.uk
Chairman *Philip Wilson*

Founded 1976. Publishes art, art history, antiques and collectibles. About 14 titles a year.

## Philip's
▷ Octopus Publishing Group

## Phillimore
▷ The History Press

## Phoenix
▷ The Orion Publishing Group Limited

## Piatkus Books

100 Victoria Embankment, London EC4Y 0DY
☎ 020 7911 8030  📠 020 8911 8100
info@littlebrown.co.uk
www.piatkus.co.uk
www.littlebrown.co.uk
Owner *Hachette Book Group*
CEO/Publisher *Ursula Mackenzie*
COO *David Kent*
Approx. Annual Turnover £10 million

Founded in 1979 by Judy Piatkus, the company was sold to Hachette in autumn 2007. Piatkus non-fiction specializes in lifestyle, health, self-help, mind, body and spirit, popular psychology, biography and business titles. Piatkus fiction

includes new novels from many of the world's bestselling international authors. DIVISIONS **Non-fiction** Publishing Director *Tim Whiting*, Editorial Director *Gill Bailey*, Commissioning Editor *Anne Lawrance* TITLES *The Holford Low GL Diet* Patrick Holford; *Women Who Think Too Much* Susan Nolan-Hoeksema; *Mind Body Bible* Mark Atkinson; **Fiction** Senior Editor *Emma Beswetherick*, Assistant Fiction Editor *Donna Condon* TITLES *High Noon* Nora Roberts; *It's in His Kiss* Julia Quinn; *Creation in Death* J.D. Robb. About 250 titles a year. Piatkus is expanding its range of books and welcomes synopses together with first three chapters.
Ⓔ Royalties twice-yearly.

✱*Authors' update* **Being part of a larger group does not seem to have dented the Piatkus enthusiasm for bright ideas. Publisher-author relations are said to be excellent.**

### Pica Press
▷ A.&C. Black Publishers Ltd

### Picador
▷ Macmillan Publishers Ltd

### Piccadilly Press
5 Castle Road, London NW1 8PR
Ⓣ 020 7267 4492    Ⓕ 020 7267 4493
books@piccadillypress.co.uk
www.piccadillypress.co.uk
Publisher/Managing Director *Brenda Gardner*
Approx. Annual Turnover £1 million

Founded 1983. Independent publisher of children's and parental books. 34 titles in 2009. Welcomes approaches from authors 'but we would like them to know the sort of books we do. It is frustrating to get inappropriate material. They should check in their local libraries, bookshops or look at our website.' No adult or cartoon-type material. Catalogue available online.
Ⓔ Royalties twice-yearly.

### Pictorial Presentations
▷ Souvenir Press Ltd

### Pimlico
▷ The Random House Group Ltd

### Pitkin
▷ The History Press

### PiXZ Books
▷ Halsgrove

### Pluto Press Ltd
345 Archway Road, London N6 5AA
Ⓣ 020 8348 2724    Ⓕ 020 8348 9133
pluto@plutobooks.com
www.plutobooks.com
Chair *Roger Van Zwanenberg*
Managing Director/Publisher *Anne Beech*

Founded 1970. Has developed a reputation for innovatory publishing in the field of non-fiction. Publishes academic and scholarly books across a range of subjects including politics and world affairs and more general titles on key aspects of currrent affairs. About 70 titles a year. Prospective authors are encouraged to consult the website for guidelines on submitting proposals.

### Pocket Books
▷ Simon & Schuster UK Limited

### Pocket Mountains Ltd
6 Church Wynd, Bo'ness EH51 0AN
Ⓣ 01506 500402    Ⓕ 01506 500405
info@pocketmountains.com
www.pocketmountains.com
Managing Directors *Robbie Porteous, April Simmons*
Approx. Annual Turnover £150,000

Founded in 2003 to publish a series of guidebooks to the Scottish hills and now publishing other guides covering the UK: mountaineering, walking, cycling and wildlife books. 30 titles in 2009. Unsolicited mss, synopses and ideas welcome; approach by email or letter. No fiction, children's or cookery.

### Poetry of Place
▷ Eland Publishing Ltd

### The Policy Press
Fourth Floor, Beacon House, Queen's Road, Bristol BS8 1QU
Ⓣ 0117 331 4054    Ⓕ 0117 331 4093
tpp-info@bristol.ac.uk
www.policypress.co.uk
Director *Alison Shaw*

Specialist social science publisher 'committed to publishing work that will impact on research, learning, policy and practice'.

### Politico's Publishing
8 Artillery Row, London SW1P 1RZ
Ⓣ 020 7802 0018    Ⓕ 020 7828 1244
sales@methuen.co.uk
www.politicospublishing.co.uk
Managing Director *Peter Tummons*
Publishing Consultant *Alan Gordon Walker*

Founded 1998. Acquired by **Methuen Publishing Ltd** in April 2003. Publishes political books. 15 titles in 2009. No unsolicited mss. Synopses and ideas welcome; telephone in the first instance.

### Polity Press
65 Bridge Street, Cambridge CB2 1UR
Ⓣ 01223 324315    Ⓕ 01223 461385
www.politybooks.com

Founded 1984. Publishes anthropology, criminology, economics, feminism, history, human geography, literature, media and cultural studies, medicine and society, philosophy, politics, psychology, religion and theology, social and political theory, sociology. 'Please do not send mss.' ⓔ Royalties annually.

## Polygon

West Newington House, 10 Newington Road, Edinburgh EH9 1QS
☏ 0131 668 4371   ℻ 0131 668 4466
info@birlinn.co.uk
www.birlinn.co.uk
Publishing Director *Neville Moir*

Bought by **Birlinn Ltd** in 2002. Publishes literary fiction, crime fiction and poetry. Does not publish children's books, science fiction/fantasy, romantic fiction or short stories. About 40 titles a year. 'We do not accept unsolicited work from authors.'
ⓔ Royalties twice-yearly.

## Pont Books
▷ Gomer Press

## Pop Universal
▷ Souvenir Press Ltd

## Portfolio Penguin
▷ Penguin Group (UK)

## Portico
▷ Anova Books

## Portland Press Ltd

Third Floor, Eagle House, 16 Procter Street, London WC1N 2JL
☏ 020 7685 2410   ℻ 020 7685 2469
editorial@portlandpress.com
www.portlandpress.com
Chairman *Professor John Clark*
Managing Director *Rhonda Oliver*
Managing Editor *Pauline Starley*
Approx. Annual Turnover £2.5 million

Founded 1990 to expand the publishing activities of the Biochemical Society (1911). Publishes biochemisty and medicine for graduate, postgraduate and research students. Expanding the list to include schools and general readership. TITLES *Biochemical Society Symposia Volume 75: Structure and Function in Cell Adhesion*; *Wenner-Gren Volume 84: The University in the Market.* Unsolicited mss, synopses and ideas welcome. No fiction.
ⓔ Royalties twice-yearly.

## Portobello Books

Twelve Addison Avenue, London W11 4QR
☏ 020 7605 1380   ℻ 020 7605 1361
mail@portobellobooks.com
www.portobellobooks.com
Owners *Sigrid Rausing, Eric Abraham, Philip Gwyn Jones*

Editorial Director *Laura Barber*
Associate Publisher *Tasja Dorkofikis*
Rights Director *Angela Rose*
Approx. Annual Turnover £800,000

Founded in 2005 by Philip Gwyn Jones, formerly publishing director of HarperCollins' literary imprint Flamingo. Owners acquired **Granta** in 2005. Publishes international literature and activist non-fiction. About 25 original titles a year. No unsolicited mss; synopses and ideas for books welcome. Approach by post to the submissions editor in the first instance. No science fiction, romance or genre fiction of any kind, no gardening, cookery, mind, body and spirit or reference.
ⓔ Royalties twice-yearly.

✴*Authors' update* **Bright books for bright people.**

## Preface Publishing
▷ The Random House Group Ltd

## Prestel Publishing Limited

4 Bloomsbury Place, London WC1A 2QA
☏ 020 7323 5004   ℻ 020 7636 8004
sales@prestel-uk.co.uk
www.prestel.com
UK Editor *Philippa Hurd*
Editor-in-Chief (Germany) *Curt Holtz*

Founded 1924. Publishes art, architecture, photography, design, fashion, children's and general illustrated books. No fiction. Unsolicited mss, synopses and ideas welcome. Approach by post or email.

## Princeton University Press
▷ University Presses of California, Columbia & Princeton Ltd

## Prion Books
▷ Carlton Publishing Group

## Profile Books

3a Exmouth House, Pine Street, Exmouth Market, London EC1R 0JH
☏ 020 7841 6300   ℻ 020 7833 3969
info@profilebooks.com
www.profilebooks.com
Publisher & Managing Director *Andrew Franklin*
Editorial Director *Stephen Brough*
Associate Publisher, Non-fiction *Daniel Crewe*
Publisher, Green Profile *Mark Ellingham*
Publisher, Serpent's Tail *Pete Ayrton*
Rights Director *Penny Daniel*

Founded 1996. Publishes general non-fiction: history, biography, current affairs, popular science, politics, business management. Also publishers of *The Economist* books. IMPRINTS **Serpent's Tail** (see entry); **Green Profile** Environmental books. No unsolicited mss; phone or send preliminary letter.

✱*Authors' update* Dedicated to 'intelligent non-fiction', Profile also has a track record of producing the oddball bestseller.

## Psychology Press (An imprint of Taylor & Francis Group, an Informa plc business)

27 Church Road, Hove BN3 2FA
☏ 020 7017 6000   🖷 020 7017 6717
www.psypress.co.uk
Managing Director *Mike Forster*
Deputy Managing Director *Rohays Perry*

Publishes psychology textbooks, monographs, professional books, tests and numerous journals which are available in both printed and online formats. Send detailed proposal plus c.v. and sample chapters. Proposal requirements and submission procedures available on the website. No poetry, fiction, travel or astrology.
£ Royalties according to contract.

## Publishing House

Trinity Place, Barnstaple EX32 9HG
☏ 01271 328892   🖷 01271 328768
info@vernoncoleman.com
www.vernoncoleman.com
Managing Director *Vernon Coleman*
Publishing Director *Sue Ward*
Approx. Annual Turnover £750,000

Founded 1989. Self-publisher of fiction, health, humour, animals, politics. Over 90 books published. TITLES *Mrs Caldicot's Cabbage War*; *England our England*; *Bodypower*; *Bilbury Chronicles*; *Second Innings*; *It's Never Too Late*; *Alice's Diary*; *Food for Thought*; *Gordon is a Moron*, all by Vernon Coleman. No submissions.

## Puffin
▷ Penguin Group (UK)

## Pushkin Press Ltd

12 Chester Terrace, London NW1 4ND
☏ 020 7730 0750   🖷 020 7730 1341
books@pushkinpress.com
www.pushkinpress.com
Chairman *Melissa Ulfane*

Publishes novels and essays in translation drawn from the best of classic and contemporary European literature.

## QED Publishing
▷ Quarto Publishing plc

## Quadrille Publishing Ltd

Alhambra House, 27–31 Charing Cross Road, London WC2H 0LS
☏ 020 7839 7117   🖷 020 7839 7118
info@quadrille.co.uk
www.quadrille.co.uk
Chairman *Sir David Cooksey*
Managing Director *Alison Cathie*

Editorial Director *Jane O'Shea*
Approx. Annual Turnover £10 million (2008)

Founded in 1994 with a view to producing a small list of top-quality illustrated books. Publishes non-fiction, including cookery, gardening, interior design and decoration, craft, health. TITLES *Feed Me Now* Bill Granger; *Life's Too Short to Drink Bad Wine* Simon Hoggart; *Wallpaper: the Ultimate Guide* Charlotte Abrahams; *Landscape Man: Making a Garden* Matthew Wilson. 40 titles in 2009. Synopses and ideas for books welcome. No fiction or children's books.
£ Royalties twice-yearly.

## Quantum Publishing
▷ Foulsham Publishers

## Quantum Publishing
▷ Quarto Publishing plc

## Quartet Books

27 Goodge Street, London W1T 2LD
☏ 020 7636 3992   🖷 020 7637 1866
david@quartetbooks.com
Chairman *Naim Attallah*

Founded 1972. Independent publisher. Publishes contemporary literary fiction including translations, popular culture, biography, music, history, politics and some photographic books. Unsolicited sample chapters with return postage welcome; no poetry, romance or science fiction. Submissions by disk or email are not accepted.
£ Royalties twice-yearly.

## The Quarto Group, Inc.

226 City Road, London EC1V 2TT
☏ 020 7700 6700   🖷 020 7700 4191
www.quarto.com
Chairman & CEO *Laurence Orbach*
CFO *Mick Mousley*
Group Creative Director *Bob Morley*
Director of Co-edition Publishing *Piers Spence*
Approx. Annual Turnover £100 million

Founded in 1976, the Quarto Group has an international portfolio of trade and co-edition publishing interests in the UK, USA, Australia and Asia. UK entities include **Aurum Press**, acquired in 2005 (see entry), **RotoVision**, acquired in 1999 (see entry), **Apple Press** (see entry); the magazine, *The World of Fine Wine* (see entry under *Magazines*); and **Quarto Publishing plc** (see entry). Other Quarto-owned concerns include the Quayside Publishing Group in the US; Lifetime Distributors, Australia's largest direct marketing bookseller; and Regent Publishing Services in Hong Kong.

## Quarto Publishing plc

The Old Brewery, 6 Blundell Street, London
N7 9BH

ⓣ 020 7700 6700    ⓕ 020 7700 4191
www.quarto.com
Chairman & CEO *Laurence Orbach*
Director of Publishing *Piers Spence*

The UK's most prolific creators of illustrated
co-editions, Quarto Publishing plc consists of
11 autonomous business units that conceive and
produce leisure, lifestyle and reference titles and
license them to publishers in 30 languages and
more than 70 countries.

DIVISIONS

**Quarto Books** Publisher *Paul Carslake*
paulc@quarto.com High quality how-to, from
crafts (knitting and crochet) to art (painting
and sculpting) to music (teach-yourself) to
high-tech (graphic design, digital arts) TITLE
*The Guitar Chord Bible*. **Quarto Children's
Books** (incorporating Design Eye) Publisher
*Sue Grabham* sueg@quarto.com Intricate
and ingenious activity and novelty books for
children TITLE *3-D Explorer: Oceans*. **Quintet
Publishing** Publisher *James Tavendale* jamest@
quarto.com Illustrated reference across the
board, specializing in food and drink TITLE *500
Cupcakes*. **QED Publishing** Associate Publisher
*Zeta Davies* zetad@quarto.com Fun and clever
list of first reading and first concept books
TITLE *Dinosaur Families*. **Marshall Editions**
(acquired 2002) Publisher *Linda Cole* linda.cole@
marshalleditions.com A big name from the past
reinvigorated and producing illustrated family
reference TITLE *Endangered Birds*. **Quintessence
Editions** Publisher *Tristan De Lancey* tristandl@
quarto.com Best known for the '1001 Series' with
sales of over three million copies in 30 languages
TITLE *1001 Movies You Must See Before You
Die*. **Quantum Publishing** Publisher *Anastasia
Cavouras* Backlist and value-priced books to
appeal to every interest and pocket. **Qu:id
Publishing** Publisher *Nigel Browning* nigelb@
quid.com Quirky, sideways take on reference
publishing: serious information presented in a
memorable and fun way TITLE *The Best Dance
Moves in the World ... Ever!*; **Global Publishing**
Publisher *Cheryl Campbell* ccampbell@
globalpub.com.au Large format, authoritative,
content rich illustrated reference TITLE *Biblica*.
**Fine Wine Editions** Publisher *Sara Morley*
saram@finewinemag.com 'The last word in wine
reference publishing', backed up by the authority
of Hugh Johnston and *The World of Fine Wine*
magazine TITLE *The Finest Wines of Champagne*.
**Iqon Editions** Publisher *David Breuer* davidb@
iqoneditions.com Highbrow pocket reference at
lowbrow prices in the areas of art, architecture
and culture generally TITLE *Isms*.

## Quercus Publishing

21 Bloomsbury Square, London WC1A 2NS
ⓣ 020 7291 7200    ⓕ 020 7291 7201
firstname.surname@quercusbooks.co.uk
www.quercusbooks.co.uk
Chief Executive *Mark Smith*
Approx. Annual Turnover £19 million

Founded in 2005 as a commercial publisher
of quality fiction and non-fiction. DIVISIONS
**Trade Publishing** Managing Director *David
North*, Editor-in-Chief *Jon Riley*, Publisher,
Non-Fiction *Richard Milner* TITLE *The
Tenderness of Wolves*; **Children's Publishing**
Editorial Head *Roisin Heycock* TITLE *The Magic
Thief*; **Contract Publishing** Managing Director
*Wayne Davies* TITLE *Speeches That Changed the
World*; **MacLehose Press** Publisher *Christopher
MacLehose* TITLE *The Girl With the Dragon
Tattoo*; **Quercus Audiobooks** *Nick Johnston*
TITLE *The Girl Who Played With Fire*. 160 titles in
2009. No unsolicited material.
ⓔ Royalties twice-yearly.

✱*Authors' update* Now that Quercus has
found security on the mammoth sales
of Stieg Larsson's *Millenium* trilogy, the
founding objective of focusing on new and
underdeveloped writers – 'refugees from big
groups who weren't high enough up the pecking
order to get into the sunshine of the big market
spends' – may come back into play.

## Quest
▷ Top That! Publishing Plc

## Qu:id Publishing
▷ Quarto Publishing plc

## Quiller Press (An imprint of Quiller Publishing Ltd)

Wykey House, Wykey, Shrewsbury SY4 1JA
ⓣ 01939 261616    ⓕ 01939 261606
info@quillerbooks.com
www.countrybooksdirect.com
Managing Director *Andrew Johnston*

Specializes in sponsored books and publications
sold through non-book trade channels as well as
bookshops. Publishes architecture, biography,
business and industry, collecting, cookery, DIY,
gardening, guidebooks, humour, reference,
sports, travel, wine and spirits. IMPRINTS
**Kenilworth Press** (see entry); **The Sportsman's
Press** (see entry); **Swan Hill Press** (see entry).
About 20 titles a year. Unsolicited mss only if the
author sees some potential for sponsorship or
guaranteed sales.
ⓔ Royalties twice-yearly.

## Quince Books Limited

209 Hackney Road, London E2 8JL
info@quincebooks.com
www.quincebooks.com
Editor *Sara Hibbard*

Founded 2005. Publishes commercial ficton.
About 15 books a year. Welcomes submissions;
send synopsis and three chapters by email.
Ⓔ Royalties annually.

## Quintessence Editions
▷ Quarto Publishing plc

## Quintet Publishing
▷ Quarto Publishing plc

## Radcliffe Press
▷ I.B. Tauris & Co. Ltd

## Radcliffe Publishing Ltd

18 Marcham Road, Abingdon OX14 1AA
Ⓣ 01235 528820   Ⓕ 01235 528830
contact.us@radcliffemed.com
www.radcliffe-oxford.com
Managing Director *Gregory Moxon*
Editorial Director *Gillian Nineham*
Approx. Annual Turnover £2 million

Founded 1987. Medical publishing house which
began by specializing in books for general
practice and health service management.
Publishes clinical, management, professional
development, health policy and medical
examination revision books, training materials,
journals, CD-ROMs, DVDs and online.
Approximately 70 titles in 2009. Unsolicited mss,
synopses and ideas welcome. No non-medical
books.
Ⓔ Royalties twice-yearly.

## The Random House Group Ltd

Random House, 20 Vauxhall Bridge Road,
London SW1V 2SA
Ⓣ 020 7840 8400   Ⓕ 020 7233 6058
enquiries@randomhouse.co.uk
www.randomhouse.co.uk
Chief Executive/Chairman *Gail Rebuck, DBE*
Deputy CEO *Ian Hudson*
Approx. Annual Turnover £295 million (UK)

The Random House Group is the UK's leading
trade publisher comprising over 40 diverse
imprints in six separate substantially autonomous
companies: CCV, Random House Enterprises,
Cornerstone, Ebury Publishing, Transworld and
Random House Children's Books.

**Random House Audio Books** Ⓣ 020 7840 8519
Commissioning Editor *Zoe Howes* (see entry
under *Audio Books*)

**Random House Digital** Ⓣ 020 7840 8400
Director of Digital *Fionnuala Duggan*

CCV
Managing Director *Richard Cable*, Publisher *Dan
Franklin*
IMPRINTS **Jonathan Cape** Ⓣ 020 7840 8836 Ⓕ
020 7233 6117 Publisher *Dan Franklin* Biography
and memoirs, current affairs, fiction, history,
photography, poetry, politics and travel. **Yellow
Jersey Press** Ⓣ 020 7840 8621 Ⓕ 020 7223 6117
Editorial Director *Matthew Phillips* Narrative
sports books. **Harvill Secker** Ⓣ 020 7840 8893
Ⓕ 020 7233 6117 Publishing Director *Liz Foley*
Principally literary fiction with some non-fiction.
Literature in translation, English literature,
quality thrillers. **Chatto & Windus** Ⓣ 020 7840
8745 Ⓕ 020 7233 6117 Publishing Director *Clara
Farmer* Memoirs, current affairs, essays, literary
fiction, history, poetry, politics, philosophy and
translations. **Pimlico** Ⓣ 020 7840 8608 Ⓕ 020
7233 6117 Publishing Director *Rachel Cugnoni*
Quality non-fiction paperbacks, specializing in
history. **Vintage** Ⓣ 020 7840 8608 Ⓕ 020 7233
6117 Publishing Director *Rachel Cugnoni* Quality
paperback fiction and non-fiction. Vintage
(founded in 1990) has been described as one of
the 'greatest literary success stories in recent
British publishing'. **The Bodley Head** Ⓣ 020 7840
8553 Ⓕ 020 7233 6117 Publishing Director *Will
Sulkin* History, science, current affairs, politics,
biography, popular culture. **Square Peg** Ⓣ 020
7840 8621 Ⓕ 020 7233 6117 Editorial Director
*Rosemary Davidson* Eclectic and commercial
non-fiction, humour, novelties and gift books.
Unsolicited mss with s.a.s.e.

CORNERSTONE
Managing Director *Susan Sandon*
IMPRINTS **Century** Ⓣ 020 7840 8569 Ⓕ 020 7233
6127 Publishing Director *Ben Dunn* General
fiction and non-fiction including commercial
fiction, autobiography, biography, sport and
humour. **William Heinemann** Ⓣ 020 7840
8400 Ⓕ 020 7233 6127 Publishing Director *Jason
Arthur* Fiction and general non-fiction: literary
fiction, quality crime and thrillers, current affairs,
translations, biography, popular science. No
unsolicited mss. **Hutchinson** Ⓣ 020 7840 8400
Ⓕ 020 7233 7870 Publishing Director *Caroline
Gascoigne* Fiction: literary and women's fiction,
crime, thrillers. Non-fiction: current affairs,
biography, memoirs, history, politics, travel,
adventure. **Random House Books** (incorporating
**Random House Business Books**) Ⓣ 020 7840
8451 Ⓕ 020 7233 6127 Publishing Director *Nigel
Wilcockson* Non-fiction: social and cultural
history, current affairs, popular culture, reference,
business and economics. **Arrow** Ⓣ 020 7840
8414 Ⓕ 020 7840 6127 Publisher *Kate Elton*
Mass-market paperback fiction and non-fiction.

**Windmill Books** ⊤ 020 7840 8414 Ⓕ 020 7840 6127 Editorial Director *Stephanie Sweeney* B-format paperback fiction and non-fiction. Unsolicited mss not accepted by Cornerstone. **Preface Publishing** ⊤ 020 7840 8443 Ⓕ 020 7840 8383 Publisher *Trevor Dolby*, Publishing Director *Rosie de Courcy* Commercial fiction and non-fiction. Unsolicited mss accepted at prefacesubmissions@randomhouse.co.uk

EBURY PUBLISHING
(Vauxhall Bridge Road address) ⊤ 020 7840 8400 Ⓕ 020 7840 8406
Managing Director *Fiona MacIntyre*, Publisher *Jake Lingwood*
IMPRINTS **Ebury Press** Publishing Director (illustrated, including cookery, family reference, history, culture and gift) *Carey Smith*, Editorial Director (fiction) *Gillian Green*, Publishing Director (popular non-fiction) *Andrew Goodfellow*. **Vermilion** Commissioning Editors *Miranda* West, *Julia Kellaway* Personal development, health, diet, relationships and parenting. **Rider** Publishing Director *Judith Kendra* Psychology, biography, philosophy, personal development, spirituality, the paranormal. **Time Out Guides** Senior Publishing Brand Manager *Luthfa Begum* Travel and walking guides, film, reference and lifestyle published in a unique partnership with *Time Out* Group. **BBC Books** Senior Publishing Director *Shirley Patton*. **Virgin Books** Editorial *Edward Faulkner, Louisa Joyner* General non-fiction: biography and autobiography, popular culture, business, popular science, cookery, travel and adventure, sport, film and TV, music, humour, true crime. IMPRINTS **Black Lace** Erotic fiction by women for women; **Nexus**.

TRANSWORLD PUBLISHERS
(See entry)

RANDOM HOUSE CHILDREN'S BOOKS DIVISION
61–63 Uxbridge Road, London W5 5SA
⊤ 020 8579 2652 Ⓕ 020 8231 6737
Managing Director *Philippa Dickinson*, Publisher (fiction) *Annie Eaton*, Editorial Director (fiction) *Kelly Hurst*, Publisher (Colour and Custom Publishing) *Fiona Macmillan*, Editorial Director (picture books) *Helen Mackenzie-Smith*, Senior Commissioning Editor (picture books) *Natascha Biebow* Picture books, fiction, poetry, non-fiction and audio cassettes under IMPRINTS **Bodley Head Children's Books; Corgi Children's Books; Doubleday Children's Books; Hutchinson Children's Books; Jonathan Cape Children's Books; Red Fox Children's Books; Tamarind Books** and **David Fickling Books** (31 Beaumont Street, Oxford OX1 2NP ⊤ 01865 339000 Ⓕ 01865 339009 dfickling@randomhouse.co.uk www.davidficklingbooks.co.uk) Publisher *David*

*Fickling*, Senior Editor *Bella Pearson* Children's fiction and picture books.
Ⓔ Royalties twice-yearly for the most part.

✱*Authors' update* Chasing hard on Hachette for the biggest share of the UK market, Random has grabbed a large share of the market for celebrity non-fiction. Restructuring has given a boost to Vintage Classics. Bodley Head has relaunched with a quality non-fiction list. Harvill Secker, another quality list, is celebrating its centenary while Cape thrives on top sales for top authors. Transworld thrives on Dan Brown. New on the street, Preface has been set up to capture readable books, fiction and non-fiction, that might otherwise escape the notice of the corporate scouts; newcomers should take heart. Gail Rebuck says that Random is keen to invest in new writers: 'There are no successes without risk.'

## Ransom Publishing Ltd
51 Southgate Street, Winchester SO23 9EH
⊤ 01962 862307    Ⓕ 05601 148881
rebecca@ransom.co.uk
www.ransom.co.uk
Managing Director *Jenny Ertle*
Creative Director *Steve Rickard*

Ransom's book programme is designed to develop reading skills (and an enthusiasm for reading) in everyone from young children through to adults. Many of Ransom's books are 'high interest age, low reading age'; both fact and fiction. The list includes well known authors such as David Orme, whose *Boffin Boy* series has proved hugely popular with reluctant readers. Another success is *Dark Man*. Aimed at adults with a very low reading age, the series won the 2006 Educational Resources Award. Approx. 188 titles in 2009. Interested in hearing from authors writing for reluctant and struggling readers. Initial enquiries by email with details of the submission and three sample chapters. 'A response cannot be guaranteed but we try to respond to all submissions within a month.' Materials returned if supplied with s.a.s.e.

## Razorbill
▷ Penguin Group (UK)

## Reader's Digest Association Ltd
11 Westferry Circus, Canary Wharf, London E14 4HE
⊤ 020 7715 8000    Ⓕ 020 7715 8181
gbeditorial@readersdigest.co.uk
www.readersdigest.co.uk
Managing Director *Chris Spratling*
Editorial Director, Books *Julian Browne*

Editorial office in the USA (see entry under *US Publishers*). Publishes health, gardening, home, natural history, cookery, history, DIY, computer, travel and word books. About 50 titles a year through trade and direct marketing channels.

*Authors' update* Now owned by Better Capital Limited who bought the UK division of Reader's Digest in April 2010.

## Reaktion Books

33 Great Sutton Street, London EC1V 0DX
☎ 020 7253 1071    🖷 020 7253 1208
info@reaktionbooks.co.uk
www.reaktionbooks.co.uk
Publisher *Michael R. Leaman*

Founded 1985. Publishes art history, architecture, animal studies, Asian studies, cultural studies, design, film, food history, geography, history, natural history, photography and travel literature (*not* travel guides or journals). TITLE *A Brief History of Nakedness* Philip Carr-Gomm; SERIES *Edible* TITLES *Caviar* Nichola Fletcher; *Cake* Nicola Humble; *Milk* Hannah Velten; *Critical Lives* TITLE *Vladimir Nabokov* Barbara Wyllie; *Animal* TITLES *Owl* Desmond Morris; *Lion* Deirdre Jackson; *Camel* Robert Irwin; *Exposures* TITLE *Photography and Africa* Erin Haney. About 45 titles a year. Written submissions only, to address above or email to ian@reaktionbooks.co.uk Please enclose s.a.s.e. if you would like to have unsuccessful submitted material returned.
💷 Royalties twice-yearly.

## Reardon Publishing

PO Box 919, Cheltenham GL50 9AN
☎ 01242 231800
reardon@bigfoot.com
www.reardon.co.uk
www.cotswoldbookshop.com
www.antarcticbookshop.com
www.nicholasreardon.com
www.chocolatemanga.com
www.limitedbooks.co.uk
Managing Editor *Nicholas Reardon*

Founded in the mid-1970s. Family-run publishing house specializing in local interest and tourism in the Cotswold area and books on Antarctica. Member of the Outdoor Writers & Photographers Guild and the National Union of Journalists. Publishes walking and driving guides. TITLES *The Cotswold Way*; *The Cotswold Way Map*; *Cotswold Walkabout*; *Cotswold Driveabout*; *The Donnington Way*; *The Haunted Cotswolds*. Also distributes for other publishers such as Ordnance Survey. 10 titles a year. Unsolicited mss, synopses and ideas welcome with return postage only.
💷 Royalties twice-yearly.

## Recollections
▷ Arnefold Publishing

## Red Arrow Books
▷ Blackstone Publishers

## Red Fox Children's Books
▷ The Random House Group Ltd

## Red Squirrel Publishing

Suite 235, 77 Beak Street, London W1F 9DB
info@redsquirrelbooks.com
www.redsquirrelbooks.com
Managing Director *Henry Dillon*

Red Squirrel Publishing specializes in citizenship test study guides. Its range of Life in the UK Test study guides has helped thousands of people achieve their goal of becoming British since the first edition was published in January 2006. Now in its third edition, the original definitive guide is expressly designed to help people pass the Life in the UK Test. Welcomes postal submissions. No fiction, poetry or children's books.
💷 Royalties annually.

## Reed Elsevier Group plc

1–3 Strand, London WC2N 5JR
☎ 020 7930 7077
www.reedelsevier.com
*Also at:* 125 Park Avenue, 23rd Floor, New York, NY 10017, USA ☎ 001 212 309 5498 🖷 001 212 309 5480
*And:* Raderweg 29, 1043 NX Amsterdam, The Netherlands ☎ 00 31 20 485 2222 🖷 00 31 20 618 0325
Chief Executive Officer *Ian Smith*
Approx. Annual Turnover £5.33 billion

Reed Elsevier Group plc is a world-leading publisher and information provider. It is owned equally by its two parent companies, Reed Elsevier Plc and Reed Elsevier NV. No unsolicited material.

## Regency House Publishing Limited

The Red House, 84 High Street, Buntingford SG9 9AG
☎ 01763 274666    🖷 01763 273501
www.regencyhousepublishing.com
Managing Director *Nicolette Trodd*

Founded 1991. Publisher and packager of mass-market non-fiction. No fiction. No unsolicited material.

## Remember When

47 Church Street, Barnsley S70 2AS
☎ 01226 734222    🖷 01226 734438
enquiries@pen-and-sword.co.uk
www.pen-and-sword.co.uk

An imprint of **Pen & Sword Books Ltd**, Remember When covers the nostalgia and

antiques collecting range of titles. No unsolicited mss; synopses and idea welcome.
Ⓔ Royalties twice-yearly.

## Reportage Press
26 Richmond Way, London W12 8LY
☎ 07971 461935
info@reportagepress.com
www.reportagepress.com
Chair *Rosie Whitehouse*

Founded in May 2007 to publish books on foreign affairs and books set in foreign countries. **Non-Fiction** TITLES *Bikila: Ethiopia's Barefoot Olympian* Tim Judah; *Passport to Enclavia* Vitali Vitaliev; *Memoirs of a Survivor* Sergei Golitsyn; *Franco's International Brigades* Christopher Othen; **Fiction** TITLE *A Girl in the Film* Charlotte Eager; **Despatches** TITLES *Red Zone: Five Bloody Days in Baghdad* Oliver Poole; *A Week in Waterloo* Magdalene De Lancey. Unsolicited mss, synopses and ideas welcome; send synopses by email and mss by post.

## Research Disclosure
▷ Kenneth Mason Publications Ltd

## Richmond House Publishing Company Ltd
70–76 Bell Street, Marylebone, London NW1 6SP
☎ 020 7224 9666    ℻ 020 7224 9688
sales@rhpco.co.uk
www.rhpco.co.uk
Managing Directors *Gloria Gordon, Spencer Block*

Publishes directories for the theatre and entertainment industries. TITLES *British Theatre Directory*; *Artistes and Agents*; *London Seating Plan Guide*. Synopses and ideas welcome.

## Rider
▷ The Random House Group Ltd

## Right Way
▷ Constable & Robinson Ltd

## Rising Stars UK Ltd
22 Grafton Street, London W1S 4EX
☎ 020 7495 6793    ℻ 020 7495 6796
info@risingstars-uk.com
www.risingstars-uk.com
Chairman *Philip Walters*
Managing Director *Andrea Carr*
*(andreacarr@risingstars-uk.com)*
Senior Publisher *Camilla Erskine*
*(camillaerskine@risingstars-uk.com)*
Approx. Annual Turnover £4 million

Founded 2002. Educational publisher of books, readers and software for schools. Winner of the ERA Supplier of the Year award for 2007 and Independent Publisher of the Year 2008. Unsolicited mss, synopses and ideas welcome; approach by email.

✴*Authors' update* Said to have a 'lively publishing programme' and a 'clear understanding of its schools market'.

## RMEP
▷ SCM – Canterbury Press Ltd

## Robert Hale Ltd
Clerkenwell House, 45–47 Clerkenwell Green, London EC1R 0HT
☎ 020 7251 2661    ℻ 020 7490 4958
enquire@halebooks.com
www.halebooks.com
Chairman/Managing Director *John Hale*

Founded 1936. Family-owned company. Publishes adult fiction (but not interested in category romance or science fiction) and non-fiction. No specialist material (education, law, medical or scientific). Acquired NAG Press Ltd in 1993 with its list of horological, gemmological, jewellery and metalwork titles, and **J.A. Allen** in 1999 with its extensive list of horse and dog books (see entry). TITLES Non-fiction: *Commercial Writing* Antonia Chitty; *Fiction Writing* Richard Skinner; *Motorhomes* David & Fiona Batten-Hill; *Sherlock Holmes's London* David Sinclair; *A Portrait of the Surrey Hills* Jane Garrett; Fiction: *Ghost Monster* Simon Clark; *Churchyard and Hawke* E.V. Thompson; *The Murder of Diana Devon and Other Mysteries* Michael Gilbert; *Girl on the Orlop Deck* Beryl Kingston; *Facing the World* Grace Thompson. Over 200 titles a year. Unsolicited mss, synopses and ideas for books welcome.
Ⓔ Royalties twice-yearly.

## Robinson
▷ Constable & Robinson Ltd

## Robson Books
▷ Anova Books

## Rodale
▷ Macmillan Publishers Ltd

## RotoVision
Sheridan House, 114 Western Road, Hove BN3 1DD
☎ 01273 727268    ℻ 01273 727269
sales@rotovision.com
www.rotovision.com
www.myspace.com/rotovision
Managing Director *Piers Spence*
Publisher *April Sankey*

Founded 1971. Acquired by **The Quarto Group, Inc.** in 1999. International publishers of graphic design, photography, digital design, advertising, film, product design and packaging design books. TITLES *World's Top Photographers: Nudes*; *What Is Graphic Design For?*; *Designs of the Times*; *First Steps in Digital Design*; *500 Digital Photography*

*Hints, Tips and Techniques*; *Print and Production Techniques for Brochures and Catalogues*. 35 titles a year. No unsolicited mss; written synopses and ideas welcome. No phone calls, please. No academic or fiction.
£ Flat fee paid.

## Rough Guides
▷ Penguin Group (UK)

## Round Hall (Ireland)
▷ Sweet & Maxwell Group

## Roundhouse Group

Maritime House, Basin Road North, Hove BN41 1WR
☎ 01273 704962   🖷 01273 704963
sales@roundhousegroup.co.uk
www.roundhousegroup.co.uk
CEO & Publisher *Alan Goodworth*
Director, Roundabout (Children's Book Division) *Bill Gills*

Founded 1991. Represents and distributes a broad range of non-fiction publishers from the UK, Europe, USA, Canada, Australia, and elsewhere. Over 1,000 titles a year. No unsolicited mss.

## Routledge (An imprint of Taylor & Francis Group, an Informa plc business)

2 Park Square, Milton Park, Abingdon OX14 4RN
☎ 020 7017 6000   🖷 020 7017 6699
www.routledge.com
CEO *Roger Horton*
Publishing Directors *Claire L'Enfant, Alan Jarvis*

Routledge is an international academic imprint publishing books primarily for university students, researchers, academics and professionals. Most Routledge authors are academics with specialist knowledge of their subject. Publishes everything from core text books to research monographs and journals in the following subject areas: humanities, including anthropology, archaeology, classical studies, communications and media studies, cultural studies, history, language and literature, music, philosophy, religious studies, theatre studies and sports studies; social sciences, including geography, politics, military and strategic studies, sociology and urban studies; business, management, law and economics; education; Asian and Middle Eastern studies. Over 2,000 titles a year. Send a detailed proposal plus c.v. and sample chapters. Proposal requirements and submission procedures can be found on the website. No poetry, fiction, travel or astrology.
£ Royalties annually or twice-yearly, according to contract.

## Rudolf Steiner Press

Hillside House, The Square, Forest Row RH18 5ES
☎ 01342 824433   🖷 01342 826437
office@rudolfsteinerpress.com
www.rudolfsteinerpress.com
Chairman *P. Martyn*
Manager *S. Gulbekian*
Approx. Annual Turnover £170,000

Founded in 1925 to publish the work of Rudolf Steiner and related materials. Publishes non-fiction: spirituality and philosophy. IMPRINT **Sophia Books** *S. Gulbekian* TITLES *From Stress to Serenity*; *Homemaking as a Social Art*. Aout 15 titles a year. No unsolicited material.
£ Royalties annually.

## RYA (Royal Yachting Association)

RYA House, Ensign Way, Hamble, Southampton SO31 4YA
☎ 023 8060 4100   🖷 023 8060 4299
phil.williamsellis@rya.org.uk
www.rya.org.uk
Chief Executive *Rod Carr*
Publishing Editor/Publications Manager *Phil Williams-Ellis*
Approx. Annual Turnover £12million+

The Royal Yachting Association, the UK governing body representing sailing, windsurfing, motorboating, powerboat racing both at sea and on inland waters, was founded in 1875. Publishes sports instruction books relating to specific training courses, technical, legal and boating advice, all written by experts in their field. TITLES *RYA Weather Handbook*; *RYA Navigation Exercises*; *RYA Powerboat Handbook*; *RYA Go Sailing!* 10 titles a year. No unsolicited mss; synopses and ideas for books welcome. Approach by email or post in the first instance.
£ Royalties twice-yearly.

## Ryelands
▷ Halsgrove

## Ryland Peters & Small Limited

20–21 Jockey's Fields, London WC1R 5BW
☎ 020 7025 2200   🖷 020 7025 2201
info@rps.co.uk
www.rylandpeters.com
Managing Director *David Peters*
Publishing Director *Alison Starling*
Art Director *Leslie Harrington*

Founded 1996. Publishes highly illustrated lifestyle books aimed at an international market, covering home and garden, food and drink, babies and kids and gift. No fiction. Acquired **Cico Books** in 2006. No unsolicited mss; synopses and ideas welcome.
£ Royalties twice-yearly.

## Sage Publications

1 Oliver's Yard, 55 City Road, London EC1Y 1SP
☎ 020 7324 8500 📠 020 7324 8600
info@sagepub.co.uk
www.sagepub.co.uk
Managing Director *Stephen Barr*
Publishing Director *Ziyad Marar*

Founded 1971. Publishes academic books and journals in humanities and the social sciences, education and science, technology and medicine. Bought academic and professional books publisher Paul Chapman Publishing Ltd in 1998 and Lucky Duck Publishing in 2004.
£ Royalties annually.

## St Pauls Publishing

187 Battersea Bridge Road, London SW11 3AS
☎ 020 7978 4300 📠 020 7978 4370
editorial@stpaulspublishing.com
www.stpaulspublishing.com
Editor *Fr. Celso Godilano*

Publishing division of the Society of St Paul. Began publishing in 1914 but activities were fairly limited until around 1948. Publishes religious material mainly: theology, scripture, catechetics, prayer books and biography. About 20 titles a year. Unsolicited mss, synopses and ideas welcome with s.a.s.e.

## Salamander
▷ Anova Books

## Salt Publishing Limited

*Head office:* Fourth Floor, 2 Tavistock Place, London WC1H 9RA
☎ 01223 882220 📠 01223 882260
sales@saltpublishing.com
www.saltpublishing.com
*Correspondence address:* 14a High Street, Fulbourn, Cambridge CB21 5DH
Chairman/Managing Director *Jen Hamilton-Emery*
Approx. Annual Turnover £170,000

Founded in 2000 to publish contemporary poetry which extends across national boundaries and a wide range of poetic practices. Since then Salt has expanded the size and range of its publishing programme, focusing on poetry, fiction, biography, translations, classics, critical companions, essays, literary criticism and text books by authors from the USA, UK, Ireland and Australia. 2010 saw the launch of a children's list, development of ebooks and audiobooks. DIVISIONS **Academic/Children's/Poetry/Non-fiction** *Chris Hamilton-Emery*; **Fiction/Gift** *Jen Hamilton-Emery*. 80 titles in 2009. No unsolicited mss except through Salt's annual debut competitions (The Crashaw Prize for Poetry and The Scott Prize for Short Stories – see website for details). Welcomes synopses and ideas for books from agents only. No religious studies.

✱*Authors' update* **Strong support for contemporary literature with a lively poetry list.**

## Samuel French Ltd

52 Fitzroy Street, London W1T 5JR
☎ 020 7387 9373 📠 020 7387 2161
theatre@samuelfrench-london.co.uk
www.samuelfrench-london.co.uk
Chairman *Leon Embry*
Managing Director *Vivien Goodwin*

Founded 1830 with the object of acquiring acting rights and publishing plays. Publishes plays only. About 50 titles a year. Unsolicited mss considered only after introductory letter addressed to the Performing Rights Department.
£ Royalties twice-yearly for books; performing royalties monthly, subject to a minimum amount.

✱*Authors' update* **Thrives on the amateur dramatic societies who are forever in need of play texts. Editorial advisers give serious attention to new material but a high proportion of the list is staged before it goes into print. New writers are advised to try one-act plays, much in demand by the amateur dramatic societies but rarely turned out by established playwrights.**

## Sandstone Press Ltd

PO Box 5725, One High Street, Dingwall IV15 9WJ
☎ 01349 862583 📠 01349 862583
info@sandstonepress.com
www.sandstonepress.com
Managing Director *Robert Davidson*

Founded 2002. Based in Highland Scotland. Publishes fiction and non-fiction. 15 titles in 2009. No unsolicited mss; synopses and ideas for books welcome. Initial approach by email, please.
£ Royalties twice-yearly.

## Saqi Books

26 Westbourne Grove, London W2 5RH
☎ 020 7221 9347 📠 020 7229 7492
www.saqibooks.com
Chairman/Managing Director *André Gaspard*
Commissioning Editor *Lynn Gaspard*

Founded in 1984, initially as a specialist publisher of books on the Middle East and the Arab world but now includes titles on Central Asia, South Asia and Europe as well as some general titles. Publishes non-fiction – academic and illustrated, including art and food and drink. IMPRINT **Telegram** (see entry). About 20 titles a year. Welcomes unsolicited material by post only.
£ Royalties annually.

✱*Authors' update* A blunt warning from the managing editor that unsolicited manuscripts are not returned. Saqi is not alone in this. Authors should retain copies of everything they send out.

## Saunders
▷ Elsevier Ltd

## Savitri Books Ltd
▷ entry under UK Packagers

## SB Publications
14 Bishopstone Road, Seaford BN25 2UB
☎ 01323 893498   🖷 01323 893860
sbpublications@tiscali.co.uk
www.sbpublications.co.uk
Owner *Mrs Lindsay Woods*

Founded 1987. Specializes in local history, including themes illustrated by old picture postcards and photographs; also travel, guides (town, walking) and railways. TITLES *Natural Kent*; *Walking the Riversides of Sussex*; *Sussex As She Was Spoke*; *The Neat and Nippy Guide to Brighton*; *Pre-Raphaelite Trail in Sussex*. Also provides marketing and distribution services for local authors. 12 titles a year.
£ Royalties annually.

## Sceptre
▷ Hachette UK

## Schiel & Denver Publishers
1 Riverside House, Heron Way, Truro TR1 2XN
☎ 0844 549 9191
enquiries@schieldenver.com
www.schieldenver.co.uk
www.schieldenver.com (USA)
Managing Director *T. Reid-Kapo*

Founded 2008. Provide a comprehensive range of publishing and promotional services including editorial, design, production, distribution and marketing paid for by the author who retains editorial and creative control. Publishes adult and children's fiction, non-fiction, biography and memoirs, current affairs, history, politics, humour, travel, health, non-fiction, spirituality, sport and academic. About 100 titles in 2009. Unsolicited material welcome; approach by email. 'No responsibility accepted for return of unsolicited mss.' OVERSEAS ASSOCIATE Schiel & Denver USA.
£ Royalties quarterly.

✱*Authors' update* Profit share deals with the author covering production costs may offend the purists but there is no denying its appeal for those who are willing to pay to see their names in print.

## Scholastic Ltd
Euston House, 24 Eversholt Street, London NW1 1DB
☎ 020 7756 7756   🖷 020 7756 7795
www.scholastic.co.uk
Chairman *D. Robinson*
Group Managing Director *Alan Hurcombe*
Approx. Annual Turnover £40 million

Founded 1964. Owned by US parent company, Scholastic Inc. who acquired **Chicken House Publishing** in May 2005. Publishes children's fiction and non-fiction and education resources for primary schools.

DIVISIONS

**Scholastic Children's Books** (Euston House address) Managing Director *Hilary Murray-Hill*

IMPRINT **Scholastic** TITLES *Horrible Histories* Terry Deary, illus. Martin Brown; *His Dark Materials Trilogy* Philip Pullman; *Mortal Engines* Philip Reeve; *Stickman* Julia Donaldson and Axel Scheffler.

**Educational Publishing** (Book End, Range Road, Witney CV32 5PR ☎ 01993 893456) Managing Director *Denise Cripps* Professional books and classroom materials for primary teachers, plus magazines such as *Child Education Plus*, *Nursery Education Plus*, *Literacy Time Plus*.

**Scholastic Book Clubs** (Euston House address) Head of Book Clubs *Julie Randles* 'SBC is the UK's number one Schools Book Club, providing the best books at the best prices and supports teachers in the process.'

**School Book Fairs** (Dolomite Avenue, Coventry Business Park, Coventry CV5 6UE) Head of Book Fairs *Steve Thompson* The Book Fair Division sells directly to children, parents and teachers in schools through 11,000 week-long book events throughout the UK.
£ Royalties twice-yearly.

## SCM – Canterbury Press
13–17 Long Lane, London EC1A 9PN
☎ 020 7776 7540   🖷 020 7776 7556
admin@hymnsam.co.uk
www.scm-canterburypress.co.uk
Publishing Director *Christine Smith*
Commissioning Editor *Natalie K. Watson (SCM)*
Approx. Annual Turnover £2.6 million

From 1 July 2009, the Church of England publishing arm, Church House Publishing, was outsourced to **Hymns Ancient & Modern Ltd** (see entry). Has three publishing imprints: **SCM Press** Text and reference books for the study of theology, philosophy of religion, ethics and religious studies; **Canterbury Press** Religious titles for the general market, liturgy, spirituality and church resources; **RMEP** Classroom resources for religious education in schools. AUTHORS include Rowan Williams, Ronald Blythe, Helen Prejean. 120 titles a year. Refer to website for guidelines on submitting proposals. £ Royalties annually.

**✱*Authors' update* Leading publisher of religious ideas with well-deserved reputation for fresh thinking. At SCM, 'questioning theology is the norm'.**

## Scottish Cultural Press/Scottish Children's Press

Unit 6, Newbattle Abbey Business Park, Newbattle Road, Dalkeith EH22 3LJ
📞 0131 660 6366    📠 0870 285 4846
info@scottishbooks.com
www.scottishbooks.com
Directors *Avril Gray, Brian Pugh*

Founded 1992. Publishes Scottish interest titles, including cultural literature, poetry, archaeology, local history. DIVISION **S.C.P. Children's Ltd** (trading as **Scottish Children's Press**) Children's fiction and non-fiction. Unsolicited mss, synopses and ideas accepted provided return postage is included, 'but *always* telephone before sending material, please. Mss sent without advance telephone call and return postage will be destroyed'.
£ Royalties paid.

## Seafarer Books Ltd

102 Redwald Road, Rendlesham, Woodbridge IP12 2TE
📞 01394 420789    📠 01394 461314
info@seafarerbooks.com
www.seafarerbooks.com
Managing Director *Patricia M. Eve*

Founded 1968. Publishes sailing titles and stories of the sea in a greatly expanded list reflecting a broader understanding of sailing narrative, maritime history and the crafts, arts and letters of the sea. AUTHORS Erskine Childers, Maldwin Drummond, Tony Groom, Tristan Lovering, Ian Tew, Peter Villiers, Ewen Southby-Tailyour. About 6 titles a year. No unsolicited mss; preliminary letter essential before making any type of submission.
£ Royalties twice-yearly.

## Seaforth Publishing (an imprint of Pen & Sword Books)

47 Church Street, Barnsley S70 2AS
📞 01226 734222    📠 01226 734438
info@seaforthpublishing.com
www.seaforthpublishing.com
Managing Director *Charles Hewitt*

Founded 2007. Specialist maritime imprint of **Pen & Sword Books Ltd**. Publishes maritime history and general maritime books. TITLES *The Wooden*

*Walls; Iron & Steel Navies; Sail & Traditional Craft; The Merchant Marine; Ship Modelling*. Unsolicited submissions welcome; send synopsis and sample chapter in printed format (double line spacing, single side). No fiction.
£ Royalties twice-yearly.

## Search Press Ltd

Wellwood, North Farm Road, Tunbridge Wells TN2 3DR
📞 01892 510850    📠 01892 515903
searchpress@searchpress.com
www.searchpress.com
Managing Director *Martin de la Bédoyère*
Commissioning Editor *Rosalind Dace*

Founded 1970. Publishes full-colour art, craft, needlecrafts – papermaking and papercrafts, painting on silk, art techniques and embroidery. No unsolicited mss; synopsis with sample chapter welcome.
£ Royalties annually.

## Seren

57 Nolton Street, Bridgend CF31 3AE
📞 01656 663018
general@seren-books.com
www.seren-books.com
Chairman *Richard Houdmont*
Publisher *Mick Felton*
Approx. Annual Turnover £200,000

Founded 1981 as a specialist poetry publisher but has now moved into general literary publishing with an emphasis on Wales. Publishes poetry, fiction, literary criticism, drama, biography, art, history and translations of fiction.

DIVISIONS **Poetry** *Amy Wack* AUTHORS Owen Sheers, Pascale Petit, Sheenagh Pugh, Kathryn Simmonds, Tiffany Atkinson, John Haynes, Carol Rumens; **Fiction** *Penny Thomas* AUTHORS Richard Collins, Lloyd Jones, Emyr Humphreys, Robert Minhinnick, Nia Wyn; **Art, Literary Criticism, History, Translations** *Mick Felton* IMPRINT **Border Lines Biographies** TITLES *Bruce Chatwin; Dennis Potter; Mary Webb; Wilfred Owen; Raymond Williams*. About 25 titles a year. Unsolicited mss, synopses and ideas for books accepted. No children's fiction.
£ Royalties annually.

## Serpent's Tail

3a Exmouth House, Pine Street, London EC1R 0JH
📞 020 7841 6300    📠 020 7833 3969
info@serpentstail.com
www.serpentstail.com
Publisher *Pete Ayrton*

Founded 1986. Since 2007, an imprint of **Profile Books**. Publishes fiction and non-fiction; literary and mainstream work, and work in translation. No unsolicited mss.

## Severn House Publishers

9–15 High Street, Sutton SM1 1DF
☎ 020 8770 3930   📠 020 8770 3850
info@severnhouse.com
www.severnhouse.com
Chairman *Edwin Buckhalter*
Editorial *Amanda Stewart*

Founded 1974. A leader in library fiction publishing. Publishes hardback fiction: romance, horror, crime. Also large print and trade paperback editions of their fiction titles. About 140 titles a year. No unsolicited material. Synopses/proposals preferred through *bona fide* literary agents only. OVERSEAS ASSOCIATE Severn House Publishers Ltd., New York.
£ Royalties twice-yearly.

## Sheldon Press

▷ Society for Promoting Christian Knowledge

## Sheldrake Press

188 Cavendish Road, London SW12 0DA
☎ 020 8675 1767   📠 020 8675 7736
enquiries@sheldrakepress.co.uk
www.sheldrakepress.co.uk
Publisher *Simon Rigge*

Founded in 1979 as a book packager and commenced publishing under its own imprint in 1991. Publishes illustrated non-fiction: history, travel, style, cookery and stationery. TITLES *The Victorian House Book*; *The Shorter Mrs Beeton*; *The Power of Steam*; *The Railway Heritage of Britain*; *Wild Britain*; *Wild France*; *Wild Spain*; *Wild Italy*; *Wild Ireland*; *Amsterdam: Portrait of a City* and Kate Greenaway stationery books. Synopses and ideas for books welcome but not interested in fiction.

## Shepheard-Walwyn (Publishers) Ltd

Suite 107, Parkway House, Sheen Lane, London SW14 8LS
☎ 020 8241 5927
books@shepheard-walwyn.co.uk
www.shepheard-walwyn.co.uk
Managing Director *Anthony Werner*
Approx. Annual Turnover £100,000

Founded 1972. 'We regard books as food for heart and mind and want to offer a wholesome diet of original ideas and fresh approaches to old subjects.' Publishes general non-fiction in four main areas: gift books in calligraphy and/or illustrated; biography, ethical economics, perennial philosophy. About 8 titles a year. Synopses and ideas for books welcome.
£ Royalties twice-yearly.

## The Shetland Times Ltd

Gremista, Lerwick ZE1 0PX
☎ 01595 693622   📠 01595 694637

publishing@shetland-times.co.uk
www.shetland-books.co.uk
Managing Director *Brian Johnston*
Publications Manager *Charlotte Black*

Founded 1872 as publishers of the local newspaper. Book publishing followed thereafter plus publication of monthly magazine, *Shetland Life*. Publishes anything with Shetland connections – local and natural history, music, crafts, maritime. Prefers material with a Shetland theme/connection. 8 titles in 2009.
£ Royalties annually.

## Shire Publications Ltd

Midland House, West Way, Botley, Oxford OX2 0PH
☎ 01865 727022   📠 01865 727017/242009
shire@shirebooks.co.uk
www.shirebooks.co.uk
Managing Director *Rebecca Smart*
Publisher *Nicholas Wright*

Founded 1962. Publishes original non-fiction paperbacks. Acquired by **Osprey Publishing Ltd** in 2007. 60 titles in 2009. No unsolicited material; send introductory letter with detailed outline of idea.
£ Royalties annually.

✱*Authors' update* You don't have to live in the country to write books for Shire but it helps. With titles like *Church Fonts*, *Haunted Inns* and *Discovering Preserved Railways* there is a distinct rural feel to the list. Another way of putting it, to quote owner John Rotheroe, Shire specializes in 'small books on all manner of obscure subjects'.

## Short Books

3A Exmouth House, Pine Street, Exmouth Market, London EC1R 0JH
☎ 020 7833 9429   📠 020 7833 9500
info@shortbooks.co.uk
www.shortbooks.co.uk
Editorial Directors *Rebecca Nicolson, Aurea Carpenter*

Founded 2001. Publishes general adult and children's non-fiction and fiction; biography, history and politics, humour and children's narrative history. No unsolicited mss.

## Sidgwick & Jackson

▷ Macmillan Publishers Ltd

## Sigma Press

Stobart House, Pontyclerc, Penybanc Road, Ammanford SA18 3HP
☎ 01269 593100   📠 01269 596116
info@sigmapress.co.uk
www.sigmapress.co.uk
Owners *Nigel Evans, Jane Evans*

Chairman/Managing Director *Nigel Evans*

Founded in 1980 as a publisher of technical books, Sigma Press now publishes mainly in the leisure area. Publishes outdoor, local heritage, adventure and biography. DIVISION **Sigma Leisure** TITLES *The Bluebird Years: Donald Campbell and the pursuit of speed* (biography); *In Search of Swallows & Amazons* (biography); *Walks in Ancient Wales*; *All-Terrain Pushchair Walks*; *Discovering Manchester*. About 10 titles a year. No unsolicited mss; synopses and ideas welcome. No poetry or novels required.
Ⓔ Royalties twice-yearly.

## Simon & Schuster UK Limited

1st Floor, 222 Gray's Inn Road, London WC1X 8HB
Ⓣ 020 7316 1900    Ⓕ 020 7316 0331
www.simonandschuster.co.uk
CEO/Managing Director *Ian S. Chapman*
Publishing Director *Suzanne Baboneau*
Publishers, Non-fiction *Mike Jones, Kerri Sharp*
Publishers, Fiction *Suzanne Baboneau, Maxine Hitchcock*
Pocket Paperback Publisher *Julie Wright*
Approx. Annual Turnover £35.9 million

Founded 1986. Owned by CBS. Sister company of the leading American company. Publishes hardback (**Simon & Schuster**) and paperback (**Pocket Books**) commercial and literary fiction, media tie-ins and non-fiction, including autobiography, biography, sport, business, history, popular science, politics, health, self-help, mind, body and spirit, humour, travel and other general interest non-fiction.

DIVISIONS **Martin Books** Editorial Director *Francine Lawrence* Specializes in bespoke publishing and branded editions, and the Simon & Schuster UK cookery list. **Simon & Schuster Children's Books** Publishing Director *Ingrid Selberg*, Editor (fiction) *Venetia Gosling*, Editor (picture books) *Emma Blackburn* Picture books, novelty and fiction titles. 440 titles a year. No unsolicited mss.
Ⓔ Royalties twice-yearly.

✱*Authors' update* Back in the top ten publishers with its mixed list of bestsellers, Simon & Schuster has the editorial nonce to detect the elusive 'good read'.

## Singing Dragon
▷ Jessica Kingsley Publishers Ltd

## Smith Gryphon Ltd
▷ John Blake Publishing Ltd

## Snowbooks Ltd
120 Pentonville Road, London N1 9JN
Ⓣ 0790 406 2414
manuscripts@snowbooks.com
www.snowbooks.com

Managing Director *Emma Barnes*

Founded 2003. Independent publisher of mainly fiction and some non-fiction. 'Keen on promoting open, friendly, productive relationships with authors and retailers'. TITLES *Adept* Robert Finn; *Needle in the Blood* Sarah Bower; *The Idle Thoughts of an Idle Fellow* J.K. Jerome; *Boxing Fitness* Ian Oliver; *Monster Trilogy* David Wellington; *City Cycling* Richard Ballantine; *Plotting for Beginners* Sue Hepworth and Jane Linfoot; *The Other Eden* and *Sand Daughter* Sarah Bryant. 15–30 titles a year. Unsolicited mss 'very welcome', by email only. Read the submission guidelines in the authors' section of the website first: www.snowbooks.com/submissions.html
Ⓔ Royalties are payable on *net* receipts.

✱*Authors' update* There are only two conditions for taking on a book: an editor must like it and believe it will sell. Snowbooks has a 'Nibbie' award as Small Publisher of the Year and was recently declared Trade Publisher of the Year by the Independent Publishers Guild.

## Society Books
17–21 Wyfold Road, London SW6 6SE
Ⓣ 020 7385 6137    Ⓕ 020 7381 8444
sbimprints@societybooks.com
www.societybooks.com
Owner *Books & Life Limited*
Managing Director *Hom Paribag*
Contact *Sarah Ismail*
Approx. Annual Turnover £500,000

Launched in 2007 by the owners of *Society Today Magazine* to publish books with a social message by victims of injustice, the disabled and ethnic minority authors, both new and established. Publishes fiction and non-fiction, arts and culture and academic titles. Will consider 60-word synopsis with introductory letter but material via literary agent preferred.
Ⓔ Royalties twice-yearly.

## Society for Promoting Christian Knowledge (SPCK)
36 Causton Street, London SW1P 4ST
Ⓣ 020 7592 3900    Ⓕ 020 7592 3939
www.spck.org.uk
www.sheldonpress.co.uk
Chief Executive Officer *Simon Kingston*

Founded 1698, SPCK is the third oldest publisher in the country. IMPRINTS **SPCK** Publishing Director *Joanna Moriarty* Theology, academic, liturgy, prayer, spirituality, biblical studies, educational resources, mission, pastoral care, gospel and culture, worldwide. **Sheldon Press** Editor *Fiona Marshall* Popular medicine, health, self-help, psychology. **Azure** Senior Editor *Alison Barr* General spirituality.

£ Royalties annually.

**Authors' update** Theology with a liberal agenda.

## Sophia Books
> Rudolf Steiner Press

## Southwater
> Anness Publishing Ltd

## Souvenir Press Ltd

43 Great Russell Street, London WC1B 3PD
☎ 020 7580 9307/7637 5711   📠 020 7580 5064
souvenirpress@ukonline.co.uk
Chairman/Managing Director *Ernest Hecht*

Independent publishing house. Founded 1951. Publishes academic and scholarly, animal care and breeding, antiques and collecting, archaeology, autobiography and biography, business and industry, children's, cookery, crafts and hobbies, crime, educational, fiction, gardening, health and beauty, history and antiquarian, humour, illustrated and fine editions, magic and the occult, medical, military, music, natural history, philosophy, poetry, psychology, religious, sociology, sports, theatre and women's studies. Souvenir's Human Horizons series for the disabled and their carers is one of the most pre-eminent in its field and recently celebrated 37 years of publishing for the disabled.

IMPRINTS/SERIES Condor; **Human Horizons; Independent Voices; Pictorial Presentations; Pop Universal; The Story-Tellers.** TITLES *Mad About the Dog* Belinda Harley; *Black Like Me* John Howard Griffin; *Ashes in the Wind* Dr Jacob Presser; *Beckett before Beckett* Brigitte Le Juez; *The Immortal Game: A History of Chess* David Shenk; *City of Night* John Rechy; *On Learning Golf* Percy Boomer; *The Secret Language of Your Face* Chi An Kuei. About 55 titles a year. Unsolicited mss considered but initial letter of enquiry and outline always required in the first instance.
£ Royalties twice-yearly.

**Authors' update** One of the last of the great independents, Souvenir covers every subject which appeals to Ernest Hecht's intellectual and commercial instincts.

## SPCK
> Society for Promoting Christian Knowledge

## Special Interest Model Books Ltd

PO Box 327, Poole BH15 2RG
☎ 01202 649930   📠 01202 649950
chrlloyd@globalnet.co.uk
www.specialinterestmodelbooks.co.uk
Contact *Chris Lloyd*

Publishes aviation, model engineering, leisure and hobbies, modelling, electronics, wine and beer

making. Send synopses rather than completed mss.
£ Royalties twice-yearly.

## Speechmark Publishing Ltd

70 Alston Drive, Bradwell Abbey, Milton Keynes MK13 9HG
☎ 01908 326944   📠 01908 326960
info@speechmark.net
www.speechmark.net
Managing Director *Liz Lane*

Speechmark publishes practical books, games and *ColorCards®* for health, education and special needs practitioners and students. Unsolicited mss, synopses and ideas welcome. Approach by letter or email.
£ Royalties twice-yearly.

## Spellmount
> The History Press

## Sphere
> Little, Brown Book Group UK

## Spon Press
> Taylor & Francis Group

## SportsBooks Limited

PO Box 422, Cheltenham GL50 2YN
☎ 01242 256755   📠 05603 108126
randall@sportsbooks.ltd.uk
www.sportsbooks.ltd.uk
Chairman & Managing Director *Randall Northam*
Approx. Annual Turnover £150,000

Founded 1995. Publishes sports books including biographies, statistical and practical. TITLES include *The Nationwide Football Annual* (formerly *The News of the World Football Annual*); *The International Track and Field Annual*; *Modern Football is Rubbish*; *A Develyshe Pastime*; *The Rebel Tours*. IMPRINT **BMM.** 15 titles in 2009. No unsolicited mss. Synopses and ideas welcome. Approach by letter or email.

## The Sportsman's Press (An imprint of Quiller Publishing Ltd)

Wykey House, Wykey, Shrewsbury SY4 1JA
☎ 01939 261616   📠 01939 261606
info@quillerbooks.com
www.countrybooksdirect.com
Managing Director *Andrew Johnston*

Specializes in books on all country subjects and general sports including fishing, fencing, shooting, equestrian, cookery, gunmaking and wildlife art. About 5 titles a year.
£ Royalties twice-yearly.

## Spring Hill
> How To Books Ltd

## Springer-Verlag London Ltd

236 Gray's Inn Road, London WC1X 8HL
☎ 020 7562 2930    📠 020 3192 2011
www.springer.com
Editorial Director *Beverley Ford*
Approx. Annual Turnover £5 million

The UK subsidiary of Springer Science & Business Media. Publishes science, technical and medical books and journals. Specializes in computing, engineering, medicine, mathematics, chemistry, biosciences. All UK published books are sold through Springer's German and US companies as well as in the UK. Globally: 4,000 books and 1,250 journals; UK: 150 titles plus 27 journals. Not interested in social sciences, fiction or school books but academic and professional science mss or synopses welcome.
💷 Royalties annually.

## Spruce
▷ Octopus Publishing Group

## Square Peg
▷ The Random House Group Ltd

## Stacey International

128 Kensington Church Street, London W8 4BH
☎ 020 7221 7166    📠 020 7792 9288
marketing@stacey-international.co.uk
www.stacey-international.co.uk
Chairman *Tom Stacey*
Managing Director *Max Scott*

Founded 1974. Part of Stacey Arts. Publishes all categories of books except textbooks. 'Through an extensive and varied list, Stacey International seeks to foster mutual understanding between contrasting peoples and cultures.' 15 titles in 2009. Unsolicited material welcome; send by email.
💷 Royalties annually.

## Stadia
▷ The History Press

## Stainer & Bell Ltd

PO Box 110, 23 Gruneisen Road, London N3 1DZ
☎ 020 8343 3303    📠 020 8343 3024
post@stainer.co.uk
www.stainer.co.uk
Managing Directors *Carol Y. Wakefield,*
*Keith M. Wakefield*
Publishing Director *Nicholas Williams*
Approx. Annual Turnover £888,000

Founded 1907 to publish sheet music. Publishes music and religious subjects related to hymnody. Unsolicited synopses/ideas for books welcome. Send letter enclosing brief précis.
💷 Royalties annually.

## Stanley Gibbons Publications

7 Parkside, Christchurch Road, Ringwood
BH24 3SH
☎ 01425 472363    📠 01425 470247
info@stanleygibbons.com
www.stanleygibbons.com
*Also at:* 399 Strand, London WC2R 0LX
Chief Executive *M. Hall*
Editorial Head *H. Jefferies*
Publishing Director *Ann-Marie Halligan*
Approx. Annual Turnover £2.9 million

Part of The Stanley Gibbons Group Limited, an AIM listed company. Long-established force in the philatelic and autograph world with 150 years in the business. Publishes philatelic reference catalogues and handbooks. Reference works relating to other areas of collecting may be considered. TITLES include *Stanley Gibbons British Commonwealth Stamp Catalogue*; *Stamps of the World*. Foreign catalogues include Japan and Korea, Portugal and Spain, Germany, Middle East, Balkans, China. Monthly publication: *Gibbons Stamp Monthly* (see entry under *Magazines*), *Philatelic Exporter*. About 20 titles a year. Unsolicited mss, synopses and ideas welcome.
💷 Royalties by negotiation.

## Stanley Thornes
▷ Nelson Thornes Ltd

## The Stationery Office
▷ TSO

## The Story-Tellers
▷ Souvenir Press Ltd

## Strand Publishing UK Ltd

Golden Cross House, 8 Duncannon Street,
Strand, London WC2N 4JF
☎ 0207 484 5088
info@strandpublishing.co.uk
www.strandpublishing.co.uk
Chairman/Managing Director *Imran Hanif*
Approx. Annual Turnover £200,000

Founded 2009. Publishes fiction, non-fiction and education. DIVISIONS **Fiction** *Romeen Malik* TITLE *The Path of Gods*; **Non-fiction** *James Castle* TITLE *Storm Over Kabul: Afghanistan, the Cockpit of Asia*; **Education** *John J. Thorpe* TITLE *Personalized Learning: Education for the 21st Century*. 12 titles in 2009. No unsolicited mss. Initial contact by email or telephone. No young children's, illustrated or erotica. 'We want all kinds of books, including non-fiction books of merit that are controversial and deal with contemporary issues as they affect the world.'
💷 Royalties annually.

## Strident Publishing Limited

22 Strathwhillan Drive, Hairmyres, East Kilbride G75 8GT

☎ 01355 220588

info@stridentpublishing.co.uk
www.stridentpublishing.co.uk
Managing Director *Keith Charters*

Founded in 2005 by author Keith Charters. Publishes fiction: children's, teenage and young adult books. Expanding list to include adult fiction. TITLES *Jessica Haggerthwaite: Witch Dispatcher* Emma Barnes; *Granny Nothing* Catherine MacPhail; *DarkIsle* D.A. Nelson (winner of the 2008 Royal Mail Award for children's books). 9 titles in 2010. Unsolicited mss (no synopses/ideas) welcome for children's, teen and young adult books only. Send mss by post or email. No non-fiction, picture books or poetry. Ⓔ Royalties twice-yearly.

## Studymates

PO Box 225, Abergele, Conwy County LL18 9AY

☎ 01745 832863   ᶠ 01745 826606

info@studymates.co.uk
www.studymates.co.uk
Managing Editor *Dr Graham Lawler*
Finance Director *Dr Judith Lawler*

Authors need to be practising or recently-retired teachers/lecturers, or published authors for the writers' guides. Ideas in writing first; unsolicited mss will not be considered. List of business books under the IMPRINT **Studymates Professional**. New list of education books planned for ABER but again Studymates will not accept unsolicited mss. 'Please approach in writing in the first instance and outline how your idea will complement the current list.'

## Summersdale Publishers Ltd

46 West Street, Chichester PO19 1RP

☎ 01243 771107   ᶠ 01243 786300

submissions@summersdale.com
www.summersdale.com
Managing Director *Alastair Williams*
Editorial Director *Jennifer Barclay*

Founded 1990. Publishes travel literature, self-help, humour and gift books, general non fiction. TITLES *Dog Heroes*; *Strictly Cat Dancing*; *Medical Murder*; *Curious Science*; *50 Things You Can Do Today to Manage IBS*. 85 titles a year. No unsolicited mss; synopses and ideas welcome by email. Ⓔ Royalties paid.

## Sutton
▷ The History Press

## Swan Hill Press (An imprint of Quiller Publishing Ltd)

Wykey House, Wykey, Shrewsbury SY4 1JA

☎ 01939 261616   ᶠ 01939 261606

info@quillerbooks.com
www.countrybooksdirect.com
Managing Director *Andrew Johnston*

Specializes in practical books on all country and field sports activities. Publishes books on fishing, shooting, gundog training, falconry, equestrian, deer, cookery, wildlife art. About 30 titles a year. Unsolicited mss must include s.a.s.e. Ⓔ Royalties twice-yearly.

## Sweet & Maxwell

100 Avenue Road, London NW3 3PF

☎ 020 7393 7000   ᶠ 020 7393 7010

www.sweetandmaxwell.co.uk
Managing Director *Peter Lake*

Founded 1799. Part of Thomson Reuters. Publishes materials in all media; looseleaf works, journals, law reports, CD-ROMs and online. The legal and professional list is varied and contains academic titles as well as treatises and reference works in the legal and related professional fields.

IMPRINTS **Sweet & Maxwell**; **W. Green (Scotland)**; **Incomes Data Services**; **Round Hall (Ireland)**. Over 1,100 products, including 180 looseleafs, 80 periodicals, more than 40 digital products and online information services and 200 new titles each year. Ideas welcome. Writers with legal/professional projects in mind are advised to contact the Legal Business Unit at the earliest possible stage in order to lay the groundwork for best design, production and marketing of a project. Ⓔ Royalties and fees according to contract.

## T&AD Poyser
▷ A.&C. Black Publishers Ltd

## T&T Clark
▷ The Continuum International Publishing Group Ltd

## T2
▷ Troubador Publishing Ltd

## Tamarind Books
▷ The Random House Group Ltd

## Tango Books Ltd

PO Box 32595, London W4 5YD

☎ 020 8996 9970   ᶠ 020 8996 9977

info@tangobooks.co.uk
www.tangobooks.co.uk
Directors *David Fielder, Sheri Safran*
Approx. Annual Turnover £1.1 million

Founded 1981. Children's novelty publisher with international co-edition potential: pop-ups, three-dimensional, cloth and board books; 700 words

maximum. About 20 titles a year. Preliminary letter or email with sample material. Material posted must include s.a.s.e. for return.
£ Primarily, fees paid.

## Taschen UK

Fifth Floor, 1 Heathcock Court, 415 Strand, London WC2R ONS
☎ 020 7845 8585    🖷 020 7836 3696
www.taschen.com

UK office of the German photographic, art and architecture publisher. Editorial office in Cologne (see entry under *European Publishers*).

## Tauris Academic Studies/Tauris Parke Paperbacks
▷ I.B. Tauris & Co. Ltd

## Taylor & Francis Group (an Informa plc business)

2 Park Square, Milton Park, Abingdon OX14 4RN
☎ 020 7017 6000    🖷 020 7017 6699
www.tandf.co.uk
CEO *Roger Horton*
Managing Director, UK (Books) *Jeremy North*
Managing Director, UK (Journals) *Ian Bannerman*
President, US (Books) *Emmett Dages*
President, US (Journals) *Kevin Bradley*

Publishes academic and reference books and journals. IMPRINTS **Routledge** (see entry); **Routledge-Cavendish**; **RoutledgeCurzon**; **RoutledgeFalmer**; **Taylor & Francis** (see entry); **Europa Publications**; **Garland Science**; **Spon Press**; **Psychology Press** (see entry); **Gordon & Breach**; **Harwood Academic**; **CRC Press**; **Frank Cass**; **BIOS Scientific Publishers**. About 3,000 titles a year.

## Taylor & Francis (An imprint of Taylor & Francis Group, an Informa plc business)

2 Park Square, Milton Park, Abingdon OX14 4RN
☎ 020 7017 6000    🖷 020 7017 6699
www.tandf.co.uk
CEO *Roger Horton*
Managing Director (UK) *Jeremy North*
President (US) *Emmett Dages*
Vice President (US) *John Lavender*

Specializes in scientific and technical books for university students and academics in the following subjects: built environment, architecture, planning and civil engineering; biology, including genetics, plant architecture, molecular biology and biotechnology, mathematics and physics, business and IT. About 600 titles a year. Send detailed proposal plus c.v. and sample chapters. Proposal requirements and submission procedures can be found on the website. No poetry, fiction, travel or astrology.
£ Royalties annually.

✱*Authors' update* An early leader in digitization, T&F has 16,000 titles online. Roger Horton sees publishing as a filtering process – searching out quality in the information overload. Very much into higher education and reference. Further expansion is predicted, particularly in the US market.

## Teach Yourself
▷ Hachette UK

## Telegram

26 Westbourne Grove, London W2 5RH
☎ 020 7229 2911    🖷 020 7229 7492
info@telegrambooks.com
www.telegrambooks.com
Chairman/Managing Director *André Gaspard*
Commissioning Editor *Shikha Sethi*

Imprint of **Saqi Books**. Independent publisher of literary fiction from around the world. 10 titles a year. Welcomes unsolicited material; approach by post.
£ Royalties annually.

## Telegraph Books

111 Buckingham Palace Road, London SW1W 0DT
☎ 020 7931 2887    🖷 020 7931 2929
www.books.telegraph.co.uk
Owner *Telegraph Media Group Ltd*
Publisher *Caroline Buckland*

Concentrates on Telegraph branded books in association/collaboration with other publishers. Publishes general non-fiction: reference, personal finance, health, gardening, cookery, heritage, guides, humour, puzzles and games. Mainly interested in books if a Telegraph link exists. About 50 titles a year. No unsolicited material.
£ Royalties twice-yearly.

## Templar Publishing

The Granary, North Street, Dorking RH4 1DN
☎ 01306 876361    🖷 01306 889097
editorial@templarco.co.uk
www.templarco.co.uk
Managing Director/Editorial Head *Amanda Wood*
Approx. Annual Turnover £17 million

Publisher and packager of quality novelty and gift books, picture books, children's illustrated non-fiction and fiction for the 8 to 12 age range. 80 titles a year. Synopses and ideas for books welcome. 'We are particularly interested in picture book mss and ideas for new novelty concepts.'
£ Royalties by arrangement.

✱*Authors' update* Templar has been praised for its 'Ology' series of books and their accompanying website, said to be 'polished and professional.'

## Temple Lodge Publishing Ltd
Hillside House, The Square, Forest Row RH18 5ES
☎ 01342 824000
office@templelodge.com
www.templelodge.com
Chairman *R. Pauli*
Chief Editor *S. Gulbekian*
Approx. Annual Turnover £65,000

Founded in 1990 to develop the work of Rudolf Steiner (see also **Rudolf Steiner Press**). Publishes non-fiction: mind, body and spirit, current affairs, health and therapy. About 12 titles a year. No unsolicited material; send initial letter of enquiry. No poetry or fiction.
£ Royalties annually.

## Tempus
▷ The History Press

## Thames and Hudson Ltd
181A High Holborn, London WC1V 7QX
☎ 020 7845 5000    ☐ 020 7845 5050
mail@thameshudson.co.uk
www.thamesandhudson.com
Chairman *Thomas Neurath*
Managing Director *Jamie Camplin*
Approx. Annual Turnover £27 million

Publishes art, archaeology, architecture and design, biography, decorative arts, fashion and textiles, garden and landscape design, graphics, history, illustrated reference, mythology, photography, popular culture, style and interior design, travel. SERIES *World of Art*; *Hip Hotels*; *StyleCity*; *New Horizons*; *Family LifeStyle*; *Earth From the Air*; *The Way We Live*; *Photofile*. TITLES *Vincent Van Gogh: The Letters*; *Art Since 1900*; *The Artist's Yearbook*; *Magnum Magnum*; *The Great LIFE Photographers*; *Hockney's Pictures*; *Grayson Perry*; *Germaine Greer's The Boy*; *Subway Art*; *The Eco-Travel Handbook*; *Fashion Makers, Fashion Shapers*; *Art Deco Complete*; *Factory Records*; *The Complete Zaha Hadid*; *World Architecture: The Masterworks*; *The Great Cities in History*; *The Iconic House*; *The Independent Design Guide*; *Secrets of the Universe*; *Reuters: Our World Now*. 200 titles a year. Send preliminary letter and outline before mss.
£ Royalties twice-yearly.

✳*Authors' update* One of the leading names in art publishing, Thames and Hudson holds its own in the production of high quality illustrated books with titles that reach beyond the specialist art market.

## Third Millennium Publishing
Third Millennium Information Ltd, 2–5 Benjamin Street, London EC1M 5QL
☎ 020 7336 0144    ☐ 020 7608 1188

info@tmiltd.com
www.tmiltd.com
Managing Director *Dr Joel Burden*
Approx. Annual Turnover £1.1 million

Founded 1999 as a highly illustrated book publisher in the international museum, heritage and art gallery markets. Also some 'coffee-table' titles in association with various universities, Oxbridge colleges, independent schools and military organizations. 20 titles annually. Unsolicited material welcome; approach in writing with c.v., book synopsis, target market and any supporting information. No fiction.
£ Royalties annually.

## Thoemmes Continuum
▷ The Continuum International Publishing Group Limited

## Thomas Cook Publishing
Thomas Cook Business Park, Coningsby Road, Peterborough PE3 8SB
☎ 01733 416477    ☐ 01733 416688
Director of Publishing *Chris Young*

Part of the Thomas Cook Group Plc, publishing commenced in 1873 with the first issue of *Cook's Continental Timetable*. Publishes guidebooks, maps and timetables. About 130 titles a year. No unsolicited mss; synopses and ideas welcome as long as they are travel-related.
£ Royalties annually.

## Thomas Nelson & Sons Ltd
▷ Nelson Thornes Limited

## Thomas Reed
▷ A.&C. Black Publishers Ltd

## Thorsons
▷ HarperCollins Publishers Ltd

## Through Writers' Eyes
▷ Eland Publishing Ltd

## Tide Mill Press
▷ Top That! Publishing Plc

## Time Out Guides
▷ The Random House Group Ltd

## Times Books
▷ HarperCollins Publishers Ltd

## Tindal Street Press Ltd
217 The Custard Factory, Gibb Street, Birmingham B9 4AA
☎ 0121 773 8157
alan@tindalstreeet.co.uk
www.tindalstreet.co.uk
Publishing Director *Alan Mahar*
Editor *Luke Brown*

Founded in 1998 to publish contemporary original fiction from the English regions. Publishes literary fiction only – novels and short story collections. No local history, memoirs or poetry. Published Man Booker 2003 shortlisted *Astonishing Splashes of Colour* and 2009 longlisted *Girl in a Blue Dress* and Costa First Novel 2007 *What Was Lost*. 8 titles in 2010. Approach with a letter, synopsis and three chapters, with s.a.s.e. if return required.
Ⓔ Royalties paid twice-yearly.

✱*Authors' update* One of the most encouraging success stories in recent publishing, Tindal has had three authors longlisted for the Man Booker Prize. Arts Council funding supports literary excellence.

## Titan Books
144 Southwark Street, London SE1 0UP
☎ 020 7620 0200     ℻ 020 7803 1990
editorial@titanemail.com
www.titanbooks.com
Managing Director *Nick Landau*
Editorial Director *Katy Wild*

Founded 1981. Now a leader in the publication of graphic novels and film and television tie-ins. Publishes comic books/graphic novels, film and television titles. IMPRINT **Titan Books** TITLES *Batman; Bones: The Official Companion; The Simpsons; Sweeney Todd: The Demon Barber of Fleet Street; Star Wars; Superman; The Winston Effect: The Art and History of Stan Winston Studio; Transformers; 24; The Official Companion Season 6; Wallace & Gromit*. About 200 titles a year. No unsolicited fiction or children's books. Ideas for film and TV titles considered; send synopsis/outline with sample chapter. No email submissions. Author guidelines available.
Ⓔ Royalties twice-yearly.

## Top That! Publishing Plc
Marine House, Tide Mill Way, Woodbridge IP12 1AP
☎ 01394 386651     ℻ 01394 386011
info@topthatpublishing.com
www.topthatpublishing.com
Managing Director *Barrie Henderson*
Editorial Director *Daniel Graham*
Approx. Annual Turnover £10 million

Founded in September 1999, Top That! Publishing is an independent children's book publisher producing novelty books that challenge and stimulate young minds. Publishes children's illustrated storybooks, fiction, reference, activity, novelty, gift and early learning books. 2010 saw the launch of an e-publishing list. DIVISIONS **Top That! Kids** TITLES *Explore, Dream, Discover; Neddy Teddy and His Phonic Friends*; **Tide Mill Press** TITLES *Cub's First Winter; Wide Mouthed Frog*; **Quest** TITLES *Orcas' Song; Vipers' Nest*; **Imagine That!** TITLES *Slow Cooking; IQ Challenge*. 60 titles in 2009. Unsolicited mss, synopses and ideas welcome. Approach by post with an email contact for response. No military or adult fiction books.
Ⓔ Royalties twice-yearly.

## Tor
▷ Macmillan Publishers Ltd

## Transworld Publishers, Random House Group Company Ltd
61–63 Uxbridge Road, London W5 5SA
☎ 020 8579 2652     ℻ 020 8579 5479
info@transworld-publishers.co.uk
www.booksattransworld.co.uk
Managing Director *Larry Finlay*
Publisher *Bill Scott-Kerr*
Approx. Annual Turnover £85 million

Founded 1950. A **Random House Group** company, New York, which in turn is a wholly-owned subsidiary of **Bertelsmann AG**, Germany. Publishes general fiction and non-fiction, gardening, sports and leisure. IMPRINTS **Bantam**; **Bantam Press** *Sally Gaminara*; **Corgi** & **Black Swan** *Linda Evans*; **Doubleday** *Marianne Velmans*; **Channel 4 Books** *Doug Young*; **Eden** and **Expert Books** *Susanna Wadeson*; **Transworld Ireland** *Eoin McHugh*. AUTHORS Monica Ali, Kate Atkinson, Dan Brown, Bill Bryson, Lee Child, Jilly Cooper, Richard Dawkins, Ben Elton, Frederick Forsyth, Tess Gerritsen, Robert Goddard, Joanne Harris, Stephen Hawking, Mo Hayder, D.G. Hessayon, John Irving, Simon Kernick, Sophie Kinsella, Kathy Lette, Paul McKenna, Andy McNab, John O'Farrell, Terry Pratchett, Patricia Scanlan, Danielle Steel, Joanna Trollope, Robert Winston. No unsolicited mss. OVERSEAS ASSOCIATES Random House Australia Pty Ltd; Random House New Zealand; Random House (Pty) Ltd (South Africa).
Ⓔ Royalties twice-yearly.

✱*Authors' update* Thriving on an enviable list of bestselling authors, Transworld has now launched its own literary festival. Aspiring authors should check for dates.

## Travel Publishing Ltd
Airport Business Centre, 10 Thornbury Road, Estover, Plymouth PL6 7PP
☎ 01752 697280     ℻ 01752 697299
info@travelpublishing.co.uk
www.travelpublishing.co.uk
Directors *Peter Robinson, Chris Day*

Founded in 1997 by two former directors of Reed Elsevier plc. Publishes travel guides covering places of interest, accommodation, food, drink

and specialist shops in Britain and Ireland. SERIES *Hidden Places*; *Hidden Inns*; *Country Pubs and Inns*; *Country Living Rural Guides* and *Country Living Garden Centres & Nurseries* (in conjunction with *Country Living* magazine); *Off the Motorway*. Over 30 titles in print. Welcomes unsolicited material; send letter in the first instance.
£ Royalties twice-yearly.

## Trentham Books Ltd

Westview House, 734 London Road, Stoke-on-Trent ST4 5NP
T 01782 745567/844699    F 01782 745553
tb@trentham-books.co.uk
www.trentham-books.co.uk
Directors *Dr Gillian Klein, Barbara Wiggins*
Approx. Annual Turnover £1 million

Publishes education (nursery, school to higher), social sciences, intercultural studies, gender studies and law for professional readers *not* for children and parents. Also academic and professional journals. No fiction, biography or poetry. Over 30 titles a year. Unsolicited mss, synopses and ideas welcome if relevant to their interests. Material only returned if adequate s.a.s.e. sent.
£ Royalties annually.

## Trident Press Ltd

Empire House, 175 Piccadilly, London W1J 9TB
T 020 7491 8770    F 020 7491 8664
admin@tridentpress.com
www.tridentpress.com
Managing Director *Peter Vine*
Approx. Annual Turnover £550,000

Founded 1997. Publishes TV tie-ins, natural history, travel, geography, underwater/marine life, history, archaeology and culture. DIVISIONS **Fiction/General Publishing** *Paula Vine*; **Natural History** *Peter Vine* TITLES *Red Sea Sharks*; *The Elysium Testament*; *BBC Wildlife Specials*; *UAE in Focus*. No unsolicited mss; synopses and ideas welcome, particularly TV tie-ins. Approach in writing or *brief* communications by email, fax or telephone.
£ Royalties annually.

## Trotman

Westminster House, Kew Road, Richmond TW9 2ND
T 020 8334 1600    F 020 8334 1601
www.trotman.co.uk
Approx. Annual Turnover £3 million

Division of **Crimson Publishing**. Publishes general careers books, higher education guides, teaching support material, employment and training resources. TITLES *Degree Course Offers*; *Guide to Student Money*; *Getting Into*; *Practice and Pass*; *Guide to Uni Life*; *Careers Uncovered*; *Real Life Guides*; *Real Life Issues*; *Careers*; *You're Hired: CV*; *You're Hired: Interview*. About 50 titles a year. Unsolicited material welcome.
£ Royalties twice-yearly.

## Troubador Publishing Ltd

5 Weir Road, Kibworth Beauchamp, Leicester LE8 0LQ
T 0116 279 2299    F 0116 279 2277
books@troubador.co.uk
www.troubador.co.uk
Chairman *Jane Rowland*
Managing Director *Jeremy Thompson*
Approx. Annual Turnover £750,000

Founded 1998. Publishes fiction, children's, academic, non-fiction, translation and poetry. Offers self-publishing agreement for fiction or non-fiction under its Matador imprint. DIVISIONS **Academic** *Jane Rowland*; **Fiction/Non-Fiction** *Jeremy Thompson*. IMPRINTS **Matador**; **Troubador Italian Studies**; **T2**; **Troubadour Storia**; **Communication and Social Justice**; **Emerging Communications Technologies**; **Communication Ethics** TITLES *Energy Beyond Oil*; *Grazia Deledda: A Biography*; *The Ethics of Teaching Practice*. 170 titles in 2009. Unsolicited mss welcome (by email only). No synopses or ideas for books.
£ Royalties quarterly.

## TSO (The Stationery Office)

St Crispins, Duke Street, Norwich NR3 1PD
T 01603 622211
www.tso.co.uk
Chief Executive *Richard Dell*
Approx. Annual Turnover £250 million

Formerly HMSO, which was founded in 1786. Became part of the private sector in October 1996. Publisher of material sponsored by Parliament, government departments and other official bodies. Also commercial publishing in the following broad categories: business and professional, environment, transport, education and law. 11,000 new titles each year with 50,000 titles in print.

## Tuckwell Press
▷ **Birlinn Ltd**

## Tusker Rock Press
▷ **Atlantic Books**

## Twenty First Century Publishers Ltd

Braunton Barn, Kiln Lane, Isfield TN22 5UE
T 01892 522802
TFCP@btinternet.com
www.twentyfirstcenturypublishers.com
Chairman *Fred Piechoczek*

Founded 2002. Publishes general fiction, financial thrillers and crime. Welcomes submissions by email (manuscripts@connectfree.co.uk). Send brief synopsis in the body of the email with extracts from the book in a file attachment ('as much as you wish to send'). No non-fiction, young adult or children's. Hard copy submissions will not necessarily be returned.
Ⓔ Royalties twice-yearly.

## Two Ravens Press Ltd
Green Willow Croft, Rhiroy, Lochbroom, Ullapool IV23 2SF
Ⓣ 01854 655307
info@tworavenspress.com
www.tworavenspress.com
Managing Director *Sharon Blackie*

Founded 2006. Publishes literary fiction, poetry and literary non-fiction, especially nature and travel writing. No commercial non-literary fiction, genre fiction (unless it's literary), anything connected with celebrities, children's books, local history, gift books. **Fiction** *Sharon Blackie* TITLES *The Falconer* Alice Thompson; *Joseph's Box* Suhayl Saadi; **Poetry** *David Knowles* TITLE *The Atlantic Forest* George Gunn; **Non-Fiction** *Sharon Blackie* TITLES *Blazing Paddles, Dances with Waves* Brian Wilson; *Fleck* Alasdair Gray; *A Wilder Vein* Linda Cracknell. 9 titles in 2010. Unsolicited material welcome subject to initial email inquiry *only* with synopsis and short biog in body of email. 'We do not accept unsolicited mss by mail and do not have time to respond to telephone enquiries (all information about how to submit is on our website)'.
Ⓔ Royalties twice-yearly.

## Ulverscroft
▷ F.A. Thorpe Publishing

## University of California Press
▷ University Presses of California, Columbia & Princeton Ltd

## University of Exeter Press
Reed Hall, Streatham Drive, Exeter EX4 4QR
Ⓣ 01392 263066   Ⓕ 01392 263064
uep@exeterpress.co.uk
www.exeterpress.co.uk
Publisher *Simon Baker*

Founded 1956. Publishes academic books: archaeology, classical studies and ancient history, history, maritime studies, English literature (especially medieval), film history, performance studies and books on Exeter and the South West. IMPRINT **Bristol Phoenix Press** Classics and ancient history. About 30 titles a year. Proposals welcomed in the above subject areas.
Ⓔ Royalties annually.

## University of Hertfordshire Press
University of Hertfordshire, College Lane, Hatfield AL10 9AB
Ⓣ 01707 284681   Ⓕ 01707 284666
UHPress@herts.ac.uk
www.herts.ac.uk/UHPress
Contact *Jane Housham*

Founded 1992. Publishes academic books on Gypsies, literature, regional and local history, education. IMPRINTS **University of Hertfordshire Press; Hertfordshire Publications**. TITLES *Here to Stay: The Gypsies and Travellers of Britain* Colin Clark and Margaret Greenfields; *Selling Shakespeare to Hollywood: The marketing of filmed Shakespeare adaptations from 1989 into the new millennium* Emma French; *Teachers' Legal Rights and Responsibilities* Jon Berry. 15 titles a year. New publishing proposals are welcome; proposal form available on the website.

## University of Wales Press
10 Columbus Walk, Brigantine Place, Cardiff CF10 4UP
Ⓣ 029 2049 6899   Ⓕ 029 2049 6108
post@uwp.co.uk
www.uwp.co.uk
Commissioning Editor *Sarah Lewis*
Assistant Commissioning Editor *Ennis Akpinar*
Approx. Annual Turnover £1 million

Founded 1922. Primarily a publisher of academic and scholarly books in English and Welsh. Subject areas include: history, political philosophy, Welsh and Celtic Studies, Literary Studies, European Studies and Medieval Studies. SERIES include *Political Philosophy Now*; *Religion and Culture in the Middle Ages*; *Iberian and Latin American Studies*; *Gender Studies in Wales*; *Gothic Literary Studies*; *Kantian Studies*; *French and Francophone Studies*; *European Crime Fiction*. 85 titles a year. Unsolicited mss considered but see website for further guidance.
Ⓔ Royalties annually.

## University Presses of California, Columbia & Princeton Ltd
1 Oldlands Way, Bognor Regis PO22 9SA
Ⓣ 01243 842165   Ⓕ 01243 842167
lois@upccp.demon.co.uk

Publishes academic titles only. US-based editorial offices. Enquiries only. Over 200 titles a year.

## Usborne Publishing Ltd
83–85 Saffron Hill, London EC1N 8RT
Ⓣ 020 7430 2800   Ⓕ 020 7430 1562
mail@usborne.co.uk
www.usborne.com
Managing Director *Peter Usborne*
Publishing Director *Jenny Tyler*
Approx. Annual Turnover £28 million

Founded 1973. Publishes non-fiction, fiction, art and activity books, puzzle books and music for children, young adults and pre-school. Some titles for parents. Up to 250 titles a year. Non-fiction books are written in-house to a specific format and therefore unsolicited mss are not normally welcome. Ideas which may be developed in-house are sometimes considered. Fiction for children will be considered. Keen to hear from new illustrators and designers.
£ Royalties twice-yearly.

❊*Authors' update* **There should be a parents' award to Peter Usborne for his marvellous books for early readers. Baby and toddler books have expanded the list.**

## Vacation Work
▷ Crimson Publishing

## Vanguard Press
▷ Pegasus Elliot Mackenzie Publishers Ltd

## Vermilion
▷ The Random House Group Ltd

## Verso

6 Meard Street, London W1F 0EG
☎ 020 7437 3546    🖷 020 7734 0059
enquiries@verso.co.uk
www.versobooks.com
Approx. Annual Turnover £2 million

Formerly New Left Books which grew out of the *New Left Review*. Publishes politics, history, sociology, economics, philosophy, cultural studies, feminism. TITLES *The Occupation* Patrick Cockburn; *Planet of Slums* Mike Davis; *Polemics* Alain Badiou; *The Soviet Century* Moshe Lewin; *NHS plc: The Privatisation of our Health Care* Allyson M. Pollock; *Street-Fighting Years* Tariq Ali; *Planet of Slums* Mike Davis. No unsolicited mss; synopses and ideas for books welcome. OVERSEAS OFFICE in New York.
£ Royalties annually.

❊*Authors' update* **Dubbed by the *Bookseller* as 'one of the most successful small independent publishers'.**

## Viking
▷ Penguin Group (UK)

## Vintage
▷ The Random House Group Ltd

## Virago Press

Little, Brown Book Group UK, 100 Victoria Embankment, London EC4Y 0DY
☎ 020 7911 8000    🖷 020 7911 8100
virago.press@littlebrown.co.uk
www.virago.co.uk
Publisher *Lennie Goodings*
Editorial Director *Ursula Doyle*

Editor, Virago Modern Classics *Donna Coonan*
Editor, Virago *Elise Dillsworth*

An imprint of **Little, Brown Book Group UK**. Founded in 1973 by Carmen Callil, Virago publishes women's literature, both fiction and non-fiction. IMPRINT **Virago Modern Classics** 19th and 20th-century fiction reprints by writers such as Daphne du Maurier, Angela Carter and Edith Wharton. Virago AUTHORS include Margaret Atwood, Maya Angelou, Sarah Dunant, Linda Grant, Michèle Roberts, Marilynne Robinson, Gillian Slovo, Talitha Stevenson, Natasha Walter, Sarah Waters, Shirley Williams. 50 titles a year. 'We do not currently accept any unsolicited submissions.'
£ Royalties twice-yearly.

❊*Authors' update* **Still the first name in women's publishing. The editorial team is highly selective. 'We are looking for something very special and there has to be a market for it,' Lennie Goodings tells *The Bookseller*.**

## Virgin Books
▷ The Random House Group Ltd

## Vision On
▷ Omnibus Press

## The Vital Spark
▷ Neil Wilson Publishing Ltd

## Voyager
▷ HarperCollins Publishers Ltd

## W. Green (Scotland)
▷ Sweet & Maxwell Group

## W. H. Freeman and Company

Palgrave Macmillan, Brunel Road, Houndmills, Basingstoke RG21 6XS
☎ 01256 329242    🖷 01256 330688
www.palgrave.com
President *Elizabeth Widdicombe (W. H. Freeman and Company, New York)*
Editorial Director *Margaret Hewinson (Palgrave Macmillan, UK)*

W. H. Freeman and Company publishes academic educational and textbooks in biochemistry, biology and zoology, chemistry, economics, mathematics and statistics, natural history, neuroscience, palaeontology, physics. The editorial offices for W. H. Freeman and Company is in New York (Basingstoke is a sales and marketing office only) but unsolicited mss can go through Basingstoke. Those which are obviously unsuitable will be sifted out; the rest will be forwarded to New York.
£ Royalties annually.

## W.W. Norton & Company Ltd

Castle House, 75–76 Wells Street, London
W1T 3QT
ⓣ 020 7323 1579    ⓕ 020 7436 4553
office@wwnorton.co.uk
www.wwnorton.co.uk
Managing Director *R. Alan Cameron*

Subsidiary of the US parent company founded in
1923 (see entry under *US Publishers*). Publishes
non-fiction and academic books.

## Walker Books Ltd

87 Vauxhall Walk, London SE11 5HJ
ⓣ 020 7793 0909    ⓕ 020 7587 1123
editorial@walker.co.uk
www.walker.co.uk
Managing Director *Helen McAleer*
Publisher *Jane Winterbotham*
Editors *Gill Evans (fiction), Deirdre McDermott
(picture books), Caroline Royds (early learning,
non-fiction & gift books), Denise Johnstone-Burt
(picture books, board & novelty)*
Approx. Annual Turnover £47.7 million (Group)

Founded 1979. Publishes illustrated children's
books, children's fiction and non-fiction. TITLES
*Maisy* Lucy Cousins; *Where's Wally?* Martin
Handford; *We're Going on a Bear Hunt* Michael
Rosen and Helen Oxenbury; *Guess How Much
I Love You* Sam McBratney and Anita Jeram;
*Alex Rider* series Anthony Horowitz; *Mortal
Instruments* series Cassandra Clare; *Chaos
Walking* series Patrick Ness. About 300 titles a
year.
ⓔ Royalties twice-yearly.

✱*Authors' update* Dedicated to
'groundbreaking books'. Noted for close
working relationships with authors.

## Wallflower Press

97 Sclater Street, London E1 6HR
ⓣ 020 7729 9533
info@wallflowerpress.co.uk
www.wallflowerpress.co.uk
Commissioning Editor *Yoram Allon*
Editorial Manager *Jackie Downs*
Approx. Annual Turnover £300,000

Founded 1999. Publishes trade and academic
books devoted to cinema and the moving image.
SERIES *Short Cuts* Introductory undergraduate
books; *Director's Cuts* Studies on significant
international filmmakers; *24 Frames* Anthologies
focusing on national and regional cinemas.
Over 25 titles a year. Unsolicited mss, synopses
and proposals welcome. No fiction or academic
material not related to the moving image.
ⓔ Royalties annually.

## Ward Lock
▷ Octopus Publishing Group

## Ward Lock Educational Co. Ltd

BIC Ling Kee House, 1 Christopher Road, East
Grinstead RH19 3BT
ⓣ 01342 318980    ⓕ 01342 410980
wle@lingkee.com
www.wardlockeducational.com
Owner *Ling Kee (UK) Ltd*

Founded 1952. Publishes educational books
(primary, middle, secondary, teaching manuals)
for all subjects, specializing in maths, science,
geography, reading and English and currently
focusing on Key Stages 1 and 2.

## Warne
▷ Penguin Group (UK)

## Waverley Books
▷ Geddes & Grosset

## Wayland
▷ Hachette Children's Books

## Weidenfeld & Nicolson
▷ The Orion Publishing Group Ltd

## Wharncliffe Books

47 Church Street, Barnsley S70 2AS
ⓣ 01226 734222    ⓕ 01226 734438
enquiries@wharncliffebooks.co.uk
www.wharncliffebooks.co.uk
Commissioning Editor *Rupert Harding*

An imprint of **Pen & Sword Books Ltd**.
Wharncliffe is the book and magazine publishing
arm of an old-established, independently owned
newspaper publishing and printing house.
Publishes local history throughout the UK,
focusing on nostalgia and old photographs.
SERIES *Foul Deeds*; *Local History Companions*;
*Pals*; *Aspects*. No unsolicited mss; synopses and
ideas welcome.
ⓔ Royalties twice-yearly.

## Which? Books

2 Marylebone Road, London NW1 4DF
ⓣ 020 7770 7000    ⓕ 020 7770 7660
rebecca.leach@which.co.uk
www.which.co.uk
Head of Book Publishing *Angela Newton*

Founded 1957. Publishing arm of Which?
Publishes non-fiction: information and reference
on personal finance, property, divorce, consumer
law. All titles offer direct value to the consumer.
IMPRINT **Which? Books** TITLES *The Good Food
Guide*; *Wills and Probate*; *Renting & Letting*.
10–12 titles a year. No unsolicited mss; send
synopses and ideas only.
ⓔ Royalties, if applicable, twice-yearly.

## White Ladder
▷ Crimson Publishing

## Wild Goose Publications

Iona Community, 4th Floor, The Savoy Centre, 140 Sauchiehall Street, Glasgow G2 3DH
☎ 0141 332 6292  🖷 0141 332 1090
admin@ionabooks.com
www.ionabooks.com
Editorial Head *Sandra Kramer*

The publishing house of the Iona Community, established in the Celtic Christian tradition of St Columba, publishes books and CDs on holistic spirituality, social justice, political and peace issues, healing, innovative approaches to worship, song and material for meditation and reflection.

## Wiley Europe Ltd

The Atrium, Southern Gate, Chichester PO19 8SQ
☎ 01243 779777  🖷 01243 775878
europe@wiley.co.uk
www.wiley.com

Founded 1807. US parent company. Publishes professional, reference trade and text books, scientific, technical and biomedical.

## Wiley Nautical

John Wiley and Sons Ltd., The Atrium, Southern Gate, Chichester PO19 8SQ
nautical@wiley.co.uk
www.wileynautical.com
Executive Editor *Miles Kendall*

Books for people who love watersports, Wiley Nautical publishes over 100 watersports and sailing books, including 'Cruising Companions' and the series of books previously published under the Fernhurst Books imprint. Recently relaunched a new digital product, the *Wiley Nautical Almanac*. Synopses and ideas welcome.

## Wiley-Blackwell

9600 Garsington Road, Oxford OX4 2DQ
☎ 01865 776868  🖷 01865 714591
www.wiley.com
Senior Vice President, Wiley-Blackwell *Eric Swanson*

Wiley-Blackwell was formed from the acquisition of Blackwell Publishing (Holdings) Ltd by John Wiley & Sons, Inc. Blackwell's publishing programme has merged with Wiley's global scientific, technical and medical business. The merged operation is now the largest of the three of John Wiley & Sons, Inc.; its other businesses are Professional/Trade and Higher Education.

## William Heinemann
▷ The Random House Group Ltd

## William Reed Directories

Broadfield Park, Crawley RH11 9RT
☎ 01293 613400  🖷 01293 610322
directories@william-reed.co.uk
www.william-reed.co.uk

William Reed Directories, a division of William Reed Business Media, was established in 1990. Its portfolio includes titles covering the food, drink, non-food, catering, hospitality, retail and export industries, including European versions. The titles are produced as directories, market research reports, exhibition catalogues and electronic publishing.

## Windgather Press
▷ Oxbow Books

## Windmill Books
▷ The Random House Group Ltd

## Wingedchariot Press

7 Court Royal, Eridge Road, Tunbridge Wells TN4 8HT
info@wingedchariot.com
www.wingedchariot.com

Founded in 2005 'to bring the best of children's books in translation to the UK for the first time'. About 4 titles a year. Will consider books in their original language with illustrations. Email synopsis and scan of drawings. No English language books.

## Wisden
▷ A.&C. Black Publishers Ltd

## Wise Publications
▷ Omnibus Press

## WIT Press

Ashurst Lodge, Ashurst, Southampton SO40 7AA
☎ 023 8029 3223  🖷 023 8029 2853
marketing@witpress.com
www.witpress.com
Owner *Computational Mechanics International Ltd, Southampton*
Chairman/Managing Director/Editorial Head *Professor C.A. Brebbia*

Founded in 1980 as Computational Mechanics Publications to publish engineering analysis titles. Changed to WIT Press to reflect the increased range of publications. Publishes scientific and technical, mainly at postgraduate level and above, including architecture, environmental engineering, bioengineering. 50 titles in 2009. Unsolicited mss, synopses and ideas welcome; approach by post or email. No non-scientific or technical material or lower level (school and college-level texts). OVERSEAS SUBSIDIARY Computational Mechanics, Inc., Billerica, USA.
£ Royalties annually.

## Wizard Books
▷ Icon Books Ltd

## The Women's Press

27 Goodge Street, London W1T 2LD
☎ 020 7636 3992   📠 020 7637 1866
www.the-womens-press.com
Chairman *Naim Attallah*

Part of the Namara Group. First title published in 1978. Publishes women only: quality fiction and non-fiction. Fiction usually has a female protagonist and a woman-centred theme. International writers and subject matter encouraged. Non-fiction: books for and about women generally; gender politics, race politics, disability, feminist theory, health and psychology, literary criticism. IMPRINTS **Women's Press Classics; Livewire Books for Teenagers** Fiction and non-fiction series for young adults. No mss without previous letter, synopsis, sample material and return postage. Submissions by disk or email not accepted.
£ Royalties twice-yearly.

## Woodhead Publishing Ltd

Abington Hall, Abington, Cambridge CB21 6AH
☎ 01223 891358   📠 01223 893694
wp@woodheadpublishing.com
www.woodheadpublishing.com
Chairman *Richard Dawes*
Managing Director *Martin Woodhead*
Approx. Annual Turnover £2.2 million

Founded 1989. Publishes engineering, materials technology, textile technology, finance and investment, food technology, environmental science. DIVISION **Woodhead Publishing** *Martin Woodhead.* About 70 titles a year. Unsolicited material welcome.
£ Royalties annually.

## Wordsworth Editions Ltd

8B East Street, Ware SG12 9HJ
☎ 01920 465167   📠 01920 462267
enquiries@wordsworth-editions.com
www.wordsworth-editions.com
Directors *E.G. Trayler, Derek Wright*
Approx. Annual Turnover £2.5 million

Founded 1987. Publishes classics of English and world literature, reference books, poetry, mystery and the supernatural. SERIES include *Children's Classics*; *Tales of Mystery and The Supernatural*; *Wordsworth Library Collection*; *Wordsworth Special Editions.* About 40 titles a year. No unsolicited mss.

## Y Lolfa Cyf

Talybont, Ceredigion SY24 5HE
☎ 01970 832304   📠 01970 832782
ylolfa@ylolfa.com
www.ylolfa.com
Managing Director *Garmon Gruffudd*
General Editor *Lefi Gruffudd*

Alcemi Editor *Gwen Davies*
Approx. Annual Turnover £1 million

Founded 1967. Publishes in Welsh and in English; has its own five-colour printing and binding facilities. Publishes mainstream Welsh language publications; language tutors; English language books for the Welsh and Celtic tourist trade; biographies. TITLES *Roberto* Roberto Martinez; *The Welsh Learner's Dictionary* Heini Gruffudd; *Half Time* Nigel Owens. IMPRINT **Alcemi** English language fiction. About 70 titles a year. Write first with synopses or ideas.
£ Royalties twice-yearly.

## Yale University Press (London)

47 Bedford Square, London WC1B 3DP
☎ 020 7079 4900   📠 020 7079 4901
sales@yaleup.co.uk
www.yalebooks.co.uk

Founded 1961. Owned by US parent company. Publishes art history, architecture, decorative arts, history, religion, music, politics, current affairs, biography and history of science. About 400 titles (worldwide) a year. Unsolicited mss and synopses welcome if within specialized subject areas.
£ Royalties annually.

✳*Authors' update* A publisher with a talent for turning out serious books which find a general market.

## Yellow Jersey Press

▷ The Random House Group Ltd

## Young Picador

▷ Macmillan Publishers Ltd

## Zambezi Publishing Ltd

PO Box 221, Plymouth PL2 2YJ
☎ 01752 367300   📠 01752 350453
info@zampub.com
www.zampub.com
Chair *Sasha Fenton*
Managing Director *Jan Budkowski*

Founded 1999. Publishes non-fiction: mind, body and spirit, self-help, finance and business, including the 'Simply' series. About 12 titles a year. Send biography, synopsis and sample chapter by mail. Brief email communication acceptable but *no* attachments, please.
£ Royalties twice-yearly.

## Zed Books Ltd

7 Cynthia Street, London N1 9JF
☎ 020 7837 4014   📠 020 7833 3960
sales@zedbooks.net
www.zedbooks.co.uk
Senior Commissioning Editor *Tamsine O'Riordan*
Marketing Director *Julian Hosie*
Approx. Annual Turnover £1.4 million

Founded 1976. Cooperatively owned and managed by the staff, Zed Books publishes international and Third World affairs, development studies, gender studies, environmental studies, human rights and specific area studies. No fiction, children's or poetry. DIVISIONS **Development & Environment, Gender Studies, Middle East** *Tamsine O'Riordan*; **Politics, Economics, Africa,**

**Latin America** *Ken Barlow* TITLES *Woman at Point Zero* Nawal El Saadawi; *The Development Dictionary* Wolfgang Sachs; *Soil Not Oil* Vandana Shiva. About 60 titles a year. No unsolicited mss; synopses and ideas welcome.
£ Royalties annually.

**Zero to Ten**
▷ Evans Brothers Ltd

# Irish Publishers

## A.&A. Farmar Ltd

78 Ranelagh Village, Dublin 6
☎ 00 353 1 496 3625
afarmar@iol.ie
www.aafarmar.ie
Editorial Director *Anna Farmar*
Production Director *Tony Farmar*

Founded 1992. Publishes non-fiction: social and
business history, biography, health, lifestyle.
About 10 titles a year. No unsolicited mss.
Synopses and ideas welcome; approach by letter
or email in the first instance.

## An Gúm

27 Sr. Fhreidric Thuaidh, Baile Átha Cliath 1
☎ 00 353 1 889 2800    📠 00 353 1 873 1140
angum@forasnagaeilge.ie
www.gaeilge.ie
Senior Editor *Seosamh Ó Murchú*
Editor *Máire Nic Mheanman*

Founded 1926. Formerly the Irish language
publications branch of the Department of
Education and Science. Has now become part
of the North/South Language Body, Foras na
Gaeilge. Publishes educational, children's, young
adult, music, lexicography and general. Little
adult fiction or poetry. About 50 titles a year.
Unsolicited mss, synopses and ideas for books
welcome. Also welcomes reading copies of first
and second level school textbooks with a view to
translating them into the Irish language.
💷 Royalties annually.

## Ashfield Press

30 Linden Grove, Blackrock, Co. Dublin
☎ 00 353 1 288 9808
susan.waine@ashfieldpress.ie
www.ashfieldpress.ie
Directors *Susan Waine, John Davey,
Gerry O'Connor*

Publishes Irish-interest non-fiction books.
TITLES *Rambling Down the Suir; Snow on the
Equator.* Unsolicited mss and synopses welcome.
ASSOCIATE COMPANY Linden Publishing Services.
💷 Royalties twice-yearly.

## Atrium
▷ Cork University Press

## Attic Press
▷ Cork University Press

## Blackhall Publishing Ltd

Lonsdale House, Avoca Avenue, Blackrock, Co.
Dublin
☎ 00 353 1 278 5090
info@blackhallpublishing.com
www.blackhallpublishing.com
Managing Director *Gerard O'Connor*
Commissioning Editor *Elizabeth Brennan*
Editor *Eileen O'Brien*

Publishes business, management, law, social
studies and life issues. Subject areas include
accounting, finance, management, social studies
and law books aimed at both students and
professionals in the industry. Also publishes
books for the general public in areas such as
health, personal finance, personal development
and current affairs. Provides legal publishing
services including Statute Law Revision, Law
Reporting, Legislative Drafting and Subscription
Services. 20 titles a year. Unsolicited mss and
synopses welcome.
💷 Royalties annually.

## Bradshaw Books

Tigh Filí, Civic Trust House, 50 Popes Quay, Cork
☎ 00 353 21 421 5175
info@tighfili.com
www.tighfili.com
Managing Director *Maire Bradshaw*
Project Manager *Fidelma Maye*

Founded 1985. Publishes poetry, women's issues,
children's books. Organizers of the annual Cork

Literary Review poetry manuscript competition.
SERIES *Cork Literary Review* and *Eurochild Anthology of Poetry and Art.*
ⓔ Royalties not generally paid.

## Brandon/Mount Eagle Publications

Dingle, Co. Kerry
ⓣ 00 353 66 915 1463    ⓕ 00 353 66 915 1234
www.brandonbooks.com
Publisher *Stephen MacDonogh*
Approx. Annual Turnover €500,000

Founded in 1997, taking over Brandon Book Publishers (founded 1982). Publishes Irish fiction, biography, memoirs and other non-fiction. About 15 titles a year. Not seeking unsolicited mss.

## Church of Ireland Publishing

Church of Ireland House, Church Avenue, Rathmines, Dublin 6
ⓣ 00 353 1 492 3979    ⓕ 00 353 1 492 4770
susan.hood@rcbdub.org
www.cip.ireland.anglican.org
Owner *Representative Church Body of the Church of Ireland*
Chairman *Dr Kenneth Milne*
Publications Officer *Dr Susan Hood*

Founded 2004. Publishes official publications of the Church of Ireland: theological, doctrinal, historical and administrative, and facilitates publication for other church-related bodies. 5 titles in 2009. No unsolicited material.

## Cló Iar-Chonnachta

Indreabhán, Connemara, Galway
ⓣ 00 353 91 593307    ⓕ 00 353 91 593362
cic@iol.ie
www.cic.ie
General Manager *Deirdre Ní Thuathail*
Approx. Annual Turnover €500,000

Founded 1985. Publishes fiction, poetry, plays, teenage fiction and children's, mostly in Irish, including translations. Also publishes cassettes of writers reading from their own works. TITLES *Seosamh Ó hÉanaí: Nár fhágha mé bás choíche* Liam Mac Con Iomaire; *Mise an Fear Ceoil: Séamus Ennis Dialann Taistil 1942–1946* Ríonach Uí Ógáin; *Ó Chósta go Cósta* Frank Reidy; *Fin de Siècle na Gaeilge* Brian Ó Conchubhair; *Salann Garbh* Joe Steve Ó Neachtain. 13 titles in 2009.
ⓔ Royalties annually.

## Cois Life

62 Rose Park, Kill Avenue, Dún Laoghaire, Co. Dublin
ⓣ 00 353 1 280 7951    ⓕ 00 353 1 280 7951
eolas@coislife.ie
www.coislife.ie
Chairman *Dr Seán Ó Cearnaigh*
Managing Director *Dr Caoilfhionn Nic Pháidín*
Approx. Annual Turnover €200,000

Founded 1996. General publishers of literature, academic and children's books, all in the Irish language. 11 titles in 2009. Mss, synopses and ideas for Irish-language books welcome; submissions by email.
ⓔ Royalties annually.

## The Collins Press

West Link Park, Doughcloyne, Wilton, Cork
ⓣ 00 353 21 434 7717
enquiries@collinspress.ie
www.collinspress.ie
Managing Director *Con Collins*
Approx. Annual Turnover €750,000

Founded 1989. Publishes general non-fiction, Irish interest and children's non-fiction books. 25–30 titles a year. No unsolicited mss; synopses and ideas welcome; approach by email. See website for submission guidelines. No technical, professional, fiction, poetry, literary criticism or short stories.

## The Columba Press

55A Spruce Avenue, Stillorgan Industrial Park, Blackrock, Co. Dublin
ⓣ 00 353 1 294 2556    ⓕ 00 353 1 294 2564
sean@columba.ie (editorial)
info@columba.ie (general)
www.columba.ie
Chairman *Neil Kluepfel*
Managing Director *Seán O'Boyle*
Approx. Annual Turnover €1.02 million

Founded 1985. Small company committed to growth. Publishes religious and counselling titles. TITLES *Already Within* Donal O'Leary; *Celtic Prayers and Reflections* Jenny Child. Backlist of 250 titles. Unsolicited ideas and synopses preferred rather than full mss.
ⓔ Royalties twice-yearly.

## Cork University Press

Youngline Industrial Estate, Pouladuff Road, Togher, Cork
ⓣ 00 353 21 490 2980    ⓕ 00 353 21 431 5329
corkuniversitypress@ucc.ie
www.corkuniversitypress.com
Owner *University College Cork*
Publications Director *Mike Collins*
Editor *Sophie Watson*
Production Editor *Maria O'Donovan*

Founded 1925. Relaunched in 1992, the Press publishes academic and some trade titles. TITLES *Ivor Browne: Music and Madness*; *Cornucopia at Home*; *The Iveragh Peninsula*; *Knock: The Virgin's Apparition in Nineteenth-Century Ireland*; *J.G. Farrell in His Own Words: Selected Letters and Diaries*. Also a biannual journal, *The Irish Review*, an interdisciplinary cultural review. No fiction. IMPRINTS **Attic Press**; **Atrium**. 15 titles in 2009. Unsolicited Irish interest related synopses

and ideas welcome for textbooks, academic monographs, trade non-fiction, illustrated histories and journals.
£ Royalties annually.

## Edmund Burke Publisher

Cloonagashel, 27 Priory Drive, Blackrock, Co. Dublin
T 00 353 1 288 2159    F 00 353 1 283 4080
deburca@indigo.ie
www.deburcararebooks.com
Managing Director *Eamonn De Búrca*
Publications Managers *Regina De Búrca, William De Búrca*
Approx. Annual Turnover €320,000

Small family-run business publishing historical and topographical and fine limited-edition books relating to Ireland. TITLES *Michael Collins and The Making of a New Ireland*, 2 vols; *The Great Book of Irish Genealogies*, 5 vols.; *The Annals of the Four Masters*, 7 vols; *Flowers of Mayo* (illus. Wendy Walsh); *O'Curry's Manners and Customs of the Ancient Irish*; *Joyce's Irish Names of Places*; *Kilroy's Fall of the Gaelic Lords*. Unsolicited mss welcome. No synopses or ideas.
£ Royalties annually.

## Four Courts Press Ltd

7 Malpas Street, Dublin 8
T 00 353 1 453 4668    F 00 353 1 453 4672
info@fourcourtspress.ie
www.fourcourtspress.ie
Director/Production Manager *Martin Healy*
Director/Senior Editor *Martin Fanning*

Founded 1969. Publishes mainly scholarly books in the humanities. About 70 titles a year. Synopses and ideas for books welcome.
£ Royalties annually.

## The Gallery Press

Loughcrew, Oldcastle, Co. Meath
T 00 353 49 854 1779    F 00 353 49 854 1779
gallery@indigo.ie
www.gallerypress.com
Managing Director *Peter Fallon*

Founded 1970. Publishes Irish poetry and drama. 13 titles in 2009. Currently, only interested in work from Irish poets and dramatists (plays must have had professional production). Unsolicited mss welcome but prefers to send submission notes to potential writers in the first instance.
£ Royalties paid.

## Gill & Macmillan

10 Hume Avenue, Park West, Dublin 12
T 00 353 1 500 9500    F 00 353 1 500 9599
www.gillmacmillan.ie
Chairman *M.H. Gill*
Managing Director *Dermot O'Dwyer*
Approx. Annual Turnover €12 million

Founded 1968 when M.H. Gill & Son Ltd and Macmillan Ltd formed a jointly owned publishing company. Publishes biography/autobiography, history, current affairs, literary criticism (all mainly of Irish interest), guidebooks and cookery. Also educational textbooks for secondary and tertiary levels. Contacts: *Anthony Murray* (educational), *Marion O'Brien* (college), *Fergal Tobin* (general). About 80 titles a year. Unsolicited synopses and ideas welcome.
£ Royalties subject to contract.

## Hachette Books Ireland

8 Castlecourt Centre, Castleknock, Dublin 15
T 00 353 1 824 6288    F 00 353 1 824 6289
info@hbgi.ie
www.hbgi.ie
Managing Director, Publishing *Breda Purdue*
Senior Commissioning Editor *Ciara Considine*
Senior Editor *Ciara Doorley*

Founded 2002. Irish division of **Hachette UK**. Publishes general fiction and non-fiction: memoirs, biography, sport, politics, current affairs, humour, food and drink. 33 titles in 2009. Consult the website for submission guidelines.
£ Royalties twice-yearly.

## Institute of Public Administration

57–61 Lansdowne Road, Dublin 4
T 00 353 1 240 3600    F 00 353 1 269 8644
sales@ipa.ie
www.ipa.ie
Director-General *Brian Cawley*
Publisher *Richard Boyle*
Approx. Annual Turnover €1.3 million

Founded in 1957 by a group of public servants, the Institute of Public Administration is the Irish public sector management development agency. The publishing arm of the organization is one of its major activities. Publishes academic and professional books and periodicals: history, law, politics, economics and Irish public administration for students and practitioners. TITLES *Economic Development 50 Years On 1958-2008*; *Education for Citizenship and Diversity in Irish Contexts*; *An Inconvenient Wait: Ireland's Quest for Membership of the EEC 1957-73*; *No Coward Soul: A Biography of Thekla Beere*; *Serving the State, The Public Sector in Ireland*; *The Corporate Governance of Commercial State-owned Enterprises in Ireland*. About 10 titles a year. No unsolicited mss; synopses and ideas welcome. No fiction or children's publishing.
£ Royalties annually.

## Irish Academic Press Ltd

2 Brookside, Dundrum Road, Dublin 14
T 00 353 1 298 9937    F 00 353 1 298 2783
info@iap.ie
www.iap.ie

Chairman *Stewart Cass (London)*
Dublin Office *Karen Donoghue*

Founded 1974. Publishes academic monographs
and humanities. Unsolicited mss, synopses and
ideas welcome.
ⓔ Royalties annually.

## Liberties Press

Guinness Enterprise Centre, Taylor's Lane,
Dublin 8
☎ 00 353 1 415 1286
info@libertiespress.com
www.libertiespress.com
Joint Managing Directors *Sean O'Keeffe,
Peter O'Connell*

Founded 2003. Publishes non-fiction: history,
sport, memoirs, biography, health, business and
finance, politics and current affairs. 25 titles
in 2009. Mss, synopses and ideas for books
welcome; submissions by email or post. No
poetry or science fiction.
ⓔ Royalties twice-yearly.

## The Liffey Press Ltd

Ashbrook House, 10 Main Street, Raheny,
Dublin 5
☎ 00 353 1 851 1458
info@theliffeypress.com
www.theliffeypress.com
Managing Director/Publisher *David Givens*

Founded 2001. Publishes general interest, Irish-
focused non-fiction. 16 titles in 2009. Synopses,
ideas and proposals with sample chapters
welcome; approach by letter or email in the first
instance. No fiction, children's or books not of
interest to Irish readers.
ⓔ Royalties twice-yearly.

## The Lilliput Press

62–63 Sitric Road, Arbour Hill, Dublin 7
☎ 00 353 1 671 1647    ☐ 00 353 1 671 1233
info@lilliputpress.ie
www.lilliputpress.ie
Chair *Kathy Gilfillan*
Managing Director *Antony Farrell*
Approx. Annual Turnover €400,000

Founded 1984. Publishes non-fiction: literature,
history, autobiography and biography, ecology,
essays; criticism; fiction. TITLES *What the Curlew
Said* John Moriaty; *Bad Day in Blackrock* Kevin
Power; *Ireland's Other Poetry: Anonymous
to Zozimus* John Wyse Jackson and Hector
McDonnell; *More Myers* Kevin Myers; *The
Dublin Edition of Ulysses* James Joyce. About 18
titles a year. Unsolicited mss, synopses and ideas
welcome. No children's or sport titles.
ⓔ Royalties annually.

## Little Island
▷ New Island

## Marino Books
▷ Mercier Press Ltd

## Maverick House Publishers

Office 19, Dunboyne Business Park, Dunboyne,
Co. Meath
☎ 00 353 1 825 5717    ☐ 00 353 1 686 5036
info@maverickhouse.com
www.maverickhouse.com
Managing Director *Jean Harrington*

Founded 2001. Publishes biography and
autobiography, history, current affairs, non-
fiction, politics, sport. TITLES *Welcome to Hell:
One Man's Fight for Life Inside the Bangkok Hilton*
Colin Martin; *The Commandant* Ian Baxter; *With
Bare Hands* Alain Robert; *Hell in Barbados* Terry
Donaldson. Unsolicited material welcome; send
by post. No fiction, children's books or poetry.
ⓔ Royalties twice-yearly.

## Mercier Press Ltd

Unit 3B, Oak House, Bessboro Road, Blackrock,
Cork
☎ 00 353 21 461 4700
info@mercierpress.ie
www.mercierpress.ie
Chairman *John Spillaner*
Managing Director *Clodagh Feehan*

Founded 1944. One of Ireland's largest publishers
with a list of approx 250 Irish interest titles.
IMPRINTS **Mercier Press; Marino Books** Editorial
Director *Mary Feehan* Children's, politics, history,
folklore, biography, mind, body and spirit, current
affairs, women's interest. TITLES *The Course of
Irish History*; all of John B. Keane's works; *Beyond
Prozac*; *It's a Long Way from Penny Apples*;
*Ireland's Master Storyteller*. Unsolicited synopses
and ideas welcome.
ⓔ Royalties annually.

## Merlin Publishing

Newmarket Hall, Cork Street, Dublin 8
☎ 00 353 1 453 5866    ☐ 00 353 1 453 5930
publishing@merlin.ie
www.merlinwolfhound.com
Managing Director/Publisher *Chenile Keogh*

Founded 2000. Member of **Clé**. Publishes true
crime, biography, lifestyle, film, music, art,
general non-fiction, history and gift books.
TITLES *Crime Wars*; *Change a Little to Change a
Lot*; *The Fraudsters*; *Colour for Living*. IMPRINT
**Wolfhound Press** Founded 1974. Non-fiction.
TITLES *Famine*; *Eyewitness Bloody Sunday*; *Father
Browne's Titanic Album*. About 12 titles a year.
Two sample chapters of unsolicited mss (with
synopses and IRCs) welcome by email (chenilek@

merlin.ie). See Merlin website for submission guidelines and proposal form.
Ⓔ Royalties annually.

## Mount Eagle
▷ Brandon/Mount Eagle Publications

## National Library of Ireland
Kildare Street, Dublin 2
Ⓣ 00 353 1 603 0200   Ⓕ 00 353 1 676 6690
info@nli.ie
www.nli.ie

Founded 1877. Publishes books and booklets based on the library's collections; folders of historical documents; academic and specialist books; reproduction folders and CD-ROMs. TITLES *The National Library of Ireland* Noel Kissane; *Ulysses Unbound: A Reader's Companion to James Joyce's Ulysses* Terence Killeen; *Cooper's Ireland: Drawings and Notes from an Eighteenth Century Gentleman* Peter Harbison; *Into the Light: An Illustrated Guide to the Photographic Collections in the National Library of Ireland* Sarah Rouse; *WB Yeats: Works and Days* James Quin, Eilis Ni Dhuibhne, Ciara McDonnell; *Librarians, Poets and Scholars: A festschrift for Donall O Luanaigh* ed. Felix M. Larkin; *Stranger to Citizens: The Irish in Europe 1600–1800* Mary Ann Lyons and Thomas O'Connor. 1 title in 2009.

## New Island Books
2 Brookside, Dundrum Road, Dundrum, Dublin 14
Ⓣ 00 353 1 298 9937/298 3411
Ⓕ 00 353 1 298 2783
sales@newisland.ie
www.newisland.ie
Managing Director/Editorial Head *Edwin Higel*
Editorial Manager *Deirdre O'Neill*

Founded 1992. Publishes fiction, Irish non-fiction, poetry and drama. IMPRINTS **New Island**; **Little Island** Children's and young adult non-illustrated (email submissions for Little Island to elaina. oneill@newisland.ie). About 25 titles a year. Unsolicited mss, synopses and ideas welcome. Send three chapters and synopsis by post.
Ⓔ Royalties twice-yearly

## The O'Brien Press Ltd
12 Terenure Road East, Rathgar, Dublin 6
Ⓣ 00 353 1 492 3333   Ⓕ 00 353 1 492 2777
books@obrien.ie
www.obrien.ie
Managing Director *Ivan O'Brien*
Publisher *Michael O'Brien*

Founded 1974. Publishes business, true crime, biography, music, travel, sport, Celtic subjects, food and drink, history, humour, politics, reference. Children's publishing – mainly fiction for every age from tiny tots to teenage. Illustrated fiction SERIES *Panda Cubs* (3 years+); *Pandas* (5 years+); *Flyers* (6 years+); *Red Flag* (8 years+). Novels (10 years+): contemporary, historical, fantasy. Some non-fiction: mainly historical, and art and craft, resource books for teachers. No poetry, adult fiction or academic. Unsolicited mss (sample chapters only), synopses and ideas for books welcome. No email submissions. Submissions will not be returned.
Ⓔ Royalties annually.

## Oak Tree Press
19 Rutland Street, Cork
Ⓣ 00 353 21 431 3855   Ⓕ 00 353 21 431 3496
info@oaktreepress.com
www.oaktreepress.com
Owners *Brian O'Kane, Rita O'Kane*
Managing Director *Brian O'Kane*

Founded 1991. Specialist publisher of business and professional books with a focus on small business start-up and development. Unsolicited mss and synopses welcome; send to the managing director, at the address above.
Ⓔ Royalties annually.

## On Stream Publications Ltd
Currabaha, Cloghroe, Co. Cork
Ⓣ 00 353 21 438 5798
info@onstream.ie
www.onstream.ie
Chairman/Managing Director *Roz Crowley*

Founded 1992. Formerly Forum Publications. Publishes academic, cookery, wine, general health and fitness, local history, railways, photography and practical guides. TITLES *While Keeping Watch* (fiction); *A Kingdom of Wine: A Celebration of Ireland's Winegeese*; *At Home in Renvyle* (recipe book); *Stop Howling at the Moon*; *101 Bedtime Stories for Managers*. About 3 titles a year. Synopses and ideas welcome. No children's books.
Ⓔ Royalties annually.

## Penguin Ireland
25 St Stephen's Green, Dublin 2
Ⓣ 00 353 1 661 7695   Ⓕ 00 353 1 661 7696
info@penguin.ie
www.penguin.ie
Managing Director *Michael McLoughlin*
Publishing Director *Patricia Deevy*
Editor *Brenda Barrington*

Founded 2002. Part of the **Penguin Group UK**. DIVISIONS **Commercial fiction/non-fiction** Editorial Director *Patricia Deevy* TITLES *In My Sister's Shoes* Sinéad Moriaty; *This Champagne Mojito is the Last Thing I Own* Ross O'Carroll Kelly; *Secret Diary of a Demented Housewife* Niamh Greene; *Hello Heartbreak* Amy Huberman; **Literary fiction/non-fiction** Editor *Brendan Barrington* TITLES *Connemara: Listening*

*to the Wind* Tim Robinson; *Notes from a Turkish Whorehouse* Philip Ó Ceallaigh; *Come What May* Dónal Óg Cusack; *Not True & Not Unkind* Ed O'Loughlin. IMPRINT **Puffin Ireland** Editor *Paddy O'Doherty*. 20 titles in 2009. Unsolicited mss, synopses and ideas for books welcome; send by post.
Ⓔ Royalties twice-yearly.

## Poolbeg Press Ltd

123 Grange Hill, Baldoyle, Dublin 13
Ⓣ 00 353 1 832 1477     Ⓕ 00 353 1 832 1430
poolbeg@poolbeg.com
www.poolbeg.com
Managing Director *Kieran Devlin*
Publisher *Paula Campbell*
Non-Fiction Commissioning Editor *Brian Langan*

Founded 1976 to publish the Irish short story and has since diversified to include all areas of fiction (literary and popular), children's fiction and non-fiction, and adult non-fiction: history, biography and topics of public interest. AUTHORS discovered and first published by Poolbeg include Maeve Binchy, Marian Keyes, Sheila O'Flanagan, Cathy Kelly, Patricia Scanlan and Melissa Hill. 'Our slogan is Poolbeg.com – The *Irish* for Bestsellers!' IMPRINTS **Poolbeg** (paperback and hardback); **Poolbeg For Children**. About 40 titles a year. Unsolicited mss, synopses and ideas welcome (mss preferred). 'Do not send original material as submissions are not acknowledged or returned.' No contact from Poolbeg within three months means the work is not suitable for their list. No drama.
Ⓔ Royalties twice-yearly.

## Puffin Ireland
▷ Penguin Ireland

## Royal Dublin Society

Science Section, Ballsbridge, Dublin 4
Ⓣ 00 353 1 240 7217     Ⓕ 00 353 1 660 4014
science@rds.ie
www.rds.ie/science
Development Executive, Science & Technology
*Dr Claire Mulhall*

Founded 1731 for the promotion of agriculture, science and the arts, and throughout its history has published books and journals towards this end. Publishes conference proceedings and books on the history of Irish science. TITLES *Agricultural Development for the 21st Century*; *The Right Trees in the Right Places*; *Agriculture & the Environment*; *Water of Life*; *Science, Blueprint for a National Irish Science Centre*; *Science, Technology and Realism*; *Sheets of Many Colours – The Mapping of Ireland's Rock's 1750–1890*; *Science in the Service of the Fishing Industry*; occasional papers in *Irish Science & Technology* series. SERIES *Science and Irish Culture* TITLES

*Science and Irish Culture: Why the History of Science Matters*; *Science and Ireland – Value for Society*; *It's Part of What We Are*.
Ⓔ Royalties not generally paid.

## Royal Irish Academy

19 Dawson Street, Dublin 2
Ⓣ 00 353 1 676 2570     Ⓕ 00 353 1 676 2346
publications@ria.ie
www.ria.ie/publications
Executive Secretary *Patrick Buckley*
Managing Editor *Ruth Hegarty*

Founded in 1785, the Academy has been publishing since 1787. Core publications are journals but more books published in recent years. Publishes academic, Irish interest and Irish language. About 8 titles a year. Limited commissioning scope. Submit mss, synopses and ideas of an academic standard. Currently encourages publications which will popularize some aspect of the sciences and humanities. Address correspondence to the managing editor of publications.

## Tír Eolas

Newtownlynch, Doorus, Kinvara, Co. Galway
Ⓣ 00 353 91 637452     Ⓕ 00 353 91 637452
info@tireolas.com
www.tireolas.com
Publisher/Managing Director *Anne Korff*

Founded 1987. Publishes books and guides on ecology, archaeology, folklore and culture. TITLES *The Book of the Burren*; *The Shores of Connemara*; *Not a Word of a Lie*; *The Book of Aran*; *Kinvara, A Seaport Town on Galway Bay*; *A Burren Journal*; *Alive, Alive-O, The Shellfish and Shellfisheries of Ireland*; *The Burren Wall*; *Corrib Country, Guide & Map*; *Shannon Valley, Guide & Map*. Unsolicited mss, synopses and ideas for books welcome. No specialist scientific and technical, fiction, plays, school textbooks or philosophy.
Ⓔ Royalties annually.

## University College Dublin Press

Newman House, 86 St Stephen's Green, Dublin 2
Ⓣ 00 353 1 477 9812/9813     Ⓕ 00 353 1 477 9821
ucdpress@ucd.ie
www.ucdpress.ie
Executive Editor *Barbara Mennell*
Assistant Editor *Noelle Moran*

Founded 1995. Academic publisher. 17 titles in 2009. Unsolicited mss, synopses and ideas welcome by post. Approach in writing. No non-academic, journals, poetry, novels or books based on academic conferences.
Ⓔ Royalties annually.

## Veritas Publications

7–8 Lower Abbey Street, Dublin 1
Ⓣ 00 353 1 878 8177     Ⓕ 00 353 1 878 6507

publications@veritas.ie
www.veritas.ie
Director *Maura Hyland*

Founded 1969 to supply religious textbooks to
schools and later introduced a wide-ranging
general list. Part of the Catholic Communications
Institute. Publishes books on religious, ethical,
moral, societal and social issues. 40 titles a year.
Unsolicited mss, synopses and ideas for books
welcome.
Ⓔ Royalties annually.

## Wolfhound Press
▷ Merlin Publishing

## The Woodfield Press
17 Jamestown Square, Dublin 8
☎ 00 353 1 454 7991
terrimcdonnell@ireland.com
www.woodfield-press.com
Managing Director *Terri McDonnell*

Founded in 1995 and since then has published
21 books in a mix of biography, women's history,
local history and industrial history. Terri
McDonnell divides her time between a full-time
job in legal publishing and managing Woodfield.
2 titles in 2009. No unsolicited mss; synopses and
ideas for books welcome. Approach by email. No
fiction.
Ⓔ Royalties annually.

# European Publishers

## Austria

### Paul Zsolnay Verlag GmbH

Prinz Eugen Strasse 30, A–1040 Vienna

℡ 00 43 1 505 7661-0  ℻ 00 43 1 505 7661-10
info@zsolnay.at
www.zsolnay.at

Founded 1923. Publishes biography, fiction, general non-fiction, history, poetry.

### Springer-Verlag GmbH

Sachsenplatz 4–6, A–1201 Vienna

℡ 00 43 1 3302415  ℻ 00 43 1 3302426
springer@springer.at
www.springer.at

Founded 1924. Austria's largest scientific publisher. Publishes academic and textbooks, journals and translations: architecture, art, cell biology, chemistry, computer science, law, medicine, neurosurgery, neurology, nursing, psychiatry, psychotherapy.

### Verlag Carl Ueberreuter GmbH

Alser Strasse 24, A–1090 Vienna

℡ 00 43 1 404440
www.ueberreuter.de

Founded 1548. Austria's largest privately-owned publishing house. Publishes children's and young adult titles; adult non-fiction includes health and nutrition, history, culture and politics.

## Belgium

### Brepols Publishers NV

Begijnhof 67, B–2300 Turnhout

℡ 00 32 14 44 80 20  ℻ 00 32 14 42 89 19
info@brepols.net
www.brepols.net

Founded 1796. Independent, academic humanities publisher of books, monographs, journals, CD-ROMs and online. Bibliography, encyclopaedias, reference, history, archaeology, language, literature, art, art history, music, philosophy, architecture.

### Standaard Uitgeverij

Mechelsesteenweg 203, B–2018 Antwerp

℡ 00 32 3 285 72 00  ℻ 00 32 3 285 72 99
info@standaarduitgeverij.be
www.standaarduitgeverij.be

Founded 1919. Publishes education, fiction, poetry, humour, maps, children's and young adult.

### Uitgeverij Lannoo NV

Kasteelstraat 97, B–8700 Tielt

℡ 00 32 51 42 42 11  ℻ 00 32 51 40 11 52
lannoo@lannoo.be
www.lannoo.be

Founded 1909. Publishes general non-fiction, art, architecture and interior design, cookery, biography, children's, gardening, health and nutrition, history, management, self-help, religion, travel, young adult fiction, photography, psychology.

### Universitaire Pers Leuven

Minderbroedersstraat 4, B–3000 Leuven

℡ 00 32 16 32 53 45  ℻ 00 32 16 32 53 52
info@upers.kuleuven.be
www.upers.kuleuven.be

Founded 1971. Academic publisher of anthropolgy, archaeology, architecture, art, arts, economics, geology, history, law, philosophy, political science, psychology, sociology, science and technology.

## Bulgaria

### Lettera

62 Rhodope Street, PO Box 802, BG–4000 Polvdiv

℡ 00 359 32 600 941  ℻ 00 359 32 600 940
office@lettera.bg
www.lettera.bg

Founded 1991. Educational publisher of textbooks and school aids, dictionaries, languages, audio and video; reference and contemporary fiction.

# Czech Republic

## Albatros Publishing House a.s.
Na Pankráci 30, CZ–140 00 Prague 4
℡ 00 420 234 633 268  📠 00 420 234 633 262
albatros@albatros.cz
www.albatros.cz

Publishes children's and educational books.

## Atlantis Ltd
PS 374 Ceská 15, CZ–602 00 Brno
℡ 00 420 542 213 221  📠 00 420 542 213 221
atlantis-brno@volny.cz
www.atlantis-brno.cz

Publishes fiction, language and linguistics, literary studies, poetry, religion, social science.

## Brána a.s.
Jankovcova 18/938, CZ–170 37 Prague 7
℡ 00 420 220 191 313  📠 00 420 220 191 313
info@brana-knihy.cz
www.brana-knihy.cz

Publishes fiction, non-fiction, art, architecture, encyclopaedias, language and linguistics, literary studies, medicine, health, social science, sport, tourism.

## Dokorán Ltd
Zborovská 40, CZ–150 00 Prague 5
℡ 00 420 257 320 803  📠 00 420 257 320 805
dokoran@dokoran.cz
www.dokoran.cz

Founded 2001. Publishes fiction, non-fiction, translations, natural sciences (physics, mathematics, nature), arts, history, social science.

## Mladá Fronta a.s.
Mezi Vodami 1952/9, CZ–143 00 Prague 4
℡ 00 420 225 276 435  📠 00 420 225 276 277
www.mladafronta.cz

Publishes translated fiction, science fiction and fantasy, poetry, non-fiction, popular science, children's books.

# Denmark

## Borgens Forlag A/S
Valbygardsvej 33, DK–2500 Valby
℡ 00 45 3615 3615
post@borgen.dk
www.borgen.dk

Founded 1948. Publishes fiction and general non-fiction: biography, children's and young adult, history, thrillers, humour, professional, psychology.

## Det Schønbergske Forlag
Landemærket 11, 5. sal, DK–1119 Copenhagen K
℡ 00 45 3373 3585  📠 00 45 3314 0115
schoenberg@nytnordiskforlag.dk
www.nytnordiskforlag.dk

Founded 1857. Part of **Nyt Nordisk Forlag Arnold Busck A/S**. Publishes directories, fiction, non-fiction, reference, textbooks.

## Forlaget Apostrof ApS
Postboks 2580, DK–2100 Copenhagen Ø
℡ 00 45 3920 8420  📠 00 45 3920 8453
apostrof@apostrof.dk
www.apostrof.dk

Founded 1980. Publishes psychology and psychiatry.

## Forum
Postbox 2252, DK–1019 Copenhagen K
℡ 00 45 3341 1800  📠 00 45 3341 1801
www.hoest.dk/forum.aspx

Founded 1940. Imprint of Rosinante & Co. Publishes fiction, bibliography, history, humour, mysteries, children's and young adults.

## Gyldendal
Klareboderne 3, DK–1001 Copenhagen K
℡ 00 45 3375 5555
gyldendal@gyldendal.dk
www.gyldendal.dk

Founded 1770. Publishes fiction, children's and young adult, directories, art, biography, education, history, how-to, medicine, music, poetry, philosophy, psychiatry, reference, general and social sciences, sociology, textbooks.

## Høst & Søn Publishers Ltd
Postbox 2252, DK–1019 Copenhagen K
℡ 00 45 3341 1800  📠 00 45 3341 1801
www.hoest.dk

Founded 1836. Imprint of Rosinante & Co. Publishes fiction, biography, children's, crafts, environmental studies, history.

## Lindhardt og Ringhof Forlag A/S
Vognmagergade 11, DK–1148 Copenhagen K
℡ 00 45 3369 5000
info@lindhardtogringhof.dk
www.lrforlag.dk

Founded 1971. Merged with Aschehoug Dansk Forlag A/S in 2008. Publishes fiction and general non-fiction.

## Nyt Nordisk Forlag Arnold Busck A/S
Landemærket 11, 5. sal, DK–1119 Copenhagen K
℡ 00 45 3373 3575  📠 00 45 3314 0115
nnf@nytnordiskforlag.dk
www.nytnordiskforlag.dk

Founded 1896. Publishes fiction, architecture, art, academic, cookery, design, education, gardening, health, natural history, philosophy, reference, religion, leisure, medicine, nursing, psychology, textbooks, travel.

## Tiderne Skifter Forlag A/S

Læderstræde 5, 1. sal, DK–1201 Copenhagen K

T 00 45 3318 6390    F 00 45 3318 6391

tiderneskifter@tiderneskifter.dk

www.tiderneskifter.dk

Founded 1973. Publishes fiction, literature and literary criticism, essays, poetry, drama, translations, history, art, politics, psychoanalysis, sexual politics, ethnicity, photography.

# Finland

## Gummerus Oy

PO Box 749, SF–00101 Helsinki

T 00 358 10 683 6200    F 00 358 9 5843 0200

info@gummerus.fi

www.gummerus.fi

Founded 1872. Publishes fiction and general non-fiction.

## Karisto Oy

PO Box 102, SF–13101 Hämeenlinna

T 00 358 3 63151    F 00 358 3 616 1565

kustannusliike@karisto.fi

www.karisto.fi

Founded 1900. Publishes fiction (including crime, thrillers and fantasy), general non-fiction (including health, history, family and childcare, self improvement, pets, fishing), juvenile and young adult.

## Otava Publishing Co. Ltd

Uudenmaankatu 8–12, SF–00120 Helsinki

T 00 358 9 19961

otava@otava.fi

www.otava.fi

Founded 1890. Publishes fiction, translations, general non-fiction, children's and young adult, multimedia, education, textbooks.

## Tammi Publishers

PO Box 410, SF–00101 Helsinki

T 00 358 9 6937 621    F 00 358 9 6937 6266

tammi@tammi.fi

www.tammi.fi

Founded 1943. Finland's third largest publisher. Publishes fiction, general non-fiction, children's and young adult, education, translations. Part of the Bonnier Group.

## WSOY (Werner Söderström Osakeyhtiö)

PO Box 222, SF–00121 Helsinki

T 00 358 9 616 81    F 00 358 9 6168 3560

firstname.lastname@wsoy.fi

www.wsoy.fi

Founded 1878. Publishes fiction, translations, general non-fiction, education, dictionaries, encyclopaedias, textbooks, children's and young adult.

# France

## Éditions Arthaud

Flammarion Groupe, 87 quai Panhard-et-Levassor, F–75647 Paris Cedex 13

T 00 33 1 40 51 31 00

contact@arthaud.fr

www.arthaud.fr

Founded 1890. Imprint of **Éditions Flammarion**. Publishes illustrated books, photography, travel.

## Éditions Belfond

12 avenue d'Italie, F–75627 Paris Cedex 13

T 00 33 1 44 16 05 00

www.belfond.fr

Founded 1963. Publishes fiction and general non-fiction.

## Editions Bordas

89 blvd Auguste-Blanqui, F–75013 Paris

T 00 33 1 44 39 54 45

www.editions-bordas.com

Founded 1946. Publishes education and general non-fiction: dictionaries, directories, encyclopaedias, reference.

## Éditions Calmann-Lévy

31 rue de Fleurus, F–75006 Paris

T 00 33 1 49 54 36 00    F 00 33 1 45 44 86 32

www.editions-calmann-levy.com

Founded 1836. Part of **Hachette Livre**. Publishes fiction, science fiction, fantasy, biography, history, humour, economics, philosophy, psychology, psychiatry, social sciences, sociology, sport.

## Éditions de la Table Ronde

14 rue Séguier, F–75006 Paris

T 00 33 1 40 46 70 70    F 00 33 1 40 46 71 01

editionslatableronde@editionslatableronde.fr

www.editionslatableronde.fr

Founded 1944. Part of Gallimard Groupe. Publishes fiction and general non-fiction, biography, history, psychology, psychiatry, religion, sport.

## Éditions Denoël

9 rue du Cherche-Midi, F–75278 Paris Cedex 06
ⓉOO 33 1 44 39 73 73    ⒻOO 33 1 44 39 73 90
denoel@denoel.fr
www.denoel.fr

Founded 1932. Part of Gallimard Groupe.
Publishes fiction, science fiction, fantasy,
thrillers, art, art history, biography, directories,
government, history, philosophy, political science,
psychology, psychiatry, reference.

## Éditions du Seuil

27 rue Jacob, F–75261 Paris Cedex 06
ⓉOO 33 1 40 46 50 50
www.seuil.com

Founded 1935. Part of the Groupe Martinière.
Publishes fiction and non-fiction, translations,
literature, literary criticism, essays, poetry,
biography, classics, illustrated, history.

## Éditions Flammarion

87 quai Panhard-et-Levassor, F–75647 Paris
Cedex 13
ⓉOO 33 1 40 51 31 00
www.editions.flammarion.com

Imprint of Groupe Flammarion. Publishes general
fiction, science fiction, thrillers, essays, adventure,
literature, literary criticism; non-fiction: art,
architecture, biography, children's, cookery,
crafts, design, gardening, history, lifestyle, natural
history, plants, popular science, reference,
medicine, nursing, wine.

## Éditions Gallimard

5 rue Sébastien-Bottin, F–75328 Paris Cedex 07
ⓉOO 33 1 49 54 42 00    ⒻOO 33 1 45 44 94 03
www.gallimard.fr

Founded 1911. Publishes fiction, poetry, art,
biography, history, music, philosophy, children's
and young adult.

## Éditions Grasset & Fasquelle

61 rue des Saints-Pères, F–75006 Paris
ⓉOO 33 1 44 39 22 00    ⒻOO 33 1 42 22 64 18
www.edition-grasset.fr

Founded 1907. Part of **Hachette Livre**. Publishes
fiction and general non-fiction, biography, essays,
literature, literary criticism, thrillers, translations,
philosophy and children's books.

## Éditions Larousse

21 rue du Montparnasse, F–75006 Paris
ⓉOO 33 1 44 39 44 00
www.larousse.fr

Founded 1852. Publishes animals, art, bilingual,
children's and young adult, childcare, cinema,
cookery, dictionaries, directories, drama,
encyclopaedias, food, games, gardening, health,
home interest, history, medicine, natural history,
nursing, music, dance, self-help, psychology,
reference, general science, language arts,
linguistics, literature, religion, sport, technology,
textbooks.

## Éditions Magnard

5 allée de la 2e DB, F–75015 Paris
ⓉOO 33 1 42 79 46 80
www.magnard.fr

Founded 1933. Part of Groupe Albin-Michel.
Publishes education, foreign language and
bilingual books, juvenile and young adults,
textbooks.

## Éditions Robert Laffont

24 avenue Marceau, F–75381 Paris Cedex 08
ⓉOO 33 1 53 67 14 00
www.laffont.fr

Founded 1987. Publishes fiction and non-fiction.

## Hachette Livre

43 quai de Grenelle, F–75905 Paris Cedex 15
ⓉOO 33 1 43 92 30 00    ⒻOO 33 1 43 92 30 30
www.hachette.com

Founded 1826. With the acquisition of Time
Warner Book Group, Hachette, already the
owner of Hodder Headline and Orion, is now
the biggest publisher in the British market and
the third largest publisher in the world behind
Pearson and Bertelsmann. Publishes fiction and
general non-fiction, bibliographies, bilingual,
children's, directories, education, encyclopaedias,
general engineering, government, language and
linguistics, reference, general science, textbooks.

## Hatier

8 rue d'Assas, F–75006 Paris
ⓉOO 33 1 49 54 48 99    ⒻOO 331 49 54 47 30
www.editions-hatier.fr

Founded 1880. Educational publishers.

## Les Éditions de Minuit SA

7 rue Bernard-Palissy, F–75006 Paris
ⓉOO 33 1 44 39 39 20    ⒻOO 33 1 44 39 39 23
www.leseditionsdeminuit.fr

Founded 1942. Publishes fiction, essays, literature,
literary criticism, philosophy, social sciences,
sociology.

## Les Éditions Nathan

25 ave Pierre de Coubertin, F–75013 Paris
ⓉOO 33 1 53 55 62 26
www.nathan.fr

Founded 1881. Publishes fiction and non-fiction
for children and young adults. Adult reference,
leisure and natural history; education, technical
and training.

## Librairie Arthème Fayard

13 due du Montparnasse, F-75006 Paris

☎ 00 33 1 45 49 82 00

www.fayard.fr

Founded 1857. Part of **Hachette Livre**. Publishes fiction, biography, directories, maps, atlases, history, dance, music, philosophy, religion, reference, sociology, general science.

## Librairie Vuibert

5 allée de la 2e DB, F–75015 Paris

☎ 00 33 1 42 79 44 00    🖷 00 33 1 42 79 46 80

www.vuibert.com

Founded 1877. Part of Groupe Albin-Michel. Publishes textbooks.

## Presses de la Cité

12 ave d'Italie, F–75627 Paris Cedex 13

☎ 00 33 1 44 16 05 00

pressesdelacite@placedesediteurs.com

www.pressesdelacite.com

Founded 1947. Imprint of **Editions Belfond**. Publishes fiction and general non-fiction, science fiction, fantasy, history, mysteries, romance, biography.

## Presses Universitaires de France (PUF)

6 ave Reille, F–75685 Paris Cedex 14

☎ 00 33 1 58 10 31 00

www.puf.com

Founded 1921. Academic publisher of scientific books, dictionaries, economics, encyclopaedias, geography, history, law, linguistics, literature, philosophy, psychology, psychiatry, political science, reference, sociology, textbooks.

# Germany

## Arena Verlag

Rottendorfer Strasse 16, D–97074 Wurzburg

☎ 00 49 9317 9644 62    🖷 00 49 9317 9644 13

www.arena-verlag.de

Publishes children's and young adult books.

## C. Bertelsmann

Neumarkter Strasse 28, D–81673 Munich

☎ 00 49 89 4136-0

Founded 1835. Imprint of the **Random House Group**. Publishes fiction and general non-fiction, art, autobiography, biography, government and politics.

## Carl Hanser Verlag GmbH & Co. KG

Postfach 86 04 20, D–81631 Munich

☎ 00 49 89 998 30 0    🖷 00 49 89 98 48 09

info@hanser.de

www.hanser.de/verlag

Founded in 1928. Publishes international and German contemporary literature; children's and juveniles; specialist books on engineering and technology, natural science, computers and computer science, economics and management, CD-ROMs, journals, academic, textbooks, translations.

## Carlsen Verlag GmbH

Völkersstrasse 14–20, D–22765 Hamburg

☎ 00 49 40 39804-0    🖷 00 49 40 39804-390

info@carlsen.de

www.carlsen.de

Founded 1953. Publishes children's and comic books.

## Deutscher Taschenbuch Verlag GmbH & Co. KG

Friedrichstrasse 1a, D–80801 Munich

☎ 00 49 89 38167-0    🖷 00 49 89 34 64 28

www.dtv.de

Founded 1961. Third largest paperback publisher in Germany. Publishes fiction and general non-fiction; biography, business, child care and development, children's, dictionaries, directories, economics, education, encyclopaedias, fantasy, government, health and nutrition, history, poetry, psychiatry, psychology, philosophy, politics, reference, religion, literature, literary criticism, essays, thrillers, translations, travel, young adult.

## Ernst Klett Verlag GmbH

Rotebühlstrasse 77, D–70178 Stuttgart

☎ 00 49 711 6672 1333

www.klett.de

Founded 1897. Educational publisher: dictionaries, encyclopaedias, textbooks.

## Heyne Verlag

Bayerstrasse 71–73, D–80335 Munich

☎ 00 49 89 4136-0

www.heyne.de

Founded 1934. Part of **Random House Group**. Publishes fiction, mystery, romance, humour, science fiction, fantasy, biography, cookery, film, history, how-to, occult, psychology, psychiatry.

## Hoffmann und Campe Verlag GmbH

Harvestehuder Weg 42, D–20149 Hamburg

☎ 00 49 40 44188-0    🖷 00 49 40 44188-202

email@hoca.de

www.hoca.de

Founded 1781. Publishes fiction and general non-fiction; art, audio, biography, dance, history, illustrated, journals, music, poetry, philosophy, psychology, psychiatry, general science, social sciences.

## Hüthig GmbH

Postfach 10 28 69, D–69018 Heidelberg
℡ 00 49 6221 489-0     🖷 00 49 6221 489-481
www.huethig.de

Founded 1925. Germany's fourth largest professional publisher: CD-ROMs, multimedia, journals, academic, textbooks, translations.

## Piper Verlag GmbH

Georgenstrasse 4, D–80799 Munich
℡ 00 49 8938 1801-0     🖷 00 49 8933 8704
info@piper.de
www.piper.de

Founded 1904. Publishes international fiction and non-fiction; thrillers, fantasy, history, politics, health, music, travel.

## Rowohlt Verlag GmbH

Hamburgerstr 17, D–21465 Reinbeck
℡ 00 49 40 72 72-0     🖷 00 49 40 72 72-319
info@rowohlt.de
www.rowohlt.de

Founded 1908. Publishes general non-fiction.

## S. Fischer Verlag GmbH

Hedderichstrasse 114, D–60596 Frankfurt am Main
℡ 00 49 69 6062-0     🖷 00 49 69 6062-214
www.fischerverlage.de

Founded 1886. Part of the **Holtzbrinck Group**. Publishes fiction, general non-fiction, biography, business, children's and young adult, history, literature, natural history, politics, psychology, reference.

## Springer-Verlag GmbH

Tiergartenstrasse 17, D–69121 Heidelberg
℡ 00 49 6221 487-0
www.springer.de

Founded 1842. Part of Springer Science+Business Media Deutschland GmbH. Leading specialist publisher of science, technology and medical books, journals, CD-ROMs, databases. Subjects include astronomy, engineering, computer science, economics, geography, law, linguistics, mathematics, philosophy, psychology, architecture, construction and transport – with eighty per cent of publications in English.

## Suhrkamp Verlag

Postfach 101945, D–60019 Frankfurt am Main
℡ 00 49 69 75601-0     🖷 00 49 69 75601-314
www.suhrkamp.de

Founded 1950. Publishes fiction, poetry, biography, cinema, film, theatre, philosophy, psychology, psychiatry, general science.

## Taschen GmbH

Hohenzollernring 53, D–50672 Cologne
℡ 00 49 221 201 80-0     🖷 00 49 221 25 49 19
contact@taschen.com
www.taschen.com

Founded 1980. Publishes photography, art, erotica, architecture and interior design.

## Ullstein Buchverlage GmbH

Friedrichstrasse 126, D–10117 Berlin
℡ 00 49 30 23456-300     🖷 00 49 30 23456-303
info@ullstein-buchverlage.de
www.ullsteinbuchverlage.de

Founded 1903. Publishes general fiction and non-fiction; literature, mystery, biography, business, economics, health, gift books, memoirs, politics.

## Verlag C.H. Beck (OHG)

Wilhelmstrasse 9, D–80801 Munich
℡ 00 49 89 38 189-0     🖷 00 49 89 38 189-398
www.beck.de

Founded 1763. Publishes general non-fiction, anthropology, archaeology, art, CD-ROMs, dance, dictionaries, directories, economics, essays, encyclopaedias, history, illustrated, journals, language arts, law, linguistics, literature, literary criticism, music, philosophy, professional, reference, social sciences, sociology, textbooks, theology.

## Verlagsgruppe Georg von Holtzbrinck GmbH

Gänsheidestrasse 26, D–70184 Stuttgart
℡ 00 49 711 2150-0     🖷 00 49 711 2150-269
info@holtzbrinck.com
www.holtzbrinck.com

Founded 1948. One of the world's largest publishing groups with 12 book publishing houses and 40 imprints. Also publishes the newspapers, *Handelsblatt* and *Die Zeit*.

## Verlagsgruppe Lübbe GmbH & Co. KG

Scheidtbachstrasse 23–31, D–51469 Bergisch Gladbach
℡ 00 49 2202 121-0     🖷 00 49 2202 121-928
www.luebbe.de

Founded 1963. Independent publisher of fiction and general non-fiction, audio, art, biography, history, how-to.

## WEKA Holding GmbH & Co. KG

Postfach 13 31, D–86438 Kissing
℡ 00 49 8233 23-0     🖷 00 49 8233 23-195
info@weka-holding.de
www.weka-holding.de

Founded 1973. Germany's largest professional publisher.

Includes FREE online access to **www.thewritershandbook.com**

# Hungary

### Akadémiai Kiadó

PO Box 245, H–1516 Budapest
℡ 00 36 1 464 8282    🖷 00 36 1 464 8251
info@akkrt.hu
www.akkrt.hu

Hungary's oldest publishing house. Founded 1828. Publishes scholarly reference, economics, education, dictionaries, languages, journals, scientific.

### Jelenkor Kiadó Szolgáltató Kft.

Munkácsy Mihály u. 30/A, H–7621 Pécs
℡ 00 36 72 314 782    🖷 00 36 72 532 047
info@jelenkor.com
www.jelenkor.com

Founded 1993. Independent publishing house. Publishes fiction, philosophy, poetry.

### Kiss József Könyvkiadó, Kereskedelmi és Reklám Kft.

Hamzsabégi út 31, H–1114 Budapest
℡ 00 36 209 9140    🖷 00 36 466 0703
konyvhet@t-online.hu
www.konyv7.hu

Publishes fiction, crime, film, history, politics, theatre, translations.

### Móra Könyvkiadó Zrt.

Váci út 19, H–1134 Budapest
℡ 00 36 1 320 4740    🖷 00 36 1 320 5382
mora@mora.hu
www.mora.hu

Founded 1950. Publishes books for children of all age groups – fiction and school books, popular science, poems, bath books and teenage horror. Recently extended range to include adult fiction and non-fiction.

# Italy

### Adelphi Edizioni S.p.A.

Via S. Giovanni sul Muro 14, I–20121 Milan
℡ 00 39 2 725731    🖷 00 39 2 89010337
info@adelphi.it
www.adelphi.it

Founded 1962. Publishes literature, literary criticism, anthropology, art, autobiography, archaeology, architecture, biography, biology, cinema, economics, history, linguistics, mathematics, medicine, music, philosophy, photography, politics, psychiatry, religion, general science, sociology, theatre, translations.

### Arnoldo Mondadori Editore S.p.A.

Via Mondadori 1, I–20090 Segrate (Milan)
℡ 00 39 2 75421    🖷 00 39 2 75422302
www.mondadori.it

Founded 1907. Italy's largest publisher. Publishes fiction and non-fiction, mystery, romance, art, biography, children's and young adult, directories, history, how-to, journals, medicine, music, poetry, philosophy, psychology, reference, religion, general science, education, textbooks.

### Bompiani

Via Mecenate 91, I–20138 Milan
℡ 00 39 2 5095 2876    🖷 00 39 2 5065 2788
www.rcslibri.it

Founded 1929. Part of RCS Libri S.p.A. publishing group. Publishes fiction and general non-fiction, autobiography, biography, art, illustrated, memoirs, philosophy.

### Bulzoni Editore S.r.L.

Via dei Liburni 14, I–00185 Rome
℡ 00 39 6 4455207    🖷 00 39 6 4450355
bulzoni@bulzoni.it
www.bulzoni.it

Founded 1969. Publisher of college textbooks.

### Cappelli Editore

Via Farini 14, I–40124 Bologna
℡ 00 39 51 239060    🖷 00 39 51 239286
info@cappellieditore.com
www.cappellieditore.com

Founded 1851. Publishes reference, textbooks, juvenile and young adult.

### Casa Editrice Longanesi S.p.A.

Via Gherardini 10, I–20145 Milan
℡ 00 39 2 34597620    🖷 00 39 2 34597212
info@longanesi.it
www.longanesi.it

Founded 1946. Mass market paperback publisher of fiction, adventure, archaeology, art, biography, essays, fantasy, history, journalism, maritime, philosophy, popular science, thrillers.

### Garzanti Libri S.p.A.

Bastioni di Porta Volta 10, I–20121 Milan
℡ 00 39 2 0062 3201
www.garzantilibri.it

Founded 1861. Publishes fiction and non-fiction, literature, literary criticism, encyclopaedias, essays, biography, cookery, crime, dictionaries, history, memoirs, philosophy, poetry, reference, textbooks.

### Giulio Einaudi Editore

Via Umberto Biancamano 2, I–10121 Turin
℡ 00 39 11 565 6341    🖷 00 39 11 565 6353
www.einaudi.it

Founded 1933. Publishes literary fiction, poetry, history, philosophy, economics, art and architecture, social sciences, science, university textbooks, reference and pocket books.

### Giunti Editoriale S.p.A.

Via Bolognese 165, I–50139 Florence
☏ 00 39 55 5062 1   📠 00 39 55 5062 398
www.giunti.it

Founded 1841. Publishes fiction and non-fiction, literature, essays, art, archaeology, alternative medicine, children's, cookery, crafts, dictionaries, education, health and beauty, history, illustrated, journals, language and linguistics, leisure, multimedia, music, psychology, reference, science, textbooks, tourism. Italian publishers of National Geographical Society books.

### Gruppo Ugo Mursia Editore S.p.A.

Via Melchiorre Gioia 45, I–20124 Milan
☏ 00 39 2 6737 8500   📠 00 39 2 6737 8601
info@mursia.com
www.mursia.com

Founded 1941. Publishes fiction, art, biography, directories, education, history, maritime, philosophy, reference, religion, sport, general science, social sciences, textbooks, juvenile and young adult.

### Rizzoli Editore

Via Mecenate 91, I–20138 Milan
☏ 00 39 2 50951
www.rcslibri.it

Founded 1945. Part of Part of RCS Libri S.p.A. publishing group. Publishes fiction and non-fiction, juvenile and young adult.

### SEI – Società Editrice Internazionale

Corso Regina Margherita 176, I–10152 Turin
☏ 00 39 11 5227 1   📠 00 39 11 5211 320
www.seieditrice.com

Founded 1908. Publishes children's, dictionaries, education, encyclopaedias, textbooks.

### Società editrice il Mulino

Strada Maggiore 37, I–40125 Bologna
☏ 00 39 51 256011   📠 00 39 51 256034
info@mulino.it
www.mulino.it

Founded 1954. Publishes economics, government, history, law, language arts, linguistics, literature, literary criticism, philosophy, political science, psychology, psychiatry, religion, social sciences, sociology, textbooks, journals.

### Sonzogno

Via Mecenate 91, I–20138 Milan
☏ 00 39 2 50951   📠 00 39 2 5065361
www.rcslibri.it

Founded 1818. Part of RCS Libri S.p.A. publishing group. Publishes fiction, mysteries, and general non-fiction.

### Sperling e Kupfer Editori S.p.A.

Via Marco D'Aviano 2, I–20131 Milan
☏ 00 39 2 28523 1   📠 00 39 2 28523 277
info@sperling.it
www.sperling.it

Founded 1899. Publishes fiction and general non-fiction; young adult, biography, health, history, science, sport, thrillers.

### Sugarco Edizioni S.r.L.

Via don Gnocchi 4, I–20148 Milan
☏ 00 39 2 407 8370   📠 00 39 2 407 8493
info@sugarcoedizioni.it
www.sugarcoedizioni.it

Founded 1956. Publishes fiction, biography, history, how-to, philosophy.

### Todariana Editrice

Via Gardone 29, I–20139 Milan
☏ 00 39 2 5521 3405   📠 00 39 2 5521 3405
toeura@tin.it
tgiutta@tin.it
www.todariana-eurapress.it

Founded 1967. Publishes fiction, poetry, science fiction, fantasy, literature, literary criticism, language arts, linguistics, psychology, psychiatry, social sciences, sociology, travel.

## Lithuania

### Eugrimas

Kalvariju, LT–8211 Vilnius
☏ 00 370 6125 3500
www.eugrimas.lt

Founded 1999. Academic and professional publisher; political science, economy, management and ethics.

## The Netherlands

### A.W. Bruna Uitgevers BV

Postbus 40203, NL–3504 AA Utrecht
☏ 00 31 30 247 0411   📠 00 31 30 241 0018
info@awbruna.nl
www.awbruna.nl

Founded 1868. Publishes fiction and general non-fiction; mysteries, thrillers, computer science, CD-ROMs, history, philosophy, psychology, psychiatry, general science, social sciences, sociology.

## Pearson Education Benelux

Postbus 75598, NL–1070 AN Amsterdam
℡ 00 31 20 575 5800    📠 00 31 20 664 5334
www.pearsoneducation.nl

Founded 1942. Publishes education, business, computer science, directories, economics, management, reference, textbooks, technology.

## Springer Science+Business Media BV

Van Godewijckstraat 30, NL–3311 GX Dordrecht
℡ 00 31 78 657 6000    📠 00 31 78 657 6555
www.springer.com

Publishes ebooks, CD-ROMs, databases and journals; biomedicine, computer science, economics, engineering, humanities, life sciences, medicine, social sciences, mathematics, physics.

## Uitgeverij Het Spectrum BV

Postbus 97, NL–3990 DB Houten
℡ 00 31 30 265 0650    📠 00 31 30 262 0850
het@spetrum.nl
www.spectrum.nl

Founded 1935. Publishes general non-fiction, children's, dictionaries, encyclopaedias, health, illustrated, language, management, parenting, reference, science, sport, travel.

## Uitgeverij J.M. Meulenhoff BV

Postbus 100, NL–1000 AC Amsterdam
℡ 00 31 20 553 3500    📠 00 31 20 535 3130
info@meulenhoff.nl
www.meulenhoff.nl

Founded 1895. Publishes international co-productions, fiction and general non-fiction. Specializes in Dutch and translated literature.

## Uitgeverij Unieboek BV

Postbus 97, NL–3990 DB Houten
℡ 00 31 30 799 8300    📠 00 31 30 799 8398
info@unieboekspectrum.nl
www.unieboek.nl

Founded 1891. Publishes fiction: mysteries, romance, thrillers; general non-fiction, business, lifestyle, travel, reference, juvenile and young adults.

## Uitgeversmaatschappij J. H. Kok BV

Postbus Box 5018, NL–8260 GA Kampen
℡ 00 31 38 339 2555
www.kok.nl

Founded 1894. Publishes fiction, history, religion, directories, encyclopaedias, reference, psychology, general science, social sciences, sociology, textbooks, juvenile and young adult.

# Norway

## Cappelen Damm AS

Postboks 350 Sentrum, N–0101 Oslo
℡ 00 47 21 616 500
www.cappelen.no

Founded 1829. Owned by the Egmont Group. Publishes fiction (including foreign fiction), general non-fiction, maps and atlases, juvenile and young adult, dictionaries, directories, encyclopaedias, reference, religion, textbooks.

## Gyldendal Norsk Forlag

Postboks 6860, St Olavs Plass, N–0130 Oslo
℡ 00 47 22 034 100    📠 00 47 22 034 105
gnf@gyldendal.no
www.gyldendal.no

Founded 1925. Publishes fiction, science fiction, children's, dictionaries, directories, encyclopaedias, journals, reference, textbooks.

## H. Aschehoug & Co (W. Nygaard)

Postboks 363 Sentrum, N–0102 Oslo
℡ 00 47 22 400 400    📠 00 47 22 206 395
epost@aschehoug.no
www.aschehoug.no

Founded 1872. Publishes fiction and general non-fiction, children's, reference, popular science, hobbies, education and textbooks, translations.

## Tiden Norsk Forlag

Postboks 6704, St Olavs Plass, N–0130 Oslo
℡ 00 47 23 327 660    📠 00 47 23 327 697
tiden@tiden.no
www.tiden.no

Founded 1933. Part of **Gyldendal Norsk Forlag** group. Publishes Norwegian and translated fiction and general non-fiction.

# Poland

## Ameet Sp. zo.o.

ul. Przybyszewskiego 176/178, PL–93–120 Lodz
℡ 00 48 42 676 27 78    📠 00 48 42 676 28 19
ameet@ameet.com.pl
www.ameet.pl

Publishes books for children.

## Nasza Ksiegarnia Publishing House

Sarabandy 24c, PL–02–868 Warsaw
℡ 00 48 22 643 93 89    📠 00 48 22 643 70 28
www.nk.com.pl

Founded 1921. Publishes books for children and young adults .

## Polish Scientific Publishers PWN

Postepu 18, PL–02–676 Warsaw
℡ 00 48 22 69 54 180    🖷 00 48 22 69 54 288
www.pwn.pl

Part of the PWN Group. Leading publisher of scientific, educational, professional and reference material.

## Prószynski i S-ka-S.A.

ul. Garazowa 7, PL–02–651 Warsaw
℡ 00 48 226 077 922    🖷 00 48 228 435 215
www.proszynski.pl

Founded 1990. Publishes foreign fiction and non-fiction; children's books, poetry, dictionaries.

## Publicat S.A.

ul. Chlebowa 24, PL–61–003 Poznan
℡ 00 48 61 652 92 52    🖷 00 48 61 652 92 00
office@publicat.pl
www.publicat.pl

Publishes fiction and non-fiction; biography, children's, crime, education, geography, history, religion.

# Portugal

## Bertrand Livreiros

Rua Professor Jorge da Silva Horta No. 1, P–1500–499 Lisbon
℡ 00 351 213 476 122
info@bertrand.pt
www.bertrand.pt

Founded 1727. Publishes bilingual, dictionaries, encyclopaedias, juvenile and young adult.

## Civilização Editora

Rua Alberto Aires de Gouveia 27, P–4050–023 Porto
℡ 00 351 22 605 0917    🖷 00 351 22 605 0999
info@civilizacao.pt
www.civilizacao.pt

Founded 1921. Publishes fiction, art, biography, cookery, dictionaries, economics, encyclopaedias, history, political science, religion, sociology, social sciences, theatre, children's and young adult.

## Editorial Caminho S.A.

Rua Cidade de Cordova No 2, P–2610–038 Alfragide
℡ 00 351 21 842 9830    🖷 00 351 21 842 9849
info@editorial-caminho.pt
www.editorial-caminho.pt

Founded 1975. Publishes fiction and non-fiction, children's and young adult; art, biography, economics, history, linguistics, music, philosophy, photography, psychology, political science, religion.

## Editorial Verbo SA

Av António Augusto de Aguiar 148, P–1069–019 Lisbon
℡ 00 351 21 380 1100    🖷 00 351 21 386 5397
www.editorialverbo.pt

Founded 1958. Publishes dictionaries, education, encyclopaedias, history, general science, juvenile and young adult.

## Gradiva–Publicações S.A.

Rua Almeida e Sousa, 21 R/C Esq, P–1399–041 Lisbon
℡ 00 351 21 397 4067    🖷 00 351 21 395 3471
geral@gradiva.mail.pt
www.gradiva.pt

Founded 1981. Publishes academic, fiction, literature, literary criticism, children's, juvenile and young adult, directories, illustrated, reference, translations.

## Livros Horizonte Lda

Rua das Chagas 17–1 Dt, P–1200–106 Lisbon
℡ 00 351 21 346 6917
info@livroshorizonte.pt
www.livroshorizonte.pt

Founded 1953. Publishes children's and young adult, education, textbooks.

## Publicações Europa-América Lda

Rua Francisco Lyon de Castro 2, Apartado 8, P–2725–354 Mem Martins
℡ 00 351 21 926 7700    🖷 00 351 21 926 7771
www.europa-america.pt

Founded 1945. Publishes fiction and non-fiction, children's, directories, reference, textbooks.

# Romania

## Curtea Veche Publishing House

11 Ion Mincu St., R–11356 Bucharest
℡ 00 40 21 222 5726    🖷 00 40 21 223 1688
www.curteaveche.ro

Publishes business, history, nutrition and diet, Romanian and foreign literature, personal development, psychology, social and political science, essays and memoirs.

## Grupul Humanitas

1 Piata Presei Libere, R–13701 Bucharest
℡ 00 402 1 408 8350    🖷 00 402 1 408 8351
www.humanitas.ro

Founded 1990. Publishes popular and literary fiction, thrillers, mystery, translations, poetry, memoirs, architecture, cookery, history, reference, music, politics, sport, medicine, anthropology, religion, psychology.

### S.C. Editura Gramar S.R.L.

Str. C-tin Radulescu Motru, R–24220 Bucharest
ⓣ 00 40 21 210 4013    ⓕ 00 40 21 211 2500
www.gramar.ro

Founded 1994. Publishes encyclopaedias,
dictionaries, fiction, poetry, short stories, essays,
social sciences, geography, history, memoirs,
travel and illustrated books for children.

## Spain

### Alianza Editorial SA

Juan Ignacio Luca de Tena 15, E–28027 Madrid
ⓣ 00 34 91 393 8888    ⓕ 00 34 91 320 7480
alianzaeditorial@alianzaeditorial.es
www.alianzaeditorial.es

Founded 1966. Part of **Grupo Anaya**. Publishes
fiction, poetry, art, history, mathematics,
philosophy, government, political science, social
sciences, sociology, general science.

### EDHASA (Editora y Distribuidora Hispano – Americana SA)

Av Diagonal 519–521, 2 piso, E–08029 Barcelona
ⓣ 00 34 93 494 9720    ⓕ 00 34 93 419 4584
info@edhasa.es
www.edhasa.es

Founded 1946. Publishes fiction, literature,
literary criticism, fantasy, essays, biography,
history.

### Ediciónes Hiperión SL

Calle de Salustiano Olózaga 14, E–28001 Madrid
ⓣ 00 34 91 577 6015    ⓕ 00 34 91 577 6016
info@hiperion.com
www.hiperion.com

Founded 1976. Publishes literature, essays, poetry,
bilingual, foreign language, children's, religion,
translations.

### Editorial Espasa-Calpe SA

Paeso Recoletos No 4 2a Planta, E–28001 Madrid
ⓣ 00 34 91 423 0370
www.espasa.com

Founded 1926. Part of Groupo Planeta.
Publishes fiction, children's, biography, essays,
education, literature, CD-ROMs, dictionaries,
encyclopaedias, reference, maps and atlases.

### Editorial Planeta SA

Avda Diagonal 662–664, E–08034 Barcelona
ⓣ 00 34 93 228 5800
www.planeta.es

Founded 1952. Part of Groupo Planeta SA.
Publishes fiction and general non-fiction.

### Editorial Seix Barral SA

Avda Diagonal 662–664, 7°, E–08034 Barcelona
ⓣ 00 349 3 496 7003    ⓕ 00 349 3 496 7004
editorial@seix-barral.es
www.seix-barral.es

Founded 1911. Part of Grupo Planeta SA. Foreign
language publisher of literature.

### Grijalbo

Travessera de Gracia 47–49, E–08021 Barcelona
ⓣ 00 34 93 366 0300    ⓕ 00 34 93 200 2219
www.grijalbo.com

Founded 1962. An imprint of **Random House
Mondadori**. Publishes fiction (historical
and thrillers) and non-fiction (self-help and
inspirational).

### Groupo Anaya

Juan Ignacio Luca de Tena 15, E–28027 Madrid
ⓣ 00 349 1 393 8800    ⓕ 00 349 1 742 6631
www.anaya.es

Founded 1959. Publishes education, reference,
literature, multimedia, children's and young adult.

### Grupo Editorial Luis Vives

Xaudaró 25, E–28034 Madrid
ⓣ 00 34 91 334 4884
dediciones@edelvives.es
www.edelvives.es

Founded 1890. Educational publisher.

### Pearson Educacion SA

C/ Ribera del Loira 28, E–28042 Madrid
ⓣ 00 34 91 382 8300    ⓕ 00 34 91 382 8327
pearson.educacion@pearson.com
www.pearsoneducacion.com

Founded 1942. Educational publishers.

### Random House Mondadori

Travessera de Gracia 47–49, E–08021 Barcelona
ⓣ 00 349 3 366 0300    ⓕ 00 349 3 200 2219
www.randomhousemondadori.es

Publishes fiction and non-fiction; children's and
young adult, art, bilingual dictionaries, biography,
economics, essays, guides, illustrated, literature,
memoirs, poetry, reference, science, self-help,
sociology, technology, thrillers,

### Tusquets Editores

Cesare Cantù 8, E–08023 Barcelona
ⓣ 00 349 3 253 0400
www.tusquets-editores.com

Founded 1968. Publishes fiction, art, autobiography,
biography, cinema, cookery, food, history, essays,
literature, memoirs, philosophy, poetry, social
sciences, theatre.

# Sweden

## Albert Bonniers Förlag

PO Box 3159, SE-103 63 Stockholm
℡ 00 46 8 696 8620
info@abforlag.bonnier.se
www.albertbonniersforlag.com

Founded 1837. Part of Bonnier Books. Publishes fiction and general non-fiction.

## Alfabeta Bokforlag

PO Box 4284, Bondegatan 21, SE–102 66 Stockholm
℡ 00 46 87 143 632    ℻ 00 46 86 432 431
info@alfabeta.se
www.alfabeta.se

Publishes literary fiction, crime, children's, young adult and non-fiction.

## B. Wåhlströms Bokförlag

PO Box 6630, SE-113 84 Stockholm
℡ 00 46 8 728 2300    ℻ 00 46 8 618 9761
info@wahlstroms.se
www.wahlstroms.se

Founded 1911. Part of Forma Publishing Group AB. Publishes fiction and general non-fiction, juvenile and young adult.

## Bokförlaget Forum AB

PO Box 3159, SE-103 63 Stockholm
℡ 00 46 8 696 8440    ℻ 00 46 8 696 8370
info@forum.se
www.forum.se

Founded 1944. Publishes fiction (including foreign) and general non-fiction.

## Bokförlaget Natur och Kultur

PO Box 706, SE-176 27 Järfälla
℡ 00 46 8 453 8500    ℻ 00 46 8 453 8520
www.nok.se

Founded 1922. Publishes fiction and general non-fiction; education and textbooks.

## Bonnier Books

PO Box 3159, SE–103 63 Stockholm
℡ 00 46 8 696 8000
info@bok.bonnier.se
www.bok.bonnier.se

Founded 2002. Part of the Bonnier AB group. Publishes fiction, non-fiction and children's books.

## Bra Böcker AB

PO Box 892, SE-201 80 Malmö
℡ 00 46 40 665 4693
www.bbb.se

Founded 1965. Publishes fiction and non-fiction.

## Brombergs Bokförlag AB

PO Box 12886, SE-112 98 Stockholm
℡ 00 46 8 562 620 80    ℻ 00 46 8 562 620 85
info@brombergs.se
www.brombergs.se

Founded 1975. Publishes literary fiction and general non-fiction.

## Norstedts Förlag

Box 2052, SE-103 12 Stockholm
℡ 00 46 8 769 88 50
info@norstedts.se
www.norstedts.se

Founded 1823. Sweden's oldest publishing house. Publishes fiction and general non-fiction.

## Rabén & Sjögren Bokförlag

PO Box 2052, SE-103 12 Stockholm
℡ 00 46 8 769 8700
info@rabensjogren.se
www.raben.se

Founded 1941. Part of **Norstedts Förlagsgrupp**. Publishes children's and young adult books.

# Switzerland

## Birkhäuser Verlag AG

Viaduktstrasse 42, CH–4051 Basel
℡ 00 41 61 205 07 37    ℻ 00 41 61 205 07 92
www.birkhauser.ch

Part of Springer Science & Business Media. Scientific and professional publisher of books and journals in the fields of architecture and design, mathematics and biosciences.

## Diogenes Verlag AG

Sprecherstrasse 8, CH-8032 Zurich
℡ 00 41 1 254 85 11    ℻ 00 41 1 252 84 07
info@diogenes.ch
www.diogenes.ch

Founded 1952. Publishes fiction, essays, literature, literary criticism, art, children's.

## Neptun-Verlag

Erlenstrasse 2, CH-8280 Kreuzlingen
℡ 00 41 71 677 96 55    ℻ 00 41 72 677 96 50
www.neptunart.ch

Founded 1946. Publishes non-fiction and children's books.

## Orell Füssli Verlag AG

Dietzingerstrasse 3, CH-8036 Zurich
℡ 00 41 1 466 77 11    ℻ 00 41 1 466 74 12
www.ofv.ch

Founded 1519. Publishes accountancy, architecture, art, art history, business, leisure, photography, politics, psychology.

## Payot Libraire

Case Postale 6730, CH-1002 Lausanne

Ⓣ 00 41 21 341 32 52   Ⓕ 00 41 21 341 32 17
www.payot-libraire.ch

Founded 1875. Publishes fiction and general non-fiction; arts and entertainment, audio, business, children's, education, language, lifestyle, professional, reference, travel.

## Sauerländer Verlage AG

Ausserfeldstrasse 9, CH-5036 Oberentfelden

Ⓣ 00 41 62 836 86 86   Ⓕ 00 41 62 836 86 95
verlag@sauerlaender.ch
www.sauerlaender.ch

Founded 1807. Publishes education and general non-fiction.

# US Publishers

## Key advice

Anyone corresponding with the big American publishers should know that they are wary of any package or letter that looks in any way suspicious. This, of course, begs the question what is or what is not suspicious which is as difficult to answer as what makes or does not make a good novel. The best policy is an initial enquiry by email.

**International Reply Coupons (IRCs)**

For return postage, send IRCs, available from post offices (£1.10 each). For current postal rates to the UK, go to the United States Postal Service website at www.usps.com and click on 'Calculate Postage'.

## ABC–CLIO

PO Box 1911, Santa Barbara CA 93116–1911
℡ 001 805 968 1911    🖷 001 805 685 9685
www.abc-clio.com
President/CEO *Ronald J. Boehm*

Founded 1953. Publishes academic and reference, focusing on history and social studies; books and multimedia. UK SUBSIDIARY ABC-Clio Ltd, Oxford. No unsolicited mss; synopses and ideas welcome.

## Abingdon Press

201 Eighth Avenue South, PO Box 801, Nashville TN 37202–0801
℡ 001 615 749 6290
www.abingdonpress.com

Founded 1789. Publishes non-fiction: religious (Protestant), reference, professional and academic texts. IMPRINTS **Dimensions for Living**; **Kingswod Books**. Over 100 titles a year.

## Academy Chicago Publishers

363 West Erie Street, 7E, Chicago IL 60654
℡ 001 312 751 7300    🖷 001 312 751 7306
info@academychicago.com
www.academychicago.com
President & Senior Editor *Anita Miller*

Founded 1975. Publishes mainstream fiction; non-fiction: art, history, mysteries. No romance, children's, young adult, religious, sexist or avant-garde. See website for submission guidelines.

## Ace Books
▷ Penguin Group (USA) Inc.

## Aladdin Paperbacks
▷ Simon & Schuster Children's Publishing

## Alfred A. Knopf
▷ The Knopf Doubleday Publishing Group

## Alpha Books
▷ Penguin Group (USA) Inc.

## Amistad
▷ HarperCollins Publishers, Inc.

## Amphoto Books
▷ Crown Publishing Group

## Anchor Books
▷ The Knopf Doubleday Publishing Group

## Anvil
▷ Krieger Publishing Co.

## Atheneum Books for Young Readers
▷ Simon & Schuster Children's Publishing

## Atlantic Monthly Press
▷ Grove/Atlantic Inc.

## Atria Books
▷ Simon & Schuster Adult Publishing Group

## Avalon Travel
▷ Perseus Books Group

## Avery
▷ Penguin Group (USA) Inc.

## Avon/Avon A/Avon Inspire/Avon Red
▷ HarperCollins Publishers, Inc.

## Back Bay Books
▷ Hachette Book Group

## Ballantine Books
▷ The Random House Publishing Group

## Bantam
▷ The Random House Publishing Group

## Barron's Educational Series, Inc.

250 Wireless Boulevard, Hauppauge NY 11788
℡ 001 631 434 3311    🖷 001 631 434 3723
barrons@barronseduc.com
www.barronseduc.com
Chairman/CEO *Manuel H. Barron*
President *Ellen Sibley*

Founded 1941. Publishes test preparation material and foreign language banks, children's fiction and non-fiction, adult reference, cookbooks, pets, hobbies, sport, photography, health, business, law, computers, art and painting. No adult fiction. Over 300 titles a year. Email queries to *Wayne Barr* (waynebarr@barronseduc.com).
🄴 Royalties twice-yearly.

## Basic Books/Basic Civitas
▷ Perseus Books Group

## Beacon Press

25 Beacon Street, Boston MA 02108
℡ 001 617 742 2110    🖷 001 617 723 3097
www.beacon.org
Director *Helene Atwan*

Founded 1854. Publishes general non-fiction. About 60 titles a year. See submission guidelines on the website. Currently, not accepting new poetry or fiction.

## Berkley
▷ Penguin Group (USA) Inc.

## Bison Books
▷ University of Nebraska Press

## Black Cat
▷ Grove/Atlantic Inc.

## Blackbirch Press
▷ Gale

## Boyds Mills Press

815 Church Street, Honesdale PA 18431
℡ 001 570 253 1164
contact@boydsmillspress.com
www.boydsmillspress.com
Publisher *Kent L. Brown Jr.*

A subsidiary of Highlights for Children, Inc. Founded 1990 as a publisher of children's picture and activity books. Publishes children's and young adult's fiction, non-fiction and poetry under five IMPRINTS: **Boyds Mills Press**; **Calkins Creek**; **Front Street**; **Lemniscaat**; **Wordsong**. About 80 titles a year. Unsolicited material welcome; submission guidelines available on the website.

## Brassey's, Inc.
▷ Potomac Books, Inc.

## Broadway Books
▷ Crown Publishing Group

## Bulfinch Press
▷ Hachette Book Group

## Business Plus
▷ Hachette Book Group

## Cadogan Guides, USA
▷ Interlink Publishing Group, Inc.

## Caedmon
▷ HarperCollins Publishers, Inc.

## Calkins Creek
▷ Boyds Mills Press

## Carolrhoda Books

241 First Avenue N, Minneapolis MN 55401
℡ 001 612 332 3344    🖷 001 612 332 7615
www.lernerbooks.com

Founded 1969. Children's book publisher. An imprint of the **Lerner Publishing Group, Inc**, Carolrhoda publishes hardcover originals. Averages 8–10 picture books each year for ages 3–8; 6 fiction titles for ages 7–18; 2–3 non-fiction titles for various ages. No longer accepting unsolicited submissions.

## Center Street
▷ Hachette Book Group

## Charles Scribner's Sons
▷ Gale

## Charlesbridge Publishing

85 Main Street, Watertown MA 02472
℡ 001 617 926 0329    🖷 001 617 926 5775
books@charlesbridge.com
www.charlesbridge.com
President/Publisher *Brent Farmer*

Publishes both picture books and transitional 'bridge books' (books ranging from early readers to middle-grade chapter books); fiction; and non-fiction with a focus on nature, science, social studies and multicultural topic.

## Christian Large Print
▷ Gale

## Chronicle Books LLC

680 Second Street, San Francisco CA 94107
℡ 001 415 537 4200    🖷 001 415 537 4460
frontdesk@chroniclebooks.com
www.chroniclebooks.com
President/Publisher *Jack Jensen*

Founded 1967. Publishes illustrated and non-illustrated adult trade and children's books as well as stationery and gift items. About 175 titles a year. Query or submit outline/synopsis and sample chapters and artwork. No adult fiction. See website for guidelines.

## Clarion Books
▷ Houghton Mifflin Harcourt

## Clarkson Potter
▷ Crown Publishing Group

## Clockroot Books
▷ Interlink Publishing Group, Inc.

## Collins/Collins Design/Collins Living
▷ HarperCollins Publishers, Inc.

## Columbia University Press
61 West 62nd Street, New York NY 10023
☎ 001 212 459 0600    ℻ 001 212 459 3678
www.cup.columbia.edu
President & Director *James D. Jordan*
Associate Director *Jennifer Crewe*

Founded 1893. Publishes scholarly and general interest non-fiction, reference, translations of Asian literature. No fiction or poetry. Welcomes unsolicited material if the subject fits their programme (see website). Approach by regular mail only.

## Crocodile Books, USA
▷ Interlink Publishing Group, Inc.

## Crown Publishing Group
1745 Broadway, New York NY 10019
☎ 001 212 782 9000
crownbiz@randomhouse.com
www.randomhouse.com/crown
President & Publisher *Jenny Frost*

Founded 1933. Division of the **Random House Publishing Group**. Publishes popular trade fiction and non-fiction.

IMPRINTS **Amphoto Books; Broadway Business; Broadway Books; Clarkson Potter; Crown Business; Crown Forum, Harmony Books; Potter Style; Potter Craft; Shaye Areheart Books; Ten Speed Press; Three Rivers Press; Tricycle Press; Watson-Guptill**. Mss submissions by agents only.

## DaCapo
▷ Perseus Books Group

## DAW Books, Inc.
375 Hudson Street, 3rd Floor, New York
NY 10014–3658
☎ 001 212 366 2096    ℻ 001 212 366 2090
daw@penguingroup.com
www.dawbooks.com
Publishers *Elizabeth R. Wollheim,*
*Sheila E. Gilbert*
Associate Editor *Peter Stampfel*

Founded 1971 by Donald and Elsie Wollheim as the first mass-market publisher devoted to science fiction and fantasy. An imprint of **Penguin Group (USA) Inc.** Publishes science

fiction/fantasy. No short stories, anthology ideas or non-fiction. About 42 titles a year. Unsolicited mss, synopses and ideas for books welcome.
💷 Royalties twice-yearly.

## Del Rey
▷ The Random House Publishing Group

## Delacorte Press
▷ The Random House Publishing Group

## Dell
▷ The Random House Publishing Group

## Dial Books for Young Readers
345 Hudson Street, New York NY 10014–3657
☎ 001 212 366 2000
Vice President & Publisher *Laura Hornik*

Founded 1961. A division of Penguin Books for Young Readers. Publishes children's books, including picture books, beginning readers, fiction and non-fiction for middle grade and young adults. 50 titles a year. Accepts unsolicited picture book mss and up to ten pages for longer works with query letter. 'Do not send self-addressed envelope with submissions. Dial will respond within four months if interested in a manuscript.'
💷 Royalties twice-yearly.

## Dial Press
▷ The Random House Publishing Group

## Dimensions for Living
▷ Abingdon Press

## Doubleday
▷ The Knopf Doubleday Publishing Group

## Dutton/Dutton's Children's Books
▷ Penguin Group (USA) Inc.

## East Gate Books
▷ M.E. Sharpe, Inc.

## Ecco
▷ HarperCollins Publishers, Inc.

## Eerdmans Publishing Company
2140 Oak Industrial Drive, NE, Grand Rapids
MI 49505
☎ 001 616 459 4591    ℻ 001 616 459 6540
info@eerdmans.com
www.eerdmans.com
President/Publisher *William B. Eerdmans Jr*

Founded in 1911 as a theological and reference publisher. Gradually began publishing in other genres with authors like C.S. Lewis, Dorothy Sayers and Malcolm Muggeridge on its lists. Publishes theology, biblical studies, ethical and social concern, social criticism and children's,

religious history, religion and literature. About
120 titles a year.
ⓔ Royalties twice-yearly.

## Eos
▷ HarperCollins Publishers, Inc.

## Exploring Community History
▷ Krieger Publishing Co.

## Faber & Faber, Inc.
▷ Farrar, Straus & Giroux, Inc.

## FaithWords
▷ Hachette Book Group

## Farrar Straus & Giroux Books for Young Readers

175 Fifth Avenue, New York NY 10010
☎ 001 646 307 5592
www.fsgkidsbooks.com

Founded 1954. Part of Macmillan Children's
Publishing Group in the US. Publishes books for
toddlers through to young adults in hardcover
and paperback. DIVISION **FSG Books for Young
Readers** *Margaret Ferguson* TITLES *The Pout-
Pout Fish*; *How Oliver Olson Changed the World*.
IMPRINT **Frances Foster Books** *Frances Foster*
TITLES *Crossing Stones*; *Last Night*. 80 titles in
2009. Will consider unsolicited submissions; mail
picture books with covering letter; for longer
works, query with the first three chapters before
sending full mss. No submissions via email, fax,
disk or CD.

## Farrar, Straus & Giroux, Inc.

18 West 18th Street, New York NY 10011
☎ 001 212 741 6900
www.fsgbooks.com

Founded 1946. Part of Macmillan in the US,
owned by German publishing group Holtzbrinck.
Publishes general and literary fiction, non-fiction,
poetry and children's books. IMPRINTS **North
Point Press**; **Hill and Wang**; **Faber & Faber, Inc.**;
**FSG Books for Young Readers** (see entry above).
No unsolicited material.

## Firebird
▷ Penguin Group (USA) Inc.

## Fireside
▷ Simon & Schuster Adult Publishing Group

## First Avenue Editions
▷ Lerner Publishing Group

## Five Star
▷ Gale

## 5-Spot
▷ Hachette Book Group

## Flux
▷ Llewellyn Worldwide Ltd

## Forever
▷ Hachette Book Group

## Frances Foster Books
▷ Farrar Straus & Giroux Books for Young Readers

## Frederick Warne
▷ Penguin Group (USA) Inc.

## Free Press
▷ Simon & Schuster Adult Publishing Group

## Front Street
▷ Boyds Mills Press

## FSG Books for Young Readers
▷ Farrar Straus & Giroux Books for Young Readers

## G.K. Hall
▷ Gale

## G.P. Putnam's Sons
▷ Penguin Group (USA) Inc.

## Gale

27500 Drake Road, Farmington Hills MI 48331
☎ 001 248 699 4253      🖷 001 248 699 8070
www.gale.com
President *Patrick C. Sommers*

Gale, part of Cengage Learning, is a world leader
in information and educational publishing.
Addresses all types of information needs,
from homework help to health questions and
business profiles, in a variety of formats: books,
electronic resources and microfilm. BRANDS/
IMPRINTS **Blackbirch Press**; **Five Star**; **Christian
Large Print**; **Greenhaven Press**; **G.K. Hall**;
**KidHaven Press**; **Large Print Press**; **Lucent
Books**; **Macmillan Reference USA**; **Primary
Source Media**; **Schirmer Reference**; **Scholarly
Resources, Inc.**; **Charles Scribner's Sons**;
**Sleeping Bear Press**; **St James Press**; **Thorndike
Press**; **Twayne Publishers**; **UXL**; **Wheeler
Publishing**.

## Gallery Books
▷ Simon & Schuster Adult Publishing Group

## Gaslight Publications

PO Box 1344, Studio City CA 91614–0344
☎ 001 818 784 8918
www.ppeps.com
Chairman *Dr William-Alan Landes*

Founded in 1952 'to keep the Holmes world alive'.
Publishes books (fiction and non-fiction) on
Sherlock Holmes only. DIVISION **Joseph W. Witt**
*Don Agey* TITLES *Shlock Holmes*; *Puzzling with
Holmes*. 10 titles in 2009. Welcomes unsolicited

mss (with IRCs). Send query letter in the first instance.

Ⓔ Royalties annually.

## The Globe Pequot Press

246 Goose Lane, PO Box 480, Guilford CT 06437
Ⓣ 001 203 458 4500    Ⓕ 001 203 458 4601
info@gobepequot.com
www.globepequot.com

Founded 1947. IMPRINTS **Footprint** Outdoor recreation guides; **Lyons** Issue-based books in the areas of current affairs, politics, true crime and history; **GPP Travel**; **GPP Life** Books for women on topics such as self-help, relationships and memoirs; **Skirt!** 'Books with attitude by dynamic women.' Unsolicited mss, synopses, and ideas welcome; growth categories include books for women and lifestyle. See website for submission guidelines.

## GPP Life/GPP Travel
▷ The Globe Pequot Press

## Grand Central Publishing
▷ Hachette Book Group

## Graphic Universe
▷ Lerner Publishing Group

## Green Willow Books
▷ HarperCollins Publishers, Inc.

## Greenhaven Press
▷ Gale

## Griffin
▷ St Martin's Press LLC

## Grosset & Dunlap
▷ Penguin Group (USA) Inc.

## Grove/Atlantic Inc.

841 Broadway, 4th Floor, New York NY 10003
Ⓣ 001 212 614 7850    Ⓕ 001 212 614 7886
editorial@groveatlantic.com
www.groveatlantic.com
President/Publisher *Morgan Entrekin*
Managing Editor *Michael Hornburg*

Founded 1952. Publishes general fiction and non-fiction. IMPRINTS **Atlantic Monthly Press**; **Black Cat**; **Grove Press**. Mss submissions by agents only.

## Hachette Book Group

237 Park Avenue, New York NY 10017
Ⓣ 001 212 364 1100
www.hachettebookgroupusa.com
CEO & Chairman *David Young*

Hachette Book Group (HBG), a wholly-owned subsidiary of Hachette Livre, is the fifth largest trade publisher in the United States. HBG's list includes adult fiction and non-fiction, illustrated, religious, children's and audio books.

DIVISIONS **Grand Central Publishing**; **FaithWords**; **Center Street**; **Little, Brown and Company**; **Little, Brown Books for Young Readers**; **Orbit**; **Hachette Digital**.

IMPRINTS **Back Bay Books**; **Bulfinch Press**; **Business Plus**; **5-Spot**; **Forever**; **Hachette Audio**; **LB Kids**; **Springboard Press**; **Poppy**; **Reagan Arthur Books**; **Twelve**; **Vision**; **Wellness Central**; **Yen Press**. About 600 adult books, 150 young adult and children's books and 65 audio books a year. No unsolicited mss.

## Harcourt Education
▷ Houghton Mifflin Harcourt

## Harmony Books
▷ Crown Publishing Group

## HarperCollins Publishers, Inc.

10 East 53rd Street, New York NY 10022
Ⓣ 001 212 207 7000
www.harpercollins.com
President/Publisher US General Books *Michael Morrison*

Founded 1817. Subsidiary of News Corporation. Publishes general and literary fiction, general non-fiction, business, children's, reference and religious books. About 1,700 titles a year.

GENERAL BOOKS GROUP IMPRINTS

**Amistad**; **Avon**; **Avon A**; **Avon Inspire**; **Avon Red**; **Caedmon**; **Collins**; **Collins Design**; **Collins Living**; **Ecco**; **Eos**; **Harper Business**; **Harper Paperbacks**; **Harper Perennial**; **HarperAudio**; **HarperCollins**; **HarperCollins e-Books**; **HarperLuxe**; **Harper One**; **ItBooks**; **Rayo**; **William Morrow**.

CHILDREN'S IMPRINTS

**Amistad**; **Eos**; **Greenwillow Books**; **HarperCollins Children's Audio**; **HarperCollins Children's Books**; **HarperFestival**; **HarperEntertainment**; **HarperTeen**; **HarperTrophy**; **Joanna Cotler Books**; **Julie Andrews Collection**; **Katherine Tegen Books**; **Laura Geringer Books**; **Rayo**. No unsolicited material with the exception of the Avon romance imprint.

## Harry N. Abrams, Inc.

115 West 18th Street, New York NY 10011
Ⓣ 001 212 206 7715    Ⓕ 001 212 229 0312
www.hnabooks.com
VP/Editor-in-Chief *Eric Himmel*

Founded 1950. Publishes illustrated books: art, natural history, photography, popular culture. No fiction. About 100 titles a year. No unsolicited material.

## Harvard University Press

79 Garden Street, Cambridge MA 02138
☎ 001 401 531 2800    ℻ 001 401 531 2801
contact_HUP@harvard.edu
www.hup.harvard.edu
Editor-in-Chief *Michael Fisher*

Founded 1913. Publishes scholarly non-fiction only; humanities, social sciences, sciences.

## Henry Holt & Company, Inc.

175 Fifth Avenue, New York NY 10010
☎ 001 646 307 5095    ℻ 001 646 307 5285
www.henryholt.com
President/Publisher *Stephen Rubin*
Editor-in-Chief, Adult Trade *Marjorie Braman*

Founded in 1866, Henry Holt is one of the oldest publishers in the United States. Part of Macmillan in the US, owned by German publishing group Holtzbrinck. Publishes fiction, by both American and international authors; biographies, science, history, politics, ecology, and psychology.

DIVISIONS/IMPRINTS **Adult Trade; Metropolitan Books; Times Books; Henry Holt Paperbacks**. About 100 titles a year. No unsolicited mss.

## Hill and Wang

▷ Farrar, Straus & Giroux, Inc.

## Hill Street Press LLC

191 E. Broad Street, Suite 216, Athens GA 30601–2848
☎ 001 706 613 7200    ℻ 001 800 621 1654
info@hillstreetpress.com
hillstreetpress.com
President *Tom Payton*

Founded 1998. Publishes limited non-fiction only, dealing with the American South as well as crossword puzzle books, college trivia and graphic novels. Some children's books. About 8 titles a year. Unsolicited submissions accepted for graphic novels (including non-fiction) only. No other unsolicited submissions. All submissions must be made electronically. See website for details.

## Hippocrene Books, Inc.

171 Madison Avenue, New York NY 10016
info@hippocrenebooks.com
www.hippocrenebooks.com
President/Editorial Director *George Blagowidow*

Founded 1970. Publishes general non-fiction and reference books. Particularly strong on foreign language dictionaries, language studies and international cookbooks, illustrated history. No fiction.

## Holiday House, Inc.

425 Madison Avenue, New York NY 10017
☎ 001 212 688 0085    ℻ 001 212 421 6134
www.holidayhouse.com
Vice President/Editor-in-Chief *Mary Cash*

Publishes children's general fiction and non-fiction (pre-school to secondary). About 50 titles a year. Send query letters only. IRCs for reply must be included. Submission guidelines available on the website.

## Houghton Mifflin Harcourt

222 Berkeley Street, Boston MA 02116
☎ 001 617 351 5000
www.hmco.com
Contact *Submissions Editor*

Founded 1832. Houghton Mifflin Company acquired Harcourt Education in 2007. Publishes fiction and non-fiction, children's books and reference. Also school textbooks and educational resources; children's fiction and non-fiction.
DIVISIONS/IMPRINTS **Houghton Mifflin Harcourt Trade & Reference; Houghton Mifflin Harcourt Children's Book Group; Clarion Books; Harcourt Children's Books; Mariner Books**. About 100 titles a year.

## Howard Books

▷ Simon & Schuster Adult Publishing Group

## HPBooks

▷ Penguin Group (USA) Inc.

## Hudson Street Press

▷ Penguin Group (USA) Inc.

## Humanity Books

▷ Prometheus Books

## Indiana University Press

601 North Morton Street, Bloomington IN 47404–3797
☎ 001 812 855 8817    ℻ 001 812 855 8507
iupress@indiana.edu
www.indiana.edu/~iupress
Director *Janet Rabinowitch*

Founded 1950. Publishes scholarly non-fiction in the following subject areas: African studies, anthropology, Asian studies, African-American studies, bioethics, environment and ecology, film, folklore, history, Jewish studies, Middle East studies, military history, music, paleontology, philanthropy, philosophy, politics, religion, Russian and East European studies, women's studies. About 140 books and 30 journals a year. Query in writing in first instance.

## Interlink Publishing Group, Inc.

46 Crosby Street, Northampton MA 01060–1804
☎ 001 413 582 7054
info@interlinkbooks.com
www.interlinkbooks.com
Owner/Publisher *Michel Moushabeck*
Editor *Pam Thompson*

Founded 1987. Independent publisher of international fiction, travel, politics, cookbooks. Specializes in Middle East titles and ethnicity. IMPRINTS **Cadogan Guides, USA** Travel guides; **Clockroot Books** International fiction; **Crocodile Books, USA** Children's books; **Olive Branch Press** Political books. About 90 titles a year. See submission guidelines on the website before making an approach.
£ Royalties annually.

## ItBooks
▷ HarperCollins Publishers, Inc.

## Joanna Cotler Books
▷ HarperCollins Publishers, Inc.

## John Wiley & Sons, Inc.
111 River Street, Hoboken NJ 07030–5774
☎ 001 201 748 6000    ℻ 001 201 748 6088
info@wiley.com
www.wiley.com
President/CEO *William J. Pesce*

Founded 1807. Wiley's core publishing programme includes scientific, technical, medical and scholarly journals, encyclopaedias, books and online products and services; professional and consumer books, and subscription and information services in all media; educational materials for undergraduate, graduate students and life-long learners. Submission guidelines available on the website under 'Resources for Authors'.

## Joseph W. Witt
▷ Gaslight Publications

## Jove
▷ Penguin Group (USA) Inc.

## Julie Andrews Collection
▷ HarperCollins Publishers, Inc.

## Kar-Ben Publishing
▷ Lerner Publishing Group

## Katherine Tegen Books
▷ HarperCollins Publishers, Inc.

## The Kent State University Press
307 Lowry Hall, PO Box 5190, Kent OH 44242
☎ 001 330 672 8099    ℻ 001 330 672 3104
jharri8@kent.edu
www.kentstateuniversitypress.com
Director *Will Underwood*
Acquiring Editor *Joyce Harrison*

Founded 1965. Publishes scholarly works in history, literary studies and general non-fiction with an emphasis on American history. 30–35 titles a year. Queries welcome; no mss.

## KidHaven Press
▷ Gale

## Kingswood Books
▷ Abingdon Press

## The Knopf Doubleday Publishing Group
1745 Broadway, New York NY 10019
☎ 001 212 572 2662 (foreign rights)
www.randomhouse.com/knopf
President *Tony Chirico*
Senior VP Publishing *Suzanne Herz*

A division of **The Random House Publishing Group**. Publishes hardcover fiction and non-fiction. IMPRINTS **Anchor Books**; **Alfred A. Knopf**; **Doubleday**; **Nan A. Talese Books**; **Pantheon Books**; **Pantheon Graphic Novels**; **Schocken Books**.

## Krieger Publishing Co.
1725 Krieger Drive, Malabar FL 32950
☎ 001 321 724 9542    ℻ 001 321 951 3671
info@krieger-publishing.com
www.krieger-publishing.com
CEO *Robert E. Krieger*
President *Donald E. Krieger*
Vice-President *Maxine D. Krieger*

Founded 1970. Publishes science, education, ecology, humanities, history, mathematics, chemistry, space science, technology and engineering.

IMPRINTS/SERIES **Anvil**; **Exploring Community History**; **Open Forum**; **Orbit**; **Professional Practices**; **Public History**. Unsolicited mss welcome. Not interested in synopses/ideas or trade type titles.

## Large Print Press
▷ Gale

## Latino Voices
▷ Northwestern University Press

## Laura Geringer Books
▷ HarperCollins Publishers, Inc.

## LB Kids
▷ Hachette Book Group

## Lemniscaat
▷ Boyds Mills Press

## Lerner Publishing Group
241 First Avenue N., Minneapolis MN 55401
☎ 001 612 332 3344    ℻ 001 612 332 7615
info@lernerbooks.com
www.lernerbooks.com

Founded 1959. Publishes children's non-fiction, fiction and curriculum material for all grade levels.

DIVISIONS/IMPRINTS **Lerner Publications**; **Carolrhoda Books** (see entry); **LernerClassroom**; **First Avenue Editions**; **Millbrook Press**; **Twenty-First Century Books**; **Kar-Ben Publishing**;

**Lerner U.K.; Graphic Universe**. No puzzle, song or alphabet books, text books, workbooks, or plays. No longer accepting unsolicited submissions.

## Little Simon
> Simon & Schuster Children's Publishing

## Little, Brown and Company/Little, Brown Books for Young Readers
> Hachette Book Group

## Llewellyn Worldwide Ltd
2143 Wooddale Drive, St. Paul MN 55125–2989
☏ 001 651 291 1970    ℻ 001 651 291 1908
www.llewellyn.com
Publisher *Bill Krause*

Division of Llewellyn Worldwide Ltd. Founded 1901. Publishes self-help and how-to: astrology, alternative health, tantra, Fortean studies, tarot, yoga, Santeria, dream studies, metaphysics, magic, witchcraft, herbalism, shamanism, organic gardening, women's spirituality, graphology, palmistry, parapsychology. Also fiction with an authentic magical or metaphysical theme. Publishes YA popular fiction under the new **Flux** imprint, and popular trade mystery fiction under the **Midnight Ink Books** imprint. About 140 titles a year. Unsolicited mss welcome; proposals preferred. IRCs essential in all cases. Submission guidelines available on the website.
£ Royalties twice-yearly.

## Louisiana State University Press
3990 West Lakeshore Drive, Baton Rouge LA 70808
☏ 001 225 578 6294    ℻ 001 225 578 6461
lsupress@lsu.edu
www.lsu.edu/lsupress
Director *MaryKatherine Callaway*

Publishes fiction and non-fiction: Southern and American history, Southern and American literary criticism, regional trade books, biography, poetry, Atlantic World studies, media studies, environmental studies, geography, political science and music (jazz). About 85 titles a year. See website for submission guidelines.

## Lucent Books
> Gale

## Lyons
> The Globe Pequot Press

## M.E. Sharpe, Inc.
80 Business Park Drive, Armonk NY 10504
☏ 001 914 273 1800    ℻ 001 914 273 2106
info@mesharpe.com
www.mesharpe.com
President/CEO *M.E. Sharpe*
Snr VP/COO *Vincent Fuentes*

Founded 1958. Publishes books and journals in the social sciences and humanities, both original works and translations in Asian and East European studies. IMPRINTS **East Gate Books**; **Sharpe Reference** *Patricia Kolb*. No unsolicited material.
£ Royalties annually.

## McFarland & Company, Inc., Publishers
PO Box 611, Jefferson NC 28640
☏ 001 336 246 4460    ℻ 001 336 246 5018
info@mcfarlandpub.com
www.mcfarlandpub.com
Executive Vice President *Rhonda Herman*
Editorial Director *Steve Wilson*
Acquisitions Editor *Gary Mitchem*

Founded 1979. A library reference and upper-end speciality market press publishing scholarly books in many fields: international studies, performing arts, popular culture, sports, automotive history, women's studies, music and fine arts, chess, history and librarianship. Specializes in general reference. Especially strong in cinema studies. Some non-fiction graphic novels. No fiction, poetry, children's, New Age or inspirational/devotional works. About 350 titles a year. No unsolicited mss; send query letter first. Synopses and ideas welcome; submissions by mail preferred.
£ Royalties annually.

## The McGraw-Hill Companies, Inc.
1221 Avenue of the Americas, New York NY 10020–1095
☏ 001 212 904 2000
www.mcgraw-hill.com
Chairman/President/CEO *Harold W. McGraw*

Founded 1888. Parent of **McGraw-Hill Education** which has offices in the UK (see entry under *UK Publishers*). Publishes a wide range of educational, professional, business, science, engineering and computing books.

## Macmillan Reference USA
> Gale

## Margaret K. McElderry Books
> Simon & Schuster Children's Publishing

## Mariner Books
> Houghton Mifflin Harcourt

## Metropolitan Books
> Henry Holt & Company, Inc.

## Midnight Ink Books
> Llewellyn Worldwide Ltd

## Millbrook Press
> Lerner Publishing Group

## Minotaur
▷ St Martin's Press LLC

## Modern Library
▷ The Random House Publishing Group

## NAL
▷ Penguin Group (USA) Inc.

## Nan A. Talese Books
▷ The Knopf Doubleday Publishing Group

## North Point Press
▷ Farrar, Straus & Giroux, Inc.

## Northwestern University Press
629 Noyes Street, Evanston IL 60208–4210
☏ 001 847 491 2046    ☏ 001 847 491 8150
nupress@northwestern.edu
www.nupress.northwestern.edu
Senior Editor/Assistant Director *Henry Carrigan Jnr*

Founded 1893. Publishes general and academic books: philosophy, drama and performance studies, history, biography, religion, Slavic studies, Latino fiction and contemporary fiction, poetry. No children's, art books. IMPRINTS **Latino Voices, TriQuarterly** Editorial Head *Henry Carrigan* Fiction and poetry. Unsolicited mss, synopses and ideas welcome; approach by mail. No unsolicited mss in fiction or poetry, please.
Ⓔ Royalties annually.

## Ohio University Press
19 Circle Drive, The Ridges, Athens
OH 45701–2979
☏ 001 740 593 1155    ☏ 001 740 593 4536
www.ohioswallow.com
Director *David Sanders*

Founded 1964. Publishes academic, regional and general trade books. IMPRINT **Swallow Press**. Synopses and ideas for books welcome. Send detailed synopsis, sample chapters and c.v. with covering letter. See website for author guidelines. No children's, how-to or genre fiction.
Ⓔ Royalties annually.

## Olive Branch Press
▷ Interlink Publishing Group, Inc.

## One World
▷ The Random House Publishing Group

## Open Forum
▷ Krieger Publishing Co.

## Orbit
▷ Krieger Publishing Co.

## Orbit
▷ Hachette Book Group

## Palgrave
▷ St Martin's Press LLC

## Pamela Dorman Books
▷ Penguin Group (USA) Inc.

## Pantheon Books/ Pantheon Graphic Novels
▷ The Knopf Doubleday Publishing Group

## Paragon House
1925 Oakcrest Avenue, Suite 7, St Paul
MN 55113–2619
☏ 001 651 644 3087    ☏ 001 651 644 0997
paragon@paragonhouse.com
www.paragonhouse.com
Executive Director *Dr Gordon L. Anderson*

Founded 1982. Publishes non-fiction: reference and academic. Subjects include history, religion, philosophy, spirituality, society, political science, psychology.
Ⓔ Royalties twice-yearly.

## Paulist Press
997 Macarthur Road, Mahwah NJ 07430–0990
☏ 001 201 825 7300    ☏ 001 201 836 3161
pmcmahon@paulistpress.com
www.paulistpress.com
Managing Editor *Paul McMahon*

Founded in 1858 and owned by Paulist Fathers. Publishes religion, spirituality, theology, pastoral resources, ecumenical and interfaith works. No fiction, memoirs or poetry. Planning to increase children's (spirituality) and catechetical list of titles. About 85 titles a year. Submissions welcome for the Paulist Press imprint only. Initial approach by email but accepts hard copy submissions only. Check the website for submissions process.
Ⓔ Royalties twice-yearly.

## Pearson Education
One Lake Street, Upper Saddle River NJ 07458
☏ 001 201-236-7000
www.pearsoned.com

The world's largest educational publisher. Publishes across a wide range of curriculum subjects from primary students to professional practitioners.

## Pelican Publishing Company, Inc.
1000 Burmaster Street, Gretna LA 70053
☏ 001 504 368 1175
www.pelicanpub.com
Editor-in-Chief *Nina Kooij*

Publishes general non-fiction: popular history, cookbooks, travel, art, business, biography, children's, architecture, collectibles guides and motivational. About 75 titles a year. Initial inquiries required for all submissions.
Ⓔ Royalties twice-yearly.

## Penguin Group (USA) Inc.

375 Hudson Street, New York NY 10014
☎ 001 212 366 2000   ℻ 001 212 366 2666
www.penguingroup.com
CEO *David Shanks*
President *Susan Petersen Kennedy*

The second-largest trade book publisher in
the world. Publishes fiction and non-fiction in
hardback and paperback; adult and children's.

ADULT DIVISION IMPRINTS Hardcover: **Avery;
Pamela Dorman Books; Dutton; G.P. Putnam's
Sons; Gotham Books; Hudson Street Press;
Tarcher; The Penguin Press; Portfolio;
Riverhead Books; Sentinel; Viking;** Trade
Paperback: **Ace Books; Alpha Books; Berkley;
HPBooks; NAL; Perigee; Penguin Books;
Plume; Prentice Hall Press; Riverhead.** Mass
Market Paperback: **Ace; Berkley; DAW Books**
(see entry); **Jove.**

YOUNG READERS DIVISION Hardcover: **Dial
Books for Young Readers** (see entry); **Dutton
Children's Books; Frederick Warne; G.P.
Putnam's Sons; Philomel Books; Viking Books
for Young Readers.** Paperback: **Firebird; Puffin
Books; Razorbill; Speak.** Mass Merchandise:
**Grosset & Dunlap; Price Stern Sloan.** No
unsolicited mss.

Ⓔ Royalties twice-yearly.

## Perennial
▷ HarperCollins Publishers, Inc.

## Perigee Books
▷ Penguin Group (USA) Inc.

## Perseus Books Group

387 Park Avenue S., 12th Floor, New York
NY 10016
☎ 001 212 340 8100   ℻ 001 212 340 8115
www.perseusbooksgroup.com
President/CEO *David Steinberger*

Founded 1997. IMPRINTS **Avalon Travel** Guide
books; **Basic Books** Psychology, science,
politics, sociology, current affairs, history; **Basic
Civitas** African and African-American studies;
**DaCapo** Art, biography, film, history, humour,
music, photography, sport; **PublicAffairs**
Current affairs, biography, history, journalism;
**Running Press** General non-fiction, children's,
illustrated books; **Seal Press** 'Books by women
for women'; **Vanguard Press** Fiction and non-
fiction; **Westview** Undergraduate and graduate
textbooks. No unsolicited mss.

## Philomel Books
▷ Penguin Group (USA) Inc.

## Picador USA
▷ St Martin's Press LLC

## Players Press

PO Box 1132, Studio City CA 91614–0132
☎ 001 818 789 4980
Vice President, Editorials *Robert W. Gordon*

Founded 1965 as a publisher of plays; now
publishes across the entire range of performing
arts: plays, musicals, theatre, film, cinema,
television, costume, puppetry, plus technical
theatre and cinema material. No unsolicited mss;
synopses/ideas welcome. Send query letter with
IRCs for response.

Ⓔ Royalties twice-yearly.

## Plume
▷ Penguin Group (USA) Inc.

## Pocket Books
▷ Simon & Schuster Adult Publishing Group

## Poppy
▷ Hachette Book Group

## Portfolio
▷ Penguin Group (USA) Inc.

## Potomac Books, Inc.

22841 Quicksilver Drive, Dulles VA 20166
☎ 001 703 661 1548   ℻ 001 703 661 1547
www.potomacbooksinc.com
Publisher *Sam Dorrance*
Senior Editors *Hilary Claggett, Elizabeth Demers*

Formerly Brassey's, Inc., founded 1983 (acquired
in 1999 by Books International of Dulles,
Virginia). Publishes non-fiction titles on topics
of history (especially military and diplomatic
history), world and US affairs, US foreign policy,
defence, intelligence, biography and sport. About
80 titles a year. No unsolicited mss; query letters/
synopses welcome.

Ⓔ Royalties annually.

## Potter Style/Potter Craft
▷ Crown Publishing Group

## Prentice Hall Press
▷ Penguin Group (USA) Inc.

## Presidio Press
▷ The Random House Publishing Group

## Price Stern Sloan
▷ Penguin Group (USA) Inc.

## Primary Source Media
▷ Gale

## Princeton University Press

41 William Street, Princeton NJ 08540–5237
☎ 001 609 258 4900   ℻ 001 609 258 6305
www.press.princeton.edu
Director *Peter J. Dougherty*

Founded 1905. Publishes academic and trade books. About 200 titles a year. No unsolicited mss. Synopses and ideas considered. Send letter in the first instance. No email submissions. No fiction. OVERSEAS ASSOCIATE Princeton University Press, 3 Market Place, Woodstock OX20 1SY, UK.
Ⓔ Royalties annually.

### Professional Practices
▷ Krieger Publishing Co.

### Prometheus Books
59 John Glenn Drive, Amherst NY 14228–2197
☎ 001 716 691 0133    🖷 001 716 691 0137
marketing@prometheusbooks.com
www.prometheusbooks.com
www.pyrsf.com
Editor-in-Chief *Steven L. Mitchell*

Founded 1969. Publishes books and journals: educational, scientific, professional, library, popular science, science fiction and fantasy, philosophy and young readers. IMPRINTS **Humanity Books; Pyr™**. About 100 titles a year. Unsolicited mss, synopses and ideas for books welcome. See website for submission guidelines.
Ⓔ Royalties twice-yearly.

### Public History
▷ Krieger Publishing Co.

### PublicAffairs
▷ Perseus Books Group

### Puffin Books
▷ Penguin Group (USA) Inc.

### Pyr™
▷ Prometheus Books

### Quayside Publishing Group
▷ The Quarto Group, Inc. under UK Publishers

### The Random House Publishing Group
1745 Broadway, New York NY 10019
☎ 001 212 782 9000
www.randomhouse.com
President/Publisher *Gina Centrello*

Publishes fiction, non-fiction, science fiction and graphic novels in hardcover, trade paperback and mass market.

IMPRINTS

**Ballantine Books; Bantam; Del Rey; Delacorte Press; Dell; Dial Press; Modern Library; One World; Presidio Press; Random House; Random House Trade Paperbacks; Spectra; Spiegel & Grau; Triumph Books; Villard**. 825 titles in 2009. No unsolicited mss.

### Rayo
▷ HarperCollins Publishers, Inc.

### Razorbill
▷ Penguin Group (USA) Inc.

### Reader's Digest Association Inc
Reader's Digest Road, Pleasantville
NY 10570–7000
☎ 001 914 238 1000
www.rd.com
President/CEO *Mary Berner*

Publishes home maintenance reference, cookery, DIY, health, gardening, children's books; videos and magazines.

### Reagan Arthur Books
▷ Hachette Book Group

### Riverhead
▷ Penguin Group (USA) Inc.

### The Rosen Publishing Group, Inc.
29 East 21st Street, New York NY 10010
☎ 001 212 777 3017
www.rosenpublishing.com
President *Roger Rosen*

Founded 1950. Publishes non-fiction books (supplementary to the curriculum, reference and self-help) for a young adult audience. Reading levels are years 7–12, 4–6 (books for teens with literacy problems), and 5–9. Subjects include conflict resolution, character building, health, safety, drug abuse prevention, self-help and multicultural titles. 100 titles a year. For all imprints, write with outline and sample chapters.

### Running Press
▷ Perseus Books Group

### Rutgers University Press
100 Joyce Kilmer Avenue, Piscataway
NJ 08854–8099
☎ 001 732 445 7762    🖷 001 732 445 7039
rutgerspress.rutgers.edu
Director *Marlie Wasserman*
Associate Director/Editor-in-Chief *Leslie Mitchner*

Founded 1936. Publishes scholarly books, regional, social sciences and humanities. About 80 titles a year. Unsolicited mss, synopses and ideas for books welcome. No original fiction or poetry.

### St James Press
▷ Gale

### St Martin's Press LLC
175 Fifth Avenue, New York NY 10010
☎ 001 646 307 5151    🖷 001 212 420 9314
inquiries@stmartins.com
www.stmartins.com
CEO (Macmillan) *John Sargent*

President/Publisher (Trade Division)
*Sally Richardson*

Founded 1952. A division of Macmillan in the US, owned by German publishing group Holtzbrinck, GmbH, St Martin's Press made its name and fortune by importing raw talent from the UK to the States and has continued to buy heavily in the UK. Publishes general fiction, especially mysteries and crime; and adult non-fiction: history, self-help, political science, travel, biography, scholarly, popular reference, college textbooks.

IMPRINTS

**Picador USA**; **Griffin** (trade paperbacks); **St Martin's Paperbacks** (mass market); **Thomas Dunne Books**; **Minotaur**; **Palgrave**; **Truman Talley Books**. All submissions via legitimate literary agents only.

## Samuel French, Inc.

45 West 25th Street, New York NY 10010–2751
☎ 001 212 206 8990   📠 001 212 206 1429
info@samuelfrench.com
www.samuelfrench.com
Managing Editor *Roxane Heinze-Bradshaw*

Founded 1830. Publishes plays in paperback: London, Broadway and off-Broadway hits, light comedies, mysteries, one-act plays and plays for young audiences. About 130 titles a year. Send online or paper query, including 10-page sample and synopsis. Visit website for more information. OVERSEAS ASSOCIATES in London and Sydney. 🅴 Royalties annually (books); twice-yearly (amateur productions); monthly (professional productions).

## Scarecrow Press Inc.

4501 Forbes Boulevard, Suite 200, Lanham MD 20706
☎ 001 301 459 3366   📠 001 301 429 5748
www.scarecrowpress.com
Publisher/Editorial Director *Edward Kurdyla*

Founded 1950 as a short-run publisher of library reference books. Acquired by **University Press of America, Inc.** in 1995 which is now part of Rowman and Littlefield Publishing Group. Publishes reference, scholarly and monographs (all levels) for libraries. Reference books in all areas except sciences, specializing in the performing arts, music, cinema and library science. About 160 titles a year. Unsolicited mss welcome but material will not be returned unless requested and accompanied by IRCs. Unsolicited synopses and ideas for books welcome.

## Schirmer Reference
▷ Gale

## Schocken Books
▷ The Knopf Doubleday Publishing Group

## Scholarly Resources, Inc.
▷ Gale

## Scribner
▷ Simon & Schuster Adult Publishing Group

## Seal Press
▷ Perseus Books Group

## Sentinel
▷ Penguin Group (USA) Inc.

## Seven Stories Press

140 Watts Street, New York NY 10013
☎ 001 212 226 8760
www.sevenstories.com
Publisher *Daniel Simon*
Subsidiary Rights *Anna Lui*

Founded 1995. Independent publisher of literature, translations, politics, media studies, popular culture, health. About 50 titles a year. No unsolicited mss; synopses and ideas from agents welcome or send query letter. No email submissions. No children's or business books. OVERSEAS SUBSIDIARIES Turnaround Publishing Services, UK and EU; Palgrave Macmillan, Australia; Macmillian Publishers NZ Ltd., New Zealand; Liberty Books (Pte.) Ltd., Pakistan; Pen International Pte. Ltd., Singapore; Stephan Phillips (Pty.) Ltd., Southern Africa; Publishers Group Canada, Canada.

## Shambhala Publications, Inc.

PO Box 308, Boston MA 02117
www.shambhala.com

Publishes contemplative religion, Buddhism, Christianity, philosophy, self-help, health, creativity and fiction. IMPRINTS **Trumpeter**; **Weatherhill**. 70 titles in 2009. Welcomes submissions – by post only – no emails. No original poetry. Check website for additional information on submission guidelines.

## Sharpe Reference
▷ M.E. Sharpe, Inc.

## Shaye Areheart Books
▷ Crown Publishing Group

## Simon & Schuster Adult Publishing Group (Division of Simon & Schuster, Inc)

1230 Avenue of the Americas, New York NY 10020
☎ 001 212 698 7000   📠 001 212 698 7007
www.simonandschuster.com
President & CEO *Carolyn K. Reidy*

Publishes fiction and non-fiction.

DIVISIONS

**Atria Books** VP & Executive Editorial Director *Emily Bestler*; **Free Press** VP & Editorial Director *Dominick Anfuso*; **Touchstone-Fireside** VP &

Editor-in-Chief *Trish Todd*; **Howard Books** VP & Editor-in-Chief *Becky Nesbitt*; **Gallery Books** VP & Editor-in-Chief *Jennifer Bergstrom*; **Scribner** VP & Editor-in-Chief *Nan Graham*; **Simon & Schuster** VP & Editorial Director *Alice Mayhew*.

IMPRINTS

**Atria Books**; **Fireside**; **Free Press**; **Howard Books**; **Pocket Books**; **Gallery Books**; **Scribner**; **Simon & Schuster**; **Threshold Editions**; **Touchstone**. Almost 2,000 titles a year. No unsolicited mss.
ⓔ Royalties twice-yearly.

## Simon & Schuster Children's Publishing

1230 Avenue of the Americas, New York NY 10020
ⓣ 001 212 698 7200
www.simonsayskids.com
President/Publisher *Jon Anderson*

A division of the Simon & Schuster Consumer Group. Publishes pre-school to young adult, picture books, hardcover and paperback fiction, non-fiction, trade, library and mass-market titles.

IMPRINTS

**Aladdin Paperbacks** Picture books, paperback fiction and non-fiction reprints and originals, and limited series for ages pre-school to young adult; **Atheneum Books for Young Readers** Picture books, hardcover fiction and non-fiction books across all genres for ages three to young adult; **Little Simon** Mass-market novelty books (pop-ups, board books, colouring & activity) and merchandise (book and audio cassette) for ages birth through eight; **Margaret K. McElderry Books** Picture books, hardcover fiction and non-fiction trade books for children ages three to young adult; **Simon & Schuster Books for Young Readers** Picture books, hardcover fiction and non-fiction for children ages three to young adult; **Simon Pulse** Paperbacks for teenagers; **Simon Scribbles** Colouring and activity books. **Simon Spotlight** Devoted exclusively to children's media tie-ins and licensed properties. About 480 titles a year. No unsolicited material.

## Simon Pulse/Simon Scribbles/Simon Spotlight
▷ Simon & Schuster Children's Publishing

## Skirt!
▷ The Globe Pequot Press

## Sleeping Bear Press
▷ Gale

## Southern Illinois University Press

1915 University Press Drive, SIUC Mail Code 6806, Carbondale IL 62901
ⓣ 001 618 453 2281    ⓕ 001 618 453 1221
www.siupress.com
Director *Lain Adkins*

Founded 1956. Publishes scholarly and general interest non-fiction books and educational materials. 40 titles a year.

## Speak
▷ Penguin Group (USA) Inc.

## Spectra
▷ The Random House Publishing Group

## Spiegel & Grau
▷ The Random House Publishing Group

## Springboard Press
▷ Hachette Book Group

## Stackpole Books

5067 Ritter Road, Mechanicsburg PA 17055
ⓣ 001 717 796 0411    ⓕ 001 717 796 0412
www.stackpolebooks.com
President *M. David Detweiler*
Vice President/Publisher *Judith Schnell*

Founded 1933. Publishes outdoor sports, fishing, hunting, nature, crafts, Pennsylvania and regional, military reference, history. About 100 titles a year.
ⓔ Royalties twice-yearly.

## Stanford University Press

1450 Page Mill Road, Palo Alto CA 94304–1124
ⓣ 001 650 723 9434    ⓕ 001 650 725 3457
info@www.sup.org
www.sup.org
Director *Dr Geoffrey R.H. Burn*
Senior Rights Manager *Ariane de Pree-Kajfez*
Editor-in-Chief *Dr Alan Harvey*

Founded 1925. Publishes non-fiction: scholarly works in all areas of the humanities and social sciences, also professional lists and college textbooks in business, economics, security studies, education and law. About 160 titles a year. No unsolicited mss; query in writing first. Editors will request a prospectus including the following: a two or three-page cover letter outlining the manuscript's argument, general content, significance to the field, fit with existing literature or competition, intended audiences, length, and plans for tables and illustrations; a detailed chapter-by-chapter description of the content of the book; table of contents; c.v.
ⓔ Royalties annually.

## State University of New York Press

194 Washington Avenue, Suite 305, Albany NY 12210–2384
ⓣ 001 518 472 5000    ⓕ 001 518 472 5038
info@sunypress.edu
www.sunypress.edu
Associate Director *James Peltz*

Founded 1966. Part of the Research Foundation of SUNY, the Press is one of the largest university presses in the USA. Publishes scholarly and trade

books in the humanities, social sciences and books of regional interest on New York State. Strengths in philosophy, religious studies and Asian studies. 189 titles in 2009. Unsolicited submissions welcome. Send material to *Jane Bunker*, Editor-in-Chief.
£ Royalties annually.

### Sterling Publishing Co. Inc.

387 Park Avenue South, New York NY 10016
☎ 001 212 532 7160
www.sterlingpub.com
VP/Publisher *Andrew Martin*
President/CEO *Charles Nurnberg*

Founded 1949. A subsidiary of Barnes & Noble, Inc. Publishes illustrated non-fiction: reference and how-to books on arts and crafts, home improvement, history, photography, children's, woodworking, pets, hobbies, gardening, games and puzzles, general non-fiction. Submission guidelines available.

### Swallow Press
▷ Ohio University Press

### Syracuse University Press

621 Skytop Road, Suite 110, Syracuse
NY 13244–5290
☎ 001 315 443 5534    ℻ 001 315 443 5545
www.SyracuseUniversityPress.syr.edu
Director *Alice R. Pfeiffer*

Founded 1943. Publishes scholarly books in the following areas: Middle East studies, Irish studies, Jewish studies, sports, gender and globalization, Iroquois studies, women and religion, medieval studies, religion and politics, disability studies, peace and conflict resolution, television and popular culture, geography and regional books. Distributes for Arlen House, Dedalus Press, Colgate University Press. About 50 titles a year. No unsolicited mss. Send query letter with IRCs.
£ Royalties annually.

### Tarcher
▷ Penguin Group (USA) Inc.

### Temple University Press

2450 West Hunting Park Avenue, Philadelphia
PA 19129–1302
☎ 001 215 926 2140    ℻ 001 215 926 2141
www.temple.edu/tempress
Editor-in-Chief *Janet M. Francendese*

Founded 1969. Publishes scholarly books. Authors generally academics. About 60 titles a year. Letter or email of inquiry with brief outline. Include fax/email address.

### Ten Speed Press
▷ Crown Publishing Group

### Thomas Dunne Books
▷ St Martin's Press LLC

### Thorndike Press
▷ Gale

### Three Rivers Press
▷ Crown Publishing Group

### Threshold Editions
▷ Simon & Schuster Adult Publishing Group

### Times Books
▷ Henry Holt & Company, Inc.

### Touchstone
▷ Simon & Schuster Adult Publishing Group

### Transaction Publishers Ltd

Rutgers – The State University of New Jersey, 35 Berrue Circle, Piscataway NJ 08854–8042
☎ 001 732 445 2280    ℻ 001 732 445 3138
trans@transactionpub.com
www.transactionpub.com
Chairman/Editorial Director *I.L. Horowitz*
President *Mary E. Curtis*

Founded 1962. Independent publisher of academic social scientific books, periodicals and serials. Publisher of record in international social research. All unsolicited material should be submitted in hard copy and full draft. Email submissions are not accepted. 'Decision-making is rapid.'

### Tricycle Press
▷ Crown Publishing Group

### TriQuarterly
▷ Northwestern University Press

### Triumph Books
▷ The Random House Publishing Group

### Truman Talley Books
▷ St Martin's Press LLC

### Trumpeter
▷ Shambhala Publications, Inc.

### Twayne Publishers
▷ Gale

### Twelve
▷ Hachette Book Group

### Twenty-First Century Books
▷ Lerner Publishing Group

### Tyndale House Publishers, Inc.

351 Executive Drive, Carol Stream IL 60188
☎ 001 630 668 8300
www.tyndale.com
President *Mark D. Taylor*

Founded 1962. Books cover a wide range of categories from non-fiction to gift books, theology, doctrine, Bibles, fiction, children's and youth. Also produces video material, calendars and audio books for the same market. No poetry. Non-denominational religious publisher of around 250 titles a year for the evangelical Christian market. No unsolicited mss. Synopses and ideas considered. Writer's guidelines available on the website.

## University of Alabama Press

Box 870380, Tuscaloosa AL 35487–0380
℡ 001 205 348 5180    🅵 001 205 348 9201
www.uapress.ua.edu
Director *Daniel J.J. Ross*
Managing Editor *Crissie Johnson*

Founded 1945. Publishes American history, Latin American history, American religious history; Caribbean archaeology; historical archaeology; ethnohistory; anthropology; Native American studies; Judaic studies, American literature and criticism; history and environmental studies; American social and cultural history. About 70 titles a year. See 'Submissions' page on the website for guidelines and editorial contacts.

## University of Alaska Press

PO Box 756240, University of Alaska, Fairbanks AK 99775
℡ 001 907 474 5831    🅵 001 907 474 5502
fypress@uaf.edu
www.uaf.edu/uapress

Publishes scholarly works about Alaska and the Pacific Rim, with a special emphasis on circumpolar regions, history, anthropology, Native studies and natural history. 25 titles a year. Queries and proposals welcome (see 'Information for Authors' page on the website for proposal guidelines).

## University of Arizona Press

355 S. Euclid Avenue, Suite. 103, Tucson AZ 85719
℡ 001 520 621 1441    🅵 001 520 621 8899
uapress@uapress.arizona.edu
www.uapress.arizona.edu
Interim Director *Kathryn Conrad*

Founded 1959. Publishes academic non-fiction, particularly with a regional/cultural link, plus Native-American and Hispanic literature. About 55 titles a year.

## University of Arkansas Press

McIlroy House, 105 N. McIlroy Avenue, Fayetteville AR 72701
℡ 001 479 575 3246    🅵 001 479 575 6044
www.uapress.com
Director *Lawrence J. Malley*

Founded 1980. Publishes scholarly monographs, poetry and general trade books. Particularly interested in scholarly works in history, African-American studies, Southern politics, Civil War, Middle East studies and civil rights. About 20 titles a year. Check website for submissions information.

## University of California Press

2120 Berkeley Way, Berkeley CA 94704–1012
℡ 001 510 642 4247    🅵 001 510 643 7127
www.ucpress.edu
Director *Lynne Withey*

Founded 1893. Publishes scholarly and scientific non-fiction; some fiction in translation. About 200 books and 40 journals annually. Preliminary letter with outline preferred.

## The University of Chicago Press

1427 East 60th Street, Chicago IL 60637
℡ 001 773 702 7700    🅵 001 773 702 9756
www.press.uchicago.edu

Founded 1891. Publishes academic non-fiction only.

## University of Hawai'i Press

2840 Kolowalu Street, Honolulu HI 96822–1888
℡ 001 808 956 8255    🅵 001 808 988 6052
uhpbooks@hawaii.edu
www.uhpress.hawaii.edu
Director *William H. Hamilton*
Executive Editor *Patricia Crosby*

Founded 1947. Publishes scholarly books pertaining to East Asia, Southeast Asia, Hawaii and the Pacific. 75 titles a year. Unsolicited synopses and ideas welcome; approach by mail. No poetry, children's books or any topics other than Asia and the Pacific.
🅴 Royalties twice-yearly.

## University of Illinois Press

1325 South Oak Street, Champaign IL 61820–6903
℡ 001 217 333 0950    🅵 001 217 244 8082
uipress@uillinois.edu
www.press.uillinois.edu
Director *Willis G. Regier*
Associate Director and Editor-in-Chief *Joan Catapano*

Founded 1918. Publishes non-fiction, scholarly and general, with special interest in Americana, women's studies, African–American studies, film, religion, American music and regional books. About 140–150 titles a year.

## University of Iowa Press

100 Kuhl House, 119 West Park Road, Iowa City IA 52242–1000
℡ 001 319 335 2000    🅵 001 319 335 2055
uipress@uiowa.edu
www.uiowapress.org

Director *Holly Carver*

Founded 1969 as a small scholarly press publishing about five books a year. Now publishing about 35 annually in a variety of scholarly fields, plus local interest, short stories, creative non-fiction and poetry anthologies. No unsolicited mss; query first. Unsolicited ideas and synopses welcome.
Ⓔ Royalties annually.

## University of Massachusetts Press

PO Box 429, Amherst MA 01004
Ⓣ 001 413 545 2217    Ⓕ 001 413 545 1226
info@umpress.umass.edu
www.umass.edu/umpress
Director *Bruce Wilcox*
Senior Editor *Clark Dougan*

Founded 1963. Publishes scholarly, general interest, African-American, ethnic, women's and gender studies, cultural criticism, economics, fiction, literary criticism, poetry, philosophy, biography, history. About 40 titles a year. Unsolicited mss considered but query letter preferred in the first instance. Synopses and ideas welcome.

## The University of Michigan Press

839 Greene Street, Ann Arbor MI 48104–3209
Ⓣ 001 734 764 4388
www.press.umich.edu

Founded 1930. Publishes scholarly and trade non-fiction, textbooks, new media studies, music, political science, English as a second language, classical studies, American history, theatre, literary criticism, gender studies, law studies and fiction.

## University of Minnesota Press

111 Third Avenue South, Suite 290, Minneapolis MN 55401
Ⓣ 001 612 627 1970    Ⓕ 001 612 627 1980
ump@umn.edu
www.upress.umn.edu
Executive Editor *Richard Morrison*

Founded 1925. Publishes academic books for scholars and selected general interest titles: American studies, anthropology, art and aesthetics, cultural theory, film and media studies, gay and lesbian studies, geography, literary theory, political and social theory, race and ethnic studies, sociology and urban studies. No original fiction or poetry. About 110 titles a year. Welcomes synopses and proposals sent by post (*no emails*). See website for submission details.

## University of Missouri Press

2910 LeMone Boulevard, Columbia MO 65201
Ⓣ 001 573 882 7641    Ⓕ 001 573 884 4498

upress@umsystem.edu
www.press.umsystem.edu
Editor in Chief *Mr Clair Willcox*

Founded 1958. Publishes academic: history, intellectual history, journalism, literary criticism and related humanities disciplines, and occasional volumes of short stories. About 70 titles a year. Best approach is by letter. Send synopses for academic work together with author c.v.

## University of Nebraska Press

1111 Lincoln Mall, Lincoln NE 68588–0630
Ⓣ 001 402 472 3581    Ⓕ 001 402 472 6214
pressmail@unl.edu
www.nebraskapress.unl.edu
Director *Donna A. Shear*

Founded 1941. Publishes scholarly, Native American studies, history of the American West, literary and cultural studies, music, Jewish studies, military history, sports history, environmental history. IMPRINT **Bison Books**. No unsolicited mss; welcomes synopses and ideas for books. Send enquiry letter in the first instance with description of project and sample material. No original fiction, children's books or poetry.
Ⓔ Royalties annually.

## University of Nevada Press

Morrill Hall Mail Stop 0166, Reno NV 89557–0166
Ⓣ 001 775 784 6573
www.nvbooks.nevada.edu
Director *Joanne O'Hare*

Founded 1961. Publishes scholarly and popular books; serious fiction, Native American studies, natural history, Western Americana, Basque studies and regional studies. Unsolicited material welcome if it fits in with areas published, or offers a 'new and exciting' direction.

## University of New Mexico Press

1717 Roma NE, Albuquerque NM 87106
Ⓣ 001 505 277 2346    Ⓕ 001 505 277 3343
unmpress@unm.edu
www.unmpress.com
Director *Luther Wilson*
Senior Acquisitions Editor *William 'Clark' Whitehorn*

Founded 1929. Publishes scholarly, regional books, biography and fiction. No how-to, humour, self-help or technical. 80 titles a year. Receives around 3,600 submissions a year.
Ⓔ Royalties annually.

## University of North Texas Press

1155 Union Circle, #311336, Denton TX 76203–5017
Ⓣ 001 940 565 2142    Ⓕ 001 940 565 4590
ronald.chrisman@unt.edu
www.unt.edu/untpress

Director *Ronald Chrisman*
Managing Editor *Karen DeVinney*

Founded 1987. Publishes regional interest, contemporary, social issues, Texas history, military history, music, women's issues, multicultural. Publishes annually the winners of the Vassar Miller Poetry Prize and the Katherine Anne Porter Prize in Short Fiction. About 15 titles a year. Synopses and ideas welcome. Do not send full ms until requested.
£ Royalties annually.

## University of Oklahoma Press

2800 Venture Drive, Norman OK 73069–8216
T 001 405 325 2000    F 001 405 325 4000
www.oupress.com
Director *B. Byron Price*
Publisher *John Drayton*

Founded 1928. Publishes general scholarly non-fiction only: American Indian studies, American West, classical studies, anthropology, natural history and political science. About 100 titles a year. See website for submissions policy.

## University of Pennsylvania Press

3905 Spruce Street, Philadelphia PA 19104–4112
T 001 215 898 6261    F 001 215 898 0404
www.pennpress.org
Director *Eric Halpern*
Editor-in-Chief *Peter Agree*

Founded 1890. Publishes serious non-fiction: scholarly, reference, professional, textbooks and semi-popular trade. No original fiction or poetry. Over 100 titles a year. No unsolicited mss but synopses and ideas for books welcome.
£ Royalties vary according to sales prospects and other relevant factors.

## University of Texas Press

PO Box 78713–7819, Austin TX 78722
T 001 512 471 7233/471 4278 (editorial)
F 001 512 232 7178
utpress@uts.cc.utexas.edu
www.utexas.edu/utpress/
www.utexaspress.com
Director *Joanna Hitchcock*
Assistant Director/Editor-in-Chief *Theresa J. May*

Publishes scholarly and regional non-fiction: anthropology, architecture, classics, environmental studies, humanities, social sciences, geography, language studies, literary modernism; Latin American/Latino/Mexican American/Middle Eastern/Native American studies, natural history and ornithology, regional books (Texas and the southwest). About 100 titles a year and 11 journals. Unsolicited material welcome in above subject areas only. Author guidelines available on the website.
£ Royalties annually.

## University of Virginia Press

PO Box 400318, Charlottesville VA 22904–4318
T 001 434 924 3468
vapress@virginia.edu
www.upress.virginia.edu
Director *Penelope J. Kaiserlian*

Founded 1963. Publishes academic books in humanities and social science with concentrations in American history, African-American studies, architecture, Victorian literature, Caribbean literature and ecocriticism. 50–60 titles a year. No unsolicited mss; will consider synopses and ideas for books if in their specific areas of concentration. Approach by mail for full proposal; email for short inquiry. No fiction, children's, poetry or academic books outside interests specified above.
£ Royalties annually.

## The University of Wisconsin Press

1930 Monroe Street, 3rd Floor, Madison WI 53711–2059
T 001 608 263 1110
uwiscpress@uwpress.wisc.edu
www.wisc.edu/wisconsinpress
Director *Sheila Leary*
Senior Acquisitions Editor *Raphael Kadushin*

Founded 1936. Publishes scholarly, general interest non-fiction and regional books about Wisconsin and the mid-west. About 50 titles a year. No unsolicited mss. Detailed author guidelines available on the website.

## University Press of America, Inc.

4501 Forbes Boulevard, Suite 200, Lanham MD 20706
T 001 301 459 3366
www.univpress.com
VP/Director *Judith L. Rothman*

Founded 1975. An imprint of Rowman & Littlefield Publishing Group. Publishes scholarly monographs, college and graduate level textbooks. No children's, elementary or high school. About 300 titles a year. Submit outline or request proposal questionnaire.

## University Press of Kansas

2502 Westbrooke Circle, Lawrence KS 66045–4444
T 001 785 864 4154    F 001 785 864 4586
upress@ku.edu
www.kansaspress.ku.edu
Director *Fred M. Woodward*

Founded 1946. Became the publishing arm for all six state universities in Kansas in 1976. Publishes scholarly books in American history, legal studies, presidential studies, American studies, political philosophy, political science, military history

and environmental history. About 55 titles a year. Proposals welcome.
ⓔ Royalties annually.

## The University Press of Kentucky

663 South Limestone Street, Lexington
KY 40508–4008
ⓣ 001 859 257 8434　ⓕ 001 859 323 1873
www.kentuckypress.com
Director *Stephen M. Wrinn*

Founded 1943. Publishes scholarly and general interest books in the humanities and social sciences. No unsolicited mss; send synopses and ideas for books by post or email. No fiction, drama, poetry, translations, memoirs or children's books.
ⓔ Royalties annually.

## University Press of Mississippi

3825 Ridgewood Road, Jackson MS 39211–6492
ⓣ 001 601 432 6205　ⓕ 001 601 432 6217
press@ihl.state.ms.us
www.upress.state.ms.us
Director *Leila W. Salisbury*
Assistant Director/Editor-in-Chief *Craig Gill*

Founded 1970. Non-profit book publisher partially supported by the eight state universities. Publishes scholarly and trade titles in literature, history, American culture, Southern culture, African-American studies, women's studies, popular culture, film and comics studies, folklife, ethnic, performance, art, architecture, photography and other liberal arts. Represented worldwide. UK REPRESENTATIVE: Roundhouse Group. About 65 titles a year. Send letter of enquiry, prospectus, table of contents and sample chapter prior to submission of full mss.
ⓔ Royalties annually.

## University Press of New England

One Court Street, Suite 250, Lebanon NH 03766
ⓣ 001 603 448 1533　ⓕ 001 603 448 7006
university.press@dartmouth.edu
www.upne.com
Director *Michael Burton*
Editor-in-Chief *Phyllis Deutsch*

Founded 1970. A scholarly books publisher sponsored by six institutions of higher education in the region: Brandeis, Dartmouth, University of Vermont, Tufts, the University of New Hampshire and Northeastern. Publishes general and scholarly non-fiction. OVERSEAS ASSOCIATES Canada: University of British Columbia; UK, Europe, Middle East: Eurospan University Press Group; Australia, New Zealand, Asia & the Pacific: East-West Export Books. About 80 titles a year. Unsolicited material welcome.
ⓔ Royalties annually.

## UXL
▷ Gale

## Vanguard Press
▷ Perseus Books Group

## Viking/Viking Books for Young Readers
▷ Penguin Group (USA) Inc.

## Villard
▷ The Random House Publishing Group

## Vision
▷ Hachette Book Group

## W.W. Norton & Company, Inc.

500 Fifth Avenue, New York NY 10110
ⓣ 001 212 354 5500　ⓕ 001 212 869 0856
www.wwnorton.com
VP/Editor-in-Chief *Starling R. Lawrence*

Founded 1923. Publishes fiction and non-fiction, college textbooks and professional books. About 400 titles a year. No unsolicited submissions.

## Walch Education

40 Walch Drive, Portland ME 04103–1286
ⓣ 001 207 772 2846　ⓕ 001 207 772 3105
www.walch.com
President *Al Noyes*

Founded 1927. Publishes supplementary educational materials for middle and secondary schools across a wide range of subjects, including English/language arts, literacy, special needs, mathematics, social studies, science and school-to-career. Always interested in ideas from secondary school teachers who develop materials in the classroom. Proposal letters, synopses and ideas welcome.

## Walker & Co.

175 Fifth Avenue, New York NY 10010
ⓣ 001 646 438 6075　ⓕ 001 212 780 0115
firstname.lastname@bloomsburyusa.com
www.walkerbooks.com (trade non-fiction)
www.walkeryoungreaders.com (BFYR)
Contact *Submissions Editor*

Founded 1959. A division of **Bloomsbury Publishing**. Publishes children's fiction and non-fiction and adult non-fiction. Please contact the following editors in advance before sending any material to be sure of their interest, then follow up as instructed: **Trade Non-fiction** *George Gibson* Permission and documentation must be available with mss. Submit prospectus first, with sample chapters and marketing analysis. **Books for Young Readers** *Emily Easton* Fiction and non-fiction for all ages. Query before sending non-fiction proposals. Especially interested in picture books – fiction and non-fiction, historical and contemporary fiction for middle grades and young adults. BFYR will consider unsolicited submissions. See website for guidelines.

## Washington State University Press

PO Box 645910, Washington State University, Pullman WA 99164–5910

Ⓣ 001 509 335 3518    Ⓕ 001 509 335 8568

wsupress@wsu.edu

wsupress.wsu.edu

Editor-in-Chief *Glen Lindeman*

Founded 1928. Publishes hardcover originals, trade paperbacks and reprints. Publishes mainly on the history, prehistory, culture, politics and natural history of the Northwest United States (Washington, Idaho, Oregon, Montana, Alaska and British Columbia). Subjects include Native American, maritime, railroad, biography, cooking/food history, nature/environment, politics. 4–6 titles a year. Submission guidelines are available on the website.

## Watson-Guptill
▷ Crown Publishing Group

## Weatherhill
▷ Shambhala Publications, Inc.

## Wellness Central
▷ Hachette Book Group

## Westview
▷ Perseus Books Group

## Wheeler Publishing
▷ Gale

## William Morrow
▷ HarperCollins Publishers, Inc.

## Wordsong
▷ Boyds Mills Press

## Yen Press
▷ Hachette Book Group

## Zondervan

5300 Patterson Avenue SE, Grand Rapids MI 49530

Ⓣ 001 616 698 6900

www.zondervan.com

President/CEO *Maureen Girkins*

Founded 1931. Subsidiary of **HarperCollins Publishers, Inc.** Publishes Protestant religion, Bibles, books, audio & video, computer software, calendars and speciality items.

# Other International Publishers

## Australia

### ACER Press

Private Bag 55, Camberwell, Victoria 3124
℡ 00 61 3 9277 5555  📠 00 61 3 9277 5500
www.acer.edu.au

Founded 1930. The publishing arm of the Australian Council for Educational Research. Publishes education, human resources, psychology, parent education and mental health, occupational therapy.

### Allen & Unwin Pty Ltd

PO Box 8500, St Leonards, NSW 1590
℡ 00 61 2 8425 0100  📠 00 61 2 9906 2218
info@allenandunwin.com
www.allenandunwin.com

Founded 1976. Independent publisher of fiction, general non-fiction, academic (social sciences, health) and children's. IMPRINTS **Arena** Fiction (adventure, crime, fantasy, thrillers, chick lit) and non-fiction (biography, DIY, health, humour, self-help, travel); **Crows Nest** Adventure, popular culture, history; **Jacana Books** Natural history; **Inspired Living** Mind, body and spirit, psychology, spirituality. About 250 titles a year.

### Arena
▷ Allen & Unwin Pty Ltd

### Cliffs Notes
▷ John Wiley & Sons Australia Ltd

### Crows Nest
▷ Allen & Unwin Pty Ltd

### Currency Press Pty Ltd

PO Box 2287, Strawberry Hills, NSW 2012
℡ 00 61 2 9319 5877  📠 00 61 2 9319 3649
enquiries@currency.com.au
www.currency.com.au

Founded 1971. Performing arts publisher – handbooks, biography, crical studies, directories and reference – drama, theatre, music, dance, film and video.

### For Dummies
▷ John Wiley & Sons Australia Ltd

### Frommer's
▷ John Wiley & Sons Australia Ltd

### Hachette Livre Australia

Level 17, 207 Kent Street, Sydney, NSW 2000
℡ 00 61 2 8248 0800  📠 00 61 2 8248 0810
www.hha.com.au

Founded 1958. Part of the **Hachette Livre Publishing Group**. Publishes popular fiction, literature, general non-fiction, autobiography, biography, current affairs, health, history, lifestyle, memoirs, self-help, sport, travel.

### HarperCollins Publishers (Australia) Pty Ltd

PO Box 321, Pymble, NSW 2073
℡ 00 61 2 9952 5000  📠 00 61 2 9952 5555
www.harpercollins.com.au

Founded 1989. Part of the News Corporation. Publishes fiction and non-fiction, children's and illustrated books.

### Inspired Living
▷ Allen & Unwin Pty Ltd

### Jacana Books
▷ Allen & Unwin Pty Ltd

### Jacaranda
▷ John Wiley & Sons Australia Ltd

### John Wiley & Sons Australia Ltd

PO Box 1226, Milton, Queensland 4064
℡ 00 61 7 3859 9755  📠 00 61 7 3859 9715
brisbane@johnwiley.com.au
www.johnwiley.com.au

Founded 1954. Owned by **John Wiley & Sons Inc.**, USA. Publishes general non-fiction and education books, dictionaries, encyclopaedias, maps and atlases, reference, textbooks. IMPRINTS **Wiley**; **Jacaranda**; **WrightBooks**; **For Dummies**; **Frommer's**; **Cliffs Notes**.

### Kangaroo Press
▷ Simon & Schuster (Australia) Pty Ltd

## Macmillan Publishers Australia Pty Ltd

1–3/15–19 Claremont Street, South Yarra, Victoria 3141

T 00 61 3 9825 1000    F 00 61 3 9825 1010

www.macmillan.com.au

Founded 1904. Part of **Macmillan Publishers**, UK. Publishes fiction and non-fiction, academic, children's, reference, education, professional, scientific, technical, medical. DIVISIONS **Macmillan Education** Primary, secondary and ELT; **Palgrave Macmillan** Academic and reference; **Pan Macmillan** (see entry); **Macquarie Dictionary Publishers** Reference (print and online);

## Macquarie Dictionary Publishers
▷ Macmillan Publishers Australia Pty Ltd

## Margaret Hamilton Books
▷ Scholastic Australia Pty Limited

## McGraw-Hill Australia & New Zealand Pty Ltd

Locked Bag 2233, Business Centre, North Ryde, NSW 1670

T 00 61 2 9900 1800

www.mcgraw-hill.com.au

Founded 1964 (Australia), 1974 (NZ). Owned by **The McGraw-Hill Companies, Inc**. Publishes textbooks and educational material for primary, secondary and higher education. Also professional (medical, general and reference).

## Melbourne University Publishing Ltd

187 Grattan Street, Carlton, Victoria 3053

T 00 61 3 9342 0300    F 00 61 3 9342 0399

mup-info@unimelb.edu.au

www.mup.unimelb.edu.au

Founded 1922. Academic publishers; general non-fiction, archaeology, architecture, art, biography, business, true crime, economics, education, finance, history, language, linguistics, law, literature, maritime, military history, natural history, philosophy, politics, psychology, psychiatry, reference, religion, science and technology, sociology, sport, travel, zoology. IMPRINTS **Melbourne University Press**; **The Miegunyah Press**; **Victory Books**. About 70 titles a year.

## The Miegunyah Press
▷ Melbourne University Publishing Ltd

## Omnibus Books
▷ Scholastic Australia Pty Limited

## Oxford University Press (Australia & New Zealand)

GPO Box 2784, Melbourne, Victoria 3001

T 00 61 3 9934 9123

www.oup.com.au

Founded 1908. Owned by **Oxford University Press**, UK. Publishes for the college, school and trade markets.

## Palgrave Macmillan
▷ Macmillan Publishers Australia Pty Ltd

## Pan Macmillan Australia Pty Ltd

Level 25, 1 Market Street, Sydney, NSW 2000

T 00 61 2 9285 9100    F 00 61 2 9285 9190

www.panmacmillan.com.au

Founded 1983. Part of **Macmillan Publishers**, UK. Publishes fiction, crime fiction, thrillers, literature; general non-fiction, biography, children's, history, self-help, travel.

## Pearson Australia Pty Ltd

Locked Bag 507, Frenchs Forest, NSW 1640

T 00 61 3 9454 2200    F 00 61 3 9453 0089

www.pearson.com.au

Part of Pearson Plc. Australia's largest educational publisher, formed in 1998 from a merger of Addison Wesley Longman Australia and Prentice Hall Australia.

## Penguin Group (Australia)

PO Box 701, Hawthorn, Victoria 3124

T 00 61 3 9811 2400    F 00 61 3 9811 2620

www.penguin.com.au

Founded 1946. Publishes adult and children's general non-fiction and fiction.

## Random House Australia

Level 3, 100 Pacific Highway, North Sydney, NSW 2060

T 00 61 2 9954 9966    F 00 61 2 9954 4562

random@randomhouse.com.au

www.randomhouse.com.au

Part of the **Random House Group**. Publishes popular and literary fiction, non-fiction, children's and illustrated.

## Scholastic Australia Pty Limited

PO Box 579, Gosford, NSW 2250

T 00 61 2 4328 3555

www.scholastic.com.au

Founded 1968. Part of **Scholastic Inc**. Educational and children's publisher of books and multimedia. IMPRINTS **Scholastic Press**; **Omnibus Books**; **Margaret Hamilton Books**. About 100 titles a year.

## Science Press

Bag 7023, Marrickville, NSW 1475

T 00 61 2 9516 1122    F 00 61 2 9550 1915

www.sciencepress.com.au

Founded 1948. Educational publisher.

## Simon & Schuster (Australia) Pty Ltd

PO Box 33, Pymble, NSW 2073
Ⓣ 00 61 2 9983 6600
Ⓕ 00 61 2 9988 4293 (editorial)
www.simonsaysaustralia.com.au

Founded 1987. Part of **Simon & Schuster Inc.**, USA. Publishes fiction and general non-fiction, children's books. IMPRINTS **Simon & Schuster Australia** Biography, health, parenting, relationships; **Kangaroo Press** Crafts, gardening, military history.

## University of Queensland Press

PO Box 6042, St Lucia, Queensland 4067
Ⓣ 00 61 7 3365 7244    Ⓕ 00 61 7 3365 7579
uqp@uqp.uq.edu.au
www.uqp.uq.edu.au

Founded 1948. Publishes academic, textbooks, children's and young adult, general non-fiction and fiction, literature, literary criticism, short stories, essays, poetry, art, architecture, biography, current affairs, history, indigenous studies, memoirs, military history, reference.

## UWA Press

The University of Western Australia (M419), 35 Stirling Highway, Perth, Crawley, WA 6009
Ⓣ 00 61 8 6488 3670    Ⓕ 00 61 8 6488 1027
admin@uwapress.uwa.edu.au
www.uwapress.uwa.edu.au

Founded 1935. Publishes academic, general non-fiction, essays, literature, literary criticism, history, natural history, children's, maritime history, reference, literary studies.

## Victory Books
▷ Melbourne University Publishing Ltd

## WrightBooks
▷ John Wiley & Sons Australia Ltd

# Canada

## Anchor Canada
▷ Random House of Canada Ltd

## Bond Street
▷ Random House of Canada Ltd

## Canadian Scholars' Press, Inc

180 Bloor Street West, Suite 801, Toronto, Ontario M5S 2V6
Ⓣ 001 416 929 2774    Ⓕ 001 416 929 1926
info@cspi.org
www.cspi.org

Founded 1986. Independent publisher of academic and trade books. IMPRINTS **Women's Press**; **Kellom Books**.

## Doubleday Canada
▷ Random House of Canada Ltd

## Douglas Gibson Books
▷ McClelland & Stewart Ltd

## Emblem
▷ McClelland & Stewart Ltd

## Firefly Books Ltd

66 Leek Crescent, Richmond Hill, Ontario L4B 1H1
Ⓣ 001 416 499 8412    Ⓕ 001 416 499 8313
www.fireflybooks.com

Founded 1977. Publishes children's and young adult non-fiction. No unsolicited submissions.

## Fitzhenry & Whiteside Limited

195 Allstate Parkway, Markham, Ontario L3R 4T8
Ⓣ 001 905 477 9700
godwit@fitzhenry.ca
www.fitzhenry.ca

Founded 1966. Publishes reference, biography, history, poetry, photography, sport, children's and young adult books. About 200 titles a year.

## H.B. Fenn & Company Ltd

34 Nixon Road, Bolton, Ontario L7E 1W2
Ⓣ 001 905 951 6600    Ⓕ 001 905 951 6601
www.hbfenn.com

Founded 1977. Publishes adult and young adult sports books.

## Harlequin Enterprises Ltd

225 Duncan Mill Road, Don Mills, Ontario M3B 3K9
Ⓣ 001 416 445 5860
www.eharlequin.com

Founded 1949. Canada's second largest publishing house. Publishes romantic and women's fiction.

## HarperCollins Canada Ltd

1995 Markham Road, Scarborough, Ontario M1B 5M8
Ⓣ 001 416 321 2241    Ⓕ 001 416 321 3033
www.harpercanada.com

Founded 1989. Publishes commerical and literary fiction, non-fiction, children's, reference, cookery and religious books.

## John Wiley & Sons Canada Ltd

5353 Dundas Street W., Suite 400, Toronto, Ontario M9B 6H8
Ⓣ 001 416 236 4433
canada@wiley.com
www.wiley.ca

Founded 1968. Subsidary of **John Wiley & Sons Inc.**, USA. Publishes professional, reference and textbooks.

## Kellom Books
▷ Canadian Scholars' Press, Inc

## Knopf Canada
▷ Random House of Canada Ltd

## LexisNexis Canada Ltd
Suite 200, 181 University Avenue, Toronto, Ontario M5H 3M7
℡ 001 416 862 7656    ℻ 001 416 862 8073
www.lexisnexis.ca

Founded 1912. Publishes law books, CD-ROMs, journals, newsletters, law reports.

## McClelland & Stewart Ltd
75 Sherbourne Street, Fifth Floor, Toronto, Ontario M5A 2P9
℡ 001 416 598 1114    ℻ 001 416 598 7764
mail@mcclelland.com
www.mcclelland.com

Founded 1906. Publishes fiction and general non-fiction, poetry. DIVISIONS/IMPRINTS **Douglas Gibson Books**; **Emblem**; **Tundra Books** (see entry). About 100 titles a year.

## McGraw-Hill Ryerson Ltd
300 Water Street, Whitby, Ontario L1N 9B6
℡ 001 905 430 5000
www.mcgrawhill.ca

Founded 1944. Subsidiary of **The McGraw-Hill Companies, Inc.** Publishes education and professional.

## Nelson Education
1120 Birchmount Road, Scarborough, Ontario M1K 5G4
℡ 001 416 752 9100    ℻ 001 416 752 8101
inquire@nelson.com
www.nelson.com

Part of the Thomson Corporation. Publishes directories, education, professional, reference, textbooks.

## New Star Books Ltd
107–3477 Commercial Street, Vancouver, British Columbia V5N 4E8
℡ 001 604 738 9429    ℻ 001 604 738 9332
info@newstarbooks.com
www.newstarbooks.com

Founded 1974. Publishes fiction and non-fiction, literature, poetry, history, sociology, social sciences.

## Oxford University Press, Canada
8 Sampson Mews, Suite 204, Don Mills, Ontario M3C 0H5
℡ 001 416 441 2941    ℻ 001 416 444 0427
www.oup.com/ca

Founded 1904. Owned by **Oxford University Press**, UK. Publishes for college, school and trade markets.

## Pearson Canada
26 Prince Andrew Place, Toronto, Ontario M3C 2T8
℡ 001 416 447 5101    ℻ 001 416 443 0948
www.pearsoncanada.ca/index.html

Founded 1966. A division of Pearson Plc and Canada's largest publisher. Publishes fiction, non-fiction and educational material in English and French.

## Penguin Group (Canada)
90 Eglinton Avenue East, Suite 700, Toronto, Ontario M4P 2Y3
℡ 001 416 925 2249    ℻ 001 416 925 0068
info@ca.penguingroup.com
www.penguin.ca

Founded 1974. Owned by Pearson Plc. Publishes fiction and non-fiction books and audio cassettes.

## Random House of Canada Ltd
One Toronto Street, Unit 300, Toronto, Ontario M5C 2V6
℡ 001 416 364 4449    ℻ 001 416 364 6863
www.randomhouse.ca

Founded 1944. Publishes fiction and non-fiction and children's. IMPRINTS **Anchor Canada** Fiction and non-fiction; **Bond Street** Fiction and non-fiction; **Doubleday Canada** Fiction (literary and commercial), business, history, memoir, journalism, young adult books; **Knopf Canada** Fiction and non-fiction; **Seal Books** Primarily reprints of fiction and some non-fiction; **Vintage Canada** Paperback fiction and non-fiction.

## Scholastic Canada Ltd
175 Hillmount Road, Markham, Ontario L6C 1Z7
℡ 001 905 887 7323    ℻ 001 905 887 3639
www.scholastic.ca

Founded 1957. Publishes (in English and French) children's books and educational material.

## Seal Books
▷ Random House of Canada Ltd

## Tundra Books
75 Sherbourne Street, 5th Floor, Toronto, Ontario M5A 2P9
℡ 001 416 598 4786
tundra@mcclelland.com
www.tundrabooks.com

Founded 1967. Division of **McClelland & Stewart Ltd**. Publishes (in English and French) children's and young adult books.

## University of Toronto Press, Inc.

10 St Mary Street, Suite 700, Toronto, Ontario M4Y 2W8

📞 001 416 978 2239
info@utpress.utoronto.ca
www.utpress.utoronto.ca

Founded 1901. Publishes scholarly books and journals. About 150 titles a year.

## Vintage Canada
▷ Random House of Canada Ltd

## Women's Press
▷ Canadian Scholars' Press, Inc

# India

## Affiliated East West Press Pvt Ltd

G–1/16 Ansari Road, Darya Ganj, New Delhi 110 002

📞 00 91 11 2327 9113
affiliate@vsnl.com
www.aewpress.com

Founded 1962. Publishes textbooks, CD-ROMs, electronic books.

## Caring
▷ Vision Books Pvt. Ltd

## HarperCollins Publishers India Ltd

A 53, Sector 57, Noida, Uttar Pradesh

📞 00 91 120 404 4800
contact@harpercollins.co.in
www.harpercollins.co.in

A joint venture between the India Today Group and **HarperCollins Publishers**. Publishes fiction and non-fiction; autobiography, astrology, biography, children's, cinema, cookery, crime, current affairs, dictionaries, fantasy, gardening, graphic novels, home interest, literature, memoirs, mind, body and spirit, poetry, politics, popular science, philosophy, psychology, puzzles, reference, religion, science fiction, short stories, sport, thrillers, travel.

## Jaico Publishing House

127 M.G. Road, Mumbai 400 023

📞 00 91 22 267 6702
www.jaicobooks.com/home.asp

Founded 1946. Publishes computer science, engineering, finance, health and nutrition, information technology, law, reference, literature, self-help, philosophy, religion, history, government, political science, sociology, textbooks.

## LexisNexis India

14th Floor, Tower B, Building No. 10, DLF Cyber City, Phase–11, Gurgaon–122002, Haryana

📞 00 91 124 477 4444     📠 00 91 124 477 4100
info.in@lexisnexis.co.in
www.lexisnexis.co.in

Part of Reed Elsevier. Publishes law, taxation, government and business books in print form, online and CD-ROM.

## Macmillan India Ltd

315–316 Raheja Chambers, 12 Museum Road, Bangalore 560 011

📞 00 91 80 2558 7878
www.macmillanindia.com

Founded 1893. Part of **Macmillan Publishers**, UK. Publishes education, non-fiction, reference, dictionaries, encyclopaedias.

## McGraw-Hill Education (India)

B-4, Sector 63, Noida, Uttar Pradesh 201 301

📞 00 91 120 438 3400
editorial_india@mcgraw-hill.com
www.tatamcgrawhill.com

Founded 1970. Educational publisher.

## Munshiram Manoharlal Publishers Pvt Ltd

PO Box 5715, 54 Rani Jhansi Road, New Delhi 110 055

📞 00 91 11 2367 1668     📠 00 91 11 2361 2745
info@mrmlonline.com
www.mrmlbooks.com

Founded 1952. Publishes academic, dictionaries, anthropology, art and art history, autobiography, architecture, archaeology, biography, environment, religion, philosophy, geography, history, humanities, language arts and linguistics, law, literature, medicine, numismatics, music, dance, theatre, Asian studies, sociology, travel.

## National Publishing House

2/35 Ansari Road, Darya Ganj, New Delhi 110 002

📞 00 91 11 2327 5267
info@nationalpublishinghouse.com
www.nationalpublishinghouse.com

Founded 1945. Publishes non-fiction, children's, history, literature, science, textbooks.

## Orient Paperbacks

5 A/8 Ansari Road, First Floor, Darya Ganj, Delhi 110 002

📞 00 91 11 2327 8877     📠 00 91 11 2327 8879
mail@orientpaperbacks.com
www.orientpaperbacks.com

Founded 1975. Division of **Vision Books Pvt. Ltd**. Publishes (in English and Hindi) fiction and general non-fiction; humour, health and nutrition, literature, New Age, puzzles, reference, religion, self-help.

## Oxford University Press India

1st Floor, YMCA Library Building, 1 Jai Singh Road, New Delhi 110 001

☎ 00 91 11 4360 0300    📠 00 91 11 2336 0897

admin.in@oup.com

www.oup.co.in

Founded 1912. Owned by **Oxford University Press**, UK. Publishes academic, education, dictionaries. Reference subjects include arts, architecture, biography, natural history, memoirs.

## Rajpal
▷ **Vision Books Pvt. Ltd**

## S. Chand & Co Ltd

7361 Ram Nagar, Qutub Road, New Delhi 110 055

☎ 00 91 11 2367 2080    📠 00 91 11 2367 7446

info@schandgroup.com

www.schandgroup.com

Founded 1917. Publishes computer science, dictionaries, encyclopaedias, engineering, reference, textbooks and educational CDs.

## Vision Books Pvt. Ltd

24, Feroze Gandhi Road, Lajpat Nagar III, New Delhi 110 024

☎ 00 91 11 2386 2201    📠 00 91 11 2983 6490

editor@visionbooksindia.com

www.visionbooksindia.com

Founded 1975. Specializes in books on business and management, finance and taxation, careers, current affairs, religion. DIVISIONS/IMPRINTS **Vision Books**; **Orient Paperbacks** (see entry); **Caring**; **Rajpal**.

# New Zealand

## Auckland University Press

University of Auckland, Private Bag 92019, Auckland

☎ 00 64 9 373 7528    📠 00 64 9 373 7465

aup@auckland.ac.nz

www.auckland.ac.nz/aup

Founded 1966. Publishes academic, biography, government, political science, history, social sciences, sociology, poetry, literature, literary criticism, Maori studies.

## Bridget Williams Books Ltd

PO Box 5482, Wellington

☎ 00 64 4 473 8128

info@bwb.co.nz

www.bwb.co.nz

Founded 1990. Independent publisher of New Zealand history, Maori experience, contemporary issues and women's studies.

## Canterbury University Press

University of Canterbury, Private Bag 4800, Christchurch

☎ 00 64 3 364 2914    📠 00 64 3 364 2044

mail@cup.canterbury.ac.nz

www.cup.canterbury.ac.nz

Founded 1964. Academic publishers; general non-fiction; history, marine biology, natural history.

## The Caxton Press

113 Victoria Street, Christchurch

☎ 00 64 3 353 0734    📠 00 64 3 365 7840

www.caxton.co.nz

Founded 1935. Publishes general non-fiction.

## Hachette Livre New Zealand Ltd

4 Whetu Place, Mairangi Bay, Auckland

☎ 00 64 9 478 1000    📠 00 64 9 478 1010

www.hachette.co.nz

Founded 1971. Part of the **Hachette Livre Publishing Group**. Publishes fiction and general non-fiction; biography, business, children's, cookery, humour, sport.

## HarperCollins Publishers (New Zealand) Ltd

PO Box 1, Shortland Street, Auckland 1140

☎ 00 64 9 443 9400    📠 00 64 9 443 9403

editors@harpercollins.co.nz

www.harpercollins.co.nz

Founded 1888. Publishes fiction, literature, children's, business, cookery, gardening, reference, religion.

## Huia (NZ) Ltd

PO Box 17-335, 39 Pipitea Street, Thorndon, Wellington, Aotearoa

☎ 00 64 4 473 9262    📠 00 64 4 473 9265

www.huia.co.nz

Founded 1991. Independent publisher of Maori cultural history and language, children's books in Maori and English, fiction.

## LexisNexis New Zealand

PO Box 472, Wellington 6140

☎ 00 64 4 385 1479    📠 00 64 4 385 1598

www.lexisnexis.co.nz

Founded 1914. Part of the LexisNexis Group. Publishes law, professional and textbooks in book and electronic formats.

## Macmillan Publishers New Zealand

6 Ride Way, Albany, Auckland

☎ 00 64 9 414 0350    📠 00 64 9 414 0351

www.macmillan.co.nz

Education publishers – general, school and academic books; fiction and non-fiction for all ages.

## McGraw-Hill New Zealand Pty Ltd
▷ McGraw-Hill Australia & New Zealand Pty Ltd

## Oxford University Press New Zealand
▷ Oxford University Press (Australia)

## Pearson New Zealand
Private Bag 102902, North Shore City 0745
℡ 00 64 9 442 7400    ℻ 00 64 9 442 7401
www.pearsoned.co.nz

Founded in 1998 from the merger of Addison
Wesley Longman New Zealand and Prentice Hall
New Zealand. Educational publishers.

## Penguin Group (NZ)
Private Bag 102 902, North Shore Mail Centre,
Auckland 0745
℡ 00 64 9 442 7400    ℻ 00 64 9 442 7401
publishing@penguin.co.nz
www.penguin.co.nz

Founded 1978. Owned by **Penguin** UK. Adult and
children's fiction and non-fiction.

## Random House New Zealand
Private Bag 102950, North Shore Mail Centre,
North Shore City 0745
℡ 00 64 9 444 7197    ℻ 00 64 9 444 7524
editor@randomhouse.co.nz
www.randomhouse.co.nz

Publishes fiction and non-fiction (cooking,
gardening, art, natural history) and children's.

## Victoria University Press
PO Box 600, Wellington
℡ 00 64 4 463 6580    ℻ 00 64 4 463 6581
victoria-press@vuw.ac.nz
www.vuw.ac.nz/vup

Founded 1979. Academic publishers; new fiction,
biography, poetry, literature, essays, New Zealand
history, Maori topics. About 25 titles a year.

# South Africa

## Heinemann International Southern Africa
PO Box 781940, Sandton 2146
℡ 00 27 11 322 8600    ℻ 00 27 11 322 8715
www.heinemann.co.za

Founded 1986. Educational publisher.

## LexisNexis South Africa
215 North Ridge Road, Morningside, Durban 4001
℡ 00 27 31 268 3111
www.lexisnexis.co.za

Publishes professional, accountancy, finance,
business, law, taxation and economics books in
print form, online and CD-ROM.

## Macmillan South Africa
Private Bag X19, Northlands 2194
℡ 00 27 11 731 3300    ℻ 00 27 11 731 3500
info@macmillan.co.za
www.macmillan.co.za

Founded 1972. Part of **Macmillan Publishers Ltd**,
UK. Educational publisher.

## Maskew Miller Longman
PO Box 396, Cape Town 8000
℡ 00 27 21 532 6000    ℻ 00 27 21 531 8103
www.mml.co.za

Founded 1893. Owned jointly by **Pearson
Education** and Caxton Publishers and Printers
Ltd. Publishes education, ELT and teacher
support books.

## Oshun
▷ Random House Struik (Pty) Ltd

## Oxford University Press Southern Africa (Pty) Ltd
PO Box 12119, N1 City, Cape Town 7463
℡ 00 27 21 596 2300    ℻ 00 27 21 596 1234
oxford.za@oup.com
www.oup.com/za

Founded 1915. Parent company: **Oxford
University Press**, UK. Publishes academic,
educational and general books.

## Random House Struik (Pty) Ltd
PO Box 1144, Cape Town 8000
℡ 00 27 21 462 4360    ℻ 00 27 21 462 4377
info@struik.co.za
www.struik.co.za

Formed by the merger of Random House SA and
Struik Publishing in September 2008. Publishes
general non-fiction, illustrated, art and culture,
lifestyle, natural history, travel. IMPRINTS **Oshun**
Fiction, biography, health, humour, parenting,
reference, self help, South African interest;
**Umuzi** Literary fiction with a South African
flavour and non-fiction; **Zebra Press** Fiction and
non-fiction.

## Shuter & Shooter (Pty) Ltd
110 CB Downes Road, Pietermaritzburg,
KwaZulu-Natal 3201
℡ 00 27 33 846 8700    ℻ 00 27 33 846 8701
www.shuters.com

Founded 1925. Publishes educational and general
books.

## Umuzi
▷ Random House Struik (Pty) Ltd

## University of Kwa-Zulu-Natal Press
Private Bag X01, Scottsville 3209
℡ 00 27 33 260 5226    ℻ 00 27 33 260 5801

books@ukzn.ac.za
www.ukznpress.co.za

Founded 1947. Publishes academic and general books; children's, African literature, poetry, economics, military history, natural sciences, social sciences.

## Wits University Press

PO Wits, Johannesburg 2050

Ⓣ 00 27 11 484 5906    Ⓕ 00 27 11 484 5971

witspress.wits.ac.za

Founded 1922. Publishes academic, biography and memoirs, economics, heritage, history, business, politics, popular science, women's writing.

## Zebra Press
▷ Random House Struik (Pty) Ltd

# Poetry Presses

## Agenda Editions

The Wheelwrights, Fletching Street, Mayfield
TN20 6TL
℡ 01435 873703
editor@agendapoetry.co.uk
www.agendapoetry.co.uk
Editor *Patricia McCarthy*

Separate collections of poetry. Very few published
in a year and not many unsolicited. Seek
publication in **Agenda Poetry Magazine** first –
see entry under *Poetry Magazines*.

## Anvil Press Poetry Ltd

Neptune House, 70 Royal Hill, London SE10 8RF
℡ 020 8469 3033    ℻ 020 8469 3363
anvil@anvilpresspoetry.com
www.anvilpresspoetry.com
Contact *Peter Jay*

Contemporary British poetry and poetry in
translation. See entry under *UK Publishers*.

## Arc Publications Ltd

Nanholme Mill, Shaw Wood Road, Todmorden
OL14 6DA
℡ 01706 812338    ℻ 01706 818948
arc.publications@btconnect.com
www.arcpublications.co.uk
Editorial Director *Tony Ward*

Contemporary poetry from new and established
writers using English as their first language
from the UK and abroad. See entry under *UK
Publishers*.

## Areté Books
▷ entry under Small Presses

## Arrowhead Press

70 Clifton Road, Darlington DL1 5DX
editor@arrowheadpress.co.uk
roger.collett@ntlworld.com
www.arrowheadpress.co.uk
Managing Editor *Roger Collett*

Quality books and pamphlets of contemporary
poetry. Not accepting unsolicited submissions at
present.

## Atlantean Publishing

4 Pierrot Steps, 71 Kursaal Way, Southend on Sea
SS1 2UY
atlanteanpublishing@hotmail.com
atlanteanpublishing.web.officelive.com
Contact *David John Tyrer*

Single author broadsheets and poetry/prose
booklets. See also **Awen**, **Monomyth**, **Bard**,
**Garbaj** and **The Supplement** magazines.

## Awen Publications
▷ entry under Small Presses

## Barque Press

26 Allerton Road, London N16 5UJ
www.barquepress.com
Contact *Andrea Brady*

Founded 1995. Poetry. 'Barque does not accept or
return unsolicited mss.'

## BB Books

Spring Bank, Longsight Road, Copster Green,
Blackburn BB1 9EU
℡ 01254 249128
Contact *Dave Cunliffe*

Post-Beat poetics and counterculture theoretic.
Iconoclastic rants and anarchic psycho-cultural
tracts. See also **Global Tapestry Journal**.

## Between the Lines
▷ entry under Small Presses

## BeWrite Books
▷ entry under Small Presses

## Beyond the Cloister Publications

74 Marina, St Leonards on Sea TN38 0BJ
beyondcloister@hotmail.co.uk
Editor *Hugh Hellicar*

Founded 1995. Anthologies of poetry and single
poet volumes. Sample submissions welcome by
post or email. Booklist on request.

## Bloodaxe Books Ltd

Highgreen, Tarset NE48 1RP
℡ 01434 240500    ℻ 01434 240505
editor@bloodaxebooks.com
www.bloodaxebooks.com

Contact *Neil Astley*

Britain's leading publisher of new poetry. No submissions by email attachments. See entry under *UK Publishers*.

## Bradshaw Books
▷ entry under Irish Publishers

## Bridge Pamphlets
PO Box 309, Aylsham, Norwich NR11 6LN
www.therialto.co.uk
Contact *Michael Mackmin*

An imprint of **Rialto Publications**.

## The Brodie Press
▷ entry under Small Presses

## Calder Wood Press
1 Beachmont Court, Dunbar EH42 1YF
☎ 01368 864953
colin.will@zen.co.uk
www.calderwoodpress.co.uk
Contact *Colin Will*

Founded 1997. Publishes poetry pamphlets and short-run publications. No unsolicited mss.

## Carcanet Press Ltd
Major poetry publisher. See entry under *UK Publishers*; also publishes **PN Review** magazine.

## Chapman Publishing
▷ entry under UK Publishers

## Cinnamon Press
Ty Meirion, Glan yr afon, Tanygrisiau, Blaenau Ffestiniog LL41 3SU
☎ 01766 832112
jan@cinnamonpress.com
www.cinnamonpress.com
Editor *Dr Jan Fortune-Wood*

Poetry, fiction, some non-fiction. Growing list from Wales, UK & international. Regular literary competitions. See website for submission guidelines. See also **Envoi** magazine.

## The Collective
c/o Penlanlas Farm, Llantilio Pertholey, Y-fenni NP7 7HN
jj@jojowales.co.uk
www.welshwriters.com
Contact *John Jones & Frank Olding*

Non-profit promoters and publishers of contemporary poetry. Hard copy submissions only. Enclose s.a.s.e.

## Day Dream Press
39 Exmouth Street, Swindon SN1 3PU
☎ 01793 523927
Contact *Kevin Bailey*

Small poetry collections. Now associated with Bluechrome Press. See also **HQ Poetry Magazine (Haiku Quarterly)**.

## The Dedalus Press
13 Moyclare Road, Baldoyle, Dublin 13, Republic of Ireland
☎ 00 353 1 839 2034    ☎ 0705 360 3342
editor@dedaluspress.com
www.dedaluspress.com
Publisher *Pat Boran*

Contemporary Irish poetry and poetry from around the world in English translation.

## Dionysia Press Ltd
127 Milton Road West, 7 Duddingston House Courtyard, Edinburgh EH15 1JG
http://dionysiapress.wordpress.com
Contact *Denise Smith*

Collections of poetry, short stories, novels and translations. See also **Understanding** magazine.

## DogHouse
PO Box 312, Tralee, Co. Kerry, Republic of Ireland
☎ 00 353 667 137547    ☎ 00 353 667 137547
info@doghousebooks.ie
www.doghousebooks.ie
Contact *Noel King*

Poetry and short story collections by individuals (Irish born only). About 4–6 titles a year.

## Dream Catcher
4 Church Street, Market Rasen, Lincoln LN8 3ET
☎ 01673 844325
paulsuther@hotmail.com
www.dreamcatchermagazine.co.uk
Contact *Paul Sutherland*

Short fiction, poetry, reviews, artwork, biographies. See also **Dream Catcher** magazine.

## Egg Box Publishing Ltd
www.eggboxpublishing.com
Publishing Director *Nathan Hamilton*

New poetry and fiction. No postal submissions.

## Enitharmon Press
26B Caversham Road, London NW5 2DU
☎ 020 7482 5967    ☎ 020 7284 1787
books@enitharmon.co.uk
www.enitharmon.co.uk
Contacts *Stephen Stuart-Smith, Isabel Brittain, Jacqueline Gabbitas*

Poetry and literary criticism. See entry under *UK Publishers*.

## Essence Press
8 Craiglea Drive, Edinburgh EH10 5PA
editor@essencepress.co.uk
www.essencepress.co.uk
Contact *Julie Johnstone*

Handbound editions of poetry and cards. Interested in minimalist writing.

## Etruscan Books

Elm House, Stowe Lane, Exbourne EX20 3RY
℡ 01837 851669
atetruscan@aol.com
www.etruscan.co.uk
Managing Editor *Nicholas Johnson*

No unsolicited submissions.

## Fal Publications

PO Box 74, Truro TR1 1XS
℡ 07887 560018
info@falpublications.co.uk
www.falpublications.co.uk
Contact *Victoria Field*

Poetry and literature from Cornwall. Please email for submissions policy.

## Feather Books
▷ entry under Small Presses

## Fineleaf
▷ entry under Small Presses

## Five Leaves Publications
▷ entry under Small Presses

## Flambard Press

16 Black Swan Court, 69 Westgate Road, Newcastle upon Tyne NE1 1SG
℡ 0191 222 1329    🖷 01434 674178
editor@flambardpress.co.uk
www.flambardpress.co.uk
Contact *Peter Lewis & Will Mackie*

Concentrates on poetry and literary fiction. Consult the website for submission information.

## Flarestack Poets Pamphlets

PO Box 14479, Birmingham B13 3GU
meria@btinternet.com
jacquirowe@hotmail.co.uk
www.flarestackpoets.co.uk
Contact *Meredith Andrea & Jacqui Rowe*

## Forward Press

Remus House, Woodston, Peterborough PE2 9JX
℡ 01733 890099    🖷 01733 313524
info@forwardpress.co.uk
www.forwardpress.co.uk
Contact *The Editorial Team*

General poetry and short fiction anthologies.

## The Frogmore Press

21 Mildmay Road, Lewes BN7 1PJ
℡ 07751 251689
J.N.Page@sussex.ac.uk
www.frogmorepress.co.uk
Managing Editor *Jeremy Page*

A forum for contemporary poetry, prose and artwork. See also **The Frogmore Papers** magazine.

## The Gallery Press

Loughcrew, Oldcastle, Co. Meath, Republic of Ireland
℡ 00 353 49 854 1779    🖷 00 353 49 854 1779
gallery@indigo.ie
www.gallerypress.com
Editor & Publisher *Peter Fallon*

Poems and plays by contemporary Irish writers. See entry under *Irish Publishers*.

## Gomer Press/Gwasg Gomer

Llandysul Enterprise Park, Llandysul SA44 4JL
℡ 01559 363090    🖷 01559 363758
gwasg@gomer.co.uk
www.gomer.co.uk
www.pontbooks.co.uk
English Books for Adults *Ceri Wyn Jones* (ceri@gomer.co.uk)
English Books for Children *Viv Sayer* (viv@gomer.co.uk)
Publishing Director *Mairwen Prys Jones* (mairwen@gomer.co.uk)

Welsh interest. See entry under *UK Publishers*.

## Green Arrow Publishing

6 Green Bank, Stacksteads, Bacup OL13 8LQ
℡ 01706 870357/07967 315270
mail@johndench.demon.co.uk
Contact *John Dench*

Collaborative publishing scheme and other services for writers. See also **Scriptor** magazine.

## HappenStance Press

21 Hatton Green, Glenrothes KY7 4SD
www.happenstancepress.com
Editor *Helena Nelson*

Poetry publishing; mainly chapbooks and usually first collections. Submission guidelines on website. Reading 'windows' December and July. See also **Sphinx** magazine.

## Headland Publications

Ty Coch, Galltegfa, Ruthin LL15 2AR
℡ 0151 625 9128
headlandpublications@hotmail.co.uk
www.headlandpublications.co.uk
Contact *Gladys Mary Coles*

Fine editions of poetry; anthologies.

## Hearing Eye

c/o 99 Torriano Avenue, London NW5 2RX
hearing_eye@torriano.org
www.torriano.org/hearing_eye/
Contact *John Rety*

Poetry/literature publishers based in Kentish Town.

## Hilltop Press

4 Nowell Place, Almondbury, Huddersfield HD5 8PD
booksmusicfilmstv.com/HilltopPress.htm
Contact *Steve Sneyd*

Specialist publisher of science fiction and dark fantasy poetry, and background material. Overseas orders to: www.bbr-online.com/catalogue

## Honno Welsh Women's Press

▷ entry under UK Publishers

## I*D Books

Connah's Quay Library, Wepre Drive, Connah's Quay, Deeside
☎ 01244 830485

Poetry, short fiction, local history – mainly self-publishing by associated writers' group.

## Indigo Dreams Publishing

132 Hinckley Road, Stoney Stanton LE9 4LN
☎ 01455 272861
idpoet@rocketmail.com
www.indigodreamsonline.com
Managing Editors *Ronnie Goodyer & Dawn Bauling*

Publishes poetry, short stories, general fiction and biography. Submissions welcome through the enquiry procedure on the website. See also *The Dawntreader*, *Reach Poetry* and *Sarasvati* magazines.

## Iron Press

5 Marden Terrace, Cullercoats, North Shields NE30 4PD
☎ 0191 253 1901
ironpress@blueyonder.co.uk
www.ironpress.co.uk
Editor *Peter Mortimer*

Poetry and fiction. Phone the editor before submitting material.

## Katabasis

10 St Martin's Close, London NW1 0HR
☎ 020 7485 3830    🖷 020 7485 3830
katabasis@katabasis.co.uk
www.katabasis.co.uk
Contact *Dinah Livingstone*

Down-to-earth and Utopian poetry and prose from home an abroad – English and Latin American. No unsolicited mss.

## Kates Hill Press

126 Watsons Green Road, Kates Hill, Dudley DY2 7LG
kateshillpress@blueyonder.co.uk
www.kateshillpress.co.uk

Fiction and social history from the West Midlands or with a West Midlands theme. Some poetry. No unsolicited mss.

## The King's England Press

Cambertown House, Commercial Road, Goldthorpe, Rotherham S63 9BL
☎ 01484 663790    🖷 01484 663790
sales@kingsengland.com
www.kingsengland.com
www.pottypoets.com
Contact *Steve Rudd*

Founded 1989. History, folklore, adult and children's poetry.

## KT Publications

16 Fane Close, Stamford PE9 1HG
☎ 01780 754193
Editor *Kevin Troop*

Always looking for new material of the highest standard. See also **The Third Half** magazine.

## Leaf Books

GTi Suite, Valleys Innovation Centre, Navigation Park, Abercynon CF45 4SN
contact@leafbooks.co.uk
www.leafbooks.co.uk

Poetry and Micro-Fiction competition anthologies: getting new writers into print.

## Leafe Press

4 Cohen Close, Chilwell, Nottingham NG9 6RW
www.leafepress.com
Contact *Alan Baker*

Pamphlets of contemporary poetry with occasional full-length collections. Currently, not accepting unsolicited submissions.

## Mare's Nest

41 Addison Gardens, London W14 0DP
☎ 020 7603 3969
www.maresnest.co.uk
Contact *Pamela Clunies-Ross*

Icelandic literature.

## Mariscat Press

10 Bell Place, Edinburgh EH3 5HT
☎ 0131 343 1070
hamish.whyte@btinternet.com
Contact *Hamish Whyte & Diana Hendry*

Currently publishing poetry pamphlets only.

## Masque Publishing

PO Box 3257, Littlehampton BN16 9AF
masque_pub@btinternet.com
http://myweb.tiscali.co.uk/masquepublishing
Contact *Lisa Stewart*

Self publishing service. See also **Decanto** poetry magazine.

## Maypole Editions
▷ entry under Small Presses

## Mudfog Press

Arts Development, The Stables, Stewart Park, The Grove, Marton, Middlesborough TS7 8AR
www.mudfog.co.uk

Community press publishing poetry and short fiction only from writers in the Tees Valley and adjoining rural areas. Email via website.

## New Island

2 Brookside, Dundrum Road, Dublin 14, Republic of Ireland
www.newisland.ie

See entry under *Irish Publishers*.

## nthposition

38 Allcroft Road, London NW5 4NE
☎ 020 7485 5002
val@nthposition.com
www.nthposition.com
Editor *Val Stevenson*
Poetry Editor *Rufo Quintavalle*
*(rquintav@gmail.com)*

Eclectic and award-winning mix of politics and opinion, travel writing, fiction and poetry, reviews and interviews and some high weirdness. Submissions: 'Poetry: two poems only plus brief biographical statement. The rest: open submission, with brief biographical statement, but 80% is rejected. Whimsy will guarantee rejection.'

## The Old Stile Press

Catchmays Court, Llandogo, Nr Monmouth
NP25 4TN
☎ 01291 689226
oldstile@dircon.co.uk
www.oldstilepress.com
Contact *Frances & Nicolas McDowall*

Fine, hand-printed books with text and images. No unsolicited mss.

## The One Time Press

Model Farm, Linstead Magna, Halesworth
IP19 0DT
☎ 01986 785422
www.onetimepress.com
Contact *Peter Wells*

Poetry of the forties and occasionally some contemporary work in limited editions. Illustrated.

## Original Plus

17 High Street, Maryport CA15 6BQ
☎ 01900 812194
smithsssj@aol.com
www.freewebs.com/thesamsmith
Contact *Sam Smith*

See also **The Journal** magazine.

## Oversteps Books Ltd

6 Halwell House, South Pool, Nr Kingsbridge
TQ7 2RX
☎ 01548 531969
alwynmarriage@overstepsbooks.com
www.overstepsbooks.com
Managing Editor *Dr Alwyn Marriage*

Publishes poetry collections by new poets who have a proven record of publishing in poetry magazines or who have won major prizes for their work. No unsolicited mss; criteria for submissions are on the website.

## Parthian

The Old Surgery, Napier Street, Aberteifi (Cardigan) SA43 1ED
☎ 01239 612059
www.parthianbooks.co.uk

New Welsh writing. See entry under *UK Publishers*.

## Partners

289 Elmwood Avenue, Feltham TW13 7QB
partners_writing_group@hotmail.com
Contact *Ian Deal*

Competitions and poetry magazines. See also **Aspire**, **A Bard Hair Day**, **Imagenation** and **Poet Tree** magazines.

## Peepal Tree Press Ltd

17 King's Avenue, Leeds LS6 1QS
☎ 0113 245 1703
contact@peepaltreepress.com
www.peepaltreepress.com
Contact *Jeremy Poynting*

Peepal Tree Press is the home of challenging and inspiring literature from the Caribbean and Black Britain. Online bookstore.

## Penniless Press

100 Waterloo Road, Ashton, Preston PR2 1EP
☎ 01772 736421
www.pennilesspress.co.uk
Contact *Alan Dent*

Novels, short fiction, poetry, plays. See also **The Penniless Press** magazine.

## Perdika Poetry

16B St Andrew's Road, Enfield EN1 3UB
☎ 020 8363 3413
editions@perdikapress.com
www.perdikapress.com
Managing Editor *Peter Brennan*

Publishes original and translated works by
contemporary poets. Unsolicited mss welcome
'but only from poets who have familiarized
themselves with the kinds of material that we
publish'. Contact by email with a sample of two or
three poems.

## Pigasus Press

13 Hazely Combe, Arreton, Isle of Wight PO30 3AJ
mail@pigasuspress.co.uk
www.pigasuspress.co.uk
Contact *Tony Lee*

Publisher of science fiction magazine and genre
poetry.

## Pighog Press

PO Box 145, Brighton BN1 6YU
info@pighog.co.uk
www.pighog.co.uk
Contact *The Editor*

'Small press. Big impact. An award-winning
press, for award winning poets. Pighog Press has
produced beautifully crafted pamphlets since
2002.' Does not accept unsolicited mss. Scouts
for unique voices all the time; look out for poetry
and short story events and competitions on the
Pighog website.

## Pikestaff Press

Ellon House, Harpford, Sidmouth EX10 0NH
☎ 01395 568941
Contact *Robert Roberts*

Contemporary poetry belonging to the English
tradition.

## Pipers' Ash Ltd
▷ entry under Small Presses

## Poems in the Waiting Room

PO Box 488, Richmond TW9 4SW
pitwr@blueyonder.co.uk
www.poemsinthewaitingroom.org
Editor *Isobel Montgomery Campbell*
Executive Editor *Cynthia Roberts*
Chairman *Michael Lee*

Arts in Health charity providing poetry cards
for NHS patients. Described in a House of
Lords debate as the most widely read national
poetry publication in the UK. Circulates some
50,000 poetry cards each quarter. 'The readers
are patients – the worried well and the worried
sick. The poems need to draw from the springs

of well-being.' Submission guidelines available on
the website and on request.

## Poetry Now

Remus House, Coltsfoot Drive, Woodston,
Peterborough PE2 8JX
☎ 01733 898101    ☐ 01733 313524
poetrynow@forwardpress.co.uk
www.forwardpress.co.uk
Editorial Manager *Michelle Afford*

Lively, personal and contemporary. Imprint of
**Forward Press**.

## Poetry Powerhouse Press
▷ Performance Poetry Society under Organizations of
Interest to Poets

## Poetry Press

26 Park Grove, Edgware HA8 7SJ
☎ 020 8958 6499
poetrypress@yahoo.co.uk
www.judyk.co.uk
www.jewishpoetrysociety.com
Contact *Judy Karbritz*

Anthologies – rhyming poetry welcome.

## Poetry Salzburg

Dept. of English, University of Salzburg,
Akademiestr. 24, A–5020 Salzburg, Austria
☎ 00 43 662 8044 4424    ☐ 00 43 662 8044 167
editor@poetrysalzburg.com
www.poetrysalzburg.com
Co-editors *Wolfgang Görtschacher & Andreas
Schachermayr*

Publishes books of poetry, poetry in translation,
poetry anthologies and literary criticism. See also
**Poetry Salzburg Review** magazine.

## Poetry Wednesbury

25 Griffiths Road, West Bromwich B71 2EH
☎ 07950 591455
ppatch66@hotmail.com
www.poetrywednesburyinternational.org.uk
Contact *Geoff Stevens*

Occasional CDs, DVDs, leaflets and booklets.
See also **Purple Patch** magazine. Public meetings
for readings, workshops, etc. monthly at St Paul's
Church, Wood Green, Wednesbury.

## Poets Anonymous

70 Aveling Close, Purley CR8 4DW
poets@poetsanon.org.uk
www.poetsanon.org.uk
Contact *Peter L. Evans*

Anthologies and collections of predominantly
south London poets. See also **Poetic Licence**
magazine.

## The Potty Poets

www.pottypoets.com

Children's poetry. See **The King's England Press**.

## Precious Pearl Press

41 Grantham Road, Manor Park, London E12 5LZ
Contact *P.G.P. Thompson*

Romantic, lyrical, spiritual, inspirational and mystical poetry; a traditionalist poetry press. See also **Rubies in the Darkness** magazine.

## PS Avalon

Box 1865, Glastonbury BA6 8YR
will@willparfitt.com
www.willparfitt.com
Contact *Will Parfitt*

Books for personal and spiritual development.

## Puppet State Press

40/1 Woodhall Road, Edinburgh EH13 0DU
☏ 0131 441 9693
richard@puppetstate.com
www.puppetstate.com
Contact *Richard Medrington*

Currently, not accepting submissions.

## QQ Press

York House, 15 Argyle Terrace, Rothesay, Isle of Bute PA20 0BD
Contact *Alan Carter*

Collections of poetry plus poetry anthologies. Enquiries should be marked 'Collections' and sent with an s.a.se. or two IRCs if from abroad. See also **Quantum Leap** magazine.

## Rack Press

The Rack, Kinnerton, Presteigne LD8 2PF
☏ 01547 560411
rackpress@nicholasmurray.co.uk
www.nicholasmurray.co.uk/RackPress.html
www.rackpress.blogspot.com
Contact *Nicholas Murray*

Welsh poetry pamphlet imprint with an international vision.

## Ragged Raven Press

1 Lodge Farm, Snitterfield, Stratford-upon-Avon CV37 0LR
☏ 01789 730320
raggedravenpress@aol.com
www.raggedraven.co.uk
Contact *Bob Mee & Janet Murch*

Poetry.

## Reality Street Editions

63 All Saints Street, Hastings TN34 3BN
☏ 07706 189253
www.realitystreet.co.uk
Contact *Ken Edwards*

New writing from Britain, Europe and America. No unsolicited mss.

## Red Candle Press

77 Homegrove House, Grove Road North, Southsea PO5 1HW
www.members.tripod.com/redcandlepress
Contact *M.L. McCarthy*

Founded 1970. Formalist poetry press. See also **Candelabrum Poetry Magazine**.

## Rialto Publications

PO Box 309, Aylsham, Norwich NR11 6LN
mail@therialto.co.uk
www.therialto.co.uk
Contact *Michael Mackmin*

'We want to publish first collections by poets of promise.' See also **The Rialto** magazine and **Bridge Pamphlets** press.

## Route Publishing

PO Box 167, Pontefract WF8 4WW
☏ 0845 158 1565
info@route-online.com
www.route-online.com
Contact *Ian Daley & Isabel Galan*

Contemporary fiction, including short stories and novels, and cultural non-fiction.

## Salt Publishing Limited
▷ entry under UK Publishers

## Second Light Publications

9 Greendale Close, London SE22 8TG
dilyswood@tiscali.co.uk
www.secondlightlive.co.uk
www.poetrypf.co.uk/secondlight.html
www.esch.dircon.co.uk/second/second.htm
Contact *Dilys Wood*
Administrator *Anne Stewart (editor@poetrypf.co.uk)*

Publishes women's poetry, mainly anthologies, occasional collections (by invitation). See **Second Light** under *Organizations of Interest to Poets*.

## Seren

57 Nolton Street, Bridgend CF31 3AE
☏ 01656 663018
seren@seren-books.com
www.seren-books.com
Contact *Mick Felton*

Poetry, fiction, lit crit, biography, essays. See entry under *UK Publishers*; also **Poetry Wales** magazine.

## Shearsman Books

58 Velwell Road, Exeter EX4 4LD
☏ 01392 434511
editor@shearsman.com
www.shearsman.com

Contact *Tony Frazer*

Publishes poetry almost exclusively. Contact editor before submitting or consult the submissions page at the Shearsman website for the most up-to-date instructions. See also **Shearsman** magazine.

## Shoestring Press

19 Devonshire Avenue, Beeston, Nottingham NG9 1BS
☎ 0115 925 1827     🖷 0115 925 1827
info@shoestringpress.co.uk
www.shoestring-press.com
Contact *John Lucas*

Poetry. No unsolicited mss.

## Sixties Press

info@sixtiespress.co.uk
www.sixtiespress.co.uk
Contact *Barry Tebb*

Small press concentrating on poetry and novellas. See also **Leeds Poetry Quarterly** and **Literature and Psychoanalysis** magazines.

## Smith/Doorstop
▷ The Poetry Business under Organizations of Interest to Poets

## Smokestack Books

PO Box 408, Middlesbrough TS5 6WA
☎ 01642 813997
info@smokestack-books.co.uk
www.smokestack-books.co.uk
Contact *Andy Croft*

Champions poets who are unconventional, unfashionable, radical or left-field.

## Spectacular Diseases

83b London Road, Peterborough PE2 9BS
Contact *Paul Green*

Innovative poetry and some prose. Translations of both.

## The Stinging Fly Press

PO Box 6016, Dublin 8, Republic of Ireland
stingingfly@gmail.com
www.stingingfly.org
Contact *Declan Meade*

Founded 2005. New Irish and international writing. No email submissions. See also **The Stinging Fly** magazine.

## tall-lighthouse

Stark Gallery, 384 Lee High Road, London SE12 8RW
☎ 020 8297 8279
info@tall-lighthouse.co.uk
www.tall-lighthouse.co.uk

Managing Editor *Les Robinson*

Publishes full collections, pamphlets, chapbooks and anthologies of poems. No unsolicited material.

## Templar Poetry

PO Box 7082, Bakewell DE45 9AF
☎ 01629 582500
info@templarpoetry.co.uk
www.templarpoetry.co.uk
Managing Editor/Publisher *Alex McMillen*

Publishes poetry pamphlets, collections and anthologies. See also **Iota** magazine.

## Two Ravens Press Ltd
▷ entry under UK Publishers

## Vane Women Press

9 Alice Row, Stockton TS18 1JU
☎ 01642 606421
anne.hine@ntlworld.com
www.vanewomen.co.uk
Contact *Anne Hine*

Poetry and short stories by women of the North East.

## Waterloo Press

95 Wick Hall, Furze Hill, Hove BN3 1NG
www.waterloopresshove.co.uk
Contact *Simon Jenner*

Poetry and periodical publisher; please phone for submissions. See also **Eratica** magazine.

## Waywiser Press
▷ entry under Small Presses

## Wendy Webb Books

9 Walnut Close, Taverham, Norwich NR8 6YN
tips4writers@yahoo.co.uk
Contact *Wendy Webb*

New forms, traditional forms, books, anthologies, Margaret Munro Gibson Memorial Poetry Competition (annual). See also **Norfolk Poets and Writers – Tips Newsletter** magazine.

## West House Books

40 Crescent Road, Nether Edge, Sheffield S7 1HN
☎ 0114 258 6035
alan@nethedge.demon.co.uk
www.westhousebooks.co.uk
Contact *Alan Halsey*

Contemporary poetry, poets' prose and related work.

## Worple Press
▷ entry under Small Presses

# Poetry Magazines

## Acumen

6 The Mount, Higher Furzeham, Brixham
TQ5 8QY
☎ 01803 851098
pwoxley@aol.com
www.acumen-poetry.co.uk
Contact *Patricia Oxley*

Good poetry, intelligent articles and wide-ranging reviews.

## Agenda Poetry Magazine

The Wheelwrights, Fletching Street, Mayfield
TN20 6TL
☎ 01435 873703
editor@agendapoetry.co.uk
www.agendapoetry.co.uk
Editor *Patricia McCarthy*

Poems, essays, reviews. Visit the website for further information and supplements to Agenda. Submissions: up to five poems, with s.a.s.e. and email address. Young poets (and young artists) and essayists (age 15 to late thirties) also encouraged in the journal and in the online Broadsheets. Subscriptions: (one vol. = 4 issues = one subscription) £28 individual (£22 OAPs/ students); £35 libraries/institutions. See **Agenda Editions** press.

## Aireings

submissions@aireings.co.uk
www.aireings.co.uk/
Contact *Lesley Quayle & Linda Marshall*

Online poetry mag: poetry, prose, reviews, b&w artwork.

## Ambit

17 Priory Gardens, London N6 5QY
☎ 020 8340 3566
info@ambitmagazine.co.uk
www.ambitmagazine.co.uk
Contact *Dr Martin Bax*

Quarterly 96-page magazine of poetry, prose and artwork. New and established writers and artists. Unsolicited contributions welcome – postal submissions: five/six poems, three short stories; greyscale versions of artwork.

## Aquarius

Flat 4, 116 Sutherland Avenue, London W9 2QP
☎ 020 7289 4338
www.geocities.com/eddielinden
Contact *Eddie S. Linden*

Literary magazine; prose and poetry. No emailed submissions.

## Areopagus

48 Cornwood Road, Plympton, Plymouth PL7 1AL
www.areopagus.org.uk
Contact *Julian Barritt*

A Christian-based arena for creative writers.

## Areté

8 New College Lane, Oxford OX1 3BN
☎ 01865 289193    ᖴ 01865 289194
craigraine@aretemagazine.co.uk
www.aretemagazine.co.uk
Editor *Craig Raine*
Assistant Editor *Ann Pasternak Slater*

Founded 1999. Fiction, poetry, reportage and reviews. No submissions via email; hard copy only. Unsolicited mss should be accompanied by s.a.s.e. See also **Areté Books** under *Small Presses*.

## ARTEMISpoetry

▷ **Second Light under Organizations of Interest to Poets**

## Aspire

289 Elmwood Avenue, Feltham TW13 7QB
partners_writing_group@hotmail.com
Contact *Ian Deal*

Launched 2009. Quarterly magazine of contemporary poetry. See also **Partners** press, **A Bard Hair Day**, **Imagenation** and **Poet Tree** magazines.

## Awen

4 Pierrot Steps, 71 Kursaal Way, Southend on Sea
SS1 2UY
atlanteanpublishing@hotmail.com
www.atlanteanpublishing.web.officelive.com
Contact *David John Tyrer*

Poetry and vignette-length fiction of any style/ genre. See also **Bard**, **Monomyth**, **Garbaj** and

The **Supplement** magazines and **Atlantean Publishing** press.

## Bard

4 Pierrot Steps, 71 Kursaal Way, Southend on Sea
SS1 2UY
atlanteanpublishing@hotmail.com
www.atlanteanpublishing.web.officelive.com
Contact *David John Tyrer*

Short poetry. See also **Awen**, **Garbaj**, **Monomyth** and **The Supplement** magazines and **Atlantean Publishing** press.

## A Bard Hair Day

289 Elmwood Avenue, Feltham TW13 7QB
Contact *Ian Deal*

Contemporary poetry magazine. See also **Partners** press, **Aspire**, **Imagenation** and **Poet Tree** magazines.

## Blithe Spirit

www.britishhaikusociety.org
Editor *Colin Blundell*

Journal of the British Haiku Society (see entry under *Organizatins of Interest to Poets*). Haiku and related forms.

## Brittle Star

PO Box 56108, London E17 0AY
℡ 0845 456 4838
magazine@brittlestar.org.uk
www.brittlestar.org.uk
Contact *Louise Hooper*

Poetry, short stories and articles on contemporary poetry.

## Candelabrum Poetry Magazine

77 Homegrove House, Grove Road North, Southsea PO5 1HW
www.members.tripod.com/redcandlepress
Contact *M.L. McCarthy*

Formalist poetry magazine for people who like poetry rhythmic and shapely. See also **Red Candle Press**.

## Carillon

19 Godric Drive, Brinsworth, Rotherham S60 5NA
℡ 01709 372875
editor@carillonmag.org.uk
www.carillonmag.org.uk
Editor *Graham Rippon*

Eclectic poetry and prose. Three issues yearly.

## Chapman

▷ entry under Magazines

## Current Accounts

Apartment 2D, Hardcastle Gardens, Bradshaw Hall, Bolton BL2 4NZ
℡ 01204 598913

fjameshartnell@aol.com
hometown.ao.co.uk/bswscribe/myhomepage/writing.html
Co-Editor *James Hartnell*

Magazine of the Bank Street Writers' Group. Poetry, short fiction, articles. Submissions also accepted from non-members (by email or by post with s.a.s.e.).

## Cyphers

3 Selskar Terrace, Ranelagh, Dublin 6, Republic of Ireland
Contact *Eiléan Ní Chuilleanáin, Macdara Woods, Pearse Hutchinson, Leland Bardwell*

Poetry, fiction, some artwork and criticism. Poetry and fiction contributions welcome. Send by post.

## Dandelion Arts Magazine

24 Frosty Hollow, East Hunsbury NN4 0SY
℡ 01604 701730  🖷 01604 701730
Editor/Publisher *Jacqueline Gonzalez-Marina, MA*

Biannual, international publication. Poetry, stories, interviews, reviews and illustrations welcome. No religious or political material. Essential to become a subscriber when seeking publication: UK, £17 p.a.; Europe, £30; RoW, US$90.

## The David Jones Journal

The David Jones Society, 22 Gower Road, Sketty, Swansea SA2 9BY
℡ 01792 206144
contact@davidjonessociety.org,
www.davidjonessociety.org/default.aspx
Contact *Anne Price-Owen*

Articles, poetry, information, reviews and inspired works.

## The Dawntreader

132 Hinckley Road, Stoney Stanton LE9 4LN
℡ 01455 272861
dawnidp@btinternet.com
www.indigodreamsonline.com
Editors *Dawn Bauling & Ronnie Goodyer*

Quarterly magazine with themes of the mystic, landscape, myth, spirituality and love/concern for the environment. Poetry, short stories, articles and legends welcome. Approach by email or post. See also *Reach Poetry* and *Sarasvati* magazines and **Indigo Dreams Publishing** press.

## Decanto

Masque Publishing, PO Box 3275, Littlehampton BN16 9AF
masque_pub@btinternet.com
http://myweb.tiscali.co.uk/masquepublishing
Contact *Lisa Stewart*

Non-conformist poetry magazine; any style considered, not just contemporary. Six issues annually; subscription details on the website. See also **Masque Publishing** press (self-publishing service).

## Dream Catcher

4 Church Street, Market Rasen, Lincoln LN8 3ET
℡ 01673 844325
www.dreamcatchermagazine.co.uk
www.inpressbooks.co.uk
Editor *Paul Sutherland*

Short fiction, poetry, reviews, artwork, biographies. See also **Dream Catcher** press.

## Earth Love

PO Box 11219, Paisley PA1 2WH
www.earthlovepoetrymagazine.co.uk
Contact *Tracy Patrick*

Poetry magazine for the environment; proceeds to conservation charities. Submissions should be accompanied by an s.a.s.e.

## Eastern Rainbow

17 Farrow Road, Whaplode Drove, Spalding PE12 0TS
p_rance@yahoo.co.uk
uk.geocities.com/p_rance/pandf.htm
Contact *Paul Rance*

Focuses on 20th century culture via poetry, prose and art. See also **Peace and Freedom** magazine.

## Envoi

Ty Meirion, Glan yr afon, Tanygrisiau, Blaenau Ffestiniog LL41 3SU
℡ 01766 832112
Editor *Jan Fortune-Wood*

Poetry, sequences, features, reviews, competitions. Now in it's 52nd year. See also **Cinnamon Press**.

## Equinox

Chemin de Cambieure, 11240 Cailhau, Aude, France
dordi.barbara@aliceadsl.fr
www.poetrymagazines.org.uk/equinox
Editor *Barbara Dordi*

Contemporary poetry invited. No email submissions; approach by letter (include s.a.s.e. with British postage stamps).

## Eratica

95 Wick Hall, Furze Hill, Hove BN3 1NG
Contact *Simon Jenner*

Biannual journal with colour plates – focus on poetry, strong on music and art. See also **Waterloo Press**.

## Esprit de Corps Literary Magazine

40 Wingfield Road, Lakenheath IP27 9HR
℡ 07790 962317
info@espritdecorpsliterarymagazine.co.uk
www.espritdecorpsliterarymagazine.co.uk
Managing Editor *James Quinton*

EDC is a left leaning literary magazine publishing short fiction, non-fiction, poetry, reviews and interviews. New writers always welcome. Previously Open Wide Magazine. See website for submission guidelines. Four issues a year. Payment: one copy.

## Fire

Field Cottage, Old White Hill, Tackley OX5 3AB
℡ 01869 331300
Editor *Jeremy Hilton*

New and little-known writers; unorthodox and unfashionable work. No unsolicited submissions; send query letter in the first instance.

## The Firing Squad

firingsquad@purplepatchpoetry.co.uk
www.the-firing-squad.com
Editor *Alex Barzdo*

Protest poetry website – 'or the grumpy old website'. Email submissions only.

## First Offense

Snails Field, Mare Hill Road, Pulborough RH20 2DS
tim@firstoffense.co.uk
www.firstoffense.co.uk
Contact *Tim Fletcher*

Magazine for contemporary poetry; not traditional but is received by most ground-breaking poets.

## The French Literary Review

Chemin de Cambieure, 11240 Cailhau, Aude, France
℡ 00 44 434 930328
dordi.barbara@aliceadsl.fr
www.barbaradordi.blogspot.com
Editor *Barbara Dordi*

Poetry, short stories, articles and b&w illustrations, all with a French connection. Unsolicited contributions welcome; approach by letter (include s.a.s.e. with British postage stamps).

## The Frogmore Papers

21 Mildmay Road, Lewes BN7 1PJ
℡ 07751 251689
J.N.Page@sussex.ac.uk
www.frogmorepress.co.uk
Editor *Jeremy Page*

Founded 1983. Biannual. Poetry, prose and artwork. See also **The Frogmore Press**.

## Gabriel: A Christian Poetry Magazine

27 Headingley Court, North Grange Road, Leeds
LS6 2QU
Editor *Thelma Laycock*

Annual Christian poetry magazine. Unsolicited
submissions by post welcome from November to
January each year.

## Garbaj

4 Pierrot Steps, 71 Kursaal Way, Southend on Sea
SS1 2UY
atlanteanpublishing@hotmail.com
www.atlanteanpublishing.web.officelive.com
Contact *D.J. Tyrer*

Humorous/non-pc poetry, vignette-length fiction,
fake news, etc. See also **Awen**, **Bard**, **Monomyth**
and **The Supplement** magazines and **Atlantean
Publishing** press.

## Global Tapestry Journal

Spring Bank, Longsight Road, Copster Green,
Blackburn BB1 9EU
☎ 01254 249128
Contact *Dave Cunliffe*

Global Bohemia, post-Beat and counterculture
orientation. See also **BB Books** press.

## Green Queen

BM Box 5700, London WC1N 3XX
eandk2@btinternet.com
Editor *Elsa Wallace*

Fiction, poetry, reviews, news, articles, opinions
– serious and humorous – dealing with issues of
humane living, ecology, veganism from a LG-BT
viewpoint. Unsolicited contributions welcome;
approach by mail.

## Haiku Scotland

2 Elizabeth Gardens, Stoneyburn EH47 8PB
haiku.scotland@btinternet.com
Contact *Frazer Henderson*

Haiku, senryu, aphorism, epigram, reviews and
articles.

## Handshake

5 Cross Farm, Station Road North, Fearnhead,
Warrington WA2 0QG
Contact *John Francis Haines*

Newsletter of **The Eight Hand Gang**, an
association of UK sci-fi poets. See entry under
*Organizations of Interest to Poets*.

## Harlequin

PO Box 23392, Edinburgh EH8 7YZ
columbine@harlequinmagazine.com
www.harlequinmagazine.com
Contact *Jim Sinclair*

Poetry and artwork of mysticism and wisdom.

## HQ Poetry Magazine (Haiku Quarterly)

39 Exmouth Street, Swindon SN1 3PU
☎ 01793 523927
Editor *Kevin Bailey*

General poetry mag with slight bias towards
imagistic/haikuesque work. Celebrates its 20th
anniversary in 2010.

## Imagenation

289 Elmwood Avenue, Feltham TW13 7QB
Contact *Ian Deal*

Poetry and artwork magazine. See also **Partners**
press, **Aspire**, **A Bard Hair Day** and **Poet Tree**
magazines.

## Inclement

White Rose House, 8 Newmarket Road, Fordham,
Ely CB7 5LL
inclement_poetry_magazine@hotmail.com
www.inclementpoetrymagazine.webs.com
Contact *Michelle Foster*

All forms and styles of poetry. Responses within
a month.

## The Interpreter's House

19 The Paddox, Oxford OX2 7PN
www.interpretershouse.org.uk
Contact *Merryn Williams*

Poems and stories up to 2,500 words; new
and established writers. Check website for
subscription details.

## Iota

PO Box 7721, Matlock DE4 9DD
☎ 01629 582500
info@iotamagazine.co.uk
www.iotamagazine.co.uk
Editor *Nigel McLoughlin*
Publisher *Alec McMillen*

Poetry, reviews, listings, interviews. See also
**Templar Poetry** press.

## Irish Pages
▷ See entry under Magazines

## The Journal

17 High Street, Maryport CA15 6BQ
☎ 01900 812194
smithsssj@aol.com
www.freewebs.com/thesamsmith/
Contact *Sam Smith*

Poems in translation alongside poetry written
in English. See also **Original Plus** press. A
maximum of six poems at a time via snailmail
or email. Poems to be embedded in email, not
attached. Reviews of collections will be published
unexpurgated.

## Krax
63 Dixon Lane, Wortley, Leeds LS12 4RR
Contact *Andy Robson*

Light-hearted, contemporary poetry, short fiction and graphics. 'No devotional writing or smutty limericks.'

## Leeds Poetry Quarterly
89 Connaught Road, Sutton SM1 3PJ
☏ 020 8286 0419
Contact *Barry Tebb*

New poems, reviews, articles on literary and psychoanalytic matters. See also **Sixties Press** and **Literature and Psychoanalysis** magazine.

## Literature and Psychoanalysis
89 Connaught Road, Sutton SM1 3PJ
☏ 020 8286 0419
Contact *Barry Tebb*

New poems, reviews, articles on literary and psychoanalytical matters. See also **Sixties Press** and **Leeds Poetry Quarterly** magazine.

## The London Magazine
▷ entry under Magazines

## Magma
43 Keslake Road, London NW6 6DH
magmapoetry@ntlworld.com
www.magmapoetry.com
Contact *David Boll*

New poetry plus poetry reviews and interviews. Submissions to: contributions@magmapoetry.com or address above.

## Markings
▷ entry under Magazines

## Modern Poetry in Translation
The Queen's College, Oxford OX1 4AW
www.mptmagazine.com
Contact *David & Helen Constantine*

Publishes and promotes poetry in English translation. Email submissions not accepted.

## Monkey Kettle
monkeykettle@hotmail.com
www.monkeykettle.co.uk
www.myspace.com/monkeykettle
Editor *Matthew M. Taylor*

Biannual poetry, prose, photos, art and articles in Milton Keynes and further afield. 'Cheerfully lo-fi with a DIY ethos. Submissons details available at www.monkeykettle.co.uk/about.html

## Monomyth
4 Pierrot Steps, 71 Kursaal Way, Southend on Sea SS1 2UY
atlanteanpublishing@hotmail.com
www.atlanteanpublishing.web.officelive.com

Contact *David John Tyrer*

Poetry, prose and articles; all genres, styles and lengths considered. New writers welcome. See also **Awen**, **Bard**, **Garbaj** and **The Supplement** magazines and **Atlantean Publishing** press.

## Mslexia
▷ entry under Magazines

## Neon Highway Poetry/Art Magazine
37 Grinshill Close, Liverpool L8 8LD
neonhighwaypoetry@yahoo.co.uk
Editor *Alice Lenkiewicz*

Established 2002. Biannual avant-garde literary journal 'for the esoteric and the visionary'. Submissions (s.a.s.e. to be included) of innovative poetry welcome but also open to formal poems with credibility. Single issue, £3 (two issues, £5.50); cheque to be made out to Alice Lenkiewicz at the address above.

## Never Bury Poetry
Bracken Clock, Troutbeck Close, Hawkshaw, Bury BL8 4LJ
☏ 01204 884080
Contact *Jean Tarry*

Founded 1989. Biannual. International reputation. Each issue has a different theme. Submissions by post with s.a.s.e.

## New Walk Magazine
c/o Nick Everett, School of English, Leicester University, University Road, Leicester LE1 7RH
newwalkmagazine@gmail.com
www.tinyurl.com/newwalkmag
Editors *Rory Waterman, Nick Everett*

A new magazine for poetry and the arts published twice a year in spring and autumn. For formalists, experimentalists and everyone in between from across the English-speaking world. Send up to six examples of your best poetry, one short story, one or two pieces of artwork (b&w preferred) or proposals for essays, reviews, etc. 'Competition is fierce. Responses within two months.'

## New Welsh Review
▷ entry under Magazines

## Norfolk Poets and Writers – Tips Newsletter
9 Walnut Close, Taverham, Norwich NR8 6YN
tips4writers@yahoo.co.uk
Editor *Wendy Webb*

Tips for Writers; six magazines per year; competitions, themes and forms. See also **Wendy Webb Books** press.

## The North
The Poetry Business, Bank Street Arts, 32–40 Bank Street, Sheffield S1 2DS
www.poetrybusiness.co.uk

Contact *Peter Sansom & Ann Sansom*

Contemporary poetry and articles, extensive reviews.

## Obsessed With Pipework

8 Abbot's Way, Pilton BA4 4BN

☏ 01749 890019

cannula.dementia@virgin.net

www.flarestack.co.uk

Contact *Meredith Andrea & Jacqui Rowe*

Quarterly magazine of new poetry 'to surprise and delight'. See also **Flarestack Publishing** press.

## Old Gothic Tales

▷ **Edgewell Publishing under Small Presses**

## Open Wide Magazine

▷ **Esprit de Corps Literary Magazine**

## Other Poetry

Fourlawshill Top, Bellingham, Hexham NE48 2EY

editors@otherpoetry.com

www.otherpoetry.com

Contact *The Editors*

Founded 1979. Twice or thrice-yearly, 60–70 poems per issue. Process of selection involves all the editors. Token payment. Submit up to four poems (with name on each sheet) plus s.a.s.e.

## Partners Annual Poetry Competition

289 Elmwood Avenue, Feltham TW13 7QB

partners_writing_group@hotmail.com

Contact *Ian Deal*

Showcases poems entered into the Partners annual open poetry competitions. Prizes: £250 (1st); £50 (2nd); year's subscription to *Aspire* poetry magazine (3rd).

## Peace and Freedom

17 Farrow Road, Whaplode Drove, Spalding PE12 0TS

☏ 01406 330242

p_rance@yahoo.co.uk

uk.geocities.com/p_rance/pandf.htm

http://pandf.booksmusicfilmstv.com

Contact *Paul Rance*

Poetry, prose, art mag; humanitarian, environmental, animal welfare. Submissions of poetry only; 32 lines maximum. All postal submissions must be accompanied by an s.a.s.e. or IRCs. See also **Eastern Rainbow** magazine.

## The Penniless Press

100 Waterloo Road, Ashton, Preston PR2 1EP

☏ 01772 736421

www.pennilesspress.co.uk

Contact *Alan Dent*

Quarterly for the poor pocket and the rich mind. Poetry, fiction, essays, reviews. See also **Penniless Press**.

## Pennine Ink Magazine

The Gallery, Mid-Pennine Arts, Yorke Street, Burnley BB11 1HD

☏ 01282 432992

sheridansdandl@yahoo.co.uk

Contact *Laura Sheridan*

Quality poetry reflecting traditional and modern trends.

## Pennine Platform

Frizingley Hall, Frizinghall Road, Bradford BD9 4LD

☏ 01274 541015

www.pennineplatform.co.uk

Contact *Nicholas Bielby*

Biannual poetry magazine; eclectic and serious-minded. Considers hard copy submissions only.

## Planet

PO Box 44, Aberystwyth SY23 3ZZ

☏ 01970 611255    ✆ 01970 611197

planet.enquiries@planetmagazine.org.uk

www.planetmagazine.org.uk

Editor *Helle Michelsen*

The Welsh Internationalist: current affairs, arts, environment.

## PN Review

▷ **Carcanet Press Ltd under UK Publishers**

## Poet Tree

289 Elmwood Avenue, Feltham TW13 7QB

Contact *Ian Deal*

Contemporary poetry magazine. See also **Partners** press, **Aspire**, **A Bard Hair Day** and **Imagenation** magazines.

## Poetic Licence

70 Aveling Close, Purley CR8 4DW

poets@poetsanon.org.uk

www.poetsanon.org.uk

Contact *Peter L. Evans*

Original unpublished poems and drawings. See also **Poets Anonymous** press.

## The Poetry Church

Eldwick Crag Farm, Otley Road, Bingley BD16 3BB

☏ 01274 563078

reavill@globalnet.co.uk

www.waddysweb.freeuk.com

Editor *Tony Reavill*

Published by Moorside Words and Music, quarterly magazine of Christian poetry and prayers. See also **Feather Books** under *Small Presses*.

## Poetry Combination Module

PEF Productions, 196 High Road, London
N22 8HH
page84direct@yahoo.co.uk
www.alienonthenet.freehosting.net
Contact *Mr PEF*

Available late spring, early autumn and mid-winter. Printable from the website. Poetry, anti-poetry, aphorism and artwork magazine by PEF and guest artists. Submissions welcome, preferably by email.

## Poetry Cornwall/Bardhonyeth Kernow

11a Penryn Street, Redruth TR15 2SP
☏ 01209 218209
poetrycornwall@yahoo.com
www.poetrycornwall.freeservers.com
Editor *Les Merton*

Publishes poets from around the world, poetry in its original language with English translation and poetry in Kernewek and Cornish dialect. Three issues a year. Submission guidelines on the website and back cover of every issue.

## Poetry Express

Biannual newsletter from **Survivors' Poetry** (see entry under *Organizations of Interest to Poets*).

## Poetry Ireland News

2 Proud's Lane, Dublin 2, Republic of Ireland
☏ 00 353 1 478 9974  🄵 00 353 1 478 0205
publications@poetryireland.ie
www.poetryireland.ie
Publications Officer *Paul Lenehan*

Bi-monthly newsletter. See also **Poetry Ireland Review** magazine.

## Poetry Ireland Review/Éigse Éireann

2 Proud's Lane, Dublin 2, Republic of Ireland
☏ 00 353 1 478 9974  🄵 00 353 1 478 0205
www.poetryireland.ie
Editor *Caitríona O'Reilly*

Quarterly journal of poetry and reviews. Email submissions not accepted. See also **Poetry Ireland News**.

## Poetry London

81 Lambeth Walk, London SE11 6DX
☏ 020 7735 8880  🄵 020 7735 8880
admin@poetrylondon.co.uk (enquiries only)
www.poetrylondon.co.uk
Poetry Editor *Colette Bryce*
Listings *Gyonghi Vegh (listings@poetrylondon. co.uk)*
Reviews *Tim Dooley*

Published three times a year, *Poetry London* includes poetry by new and established writers, reviews of recent collections and anthologies, articles on issues relating to poetry, and an encyclopædic listings section of virtually everything to do with poetry in the capital and elsewhere in the UK. Email submissions not accepted.

## Poetry Nottingham

11 Orkney Close, Stenson Fields, Derby DE24 3LW
Editor *Adrian Buckner*

Poetry, articles, reviews, published thrice-yearly. Subscription: £12 p.a. (£4 individual issues).

## Poetry Reader

Scottish Poetry Library, 5 Crichton's Close, Canongate, Edinburgh EH8 8DT
☏ 0131 557 2876  🄵 0131 557 8393
inquiries@spl.org.uk
www.spl.org.uk
www.readingroom.spl.org.uk
Marketing Officer *Jane Alexander*

Newsletter of the **Scottish Poetry Library**.

## Poetry Review

Poetry Society, 22 Betterton Street, London WC2H 9BU
☏ 020 7420 9883  🄵 020 7240 4818
poetryreview@poetrysociety.org.uk
www.poetrysoc.org.uk
Editor *Fiona Sampson*

The senior review of contemporary poetry. Quarterly.

## Poetry Salzburg Review

Dept. of English, University of Salzburg, Akademiestr. 24, A–5020 Salzburg, Austria
☏ 00 43 662 8044 4424  🄵 00 43 662 8044 167
editor@poetrysalzburg.com
www.poetrysalzburg.com
Co-editors *Wolfgang Görtschacher & Andreas Schachermayr*

Poetry magazine, formerly *The Poet's Voice*. Published at the University of Salzburg. Publishes new poetry, translations, interviews, review-essays, essays on poetics and poetry, artwork.

## Poetry Scotland

91–93 Main Street, Callander FK17 8BQ
www.poetryscotland.co.uk
Contact *Sally Evans*

All-poetry broadsheet with Scottish emphasis.

## Poetry Wales

57 Nolton Street, Bridgend CF31 3AE
☏ 01656 663018
www.poetrywales.co.uk
Editor *Zöe Skoulding (School of English, Bangor University, Bangor LL57 2DG)*

An international magazine with a reputation for fine writing and criticism. Email submissions not accepted. See also **Seren Books** under *UK Publishers*.

## Premonitions

13 Hazely Combe, Arreton, Isle of Wight PO30 3AJ

☎ 01983 865668

mail@pigasus.press.co.uk

www.pigasuspress.co.uk

Contact *Tony Lee*

Magazine of science fiction and horror stories, with genre poetry and artwork. See also **Pigasus Press**.

## Presence

90D Fishergate Hill, Preston PR1 8JD

m.lucas27@btinternet.com

http://haiku-presence.50webs.com

Contact *Martin Lucas*

Haiku, senryu, tanka, renku and related poetry in English.

## Pulsar Poetry Webzine

34 Lineacre Close, Grange Park, Swindon SN5 6DA

☎ 01793 875941

pulsar.ed@btopenworld.com

www.pulsarpoetry.com

Editor *David Pike*

Hard-hitting/inspirational poetry with a message and meaning. Now a webzine only, came into effect after the printed publication of the September 2009 edition #52. Selected poems are posted on a quarterly basis.

## Purple Patch

25 Griffiths Road, West Bromwich B71 2EH

ppatch66@hotmail.com

www.purplepatchpoetry.co.uk

www.poetrywednesburyinternational.org.uk

www.geoffstevens.co.uk

Contact *Geoff Stevens*

Poetry mag founded 1976. Includes reviews and gossip column. See also **Poetry Wednesbury** press.

## Quantum Leap

York House, 15 Argyle Terrace, Rothesay, Isle of Bute PA20 0BD

Contact *Alan Carter*

User-friendly magazine – encourages new writers; all types of poetry – pays. Mark envelope 'Guidelines' and enclose s.a.s.e. or two IRCs if from abroad. See also **QQ Press**.

## The Radiator: A Journal of Contemporary Poetics

Flat 10, 21 Greenheys Road, Liverpool L8 0SX

☎ 0151 727 2681

scottthurston@btinternet.com

Editor *Scott Thurston*

Commissioned essays on contemporary innovative poetry by practitioners. No unsolicited contributions. Query by email.

## Rainbow Poetry News

74 Marina, St Leonards-on-Sea TN38 0BJ

☎ 01424 444072

beyondcloister@hotmail.co.uk

Contact *Hugh C. Hellicar*

Quarterly with poems, articles and news of poetry recitals in London and the South East but with overseas readership.

## Reach Poetry

132 Hinckley Road, Stoney Stanton LE9 4LN

☎ 01455 272861

idpoet@rocketmail.com

www.indigodreamsonline.com

Editor *Ronnie Goodyer*

Long established monthly poetry magazine. Accepts new and experienced poets. Cash prizes each month for best poems voted for by readers. Reviews, lively letters pages. Send contributions by email or post. See also *The Dawntreader* and *Sarasvati* magazines and **Indigo Dreams Publishing** press.

## The Reater

Wrecking Ball Press, 24 Cavendish Square, Hull HU3 1SS

editor@wreckingballpress.com

www.wreckingballpress.com

Contact *Shane Rhodes*

No flowers, just blunt, chiselled poetry.

## The Recusant

www.therecusant.moonfruit.com

Editor *Alan Morrison*

Non-conformist poetry, prose, polemic, reviews and articles. Seeks to provide a home for contemporary writing that goes against the grain, in subject more than style. Socially-inclined work with a left-field slant particularly welcome. Visit the website for submission guidelines.

## Red Poets – Y Beirdd Coch

26 Andrew's Close, Heolgerrig, Merthyr Tydfil CF48 1SS

☎ 01685 376726

mjenkins1927@gmail.com

www.redpoets.org

Contact *Mike Jenkins & Marc Jones*

A magazine of left-wing poetry from Wales and beyond: socialist, republican and anarchist.

## The Rialto

PO Box 309, Aylsham, Norwich NR11 6LN

mail@therialto.co.uk

www.therialto.co.uk

Contact *Michael Mackmin*

Excellent poetry in a clear environment. Buy online at www.impressbooks.co.uk See also **Rialto Publications** press.

## Rubies In the Darkness

41 Grantham Road, Manor Park, London E12 5LZ
Contact *P.G.P. Thompson*

Romantic, lyrical, spiritual, inspirational, traditional and mystical poetry. See also **Precious Pearl Press**.

## Sable

PO Box 33504, London E9 7YE
editorial@sablelitmag.org
www.sablelitmag.org
Contact *Kadija Sesay*
*Also at:* PO Box 2803, Upper Darby, PA 19082, USA

Literary magazine for writers of African, Caribbean and Asian descent. Poetry, prose, memoirs, travel, etc.

## Sarasvati

132 Hinckley Road, Stoney Stanton LE9 4LN
☎ 01455 272861
dawnidp@btinternet.com
www.indigodreamsonline.com
Editors *Dawn Bauling & Ronnie Goodyer*

Bi-monthly magazine of poetry, short stories and articles that allow each writer to showcase their work over 3–4 pages. Biographies included. Contributions welcome by email or post. See also *The Dawntreader* and *Reach Poetry* magazines and **Indigo Dreams Publishing** press.

## Scar Tissue

Pigasus Press, 13 Hazley Combe, Arreton, Isle of Wight PO30 3AJ
Contact *Tony Lee*

SF genre poetry, short fiction, cartoons and artwork.

## Scribbler!

Remus House, Coltsfoot Drive, Woodston, Peterborough PE2 9JX
☎ 01733 890066   📠 01733 313524
admin@youngwriters.co.uk
www.youngwriters.co.uk
Editor *Donna Samworth*

Thrice-yearly magazine by 7–11 year-olds: poetry, stories, guest authors, workshops, features.

## Scriptor

6 Green Bank, Stacksteads, Bacup OL13 8LQ
☎ 01706 870357/07967 315270
mail@johndench.demon.co.uk
Contact *John Dench*

Poetry, short stories, essays from the UK; published biennially (next issue 2010); guidelines

available from July 2009. See also **Green Arrow Publishing** press.

## The Seventh Quarry – Swansea Poetry Magazine

Dan-Y-Bryn, 74 Cwm Level Road, Brynhyfryd, Swansea SA5 9DY
info@peterthabitjones.com
www.peterthabitjones.com
www.myspace.com/theseventhquarry
Contact *Peter Thabit Jones*
Consulting Editor, America *Vince Clemente*

Quality poetry from Swansea and beyond. Submission details available on the website. No more than *four* poems. S.a.s.e required if a postal submission.

## Shearsman

58 Velwell Road, Exeter EX4 4LD
☎ 01392 434511
editor@shearsman.com
www.shearsman.com
Contact *Tony Frazer*

Mainly poetry, some prose, some reviews. Poetry in the modernist tradition. No fiction. See also **Shearsman Books** press.

## The SHOp: A Magazine Of Poetry

Skeagh, Schull, Co. Cork, Republic of Ireland
theshop@theshop-poetry-magazine.ie
(not for submissions)
www.theshop-poetry-magazine.ie
Editors *John Wakeman & Hilary Wakeman*

International but with emphasis on Irish poetry.

## Smoke

The Windows Project, Liver House, 96 Bold Street, Liverpool L1 4HY
☎ 0151 709 3688
www.windowsproject.co.uk
Contact *Dave Ward*

Poetry, graphics, short prose. Biannual; 24pp.

## South

PO Box 3744, Cookham, Maidenhead SL6 9UY
www.southpoetry.org
Contact *Chrissie Williams*

Poetry magazine from the southern counties of England that welcomes poets from across the world. Poems judged anonymously. Visit the website for full submission details.

## Southword: A Literary Journal Online
▷ entry under Magazines

## Sphinx

21 Hatton Green, Glenrothes KY7 4SD
www.happenstancepress.com
Editor *Helena Nelson*

A feature and review-based magazine focusing on poetry publishing: who, how and why. No unsolicited submissions. See also **Happenstance Press**.

## Springboard

Corrimbla, Ballina, Co. Mayo, Republic of Ireland
bobgroom@eircom.net
Contact *Robert Groom*

Short fiction, poetry and articles.

## The Stinging Fly

PO Box 6016, Dublin 8, Republic of Ireland
stingingfly@gmail.com
www.stingingfly.org
Publisher/Editor *Declan Meade*

New Irish and international writing. Poetry, short fiction, author interviews, essays and book reviews. See also **The Stinging Fly Press**. Submissions are considered from January to March each year. No email submissions.

## The Supplement

4 Pierrot Steps, 71 Kursaal Way, Southend on Sea
SS1 2UY
atlanteanpublishing@hotmail.com
www.atlanteanpublishing.web.officelive.com
Contact *D.J. Tyrer*

Non-fiction: articles, news, reviews, letters, etc. See also **Awen**, **Bard**, **Garbaj** and **Monomyth** magazines and **Atlantean Publishing** press.

## Taliesin

Academi, Mount Stuart House, Mount Stuart Square, Cardiff CF10 5FQ
☏ 029 2047 2266  🖷 029 2049 2930
taliesin@academi.org
www.academi.org
Joint Editors *Angharad Elen & Siân Melangell Dafydd*

Wales' leading Welsh language literary journal, published three times a year.

## Tears in the Fence

38 Hod View, Stourpaine, Nr Blandford Forum
DT11 8TN
☏ 01258 456803  🖷 01258 454026
david@davidcaddy.wanadoo.co.uk
www.myspace.com/tearsinthefence
Contact *David Caddy*

'An international magazine that appreciates social and poetic awareness and writing that prompts close and divergent readings.'

## 10th Muse

33 Hartington Road, Southampton SO14 0EW
a.jordan@surfree.co.uk
www.nonism.org.uk/muse.html

Editor *Andrew Jordan*

Publishes poetry and reviews as well as short prose and graphics. 10th Muse is an occasional publication. £3.50 each/three-issue subscription, £9/back issues of most issues available for £2. Send up to six poems or 2,000 words of prose. Enclose an s.a.e. or International Reply Coupons) if reply required (no need to send IRCs if your email address is in covering letter). Email for information only; no submissions by email. 'It is quite a good idea to read a copy of the magazine before sending your poems.' No payment for publication; contributors receive a complimentary copy of the magazine.

## The Third Half

16 Fane Close, Stamford PE9 1HG
☏ 01780 754193
Editor *Kevin Troop*

Searches for good poetry and fiction. See also **KT Publications** press.

## Time Haiku

Basho-an, 105 King's Head Hill, London E4 7JG
☏ 020 8529 6478
facey@aol.com
Contact *Erica Facey*
Editor *Diana Webb*

Founded 1994. Promotes haiku, tanka and haiku-related forms. Aims to create accessibility to all people interested in this form of poetry as well as established poets. A biannual journal and newsletter are available to subscribers.

## Understanding

127 Milton Road West, 7 Duddingston House Courtyard, Edinburgh EH15 1JG
Editor *Denise Smith*

Original poetry, short stories, parts of plays, reviews and articles. See also **Dionysia Press**.

## Wasafiri

The Open University in London, 1–11 Hawley Crescent, London NW1 8NP
☏ 020 7556 6110  🖷 020 7556 6187
wasafiri@open.ac.uk
www.wasafiri.org
Editor *Susheila Nasta*
Deputy Editor *Sharmilla Beezmohun*
Assistant Editor *Nisha Jones*
Editorial Manager *Teresa Palmiero*
Reviews Editor *Mark Stein*

The magazine of international contemporary writing.

# Organizations of Interest to Poets

## Academi (Yr Academi Gymreig)

3rd Floor, Mount Stuart House, Mount Stuart
Square, Cardiff Bay CF10 5FQ
℡ 029 2047 2266    🖷 029 2049 2930
post@academi.org
www.academi.org
*Glyn Jones Centre:* Wales Millennium Centre,
Bute Place Cardiff CF10 5AL
*North Wales Office:* Tŷ Newydd, Llanystumdwy,
Cricieth, Gwynedd LL52 0LW
*West Wales Office:* Dylan Thomas Centre,
Somerset Place, Swansea SA1 1RR
Chief Executive *Peter Finch*

Academi is the Welsh National Literature
Promotion Agency and Society for Authors with
special responsibility for literary activity, writers'
residencies, writers on tour, festivals, writers'
groups, readings, tours, exchanges and other
development work. Academi awards financial
bursaries annually, runs a criticism service and
organizes the **Wales Book of the Year Award**.
**Yr Academi Gymreig/The Welsh Academy**
operates the Arts Council of Wales franchise for
Wales-wide literature development. The Academi
sponsors a range of annual contests including
the John Tripp Award For Spoken Poetry and
the prestigious **Academi Cardiff International
Poetry Competition**. Publishes *A470*, a quarterly
literary information magazine. Academi is
also responsible for the National Poet of Wales
project and the Welsh Academy Encyclopaedia of
Wales. For further information, see entry under
*Professional Associations and Societies.*

## Apples & Snakes

The Albany, Deptford, London SE8 4AG
℡ 0845 521 3460    🖷 0845 521 3461
info@applesandsnakes.org
www.myspace.com/applesandsnakespoetry
www.applesandsnakes.org
Director *Lucy Crompton-Reid*

Works nationwide to promote popular,
high-quality and cross-cultural poetry;
programmes live events for new and established
poets including open mic events; operates a
Poets in Education scheme where poets run
workshops and perform in schools, prisons and
community settings; coordinates the professional
development of poets, with opportunities
for training mentoring, national touring and
residencies.

## The Arvon Foundation
▷ entry under UK and Irish Writers' Courses

## The British Haiku Society

www.britishhaikusociety.org
General Secretary *Jon Baldwin*

Formed in 1990. Promotes the appreciation and
writing of haiku and related forms; welcomes
overseas members. Provides tutorials, workshops,
readings, critical comment and information.
Specialist advisers are available. Runs a haiku
library and administers a haiku contest. The
journal, *Blithe Spirit*, is produced quarterly.
The Society has active local groups and issues a
regular newsletter. Current membership details
can be obtained from the website.

## Contemporary Poetics Research Centre

School of English and Humanities, Birkbeck
College, Malet Street, London WC1E 7HX
w.rowe@bbk.ac.uk
c.watts@bbk.ac.uk
www.bbk.ac.uk/cprc
Contact *Will Rowe & Carol Watts*

Dedicated to fostering research, performance
and practice in all modes and formats of
contemporary poetry. Runs a Creative Reading
Workshop, holds readings, performances,
talks and debates and organizes conferences.
Collaborates closely with Royal Holloway
University of London and the Centre for Cultural
Poetics at the University of Southampton (e.g.
on the British Electronic Poetry Centre at www.
soton.ac.uk/~bepc). Poets associated with the
Centre include Ulli Freer, Caroline Bergvall
and Sean Bonney. The Centre publishes the
web journals, *Pores* (www.pores.bbk.ac.uk) and
*Readings* (www.bbk.ac.uk/readings/).

## The Eight Hand Gang

5 Cross Farm, Station Road North, Fearnhead,
Warrington WA2 0QG
Secretary *John F. Haines*

An association of SF poets. Publishes *Handshake*, a single-sheet newsletter of SF poetry and information available free in exchange for an s.a.s.e.

## 57 Productions

57 Effingham Road, Lee Green, London SE12 8NT
☎ 020 8463 0866    🖷 020 8463 0866
info@57productions.com
www.57productions.com
Contact *Paul Beasley*

Specializes in the promotion of poetry and its production through an agency service, a programme of events and a series of audio publications. Services are available to event promoters, festivals, education institutions and the media. Poets represented include Jean 'Binta' Breeze, Michael Rosen, Patience Agbabi and many others. 57 Productions' series of audio cassettes, CDs and Poetry in Performance compilations offer access to some of the most exciting poets working in Britain today – check their Poetry Jukebox on the website.

## The Football Poets

4 The Retreat, Butterow, Stroud GL5 2LS
☎ 01453 757376
editors@footballpoets.org
www.footballpoets.org
Editor & Performance Poet *Crispin Thomas*
Co-Editors *Simon Williams, Peter Goulding*

The Football Poets exist to promote and encourage the writing, reading and performing of football poetry. Formed collectively in 1995, they perform extensively and also provide comprehensive football poetry workshops in schools, football clubs, prisons and communities. Their website, which was launched in 2000, is a fast, free and entertaining mix of literature and soccer poetry from around the world. Anyone may contribute. The site has been archived by the British Library and currently hosts over 11,000 football poems.

## The Indian King Poets

Garmoe Cottage, 2 Trefrew Road, Camelford PL32 9TP
☎ 01840 212161
indianking@btconnect.com
Director *Helen Jagger Wood*

A weekly half-day workshop where regular participants write together, bring work in progress for criticism, share experience of publication and invite visiting poets who run occasional day-long workshops. This is also the home of the Poetry Society's Camelford Poetry Stanza. Quarterly evening readings are held at the Camelford Gallery with an Open Mic slot. A Novel Writing Surgery, led by Karen Hayes, takes place on a Tuesday every other month.

## Kent & Sussex Poetry Society

Broomhill Farm, Beneden TN17 4JT
☎ 01892 662781
info@kentandsussexpoetrysociety.org
www.kentandsussexpoetrysociety.org
Honorary Secretary *Keith Francis*

Founded 1946. Monthly guest readers, monthly workshops, open competition, members' competition, writers' week and an annual publication, *The Folio*, now in its 64th year.

## Performance Poetry Society

PO Box 11178, Birmingham B11 4WP
☎ 0121 242 6644
performancepoetry@yahoo.com
*Branch contact & rehearsal room:* The Old Meeting House, behind 20/22 Wolverhampton Street, Dudley
Contact *Sandra Dennis*
Dudley Branch Contact *Jim MacCool (01384 258191)*

PPS was founded in 1999 by members of the Birmingham Midland Institute to serve the interests of performance poets and poetry in performance throughout the UK and Ireland. Initiated October as National Poetry Month in 2000. Organizes an extensive national tour each autumn, visiting small halls, colleges, schools and prisons. The Society has an in-house publishing company, Poetry Powerhouse Press.

## The Poetry Archive

PO Box 286, Stroud GL6 1AL
☎ 01453 832090    🖷 01453 836450
info@poetryarchive.org
www.poetryarchive.org
Directors *Richard Carrington, Andrew Motion*

A permanent and continually expanding online archive of audio recordings of work by poets who write in English. New 60-minute recordings are made for the Archive by a wide-ranging list of contemporary poets; extracts from those recordings are available on the website launched in 2005 at www.poetryarchive.org and the full-length recordings are available on CD for purchase by mail order. The site also contains historic recordings by now-dead poets, a separate archive of recordings for young children, and a wealth of educational and intepretative information. A registered charity, the Archive intends to ensure that all significant poets are recorded for posterity and that their recordings are widely valued and enjoyed.

## The Poetry Book Society

Dutch House, 307–308 High Holborn, London
WC1V 7LL
☎ 020 7833 9247   🖷 020 7833 5990
info@poetrybooks.co.uk
www.poetrybooks.co.uk
www.poetrybookshoponline.com
www.childrenspoetrybookshelf.co.uk
Director *Chris Holifield*
Editorial/Marketing Officer *Davd Isaac*
Sales/Membership Officer *David McDonagh*

For readers, writers, students and teachers of poetry. Founded in 1953 by T.S. Eliot and friends and funded by the Arts Council, the PBS is a unique membership organization and book club providing up-to-date and comprehensive information about poetry from publishers in the UK and Ireland. Members receive the quarterly *Selectors' Choice* and the quarterly *PBS Bulletin* packed with articles by poets, poems, news, listings and access to discounts of at least 25% off featured titles. Subscriptions start at £12. The PBS website (www.poetrybooks.co.uk) has a recruitment area and a closed section for members. The PBS also offers a range of nearly 90,000 poetry titles at www.poetrybookshoponline.com, sells the Poetry Archive CDs and has a new section called SoundBlast for performance poets' CDs. During 2005 the PBS relaunched the Children's Poetry Bookshelf (www.childrenspoetrybookshelf.co.uk) for 7–11-year-olds with new membership schemes for parents, grandparents and libraries, adding an annual competition for child poets, **Old Possum's Children's Poetry Competition**, a year later. Also runs the annual **T.S. Eliot Prize** for the best collection of new poetry, which has a Shadowing Scheme for schools and a new reading groups scheme. In 2004 the PBS ran the Next Generation Poets promotion. In 2009 the PBS worked in partnership with the British Library to run the inaugural **Michael Marks Awards for Poetry Pamphlets**.

## The Poetry Business

Bank Street Arts, 32–40 Bank Street, Sheffield
S1 2DS
office@poetrybusiness.co.uk
www.poetrybusiness.co.uk
Directors *Peter Sansom, Ann Sansom*

Founded in 1986, the Business publishes *The North* magazine and books and pamphlets under the Smith/Doorstop imprint. It runs an annual competition and organizes writing days and a Writing School. See the website for full details.

## Poetry Can

12 Great George Street, Bristol BS1 5RH
☎ 0117 933 0900
admin@poetrycan.co.uk
www.poetrycan.co.uk
Director *Colin Brown*
Administrator *Peter Hunter*

Based in Bristol and founded in 1995, Poetry Can is a poetry development agency working across Bristol and the South West Region. It organizes events such as the annual Bristol Poetry Festival, an education programme and provides information and advice concerning all aspects of poetry to individuals, agencies and organizations. It also hosts the Literature South West website.

## Poetry Ireland/Éigse Éireann

2 Proud's Lane, Dublin 2, Republic of Ireland
☎ 00 353 1 478 9974   🖷 00 353 1 478 0205
poetry@iol.ie
www.poetryireland.ie
Education: 00 353 1 475 8605
Writers in Schools scheme: 00 353 1 475 8601
Director *Joseph Woods*

Poetry Ireland is the national organization for poetry in Ireland, with its four core activities being readings, publications, education and an information and resource service. Organizes readings by Irish and international poets countrywide. Through its website, telephone, post and public enquiries, Poetry Ireland operates as a clearing house for everything pertaining to poetry in Ireland. Operates the Writers in Schools scheme. Publishes *Poetry Ireland News*, a bi-monthly newsletter containing information on events, competitions and opportunities. *Poetry Ireland Review* appears quarterly and is the journal of record for poetry in Ireland; current editor: *Caitríona O'Reilly*.

## The Poetry Library

Royal Festival Hall, Level 5, London SE1 8XX
☎ 020 7921 0943/0664   🖷 020 7921 0607
info@poetrylibrary.org.uk
www.poetrylibrary.org.uk
www.poetrymagazines.org.uk
Librarians *Chris McCabe, Miriam Valencia*

Founded by the Arts Council in 1953. A collection of 100,000 books, pamphlets, CDs, DVDs, videos and magazines of modern poetry since 1912, from Georgian to Rap, representing all English-speaking countries and including translations into English by contemporary poets. Two copies of each title are held, one for loan and one for reference. A wide range of poetry magazines and ephemera from all over the world are kept along with cassettes, CDs and videos for consultation, with many available for loan. There is a children's poetry section with teacher's resource collection.

Library staff can help with identifying 'lost quotations', suggesting poems by theme, or providing information about poets. They can also help with enquiries about publishing poetry, and the website is regularly updated with new poetry competitions, events, magazines and publishers.

## poetry pf

20 Clovelly Way, Orpington BR6 0WD
☏ 01689 811394
editor@poetrypf.co.uk
www.poetrypf.co.uk
Contact *Anne Stewart*

Poet showcase site. Creates and manages searchable and contactable internet presence for poets. Publishes Poem Cards and occasional books for public sale. Poetry-related print and design services and project work.

## The Poetry School

81 Lambeth Walk, London SE11 6DX
☏ 020 7582 1679   ℻ 020 7625 5391
administration@poetryschool.com
www.poetryschool.com
Contact *The Administrator*

Founded in 1997 and funded by Arts Council England, London, The Poetry School offers a wide range of courses and workshops, tutorials and seminars on writing and reading poetry and is open to all levels of writer, regardless of experience or formal qualifications. Tutors include Mimi Khalvati, Graham Fawcett, Roddy Lumsden, Tamar Yoseloff, Don Paterson, Daljit Nagra. Activities are offered throughout the country as well as increasingly online. The School provides a forum for (aspiring) poets to share experiences, develop skills and extend appreciation of the traditional and innovative aspects of their art. Visit the website for more information and to make a booking. The Poetry School is a Registered Charity No. 1069314.

## The Poetry Society

22 Betterton Street, London WC2H 9BX
☏ 020 7420 9880   ℻ 020 7240 4818
info@poetrysociety.org.uk
www.poetrysociety.org.uk
Director *Judith Palmer*

Founded in 1909, the Society exists to help poets and poetry thrive in Britain. Reaching out from its Covent Garden base, it promotes the national health of poetry in a range of imaginative ways. Membership costs £40 for individuals (concessions available); youth memberships from £18. *Poetry News* membership is from £18. Current activities include:

* Quarterly magazine of new verse, views and criticism, *Poetry Review*.
* Quarterly newsletter, *Poetry News*, with lively relevant articles for members and competitions.

* Promotions, events and cooperation with Britain's many literature festivals, poetry venues and poetry publishers.
* Competitions and awards, including the annual **National Poetry Competition** with a £5,000 first prize.
* Improve your poetry with *Poetry Prescription*. Reduced rates for members.
* Members' web pages including photos and profiles.
* Stanzas – poetry groups run by Poetry Society members in your area.
* Monthly e-bulletins – latest news, competitions and events – sent to your inbox.
* Provides information and advice, publishes books, posters and resources for schools and libraries. Education membership costs £50 (secondary) or £30 (primary) which includes free poetry anthologies, lesson plans and a subscription to Poems on the Underground. Publications include *The Poetry Book for Primary Schools* and *Jumpstart Poetry for the Secondary School*, colourful poetry posters for Keystages 1, 2, 3 and 4. Many of Britain's most popular poets – including Michael Rosen, Roger McGough and Jackie Kay – contribute, offering advice and inspiration.
* The Society's online poetry classroom is at www.poetryclass.net
* The Poetry Café serving snacks and drinks to members, friends and guests, part of The Poetry Place, a venue for many poetry activities – readings, workshops and poetry launches. This space is available for bookings (020 7420 9887).
* *Poetry Landmarks of Britain*, a free resource on the Society's website, packed with regional poetry places, publishers, venues and regular events.

## The Poetry Trust

The Cut, 9 New Cut, Halesworth IP19 8BY
☏ 01986 835950   ℻ 01986 874524
info@thepoetrytrust.org
www.thepoetrytrust.org
Director *Naomi Jaffa*

One of the UK's flagship poetry organizations, delivering a year-round programme of live events, creative education opportunities, courses, prizes and publications. Major events include the international **Aldeburgh Poetry Festival** held annually in early November (5th–7th in 2010) and the **Aldeburgh First Collection Prize**. Registered Charity No. 1102893.

## Poets Anonymous
▷ entry under Poetry Presses

## Point Editions

Ithaca, Apdo. 125, E–03590 Altea, Spain
℡ 00 34 96 584 2350   🖷 00 34 96 584 2350
elpoeta@point-editions.com
www.point-editions.com
*Also at:* Rekkemsestraat 167, B–8510 Marke,
Belgium
Director *Germain Droogenbroodt*

Founded as POetry INTernational in 1984, Point
is based in Spain and Belgium. A multilingual
publisher of contemporary verse from *established*
poets, the organization has brought out more
than 80 titles in at least eight languages, including
English. Point Editions runs the original work
alongside a verse translation into Dutch made
in cooperation with the poet. The organization's
website features the world's best known and
unknown poets in English, Spanish and Dutch.
Point also co-organizes several international
poetry festivals.

## Regional Arts Councils

See **Arts Councils and Regional Offices**, page
565.

## Scottish Pamphlet Poetry

25 Market Square, Masham, Ripon HG4 4EG
hazel@scottish-pamphlet-poetry.biz
www.scottish-pamphlet-poetry.com
Administrator *Hazel Cameron*

A publisher collective of mainly Scottish-based
presses promoting poetry published in pamphlet
form. A pamphlet is defined as not less than six
pages but no more than 36. The organization has
an excellent website with online sales and audio.
It runs book fairs and events, and is supported by
the Scottish National Library of Scotland which
administers the Annual Memorial Award created
to recognize publishing skill and effort in the
pamphlet form.

## Scottish Poetry Library

5 Crichton's Close, Canongate, Edinburgh
EH8 8DT
℡ 0131 557 2876   🖷 0131 557 8393
inquiries@spl.org.uk
www.spl.org.uk
Director *Robyn Marsack*
Librarian *Julie Johnstone*

A comprehensive reference and lending collection
of work by Scottish poets in Gaelic, Scots and
English, plus the work of British, European and
international poets. Stock includes books, audio,
videos, news cuttings and magazines. Borrowing
is free to all. Services include a postal lending
scheme, for which there is a small fee; enquiries;
schools workshops throughout Scotland
on application; exhibitions; bibliographies;
publications; general information in the field of
poetry. Also available is an online catalogue and
index to Scottish poetry periodicals. There is a
Friends' scheme costing £25 annually. Friends
receive a newsletter and other benefits and
support the library.

## Second Light

9 Greendale Close, London SE22 8TG
dilyswood@tiscali.co.uk
www.esch.dircon.co.uk/second/second.htm
Director *Dilys Wood*
Administrator *Anne Stewart (editor@poetrypf.
co.uk)*

A network of over 350 women poets, major
names and less well-known. With its publishing
arm, Second Light Publications, the network aims
to develop and promote women's poetry. Issues
a twice-yearly poetry journal, *ARTEMISpoetry*,
open to submissions for non-members; runs an
annual poetry competition; holds residential
workshops, readings; has published three
anthologies of women's poetry.

## Spiel Unlimited

20 Coxwell Street, Cirencester GL7 2BH
℡ 01285 640470/07814 830031
spiel@arbury.freeserve.co.uk
info@spiel.wanadoo.co.uk
Directors *Marcus Moore, Sara-Jane Arbury*

'Spoken word, written word, anywhere,
everywhere' with two writers who put a positive
charge in live literature by organizing quirky and
original events such as Slam!Fests, Spontaneity
Days and Living Room Poetry performances,
hosting UK poetry slams and running workshops
for schools and adults. Also produce *SPIEL*, a
monthly email newsletter. Specialists in breathing
new life into literature.

## Survivors' Poetry

Studio 11, Bickerton House, 25–27 Bickerton
Road, London N19 5JT
℡ 020 7281 4654   🖷 020 7281 7894
info@survivorspoetry.org.uk
www.survivorspoetry.com
Director *Dr Simon Jenner
(simon@survivorspoetry.org.uk)*
National Outreach & National Mentoring Scheme
*Roy Birch (royb@survivorspoetry.org.uk)*
Administration & Marketing, Poetry Express
*Blanche Donnery (blanche@survivorspoetry.org.
uk)*

A unique national literature organization
promoting poetry by survivors of mental distress
through workshops, readings and performances
to audiences all over the UK. It was founded in
1991 by four poets with first-hand experience
of the mental health system. Survivors'
community outreach work provides training and

performance workshops and publishing projects. For the organization's quarterly magazine, see *Poetry Express*. Survivors work with those who have survived mental distress and those who empathize with their experience.

## Tŷ Newydd Writers' Centre

Llanystumdwy, Cricieth LL52 0LW
℡ 01766 522811    ℻ 01766 523095
post@tynewydd.org
www.tynewydd.org
Executive Director *Sally Baker*

Run by the Taliesin Trust, an independent, Arvon-style residential writers' centre established in the one-time home of Lloyd George in North Wales. The programme (in both Welsh and English) has a regular poetry content. (See also entry under *UK and Irish Writers' Courses*.) Fees start at £230 for weekends and £465 for week-long courses. Bursaries available. Among the many tutors to-date have been: Gillian Clarke, Carol Ann Duffy, Owen Sheers, Robert Minhinnick, Liz Lochhead, Michael Longley, Ian Macmillan and Jo Shapcott. Send for the centre's descriptive leaflets and programme of courses.

## The Western Writers' Centre/Ionad Scriobhneoirí Chaitlin Maud

27 Nuns Island, Galway, Republic of Ireland
℡ 00 353 91 533594
westernwriters@eircom.net
www.twwc.ie
Director *Fred Johnston*
Administration *Marvelle Maguire*

Founded 2001. Based in the West Shannon region of Ireland and named after the late Caitlin Maud, prominent Irish-language activist, poet and singer. Courses, organized readings and workshops are held in the Galway City and County area and the West of Ireland generally. Small library of literary periodicals. News and reviews of books in the 'Kiosque!' section of the website. Produces a monthly literary newsletter, *The Word Tree*. Accepts original work for online publication in Gaelic, English and French. No payment for publication.

# Small Presses

## Aard Press

c/o Aardverx, 31 Mountearl Gardens, London
SW16 2NL
Managing Editors *Dawn Redwood, D. Jarvis*

Founded 1971. Publishes artists' bookworks, experimental/visual poetry and art theory, topographics, ephemera, international mail-art and performance art documentation. Very small editions. No unsolicited material or proposals.
£ Royalties not paid. No sale-or-return deals.

## Accent Press

The Old School, Upper High Street, Bedlinog
CF46 6RY
T 01443 710930    F 01443 710940
info@accentpress.co.uk
www.accentpress.co.uk
Managing Editor *Hazel Cushion*

Founded 2003. Publishes general fiction, erotic fiction, health, humour and biography. 40 titles a year. Visit the website for submission guidelines.
£ Royalties twice-yearly.

## Allardyce, Barnett, Publishers

14 Mount Street, Lewes BN7 1HL
T 01273 479393
www.abar.net
Publisher *Fiona Allardyce*
Managing Editor *Anthony Barnett*

Founded 1981. Publishes art, literature and music. IMPRINT **Allardyce Book**. About 3 titles a year. Unsolicited mss and synopses cannot be considered.

## Alloway Publishing Limited
▷ Stenlake Publishing Limited

## Anglo-Saxon Books

25 Brocks Road, Swaffham PE37 7XG
T 0845 430 4200
www.asbooks.co.uk
Managing Editor *Tony Linsell*

Founded 1990 to promote a greater awareness of and interest in early English history, language and culture. Publishes Anglo-Saxon history, culture,
language. About 5 titles a year. Please phone before sending mss.
£ Royalties at standard rate.

## Areté Books

8 New College Lane, Oxford OX1 3BN
T 01865 289193    F 01865 289194
craigraine@aretemagazine.co.uk
www.aretemagazine.co.uk
Editor *Craig Raine*
Deputy Editor *Ann Pasternak Slater*

Founded in 2009 as an offshoot of the tri-quarterly literary journal, *Areté* (see entry under *Poetry Magazines*). Areté Books' first publication, *A Scattering*, Christopher Reid's collection of poems, won the 2009 Costa Book of the Year award.

## Awen Publications

7 Dunsford Place, Bath BA2 6HF
T 01225 334204
publisher@awenpublications.co.uk
www.awenpublications.co.uk
Managing Editor *Kevan Manwaring*

Founded 2003. Publishes eco-spiritual poetry, fiction and creative non-fiction 6 titles in 2009. Unsolicited submissions not welcome at present; check website for details and updates.
£ Royalties by arrangement with the author.

## Barny Books

The Cottage, Hough on the Hill, near Grantham
NG32 2BB
T 01400 250246    F 01400 251737
www.barnybooks.biz
Managing Director/Editorial Head *Molly Burkett*
Business Manager *Jayne Thompson*

Founded with the aim of encouraging new writers and illustrators. Publishes a wide variety of books including adult fiction and non-fiction. Offers schools' projects where students help to produce books. About 12–15 titles a year. Too small a concern to have the staff/resources to deal with unsolicited mss. Writers with strong ideas should approach Molly Burkett by letter in the first instance. Also runs a readership and advisory

service for new writers (£20 fee for short stories or illustrations; £50 for full-length stories).
Ⓔ Division of profits 50/50.

## BB Books
▷ entry under Poetry Presses

## Between the Lines
Bench House, 82 London Road, Chipping Norton OX7 5FN
☎ 01608 644755
btluk@aol.com
waywiser-press.com/imprints/betweenthelines.html
Editorial Board *Peter Dale, Philip Hoy, J.D. McClatchy*
Editorial Assistant *Ryan Roberts*

Founded in 1998 and since 2002 an imprint of the **Waywiser Press**. Publishes in book form extended interviews with leading contemporary poets, amongst them Peter Dale, Dick Davis, Rachel Hadas, Timothy Steele, John Ashbery, Charles Simic, Ian Hamilton, Donald Justice, Seamus Heaney, Richard Wilbur, Thom Gunn, Donald Hall, Anthony Hecht, Anthony Thwaite, Michael Hamburger and W.D. Snodgrass. Most volumes include a career sketch, a comprehensive bibliography and a representative selection of quotations from the poets' critics and reviewers; some also include photographs as well as previously uncollected poems.
Ⓔ Royalties not paid.

## BeWrite Books
32 Bryn Road South, Ashton-in-Makerfield, Wigan WN4 8QR
contact@bewrite.net
www.bewrite.net
Managing Editor *Neil Marr*

Founded 2002. Publishes fiction and poetry in paperback and ebook form. 12–24 titles a year. Unsolicited mss and synopses welcome using the online form on the website (www.bewrite.net/bookshop/submission_requirements.htm).
Ⓔ Royalties quarterly.

## Bluemoose Books Limited
25 Sackville Street, Hebden Bridge HX7 7DJ
☎ 01422 842731
dufk@aol.com
www.bluemoosebooks.com
Managing Editor *Hetha Duffy*
Publisher *Kevin Duffy*

Founded 2006. Publishes fiction and non-fiction. 4 titles in 2009. No unsolicited mss; send first three chapters and synopsis either by post (include s.a.s.e.) or email. No poetry or children's books.
Ⓔ Royalties annually.

## Book Castle Publishing
2a Sycamore Business Park, Dishforth Road, Copt Hewick HG4 5DF
☎ 07837 217893
info@book-castle.co.uk
www.book-castle.co.uk
Managing Editor *Paul Bowes*
Assistant Editor *Sally Siddons*

Founded 1986. Publishes non-fiction of local interest (Bedfordshire, Hertfordshire, Buckinghamshire, Oxfordshire, the Chilterns). Over 120 titles in print. 5 titles a year. Unsolicited mss, synopses and ideas for books welcome.
Ⓔ Royalties paid.

## Brilliant Publications
Unit 10, Sparrow Hall Farm, Edlesborough, Dunstable LU6 2ES
☎ 01525 222292    🖷 01525 222720
info@brilliantpublications.co.uk
www.brilliantpublications.co.uk
Publisher *Priscilla Hannaford*

Founded 1993. Publishes practical resource books for teachers, parents and others working with 3–14 year-olds. Their vision is to make learning a fun, rewarding and creative experience for both teachers and pupils. About 10–15 titles a year. See website for author guidelines. Potential authors are strongly advised to look at the format of existing books before submitting synopses. 'We do not publish children's picture books.'
Ⓔ Royalties twice-yearly.

## The Brodie Press
thebrodiepress@hotmail.com
www.brodiepress.co.uk
Project Editor *Penny Price*
Managing Editors *Hannah Sheppard, Tom Sperlinger*

Founded 2002. Publishes poetry and anthologies. SERIES *The Brodie Poets*, launched in 2003, including Julie-ann Rowell, Poetry Book Society Pamphlet Choice, winter 2003. 1 title a year. Send synopses and proposals by email in the first instance.

## Charlewood Press
7 Weavers Place, Chandlers Ford SO53 1TU
☎ 023 8026 1192
gponting@clara.net
www.home.clara.net/gponting/index-page11.html
Managing Editors *Gerald Ponting, Anthony Light*

Founded 1987. Publishes local history books and walks booklets on Hampshire and adjacent counties, especially around the Fordingbridge area. Publishes work by the managing editors only.

## CNP Publications
> Lyfrow Trelyspen

## Contact Publishing Ltd
☎ 020 7193 1782
info@contact-publishing.co.uk
www.contact-publishing.co.uk
Managing Director *Anne Kontoyannis*
Submissions Editor *Grant Cooper*

Founded 2003. Publishes general fiction,
non-fiction, self help and New Age, thrillers,
historical fiction, biography; 'anything original'.
No children's, sports or westerns. 'Interested in
seeing mss by new authors with original angles
and new ideas'. See website for author guidelines.
Send synopsis, three sample chapters and outline
in the first instance as email submission only (to
manuscript@contact-publishing.co.uk) as doc or
pdf files.
£ Royalties twice-yearly.

## The Cosmic Elk
68 Elsham Crescent, Lincoln LN6 3YS
☎ 01522 820922
post@cosmicelk.net
www.cosmicelk.net
Contact *Heather Hobden*

Founded 1988 for 'desk-top publishing' easily
updated print-on-demand books/booklets
on science, history and the history of science;
also websites designed and maintained. 'Please
check the website first where you will find the
information you need and contact us by the email
address above'.

## Crescent Moon Publishing and Joe's Press
PO Box 393, Maidstone ME14 5XU
☎ 01622 729593
cresmopub@yahoo.co.uk
www.crescentmoon.org.uk
Managing Editor *Jeremy Robinson*

Founded 1988 to publish critical studies of
figures such as D.H. Lawrence, Thomas Hardy,
André Gide, Walt Disney, Rilke, Leonardo da
Vinci, Mark Rothko, C.P. Cavafy and Hélène
Cixous. Publishes literature, criticism, media, art,
feminism, painting, poetry, travel, guidebooks,
cinema and some fiction. Literary magazine,
*Passion* (quarterly) and *Pagan America*, biannual
anthology of American poetry. About 15–20 titles
per year. Unsolicited synopses and ideas welcome
but approach in writing first and send an s.a.s.e.
Do not send whole mss.
£ Royalties negotiable.

## Crime Express
> Five Leaves Publications

## David Porteous Editions
PO Box 5, Chudleigh, Newton Abbot TQ13 0YZ
☎ 01626 853310
editorial@davidporteous.com
www.davidporteous.com
Publisher *David Porteous*

Founded 1992 to produce colour illustrated
books on hobbies and leisure for the UK and
international markets. Publishes crafts, hobbies,
art techniques and needlecrafts. No poetry
or fiction. 3–4 titles a year. Unsolicited mss,
synopses and ideas welcome if return postage
included.
£ Royalties twice-yearly.

## Day Books
Orchard Piece, Crawborough, Charlbury OX7 3TX
☎ 01608 811196
diaries@day-books.com
www.day-books.com
Managing Editor *James Sanderson*

Founded in 1997 to publish a series of great
diaries from around the world. Unsolicited mss,
synopses and ideas about diaries only, please.
Include postage if return of material is required.

## The Dragonby Press
30 King Street, Winterton DN15 9TP
☎ 01724 737254
rich@rahwilliams.orangehome.co.uk
Managing Editor *Richard Williams*

Founded 1987 to publish affordable bibliography
for reader, collector and dealer. About 3 titles
a year. Unsolicited mss, synopses and ideas
welcome for bibliographical projects only.
£ Royalties paid.

## Dramatic Lines
PO Box 201, Twickenham TW2 5RQ
☎ 020 8296 9502    ᶠ 020 8296 9503
mail@dramaticlinespublishers.co.uk
www.dramaticlines.co.uk
Managing Editor *John Nicholas*

Founded to promote drama for all. Publications
with a wide variety of theatrical applications
including classroom use and school assemblies,
drama examinations, auditions, festivals, theatre
group performance, musicals and drama resource
handbooks. Unsolicited drama-related mss,
proposals and synopses welcome; enclose s.a.s.e.
£ Royalties paid.

## Dreamstairway Books
8 Yew Tree Grove, Highley, Bridgenorth
WV16 6DG
☎ 01746 861330    ᶠ 01746 861330
robert@dreamstairway.co.uk
www.dreamstairway.co.uk
Managing Editor *Robert Foster*

Founded 2009. Publishes mind, body and spirit, crafts and practical guides. 8 titles in 2009. No unsolicited mss; send synopses and ideas for books by email or post.
£ Royalties twice-yearly.

## Edgewell Publishing

5A Front Street, Prudhoe NE42 5HJ
☎ 01661 835330   📠 01661 835330
keithminton@btconnect.com
www.edgewell-publishing.co.uk
Editor *Keith Minton*
Co-Editor *Stephen Appleby*

Quarterly short story and poetry magazine (*Old Gothic Tales* – four issues annually) to encourage writing from new and established writers, also children. Writers from the North East especially welcome. Offers a reading and literary agency service 'especially for new writers'. Contact for charges, by email, where possible..
£ No royalties; writers maintain copyright.

## Educational Heretics Press

113 Arundel Drive, Bramcote, Nottingham NG9 3FQ
☎ 0115 925 7261   📠 0115 925 7261
www.edheretics.gn.apc.org
Directors *Janet & Roland Meighan*

Non-profit venture which aims to question the dogmas of schooling in particular and education in general, and establish the logistics of the next personalized learning system. No unsolicited material. Enquiries only.
£ Royalties paid on recent contracts.

## Eilish Press

4 Collegiate Crescent, Broomhall Park, Sheffield S10 2BA
☎ 07973 353964
eilishpress@hotmail.co.uk
eilishpress.tripod.com
Head of Marketing *Suzi Kapadia*

Specializes in academic work and popular non-fiction in the areas of women's studies, human rights and anti-racism. Also produces children's literature with humanitarian themes. No adult fiction. Accepting no new projects.
£ Royalties annually.

## Fand Music Press

Glenelg, 10 Avon Close, Petersfield GU31 4LG
☎ 01730 267341   📠 01730 267341
paul@fandmusic.com
www.fandmusic.com
Managing Editor *Peter Thompson*

Founded in 1989 as a sheet music publisher, Fand Music Press has expanded its range to include CD recordings and books on music. Also publishes poetry and short stories. 10 titles a year. No

unsolicited mss. Write with ideas in the first instance.

## Feather Books

PO Box 438, Shrewsbury SY3 0WN
☎ 01743 872177   📠 01743 872177
john@waddysweb.freeuk.com
www.waddysweb.freeuk.com
Managing Director *Rev. John Waddington-Feather*
Music Director *David Grundy*
Drama Director/Recordings Manager *Tony Reavill*

Founded in 1980 to publish writers' group work. All material has a strong Christian ethos. Publishes poetry (mainly, but not exclusively, religious); Christian mystery novels (the Revd. D.I. Blake Hartley series); Christian children's *Quill Hedgehog* novels; seasonal poetry collections and *Feather's Miscellany*. Produces poetry, drama and hymn CD/cassettes; also the anthology 'Pilgrimages' ed. Professor Walter Nash ('a milestone in Christian poetry publishing'). About 20 titles a year. All correspondence to include s.a.e., please.

## Fern House Personal Publishing

19 High Street, Haddenham, Ely CB6 3XA
☎ 01353 740222
info@fernhouse.com
www.fernhouse.com
Managing Director *Rodney Dale*

Founded 1995. Has published mainly non-fiction (18 titles in print). Now specializes in helping writers turn their works into book form.

## Fidra Books

219 Bruntsfield Place, Edinburgh EH10 4DH
☎ 0131 447 1917
vanessa@fidrabooks.com
www.fidrabooks.com
Managing Editor *Vanessa Robertson*

Founded 2005. Publishes children's fiction, specializing in reprints of sought-after 20th century children's fiction. No unsolicited material; initial approach by letter or email.
£ Royalties quarterly.

## Fineleaf

Moss Cottage, Pontshill, Ross-on-Wye HR9 5TB
☎ 01989 750369
books@fineleaf.co.uk
www.fineleaf.co.uk
Managing Editor *Philip Gray*

Founded 2001. Publishes landscape, history, fine art and poetry. 8 titles in 2009. Unsolicited mss, snopses and ideas welcome; approach by email.
£ Royalties annually.

## Five Leaves Publications

PO Box 8786, Nottingham NG1 9AW
☎ 0115 969 3597

info@fiveleaves.co.uk
www.fiveleaves.co.uk
Contact *Ross Bradshaw*

Founded 1995 (taking over the publishing programme of Mushroom Bookshop). Publishes fiction, poetry, social history, young adult fiction and Jewish interest. Also publishes for the Anglo-Catalan Society. IMPRINTS **Crime Express** Crime fiction novellas; **New London Editions** Classic London fiction; **Roger Hollis** Poetry and memoirs. About 15 titles a year. Titles normally commissioned.
Ⓔ Royalties paid.

## Frontier Publishing

Windetts, Kirstead NR15 1EG
Ⓣ 01508 558174
frontierpublishing@ymail.com
www.frontierpublishing.co.uk
Managing Editor *John Black*

Founded 1983. Publishes non-fiction only: travel, photography, sculptural history and literature. 2–3 titles a year. No unsolicited mss; synopses and ideas welcome.
Ⓔ Royalties paid.

## Galactic Central Publications

25A Copgrove Road, Leeds LS8 2SP
gcp@philsp.com
www.philsp.com
Managing Editor *Phil Stephenson-Paine*

Founded 1980. Publishes science fiction biographies only. No unsolicited mss; email synopses or ideas for books.

## Glosa Education Organisation

PO Box 18, Richmond TW9 2GE
www.glosa.org
Managing Editor *Wendy Ashby*

Founded 1981. Publishes textbooks, dictionaries and translations for the teaching, speaking and promotion of Glosa (an international, auxiliary language); also a newsletter, *Plu Glosa Nota* and journal, *PGN*. Unsolicited mss and ideas for Glosa books welcome.

## Godstow Press

60 Godstow Road, Wolvercote, Oxford OX2 8NY
Ⓣ 01865 556215
info@godstowpress.co.uk
www.godstowpress.co.uk
Managing Editors *Linda Smith, David Smith*

Founded in 2003 for the publication of creative work with a philosophical/spiritual content. 2–3 titles a year. Considers works self-published by the author for inclusion in the catalogue. The press works on direct sales rather than through usual trade outlets. Synopses and ideas welcome but no unsolicited mss.

## Graffeg

2 Radnor Court, 256 Cowbridge Road East, Cardiff CF5 1GZ
Ⓣ 029 2037 7312    Ⓕ 029 2039 8101
info@graffeg.com
www.graffeg.com
Managing Editor *Peter Gill*

Publisher of photographic books about Wales: food, travel, landscapes, gardens and architecture. 23 titles annually. Unsolicited material welcome by post.
Ⓔ Royalties twice-yearly.

## Grant Books

The Coach House, New Road, Cutnall Green, Droitwich WR9 0PQ
Ⓣ 01299 851588    Ⓕ 01299 851446
golf@grantbooks.co.uk
www.grantbooks.co.uk
Managing Editor *H.R.J. Grant*

Founded 1978. Publishes golf-related titles only: course architecture, history, biography, etc., but not instructional, humour or fiction material. About 3 titles a year. Mss, synopses and ideas welcome.
Ⓔ Royalties paid.

## Great Northern Publishing

PO Box 202, Scarborough YO11 3GE
Ⓣ 01723 581329    Ⓕ 01723 581329
books@greatnorthernpublishing.co.uk
www.greatnorthernpublishing.co.uk
Production Manager/Senior Editor *Mark Marsay*

Small, independent, award-winning, family-owned company founded in 1999. Publishers of books, magazines and journals. Also provides full book, magazine and print production services to individuals, businesses, charities, museums and other small publishers. Publishes non-fiction (mainly military), adult erotic material and, occasionally, fiction. No romance, religious, political, feminist, New Age, medical or children's books. Publishers of the bi-monthly military history magazine, *The Great War* and monthly erotic art magazine, *Jade* (see entries under *Magazines*). Mail order and internet-based bookshop stocking selected titles alongside its own. Website provides submission guidelines and current publishing requirements. No unsolicited phone calls or mss; send letter or text-only email in the first instance.
Ⓔ Royalties twice-yearly.

## Griffin Publishing Associates

61 Grove Avenue, London N10 2AL
Ⓣ 020 8444 0815/0778 842 5262
griffin.publishing.associates@googlemail.com
Managing Editor *Ali Ismail*

Founded 2007. Publishes all kinds of fiction (plays, poetry and prose) with an emphasis on science fiction, fantasy and horror. Unsolicited mss, synopses and ideas for books welcome. Approach by post (return postage must be included) or telephone. (For Australian and New Zealand writers and literary agents, postal contact address: Seeham Moheed, GPA local agent, 58 Challenger Street, St Helier, Auckland, New Zealand.
£ Royalties annually.

## GSSE

11 Malford Grove, Gilwern, Abergavenny NP7 0RN
☎ 01873 830872
GSSE@zoo.co.uk
www.gsse.org.uk
Owner/Manager *David P. Bosworth*

Publishes books and booklets describing classroom practice (at all levels of education and training). Ideas welcome – particularly from practising teachers, lecturers and trainers describing how they use technology in their teaching.
£ Royalties by arrangement.

## Haunted Library

Flat 1, 36 Hamilton Street, Hoole, Chester CH2 3JQ
☎ 01244 313685
pardos@globalnet.co.uk
www.users.globalnet.co.uk/~pardos/GS.html
Managing Editor *Rosemary Pardoe*

Founded 1979. Publishes the *Ghosts and Scholars M.R. James Newsletter* twice a year, featuring articles, news and reviews (no fiction).
£ Royalties not paid.

## Hawkwood Books

3 Turner Street, Lincoln LN1 3JL
☎ 01522 831603
enquiries@hawkwoodbooks.co.uk
www.hawkwoodbooks.co.uk
Managing Editor *Ellis Delmonte*

Founded 2006. Publishes children's and young adult fiction. No picture books or non-fiction. 2 titles in 2009. 'We are a small venture with a limited list but are not a vanity press and do not charge for any aspect of the publishing process. However, there is no guarantee that published books will be widely distributed through retailers.' See website for submission details.

## Headpress

Suite 306, The Colourworks, 2a Abbot Street, London E8 3DP
☎ 0845 330 1844
headoffice@headpress.com
www.headpress.com
Managing Editor *David Kerekes*

Founded 1991. Publishes *Headpress* journal and books devoted to film, music, popular culture, the strange and esoteric. No fiction or poetry. Unsolicited material welcome; send letter and outline in the first instance.
£ Royalties/flat fees vary.

## Heart of Albion Press

62 Wartnaby Street, Market Harborough LE16 9BE
☎ 01858 431717
albion@indigogroup.co.uk
www.hoap.co.uk
Managing Editor *R.N. Trubshaw*

Founded 1990 to publish local history. Publishes folklore, local history and mythology. Synopses relating to these subjects welcome.
£ Royalties negotiable.

## Ignotus Press

BCM–Writer, London WC1N 3XX
☎ 00 353 625 6582
ignotuspress@eircom.net
suzanneruthvenatignotuspress.blogspot.com
Commissioning Editor *Suzanne Ruthven*

Founded 1994. Publishes metaphysical non-fiction. No unsolicited mss; send outline by email or post after reading submission guidelines on the website.
£ Royalties twice-yearly.

## Immanion Press

8 Rowley Grove, Stafford ST17 9BJ
☎ 01785 613299
editorial@immanion-press.com
www.immanion-press.com
Managing Editor *Storm Constantine*

Founded 2003. Publishes genre fiction (science fiction, fantasy, horror), slipstream fiction; esoteric non-fiction. IMPRINT **Megalithica Books** Commissioning Editor *Taylor Ellwood* (info@immanion-press.com) Non-fiction, specifically cutting edge esoteric work. Submissions by email or post (enclose s.a.s.e.). Website gives full submission details.
£ Royalties paid.

## Indepenpress

▷ Pen Press & Indepenpress Publishing Ltd

## Indigo Dreams Publishing

▷ entry under Poetry Presses

## Infinity Junction

PO Box 64, Neston CH64 0WB
infin-info@infinityjunction.com
www.infinityjunction.com
Managing Editor *Neil Gee*

Established originally as a self-publishing, self-help organization in 1999, Infinity Junction became a publisher in 2001. Currently offers a variety of services to authors, including advice, editing, book layout and full commercial publication. Authors are strongly advised to read the detailed information on the website before making contact. Email communication preferred. 'We cannot undertake to read unsolicited whole mss.'

## Iolo

38 Chaucer Road, Bedford MK40 2AJ
☎ 01234 301718/07909 934866    📠 01234 301718
dedwydd1@ntlworld.com
www.dedwyddjones.co.uk
Managing Director *Dedwydd Jones*

Publishes Welsh theatre-related material and campaigns for a Welsh National Theatre. Ideas on Welsh themes welcome; approach in writing.

## Ivy Publications

72 Hyperion House, Somers Road, London SW2 1HZ
☎ 020 8671 6872
Proprietor *Ian Bruton-Simmonds*

Founded 1989. Publishes educational, science, fiction, philosophy, children's, travel, literary criticism, history, film scripts. No unsolicited mss; send two pages, one from the beginning and one from the body of the book, together with synopsis (one paragraph) and s.a.s.e. No cookery, gardening or science fiction.
£ Royalties annually.

## Jane Nissen Books

Swan House, Chiswick Mall, London W4 2PS
☎ 020 8994 8203    📠 020 8742 8198
jane@nissen.demon.co.uk
www.janenissenbooks.co.uk
Publisher *Jane Nissen*

Founded 2000. Publishes reprints of children's fiction only. Winner of the 2007 Eleanor Farjeon Award. Does not consider submitted material as all titles published are reprints of classics or forgotten children's books.

## The Jupiter Press

The Hundred House, The Ludlow Road, Bridgnorth WV16 5NL
☎ 01746 763795
gordonthomasdrury@btinternet.com
Managing Editor *Gordon Thomas Drury*

Founded 1995. Looking for niche market and information publications, also quiz and game subjects. Interested in holistic, clairvoyance and esoteric subjects. Synopses and ideas for books welcome. Send idea and sample chapter (include contact telephone number and return postage). Unsolicited mss welcome 'but may request a reader's fee'. Currently seeking books and ideas for SERIES entitled *How To Make Money At ....*
£ Royalties paid.

## Kittiwake

3 Glantwymyn Village Workshops, Glantwymyn, Nr Machynlleth SY20 8LY
☎ 01650 511314    📠 01650 511314
perrographics@btconnect.com
www.kittiwake-books.com
Managing Editor *David Perrott*

Founded 1986. Publishes guidebooks only, with an emphasis on careful design/production. Specialist research, writing, cartographic and electronic publishing services available. Unsolicited mss, synopses and ideas for guidebooks welcome.
£ Royalties paid.

## The Lindsey Press

Unitarian Headquarters, 1–6 Essex Street, London WC2R 3HY
☎ 020 7240 2384
info@unitarian.org.uk
Convenor *Kate Taylor*

Established at the end of the 18th century as a vehicle for disseminating liberal religion. Adopted the name of The Lindsey Press at the beginning of the 20th century (after Theophilus Lindsey, the great Unitarian Theologian). Publishes books reflecting liberal religious thought or Unitarian denominational history. Also worship material – hymn books, collections of prayers, etc. No unsolicited mss; synopses and ideas welcome.
£ Royalties not paid.

## The Linen Press

75c (13) South Oswald Road, Edinburgh EH9 2HH
lynnmichell@googlemail.com
www.linenpressbooks.co.uk
Managing Director *Lynn Michell*

Founded 2006. Small, independent publishing house based in Edinburgh run by women for women. 'We are keen to champion new literary fiction, international writers, innovative prose and women writers from minority groups. We will consider fiction and non-fiction but not short stories, children's books, science fiction, romance or poetry.' 4 titles a year. Happy to receive submissions from emerging as well as established writers as long as they fulfil the criteria and ethos of the Linen Press. Please read the submission guidelines before sending in a proposal.

## Logaston Press

Little Logaston, Logaston, Woonton, Almeley HR3 6QH
☎ 01544 327344
logastonpress@btinternet.com
www.logastonpress.co.uk
Managing Editors *Andy Johnson, Karen Stout*

---

Founded 1985. Publishes history, social history, archaeology, art, architecture, guides to rural West Midlands, Central and South Wales. 20 titles a year. Unsolicited material welcome but prefers initial phone call or email to ensure subject is one they might consider.
£ Royalties paid on publication.

## Lyfrow Trelyspen

The Roseland Institute, Gorran, St Austell PL26 6NT
☎ 01726 843501   📠 01726 843501
trelispen@care4free.net
www.theroselandinstitute.co.uk
Managing Editor *Dr James Whetter*

Founded 1975. Publishes works on Cornish history, biography, essays, etc. Also **CNP Publications** which publishes the quarterly journal *The Cornish Banner/An Baner Kernewek*. 1–2 titles a year. Unsolicited mss, synopses and ideas welcome.
£ Royalties not paid.

## M&M Baldwin

24 High Street, Cleobury Mortimer, Kidderminster DY14 8BY
☎ 01299 270110
mb@mbaldwin.free-online.co.uk
Managing Editor *Dr Mark Baldwin*

Founded 1978. Publishes local interest/history and WW2 codebreaking books. Unsolicited mss, synopses and ideas for books welcome (not general fiction).
£ Royalties paid annually.

## The Maia Press

15–16 Nassau Street, London W1W 7AB
☎ 020 7743 9898
angeline@arcadiabooks.co.uk
www.arcadiabooks.co.uk
Contact *Angeline Rothermundt*

Founded 2002. Publishes fiction by new writers. 6 titles a year. No unsolicited mss, write with c.v. and synopsis only. S.a.s.e. essential.
£ Royalties twice-yearly.

## Maypole Editions

18 The Lindens, Shifnal TF11 8AB
Managing Editor *Barry Taylor*

Founded 1987. Publishes fiction, drama, film scripts and poetry. No submissions until 2011 at which time an s.a.s.e. will be required. Photocopies only. 'No reply means no. Sounds harsh but I was drowning in poets.'

## Megalithica Books

▷ Immanion Press

## Mercia Cinema Society

29 Blackbrook Court, Durham Road, Loughborough LE11 5UA
☎ 01509 218393
admin@merciacinema.org
http://merciacinema.org
Chairman & Editor-in-Chief *Kate Taylor*
Series Editor/Administrator *Mervyn Gould*

Founded in 1980 to foster research into the history of picture houses. Publishes books and booklets on the subject, including cinema circuits and chains. Books are often tied in with specific geographical areas. Quarterly journal: *Mercia Bioscope*, editor *Paul Smith* (editor@merciacinema.org). 63 titles to date. Unsolicited mss (preferably on disk); synopses and ideas.
£ Royalties annually where paid. Six free copies to the author.

## Meridian Books

40 Hadzor Road, Oldbury B68 9LA
☎ 0121 429 4397
meridian.books@btopenworld.com
Managing Editor *Peter Groves*

Founded in 1985 as a small home-based enterprise following the acquisition of titles from Tetradon Publications Ltd. Publishes walking and regional guides. 4–5 titles a year. Unsolicited mss, synopses and ideas welcome if relevant. Send s.a.s.e. if mss to be returned.
£ Royalties paid.

## Millers Dale Publications

7 Weavers Place, Chandlers Ford, Eastleigh SO53 1TU
☎ 023 8026 1192
home.clara.net/gponting/index-page11.html
Managing Editor *Gerald Ponting*

Founded 1987. Self-publisher of local history and walks.

## Mirage Publishing

PO Box 161, Gateshead NE8 1WY
☎ 0870 720 9498
sales@miragepublishing.com
www.miragepublishing.com
Artistic Director *Sharon Anderson*

Founded 1998. Publishes mind, body and spirit, autobiography, biography. 15 titles in 2009. No unsolicited mss; send synopses and ideas by post or email.
£ Royalties twice-yearly.

## Need2Know

Remus House, Coltsfoot Drive, Woodston PE2 9JX
☎ 01733 898103   📠 01733 313524
sales@n2kbooks.com
www.need2knowbooks.co.uk

Founded in 1995 'to fill a gap in the market for self-help books'. Need2Know is an imprint of **Forward Press** (see entry under *Poetry Presses*). Publishes contemporary health and lifestyle issues. No unsolicited mss. Call in the first instance.
€ Advance plus 15% royalties.

## Neil Miller Publications

c/o Ormonde House, 49 Ormonde Road, Hythe CT21 6DW
contact@neilmillersbooks.co.uk
www.neilmillersbooks.co.uk
*Also at:* 4 Rue D'Equirre Bergueneuse, Pas de Calais, France 62134
Managing Editor *Neil Miller*

Founded 1994. Publishes novels and collections of short stories on any subject, fact or fiction: memoirs, autobiography, mystery, horror, war, comedy, science fiction. Interested in working with new and published writers. Evaluation and critique service available. 'Send your novel or selection of short stories for consideration plus our reader's fee of £45. For this you receive a brief critique plus two of our latest authors' books and have the opportunity of being published by us. European authors receive one novel.' Enclose s.a.s.e. for return of work. No unrequested tales.
€ *Authors' Update* Liable to ask authors to contribute to production costs.

## New London Editions
▷ Five Leaves Publications

## North Staffordshire Press Ltd

Staffordshire University, Business Village, 72 Leek Road, Stoke on Trent ST4 2AR
☎ 01782 442831
malcolm.henson@editorial.staffs.com
www.northstaffordshirepress.com
Managing Editor *Malcolm Henson*
E-learning Consultant *Maria De Marks*
Operations Manager *Christopher Bailey*

Founded 2006. Publishes sociology, history, sport, poetry, fiction and academic books. 3 titles in 2009. Unsolicited mss, synopses and ideas welcome; approach by letter, email or phone call.
€ Royalties annually.

## The Nostalgia Collection

Silver Link Publishing Ltd, The Trundle, Ringstead Road, Great Addington, Kettering NN14 4BW
☎ 01536 330588    € 01536 330588
sales@nostalgiacollection.com
www.nostalgiacollection.com
Managing Editor *Will Adams*

Founded 1985. Small independent company specializing in illustrated post-war nostalgia titles including railways, trams, ships and other transport subjects, towns and cities, villages and rural life, rivers and inland waterways, and industrial heritage under the **Silver Link** and **Past and Present** imprints.
€ Fees/royalties paid.

## Nyala Publishing

4 Christian Fields, London SW16 3JZ
☎ 020 8764 6292/0115 981 9418
€ 020 8764 6292/0115 981 9418
nyala.publishing@geo-group.co.uk
www.geo-group.co.uk
Editorial Director *J.F.J. Douglas*

Founded 1996. Publishing arm of Geo Group. Publishes a limited range of biography, travel and general non-fiction. Offers a wide range of printing and publishing services. Reading, editing, proofreading and advice to self-publishers. No unsolicited mss; synopses and ideas considered.
€ Royalties annually.

## Orpheus Publishing House

4 Dunsborough Park, Ripley Green, Ripley, Guildford GU23 6AL
☎ 01483 225777    € 01483 225776
orpheuspubl.ho@btinternet.com
Managing Editor *J.S. Gordon*

Founded 1996. Publishes 'well-researched and properly argued' books in the fields of occult science, esotericism and comparative philosophy/religion. 'Keen to encourage good (but sensible) new authors.' In the first instance, send maximum three-page synopsis with s.a.s.e.
€ Royalties by agreement.

## Packard Publishing Limited

Forum House, Stirling Road, Chichester PO19 7DN
☎ 01243 537977    € 01243 537977
info@packardpublishing.co.uk
www.packardpublishing.co.uk
Chairman/Managing Director *Michael Packard*

Founded in 1977 to distribute overseas publishers' lists in biology, biochemistry and ecology. Publishes academic & professional: school/university interface, postgraduate – mainly in land management and applied ecology; agriculture, forestry, nature conservation, rural studies, landscape architecture and garden design; some languages (French). SERIES Instructional books on garden and landscape design, and monographs or critical biographies in garden and landscape design and history. 36 titles to date. No unsolicited mss. Synopses and ideas in relevant areas welcome. Telephone first.
€ Royalties twice-yearly in first year, then annually.

## Panacea Press Limited

86 North Gate, Prince Albert Road, London
NW8 7EJ
☎ 020 7722 8464   ℻ 020 7586 8187
ebrecher@panaceapress.net
www.panaceapress.net
Managing Editor *Erwin Brecher, PhD*

Founded as a self-publisher but now open for
fiction and non-fiction from other authors.
Material of academic value considered provided it
commands a wide general market. No unsolicited
mss; synopses and ideas welcome. No telephone
calls. Approach by email, fax or letter.
£ Royalties annually.

## Parapress

The Basement, 9 Frant Road, Tunbridge Wells
TN2 5SD
☎ 01892 512118   ℻ 01892 512118
office@parapress.myzen.co.uk
www.parapress.co.uk
Managing Editor *Elizabeth Imlay*

Founded 1993. Publishes animals, autobiography,
biography, history, literary criticism, military and
naval, music, self-help. Some self-publishing.
About 2 titles a year.

## Past and Present
▷ The Nostalgia Collection

## Paupers' Press

37 Quayside Close, Turney's Quay, Trent Bridge,
Nottingham NG2 3BP
☎ 0115 986 3334   ℻ 0115 986 3334
books@pauperspress.com
www.pauperspress.com
Managing Editor *Colin Stanley*

Founded 1983. Publishes extended essays in
booklet form (about 15,000 words) on literary
criticism and philosophy. 'Sometimes we stray
from these criteria and produce full-length
books, but only to accommodate an exceptional
manuscript.' Limited hardback editions of
bestselling titles. Centre for Colin Wilson Studies.
Critical essays on his work encouraged. About 6
titles a year. No unsolicited mss but synopses and
ideas for books welcome.
£ Royalties annually.

## Peepal Tree Press Ltd
▷ entry under Poetry Presses

## Pen Press & Indepenpress Publishing Ltd

25 Eastern Place, Brighton BN2 1GJ
☎ 0845 108 0530   ℻ 01273 261434
info@penpress.co.uk
www.penpress.co.uk
Managing Director *Lynn Ashman*

Founded 1996. Publishing across a wide range of
categories, Pen Press helps new authors to self-
publish. Distribution, promotion and marketing
included in self-publishing deal. Publisher, not
author, pays for reprints and shares in sales
revenue. Now offering traditional publishing
under the **Indepenpress** and **Pulp Press**
IMPRINTS. Over 1,000 titles to date. Write, phone
or email for submission form and full details.
Return form with full ms (digital copy preferred).
£ Royalties twice-yearly.

## Perfect Publishers Ltd

23 Maitland Avenue, Cambridge CB4 1TA
☎ 01223 424422   ℻ 01223 424414
editor@perfectpublishers.co.uk
www.perfectpublishers.co.uk
Managing Director *S.N. Rahman*

A print on demand book publishing company,
founded 2005. Offers self-publishing services.
Publishes books of all genres. Submissions by
email or letter.
£ 'We pay 100% royalties twice-yearly.'

## Pipers' Ash Ltd

'Pipers' Ash', Church Road, Christian Malford,
Chippenham SN15 4BW
☎ 01249 720563
pipersash@supamasu.com
www.supamasu.com
Managing Editor *Mr A. Tyson*

Founded 1976. The company's publishing
activities include individual collections of
contemporary short stories, science fiction short
stories, poetry, plays, short novels, local histories,
children's fiction, philosophy, biographies,
translations and general non-fiction. 12 titles a
year. Synopses and ideas welcome; 'new authors
with potential will be actively encouraged'. Offices
in Indonesia, New Zealand and Australia.
£ Royalties annually.

## Playwrights Publishing Co.

70 Nottingham Road, Burton Joyce, Nottingham
NG14 5AL
☎ 0115 931 3356
playwrightspublishingco@yahoo.com
www.playwrightspublishing.co.uk
www.playwrightspublishing.com
Managing Editors *Liz Breeze, Tony Breeze*

Founded 1990. Publishes one-act and full-length
plays online. Unsolicited scripts welcome. No
synopses or ideas. Reading fees: £15 one act; £30
full length (waived if evidence of professional
performance or if writer is unwaged). S.a.s.e.
required.
£ Royalties paid.

## Pomegranate Press

Dolphin House, 51 St Nicholas Lane, Lewes BN7 2JZ
☎ 01273 470100   📠 01273 470100
pomegranatepress@aol.com
www.pomegranate-press.co.uk
Publisher *David Arscott*

Founded in 1992 by writer/broadcaster David Arscott, who also administers the **Sussex Book Club**. Specializes in books about Sussex and self publishing. IMPRINT **Pomegranate Practicals** How-to books. 6 titles a year.
💷 Royalties twice-yearly.

## Praxis Books

Crossways Cottage, Walterstone HR2 0DX
☎ 01873 890695
rebeccatope@btinternet.com
www.rebeccatope.com
Proprietor *Rebecca Smith*

Founded 1992. Publishes reissues of the works of Sabine Baring-Gould, memoirs, diaries and local interest. 24 titles to date. Unsolicited mss accepted with s.a.s.e. No fiction. Editing and advisory service available. Funding negotiable. 'I am most likely to accept work with a clearly identifiable market.'

## Pulp Press

▷ Pen Press & Indepenpress Publishing Ltd

## Punk Publishing Ltd

3 The Yard, Pegasus Place, London SE11 5SD
☎ 020 7820 9333
enquiries@punkpublishing.co.uk
www.punkpublishing.co.uk
Managing Editor *Jonathan Knight*

Founded in 2006 by Jonathan Knight to publish his *Cool Camping: England*, which became a bestseller. Since then the company has expanded to publish a series of travel guides: *Cool Camping*; *Wild Swimming*. See website for areas of interest and submission guidelines.

## Robinswood Press

30 South Avenue, Stourbridge DY8 3XY
☎ 01384 397475   📠 01384 440443
info@robinswoodpress.com
www.robinswoodpress.com
Managing Editor *Christopher J. Marshall*

Founded 1985. Publishes children's, educational, teaching resources, SEN. Also collaborative publishing, e.g., with Camphill Foundation. About 30–50 titles a year. Unsolicited mss, synopses and ideas welcome. See website.
💷 Royalties paid.

## Roger Hollis

▷ Five Leaves Publications

## St James Publishing

St James's School, Earsby Street, London W14 8SH
☎ 020 7348 1799
stjamespublishing@stjamesschools.co.uk
www.stjamespublishing.co.uk
Managing Editor *Paul Palmarozza*

Publishes children's educational books. Does not consider unsolicited material.

## Searle Publishing

116 Commercial Road, Swindon SN1 5BD
info@searlepublishing.com
www.searlepublishing.com
Managing Editor *David James Searle*

Founded 2009. Books published include poetry (traditional and experimental), experimental fiction, art, English history, guides to English seaside towns. 9 titles in 2009. Welcomes unsolicited material; will consider all subjects. See website for FAQs and submission guidelines. Approach by email. Profits from sales are shared equally between the author and Searle Publishing. Production costs are met by the author. At the time of going to print, the formula is stated as: 'The cost per publication is approximately equal to the profit made from the sale of 15 books, so we keep all profits from the sale of the first 15 of each publication.'

## Serif

47 Strahan Road, London E3 5DA
☎ 020 8981 3990   📠 020 8981 3990
stephen@serif.books.co.uk
www.serifbooks.co.uk
Managing Editor *Stephen Hayward*

Founded 1993. Publishes cookery, Irish and African studies, travel writing and modern history; no fiction.
💷 Royalties paid.

## Silver Link

▷ The Nostalgia Collection

## Smoking Gun Books Ltd

1 Golfside Close, Whetstone, London N20 0RD
☎ 020 8446 8070   📠 020 8446 8070
publisher@smokinggunbooks.com
www.smokinggunbooks.com
Managing Director *Janet Weitz*

Founded in 2006, Smoking Gun Books is a niche print-on-demand publisher offering a full publishing service including editing, page layout, ISBN registration, printing, cover design, marketing and international distribution. Full details available on the website. 4 titles in 2009. Submissions by email.
💷 Royalties twice-yearly.

## Spacelink Books

115 Hollybush Lane, Hampton TW12 2QY
☎ 020 8979 3148
www.spacelink.fsworld.co.uk
Managing Director *Lionel Beer*

Founded 1967. Named after a UFO magazine published in the 1960/70s. Publishes non-fiction titles connected with UFOs, Fortean phenomena and paranormal events. Publishers of *TEMS News* for the Travel and Earth Mysteries Society. Distributors of a wide range of related titles and magazines. No unsolicited mss; send synopses and ideas – non-fiction.

## Stenlake Publishing Limited

54–58 Mill Square, Catrine KA5 6RD
☎ 01290 552233
enquiries@stenlake.co.uk
www.stenlake.co.uk

Publishes illustrated local history, railways, shipping, aviation and industrial. IMPRINT **Alloway Publishing Limited** acquired in January 2008. Around 30 titles in 2009. Unsolicited mss, synopses and ideas welcome if accompanied by s.a.s.e. or sent by email. Freelance writers with experience in above fields also sought for specific commissions.
Ⓔ Royalties or fixed fee paid.

## Stone Flower Limited

PO Box 1513, Ilford IG1 3QU
renalomen@hotmail.com
Managing Editor *L.G. Norman*

Founded 1989. Publishes humour and general fiction. No submissions. All material commissioned or generated in-house.

## Superscript

404 Robin Square, Newtown SY16 1HP
☎ 01588 650452
drjbford@yahoo.co.uk
www.superscript.org.uk
www.dubsolution.org
Editor *Polly Kaan*
Chair *Anand Somesh*
Secretary *Ray Pahl*

Founded 2002. Publishes literary, philosophical and political fiction, humanities and social sciences. 15 titles in 2008. Initial enquiries by letter, telephone or email to the editor.
Ⓔ Royalties usually 10% RRP but vary with each contract.

## Sylph Editions

5 St Swithun's Terrace, Lewes BN7 1UJ
☎ 01273 471706　Ⓕ 01273 808247
or@sylpheditions.com
www.sylpheditions.com
Publisher *Ornan Rotem*

Associate Publisher *Num Stibbe*

Founded 2006. Publishes art and photography books that centre on marriage between text and image, monographs, theoretical essays and critical writing. 6 titles in 2009. Ideas welcome; approach by email. Ⓔ Royalties twice-yearly.

## Tartarus Press

Coverley House, Carlton-in-Coverdale, Leyburn DL8 4AY
☎ 01969 640399　Ⓕ 01969 640399
tartarus@pavilion.co.uk
www.tartaruspress.com
Proprietor *Raymond Russell*
Editor *Rosalie Parker*

Founded 1987. Publishes fiction, short stories, reprinted classic supernatural fiction and reference books. About 10 titles a year. 'Please do not send submissions. We cater to a small, collectible market and commission the fiction we publish.'

## Tlön Books Publishing Ltd

64 Arthurdon Road, London SE4 1JU
☎ 05603 442122　Ⓕ 020 8690 0642
mark.reid@tlon.co.uk
www.tlon.co.uk
Managing Editor *Jason Shelley*

Founded 2002. Publishes fiction, art, photography, poetry and children's books. Unsolicited mss, synopses and ideas welcome; approach by email. Ⓔ Royalties twice-yearly.

## To Hell With Publishing

10 Woburn Walk, London WC1H 0JL
☎ 020 7387 8417
info@tohellwithpublishing.com
www.tohellwithpublishing.com
Managing Editor *Laurence Johns*

Founded 2006. Publishes literary fiction. Unsolicited mss welcome.

## Tonto Books Limited

71 Thornton Crescent, Blaydon on Tyne NE21 4BA
☎ 0191 442 1857
contact@tontobooks.com
www.tontobooks.com
Managing Editor *Stuart Wheatman*

Founded 2005. Publishes fiction, non-fiction and biography. 5 titles in 2009. Unsolicited mss, synopses and ideas for books welcome. Send initial query letter. Ⓔ Royalties twice-yearly.

## Wakefield Historical Publications

19 Pinder's Grove, Wakefield WF1 4AH
☎ 01924 372748
kate@airtime.co.uk
Managing Editor *Kate Taylor*

Founded in 1977 by the Wakefield Historical Society to publish well-researched, scholarly works of regional (namely West Riding) historical significance. 1–2 titles a year. Unsolicited mss, synopses and ideas for books welcome. Ⓔ Royalties not paid.

## Watling Street Publishing Ltd

33 Hatherop, Nr Cirencester GL7 3NA
Ⓣ 01285 750212
chris.mclaren@saltwaypublishing.co.uk
Managing Director *Chris McLaren*
Editorial Head *Christine Kidney*

Founded 2001. Publishes non-fiction titles on London and its history; children's and adult. No unsolicited mss. Ideas and synopses welcome. Approach in writing or by email with a one-page proposal. 'Not interested in anything not related to London.'

## Waywiser Press

Bench House, 82 London Road, Chipping Norton OX7 5FN
Ⓣ 01608 644755
waywiser-press@aol.com
waywiser-press.com/
Managing Editor *Philip Hoy*
Editorial Advisers *Joseph Harrison, Clive Watkins, Greg Williamson*

Founded 2002. An independent press, publishing poetry, fiction, criticism and other kinds of literary work by new as well as established writers, as well as a series of in-depth interviews with senior contemporary poets (under the press's IMPRINT **Between The Lines**). The press also runs an annual literary award, the **Anthony Hecht Poetry Prize** Submission details available on the website. Ⓔ Royalties paid.

## Whitchurch Books Ltd

67 Merthyr Road, Whitchurch, Cardiff CF14 1DD
Ⓣ 029 2052 1956
whitchurchbooks@btconnect.com
Managing Director *Gale Canvin*

Founded 1994. Publishes local interest, particularly local history. Welcomes unsolicited mss, synopses and ideas on relevant subjects. Initial approach by phone or in writing. Ⓔ Royalties twice-yearly.

## Whittles Publishing

Dunbeath Mains Cottages, Dunbeath KW6 6EY
Ⓣ 01593 731333   Ⓕ 01593 731400
info@whittlespublishing.com
www.whittlespublishing.com
Publisher *Dr Keith Whittles*

Publisher in geomatics, civil and structural engineering, geotechnics, manufacturing and materials technology and fuel and energy science. Also publishes non-technical books within the following areas: Asian connections, maritime,

pharology, military history, landscape and nature writing. Unsolicited mss, synopses and ideas welcome on appropriate themes. Ⓔ Royalties annually.

## Willow Bank Publishers Ltd

E–Space North, 181 Wisbech Road, Littleport, Ely CB6 1RA
Ⓣ 01353 687934
editorial@willowbankpublishers.co.uk
www.willowbankpublishers.co.uk
Managing Editor *Christopher Sims*

Founded 2005. Publishes all genres, fiction and non-fiction. Unsolicited mss, synopses and ideas for books welcome; send via post . Ⓔ Royalties annually.

## Witan Books

Cherry Tree House, 8 Nelson Crescent, Cotes Heath, via Stafford ST21 6ST
Ⓣ 01782 791673
witan@mail.com
Director & Managing Editor *Jeff Kent*

Founded in 1980 for self-publishing and commenced publishing other writers in 1991. Publishes general books, including biography, education, environment, geography, history, politics, popular music and sport. 1 or 2 titles a year. Unsolicited mss, synopses and ideas welcome (include s.a.s.e.). Ⓔ Royalties paid.

## Worple Press

PO Box 328, Tonbridge TN9 1WR
Ⓣ 01732 368958
theworpleco@aol.com
www.theworplepress.co.uk
Managing Editors *Peter Carpenter, Amanda Knight*

Founded 1997. Independent publisher specializing in poetry, art and alternative titles. Write or phone for catalogue and flyers. 4 titles a year. No unsolicited mss.

## Zymurgy Publishing

Hoults Estate, Walker Road, Newcastle upon Tyne NE6 2HL
Ⓣ 0191 276 2425   Ⓕ 0191 276 2425
zymurgypublishing@googlemail.com
Chairman *Martin Ellis*

Founded 2000. Publishes adult non-fiction, ranging from full colour illustrated hardbacks to mass market paperbacks and children's non-fiction. Synopses and ideas for books welcome; initial contact by telephone or email. No unsolicited mss. Ⓔ Royalties twice-yearly.

# Audio Books

## BBC Audiobooks Ltd

St James House, The Square, Lower Bristol Road, Bath BA2 3BH

☎ 01225 878000    🖷 01225 310771

bbcaudiobooks@bbc.com

www.bbcworldwide.com

Managing Director *Mike Bowen*

Publishing Director *Jan Paterson*

BBC Audiobooks Ltd was established in 2003 with the integration of BBC Radio Collection, Cover to Cover and Chivers Audio Books. Since then it has become a leading trade and library publisher in the UK, publishing a wide range of entertainment on a variety of formats including CD, cassette, MP3-CD and downloads, podcasts, etc.

DIVISIONS

**Trade** Titles range from original books and full-cast radio dramas through to abridged readings and full, unabridged recordings TITLES *War and Peace*; *The Odyssey*; *Captain Corelli's Mandolin*; the complete Sherlock Holmes canon. IMPRINTS **BBC Audio – Children's** From pre-school nursery rhymes to modern classics TITLES *The Chronicles of Narnia*; *His Dark Materials* trilogy. **BBC Audio – Radio Collection** Comedy classics from TV and radio TITLES *The Goons*; *Hancock's Half Hour*; *Fawlty Towers*; *The News Quiz*; *Just a Minute*. **BBC Audio** Contemporary comedy. TITLES *The Mighty Boosh*; *Have I Got News For You*; *Little Britain*. **BBC Audio – Lifestyle** A range of titles to help the listener improve or learn TITLES *Glenn Harrold's Ultimate Guide to ...* series. **BBC Audio – Radio 4** TITLES *This Sceptred Isle*; *Letter from America*; *Churchill Remembered*. **Science Fiction & Fantasy** TITLES *The Hitchhiker's Guide to the Galaxy*; *Doctor Who*; *Journey into Space*; *The Lord of the Rings*.

**Library** IMPRINTS **Chivers Audio Books**; **Chivers Children's Audio Books** Founded 1980. Publishes over 500 titles annually. Complete and unabridged books for adults and children are sold to libraries and to the public by direct mail through **The Audiobook Collection**. CDs and cassettes covering bestselling fiction and popular non-fiction.

## Bloomsbury Publishing

▷ entry under UK Publishers

## Chivers Audio Books/Chivers Children's Audio Books

▷ BBC Audiobooks Ltd

## Chrome Dreams

12 Seaforth Avenue, New Malden KT3 6JP

☎ 020 8715 9781    🖷 020 8241 1426

mail@chromedreams.co.uk

www.chromedreams.co.uk

Managing Director *Rob Johnstone*

A small record company and publisher founded 1998 to produce audio-biographies of current rock and pop artists and legendary performers on CD and, more recently, books on the same subjects. Ideas for biographies welcome.

## Creative Content Ltd

8 The Cornfields, Hatch Warren, Basingstoke RG22 4QB

☎ 01256 466641

ali@creativecontentdigital.com

www.creativecontentdigital.com

Owners *Alison Muirden, Lorelei King*

Managing Director *Alison Muirden*

Established in 2008, Creative Content is a joint partnership company owned by Ali Muirden, former Audio Publisher at Macmillan and Lorelei King, a multi award winning voice artist and actor. Publishes specially commissioned worldrights content in audio download, ebooks, on demand CD and preloaded MP3 players. Genres published include speech improvement, self help/self improvement, lifestyle and business etiquette. 12 titles in 2010. Ideas welcome from agents and authors.

🄴 Royalties annually.

## Crimson Cats Audio Books

The Red Cottage, The Street, Starston IP20 9NN

☎ 01379 854888

editor@crimsoncats.co.uk

www.crimsoncats.co.uk

Managing Directors *Michael Bartlett, Dee Palmer*

Founded in 2005 by two ex-BBC radio producers. Produces unusual/quirky (mostly out of copyright) books in any genre. TITLES *My Life and Times* Jerome K. Jerome; *The Beautiful Cassandra and Other Stories* Jane Austen; *My Life As a Spy* Robert Baden-Powell. 3 titles in 2009. Submissions not accepted.

## CSA Word

6a Archway Mews, 241a Putney Bridge Road, London SW15 2PE
☎ 020 8871 0220   📠 020 8877 0712
info@csaword.co.uk
www.csaword.co.uk
Managing Director *Clive Stanhope*
Audio Director *Victoria Williams*
Editorial and Production Assistant *Vanessa Brown*

Founded 1989. Publishes fiction, non-fiction, children's, short stories, poetry, travel, biographies, classics. Over 170 titles to date on CD and available as downloads. Tends to favour quality/classic/nostalgic/timeless literature. TITLES include *Carry on Jeeves* P.G. Wodehouse; *Room with a View* E.M. Forster; *Brideshead Revisited* Evelyn Waugh; *William's Happy Days* Richmal Crompton; *Lady Chatterley's Lover* D.H. Lawrence; *Bunter Does His Best* Frank Richards; *The Prince* Nicolo Machiavelli; *Seven Pillars of Wisdom* T.E. Lawrence. Also produces radio programmes for the BBC – ideas welcome – see entry under *Film, TV and Radio Producers*.

## CYP

The Fairway, Bush Fair, Harlow CM18 6LY
☎ 01279 444707   📠 01279 445570
enquiries@cyp.co.uk
www.kidsmusic.co.uk
Operations Director *Mike Kitson*

Founded 1978. Publishes music DVDs and audiobooks for young children; educational, entertainment. TV music and soundtrack production.

## 57 Productions

▷ entry under Organizations of Interest to Poets

## Greenpark Productions Ltd

▷ see entry under Library Services

## Hachette Digital

100 Victoria Embankment, London EC4Y 0DY
☎ 020 7911 8044   📠 020 7911 8100
sarah.shrubb@littlebrown.co.uk
www.littlebrown.co.uk
Editorial Director *Sarah Shrubb*

Launched in 2003 and now incorporating audiobooks from Little, Brown, Headline and John Murray, along with ebooks from Little, Brown. Publishes fiction, humour, poetry and non-fiction. TITLES include new fiction from authors such as Alexander McCall Smith, Patricia Cornwell and Sarah Waters, non-fiction such as *The Last Fighting Tommy* by Harry Patch with Richard van Emden and unabridged classics such as Joseph Heller's *Catch 22*. A list of over 250 titles.

## HarperCollins AudioBooks

77–85 Fulham Palace Road, London W6 8JB
☎ 020 8307 4444
jo.forshaw@harpercollins.co.uk
www.harpercollins.co.uk
Head of Audio *Jo Forshaw*

The HarperCollins audio list was launched in 1990. Publishes a wide range including popular and classic fiction, non-fiction, children's, Shakespeare, poetry and self help on cassette, CD and digital download. AUTHORS Agatha Christie, Ian McEwan, C.S. Lewis, Roald Dahl, Dr Seuss, Lemony Snicket, J.R.R. Tolkien, Bernard Cornwell, Enid Blyton, Nick Butterworth, Judith Kerr.

## Hodder & Stoughton Audiobooks

338 Euston Road, London NW1 3BH
☎ 020 7873 6000   📠 020 7873 6194
rupert.lancaster@hodder.co.uk
www.hodder.co.uk
Publisher *Rupert Lancaster*

Launched in 1994 with the aim of publishing outstanding authors from within the Hodder group and commissioning independent audio titles. Publishes fiction and non-fiction. AUTHORS include Stephen King, John le Carré, Michael Parkinson, Al Murray, Alan Titchmarsh, Melvyn Bragg, Jeffery Deaver, Peter Robinson, Jodi Picoult, David Mitchell and Pam Ayres.

## Macmillan Digital Audio

20 New Wharf Road, London N1 9RR
☎ 020 7014 6000   📠 020 7014 6141
www.panmacmillan.com
Owner *Macmillan Publishers Ltd*
Digital Director *Sara Lloyd*

Founded 1995. Publishes adult fiction, non-fiction and autobiography, focusing mainly on lead book titles and releasing audio simultaneously with hard or paperback publication. Also publishes children's audio titles. Won Audio Publisher of the Year at the 2003 Spoken Word Awards and has won many other Spoken Word Awards for individual audio titles in previous years About 40–60 titles a year. Submissions not accepted.

## Naxos AudioBooks

40a High Street, Welwyn AL6 9EQ
☎ 01438 717808   📠 01438 717809
naxos_audiobooks@compuserve.com
www.naxosaudiobooks.com

Owner *HNH International, Hong Kong/Nicolas Soames*
Managing Director *Nicolas Soames*

Founded 1994. Part of Naxos, the classical budget CD company. Publisher of the Year in the 2001 Spoken Word Awards. Publishes classic and modern fiction, non-fiction, young adult and junior classics, drama and poetry. TITLES *Ulysses* Joyce; *King Lear* Shakespeare; *History of the Musical* Fawkes; *Just So Stories* Kipling. A list of 400 titles.

## Oakhill Publishing Ltd

PO Box 3855, Bath BA1 3WW
☎ 01225 874355    🖷 01225 874442
info@oakhillpublishing.com
Managing Director *Julian Batson*

Founded in 2005 by Julian Batson, formerly with Chivers Press, to offer a wider selection of titles for the library market. Produces complete and unabridged audiobooks of popular fiction and non-fiction for adults and children. TITLES *The Gathering* Anne Enright; *The Visible World* Mark Slouka; *The Sixth Wife* Suzannah Dunn; *Dirty Bertie* David Roberts and Alan Macdonald; *Free Lance* Paul Stewart and Chris Riddell; *My Swordhand is Singing* Marcus Sedgwick; *The White Tiger* Aravind Adiga. 156 titles in 2009. Suggestions for titles from literary agents only.

## Orion Audiobooks (Division of the Orion Publishing Group Ltd)

Orion House, 5 Upper St Martin's Lane, London WC2H 9EA
☎ 020 7520 4425    🖷 020 7240 4822
pandora.white@orionbooks.co.uk
www.orionbooks.co.uk
Publisher *Pandora White*

Orion Audio has released over 600 titles since it was founded in 1998 and over the years has won many audio awards. It publishes fiction, non-fiction, humour, autobiography, children's, science fiction, crime and thrillers. AUTHORS include Maeve Binchy, Ian Rankin, Robert Crais, Michael Connelly, Francesca Simon (*Horrid Henry* series), Harlan Coben, Kate Mosse, Dan Brown, Sally Gardner, Michelle Paver, Linwood Barclay and Richard Hammond. 60 titles a year. In 2006, Orion expanded into the digital download market via www.audible.co.uk

## Penguin Audiobooks

80 Strand, London WC2R ORL
☎ 020 7010 3000    🖷 020 7010 6060
audio@penguin.co.uk
www.penguin.co.uk
Audio Books Editor *Ravina Bajwa*

Launched in November 1993 and has expanded rapidly since then to reflect the diversity of Penguin's publishing. Publishes classics and contemporary fiction, non-fiction, autobiography, an increasing range of digital audiobooks as well as children's titles under the **Puffin Audiobooks** imprint. AUTHORS include: Nick Hornby, Zadie Smith, Marian Keyes, Gervase Phinn, Antony Beevor, Niall Fergusson, Roald Dahl, Eoin Colfer and Charlie Higson. About 30 titles a year.

## Puffin Audiobooks
▷ Penguin Audiobooks

## Quercus Audiobooks
▷ Quercus Publishing under UK Publishers

## Random House Audio Books

20 Vauxhall Bridge Road, London SW1V 2SA
☎ 020 7840 8529
Owner *The Random House Group Ltd.*
Editorial Director *Zoe Howes*

The audiobooks division of the Random House Group Ltd started in 1991. Acquired the Reed Audio list in 1997. Publishes fiction and non-fiction. AUTHORS include James Patterson, Donna Leon, Sebastian Faulks, John Grisham, Mark Haddon, Robert Harris, Terry Pratchett, Ian McEwan, Andy McNab, Ruth Rendell, Kathy Reichs, Richard Dawkins.

## Salt Publishing Limited
▷ entry under UK Publishers

## Shakespeare Appreciated
▷ SmartPass Ltd

## Simon & Schuster Audio

1st Floor, 222 Gray's Inn Road, London WC1X 8HB
☎ 020 7316 1900    🖷 020 7316 0332
info@simonandschuster.co.uk
www.simonandschuster.co.uk
Audio Manager *Kirsty McNeil*

Simon & Schuster Audio began by distributing their American parent company's audio products. Moved on to repackaging products specifically for the UK market and in 1994 became more firmly established in this market with a huge rise in turnover. Publishes adult fiction, self help and business titles. TITLES *Secret Scream* Lynda la Plante; *The White Queen* Philippa Gregory; *The 7 Habits of Highly Effective People* Stephen R. Covey; *Chronicles* Bob Dylan; *Angels and Demons* and *Deception Point* Dan Brown; *The Kite Runner* and *A Thousand Splendid Suns* Khaled Hosseini; *The Secret* Rhonda Byrne.

## SmartPass Ltd

15 Park Road, Rottingdean, Brighton BN2 7HL
☎ 01273 300742
info@smartpass.co.uk
www.smartpass.co.uk
www.shakespeareappreciated.com

www.spaudiobooks.com
Managing Director *Phil Viner*
Creative Director *Jools Viner*

Founded 1999. SmartPass produces audio education study resources, full-cast drama with commentary analysis and study strategies for English Literature set texts – novels, plays and poetry. Available in student and teacher formats, including audio linked study materials. TITLES include *A Kestrel for a Knave*; *Animal Farm*; *Pride and Prejudice*; *Great Expectations*; *The Mayor of Casterbridge*; *Shakespeare: the works*; *War Poetry*; *An Inspector Calls*; *Twelfth Night*; *Hamlet*; *Julius Caesar*. IMPRINTS **Shakespeare Appreciated** Full-cast unabridged dramas with commentary explaining who's who and what's going on plus historical insights and background information; created for adults rather than students, for enjoyment rather than education. TITLES *Twelfth Night*; *Othello*; *King Lear*. **SPAudiobooks** Full-cast unabridged dramas of classic and cult texts

TITLES *The Antipope* Robert Rankin; *Macbeth*; *Henry V*; *Romeo and Juliet*. Welcomes ideas from authors who are teachers and from agents who represent the authors of studied texts.

## Soundings Audio Books

Isis House, Kings Drive, Whitley Bay NE26 2JT
☎ 0191 253 4155   🖷 0191 251 0662
www.isis-publishing.co.uk

Founded in 1982. Part of the Ulverscroft Group Ltd. Together with **Isis Publishing** publishes fiction and non-fiction; crime, romance. AUTHORS include Lyn Andrews, Rita Bradshaw, Lee Child, Catherine Cookson, Alexander Fullerton, Joanne Harris, Anna Jacobs, Robert Ludlum, Patrick O'Brian, Pamela Oldfield, Susan Sallis, Judith Saxton, Mary Jane Staples, Sally Worboyes. About 190 titles a year. Many also available on CD and MP3.

## SPAudiobooks
▷ SmartPass Ltd

# UK Packagers

## Aladdin Books Ltd

PO Box 53987, London SW15 2SF
☏ 020 3174 3090    ⊞ 020 8780 3939
sales@aladdinbooks.co.uk
www.aladdinbooks.co.uk
Managing Director *Charles Nicholas*

Founded in 1979 as a packaging company but with joint publishing ventures in the UK and USA. Commissions children's fully illustrated, non-fiction reference books. IMPRINTS **Aladdin Books** *Bibby Whittaker* Children's reference; **Nicholas Enterprises** *Charles Nicholas* Adult non-fiction; **The Learning Factory** *Charles Nicholas* Early learning concepts 0–4 years. TITLES *World Issues*; *Science Readers*; *The Atlas of Animals*. About 40 titles a year. Will consider synopses and ideas for children's non-fiction with international sales potential only. No fiction.
⊞ Fees usually paid instead of royalties.

## The Albion Press Ltd

Spring Hill, Idbury OX7 6RU
☏ 01993 831094
Chairman/Managing Director *Emma Bradford*

Founded 1984. Commissions illustrated trade titles, particularly children's. TITLES *The Great Circle: A History of the First Nations* Neil Philip; *Fairy Tales of Hans Christian Andersen* Isabelle Brent. About 2 titles a year. Unsolicited synopses and ideas for books not welcome.
⊞ Royalties paid; fees paid for introductions and partial contributions.

## Amber Books Ltd

Bradleys Close, 74–77 White Lion Street, London N1 9PF
☏ 020 7520 7600    ⊞ 020 7520 7606/7
enquiries@amberbooks.co.uk
www.amberbooks.co.uk
Managing Director *Stasz Gnych*
Rights Director *Sara Ballard*
Publishing Manager *Charles Catton*

Founded 1989. Commissions history, military, aviation, transport, sport, combat, survival and fitness, cookery, lifestyle, naval history, crime, childrens, encyclopaedias, schools and library sets and general reference. No fiction, poetry or biography. IMPRINT **Tiptoe Books** Picture books, sticker books, calendars, reference and educational for children of all ages. 80 titles in 2009. No unsolicited material outside subject areas listed above.
⊞ Fees paid.

## BCS Publishing Ltd

2nd Floor, Temple Court, 109 Oxford Road, Cowley OX4 2ER
☏ 01865 770099    ⊞ 01865 770050
bcs-publishing@dsl.pipex.com
Managing Director *Steve McCurdy*
Approx. Annual Turnover £150,000

Commissions general interest non-fiction for the international co-edition market.

## Bender Richardson White

PO Box 266, Uxbridge UB9 5NX
☏ 01895 832444    ⊞ 01895 835213
brw@brw.co.uk
www.brw.co.uk
Partners *Lionel Bender, Kim Richardson, Ben White*

Founded 1990 to produce illustrated non-fiction for children, adults and family reference for publishers in the UK and USA. 60–70 titles a year. Unsolicited material not welcome. Also offers professional advice to authors, editors and illustrators. ⊞ Fees paid.

## Book House

▷ Salariya Book Company Ltd

## Breslich & Foss Ltd

Unit 2A, Union Court, 20–22 Union Road, Clapham, London SW4 6JP
☏ 020 7819 3990    ⊞ 020 7819 3998
sales@breslichfoss.com
www.breslichfoss.co.uk
Directors *Paula Breslich, K.B. Dunning*
Approx. Annual Turnover £650,000

Packagers of adult non-fiction titles, including interior design, crafts, gardening, health and children's non-fiction and picturebooks. Unsolicited mss welcome but synopses preferred. Include s.a.s.e. with all submissions.
⊞ Royalties paid twice-yearly.

## Brown Wells and Jacobs Ltd

2 Vermont Road, London SE19 3SR
☎ 020 8653 7670    🖷 020 8653 7774
graham@bwj-ltd.com
www.bwj.org
Managing Director *Graham Brown*
Production Manager & PA *Jenny Broom*

Founded 1979. Commissions non-fiction, novelty, pre-school and first readers. About 40 titles a year. Unsolicited synopses and ideas for books welcome.
💷 Fees paid.

## Cameron & Hollis

PO Box 1, Moffat DG10 9SU
☎ 01683 220808    🖷 01683 220012
cameronhollis@cameronbooks.co.uk
www.cameronbooks.co.uk
Director *Jill Hollis*
Approx. Annual Turnover £400,000

Commissions contemporary art, design, collectors' reference, decorative arts, architecture. About 3 titles a year.
💷 Payment varies with each contract.

## Compendium Publishing Ltd
▷ entry under UK Publishers

## David West Children's Books

7 Princeton Court, 55 Felsham Road, London SW15 1AZ
☎ 020 8780 3836    🖷 020 8780 9313
dww@btinternet.com
www.davidwestchildrensbooks.com

Founded 1992. Commissions children's illustrated reference books. No fiction or adult books. Unsolicited ideas and synopses welcome; approach in writing in the first instance
💷 Fees/royalties paid annually.

## Diagram Visual Information Ltd

34 Elaine Grove, London NW5 4QH
☎ 020 7485 5941    🖷 020 7485 5941
brucerobertson@diagramgroup.com
Managing Director *Bruce Robertson*

Founded 1967. Producer of library, school, academic and trade reference books. About 10 titles a year. 'We no longer seek submissions.'
💷 Fees paid; no payment for sample material.

## Eddison Sadd Editions

St Chad's House, 148 King's Cross Road, London WC1X 9DH
☎ 020 7837 1968    🖷 020 7837 2025
info@eddisonsadd.com
www.eddisonsadd.com
Managing Director *Nick Eddison*
Editorial Director *Ian Jackson*
Approx. Annual Turnover £2.5 million

Founded 1982. Produces a wide range of popular illustrated non-fiction with books and interactive kits published in 30 countries. Many titles suit gift markets. Incorporates Bookinabox Ltd. Ideas and synopses, rather than mss, are welcome but titles must have international appeal.
💷 Royalties paid twice-yearly; fees paid when appropriate.

## Elwin Street Productions

144 Liverpool Road, London N1 1LA
☎ 020 7700 6785    🖷 020 7700 6031
info@elwinstreet.com
www.elwinstreet.com
Managing Director *Silvia Langford*
Approx. Annual Turnover £1 million

Founded 2003. Produces illustrated non-fiction in quirky humour, gift, reference and practical life including self-help and parenting. About 20 titles a year. Synopses and ideas for books welcome; approach by post only with s.a.s.e. No cookery or gardening. All books must have an international angle and not of UK interest only.
💷 Fees paid.

## Erskine Press

The White House, Sandfield Lane, Eccles, Norwich NR16 2PB
☎ 01953 887277    🖷 01953 888361
erskpres@aol.com
www.erskine-press.com
Chief Executive *Crispin de Boos*
Approx. Annual Turnover £65,000

Specialist publisher of books on Antarctic exploration – facsimiles, diaries, previously unpublished works and first English translations of expeditions of the late 19th and early 20th centuries. Publisher of general interest biographies with special reference to Norfolk. Recent publications include a number of Second World War related subjects. Also produces scholarly reprints and limited edition publications for academic/business organizations ranging from period print reproductions to facsimiles of rare and important books no longer available. 6 titles in 2009. No unsolicited mss. Ideas welcome.
💷 Royalties paid twice-yearly.

## Expert Publications Ltd

Sloe House, Halstead CO9 1PA
☎ 01787 474744    🖷 01787 474700
expert@lineone.net
Chairman *Dr. D.G. Hessayon*

Founded 1993. Produces the *Expert* series of books by Dr. D.G. Hessayon. Currently 24 titles in the series, including *The Flower Expert*; *The Evergreen Expert*; *The Vegetable & Herb Expert*; *The Flowering Shrub Expert*; *The Container*

*Expert*; *The Orchid Expert*. No unsolicited material.

## Graham-Cameron Publishing & Illustration

The Studio, 23 Holt Road, Sheringham NR26 8NB
☎ 01263 821333   🖷 01263 821334

*Also at:* 59 Hertford Road, Brighton BN1 7GG
☎ 01273 385890
enquiry@gciforillustration.com
www.graham-cameron-illustration.com
Editorial Director *Mike Graham-Cameron*
Art Director *Helen Graham-Cameron*

Founded 1984. Packages illustrated information and educational books. TITLES *Up From the Country*; *In All Directions*; *The Holywell Story*; *Flashbacks*; *Let's Look at Dairying*. Has over 37 contracted book illustrators concentrating on information, educational and children's books. 'Absolutely *no* unsolicited mss, please.'
Ⓔ Royalties annually.

## Haldane Mason Ltd

PO Box 34196, London NW10 3YB
☎ 020 8459 2131   🖷 020 8728 1216
info@haldanemason.com
www.haldanemason.com
Editorial Director *Sydney Francis*
Art Director *Ron Samuel*

Founded 1995. Commissions mainly children's illustrated non-fiction. Children's books are published under the **Red Kite Books** imprint. Adult list consists mainly of mind, body and spirit plus alternative health books under the Neal's Yard Remedies banner; children's age range 0–15. Unsolicited synopses and ideas welcome but approach by phone or email first. No fiction, please.
Ⓔ Fees paid.

## The Ilex Press Limited

210 High Street, Lewes BN7 2NS
☎ 01273 403124   🖷 01273 487441
surname@ilex-press.com
www.ilex-press.com
Managing Director *Stephen Paul*
Rights Director *Roly Allen*

Founded 1999. Sister company of **The Ivy Press Limited**. IMPRINTS **Ilex Photo** Commissions titles on digital photography and graphics; **Ilex Press** Publishes reference works on popular culture, comics, some crafts and practical arts. No fiction. Unsolicited synopses and ideas welcome; send a *brief* idea outline (3 or 4 pages) and a letter.
Ⓔ Fees paid.

## The Ivy Press Limited

210 High Street, Lewes BN7 2NS
☎ 01273 487440   🖷 01273 487441

surname@ivy-group.co.uk
www.ivy-group.co.uk
Managing Director *Stephen Paul*
Publisher *Jason Hook*

Founded 1996. Sister company of **The Ilex Press Limited**. Commissions illustrated non-fiction books covering subjects such as art, lifestyle, popular culture, health, self-help and humour. No fiction. 40 titles a year. Unsolicited synopses and ideas welcome; send a *brief* idea outline (3 or 4 pages) and a letter.
Ⓔ Fees paid.

## The Learning Factory
▷ Aladdin Books Ltd

## Lennard Associates Ltd

Windmill Cottage, Mackerye End, Harpenden AL5 5DR
☎ 01582 715866   🖷 01582 715866
stephenson@lennardqap.co.uk
Chairman/Managing Director *Adrian Stephenson*

Founded 1979. Now operating as a packager/production company only with no trade distribution. 3 title in 2009. No unsolicited mss.
Ⓔ Both fees and royalties by arrangement.

## Lexus Ltd

60 Brook Street, Glasgow G40 2AB
☎ 0141 556 0440   🖷 0141 556 2202
peterterrell@lexusforlanguages.co.uk
www.lexusforlanguages.co.uk
Managing/Editorial Director *P.M. Terrell*

Founded 1980. Compiles bilingual reference, language and phrase books. TITLES Rough Guide phrasebooks; Langenscheidt dictionaries; *HarperCollins English-Chinese*; *Oxford Italian Pocket Dictionary*. Own series of *Travelmates* published in 2004; launched a new beginner's course for learning Chinese, *The Chinese Classroom*, in 2007; *Insider China – cultural background for Chinese studies* and Read, Listen and Speak (interactive CD-Rom for learners of Chinese). About 5 titles a year. Unsolicited material considered although books are mostly commissioned. Freelance contributors employed for a wide range of languages.
Ⓔ Flat fee.

## Lionheart Books

10 Chelmsford Square, London NW10 3AR
☎ 020 8459 0453
lionheart.brw@btinternet.com
Senior Partner *Lionel Bender*
Partner *Madeleine Samuel*

A design/editorial packaging team. Titles are primarily commissioned from publishers. Highly illustrated non-fiction for children aged 8+, mostly natural history, history and general

science. See also **Bender Richardson White**. About 20 titles a year.
Ⓔ Flat fee.

## M&M Publishing Services

33 Warner Road, Ware SG12 9JL
☎ 01920 466003
mikemoran@moran01.wanadoo.co.uk
www.photography-london.co.uk
Proprietors *Mike Moran, Maggie Copeland*

Project managers, production and editorial, packagers and publishers. TITLES *MM Publisher Database*; *MM Printer Database* (available in UK, European and international editions).

## Market House Books Ltd

Suite B, Elsinore House, 43 Buckingham Street, Aylesbury HP20 2NQ
☎ 01296 484911
books@mhbref.com
www.markethousebooks.com
Directors *Anne Kerr, Jonathan Law*

Founded 1970. Compiles dictionaries, encyclopaedias and other reference books. TITLES *APA Dictionary of Psychology*; *Collins English Dictionary*; *Collins Encyclopedia* (10 vols.); Facts on File Dictionaries of: *Astronomy*, *Chemistry*, etc.; *Larousse Thematica* (6 vols.); *The Macmillan Dictionary of Philosophy*, etc.; Oxford Dictionaries of: *Business*, *Science*, *Medicine*, etc.; *Penguin Dictionary of Electronics*; *Penguin Rhyming Dictionary*, etc.; Taylor & Francis *Biographical Encyclopedia of Scientists*. About 15 titles a year. Unsolicited material not welcome as most books are compiled in-house.
Ⓔ Fees paid.

## Mathew Price Ltd

Albury Court, Albury, Thame OX9 2LP
mathewp@mathewprice.com
www.mathewprice.com

*Main office:* 12300 Ford Road, Suite 455, Dallas, TX 75234, USA
Chairman/Managing Director *Mathew Price*

Founded 1983. Commissions full-colour novelty and picture books and fiction for young children. 'We accept submissions by email, only. We are looking for stories for the very young and intelligent novelties, whether fiction or non-fiction. We do not accept stories in rhyme. Please understand that we cannot reply at length to all submissions, so if we can't publish it, our response will probably be brief and impersonal!'

## Monkey Puzzle Media Ltd

48 York Avenue, Hove BN3 1PJ
☎ 01273 279928

info@monkeypuzzlemedia.com
Chairman/Managing Director *Roger Goddard-Coote*
Commissioning and List Development *Paul Mason*

Founded 1998. Packager of adult and children's non-fiction for trade, school, library and mass markets. No fiction or textbooks. About 80 titles a year. Synopses and ideas welcome. Include s.a.s.e. for return.
Ⓔ Fees paid; no royalties.

## Nicholas Enterprises
▷ Aladdin Books Ltd

## Orpheus Books Limited

6 Church Green, Witney OX28 4AW
☎ 01993 774949    ℻ 01993 700330
info@orpheusbooks.com
www.orpheusbooks.com
Directors *Nicholas Harris, Sarah Hartley*

Founded 1993. Commissions children's non-fiction. About 20 titles a year. No unsolicited material.
Ⓔ Fees paid.

## Red Kite Books
▷ Haldane Mason Ltd

## Regency House Publishing Limited
▷ entry under UK Publishers

## Salariya Book Company Ltd

25 Marlborough Place, Brighton BN1 1UB
☎ 01273 603306    ℻ 01273 693857
salariya@salariya.com
www.salariya.com
www.book-house.co.uk
www.scribblersbook.com
Managing Director *David Salariya*
Editor *Stephen Haynes*
Art Director *Carolyn Franklin*

Founded 1989. Books for children: history, art, music, science, architecture, education, nature, environment, fantasy, folktales, multicultural, fiction, young readers and picture books. IMPRINTS **Book House**; **Scribblers** Highly illustrated books in all subjects for babies and children, from pre-school board books to teenage graphic novels. For fiction, submit complete c.v., ms for picture books; outline/synopsis and one sample story or chapter for collections, novels or graphic novels. Response in four months if s.a.s.e. included. Illustrations: no original artwork. Send c.v., promotion sheet to be kept on file. Samples returned if s.a.s.e. included.
Ⓔ Payment by arrangement.

## Savitri Books Ltd

4 Nutholt Lane, Ely CB7 4PL
☎ 01353 654327    ℻ 01353 654327

munni.srivastava@virginmedia.com
Managing Director *Mrinalini S. Srivastava*
Approx. Annual Turnover £200,000

Founded 1983 and since 1998, Savitri Books has also become a publisher in its own right (textile crafts). Keen to work 'very closely with authors/ illustrators and try to establish long-term relationships with them, doing more books with the same team of people'. Commissions illustrated non-fiction: biography, history, travel. About 7 titles a year. Unsolicited synopses and ideas for books 'very welcome'.

## Scribblers
▷ Salariya Book Company Ltd

## Stonecastle Graphics Ltd
Highlands Lodge, Chartway Street, Sutton Valence ME17 3HZ
☎ 01622 844414
paul@stonecastle-graphics.com
sue@stonecastle-graphics.com
www.stonecastle-graphics.com
Director *Paul Turner*
Editorial Head *Sue Pressley*
Approx. Annual Turnover £200,000

Founded 1976. Formed additional design/ packaging partnership, **Touchstone**, in 1983 (see entry). Commissions illustrated non-fiction general books: motoring, health, sport, leisure, lifestyle and children's non-fiction. TITLES *Red Arrows – The Royal Air Force Aerobatic Team in Action*; *The Illustrated Surgery Guide*; *A Book of the Sea*; *Guitar Chords*; *Spin-A-Quiz Series*; *Presenting the Philippines*; *Tropical Dive Destinations*. 10 titles in 2009. Unsolicited synopses and ideas for books welcome
£ Fees paid.

## Templar Publishing
▷ entry under UK Publishers

## Tiptoe Books
▷ Amber Books Ltd

## Toucan Books Ltd
Third Floor, 89 Charterhouse Street, London EC1M 6HR
☎ 020 7250 3388    📠 020 7250 3123
Managing Director *Ellen Dupont*

Founded 1985. Specializes in international co-editions and fee-based editorial, design and production services. Commissions illustrated non-fiction for children and adults. No fiction or non-illustrated titles. About 15 titles a year.
£ Royalties paid twice-yearly; fees paid in addition to or instead of royalties.

## Touchstone Books Ltd
Highlands Lodge, Chartway Street, Sutton Valence ME17 3HZ
☎ 01622 844414
paul@touchstone-books.net
sue@touchstone-books.net
www.touchstone-books.net
Directors *Paul Turner, Sue Pressley, Nick Wigley, Trevor Legate*

Founded in 2006, Touchstone Books Ltd is an independent publisher of lavishly-illustrated books on motoring, aviation, art, photography and general-interest subjects. TITLES *Cobra: The First 40 Years*; *100 Years of Grand Prix*; *The Very Best of Volkswagen*; *In the Company of Dali*; *100 Years of Brooklands: The Birthplace of British Motorsport and Aviation*; *Fighting Force: The 90th Anniversary of the Royal Air Force* (officially endorsed by the RAF); *The Red Arrows*. 6 titles in 2009. Unsolicited synopses and ideas for books welcome.
£ Fees paid.

## User Design
31 Dover Street, Leicester LE1 6PW
☎ 07790 924159
info@userdesign.co.uk
www.userdesign.co.uk
Managing Director *Thomas Bohm*

Founded 2005. Offers a range of services: book design, book cover design, illustration (freehand, technical) and production (corrections, scanning, typesetting) for academic, children's, educational and general trade book publishers. 10 titles in 2009. No unsolicited material.
£ Fees paid.

## Working Partners Ltd
Stanley House, St Chad's Place, London WC1X 9HH
☎ 020 7841 3939    📠 020 7841 3940
enquiries@workingpartnersltd.co.uk
www.workingpartnersltd.co.uk
Chairman *Ben Baglio*
Managing Director *Chris Snowdon*
Contact *James Noble*

Specializes in quality mass-market fiction for leading children's publishers including Bloomsbury, HarperCollins, Hodder, Macmillan, Orchard, OUP, Random House and Scholastic. First chapter books to young adult. No picture books. **Working Partners Two** handles adult fiction across all popular genres on a similar basis. Welcomes approaches from interested writers.
£ Both fees and royalties by arrangement.

# Electronic Publishing and Other Services

## www.ABCtales.com

8 Clifton Road, Brighton BN1 3HP

☏ 07899 901023

tcook@abctales.com

www.abctales.com

Owner *Burgeon Creative Ideas Ltd*

Editor *Tony Cook*

Founded by A. John Bird, MBE, co-founder of the *Big Issue* magazine, Tony Cook and Gordon Roddick. ABCtales is a free website dedicated to publishing and developing new writing. Content is predominantly short stories, autobiography and poetry. Anyone can upload creative writing to the website for free. Authors such as Joe Dunthorne and Drew Gummerson started out on ABCtales and still write regularly on the site, testing out new work. There are frequent ABCtales evenings around the country, competitions and excellent feedback on work from fellow writers.

## Authors OnLine

▷ entry under UK Publishers

## Books 4 Publishing

Unit 10, Alliance Court, Eco Park Road, Ludlow SY8 1FB

☏ 0870 777 1113

editor@books4publishing.com

www.books4publishing.com

Owner *Corvedale Media Ltd*

Managing Director *Mark Oliver*

Founded 2000. Specializes in showcasing unpublished books and helping new authors gain recognition for their work by displaying synopses and up to 5,000 words on the Books 4 Publishing website and promoting titles to publishers and literary agents. Mss, synopses and ideas welcome; initial enquiries by email, post or submission form on website.

## 50connect – www.50connect.co.uk

Morley House, Badminton Court, Church Street, Old Amersham HP7 0DD

☏ 01494 736130    ℻ 0800 634 2250

admin@50connect.com

www.50connect.co.uk

Editors *Mark O'Haire (mark@50connect.com)*, *(Gareth Hargreaves (gareth@50connect.com)*

Launched in 2000, and awarded 'esuperbrand status' in 2006, www.50connect.co.uk is an editorial-led lifestyle website for today's over-45s. Covers finance, health, travel, overseas retirement, living abroad, retirement, food, wine, theatre, music, gardening, pets and much more. Also strong community element including forums and a chat room. Submissions welcome but please contact via email in the first instance with some previous examples of work published.

## Fledgling Press

7 Lennox Street, Edinburgh EH4 1QB

☏ 0131 343 2367

info@fledglingpress.co.uk

www.fledglingpress.co.uk

Director *Zander Wedderburn*

Founded 2000. Internet publisher, aiming to be a launching pad for new authors. Special interest in authentic writing about the human condition, including autobiography, diaries, poetry and fictionalized variations on these. Monthly online competition (www.canyouwrite.com) with small prizes for short pieces. Also free books and reports in the areas of shiftwork and working time. Send mss and other details by email from the website or by post. Links into short-run book production. 24 titles to date.

## Justis Publishing Ltd

Grand Union House, 20 Kentish Town Road, London NW1 9NR

☏ 020 7267 8989    ℻ 020 7267 1133

communications@justis.com

www.justis.com

www.justcite.com

Founded 1986. Electronic publisher of UK and European legal and offical information on CD-ROM, online and the internet.

## Lulu

▷ entry under UK Publishers

## M-Y Ebooks

M-Y Books Ltd, 187 Ware Road, Hertford
SG13 7EQ
☎ 01992 586279
jonathan@m-yebooks.co.uk
www.m-yebooks.co.uk
Contact *Jonathan Miller*

M-Y Ebooks offers a one stop shop for ebook publishing, marketing and distribution. Services cover everything for successful ebook creation through to sales, including formatting, optional typesetting and design. Sells through all of the established and developing channels for ebooks sales worldwide including the internet, mobile phone and ebook readers including, Kindle (Amazon) as well as elibraries in both the UK and USA.

## New Authors Showcase

▷ Barrie James Literary Agency under UK Literary Agents

## Online Originals

Priory Cottage, Wordsworth Place, London
NW5 4HG
☎ 07754 095074
editor@onlineoriginals.com
www.onlineoriginals.com
Managing Director *David Gettman*
Technical Director *Neill Sanders*

Publishes book-length works as ebooks on the web and occasionally as print on demand or short-run library editions. Acquires global electronic rights in literary fiction, intellectual non-fiction, drama, and youth fiction (ages 8–16). No poetry, fantasy, how-to, self-help, picture books, cookery, hobbies, crafts or local interest. TITLES *Being and Becoming* Christopher Macann (4 vols.); *The Jealous God* Martin Boyland; *Lard and Speck* Andrew Cogan; *Quintet* Frederick Forsyth; *The Angels of Russia* Patricia Leroy. 3 titles in 2009. 'Unsolicited mss, synopses and proposals for books welcome via our unique, peer-review, automated, online submissions system, accessed via the website address above. No submissions by email or on paper, please.' € Royalties paid annually (50% royalties on ebook price of £6).

## Wuggles Publishing

185 Kingrosia Park, Clydach, Swansea SA6 5PF
☎ 01792 844445
wuggles@ic24.net
www.wugglespublishing.co.uk
Contact *Chris Thomas*

Established in 2001 to assist unpublished writers 'ignored by the big boys to get off the ground'. Only interested in non-PC work of any genre except children's, religious, gay, academic or porn. 2–3 titles a year. No mss; submit by phone or email in first instance. See website for further details.

# UK Literary Agents

\* = Members of the **Association of Authors' Agents (AAA)**
\*\* = Members of the **Association of Scottish Literary Agents (ASLA)**

## Aarau Literary Agency

106 Mansfield Drive, Merstham RH1 3JN
submission@aaraulit.com
www.aaraulit.com
Senior Partner & Editor *Paul M. Muller, PhD*
Correspondence Secretary *David Sherriff*

Founded 2002. Handles full-length fiction, especially thrillers (techno a plus); literary fiction, especially political and/or social satire (controversy a plus); quality erotica, in any of the above genres. No poetry, screenplays, non-fiction, short fiction, 'who-dunnits' or autobiography. COMMISSION 15% (all rights all genres including international). 'We only consider work from *unpublished* authors (or major changes of genre).' Electronic submissions only: first contact via the website then email only. All requirements listed on the web page must be fulfilled before email contact is made. No fees charged.

## Abner Stein*

10 Roland Gardens, London SW7 3PH
☎ 020 7373 0456    🖷 020 7370 6316
Contact *Arabella Stein*

Founded 1971. Mainly represents US agents and authors but handles some full-length fiction and general non-fiction. No scientific, technical, etc. No scripts. COMMISSION Home 10%; US & Translation 20%.

## The Agency (London) Ltd*

24 Pottery Lane, London W11 4LZ
☎ 020 7727 1346    🖷 020 7727 9037
info@theagency.co.uk
Contacts *Stephen Durbridge, Leah Schmidt, Norman North, Julia Kreitman, Bethan Evans, Hilary Delamere (children's books), Katie Haines, Faye Webber, Nick Quinn, Fay Davies, Ian Benson, Jago Irwin*

Founded 1995. Deals with writers and rights for TV, film, theatre, radio scripts and children's books: picture books and fiction, 0–teen. Also handles directors. Only existing clients for adult fiction or non-fiction. COMMISSION Home 10%; US by arrangement; Children's books: Home 15%; Overseas 20%. Send letter with s.a.s.e. No unsolicited mss. No reading fee.

## Aitken Alexander Associates Ltd*

18–21 Cavaye Place, London SW10 9PT
☎ 020 7373 8672    🖷 020 7373 6002
reception@aitkenalexander.co.uk
www.aitkenalexander.co.uk
Contacts *Gillon Aitken, Clare Alexander, Matthew Hamilton, Ayesha Karim, Andrew Kidd, Lesley Thorne (film/TV)*
Associated Agents *Anthony Sheil, Mary Pachnos, Lucy Luck*

Founded 1977. Handles fiction and non-fiction. No plays, scripts or children's fiction unless by existing clients. CLIENTS include Clare Allan, Pat Barker, Sarah Bradford, Jung Chang, John Cornwell, Sarah Dunant, Susan Elderkin, Diana Evans, Roopa Farooki, Sebastian Faulks, Helen Fielding, Germaine Greer, Julia Gregson, Mark Haddon, Mohammed Hanif, Nicky Haslam, Susan Howatch, Liz Jensen, Pankaj Mishra, Charles Moore, Jonathan Raban, Piers Paul Read, Louise Rennison, Michèle Roberts, Jennie Rooney, James Scudamore, Nicholas Shakespeare, Gillian Slovo, Edward St. Aubyn, Rory Stewart, Nick Stone, Colin Thubron, Penny Vincenzi, Willy Vlautin, A.N. Wilson, Robert Wilson. COMMISSION Home 15%; US & Translation 20%; Film/TV 15%. Send preliminary letter, with half-page synopsis and first 30pp of sample material, and adequate return postage, in the first instance. 'We do not accept poetry, self-help/how-to or unsolicited dramatic screenplays/scripts.' No reading fee.

## Alan Brodie Representation Ltd

Paddock Suite, The Courtyard, 55 Charterhouse Street, London EC1M 6HA
☎ 020 7253 6226    🖷 020 7183 7999
info@alanbrodie.com
www.alanbrodie.com
Contacts *Alan Brodie, Sarah McNair, Lisa Foster, Harriet Pennington Legh*

Founded 1989. Handles theatre, film and TV scripts. No books. COMMISSION Home 10%; Overseas 15%. Preliminary letter plus professional recommendation and c.v. essential. No reading fee but s.a.s.e. required.

## A.M. Heath & Co. Ltd*

6 Warwick Court, London WC1R 5DJ
☎ 020 7242 2811   📠 020 7242 2711
www.amheath.com
Contacts *Bill Hamilton, Victoria Hobbs,*
*Euan Thorneycroft, Sarah Molloy (children's),*
*Jennifer Custer (foreign rights)*

Founded 1919. Handles fiction, general non-
fiction and children's. No dramatic scripts,
poetry, picture books or short stories. CLIENTS
include Christopher Andrew, Nadeem Aslam,
Bella Bathurst, Anita Brookner, Geoff Dyer, R.J.
Ellory, Katie Fforde, Graham Hancock, Conn
Iggulden, Marina Lewycka, Lucy Mangan, Hilary
Mantel, Joanna Nadin, Maggie O'Farrell, Susan
Price, Helen Simpson, John Sutherland, Barbara
Trapido. COMMISSION Home 10–15%; US &
Translation 20%; Film & TV 20%. OVERSEAS
ASSOCIATES in the US, Europe, South America,
Japan and the Far East. Preliminary letter and
synopsis essential. No reading fee.

## The Ampersand Agency Ltd*

Ryman's Cottages, Little Tew OX7 4JJ
☎ 01608 683677   📠 01608 683449
info@theampersandagency.co.uk
www.theampersandagency.co.uk
www.theampersandagency.co.uk/blog
Contacts *Peter Buckman, Anne-Marie Doulton,*
*Melanie Michael-Greer, Amy Wigelsworth*
Consultants *Peter Janson-Smith, Patrick Neale*

Founded 2003. Handles literary and commercial
fiction and non-fiction; contemporary and
historical novels, crime, thrillers, biography,
women's fiction, history, memoirs. No scripts
unless by existing clients. No poetry, science
fiction or fantasy. CLIENTS include Quentin Bates,
Helen Black, S.J. Bolton, Georgette Heyer estate,
Druin Burch, Martin Conway, Vanessa Curtis,
Will Davis, Tracy Gilpin, Cora Harrison, Michael
Hutchinson, Beryl Kingston, Miriam Morrison,
Philip Purser, Niamh Shaw, P. Robert Smith, Ivo
Stourton, Vikas Swarup, Nick van Bloss, Michael
Walters. COMMISSION Home 15%; US & Translation
20%. Translation rights handled by **The Buckman
Agency**. Unsolicited mss, synopses and ideas
welcome, as are email enquiries, but send sample
chapters and synopsis by post (s.a.s.e. required if
material is to be returned). No reading fee.

## The Andrew Lownie Literary Agency*

36 Great Smith Street, London SW1P 3BU
☎ 020 7222 7574   📠 020 7222 7576
lownie@globalnet.co.uk
www.andrewlownie.co.uk
Contact *Andrew Lownie*

Founded 1988. Specializes in non-fiction,
especially history, biography, current affairs,
military history, reference and packaging

celebrities and journalists for the book market.
No poetry, short stories or science fiction.
Formerly a journalist, publisher and himself the
author of 12 non-fiction books, Andrew Lownie's
CLIENTS include Richard Aldrich, Juliet Barker,
Doug Beattie, Guy Bellamy, the Joyce Cary
estate, Roger Crowley, Warwick Davis, Tom
Devine, Duncan Falconer, Laurence Gardner,
Cathy Glass, Timothy Good, David Hasselhoff,
Robert Hutchinson, Lawrence James, Suzannah
Lipscomb, Christopher Lloyd, Sean Longden,
Julian Maclaren-Ross estate, Michael Perham,
Martin Pugh, Sian Rees, Desmond Seward, David
Stafford and Peter Thompson. COMMISSION
Worldwide 15%. Preferred approach by email in
format suggested on website.

## Andrew Mann Ltd*

1 Old Compton Street, London W1D 5JA
☎ 020 7734 4751
info@andrewmann.co.uk
www.andrewmann.co.uk
Director/Agent *Anne Dewe*
Director/Agent *Tina Betts*
Agent *Louise Burns*

Founded 1975. Handles fiction, general
non-fiction, children's books and TV, theatre,
radio scripts. COMMISSION Home 15%; US &
Translation 20%. OVERSEAS ASSOCIATES various.
Unsolicited mss accepted but only first three
chapters together with preliminary letter, and
synopsis (s.a.s.e. essential). Email submissions for
synopses only; 'we do not open attachments'. No
reading fee.

## Andrew Nurnberg Associates Ltd*

Clerkenwell House, 45–47 Clerkenwell Green,
London EC1R 0QX
☎ 020 3327 0400   📠 020 7253 4851
www.andrewnurnberg.com
Directors *Andrew Nurnberg, Sarah Nundy,*
*D. Roger Seaton, Vicky Mark*

Agent for UK and international authors, including
children's writers. Sells rights internationally
on behalf of UK/US agencies and publishers.
Branches in Moscow, Budapest, Prague, Sofia,
Warsaw, Riga, Beijing and Taipei. COMMISSION
Home 15%; US & Translation 20%. See website
for full information, contact and submission
guidelines.

## Annette Green Authors' Agency*

1 East Cliff Road, Tunbridge Wells TN4 9AD
☎ 01892 514275
david@annettegreenagency.co.uk
www.annettegreenagency.co.uk
Contact *Address material to the Agency*

Founded 1998. Handles literary and general
fiction and non-fiction, popular culture and

current affairs, science, music, film, history, biography, older children's and teenage fiction. No dramatic scripts or poetry. CLIENTS include Andrew Baker, Tim Bradford, Bill Broady, Meg Cabot, Simon Conway, Mary Hogan, Anvar Khan, Maria McCann, Adam Macqueen, Ian Marchant, Professor Charles Pasternak, Lilian Pizzichini, Kirsty Scott, Bernadette Strachan, Elizabeth Woodcraft. COMMISSION Home 15%; US & Translation 20%. Letter, synopsis, sample chapters and s.a.s.e. essential. No reading fee.

## Antony Harwood Limited*

103 Walton Street, Oxford OX2 6EB
☎ 01865 559615 📠 01865 310660
mail@antonyharwood.com
www.antonyharwood.com
Contacts *Antony Harwood,*
*James Macdonald Lockhart*

Founded 2000. Handles fiction and non-fiction. CLIENTS Amanda Craig, Peter F. Hamilton, Alan Hollinghurst, A.L. Kennedy, Douglas Kennedy, Dorothy Koomson, Chris Manby, George Monbiot, Garth Nix, Tim Parks. COMMISSION Home 15%; US & Translation 20%. Send letter and synopsis with return postage in the first instance. No reading fee.

## A.P. Watt Ltd*

20 John Street, London WC1N 2DR
☎ 020 7405 6774 📠 020 7831 2154
apw@apwatt.co.uk
www.apwatt.co.uk
Directors *Caradoc King, Linda Shaughnessy,*
*Derek Johns, Georgia Garrett,*
*Natasha Fairweather*

Founded 1875. The oldest-established literary agency in the world. Handles full-length typescripts, including children's books, screenplays for film and TV. No poetry, academic or specialist works. CLIENTS include Monica Ali, Trezza Azzopardi, David Baddiel, Sebastian Barry, Quentin Blake, Melvin Burgess, Marika Cobbold, Helen Dunmore, Nicholas Evans, Giles Foden, Esther Freud, Janice Galloway, Martin Gilbert, Nadine Gordimer, Linda Grant, Philip Hensher, Reginald Hill, Michael Holroyd, Michael Ignatieff, Mick Jackson, Philip Kerr, Dick King-Smith, India Knight, John Lanchester, Alison Lurie, Jan Morris, Andrew O'Hagan, Caryl Phillips, Philip Pullman, James Robertson, Jancis Robinson, Jon Ronson, Elaine Showalter, Zadie Smith, Graham Swift, Salley Vickers and the estates of Graves and Maugham. COMMISSION Home 15%; US & Translation 20%. No unsolicited mss accepted.

## Artellus Limited

30 Dorset House, Gloucester Place, London
NW1 5AD
☎ 020 7935 6972 📠 020 8609 0347

www.artellusltd.co.uk
Chairman *Gabriele Pantucci*
Director *Leslie Gardner*
Associates *Darryl Samaraweera, Andrew Marszal*

Founded 1986. Full-length and short mss. Handles crime, science fiction, historical, contemporary and literary fiction; non-fiction: art history, current affairs, biography, general history, science. Works directly in the USA and with agencies internationally. COMMISSION Home 10%; Overseas 20%. Sample chapters in the first instance. No reading fee. Return postage essential.

## Author Literary Agents

53 Talbot Road, London N6 4QX
☎ 020 8341 0442/07767 022659
📠 020 8341 0442
agile@authors.co.uk
Contact *John Ridley Havergal*

Founded 1997. 'We put to leading publishers and producers strong new works in all genres.' COMMISSION (VAT extra) Book, Screen & Internet production rights: Home 15%; Overseas & Translation 25%. Send s.a.s.e. with first chapter, scene or section writing sample, plus half-to-one page outline. Please include graphics samples, if applicable. No reading fee.

## Barbara Levy Literary Agency*

64 Greenhill, Hampstead High Street, London
NW3 5TZ
☎ 020 7435 9046 📠 020 7431 2063
Contacts *Barbara Levy, John Selby*

Founded 1986. Handles general fiction, non-fiction, and film and TV rights. COMMISSION Home 15%; US 20%; Translation by arrangement, in conjunction with **The Buckman Agency**. US ASSOCIATE Marshall Rights. No unsolicited mss. No reading fee.

## Barrie James Literary Agency

Rivendell, Kingsgate Close, Torquay TQ2 8QA
☎ 01803 326617
mail@newauthors.org.uk
www.newauthors.org.uk
Contact *Barrie E. James*

Founded 1997. Includes New Authors Showcase an internet site for new writers and poets to display their work to publishers and others. *No unsolicited mss.* First contact: s.a.s.e. or email.

## The Bell Lomax Moreton Agency

James House, 1 Babmaes Street, London
SW1Y 6HF
☎ 020 7930 4447 📠 020 7925 0118
agency@bell-lomax.co.uk
Executives *Eddie Bell, Pat Lomax, Paul Moreton,*
*June Bell*

Established 2002. Handles quality fiction and non-fiction, biography, children's, business and sport. No unsolicited mss without preliminary letter. No scripts. No reading fee.

## Berlin Associates

7 Tyers Gate, London SE1 3HX
☎ 020 7836 1112   🅕 020 7632 5296
agents@berlinassociates.com
submissions@berlinassociates.com
www.berlinassociates.com
Contacts *Lindsey Bender, Marc Berlin, Mat Connell, Rachel Daniels, Stacy Hawkes (technicians), Julia Mills, Laura Reeve (technicians), Emily Summerscale, Fiona Williams*

Formerly London Management, founded in 1970. Represents writers for TV, theatre and film. Also directors, composers, producers and technicians. COMMISSION Home 12½%. Unsolicited material accepted with professional recommendation or preliminary letter; approach via the submissions email above. No reading fee.

## Bill McLean Personal Management

23B Deodar Road, London SW15 2NP
☎ 020 8789 8191
Contact *Bill McLean*

Founded 1972. Handles scripts for all media. No books. CLIENTS include Dwynwen Berry, Graham Carlisle, Phil Clark, Pat Cumper, Jane Galletly, Patrick Jones, Tony Jordan, Bill Lyons, John Maynard, Michael McStay, Les Miller, Ian Rowlands, Jeffrey Segal, Richard Shannon, Ronnie Smith, Barry Thomas, Garry Tyler, Frank Vickery, Laura Watson, Mark Wheatley. COMMISSION Home 10%. No unsolicited mss. Phone call or introductory letter essential. No reading fee.

## Blake Friedmann Literary Agency Ltd*

122 Arlington Road, London NW1 7HP
☎ 020 7284 0408   🅕 020 7284 0442
firstname@blakefriedmann.co.uk
www.blakefriedmann.co.uk
Contacts *Carole Blake (books), Katie Williams (scripts), Conrad Williams (original scripts/radio), Isobel Dixon (books), Oli Munson (books)*

Founded 1977. Handles all kinds of fiction from genre to literary; a varied range of specialist and general non-fiction, plus scripts for TV, radio and film. No poetry, science fiction or short stories (unless from existing clients). Special interests: commercial women's fiction, intelligent thrillers, literary fiction, upmarket non-fiction, teenage fiction. CLIENTS include Gilbert Adair, Jane Asher, Andy Briggs, Edward Carey, Elizabeth Chadwick, Anne de Courcy, Anna Davis, Barbara Erskine, David Gilman, Ann Granger, Billy Hopkins, Peter James, Deon Meyer, Lawrence Norfolk,

Gregory Norminton, Joseph O'Connor, Sheila O'Flanagan, Michael Ridpath, Craig Russell, Tess Stimson, Michael White. COMMISSION Books: Home 15%; US & Translation 20%. Radio/TV/Film: 15%. OVERSEAS ASSOCIATES 24 worldwide. Unsolicited mss welcome but initial letter with synopsis and first two chapters preferred. Letters should contain as much information as possible on previous writing or relevant experience, aims for the future, etc. No reading fee.

## BookBlast Ltd

PO Box 20184, London W10 5AU
☎ 020 8968 3089
www.bookblast.com
Contact *Address material to the Company*

Handles fiction and non-fiction. Memoirs, travel, popular culture, multicultural writing only. Reading very selectively. No new authors taken on except by recommendation. Editorial advice given to own authors. Initiates in-house projects. Also offers translation consultancy. Film, TV and radio rights sold in works by existing clients. COMMISSION Home 12%; US & Translation 20%; Film, TV & Radio 20%. No reading fee.

## Bookseeker Agency

PO Box 7535, Perth PH2 1AF
☎ 01738 620688
bookseeker@blueyonder.co.uk
http://book-seeker.co.uk
http://bookseeker.webs.com
Contact *Paul Thompson*

The agency was formed originally to help a handful of writers in the East of Scotland get a foothold in publishing. Keeps its client base small in order to concentrate its efforts but keeps details of enquirers on file in case an opening comes along. COMMISSION 15%. 'Enquiries from writers looking for a publisher, or publishers looking for a writer, are always welcome.' Approach in writing, giving a personal introduction, a summary of work so far and any current projects. No mss or sample chapters; synopsis or half-a-dozen poems acceptable. No reading fee.

## Brie Burkeman & Serafina Clarke Ltd*

14 Neville Court, Abbey Road, London NW8 9DD
☎ 0870 199 5002   🅕 0870 199 1029
info@burkemanandclarke.com
www.burkemanandclarke.com
Contact *Brie Burkeman*

Founded 2000. Handles commercial and literary full-length fiction and non-fiction; full length, commercial film, TV and theatre material but no musicals, short films, community projects. Independent film and TV consultant. No text book, academic, poetry, short stories. CLIENTS include Richard Askwith, Alastair Chisholm,

Kitty Ferguson, Shaun Hutson, Robin Norwood, Steven Sivell. COMMISSION Home 15% US & Translation 20%. Approach in writing in the first instance – email or letter – but see full submission guidelines on the website. Unsolicited email attachments will be deleted without opening. Return postage essential. No reading fee. OVERSEAS ASSOCIATE The Buckman Agency and Grandi & Associati.

## The Buckman Agency

Ryman's Cottages, Little Tew OX7 4JJ
℡ 01608 683677/020 8544 2674
℻ 01608 683449/020 8543 9653
r.buckman@talk21.com
j.buckman@talk21.com
Partners *Rosie Buckman, Jessica Buckman*

Founded in the early 1970s, the agency specializes in foreign rights and represents leading authors and agents from the UK and US. COMMISSON 20% (including sub-agent's commission). No unsolicited mss.

## Campbell Thomson & McLaughlin Ltd*

50 Albemarle Street, London W1S 4BD
℡ 020 7493 4361    ℻ 020 7495 8961
submissions@ctmcl.co.uk
Contact *Charlotte Bruton*

Founded 1931. Handles fiction and general non-fiction, excluding children's. No plays, film or TV scripts, articles, short stories or poetry. Translation rights handled by **The Marsh Agency**. No unsolicited mss or synopses. Preliminary enquiry essential, by letter or email. No reading fee.

## Capel & Land Ltd*

29 Wardour Street, London W1D 6PS
℡ 020 7734 2414    ℻ 020 7734 8101
rosie@capelland.co.uk
www.capelland.com
Contact *Georgina Capel*

Handles fiction and non-fiction. Also film and TV rights and broadcasters; not scripts. CLIENTS Kunal Basu, Vince Cable, John Gimlette, Andrew Greig, Dr Tristram Hunt, Tobias Jones, Adam Nicolson, Andrew Roberts, Simon Sebag Montefiore, Stella Rimington, Diana Souhami, Louis Theroux, Fay Weldon. COMMISSION Home, US & Translation 15%. Send sample chapters and synopsis with covering letter and email address in the first instance. 'We are unable to return material so be sure to keep a hard copy. We will contact you within six weeks if we are interested in taking things further.' No reading fee.

## Caroline Davidson Literary Agency*

5 Queen Anne's Gardens, London W4 1TU
℡ 020 8995 5768    ℻ 020 8994 2770
cdla@ukgateway.net
www.cdla.co.uk
Contact *Caroline Davidson*

Founded 1988. Handles a wide range of original fiction and non-fiction. Non-fiction includes archaeology, architecture, art, biography, climate change, current affairs, design, gardening, health, history, medicine, natural history, reference, science. Finished, polished first novels positively welcomed. No thrillers, crime, fantasy, science fiction. Short stories, children's, plays, poetry not considered. CLIENTS Peter Barham, Andrew Beatty, Andrew Dalby, Emma Donoghue, Chris Greenhalgh, Ed Husain, Tom Jaine, Simon Unwin. COMMISSION Home & Commonwealth, US, Translation 12½%; 20% if sub-agents are involved in foreign sales or films. Send an initial letter giving details of the project, plus c.v. and s.a.s.e. Non-fiction writers should include a detailed proposal with chapter synopsis. Fiction writers should submit the first 50 and last ten pages of their novel. No response to submissions sent by fax or email and those without return postage/s.a.s.e. Absolutely no telephone enquiries.

## Caroline Sheldon Literary Agency Ltd*

71 Hillgate Place, London W8 7SS
℡ 020 7727 9102
carolinesheldon@carolinesheldon.co.uk
pennyholroyde@carolinesheldon.co.uk
www.carolinesheldon.co.uk
Literary Agents *Caroline Sheldon, Penny Holroyde*

Founded 1985. Handles fiction and children's books. Special interests: all fiction for women, sagas, suspense, contemporary, chick-lit, historical fiction, horror, fantasy and comic novels. Non-fiction: true life stories, memoirs, humour, quirky, animal interest. Children's books: fiction for all age groups, contemporary, comic, fantasy and all major genres including fiction series. Picture book stories. Non-fiction. Children's illustration: all quality illustrations. COMMISSION Home 15%; US & Translation 20%; Film & TV 15%. Send submissions by email only with subject line 'submissions/title of work/author's name'. Open submissions with full email about yourself and your work, and attach three chapters of ms or other representative sample material. 'Choose agent Caroline Sheldon or Penny Holroyde. Do not submit to both. We do not represent TV or film scripts except for current clients. Editorial advice only given on work of exceptional promise.'

## Casarotto Ramsay and Associates Ltd

Waverley House, 7–12 Noel Street, London W1F 8GQ
℡ 020 7287 4450    ℻ 020 7287 9128

info@casarotto.co.uk
www.casarotto.uk.co.uk
Film/TV/Radio *Jenne Casarotto, Rachel Holroyd, Charlotte Kelly, Jodi Shields, Elinor Burns, Mark Casarotto, Abby Singer, Lucinda Prain, Sophie Dolan*
Stage *Tom Erhardt, Mel Kenyon*

Handles scripts for TV, theatre, film and radio. CLIENTS include: **Theatre** Alan Ayckbourn, Caryl Churchill, Christopher Hampton, David Hare, Sarah Kane estate, Mark Ravenhill. **Film** Laura Jones, Neil Jordan, Nick Hornby, Shane Meadows, Purvis & Wade, Lynne Ramsay. **TV** Howard Brenton, Amy Jenkins, Susan Nickson, Jessica Hynes. COMMISSION Home 10%. OVERSEAS ASSOCIATES worldwide. No unsolicited material without preliminary letter.

## Cecily Ware Literary Agents

19C John Spencer Square, London N1 2LZ
☏ 020 7359 3787    🖷 020 7226 9828
info@cecilyware.com
Contacts *Cecily Ware, Gilly Schuster, Warren Sherman*

Founded 1972. Primarily a film and TV script agency representing work in all areas: drama, children's, series/serials, adaptations, comedies, etc. No books or theatre. COMMISSION Home 10%; US 10–20% by arrangement. No unsolicited mss or phone calls. Approach in writing only. No reading fee.

## Celia Catchpole

56 Gilpin Avenue, London SW14 8QY
☏ 020 8255 4835
celiacatchpole@yahoo.co.uk
www.celiacatchpole.co.uk
Contact *Celia Catchpole*

Founded 1996. Handles children's books – artists and writers. No TV, film, radio or theatre scripts. No poetry. COMMISSION Home 10% (writers) 15% (artists); US & Translation 20%. Works with associate agents abroad. No unsolicited mss.

## Chapman & Vincent*

7 Dilke Street, London SW3 4JE
☏ 020 7352 5582    🖷 020 7352 5582
chapmanvincent@hotmail.co.uk
Directors *Jennifer Chapman, Gilly Vincent*

A small agency handling non-fiction work only. Clients come mainly from personal recommendation and the agency is not actively seeking clients but will consider original work, although not travel, domestic tragedies or academic books. CLIENTS include George Carter, Leslie Geddes-Brown, Lucinda Lambton, Rowley Leigh and Eve Pollard. COMMISSION Home 15%; US & Europe 20%. Works with The Elaine Markson Agency. Submissions require synopsis,

two sample chapters and s.a.s.e. Will consider email enquiries without attachments. No reading fee.

## Christine Green Authors' Agent*

6 Whitehorse Mews, Westminster Bridge Road, London SE1 7QD
☏ 020 7401 8844    🖷 020 7401 8860
info@christinegreen.co.uk
www.christinegreen.co.uk
Contact *Christine Green*

Founded 1984. Handles fiction (general and literary) and general non-fiction. No scripts, fantasy, sci-fi, poetry or children's. COMMISSION Home 10%; US & Translation 20%. Initial letter, synopsis and first three chapters preferred. No reading fee but return postage essential.

## The Christopher Little Literary Agency*

Eel Brook Studios, 125 Moore Park Road, London SW6 4PS
☏ 020 7736 4455    🖷 020 7736 4490
info@christopherlittle.net
firstname@christopherlittle.net
www.christopherlittle.net
Contact *Christopher Little*

Founded 1979. Handles commercial and literary full-length fiction and non-fiction. No poetry, plays, science fiction, fantasy, textbooks, illustrated children's or short stories. Film scripts for established clients only. AUTHORS include Paul Bajoria, Sophia Bennett, Janet Gleeson, Cathy Hopkins, Carol Hughes, Ishani Kar-Pukayastha, Alastair MacNeill, Robert Mawson, Haydn Middleton, A.J. Quinnell, Christopher Matthew, Robert Radcliffe, J.K. Rowling, Darren Shan, D.B. Shan, Wladyslaw Szpilman, John Watson, Pip Vaughan-Hughes, Christopher Hale, Peter Howells, Gen. Sir Mike Jackson, Shiromi Pinto, Anne Zouroudi. COMMISSION UK 15%; US, Canada, Translation, Audio, Motion Picture, TV & Electronic 20%.

## Conville & Walsh Limited*

2 Ganton Street, London W1F 7QL
☏ 020 7287 3030    🖷 020 7287 4545
info@convilleandwalsh.com
www.convilleandwalsh.com
Directors/Literary Agents *Clare Conville, Patrick Walsh, Jake Smith-Bosanquet*
Literary Agents *Ben Mason, Susan Armstrong*
Children's/Comedy Agent *Jo Unwin*
Finance Director  *Alan Oliver*
Foreign Rights Director *Jake Smith-Bosanquet*

Established in 2000 by Clare Conville (ex-A.P. Watt) and Patrick Walsh (ex-Christopher Little Literary Agency). Handles all genres of fiction, non-fiction and children's worldwide. 'Agency taste is generally upmarket to the intelligent end of the mass-market.' Fiction CLIENTS range from the Booker winner DBC Pierre to John Llewellyn Rhys prize-winner Sarah Hall and the Orwell Prize winner Delia Jarrett-Macauley. Commercial novelists include Isabel Wolff, Nick Harkaway, John Niven, Ali Shaw, Howard Marks, Adam Creed, Matt Dunn and Michael Cordy. Non-fiction CLIENTS such as Tom Holland, Helen Castor, Simon Singh, Ben Wilson, Arthur Potts Dawson, Dr Christian Jessen, Gavin Pretor-Pinney, Belle de Jour, Kate Spicer, Richard Wiseman, Kate Rew, Natalie Haynes, Clive Stafford-Smith, Edward Vallance, Ian Stewart, Ruth Padel, Misha Glenny, Michael Bywater, Brett Kahr and Michael Scott. Artists represented for books include David Shrigley and Steven Appleby and the estate of Francis Bacon. Comedians and satirists represented include Vic Reeves, Charlie Brooker and Jon Hegley. Prestigious children's and young adult list that includes John Burningham, Steve Voake, David Bedford, Philip Caveney, P.J. Lynch and the estate of Astrid Lindgren. COMMISSION Home 15%; US & Translation 20%. Send by post first three chapters, one-page synopsis, covering letter and s.a.s.e. No reading fee.

## Creative Authors Ltd Literary Agency

11A Woodlawn Street, Whitstable CT5 1HQ
write@creativeauthors.co.uk
www.creativeauthors.co.uk
Director/Literary Agent *Isabel Atherton*

Founded 2008. Handles fiction and non-fiction; academic, children's, journalism, crime, mysteries and thrillers. Specializes in lifestyle, crafts, mind, body and spirit and music titles. Scripts on behalf of established agency clients only. CLIENTS include Adele Nozedar, John Robb, E.A. Hanks, Fiona McDonald, Dr Keith Souter, Lee Bullman, Clare Gee, Amanda Hallay. COMMISSION Home 15%; US 20%. OVERSEAS ASSOCIATES Uses a variety of trusted sub-agents. Unsolicited submissions welcome. Please do not telephone; prefers email submissions. No novellas, short stories or poetry. No reading fee.

## Curtis Brown Group Ltd*

Haymarket House, 28/29 Haymarket, London SW1Y 4SP
☎ 020 7393 4400    🖷 020 7393 4401
cb@curtisbrown.co.uk
www.curtisbrown.co.uk
CEO *Jonathan Lloyd*
Director of Operations *Ben Hall*
Directors *Jacquie Drewe, Jonny Geller, Nick Marston, Sarah Spear*
Books *Jonny Geller (MD, Book Division), Jonathan Lloyd, Camilla Hornby, Vivienne Schuster, Elizabeth Sheinkman,*

*Gordon Wise, Karolina Sutton, Stephanie Thwaites*
Joint Head of Foreign Rights *Betsy Robbins, Kate Cooper*
Foreign Rights Agents *Carol Jackson, Katie McGowan, Elizabeth Iveson (on behalf of ICM books), Daisy Meyrick (on behalf of ICM books)*
Film/TV/Theatre *Nick Marston (MD, Media Division), Tally Garner, Ben Hall, Joe Phillips, Amanda Davis*
Actors *Grace Clissold, Maxine Hoffman, Sarah MacCormick, Sarah Spear, Kate Staddon, Lucy Johnson, Mary Fitzgerald*
Presenters *Jacquie Drewe*

Founded 1899. Agents for the negotiation in all markets of novels, general non-fiction, children's books and associated rights (including multimedia) as well as plays, film, theatre, TV and radio scripts. OVERSEAS ASSOCIATES in Australia and the US. Send outline for non-fiction and short synopsis for fiction with the first two/three chapters and autobiographical note. Submissions accepted only via the website. No reading fee. Also represents playwrights, film and TV writers and directors, theatre directors and designers, TV and radio presenters and actors.

## Darley Anderson Children's Literary Agency Ltd*

Estelle House, 11 Eustace Road, London SW6 1JB
☎ 020 7386 2674    🖷 020 7386 5571
enquiries@darleyanderson.com
www.darleyandersonchildrens.com
Proprietor *Darley Anderson*
Director *Peter Colegrove*
Children's Agent *Becky Stradwick*
Head of Rights *Madeleine Buston*

Handles children's fiction and young adult non-fiction. No poetry or academic books. CLIENTS Cathy Cassidy, Lisa Clark, John Connolly, Helen Grant, Michelle Harrison, Adrienne Kress, Carmen Reid, Rob Stevens, Kate Wild. COMMISSION Home 15%; US & Translation 20%. OVERSEAS ASSOCIATE APA Talent and Literary Agency (LA/Hollywood); and leading foreign agents worldwide. Postal queries preferred. Send preliminary letter, synopsis and first three chapters. Return postage/s.a.s.e. essential for reply. No reading fee.

## Darley Anderson Literary, TV & Film Agency*

Estelle House, 11 Eustace Road, London SW6 1JB
☎ 020 7385 6652    🖷 020 7386 5571
enquiries@darleyanderson.com
www.darleyanderson.com
Contacts *Darley Anderson (thrillers), Zoe King (non-fiction), Madeleine Buston (Head of Rights), general and women's fiction), Camilla Bolton*

(crime and mystery), Kasia Benke (Rights Manager), Rosanna Bellingham (Finance)

Founded 1988. Run by an ex-publisher with a knack for spotting talent and making great deals – many for six and some for seven and eight figure advances. Handles commercial fiction and non-fiction, children's fiction and non-fiction; also selected scripts for film and TV. No academic books or poetry. Special fiction interests: all types of thrillers and crime (American/hard boiled/cosy/historical); women's fiction (sagas, chick-lit, contemporary, love stories, 'tear jerkers', women in jeopardy) and all types of American and Irish novels. Non-fiction interests: autobiographies, memoirs, biographies, sports books, 'true life' women in jeopardy, revelatory history and science, popular psychology, self improvement, diet, beauty, health, finance, fashion, animals, humour/cartoon, gardening, cookery, inspirational and religious. CLIENTS include Anne Baker, Alex Barclay, Constance Briscoe, Chris Carter, Lee Child, Martina Cole, John Connolly, Jane Costello, Margaret Dickinson, Clare Dowling, Tana French, Dr Daniel & Jason Freeman, Helen Grant, Tara Hyland, Milly Johnson, Chris Mooney, Lesley Pearse, Lynda Page, Adrian Plass, Carmen Reid, Rebecca Shaw, Graeme Sims, Cally Taylor, Lee Weeks. COMMISSION Home 15%; US & Translation 20%; Film/TV/Radio 20%. Established the **Darley Anderson Children's Literary Agency Ltd** in 2008 (see entry). OVERSEAS ASSOCIATES APA Talent and Literary Agency (LA/Hollywood); and leading foreign agents throughout the world. Send letter, synopsis and first three chapters; plus s.a.s.e. for return. No reading fee.

## David Godwin Associates

55 Monmouth Street, London WC2H 9DG
020 7240 9992    020 7395 6110
assistant@davidgodwinassociates.co.uk
www.davidgodwinassociates.co.uk
Contact *Charlotte Knight, Assistant Agent*

Founded 1996. Handles literary and general fiction, non-fiction, biography. No scripts, science fiction or children's. No reading fee. COMMISSION Home 15%; Overseas 15%. 'Please contact the office before making a submission.'

## David Grossman Literary Agency Ltd

118b Holland Park Avenue, London W11 4UA
020 7221 2770    020 7221 1445
Contact *Submissions Dept.*

Founded 1976. Handles full-length fiction and general non-fiction – good writing of all kinds and anything healthily controversial. No verse or technical books for students. No original screenplays or teleplays (only works existing in volume form are sold for performance rights).

Generally works with published writers of fiction only but 'truly original, well-written novels from beginners' will be considered. COMMISSION Rates vary for different markets. OVERSEAS ASSOCIATES throughout Europe, Asia, Brazil and the US. Best approach by preliminary letter giving full description of the work and, in the case of fiction, with the first 50 pages. 'Please ensure material is printed in at least 1½ spacing on one side of paper only; that all pages are numbered consecutively throughout and in minimum 11pt typeface.' All material must be accompanied by return postage. No submissions by fax or email. No unsolicited mss. No reading fee.

## David Higham Associates Ltd*

5–8 Lower John Street, Golden Square, London W1F 9HA
020 7434 5900    020 7437 1072
dha@davidhigham.co.uk
www.davidhigham.co.uk
Books *Anthony Goff, Bruce Hunter, Veronique Baxter, Lizzy Kremer, Caroline Walsh, Andrew Gordon, Georgia Glover, Alice Williams*
Scripts *Nicky Lund, Georgina Ruffhead*

Founded 1935. Handles fiction, general non-fiction (biography, history, current affairs, etc.) and children's books. Also scripts. CLIENTS include J.M. Coetzee, Stephen Fry, Jane Green, James Herbert, Alexander McCall Smith, Lynne Truss, Jacqueline Wilson. COMMISSION Home 15%; US & Translation 20%; Scripts 10%. Preliminary letter with synopsis essential in first instance. See website for submission guidelines. No reading fee.

## Deborah Owen Ltd*

78 Narrow Street, Limehouse, London E14 8BP
020 7987 5119    020 7538 4004
do@deborahowen.co.uk
Contact *Deborah Owen*

Founded 1971. Small agency specializing in only representing three authors direct around the world. *No new authors.* CLIENTS Delia Smith, Amos Oz and David Owen. COMMISSION Home 15%; US & Translation 15%.

## The Dench Arnold Agency

10 Newburgh Street, London W1F 7RN
020 7437 4551    020 7439 1355
contact@denngarnold.co.uk
www.denngarnold.co.uk
Contacts *Elizabeth Dench, Michelle Arnold*

Founded 1972. Handles scripts for TV and film. CLIENTS include Peter Chelsom. COMMISSION Home 10–15%. OVERSEAS ASSOCIATES include William Morris/Sanford Gross and C.A.A., Los Angeles. Unsolicited mss will be read, but a letter with sample of work and c.v. (plus s.a.s.e.) is required.

## Diane Banks Associates Ltd*

submissions@dianebanks.co.uk
www.dianebanks.co.uk
Contact *Diane Banks*

Founded 2006. Handles commercial fiction and non-fiction. Fiction: women's, crime, thrillers, literary fiction with a strong storyline; non-fiction: memoirs, real-life stories, celebrity, autobiography, biography, popular history, popular science, self-help, popular psychology, fashion, health and beauty. No poetry, children's, academic books, plays, scripts or short stories. CLIENTS include Miss S, Andy Taylor (Duran Duran), Alex Higgins, Elise Lindsay, Elizabeth Burton-Phillips, Hannah Sandling, Narinder Kaur, Emily Winterburn, Sasha Wagstaff and Polly Courtney. COMMISSION Home 15%; Overseas 20%. Send brief c.v., synopsis and first three chapters as Word or Open Document attachments to the email address above. Aims to give initial response within two weeks. No reading fee. 'We do not accept hard copy submissions and will neither look at nor return any submissions sent in this format. Unfortunately we are not able to offer any kind of analysis of mss which we choose to reject.'

## Dinah Wiener Ltd*

12 Cornwall Grove, Chiswick, London W4 2LB
℡ 020 8994 6011    ℻ 020 8994 6044
Contact *Dinah Wiener*

Founded 1985. Handles fiction and general non-fiction: biography, popular science. No scripts, children's or poetry. CLIENTS include Valerie-Anne Baglietto, Malcolm Billings, Guy Burt, David Deutsch, Sandy Gall, Wendy K. Harris, Jenny Hobbs, Daniel Snowman, Marcia Willett. COMMISSION Home 15%; US & Translation 20%. Approach with preliminary letter in first instance, giving full but brief c.v. of past work and future plans. Mss submitted must include s.a.s.e. and be typed in double-spacing.

## Dorian Literary Agency (DLA)*

Upper Thornehill, 27 Church Road, St Marychurch, Torquay TQ1 4QY
℡ 01803 312095
Contact *Dorothy Lumley*

Founded 1986. Handles popular genre fiction for adults: romance, historicals, sagas; crime & thrillers; science fiction, fantasy, horror. No short stories, poetry, autobiography, non-fiction, film/TV scripts or plays. CLIENTS include Gillian Bradshaw, Gary Gibson, Kate Hardy, Stephen Jones, Brian Lumley, Carol Rivers, Lyndon Stacey, Julia Williams. COMMISSION Home 12½%; US 15%;

Translation 20–25%. Works with agents in most countries for translation. 'Regret only reading very selectively; please approach by letter only with c.v., writing history and stamp or email address for reply. If interested, will request sample chapters.' No reading fee.

## Dorie Simmonds Agency*

Riverbank House, 1 Putney Bridge Approach, London SW6 3JD
℡ 020 7736 0002    ℻ 020 7736 0010
dorie@doriesimmonds.com
Contact *Dorie Simmonds*

Handles a wide range of subjects including commercial fiction and non-fiction, children's books and associated rights. Specialities include contemporary personalities and historical biographies, commercial women's fiction, historical fiction, crime fiction and children's books. COMMISSION Home & US 15%; Translation 20%. Outline required for non-fiction; a short synopsis for fiction with 2–3 sample chapters, and a c.v. with writing experience/publishing history. No reading fee. Return postage essential.

## Duncan McAra**

28 Beresford Gardens, Edinburgh EH5 3ES
℡ 0131 552 1558
duncanmcara@mac.com
Contact *Duncan McAra*

Founded 1988. Handles fiction (literary fiction) and non-fiction, including art, architecture, archaeology, biography, history, military, travel and books of Scottish interest. COMMISSION Home 15%; US & Translation/Film & TV 20%. Preliminary letter, synopsis and sample chapters (including return postage) essential. No reading fee.

## Ed Victor Ltd*

6 Bayley Street, Bedford Square, London WC1B 3HE
℡ 020 7304 4100    ℻ 020 7304 4111
mary@edvictor.com
Contacts *Ed Victor, Maggie Phillips, Sophie Hicks, Charlie Campbell*
Foreign Rights Manager *Morag O'Brien*

Founded 1976. Handles a broad range of material but leans towards the more commercial ends of the fiction and non-fiction spectrums. Excellent children's book list. No poetry, scripts or academic. Takes on very few new writers and does not accept unsolicited submissions. After trying his hand at book publishing and literary magazines, Ed Victor, an ebullient American, found his true vocation. Strong opinions, very pushy and works hard for those whose intelligence he respects. Loves nothing more than a good title auction. CLIENTS include

John Banville, Herbie Brennan, Eoin Colfer, D.J. Connell, Frederick Forsyth, A.A. Gill, Josephine Hart, Jack Higgins, Nigella Lawson, Kathy Lette, Allan Mallinson, Andrew Marr, Danny Scheinmann, Janet Street-Porter and the estates of Douglas Adams, Raymond Chandler, Dame Iris Murdoch, Sir Stephen Spender and Irving Wallace. COMMISSION Home & US 15%; Translation 20%. No unsolicited mss.

## Eddison Pearson Ltd*

West Hill House, 6 Swains Lane, London N6 6QS
📞 020 7700 7763  📠 020 7700 7866
mail@eddisonpearson.com
www.eddisonpearson.com
Contact *Clare Pearson*

Founded 1996. Small, personally-run agency. Handles children's books, fiction and non-fiction, poetry. CLIENTS include Sue Heap, Robert Muchamore, Valerie Bloom. COMMISSION Home 10%; US & Translation 15–20%. Enquiries and submissions by email only. Please email for up-to-date submission guidelines by return. No reading fee.

## Edwards Fuglewicz*

49 Great Ormond Street, London WC1N 3HZ
📞 020 7405 6725  📠 020 7405 6726
Contacts *Ros Edwards, Helenka Fuglewicz, Julia Forrest*

Founded 1996. Handles literary and commercial fiction (not children's, science fiction, horror or fantasy); non-fiction: biography, history, popular culture. COMMISSION Home 15%; US & Translation 20%. No unsolicited mss or email submissions.

## Elaine Steel*

110 Gloucester Avenue, London NW1 8HX
📞 020 7483 2681
ecmsteel@aol.com
Contact *Elaine Steel*

Founded 1986. Handles scripts, screenplays and books. No technical or academic. CLIENTS include Les Blair, Anna Campion, Michael Eaton, Gwyneth Hughes, Brian Keenan, Troy Kennedy Martin, James Lovelock, Vince O'Connell, Albie Sachs, Ben Steiner. COMMISSION Home 10%; US & Translation 20%. Initial phone call preferred.

## Eric Glass Ltd

25 Ladbroke Crescent, London W11 1PS
📞 020 7229 9500  📠 020 7229 6220
eglassltd@aol.com
Contact *Janet Glass*

Founded 1934. Handles fiction, non-fiction and scripts for publication or production in all media. No poetry, short stories or children's works. CLIENTS include Pierre Chesnot, Charles Dyer,

Henry Fleet, Tudor Gates, Philip Goulding, Pauline Macaulay, Sebastian Beaumont and the estates of Rodney Ackland, Marc Camoletti, Jean Cocteau, Warwick Deeping, William Douglas Home, Philip King, Robin Maugham, Alan Melville, Beverley Nichols, Jean-Paul Sartre. COMMISSION Home 15%; US & Translation 20%. OVERSEAS ASSOCIATES in the US, Australia, Czech Republic, France, Germany, Greece, Holland, Italy, Japan, Poland, Scandinavia, Slovakia, South Africa, Spain. No unsolicited mss. Return postage required. No reading fee.

## Eunice McMullen Ltd

Low Ibbotsholme Cottage, Off Bridge Lane, Troutbeck Bridge, Windermere LA23 1HU
📞 01539 448551
eunicemcmullen@totalise.co.uk
www.eunicemcmullen.co.uk
Contact *Eunice McMullen*

Founded 1992. Handles all types of children's fiction. Especially interested in 9plus and has 'an excellent' list of picture book authors and illustrators. CLIENTS include Wayne Anderson, Jon Berkeley, Sam Childs, Caroline Jayne Church, Jason Cockcroft, Ross Collins, Charles Fuge, Angela McAllister, David Melling, Angie Sage, Gillian Shields, David Wood. COMMISSION Home 10%; US 15%; Translation 20%. *No unsolicited scripts.* Telephone or email enquiries only.

## Eve White Literary Agent*

54 Gloucester Street, London SW1V 4EG
📞 020 7630 1155
eve@evewhite.co.uk
www.evewhite.co.uk
Contact *Eve White*

Founded 2003. Handles full-length adult and children's fiction and non-fiction. No poetry, short stories or textbooks. CLIENTS Susannah Corbett, Carolyn Ching, Tracey Corduroy, Jimmy Docherty, Rae Earl, David Flavell, Abie Longstaff, Kate Maryon, Rachel Mortimer, Cairan Murtagh, Chris Pascoe, Gillian Rogerson, Ruth Saberton, Andy Stanton, Alex Stobbs, Tabitha Suzuma. COMMISSION Home 15%; US & Translation 20%. 'Authors must see the website for up-to-date submission requirements.' No initial approach by email necessary. No reading fee.

## Faith Evans Associates*

27 Park Avenue North, London N8 7RU
📞 020 8340 9920  📠 020 8340 9410
faith@faith-evans.co.uk

Founded 1987. Small agency. CLIENTS include Melissa Benn, the estate of Madeleine Bourdouxhe, Eleanor Bron, Caroline Conran, Alicia Foster, Midge Gillies, Ed Glinert, Vesna Goldsworthy, Cate Haste, Jim Kelly, Helena

Kennedy, Seumas Milne, Tom Paulin, Sheila Rowbotham, the estate of Lorna Sage, Rebecca Stott, Harriet Walter, Elizabeth Wilson. COMMISSION Home 15%; US & Translation 20%. OVERSEAS ASSOCIATES worldwide. List full; no submissions, please.

## The Feldstein Agency

123–125 Main Street, 2nd Floor, Bangor BT20 4AE
☏ 028 9147 2823
paul@thefeldsteinagency.co.uk
www.thefeldsteinagency.co.uk
Contacts *Paul Feldstein, Susan Feldstein*

Founded 2007. Handles commercial fiction, literary fiction and criticism, romantic fiction, crime fiction, mysteries and thrillers, essays, adventure, autobiography/biography, cookery, current affairs, football, golf, guidebooks, heritage, history, humanities, humour, Celtic and Irish interest, journalism, leisure, lifestyle, local history, media, memoirs, military, music, politics, sociology, travel and true crime. CLIENTS inclue Mary O'Sullivan, Ray Strobel, Adrian White, Stephen Walker, Darragh McIntyre, John Richardson, Patrick Greg, Richard Killeen and William Sheehan. COMMISSION Home 15%; US & Translation 20%. Send synopsis first; approach by email. No children's books. No reading fee.

## Felicity Bryan*

2A North Parade, Banbury Road, Oxford OX2 6LX
☏ 01865 513816   📠 01865 310055
agency@felicitybryan.com
www.felicitybryan.com
Agents *Felicity Bryan, Catherine Clarke, Caroline Wood*
Associate Agent *Sally Holloway*

Founded 1988. Handles fiction of various types and non-fiction with emphasis on history, biography, science and current affairs. No scripts for TV, radio or theatre. No crafts, picture books, how-to, science fiction or light romance. CLIENTS include Carlos Acosta, David Almond, Karen Armstrong, Simon Blackburn, Archie Brown, Isla Dewar, John Dickie, Jenny Downham, Sally Gardner, A.C. Grayling, Tim Harford, Sadie Jones, Phyllida Law, Diarmaid MacCulloch, John Man, Martin Meredith, James Naughtie, Linda Newbery, John Julius Norwich, Iain Pears, Robin Pilcher, Rosamunde Pilcher, Matt Ridley, Meg Rosoff, Miriam Stoppard, Roy Strong, Adrian Tinniswood, Colin Tudge, Eleanor Updale, Martin Wolf, Lucy Worsley and the estates of Robertson Davies and Humphrey Carpenter. COMMISSION Home 15%; US & Translation 20%. OVERSEAS ASSOCIATES **Andrew Nurnberg**, Europe; several agencies in US. See website for submission guidelines. No email submissions. No reading fee.

## Felix de Wolfe Limited

Kingsway House, 103 Kingsway, London WC2B 6QX
☏ 020 7242 5066   📠 020 7242 8119
info@felixdewolfe.com
Contact *Felix de Wolfe*

Founded 1938. Handles quality fiction only, and scripts. No non-fiction or children's. CLIENTS include Jan Butlin, Jeff Dowson, Brian Glover, Sheila Goff, Aileen Gonsalves, John Kershaw, Ray Kilby, Bill MacIlwraith, Angus Mackay, Gerard McLarnon, Malcolm Taylor, David Thompson, Paul Todd, Dolores Walshe. COMMISSION Home 12½%; US 20%. No unsolicited mss. No reading fee.

## Fox & Howard Literary Agency*

4 Bramerton Street, London SW3 5JX
☏ 020 7352 8691   📠 020 7352 8691
Contacts *Chelsey Fox, Charlotte Howard*

Founded 1992. A small agency, specializing in non-fiction, that prides itself on 'working closely with its authors'. Handles biography, history and popular culture, reference, business, mind, body and spirit, health and personal development, popular psychology. COMMISSION Home 15%; US & Translation 20%. No unsolicited mss; send letter and synopsis with s.a.se. for response. No reading fee.

## Frances Kelly*

111 Clifton Road, Kingston upon Thames KT2 6PL
☏ 020 8549 7830   📠 020 8547 0051
Contact *Frances Kelly*

Founded 1978. Handles non-fiction, including illustrated: biography, history, art, self-help, food & wine, complementary medicine and therapies, finance and business books; and academic non-fiction in all disciplines. No scripts except for existing clients. COMMISSION Home 10%; US & Translation 20%. No unsolicited mss. Approach by letter with brief description of work or synopsis, together with c.v. and return postage.

## Fraser Ross Associates**

6 Wellington Place, Edinburgh EH6 7EQ
☏ 0131 657 4412/0131 553 2759
kjross@tiscali.co.uk
lindsey.fraser@tiscali.co.uk
www.fraserross.co.uk
Contacts *Lindsey Fraser, Kathryn Ross*

Founded 2002. Handles children's books, adult literary and mainstream fiction. No poetry, short stories, adult fantasy and science fiction, academic, scripts. CLIENTS include Gill Arbuthnott, Thomas Bloor, Ella Burfoot, Vivian French, Chris Higgins, Barry Hutchison, Ann Kelley, Joan Lennon, Joan Lingard, Lynne Rickard, Jamie Rix, Dugald Steer. COMMISION Home 12½%;

US & Translation 20%. Send preliminary letter, synopsis, first three chapters or equivalent, c.v. and s.a.s.e. No reading fee.

### Freeze Creative Limited

Alicia Cottage, The Street, Charmouth DT6 6QE
☎ 01279 450324
freezecreative@btinternet.com
www.freezecreative.com
Contact *Sammy Joe Hansworth*

Founded 2009. Represents authors of screenplays for film and TV only as well as writer/directors. Specializes in strong British and international character-led dramas. No horror or science fiction. COMMISSION Home 15%; US 20%. Unsolicited submissions should consist of a one-page synopsis, pdf file attachment, via email only. No reading fee.

### Futerman, Rose & Associates*

91 St Leonards Road, London SW14 7BL
☎ 020 8255 7755    ☒ 020 8286 4860
enquiries@futermanrose.co.uk
www.futermanrose.co.uk
Contact *Guy Rose*

Founded 1984. Handles fiction, biography, show business, current affairs, teen fiction and scripts for TV and film. No children's, science fiction or fantasy. CLIENTS include Larry Barker, Christian Piers Betley, David Bret, John Clive, Tom Conti, Iain Duncan Smith, Sir Martin Ewans, Susan George, Paul Hendy, Russell Warren Howe, Keith R. Lindsay, Eric MacInnes, Paul Marx, Max Morgan-Witts, Antonia Owen, Liz Rettig, Valerie Rossmore, Peter Sallis, Paul Stinchcombe, Gordon Thomas, Dr Mark White, Toyah Willcox, Simon Woodham, Allen Zeleski. No unsolicited mss. Send preliminary letter with brief resumé, synopsis, first 20pp and s.a.s.e.

### Graham Maw Christie Literary Agency

19 Thornhill Crescent, London N1 1BJ
☎ 020 7737 4766
enquiries@grahammawchristie.com
www.grahammawchristie.com
Contacts *Jane Graham Maw, Jennifer Christie*

Founded 2005. Handles general non-fiction. No children's or poetry.

### Greene & Heaton Ltd*

37 Goldhawk Road, London W12 8QQ
☎ 020 8749 0315    ☒ 020 8749 0318
info@greeneheaton.co.uk
www.greeneheaton.co.uk
Contacts *Carol Heaton, Judith Murray, Antony Topping, Linda Davis (children's)*

A medium-sized agency with a broad range of clients. Handles all types of fiction and non-fiction. No original scripts for theatre, film or TV. CLIENTS include Poppy Adams, Mark Barrowcliffe, Bill Bryson, Jan Dalley, Marcus du Sautoy, Suzannah Dunn, Hugh Fearnley-Whittingstall, Jane Fearnley-Whittingstall, Michael Frayn, P.D. James, William Leith, James McGee, Shona MacLean, Thomasina Miers, The Royal Society, C.J. Sansom, William Shawcross, Sarah Waters. COMMISSION Home 15%; US & Translation 20%. OVERSEAS ASSOCIATES worldwide. No reply to unsolicited submissions without email address and/or return postage.

### The Greenhouse Literary Agency Ltd

Stanley House, St Chad's Place, London WC1X 9HH
☎ 020 7841 3959    ☒ 020 7841 3940
submissions@greenhouseliterary.com
www.greenhouseliterary.com
Contacts *Julia Churchill, Sarah Davies*

A transatlantic agency with an international outlook. Handles children's and teen fiction: high-concept young series; 8–12 and young adult fiction. No picture books, pre-school texts or illustrators, non-fiction, inspirational/religious work, short stories or writing for adults. Clients include Sarwat Chadda, Harriet Goodwin, Lindsey Leavitt, Brenna Yovanoff, Jon Mayhew, Michael Ford, Anne-Marie Conway, Alexandra Diaz, Tricia Springstubb, Sarah Aronson, Tami Lewis Brown, Teresa Harris, Sue Cowing, Winifred Conkling, Valerie Patterson. COMMISSION UK & US 15%; Translation 25%. OVERSEAS ASSOCIATE **The Greenhouse Literary Agency Ltd**, USA. See website for submission details. Welcomes query emails (no snailmail) with up to five pages (first five) pasted into the body of the email. No reading fee.

### Gregory & Company Authors' Agents* (formerly Gregory & Radice)

3 Barb Mews, London W6 7PA
☎ 020 7610 4676    ☒ 020 7610 4686
info@gregoryandcompany.co.uk (general enquiries)
www.gregoryandcompany.co.uk
Contact *Jane Gregory*
Editor *Stephanie Glencross*
Rights Manager *Claire Morris*
Rights Executive *Jemma McDonagh*

Founded 1987. Handles all kinds of fiction and general non-fiction. Special interest fiction – literary, commercial, women's fiction, crime, suspense and thrillers. 'We are particularly interested in books which will also sell to publishers abroad.' No original plays, film or TV scripts (only published books are sold to film and TV). No science fiction, fantasy, poetry, academic or children's books. No reading fee. Editorial advice given to own authors. COMMISSION

Home 15%; US, Translation, Radio/TV/Film 20%. OVERSEAS ASSOCIATES Is well represented throughout Europe, Asia and US. No unsolicited mss; send a preliminary letter with c.v., synopsis, first ten pages and future writing plans (plus return postage) by post or email (maryjones@ gregoryandcompany.co.uk).

## Gunnmedia Ltd

50 Albemarle Street, London W1S 4BD
℡ 020 7529 3745
ali@gunnmedia.co.uk
sarah@gunnmedia.co.uk
www.gunnmedia.co.uk
Contacts *Ali Gunn, Sarah McFadden, Georgina Bean, Jonny McCune*

Founded 2005. Handles commercial fiction and non-fiction; TV and film scripts. CLIENTS include Jenny Colgan, Brian Freeman, Mil Millington, Michael Bilton, Carole Malone, Carol Thatcher, Martina Navratilova, Lowri Turner, Fiona McIntosh, Nabeel Yasin. COMMISSION Home 15%; US & Translation 20%. OVERSEAS ASSOCIATES **Gelfman Schneider**; Fletcher Parry. Unsolicited mss, sample chapters and synopses welcome; approach in writing. No reading fee.

## The Hanbury Agency*

28 Moreton Street, London SW1V 2PE
℡ 020 7630 6768
enquiries@hanburyagency.com
www.hanburyagency.com
Contact *Margaret Hanbury*

Personally run agency representing quality fiction and non-fiction. No plays, scripts, poetry, children's books, fantasy, horror. CLIENTS include George Alagiah, the estate of J.G. Ballard, Simon Callow, Jane Glover, Bernard Hare, Judith Lennox, Katie Price. COMMISSION Home 15%; Overseas 20%. No unsolicited approaches via email.

## Henser Literary Agency

174 Pennant Road, Llanelli SA14 8HN
℡ 01554 753520   ℻ 01554 753520
Contact *Steve Henser*

Founded 2002. Handles fiction: mystery and general; science fiction and fantasy; translation between English and Japanese; non-fiction book on Japan. Also TV, film, radio and theatre scripts. No horror or sadism. No reading fee. COMMISSION Home 15%; US 20%. No unsolicited material. Not looking for additional clients at present. No reading fee.

## hhb agency ltd*

6 Warwick Court, London WC1R 5DJ
℡ 020 7405 5525
heather@hhbagency.com
elly@hhbagency.com
www.hhbagency.com

Contacts *Heather Holden-Brown, Elly James*

Founded in 2005 by Heather Holden-Brown, a publishing editor for 20 years with Waterstone's, Harrap, BBC Books and Headline. Handles non-fiction: history and politics, contemporary autobiography/biography, popular culture, entertainment and TV, business, family memoir, food and cookery a speciality. No scripts. COMMISSION Home 15%. No unsolicited mss; please contact by email or telephone before sending any material. No reading fee.

## Hilary Churchley, Literary Agent

23 Beech Road, Wheatley OX33 1UP
℡ 07768 353082
hilarychurchley@live.co.uk
hilarychurchley.co.uk
Contact *Hilary Churchley*

Founded 2008. Handles adult fiction, crossover and teenage fiction, historical biographies, historical non-fiction. Specializes in crossover fiction, historical fiction and non-fiction, new writers. COMMISSION Home 10%; Translation 20%. Welcomes submissions; see guidelines on the website. No reading fee. No science fiction/ fantasy.

## Independent Talent Group Limited

Oxford House, 76 Oxford Street, London W1D 1BS
℡ 020 7636 6565   ℻ 020 7323 0101
Contacts *Sue Rodgers, Jessica Sykes, Cathy King, Greg Hunt, Hugo Young, Michael McCoy, Duncan Heath, Paul Lyon-Maris*

Founded 1973. Specializes in scripts for film, theatre, TV and radio. No unsolicited scripts.

## The Inspira Group

5 Bradley Road, Enfield EN3 6ES
℡ 020 8292 5163   ℻ 0870 139 3057
darin@theinspiragroup.com
www.theinspiragroup.com
Managing Director *Darin Jewell*

Founded 2001. Handles children's books, crime fiction, lifestyle/relationships, science fiction/ fantasy, humour, non-fiction. No scripts. CLIENTS Stephanie Baudet, Fergus O'Connell, Michael G.R. Tolkien, Simon Brown, Simon Hall. COMMISSION Home & US 15%. Unsolicited mss and synopses welcome; approach by email or telephone. No reading fee but may charge admin fees.

## Intercontinental Literary Agency*

Centric House, 390–391 Strand, London WC2R 0LT
℡ 020 7379 6611   ℻ 020 7240 4724
ila@ila-agency.co.uk
www.ila-agency.co.uk

Contacts *Nicki Kennedy, Sam Edenborough, Tessa Girvan, Katherine West, Mary Esdaile, Jenny Robson*

Founded 1965. Represents translation rights only on behalf of Lucas Alexander Whitley; Luigi Bonomi Associates; Mulcahy Conway Associates; The Viney Agency; Wade & Doherty; John Beaton Writers' Agent; Jane Conway-Gordon; Short Books; Free Agents; Bell Lomax Moreton; The Agency (Hilary Delamere clients only); Stimola Literary Studio (New York); Elyse Cheney Literary Associates (New York); Barer Literary (New York); Park Literary Group (New York); Harold Matson Company (New York); The Steinberg Agency (New York); The Turnbull Agency (John Irving); and selected authors on behalf of various other agencies. No unsolicited submissions accepted.

## J.M. Thurley Management

Archery House, 33 Archery Square, Walmer
CT14 7JA
☎ 01304 371721   🖷 01304 371416
jmthurley@aol.com
www.thecuttingedge.biz
Contacts *Jon Thurley, Patricia Preece*

Founded 1976. Handles fiction and non-fiction, including academic, adventure, anthropology, archaeology, art and art history, autobiography, biotechnology, China, cinema, crime, current affairs, defence, documentary, economics, film, history, journalism, law, medical, memoirs, mysteries, mythology, occult, politics, psychology, romantic and science fiction, television, theatre, thrillers, world affairs. Also offers a paid manuscript consultancy service for assessment and creative input (see website for details).

## Jane Conway-Gordon Ltd*

213 Westbourne Grove, London W11 2SE
☎ 020 7792 3718
jane@conway-gordon.co.uk
Contact *Jane Conway-Gordon*

Founded 1982. Handles fiction and general non-fiction. No poetry, science fiction, children's or short stories. COMMISSION Home 15%; US & Translation 20%. OVERSEAS ASSOCIATES Lyons Literary LLC, New York; plus agencies throughout Europe and Japan. Unsolicited mss welcome; preliminary letter and return postage essential. No reading fee.

## Jane Judd Literary Agency*

18 Belitha Villas, London N1 1PD
☎ 020 7607 0273   🖷 020 7607 0623
www.janejudd.com
Contact *Jane Judd*

Founded 1986. Handles general fiction and non-fiction: women's fiction, crime, thrillers, literary fiction, humour, biography, investigative journalism, health, women's interests and travel. 'Looking for good contemporary women's fiction but not Mills & Boon-type; also historical fiction and non-fiction.' No scripts, academic, gardening, short stories or DIY. CLIENTS include Andy Dougan, Cliff Goodwin, Jill Mansell, Jonathon Porritt, Rosie Rushton, Manda Scott, David Winner. COMMISSION Home 10%; US & Translation 20%. Approach with letter, including synopsis, first chapter and s.a.s.e. Initial telephone call helpful in the case of non-fiction.

## Jane Turnbull*

*Mailing address:* Barn Cottage, Veryan, Truro
TR2 5QA
☎ 01872 501317
jane@janeturnbull.co.uk
www.janeturnbull.co.uk
*London office:* 58 Elgin Crescent, London W11 2JJ
☎ 020 7727 9409
Contact *Jane Turnbull*

Founded 1986. Handles fiction and non-fiction but specializes in biography, history, current affairs, self-help and humour. Translation rights handled by **Aitken Alexander Associates Ltd**. COMMISSION Home 15%; US & Foreign 20%. No unsolicited mss. Approach with letter in the first instance.

## Janklow & Nesbit (UK) Ltd*

33 Drayson Mews, London W8 4LY
☎ 020 7376 2733   🖷 020 7376 2915
queries@janklow.co.uk
www.janklowandnesbit.co.uk
Agents *Tif Loehnis, Claire Paterson, Will Francis*
Head of Rights *Rebecca Folland*

Founded 2000. Handles fiction and non-fiction; commercial and literary. Send full outline (non-fiction), synopsis and three sample chapters (fiction) plus informative covering letter and return postage. No reading fee. No poetry, plays or film/TV scripts. US rights handled by **Janklow & Nesbit Associates** in New York.

## Jeffrey Simmons

15 Penn House, Mallory Street, London NW8 8SX
☎ 020 7224 8917
jasimmons@unicombox.co.uk
Contact *Jeffrey Simmons*

Founded 1978. Handles biography and autobiography, cinema and theatre, fiction (both quality and commercial), history, law and crime, politics and world affairs, parapsychology and sport (but not exclusively). No science fiction/fantasy, children's books, cookery, crafts, hobbies or gardening. Film scripts handled only if by book-writing clients. Special interest in personality books of all sorts and fiction from young writers (i.e. under 40) with a future. COMMISSION Home 10–15%; US & Foreign 15%. Writers become clients by personal introduction or by letter, enclosing a synopsis if possible, a brief biography, a note of any previously published books, plus a list of any publishers and agents who have already seen the mss.

## Jenny Brown Associates**

33 Argyle Place, Edinburgh EH9 1JT
☎ 0131 229 5334
info@jennybrownassociates.com
www.jennybrownassociates.com
Contacts *Jenny Brown, Mark Stanton*
Children's *Lucy Juckes*
Crime *Allan Guthrie*
Foreign Rights *Kevin Pockington*

Founded 2002. Handles literary fiction, crime writing, writing for children; non-fiction: memoirs, sport, music, nature, popular culture. No poetry, short stories, science fiction, fantasy or academic. CLIENTS include Lin Anderson, David Belbin, Stona Fitch, Alasdair Gray, Alex Gray, Keith Gray, Gaby Halberstam, Roger Hutchinson, Sara Maitland, Richard Moore, Natasha Solomons, Paul Torday, Esther Woolfson. COMMISSION Home 12½%; US & Translation 20%. See website for submission guidelines.

## Jill Foster Ltd

9 Barb Mews, Brook Green, London W6 7PA
☎ 020 7602 1263   🖷 020 7602 9336
www.jflagency.com
Contacts *Alison Finch, Simon Williamson, Dominic Lord, Gary Wild, Jill Foster*

Founded 1976. Handles scripts for TV, radio, film and theatre. No fiction, short stories or poetry. CLIENTS include Ian Brown, Grant Cathro, Phil Ford, Rob Gittins, Wayne Jackman, David Lane, Tony Millan and Mike Walling, Jim Pullin, Fraser Steele, Pete Sinclair, Paul Smith, Peter Tilbury, Roger Williams, Susan Wilkins. COMMISSION Home 12½%; US & Translation 15%. No unsolicited mss; approach by letter in the first instance. No approaches by email. No reading fee.

## John Pawsey

8 Snowshill Court, Giffard Park, Milton Keynes
MK14 5QG
☎ 01908 217179
Contact *John Pawsey*

Founded 1981. Handles non-fiction only: biography, current affairs, popular culture, sport. Special interests: sport and biography. No fiction, children's, drama scripts, poetry, short stories, journalism, academic or submissions from outside the UK. CLIENTS include David Rayvern Allen, William Fotheringham, Don Hale, Patricia Hall, Roy Hudd OBE, Gary Imlach, Anne Mustoe,

Gavin Newsham, Matt Rendell. COMMISSION Home 12½%; US & Translation 19–25%. OVERSEAS ASSOCIATES in the US, Japan, South America and throughout Europe. Preliminary letter with s.a.s.e. essential (no email submissions). No reading fee.

## Johnson & Alcock*

Clerkenwell House, 45–47 Clerkenwell Green, London EC1R 0HT
☎ 020 7251 0125   🖷 020 7251 2172
info@johnsonandalcock.co.uk
www.johnsonandalcock.co.uk
Contacts *Andrew Hewson, Michael Alcock, Anna Power, Ed Wilson*

Founded 1956. Handles literary and commercial fiction, children's 9+ fiction; general non-fiction including current affairs, biography, memoirs, history, culture, design, lifestyle and health, film and music, sci-fi and graphic novels. No poetry, screenplays, technical or academic material. COMMISSION Home 15%; US & Translation 20%. For fiction, send first three chapters, full synopsis and covering letter. For non-fiction, send covering letter with details of writing and other relevant experience, plus full synopsis. No response to email submissions. No reading fee but return postage essential.

## Jonathan Clowes Ltd*

10 Iron Bridge House, Bridge Approach, London NW1 8BD
☎ 020 7722 7674   🖷 020 7722 7677
www.jonathanclowes.co.uk
Contacts *Ann Evans, Nemonie Craven Roderick*

Founded 1960. Pronounced 'clewes'. Now one of the biggest fish in the pond and not really for the untried unless they are true high-flyers. Fiction and non-fiction. Special interests: situation comedy, film and television rights. CLIENTS include David Bellamy, Len Deighton, Elizabeth Jane Howard, Doris Lessing, David Nobbs, Barbara Voors, Gillian White and the estate of Kingsley Amis. COMMISSION Home & US 15%; Translation 19%. OVERSEAS ASSOCIATES **Andrew Nurnberg Associates**; Sane Töregard Agency. No unsolicited mss; authors come by recommendation or by successful follow-ups to preliminary letters.

## Jonathan Pegg Literary Agency*

32 Batoum Gardens, London W6 7QD
☎ 020 7603 6830   🖷 020 7348 0629
info@jonathanpegg.com
www.jonathanpegg.com
Contact *Jonathan Pegg*

Founded 2008. Handles fiction: literary, thrillers, historical and quality commercial; non-fiction: current affairs, memoir and biography, history, popular science, nature, arts and culture, lifestyle, popular psychology. No film, TV, theatre or radio scripts; no children's illustrated books. CLIENTS Nick Davies, Zac Goldsmith, Robert Lacey, Tom McCarthy, Stephen Pollard, General Sir Michael Rose, Mira Stout. COMMISSION Home 15%; US 20%. Accepts submissions by email (preferably) or post with s.a.s.e. Include a one-page mini-synopsis, a half-page c.v., a longer synopsis (for non-fiction) and two sample chapters. If sending material by email, please ensure it is via 'Word document' attachments, 1½ line spacing. No reading fee.

## Josef Weinberger Plays

12–14 Mortimer Street, London W1T 3JJ
📞 020 7580 2827    📠 020 7436 9616
general.info@jwmail.co.uk
www.josef-weinberger.com
Contact *Michael Callahan*

Josef Weinberger is both agent and publisher of scripts for the theatre. CLIENTS include Ray Cooney, John Godber, Peter Gordon, Debbie Isitt, Arthur Miller, Sam Shepard, John Steinbeck. OVERSEAS REPRESENTATIVES in the USA, Canada, Australia, New Zealand and South Africa. No unsolicited mss; introductory letter essential. No reading fee.

## Judith Chilcote Agency*

8 Wentworth Mansions, Keats Grove, London NW3 2RL
📞 020 7794 3717
judybks@aol.com
Contact *Judith Chilcote*

Founded 1990. Handles commercial fiction, TV tie-ins, health and nutrition, popular psychology, biography and celebrity autobiography and current affairs. COMMISSION Home 15%; Overseas 20–25%. *No* academic, science fiction, short stories, film scripts or poetry. *No approaches by email.* Send letter with c.v., synopsis, three chapters and s.a.s.e. for return. No reading fee. No cheques or postal vouchers accepted.

## Judith Murdoch Literary Agency*

19 Chalcot Square, London NW1 8YA
📞 020 7722 4197
Contact *Judith Murdoch*

Founded 1993. Handles full-length fiction only, especially accessible literary, crime and popular women's fiction. No science fiction/fantasy, children's, poetry or short stories. CLIENTS include Anne Bennett, Anne Berry, Tony Black, Alison Bond, Frances Brody, Leah Fleming, Meg Hutchinson, Lola Jaye, Pamela Jooste, Jessie Keane, Catherine King, Jill McGivering, Eve Makis, Jaishree Misra, Kitty Neale, James Steel. Translation rights handled by **The Marsh Agency Ltd**. COMMISSION Home 15%; US & Translation 20%. No unsolicited mss; approach in writing only

enclosing first two chapters and brief synopsis. Submissions by email cannot be considered. Return postage/s.a.s.e. essential. Editorial advice given. No reading fee.

## Judy Daish Associates Ltd

2 St Charles Place, London W10 6EG
📞 020 8964 8811    📠 020 8964 8966
Contacts *Judy Daish, Tracey Elliston, Howard Gooding*

Founded 1978. Theatrical literary agent. Handles scripts for film, TV, theatre and radio. No books. Preliminary letter essential. No unsolicited mss.

## Juliet Burton Literary Agency*

2 Clifton Avenue, London W12 9DR
📞 020 8762 0148    📠 020 8743 8765
juliet.burton@btinternet.com
Contact *Juliet Burton*

Founded 1999. Handles fiction and non-fiction. Special interests crime and women's fiction. No plays, film scripts, articles, poetry or academic material. COMMISSION Home 15%; US & Translation 20%. Approach in writing in the first instance; send synopsis and two sample chapters with s.a.s.e. No email submissions. No unsolicited mss. No reading fee.

## Jüri Gabriel

35 Camberwell Grove, London SE5 8JA
📞 020 7703 6186
Contact *Jüri Gabriel*

Handles quality fiction and non-fiction and (almost exclusively for existing clients) film, TV and radio rights. Jüri Gabriel worked in television, wrote books for 20 years and is chairman of Dedalus publishers (see entry under *UK Publishers*). No short stories, articles, verse or books for children. CLIENTS include Maurice Caldera, Diana Constance, Miriam Dunne, Matt Fox, Paul Genney, Pat Gray, Mikka Haugaard, Robert Irwin, Andrew Killeen, John Lucas, David Madsen, Richard Mankiewicz, David Miller, Andy Oakes, John Outram, Philip Roberts, Dr Stefan Szymanski, Frances Treanor, Jeremy Weingard, Dr Terence White. COMMISSION Home 10%; US & Translation 20%. Unsolicited mss ('two-page synopsis and three sample chapters in first instance, please') welcome if accompanied by return postage and letter giving sufficient information about author's writing experience, aims, etc.

## Kate Hordern Literary Agency

18 Mortimer Road, Clifton, Bristol BS8 4EY
katehordern@blueyonder.co.uk
annewilliamskhla@googlemail.com
Contact *Kate Hordern (0117 923 9368)*
Associate Agent *Anne Williams (07768 518897)*

Agency founded 1999. Anne Williams joined as associate agent in 2009 to expand commercial fiction. Handles fiction, general non-fiction and children's. CLIENTS Richard Bassett, Cheryl Browne, Jeff Dawson, Kylie Fitzpatrick, Sven Hassel, Duncan Hewitt, J.T. Lees, Will Randall, Dave Roberts, John Sadler. COMMISSION Home 15%; US & Translation 20%. OVERSEAS ASSOCIATES Carmen Balcells Agency, Spain; Synopsis Agency, Russia and various agencies in Asia. Submission by email preferred. 'We are looking for commercial women's fiction, including historical, crime, thrillers, quality and literary fiction, general non-fiction, children's for 8+.' New clients taken on very selectively. Pitch, synopsis and sample chapters required for fiction; pitch and proposal/chapter breakdown required for non-fiction. Postal submissions: include email address for response, otherwise s.a.s.e. essential. No reading fee.

## Knight Features

20 Crescent Grove, London SW4 7AH
☎ 020 7622 1467     ☎ 020 7622 1522
peter@knightfeatures.co.uk
Contacts *Peter Knight, Samantha Ferris, Gaby Martin, Andrew Knight*

Founded 1985. Handles puzzle, cartoon, business, history, biographical and autobiographical books. No poetry, science fiction or cookery. CLIENTS include Barbara Minto, Frank Dickens, Gray Jolliffe, David J. Bodycombe. ASSOCIATES UnitedMedia Inc., US; Creators Syndicate, US; Auspac Media, Australia; Puzzle Company, New Zealand; Asia Features, India. No unsolicited mss and no email submissions. Send letter accompanied by c.v. and s.a.s.e. with synopsis of proposed work.

## Laurence Fitch Ltd (incorporating The London Play Co. 1922)

258 Belsize Road, London NW6 4BT
☎ 020 7316 1837
information@laurencefitch.com
www.laurencefitch.com
Contact *Brendan Davis*

Founded 1952, incorporating the London Play Company (1922) and in association with Film Rights Ltd (1932). Handles children's and horror books, scripts for theatre, film, TV and radio only. CLIENTS include Carlo Ardito, Hindi Brooks, John Chapman & Ray Cooney, Jeremy Lloyd, Dave Freeman, John Graham, Robin Hawdon, Glyn Robbins, Lawrence Roman, Gene Stone, the estate of Dodie Smith, Edward Taylor. COMMISSION UK 10%; Overseas 15%. OVERSEAS ASSOCIATES worldwide. No unsolicited mss. Send synopsis with sample scene(s)/first chapters in the first instance. No reading fee.

## Lavinia Trevor Literary Agency*

29 Addison Place, London W11 4RJ
☎ 020 7603 5254     ☎ 0870 129 0838
www.laviniatrevor.co.uk
Contact *Lavinia Trevor*

Founded 1993. Handles general fiction (literary and commercial) and non-fiction, including popular science. No fantasy, science-fiction, poetry, academic, technical or children's books. No TV, film, radio, theatre scripts. COMMISSION rate by agreement with author. No unsolicited submissions.

## LAW*

14 Vernon Street, London W14 0RJ
☎ 020 7471 7900     ☎ 020 7471 7910
Contacts *Mark Lucas, Julian Alexander, Araminta Whitley, Alice Saunders, Peta Nightingale, Philippa Milnes-Smith (children's), Holly Vitow (children's)*

Founded 1996. Handles full-length commercial and literary fiction, non-fiction and children's books. No fantasy (except children's), plays, poetry or textbooks. Film and TV scripts handled for established clients only. COMMISSION Home 15%; US & Translation 20%. OVERSEAS ASSOCIATES worldwide. Unsolicited mss considered; send brief covering letter, short synopsis and two sample chapters. S.a.s.e. essential. No emailed or disk submissions.

## LBA

▷ Luigi Bonomi Associates Limited

## The Leo Media & Entertainment Group

150 Minories, London EC3N 1LS
☎ 020 7183 3177     ☎ 0700 605 7893
info@leomediagroup.com
www.leomediagroup.com
Contacts *Beth Newton, Alex Sullivan, Jennifer Peters*

Founded 2003. Handles adventure and exploration, children's, crime, fiction, humour, lifestyle, literature, memoirs, mysteries, romantic fiction, science fiction/fantasy, teenage books. Also TV, film, radio and theatre scripts. Specializes in fiction, science fiction, children's and teenage books. No academic, poetry or sport. CLIENTS Buddy Bregman, Chuck Kelly, Mia Sperber, John Hardman, Steve Day, E.M. Lake, Nigel Pascoe QC, Mark Tepsich, Jim Curtis, Nik Goldman, Jay Milner, Brenda Lee. COMMISSION Home 15%; US 20%. No unsolicited mss. Send sample chapters and/or synopses via email only. No reading fee.

## Limelight Management Limited*

Unit 10, Filmer Mews, 75 Filmer Road, London
SW6 7JS
☎ 020 7637 2529    ☏ 020 7637 2538
limelight.management@virgin.net
www.limelightmanagement.com
Contacts *Fiona Lindsay, Mary Bekhait*

Founded 1991. Specialists in general non-fiction in
the areas of cookery, gardening, antiques, interior
design, wine, art and crafts, fashion, beauty and
health. No academic text, no poetry, short stories,
plays, musicals or short films. COMMISSION
Home 15%; US & Translation 20%. Unsolicited
mss welcome; send preliminary letter (s.a.s.e.
essential). No reading fee.

## The Lindsay Literary Agency

East Worldham House, East Worldham, Alton
GU34 3AT
☎ 01420 83143
info@lindsayliteraryagency.co.uk
Contact *Becky Bagnell*

Founded 2005. Handles literary fiction,
non-fiction and children's fiction. No scripts.
AUTHORS include Sue Lloyd Roberts, Mike
Lancaster, Stella Wiseman. COMMISSION Home
15%. Send letter with first three chapters and
s.a.s.e. No reading fee.

## London Independent Books

26 Chalcot Crescent, London NW1 8YD
☎ 020 7706 0486    ☏ 020 7724 3122
Proprietor *Carolyn Whitaker*

Founded 1971. A self-styled 'small and
idiosyncratic' agency. Handles fiction and
non-fiction reflecting the tastes of the proprietor.
All subjects considered (except computer books
and young children's), providing the treatment
is strong and saleable. Scripts handled only
if by existing clients. Special interests: boats,
travelogues, commercial fiction, science fiction
and fantasy, teenage fiction. COMMISSION Home
15%; US & Translation 20%. No unsolicited mss;
letter, synopsis and first two chapters with return
postage the best approach. No reading fee.

## London Management
▷ Berlin Associates

## The London Play Co. 1922
▷ Laurence Fitch Ltd

## Lorella Belli Literary Agency (LBLA)*

54 Hartford House, 35 Tavistock Crescent,
Notting Hill, London W11 1AY
☎ 020 7727 8547    ☏ 0870 787 4194
info@lorellabelliagency.com
www.lorellabelliagency.com
Contact *Lorella Belli*

Founded 2002. Handles full-length fiction
(from literary to genre – women's fiction, crime,
thrillers, historical) and general non-fiction.
Particularly interested in first-time novelists,
journalists, multi-cultural and international
writing, books on or about Italy. No children's,
fantasy, science fiction, short stories, poetry,
plays, screenplays or academic books.
CLIENTS include Susan Brackney, Zöe Brân,
Gesine Bullock-Prado, Diablo Cody, Annalisa
Coppolaro-Nowell, Sally Corner, Emily Giffin,
Linda Kavanagh, Edward Kritzler, William
Little, Nisha Minhas, Alanna Mitchell, Rick
Mofina, Sandro Monetti, Angela Murrills, Ingrid
Newkirk, Judy Nunn, Robyn Okrant, Jennifer
Ouellette, Robert Ray, Anneli Rufus, Sheldon
Rusch, Grace Saunders, Dave Singleton, Rupert
Steiner, Justine Trueman, Diana Winston, Carol
Wright. COMMISSION Home 15%; US, Translation
& Dramatic Rights 20%. Works in conjunction
with leading associate agencies abroad and a film/
TV agency in London. Also represents American,
Canadian, Australian and European agencies in
the UK. Welcomes approaches from new and
established authors and journalists by letter or
email in the first instance. Return postage and
s.a.s.e. essential for reply. No reading fee. May
suggest revision.

## Louise Greenberg Books Ltd*

The End House, Church Crescent, London N3 1BG
☎ 020 8349 1179    ☏ 020 8343 4559
louisegreenberg@msn.com
Contact *Louise Greenberg*

Founded 1997. Handles full-length literary fiction
and serious non-fiction only. COMMISSION
Home 15%; US & Translation 20%. CHILDREN'S
ASSOCIATE **Sarah Manson Literary Agent**. *No
telephone approaches.* No reading fee. S.a.s.e.
essential.

## Lucas Alexander Whitley
▷ LAW

## Lucy Luck Associates (In association with Aitken Alexander Associates Ltd)

18–21 Cavaye Place, London SW10 9PT
☎ 020 7373 8672
lucy@lucyluck.com
www.lucyluck.com
Contact *Lucy Luck*

Founded 2006. Handles quality fiction and non-
fiction. No TV, film, radio or theatre scripts; no
illustrated or children's books. CLIENTS include
Kevin Barry, Jon Hotten, Ewan Morrison, Philip
Ó Ceallaigh, Catherine O'Flynn, Rebbecca Ray,
Amanda Smyth, Adam Thorpe, Rupert Wright.
COMMISSION Home 15%; US & Translation 20%.
Send sample chapters by post. No reading fee.

## Luigi Bonomi Associates Limited (LBA)*

91 Great Russell Street, London WC1B 3PS
☎ 020 7637 1234    ℻ 020 7637 2111
info@bonomiassociates.co.uk
www.bonomiassociates.co.uk
Contacts *Luigi Bonomi, Amanda Preston*

Handles commercial and literary fiction, thrillers, crime and women's fiction; non-fiction: history, science, parenting, lifestyle, diet, health; teen and young adult fiction. CLIENTS include James Barrington, James Becker, Sean Black, Chris Beardshaw, Linda Blair, Jo Carnegie, Gennaro Contaldo, Josephine Cox, Nick Foulkes, David Gibbins, Richard Hammond, Bruno Hare, Jane Hill, Matt Hilton, John Humphrys, Graham Joyce, Simon Kernick, James May, Gavin Menzies, Niki Monaghan, Mike Morley, Gillian McKeith, Richard Madeley & Judy Finnigan, Sue Palmer, William Petre, Melanie Phillips, Gervaise Phinn, Jem Poster, Mitch Symons, Alan Titchmarsh, Jon Trigell, Sarah Tucker, Charlotte Ward, Sally Worboyes, Sir Terry Wogan. COMMISSION Home 15%; US & Translation 20-25%. 'We are very keen to find new authors and help them develop their careers.' No scripts, poetry, science fiction/fantasy, children's storybooks. *Send synopsis and first chapter only.* 'We do *not* reply by email so return postage/s.a.s.e. essential for reply.' No reading fee.

## Lutyens and Rubinstein*

21 Kensington Park Road, London W11 2EU
☎ 020 7792 4855
submissions@lutyensrubinstein.co.uk
www.lutyensrubinstein.co.uk
Partners *Sarah Lutyens, Felicity Rubinstein*

Founded 1993. Handles adult fiction and non-fiction books. No TV, film, radio or theatre scripts. COMMISSION Home 15%; US & Translation 20%. Unsolicited mss accepted by email; send introductory letter, c.v. and two chapters. No reading fee.

## Luxton Harris Ltd*

104a Park Street, London W1K 6NG
☎ 020 7318 1248
rebecca.winfield@btopenworld.com
Contacts *Rebecca Winfield, Jonathan Harris, David Luxton*

Founded 2003. Specializes in sport and general non-fiction; biography, popular culture, history. No children's books, science fiction, fantasy, horror or poetry. COMMISSION Home 15%; US & Translation 20%. No unsolicited mss. Please send a letter of enquiry or email in the first instance.

## McKernan Literary Agency & Consultancy**

5 Gayfield Square, Edinburgh EH1 3NW
☎ 0131 557 1771
info@mckernanagency.co.uk
www.mckernanagency.co.uk
Agents *Maggie McKernan, Edwin Hawkes*

Founded 2005. Maggie McKernan handles literary and commercial fiction and non-fiction: Scottish, biography, history, current affairs, memoirs; Edwin Hawkes handles historical fiction, science fiction, fantasy and non-fiction (history, current affairs, travel and popular science.) Sells to the UK market, and works with **Capel & Land Ltd** for translation, US, film and TV sales. CLIENTS include Michael Collins, Alan Taylor, Belinda Seaward, Michael Fry, Michael Schmidt, Carlos Alba, Steven Mithen, Josie Long and Christopher New. No reading fee. See website for submission guidelines. No reading fee.

## The Maggie Noach Literary Agency

7 Peacock Yard, Iliffe Street, London SE17 3LH
☎ 020 7708 3073
info@mnla.co.uk
www.mnla.co.uk
Contact *Address material to the Company.*

Pronounced 'no-ack'. Handles fiction, general non-fiction and children's. COMMISSION Home 15%; US & Translation 20%. No unsolicited mss; submissions by arrangement only.

## Maggie Pearlstine Associates*

31 Ashley Gardens, Ambrosden Avenue, London SW1P 1QE
☎ 020 7828 4212    ℻ 020 7834 5546
maggie@pearlstine.co.uk
Contact *Maggie Pearlstine*

Founded 1989. Small agency representing a select few authors. CLIENTS include Matthew Baylis, Toby Green, Roy Hattersley, Dr Paul Keedwell, Charles Kennedy, Quentin Letts, Claire Macdonald, Prof. Raj Persaud, Prof. Lesley Regan, Malcolm Rifkind, Winifred Robinson, Henrietta Spencer-Churchill, Prof. Kathy Sykes, Prof. Robert Winston. No new authors. Translation rights handled by **Aitken Alexander Associates Ltd**.

## Marjacq Scripts Ltd

34 Devonshire Place, London W1G 6JW
☎ 020 7935 9499    ℻ 020 7935 9115
subs@marjacq.com
www.marjacq.com
Books *Philip Patterson*
Film/TV *Luke Speed*

Handles fiction and non-fiction, literary and commercial, graphic novels and comic books as well as film, TV, screenwriters and directors. No poetry. COMMISSION Home 15%; Overseas 20%. New work welcome; send brief letter, synopsis and approx. first 50 pages plus s.a.s.e. No reading fee.

## The Marsh Agency Ltd*

50 Albemarle Street, London W1S 4BD
☎ 020 7493 4361   ℻ 020 7495 8961
enquiries@marsh-agency.co.uk
www.marsh-agency.co.uk
Chair *Susie Nicklin*
Head of English Sales *Jessica Woollard*
Foreign Rights Department *Camilla Ferrier*

Founded in 1994 as international rights specialist for British, American and Canadian agencies. Expanded to act as agents handling fiction and non-fiction. Specializes in authors with international potential. See website for further information on individual agents' areas of interest. COMMISSION Home 15%; Elsewhere 20%; Film & TV 15%. See also **Paterson Marsh Ltd**. No plays, scripts or poetry. Unsolicited mss considered. Electronic submissions to go through the website.

## Mary Clemmey Literary Agency*

6 Dunollie Road, London NW5 2XP
☎ 020 7267 1290   ℻ 020 7813 9757
mcwords@googlemail.com
Contact *Mary Clemmey*

Founded 1992. Handles fiction and non-fiction – high-quality work with an international market. No science fiction, fantasy or children's books. TV, film, radio and theatre scripts from existing clients only. US clients: Frederick Hill Bonnie Nadell Inc., Lynn C. Franklin Associates Ltd, The Miller Agency, Roslyn Targ, Weingel-Fidel Agency Inc, Betsy Amster Literary Enterprises. COMMISSION Home 15%; US & Translation 20%. OVERSEAS ASSOCIATE Elaine Markson Literary Agency, New York. No unsolicited mss and no email submissions. Approach by letter only in the first instance giving a description of the work (include s.a.s.e.). No reading fee.

## MBA Literary Agents Ltd*

62 Grafton Way, London W1T 5DW
☎ 020 7387 2076   ℻ 020 7387 2042
firstname@mbalit.co.uk
www.mbalit.co.uk
Contacts *Diana Tyler, Meg Davis, Laura Longrigg, David Riding, Sophie Gorell Barnes, Susan Smith, Jean Kitson*

Founded 1971. Handles fiction and non-fiction, TV, film, radio and theatre scripts. Works in conjunction with agents in most countries. Also UK representative for Donald Maass Agency, **Frances Collin Literary Agency**, Martha Millard Literary Agency. In November 2006, the Merric Davidson Literary Agency became part of MBA. COMMISSION Home 15%; Overseas 20%; Theatre/TV/Radio 10%; Film 10–20%. See website for submission details.

## Mic Cheetham Associates

50 Albemarle Street, London W1S 4BD
☎ 020 7495 2002
info@miccheetham.com
www.miccheetham.com
Contacts *Mic Cheetham, Simon Kavanagh*

Established 1994. Handles general and literary fiction, crime and science fiction, and some specific non-fiction. No film/TV scripts apart from existing clients. No children's, illustrated books or poetry. CLIENTS include Iain Banks, Simon Beckett, Carol Birch, Alan Campbell, Paul Cornell, Laurie Graham, M. John Harrison, Toby Litt, Ken MacLeod, China Miéville, Antony Sher. COMMISSION Home 15%; US & Translation 20%. New clients by invitation only.

## Michael Motley Ltd

The Old Vicarage, Tredington, Tewkesbury GL20 7BP
☎ 01684 276390   ℻ 01684 297355
Contact *Michael Motley*

Founded 1973. Handles only full-length mss (adult: 60,000+; children's – all ages – and humour: length variable). No short stories or journalism. No science fiction, horror, poetry or original dramatic material. COMMISSION Home 10–15%; US 15%; Translation 20%. New clients by referral only: no unsolicited material considered. No reading fee.

## Micheline Steinberg Associates

104 Great Portland Street, London W1W 6PE
☎ 020 7631 1310
info@steinplays.com
www.steinplays.com
Contacts *Micheline Steinberg, Helen MacAuley*

Founded 1988. Specializes in drama for stage, TV, film, animation and radio. Represents writers for film and TV rights in fiction and non-fiction on behalf of book agents. COMMISSION Home 12½%; Overseas 15–20%. Works in association with agents in the USA and overseas. Return postage essential.

## Michelle Kass Associates*

85 Charing Cross Road, London WC2H 0AA
☎ 020 7439 1624   ℻ 020 7734 3394
office@michellekass.co.uk
Contacts *Michelle Kass, Andrew Mills*

Founded 1991. Handles literary fiction, film and television primarily. COMMISSION Home 10%; US & Translation 10–20%. No mss accepted without preliminary phone call. No email submissions. No reading fee.

## Miles Stott Children's Literary Agency*
East Hook Farm, Lower Quay Road, Hook,
Haverfordwest SA62 4LR
☏ 01437 890570
nancy@milesstottagency.co.uk
www.milesstottagency.co.uk
Director *Nancy Miles*
Associate Agent *Victoria Birkett* (☏ 01789 488142;
*victoriabirkett@tiscali.co.uk)*

Established in 2003. Handles children's novelty
books, fiction for all ages including teenage fiction
and series fiction. No non-fiction or poetry.
CLIENTS include Ronda Armitage, Dominic
Barker, Sebastien Braun, Jan Fearnley, Stacy
Gregg, Frances Hardinge, Hiawyn Oram, Justin
Richards. Publisher clients include Roaring Brook
Press (USA) and Allen & Unwin (Australia).
COMMISSION Home 15%; USA & Translation 20%.
Welcomes unsolicited submissions by post or
email; send covering letter, brief synopsis and first
three chapters (s.a.s.e. essential). No reading fee.
OVERSEAS ASSOCIATE Barry Goldblatt Literary
LLC, USA.

## Mulcahy Conway Associates*
15 Canning Passage, London W8 5AA
enquiries@mca-agency.com
www.mulcahyconwayassociates.com
Contacts *Ivan Mulcahy, Jonathan Conway,
Laetitia Rutherford*

Handles literary and commercial fiction and
non-fiction; children's books. No film, TV, radio
or theatre scripts. CLIENTS include Clare Brown,
Ha-Joon Chang, Noah Charney, Josie Curran,
Tim Etchells, Henry Hemming, Jon Henderson,
Ian Kelly, David Mitchell, Mark Robson, Phil
Spencer, Bill Turnbull, Robert Webb, Mike Wilks,
Evie Wyld, Jo Whiley, Wagamama, TalkSPORT.
COMMISSION Home 15%; US & Translation 20%.
Unsolicited mss, synopses and ideas welcome;
approach with sample material, covering
letter and s.a.s.e. Refer to website for more
details. No reading fee. OVERSEAS ASSOCIATE
Intercontinental Agency (translation rights).

## The Narrow Road Company
182 Brighton Road, Coulsdon CR5 2NF
☏ 020 8763 9895　🅵 020 8763 2558
richardireson@narrowroad.co.uk
www.narrowroad.co.uk
*Also at:* 3rd Floor, 76 Neal Street, London
WC2H 9PL
Agent/Director *Richard Ireson*
Agent *Amy Ireson* (☏ 020 7379 9598)

Founded 1986. Part of the Narrow Road Group.
Theatrical literary agency. Handles scripts for
TV, theatre, film and radio. No novels or poetry.
CLIENTS include Geoff Aymer, Steve Gooch, Joe
Graham, Renny Krupinski, Simon McAllum. No
unsolicited mss or email attachments; approach
by letter with c.v. only. Interested in writers with
some experience in television, theatre and radio.

## Paterson Marsh Ltd*
50 Albemarle Street, London W1S 4BD
☏ 020 7297 4312　🅵 020 7495 8961
www.patersonmarsh.co.uk
Contact *Stephanie Ebdon*

World rights representatives of authors
and publishers primarily in the UK and
US, specializing in books for mental health
professionals. CLIENTS include the estates of
Sigmund and Anna Freud, Donald Winnicott,
Michael Balint and Wilfred Bion. Also interested
in serious and popular non-fiction. Visit website
for further information and submissions.
COMMISSION 20% (including sub-agent's
commission). See also **The Marsh Agency Ltd**.
No fiction, scripts, poetry, children's, articles or
short stories. Unsolicited mss considered. Send
outline and sample chapter.

## PBJ Management Ltd
5 Soho Square, London W1O 3QA
☏ 020 7287 1112　🅵 020 7287 1191
suzanne@pbjmgt.co.uk
www.pbjmgt.co.uk
Managing Director *Caroline Chignell*
Chairman *Peter Bennet-Jones*
Writer's Agent *Suzanne Milligan*

Founded 1987. Represents writers, performers,
presenters, composers, directors, producers, DJs.
Particularly interested in film and TV comedy.
Handles film, TV, radio and theatre scripts.
No novels or short stories. CLIENTS Armando
Iannucci, Barry Humphries, Chris Morris, Drew
Pearce, Dylan Moran, Eddie Izzard, Harry Enfield,
Lenny Henry, The Mighty Boosh, Jon Canter,
Rowan Atkinson, Sean Hughes. COMMISSION
Home 12½%. Welcomes unsolicited material; send
completed script in Word format with covering
letter and c.v.

## Peters Fraser and Dunlop*
Drury House, 34–43 Russell Street, London
WC2B 5HA
☏ 020 7344 1000　🅵 020 7836 9539/7836 9541
postmaster@pfd.co.uk
www.pfd.co.uk
Chief Executive Officer *Caroline Michel*
Managing Director *Lesley Davey*
Books *Caroline Michel, Michael Sissons,
Annabel Merullo, Tom Williams*
Foreign Rights *Louisa Pritchard*
Film, TV, Theatre, Presenters, Public
Speakers *Gemma Hirst, Jessica Cooper,
Alexandra Henderson*
Children's Books *Suzy Jenvey*
Journalism *Robert Caskie*

Peters Fraser and Dunlop represents authors of fiction and non-fiction, children's writers, screenwriters, playwrights, documentary makers, technicians, presenters and public speakers throughout the world. The agency has 85 years of international experience in all media. Outline sample chapters and author biographies should be addressed to the books department. Material should be submitted on an exclusive basis; or in any event disclose if material is being submitted to other agencies or publishers. Return postage essential. No reading fee. Response to email submissions cannot be guaranteed. See website for submission guidelines.

## PFD
▷ Peters Fraser and Dunlop

## Pollinger Limited*

9 Staple Inn, London WC1V 7QH
☎ 020 7404 0342    🖷 020 7242 5737
info@pollingerltd.com
www.pollingerltd.com
Managing Director *Lesley Pollinger*
Agents *Lesley Pollinger, Joanna Devereux, Tim Bates*
Permissions *Electronic form available on website*
Rights Manager *rightsmanager@pollingerltd.com*
Consultants *Leigh Pollinger, James Fox*

CLIENTS include Michael Coleman, Catherine Fisher, Philip Gross, Kelly McKain, Jeremy Poolman, Nicholas Rhea. Also the estates of H.E. Bates, Erskine Caldwell, Rachel Carson, D.H. Lawrence, John Masters, Carson McCullers, Alan Moorehead, W. Heath Robinson, Eric Frank Russell, Clifford D. Simak and other notables. COMMISSION Home 15%; Translation 20%. Overseas, theatrical and media associates. Submissions policy outlined on the website.

## PVA Management Limited

County House, St Mary's Street, Worcester WR1 1HB
☎ 01905 616100    🖷 01905 610709
maggie@pva.co.uk
www.pva.co.uk
Managing Director *Paul Vaughan*

Founded 1978. Handles non-fiction only. COMMISSION 15%. Send synopsis and sample chapters together with return postage.

## Redhammer Management Ltd*

186 Bickenhall Mansions, Bickenhall Street, London W1U 6BX
☎ 020 7486 3465    🖷 020 7000 1249
info@redhammer.info
www.redhammer.info
Vice President *Peter Cox*

'Provides in-depth management for a small number of highly talented authors.' Willing to take on unpublished authors who are professional in their approach and who have major international potential, ideally, for book, film and/or TV. CLIENTS Martin Bell OBE, Mark Borkowski, Brian Clegg, Audrey Eyton, Rebecca Hardy, Maria Harris, Senator Orrin Hatch, Amanda Lees, Saah Mussi, Michelle Paver, Carolyn Soutar, Carole Stone, Donald Trelford, David Yelland. Submissions considered only if the guidelines given on the website have been followed. Do not send unsolicited mss by post. No radio or theatre scripts. No reading fee.

## RLA Literary Agency
▷ Rogers, Coleridge & White Ltd

## Robert Smith Literary Agency Ltd*

12 Bridge Wharf, 156 Caledonian Road, London N1 9UU
☎ 020 7278 2444    🖷 020 7833 5680
robertsmith.literaryagency@virgin.net
Contact *Robert Smith*

Founded 1997. Handles non-fiction; biography, memoirs, current affairs, history, entertainment, sport, true crime, cookery and lifestyle. No scripts, fiction, poetry, academic or children's books. CLIENTS Kate Adie (serializations), Martin Allen, Peta Bee, Jamie Cameron Stewart, Martyn and Michelle Compton, Judy Cook, Stewart Evans, Neil and Christine Hamilton, Bob Harris, James Haspiel, Albert Jack, Lois Jenkins, Siobhan Kennedy-McGuinness, Heidi Kingstone, Roberta Kray, Jean MacColl, Ann Ming, Michelle Morgan, Theo Paphitis, Howard Raymond, Frances Reilly, Keith Skinner, Jayne Sterne. COMMISSION Home 15%; US & Translation 20%. No unsolicited mss. Send a letter and synopsis in the first instance. No reading fee.

## The Rod Hall Agency Limited

6th Floor, Fairgate House, 78 New Oxford Street, London WC1A 1HB
☎ 020 7079 7987    🖷 0845 638 4094
office@rodhallagency.com
www.rodhallagency.com
Contact *Charlotte Knight*

Founded 1997. Handles drama for film, TV and theatre and writers-directors. Does not represent writers of episodes for TV series where the format is provided but represents originators of series. CLIENTS include Simon Beaufoy (*The Full Monty*), Jeremy Brock (*Mrs Brown*), Liz Lochhead (*Perfect Days*), Martin McDonagh (*The Pillowman*), Simon Nye (*Men Behaving Badly*). COMMISSION Home 10%; US & Translation 15%. No reading fee.

## Roger Hancock Ltd

7 Broadbent Close, Highgate Village, London
N6 5JW
℡ 020 8341 7243
Contact *Material should be addressed to the
Submissions Department*

Founded 1960. Special interests: comedy drama
and light entertainment. COMMISSION Home
10%; Overseas 15%. Unsolicited mss not welcome.
Initial phone call required.

## Rogers, Coleridge & White Ltd*

20 Powis Mews, London W11 1JN
℡ 020 7221 3717    ℻ 020 7229 9084
www.rcwlitagency.com
Chairman *Deborah Rogers*
Managing Director *Peter Straus*
Directors *Gill Coleridge, Peter Robinson,
Pat White, David Miller, Zoe Waldie,
Laurence Laluyaux, Stephen Edwards*
Other Agents *Catherine Pellegrino,
Hannah Westland, Sam Copeland*

Founded 1967. Since April incorporates the
RLA Literary Agency. Handles full-length
mss fiction, non-fiction and children's books.
COMMISSION Home 15%; US & Translation 20%.
No submissions by fax or email.

## Rosica Colin Ltd

1 Clareville Grove Mews, London SW7 5AH
℡ 020 7370 1080    ℻ 020 7244 6441
Contact *Joanna Marston*

Founded 1949. Handles all full-length mss, plus
theatre, film, television and sound broadcasting
but few new writers being accepted. COMMISSION
Home 10%; US & Translation 20%. Preliminary
letter with return postage essential; writers
should outline their writing credits and whether
their mss have previously been submitted
elsewhere. May take 3–4 months to consider full
mss; synopsis preferred in the first instance. No
reading fee.

## Rupert Crew Ltd*

1A King's Mews, London WC1N 2JA
℡ 020 7242 8586    ℻ 020 7831 7914
info@rupertcrew.co.uk (correspondence only)
www.rupertcrew.co.uk
Contacts *Doreen Montgomery, Caroline
Montgomery*

Founded 1927. International representation,
handling volume and subsidiary rights in fiction
and non-fiction properties. No screenplays, plays
or poetry, journalism or short stories, science
fiction or fantasy. COMMISSION Home 15%;
Elsewhere 20%. Preliminary letter and return
postage essential. No reading fee.

## Rupert Heath Literary Agency

50 Albemarle Street, London W1S 4BD
℡ 020 7788 7807    ℻ 020 7691 9331
emailagency@rupertheath.com
www.rupertheath.com
Contact *Rupert Heath*

Founded 2000. Handles literary and general
fiction and non-fiction, including history,
biography and autobiography, current affairs,
popular science, the arts and popular culture.
No scripts or poetry. COMMISSION Home 15%;
US & Translation 20%. OVERSEAS ASSOCIATES
worldwide. Visit the website first to familiarize
yourself with the agency's list. 'Approach by email
telling us a bit about yourself and the book you
wish to submit.' No reading fee.

## Sarah Manson Literary Agent

6 Totnes Walk, London N2 0AD
℡ 020 8442 0396
info@sarahmanson.com
www.sarahmanson.com
Contact *Sarah Manson*

Founded 2002. Handles quality fiction for
children and young adults. COMMISSION Home
10%; Overseas & Translation 20%. Please consult
website for submission guidelines. No reading fee.

## Sarah Such Literary Agency

81 Arabella Drive, London SW15 5LL
℡ 020 8876 4228    ℻ 020 8878 8705
sarah@sarahsuch.com
www.sarahsuch.com
www.twitter.com/sarahsuch
Director *Sarah Such*

Founded 2006. Former publisher with 19 years'
experience. Handles high-quality literary and
commercial non-fiction and fiction including
children's books. 'Always looking for original
work and new talented writers.' COMMISSION
Home 15%; US & Translation 20%. OVERSEAS
ASSOCIATES worldwide. Film and TV
representation: **Aitken Alexander Associates
Ltd**. Works by recommendation but welcomes
unsolicited synopses and sample chapter
submissions. TV/film scripts for established
clients only. No radio or theatre scripts, poetry,
fantasy, self-help or short stories. Contact by
email with brief synopsis, author biog and first
two chapters (as Word attachment). S.a.s.e.
essential if postal submission. No unsolicited mss
and no telephone submissions (hand delivered
submissions not accepted). No reading fee. Will
suggest revision.

## The Sayle Literary Agency*

1 Petersfield, Cambridge CB1 1BB
℡ 01223 303035    ℻ 01223 301638
www.sayleliteraryagency.com

Contact *Rachel Calder*

Handles literary, crime and general fiction, current affairs, social issues, travel, biography, history, general non-fiction. No plays, poetry, textbooks, technical, legal or medical books. COMMISSION Home 15%; US & Translation 20%. OVERSEAS ASSOCIATES Dunow & Carlson & Lerner Literary Agency; Darhansoff, Verrill Feldman; New England Publishing Associates, USA; translation rights handled by **The Marsh Agency Ltd**; film rights by **Sayle Screen Ltd**. No unsolicited mss. Preliminary letter essential, including a brief biographical note and a synopsis plus two or three sample chapters. 'Can only respond to successful submissions.' No reading fee.

## Sayle Screen Ltd

11 Jubilee Place, London SW3 3TD
☎ 020 7823 3883   🖷 020 7823 3363
info@saylescreen.com
www.saylescreen.com
Agents *Jane Villiers, Matthew Bates, Toby Moorcroft*

Specializes in writers and directors for film and television. Also deals with theatre and radio. Works in association with **The Sayle Literary Agency**, **Greene & Heaton Ltd**, **Robinson Literary Agency Ltd** representing film and TV rights in novels and non-fiction. CLIENTS include Andrea Arnold, Shelagh Delaney, Marc Evans, Margaret Forster, John Forte, Rob Green, Mark Haddon, Christopher Monger, Paul Morrison, Gitta Sereny, Sue Townsend. No unsolicited material without preliminary letter. No email submissions.

## The Science Factory Limited

2 Twyford Place, Tiverton EX16 6AP
☎ 020 7193 7296
info@sciencefactory.co.uk
www.sciencefactory.co.uk
Contact *Peter Tallack*

Established in 2008 by Peter Tallack, ex-**Conville & Walsh** director and **Weidenfeld & Nicolson** publishing director. Handles all areas of non-fiction, including history, biography, memoirs, politics, current affairs, travel and science. Specializes in science, technology, medicine and natural history. No poetry, science fiction and fantasy; no TV, film, theatre or radio scripts. COMMISSION Home 15%; US & Translation 20%. Unsolicited mss, sample chapters or synopses welcome; approach by email with covering letter and proposal, including c.v., chapter summaries and sample chapter. No reading fee.

## The Sharland Organisation Ltd

The Manor House, Manor Street, Raunds
NN9 6JW
☎ 01933 626600
tso@btconnect.com
www.sharlandorganisation.co.uk
Contacts *Mike Sharland, Alice Sharland*

Founded 1988. Specializes in national and international film, theatre and TV negotiations. Also negotiates multimedia, interactive TV deals and computer game contracts. Handles scripts for film, TV, radio and theatre; also non-fiction. Markets books for film and handles stage, radio, film and TV rights for authors. No scientific, technical or poetry. COMMISSION Home 15%; US & Translation 20%. OVERSEAS ASSOCIATES various. No unsolicited mss. Preliminary enquiry by letter or phone essential.

## Sheil Land Associates Ltd*

52 Doughty Street, London WC1N 2LS
☎ 020 7405 9351   🖷 020 7831 2127
www.sheilland.co.uk
Agents, UK & US *Sonia Land, Vivien Green, Ian Drury, Piers Blofeld*
Film/Theatrical/TV *Sophie Janson, Lucy Fawcett*
Foreign *Gaia Banks, Emily Dyson*

Founded 1962. Handles full-length, literary and commercial fiction and non-fiction, including: biography, politics, history, military history, gardening, science, thrillers, crime, romance, drama, science fiction, fantasy, business, travel, cookery, memoirs, humour, UK and foreign estates. Also theatre, film, radio and TV scripts. CLIENTS include Sally Abbott, Peter Ackroyd, Benedict Allen, Charles Allen, Roy Apps, Pam Ayres, Melvyn Bragg, Steven Carroll, Mark Chadbourn, David Cohen, Frank Coles, Anna del Conte, Judy Corbalis, Elizabeth Corley, Seamus Deane, Angus Donald, Alex Dryden, Greg Dyke, Rosie Goodwin, Chris Ewan, Alan Gilbey, Jean Goodhind, Robert Green, Peter Hart, Susan Hill, Richard Holmes, Nicholas Hopkins, HRH The Prince of Wales, Iain Johnstone, Irene Karafilly, Brooke Kinsella, Adam Long, Richard Mabey, The Brothers Macleod, Mary Morris, Rachel Murrell, Graham Rice, Robert Rigby, Steve Rider, Martin Riley, Robert Scott, Paul Seller, Diane Setterfield, Tom Sharpe, Dom Shaw, Martin Stephen, Jeffrey Tayler, Andrew Taylor, Sue Teddern, Rose Tremain, Barry Unsworth, Kevin Wells, Prof. Stanley Wells, Neil White, John Wilsher, Paul Wilson, James Wyllie and the estates of Catherine Cookson, Patrick O'Brian, Penelope Mortimer, Jean Rhys and F.A. Worsley. COMMISSION Home 15%; US & Translation 20%. OVERSEAS ASSOCIATES **Georges Borchardt, Inc.** (Richard Scott Simon). US film and TV representation: CAA, APA, and others. Welcomes approaches from new clients either to start or to develop their careers. Preliminary letter with s.a.s.e. essential. No reading fee.

## Sheila Ableman Literary Agency*

48–56 Bayham Place, London NW1 0EU
☏ 020 7388 7222
sheila@sheilaableman.co.uk
www.sheilaableman.com
Contact *Sheila Ableman*

Founded 1999. Handles non-fiction including history, science, biography and autobiography. Specializes in celebrity ghostwriting and TV tie-ins. No poetry, children's, gardening or sport. COMMISSION Home 15%; US & Translation 20%. Unsolicited mss welcome. Approach in writing with publishing history, c.v., synopsis, three chapters and s.a.s.e. for return. No reading fee.

## Shelley Power Literary Agency Ltd*

13 rue du Pré Saint Gervais, 75019 Paris, France
☏ 00 33 1 42 38 36 49    ☏ 00 33 1 40 40 70 08
shelley.power@wanadoo.fr
Contact *Shelley Power*

Founded 1976. Shelley Power works between London and Paris. This is an English agency with London-based administration/accounts office and the editorial office in Paris. Handles general commercial fiction, quality fiction, business books, self-help, true crime, investigative exposés, film and entertainment. No scripts, short stories, children's or poetry. COMMISSION Home 12½%; US & Translation 20%. Preliminary letter with brief outline of project (plus return postage as from UK or France) essential. 'We do not consider submissions by email.' No reading fee.

## Shirley Stewart Literary Agency*

3rd Floor, 4A Nelson Road, London SE10 9JB
☏ 020 8293 3000
Director *Shirley Stewart*

Founded 1993. Handles fiction and non-fiction. No scripts, children's, science fiction, fantasy or poetry. COMMISSION Home 10–15%; US & Translation 20%. OVERSEAS ASSOCIATE **Curtis Brown Ltd**, New York. Will consider unsolicited material; send letter with two or three sample chapters in the first instance. S.a.s.e. essential. Submissions on disk not accepted. No reading fee.

## Sinclair-Stevenson

3 South Terrace, London SW7 2TB
☏ 020 7581 2550    ☏ 020 7581 2550
Contact *Christopher Sinclair-Stevenson*

Founded 1995. Handles biography, current affairs, travel, history, fiction, the arts. No scripts, children's, academic, science fiction/fantasy. CLIENTS include Jennifer Johnston, J.D.F. Jones, Ross King, Christopher Lee, Andrew Sinclair and the estates of Alec Guinness, John Cowper Powys and John Galsworthy. COMMISSION Home 10%; US 15%; Translation 20%. OVERSEAS ASSOCIATE T.C. Wallace Ltd, New York. Translation rights handled by **David Higham Associates Ltd**. Send synopsis with s.a.s.e. in the first instance. No reading fee.

## The Standen Literary Agency

53 Hardwicke Road, London N13 4SL
☏ 020 8889 1167    ☏ 020 8889 1167
www.standenliteraryagency.com
Contact *Mrs Yasmin Standen*

Founded 2004. Handles fiction, both adult and children's. 'We are interested in first time writers.' No TV/radio scripts. CLIENTS include Jonathan Yeatman Biggs, Lisa Bratby, John Brindley, Zara Kane, Zoe Marriott, D.M. Mullan, Andrew Murray, Leonora Rustamova. COMMISSION Home 15%; Overseas 20%. Follow the submissions procedure on the website. 'With regard to non-fiction, please contact us in the first instance.' No email submissions. No reading fee.

## Sunflower Literary Agency
▷ Aarau Literary Agency

## Susan Yearwood Literary Agency

2 Knebworth House, Londesborough Road, London N16 8RL
☏ 020 7503 0954
susan@susanyearwood.com
www.susanyearwood.com
Contact *Susan Yearwood*

Founded May 2007. Handles adult commercial and literary fiction and non-fiction, children's fiction (not picture books). COMMISSION Home 15%. Unsolicited mss, sample chapters or synopses welcome. No reading fee.

## The Susijn Agency Ltd

3rd Floor, 64 Great Titchfield Street, London W1W 7QH
☏ 020 7580 6341    ☏ 020 7580 8626
info@thesusijnagency.com
www.thesusijnagency.com
Agents *Laura Susijn, Nicola Barr*

Founded April 1998. Specializes in selling rights worldwide in literary fiction and non-fiction. Also represents non-English language authors and publishers for UK, US and translation rights worldwide. COMMISSION Home 15%; US & Translation 15–20%. Send preliminary letter, synopsis and first two chapters by post (plus s.a.s.e. if material is to be returned). No reading fee.

## The Tennyson Agency

10 Cleveland Avenue, Wimbledon Chase, London
SW20 9EW
☎ 020 8543 5939
submissions@tenagy.co.uk
www.tenagy.co.uk
Contacts *Christopher Oxford, Adam Sheldon*

Founded 2001. Specializes in theatre, radio,
television and film scripts. Related material
considered on an ad hoc basis. No short stories,
children's, poetry, travel, military/historical,
academic, fantasy, science fiction or sport.
CLIENTS include Tony Bagley, Kristina Bedford,
A.D. Cooper, Alastair Cording, Caroline Coxon,
Iain Grant, Jonathan Holloway, Philip Hurd-
Wood, Joanna Leigh, Steve MacGregor, Antony
Mann, Elizabeth Moynihan, Kathryn Radmall,
Ken Ross, John Ryan, Matthew Salkald, Walter
Saunders, Graeme Scarfe, Diane Speakman,
Diana Ward and the estate of Julian Howell.
COMMISSION Home 12½–15%; Overseas 20%. No
unsolicited material; send introductory letter with
author's résumé and proposal/outline of work. No
reading fee.

## Teresa Chris Literary Agency Ltd*

43 Musard Road, London W6 8NR
☎ 020 7386 0633
Contacts *Teresa Chris, Charles Brudenell-Bruce*

Founded 1989. Handles crime, general, women's,
commercial and literary fiction, and non-fiction:
history, biography, cookery, lifestyle, gardening,
etc. Specializes in crime fiction and commercial
women's fiction. No scripts. Film and TV rights
handled by co-agent. No poetry, short stories,
fantasy, science fiction or horror. CLIENTS include
Stephen Booth, Martin Davies, Tamara McKinley,
Marguerite Patten, Eileen Ramsay, Debby Holt.
COMMISSION Home 10%; US & Translation 20%.
OVERSEAS ASSOCIATES Patty Moosbrugger
Literary Agency, USA; representatives in all
other countries. Unsolicited mss welcome. Send
query letter with first two chapters plus two-page
synopsis (s.a.s.e. *essential*) in first instance. No
reading fee.

## Tibor Jones & Associates

Unit 12b, Piano House, 9 Brighton Terrace,
London SW9 8DJ
☎ 020 7733 0555
www.tiborjones.com
Contact *Kevin Conroy Scott*

Founded 2008. Handles fiction, literary
biography, black writing/issues, Commonwealth,
EU, Far East and Latin American literature,
autobiography, commercial fiction. Unsolicited
material welcome; send covering letter, first five
pages and a synopsis.

## Toby Eady Associates

9 Orme Court, London W2 4RL
☎ 020 7792 0092   ☒ 020 7792 0879
toby@tobyeady.demon.co.uk
Contacts *Toby Eady, Jamie Coleman,
Samar Hammam*

Established in 1968. Handles fiction, and
non-fiction. Special interests: global stories with
international appeal. CLIENTS include Bernard
Cornwell, Susan Lewis, Mark Burnell, Ching
He Huang, Mary Wesley, Anna Politkovskaya,
John Carey, Alison Wong, Yasmin Crowther,
Gavin Esler, Sophie Gee, Richard Lloyd Parry,
Julia Lovell, Francesca Marciano, Deborah
Scroggins, Rachel Seiffert, Diane Wei Liang,
Robert Winder, Fan Wu, Xinran Xue, John
Stubbs. COMMISSION Home 15%; Elsewhere 20%.
OVERSEAS ASSOCIATES USA: ICM; France: La
Nouvelle Agence; Holland & Scandinavia: Jan
Michael; Italy & Germany: Marco Vigevani; Spain
& Portugal: MB Agenzia; Korea: The Eric Yang
Agency; Turkey: Akcali Copyright; Hungary:
Katai & Bolza; Czech Republic: Kristin Olson;
Brazil: Tassy Barham; Japan: The English Agency;
Eastern Europe: Prava I Prevodi; Greece: JLM.
No film/TV scripts or poetry. Submissions by
email only. Send first 50 pages to submissions@
tobyeady.demon.co.uk

## TVmyworld

14 Dean Street, London W1D 3RS
☎ 020 7437 4188
business@10muses.com
www.10muses.com
Contacts *Mark Maco, Malcolm Rasala*

Specializes in books, television, movies, internet
tv, games, second life. Has a production arm
making motion pictures, television, commercials
and internet television. Send first 30 pages only
initially.

## Uli Rushby-Smith Literary Agency

72 Plimsoll Road, London N4 2EE
☎ 020 7354 2718   ☒ 020 7354 2718
Contact *Uli Rushby-Smith*

Founded 1993. Handles fiction and non-fiction,
commercial and literary, both adult and children's.
Film and TV rights handled in conjunction with
a sub-agent. No plays or poetry. COMMISSION
Home 15%; US & Translation 20%. Represents UK
rights for Columbia University Press. Approach
with an outline, two or three sample chapters and
explanatory letter in the first instance (s.a.s.e.
essential). No disks. No reading fee.

## United Agents Limited*

12–26 Lexington Street, London W1F 0LE
www.unitedagents.co.uk

Contacts *Simon Trewin, Caroline Dawnay, Jim Gill, Sarah Ballard, Rosemary Scoular, Rosemary Canter, Anna Webber, Jane Willis, Jessica Craig, Robert Kirby*

Founded 2007. Handles fiction, non-fiction, biography, children's. Also TV, film, radio and theatre scripts. Unsolicited material welcome; see submissions policy on the website. No reading fee.

## Valerie Hoskins Associates Limited

20 Charlotte Street, London W1T 2NA
☎ 020 7637 4490   📠 020 7637 4493
vha@vhassociates.co.uk
Contacts *Valerie Hoskins, Rebecca Watson*

Founded 1983. Handles scripts for film, TV and radio. Special interests feature films, animation and TV. COMMISSION Home 12½%; US 20% (maximum). No unsolicited scripts; preliminary letter of introduction essential. No reading fee.

## Vanessa Holt Ltd*

59 Crescent Road, Leigh-on-Sea SS9 2PF
☎ 01702 473787
info@vanessaholt.eclipse.co.uk
Contact *Vanessa Holt*

Founded 1989. Handles general fiction, non-fiction and non-illustrated children's books. No scripts, poetry, academic or technical. Specializes in crime fiction, commercial and literary fiction, and particularly interested in books with potential for sales abroad and/or to TV. COMMISSION Home 15%; US & Translation 20%; Radio/TV/Film 15%. Represented in all foreign markets. Approach by letter or telephone only; no unsolicited mss.

## Wade & Doherty Literary Agency Ltd

33 Cormorant Lodge, Thomas More Street, London E1W 1AU
☎ 020 7488 4171   📠 020 7488 4172
rw@rwla.com
bd@rwla.com
www.rwla.com
Contacts *Robin Wade, Broo Doherty*

Founded 2001. Handles general fiction and non-fiction including children's books. Specializes in military history and crime books. No scripts, poetry, plays or short stories. COMMISSION Home 10%; Overseas & Translation 20%. 'Fees negotiable if a contract has already been offered.' Send detailed synopsis and first 10,000 words by email with a brief biography. No reading fee.

## Watson, Little Ltd*

48–56 Bayham Place, London NW1 0EU
☎ 020 7388 7529   📠 020 7388 8501
office@watsonlittle.com
Contacts *Mandy Little, James Wills, Sallyanne Sweeney*

Handles fiction, commercial women's fiction, crime and literary fiction. Non-fiction special interests include history, science, popular psychology, self-help and general leisure books. Also children's fiction and non-fiction. No short stories, poetry, TV, play or film scripts. Not interested in purely academic writers. COMMISSION Home 15%; US & Translation 20%. OVERSEAS ASSOCIATE The Marsh Agency Ltd; FILM & TV ASSOCIATES The Sharland Organisation Ltd; MBA Literary Agents Ltd; USA: Howard Morhaim (adult); The Chudney Agency (children). Preliminary letter, synopsis and sample chapters. Return postage essential.

## Whispering Buffalo Literary Agency Ltd

97 Chesson Road, London W14 9QS
☎ 020 7565 4737
mariam@whisperingbuffalo.com
www.whisperingbuffalo.com
Director *Mariam Keen*

Founded 2007. Handles commercial and literary fiction and non-fiction including children's and teenage fiction. Subjects include adventure, anthropology, art, arts and entertainment, astronomy, autobiography, cartoons, classical studies, etiquette, fashion, film, graphic novels, health and beauty, humour, interior design, lifestyle, magic, memoirs, music, mythology, natural history, occult, politics, romantic and science fiction, self help, thrillers. Special interest: book to film adaptations. No TV, film, radio or theatre scripts and no poetry or academic material. COMMISSION Home 12½%; Film, US & Translation 20%. Welcomes unsolicited material; send preliminary letter, c.v., synopsis and three sample chapters together with s.a.s.e. for return. No reading fee.

## William Morris Agency Endeavor Entertainment*

Centre Point, 103 New Oxford Street, London WC1A 1DD
☎ 020 7534 6800   📠 020 7534 6900
www.wma.com
Books *Eugenie Furniss (Head of Literary UK), Rowan Lawton, Cathryn Summerhayes*
TV *Holly Pye (Head of TV UK), Isabella Zoltowski*

Worldwide talent and literary agency with offices in New York, Beverly Hills and Nashville. UK office handles fiction and general non-fiction and TV formats. COMMISSION Film & TV 10%; UK Books 15%; USA Books & Translation 20%. Accepts *postal* submissions only. Send query letter, synopsis and three chapters (maximum 50 pages). No poetry, plays, screenplays or picture books. No reading fee.

## The Wylie Agency (UK) Ltd

17 Bedford Square, London WC1B 3JA
Ⓣ 020 7908 5900    Ⓕ 020 7908 5901
mail@wylieagency.co.uk

Handles fiction and non-fiction. No scripts or children's books. COMMISSION Home 10%; US 15%; Translation 20%. The Wylie Agency does not accept unsolicited submissions. Enquire by letter or email before submitting. Any submission must be accompanied by return postage/s.a.s.e.

## Agency Consultants

## Agent Research & Evaluation, Inc. (AR&E)

425 North 20th Street, Philadelphia, PA 19130 USA
Ⓣ 001 215 563 1867    Ⓕ 001 215 563 6797
info@agentresearch.com
www.agentresearch.uk.net
Contact *Bill Martin*

US consultancy founded in 1996. Works with authors seeking a new or first literary agent in the USA, UK, or Canada. 'A hands-on, customized service with hard data – who sells what to whom for how much – presented from a writer's point of view. An individually designed approach to the right agents for what you're doing, including our Very Special Service: personal recommendations to top level agents for work that is "ready for its close-up". See the website for 'Who We Are', details of pricing on the 'Our Services' page and some of the authors worked with on the 'Links' page.

# UK Literary Scouts

Literary scouts gather information from UK agents, publishers and editors on behalf of foreign clients. They are not literary agents and work only with material that is already commissioned or being handled by an agent. They do *not* accept unsolicited material and do *not* deal directly with writers.

## Anne Louise Fisher Associates

29 D'Arblay Street, London W1F 8EP
☎ 020 7494 4609    🖷 020 7494 4611
annelouise@alfisher.co.uk
Contacts *Anne Louise Fisher, Catherine Eccles*

Scouts for Doubleday/Nan Talese, US; Doubleday/Bond Street Books, Canada; Univers Poche/Fleuve Noir, 10/18, Pocket Jeunesse, France; C. Bertelsmann, Knaus, Blanvalet, DVA/Siedler/Pantheon, C. Bertelsmann Jugenbuch, Germany; Arnoldo Mondadori Editore, Oscar and Mondadori Ragazzi, Italy; Albert Bonniers Forlag, Sweden; Otava, Finland; Gyldendal Norsk Forlag, Norway; Gyldendal, Denmark; Mouria, Holland; Random House Mondadori and Montena, Spain; Editora Nova Fronteira, Brazil; Patakis Publications, Greece; Heyday Films, UK.

## Badcock & Rozycki Literary Scouts

1 Old Compton Street, London W1D 5JA
☎ 020 7734 7997    🖷 020 7734 6886
Contacts *June Badcock (june@litscouts. co.uk), Barbara Rozycki (barbara@litscouts. co.uk), Claire Holt (claire@litscouts.co.uk), Hollie Paterson (hollie@litscouts.co.uk)*

Scouts for Wilhelm Heyne Verlag and Diana Verlag, Germany; Unieboek BV and Het Spectrum, The Netherlands; RCS Libri Group, Italy; Editions Jean-Claude Lattès, France; Ediciones Salamandra, Spain; Ellinika Grammata, Greece; Werner Söderström Osakeyhtiö, Finland; Forum, Sweden; Cappelen Damm, Norway; Lindhardt & Ringhof Forlag, Denmark; Verold division of Bjartur-Verold, Iceland; Blueprint Pictures, UK.

## Folly Marland, Literary Scout

☎ 020 8986 0111    🖷 020 8986 0111
fmarland@pobox.com
Contact *Folly Marland*

Scout for S. Fischer, Scherz Verlag, Krüger Verlag, Germany; Livani, Greece; FMG: Truth & Dare, Trademark, Mistral, Pimento, The Netherlands; Village Books, Japan; Kowalski, Italy.

## Heather Schiller

heather.schiller@virgin.net
Contact *Heather Schiller*

Scouts for Prometheus, The Netherlands; Metaichmio, Greece; Piper, Germany; Bompiani, Italy; Urano, Spain; Pax, Norway.

## Jane Southern

11 Russell Avenue, Bedford MK40 3TE
☎ 01234 400147
jane@janesouthern.com
Contact *Jane Southern*

UK Scout for Der Club, Germany; The House of Books, The Netherlands; Belfond, Presses de la Cité and France Loisirs, France; Newton Compton, Italy; Damm Forlag, Sweden; Planeta Group, Spain; Grayhawk Agency, Taiwan; Minoas, Greece; Owls Agency, Japan.

## Louise Allen-Jones Literary Scouts

5c Old Town, London SW4 0JT
☎ 020 7720 2453    🖷 020 7627 3510
Contacts *Louise Allen-Jones (louise@ louiseallenjones.com), Sally Page (sally@ louiseallenjones.com), Naomi Tongue (naomi@ louiseallenjones.com)*

Scouts for Ullstein Buchverlage (Ullstein hb/ppbk, List, Claassen, Marion von Schröder, Econ, Propyläen, Graf Verlag), Germany; AW Bruna (Bruna, Orlando, Signatuur, VIP, Signature) and Meulenhoff Boekerij (Arena, Boekerij, Mynx, J.M. Meulenhoff), The Netherlands; Editora Record (Galera, Best Seller, Record), Brazil; Sperling & Kupfer (Sperling & Kupfer, Frassinelli), Italy; Keter Books, Israel; Oceanida, Greece; The English Agency, Japan; Jentas, Denmark; Origin Pictures, UK.

## Lucy Abrahams Literary Scouting

℡ 0781 000 8243
lucy@literaryscout.co.uk
www.literaryscout.co.uk
Contact *Lucy Abrahams*

Scouts for Books in the Attic and Yedioth Books/
Miskal, Israel; Carrera, The Flying Dutchman and
Lebowski/ Dutch Media, The Netherlands; Sonia
Draga, Poland.

## Petra Sluka

3 Dry Bank Road, Tonbridge TN10 3BS
℡ 01732 353694
petrasluka@petrasluka.plus.com
Contact *Petra Sluka*

Scouts for Verlagsgruppe Lübbe, Germany; De
Fontein/De Kern, Holland; Ediciones B, Spain;
Bra Böcker, Sweden; Japan Uni Agency, Inc.,
Japan; Harlenic Hallas, Greece.

## Rosalind Ramsay Limited

Third Floor, 23 Monmouth Street, London
WC2H 9DD
℡ 020 7836 0054
ros@rosalindramsay.com
www.rosalindramsay.com
Contacts *Rosalind Ramsay, Françoise Higson,
Pippa Wright*

Scouts for Ambo Anthos and Uitgeverij Artemis,
The Netherlands; Kadokawa Shoten, Japan;
Kiepenheuer and Witsch, Germany; Droemer
Knaur, Germany; Kinneret Zmora Bitan Dvir,
Israel; Norstedts Forlagsgrupp, Sweden; Il
Saggiatore, Italy; Santillana Ediciones General,
Spain; Editora Objectiva, Portugal; Editora
Objetiva, Brazil; Channel 4 and FilmFour, UK.

## Sylvie Zannier-Betts

114 Springfield Road, Brighton BN1 6DE
℡ 01273 557370    ℻ 01273 557370
szannier@lineone.net
Contact *Sylvie Zannier-Betts*

Scouts for Uitgeverij De Geus, The Netherlands;
S. Fischer Verlag, Germany; Gummerus
Publishers, Finland; Alfabeta Bokfoerlag, Sweden;
Garzanti Libri, Italy. Exclusive agent in the UK for
S. Fischer Verlag, Germany.

# PR Consultants

## Amanda Johnson PR

enquiries@amandajohnsonpr.com
www.amandajohnsonpr.com
Contacts *Amanda Johnson, Jessica Jackson*

Represents literary prizes, festivals and events; author publicity campaigns. CLIENTS include the Orange Prize; Costa Book Awards; International PEN Free the Word! Festival; Puffin Books.

## Andrea Marks Public Relations

27 London House, Canon's Corner, Edgware HA8 8AX
☏ 020 8958 5128   📠 020 8958 5128
info@andreamarks.co.uk
www.andreamarks.co.uk
Contact *Andrea Marks*

Corporate PR for publishers, booksellers, publishers' lists (not individual titles), prizes and awards, events, projects and initiatives, professional associations.

## Booked PR

Marden House, Penns Road, Petersfield GU32 2EW
☏ 01730 233885   📠 01730 233880
helen@bookedpr.com
www.bookedpr.com
Head of Publicity *Helen McCusker*

Launched by a trained journalist, Booked PR handles non-fiction authors, publishers' lists and PR support for agencies. With experience in promoting finance, political, business, lifestyle and biographies, services can be adapted to suit most titles. Clients include Harriman House, Infinite Ideas, Ovolo Publishing, Foulsham, Matador, Giles de la Mare, Brambleby Books.

## Cameron Publicity and Marketing

6 St Thomas Close, Farnham GU9 8AT
☏ 07903 951957
ben@cameronpm.co.uk
www.cameronpm.co.uk
Contact *Ben Cameron*

Represents publishers and authors for both adult and children's books. Recent CLIENTS include Anova Books; Hachette Children's Books (Hodder and Orchard); The National Maritime Museum; Pavilion Books; AuthorHouse.

## Catherine Stokes Consultancy Ltd

Woolstone, 27 Blacksmiths Hill, Aynho, Banbury OX17 3AH
☏ 07875 567110
stokes.catherine@gmail.com
Director *Catherine Stokes*

Specializes in marketing and PR for children's books. Represents authors, publishers' lists, series and individual titles, prizes, professional associations. CLIENTS include Random House; Egmont; Oxford University Press; Letterland International; National Literacy Trust.

## Cathy Frazer PR

Old Hayle Barn, Hayle Farm, Marle Place Road, Horsmonden TN12 8DZ
☏ 01892 724156   📠 01892 724155
cathyfrazer@hotmail.com
Contact *Cathy Frazer*

Handles illustrated non-reference, consumer education, children's books; subjects include wine, food, antiques, sport, languages, health, parenting. Individual authors also represented. Over 20 years of publishing experience. CLIENTS include Hodder Education; Hodder Children's Books; Virgin; Mitchell Beazley; Millers; DBP; Kingfisher; Carol Vorderman; Uri Geller; Viscount Linley; Sir Patrick Moore; Carol Smillie; Lorraine Kelly; Peter Allis; Jonty Hearnden; Eric Knowles; Judith Miller; Marguerite Patten; Sarah Kennedy.

## Claire Bowles Publicity

Ashley Court, Ashley, Market Harborough LE16 8HF
☏ 01858 565800   📠 01858 565811
www.clairebowlespr.co.uk
Contact *Claire Bowles*

Represents non-fiction: The Good Pub Guide; Oz Clarke. CLIENTS include Workman (US); Artisan (US); Ebury Press and Anova Books

## Claire Sawford PR

36 Gloucester Avenue, London NW1 7BB
☏ 020 7722 4114
claire@clairesawford.net
Contact *Claire Sawford*

Specializes in non-fiction: art, architecture, fashion, design, biography. Recent CLIENTS include V&A Publications; RIBA Bookshops; Royal Academy of Arts; National Portrait Gallery; Gill Hicks.

## Colbert Macalister PR

7 Killieser Avenue, London SW2 4NU
☎ 020 8671 6615
ailsa@colmacpr.co.uk
www.colbertmacalister.co.uk
Contact *Ailsa Macalister*

Represents publishers' lists and authors. CLIENTS include John Blake Publishing; Random House; Faber & Faber; Hay House; Bonnie; Robert Hale; Absolute Press.

## Colman Getty

28 Windmill Street, London W1T 2JJ
☎ 020 7631 2666   📠 020 7631 2699
info@colmangetty.co.uk
www.colmangetty.co.uk
Chief Executive *Dotti Irving*
Managing Director *Liz Sich*

Handles prizes, book campaigns, authors, publishers' lists, professional associations. CLIENTS include The Man Booker Prize for Fiction; World Book Day; National Poetry Day; Little, Brown; Pan Macmillan; Penguin; Quick Reads; *The Times* Cheltenham Festival of Literature.

## Day Five Ltd

1 Archibald Road, London N7 0AN
☎ 020 7619 0098
eobr@blueyonder.co.uk
Contact *Emma O'Bryen*

Represents publishers' lists and individual book projects. CLIENTS include Frances Lincoln Publishers; Macmillan; Natural History Museum.

## The Farley Partnership

Friars Cote Farm, Crockers Lane, Northiam, Rye TN31 6PY
☎ 01797 253668   📠 01797 253565
carol.farley@farleypart.com
www.thefarleybooklist.com
Contacts *Carol Farley, Nicholas Farley*

Represents authors and publishers (travel, children's/parenting, business, non-fiction, fiction, food and drink, gardening). 'Extensive experience of all aspects of publishing industry – publicity, marketing, sales and distribution, self-publishing consultancy.' CLIENTS include Punk Publishing (*Cool Camping*; *Wild Swimming*); Chastleton; Survival Books; Franck Pontais; Mediterranean Islands; Oxygen Books (city-lit series); Crimson Publishing (travel series, business and children's/parenting titles); Pallas Athene; Veloce Publishing (automotive and pets series); Trailblazer.

## FMcM Associates

3rd Floor, Colonial Buildings, 59–61 Hatton Garden, London EC1N 8LS
☎ 020 7405 7422   📠 020 7405 7424
publicity@fmcm.co.uk
www.fmcm.co.uk
Director *Fiona McMorrough*

'FMcM is an award-winning PR and marketing consultancy specializing in publishing, literature, the arts and events.' The agency works closely with new and established authors and artists and across the breadth of UK publishing. CLIENTS include: Penguin; Hachette; Random House; HarperCollins; British Council; English PEN; Althorp Literary Festival and London Review of Books. Author campaigns include Marian Keyes; Alastair Campbell; Margaret Atwood; Bernard Cornwell; Amanda Foreman and Lionel Shriver.

## Giant Rooster PR

9 Hereford Road, London W2 4AB
☎ 020 7792 4701
gina@giantroosterpr.co.uk
Managing Director *Gina Rozner*

Established 1997. Handles individual authors, publishers' lists and prizes. CLIENTS include Fourth Estate; Harper Press; Simon & Schuster; Weidenfeld & Nicolson; Poetry Book Society (T.S. Eliot Prize); Constable & Robinson; Gallic Books.

## Idea Generation

11 Chance Street, London E2 7JB
☎ 020 7749 6850
hector@ideageneration.co.uk
www.ideageneration.co.uk
Managing Director *Hector Proud*

Arts and entertainment PR consultancy, working across the arts, photography, publishing, museums, film, TV, media and lifestyle sectors. Specializes in PR for illustrated, arts and photographic books. CLIENTS include Gloria Books; Genesis Publications; Vision On Publishing; Dazed Publishing; Waddell Publishing.

## Louise Page PR

☎ 020 8741 5663   📠 020 8741 5663
louise@lpagepr.co.uk
Contact *Louise Page*

PR for fiction and non-fiction authors. CLIENTS include Maeve Binchy; Martina Cole; Wendy Holden.

## Maria Boyle Communications Limited

36 Shalstone Road, Mortlake, London SW14 7HR
☎ 020 8876 8444
maria@mbcomms.co.uk
www.mbcomms.co.uk
Director *Maria Boyle*

Handles PR projects for some of the UK's leading publishers and bestselling authors. Named as one of the top three independent PR consultants in the UK by *PR Week*. Works for a range of consumer and corporate clients. 'One area of specialism includes developing and delivering media campaigns that jump off the books pages and feature in the main body of newspapers and magazines, on national high profile TV and radio programmes and online.' CLIENTS include Penguin Classics; Pearson Booktime; DK Rough Guides; HarperCollins; The Book People and a number of high-profile bestselling authors, e.g. Barbara Taylor Bradford.

## MGA (Macdougall Gabriel Associates)

190 Shaftesbury Avenue, London WC2H 8JL
☎ 020 7836 4774   📠 020 7836 4775
mga@mga-pr.com
www.mga-pr.com
Managing Director *Beth Macdougall*
Director *Bethan Jones*

Publishing consultancy which specializes in promoting publishers' lists and authors. CLIENTS include (publishers' lists) Arris Books; English Heritage; Good Life Press; Inside Pocket Publishing; Psychology News Press; Sibling Press; Souvenir Press; Welsh Books Council (Gomer; Graffeg; Parthian; Pont; Seren, etc). Selected books for Alastair Sawday; Constable & Robinson; HarperCollins; Malavan Books; Particular Books; Penguin Press; Quirk Books; Transatlantic Press. Authors include Brian Aldiss, L.P. Berger, Bowvayne.

## Midas Public Relations

10–14 Old Court Place, London W8 4PL
☎ 020 7361 7860   📠 020 7938 1268
info@midaspr.co.uk
www.midaspr.co.uk
Chairman *Tony Mulliken*
Joint Managing Directors *Steven Williams, Jacks Thomas*

Publicity specialists for book and magazine publishing, the arts, online, event and entertainment industries, including crisis management. Creative and original PR campaigns across news, consumer, regional, trade, children's, arts, business and online press. Named in *The Bookseller*'s Top 100 'most influential in publishing'. CLIENTS include authors (fiction, non fiction, children's, religious, business, music), publishers, prizes/awards, literary events, magazines, newspapers, musicians, actors and artists – The London Book Fair; Galaxy British Book Awards; Harlequin, Mills & Boon Brand PR; Bloomsbury; Penguin; Rebecca Farnworth; Bruce Parry; Encyclopaedia Britannica; Private Eye; Disney.

## Nicky Potter

181 Alexandra Park Road, London N22 7UL
☎ 020 8889 9735
nicpot@dircon.co.uk
Contact *Nicky Potter*

Represents publishers' lists, prizes and professional organizations. CLIENTS include School Library Association (SLA School Librarian of the Year Award); Marsh Award for Children's Literature in Translation; Frances Lincoln; Barrington Stoke; The Children's Book Show and Frances Lincoln Diverse Voices Children's Book Award.

## Platypus PR

2 Tidy Street, Brighton BN1 4EL
☎ 01273 692215
info@platypuspr.com
www.platypuspr.com
Contact *Jeff Scott*

Represents publishers, authors, including self-published authors. CLIENTS include Cambridge University Press; Simon & Schuster; Dorling Kindersley; Pearson Education; Continuum; Clairview Books; Gibson Square; John Wiley & Sons; various independent publishers. Books handled have won the Economist Biography Book of the Year 2008, Economist Economics and Business Book of the Year 2008 and Financial Times Best Business Book of 2007.

## Riot Communications

info@riotcommunciations.com
www.riotcommunciations.com
Founders/Co-Directors *Anwen Hooson, Preena Gadher*

Handles authors, publishers, prizes, exhibitions, brand campaigns, general arts projects. CLIENTS include Penguin Books; Waterstone's; Rough Guides; Children's Laureate; William Hill Sports Book of the Year; Little, Brown.

## Sally Randall PR

Forty Acre Oast, Woodchurch, Ashford TN26 3PW
☎ 01233 860670
sallyrandall@reynoldsm.f2s.com
Contact *Sally Randall*

Handles general fiction and non-fiction, independent publishers' lists, individual authors/titles, arts. CLIENTS include Meet the Author; Galore Park; Simon & Schuster; Michael O'Mara.

## Sonia Pugh PR

☎ 01375 891063
sonia.pugh@virgin.net
Contact *Sonia Pugh*

Independent PR consultancy which handles illustrated non-fiction books in all areas including launches, etc.

# Irish Literary Agents

## The Book Bureau Literary Agency

7 Duncairn Avenue, Bray, Co. Wicklow
☎ 00 353 1 276 4996 📠 00 353 1 276 4834
thebookbureau@oceanfree.net
Contact *Ger Nichol*

Handles general and literary fiction. Special interest in women's fiction, crime and thrillers and some non-fiction. COMMISSION Home 10%; Overseas 20%. Works with foreign associates. Will suggest revision. Send preliminary letter, synopsis and first three chapters (single line spacing preferred); return postage essential (IRCs only from UK and abroad). No reading fee.

## Font International Literary Agency

Hollyville House, Hollybrook Road, Clontarf, Dublin 3
☎ 00 353 1 853 2356
info@fontlitagency.com
Contact *Ita O'Driscoll*

Founded 2003. Handles book-length adult fiction and non-fiction from previously published writers (no children's, drama, sci-fi, erotic, technical or poetry). COMMISSION Home 15–20%; Translation 20–25%. No unsolicited mss. 'As an initial contact, please query our interest in your property either by email or post only.' Include details of writing and other media experience. IRCs required. 'We regret that we cannot discuss queries over the phone.' No reading fee.

## Jonathan Williams Literary Agency

Rosney Mews, Upper Glenageary Road, Glenageary, Co. Dublin
☎ 00 353 1 280 3482 📠 00 353 1 280 3482
Contact *Jonathan Williams*

Founded 1980. Handles general trade books: fiction, auto/biography, travel, politics, history, music, literature and criticism, gardening, cookery, sport and leisure, humour, reference, social questions, photography. Some poetry. No plays, science fiction, children's books, mind, body and spirit, computer books, theology, multimedia, motoring, aviation. COMMISSION Home 10%; US & Translation 20%. OVERSEAS ASSOCIATES Piergiorgio Nicolazzini Literary Agency, Italy; Linda Kohn, International Literature Bureau, Holland and Germany; Antonia Kerrigan Literary Agency, Spain; Tuttle-Mori Agency Inc., Japan. No reading fee 'unless the author wants a very fast opinion'. Initial approach by phone or letter. UK submissions should be accompanied by Irish postage stamps or International Reply Coupons (IRCs).

## The Lisa Richards Agency

108 Upper Leeson Street, Dublin 4
☎ 00 353 1 637 5000 📠 00 353 1 667 1256
info@lisarichards.ie
www.lisarichards.ie
Literary Agent *Faith O'Grady*

Founded 1998. Handles fiction and general non-fiction. CLIENTS include Laura Cassidy, Helena Close, Susan Connolly, June Considine (aka Laura Elliot), Matt Cooper, Damian Corless, Denise Deegan, Christine Dwyer Hickey, Robert Fannin, Karen Gillece, Tara Heavey, Paul Howard (aka Ross O'Carroll-Kelly), Amy Huberman, Arlene Hunt, Roisin Ingle, Alison Jameson, Declan Lynch, Roisin Meaney, Pauline McLynn, Aisling McDermott, Anna McPartlin, David O'Doherty, Damien Owens, Kevin Rafter, Eirin Thompson. COMMISSION Home 10%; UK 15%; US & Translation 20%; Film & TV 15%. OVERSEAS ASSOCIATE **The Marsh Agency Ltd** for translation rights. Approach with proposal and sample chapter for non-fiction, and 3–4 chapters and synopsis for fiction (s.a.s.e. essential). No reading fee.

## Marianne Gunn O'Connor Literary Agency

Morrison Chambers, Suite 17, 32 Nassau Street, Dublin 2
☎ 00 353 1 677 9100 📠 00 353 1 677 9101

mgoclitagency@eircom.net
Contact *Marianne Gunn O'Connor*

Founded 1996. Handles literary and commercial fiction, non-fiction: biography; film and TV rights. CLIENTS include Patrick McCabe, Cecelia Ahern, Claudia Carroll, Morag Prunty, Chris Binchy, Anita Notaro, Noelle Harrison, Sinead Moriarty, David McWilliams, Alan Gilsenan, John Kelly, King Adz, Mike McCormack, Peter Murphy, Paddy McMahon, Thrity Engineer, John Lynch, Louise Douglas, Abbie Taylor, Julia Kelly, Kevin Power, Alison Walsh, Christy Lefteri. COMMISSION UK 15%; Overseas 20%; Film & TV 20%. Translation rights handled by Vicki Satlow

Literary Agency, Milan. No unsolicited mss; send preliminary enquiry letter plus half-page synopsis per email.

## Walsh Communications

info@walshcommunications.ie
www.walshcommunications.ie
Contact *Emma Walsh*

Founded 2006. Handles fiction, non-fiction, children's literature. Also TV, film, theatre and radio scripts. No poetry, science fiction or short stories. COMMISSION Home 15%; US 20%. Email query in the first instance; see website for author guidelines. No reading fee.

# US Literary Agents

\* = Members of the **Association of Authors' Representatives, Inc.**

## Key advice

**International Reply Coupons (IRCs)** For return postage, IRCs are required (*not* UK postage stamps). These are available from post offices (letters £1.10 each). To calculate postage from the Republic of Ireland to the UK go to www.anpost.ie/AnPost and click on 'Calculate the Postage'.

## The Aaron M. Priest Literary Agency\*

708 Third Avenue, 23rd Floor, New York NY 10017–4201
℡ 001 212 818 0344
www.aaronpriest.com

Founded 1974. Handles literary and commercial fiction. No scripts, children's fantasy or sci-fi. Submissions by email only; refer to the website for full details. No reading fee.

## Alison J. Picard Literary Agent

PO Box 2000, Cotuit MA 02635
℡ 001 508 477 7192
℻ 001 508 477 7192 (notify before faxing)
ajpicard@aol.com
Contact *Alison Picard*

Founded 1985. Handles mainstream and literary fiction, contemporary and historical romance, children's and young adult, mysteries and thrillers; plus non-fiction. No short stories or poetry. Rarely any science fiction and fantasy. Particularly interested in expanding non-fiction titles. COMMISSION 15%. OVERSEAS ASSOCIATE **John Pawsey**, UK. Approach with written query. No reading fee.

## The Angela Rinaldi Literary Agency\*

PO Box 7877, Beverly Hills CA 90212–7877
℡ 001 310 842 7665    ℻ 001 310 837 8143
amr@rinaldiliterary.com
www.rinaldiliterary.com
Contact *Angela Rinaldi*

Founded 1995. Handles fiction: commercial, literary, upmarket contemporary women's, suspense, literary historical thrillers, gothic suspense, children's and young adult fiction. Non-fiction: memoirs, women's issues/studies, current issues, biography, psychology, popular reference, self-help, health books on specific issues, business, career, personal finance. No scripts, humour, techno/espionage/drug thrillers, category romance, science fiction, fantasy, horror, westerns, cookbooks, poetry, film scripts, religion or occult. COMMISSION Home 15%; Foreign 25%. Email enquiry (no attachments). Novels: send brief synopsis and first three chapters only; non-fiction queries: send detailed covering letter or proposal. 'Please tell me if I have your work exclusively or if it is a multiple submission. Allow 4–6 weeks for response.' No reading fee.

## Ann Rittenberg Literary Agency, Inc.\*

30 Bond Street, New York NY 10012
℡ 001 212 684 6936    ℻ 001 212 684 6929
www.rittlit.com
Contacts *Ann Rittenberg, Penn Whaling*

Founded 1992. Handles literary fiction, serious narrative non-fiction, biography/memoirs, cultural history, upmarket women's fiction and thrillers. No scripts, romance, science fiction, self-help, inspirational and nothing at genre level. COMMISSION Home 15%; Foreign 20%. Send query letter, sample chapters and synopses by post; no emails. Enclose s.a.s.e. No reading fee.

## The Axelrod Agency, Inc.\*

55 Main Street, PO Box 357, Chatham NY 12037
℡ 001 518 392 2100
steve@axelrodagency.com
Contact *Steven Axelrod*

Founded 1983. Handles commercial women's fiction, romantic fiction and mysteries. No non-fiction or scripts. COMMISSION Home 15%; Translation 20%. Welcomes submissions by post (include IRCs) or email. No reading fee.

## The Balkin Agency, Inc.\*

PO Box 222, Amherst MA 01004
℡ 001 413 322 8697    ℻ 001 413 322 8697
rick62838@crocker.com
Contact *Richard Balkin*

Founded 1973. Handles adult non-fiction only. COMMISSION Home 15%; Foreign 20%. No reading fee for outlines and synopses.

## Barbara Braun Associates, Inc.*

151 West 19th Street, 4th Floor, New York
NY 10011

☎ 001 212 604 9023 📠 001 212 871 9530
barbara@barbarabraunagency.com
www.barbarabraunagency.com
Contacts *Barbara Braun, John Baker*

Founded 1995. Handles literary and mainstream, women's and historical fiction, also serious non-fiction, including psychology and biography. Also young adult and mystery books. Specializes in art history, archaeology, architecture and cultural history. No scripts, poetry, science fiction. COMMISSION Home 15%; Foreign 20%. No unsolicited mss; send query letter, synopsis and first five chapters of project pasted into body of email (to: bbasubmissions@gmail.com). No reading fee.

## The Barbara Hogenson Agency, Inc.*

165 West End Avenue, Suite 19–C, New York
NY 10023

☎ 001 212 874 8084 📠 001 212 362 3011
Owner *Barbara Hogenson*

Founded 1994. Handles non-fiction, literary fiction and plays. No science fiction, romance or children's books. Specializes in theatre-related books and literary fiction. COMMISSION Home 15%; Foreign & Translation 20%. OVERSEAS ASSOCIATES Lora Fountain, France; Andrew Nurnberg, Eastern Europe; Japan Uni, Japan. No unsolicited mss; query letter only. Recommendation from current clients preferred or reference to a client's work. No reading fee.

## Barbara W. Stuhlmann Author's Representative

PO Box 276, Becket MA 01223–0276
☎ 001 413 623 5170
Contact *Barbara Ward Stuhlmann*

Founded 1954. COMMISSION Home 10%; Foreign 15%; Translation 20%. 'No new clients at this time.'

## B.J. Robbins Literary Agency*

5130 Bellaire Avenue, North Hollywood CA 91607
☎ 001 818 760 6602
robbinsliterary@aol.com
Contact *B.J. Robbins*

Founded 1992. Handles literary fiction, narrative and general non-fiction. No scripts, genre fiction, romance, horror, science fiction or children's books. COMMISSION Home 15%; Foreign & Translation 20%. OVERSEAS ASSOCIATES **Abner Stein** and **The Marsh Agency Ltd**, UK. Send covering letter with first three chapters or email query in the first instance. No reading fee.

## B.K. Nelson Literary Agency

84 Woodland Road, Pleasantville NY 10570
☎ 001 914 741 1322 📠 001 914 741 1324
bknelson4@cs.com
www.bknelson.com
*Also at:* 1565 Paseo Vida, Palm Springs, CA 92264
☎ 001 760 778 8800 📠 001 914 778 0034
President *Bonita K. Nelson*
Vice President *Leonard 'Chip' Ashbach*
Editorial Director *John W. Benson*

Founded 1979. Specializes in novels, business, self-help, how-to, political, autobiography, celebrity biography. Major motion picture and TV documentary success. COMMISSION 20%. Lecture Bureau for Authors founded 1994; Foreign Rights Catalogue established 1995; BK Nelson Infomercial Marketing Co. 1996, primarily for authors and endorsements, and BKNelson, Inc. for motion picture production in 1998. Signatory to Writers Guild of America, West (WGAW). No unsolicited mss. Letter of inquiry. Reading fee charged.

## Browne & Miller Literary Associates*

410 S. Michigan Avenue, Suite 460, Chicago
IL 60605
mail@browneandmiller.com
www.browneandmiller.com
President *Danielle Egan-Miller*

Founded 1971. Formerly known as Multimedia Product Development. Handles commercial and literary fiction, and practical non-fiction with wide appeal. No scripts, juvenile, science fiction, horror or poetry. COMMISSION Home 15%; Foreign & Translation 20%. OVERSEAS ASSOCIATES in Europe, Latin America, Japan and Asia. In the first instance, send query email or letter with IRCs. No unsolicited material. No reading fee.

## Carol Mann Agency*

55 Fifth Avenue, New York NY 10003
☎ 001 212 206 5635 📠 001 212 675 4809
www.carolmannagency.com
Contacts *Carol Mann*

Founded 1977. Handles literary and commercial fiction and non-fiction. No scripts or genre fiction. COMMISSION Home 15%; Foreign 20%. No unsolicited material; send query letter with return postage. No reading fee.

## Castiglia Literary Agency

1155 Camino del mar, Suite 510, Del Mar CA 92014
☎ 001 858 755 8761 📠 001 858 755 7063
www.castigliaagency.com
Contacts *Julie Castiglia, Winifred Golden, Sally Van Haitsma, Deborah Ritchken*

Founded 1993. Handles fiction: literary, mainstream, ethnic; non-fiction: narrative, biography, business, science, health, parenting, memoirs, psychology, women's and contemporary issues. Specializes in science, biography and literary fiction. No scripts, poetry, horror or fantasy. COMMISSION Home 15%; Foreign & Translation 25%. No unsolicited material; send query letter only in the first instance. No reading fee.

## The Charlotte Gusay Literary Agency

10532 Blythe Avenue, Los Angeles CA 90064
☏ 001 310 559 0831     🖷 001 310 559 2639
gusay1@ca.rr.com (queries only)
www.gusay.com
Owner/President *Charlotte Gusay*

Founded 1989. Handles fiction, non-fiction, young adult/teen and screenplays (feature scripts/plays to film). No short stories, poetry. Enjoys representing novels, narrative non-fiction, gardening and travel. Queries for most categories and genres are welcome. COMMISSION Home 15%; Foreign & Translation 25%. No unsolicited mss; send one-page query letter and s.a.s.e. in the first instance. No reading fee.

## Cherry Weiner Literary Agency

28 Kipling Way, Manalapan NJ 07726
☏ 001 732 446 2096     🖷 001 732 792 0506
Cherry8486@aol.com
Contact *Cherry Weiner*

Founded 1977. Handles all types of fiction: science fiction and fantasy, mainstream, romance, mystery, westerns and some non-fiction. COMMISSION 15%. No submissions except through referral. No reading fee.

## Curtis Brown Ltd*

10 Astor Place, New York NY 10003
☏ 001 212 473 5400
www.curtisbrown.com
President *Peter Ginsberg*
Book Rights *Laura Blake Peterson, Katherine Fausset, Peter Ginsberg, Emilie Jacobson, Ginger Knowlton, Maureen Walters, Mitchell Waters, Elizabeth Harding, Ginger Clark*
Film & TV Rights *Timothy Knowlton, Holly Frederick*
Translation *Dave Barbor*

Founded 1914. Handles general fiction and non-fiction. Also some scripts for film, TV and theatre. OVERSEAS ASSOCIATES Representatives in all major foreign countries. No unsolicited mss; queries only, with IRCs for reply. No reading fee.

## Dominick Abel Literary Agency, Inc.*

146 West 82nd Street, Suite 1A, New York NY 10024
☏ 001 212 877 0710
agency@dalainc.com
Contact *Dominick Abel*

Established 1975. No scripts, children's books or poetry. COMMISSION Home 15%; Translation 20%. No unsolicited material. Query by email (no attachments, please). No reading fee.

## Don Congdon Associates, Inc.*

156 Fifth Avenue, Suite 625, New York NY 10010
☏ 001 212 645 1229     🖷 001 212 727 2688
Contacts *Michael Congdon, Susan Ramer, Cristina Concepcion, Maura Kye-Casella, Katie Kotchman*

Founded 1983. Handles fiction and non-fiction. No academic, technical, romantic fiction or scripts. COMMISSION Home 15%; UK & Translation 19%. OVERSEAS ASSOCIATES worldwide. No unsolicited mss. Queries via email or snailmail letter with return postage. No reading fee.

## Doris S. Michaels Literary Agency Inc.*

1841 Broadway, Suite 903, New York NY 10023
☏ 001 212 265 9474     🖷 001 212 265 9480
query@dsmagency.com
www.dsmagency.com
Contact *Doris Michaels*

Handles literary fiction that has a commercial appeal and strong screen potential and women's fiction; non-fiction: business, current affairs, biography and memoirs, self-help, humour, history, health, classical music, sports, women's issues, social sciences and pop culture. No action/adventure, suspense, science fiction, romance, New Age, religion/spirituality, gift books, art books, fantasy, thrillers, mysteries, westerns, occult and supernatural, horror, historical fiction, poetry, children's literature, humour or travel books. COMMISSION Home 15% Send query letter via email with a one-page synopsis and include a short paragraph detailing credentials. No reading fee. No unsolicited calls.

## Dunham Literary Inc.*

156 Fifth Avenue, Suite 625, New York NY 10010
☏ 001 212 929 0994     🖷 001 212 929 0904
www.dunhamlit.com
Contacts *Jennie Dunham, Blair Hewes*

Founded 2000. Handles literary fiction, non-fiction and children's (from picture books through young adult). No scripts, romance, westerns, horror, science-fiction/fantasy or poetry. COMMISSION Home 15%; Foreign & Translation 20%. OVERSEAS ASSOCIATE UK & Europe: **A.M. Heath & Co. Ltd**. The Rhoda Weyr Agency is now a division of Dunham Literary Inc. No unsolicited material. Send query letter (with return postage) giving information on the author and ms. No email or faxed queries; see the website for further contact information. No reading fee.

## Dystel & Goderich Literary Management*

One Union Square West, Suite 904, New York
NY 10003
℡ 001 212 627 9100   (F) 001 212 627 9313
www.dystel.com
Contacts *Jane Dystel, Miriam Goderich,
Stacey Glick, Michael Bourret, Jim McCarthy,
Lauren E. Abramo, Chasya Milgrom*

Founded 1994. Handles non-fiction and fiction.
Specializes in politics, history, biography,
cookbooks, current affairs, celebrities,
commercial and literary fiction. No reading fee.
See website for submissions guide.

## Educational Design Services, LLC

5750 Bou Avenue, Ste. 1508, N. Bethesda
MD 20852
℡ 001 301 881 8611
blinder@educationaldesignservices.com
www.educationaldesignservices.com
President *Bertram Linder*

Founded 1979. Specializes in texts and
professional development materials for the
education and school market. COMMISSION
Home 15%; Foreign 25%. 'E-submissions' welcome.
IRCs must accompany mail submissions.

## Ethan Ellenberg Literary Agency*

548 Broadway, Suite 5E, New York NY 10012
℡ 001 212 431 4554   (F) 001 212 941 4652
agent@ethanellenberg.com
www.ethanellenberg.com
Contact *Ethan Ellenberg*

Founded 1984. Handles fiction: commercial,
genre, literary and children's; non-fiction:
history, biography, science, health, cooking,
current affairs. Specializes in commercial fiction,
thrillers, suspense and romance. No scripts,
poetry or short stories. COMMISSION Home 15%;
Translation 20%. Prefers submissions via email.
Will accept submissions by post but does not
return submission material from outside the USA.
For fiction send synopsis and first 50 pages; for
non-fiction send proposal and sample chapters, if
available. No reading fee. For email submissions
send query letter only; no attachments.

## The Evan Marshall Agency*

Six Tristam Place, Pine Brook NJ 07058–9445
℡ 001 973 882 1122   (F) 001 973 822 3099
evanmarshall@optonline.net
www.theevanmarshallagency.com
President *Evan Marshall*

Founded 1987. Handles fiction (adult and young
adult). No scripts, children's books or non-fiction.
COMMISSION Home 15%; Foreign & Translation
20%. No unsolicited material. 'We are currently
taking clients by professional referral only. Do not
query us at this time.'

## Fine Print Literary Management*

240 West 35th Street, Suite 500, New York
NY 10001   (F) 001 212 279 0927
www.fineprintlit.com
CEO *Peter Rubie*
President *Stephany Evans*
Contacts *Janet Reid, Brendan Deneen,
Colleen Lindsay, Amy Tipton, Diane Freed,
Meredith Hayes*

Founded in 2007 following a merger of Peter
Rubie Agency and Imprint. Handles fiction,
non-fiction, children's and adult. No poetry or
short stories. COMMISSION Home 15%; Foreign &
Translation 15–30%. Send query letter plus first
chapter by email or post (no email attachments).
No reading fee.

## Frances Collin Literary Agent*

PO Box 33, Wayne PA 19087–0033
℡ 001 610 254 0555   (F) 001 610 254 5029
queries@francescollin.com
www.francescollin.com
Contact *Frances Collin*

Founded 1948. Successor to Marie Rodell.
Handles general fiction and non-fiction. No
scripts. OVERSEAS ASSOCIATES worldwide. No
unsolicited mss. Email queries preferred but
with no attachments. Or send query letter only,
with IRCs for reply. No fax or telephone queries,
please. No reading fee. Rarely accepts non-
professional writers or writers not represented in
the UK.

## Gelfman Schneider Literary Agents, Inc.*

250 West 57th Street, Suite 2122, New York
NY 10107
℡ 001 212 245 1993   (F) 001 212 245 8678
Contacts *Deborah Schneider, Jane Gelfman*

Founded 1919 (London), 1980 (New York).
Formerly John Farquharson Ltd. Works mostly
with established/published authors. Specializes in
general trade fiction and non-fiction. No poetry,
short stories or screenplays. COMMISSION Home
15%; Dramatic 15%; Foreign 20%. OVERSEAS
ASSOCIATE **Curtis Brown Group Ltd**, UK. No
reading fee for outlines. Submissions must be
accompanied by IRCs. No email queries please.

## Georges Borchardt, Inc.*

136 East 57th Street, New York NY 10022
℡ 001 212 753 5785   (F) 001 212 838 6518

Founded 1967. Works mostly with established/
published authors. Specializes in fiction,
biography and general non-fiction of unusual
interest. COMMISSION Home, UK & Dramatic
15%; Translation 20%. UK ASSOCIATE **Sheil Land
Associates Ltd (Richard Scott Simon)**, London.
Unsolicited mss not read.

## The Greenhouse Literary Agency Ltd

11308 Lapham Drive, Oakton VA 22124
℡ 001 703 865 4990
submissions@greenhouseliterary.com
www.greenhouseliterary.com
Vice President *Sarah Davies*

A transatlantic agency with an international
outlook. Handles children's and teen fiction:
high-concept chapter book series; middle grade
and YA fiction. No picture books, pre-school
texts or illustrators, stand-alone very short
chapter books of less than 10,000 words,
non-fiction, inspirational/religious work, short
stories or writing for adults. CLIENTS Sarwat
Chadda, Lindsey Leavitt, Brenna Yovanoff, Valerie
Patterson, Alexandra Diaz, Tricia Springstubb,
Sarah Aronson,Tami Lewis Brown, Teresa
Harris, Jon Mayhew, Sue Cowing, Anne-Marie
Conway, Winifred Conkling, Michael Ford,
Harriet Goodwin. COMMISSION US & UK 15%;
Translation 25%. OVERSEAS ASSOCIATE **The
Greenhouse Literary Agency Ltd**, UK. See
website for submission details. Welcomes query
emails (no snailmail) with up to five pages (first
five) pasted into the body of the email. No reading
fee.

## Harvey Klinger, Inc.*

300 West 55th Street, Suite 11V, New York
NY 10019
queries@harveyklinger.com
www.harveyklinger.com
Contact *Harvey Klinger*

Founded 1977. Handles mainstream fiction
and non-fiction. Specializes in commercial and
literary fiction, psychology, health, science and
children's books. No scripts, poetry or computer
books. COMMISSION Home 15%; Foreign 25%.
OVERSEAS ASSOCIATES in all principal countries.
Welcomes unsolicited material; send by email or
post (no faxes). No reading fee.

## Helen Rees Literary Agency*

376 North Street, Boston MA 02113–2103
℡ 001 617 227 9014    ℻ 001 617 227 8762
reesagency@reesagency.com
Contact *Helen Rees*
Associates *Ann Collette, Lorin Rees*

Founded 1982. Specializes in books on health and
business; also handles biography, autobiography
and history; quality fiction. No scholarly or
technical books. No scripts, science fiction,
children's, poetry, photography, short stories,
cookery. COMMISSION Home 15%; Foreign 20%.
No email queries or attachments. Send query
letter with IRCs. No reading fee.

## Howard Morhaim Literary Agency*

30 Pierrepont Street, Brooklyn NY 11201
℡ 001 718 222 8400    ℻ 001 718 222 5056
www.morhaimliterary.com
Contact *Howard Morhaim*

Founded 1979. Handles general adult and
young adult fiction and non-fiction. No scripts
poetry or religious. COMMISSION Home 15%;
UK & Translation 20%. OVERSEAS ASSOCIATES
worldwide. Send query letter with synopsis and
sample chapters for fiction; query letter with
outline or proposal for non-fiction. Include return
postage. No unsolicited mss. No reading fee.

## Jack Scagnetti Talent & Literary Agency

5118 Vineland Avenue, Suite 106, North
Hollywood CA 91601
℡ 001 818 762 3871/761 0580
www.jackscagnetti.com
Contact *Jack Scagnetti*

Founded 1974. Works mostly with established/
published authors. Handles non-fiction, fiction,
film and TV scripts. No reading fee. COMMISSION
Home & Dramatic 10% (scripts), 15% (books);
Foreign 15%.

## Jane Chelius Literary Agency, Inc.*

548 Second Street, Brooklyn, New York NY 11215
℡ 001 718 499 0236    ℻ 001 718 832 7335
queries@janechelius.com
www.janechelius.com
Contact *Jane Chelius*

Founded 1995. Handles popular and literary
fiction; narrative non-fiction. No children's, young
adult, screenplays, scripts or poetry. COMMISSION
Home 15% Foreign 20%. Submission guidelines
available on the website.

## Janklow & Nesbit Associates

445 Park Avenue, 13th Floor, New York
NY 10022–2606
℡ 001 212 421 1700    ℻ 001 212 980 3671
info@janklow.com
foreignrights@janklow.com
Partners *Morton L. Janklow, Lynn Nesbit*
Senior Vice-President *Anne Sibbald*
Director *Tina Bennett*
Agents *Priscilla Gilman, Luke Janklow,
Richard Morris*
Foreign Rights Director *Cullen Stanley*

Founded 1989. Handles fiction and non-fiction;
commercial and literary. See also **Janklow &
Nesbit (UK) Ltd** under *UK Literary Agents*. No
unsolicited mss.

## Jeanne Fredericks Literary Agency, Inc.*

221 Benedict Hill Road, New Canaan CT 06840
℡ 001 203 972 3011    ℻ 001 203 972 3011

jeanne.fredericks@gmail.com
www.jeannefredericks.com
Contact *Jeanne Fredericks*

Founded 1997. Handles quality adult non-fiction, usually of a practical and popular nature by authorities in their fields. Specializes in health, gardening, science/nature/environment, business, self-help, reference. No fiction, juvenile, poetry, essays, politics, academic or textbooks. COMMISSION Home 15%; Foreign 25% (with co-agent) or 20% (direct). No unsolicited material; send query by email (no attachments) or letter with return postage in the first instance. No reading fee.

## The Jeff Herman Agency, LLC

PO Box 1522, Stockbridge MA 01262
℡ 001 413 298 0077    🖷 001 413 298 8188
jeff@jeffherman.com
www.jeffherman.com
Contact *Jeffrey H. Herman*

Handles all areas of non-fiction, textbooks and reference, business, spiritual and psychology. No scripts. COMMISSION Home 15%; Translation 10%. No unsolicited mss. Query by letter with IRCs in the first instance or by email. No reading fee. Jeff Herman publishes a useful reference guide to the book trade called *Jeff Herman's Guide to Book Editors, Publishers & Literary Agents* (Three Dog Press).

## John A. Ware Literary Agency

392 Central Park West, New York NY 10025
℡ 001 212 866 4733    🖷 001 212 866 4734
Contact *John Ware*

Founded 1978. Specializes in non-fiction: biography, history, current affairs, investigative journalism, science, nature, inside looks at phenomena, medicine and psychology (academic credentials required). No personal memoirs. Also handles literary fiction, mysteries/thrillers, sport, oral history, Americana and folklore. COMMISSION Home & Dramatic 15%; Foreign 20%. Unsolicited mss not read. Send query letter first with IRCs to cover return postage. No reading fee.

## John Hawkins & Associates, Inc.*

71 West 23rd Street, Suite 1600, New York NY 10010
℡ 001 212 807 7040    🖷 001 212 807 9555
www.jhalit.com
Contacts *John Hawkins, William Reiss, Anne Hawkins, Warren Frazier, Moses Cardona*

Founded 1893. Handles film and TV rights. COMMISSION Apply for rates. No unsolicited mss; welcome queries by email or post (IRCs necessary for response). No reading fee.

## Joy Harris Literary Agency, Inc.*

156 Fifth Avenue, Suite 617, New York NY 10010
℡ 001 212 924 6269    🖷 001 212 924 6609
Contact *Joy Harris*

Handles adult non-fiction and fiction. COMMISSION Home 15%; Foreign 20%. Query letter in the first instance (s.a.s.e. required). No reading fee.

## The Karpfinger Agency

357 West 20th Street, New York NY 10011
℡ 001 212 691 2690    🖷 010 212 691 7129
info@karpfinger.com
www.karpfinger.com
Contact *Barney Karpfinger*

Founded 1985. Handles quality fiction and non-fiction. No scripts, poetry, drama, romance, horror, science fiction, fantasy or children's picture books. See website for submission guidelines; approach by snailmail. No reading fee.

## Kimberley Cameron & Associates
▷ Reece Halsey North Literary Agency

## Lescher & Lescher Ltd*

346 East 84th Street, New York NY 10028
℡ 001 212 396 1999    🖷 001 212 396 1991
Contacts *Robert Lescher, Susan Lescher*

Founded 1964. Handles a broad range of serious non-fiction including current affairs, history, biography, memoirs, politics, law, contemporary issues, popular culture, food and wine, literary and commercial fiction including mysteries and thrillers; some children's books. Specializes in wine books and cookbooks. No poetry, science fiction, New Age, spiritual, romance. COMMISSION Home 15%; Foreign 20%. No unsolicited mss – please query first. No reading fee.

## Linda Chester & Associates*

Rockefeller Center, 630 Fifth Avenue, Suite 2036, New York NY 10111
www.lindachester.com
Contact *Linda Chester*

Founded 1978. Handles literary and commercial fiction and non-fiction in all subjects. No scripts, children's or textbooks. COMMISSION Home & Dramatic 15%; Translation 25%. No unsolicited mss or queries. No reading fee for solicited material.

## Linda Konner Literary Agency*

10 West 15 Street, Suite 1918, New York NY 10011
ldkonner@cs.com
www.lindakonnerliteraryagency.com
President *Linda Konner*

Founded 1996. Handles non-fiction only, specializing in health, self-help, diet/fitness, pop psychology, relationships, parenting, personal finance. Also some pop culture/celebrities. COMMISSION Home 15%; Foreign 25%. Books must be written by or with established experts in their field. No scripts, fiction, poetry or children's books. No unsolicited material; send one-page query with return postage or via email. No reading fee.

## Linda Roghaar Literary Agency, LLC*
133 High Point Drive, Amherst MA 01002
☎ 001 413 256 1921
contact@LindaRoghaar.com
www.lindaroghaar.com
Contact *Linda L. Roghaar*

Founded 1997. Handles fiction and specializes in non-fiction titles. No romance, science fiction or horror. COMMISSION Home 15%; Foreign & Translation rate varies. Send query by email or letter with return postage (for fiction, include the first five pages). No reading fee.

## Loretta Barrett Books, Inc.*
220 East 23rd Street, New York NY 10010
☎ 001 212 242 3420
www.lorettabarrettbooks.com
Contact *Loretta Barrett*

Founded 1990. Handles all non-fiction and fiction genres except children's, poetry, science fiction/fantasy and historical romance. Specializes in women's fiction, history, spirituality. No scripts. COMMISSION Home 15%; Foreign 20%. Send query letter with biography and return postage only in the first instance. Email submissions to query@lorettabarrettbooks.com

## Lowenstein Associates Inc.*
121 West 27th Street, Suite 601, New York NY 10001
☎ 001 212 206 1630    ✆ 001 212 727 0280
www.lowensteinyost.com
Agent *Barbara Lowenstein*

Founded 1976. Handles fiction: upmarket, commercial and multicultural, thrillers, mysteries, women's fiction and young adult. Non-fiction: narrative, business, health, spirituality, psychology, relationships, politics, history, memoirs, natural science, women's issues, social issues, pop culture, personal finance. COMMISSION Home 15%; Foreign & Translation 20%. OVERSEAS ASSOCIATES in all major countries. Fiction: send query letter with short synopsis, first chapter and s.a.s.e. Include any prior literary credits (previously published titles and reviews, writing courses, awards/grants); non-fiction: send query letter, project overview, list of credentials, media appearances, previous titles, reviews and s.a.s.e. No reading fee.

## Malaga Baldi Literary Agency
233 West 99th Street, Suite 19C, New York NY 10025
☎ 001 212 222 3213
baldibooks@gmail.com
Contact *Malaga Baldi*

Founded 1986. Handles quality fiction and non-fiction. No scripts. No westerns, men's adventure, science fiction/fantasy, romance, how-to, young adult or children's. COMMISSION 15%. OVERSEAS ASSOCIATES **Abner Stein**, UK; Owl Agency; Eliane Benisti, France; Marsh Agency. Writers of fiction should send query letter describing the novel plus IRCs. For non-fiction, approach in writing with a proposal, table of contents and two sample chapters. No reading fee.

## Manus & Associates Literary Agency, Inc.*
425 Sherman Avenue, Suite 200, Palo Alto CA 94306
ManusLit@ManusLit.com
www.ManusLit.com
Contacts *Jillian Manus, Jandy Nelson, Penny Nelson, Dena Fischer*

Handles commercial and literary fiction, also young adult and middle grade; non-fiction, including true crime, self-help, memoirs, history, pop culture and popular science. *No* scripts, science fiction/fantasy, westerns, romance, horror, poetry or children's books. COMMISSION Home 15%; Foreign 25%. No unsolicited mss. Fiction: send query letter and first 30 pages; non-fiction: query letter and proposal. Include return postage. No reading fee.

## The Margret McBride Literary Agency*
7744 Fay Avenue, Suite 200, La Jolla CA 92037
☎ 001 858 454 1550    ✆ 001 858 454 2156
staff@mcbridelit.com
www.mcbrideliterary.com
Submissions Manager *Michael Daley*
Assistant to Margret McBride *Faye Atchison*
Vice President/Associate Agent *Donna DeGutis*

Founded 1981. Handles non-fiction mostly, plus business and some fiction. No scripts, science fiction, fantasy, romance or children's books. Specializes in business and management. COMMISSION Home 15%; Overseas 15–25%; Translation 25%. No unsolicited mss. Send query letter with brief synopsis and s.a.s.e. only. Check the submission guidelines on the website before making contact.

## Maria Carvainis Agency, Inc.*
Rockefeller Center, 1270 Avenue of the Americas, Suite 2320, New York NY 10020
☎ 001 212 245 6365    ✆ 001 212 245 7196
mca@mariacarvainisagency.com
President *Maria Carvainis*

Founded 1977. Handles fiction: literary and mainstream, contemporary women's, mystery, suspense, historical, young adult novels; non-fiction: business, women's issues, memoirs, health, biography, medicine. No film scripts. No science fiction. Clients include Mary Balogh, Sandra Brown, Candace Camp, Phillip Depoy, Cindy Gerard, Kristan Higgins, P.J. Parrish, and Will Thomas. COMMISSION Home 15%; Translation 20%. No faxed or emailed queries. No unsolicited mss; they will be returned unread. Queries only, with IRCs for response. No reading fee. Member of AAR.

## Marie Rodell
▷ Francis Collin Literary Agent

## Meredith Bernstein Literary Agency, Inc.*
2095 Broadway, Suite 505, New York NY 10023
☎ 001 212 799 1007   🖷 001 212 799 1145
Contacts *Meredith Bernstein*

Founded 1981. Handles commercial and literary fiction, mysteries and non-fiction (women's and young adult issues, biography, memoirs, health, current affairs, crafts). COMMISSION Home & Dramatic 15%; Translation 20%. OVERSEAS ASSOCIATES **Abner Stein**, UK; Lennart Sane, Holland, Scandinavia and Spanish language; Thomas Schluck, Germany; Bardon Chinese Media Agency; William Miller, Japan; Frederique Porretta, France; Agenzia Letteraria, Italy.

## Mews Books Ltd
c/o Sidney B. Kramer, 20 Bluewater Hill, Westport CT 06880
☎ 001 203 227 1836   🖷 001 203 227 1144
mewsbooks@aol.com (initial contact only; submission by regular mail)
Contacts *Sidney B. Kramer, Fran Pollak*

Founded 1970. Handles adult fiction and non-fiction, children's, pre-school and young adult. No scripts, short stories or novellas (unless by established authors). Specializes in cookery, medical, health and nutrition, scientific non-fiction, children's and young adult. COMMISSION Home 15%; Film & Translation 20%. Unsolicited material welcome. Presentation must be professional and should include brief summary of plot/characters ('avoid a reviewer's point of view when describing plot'), one or two sample chapters, personal credentials and targeted market, all suitable for forwarding to a publisher. Send by regular mail, not email. No reading fee. Requests exclusivity while reading and information if material has been circulated. Charges for photocopying, postage expenses, telephone calls and other direct costs. Principal agent is an attorney and former publisher (a founder of Bantam Books, Corgi Books, London). Offers consultation service through which writers

can get advice on a contract or on publishing problems.

## Michael Larsen/Elizabeth Pomada Literary Agency*
1029 Jones Street, San Francisco CA 94109
☎ 001 415 673 0939
larsenpoma@aol.com
laurie@agentsavant.com
www.larsenpomada.com
www.agentsavant.com
Non-Fiction *Michael Larsen*
Fiction *Elizabeth Pomada*
Genre Fiction/Young Adult *Laurie McLean*

Founded 1972. Handles adult fiction and non-fiction – literary and commercial. No scripts, poetry, children's, science fiction. COMMISSION Home 15%; Foreign 20–30%. ASSOCIATE Laurie McLean (represents romance, science fiction and fantasy and young adult books). OVERSEAS ASSOCIATES **David Grossman Literary Agency Ltd**, British rights; Chandler Crawford (foreign rights). Fiction: send first ten pages with two-page synopsis and s.a.s.e.; non-fiction: email first ten pages with two-page synopsis in the body of the email (no attachments) to larsenpoma@aol.com; genre fiction and YA: email first ten pages and two-page synopsis in the body of the email to laurie@agentsavant.com (consult website for guidelines).

## Michael Snell Literary Agency
PO Box 1206, Truro MA 02666–1206
☎ 001 508 349 3718
snellliteraryagency@yahoo.com
President *Michael Snell*
Vice President *Patricia Smith*

Founded 1978. Adult non-fiction, especially business; leadership; entrepreneurship and psychology, health, parenting, science, relationships and women's issues. Specializes in business (professional and reference to popular trade how-to); general how-to and self-help on all topics, from diet and exercise to parenting, relationships, health, sex, psychology, personal finance, and dogs, cats and horses, plus literary and suspense fiction. COMMISSION Home 15%. No unsolicited mss. Send outline and sample chapter with return postage for reply. No reading fee for outlines. Brochure available on how to write a book proposal. Model proposal also for sale directly to prospective clients. Author of *From Book Idea to Bestseller*, published by Prima. Rewriting, developmental editing, collaborating and ghost-writing services available on a fee basis. Send IRCs.

## Miriam Altshuler Literary Agency*
53 Old Post Road North, Red Hook NY 12571
☎ 001 845 758 9408

www.miriamaltshulerliteraryagency.com/

Founded 1994. Handles literary and commercial fiction; general non-fiction, narrative non-fiction, memoirs, psychology, biography, young adult fiction. No romance, science fiction, mysteries, poetry, westerns, fantasy, how-to, techno thrillers or self-help. COMMISSION Home 15%; Foreign & Translation 20%. Send query letter and synopsis in the first instance with an email address. 'Should we wish to contact you to see material we can no longer respond without an s.a.s.e.' (no email or fax queries). No reading fee.

## Mollie Glick – Foundry Literary + Media*

33 West 17th Street, PH, New York City NY 10011
submissions@foundrymedia.com
www.foundrymedia.com
Literary Agent *Mollie Glick*

Represents literary fiction, narrative non-fiction and some practical non- fiction. Particularly interested in fiction that bridges the literary/commercial divide, combining strong writing with a great plot. Also, non-fiction dealing with popular science, medicine, psychology, cultural history, memoir and current events. Send email query. No reading fee.

## Nancy Love Literary Agency*

250 East 65th Street, Suite 4A, New York NY 10065
☏ 001 215 980 3499    🖷 001 215 308 6405
nloveag@aol.com
Contact *Nancy Love*

Founded 1985. Handles adult non-fiction. Specializes in health, women's issues, politics, foreign affairs and parenting. No fiction, scripts or children's. COMMISSION Home 15%; Foreign 20%. Send query letter or email. No reading fee.

## The Ned Leavitt Agency*

70 Wooster Street, Suite 4F, New York NY 10012
www.nedleavittagency.com
President *Ned Leavitt*
Agent *Britta Steiner Alexander*

Ned Leavitt specializes in creativity, spirituality, health and literary fiction; Britta Alexander specializes in smart non-fiction for 20 and 30-somethings (dating, relationships, business) and commercial fiction. No screenplays or genre fiction. COMMISSION Home 15%. Submissions by recommendation only; see guidelines on the website. No reading fee.

## Patricia Teal Literary Agency*

2036 Vista del Rosa, Fullerton CA 92831
☏ 001 714 738 8333    🖷 001 714 738 8333
Contact *Patricial Teal*

Founded 1978. Handles women's fiction: series and single-title works; commercial and popular

non-fiction. Specializes in romantic fiction. No scripts, science fiction, fantasy, horror, academic texts. COMMISSION Home 15%; Foreign 20%. No unsolicited material. Send query letter with s.a.s.e. in the first instance. No reading fee. 'Accepting queries in fiction from published novelists only.'

## Pema Browne Ltd, Literary Agents

11 Tena Place, Valley Cottage NY 10989
ppbltd@optonline.net
www.pemabrowneltd.com
Contact *Pema Browne*

Founded 1966. ('Pema rhymes with Emma.') Handles mass-market mainstream and hardcover fiction: romance, business, children's picture books and young adult; non-fiction: how-to and reference. COMMISSION Home 20%; Translation 20%; Overseas authors 20%. No unsolicited mss; send query letter with IRCs. No fax or email queries. No longer accepting screenplays. 'We are accepting very few new clients at this time.'

## Peter Lampack Agency, Inc.

551 Fifth Avenue, Suite 1613, New York NY 10176
☏ 001 212 687 9106    🖷 001 212 687 9109
alampack@verizon.net
Agent *Andrew Lampack*
International Rights *Ripsime Dilanyan*

Founded in 1977. Handles commercial fiction: male action and adventure, contemporary relationships, mysteries and suspense, literary fiction; also non-fiction from recognized experts in a given field, plus biographies. Handles theatrical, motion picture and TV rights from book properties. No original scripts or screenplays, series or episodic material. COMMISSION Home & Dramatic 15%; Translation & UK 20%. Cover letter and 1–2 page synopsis via email only. 'We will respond within four weeks and invite the submission of mss which we would like to examine.' No reading fee. No unsolicited mss.

## Philip G. Spitzer Literary Agency, Inc.*

50 Talmage Farm Lane, East Hampton NY 11937
☏ 001 631 329 3650
www.spitzeragency.com/
Contact *Philip Spitzer*

Founded 1969. Works mostly with established/published authors. Specializes in general non-fiction and fiction – thrillers. COMMISSION Home & Dramatic 15%; Foreign 20%. No reading fee for outlines.

## Pinder Lane & Garon-Brooke Associates Ltd*

159 West 53rd Street, Suite 14–C, New York NY 10019
☏ 001 212 489 0880

pinderl@rcn.com
Owner Agents *Dick Duane, Robert Thixton*

Founded 1951. Fiction and non-fiction. No category romance, westerns or mysteries. COMMISSION Home 15%; Dramatic 10–15%; Foreign 30%. No unsolicited mss. First approach by query letter. No reading fee.

## PMA Literary & Film Management, Inc.

PO Box 1817, Old Chelsea Station, New York NY 10113
☎ 001 212 929 1222   📠 001 212 206 0238
peter@pmalitfilm.com
www.pmalitfilm.com
President *Peter Miller*

Founded 1976. Commercial fiction and non-fiction. Specializes in books with motion picture and television potential and in true crime. No poetry, pornography, non-commercial or academic. COMMISSION Home 15%; Dramatic 10–15%; Foreign 20–25%. No unsolicited mss. Approach by query letter (with s.a.s.e.) or, preferred, email (no attachments).

## Quicksilver Books, Literary Agents

508 Central Park Avenue, Suite 5101, Scarsdale NY 10583
☎ 001 914 722 4664   📠 001 914 722 4664
quickbooks@optonline.net
President *Bob Silverstein*

Founded 1973. Handles literary fiction and mainstream commercial fiction: blockbuster, suspense, thriller, contemporary, mystery and historical; and general non-fiction, including self-help, psychology, holistic healing, ecology, environmental, biography, fact crime, New Age, health, nutrition, cookery, enlightened wisdom and spirituality. No scripts, science fiction and fantasy, pornography, children's or romance. COMMISSION Home & Dramatic 15%; Translation 20%. UK material being submitted must have universal appeal for the US market. Unsolicited material welcome but must be accompanied by IRCs for response, together with biographical details, covering letter, etc. No reading fee.

## Raines & Raines*

103 Kenyon Road, Medusa NY 12120
☎ 001 518 239 8311   📠 001 518 239 6029
Contacts *Theron Raines, Joan Raines, Keith Korman*

Founded 1961. Handles general non-fiction. No scripts. COMMISSION Home 15%; Foreign & Translation 20%. No unsolicited material; send one-page letter in the first instance. No reading fee.

## Reece Halsey North Literary Agency/ Kimberley Cameron & Associates*

1550 Tiburon Blvd., #704, Tiburon CA 94920
☎ 001 415 789 9191   📠 001 415 789 9177
info@reecehalseynorth.com (Amy Burkhardt)
www.reecehalseynorth.com
www.kimberleycameron.com
Contacts *Kimberley Cameron, Elizabeth Evans, April Eberhardt*

The Reece Halsey Agency was founded 1957. Aldous Huxley, Upton Sinclair and William Faulkner have been among their clients. Represents literary and mainstream fiction, non-fiction. No scripts, poetry or children's books. COMMISSION Home 15%; Foreign 20%. Send query letter with first 10–50 pages by email. No reading fee.

## Rhoda Weyr Agency, The
▷ Dunham Literary Inc.

## Richard Parks Agency*

PO Box 693, Salem NY 12865
☎ 001 518 854 9466   📠 001 518 854 9466
rp@richardparksagency.com
www.richardparksagency.com
Contact *Richard Parks*

Founded 1989. Handles general trade fiction and non-fiction: literary novels, mysteries and thrillers, commercial fiction, science fiction, biography, pop culture, psychology, self-help, parenting, medical, cooking, gardening, history, etc. No scripts. No technical or academic. COMMISSION Home 15%; UK & Translation 20%. OVERSEAS ASSOCIATES **The Marsh Agency Ltd**; **Barbara Levy Literary Agency**. No unsolicited mss. Fiction read by referral only. Non-fiction query with s.a.s.e. Faxed or email queries will *not* be considered. No reading fee.

## Robert A. Freedman Dramatic Agency, Inc.*

Suite 2310, 1501 Broadway, New York NY 10036
☎ 001 212 840 5760
President *Robert A. Freedman*
Senior Vice-President *Selma Luttinger*
Vice-President *Marta Praeger*

Founded 1928 as Brandt & Brandt Dramatic Department, Inc. Took its present name in 1984. Works mostly with established authors. Send letter of enquiry first with s.a.s.e. Specializes in plays, film and TV scripts. COMMISSION Dramatic 10%. Unsolicited mss not read.

## Rosalie Siegel, International Literary Agent, Inc.*

1 Abey Drive, Pennington NJ 08534
☎ 001 609 737 1007   📠 001 609 737 3708
Contact *Rosalie Siegel*

Founded 1978. A one-woman, highly selective agency that takes on only a limited number of new projects. Handles fiction and non-fiction (especially narrative). Specializes in French history and literature; Europe in general; American history, social history. No science fiction, photography, illustrated art books or children's. COMMISSION Home 15%; Foreign 20%. No unsolicited material.

## The Rosenberg Group*

23 Lincoln Avenue, Marblehead MA 01945
☎ 001 781 990 1341　📠 001 781 990 1344
www.rosenberggroup.com
Contact *Barbara Collins Rosenberg*

Founded 1998. Handles non-fiction (please see website for areas of interest), fiction, specializing in romance (single title and category) and women's; trade non-fiction with a special interest in sports, also handles college textbooks for the first and second year courses. No scripts, science fiction, true crime, inspirational fiction, children's and young adult. COMMISSION Home 15%; Foreign 25%. No unsolicited material; send query letter by post only. No reading fee.

## Russell & Volkening, Inc.*

50 West 29th Street, #7E, New York NY 10001
☎ 001 212 684 6050　📠 001 212 889 3026
Contacts *Tim Seldes, Jesseca Salky (adult fiction & memoirs), Carrie Hannigan (children's & young adult fiction), Joy Azmitia (humour & fiction), Josh Getzler (fiction, thrillers, crime)*

Founded 1940. Handles literary fiction and non-fiction; science, history, current affairs, politics, mystery, thriller, true crime, children's. Specializes in politics and current affairs. No scripts, science fiction/fantasy or very prescriptive non-fiction. COMMISSION Home 15%; Foreign 20%. No unsolicited mss; sample chapters or synopsis with query and covering letter in the first instance (include s.a.s.e.). No reading fee.

## The Sagalyn Literary Agency*

4922 Fairmont Avenue, Suite 200, Bethesda MD 20814
☎ 001 301 718 6440
query@sagalyn.com
www.sagalyn.com

Founded 1980. Handles mostly upmarket nonfiction with some fiction. No screenplays, romance, science fiction/fantasy, children's literature. No unsolicited material. See website for submissions procedure. No reading fee.

## Sandra Dijkstra Literary Agency*

1155 Camino Del Mar, PMB 515, Del Mar CA 92014
☎ 001 858 755 3115　📠 001 858 794 2822
www.dijkstraagency.com

Founded 1981. Handles quality and commercial non-fiction and fiction, including some genre fiction. No scripts. No westerns, science fiction or poetry. Specializes in quality fiction including women's and multicultural fiction, mystery/thrillers, romance, suspense, children's literature, narrative non-fiction, psychology, self-help, science, health, business, memoirs, biography, current affairs and history. 'Dedicated to promoting new and original voices and ideas.' COMMISSION Home 15%; Translation 20%. OVERSEAS ASSOCIATES **Abner Stein**, UK; Agence Hoffman, Germany; Licht & Burr, Scandinavia; Luigi Bernabo, Italy; Sandra Bruna, Spain/Portugal; Caroline Van Gelderen, Netherlands; La Nouvelle Agence, France; The English Agency, Japan; Bardon-Chinese Media Agency, China/Taiwan; Prava I Prevodi, Eastern Europe; Tuttle Mori, Thailand; Synopsis, Russia/Baltic States; Maxima Creative Agency, Indonesia; Graal, Poland. For fiction send brief synopsis (one page) and first 50 pages; for non-fiction send proposal with overview, chapter outline, author biog, 1–2 sample chapters and profile of competition. All submissions should be accompanied by IRCs. No reading fee.

## Sanford J. Greenburger Associates, Inc.*

55 Fifth Avenue, 15th Floor, New York NY 10003
☎ 001 212 206 5600　📠 001 212 463 8718
www.greenburger.com
Contacts *Heide Lange, Faith Hamlin, Daniel Mandel, Matt Bialer, Brenda Bowen, Michael Harriot*

Handles fiction and non-fiction. First approach with query letter, sample chapter and synopsis. No reading fee.

## Schiavone Literary Agency, Inc.

*Corporate Offices:* 236 Trails End, West Palm Beach FL 33413–2135
☎ 001 561 966 9294　📠 001 561 966 9294
profschia@aol.com
www.publishersmarketplace.com/members/profschia
*New York branch office:* 3671 Hudson Manor Terrace, Suite 11H, Bronx, NY 10463
☎/📠 001 718 548 5332
CEO *James Schiavone (Florida office)*
President *Jennifer Duvall (New York; jendu77@aol.com)*
Executive VP *Kevin McAdams (New York; kvn.mcadams@yahoo.com)*

Founded 1996. Handles fiction and non-fiction (all genres). Specializes in biography, autobiography, celebrity memoirs. No poetry. COMMISSION Home 15%; Foreign & Translation 20%. OVERSEAS ASSOCIATES in Europe and Asia. No unsolicited mss; send query letter only with s.a.s.e. and IRCs.

No response without s.a.s.e. For fastest response, email queries of one page (no attachments) are acceptable and encouraged. No reading fee.

## S©ott Treimel NY*

434 Lafayette Street, New York NY 10003
☎ 001 212 505 0664
jmc.st.ny@verizon.net
www.ScottTreimelNY.com
Agent *John M. Cusick*

Founded 1995. Handles children's books only, from concept/board books to teen fiction. No picture book texts or film scripts. COMMISSION Home 15%; Foreign 20%. No unsolicited material. All queries considered via the website.

## Scovil Galen Ghosh Literary Agency, Inc.*

276 Fifth Avenue, Suite 708, New York NY 10001
☎ 001 212 679 8686    🖷 001 212 679 6710
info@sgglit.com
www.sgglit.com
Contacts *Russell Galen, Anna M. Ghosh, Jack Scovil, Ann Behar*

Founded 1993. Handles all categories of books. No scripts. COMMISSION Home 15%; Foreign 20%. No unsolicited material; query by email only; do not send attachments to emails unless invited to do so. No reading fee.

## Sheree Bykofsky Associates, Inc.*

4326 Beach View Blvd., PO Box 706, Brigantine NJ 08203
shereebee@aol.com
www.shereebee.com
Contact *submitbee@aol.com*

Founded 1991. Handles adult fiction and non-fiction. No scripts. No children's, young adult, horror, science fiction, romance, westerns, occult or supernatural. COMMISSION Home 15%; UK (including sub-agent's fee) 20%. No unsolicited mss. Send query letter first with brief synopsis or outline and writing sample (1–3 pp) for fiction. IRCs essential for reply or return of material. 'Please do not send material via methods that require signature, such as FedEx, etc.' No phone calls. See website for submission guidelines. No reading fee.

## Spectrum Literary Agency

320 Central Park West, Suite 1-D, New York NY 10025
☎ 001 212 362 4323    🖷 001 212 362 4562
www.spectrumliteraryagency.com
President and Agent *Eleanor Wood*
Agent *Justin Bell*

Founded 1976. Handles science fiction, fantasy, mystery, suspense and romance. No scripts. Not interested in self-help, New Age, religious fiction/non-fiction, children's books, poetry, short stories, gift books or memoirs. COMMISSION Home 15%; Translation 20%. Send query letter in first instance with s.a.s.e. Faxed or electronic submissions are not accepted. No reading fee.

## Sterling Lord Literistic, Inc.

65 Bleecker Street, New York NY 10012
☎ 001 212 780 6050    🖷 001 212 780 6095
info@sll.com
www.sll.com
Contacts *Philippa Brophy, Chris Calhoun, Laurie Liss*

Founded 1979. Handles all genres, fiction and non-fiction. COMMISSION Home 15%; UK & Translation 20%. No unsolicited mss. Prefers letter outlining all non-fiction. No reading fee.

## Stimola Literary Studio, LLC*

306 Chase Court, Edgewater NJ 07020
☎ 001 201 945 9353
info@stimolaliterarystudio.com
www.stimolaliterarystudio.com
Contact *Rosemary B. Stimola*

Founded 1997. Handles children's books – pre-school through young adult – fiction and non fiction. Specializes in picture books, middle/young adult novels. No adult fiction. COMMISSION Home 15%; Foreign 20%. No unsolicited material; send query email (no attachments). No reading fee.

## Susan Ann Protter Literary Agent*

320 Central Park West, Suite #12E, New York NY 10025
☎ 001 212 362 7920    🖷 001 212 362 4562
sapla@aol.com
www.susanannprotter.com
Contact *Susan Ann Protter*

Founded 1971. Handles general fiction, mysteries, thrillers, science fiction and fantasy; non-fiction: history, general reference, biography, science, health, current affairs and parenting. No romance, poetry, westerns, religious, children's or sport manuals. No scripts. COMMISSION Home & Dramatic 15%. OVERSEAS ASSOCIATES **Abner Stein**, UK; agents in all major markets. No unsolicited queries. By referral only.

## Susan Schulman, A Literary Agency*

454 West 44th Street, New York NY 10036
☎ 001 212 713 1633    🖷 001 212 581 8830
schulman@aol.com
Submissions Editor (books) *Emily Uhry*
Submissions Editor (plays) *Linda Kiss*
Rights & Permissions Editor *Eleanora Tevís*

Specializes in non-fiction of all types but particularly in health and psychology-based self-help for men, women and families. Other interests include business, memoirs, the social sciences, biography, language and international law. Fiction interests include contemporary fiction, including women's, mysteries, historical and thrillers with a cutting edge. 'Always looking for something original and fresh.' Represents properties for film and theatre, and works with agents in appropriate territories for translation rights. COMMISSION Home & Dramatic 15%; Translation 20%. No unsolicited mss. Query first, including outline and three sample chapters with IRCs. No reading fee.

## 2M Communications Ltd*

121 West 27th Street, Suite 601, New York NY 10001
📞 001 212 741 1509    📠 001 212 691 4460
www.2mcommunications.com
Contact *Madeleine Morel*

Founded 1982. Specializes in representing ghostwriters. Represents ghostwriters only. No unsolicited mss.

## Victoria Sanders & Associates LLC*

241 Avenue of the Americas, Suite 11H, New York NY 10014
📞 001 212 633 8811    📠 001 212 633 0525
queriesvsa@hotmail.com
www.victoriasanders.com
Contacts *Victoria Sanders, Diane Dickensheid*

Founded 1992. Handles general trade fiction and non-fiction, plus ancillary film and television rights. COMMISSION Home & Dramatic 15%; Translation 20%. Please send all queries via email.

## Wales Literary Agency, Inc.*

PO Box 9428, Seattle WA 98109–0428
📞 001 206 284 7114
waleslit@waleslit.com
www.waleslit.com
Contacts *Elizabeth Wales, Neal Swain*

Founded 1988. Handles quality fiction and non-fiction. No westerns, romance, self help, science fiction or horror. Special interest in 'Pacific Rim', West Coast and Pacific Northwest stories. COMMISSION Home 15%; Dramatic & Translation/Foreign 20%. No unsolicited mss; send query letter. No email queries longer than one page and no attachments. No reading fee.

## The Wendy Weil Agency, Inc.*

232 Madison Avenue, Suite 1300, New York NY 10016
📞 001 212 685 0030    📠 001 212 685 0765
info@wendyweil.com
www.wendyweil.com
Agents *Wendy Weil, Emily Forland*
Rights *Emma Patterson*

Founded 1987. Handles fiction (commercial and literary), non-fiction (journalism, memoirs, etc.). No scripts, science fiction, romance, visual

books, self-help, cookery. OVERSEAS ASSOCIATES **David Higham Associates Ltd**; Paul & Peter Fritz Agency. Send query letter with synopsis. No reading fee.

## William Clark Associates*

186 Fifth Avenue, 2nd Floor, New York NY 10010
📞 001 212 675 2784
www.wmclark.com
Contact *William Clark*

Founded 1997. Handles non-fiction, mainstream literary fiction and some young adult books. No scripts, horror, science fiction, fantasy, diet or mystery. COMMISSION Home 15%; Foreign & Translation 20%. OVERSEAS ASSOCIATES **Ed Victor Ltd; Andrew Nurnberg Associates Ltd** (translation rights). 'William Clark began accepting representation queries via email in 1993. Since that time the query mailbox address has propagated across the net and receives an overwhelming amount of spam email daily. The agency no longer accepts queries via any other means than the form on our Query Guidelines page at www.wmclark.com/queryguidelines.html and queries sent any other way will be deleted unread. The agency responds to all queries sent according to its guidelines, whether or not it is interested in the work or idea.'

## Writers House, LLC.*

21 West 26th Street, New York NY 10010
📞 001 212 685 2400    📠 001 212 685 1781
www.writershouse.com
Contacts *Albert Zuckerman, Amy Berkower, Merrilee Heifetz, Susan Cohen, Susan Ginsburg, Robin Rue, Simon Lipskar, Steven Malk, Jodi Reamer, Dan Lazar, Rebecca Sherman, Ken Wright, Dan Conaway, Michele Rubin*

Founded 1973. Handles all types of fiction, including children's and young adult, plus narrative non-fiction: history, biography, popular science, pop and rock culture as well as how-to, business and finance, and New Age. Specializes in popular and literary fiction, women's novels, thrillers and children's. No scripts. No professional or scholarly. COMMISSION Home & Dramatic 15%; Foreign 20%. Albert Zuckerman is author of *Writing the Blockbuster Novel*, published by Little, Brown & Co. and Warner Paperbacks. For consideration of unsolicited mss, send a one-page letter of enquiry, 'explaining why your book is wonderful, briefly what it's about and outlining your writing background'. No reading fee.

## Agency Consultants

## Agent Research & Evaluation, Inc. (AR&E)
▷ entry under Agency Consultants (p.259)

# Book Clubs

## Artists' Choice

PO Box 3, Huntingdon PE28 0QX
☏ 01832 710201  🖷 01832 710488
www.artists-choice.co.uk

Specializes in books for the amateur artist at all levels of ability.

## Baker Books

Manfield Park, Cranleigh GU6 8NU
☏ 01483 267888  🖷 01483 267409
enquiries@bakerbooks.co.u
www.bakerbooks.co.uk

Book clubs for schools: Funfare for ages 3–8, Bookzone, ages 9–12 and My Word, ages 13–16. Four issues per year operated in the UK and overseas.

## BCA (Book Club Associates)

Hargreaves Road, Groundwell, Swindon SN25 5BG
☏ 01793 723547
www.booksdirect.co.uk

BCA was acquired by Aureliuis GMBh in January 2009 and is Britain's largest book club organization. Clubs include: Books For Children (BFC), BooksDirect, Fantasy & SF, World of Mystery & Thriller, Railway Book Club, History Guild, Military & Aviation Book Society.

## Bibliophile Books

Unit 5, Datapoint, 6 South Crescent, London E16 4TL
☏ 020 7474 2474  🖷 020 7474 8589
orders@bibliophilebooks.com
www.bibliophilebooks.com

New books covering a wide range of subjects at discount prices. Write, phone, fax or email for free catalogue of 3,000 titles issued 10 times a year.

## Book Club Associates
▷ BCA

## Cygnus Books

PO Box 15, Llandeilo SA19 6YX
☏ 01558 825500  🖷 01558 825517
info@cygnus-books.co.uk
www.cygnus-books.co.uk

Bookseller offering books 'for your next step in spirituality and complementary health care.' Over 1,000 hand-picked titles. Publishes *The Cygnus Review* magazine (free of charge) which features 30–40 reviews on new mind, body and spirit titles each month.

## David Arscott's Sussex Book Club

Dolphin House, 51 St Nicholas Lane, Lewes BN7 2JZ
☏ 01273 470100  🖷 01273 470100
sussexbooks@aol.com
www.pomegranate-press.co.uk

Founded January 1998. Specializes in books about the county of Sussex. Represents all the major publishers of Sussex books and offers a wide range of titles. Free membership without obligation to buy.

## The Folio Society

44 Eagle Street, London WC1R 4FS
☏ 020 7400 4222  🖷 020 7400 4242
enquiries@foliosociety.com
www.foliosociety.com

Fine editions of classic fiction, history and memoirs; also some children's classics.

## Letterbox Library

71–73 Allen Road, London N16 8RY
☏ 020 7503 4801  🖷 020 7503 4800
info@letterboxlibrary.com
www.letterboxlibrary.com

Children's book cooperative. Hard and softcover specialist in multicultural and inclusive books for children from one to 12 years.

## Poetry Book Society
▷ entry under Organizations of Interest to Poets

## Scholastic Book Clubs
▷ Scholastic Ltd under UK Publishers

# Magazines

## Acclaim
> The New Writer

## Accountancy
145 London Road, Kingston Upon Thames
KT2 6SR
☎ 020 8247 1379　📠 020 8247 1424
accountancynews@cch.co.uk
www.accountancymagazine.com
Owner *Wolters Kluwer UK*
Editor *Sally Percy*
Circulation 150,456

Founded 1889. MONTHLY. Written ideas welcome.
FEATURES *Sally Percy* Accounting, tax, business-
related articles of high technical content aimed at
professional/managerial readers. Maximum 2,000
words. 💷 Payment by arrangement.

## Accountancy Age
32–34 Broadwick Street, London W1A 2HG
☎ 020 7316 9000　📠 020 7316 9250
accountancy.age@incisivemedia.com
www.accountancyage.com
Owner *Incisive Media*
Editor-in-Chief *Lem Bingley*
Editor *Gavin Hinks* (020 7316 9608)
Circulation 44,945

Founded 1969. WEEKLY. Unsolicited mss welcome.
Ideas may be suggested in writing provided they
are clearly thought out.
FEATURES *Kevin Reed* Topics right across the
accountancy, business and financial world. Max.
1,400 words. 💷 Payment negotiable.

## Accounting and Business
ACCA, 29 Lincoln's Inn Fields, London WC2A 3EE
☎ 020 7059 5946　📠 020 7059 5771
chris.quick@accaglobal.com
www.accaglobal.com
Owner *ACCA*
Editor *Chris Quick*
Circulation 134,449

Founded 1998. Monthly journal of the Association
of Chartered Certified Accountants (July/
August and November/December are combined
issues). Features on finance and business, world

accountancy news and other matters of interest to
accountants.
FEATURES *Chris Quick* (at email above) UK &
international; *Colette Steckel* (colette.steckel@
accaglobal.com) Asia-Pacific related. Welcomes
ideas by email. Max. 1,500 words. 💷 Approx.
£320 per £1,000 words.

## ACE Tennis Magazine
Seven Squared, Sea Containers House, 20 Upper
Ground, London SE1 9PD
☎ 020 7775 7775
editor@acetennismagazine.com
Owner *LTA*
Editor *Scott Manson*
Circulation 25,000

Founded 1996. MONTHLY specialist tennis
magazine. News and features. No unsolicited
mss; send feature synopses by email in the first
instance. No tournament reports.

## Acoustic
Oyster House Media Ltd, Oyster House, Hunter's
Lodge, Kentisbeare EX15 2DY
☎ 01884 266100　📠 01884 266101
editor@acousticmagazine.com
www.acousticmagazine.com
Owner *Oyster House Media Ltd*
Assistant Editor *Ben Cooper*
Consulting Editor *Mark Tucker*
Circulation 25,000

Launched 2004. MONTHLY. Reviews, news,
features and interviews related to acoustic guitars,
acoustic instruments and acoustic instrument
manufacturing. Approach by email.

## Acumen
> under Poetry Magazines

## Aeroplane
IPC Media Ltd, Blue Fin Building, 110 Southwark
Street, London SE1 0SU
☎ 020 3148 4100
editoraero@ipcmedia.com
www.aeroplanemonthly.com
Owner *IPC Inspire*
Editor *Michael Oakey*

Circulation 35,169

Founded 1973. MONTHLY. Historic aviation and aircraft preservation from its beginnings to the 1960s. No poetry. Will consider news items and features written with authoritative knowledge of the subject, illustrated with good quality photographs.
NEWS *Tony Harmsworth* Max. 500 words.
FEATURES *Nick Stroud* Max. 2,000 words.
Approach by letter or email in the first instance.

£ £60 per 1,000 words; £10–40 per picture used.

## Aesthetica Magazine

PO Box 371, York YO23 1WL
☎ 01904 479168    ℻ 01904 479749
info@aestheticamagazine.com
www.aestheticamagazine.com
Managing Editor *Cherie Federico*
Director *Dale Donley*

Founded 2002. Arts and culture publication which explores the varied nature of the arts and recognizes the dynamics of contemporary culture. 'Bringing a fresh perspective to the national forum, *Aesthetica* is at the forefront of contemporary arts by critically engaging with visual art, film, music, literature and theatre.' Submissions not accepted.

## AIR International

PO Box 100, Stamford PE9 1XQ
☎ 01780 755131    ℻ 01780 757261
airint@keypublishing.com
www.airinternational.com
Owner *Key Publishing Ltd*
Editor *Mark Eyton*
Circulation 25,000

Founded 1971. MONTHLY. Civil and military aircraft magazine. Unsolicited mss welcome but initial approach by phone or in writing preferred.

## AirForces Monthly

PO Box 100, Stamford PE9 1XQ
☎ 01780 755131    ℻ 01780 757261
edafm@keypublishing.com
Owner *Key Publishing Ltd*
Editor *Alan Warnes*
Circulation 20,439

Founded 1988. MONTHLY. Modern military aviation magazine. Unsolicited contributions welcome but initial approach by phone, email or post preferred.

## All About Soap

Hachette Filipacchi (UK) Ltd, 64 North Row, London W1K 7LL
☎ 020 7150 7000    ℻ 020 7150 7681
allaboutsoap@hf-uk.com
www.allaboutsoap.co.uk
Editor *Johnathon Hughes*

Circulation 96,341

Founded 1999. FORTNIGHTLY. Soap opera storylines, television gossip and celebrity features. No unsolicited contributions; make initial contact by phone or email.

## Amateur Gardening

Westover House, West Quay Road, Poole BH15 1JG
☎ 01202 440840    ℻ 01202 440860
Owner *IPC Media (A Time Warner Company)*
Editor *Tim Rumball*
Circulation 35,954

Founded 1884. WEEKLY. Topical and practical gardening articles. News items are compiled and edited in-house, generally. Approach with news stories by email in the first instance.

£ Payment negotiable.

## Amateur Photographer

IPC Media Ltd, Blue Fin Building, 110 Southwark Street, London SE1 0SU
☎ 020 3148 4138
amateurphotographer@ipcmedia.com
Owner *IPC Media (A Time Warner Company)*
Editor *Damien Demolder*
Circulation 21,146

Founded 1884. WEEKLY. For the competent amateur with a technical interest. Freelancers are used but writers should be aware that there is ordinarily no use for words without pictures.

## Angler's Mail

IPC Media Ltd, Blue Fin Building, 110 Southwark Street, London SE1 0SU
☎ 020 3148 5000
anglersmail@ipcmedia.com
www.anglersmail.com
Owner *IPC Media (A Time Warner Company)*
Editor *Tim Knight*
Circulation 34,413

Founded 1965. WEEKLY. Angling news and matches. Interested in pictures, stories, tip-offs and features. Approach the news desk by telephone. £ £10–40 per 200 words; pictures, £20–40.

## Animal Action

Wilberforce Way, Southwater, Horsham RH13 9RS
☎ 0300 123 0392    ℻ 0303 123 0392
publications@rspca.org.uk
www.rspca.org.uk
Owner *RSPCA*
Editor *Sarah Evans*
Circulation 30,655

BI-MONTHLY (SEVEN ISSUES YEARLY). RSPCA youth membership magazine. Articles (pet care, etc.) are written in-house. Good-quality animal photographs welcome.

## Animals and You

D.C. Thomson & Co. Ltd, 2 Albert Square,
Dundee DD1 9QJ
☏ 01382 223131    ℻ 01382 322214
animalsandyou@dcthomson.co.uk
Owner *D.C. Thomson & Co. Ltd*
Editor-in-Chief *Lowaine Wilson (lowilson@ dcthomson.co.uk)*
Circulation 38,503

Founded 1998. THREE-WEEKLY magazine aimed at girls aged 7–11 years who love animals. Contains cute 'n' cuddly pin-ups, pet tips and advice, animal charity information, photo stories, puzzles, readers' pictures and drawings. Most work done in-house but will consider short features, photographic or with good illustrations. Approach by email.

## Antique Collecting
▷ ACC Publishing Group under UK Publishers

## Antiquesnews

PO BOX 3369, Chippenham SN15 9DU
☏ 01225 742240
mail@antiquesnews.co.uk
www.antiquesnews.co.uk
Owner/Editor *Gail McLeod*

Founded 1997. WEEKLY. Up-to-date information online for the British antiques and art trade. News, photographs, gossip and controversial views on all aspects of the fine art and antiques world welcome. Articles on antiques and fine arts themselves are also featured. Approach in writing with ideas.

## Apex Advertiser

PO Box 7086, Clacton-on-Sea CO15 5WN
☏ 01255 428500
mail@apexpublishing.co.uk
www.apexadvertiser.co.uk
Owner *Apex Publishing Ltd*
Editor *Chris Cowlin*
Circulation 12,000

Founded 2002. MONTHLY advertising magazine including guides and information, features, listings and reviews. Approach by post, only.

## Apollo

22 Old Queen Street, London SW1H 9HP
☏ 020 7961 0107
editorial@apollomag.com
www.apollo-magazine.com
Owner *Press Holdings Ltd*
Chief Executive *Andrew Neil*
Editor *Michael Hall*

Founded 1925. MONTHLY. Specialist articles on art and antiques, collectors and collecting, exhibition and book reviews, information on dealers and auction houses. Unsolicited mss welcome but

please make initial contact by email. Interested in new research on fine and decorative arts of all eras, and architecture.

## The Architects' Journal

Greater London House, Hampstead Road, London NW1 7EJ
☏ 020 7728 4574    ℻ 020 7728 4666
firstname.surname@emap.com
www.architectsjournal.co.uk
Owner *Emap*
Editor-in-Chief *Kieran Long*
Circulation 9,088

WEEKLY trade magazine dealing with all aspects of the industry. Approach in writing with ideas or synopses.

## Architectural Design

John Wiley & Sons, 3rd Floor, International House, Ealing Broadway Centre, London W5 5DB
☏ 020 8326 3800    ℻ 020 8326 3801
Owner *John Wiley & Sons Ltd*
Editor *Helen Castle*
Circulation 5,000

Founded 1930. BI-MONTHLY. Sold as a book and journal as well as online, *AD* charts theoretical and topical developments in architecture. Format consists of a minimum of 128pp, the first part dedicated to a theme compiled by a specially commissioned guest editor; the back section (AD+) carries series and more current one-off articles. Unsolicited mss not welcome generally, though journalistic contributions will be considered for the back section.

## The Architectural Review

Greater London House, Hampstead Road, London NW1 7EJ
☏ 020 7728 4589
www.arplus.com
Editor-in-Chief *Kieran Long*
Circulation 14,869

MONTHLY international professional magazine dealing with architecture and all aspects of design. Approach in writing with ideas or synopses.

## The Armourer Magazine

Warners Group Publications, The Maltings, West Street, Bourne PE10 9PH
☏ 01260 278044    ℻ 01260 278044
editor.armourer@btinternet.com
www.armourer.co.uk
Owner *Warners Group Publications*
Editor *Irene Moore*
Circulation 15,000

Founded 1994. BI-MONTHLY. Militaria, military antiques, military history, events and auctions, with an emphasis on the First and Second World

Wars. Unsolicited military-related contributions are welcome. Approach by email in the first instance.

## Art Monthly

4th Floor, 28 Charing Cross Road, London
WC2H 0DB
☎ 020 7240 0389   🖷 020 7497 0726
info@artmonthly.co.uk
www.artmonthly.co.uk
Owner *Britannia Art Publications Ltd*
Editor *Patricia Bickers*
Circulation 6,000

Founded 1976. TEN ISSUES YEARLY. News and features of relevance to those interested in modern and contemporary visual art. Unsolicited mss welcome. Contributions should be addressed to the associate editor, accompanied by s.a.s.e.
FEATURES Always commissioned. Interviews and articles of up to 2,000 words on art theory, individual artists, contemporary art history and issues affecting the arts (e.g. funding and arts education). Exhibition and book reviews: 750–1,000 words.
NEWS Brief reports (250–300 words) on art issues.
£ Payment negotiable.

## The Art Newspaper

70 South Lambeth Road, London SW8 1RL
☎ 020 7735 3331   🖷 020 7735 3332
contact@theartnewspaper.com
www.theartnewspaper.com
www.theartnewspaper.tv
Owner *Umberto Allemandi & Co. Publishing*
Editor *Jane Morris*
Circulation 22,600

Founded 1990. MONTHLY. Tabloid format with hard news on the international art market, news, museums, exhibitions, archaeology, conservation, books and current debate topics. Length 250–2,000 words. No unsolicited mss. Approach with ideas in writing. Commissions only.

## Art Review

1 Sekforde Street, London EC1R 0BE
☎ 020 7107 2760   🖷 020 7107 2761
office@artreview.com
www.artreview.com
Owner *Dennis Hotz*
Editor *Mark Rappolt*
Executive Editor *David Terrien*
Circulation 35,000

Founded 1949. MONTHLY magazine of international contemporary art. Features, profiles, previews and reviews of artists and exhibitions worldwide.
NEWS All items written in-house.

FEATURES *Mark Rappolt* Max. 1,500 words. Email pitch and c.v. at least three months in advance of related show/publication. £ 35p. per word.

## The Artist

Caxton House, 63–65 High Street, Tenterden TN30 6BD
☎ 0158076 3673   🖷 0158076 5411
www.painters-online.co.uk
Owner/Editor *Sally Bulgin*
Circulation 25,000

Founded 1931. MONTHLY. Art journalists, artists, art tutors and writers with a good knowledge of art materials are invited to write to the editor with ideas for practical and informative features about art, techniques and artists.

## Astronomy Now

PO Box 175, Tonbridge TN10 4ZY
☎ 01732 446110   🖷 01732 300148
editorial2010@astronomynow.com
www.astronomynow.com
Owner *Pole Star Publications*
Editor *Keith Cooper*
Circulation 25,000

Founded 1987. MONTHLY astronomy magazine containing the latest astro news, discoveries, images, amateur observing and equipment guides. Unsolicited contributions on amateur observing equipment and astrophotography welcome. 'No pet theories on the universe.' Approach the editor in writing or via email.
FEATURES Max. 1,000/1,500 or 2,000 words.
£ £75 per 500 words.
NEWS All news is compiled in-house.

## Athletics Weekly

PO Box 1026, Peterborough PE1 9GZ
☎ 01733 808550   🖷 01733 808530
results@athletics-weekly.com
www.athleticsweekly.com
Owner *Descartes Publishing*
Editor *Jason Henderson*
Circulation 15,000

Founded 1945. WEEKLY. Covers track and field, road, fell, cross-country, race walking. Results, fixtures, interviews, technical coaching articles and news surrounding build up to 2012 Olympics.
NEWS *Paul Halford* Max. 400 words.
FEATURES *Jason Henderson* Max. 2,000 words. Approach in writing.
£ Payment negotiable.

## Attitude Magazine

207 Old Street, London EC1V 9NR
attitude@attitude.co.uk
www.attitude.co.uk
Owner *Trojan Publishing*
Editor *Matthew Todd*

Circulation 90,000

Launched 1994. MONTHLY. Top selling gay magazine with global cover interview exclusives including Madonna, David Beckham, Heath Ledger. Welcomes feature suggestions and opinion pieces. 'No sex and the city style columns.' Approach by email.

## The Author

84 Drayton Gardens, London SW10 9SB
☎ 020 7373 6642
theauthor@societyofauthors.org
Owner *The Society of Authors*
Editor *Andrew Rosenheim*
Manager *Kate Pool*
Circulation 9,500

Founded 1890. QUARTERLY journal of the **Society of Authors**. Most articles are commissioned.

## Auto Express

30 Cleveland Street, London W1T 4JD
☎ 020 7907 6000   📠 020 7907 6234
editorial@autoexpress.co.uk
www.autoexpress.co.uk
Owner *Dennis Publishing*
Editor *David Johns*
Deputy Editor *Graham Hope*
News and Features Editor *Julie Sinclair*
Circulation 64,114

Founded 1988. WEEKLY consumer motoring title with news, drives, tests, investigations, etc.
NEWS News stories and tip-offs welcome. Fillers 100 words max.; leads 300 words.
FEATURES Ideas welcome. No fully-written articles – features will be commissioned if appropriate. Max. 2,000 words.

## Autocar

Haymarket Consumer Media, Teddington Studios, Broom Road, Teddington TW11 9BE
☎ 020 8267 5630   📠 020 8267 5759
autocar@haymarket.com
www.autocar.co.uk
Owner *Haymarket Consumer Media,*
Editor *Chas Hallett*
News Editor *Dan Stevens*
Circulation 47,371

Founded 1895. WEEKLY. All news stories, features, interviews, scoops, ideas, tip-offs and photographs welcome. 💷 Payment negotiable.

## Aviation News

HPC, Drury Lane, St Leonards on Sea TN38 9BJ
☎ 01424 205537   📠 01424 443693
editor@aviation-news.co.uk
www.aviation-news.co.uk
Owner *Hastings Printing Company*
Editor *David Baker*

Circulation 21,000

Founded 1939. MONTHLY review of aviation for those interested in military, commercial and business aircraft and equipment – old and new. Will consider articles on military and civil aircraft, airports, air forces, current and historical subjects. No fiction.
FEATURES 'New writers always welcome and if the copy is not good enough it is returned with guidance attached.' Max. 5,000 words.
NEWS Items are always considered from new sources. Max. 300 words.
💷 Payment negotiable.

## Balance

Diabetes UK, Macleod House, 10 Parkway, London NW1 7AA
☎ 020 7424 1000   📠 020 7424 1001
balance@diabetes.org.uk
Owner *Diabetes UK*
Editor *Martin Cullen*
Circulation 250,000

Founded 1935. BI-MONTHLY. Unsolicited mss are not accepted. Writers may submit a brief proposal in writing. Only topics relevant to diabetes will be considered.
NEWS Short pieces about activities relating to diabetes and the lifestyle of people with diabetes. Max. 150 words.
FEATURES Medical, diet and lifestyle features written by people with diabetes or with an interest and expert knowledge in the field. General features are mostly based on experience or personal observation. Max. 1,500 words.
💷 NUJ rates.

## The Banker

Number One Southwark Bridge, London SE1 9HL
☎ 020 7873 3000
brian.caplen@ft.com
www.thebanker.com
Owner *FT Business*
Editor *Brian Caplen*
Editor Emeritus *Stephen Timewell*
Circulation 28,022

Founded 1926. MONTHLY. News and features on banking, finance and capital markets worldwide and technology.

## BBC Gardeners' World Magazine

Media Centre, 201 Wood Lane, London W12 7TQ
☎ 020 8433 3959   📠 020 8433 3986
adam.pasco@bbc.com
www.gardenersworld.com
Owner *BBC Worldwide Publishing Ltd*
Editor *Adam Pasco*
Deputy Editor *Lucy Hall*
Commissioning Editor *Kevin Smith*

Circulation 216,936

Founded 1991. MONTHLY. Gardening advice, ideas and inspiration. No unsolicited mss. Approach by phone or in writing with ideas – interested in features about exceptional small gardens. Also interested in any exciting new gardens showing good design and planting ideas as well as new plants and the stories behind them. Includes sections on Grow & Eat, problem solving, wildlife gardening, shopping and buyers' guides, plus practical projects.

## BBC Good Food

Media Centre, 201 Wood Lane, London W12 7TQ
🕾 020 8433 2000
www.bbcgoodfood.com
Owner *BBC Worldwide*
Editor *Gillian Carter*
Circulation 351,430

Founded 1989. MONTHLY recipe magazine featuring celebrity chefs. No unsolicited mss. Features enquiries should go to elaine.stocks@bbc.com

## BBC History Magazine

Tower House, Fairfax Street, Bristol BS1 3BN
🕾 0117 927 9009
historymagazine@bbcmagazinesbristol.com
www.bbchistorymagazine.com
Owner *Bristol Magazines Ltd*
Editor *Dave Musgrove*
Circulation 64,712

Founded 2000. MONTHLY. General news and features on British and international history, with book reviews, listings of history events, TV and radio history programmes and regular features for those interested in history and current affairs. Will consider submissions from academic or otherwise expert historians/archaeologists as well as, occasionally, from historically literate journalists who include expert analysis and historiography with a well-told narrative. Ideas for regular features are welcome. Also publishes cartoons, quizzes and crosswords. 'We cannot guarantee to acknowledge all unsolicited mss.' FEATURES should be pegged to anniversaries or forthcoming books/TV programmes, current affairs topics, etc. 750–3,000 words. NEWS 400–500 words. Send short letter or email with synopsis, giving appropriate sources, pegs for publication dates, etc.
£ Payment negotiable.

## BBC Music Magazine

9th Floor, Tower House, Fairfax Street, Bristol BS1 3BN
🕾 0117 927 9009    🖷 0117 934 9008
music@bbcmagazinesbristol.com
www.bbcmusicmagazine.com

Owner *BBC Worldwide Publishing Ltd*
Editor *Oliver Condy*
Circulation 42,810

Founded 1992. MONTHLY. All areas of classical music. Not interested in unsolicited material. Approach with ideas by email or fax.

## BBC Top Gear Magazine

Second Floor A, Energy Centre, Media Village, 201 Wood Lane, London W12 7TQ
🕾 020 8433 2000
www.topgear.com
Owner *BBC Magazines*
Editorial Director *Conor McNicholas*
Circulation 200,796

Founded 1993. MONTHLY general interest motoring magazine. No unsolicited material as most features are commissioned.

## BBC Who Do You Think You Are? Magazine

9th Floor, Tower House, Fairfax Street, Bristol BS8 3BN
🕾 0117 314 7400
editorial@bbcwhodoyouthinkyouare.com
www.bbcwhodoyouthinkyouare.com
Owner *Bristol Magazines Ltd*
Editor *Sarah Williams*
Circulation 20,266

Founded 2007. THIRTEEN ISSUES YEARLY. Articles relating to family history and social history. Submissions welcomed from experienced family history writers and social and military historians with an understanding of primary records. Send suggestions for articles to the editor.

## BBC Wildlife Magazine

14th Floor, Tower House, Fairfax Street, Bristol BS1 3BN
🕾 0117 927 9009    🖷 0117 927 9008
wildlifemagazine@bbcmagazines.com
www.bbcwildlifemagazine.com
Owner *Bristol Magazines Ltd*
Editor *Sophie Stafford*
Features Editor *Ben Hoare*
News & Travel Editor *James Fair*
Circulation 44,560

Founded 1963. Formerly *Wildlife*; became *BBC Wildlife Magazine* in 1983. THIRTEEN ISSUES YEARLY. Welcomes idea pitches but not unsolicited mss. Read guidelines on website to improve chances of success.
FEATURES Most features commissioned from excellent writers with expert knowledge of wildlife or conservation subjects. Max. 2,000 words.
NEWS Most news stories commissioned from known freelancers. Max. 400 words.
£ Features, £200–450; news, £80–120.

## Bee World
> Journal of Apicultural Research

## Bella

H. Bauer Publishing, Academic House, 24–28 Oval Road, London NW1 7DT

℡ 020 7241 8000    🅕 020 7241 8056

Owner *H. Bauer Publishing*
Editor *Julia Davis*
Circulation 253,001

Founded 1987. WEEKLY. Women's magazine specializing in real-life, human interest stories plus celebrity, fashion, beauty, home and health. No fiction. Does not accept unsolicited submissions.

## Best

33 Broadwick Street, London W1F 0DQ

℡ 020 7339 4500    🅕 020 7339 4580

Owner *National Magazine Company*
Editor *Jackie Hatton*
Circulation 301,440

Founded 1987. WEEKLY women's magazine. Multiple features, news, short stories on all topics of interest to women. No unsolicited material.

## Best of British

Church Lane Publishing Ltd, Bank Chambers, 6 Market Place, Market Deeping PE6 8DL

℡ 01778 342814
info@bestofbritishmag.co.uk
www.bestofbritishmag.co.uk

Owner *Church Lane Publishing Ltd*
Editor-in-Chief *Ian Beacham*
Editor *Linne Matthews*

Founded 1994. MONTHLY magazine celebrating all things British, both past and present. Emphasis on nostalgia – memories from the 1930s, 1940s, 1950s, 1960s and 1970s. Study of the magazine is advised in the first instance. All preliminary approaches should be made in writing or by email.

## The Big Issue and The Big Issue South West

1–5 Wandsworth Road, London SW8 2LN

℡ 020 7526 3200    🅕 020 7526 3201

helena.drakakis@bigissue.com
www.bigissue.com

Editor-in-Chief *A. John Bird*
Editor *Charles Howgego*
Associate Editor *Helena Drakakis*
 Circulation 75,027; 15,011 (South West); 146,723 (group)

Founded 1991. WEEKLY. An award-winning campaigning and street-wise general interest magazine sold in London, the Midlands, the North East and South of England. Separate regional editions sold in Manchester, Scotland, Wales and the South West.

NEWS *Helena Drakakis* Hard-hitting exclusive stories with emphasis on social injustice aimed at national leaders.

ARTS *Charles Howgego* Interested in interviews and analysis ideas. Reviews written in-house. Send synopses to arts editor.

FEATURES Interviews, campaigns, comment, opinion and social issues reflecting a varied and informed audience. Balance includes social issues but mixed with arts and cultural features. Freelance writers used each week commissioned from a variety of contributors. Best approach in the first instance is to email or post synopses to Editor, *Charles Howgego* with examples of work. Max. 1,500 words.

## The Big Issue Cymru

55 Charles Street, Cardiff CF10 2GD

℡ 029 2025 5670    🅕 029 2025 5672

alex.donohue@bigissuecymru.co.uk
www.bigissuecymru.co.uk

Editor *Paul McNamee*
Circulation 9,250

Founded 1993. WEEKLY. Hard-hitting social and news features plus arts and entertainment.

FEATURES/NEWS *Paul McNamee* Welcomes contributions but must be relevant to a Welsh readership. Approach by email. Lead in of four to six weeks. Max. 1,000 words (features), 350 (news). No poetry or fiction.

## The Big Issue in Scotland

2nd Floor, 43 Bath Street, Glasgow G2 1HW

℡ 0141 352 7260    🅕 0141 333 9049

editorial@bigissuescotland.com
www.bigissuescotland.com

Editor *Paul McNamee*
Circulation 21,846

WEEKLY. Coverage of social issues, arts and culture, sport, interviews; news and hard-hitting journalism.

## The Big Issue in the North

10 Swan Street, Manchester M4 5JN

℡ 0161 831 5550    🅕 0161 831 5577

kevin.gopal@bigissuenorth.co.uk
www.thebiglifecompany.com

Owner *The Big Life Company*
Editor *Kevin Gopal*
Circulation 25,589

Launched 1992. WEEKLY. Entertainment, celebrity interviews, political and environmental issues and issues from the social sector as well as homelessness. Welcomes news stories not covered by the national press; approach by email.

## The Big Issue South West

Founded 1994. WEEKLY. Hard-hitting, socially aware news and features, local arts and entertainment. Welcomes news features with a regional relevance. No poetry or fiction. Editorial dealt with by London office – see entry for *The Big Issue*.

## Bike

Bauer Automotive Ltd, Media House, Lynch Wood, Peterborough PE2 6EA
☎ 01733 468099    ℻ 01733 468290
bike@bauermedia.co.uk
www.bikemagazine.co.uk
Owner *Bauer Automotive Ltd*
Editor *Tim Thompson*
Circulation 65,147

Founded 1971. MONTHLY. Broad-based motorcycle magazine. Approach by email.

## Bird Life Magazine

The RSPB, UK Headquarters, The Lodge, Sandy SG19 2DL
☎ 01767 680551    ℻ 01767 683262
derek.niemann@rspb.org.uk
Owner *Royal Society for the Protection of Birds*
Editor *Derek Niemann*
Circulation 90,000

Founded 1965. BI-MONTHLY. Bird, wildlife and nature conservation for 8–12-year-olds (RSPB Wildlife Explorers members). No unsolicited mss. No 'captive/animal welfare' articles.
FEATURES Unsolicited material rarely used.
NEWS News releases welcome. Approach in writing in the first instance.

## Bird Watching

Media House, Lynchwood, Peterborough PE2 6EA
☎ 01733 468000    ℻ 01733 468387
birdwatching@bauermedia.co.uk
www.birdwatching.co.uk
Editor *Sheena Harvey*
Circulation 17,663

Founded 1986. MONTHLY magazine for birdwatchers at all levels of experience. Offers practical advice of where and when to see birds, how to identify them, product reviews (particularly binoculars, telescopes and digiscoping equipment), extensive bird sightings, 'Go Birding' walks (10 each issue), news and general articles.
FEATURES Interested in authoratative articles on bird behaviour and bird watching/sites in the UK. Please send synopsis, not the whole piece. Max. 1,000 words.
NEWS UK news pre-eminent, preferably with pictures. Max. 350 words.
£ Fees negotiable.

## Birds

The RSPB, UK Headquarters, The Lodge, Sandy SG19 2DL
☎ 01767 680551    ℻ 01767 683262
birdsmagazine@rspb.org.uk
www.rspb.org.uk
Owner *Royal Society for the Protection of Birds*
Editor *Sarah Brennan*
Circulation 622,322

QUARTERLY magazine which covers not only wild birds but also other wildlife and related conservation topics. No interest in features on pet birds or 'rescued' sick/injured/orphaned ones. Content refers mostly to RSPB work so opportunities for freelance work on other subjects are limited but some freelance submissions are used in most issues. Phone or email to discuss.

## Birdwatch

The Chocolate Factory, 5 Clarendon Road, London N22 6XJ
☎ 020 8881 0550    ℻ 020 8881 0990
editorial@birdwatch.co.uk
www.birdwatch.co.uk
Owner *Solo Publishing*
Editor *Dominic Mitchell*
Circulation 14,000

Founded 1992. MONTHLY magazine featuring illustrated articles on all aspects of birds and birdwatching, especially in Britain. No unsolicited mss. Approach in writing with synopsis of 100 words max. Annual **Birdwatch Bird Book of the Year** award (see entry under *Prizes*).
FEATURES Unusual angles/personal accounts, if well-written. Articles of an educative or practical nature suited to the readership. Max. 2,000 words.
FICTION Very little opportunity although occasional short story published. Max. 1,500 words.
NEWS Very rarely use external material.

## Bizarre

30 Cleveland Street, London W1T 4JD
☎ 020 7907 6000
bizarre@dennis.co.uk
www.bizarremag.com
Owner *Dennis Publishing*
Editor *David McComb*
Deputy Editor *Kate Hodges*
Circulation 28,979

Founded 1997. THIRTEEN ISSUES YEARLY. Amazing stories and images from around the world. No fiction, poetry, illustrations, short snippets.
FEATURES *Kate Hodges* Send synopsis by post or email.
£ Payment negotiable.

## Black Beauty & Hair

2nd Floor, Culvert House, Culvert Road, London SW11 5DH
☎ 020 7720 2108    ℻ 020 7498 3023

info@blackbeautyandhair.com
www.blackbeautyandhair.com
Owner *Hawker Consumer Publications Ltd*
Publisher *Pat Petker (ext. 207)*
Editor *Irene Shelley (ext. 215)*
Circulation 19,697

BI-MONTHLY with one annual special: *The Hairstyle Book* in October, and a *Bridal Supplement* in the April/May issue. Black hair and beauty magazine with emphasis on authoritative articles relating to hair, beauty, fashion, health and lifestyle. Unsolicited contributions welcome. FEATURES Beauty and fashion pieces welcome from writers with a sound knowledge of the Afro-Caribbean beauty scene plus bridal features. Minimum 1,000 words.
£ £100 per 1,000 words.

## Black Static

TTA Press, 5 Martins Lane, Witcham, Ely CB6 2LB
blackstatic@ttapress.com
www.ttapress.com
Owner *TTA Press*
Editor *Andy Cox*

Formerly *The Third Alternative*, founded 1993. BIMONTHLY colour horror magazine: short fiction, news and gossip, film and television, manga and anime books. Publishes talented newcomers alongside famous authors. Unsolicited mss welcome if accompanied by s.a.s.e. or email address for overseas submissions (max. 8,000 words). Fiction submissions as hard copy only; feature suggestions can be emailed. Potential contributors are advised to study the magazine. Contracts are exchanged and payment made upon acceptance. Winner of British Fantasy Awards, International Horror Guild Award.

## Bliss Magazine

Panini UK Ltd, Brockbourne House, 77 Mount Ephraim, Tunbridge Wells TN4 8BS
☎ 01892 500100   📠 01892 545666
bliss@panini.co.uk
www.mybliss.co.uk
Owner *Panini UK Ltd*
Features Editor *Angeli Milburn*
Circulation 88,801

Founded 1995. MONTHLY teenage lifestyle magazine for girls. No unsolicited mss; email the Features Editor (amilburn@panini.co.uk) with an idea and then follow up with a call.
NEWS Worldwide teenage news. Max. 200 words.
FEATURES Real life teenage stories with subjects willing to be photographed. Reports on teenage issues. Max. 1,500 words.

## Blueprint

Progressive Media Publishing, John Carpenter House, John Carpenter Street, London EC4Y 0AN
☎ 020 7936 6400   📠 020 7936 6813

vrichardson@blueprintmagazine.co.uk
www.blueprintmagazine.co.uk
Owner *Progressive Media Publishing*
Editor *Vicky Richardson*
Associate Editor *Tim Abrahams*
Assistant Editor *Peter Kelly*
Circulation 6,453

Founded 1983. MONTHLY. 'Crossing the boundaries of design and architecture (from graphics and product design to interiors and urban development), Blueprint examines their impact on society in general.' Features, news and reviews. Welcomes ideas and pitches for articles on topics relevant to the magazine; no completed features. Approach by email.

## Book & Magazine Collector

Warners Group Publications, The Chocolate Factory, 5 Clarendon Road, London N22 6XJ
☎ 020 8889 0583   📠 020 8889 0990
editor@bookandmagazinecollector.com
www.bookandmagazinecollector.com
Owner *Warners Group Publications*
Editor *Chris Peachment*
Circulation 10,000

Founded 1984. THIRTEEN ISSUES YEARLY. News and reviews relating to the hobby of book and magazine collecting. Over 40 pages of for-sale or wanted books (free listings for subscribers). Welcomes items on famous authors and their works and articles on collecting; max. 2,500 words. Contact the editor via email.

## The Book Collector

PO Box 12426, London W11 3GW
☎ 020 7792 3492   📠 020 7792 3492
editor@thebookcollector.co.uk (editorial)
www.thebookcollector.co.uk
Owner *The Collector Ltd*
Editor *Nicolas J. Barker*

Founded 1950. QUARTERLY magazine on bibliography and the history of books, book-collecting, libraries and the book trade.

## Book World Magazine

2 Caversham Street, London SW3 4AH
☎ 020 7351 4995   📠 020 7351 4995
leonard.holdsworth@btinternet.com
Owner *Christchurch Publishers Ltd*
Editor *James Hughes*
Circulation 5,500

Founded 1980. MONTHLY news and reviews for serious book collectors, librarians, antiquarian and other booksellers. No unsolicited mss. Interested in material relevant to literature, art and book collecting. Send letter in the first instance.

## The Bookseller

5th Floor, Endeavour House, 189 Shaftesbury Avenue, London WC2H 8TJ

☎ 020 7420 6006　📠 020 7420 6103

www.theBookseller.com

Owner *Nielsen Business Media*

Editor-in-Chief *Neill Denny*

Trade journal of the book trade, publishing, retail and libraries – the essential guide to the book business since 1858. Trade news and features, including special features, company news, publishing trends, bestseller data, etc. Unsolicited mss rarely used as most writing is either done in-house or commissioned from experts within the trade. Approach in writing first.

## Boxing Monthly

40 Morpeth Road, London E9 7LD

☎ 020 8986 4141　📠 020 8986 4145

mail@boxing-monthly.co.uk

www.boxing-monthly.co.uk

Owner *Topwave Ltd*

Editor *Glyn Leach*

Circulation 25,000

Founded 1989. MONTHLY. International coverage of professional boxing; previews, reports and interviews. Unsolicited material welcome. Interested in small hall shows and grass-roots knowledge. No big fight reports. Approach in writing in the first instance.

## Boyz

18 Brewer Street, London W1F 0SH

☎ 020 7025 6100　📠 020 7025 6109

davidb@boyz.co.uk

Managing Director *David Bridle*

Editor *Stuart Brumfitt*

Circulation 55,000

Founded 1991. WEEKLY entertainment and features magazine aimed at a gay readership covering clubs, fashion, TV, films, music, theatre, celebrities and the London gay scene in general. Unsolicited mss welcome.

## Brand Literary Magazine

King William Court, University of Greenwich, Maritime Campus, Park Row, London SE10 9LS

☎ 020 8331 8952/8940

info@brandliterarymagazine.co.uk

www.brandliterarymagazine.co.uk

Editor-in-Chief *Nina Rapi*

Circulation 1,000

BIANNUAL literary magazine, launched in 2007, focusing on the short form; left of field work and international writing. Welcomes stories, plays, poems and creative non-fiction; no genre or historical material. Approach by post; see submission guidelines on the website.

FICTION *Nina Rapi* 'Looking for writing that takes risks and has a strong voice.' Max. 2,500 words.

POETRY *Cherry Smyth* Max. 40 lines.

PLAYS *Nina Rapi, Harry Derbyshire* (Advisor) Bold, short plays and performance text. Max. ten minutes' performance time.

💷 Two free copies of the magazine.

## Brides

Vogue House, Hanover Square, London W1S 1JU

☎ 020 7499 9080

bridseditor@condenast.co.uk

www.bridesmagazine.co.uk

Owner *Condé Nast Publications Ltd*

Editor *Deborah Joseph*

Circulation 68,586

Founded 1955. BI-MONTHLY. Much of the magazine is produced in-house, but good, relevant emotional and wedding planning features always welcome. Max. 1,000 words. Prospective contributors should make initial contact by email.

## Britain

The Chelsea Magazine Co. Ltd, 30 Old Church Street, London SW3 5BY

☎ 020 7349 3150　📠 020 7349 3160

andrea.spain@britain-magazine.com

Editor *Andrea Spain*

Circulation 27,882

Founded in the 1930s. BI-MONTHLY. Travel magazine of 'VisitBritain'. Not much opportunity for unsolicited work – approach with ideas and samples (by email only). Also interested in article ideas with good quality digital images or transparencies.

## British Birds

4 Harlequin Gardens, St Leonards-on-Sea TN37 7PF

☎ 01424 755155　📠 01424 755155

editor@britishbirds.co.uk

Editor *Dr R. Riddington*

Circulation 6,000

Founded 1907. MONTHLY ornithological journal. Features main papers on topics such as behaviour, distribution, ecology, identification and taxonomy; annual *Report on Rare Birds in Great Britain* and *Rare Breeding Birds in the UK*; sponsored competition for Bird Photograph of the Year. Unsolicited mss welcome from ornithologists.

FEATURES Well-researched, original material relating to Western Palearctic birds welcome.

NEWS *Adrian Pitches* (adrian.pitches@ blueyonder.co.uk) Items ranging from conservation to humour. Max. 200 words.

💷 Payment for photographs, drawings, paintings and main papers.

## British Chess Magazine

The Chess Shop, 44 Baker Street, London
W1U 7RT
☏ 020 7486 8222   🖷 020 7486 3355
editor@bcmchess.co.uk
www.bcmchess.co.uk
Director/Editor *John Saunders*

Founded 1881. MONTHLY. Emphasis on
tournaments, the history of chess and chess-
related literature. Approach in writing with
ideas. Unsolicited mss not welcome unless from
qualified chess experts and players.

## British Journalism Review

Sage Publications Ltd, 1 Oliver's Yard, 55 City
Road, London EC1Y 1SP
☏ 020 7324 8500   🖷 020 7324 8756
bill.hagerty@btinternet.com
www.bjr.org.uk
www.bjr.sagepub.com
Owner *BJR Publishing Ltd*
Editor *Bill Hagerty*

Founded 1989. QUARTERLY. Media magazine
providing a forum for the news media. Freelance
contributions welcome: non-academic articles on
all aspects of the news media only. No academic
papers. Approach by email. Max. 2,500 words.

## British Medical Journal

BMA House, Tavistock Square, London WC1H 9JR
☏ 020 7387 4499   🖷 020 7383 6418
editor@bmj.com
www.bmj.com
Owner *British Medical Association*
Editor *Dr Fiona Godlee*
Circulation 120,071

One of the world's leading general medical
journals.

## British Philatelic Bulletin

35–50 Rathbone Place, London W1T 1HQ
☏ 020 7441 4744
www.royalmail.com/stamps
Owner *Royal Mail*
Editor *John Holman*
Circulation 15,000

Founded 1963. MONTHLY bulletin giving details
of forthcoming British stamps, features on older
stamps and postal history, and book reviews.
Welcomes photographs of interesting, unusual or
historic letter boxes.
FEATURES Articles on all aspects of British
philately. Max. 1,500 words.
NEWS Reports on exhibitions and philatelic
events. Max. 500 words. Approach in writing in
the first instance.
💷 £75 per 1,000 words.

## British Railway Modelling

The Maltings, West Street, Bourne PE10 9PH
☏ 01778 391027   🖷 01778 425437
johne@warnersgroup.co.uk
www.brmodelling.co.uk
Owner *Warners Group Publications Plc*
Editor *John Emerson*
Assistant Editor *Tony Wright*
Circulation 26,000

Founded 1993. MONTHLY. A general magazine
for the practising modeller. Ideas are welcome.
Interested in features on quality models, from
individual items to complete layouts. Approach in
writing first.
FEATURES Articles on practical elements of
the hobby, e.g. locomotive construction, kit
conversions, etc. Layout features and articles on
individual items which represent high standards
of the railway modelling art. Max. length 6,000
words (single feature).
NEWS News and reviews containing the model
railway trade, new products, etc. Max. length
1,000 words.

## Broadcast

Greater London House, Hampstead Road,
London NW1 7EJ
☏ 020 7728 5000
www.broadcastnow.co.uk
Owner *Emap Ltd*
Editor *Lisa Campbell*
Deputy Editor *Emily Booth*
Circulation 8,082

Founded 1959. WEEKLY. Some opportunities for
freelance contributions. Write to the relevant
editor in the first instance.
FEATURES *Emily Booth* Any broadcasting issue.
NEWS *Chris Curtis* Broadcasting news.
💷 £270 per 1,000 words.

## Build It

Ocean Media Group Ltd, 9th Floor, 1 Canada
Square, Canary Wharf, London E14 5AP
☏ 020 7772 8452   🖷 020 7772 8599
buildit@oceanmedia.co.uk
www.buildit-online.co.uk
Owner *Ocean Media*
Acting Editor *Duncan Hayes*
Circulation 12,927

Founded 1990. MONTHLY magazine covering
self-build, conversion and renovation. Unsolicited
material welcome on self-build case studies
as well as articles on technical construction,
architecture and design, home improvements and
dealing with builders. Max. length 2,500 words.
Approach by post or email.

## Building Magazine

3rd Floor, Ludgate House, 245 Blackfriars Road,
London SE1 9UY
☏ 020 7560 4141    ⊞ 020 7560 4080
building@ubm.co.uk
www.building.co.uk
Owner *UBM Information*
Editor *Denise Chevin*
Circulation 24,241

WEEKLY magazine for the whole construction
industry, from architects and engineers to
contractors and house builders. News and
features. No unsolicited contributions as all
material is commissioned, although legal items
welcome.

## The Burlington Magazine

14–16 Duke's Road, London WC1H 9SZ
☏ 020 7388 1228    ⊞ 020 7388 1230
editorial@burlington.org.uk
www.burlington.org.uk
Owner *The Burlington Magazine Publications Ltd*
Managing Director *Kate Trevelyan*
Editor *Richard Shone*
Deputy Editor *Bart Cornelis*
Associate Editor *Jane Martineau*

Founded 1903. MONTHLY. Unsolicited
contributions welcome on the subject of art
history provided they are previously unpublished.
All preliminary approaches should be made in
writing.
EXHIBITION & BOOK REVIEWS Usually
commissioned, but occasionally unsolicited
reviews are published if appropriate. Max. 1,000
words. ⓔ Max. £140.
ARTICLES Max. 4,500 words. ⓔ Max. £150.
SHORTER NOTICES Max. 2,000 words. ⓔ Max.
£80.

## Buses

PO Box 14644, Leven KY9 1WX
☏ 01333 340637    ⊞ 01333 340608
buseseditor@btconnect.com
www.busesmag.com
Owner *Ian Allan Publishing*
Editor *Alan Millar*
Circulation 13,000

Founded 1949. MONTHLY magazine of the bus and
coach industry; vehicles, companies and people.
Specialist articles welcome; no general articles
with no specific bus content. Email enquiry
welcome.

## Business Traveller

2nd Floor, Cardinal House, Albemarle Street,
London W1S 4TE
☏ 020 7647 6330    ⊞ 020 7647 6331
editorial@businesstraveller.com
www.businesstraveller.com

Owner *Panacea Publishing*
Editorial Director *Tom Otley*
Circulation 54,969

MONTHLY consumer publication. Opportunities
exist for freelance writers. Would-be contributors
are strongly advised to study the magazine or the
website first. Approach in writing with ideas.
ⓔ Payment varies.

## Buzz Extra

IBRA, 16 North Road, Cardiff CF10 3DY
☏ 029 2037 2409    ⊞ 05601 135640
mail@ibra.org.uk
www.ibra.org.uk
Owner *International Bee Research Association*
Editor *Richard Jones*

Informative, full-colour newsletter of the
International Bee Research Association.
QUARTERLY (March, June, September and
December).

## Cage & Aviary Birds

IPC Media Ltd, Blue Fin Building, 110 Southwark
Street, London SE1 0SU
☏ 020 3148 4171    ⊞ 020 3148 8129
birds@ipcmedia.com
Owner *IPC Media (A Time Warner Company)*
Editor *Kim Forrester*
Circulation 13,918

Launched 1902. WEEKLY. 'Everything you need
to know to keep, breed and show captive birds,
including news and opinion.' No wild birds, bird
watching, holidays/travel trips about birds.
FEATURES Happy to consider features based on
first-hand experience of keeping birds or profiles
of successful birdkeepers. Approach by email.
Max. 1,200 words. ⓔ Payment varies but from
£40 per 1,000 words.

## The Cambridgeshire Journal Magazine

Cambridge Newspapers Ltd, Winship Road,
Milton CB24 6PP
☏ 01223 434419
Owner *Cambridge Newspapers Ltd*
Editor *Alice Ryan*
Circulation 11,000

Founded 1990. MONTHLY magazine covering
items of local interest – history, people,
conservation, business, places, food and wine,
fashion, homes and gardens.
FEATURES 750–1,500 words max., plus pictures.
Approach the editor with ideas in the first
instance.

## Campaign

174 Hammersmith Road, London W6 7JP
☏ 020 8267 4683    ⊞ 020 8267 4927
campaign@haymarket.com
www.campaignlive.co.uk

Owner *Haymarket Media Group*
Editor *Claire Beale*
Circulation 9,005

Founded 1968. WEEKLY. Lively magazine serving the advertising and related industries. Freelance contributors are best advised to write in the first instance.
FEATURES Ideas welcome.
NEWS No submissions as all news handled in-house.
£ Payment negotiable.

## Camping and Caravanning

Greenfields House, Westwood Way, Coventry CV4 8JH
☎ 0845 130 7631    ℻ 024 7647 5413
www.campingandcaravanningclub.co.uk
Owner *The Camping and Caravanning Club*
Features Editor *Sue Taylor*
Circulation 237,168
Founded 1901. MONTHLY. Interested in journalists with camping and caravanning knowledge. Write with ideas for features in the first instance.
FEATURES Outdoor pieces in general, plus items on specific regions of Britain. Max. 1,200 words. Illustrations to support text essential.

## Canals and Rivers

PO Box 618, Norwich NR7 0QT
☎ 01603 708930
chris@themag.fsnet.co.uk
www.canalsandrivers.co.uk
Owner *A.E. Morgan Publications Ltd*
Editor *Chris Cattrall*
Circulation 12,000

Covers all aspects of waterways, narrow boats and cruisers. Contributions welcome. Make initial approach by post or email.
FEATURES Waterways, narrow boats and motor cruisers, cruising reports, practical advice, etc. Unusual ideas and personal comments are particularly welcome. Max. 2,000 words. Articles should be supplied in PC Windows format disk or emailed with JPEG images.
NEWS Items of up to 200 words welcome on the Inland Waterways System, plus photographs if possible.
£ Features, around £60 per page; news, £20.

## Candis

Newhall Lane, Hoylake CH47 4BQ
☎ 0151 632 3232    ℻ 0844 545 8103
www.candis.co.uk
Owner *Newhall Publications*
Editor *Debbie Atwell*
Circulation 263,754
MONTHLY. General features on health and family lifestyle including makeover and celebrity. No unsolicited contributions as most features are

individually commissioned. Ideas for features should be emailed to helen@candis.co.uk
FICTION Requests for fiction guidelines can be sent to fiction@candis.co.uk (2,000 words, plus or minus 10%, only).

## Car Mechanics

Kelsey Publishing Group, Cudham Tithe Barn, Berry's Hill, Cudham TN16 3AG
☎ 01959 541444
cm.ed@kelsey.co.uk
www.carmechanicsmag.co.uk
Editor *Martyn Knowles*
Circulation 32,000

MONTHLY. Practical guide to maintenance and repair of post-1983 cars for DIY and the motor trade. Initial approach by email, letter or phone strongly recommended, 'but please study a recent copy for style and content first'.
FEATURES Good, entertaining and well-researched technical material welcome, especially anything presenting complex matters clearly and simply.
£ Payment negotiable 'but generous for the right material. We normally require words and pictures and can rarely use words alone'.

## Caravan Magazine

IPC Media Ltd, Leon House, 233 High Street, Croydon CR9 1HZ
☎ 020 8726 8000    ℻ 020 8726 8299
caravan@ipcmedia.com
www.caravanmagazine.co.uk
Owner *IPC Media (A Time Warner Company)*
Editor *Victoria Bentley*
Circulation 16,057

Founded 1933. MONTHLY. Unsolicited mss welcome. Approach in writing with ideas. All correspondence should go direct to the editor.
FEATURES Touring with strong caravan bias, technical/DIY features and how-to section. Max. 1,500 words.
£ Payment by arrangement.

## Cat World

Ancient Lights, 19 River Road, Arundel BN18 9EY
☎ 01903 884988    ℻ 01903 885514
laura@ashdown.co.uk
www.catworld.co.uk
Owner *Ashdown Publishing Ltd*
Editor *Laura Quiggan*

Founded 1981. MONTHLY. Unsolicited mss welcome but initial approach in writing preferred. No poems or fiction.
FEATURES Lively, first-hand experience features on every aspect of the cat. Breeding features and veterinary articles by acknowledged experts only. Preferred length 750 or 1,700 words. Accompanying pictures should be good

quality and sharp. Submissions by email (MS Word attachment) if possible, or by disk with accompanying hard copy and s.a.s.e. for return.

## The Catholic Herald

Lamb's Passage, Bunhill Row, London EC1Y 8TQ
☏ 020 7448 3603   🖷 020 7256 9728
editorial@catholicherald.co.uk
www.catholicherald.co.uk
Editor *Luke Coppen*
Features Editor *Ed West*
Literary Editor *Stav Sherez*
Circulation 22,000

WEEKLY. Interested mainly in straight Catholic issues but also in general humanitarian matters, social policies, the Third World, the arts and books. £ Payment by arrangement.

## Celebs on Sunday

▷ Sunday Mirror under National Newspapers

## Chapman

4 Broughton Place, Edinburgh EH1 3RX
☏ 0131 557 2207
chapman-pub@blueyonder.co.uk
www.chapman-pub.co.uk
Owner/Editor *Dr Joy Hendry*
Circulation 2,000

Founded 1970. Scotland's quality literary magazine. Features poetry, short works of fiction, criticism, reviews and articles on theatre, politics, language and the arts. Gives priority to Scottish writers but also has work from UK and international authors. Submissions welcome from anywhere in the world. Snailmail only – no email submissions. S.a.s.e., IRC or reply email address required. Priority is given to full-time writers.
FEATURES Topics of literary interest, especially Scottish literature, theatre, culture or politics. Max. 5,000 words.
FICTION Short stories, occasional novel extracts if self-contained. Shortage of space means that longer work is less likely to be used.
SPECIAL PAGES Poetry, both UK and non-UK in translation (mainly, but not necessarily, European). Featured artists in every issue.
£ Payment by negotiation.

## Chat

IPC Media Ltd, Blue Fin Building, 110 Southwark Street, London SE1 0SU
☏ 020 3148 5000
www.ipcmedia.com/pubs/chat.htm
Owner *IPC Media (A Time Warner Company)*
Editor *Gilly Sinclair*
Circulation 440,093

Founded 1985. WEEKLY general interest women's magazine. Unsolicited mss considered; approach in writing with ideas. Not interested in contributors 'who have never bothered to read *Chat* and therefore don't know what type of magazine it is. *Chat* does not publish fiction.'
FEATURES *Devinder Bains* Human interest and humour. Max. 1,000 words.
£ Up to £500 maximum.

## Chemist & Druggist

Riverbank House, Angel Lane, Tonbridge TN9 1SE
☏ 01732 377487   🖷 01732 367065
chemdrug@cmpmedica.com
www.chemistanddruggist.co.uk
Owner *CMP Medica*
Editor *Gary Paragpuri*
Circulation 13,263

Founded 1859. WEEKLY news magazine for the UK community pharmacy.
FEATURES Practice, politics, professional matters, clinical articles and business advice. Contact by phone or email to discuss ideas.
NEWS Contact news editor with articles relating to local pharmacy matters, local pharmaceutical industry events and pharmacists in the news. Max. 200 words.
£ Payment at standard rates.

## Choice

First Floor, 2 King Street, Peterborough PE1 1LT
☏ 01733 555123   🖷 01733 427500
editorial@choicemag.co.uk
www.choicemag.co.uk
Owner *Choice Publishing Ltd*
Editor *Norman Wright*
Circulation 85,000

Monthly full-colour, lively and informative magazine for people aged 50 plus which helps them get the most out of their lives, time and money after full-time work.
FEATURES Real-life stories, hobbies, interesting (older) people, British heritage and countryside, involving activities for active bodies and minds, health, relationships, book/entertainment reviews. Unsolicited mss read (s.a.s.e. for return of material); write with ideas and copies of cuttings if new contributor. No phone calls, please.
RIGHTS/MONEY All items affecting the magazine's readership are written by experts. Areas of interest include pensions, state benefits, health, finance, property, legal.
£ Payment by arrangement.

## Church Music Quarterly

19 The Close, Salisbury SP1 2EB
☏ 01722 424848   🖷 01722 424849
cmq@rscm.com
www.rscm.com
Owner *Royal School of Church Music*
Communications Officer *Cathy Markall*

Circulation 14,000

QUARTERLY membership magazine.
Contributions welcome two months before
publication. Phone or email.

FEATURES Articles on church music or related
subjects considered. Max. 1,400 words.

## Church of England Newspaper

14 Great College Street, London SW1P 3RX
☎ 020 7878 1001    🖷 020 7878 1031
cen@churchnewspaper.com
www.churchnewspaper.com
Owner *Religious Intelligence*
Editor *Colin Blakely*
Circulation 11,000

Founded 1828. WEEKLY. Almost all material
is commissioned but unsolicited mss are
considered.

FEATURES Preliminary enquiry essential. Max.
1,200 words.

NEWS Items must be sent promptly to the
email address above and should have a church/
Christian relevance. Max. 200–400 words.

💷 Payment negotiable.

## Church Times

13–17 Long Lane, Smithfield, London EC1A 9PN
☎ 020 7776 1060    🖷 020 7776 1086
news@churchtimes.co.uk
features@churchtimes.co.uk
www.churchtimes.co.uk
Owner *Hymns Ancient & Modern*
Editor *Paul Handley*
Circulation 27,000

Founded 1863. WEEKLY. Unsolicited mss
considered (no poetry).

FEATURES *Christine Miles* Articles and pictures
(any format) on religious topics. Max. 1,500
words.

NEWS *Helen Saxbee* Occasional reports
(commissions only) and up-to-date photographs.

💷 Features, £100 per 1,000 words; news, by
arrangement.

## Classic & Sports Car

Teddington Studios, Broom Road, Teddington
TW11 9BE
☎ 020 8267 5399    🖷 020 8267 5318
letters.c&sc@haymarket.com
www.classicandsportscar.com
Owner *Haymarket Consumer Media*
Editor *James Elliott*
Circulation 73,086

Founded 1982. MONTHLY. Best-selling classic car
magazine covering all aspects of buying, selling,
owning and maintaining historic vehicles. Will
consider news stories and features involving
classic cars but 'be patient because hundreds of

unsolicited material received monthly'. Initial
contact by email (james.elliott@haymarket.com).

## Classic Bike

Bauer Automotive, Media House, Lynchwood,
Peterborough Business Park, Peterborough
PE2 6EA
☎ 01733 468000    🖷 01733 468290
classic.bike@bauermedia.co.uk
www.classicbike.co.uk
Owner *Bauer Automotive*
Editor *Hugo Wilson*
Deputy Editor *Jim Moore*
Features Editor *Ben Miller*
Circulation 43,178

Founded 1978. MONTHLY. Mainly pre-1990 classic
motorcycles. Freelancers must contact the editor
before making submissions. Approach in writing
or by email.

NEWS Genuine news with good illustrations, if
possible, suitable for a global audience. Max. 400
words.

FEATURES British motorcycle industry inside
stories, technical features 'that can be understood
by all', German, Spanish and French machine
features, people and classic Japanese bikes from
the 1950s to the 1990s. Restoration features, riding
features; 'good reads'. Max. 2,000 words.

SPECIAL PAGES How-to features, oddball
machines, stunning pictures, features with a fresh
slant.

💷 Payment negotiable.

## Classic Boat

IPC Media Ltd, Leon House, 233 High Street,
Croydon CR9 1HZ
☎ 020 8726 8000    🖷 020 8726 8195
cb@ipcmedia.com
www.classicboat.co.uk
Owner *IPC Media (A Time Warner Company)*
Editor *Dan Houston*
Circulation 12,012

Founded 1987. MONTHLY. Traditional boats and
classic yachts, old and new; maritime history.
Unsolicited mss, particularly if supported by
good photos, are welcome. Sail and power boat
pieces considered. Approach in writing with
ideas. Interested in well-researched stories on
all nautical matters. News reports welcome.
Contributor's notes available online.

FEATURES Boatbuilding, boat history and design,
events, yachts and working boats. Material must
be well-informed and supported where possible
by good-quality or historic photos. Max. 3,000
words. Classic Boat is defined by excellence of
design and construction – the boat need not be
old and wooden!

NEWS New boats, restorations, events,
boatbuilders, etc. Max. 500 words.

Ⓔ Features, £75–100 per published page; news, according to merit.

## Classic Cars

Media House, Peterborough Business Park, Lynchwood, Peterborough PE2 6EA
Ⓣ 01733 468582    Ⓕ 01733 468888
classic.cars@bauermedia.co.uk
www.classiccarsmagazine.co.uk
Owner *Bauer*
Editor *Phil Bell*
Circulation 35,564

Founded 1973. TWELVE ISSUES YEARLY. International classic car magazine containing entertaining and informative articles about classic cars, events and associated personalities. Contributions welcome. Ⓔ Payment negotiable.

## Classic Land Rover World
▷ Land Rover World

## Classic Rock

Future Publishing Ltd, 2 Balcombe Street, London NW1 6NW
Ⓣ 020 7042 4000
classicrock@futurenet.co.uk
www.classicrockmagazine.com
Owner *Future Publishing Ltd*
Editor *Sian Llewellyn*
Circulation 70,301

Launched 1998. MONTHLY rock music magazine, both old and new.
FEATURES Welcomes star interviews, well-written and relevant features. Email the editor. 'Read the magazine first.' Ⓔ Payment varies.

## Classic Stitches

D.C. Thomson & Co. Ltd, 80 Kingsway East, Dundee DD4 8SL
Ⓣ 01382 223131    Ⓕ 01382 452491
editorial@classicstitches.com
www.classicstitches.com
Owner *D.C. Thomson & Co. Ltd*
Editor *Mrs Bea Neilson*
Circulation 7,500

Founded 1994. BI-MONTHLY magazine with exclusive designs to stitch, textile-based features, designer profiles, book reviews and 'what's new shopping' information. Will consider features covering textiles – historical and modern, designer profiles and original designs in context with the style of the magazine (no items on knitting or crochet). Approach by email or letter with short synopsis to avoid clashes. For designs, send sketches, swatches and brief description of the idea.
FEATURES *Bea Neilson* Three features per issue with accompanying photography. Max. 2,000 words.
NEWS *Liz O'Rourke* Max. 300 words.

SPECIAL PAGES *Liz O'Rourke* Details only.
WEBSITE Sub-editor, *Allison Hay* Short, craft-orientated features. Max 500 words.

## Classical Guitar

1 & 2 Vance Court, Trans Britannia Enterprise Park, Blaydon on Tyne NE21 5NH
Ⓣ 0191 414 9000    Ⓕ 0191 414 9001
classicalguitar@ashleymark.co.uk
www.classicalguitarmagazine.com
Owner *Ashley Mark Publishing Co.*
Editor *Guy Traviss*

Founded 1982. MONTHLY.
FEATURES *Colin Cooper* Usually written by staff writers. Max. 1,500 words.
NEWS *Thérèse Wassily Saba* Small paragraphs and festival concert reports welcome.
REVIEWS *Tim Panting* Concert reviews of up to 250 words are usually written by staff reviewers. Ⓔ Features, by arrangement; no payment for news.

## Classical Music

241 Shaftesbury Avenue, London WC2H 8TF
Ⓣ 020 7333 1742    Ⓕ 020 7333 1769
classical.music@rhinegold.co.uk
www.rhinegold.co.uk
Owner *Rhinegold Publishing Ltd*
Editor *Keith Clarke*

Founded 1976. FORTNIGHTLY. A specialist magazine using precisely targeted news and feature articles aimed at the music business. Most material is commissioned but professionally written unsolicited mss are occasionally published. Freelance contributors may approach in writing with an idea but should familiarize themselves beforehand with the style and market of the magazine. Ⓔ Payment negotiable.

## Climber

Warners Group Publications plc, West Street, Bourne PE10 9PH
Ⓣ 01778 391000
climberscomments@warnersgroup.co.uk
www.climber.co.uk
Owner *Warners Group Publications plc*
Editor *Andy McCue*

Founded 1962. MONTHLY. Features articles and news on climbing and mountaineering in the UK and abroad. No unsolicited mss.

## Closer

Endeavour House, 189 Shaftesbury Avenue, London WC2H 8JG
Ⓣ 020 7437 9011    Ⓕ 020 7859 8600
closer@closermag.co.uk
www.closeronline.co.uk
Owner *Bauer Media*
Editor *Lisa Burrow*

Circulation 539,135

Launched 2002. WEEKLY celebrity and real life magazine. No unsolicited material; approach by email with idea in the first instance.

## Club International

3rd Floor, Trojan Publishing, 207 Old Street, London EC1V 9NR

☎ 020 7608 6300
andrewe@paulraymond.com
Editor *Andrew Emery*
Circulation 80,000

Founded 1972. MONTHLY. Features and short humorous items aimed at young male readership aged 18–30.
FEATURES Max. 1,000 words.
SHORTS 200–750 words.
£ Payment negotiable.

## Coach and Bus Week

Rouncy Media Ltd, 3 The Office Village, Forder Way, Cygnet Park, Hampton, Peterborough PE7 8GX

☎ 01733 293240    ℻ 0845 280 2927
jacqui.grobler@rouncymedia.co.uk
www.CBWonline.com
Owner *Rouncy Media Ltd*
Editorial Director *Andrew Sutcliffe*
Circulation 4,000

WEEKLY magazine, aimed at coach and bus operators. Interested in coach and bus industry-related items only: business, financial and legal. Contact by letter, email, telephone or via website.

## Coin News

Token Publishing Ltd, Orchard House, Duchy Road, Heathpark, Honiton EX14 1YD

☎ 01404 46972    ℻ 01404 44788
info@tokenpublishing.com
www.tokenpublishing.com
Owner *P.C. Mussell*
Owners *J.W. Mussell, Carol Hartman*
Editor *J.W. Mussell*
Circulation 10,000

Founded 1964. MONTHLY. Contributions welcome. Approach by phone or email in the first instance.

FEATURES Opportunity exists for well-informed authors 'who know the subject and do their homework'. Max. 2,000 words.
£ Payment by arrangement.

## Company

National Magazine House, 72 Broadwick Street, London W1F 9EP

☎ 020 7312 3775    ℻ 020 7312 3797
company.mail@natmags.co.uk
www.getlippy.com
Owner *National Magazine Co. Ltd*

Editor *Victoria White*
Circulation 240,035

MONTHLY. Glossy women's magazine appealing to the independent and intelligent young woman. A good market for freelancers: 'We look for great newsy features relevant to young British women'. Keen to encourage bright, new, young talent, but uncommissioned material is rarely accepted. Feature outlines are the only sensible approach in the first instance. Max. 1,500–2,000 words. Send to *Emma Justice*, Features Editor. £ £250 per 1,000 words.

## Computer Arts

30 Monmouth Street, Bath BA1 2BW

☎ 01225 442244    ℻ 01225 732295
computerarts@futurernet.com
www.computerarts.co.uk
Owner *Future Publishing*
Publisher *Matthew Pierce*
Editor *Rob Carney*
Circulation 19,391

Founded 1995. MONTHLY. The world of computer arts and digital creative design. 3D, web design, Photoshop, digital video. No unsolicited mss. Interested in tutorials, profiles, tips, software and hardware reviews. Approach by post or email.

## Computer Weekly

Quadrant House, The Quadrant, Sutton SM2 5AS

☎ 020 8652 3122    ℻ 020 8652 8979
computer.weekly@rbi.co.uk
www.computerweekly.com
Owner *Reed Business Information*
Editor-in-Chief *Bryan Glick*
Circulation 120,970

Founded 1966. Freelance contributions welcome.
FEATURES Always looking for good new writers with specialized industry knowledge. Max. 1,800 words.
NEWS *Bill Goodwin* Some openings for regional or foreign news items. Max. 300 words.
£ Payment negotiable.

## Computeractive

Incisive Media Ltd, 32–34 Broadwick Street, London W1A 2HG

☎ 020 7316 9000    ℻ 020 7316 9520
letters@computeractive.co.uk
www.computeractive.co.uk
Owner *Incisive Media Ltd*
Editor *Paul Allen*
Circulation 200,307

Launched 1998. FORTNIGHTLY. Consumer and computer technology magazine. No unsolicited material.

## Computing

Incisive Media Ltd, 32–34 Broadwick Street,
London W1A 2HG
☏ 020 7316 9000   📠 020 7316 9160
feedback@computing.co.uk
www.computing.co.uk
Owner *Incisive Media Ltd*
Editor *Bryan Glick*
Circulation 86,000

Founded 1973. WEEKLY newspaper for IT
professionals. Unsolicited articles welcome.
Please enclose s.a.s.e. for return.
💷 Up to £300 per 1,000 words.

## Condé Nast Traveller

Vogue House, Hanover Square, London W1S 1JU
☏ 020 7499 9080   📠 020 7493 3758
editorcntraveller@condenast.co.uk
www.cntraveller.com
Owner *Condé Nast Publications*
Editor *Sarah Miller*
Circulation 81,031

Founded 1997. MONTHLY travel magazine.
Proposals rather than completed mss preferred.
Approach in writing in the first instance. No
unsolicited photographs. 'The magazine has a no
freebie policy and no writing can be accepted on
the basis of a press or paid-for trip.'

## Construction News

Greater London House, Hampstead Road,
London NW1 7EJ
☏ 020 7728 4632   📠 020 7391 3435
cneditorial@emap.com
www.cnplus.co.uk
Owner *Emap*
Editor *Nick Edwards*
Associate Editor (News) *Alex Hawkes*
Circulation 16,523

WEEKLY. Since 1871, *Construction News* has
provided news to the industry. Readership ranges
from construction site to the boardroom, from
building and specialist construction through to
civil engineering. Initial approach by email.

## Contemporary Review

PO Box 1242, Oxford OX1 4FJ
☏ 01865 201529   📠 01865 201529
editorial@contemporaryreview.co.uk
www.contemporaryreview.co.uk
Owner *Contemporary Review Co. Ltd*
Editor *Dr Richard Mullen*

Founded 1866. QUARTERLY. Independent review
dealing with questions of the day, chiefly politics,
international affairs, religion, literature and the
arts.
FEATURES Limited scope for freelancers.
Authoritative articles of 2,000–3,000 words. Send
by email. 💷 Payment £5 per page.

## Coop Traveller

River Publishing, 3rd Floor, One Neal Street,
London WC2H 9QL
☏ 020 7306 0304   📠 020 7306 0314
travellereditorial@riverltd.co.uk
www.riverltd.co.uk
Owner *River Publishing*
Editor *Heather Farmbrough*
Circulation 220,000

Founded 2001. THREE ISSUES YEARLY. Magazine
for Co-op Travel customers. Travel articles and
travel-related news for the independent traveller.
Approach by telephone, fax, letter or email in the
first instance.
FEATURES Ideas must reflect destinations that
Co-op Travel customers go to. Design and tone
are aspirational. Max. 1,000 words.
💷 Payment negotiable.

## Cosmopolitan

National Magazine House, 72 Broadwick Street,
London W1F 9EP
☏ 020 7439 5000   📠 020 7439 5016
cosmo.mail@natmags.co.uk
www.cosmopolitan.co.uk
Owner *National Magazine Co. Ltd*
Editor *Louise Court*
Deputy Editor *Claire Askew*
Circulation 430,353

MONTHLY. Designed to appeal to the mid-
twenties, modern-minded female. Popular mix
of articles, with emphasis on relationships and
careers, and hard news. No fiction. Will rarely
use unsolicited mss but always on the look-out
for 'new writers with original and relevant ideas
and a strong voice'. All would-be writers should be
familiar with the magazine. Email for submission
guidelines.

## Counselling at Work

BACP Workplace, BACP, BACP House, 15 St
John's Business Park, Lutterworth LE17 4HB
☏ 01455 883300   📠 01455 550243
acw@bacp.co.uk
www.bacpworkplace.org.uk
Owner *British Association for Counselling and
Psychotherapy*
Editor *Rick Hughes*
Circulation 1,600
Founded 1993. QUARTERLY official journal of
BACP Workplace, a Division of BACP. Looking
for well-researched articles (500–2,400 words)
about *any* aspect of workplace counselling and
employee support. Mss from those employed as
counsellors or in welfare posts are particularly
welcome. Photographs accepted. No fiction or
poetry. Contact the editor for submission details
💷 No payment.

## Country Homes and Interiors

IPC Media Ltd, Blue Fin Building, 110 Southwark
Street, London SE1 0SU
☎ 020 3148 5000
countryhomes@ipcmedia.com
Owner *IPC Media (A Time Warner Company)*
Editor *Rhoda Parry*
Circulation 87,326

Founded 1986. MONTHLY. The best approach for
prospective contributors is with an idea in writing
as unsolicited mss are not welcome.
FEATURES *Rhoda Parry* Monthly personality
interviews of interest to an intelligent, affluent
readership (women and men), aged 25–44. Max.
1,200 words.
HOUSES *Vivienne Ayers* Country-style homes with
excellent design ideas. Length 1,000 words.
£ Payment negotiable.

## Country Life

IPC Media Ltd, Blue Fin Building, 110 Southwark
Street, London SE1 0SU
☎ 020 3148 4444
www.countrylife.co.uk
Owner *IPC Media (A Time Warner Company)*
Editor-in-Chief *Mark Hedges*
Circulation 36,836

Established 1897. WEEKLY. *Country Life* features
articles which relate to architecture, countryside,
wildlife, rural events, sports, arts, exhibitions,
current events, property and news articles
of interest to town and country dwellers.
Strong informed material rather than amateur
enthusiasm. 'Uncommissioned material is rarely
accepted. We regret we cannot be liable for
the safe custody or return of any solicited or
unsolicited materials.'

## Country Living

National Magazine House, 72 Broadwick Street,
London W1F 9EP
☎ 020 7439 5000   📠 020 7439 5093/5077
www.allaboutyou.com
Owner *National Magazine Co. Ltd*
Editor *Susy Smith*
Circulation 197,891

Magazine aimed at both country dwellers
and town dwellers who love the countryside.
Covers people, conservation, wildlife, houses
(gardens and interiors) and rural businesses. No
unsolicited mss

## Country Smallholding

Archant Publishing, Fair Oak Close, Exeter
Airport Business Park, Clysts Honiton, Nr Exeter
EX5 2UL
☎ 01392 888481   📠 01392 888499
editorial.csh@archant.co.uk
www.countrysmallholding.com

Owner *Archant Publishing*
Editor *Simon McEwan*
Circulation 20,181

Founded 1975. MONTHLY magazine for small
farmers, smallholders, practical landowners and
for anyone interested in self sufficiency at all
levels both in town and in the country. Articles
welcome on organic growing, keeping poultry,
livestock and other animals, crafts, cookery,
herbs, building and energy. Articles should
be detailed and practical, based on first-hand
knowledge and experience. Approach in writing
or by email.

## Country Walking

Media House, Lynchwood, Peterborough PE2 6EA
☎ 01733 468000
country.walking@bauermedia.co.uk
www.livefortheoutdoors.com
Owner *Bauer Media*
Editor *Jonathan Manning*
Circulation 42,773

Founded 1987. MONTHLY magazine focused on
walking in the finest landscapes of the UK and
Europe. News, gear tests and profiles of celebrity
walkers also feature. Special monthly pull-out
section of detailed walking routes. Original
high quality photography and ideas for features
considered, although very few unsolicited mss
accepted. Approach by letter or email.
SPECIAL PAGES Routes section of magazine,
including outline Ordnance Survey map
and step-by-step directions. Accurately and
recently researched walk and fact file including
photographs and points of interest en route.
Please contact *Ruth Addison* for guidelines
(unsolicited submissions generally not accepted
for this section).
£ Payment not negotiable.

## The Countryman

The Water Mill, Broughton Hall, Skipton
BD23 3AG
☎ 01756 701381   📠 01756 701326
editorial@thecountryman.co.uk
www.thecountryman.co.uk
Owner *Country Publications Ltd*
Editor *Paul Jackson*
Circulation 18,240

Founded 1927. Monthly. Countryside feature ideas
welcome by email, phone or in writing. Articles
supplied with top quality illustrations (digital,
colour transparencies, archive b&w prints and
line drawings) are far more likely to be used. Max.
article length 1,200 words.

## The Countryman's Weekly (incorporating Gamekeeper and Sporting Dog)

Unit 2, Lynher House, 3 Bush Park, Estover, Plymouth PL6 7RG

☎ 01752 762990   📠 01752 771715

editorial@countrymansweekly.com

www.countrymansweekly.com

Owner *Diamond Publishing Ltd*

Editor *David Venner*

Founded 1895. WEEKLY. Unsolicited material welcome.

FEATURES on any country sports topic. Max. 1,000 words.

💷 Payment rates available on request.

## The Cricketer International
▷ The Wisden Cricketer

## Crimewave

TTA Press, 5 Martins Lane, Witcham, Ely CB6 2LB

www.ttapress.com

Owner *TTA Press*

Editor *Andy Cox*

Circulation 8,000

Founded 1998. BIANNUAL large paperback format magazine of cutting edge crime and mystery fiction. 'The UK's only magazine specializing in crime short stories, publishing the very best from across the spectrum.' Every issue contains stories by authors who are household names in the crime fiction world but room is found for lesser known and unknown writers. Submissions welcome (not via email) with appropriate return postage. Potential contributors are advised to study the magazine. Contracts exchanged upon acceptance.

💷 Payment on publication.

## CrossStitcher

30 Monmouth Street, Bath BA1 2BW

☎ 01225 442244

www.crossstitchermagazine.co.uk

Owner *Future Publishing*

Editor *Cathy Lewis*

Circulation 53,303

Founded 1992. MONTHLY. Charts, interviews with designers and occasional features about stitching. All writing is done in-house but there is a bank of regular freelance designers. Will consider submissions on cross-stitch designs, 'preferably to a very high quality and using up-to-date and new techniques'. No feature material. Approach by email or letter.

## Crucible
▷ Hymns Ancient & Modern Ltd under UK Publishers

## Cumbria

The Water Mill, Broughton Hall, Skipton BD23 3AG

☎ 01535 637983

editorial@cumbriamagazine.co.uk

www.cumbriamagazine.co.uk

Owner *Country Publications Ltd*

Features Editor *Kevin Hopkinson*

Circulation 12,183

Founded 1951. MONTHLY. County magazine of strong regional and countryside interest, focusing on the Lake District. Unsolicited mss welcome. Max. 1,200 words. Approach in writing, by phone or email with feature ideas.

💷 £70 per 1,000 words plus extra for photos.

## Cycle Sport

IPC Media Ltd, Leon House, 233 High Street, Croydon CR9 1HZ

☎ 020 8726 8460

cyclesport@ipcmedia.com

www.cyclesport.co.uk

Owner *IPC Media (A Time Warner Company)*

Managing Editor *Robert Garbutt*

Circulation 15,871

Founded 1993. MONTHLY magazine dedicated to professional cycle racing. Unsolicited ideas for features welcome.

## Cycling Weekly

IPC Media Ltd, Leon House, 233 High Street, Croydon CR9 1HZ

☎ 020 8726 8453   📠 020 8726 8499

cycling@ipcmedia.com

www.cyclingweekly.co.uk

Owner *IPC Media (A Time Warner Company)*

Publishing Director *Keith Foster*

Editor *Robert Garbutt*

Circulation 29,029

Founded 1891. WEEKLY. All aspects of cycle sport covered. Unsolicited mss and ideas for features welcome. Approach in writing with ideas. Fiction rarely used. 'It is important that you are familiar with the format so read the magazine first.'

FEATURES Cycle racing, coaching, technical, health and fitness material and related areas. Max. 2,000 words. Most work commissioned but interested in seeing new work.

NEWS Short news pieces, local news, etc. Max. 300 words.

💷 Features, around £100–120 per 1,000 words (quality permitting); news, £15 per story.

## Dalesman

The Water Mill, Broughton Hall, Skipton BD23 3AG

☎ 01756 701381   📠 01756 701326

editorial@dalesman.co.uk

www.dalesman.co.uk

Owner *Country Publications Ltd*

Editor *Paul Jackson*

Circulation 37,463

Founded 1939. Now the biggest-selling regional publication of its kind in the country. MONTHLY magazine with articles of specific Yorkshire interest. Unsolicited mss welcome; receives approximately ten per day. Initial approach in writing, by phone or email. Max. 1,000 words. £ £70 per 1,000 words plus extra for photos.

## Dance Today

Dancing Times Ltd, Clerkenwell House, 45–47 Clerkenwell Green, London EC1R 0EB
☎ 020 7250 3006    🖷 020 7253 6679
dancetoday@dance-today.co.uk
www.dance-today.co.uk
Owner *The Dancing Times Ltd*
Editor *Nicola Rayner*
Editorial Assistant *Jon Gray*
Circulation 8,000

Founded in 1956 as the *Ballroom Dancing Times* and relaunched as *Dance Today* in 2001. MONTHLY magazine for anyone interested in social and competitive dance. Styles include ballroom and Latin American, salsa, tango, flamenco, tea dancing, sequence and all social dances. Related topics covered include health and fitness and dance fashion. Contributions welcome; send cuttings and short pitch with sample paragraph by post or email to the editor.

## The Dancing Times

The Dancing Times Ltd, 45–47 Clerkenwell Green, London EC1R 0EB
☎ 020 7250 3006    🖷 020 7253 6679
editorial@dancing-times.co.uk
www.dancing-times.co.uk
Owner *The Dancing Times Ltd*
Editor *Jonathan Gray*

Founded 1910. MONTHLY. Freelance suggestions welcome from specialist dance writers and photographers only. Approach in writing.

## The Dandy

D.C. Thomson & Co. Ltd, Albert Square, Dundee DD1 8QJ
☎ 01382 575743
editor@dandy.co.uk
www.dandy.com
Owner *D.C. Thomson & Co. Ltd*
Editor *Craig Graham*
Circulation 18,938

Founded 1937. WEEKLY comic. Original, humorous comic strips for 7–12-year-olds, featuring established characters such as Desperate Dan as well as new ones like Jak. Generally, unsolicited contributions are welcome but 'please read the comic before sending any material. Try a one-off script for an established, current *Dandy* character. Send by email or post but never send

your only copy.' Include s.a.s.e. for reply. No non-script material such as puzzles, poetry, etc.

## Darts World

25 Orlestone View, Ham Street, Ashford TN26 2LB
☎ 01233 733558
mb.graphics@virgin.net
www.dartsworld.com
Owner *World Magazines Ltd*
Editor *M. Beeken*
Circulation 12,263

FEATURES Single articles or series on technique and instruction. Max. 1,200 words.
FICTION Short stories with darts theme. Max. 1,000 words.
NEWS Tournament reports and general or personality news required. Max. 800 words.
£ Payment negotiable.

## Dazed & Confused

112–116 Old Street, London EC1V 9BG
☎ 020 7336 0766    🖷 020 7336 0966
rod@dazedgroup.com
www.dazeddigital.com
Owner *Waddell Ltd*
Editor *Rod Stanley*
Circulation 93,000

Founded 1992. MONTHLY. Cutting edge fashion, music, film, art, interviews and features. No unsolicited material. Approach by email with ideas in the first instance.

## Decanter

Blue Fin Building, 110 Southwark Street, London SE1 0SU
☎ 020 3148 4488    🖷 020 3148 8524
editor@decanter.com
www.decanter.com
Editor *Guy Woodward*
Circulation 40,000

FOUNDED 1975. Glossy consumer wines magazine. Feature ideas welcome; send to the editor by post, fax or email. No fiction.
NEWS/FEATURES All items and articles should concern wines, food and related subjects. £ £250 per 1,000 words.

## Delicious

Seven Publishing Group, Sea Container's House, 20 Upper Ground, London SE1 9PD
☎ 020 7775 7757    🖷 020 7775 7705
laura.grossman@eyetoeyemedia.co.uk
www.deliciousmagazine.co.uk
Owner *Eye to Eye Media*
Editor *Matthew Drennan*
Circulation 104,178

Launched 2003. MONTHLY publication for people who love food and cooking. Recipes – from the simple to the sophisticated – plus features on food and where it comes from. Welcomes freelance contributions; approach by email.

## Derbyshire Life and Countryside

61 Friar Gate, Derby DE1 1DJ
☏ 01332 227850   🖷 01332 227860
www.derbyshirelife.co.uk
Owner *Archant Life*
Editor *Joy Hales*
Circulation 16,106

Founded 1931. MONTHLY county magazine for Derbyshire. Unsolicited mss and photographs of Derbyshire welcome, but written approach with ideas preferred.

## Descent

Wild Places Publishing, PO Box 100, Abergavenny NP7 9WY
☏ 01873 737707
descent@wildplaces.co.uk
www.caving.uk.com
Owner *Wild Places Publishing*
Editor *Chris Howes*
Assistant Editor *Judith Calford*

Founded 1969. BI-MONTHLY magazine for cavers and mine enthusiasts. Submissions welcome from freelance contributors who can write accurately and knowledgeably and in the style of existing content on any aspect of caves, mines or underground structures.
FEATURES General interest articles of under 1,000 words welcome, as well as short foreign news reports, especially if supported by photographs/illustrations. Suitable topics include exploration (particularly British, both historical and modern), expeditions, equipment, techniques and regional British news. Max. 2,000 words.
£ Payment on publication according to page area filled.

## Destination France

151 Station Street, Burton on Trent DE14 1BG
☏ 01283 742950   🖷 01283 742957
carmen.konopka@wwonline.co.uk
www.destinationfrancemagazine.co.uk
Owner *Waterways World*
Editor *Carmen Konopka*
Circulation 15,000

A stylish, contemporary, QUARTERLY magazine aimed at Francophiles. Covers food, wine, art, architecture, books and lifestyle as well as travel within France. Features include sports, general tourism, property and regional information. Photographs to accompany articles. Approach in writing, preferably by email, in the first instance.
£ Payment negotiable.

## Devon Life

Archant House, Babbage Road, Totnes TQ9 5JA
☏ 01803 860910   🖷 01803 860926
devonlife@archant.co.uk
www.devon.greatbritishlife.co.uk
Owner *Archant*
Editor *Jane Fitzgerald*
Circulation 15,625

Originally launched in 1963; relaunched 1996. MONTHLY. Articles on all aspects of Devon – places of interest, personalities, events, arts, history and food. Contributions welcome; send by email to the editor.
FEATURES Words/picture package preferred on topics listed above. Max. 1,300 words.
£ Payment by arrangement.

## Director

116 Pall Mall, London SW1Y 5ED
☏ 020 7766 8950   🖷 020 7766 8840
director-ed@iod.com
Group Editor *Richard Cree*
Circulation 56,701

1991 Business Magazine of the Year. Published by Director Publications Ltd for members of the Institute of Directors and subscribers. Wide range of features from political and business profiles and management thinking to employment and financial issues. Also book reviews. Regular contributors used. Send letter with synopsis/published samples rather than unsolicited mss.
£ £350 per 1,000 words.

## Disability Now

6 Market Road, London N7 9PW
☏ 020 7619 7323   🖷 020 7619 7331
editor@disabilitynow.org.uk
www.disabilitynow.org.uk
Owner *Scope*
Editor *Ian Macrae*
Circulation 18,514

Founded 1984. Leading MONTHLY magazine for disabled people. Freelance contributions welcome. No fiction. Approach in writing or by email.
FEATURES cover new initiatives and services, personal experiences and general issues of interest to a wide national readership. Max. 900 words. Disabled contributors welcome.
NEWS Max. 300 words.
£ Payment by arrangement.

## Diva, lesbian life and style

Spectrum House, 32–34 Gordon House Road, London NW5 1LP
☏ 020 7424 7400   🖷 020 7424 7401
edit@divamag.co.uk
www.divamag.co.uk
Owner *Millivres-Prowler Group*
Editor *Jane Czyzselska*

Founded 1994. MONTHLY glossy magazine featuring lesbian news and culture including fashion, lifestyle and satire. Welcomes news, features and photographs. No poetry. Contact the news editor at the email address above with news items and feature ideas and photo/fashion photographs.

## DIY Week

Faversham House, 232a Addington Road, Selsdon CR2 8LE
☎ 020 8651 7100   ℻ 020 8651 7117
diyweek@fav-house.com
www.diyweek.net
Owner *Faversham House Group*
Editor *Will Parsons*
Circulation 7,900

Founded 1874. FORTNIGHTLY B2B (business-to-business) magazine covering the entire home improvement market. Welcomes exclusive, timely, sector-specific feature content but 'very rarely work outside of our freelance database'. Approach by email.

## Dog World

Somerfield House, Wotton Road, Ashford TN23 6LW
☎ 01233 621877   ℻ 01233 645669
info@dogworld.co.uk
www.dogworld.co.uk
Owner *DW Media Holdings Ltd*
Editor *Stuart Baillie*
Circulation 155,448

Founded 1902. WEEKLY newspaper for people who are seriously interested in pedigree dogs. Unsolicited mss occasionally considered but initial approach in writing preferred. FEATURES Well-researched historical items or items of unusual interest concerning dogs. Max. 2,000 words. Photographs of unusual 'doggy' situations occasionally of interest. NEWS Freelance reports welcome on court cases and local government issues involving dogs. ℻ Features, up to £75; photos, £15.

## Dogs Monthly

ABM Publishing Ltd, 61 Great Whyte, Ramsey, Huntingdon PE26 1HJ
☎ 0845 094 8958   ℻ 0845 519 0229
info@dogsmonthly.co.uk
www.dogsmonthly.co.uk
Owner *ABM Publishing Ltd*
Editor *Caroline Davis*
Assistant Editor *Hannah Roche*
Sub-editor/Writer *Jenny Chafer*
Sub-editor *Gareth Salter*
Circulation 20,000

MONTHLY magazine containing expert, practical advice on caring for and training pet dogs by 'the best of the canine world's writers, vets and trainers'. Submissions by post or email; may be up to three months before a reply is given.

## Dorset Life – The Dorset Magazine

7 The Leanne, Sandford Lane, Wareham BH20 4DY
☎ 01929 551264   ℻ 01929 552099
editor@dorsetlife.co.uk
www.dorsetlife.co.uk
Owner/Editor *John Newth*
Circulation 10,000

Founded 1968. MONTHLY magazine about Dorset, both present and past. Unsolicited contributions welcome if specifically about Dorset.

## Early Music Today

Rhinegold Publishing, 241 Shaftesbury Avenue, London WC2H 8TF
☎ 020 7333 1744   ℻ 020 1733 1769
emt@rhinegold.co.uk
www.rhinegold.co.uk
Owner *Rhinegold Publishing*
Editor *Jonathan Wikeley*
Circulation 10,000

Launched 1993. BI-MONTHLY. Articles on early music, concerts, CD and book reviews, comprehensive concert listings, early music news. FEATURES All contributors are experts in the early music field (pre-1800). Relevant articles and news welcome. Approach by email or post.

## Eastern Eye

Asian Media & Marketing Group, Garavi Gujarat House, No. 1 Silex Street, London SE1 0DW
☎ 020 7928 1234   ℻ 020 7261 0055
editor@easterneye.eu
www.easterneye.eu
Owner *Asian Media & Marketing Group*
Editor *Hamant Verma*
Circulation 20,844

Founded 1989. WEEKLY community paper for the Asian community in Britain. Interested in relevant general, local and international issues. Approach in writing with ideas for submission.

## Easy Living

6–8 Old Bond Street, London W1S 4PH
☎ 020 7499 9080   ℻ 020 7399 2625
easylivingeditorial@condenast.co.uk
www.easylivingmagazine.com
Owner *Condé Nast*
Editor *Susie Forbes*
Health and Beauty Director *Catherine Turner*
Fashion Director *Liz Thody*
Food Editor *David Herbert*
Circulation 170,033

Launched 2005. MONTHLY magazine aimed at women aged 30–50-plus. Focuses on aspects of fashion, home life, beauty and health, food, relationships; stylish and glossy with real-life, practical and useful items.

## The Ecologist

Unit 102, Lana House Studios, 116–118 Commercial Street, London E1 6NF

☏ 020 7422 8100　🖷 020 7422 8101

mark@theecologist.org

www.theecologist.org

Owner *Ecosystems Ltd*

Editor *Mark Anslow*

Director *Zac Goldsmith*

Founded 1970. Formerly a magazine, now a website. Unsolicited mss welcome but best approach is a brief proposal to the editor by email (address above), outlining experience and background and summarizing suggested article. FEATURES Radical approach to political, economic, social and environmental issues, with an emphasis on rethinking the basic assumptions that underpin modern society. Articles of between 500 and 2,000 words.
💷 Payment negotiable.

## The Economist

25 St James's Street, London SW1A 1HG

☏ 020 7830 7000　🖷 020 7839 2968

www.economist.com

Owner *The Economist Group*

Editor *John Micklethwait*

Circulation 189,201 (UK)

Founded 1843. WEEKLY. Worldwide circulation. Approaches should be made in writing to the editor. No unsolicited mss.

## The Edge

Unit 138, 22 Notting Hill Gate, London W11 3JE

☏ 0845 456 9337

www.theedge.abelgratis.co.uk

Editor *David Clark*

QUARTERLY. Reviews and features: film (indie, arts), books, popular culture.
FICTION 2,000+ words, almost any 'type' of story, but see website or magazine to email for examples/tone. Sample issue £5 post free (cheques payable to 'The Edge'). Guidelines on website or send s.a.s.e.
💷 Payment 'negotiable after we express interest'.

## Edinburgh Review

22A Buccleugh Place, Edinburgh EH8 9LN

☏ 0131 651 1415

edinburgh.review@ed.ac.uk

www.edinburghreview.org.uk

Owner *English Literature, University of Edinburgh*

Editor *Brian McCabe*

Founded 1802. THREE ISSUES YEARLY. Cultural and literary magazine. Articles on Scottish and international culture, history, art, literature and

politics. Each issue is broadly geographically themed; recent issues include the Carribean, Poland and Northern Ireland. Unsolicited contributions welcome, particularly articles, fiction and poetry; send by post or email. 'Writers offering unsolicited essays should contact us by email in the first instance.'

## Elle

64 North Row, London W1K 7LL

☏ 020 7150 7000　🖷 020 7150 7670

Owner *Hachette Filipacchi*

Editor *Lorraine Candy*

Features Director *Kerry Potter*

Circulation 195,455

Founded 1985. MONTHLY fashion glossy. Prospective contributors should approach the relevant editor in writing in the first instance, including cuttings.
FEATURES Max. 2,000 words.
PREVIEW Short articles on hot trends, events, fashion and beauty. Max. 500 words.
💷 Payment negotiable.

## Elle Decoration

64 North Row, London W1K 7LL

☏ 020 7150 7000　🖷 020 7150 7671

Owner *Hachette Filipacchi*

Editor-in-Chief *Michelle Ogundehin*

Circulation 60,056

Launched 1989. MONTHLY style magazine for the home – 'beautiful homes, great designs, inspiration and ideas'. No unsolicited contributions; approach by email to the appropriate department.

## Embroidery

51 Howey Hill, Congleton CW12 4AF

☏ 01260 273891

jo.editor@btinternet.com

www.embroiderersguild.com

Owner *The Embroiderers' Guild*

Editor *Joanne Hall*

Circulation 6,000

Founded 1933. BI-MONTHLY. Features articles on contemporary practice within art textiles and embroidery. Insights into historical and world embroidery. Also includes news, exhibition and book reviews. Unsolicited material welcome. Max. 1,000 words. Authors must source or identify photography/images.
💷 Payment negotiable.

## Empire

4th Floor, Mappin House, 4 Winsley Street, London W1W 8HF

☏ 020 7182 8781　🖷 020 7182 8703

empire@bauermedia.com

www.empireonline.com

Owner *Bauer Media*

Editor *Mark Dinning*
Circulation 194,239

Founded 1989. Launched at the Cannes Film Festival. MONTHLY guide to the movies which aims to cover the world of films in a 'comprehensive, adult, intelligent and witty package'. Although most of *Empire* is devoted to films and the people behind them, it also looks at the developments and technology behind television and DVDs plus music, multimedia and books. Wide selection of in-depth features and stories on all the main releases of the month, and reviews of over 100 films and videos. Contributions welcome but approach in writing first.

FEATURES Behind-the-scenes features on films, humorous and factual features.
£ Payment by agreement.

## The Engineer

50 Poland Street, London W1F 7AX
℡ 020 7970 4154    📠 0870 600 8399
tepr@centaur.co.uk
www.theengineer.co.uk
Owner *Centaur Media*
Editor *Jon Excell*
Circulation 32,546

FOUNDED 1856. FORTNIGHTLY news magazine for technology and innovation.

FEATURES Most outside contributions are commissioned but good ideas are always welcome. Max. 2,000 words.
NEWS Scope for specialist regional freelancers, and for tip-offs. Max. 500 words.
TECHNOLOGY Technology news from specialists. Max. 500 words.
£ Payment by arrangement.

## The English Garden

Archant House, Oriel Road, Cheltenham GL50 1BB
℡ 01242 211080    📠 01242 211081
theenglishgarden@archant.co.uk
www.theenglishgarden.co.uk
Owner *Archant Specialist*
Editor *Tamsin Westhorpe*
Circulation 32,182 (UK), 58,043 (worldwide)

Founded 1996. MONTHLY. Features on beautiful gardens with practical ideas on design and planting. No unsolicited mss.
FEATURES Max. 800 words. Approach in writing in the first instance; send synopsis of 150 words with strong design and planting ideas, feature proposals, or sets of photographs of interesting gardens.

## ES

▷ Evening Standard under Regional Newspapers

## Esquire

National Magazine House, 72 Broadwick Street, London W1F 9EP
℡ 020 7439 5000    📠 020 7439 5675
www.esquire.co.uk
Owner *National Magazine Co. Ltd*
Editor *Jeremy Langmead*
Circulation 59,160

Founded 1991. MONTHLY. Quality men's general interest magazine. No unsolicited mss or short stories.

## Essentials

IPC Media Ltd, Blue Fin Building, 110 Southwark Street, London SE1 0SU
℡ 020 3148 7211
Owner *IPC Media (A Time Warner Company)*
Editor *Jules Barton-Breck*
Circulation 112,135

Founded 1988. MONTHLY women's lifestyle magazine. Initial approach by email or telephone preferred. Prospective contributors should study the magazine thoroughly before submitting anything. No fiction.
FEATURES Max. 2,000 words (A4/double-spaced).
£ Payment negotiable.

## Eventing

▷ Horse and Hound

## Evergreen

PO Box 52, Cheltenham GL50 1YQ
℡ 01242 537900    📠 01242 537901
Editor *Stephen Garnett*
Circulation 48,000

Founded 1985. QUARTERLY magazine featuring articles and poems about Britain. Unsolicited contributions welcome.
FEATURES Britain's natural beauty, towns and villages, nostalgia, wildlife, traditions, odd customs, legends, folklore, crafts, etc. Length 250–2,000 words.
£ £15 per 1,000 words; poems £4.

## Eye: The international review of graphic design

Eye Magazine Ltd, Studio 6, The Lux Building, 2–4 Hoxton Square, London N1 6NU
℡ 020 7684 6531
john.walters@eyemagazine.com
www.eyemagazine.com
www.blog.eyemagazine.com
Owner *Eye Magazine Ltd*
Editor *John L. Walters*
Circulation 9,000

Founded 1990. QUARTERLY. 'The world's most beautiful and authoritative magazine about graphic design and visual culture.' Unsolicited contributions not generally welcomed.

FEATURES Specialist articles about graphic design and designers. Design history by experts. Max. 3,000 words.
REVIEWS Max. 200–1,000 words.
£ Payment negotiable.

## Fabulous
▷ News of the World under National Newspapers

## Fairgame Magazine
▷ She Kicks

## Families First

Mary Sumner House, 24 Tufton Street, London SW1P 3RB
☎ 020 7222 5533    ⊠ 020 7227 9731
publications@themothersunion.com
www.familiesfirstmagazine.com
Owner *MU Enterprises Ltd*
Editor *Catherine Butcher*
Circulation 42,082

Founded 1976. Formerly *Home & Family*.
BI-MONTHLY. Unsolicited mss very rarely considered. No fiction or poetry. Features on family life, social issues, marriage, Christian faith, etc. Max. 1,000 words. £ Payment 'modest'.

## Family Tree

61 Great Whyte, Ramsey, Huntingdon PE26 1HJ
☎ 01487 814050    ⊠ 01487 711361
sue.f@abmpublishing.co.uk
www.family-tree.co.uk
Owner *ABM Publishing Ltd*
Editor *Helen Tovey*
Circulation 35,000

Founded 1984. THIRTEEN ISSUES YEARLY. News and features on matters of genealogy. Articles with a social or military history angle, or with useful research steps for fellow readers favoured. Approach in writing with ideas. All material should be addressed to *Helen Tovey*.
FEATURES Any genealogically related subject. Max. 2,000 words.
£ Features payment negotiable.

## Fancy Fowl

The Publishing House, Station Road, Framlingham IP13 9EE
☎ 01728 622030    ⊠ 01728 622031
fancyfowl@todaymagazines.co.uk
www.fancyfowl.com
Owner *Today Magazines Ltd*
Editor *Kevin Davis*
Circulation 3,000

Founded 1979. MONTHLY. Devoted entirely to poultry, waterfowl, turkeys, geese, pea fowl, etc. – management, breeding, rearing and exhibition. Outside contributions of knowledgeable, technical poultry-related articles and news welcome. Max. 1,000 words. Approach by letter or email.
£ Payment negotiable.

## Farmers Weekly

Quadrant House, Sutton SM2 5AS
☎ 020 8652 4911    ⊠ 020 8652 4005
farmers.weekly@rbi.co.uk
www.fwi.co.uk
Owner *Reed Business Information*
Editor *Jane King*
Online Editor *Julian Gairdner*
Circulation 68,461

WEEKLY B2B (business-to-business) title with website. PPA Business Magazine of the Year 2006. For practising farmers and those in the ancillary industries. Increasingly strong interactive/community focus to content. Unsolicited mss considered.
FEATURES A wide range of material in print and online relating to running a farm business: specific sections on arable and livestock farming, farm life, practical and general interest, machinery and business.
NEWS General farming news, technical stories and case studies.
£ Payment negotiable.

## Fast Car

Future Publishing Ltd, 30 Monmouth Street, Bath BA1 2BW
☎ 01225 442244
www.fastcar.co.uk
Owner *Future Publishing Ltd*
Editor *Steve Chalmers*
Circulation 26,120

Founded 1987. THIRTEEN ISSUES YEARLY. Lad's magazine about performance tuning and modifying cars. Covers all aspects of this youth culture including the latest street styles and music. Features cars and their owners, product tests and in-car entertainment. Also includes a free reader ads section.
NEWS Any item in line with the above.
FEATURES Innovative ideas in line with the above and in the *Fast Car* writing style. Generally two to four pages in length. No Kit-car features, race reports or road test reports of standard cars. Copy should be as concise as possible.
£ Payment negotiable.

## FHM

Mappin House, 4 Winsley Street, London W1W 8HF
☎ 020 7182 8000    ⊠ 020 7182 8021
general@fhm.com
www.fhm.com
Owner *Bauer Media*
Editor-in-Chief *Colin Kennedy*

Circulation 231,235

Founded in 1986 as a free fashion magazine, FHM evolved to become more male oriented. Best-selling men's magazine covering all areas of men's lifestyle. Published MONTHLY with editions worldwide. No unsolicited mss. Synopses and ideas welcome by email. £ Payment negotiable.

## The Field

IPC Media Ltd, Blue Fin Building, 110 Southwark Street, London SE1 0SU
☎ 020 3148 4772   📠 020 3148 8179
field_secretary@ipcmedia.com
www.thefield.co.uk
Owner *IPC Media (A Time Warner Company)*
Editor-in-Chief *Jonathan Young*
Circulation 30,428

Founded 1853. MONTHLY magazine for those who are serious about the British countryside and its pleasures. Unsolicited mss welcome but initial approach should be made in writing or by phone.
FEATURES Exceptional work on any subject concerning the countryside. Most work tends to be commissioned.
£ Payment varies.

## 50 Connect – www.50connect.co.uk
▷ entry under Electronic Publishing and Other Services

## Fishing News

IntraFish, 6th Floor, Eldon House, 2 Eldon Street, London EC2M 7UA
☎ 020 7650 1034
tim.oliver@fishingnews.co.uk
www.fishingnews.co.uk
Owner *IntraFish Media*
Editor *Tim Oliver*
Circulation 8,320

Founded 1913. WEEKLY. All aspects of the commercial fishing industry in the UK and Ireland. No unsolicited mss; email enquiry in the first instance. Max. 600 words for news and 1,500 words for features. £ £100 per 1,000 words.

## Flash: The International Short-Short Story Magazine

Department of English, University of Chester, Parkgate Road, Chester CH1 4BJ
☎ 01244 513152   📠 01244 511330
flash.magazine@chester.ac.uk
www.chester.ac.uk/flash.magazine
Editors *Dr Peter Blair, Dr Ashley Chantler*

Launched 2008. BIANNUAL magazine of short stories of up to 360 words (title included). Articles and reviews considered. Full details on the website.

## Flight International

Quadrant House, The Quadrant, Sutton SM2 5AS
☎ 020 8652 3842   📠 020 8652 3840

flight.international@flightglobal.com
www.flightglobal.com
Owner *Reed Business Information*
Editor *Murdo Morrison*
Circulation 41,000

Founded 1909. WEEKLY. International trade magazine for the aerospace industry, including civil, military and space. Unsolicited mss considered. Commissions preferred – phone with ideas and follow up with letter. Email submissions encouraged.
FEATURES *Murdo Morrison* Technically informed articles and pieces on specific geographical areas with international appeal. Analytical, in-depth coverage required, preferably supported by interviews. Max. 1,800 words.
NEWS *Dan Thisdell* Opportunities exist for news pieces from particular geographical areas on specific technical developments. Max. 350 words.
£ NUJ rates.

## Flora International

The Fishing Lodge Studio, 77 Bulbridge Road, Wilton, Salisbury SP2 0LE
☎ 01722 743207   📠 01722 743207
floramag@aol.com
www.flora-magazine.co.uk
Publisher/Editor *Maureen Foster*
Circulation 16,000

Founded 1974. BI-MONTHLY magazine for flower arrangers and florists. Unsolicited mss welcome. Approach in writing with ideas. Not interested in general gardening articles.
FEATURES Fully illustrated: colour photographs, transparencies or CD. Flower arranging, flower arrangers' gardens and flower crafts. Floristry items written with practical knowledge and well illustrated are particularly welcome. Average 1,000 words.
£ £60 per 1,000 words plus additional payment for suitable photographs, by arrangement.

## FlyPast

PO Box 100, Stamford PE9 1XQ
☎ 01780 755131   📠 01780 757261
flypast@keypublishing.com
www.flypast.com
Owner *Key Publishing Ltd*
Editor *Ken Ellis*
Circulation 40,106

Founded 1981. MONTHLY. Historic aviation and aviation heritage, mainly military, First and Second World War period 1914–1970. Unsolicited mss welcome.

## Focus

Bristol Magazines Ltd, Tower House, Fairfax Street, Bristol BS1 3BN
☎ 0117 927 9009 (switchboard)   📠 0117 934 9008

focus@bbcmagazinesbristol.com
www.bbcfocusmagazine.com
Owner *Bristol Magazines Ltd, a subsidiary of BBC Worldwide*
Editor *Jheni Osman*
Deputy Editor *Andy Ridgway*
Circulation 70,122

Founded 1992. FOUR-WEEKLY. Popular science, technology, the future and discovery. Welcomes *relevant* summaries of original feature ideas by email.

## Fortean Times: The Journal of Strange Phenomena

Dennis Publishing, 30 Cleveland Street, London W1T 4JD
☎ 020 7907 6235    ℻ 020 7907 6139
david_sutton@dennis.co.uk
www.forteantimes.com
Editor *David Sutton*
Circulation 20,941

Founded 1973. THIRTEEN ISSUES YEARLY. Accounts of strange phenomena and experiences, curiosities, mysteries, prodigies and portents. Unsolicited mss welcome. Approach by email with ideas. No fiction, poetry, rehashes or politics. FEATURES Well-researched and referenced material on current or historical mysteries, or first-hand accounts of oddities. Max. 4,000 words, preferably with good relevant photos/illustrations.
NEWS Concise copy with full source references essential.
£ Payment negotiable.

## Foundation: the international review of science fiction

Journal of the **Science Fiction Foundation** (see entry under *Professional Associations and Societies*).

## FQ Magazine

3–5 Spafield Street, London EC1R 4QB
☎ 020 7841 0344
tom@touchline.com
www.fqmagazine.co.uk
Owner *3D Media Ltd*
Editor *Andy Tongue*

Launched 2003. BI-MONTHLY. Lifestyle magazine for young fathers. Interviews, reviews, finance, kids' toys, child issues, parenting tips and general fatherhood advice. Ideas welcome; contact the editor by email.

## France Magazine

Archant House, Oriel Road, Cheltenham GL50 1BB
☎ 01242 216050    ℻ 01242 216094

editorial@francemag.com
www.francemag.com
Owner *Archant Life France*
Editor *Carolyn Boyd*
Circulation 37,082 (group, including US edition); 19,351 (UK)

FOUNDED 1989. MONTHLY magazine containing all things of interest to Francophiles – in English. Does not cover expat-related subjects or property. Approach by email in the first instance.

## Freelance Market News

Sevendale House, 7 Dale Street, Manchester M1 1JB
☎ 0161 228 2362    ℻ 0161 228 3533
fmn@writersbureau.com
www.freelancemarketnews.com
Editor *Angela Cox*

MONTHLY. News and information on the freelance writers' market, both inland and overseas. Includes market information, competitions, seminars, courses, overseas openings, etc. Short articles (700–1,500 words). Unsolicited contributions welcome.
£ Payment by negotiation.

## The Freelance
▷ National Union of Journalists under Professional Associations and Societies

## Front Magazine

2nd Floor, 2–4 Noel Street, London W1F 8GB
☎ 020 3358 3301
front@frontarmy.co.uk
Owner *The Kane Corporation*
Editor *Joe Barnes*
Circulation 41,946

Founded 1998. MONTHLY men's interest magazine. Includes features on sport (including extreme sport), music, crime, war, humour, sex, alternative culture, real life stories and irreverent celebrity interviews. Welcomes contributions. Approach by email.

## Garden Answers

Media House, Lynchwood, Peterborough PE2 6EA
☎ 01733 46800
Owner *Bauer Consumer Media*
Editor *Geoff Stebbings*
Circulation 20,319

Founded 1982. MONTHLY. Interested in hearing from gardening writers on any subject, whether flowers, fruit, vegetables, houseplants or greenhouse gardening. 'It is unlikely that unsolicited manuscripts will be used, as articles are usually commissioned and must be in the magazine style.' Prospective contributors should approach the editor in writing.

## Garden News

Media House, Lynchwood, Peterborough PE2 6EA

☎ 01733 468000

Owner *Bauer Consumer Media*
Editor *Clare Foggett*
Circulation 31,455

Founded 1958. Britain's biggest-selling, full-colour gardening WEEKLY. News and advice on growing flowers, fruit and vegetables, plus colourful features on all aspects of gardening especially for the committed gardener. News and features welcome, especially if accompanied by top-quality photos or illustrations. Contact the editor before submitting any material.

## The Garden, Journal of the Royal Horticultural Society

RHS Media, 4th Floor, Churchgate New Road, Peterborough PE1 1TT

☎ 0845 260 0909  📠 01733 341633

thegarden@rhs.org.uk
www.rhs.org.uk

Owner *The Royal Horticultural Society*
Editor *Ian Hodgson*
Circulation 340,781

Founded 1866. MONTHLY journal of the Royal Horticultural Society. Covers all aspects of the art, science and practice of horticulture and garden making. 'Articles must have depth and substance.' Approach by letter or email with a synopsis in the first instance.

## Gardens Illustrated

Bristol Magazines Ltd, Tower House, Fairfax Street, Bristol BS1 3BN

☎ 0117 314 8774  📠 0117 933 8032

gardens@bbcmagazinesbristol.com
www.gardensillustrated.com

Owner *Bristol Magazines Ltd*
Editor *Juliet Roberts*
Circulation 32,101

Founded 1993. MONTHLY. Dubbed the 'World of Exteriors'. 'Internationally regarded as the most inspiring, authoritative and thought-provoking guide to gardens, plants and design.' Unsolicited mss are rarely used and it is best that prospective contributors approach the editor with ideas in writing, supported by photographs.

## Gay Times

🔁 GT

## Gibbons Stamp Monthly

Stanley Gibbons, 7 Parkside, Christchurch Road, Ringwood BH24 3SH

☎ 01425 472363  📠 01425 470247

hjefferies@stanleygibbons.co.uk
www.gibbonsstampmonthly.com

Owner *The Stanley Gibbons Group Plc*

Editor *Hugh Jefferies*
News & Art Editor *John Moody (jmoody@stanleygibbons.co.uk)*
Circulation 20,000

Founded 1890. MONTHLY. News and features. Unsolicited mss welcome. Make initial approach in writing by telephone or email to avoid disappointment.

FEATURES *Hugh Jefferies* Unsolicited material of specialized nature and general stamp features welcome. Max. 3,000 words but longer pieces can be serialized.

NEWS *John Moody* Any philatelic news item. Max. 500 words.

💷 Features, £40–50 per 1,000 words; news, no payment.

## Girl Talk

Media Centre, 201 Wood Lane, London W12 7TQ

☎ 020 8433 1010

girltalk.magazine@bbc.co.uk
www.bbcgirltalk.com

Owner *BBC Worldwide*
Editor *Sam Robinson*
Circulation 68,026

Founded 1995. FORTNIGHTLY. Lifestyle magazine for girls aged 8–12. Friendship and belonging, celebrity. No unsolicited submissions; send c.v. by email.

## Glamour

6–8 Old Bond Street, London W1S 4PH

☎ 020 7499 9080  📠 020 7491 2551

letters@glamourmagazine.co.uk
www.glamourmagazine.co.uk

Owner *Condé Nast*
Editor *Jo Elvin*
Circulation 515,281

Founded 2001. Handbag-size glossy women's magazine – fashion, beauty and celebrities. No unsolicited mss; send ideas for features in synopsis form to the Features Editor, *Corrie Jackson*.

## Go Caravan

Warners Group Publications plc, The Maltings, West Street, Bourne PE10 9PH

☎ 01778 391111  📠 01778 425437

sallyp@warnersgroup.co.uk
www.outandaboutlive.co.uk

Editor *Sally Pepper*

Founded in 2008 as an annual and MONTHLY from April 2010. 'Holiday ideas and inspiration magazine with high production values and perfect bound for caravanners.' Opportunities for UK travel writing particularly from those who have caravanning experience.

## Go Girl

Egmont UK Ltd, 239 Kensington High Street,
London w8 6sa
☏ 020 7761 3719
gogirlmag@euk.egmont.com
www.gogirlmag.co.uk
Owner *Egmont UK Ltd*
Editor *Emma Prosser*
Circulation 45,005

THREE-WEEKLY magazine covering everything in
a girl's world from fashion to friendship to celebs.
No unsolicited submissions but occasionally hire
freelance writers to work in-house. Approach by
email.

## Gold Dust Magazine

55 Elmdale Road, London E17 6PN
☏ 020 8765 4665
davidgardiner@worldonline.co.uk
www.golddustmagazine.co.uk
Owner *Omma Velada*
Editors *David Gardiner, Claire Tyne*
Circulation 500

Launched 2004. Biannual magazine of short
stories, flash fiction, short plays, film scripts and
novel extracts. Articles about writing, book and
film reviews. Unsolicited submissions welcome;
approach by email.
FICTION *David Gardiner* 'We are always looking
for quality work from new or experienced writers.
Please read all our submission guidelines before
submitting work.' Max. 3,000 words.
POETRY *Claire Tyne* Max. 50 lines.
FEATURES *David Gardiner* Book reviews. Max.
2,000 words.

## Golf Monthly

IPC Media Ltd, Blue Fin Building, 110 Southwark
Street, London SE1 0SU
☏ 020 3148 4530    ☏ 020 3148 8130
golfmonthly@ipcmedia.com
www.golf-monthly.co.uk
Owner *IPC Media (A Time Warner Company)*
Editor *Michael Harris*
Assistant Editor *Alex Narey*
Deputy Editor *Neil Tappin*
Associate Editor *Jeremy Elwood*
Circulation 65,133

Founded 1911. MONTHLY. Player profiles, golf
instruction, general golf features and columns.
Not interested in instruction material from
outside contributors. Approach in writing with
ideas.
FEATURES Max. 1,500–2,000 words.
£ Payment by arrangement.

## Golf World

Media House, Lynchwood, Peterborough
Business Park, Peterborough PE2 6EA
☏ 01733 468000    ☏ 01733 468843
chris.jones@bauermedia.co.uk
www.golf-world.co.uk
Owner *Bauer Media*
Editor *Chris Jones*
Commissioning Editor *Jock Howard*
Editorial Assistant *Linda Manigan*
Circulation 39,146

Founded 1962. MONTHLY. No unsolicited mss.
Approach in writing with ideas.

## Good Holiday Magazine

Parman House, 30–36 Fife Road, Kingston-upon-
Thames KT1 1SY
☏ 020 8547 9822    ☏ 020 8546 0984
edit@goodholidayideas.com
www.goodskiguide.com
Editor *John Hill*
Circulation 100,000

Founded 1985. QUARTERLY aimed at affluent
holiday-makers (income of £120,000+) rather
than travellers/backpackers. Worldwide
destinations including Europe and domestic. No
unsolicited material but approach in writing or by
email with ideas and/or synopsis.
£ Payment negotiable.

## Good Housekeeping

National Magazine House, 72 Broadwick Street,
London W1F 9EP
☏ 020 7439 5000    ☏ 020 7439 5616
firstname.lastname@natmags.co.uk
www.allaboutyou.com/goodhousekeeping
Owner *National Magazine Co. Ltd*
Editor *Lindsay Nicholson*
Circulation 430,089

Founded 1922. MONTHLY glossy. No unsolicited
mss. Write with ideas in the first instance to the
appropriate editor.
FEATURES *Lucy Moore* Most work is
commissioned but original ideas are always
welcome. No ideas are discussed on the
telephone. Send short synopsis, plus relevant
cuttings, showing previous examples of work
published. No unsolicited mss.
HEALTH *Julie Powell.* Submission guidelines as for
features; no unsolicited mss.

## Good Motoring

Station Road, Forest Row RH18 5EN
☏ 01342 825676    ☏ 01342 824847
editor@motoringassist.com
www.roadsafety.org.uk
Owner *Gem Motoring Assist*
Editor *James Luckhurst*

Circulation 62,000

Founded 1932. QUARTERLY motoring, road safety, travel and general features magazine. 1,500 words max. Prospective contributors should approach in writing only.

## Good News

PO Box 9831, Nottingham NG2 9JN
☎ 0115 923 3424
goodnewseditor@ntlworld.com
www.goodnews-paper.org.uk
Owner *Good News Fellowship*
Editor *Andrew Halloway*
Circulation 60,000

Founded 2001. MONTHLY Christian evangelistic newspaper which welcomes contributions. No fiction. Send for sample copy of writers' guidelines in the first instance. No poetry or children's stories.

NEWS Items of up to 500 words (preferably with pictures) showing the Christian faith in action. 'Churchy' items not wanted. Testimonies of how people have come to personal faith in Jesus Christ and the difference it has made are always welcome.
£ Payment negotiable.

## The Good Ski Guide

Parman House, 30–36 Fife Road, Kingston-upon-Thames KT1 1SY
☎ 020 8547 9822    ☐ 020 8546 0984
edit@goodholidayideas.com
www.goodskiguide.com
Owner *Good Holiday Group*
Editors *John Hill, Nick Dalton*
Circulation 250,000

Founded 1976. SIX ISSUES (October–Easter). Unsolicited mss welcome from writers with a knowledge of skiing and ski resorts. Prospective contributors are best advised to make initial contact in writing as ideas and work need to be seen before any discussion can take place. Good Ski Guide is distributed worldwide by British Airways, so features must have international flavour, e.g. not British interest only.
£ Payment negotiable.

## GQ

Vogue House, Hanover Square, London W1S 1JU
☎ 020 7499 9080    ☐ 020 7495 1679
www.gq.com
Owner *Condé Nast Publications Ltd*
Editor *Dylan Jones*
Circulation 120,057

Founded 1988. MONTHLY. Men's style magazine. No unsolicited material. Write or fax with an idea in the first instance.

## Granta

12 Addison Avenue, London W11 4QR
☎ 020 7605 1360    ☐ 020 7605 1361
editorial@granta.com
www.granta.com
Editor *John Freeman*
Deputy Editor *Ellah Allfrey*

QUARTERLY magazine of new writing, including fiction, poetry, memoirs, reportage and photography published in paperback book form. Highbrow, diverse and contemporary, often with a thematic approach. Unsolicited mss (including fiction) considered. A lot of material is commissioned. Vital to read the magazine first to appreciate its very particular fusion of cultural and political interests. No reviews or news articles. Access the website for submission guidelines.
£ Payment negotiable.

## Grazia

Bauer Media, Endeavour House, 189 Shaftesbury Avenue, London WC2H 8JG
☎ 020 7437 9011    ☐ 020 7520 6589
www.graziamagazine.co.uk
Owner *Bauer Consumer Media*
Editor *Jane Bruton*
Circulation 229,732

Launched February 2005. Britain's first WEEKLY glossy aimed at women aged 25 to 45. Features articles on fashion, celebrity, news, beauty and lifestyle.

## The Great Outdoors
> TGO

## The Great War (1914–1918)

PO Box 202, Scarborough YO11 3GE
☎ 01723 581329    ☐ 01723 581329
books@greatnorthernpublishing.co.uk
www.greatnorthernpublishing.co.uk
Owner *Great Northern Publishing*
Editor *Mark Marsay*

Founded 2001. BI-MONTHLY subscription only, non-academic magazine published in A5 format. 'The little magazine dedicated to the Great War (1914–18) and to those who perished and those who returned.' Articles, personal stories and accounts of those who served (men and women of all nationalities) and their families: diaries, anecdotes, letters, postcards, poetry, unit histories, events and memorials, etc. Absolutely no fiction or long academic works expounding historian's personal views. New material welcome but contact editor prior to sending. See website for submission guidelines and editorial content. 'Open door policy: all welcome regardless of ability to write to high standard as all work is

carefully edited. No subject or topic excluded.'
Sample copy £5.

## Grow Your Own

25 Phoenix Court, Hawkins Road, Colchester
CO2 8JY
℡ 01206 505979    🖷 01206 505945
lucy.halsall@aceville.co.uk
www.growfruitandveg.co.uk
Owner *Helen Tudor*
Editor *Lucy Halsall*

Launched 2005. MONTHLY magazine aimed at
aspiring self-sufficients giving information on the
best ways to grow (and cook) seasonal produce.
Contributions (features and news) are welcome
on anything relevant to fruit, vegetables and
herbs. Approach the editor by email.

## GT (Gay Times)

Unit M, Spectrum House, 32–34 Gordon House
Road, London NW5 1LP
℡ 020 7424 7400    🖷 020 7424 7401
edit@gaytimes.co.uk
www.gaytimes.co.uk
Owner *Millivres Prowler Group*
Editor *Tris Reid-Smith*
Circulation 68,000

Covers all aspects of gay life, plus general interest
likely to appeal to the gay community, art reviews,
fashion, style and news. Regular freelance writers
used.
💷 Payment negotiable.

## Guardian Weekend

▷ The Guardian under National Newspapers

## Guiding magazine

17–19 Buckingham Palace Road, London
SW1W 0PT
℡ 020 7834 6242    🖷 020 7828 8317
guiding@girlguiding.org.uk
www.girlguiding.org.uk
Owner *Girlguiding UK*
Editor *Jane Yettram*
Circulation 79,000

Founded 1914. UNSOLICITED mss welcome provided
topics relate to guiding and/or women's role
in society. Ideas in writing appreciated in first
instance. No nostalgic, 'when I was a Guide',
pieces, please.
ACTIVITY IDEAS Interesting, contemporary ideas
and instructions for activities for girls aged 5 to
18+ to do during unit meetings – crafts, games
(indoor/outdoor), etc.
FEATURES Topics relevant to today's women. 500
words.
NEWS Items likely to be of interest to members.
Max. 100–150 words.
💷 Payment negotiable.

## H&E Naturist

Burlington Court, Carlisle Street, Goole DN14 5EG
℡ 01405 760298/766769    🖷 01405 763815
editor@henaturist.net
www.henaturist.net
Owner *New Freedom Publications Ltd*
Editor *Sam Hawcroft*
Circulation 20,000

Founded 1898. MONTHLY naturist magazine.
FEATURES Will consider short features on social
nudism, longer features on nudist holidays
and nudist philosophy. 90% of every issue is by
freelance contributors. 800–1,500 words.
NEWS 'We are always on the lookout for
national and international nudist news stories.'
250–500 words. No soft porn, 'sexy' stories or
sleazy photographs. Approach by post, email or
telephone.

## Hair

IPC Media Ltd, Blue Fin Building, 110 Southwark
Street, London SE1 0SU
℡ 020 3148 7274
Owner *IPC Media (A Time Warner Company)*
Editor *Louise White*
Circulation 42,298

Founded 1977. TWELVE ISSUES YEARLY. Hair and
beauty magazine. No unsolicited mss, but always
interested in good photographs. Approach with
ideas in writing.
FEATURES Fashion pieces on hair trends and
styling advice.
💷 Payment negotiable.

## Hair Ideas

Origin Publishing, 9th Floor, Tower House,
Fairfax Street, Bristol BS1 3BN
℡ 0117 314 8805
michelletiernan@originpublishing.co.uk
www.loveyourhair.com
Owner *Origin Publishing*
Editor *Michelle Tiernan*
Commissioning Editor *Sophie Jordan*
Circulation 64,018

Launched 2003. MONTHLY guide for women
looking for a new hairstyle – hundreds of new
ideas, celebrity features and product tests. No
unsolicited material.

## Hairflair & Beauty

Haversham Publications Ltd, Freebournes House,
Freebournes Road, Witham CM8 3US
℡ 01376 534540    🖷 01376 534546
Owner *Haversham Publications Ltd*
Editor *Angela Barnes*
Circulation 27,628

Founded 1982. BI-MONTHLY. Original and
interesting hair and beauty-related features

written in a young, lively style to appeal to a readership aged 16–35 years.

## Harper's Bazaar

National Magazine House, 72 Broadwick Street, London W1F 9EP

☎ 020 7439 5000    🖷 020 7439 5506

www.harpersbazaar.co.uk

Owner *National Magazine Co. Ltd*
Editor *Lucy Yeomans*
Circulation 110,638

MONTHLY. Up-market glossy combining the stylish and the streetwise. Approach in writing (not by phone) with ideas.

FEATURES *Ajesh Patalay* Ideas only in the first instance.

NEWS Snippets welcome if very original.

💷 Payment negotiable.

## Harpers Wine and Spirit Magazine

William Reed Publishing, Broadfield Park, Crawley RH11 9RT

☎ 01293 613400    🖷 01293 610317

editorial@harpers.co.u
www.wine-spirit.com

Owner *William Reed Business Media*
Editor *Richard Siddle*
Circulation 11,275

FORTNIGHTLY. Business to business trades review magazine covering wine, spirits and beer. No unsolicited mss. Prospective contributors should approach in writing or by email.

## Health & Fitness Magazine

Swan House, 37–39 High Holborn, London WC1V 6AA

☎ 020 7421 5403

mary.comber@burdamagazines.co.uk
www.healthandfitnessonline.co.uk

Owner *Hubert Burda Media*
Editor *Mary Comber*
Circulation 32,580

Founded 1983. MONTHLY. Target reader: active, health-conscious women aged 25–40. Features news and articles on nutrition, exercise, healthy eating, holistic health and well-being. Will consider ideas; approach in writing in the first instance.

## Healthy

River Publishing, 3rd Floor, One Neal Street, London WC2H 9QL

☎ 020 7413 9447    🖷 020 7306 0314

healthy@riverltd.co.uk
www.healthy-magazine.co.uk

Owner *River Publishing*
Editor *Jane Druker*

Circulation 142,638

EIGHT ISSUES YEARLY. Customer health magazine for Holland & Barrett health food shops. Only mentions products sold in store and will *not* mention any products by brand name, even those sold in store. Authoritative and accessible information and advice. Will consider holistic health and lifestyle features; approach by email or post.

## Heat

Endeavour House, 189 Shaftesbury Avenue, London WC2H 8JG

☎ 020 7859 8657    🖷 020 7859 8670

heat@heatmag.com
www.heatworld.com

Owner *Bauer Consumer Media*
Editor *Sam Delaney*
Circulation 458,858

Founded 1999. WEEKLY entertainment magazine dealing with TV, film and radio information, fashion and features, with an emphasis on celebrity interviews and news. Targets 18 to 40-year-old readership, male and female. Articles written both in-house and by trusted freelancers. No unsolicited mss.

## Hello!

Wellington House, 69–71 Upper Ground, London SE1 9PQ

☎ 020 7667 8700    🖷 020 7667 8716

www.hellomagazine.com

Owner *Hola! (Spain)*
Deputy Editor *Ruth Sullivan*
Assistant Editor, Features *Rosie Nixon*
News Editor *Thomas Whitaker*
Circulation 409,043

WEEKLY. Owned by a Madrid-based publishing family, *Hello!* has grown faster than any other British magazine since its launch here in 1988. The magazine has editorial offices both in Madrid and London. Major colour features plus regular news pages. Although much of the material is provided by regulars, good proposals do stand a chance. Approach the commissioning editor with ideas in the first instance. No unsolicited mss.

FEATURES Interested in celebrity-based features, with a newsy angle, and exclusive interviews from generally unapproachable personalities.

💷 Payment by arrangement.

## Hi-Fi News

IPC Media Ltd, Leon House, 233 High Street, Croydon CR9 1HZ

☎ 020 8726 8311    🖷 020 8726 8397

hi-finews@ipcmedia.com
www.hifinews.co.uk

Owner *IPC Media (A Time Warner Company)*
Editor *Paul Miller*

Circulation 9,438

Founded 1956. MONTHLY. Write in the first instance with suggestions based on knowledge of the magazine's style and subject. All articles must be written from an informed technical or enthusiast viewpoint. £ Payment negotiable, according to technical content.

## High Life

85 Strand, London WC2R 0DW
☎ 020 7550 8000   📠 020 7550 8250
high.life@cedarcom.co.uk
www.cedarcom.co.uk
Owner *Cedar Communications*
Editor *Kerry Smith*
Circulation 190,257

Founded 1973. MONTHLY glossy. British Airways in-flight magazine. Almost all the content is commissioned. No unsolicited mss. Few opportunities for freelancers.

## History Today

20 Old Compton Street, London W1D 4TW
☎ 020 7534 8000
p.lay@historytoday.com
www.historytoday.com
Owner *History Today Trust for the Advancement of Education*
Editor *Paul Lay*
Circulation 25,274

Founded 1951. MONTHLY. General history and archaeology worldwide, history behind the headlines. Serious submissions only. Approach by email.

## Home & Family
▷ Families First

## Homes & Antiques

9th Floor, Tower House, Fairfax Street, Bristol BS1 3BN
☎ 0117 927 9009
natashagoodfellow@bbcmagazines.co
www.bbchomesandantiques.com
Owner *BBC Magazines Bristol*
Editor *Angela Linforth*
Circulation 60,222

Founded 1993. MONTHLY traditional home interest magazine with a strong bias towards antiques and collectibles. 'There are plenty of opportunities for freelancers and *Homes & Antiques* is always looking for interesting people and stories to feature in the fields of heritage, collecting, crafts and home design.' No fiction or health and beauty. Approach with ideas by phone or in writing.
FEATURES *Natasha Goodfellow* Pieces commissioned on recce shots and cuttings. Guidelines available on request. Send cuttings of

relevant work published. Max. 1,500 words.
£ Payment negotiable.

## Homes & Gardens

IPC Media Ltd, Blue Fin Building, 110 Southwark Street, London SE1 0SU
☎ 020 3148 7311   📠 020 3148 8165
www.homesandgardens.com
Owner *IPC Media (A Time Warner Company)*
Editor *Deborah Barker*
Circulation 132,117

Founded 1919. MONTHLY. Almost all published articles are specially commissioned. No fiction or poetry. Best to approach in writing with an idea, enclosing snapshots, if appropriate.

## Horse

IPC Media Ltd, Blue Fin Building, 110 Southwark Street, London SE1 0SU
☎ 020 3148 4609
joanna_browne@ipcmedia.com
www.horsemagazine.co.uk
Owner *IPC Media (A Time Warner Company)*
Editor *Joanna Browne*
Circulation 17,113

Founded 1997. MONTHLY magazine aimed at serious leisure riders and keen competitors. Each month the magazine includes easy-to-follow training tips from some of the world's top riders, alongside horsecare features and the latest news and gossip from the horseworld. Send feature ideas with a short synopsis to the editor.

## Horse and Hound

IPC Media Ltd, 9th Floor, Blue Fin Building, 110 Southwark Street, London SE1 0SU
☎ 020 3148 4562
jenny_sims@ipcmedia.com
www.horseandhound.co.uk
Owner *IPC Media (A Time Warner Company)*
Editor *Lucy Higginson*
Circulation 55,489

Founded 1884. WEEKLY. The oldest equestrian magazine on the market. Contains regular veterinary advice and instructional articles, as well as authoritative news and comment on fox hunting, international and national showjumping, horse trials, dressage, driving and endurance riding. Also weekly racing and point-to-points, breeding reports and articles. Regular books and art reviews, and humorous articles and cartoons are frequently published. Plenty of opportunities for freelancers. Unsolicited contributions welcome.
    Also publishes a sister monthly publication, *Eventing*, which covers the sport of horse trials comprehensively; Editor *Julie Harding*.
£ NUJ rates.

## Horse and Rider

Headley House, Headley Road, Grayshott
GU26 6TU
ⓣ 01428 601020    ⓕ 01428 601030
djm@djmurphy.co.uk
www.horseandrideruk.com
Owner *D.J. Murphy (Publishers) Ltd*
Editor *Nicky Moffatt*
Deputy Editor *Jane Gazzard*
Circulation 45,327

Founded 1949. MONTHLY. Adult readership,
largely horse-owning. News and instructional
features, which make up the bulk of the magazine,
are almost all written in-house or commissioned.
New contributors and unsolicited articles are
occasionally used. Approach the editor in writing
with ideas.

## Hotline

Ink Publishing, 141–143, Shoreditch High Street,
London E1 6JE
ⓣ 020 7613 8777
firstname.lastname@ink-publishing.co.uk
www.ink-publishing.com
www.ink-live.com/hotline/
Editors *Sophy Grimshaw, Steven Watson*

Published by INK since 2008. MONTHLY events-
led entertainment magazine reaching all Virgin
Trains passengers. Covers the 'best of the best'
across the Virgin Trains network including food
& drink, bars & clubs, music, film, comedy, art,
theatre, family entertainment and travel.

## House & Garden

Vogue House, Hanover Square, London W1S 1JU
ⓣ 020 7499 9080    ⓕ 020 7629 2907
www.houseandgarden.co.uk
Owner *The Condé Nast Publications Ltd*
Editor *Susan Crewe*
Circulation 130,777

Founded 1947. MONTHLY. Most feature material
is produced in-house but occasional specialist
features are commissioned from qualified
freelancers, mainly for the interiors, wine and
food sections and travel.
FEATURES *Hatta Byng* Suggestions for features,
preferably in the form of brief outlines of
proposed subjects, will be considered. Garden
features should be sent to *Clare Foster*.

## House Beautiful

National Magazine House, 72 Broadwick Street,
London W1F 9EP
ⓣ 020 7439 5000
Owner *National Magazine Co. Ltd*
Editor *Julia Goodwin*
Circulation 168,035

Founded 1989 and relaunched in November
2003. MONTHLY. Lively magazine offering sound,
practical information and plenty of inspiration for
those who want to make the most of where they
live. Over 100 pages of easy-reading editorial.
Regular features about decoration, DIY, food,
gardening and occasionally property, plus topical
features with a newsy slant to fit the 'Hot Topics'
slot. Approach in writing with synopses in the
first instance including a brief outline of areas of
specialism and examples of previously published
articles in a similar vein.

## How to Spend It
▷ Financial Times under National Newspapers

## i-D Magazine

124 Tabernacle Street, London EC2A 4SA
ⓣ 020 7490 9710    ⓕ 020 7490 9737
editor@i-dmagazine.co.uk
www.i-dmagazine.co.uk
Owner *Levelprint*
Editor *Ben Reardon*
Circulation 77,985

Founded 1980. MONTHLY lifestyle magazine for
both sexes with a fashion bias. International. Very
hip. Does not accept unsolicited contributions
but welcomes new ideas from the fields of
fashion, music, clubs, art, film, technology, books,
sport, etc. No fiction or poetry. 'We are always
looking for freelance non-fiction writers with new
or unusual ideas'. A different theme each issue –
past themes have included Green politics, taste,
films, sex, love and loud dance music – means
it is advisable to discuss feature ideas in the first
instance.

## Ideal Home

IPC Media Ltd, Blue Fin Building, 110 Southwark
Street, London SE1 0SU
ⓣ 020 3148 7357
Owner *IPC Media (A Time Warner Company)*
Editorial Director *Isobel McKenzie-Price*
Circulation 187,322

Founded 1920. MONTHLY glossy. Unsolicited
feature ideas are welcome only if appropriate to
the magazine. Prospective contributors wishing to
submit ideas should do so in writing to the editor.
No fiction.
FEATURES Furnishing and decoration of houses,
kitchens or bathrooms; interior design, soft
furnishings, furniture and home improvements,
lifestyle, consumer, gardening, property, food
and readers' homes. Length to be discussed with
editor.
ⓔ Payment negotiable.

## Image Magazine

Upper Mounts, Northampton NN1 3HR
ⓣ 01604 467043/467044    ⓕ 01604 467190
image@northantsnews.co.uk
www.northamptontoday.co.uk

Owner *Northamptonshire Newspapers Ltd*
Editor *Ruth Supple*
Circulation 12,000

Founded 1905. MONTHLY general interest regional magazine. No unsolicited mss.
FEATURES Local issues, personalities, businesses, etc., of Northamptonshire, Bedfordshire, Buckinghamshire interest. Max. 500 words.
NEWS No hard news as such, just monthly diary column.
OTHER Regulars on motoring, fashion, beauty, lifestyle and travel. Max. 500 words.
£ Payment for features negotiable.

## Independent Magazine
▷ The Independent under National Newspapers

## Inspire
CPO, Garcia Estate, Canterbury Road, Worthing
BN13 1BW
☎ 01903 264556
russbravo@cpo.org.uk
www.inspiremagazine.org.uk
Owner *Christian Publishing & Outreach Ltd*
Editor *Russ Bravo*
Sub-editor *Sharon Barnard*
Circulation 85,000

MONTHLY. Upbeat, good news Christian magazine featuring human interest stories, growing churches and community transformation. Limited freelance articles used. Contributor's guidelines available. Contact the editor by email in the first instance.

## InStyle
IPC Media Ltd, Blue Fin Building, 110 Southwark Street, London SE1 0SU
☎ 020 3148 5000   🖷 020 3148 8166
firstname_lastname@instyleuk.com
Owner *IPC Media (A Time Warner Company)*
Editor *Eilidh Macaskill*
Fashion Director *Chloe Beeney*
Circulation 184,141

Launched March 2001. MONTHLY. UK edition of US fashion, beauty, celebrity and lifestyle magazine. No unsolicited material.

## Interzone: Science Fiction & Fantasy
TTA Press, 5 Martins Lane, Witcham, Ely CB6 2LB
interzone@ttapress.com
www.ttapress.com
Owner *TTA Press*
Editor *Andy Cox*
Circulation 15,000

Founded 1982. BI-MONTHLY magazine of science fiction and fantasy. Unsolicited mss are welcome 'from writers who have a knowledge of the magazine and its contents'. S.a.s.e. essential for return.

FICTION 2,000–8,000 words.
FEATURES Book/film reviews, interviews with writers and occasional short articles. Length by arrangement.
£ Fiction, £30 per 1,000 words; features, negotiable.

## Investors Chronicle
Number One, Southwark Bridge, London SE1 9HL
☎ 020 7873 3000   🖷 020 7382 8105
www.investorschronicle.co.uk
Owner *Pearson*
Editor *Jonathan Eley*
Deputy Editor *Rosie Carr*
Circulation 30,006

Founded 1860. WEEKLY. Opportunities for freelance contributors in the survey section only. All approaches should be made in writing. Over forty surveys are published each year on a wide variety of subjects, generally with a financial, business or investment emphasis. Copies of survey list and synopses of individual surveys are obtainable from the surveys editor.
£ Payment negotiable.

## Irish Pages
The Linen Hall Library, 17 Donegall Square North, Belfast BT1 5GB
☎ 028 9043 4800
irishpages@yahoo.co.uk
www.irishpages.org
Editor *Chris Agee*
Circulation 2,800

Founded 2002. BIANNUAL, non-partisan, non-sectarian journal publishing writing from Ireland and abroad. 'The most important cultural journal in Ireland at the present moment' (Jonathan Allison, Director of the Yeats Summer School).
FEATURES Poetry, short fiction, essays, non-fiction, memoirs, nature-writing, translated work, literary journalism and other autobiographical, historical and scientific writing of literary distinction. Irish language and Ulster Scots writing are published in the original, with English translations. Equal editorial attention is given to established, emergent and new writers. Send submissions to the editor by post.

## Jade – The International Erotic Art and Literature Magazine
PO Box 202, Scarborough YO11 3GE
☎ 01723 581329   🖷 01723 581329
books@greatnorthernpublishing.co.uk
www.greatnorthernpublishing.co.uk
Owner *Great Northern Publishing*
Editor *Mark Marsay*

Founded 2002. MONTHLY adult subscription-only magazine. Uncensored, high-res PDF download format since April 2008. Official magazine of

The Guild of Erotic Artists. Features new and established international photographers, artists, sculptors and writers in all erotic genres from around the world. Some editorial articles and features within the genre. No advice columns or non-erotica related submissions. Contributors (photographers, artists, sculptors and writers – articles and fiction) should consult the website for submission guidelines, style and current requirements; contact the editor prior to sending material. Sample download copy £5 (contains strong adult content).

## Jane's Defence Weekly

Sentinel House, 163 Brighton Road, Coulsdon CR5 2YH
☏ 020 8700 3700    🅕 020 8763 1007
jdw@janes.com
www.janes.com
Owner *IHS Jane's*
Editor *Peter Felstead*
Circulation 28,100

Founded 1984. WEEKLY. No unsolicited mss. Approach in writing with ideas in the first instance.
FEATURES Current military and defence industry topics of worldwide interest. No history pieces. Max. 4,000 words.

## Jersey Now

PO Box 582, Five Oaks, St Saviour JE4 8XQ
☏ 01534 611743    🅕 01534 611610
eperchard@msppublishing.com
Owner *MSP Publishing*
Editor *Elisabeth Perchard*
Circulation 26,000

Founded 1987. QUARTERLY lifestyle magazine for Jersey covering homes, gardens, the arts, Jersey heritage, motoring, boating, fashion, beauty, food and drink, health, travel and technology. Upmarket glossy aimed at an informed and discerning readership. Interested in Jersey-orientated articles only; 1,200 words max. Approach the editor initially. 🅔 Payment negotiable.

## Jewish Chronicle

25 Furnival Street, London EC4A 1JT
☏ 020 7415 1500    🅕 020 7405 9040
editorial@thejc.com
www.thejc.com
Owner *Kessler Foundation*
Editor *Stephen Pollard*
Circulation 31,224

WEEKLY. Unsolicited mss welcome if 'the specific interests of our readership are borne in mind by writers'. Approach in writing, except for urgent current news items. No fiction. Max. 1,500 words for all material.

FEATURES *Alan Montague*
NEWS EDITOR *Jenni Frazer*
FOREIGN EDITOR *Miriam Shaviv*
POLITICAL EDITOR *Martin Bright*
SUPPLEMENTS *Angela Kiverstein*
🅔 Payment negotiable.

## Jewish Telegraph Group of Newspapers

Telegraph House, 11 Park Hill, Bury Old Road, Prestwich, Manchester M25 0HH
☏ 0161 740 9321    🅕 0161 740 9325
manchester@jewishtelegraph.com
www.jewishtelegraph.com
Owner *Jewish Telegraph Ltd*
Editor *Paul Harris (pharris@jewishtelegraph.com*
☏*0161 741 2633* 🅔 *0161 740 5555 )*
Circulation 16,000

Founded 1950. WEEKLY publication with local, national and international news and features. (Separate editions published for Manchester, Leeds, Liverpool and Glasgow.) Unsolicited features on Jewish humour and history welcome. NEWS Contact the newsdesk at newsdesk@jewishtelegraph.com or 0161 741 2631.

## Journal of Apicultural Research (incorporating Bee World)

IBRA, 16 North Road, Cardiff CF10 3DY
☏ 029 2037 2409    🅕 05601 135640
mail@ibra.org.uk
www.ibra.org.uk
Owner *International Bee Research Association*
Editor *Norman Carreck*
Circulation 1,700

Bee World, founded 1919; Journal of Apicultural Research, 1962. QUARTERLY. High-quality factual journal, including peer-reviewed articles, with international readership. Features on apicultural science and technology. Unsolicited mss welcome but authors should write to the editor for guidelines before submitting material.

## Journal of ApiProduct and ApiMedical Science

IBRA, 16 North Road, Cardiff CF10 3DY
☏ 029 2037 2409    🅕 05601 135640
mail@ibra.org.uk
www.ibra.org.uk
Owner *International Bee Research Association*
Editor *Professor Rose Cooper*

E-journal launched January 2009. Peer reviewed papers of highest standards published as original research covering the biologically relevant properties and substances of the six main hive products: honey, propolis, pollen, royal jelly, wax and bee venom.

## Kerrang!

Mappin House, 4 Winsley Street, London
W1W 8HF
☎ 020 7182 8000    ✆ 020 7182 8910
feedback@kerrang.com
www.kerrang.com
Owner *Bauer Consumer Media*
Editor *Nichola Browne*
Circulation 41,125

Founded 1981. WEEKLY rock, punk and metal
magazine. 'Written by fans for fans.' Will consider
ideas for features but not actively seeking new
contributors unless expert in specialist fields such
as black metal and hardcore, emo, etc.

## Koi Magazine

Origin Publishing, Tower House, Fairfax Street,
Bristol BS1 3BN
☎ 0117 927 9009    ✆ 0117 934 9008
koi@originpublishing.co.uk
www.koimag.co.uk
Owner *Origin Publishing*
Editor *Beckie Rodgers*
Circulation 15,000

Launched 1999. FOUR WEEKLY. Practical guide
to koi-keeping featuring everything you need to
know about the hobby, including koi appreciation
and pond construction. Contributions welcome;
send press releases, articles and photographs by
email.
FEATURES *Beckie Rodgers* Max. 4,000 words.
£ £150–£400.
NEWS *Amy Schroeter* Max. 200 words.

## Lads Mag
▷ Daily Sport under National Newspapers

## The Lady

39–40 Bedford Street, London WC2E 9ER
☎ 020 7379 4717    ✆ 020 7836 4620
www.lady.co.uk
Editor *Rachel Johnson*
Circulation 28,782

Founded 1885. WEEKLY. Unsolicited mss are
accepted provided they are not on the subject of
politics or religion, or on topics covered by staff
writers or special correspondents, i.e. fashion and
beauty, health, cookery, household, gardening,
finance and shopping.
FEATURES Well-researched pieces on British
and foreign travel, historical subjects or events;
interviews and profiles and other general interest
topics. Max. 1,100 words for illustrated two-page
articles; 800 words for one-page features.
Contributions from readers on how *The Lady*
changed their lives in some way are encouraged.
All material should be addressed to the editor
with s.a.s.e. enclosed. Photographs supporting
features may be supplied as colour transparencies,
b&w prints or on disk. Telephone enquiries about
features are not encouraged.

## Lancashire Life

3 Tustin Court, Port Way, Preston PR2 2YQ
☎ 01772 722022    ✆ 01772 736496
roger.borrell@lancashirelife.co.uk
www.lancashirelife.co.uk
Owner *Archant Life*
Editor *Roger Borrell*
Circulation 23,496

Founded 1947. MONTHLY county magazine.
Features and pictures about Lancashire.
Unsolicited contributions welcome; approach the
editor by email.

## Land Rover World

Kelsey Publishing Ltd, PO Box 978, Peterborough
PE1 9FL
www.landroverworld.co.uk
Owner *Kelsey Publishing Group*
Editor *Mike Gould*
Circulation 30,000

Founded 1994. MONTHLY. Incorporates *Practical
Land Rover World* and *Classic Land Rover World*.
Unsolicited material welcome, especially if
supported by high-quality illustrations.
FEATURES All articles with a Land Rover theme
of interest. Potential contributors are strongly
advised to examine previous issues before starting
work.
£ Payment negotiable.

## Lawyer 2B

50 Poland Street, London W1F 7AX
☎ 020 7970 4000
www.lawyer2b.com
Owner *Centaur Media plc*
Editor *Husnara Begum*
Circulation 25,000

Launched in 2001, *Lawyer 2B* is a sister
publication of *The Lawyer*, circulated to graduate
recruitment professionals, law students and
trainees. FOUR ISSUES PER ACADEMIC YEAR.
Regular opportunities for profiles/features
relating to legal education and law students.

## The Lawyer

50 Poland Street, London W1F 7AX
☎ 020 7970 4000
editorial@thelawyer.com
www.thelawyer.com
Owner *Centaur Media plc*
Editor *Catrin Griffiths*
News Editor *Margaret Taylor*
Special Reports Editor *Tom Phillips*
Associate Editor (New York) *Matt Byrne*

Circulation 28,221 (print); 182,633 (online)

Launched 1987. WEEKLY plus daily news email alerts and commentaries. Leading news magazine for City and commercial lawyers. Strong international coverage and leading-edge research. FEATURES Technical legal content, written in an accessible style. No unsolicited contributions; initial approach by email with specific idea.

## Lincolnshire Life

County House, 9 Checkpoint Court, Sadler Road, Lincoln LN6 3PW
☏ 01522 527127    🆔 01522 842000
editorial@lincolnshirelife.co.uk
www.lincolnshirelife.co.uk
Publisher *C. Bingham*
Executive Editor *Josie Thurston*
Circulation 10,000

Founded 1961. MONTHLY county magazine featuring geographically relevant articles on local culture, history, personalities, etc. Max. 1,500 words. Contributions supported by three or four good-quality photographs welcome. Approach in writing. 🖆 Payment varies.

## The Lincolnshire Poacher

County House, 9 Checkpoint Court, Sadler Road, Lincoln LN6 3PW
☏ 01522 527127    🆔 01522 842000
editorial@lincolnshirelife.co.uk
www.lincolnshirelife.co.uk
Publisher *C. Bingham*
Executive Editor *Josie Thurston*
Circulation 5,000

QUARTERLY county magazine featuring geographically relevant but nostalgic articles on history, culture and personalities of Lincolnshire. Max. 2,000 words. Contributions supported by three or four good-quality photographs/illustrations appreciated. Approach in writing. 🖆 Payment varies.

## The List

14 High Street, Edinburgh EH1 1TE
☏ 0131 550 3050    🆔 0131 557 8500
jonathan.ensall@list.co.uk
www.list.co.uk
Owner *The List Ltd*
Publisher *Robin Hodge*
Editor *Jonathan Ensall*
Circulation 8,493

Founded 1985. FORTNIGHTLY. Events guide covering Glasgow and Edinburgh. Interviews and profiles of people working in film, theatre, music and the arts. Max. 1,200 words. No unsolicited mss. News material tends to be handled in-house

## Literary Review

44 Lexington Street, London W1F 0LW
☏ 020 7437 9392    🆔 020 7734 1844
editorial@literaryreview.co.uk
www.literaryreview.co.uk
Editor *Nancy Sladek*
Circulation 15,000

Founded 1979. MONTHLY. Publishes book reviews (commissioned), features and articles on literary subjects. Prospective contributors are best advised to contact the editor in writing. Unsolicited mss not welcome. Runs a monthly competition for 'poems which rhyme, scan and make sense'. 🖆 Payment 'minuscule'.

## Litro Magazine

☏ 020 7792 0750
editor@litro.co.uk
www.litro.co.uk
Owner *Ocean Media Books Ltd*
Editor *Tom Chivers*

A free MONTHLY magazine distributed online, in libraries, galleries, bars, cafes and other venues in London and beyond. Publishes new, original short fiction. Contributors have included Irvine Welsh, Yiyun Li, Glyn Maxwell, Benjamin Zephaniah and Andrew Crumey. Unsolicited contributions are welcome; send to the editor by email. 🖆 No payment.

## Living etc

IPC Media Ltd, Blue Fin Building, 110 Southwark Street, London SE1 0SU
☏ 020 3148 7443
livingetc@ipcmedia.com
www.livingetc.co.uk
Owner *IPC Media (A Time Warner Company)*
Editor *Suzanne Imre*
Circulation 91,543

Founded 1997. MONTHLY homes magazine for modern living. Approach by email.
FEATURES *Mary Weaver/Claudia Bailie* House features – readers' homes.
NEWS *Bethan Ryder* Special reports on international design.

## Living France

Archant House, Oriel Road, Cheltenham GL50 1BB
☏ 01242 216086    🆔 01242 211081
editorial@livingfrance.com
www.livingfrance.com
Owner *Archant Life*
Editor *Eleanor O'Kane*

Founded 1989. FOUR-WEEKLY. A magazine for people who are considering or actively purchasing a home in France. Editorial consists of travel features as well as practical, friendly information and advice on living in France or becoming an owner of French property. Will consider articles describing different regions of France, interviews with owners of property or new-build and aspects of living in France. Also interested in submissions on interiors in France and gardens and gardening in France. Welcomes contact from professional writers. No unsolicited mss; approach by post or by email with an idea.

## Loaded

IPC Media Ltd, Blue Fin Building, 110 Southwark Street, London SE1 0SU
☎ 020 3148 5000    📠 020 3148 8107
andy_sherwood@ipcmedia.com
Owner *IPC Media (A Time Warner Company)*
Editor *Martin Daubney*
Deputy Editor *Andy Sherwood*
Circulation 71,251

Founded 1994. MONTHLY men's lifestyle magazine featuring music, sport, sex, humour, travel, fashion, hard news and popular culture. Will consider material which comes into these categories; approach the features editor in writing or by email in the first instance. No fiction or poetry.

## The London Magazine

*Editorial office:* Flat 5, 11 Queen's Gate, London SW7 5EL
☎ 0207 584 5977 (admin)    📠 020 7225 3273
admin@thelondonmagazine.net
www.thelondonmagazine.net
Publisher *Dr Burhan Al-Chalabi*
Editor *Sara-Mae Tuson*
Circulation 3,000

Founded originally in 1732. BI-MONTHLY review of literature and the arts. Publishes poems, short stories, features, memoirs, and book and performance reviews. Not interested in overtly political material. Postal submissions should include an s.a.s.e. (plus IRCs if overseas) to the editorial office.
FEATURES Max. 3,000 words.
FICTION Max. 3,000 words.
POEMS 'Any length within reason.'
£ No payment.

## London Review of Books

28 Little Russell Street, London WC1A 2HN
☎ 020 7209 1101    📠 020 7209 1102
edit@lrb.co.uk
www.lrb.co.uk
Owner *LRB Ltd*
Editor *Mary-Kay Wilmers*
Circulation 48,269

Founded 1979. FORTNIGHTLY. Reviews, essays and articles on political, literary, cultural and scientific subjects. Also poetry. Unsolicited contributions

welcome (approximately 50 received each week). No pieces under 2,000 words. Contact the editor in writing. Please include s.a.s.e.
£ £300 per 1,000 words; poems, £125.

## Lothian Life

Larick House, Whitehouse, Tarbert PA29 6XR
☎ 01880 730360/07734 699607
info@lothianlife.co.uk
www.lothianlife.co.uk
Owner *Pages Editorial & Publishing Services*
Editor *Suse Coon*

Online county magazine for people who live, work or have an interest in the Lothians. Features on successful people, businesses or initiatives. Regular articles by experts in the Lifestyle section including Homes & Gardens, Tastebuds, Out & About, The Arts and Health & Fitness. Phone first to discuss content and timing.
£ Payment: see website.

## Machine Knitting Monthly

PO Box 1479, Maidenhead SL6 8YX
☎ 01628 783080    📠 01628 633250
mail@machineknittingmonthly.net
www.machineknittingmonthly.net
Owner *RPA Publishing Ltd*
Editor *Anne Smith*

Founded 1986. MONTHLY. Unsolicited mss considered 'as long as they are applicable to this specialist publication. We have our own regular contributors each month but we're always willing to look at new ideas from other writers.' Approach in writing in the first instance.

## MacUser

30 Cleveland Street, London W1T 4JD
☎ 020 7907 6000
mailbox@macuser.co.uk
www.macuser.co.uk
Owner *Dennis Publishing*
Editor *Nik Rawlinson*
Circulation 10,675

Launched 1985. FORTNIGHTLY computer magazine.

## Management Today

174 Hammersmith Road, London W6 7JP
☎ 020 8267 4610
editorial@managementtoday.com
Owner *Haymarket Management Group*
Managing Director *Nick Farish*
Editor *Matthew Gwyther*
Circulation 100,016

Launched in 1966. General business topics and features. Ideas welcome. Send brief synopsis to the deputy editor.

## marie claire

IPC Media Ltd, Blue Fin Building, 110 Southwark Street, London SE1 0SU
☎ 020 3148 7513    🖷 020 3148 8120
marieclaire@ipcmedia.com
www.marieclaire.co.uk
Owner *European Magazines Ltd*
Editor *Trish Halpin*
Circulation 283,025

Founded 1988. MONTHLY. An intelligent glossy magazine for women, with strong international features and fashion. No unsolicited mss. Approach with ideas in writing or by email (marieclaireideas@ipcmedia.com). No fiction.
FEATURES *Miranda McMinn* Detailed proposals for feature ideas should be accompanied by samples of previous work.

## Marketing Week

St Giles House, 50 Poland Street, London W1F 7AX
☎ 020 7970 4000    🖷 020 7970 6722
mw.editorial@centaur.co.uk
www.marketingweek.co.uk
Owner *Centaur Communications*
Editor *Mark Shoueke*
Circulation 35,806

WEEKLY trade magazine of the marketing industry. Features on all aspects of the business, written in a newsy and up-to-the-minute style. Approach with ideas in the first instance.
FEATURES *Ruth Mortimer*
🖻 Payment negotiable.

## Markings

The Bakehouse, 44 High Street, Gatehouse of Fleet DG7 2HP
☎ 01557 814175
info@markings.org.uk
www.markings.org.uk
Editors *John Hudson, Chrys Salt*
Circulation 500+

Launched 1995. BIANNUAL Scottish literary magazine with an international readership. Poetry, contemporary art, short stories, criticism and reviews. Occasionally publishes special features on established or emergent literary figures. Unsolicited submissions are invited in any language but must be accompanied by English translation. 'Markings wishes to be challenging, controversial and loyal to honest and humane writing.' Black and white artwork appears in every issue; artists and illustrators are encouraged to submit work. Electronic submissions preferred but if sending by post enclose an s.a.se.
FEATURES Any subject relating to the arts. Max. 2,500 words.
FICTION Short stories on any subject. Max. 3,000 words.
POETRY Max. six submissions.

## Match!

Media House, Peterborough Business Park, Lynchwood, Peterborough PE2 6EA
☎ 01733 468040    🖷 01733 468724
match.magazine@bauermedia.co.uk
www.matchmag.co.uk
Owner *Bauer Media*
Editor *James Bandy*
Circulation 72,881

Founded 1979. WEEKLY. The UK's biggest-selling football magazine aimed at 8–14-year-olds. All material is generated in-house by a strong news and features team. Work experience placements often given to trainee journalists and students; several staff have been recruited through this route. Approach in writing or by phone.

## Maxim

30 Cleveland Street, London W1T 4JD
☎ 020 7907 6000    🖷 020 7907 6439
stuart_messham@dennis.co.uk
www.maxim.co.uk
Owner *Dennis Publishing*
Editor *Ben Raworth*
Deputy Editor *Stuart Messham*

Established in 1995, the print edition ceased publication in April 2009. Available as an online men's lifestyle magazine featuring sex, grooming, girls, fun-stuff, cars and fashion. No fiction or poetry. Send outlines of ideas only together with examples of published work.

## Mayfair

3rd Floor, Trojan Publishing, 207 Old Street, London EC1V 9NR
☎ 020 7608 6300
mayfair@paulraymond.com
www.paulraymond.com
Owner *Paul Raymond Publications*
Editor *Matt Berry*
Circulation 40,000

Founded 1966. THIRTEEN ISSUES YEARLY. Unsolicited material accepted if pertinent to the magazine and if accompanied by suitable illustrative material. 'We will *only* publish work if we can illustrate it.' Interested in features and humour aimed at men aged 18 to 80; 800–1,000 words. Punchy, bite-sized humour, top ten features, etc. 'Must make the editor laugh.' Also considers erotica.

## Mayfair Times

Blandel Bridge House, 56 Sloane Square, London SW1W 8AX
☎ 020 7259 1050    🖷 020 7901 9042
mayfair.times@pubbiz.com
www.pubbiz.com
Owner *Publishing Business Ltd*
Editor *Selma Day*

Circulation 21,000

Founded 1985. MONTHLY. Features on Mayfair of interest to residents, local workers, visitors and shoppers.

## Medal News

Token Publishing, Orchard House, Duchy Road, Heathpark, Honiton EX14 1YD

☎ 01404 46972  🖷 01404 44788

info@tokenpublishing.com

www.tokenpublishing.com

Owners *J.W. Mussell, Carol Hartman*

Editor *J.W. Mussell*

Circulation 6,000

Founded 1989. TEN ISSUES YEARLY. Unsolicited material welcome but initial approach by phone or in writing preferred.

FEATURES 'Opportunities exist for well-informed authors who know the subject and do their homework. It would help if a digital copy of the contribution is provided in the form of an email or CD. Relevant illustrations welcome.' Max. 2,500 words.

£ £30 per 1,000 words.

## Melody Maker

▷ New Musical Express

## Men's Health

National Magazine House, 33 Broadwick Street, London W1F 0DQ

☎ 020 7339 4400  🖷 020 7339 4444

www.menshealth.co.uk

Owner *Natmag-Rodale Publishing*

Editor *Morgan Rees*

Circulation 250,577

Founded 1994. MONTHLY men's healthy lifestyle magazine covering health, fitness, nutrition, stress and sex issues. No unsolicited mss; will consider ideas and synopses tailored to men's health. No fiction or extreme sports. Approach in writing in the first instance.

## MiniWorld Magazine

IPC Media Ltd, Leon House, 233 High Street, Croydon CR9 1HZ

☎ 020 8726 8000  🖷 020 8726 8398

miniworld@ipcmedia.com

www.miniworld.co.uk

Owner *IPC Media (A Time Warner Company)*

Editor *Monty Watkins*

Circulation 25,000

Founded 1991. THIRTEEN ISSUES YEARLY. Car magazine devoted to the Mini. Unsolicited material welcome but prospective contributors are advised to contact the editor.

FEATURES Maintenance, tuning, restoration, technical advice, classified, sport, readers' cars and social history of this cult car.

£ Payment negotiable.

## Mixmag

Development Hell Ltd, 90–92 Pentonville Road, London N1 9HS

☎ 020 7078 8400

www.mixmag.net

Owner *Development Hell Ltd*

Editor *Nick DeCosemo*

Circulation 26,116

Originally launched in 1982 and relaunched by new owners, Development Hell Ltd in 2006. Bestselling MONTHLY dance music magazine and 'leading authority on club culture'. News, features, night-life reviews, fashion, club listings and free mix CDs.

## Mizz

Panini UK Ltd, Broxbourne House, 77 Mount Ephraim, Tunbridge Wells TN4 8BS

☎ 01892 500100  🖷 01892 545666

mizz@panini.co.uk

www.mizz.com

Owner *Panini UK Ltd*

Editor *Karen O'Brien*

Circulation 49,230

Founded 1985. FORTNIGHTLY magazine for the 10–14-year-old girl.

FEATURES 'We have a full features team and thus do not accept freelance features.'

## Mobilise

Mobilise Organisation, Ashwellthorpe, Norwich NR16 1EX

☎ 01508 489449  🖷 01508 488173

editor@mobilise.info

www.mobilise.info

Owner *Mobilise Organisation*

Editor *Helen Smith*

Circulation 20,000

MONTHLY publication of the Mobilise Organisation, which aims to promote and protect the interests and welfare of disabled people and help and encourage them in gaining increased mobility. Various discounts available for members; membership costs £16 p.a. (single), £21 (joint). The magazine includes information for members plus members' letters. Approach in writing with ideas. Unsolicited mss welcome.

## Model Collector

IPC Focus Network, Leon House, 233 High Street, Croydon CR9 1HZ

☎ 020 8726 8000  🖷 020 8726 8299

modelcollector@ipcmedia.com

www.modelcollector.com

Owner *AOL Time Warner/IPC Media*

Editor *Lindsey Amrani*
Circulation 8,795

Founded 1987. THIRTEEN ISSUES YEARLY. Britain's best selling die cast magazine. From the latest models to classic 1930's Dinkys. Interested in historical articles about particular models or ranges and reviews of the latest products. Photographs welcome. Not interested in radio controlled models or model railways.
FEATURES Freelancers should contact the editor. Unsolicited material may be considered.
NEWS Trade news welcome. Call *Lindsey Armrani* (020 8726 8238) to discuss details.

## Mojo

Mappin House, 4 Winsley Street, London W1W 8HF
☎ 020 7182 8616   🖷 020 7182 8596
mojo@bauermedia.co.uk
www.mojo4music.com
Owner *Bauer Consumer Media*
Editor-in-Chief *Phil Alexander*
Circulation 98,484

Founded 1993. MONTHLY magazine containing features, reviews and news stories about all types of music and its influences. Receives about five mss per day. No poetry, think-pieces on dead rock stars or similar fan worship.
FEATURES Amateur writers discouraged except as providers of source material, contacts, etc.
NEWS All verifiable, relevant stories considered.
REVIEWS Write to Reviews Editor, *Jenny Bulley* with relevant specimen material.
£ Payment negotiable.

## Moneywise

1st Floor, Standon House, Mansell Street, London E1 8AA
☎ 020 7680 3600   🖷 020 7702 0710
rachel.lacey@moneywise.co.uk
johanna.gornitzki@moneywise.co.uk
www.moneywise.co.uk
Owner *Moneywise Publishing Ltd*
Editors *Rachel Lacey, Johanna Gornitzki*
Circulation 23,028

Founded 1990. MONTHLY. No unsolicited mss; ideas welcome. Make initial approach in writing to the editorial department (c.v. preferred).

## More

Endeavour House, 189 Shaftesbury Avenue, London WC2H 8JG
☎ 020 7208 3165   🖷 020 7208 3595
www.moremagazine.co.uk
Owner *Bauer Publishing Consumer Media*
Editor *Chantelle Horton*
Circulation 192,860

Founded 1988. WEEKLY women's magazine aimed at the working woman aged 18–25. Useful features on men, sex and relationships plus celebrity relationships gossip, and highstreet fashion, shocking real lives and practical beauty. Most items are commissioned; approach features editor and news editor with ideas. Prospective contributors are strongly advised to study the magazine's style before submitting anything.

## Mother and Baby

Endeavour House, 189 Shaftesbury Avenue, London WC2H 8JG
☎ 020 7295 5560
mother&baby@bauermedia.co.uk
Owner *Bauer Media*
Editor *Miranda Levy*
Circulation 52,416

Founded 1956. MONTHLY. Welcomes suggestions for feature ideas from national magazine and newspaper writers about pregnancy, newborn basics and baby development. Approaches may be made by email or in writing to the Senior Writer, *Jemma Walton.*

## Motor Boat & Yachting

IPC Media Ltd, Blue Fin Building, 110 Southwark Street, London SE1 0SU
☎ 020 3148 4651
mby@ipcmedia.com
www.mby.com
Owner *Time Warner*
Editor *Hugo Andreae*
Deputy Editor *Rob Peake*
Technical Editor *David Marsh*
Circulation 14,414

Founded 1904. MONTHLY for those interested in motor boats and motor cruising.
FEATURES *Hugo Andreae* Cruising features and practical features especially welcome. Illustrations/photographs (colour) are just as important as text. £ Payment by arrangement.
NEWS *Rob Peake* Factual pieces. Max. 200 words.
£ Up to £50 per item.

## Motor Boats Monthly

IPC Media Ltd, 9th Floor, Blue Fin Building, 110 Southwark Street, London SE1 0SU
☎ 020 3148 4664   🖷 020 3148 8128
carl_richardson@ipcmedia.com
Owner *IPC Media (A Time Warner Company)*
Editor *Carl Richardson*
Circulation 14,039

Founded 1987. MONTHLY magazine devoted to motor boating. Unsolicited contributions welcome: cruising articles, DIY and human interest. No powerboat racing or any sailing-related material. Approach by email.

## Motor Caravan Magazine

IPC Media Ltd, Leon House, 233 High Street,
Croydon CR9 1HZ
☏ 020 8726 8000
motorcaravan@ipcmedia.com
www.motorcaravanmagazine.co.uk
Owner *IPC Media (A Time Warner Company)*
Editor *Victoria Bentley*
Assistant Editor *Naomi Leach*
Circulation 10,000

Founded 1986. MONTHLY magazine with ideas
about where to go in your motor caravan, expert
advice to keep trips fun and practical advice on
vans and kit. Interested in features on touring,
unusual motorhomes, and activities people
pursue while out and about in their van. Email the
editor with ideas.

## Motor Cycle News

Media House, Lynchwood, Peterborough
Business Park, Peterborough PE2 6EA
☏ 01733 468000    ⅎ 01733 468028
MCN@motorcylenews.com
www.motorcylenews.com
Owner *Bauer Consumer Media*
Editor *Marc Potter*
Circulation 114,304

Founded 1955. WEEKLY. Interested in short
news stories and features on motorcyles and
motorcycle racing. Contact relevant desks direct.

## Motorcaravan Motorhome Monthly (MMM)

PO Box 88, Tiverton EX16 7ZN
mmmeditor@warnersgroup.co.uk
www.outandaboutlive.co.uk
Owner *Warners Group Publications Plc*
Editor *Mike Jago*
Circulation 39,001

Founded 1966. MONTHLY. 'There's no money in
motorcaravan journalism but for those wishing
to cut their first teeth ...' Unsolicited mss welcome
if relevant, but ideas in writing preferred in first
instance.
FEATURES Caravan site reports. Max. 500 words.
TRAVEL Motorcaravanning trips (home and
overseas). Max. 2,000 words.
NEWS Short news items for miscellaneous pages.
Max. 200 words.
SPECIAL PAGES DIY – modifications to
motorcaravans. Max. 1,500 words.
OWNER REPORTS Contributions welcome from
motorcaravan owners. Contact the editor for
requirements. Max. 2,000 words.
£ Payment varies but starts at £70 per published
page (incl.).

## Mountain Magic

Parman House, 30–36 Fife Road, Kingston-upon-
Thames KT1 1ST
☏ 020 8547 9822    ⅎ 020 8546 0984
info@goodholidayideas.com
Editor *John Hill*
Circulation 150,000

QUARTERLY magazine for those seeking healthy
holidays away from beaches and cities, such as hill
walking, trekking, fishing, windsurfing, etc. No
backpacking or serious climbing. No unsolicited
material but approach in writing or by email with
ideas and/or synopsis. £ Payment negotiable.

## Mslexia (For Women Who Write)

PO Box 656, Newcastle upon Tyne NE99 1PZ
☏ 0191 233 3860
postbag@mslexia.co.uk
www.mslexia.co.uk
Owner *Mslexia Publications Limited*
Editor *Daneet Steffens*
Circulation 10,700

Founded 1997. QUARTERLY. Articles, advice,
reviews, interviews, events for women writers
plus new poetry and prose. Will consider fiction,
poetry, features and letters but contributors *must*
send for guidelines first or see website for details.

## Music Week

Ludgate House, 245 Blackfriars Road, London
SE1 9UY
☏ 020 7921 5000
ben@musicweek.com
www.musicweek.com
Owner *United Business Media*
Publisher *Joe Hosken*
Editor *Paul Williams*
Circulation 5,292

Britain's only WEEKLY music business magazine.
Also produces a news and data website. Annual
music industry contacts book – the *Music Week
Directory*. Free daily news email. No unsolicited
mss. Approach in writing with ideas.
FEATURES Analysis of specific music business
events and trends.
NEWS Music industry news only.

## Musical Opinion

1 Exford Road, London SE12 9HD
☏ 020 8857 1582
musicalopinion@hotmail.co.uk
www.musicalopinion.com
Owner *Musical Opinion Ltd*
Editor *Robert Matthew-Walker*
Editor Emeritus *Denby Richards*
Circulation 8,500

Founded 1877. Glossy, full-colour BI-MONTHLY
magazine. International classical music content,
with topical features on music, musicians,

festivals, etc., and reviews (concerts, festivals, opera, ballet, jazz, CDs, DVDs, videos, books and printed music). International readership. No unsolicited mss; commissions only. Ideas always welcome though; approach by phone or email, giving telephone number. Visit the website for full information.
£ Payment negotiable.

## My Weekly
80 Kingsway East, Dundee DD4 8SL
☎ 01382 223131  📠 01382 452491
myweekly@dcthomson.co.uk
www.dcthomson.co.uk
Owner *D.C. Thomson & Co. Ltd*
Editor *Sally Hampton*
Features Editors *Sally Rodger (srodger@dcthomson.co.uk), Jennifer McEwan (jmcewan@dcthomson.co.uk)*
Fiction Editor *Liz Smith (lsmith@dcthomson.co.uk)*
Circulation 145,750

WEEKLY title aimed at women of 50-plus. Mix of health, fashion, food, celebrity news and fiction.
FEATURES Accepted on commission only basis. Pitch ideas in writing or via email.
FICTION Email for current guidelines. 'Unsuccessful mss will not be returned. Do not send an s.a.s.e. or the only copy of your work.'
£ Payment negotiable.

## My Weekly Pocket Novels
D.C. Thomson & Co. Ltd, 80 Kingway East, Dundee DD4 8SL
☎ 01382 575540  📠 01382 452491
mseed@dcthomson.co.uk
Owner *D.C. Thomson & Co. Ltd*
Editor *Maggie Seed*

WEEKLY. Publishes 25,000–30,000-word romantic stories aimed at the adult market. Send first three chapters and synopsis by post.

## National Trust Magazine
Heelis, Kemble Drive, Swindon SN2 2NA
☎ 01793 817716  📠 01793 817401
magazine@nationaltrust.org.uk
Owner *National Trust*
Editor *Sue Herdman*
Circulation 1.75 million

Founded 1968. THREE ISSUES YEARLY. Conservation of historic houses, coast and countryside in England, Northern Ireland and Wales. No unsolicited mss. Approach in writing with ideas.

## The Naturalist
c/o University of Bradford, Bradford BD7 1DP
☎ 01274 234212  📠 01274 234231
m.r.d.seaward@bradford.ac.uk
Owner *Yorkshire Naturalists' Union*

Editor *Prof. M.R.D. Seaward*
Circulation 5,000

Founded 1875. QUARTERLY. Natural history, biological and environmental sciences for a professional and amateur readership. Unsolicited mss and b&w illustrations welcome. Particularly interested in material – scientific papers – relating to the north of England. £ No payment.

## Nature
The Macmillan Building, 4–6 Crinan Street, London N1 9XW
☎ 020 7833 4000  📠 020 7843 4596
nature@nature.com
www.nature.com/nature
Owner *Nature Publishing Group*
Editor-in-Chief *Philip Campbell*
Circulation 55,000

Covers all fields of science, with articles and news on science and science policy only. Scope only for freelance writers with specialist knowledge in these areas.

## Nautical Magazine
4–10 Darnley Street, Glasgow G41 2SD
☎ 0141 429 1234  📠 0141 420 1694
info@skipper.co.uk
www.skipper.co.uk
Editor *Richard Brown*
Circulation 700

Founded 1832. MONTHLY. Nautical news and book reviews. Merchant navy related features and fiction. Max. 2,300 words. No political or 'hate' articles (e.g. naming of bad captains).

## NB
105 Judd Street, London WC1H 9NE
☎ 020 7391 2128
www.rnib.org.uk/nbmagazine
Owner *Royal National Institute of Blind People*
Editor *Ann Lee*
Circulation 10,000

Founded 1917. MONTHLY. Published in print, braille and on audio CD, DAISy CD and email. Unsolicited mss accepted. Authoritative items by professionals or volunteers working in the field of eye health and sight loss welcome. Max. 1,500 words.
£ Payment negotiable.

## New!
The Northern & Shell Building, 10 Lower Thames Street, London EC3R 6EN
☎ 020 8612 7016
karmel.doughty@express.co.uk
Owner *Richard Desmond*
Editor *Kirsty Tyler*

Circulation 600,741

Founded 2003. Celebrity WEEKLY with true life content plus fashion and beauty. Celebrity interviews and true life submissions welcome. Approach by email.

## New Humanist

1 Gower Street, London WC1E 6HD
☎ 020 7436 1151    🖷 020 7079 3588
editor@newhumanist.org.uk
www.newhumanist.org.uk
Owner *Rationalist Association*
Editor *Caspar Melville*
Circulation 5,000

Founded 1885. BI-MONTHY. Unsolicited mss welcome. No fiction.
FEATURES Articles with a humanist perspective welcome in the following fields: religion (critical), humanism, human rights, philosophy, current events, literature, history and science. 2,000 words.
BOOK REVIEWS 750–1,000 words, by arrangement with the editor.
£ Payment for features is nominal, but negotiable.

## New Internationalist

55 Rectory Road, Oxford OX4 1BW
☎ 01865 811400    🖷 01865 793152
ni@newint.org
www.newint.org/
Owner *New Internationalist Trust*
Co-Editors *Vanessa Baird, David Ransom, Adam Ma'anit, Jess Worth, Chris Brazier*
Editorial Coordinator *Jo Lateu*
Circulation 75,000

Radical and broadly leftist in approach, but unaligned. Concerned with world poverty and global issues of peace and politics, feminism and environmentalism, with emphasis on the Third World. Difficult to use unsolicited material as they work to a theme each month and features are commissioned by the editor on that basis. The way in is to send examples of published or unpublished work; writers of interest are taken up.

## New Musical Express

IPC Media Ltd, Blue Fin Building, 110 Southwark Street, London SE1 0SU
☎ 020 3148 5000    🖷 020 3148 8107
www.nme.com
Owner *IPC Media (A Time Warner Company)*
Editor *Krissi Murison*
Circulation 38,486

Britain's best-selling musical WEEKLY, incorporating *Melody Maker*. Freelancers used, but always for reviews in the first instance. Specialization in areas of music is a help.

REVIEWS: ALBUMS *Julian Marshall* Send in examples of work, either published or specially written samples.

## New Scientist

8th Floor, Lacon House, 84 Theobalds Road, London WC1X 8NS
☎ 020 8652 3500    🖷 020 7611 1250 (news)
www.newscientist.com
Owner *Reed Business Information Ltd*
Editor *Jeremy Webb*
Circulation 151,324 (worldwide)

Founded 1956. WEEKLY. No unsolicited mss. Approach with ideas (one A4-page synopsis) by fax or email.
FEATURES Commissions only, but good ideas welcome. Max. 3,500 words.
NEWS *Shaoni Bhattacharya* Mostly commissions, but ideas for specialist news welcome. Max. 1,000 words.
REVIEWS are commissioned.
OPINION Unsolicited material welcome if of general/humorous interest and related to science. Max. 1,000 words.
£ Payment negotiable.

## The New Shetlander

Market House, 14 Market Street, Lerwick ZE1 0JP
☎ 01595 743902    🖷 01595 696787
vas@shetland.org
www.va-shetland.co.uk
Owner *Voluntary Action Shetland*
Editors *Brian Smith, Laureen Johnson*
Circulation 1,400

Founded 1947. QUARTERLY literary magazine containing short stories, essays, poetry, historical articles, literary criticism, political comment, arts and books. The magazine has two editors and an editorial committee who all look at submitted material. Interested in considering short stories, poetry, historical articles with a northern Scottish or Scandinavian flavour, literary pieces and articles on Shetland. As a rough guide, items should be between 1,000 and 2,000 words although longer mss are considered. Initial approach in writing, please.
£ Complimentary copy.

## New Statesman

7th Floor, John Carpenter House, 7 Carmelite Street, London EC4Y 0BS
☎ 020 7936 6400    🖷 020 7936 6501
info@newstatesman.co.uk
www.newstatesman.com
Publisher *Spencer Neal*
Editor *Jason Cowley*
Culture Editor *Jonathan Derbyshire*
Deputy Culture Editor *Daniel Trilling*

Circulation 23,128

WEEKLY magazine, the result of a merger (1988) of *New Statesman* and *New Society*. Coverage of news, book reviews, arts, current affairs, politics and social reportage. Unsolicited contributions with s.a.s.e. will be considered.

## New Theatre Quarterly

Oldstairs, Kingsdown, Deal CT14 8ES
☎ 01304 373448
simontrussler@btinternet.com
www.uk.cambridge.org
Publisher *Cambridge University Press*
Editors *Simon Trussler, Maria Shevtsova*

Founded 1985 (originally launched in 1971 as *Theatre Quarterly*). Articles, interviews, documentation and reference material covering all aspects of live theatre. Recommend preliminary email enquiry before sending contributions. No theatre reviews or anecdotal material. Online submission preferred.

## New Welsh Review

PO Box 170, Aberystwyth SY23 1WZ
☎ 01970 628410    🖷 01970 628410
editor@newwelshreview.com
www.newwelshreview.com
Owner *New Welsh Review Ltd*
Editor *Kathryn Gray*
Circulation 3,500

Founded 1988. QUARTERLY Welsh literary magazine in the English language. Welcomes material of literary and cultural relevance to Welsh readers and those with an interest in Wales. Approach in writing in the first instance.
FEATURES Max. 3,000 words.
FICTION Max. 3,000 words.
REVIEWS Max. 800 words.
💷 Average of £165 (features); £80 (fiction); £45 (reviews); £27 per poem.

## The New Writer

PO Box 60, Cranbrook TN17 2ZR
☎ 01580 212626
editor@thenewwriter.com
www.thenewwriter.com
Publisher *Merric Davidson*
Editor *Suzanne Ruthven*

Founded 1996. Published BI-MONTHLY following the merger between *Acclaim* and *Quartos* magazines. Available by subscription only, TNW continues to offer practical 'nuts and bolts' advice on creative writing but with the emphasis on *forward-looking* articles and features on all aspects of the written word that demonstrate the writer's grasp of contemporary writing and current editorial/publishing policies. Plenty of news, views, competitions, reviews; writers' guidelines available with s.a.s.e. Monthly email

news bulletin is included free of charge in the subscription package.
FEATURES Unsolicited mss welcome. Interested in lively, original articles on writing in its broadest sense. Approach with ideas in writing in the first instance. No material is returned unless accompanied by s.a.s.e.
FICTION Publishes short-listed entries from competitions and subscriber-only submissions.
POETRY Unsolicited poetry welcome. Both short and long unpublished poems, providing they are original and interesting.
💷 Features, £20 per 1,000 words; fiction, £10 per story; poetry, £5 per poem.
Full guidelines and details of annual Prose & Poetry Prizes at the website.

## New Writing Scotland

Association for Scottish Literary Studies, c/o Department of Scottish Literature, 7 University Gardens, University of Glasgow G12 8QH
☎ 0141 330 5309    🖷 0141 330 5309
nws@asls.org.uk
www.asls.org.uk
Owner *Association for Scottish Literary Studies*
Managing Editor *Duncan Jones*
Editors *Alan Bissett, Carl MacDougall*

ANNUAL anthology of contemporary poetry and prose in English, Gaelic and Scots, produced by the **Association for Scottish Literary Studies** (see entry under *Professional Associations and Societies*). Will consider poetry, drama, short fiction or other creative prose but not full-length plays or novels, though self-contained extracts are acceptable. Contributors should be Scottish by birth, upbringing, or inclination, or resident in Scotland. Max. 3,500 words is suggested. Send no more than one short story and no more than four poems. Submissions should be accompanied by two s.a.s.e.s (one for receipt, the other for return of mss). Mss, which must be sent by 30th September, should be typed on one side of the paper only with the sheets secured at top left-hand corner. Prose pieces should be double-spaced. Provide covering letter with full contact details but do not put name or address on individual work(s). Prose pieces should carry an approximate word count. 💷 £20 per printed page.

## newbooks

4 Froxfield Close, Winchester SO22 6JW
guy@newbooksmag.com
www.newbooksmag.com
Owner/Editor *Guy Pringle*
Circulation 80,000

Founded 2000. BI-MONTHLY magazine for readers and reading groups with extracts from the best new fiction and free copies to be claimed. No unsolicited contributions. Also publishes *tBkmag* for 8–12-year-olds.

## The North
▷ under Poetry Magazines

## Notes from the Underground
23 Sutherland Square, London SE17 3EQ
info@notesfromtheunderground.co.uk
www.notesfromtheunderground.co.uk
Joint Editors *Christopher Vernon,*
*Tristan Summerscale*
Commissioning Editors *Hannah Gilkes,*
*Sunita Soliar*

Free creative writing paper, aimed at bridging the gap between niche literary journals and those free newspapers that have a low quality of editorial content. Publishes short stories, cartoons and original illustrations as well as arts-based non-fiction. A print run of 100,000 copies. All the content from the print edition is available online along with web exclusive stories, videos and podcasts. Welcomes unsolicited contributions. Contributor information available on the website.

## Now
IPC Media Ltd, Blue Fin Building, 110 Southwark Street, London SE1 0SU
☏ 020 3148 6373
Owner *IPC Media (A Time Warner Company)*
Editor *Abigail Blackburn*
Circulation 394,130

Founded 1996. WEEKLY magazine of celebrity gossip, news and topical features aimed at the working woman. Unlikely to use freelance contributions due to specialist content – e.g. exclusive showbiz interviews – but ideas will be considered. Approach in writing; no faxes.

## Nursery Education
Scholastic Ltd, Villiers House, Clarendon Avenue, Leamington Spa CV32 5PR
☏ 01926 887799    ⊕ 01926 883331
earlyyears@scholastic.co.uk
www.scholastic.co.uk
Owner *Scholastic Ltd*
Publishing Director *Helen Freeman*
Circulation 14,506

Founded 1997. MONTHLY print and online magazine for early years professionals. Welcomes contributions from freelance journalists specializing in education (print and online).
FEATURES Early years issues explored. Max. 1,500 words.
NEWS Relevant and timely news for the early years sector. Max. 400 words.
PRACTICAL IDEAS Themed activities that support the Foundation Stage curriculum and Birth-to-Three activites. Max. 800 words. Approach by letter or email.

## Nursery World
174 Hammersmith Road, London W6 7JP
☏ 020 8267 8409
news.nw@haymarket.com
www.nurseryworld.co.uk
Owner *Haymarket Publishing*
Editor *Liz Roberts*
Circulation 18,000

Founded 1925. WEEKLY publication for professionals working with children from birth to five in nurseries, schools, pre-schools, etc. No articles aimed at parents. Contact by email.
FEATURES *Ruth Thomson* Case studies, good practice and management. ⓔ £200 per 1,000 words.
NEWS *Catherine Gaunt* Occasional freelance commissions or shifts by people with some relevant experience. ⓔ £200 per 1,000 words.

## Nursing Times
Greater London House, Hampstead Road, London NW1 7EJ
☏ 020 7728 3702    ⊕ 020 7728 3700
nt@emap.com
www.nursingtimes.net
Owner *Emap Public Sector*
Editor *Alastair McLellan*
Circulation 30,923

Some of *Nursing Times'* feature content is from unsolicited contributions sent on spec. Articles on all aspects of nursing and health care, including clinical information, written in a contemporary style, are welcome.

## Nuts
IPC Media Ltd, Blue Fin Building, 110 Southwark Street, London SE1 0SU
☏ 020 3148 5000
nutsmagazine@ipcmedia.com
www.nuts.co.uk
Owner *IPC Media (A Time Warner Company)*
Editor *Dominic Smith*
Circulation 176,835

Launched 2004. WEEKLY magazine for men aged 16–45 'who like chat, women, gossip, cars and football'. Welcomes true life stories and items on gadgets. Approach by email in the first instance.

## OK! Magazine
The Northern & Shell Building, 10 Lower Thames Street, London EC3R 6EN
☏ 020 8612 7000    ⊕ 020 8612 7005
firstname.lastname@express.co.uk
Owner *Northern & Shell Media/Richard Desmond*
Editor *Lisa Byrne*
Circulation 588,546

Founded 1996. WEEKLY celebrity-based magazine. Welcomes interviews and pictures on well known personalities, and ideas for general features.

Approach the editor by phone or fax in the first instance.

## Old Glory

Mortons Heritage Media, Mortons Way, Horncastle LN9 6JR
☎ 01507 529306    🖷 01507 529495
editor@oldglory.co.uk
www.oldglory.co.uk
Owner *Mortons Media Ltd*
Editor *Colin Tyson*

Founded 1988. MONTHLY magazine covering all aspects of transport and industrial heritage both in the UK and overseas. Specializes in vintage vehicles and steam preservation. No unsolicited articles. Approach with idea by email to the editor.
FEATURES Restoration projects from steam rollers to windmills. Max. 2,000 words.
NEWS Steam and transport rally reports. News concerning industrial and heritage sites and buildings.

## The Oldie

65 Newman Street, London W1T 3EG
☎ 020 7436 8801    🖷 020 7436 8804
editorial@theoldie.co.uk
www.theoldie.co.uk
Owner *Oldie Publications Ltd*
Editor *Richard Ingrams*
Circulation 35,965

Founded 1992. MONTHLY general interest magazine with a strong humorous slant for the older person. Submissions welcome; enclose s.a.s.e. No poetry.

## Olive

Media Centre, 201 Wood Lane, London W12 7TQ
☎ 020 8433 1402
firstname.surname@bbc.com
www.bbcgoodfood.com
Owner *BBC Worldwide Publishing Ltd*
Editor *Christine Hayes*
Editorial Coordinator *Danielle Theunissen*
Features Editor *Jessica Gunn (jess.gunn@bbc.com)*
Food Editor *Janine Ratcliffe*
Travel Editor *Alison Bowles*
Circulation 93,300

Launched 2003. MONTHLY food magazine. Recipes, restaurants and food lovers' travel. Unsolicited pitches are welcome 'but please note that we will reply *only* to those we plan to commission'.

## Opera

36 Black Lion Lane, London W6 9BE
☎ 020 8563 8893    🖷 020 8563 8635
editor@opera.co.uk
www.opera.co.uk
Owner *Opera Magazine Ltd*

Editor *John Allison*
Circulation 6,500

Founded 1950. MONTHLY review of the current opera scene. Almost all articles are commissioned and unsolicited mss are not welcome. All approaches should be made in writing.

## Opera Now

241 Shaftesbury Avenue, London WC2H 8TF
☎ 020 7333 1740    🖷 020 7333 1769
opera.now@rhinegold.co.uk
www.rhinegold.co.uk/operanow
Publisher *Rhinegold Publishing Ltd*
Editor-in-Chief *Ashutosh Khandekar*
Deputy Editor *Antonia Couling*
Assistant Editor *Hannah Beynon*

Founded 1989. BI-MONTHLY. News, features and reviews aimed at those involved as well as those interested in opera. No unsolicited mss. All work is commissioned. Approach with ideas in writing.

## OS (Office Secretary) Magazine

15 Grangers Place, Witney OX28 4BS
☎ 0141 567 6028
kim.mcallister@peeblesmedia.com
www.peeblesmedia.com
Owner *Peebles Media Group Ltd*
Editor *Kim McAllister*
Circulation 22,520

Founded 1986. BI-MONTHLY. Features articles of interest to secretaries and personal assistants aged 25–60. No unsolicited mss.
FEATURES Informative pieces on technology and practices, office and employment-related topics. Length 1,000 words. Approach with ideas by telephone or in writing.
💷 Payment by negotiation.

## Park Home & Holiday Caravan

IPC Media Ltd, Leon House, 233 High Street, Croydon CR9 1HZ
☎ 020 8726 8252    🖷 020 8726 8299
alex_melvin@ipcmedia.com
www.phhc.co.uk
Owner *IPC Inspire*
Editor *Alex Melvin*
Circulation 5,000

Founded 1960. THIRTEEN ISSUES YEARLY. News of parks, models, legislation, accessories, lifestyle for those living in residential park homes or owning holiday caravans. Welcomes material specifically related to park home and holiday caravan sites; no general touring features. Approach by email.

## PC Format

Future Publishing, 30 Monmouth Street, Bath BA1 2BW
☎ 01225 442244    🖷 01225 732295

pcfmail@futurenet.com
www.pcformat.co.uk
Owner *Future Publishing*
Publisher *Stuart Anderton*
Editor *Alan Dexter*
Circulation 11,914

Founded 1991. FOUR-WEEKLY magazine covering performance hardware and gaming. Welcomes feature and interview ideas in the first instance; approach by email only.

## Pensions World

2 Addiscombe Road, Croydon CR9 5AF
☎ 020 8212 1946    📠 020 8212 1920
stephanie.hawthorne@lexisnexis.co.uk
www.pensionsworld.co.uk
Owner *Reed Elsevier*
Editor *Stephanie Hawthorne*
Circulation 8,429

Launched 1972. MONTHLY magazine for pensions professionals. Pensions, law, investment and retirement issues. Specialist contributors only. No consumer pension stories. Approach by email.
FEATURES Max. 1,500 words.
NEWS Max. 400 words.
£ Payment negotiable.

## People Management

Personnel Publications Ltd., 17–18 Britton Street, London EC1M 5TP
☎ 020 7324 2729    📠 020 7324 2791
editorial@peoplemanagement.co.uk
www.peoplemanagement.co.uk
Editor *Rima Evans*
Circulation 132,168

FORTNIGHTLY magazine on human resources, industrial relations, employment issues, etc. Welcomes submissions but apply for 'Guidelines for Contributors' in the first instance.
FEATURES *Jane Pickard/Claire Warren*
NEWS *James Brockett*
LAW AT WORK *Jill Evans*

## People's Friend Pocket Novels

D.C. Thomson & Co. Ltd, 80 Kingway East, Dundee DD4 8SL
☎ 01382 575540    📠 01382 452491
tsteel@dcthomson.co.uk
Owner *D.C. Thomson & Co. Ltd*
Editor *Tracey Steel*

Founded 1938. FORTNIGHTLY. Publishes 50,000-word romantic stories. Unsolicited contributions welcome. Phone, write or email for guidelines; synopsis and first three chapters to be sent initially.

## The People's Friend

80 Kingsway East, Dundee DD4 8SL
☎ 01382 223131    📠 01382 452491
peoplesfriend@dcthomson.co.uk
Owner *D.C. Thomson & Co. Ltd*
Editor *Angela Gilchrist*
Circulation 306,591

The *Friend* is a fiction magazine, with two serials and several short stories each week. Founded in 1869, it has always prided itself on providing 'a good read for all the family'. Stories should be about ordinary, identifiable characters with the kind of problems the average reader can understand and sympathize with. 'We look for the romantic and emotional developments of characters, rather than an over-complicated or contrived plot. We regularly use period serials and, occasionally, mystery/adventure.' Guidelines on request with s.a.s.e.
SHORT STORIES Can vary in length from 1,000 words or less to 4,000.
SERIALS Serials of 8–12 instalments.
ARTICLES Short fillers welcome.
£ Payment on acceptance.

## Period Living

St Giles House, 50 Poland Street, London W1F 7AX
☎ 020 7970 4433    📠 020 7970 4438
period.living@centaur.co.uk
www.periodliving.co.uk
Owner *Centaur*
Editor-in-Chief *Michael Holmes*
Editor *Sarah Whelan*
Circulation 45,223

Founded 1992. Formed from the merger of *Period Living* and *Traditional Homes*. Covers interior decoration in a period style, period house profiles, traditional crafts, renovation of period properties.
£ Payment varies according to length/type of article.

## The Philosopher

6 Craghall Dene Avenue, Newcastle upon Tyne NE3 1QR
papers@the-philosopher.co.uk
www.the-philosopher.co.uk
Owner *The Philosophical Society of England*
Editor *Martin Cohen*

Founded 1913. BIANNUAL journal of the Philosophical Society of Great Britain with an international readership made up of members, libraries and specialist booksellers. Wide range of philosophical interests, but leaning towards articles that present philosophical investigation which is relevant to the individual and to society in our modern era. Accessible to the non-specialist. Will consider articles and book reviews. 'Notes for Contributors' available; see website. As well as short philosophical papers, will accept:

NEWS about lectures, conventions, philosophy groups. Ethical issues in the news. Max. 1,000 words.

REVIEWS of philosophy books (max. 600 words); discussion articles of individual philosophers and their published works (max. 2,000 words).

MISCELLANEOUS items, including graphics, of philosophical interest and/or merit.

£ Free copies.

## Piano

241 Shaftesbury Avenue, London WC2H 8TF

T 020 7333 1744     F 020 7333 1766

piano@rhinegold.co.uk

www.rhinegold.co.uk

Owner *Rhinegold Publishing*

Editor *Jeremy Siepmann*

Circulation 11,000

Founded 1993. BI-MONTHLY magazine containing features, profiles, technical information, news, reviews of interest to those with a serious amateur or professional concern with pianos or their playing. No unsolicited material. Approach with ideas in writing only.

## Picture Postcard Monthly

15 Debdale Lane, Keyworth, Nottingham NG12 5HT

T 0115 937 4079     F 0115 937 6197

reflections@postcardcollecting.co.uk

www.postcardcollecting.co.uk

Owners *Brian & Mary Lund*

Editor *Brian Lund*

Circulation 4,000

Founded 1978. MONTHLY. News, views, clubs, diary of fairs, sales, auctions, and well-researched postcard-related articles. Might be interested in general articles supported by postcards. Unsolicited mss welcome. Approach by phone, email or post with ideas.

## Pilot

Archant Specialist, 3 The Courtyard, Denmark Street, Wokingham RG40 2AZ

T 0118 989 7246

nick.bloom@pilotweb.aero

www.pilotweb.co.uk

Publisher *Archant Specialist*

Editor *Nick Bloom*

Circulation 15,754

Founded 1966. MONTHLY magazine for private plane pilots. Much of the magazine is written by invitation from regular contributors. Unsolicited mss welcome but ideas in writing preferred. Non-pilot authors unlikley to be considered. Perusal of any issue of the magazine will reveal the type of material bought. See the website for advice to would-be contributors.

NEWS Contributions, preferably with photographs, need to be as short as possible. See *Pilot Notes* and *Old-Timers* in the magazine.

FEATURES Many articles are unsolicited personal experiences/travel accounts from pilots of private planes; good photo coverage is very important. Max. 5,000 words.

## Pink Paper

Unit M, Spectrum House, 32–34 Gordon House Road, London NW5 1LP

T 020 7424 7400     F 020 7424 7401

news@pinkpaper.com

www.pinkpaper.com

Owner *Millivres Prowler Group*

Editor *Tris Reid-Smith*

Founded 1987. Having suspended its print run in 2009, *Pink Paper* is now online only. 'Britain's premiere website' for lesbians and gay men covering politics, social issues, health, the arts, celebrity interviews and all areas of concern to lesbian/gay people. Unsolicited mss welcome. Initial approach by post with an idea preferred. Interested in profiles, reviews, in-depth features and short news pieces.

£ Payment by arrangement.

## Poetry Ireland Review
▷ under Poetry Magazines

## Poetry Review
▷ under Poetry Magazines

## Poetry Scotland
▷ under Poetry Magazines

## Poetry Wales
▷ under Poetry Magazines

## PONY

D.J. Murphy (Publishers) Ltd, Headley House, Headley Road, Grayshott GU26 6TU

T 01428 601020     F 01428 601030

djm@djmurphy.co.uk (text only)

www.ponymag.com

Owner *D.J. Murphy (Publishers) Ltd*

Editor *Janet Rising*

Assistant Editor *Penny Rendall*

Circulation 42,219

Founded 1948. Lively MONTHLY aimed at 10–16-year-olds. News, instruction on riding, stable management, veterinary care, interviews. Approach in writing with an idea.

FEATURES welcome. Max. 900 words.

NEWS Written in-house. Photographs and illustrations (serious/cartoon) welcome.

£ £65 per 1,000 words.

## post plus
▷ Sunday Post under National Newspapers

## PR Week

174 Hammersmith Road, London W6 7JP
☎ 020 8267 4429   📠 020 8267 4509
newsdesk.prweek@haymarket.com
www.prweek.com/uk
Owner *Haymarket Business Publications Ltd*
Editor *Daniel Rogers*
Circulation 14,818

Founded 1984. WEEKLY. Contributions accepted
from experienced journalists. Approach in
writing with an idea.
💷 Payment negotiable.

## Practical Boat Owner

IPC Media Ltd, Westover House, West Quay
Road, Poole BH15 1JG
☎ 01202 440820   📠 01202 440860
pbo@ipcmedia.com
www.pbo.co.uk
Owner *IPC Media (A Time Warner Company)*
Editor *Sarah Norbury*
Circulation 42,096

Britain's biggest selling boating magazine.
MONTHLY magazine of practical information for
cruising boat owners. Receives about 1,500 mss
per year.
FEATURES Technical articles about maintenance,
restoration, modifications to cruising boats,
power and sail up to 45ft. European and British
regional pilotage articles and cruising guides.
Approach in writing with synopsis in the first
instance.
💷 Payment negotiable.

## Practical Caravan

Teddington Studios, Broom Road, Teddington
TW11 9BE
☎ 020 8267 5629   📠 020 8267 5725
practical.caravan@haymarket.com
www.practicalcaravan.com
Owner *Haymarket Consumer Media*
Editor *Nigel Donnelly*
Circulation 40,242

Founded 1967. MONTHLY. Contains caravan
reviews, travel features, investigations, products,
park reviews. Unsolicited mss welcome on travel
relevant only to caravanning/touring vans. No
motorcaravan or static van stories. Approach with
ideas by phone, letter or email.
FEATURES Must refer to caravanning. Written in
friendly, chatty manner. Pictures essential. Max.
length 2,000 words.
💷 Payment negotiable.

## Practical Fishkeeping

Bauer Media, Media House, Lynchwood,
Peterborough PE2 6EA
☎ 01733 468000
firstname.lastname@bauermedia.co.uk
www.practicalfishkeeping.co.uk
Owner *Bauer Media*
Editor *Karen Youngs*
Circulation 14,558

THIRTEEN ISSUES YEARLY. Practical articles
on all aspects of fishkeeping. Unsolicited mss
welcome; approach in writing with ideas. Quality
photographs of fish always welcome. No fiction
or verse.

## Practical Land Rover World
▷ Land Rover World

## Practical Parenting

Magicalia Ltd, 15–18 White Lion Street, London
N1 9PD
☎ 020 7843 8871
Owner *Magicalia Ltd*
Editor *Daniella Delaney*
Circulation 24,022

Founded 1987. MONTHLY. Practical advice on
pregnancy, birth, babycare and childcare, 0–4
years. Submit ideas/synopses by email. Interested
in feature articles of up to 1,500 words in length,
and in readers' experiences/personal viewpoint
pieces of between 750–1,000 words. All material
must be written for the magazine's specifically
targeted audience and in-house style. Submissions
to: *Jenny Stallard* (jenny.stallard@magicalia.com)
💷 Payment negotiable.

## Practical Photography

Bauer Media, Media House, Lynchwood,
Peterborough PE2 6EA
☎ 01733 468000   📠 01733 468387
practical.photography@bauermedia.co.uk
www.photoanswers.co.uk
Owner *Bauer Media*
Editor-in-Chief *Andrew James*
Circulation 50,654

MONTHLY All types of photography, particularly
technique-orientated pictures. No unsolicited
mss. Preliminary approach may be made by email.
Always interested in new, especially unusual,
ideas.
FEATURES Anything relevant to the world of
photography, but 'not the sort of feature produced
by staff writers. Features on technology, digital
imaging techniques and humour are three areas
worth exploring. Bear in mind that there is a
three-month lead-in time.' Max. 2,000 words.
💷 Payment varies.

## Practical Wireless

Arrowsmith Court, Station Approach, Broadstone
BH18 8PW
☎ 0845 803 1979   📠 01202 659950
rob@pwpublishing.ltd.uk
www.pwpublishing.ltd.uk

Owner *PW Publishing Ltd*
Editor *Rob Mannion (G3XFD/EI5IW)*
Circulation 20,000

Founded 1932. MONTHLY. News and features relating to Amateur Radio, radio construction and radio communications. Author's guidelines available on request and are essential reading (send s.a.s.e. or email). Approach by phone or email with ideas in the first instance. Only interested in hearing from people with an interest in and knowledge of Amateur Radio. Copy (on disk) should be supported where possible by artwork, either illustrations, diagrams or photographs.
Ⓔ £54–70 per page.

## The Practitioner

10 Fernthorpe Road, London SW16 6DR
Ⓣ 020 8677 3508
cshort@thepractitioner.co.uk
www.thepractitioner.co.uk
Owner *Practitioner Medical Publilshing Ltd*
Editor *Corinne Short*
Circulation 35,656

Founded 1868. MONTHLY publication that keeps General Practitioners up to date on clinical issues. All articles are independently commissioned.

## Prediction

IPC Focus Network, Leon House, 233 High Street, Croydon CR9 1HZ
Ⓣ 020 8726 8000
prediction@ipcmedia.com
www.predictionmagazine.co.uk
Owner *IPC Media (A Time Warner Company)*
Editor *Marion Williamson*
Circulation 12,112

Founded 1936. MONTHLY. Covering astrology and mind, body, spirit topics. Unsolicited material in these areas welcome (about 200–300 mss received every year). Writers' guidelines available on request.
ASTROLOGY Pieces should be practical and of general interest.
FEATURES Articles on divination, shamanism, alternative healing, psychics and other supernatural phenonema considered. Please read a recent copy of the magazine before sending unsolicited material.

## Pregnancy & Birth

Endeavour House, 189 Shaftebury Avenue, London WC2H 8JG
Ⓣ 020 7295 5563
p.andb@bauermedia.co.uk
Owner *Bauer London Lifestyle*
Editor *Ellie Hughes*

Circulation 36,772

MONTHLY magazine covering all aspects of pregnancy from health to fashion. Regularly commissions features from health, real life and features writers. Freelancers should approach by email in the first instance.
Ⓔ Payment varies.

## Press Gazette

John Carpenter House, John Carpenter Street, London EC4Y 0AN
Ⓣ 020 7936 6433
pged@pressgazette.co.uk
www.pressgazette.co.uk
Owner *Progressive Media Group*
Editor *Dominic Ponsford*
Circulation 5,010

MONTHLY multi-media magazine and news website for all journalists – in regional and national newspapers, magazines, broadcasting and online – containing news, features and analysis of all areas of journalism, print and broadcasting. Unsolicited mss welcome; interested in profiles of magazines, broadcasting companies and news agencies, personality profiles, technical and current affairs relating to the world of journalism. Also welcomes gossip for diary page. Approach with ideas by phone, email, fax or post.

## Pride

Pride House, 55 Battersea Bridge Road, London SW11 3AX
Ⓣ 020 7228 3110    Ⓕ 020 7801 6717
info@pridemagazine.com
Owner/Editor *Carl Cushnie Junior*
Circulation 40,000

Founded 1991. MONTHLY lifestyle magazine for black women with features, beauty, arts and fashion. Approach in writing or by email with ideas.
FEATURES Issues pertaining to the black community. 'Ideas and solicited mss are welcomed from new freelancers.' Max. 2,000 words.
FASHION & BEAUTY *Shevelle Rhule* Freelancers used for short features. Max. 1,000 words.

## Prima

National Magazine House, 72 Broadwick Street, London W1F 9EP
Ⓣ 020 7439 5000
prima@natmags.co.uk
www.natmags.co.uk
Owner *National Magazine Company*
Editor *Maire Fahey*
Circulation 288,301

Founded 1986. MONTHLY women's magazine.

HEALTH & FEATURES Coordinator *Karen Swayn* Mostly practical and real life, written by specialists or commissioned from known freelancers. Unsolicited mss not welcome.

## Private Eye

6 Carlisle Street, London W1D 3BN
☏ 020 7437 4017    🖷 020 7437 0705
strobes@private-eye.co.uk
www.private-eye.co.uk
Owner *Pressdram*
Editor *Ian Hislop*
Circulation 210,218

Founded 1961. FORTNIGHTLY satirical and investigative magazine. Prospective contributors are best advised to approach the editor in writing. News stories and feature ideas are always welcome, as are cartoons.

## Prospect

2 Bloomsbury Place, London WC1A 2QA
☏ 020 7255 1344 (editorial)/1281 (publishing)
🖷 020 7255 1279
editorial@prospect-magazine.co.uk
publishing@prospect-magazine.co.uk
www.prospect-magazine.co.uk
Owner *Prospect Publishing Limited*
Editor *David Goodhart*
Circulation 28,446

Founded 1995. MONTHLY. Covers politics, international affairs, arts, culture, science, technology and economics. Email proposals or articles to submissions@prospect-magazine.co.uk Please read the magazine before submitting material.

## Psychic News

The Coach House, Stansted Hall, Stansted CM24 8UD
☏ 01279 817050    🖷 01279 817051
pnadverts@btconnect.com
www.psychicnewsbookshop.co.uk
Owner *Psychic Press 1995 Ltd*
Editor *Susan Farrow*
Circulation 40,000

Founded 1932. *Psychic News* is the world's only WEEKLY Spiritualist newspaper. It covers subjects such as psychic research, hauntings, ghosts, poltergeists, spiritual healing, survival after death and paranormal gifts. Unsolicited material considered. Ⓔ No payment.

## Psychologies

64 North Row, London W1K 7LL
☏ 020 7150 7000
psychologies@hf-uk.com
www.psychologies.co.uk
Owner *Hachette Filipacchi UK*
Editor *Louise Chunn*

Features Editors *Rebecca Alexander (rebecca.alexander@hf-uk.com), Catherine Jones*
Deputy Editor *Clare Longrigg*
Circulation 130,860

Founded 2005. MONTHLY. Women's lifestyle magazine covering relationships, family and parenting, health, beauty, social trends, travel, personality and behaviour, spirituality and sex. No fashion items. Consult the submissions guidelines on the website before submitting ideas for features.

## Q

Mappin House, 4 Winsley Street, London W1W 8HF
☏ 020 7182 8482    🖷 020 7182 8547
firstname.surname@qthemusic.com
www.qthemusic.com
Owner *Bauer Media*
Editor-in-Chief *Paul Rees*
Senior Editor *Matt Mason*
Circulation 94,811

Founded 1986. MONTHLY. Glossy aimed at educated popular music enthusiasts. Few opportunities for freelance writers. Prospective contributors should approach in writing only to the relevant section editor.

## Quality Fiction (QF)

c/o AllWriters' Workplace & Workshop, 234 Brook Street, Unit 2, Waukesha, WI 53188, USA
qfsubmissions@yahoo.com
www.allwriters.org (click on QWF)

Quality Fiction magazine, formerly QWF (Quality Women's Fiction), is a semi-annual literary magazine published in January and July through AllWriters' Workplace & Workshop in the USA. 'Evoking emotion is the most important characteristic of a QF story. Any genre, written by any gender, about any situation, is welcome as long as it is quality writing.' Submissions by email only. Submit only during reading periods: September 1st to November 31st and March 1st to May 31st. Submissions received outside of the reading period will be returned unread. For further information and detailed guidelines, please access the website or contact the editor at the email address above.

## Quarterly Review

26 Meadow Lane, Sudbury CO10 2TD
☏ 01507 339056
editor@quarterly-review.org
www.quarterly-review.org
Owner *Quarterly Review*
Editor *Derek Turner*

Founded 2007. A revival of the classic journal founded by Sir Walter Scott, Robert Southey and George Canning in 1809. QUARTERLY journal of

politics, ideas and culture, with an emphasis on regionalism, ecology and socio-biology. Approach in writing in the first instance. Articles from 1,000–8,000 words.
£ No payment.

## Quartos
▷ The New Writer

## QWF Magazine
▷ Quality Fiction

## RA Magazine (Royal Academy of Arts Magazine)
Royal Academy of Arts, Burlington House, Piccadilly, London W1J 0BD
☎ 020 7300 5820    🖷 020 7300 8032
ramagazine@royalacademy.org.uk
www.ramagazine.org.uk
Owner *Royal Academy of Arts*
Editor *Sarah Greenberg*
Circulation 100,000

Founded 1983. QUARTERLY visual arts, architecture and culture magazine distributed to Friends of the RA, promoting exhibitions and publishing articles on art shows, books and news worldwide. 'There are opportunities for freelancers who can write about art in an accessible way. Most articles need to be tied in to forthcoming exhibitions, books and projects.' No unsolicited contributions. Approach by telephone or email.

## Racecar Engineering
IPC Media Ltd, Leon House, 233 High Street, Croydon CR9 1HZ
☎ 020 8726 8362    🖷 020 8726 8398
racecar@ipcmedia.com
www.racecar-engineering.com
Owner *IPC Media (A Time Warner Company)*
Editor *Graham Jones*
Circulation 15,000

Founded 1990. MONTHLY. In-depth features on motorsport technology plus news and products. Interested in receiving news and features from freelancers.
FEATURES Informed insight into current motorsport technology. Max. 3,000 words.
NEWS New cars, products or business news relevant to motorsport. No items on road cars or racing drivers. Max. 500 words. Call or email to discuss proposal.

## Racing Post
1 Canada Square, Canary Wharf, London E14 5AP
☎ 020 7293 3000    🖷 020 7293 3758
editor@racingpost.com
www.racingpost.com
Owner *Centurycomm Ltd*
Editor *Bruce Millington*

Circulation 47,976
Founded 1986. DAILY horse racing paper with some general sport. In 1998, following an agreement between the owners of *The Sporting Life* and the *Racing Post*, the two papers merged.

## Radio Times
Media Centre, 201 Wood Lane, London W12 7TQ
☎ 020 8433 2000    🖷 020 8433 3160
radio.times@bbc.com
www.radiotimes.com
Owner *BBC Worldwide Limited*
Editor *Ben Preston*
Circulation 1.1 million

WEEKLY. UK's leading broadcast listings magazine. The majority of material is provided by freelance and retained writers, but the topicality of the pieces means close consultation with editors is essential. Very unlikely to use unsolicited material. Detailed BBC, ITV, Channel 4, Channel 5 and satellite television and radio listings are accompanied by feature material relevant to the week's output.
£ Payment by arrangement.

## Radio User
Arrowsmith Court, Station Approach, Broadstone BH18 8PW
☎ 0845 803 1979    🖷 01202 659950
roger@pwpublishing.ltd.uk
www.pwpublishing.ltd.uk
Owner *PW Publishing Ltd*
Editor *Roger Hall*
Circulation 19,000

Founded 1937. In 2006, *Shortwave Magazine* merged with *Radio Active* magazine to become *Radio User*. MONTHLY. Specialist electronics and radio enthusiasts' magazine covering all aspects of listening to radio signals. Features, news, regular columns and projects. Will consider contributions but recommends that contact be made prior to submission to check publishing plans.

## Rail Express
Foursight Publications Ltd, 20 Park Street, King's Cliffe, Peterborough PE8 6XN
☎ 01780 470086    🖷 01780 470060
editor@railexpress.co.uk
www.railexpress.co.uk
Owner/Editor *P.A. Sutton*
Circulation c.16,500

Launched 1996. MONTHLY glossy dealing with contemporary railway operations in the UK. Well researched and illustrated news (max. 500 words) and features (max. 4,000 words) that appeal to the core enthusiast readership are considered. Contact with brief in the first instance.

## Rail Magazine

Media House, Peterborough Business Park, Lynchwood,, Peterborough PE2 6EA
☎ 01733 468000   ℻ 01733 288163
www.rail-magazine.co.uk
Owner *Bauer Media*
Managing Editor *Nigel Harris*
Circulation 34,715

Founded 1981. FORTNIGHTLY magazine dedicated to modern railway. News and features, and topical newsworthy events. Unsolicited mss welcome. Approach by phone with ideas. Not interested in personal journey reminiscences. No fiction.
FEATURES By arrangement with the editor. All modern railway British subjects considered. Max. 2,000 words.
NEWS Any news item welcome. Max. 500 words.
€ Features, varies/negotiable; news, up to £100 per 1,000 words.

## Railway Gazette International

NINE Sutton Court Road, Sutton SM1 4SZ
☎ 020 8652 5200   ℻ 020 8652 5210
editor@railwaygazette.com
www.railwaygazette.com
Owner *DVV Media UK*
Editor *Chris Jackson*
News Editor *Andrew Grantham*
Industry Editor *Nick Kingsley*
Circulation 9,945

Founded 1835. MONTHLY magazine for senior railway, metro and tramway managers, engineers and suppliers worldwide. 'No material for railway enthusiast publications.' Email to discuss ideas in the first instance.

## The Railway Magazine

IPC Media Ltd, Blue Fin Building, 110 Southwark Street, London SE1 0SU
☎ 020 3148 4683   ℻ 020 3148 8521
railway@ipcmedia.com
www.ipcmedia.com
Owner *IPC Media (A Time Warner Company)*
Editor *Nick Pigott*
Circulation 34,715

Founded 1897. MONTHLY. Articles, photos and short news stories of a topical nature, covering modern railways, steam preservation and railway history, welcome. Max. 2,000 words, with sketch maps of routes, etc., where appropriate. Unsolicited mss welcome. No poetry.
€ Payment negotiable.

## Random Acts of Writing

11 Alexander Place, Inverness IV3 5BX
info@randomactsofwriting.co.uk
www.randomactsofwriting.co.uk
Owners *Jennifer Thomson, Vikki Trelfer*
Editor *Jennifer Thomson*
Sub-editor *Vikki Trelfer*
Circulation 200

Founded 2005. THREE ISSUES YEARLY. Promotes the short story. 'Unsolicited short fiction (max. 3,000 words) welcomed from all, particularly new writers or those based in or writing about the north of Scotland. Will give advice and feedback.' Approach by email. € Complimentary copy of the magazine plus discounts on further copies.

## Reader's Digest

11 Westferry Circus, Canary Wharf, London E14 4HE
☎ 020 7715 8000   ℻ 020 7715 8712
www.readersdigest.co.uk
Editor *Gill Hudson*
Circulation 465,028

Founded 1922. Monthly general interest magazine. Open to pitches from freelancers. Opportunities exist for short, humorous contributions. Allow three months' lead time.

## The Reader

Reader Office, 19 Abercromby Square, Liverpool L69 7ZG
☎ 0151 794 2830
readers@thereader.org.uk
http://magazine.thereader.org.uk/
Editor *Philip Davis*
Circulation 1,200

Founded 1997. QUARTERLY. Poetry, short fiction, literary articles and essays, thought, reviews, recommendations for individual readers and reading groups. Contributions from internationally lauded and new voices. Welcomes articles/essays about reading and about issues in modern life, max. 2,000 words. Recommendations for good reading, max. 1,000 words. Short stories, max. 2,500 words. The printed voice of The Reader Organization – literary outreach. Aims to reach serious readers in non-academic but thoughtful style. Approach in writing.
€ Payment negotiable.

## Readers' Review Magazine
▷ The Self Publishing Magazine

## Real People

33 Broadwick Street, London W1F 0DQ
☎ 020 7439 5000
Owner *National Magazine Co.*
Editor *Samm Taylor*
Circulation 216,038

WEEKLY magazine of real-life stories in a popular women's weekly, first-person format.
FEATURES Real-life stories for specific sections and tailored to the magazine. No fiction. Max. 1,800 words. € Payment varies; £300+ for spread.

## Record Collector

Room 101, The Perfume Factory, 140 Wales Farm Road, Acton, London W3 6UG

⊤ 020 8752 8181    Ⓕ 0870 732 6060

alan.lewis@metropolis.co.uk

www.recordcollectormag.com

Owner *Metropolis*

Editor *Alan Lewis*

Founded 1979. MONTHLY. Detailed, well-researched articles welcome on any aspect of record collecting or any collectible artist in the field of popular music (1950s to present day), with complete discographies where appropriate. Unsolicited mss welcome. Approach the editor with ideas by phone or email.

Ⓔ Payment negotiable.

## Red

64 North Row, London W1K 7LL

⊤ 020 7150 7000    Ⓕ 020 7150 7684

may.frost@redmagazine.co.uk

www.redmagazine.co.uk

Owner *Hachette Filipacchi UK Ltd*

Editor *Sam Baker*

Features Director *Lindsay Frankel*

Features & Lifestyle Assistant *May Frost*

Circulation 226,502

Founded 1998. MONTHLY magazine aimed at the 30-something woman. Will consider ideas sent in 'on spec' but tends to rely on regular contributors.

## Report

ATL, 7 Northumberland Street, London WC2N 5RD

⊤ 020 7930 6441    Ⓕ 020 7782 1618

www.atl.org.uk

Owner *Association of Teachers and Lecturers*

Editor *Alex Tomlin*

Managing Editor *Victoria Poskitt*

Circulation 160,000

Founded 1978. NINE ISSUES YEARLY during academic terms. Contributions welcome. All submissions should go directly to the editor. Articles should be no more than 800 words and must be of practical interest to the classroom teacher, F.E. lecturers and all other education staff.

## Retail Week

Emap, Greater London House, Hampstead Road, London NW1 7EJ

⊤ 020 7728 5000

editorial@retail-week.com

www.retail-week.com

Owner *Emap*

Editor *Tim Danaher*

Circulation 9,128

Founded 1988. WEEKLY. Leading news publication for the multiple retail sector. No unsolicited contributions. Limited opportunities for both features and news. Approach by email.

FEATURES *Joanna Perry* Max. 1,600 words.

NEWS *Jennifer Creevy*

## Reveal

National Magazine Co., National Magazine House, 33 Broadwick Street, London W1F 0DQ

⊤ 020 7339 4500    Ⓕ 020 7339 4529

michael@natmags.co.uk

www.natmags.co.uk

Owner *National Magazine Co.*

Editor *Jane Ennis*

Circulation 330,911

Launched 2004. WEEKLY glossy lifestyle magazine with celebrities, real-life stories, cookery, beauty and seven-day TV guide. No unsolicited contributions.

## Romance Matters

▷ The Romantic Novelists' Association under Professional Associations and Societies

## Rugby World

IPC Media Ltd, Blue Fin Building, 110 Southwark Street, London SE1 0SU

⊤ 020 3148 4708

paul_morgan@ipcmedia.com

www.rugbyworld.com

Owner *IPC Media (A Time Warner Company)*

Editor *Paul Morgan*

Circulation 45,677

Founded 1960. MONTHLY. Features of special rugby interest only. Unsolicited contributions welcome but s.a.s.e. essential for return of material. Prior approach by phone or in writing preferred.

## Runner's World

33 Broadwick Street, London W1F 0DQ

⊤ 020 7339 4400    Ⓕ 020 7339 4420

rwedit@natmag-rodale.co.uk

www.runnersworld.co.uk

Owner *Natmag – Rodale Ltd*

Editor *Andy Dixon*

Circulation 96,352

Founded 1979. MONTHLY magazine giving practical advice on all areas of recreational running including products and training, travel features, health, news and cross-training advice. Innovative ideas for training and nutrition features, as well as longer 'read' pieces, are welcome. Approach with ideas in writing in the first instance.

## Running Fitness

PO Box 13, Westerham TN16 3WT
rf.ed@kelsey.co.uk
www.runningfitnessmag.co.uk
Owner *Kelsey Publishing*
Editor *David Castle*
Circulation 25,000

Founded 1985. MONTHLY. Instructional articles on running, fitness, and lifestyle, plus running-related activities and health.
FEATURES Specialist knowledge an advantage. Opportunities are wide, but approach with ideas in the first instance.
NEWS Opportunities for people stories, especially if backed up by photographs.

## Saga Magazine

Saga Publishing Ltd, The Saga Building, Enbrook Park, Sandgate, Folkstone CT20 3SE
☏ 01303 771523    🖷 01303 776699
editor@saga.co.uk
www.saga.co.uk
Owner *Saga Publishing Ltd*
Editor *Kate Bravery*
Circulation 655,728

Founded 1984. MONTHLY magazine that sets out to celebrate the role of older people in society, reflecting their achievements, promoting their skills, protecting their interests, and campaigning on their behalf. A warm personal approach, addressing the readership in an up-beat and positive manner. It has a hard core of celebrated commentators/writers (e.g. Paul Lewis, Katharine Whitehorn, Sally Brampton) as regular contributors. Articles mostly commissioned or written in-house but exclusive celebrity interviews welcome if appropriate/relevant. Length 1,000–1,200 words (max. 1,600). No short stories, travel, fiction or poems, please.

## Sailing Today

Swanwick Marina, Lower Swanwick, Southampton SO31 1ZL
☏ 01489 585224    🖷 01489 565054
info@sailingtoday.co.uk
www.sailingtoday.co.uk
Owner *Marine Media Ltd*
Managing Editor *Rodger Witt*

Founded 1997. MONTHLY practical magazine for cruising sailors. *Sailing Today* covers owning and buying a boat, equipment and products for sailing and is about improving readers' skills, boat maintenance and product tests. Most articles are commissioned but will consider practical features and cruising stories with photos. Approach by telephone or in writing in the first instance.

## Sainsbury's Magazine

20 Upper Ground, London SE1 9PD
☏ 020 7775 7775    🖷 020 7775 7705
edit@sainsburysmagazine.co.uk
Owner *Seven Publishing Ltd*
Editor *Helena Lang*
Circulation 335,515

Founded 1993. MONTHLY featuring a main core of food and cookery with features, health, beauty, travel, home and gardening. No unsolicited mss. Approach in writing with ideas only in the first instance.

## The Salisbury Review

33 Canonbury Park South, London N1 2JW
☏ 020 7226 7791    🖷 020 7354 0383
info@salisburyreview.co.uk
www.salisburyreview.com
Managing Editor *Merrie Cave*
Consulting Editors *Roger Scruton, Myles Harris, Lord Charles Cecil, Mark Baillie, Christie Davies*
Literary Editor *Ian Crowther*
Circulation 2,000

Founded 1982. QUARTERLY magazine of conservative thought. Editorials and features from a right-wing viewpoint. Unsolicited articles, reviews by arrangement. No fiction or poetry.
FEATURES Max. 2,000 words.
REVIEWS Max. 1,000 words.
💷 Small payment.

## Scotland in Trust

91 East London Street, Edinburgh EH7 4BQ
☏ 0131 556 2220    🖷 0131 556 3300
editor@scotlandintrust.co.uk
www.scotlandintrust.co.uk
Owner *National Trust for Scotland*
Editor *Iain Gale*
Contact *Neil Braidwood*
Circulation 179,576

Founded 1983. THREE ISSUES YEARLY. Membership magazine of the National Trust for Scotland. Contains heritage/conservation features relating to the Trust, their properties and work. No unsolicited mss.

## Scotland Means Business
▷ Daily Record under National Newspapers

## The Scots Magazine

D.C. Thomson & Co., 80 Kingsway East, Dundee DD4 8SL
☏ 01382 575137
mail@scotsmagazine.com
www.scotsmagazine.com
Owner *D.C. Thomson & Co. Ltd*
Editor *Karen Byrom*

Circulation 33,438

Founded 1739. MONTHLY. Covers a wide field of Scottish interests ranging from personalities to wildlife, climbing, reminiscence, history and folklore. Outside contributions welcome; 'staff delighted to discuss in advance by letter or email'.
EVENTS LISTING *Karen Byrom*
PHOTOGRAPHY/MUSIC *Ian Neilson*
BOOKS *Alison Cook*

## The Scottish Farmer

Newsquest Magazines, 200 Renfield Street, Glasgow G2 3QB
℡ 0141 302 7727    🖷 0141 302 7799
ken.fletcher@thescottishfarmer.co.uk
www.thescottishfarmer.co.uk
Owner *Newsquest Magazines*
Editor *Alasdair Fletcher*
Circulation 18,755

Founded 1893. WEEKLY. Farmer's magazine covering most aspects of Scottish agriculture. Unsolicited mss welcome. Approach with ideas in writing, by fax or email.
FEATURES *Ken Fletcher* Technical articles on agriculture or farming units. 1,000–2,000 words.
NEWS *Ken Fletcher* Factual news about farming developments, political, personal and technological. Max. 800 words.
WEEKEND FAMILY PAGES Rural and craft topics.

## Scottish Field

Special Publications, Craigcrook Castle, Craigcrook Road, Edinburgh EH4 3PE
℡ 0131 312 4550    🖷 0131 312 4551
editor@scottishfield.co.uk
www.scottishfield.co.uk
Owner *Wyvex Media Ltd*
Editor *Archie Mackenzie*
Circulation 13,610

Founded 1903. MONTHLY. Scotland's quality lifestyle magazine. Unsolicited mss welcome but writers should study the magazine first.
FEATURES Articles of general interest on Scotland and Scots abroad with good photographs or, preferably, emailed images at 300 DPI. Approx. 1,000 words.
💷 Payment negotiable.

## Scottish Home & Country

42 Heriot Row, Edinburgh EH3 6ES
℡ 0131 225 1724    🖷 0131 225 8129
magazine@swri.demon.co.uk
www.swri.org.uk
Owner *Scottish Women's Rural Institutes*
Editor *Liz Ferguson*
Circulation 8,000

Founded 1924. MONTHLY. Scottish or rural-related issues, health, travel, women's issues and general interest. Unsolicited mss welcome.

Commissions are rare and tend to go to established contributors only.

## Scouting Magazine

Gilwell House, Gilwell Park, Chingford E4 7QW
℡ 020 8433 7100    🖷 020 8433 7103
scouting.magazine@scout.org.uk
www.scouts.org.uk
Owner *The Scout Association*
Editors *Chris James, Hilary Galloway, Ellis Matthews, David O'Carroll*
Circulation 95,000

BI-MONTHLY magazine for adults connected to or interested in the Scout Movement. Interested in Scouting-related submissions only. Sent to Members of the Association; not available on subscription
💷 Payment by negotiation.

## Screen

Gilmorehill Centre for Theatre, Film and Television, University of Glasgow, Glasgow G12 8QQ
℡ 0141 330 5035    🖷 0141 330 3515
screen@arts.gla.ac.uk
www.screen.arts.gla.ac.uk
screen.oxfordjournals.org
Publisher *Oxford University Press*
Editorial Office *Caroline Beven, Heather Middleton*
Editors *Annette Kuhn, John Caughie, Karen Lury, Jackie Stacey, Sarah Street*

QUARTERLY academic journal of film and television studies for a readership ranging from undergraduates to screen studies academics and media professionals. There are no specific qualifications for acceptance of articles. Straightforward film reviews are not normally published. Check the magazine's style and market in the first instance.

## Screen International

Greater London House, Hampstead Road, London NW1 7EJ
℡ 020 7728 5605    🖷 020 7728 5555
mike.goodridge@emap.com
www.screendaily.com
Owner *Emap*
Editor *Mike Goodridge*
Circulation 6,306

'The international voice of the international film industry'. Expert freelance writers used in all areas. No unsolicited mss. Approach by email.
NEWS *Caroline Parry*
FEATURES *Louise Tutt, Sarah Cooper*
REVIEWS Editor *Finn Halligan*
BOX OFFICE *Jack Warner*
💷 Negotiable on NUJ basis.

## Screentrade Magazine

Screentrade Media Ltd, PO Box 144, Orpington
BR6 6LZ
☎ 01689 833117   🖷 01689 833117
philip@screentrademagazine.com
philip45other@yahoo.co.uk
www.screentrademagazine.com
Owner *Screentrade Media Ltd*
Editor *Philip Turner*
US Representative, Screentrade Media Ltd
*Pamala Stanton (001 616 847 0144/616 405 243;
pam@screentrademagazine.com)*
Circulation 3,000–5,000

Founded 2000. QUARTERLY journal for British,
European and North American exhibitors and
film distributors. Now also serving Russia, China,
India and Pacific Rim territories.
FEATURES Dedicated sections on the following:
cinema management, digital cinema & alternative
content, 3D cinema, cinema architecture,
seating, point-of-sale/ticketing, concessions,
screen technology, health & safety, nostalgia,
showmanship, book reviews, box office figures
and film columns, career tips and other
managerial matters. Some film reviews but no
'film star' interviews. Most contributions are
from within the exhibition industry. Items on
the state of cinema exhibition, film distribution,
cinema architecture, interviews with key industry
personnel frequently undertaken. 1,500–3,000
words.
NEWS Topical items (if substantiated) welcome.
Events coverage (e.g. festivals, premières) from an
exhibitor's viewpoint preferred.

## ScriptWriter
▷ TwelvePoint.com

## Sea Breezes

Sea Breezes Publications Ltd, Media House,
Cronkbourne, Douglas IM4 4SB
☎ 01624 696573   🖷 01624 661655
sb@seabreezes.co.im
www.seabreezes.co.im
Owner *Sea Breezes Publications Ltd*
Editor *Captain A.C. Douglas*
Circulation 17,000

Founded 1919. MONTHLY. Covers virtually
everything relating to ships and seamen.
Unsolicited mss welcome; they should be
thoroughly researched and accompanied by
relevant photographs. No fiction, poetry, or
anything which 'smacks of the romance of the sea'.
FEATURES Factual tales of ships, seamen and
the sea, Royal or Merchant Navy, sail or power,
nautical history, shipping company histories,
epic voyages, etc. Length 1,000–4,000 words.
'The most readily acceptable work will be that
which shows it is clearly the result of first-hand
experience or the product of extensive and
accurate research.' 🖻 £14 per page (about 800
words).

## The Self Publishing Magazine (incorporating Readers' Review Magazine)

5 Weir Road, Kibworth Beauchamp, Leicester
LE8 0LQ
☎ 0116 279 2299   🖷 0116 279 2277
selfpublishing@troubador.co.uk
www.selfpublishingmagazine.co.uk
Owner *Troubador Publishing Ltd*
Editor *Jane Rowland*
Circulation 2,000

Launched 2006. THREE ISSUES YEARLY. Articles
on self publishing. Submissions welcome on
aspects of authors publishing their own books
(design, production, editing, print, marketing).
Email the editor in the first instance. Max. 2,000
words.
BOOK REVIEWS Self-published books, only. Max.
300 words.

## She Kicks

The Baltic Business Centre, Gateshead NE8 3DA
☎ 0191 442 4003   🖷 0191 442 4002
jen@fgmag.com
www.fgmag.com
Editor *Jennifer O'Neill*
Circulation 10,000

Founded in 2003 as *Fairgame Magazine*. BI-
MONTHLY magazine for women football players.
Contributions welcome.
FEATURES *Jennifer O'Neill* International reports,
player and team profiles, interviews, diet,
health and fitness, tactics, training advice, play
improvement, fund-raising. Max. 1,500 words.
NEWS *Wilf Frith* (wilf@fgmag.com) Match
reports, team news, transfers, injuries, results and
fixtures. Max. 600 words.

## She Magazine

National Magazine House, 72 Broadwick Street,
London W1F 9EP
☎ 020 7439 5000   🖷 020 7312 3940
www.allaboutyou.com
Owner *National Magazine Company Ltd*
Editor *Claire Irvin*
Circulation 150,074

Glossy MONTHLY for the 35-plus woman,
addressing her needs as a modern individual, a
parent and a homemaker. Talks to its readers in
an intelligent, humorous and informative way.
FEATURES Approach with feature ideas in writing
or email to Deputy Editor, *Carmen Bruegmann*
(carmen.bruegmann@natmags.co.uk). No
unsolicited material.
🖻 Payment negotiable.

## Ships Monthly

IPC Inspire, 222 Branston Road, Burton-upon-Trent DE14 3BT
☎ 01283 542721   🖷 01283 546436
shipsmonthly@ipcmedia.com
www.shipsmonthly.com
Owner *IPC Inspire*
Editor *Iain Wakefield*
Deputy Editor *Nicholas Leach*
Circulation 22,000

Founded 1966. MONTHLY A4 format magazine for ship enthusiasts and maritime professionals. News, photographs and illustrated articles on all kinds of ships – mercantile and naval, past and present. No yachting. Welcomes contributions on company histories, classic and modern ships, ports and harbours. Most articles are commissioned; prospective contributors should telephone or email in the first instance.

## Shooting and Conservation

BASC, Marford Mill, Rossett, Wrexham LL12 0HL
☎ 01244 573000   🖷 01244 573001
jeffrey.olstead@basc.org.uk
www.basc.org.uk
Owner *The British Association for Shooting and Conservation (BASC)*
Editor *Jeffrey Olstead*
Circulation 130,000

SIX ISSUES YEARLY. Good articles and stories on shooting, conservation and related areas may be considered although most material is produced in-house. Max. 1,500 words.
£ Payment negotiable.

## The Shooting Gazette

PO Box 225, Stamford PE9 2HS
☎ 01780 485350   🖷 01780 485390
will_hetherington@ipcmedia.com
Owner *IPC Media (A Time Warner Company)*
Editor *Will Hetherington*
Circulation 14,765

Launched 1989. MONTHLY aimed at those who enjoy driven game shooting in the UK and around the world.
FEATURES *Will Hetherington* Welcomes articles directly related to interesting or unusual game shooting stories; max. 2,000 words. Approach by email.
NEWS *Martin Puddifer* Max. 500 words.

## Shooting Times & Country Magazine

IPC Media Ltd, Blue Fin Building, 110 Southwark Street, London SE1 0SU
☎ 020 3148 4741
steditorial@ipcmedia.com
www.shootingtimes.co.uk
Owner *IPC Media (A Time Warner Company)*
Editor *Camilla Clark*

Circulation 24,348
Founded 1882. WEEKLY. Covers shooting, fishing and related countryside topics. Unsolicited contributions considered.
£ Payment negotiable.

## Shout Magazine

D.C. Thomson & Co., Albert Square, Dundee DD1 9QJ
☎ 01382 223131   🖷 01382 200880
shout@dcthomson.co.uk
www.shoutmag.co.uk
Owner *D.C. Thomson Publishers*
Editor *Maria T. Welch*
Circulation 72,008

Founded 1993. FORTNIGHTLY. Pop music, quizzes, emotional, beauty, fashion, soap features.

## Shout!

PO Box YR46, Leeds LS9 6XG
☎ 0113 248 5700
shout.magazine@ntlworld.com
www.shoutweb.co.uk
Owner/Editor *Mark Michalowski*
Circulation 7,000

Founded 1995. MONTHLY lesbian/gay and bisexual news, views, arts and scene for Yorkshire; lgb health and politics. Interested in reviews of Yorkshire lgb events, happenings, news, analysis – 300 to 1,000 words max. No fiction, fashion or items with no reasonable relevance to Yorkshire and the north.

## Showing World

Robin Aldwood Publications Ltd, PO Box 793, Needham IP6 8WN
☎ 01449 722505
sandy@showingworldonline.co.uk
www.showingworldonline.co.uk
Owner *Robin Aldwood Publications Ltd*
Editor *Sandy Wooderson*
Circulation 10,000

Founded 1991. SIX ISSUES YEARLY Features every aspect of showing horses and ponies plus natives, miniatures and donkeys. Knowledgeable and how-to-do-it articles welcome. Photos essential.
£ Payment negotiable.

## Shropshire Magazine

Waterloo Road, Telford TF1 5HU
☎ 01952 241455   🖷 01952 254605
Owner *Midland News Association*
Editor *Neil Thomas*

Founded 1950. MONTHLY.

FEATURES Personalities, topical items, historical (e.g. family) of Shropshire; also general interest: homes, weddings, antiques, etc. Max. 1,000 words. Unsolicited mss welcome but ideas in writing preferred.

£ Payment negotiable 'but modest'.

## Sight & Sound

BFI, 21 Stephen Street, London W1T 1LN

☎ 020 7255 1444   🖷 020 7436 2327

s&s@bfi.org.uk

www.bfi.org.uk/sightandsound

Owner *BFI*

Editor *Nick James*

Circulation 19,842

Founded 1932. MONTHLY. Topical and critical articles on international cinema, with regular columns from the USA and Europe. Length 1,000–5,000 words. Relevant photographs appreciated. Also book, film and DVD release reviews. Approach in writing with ideas.

£ Payment by arrangement.

## The Skier and Snowboarder Magazine

Mountain Marketing Ltd, PO Box 386, Sevenoaks TN13 1AQ

☎ 0845 310 8303

frank.baldwin@skierandsnowboarder.co.uk

www.skierandsnowboarder.co.uk

Publisher *Mountain Marketing Ltd*

Editor *Frank Baldwin*

Circulation 30,000

SEASONAL. FIVE ISSUES YEARLY, from July to May. Outside contributions welcome.

FEATURES Various topics covered, including race and resort reports, fashion, equipment update, artificial slope, club news, new products, health and safety. Crisp, tight, informative copy of 800 words or less preferred.

NEWS All aspects of skiing news covered.

£ Payment negotiable.

## Slightly Foxed: The Real Reader's Quarterly

67 Dickinson Court, 15 Brewhouse Yard, London EC1V 4JX

☎ 020 7549 2121/2111   🖷 0870 199 1245

all@foxedquarterly.com

www.foxedquarterly.com

Owner *Slightly Foxed Ltd*

Editors *Gail Pirkis, Hazel Wood*

Circulation 5,500

Founded 2004. 'Reviews of books (fiction and non-fiction) that have stood the test of time or books that have been published recently and are of real quality but which have been overlooked by reviewers and bookshops.' Unsolicited contributions of 'lively, personal, idiosyncratic writing of real quality' are welcome but it is recommended that would-be contributors read the magazine first to gauge its approach. Send email with sample work in the first instance. Not interested in anything that is not actually a review of a book or author.

## Slim at Home

25 Phoenix Court, Hawkins Road, Colchester CO2 8JY

☎ 01206 505987

laura@aceville.co.uk

www.slimathome.co.uk

Owner *Aceville Publications (2001) Ltd*

Editor *Naomi Abeykoon*

Circulation 110,000

Launched 2007. MONTHLY. National women's magazine for dieters who want to lose weight in their own way.

FEATURES *Naomi Abeykoon* Targeted weight-loss articles must be highly researched and of use to people who want to lose weight. No unrelated health features, quick tips or reviews of products and services. Approach by email.

CELEBS *Naomi Abeykoon* Current celebrity weight-loss stories commissioned. Pictures or shoot must be available.

## Smallholder

*Editorial:* Hook House, Wimblington March PE15 0QL

☎ 01354 741538   🖷 01354 741182

liz.wright1@btconnect.com

www.smallholder.co.uk

Owner *Newsquest Plc*

Editor *Liz Wright*

Circulation 20,000

Founded 1882. MONTHLY. Outside contributions welcome. Send for sample magazine and editorial schedule before submitting anything. Follow up with samples of work to the editor so that style can be assessed for suitability. No poetry or humorous, unfocused personal tales; no puzzles.

FEATURES *Stephen Ivall, Jonathan Millar* (☎ 01326 213337) New writers always welcome, but must have high level of technical expertise – 'not textbook stuff'. 'How to do it' articles with photos welcome if relevant to smallholding. Illustrations and photos paid for. Length 750–1,500 words.

NEWS All agricultural and rural news welcome. Length 200–500 words.

£ Payment negotiable.

## SmartLife International

25 Phoenix Court, Hawkins Road, Colchester CO2 8JY

☎ 01206 505924   🖷 01206 505929

stuart@smartlifeint.com

www.smartlifeint.com

Owner *Aceville Publications Ltd*

Editor *Stuart Pritchard*

Circulation 18,933

Launched 2000. TEN ISSUES YEARLY. High-end lifestyle magazine: consumer technology, interior design, luxury cars, travel, fashion and grooming. Welcomes proposals by email.
FEATURES 'No good feature ideas that fit the remit are ruled out.' Max. 2,000 words.
NEWS Generally compiled in-house.

## Snooker Scene

Hayley Green Court, 130 Hagley Road, Hayley Green, Halesowen, Birmingham B63 1DY
☎ 0121 585 9188    🖷 0121 585 7117
clive.everton@talk21.com
www.snookersceneonline.com
Owner *Everton's News Agency*
Editor *Clive Everton*
Circulation 8,000

Founded 1971. MONTHLY. No unsolicited mss. Approach in writing with an idea.

## Somerset Life

Archant House, Babbage Road, Totnes TQ9 5JA
☎ 01803 860910    🖷 01803 860926
www.somerset.greatbritishlife.co.uk
Owner *Archant Life*
Editor *Natalie Vizzard*
Circulation 8,000

Launched 1994. MONTHLY magazine that celebrates Somerset with coverage of towns, villages, the county's history, property, antiques, fashion, interior design, gardens and motoring.
FEATURES *Natalie Vizzard* Prefers picture/word packages, particularly on food and drink, celebrity interviews, countryside, heritage, health and beauty, people, business and the arts. 'Always on the lookout for new voices (particularly if also a competent photographer) on a regular or one-off freelance basis.' Max. 1,500 words. Approach by email. 🗈 £75 for a 1,000-word/picture package.

## Southword: A Literary Journal Online

The Munster Literature Centre, Frank O'Connor House, 84 Douglas Street, Cork, Republic of Ireland
☎ 00 353 21 431 2955
administrator@munsterlit.ie
www.munsterlit.ie
Editor *Patrick Cotter*

Founded 2001. Published biannually by the Munster Literature Centre, *Southword* is a literary journal featuring poems, fiction and reviews. The magazine has featured Haruki Murakami, James Lasdun and Colm Toibín but also welcomes work from new writers. Both poetry and fiction are considered between January and March 15th each year for the summer issue. Poetry alone is considered between July and September 15th

for the winter issue. See the website for further submission guidelines.

## Spear's Wealth Management Survey

*Registered address:* 4 Croxted Mews, 286–288 Croxted Road, London SW24 9DA
☎ 020 7602 7095
josh.spero@luxurypublishing.com
www.spearswms.com
Editor-in-Chief *William Cash*
Senior Editor/Website Editor *Josh Spero*
Managing Editor *Penelope Bennett*
Circulation 30,000

Launched 2005. Financial lifestyle and cultural quarterly for high net worth individuals, entrepreneurs and wealth management professionals, including decision-makers and wealth creators. Subscription only. Email address above to offer contributions.

## Speciality Food

25 Phoenix Court, Hawkins Road, Colchester CO2 8JY
☎ 01206 505971
carolyn.wilson@aceville.co.uk
www.specialityfoodmagazine.co.uk
Owner *Aceville Publications Ltd*
Editor *Carolyn Wilson*
Circulation 8,699

Founded 2002. NINE ISSUES YEARLY. Trade magazine focusing on premium food and drink relevant for the fine food retail sector. Unsolicited contributions not generally welcome.
FEATURES Subjects such as cheese, dairy and chocolate; also organic food. Approach by email. Max. 1,800 words. 🗈 Payment negotiable.

## The Spectator

22 Old Queen Street, London SW1H 9HP
☎ 020 7961 0200    🖷 020 7961 0058
editor@spectator.co.uk
www.spectator.co.uk
Owner *The Spectator (1828) Ltd*
Editor *Fraser Nelson*
Books *Mark Amory*
Business Editor *Martin Vander Weyer*
Art Editor *Liz Anderson*
Circulation 70,300

Founded 1828. WEEKLY political and literary magazine. Prospective contributors should write in the first instance to the relevant editor. Unsolicited mss welcome, but no 'follow up' phone calls, please. 🗈 Payment nominal.

## Staffordshire Life

The Publishing Centre, Derby Street, Stafford ST16 2DT
☎ 01785 257700    🖷 01785 253287

louise.elliott@staffordshirenewspapers.co.uk
www.staffordshirelife.co.uk
Owner *Staffordshire Newspapers Ltd*
Editor *Louise Elliott*
Circulation 15,500

Founded 1982. MONTHLY. Full-colour county magazine devoted to Staffordshire, its surroundings and people. Contributions welcome. Approach in writing with ideas.
FEATURES Max. 1,200 words.
Ⓔ NUJ rates.

## Stage

Stage House, 47 Bermondsey Street, London SE1 3XT
Ⓣ 020 7403 1818 Ⓕ 020 7939 8478
editor@thestage.co.uk
www.thestage.co.uk
Owner *The Stage Newspaper Ltd*
Editor *Brian Attwood*
Circulation 18,611

Founded 1880. WEEKLY. No unsolicited mss. Prospective contributors should write with ideas in the first instance.
FEATURES Preference for middle-market, tabloid-style articles. 'Puff pieces', PR plugs and extended production notes will not be considered. Max. 800 words. Profiles: 1,200 words.
NEWS News stories from outside London are always welcome. Max. 300 words.
Ⓔ £100 per 1,000 words.

## Stamp Lover

Harvard House, 621 London Road, Isleworth TW7 4ER
Ⓣ 020 8568 2433
stamplover@ukphilately.org.uk
www.ukphilately.org.uk/nps
Owner *National Philatelic Society*
Editor *Michael L. Goodman*
Circulation 800

Founded 1908. SIX ISSUES YEARLY. Magazine of the National Philatelic Society. Welcomes articles about the hobby; stamps, old and new. Approach by email.
NEWS Information about worldwide philately. Max. 1,000 words. Ⓔ No payment.

## Stamp Magazine

IPC Media Ltd, Leon House, 233 High Street, Croydon CR9 1HZ
Ⓣ 020 8726 8241 Ⓕ 020 8726 8299
guy_thomas@ipcmedia.com
Owner *IPC Media (A Time Warner Company)*
Editor *Guy Thomas*
Circulation 12,000

Founded 1934. MONTHLY news and features on the world of stamp collecting from the past to the present day. Interested in articles by experts on particular countries or themes such as subject matter illustrated on stamps. Approach in writing.
NEWS *Julia Lee* News of latest stamp issues or industry news. Max. 500 words.
FEATURES *Guy Thomas* Any features welcome on famous stamps, rarities, postmarks, postal history, postcards, personal collections. Must be illustrated with colour images ('we can arrange for photography of original stamps').
Ⓔ Payment negotiable.

## Standpoint

11 Manchester Square, London W1V 3PW
Ⓣ 020 7563 9840 Ⓕ 020 7563 9841
editor@standpointmag.co.uk
www.standpointmag.co.uk
Owner *The Social Affairs Unit*
Editor *Daniel Johnson*
Circulation 11,000

Launched 2008. MONTHLY. International politics and current affairs, cultural issues, essays, reviews, short fiction and poetry. Most articles are commissioned but will look at unsolicited contributions. Approach by email.

## Star

The Northern & Shell Building, Number 10 Lower Thames Street, London EC3R 6EN
Ⓣ 020 8612 7000 Ⓕ 020 8612 7505
starmagazine@express.co.uk
www.star-magazine.co.uk
Owner *Richard Desmond*
Editor *Busola Odulate*
Circulation 317,940

Founded 2003. WEEKLY celebrity magazine with diet pages, travel and TV listings. No unsolicited contributions. Contact by email.

## Steam Railway Magazine

Bauer Media, Media House, Lynchwood Business Park, Peterborough PE2 6EA
Ⓣ 01733 288026
steam.railway@bauermedia.co.uk
Owner *Bauer Media*
Editor *Danny Hopkins*
News Editor *Gary Boydhope*
Circulation 32,842

Founded 1979. FOUR-WEEKLY magazine targeted at all steam enthusiasts interested in the modern preservation movement. Unsolicited material welcome. News reports, photographs, steam-age reminiscences. Approach in writing or by email.

## Stella Magazine

▷ The Sunday Telegraph under National Newspapers

## The Strad

Newsquest Specialist Media Ltd, 2nd Floor, 30 Cannon Street, London EC4M 6YJ
☎ 020 7618 3095    📠 020 7618 3483
thestrad@thestrad.com
www.thestrad.com
Owner *Newsquest Specialist Media Group*
Editor *Ariane Todes*
Circulation 17,500

Founded 1890. MONTHLY for classical string musicians, makers and enthusiasts. Unsolicited mss accepted occasionally 'though acknowledgement/return not guaranteed'. FEATURES Profiles of string players, teachers, luthiers and musical instruments, also relevant research. Max. 2,000 words.
REVIEWS *Matthew Rye*
💷 £150 per 1,000 words.

## Stuff

Teddington Studios, Broom Road, Teddington TW11 9BE
☎ 020 8267 5036
stuff@haymarket.com
www.stuff.tv
Owner *Haymarket Publishing Ltd*
Editor *Fraser Macdonald*
Circulation 95,695

Launched 1998. MONTHLY guide to new technology with information on all the latest gadgets and gear. Includes a 'Buyers' Guide' with information about the top tried and tested items. No unsolicited contributions.

## Suffolk and Norfolk Life

The Publishing House, Framlingham IP13 9EE
☎ 01728 622030    📠 01728 622031
todaymagazines@btopenworld.com
www.suffolknorfolklife.com
Owner *Today Magazines Ltd*
Editor *Kevin Davis*
Circulation 17,000

Founded 1989. MONTHLY. General interest, local stories, historical, personalities, wine, travel, food. Unsolicited mss welcome. Approach by phone or in writing with ideas. Not interested in anything which does not relate specifically to East Anglia.
FEATURES *Kevin Davis* Max. 1,500 words, with photos.
NEWS *Kevin Davis* Max. 1,000 words, with photos.
SPECIAL PAGES *William Locks* Study the magazine for guidelines. Max. 1,500 words.
💷 £45–£80.

## Sugar Magazine

64 North Row, London W1K 7LL
☎ 020 7150 7087
www.sugarscape.com

Owner *Hachette Filipacchi (UK)*
Editor *Annabel Brog*
Circulation 140,980

Founded 1994. MONTHLY. Everything that might interest the teenage girl. No unsolicited mss. Will consider ideas or contacts for real-life features. No fiction. Approach in writing in the first instance.

## SuperBike Magazine

IPC Media Ltd, Leon House, 233 High Street, Croydon CR9 1HZ
☎ 020 8726 8455    📠 020 8726 8499
kenny_pryde@ipcmedia.com
www.superbike.co.uk
Owner *IPC Media (A Time Warner Company)*
Publishing Director *Keith Foster*
Editor *Kenny Pryde*
Circulation 33,269

Founded 1977. FOUR-WEEKLY. Dedicated to all that is best and most exciting in the world of motorcycling. Unsolicited mss, synopses and ideas by email are welcome.

## Surrey Life

Holmesdale House, 46 Croydon Road, Reigate RH2 0NH
☎ 01737 247188    📠 01737 246596
editor@surreylife.co.uk
www.surreylife.co.uk
Owner *Archant Life*
Editor *Caroline Harrap*
Circulation 15,000

Launched 2002. MONTHLY county magazine featuring the best property, food and drinks and celebrity interviews. Unsolicited submissions of Surrey-based feature ideas are welcome; send by email.

## Sussex Life

Baskerville Place, 28 Teville Road, Worthing BN11 1UG
☎ 01903 703730    📠 01903 703770
simon.irwin@archant.co.uk
www.susssexlife.co.uk
Owner *Archant Life*
Editor *Simon Irwin*
Circulation 11,000

Founded 1965. MONTHLY. Sussex and general interest magazine. Regular supplements on education, fashion, homes and gardens. Interested in investigative, journalistic pieces relevant to the area and celebrity profiles. Unsolicited mss, synopses and ideas welcome; approach by email.

## Swimming Times Magazine

41 Granby Street, Loughborough LE11 3DU
☎ 01509 632230    📠 01509 618701

swimmingtimes@swimming.org
www.britishswimming.org
Owner *Amateur Swimming Association*
Editor *P. Hassall*
Circulation 20,000

Founded 1923. MONTHLY. Competitive swimming and associated subjects. Unsolicited mss welcome.
FEATURES Technical articles on swimming, water polo, diving or synchronized swimming. Length and payment negotiable.

## The Tablet

1 King Street Cloisters, Clifton Walk, London W6 0GY
📞 020 8748 8484    📠 020 8748 1550
thetablet@tablet.co.uk
www.thetablet.co.uk
Owner *The Tablet Publishing Co Ltd*
Publisher *Ignatius Kusiak*
Editor *Catherine Pepinster*
Circulation 21,978

Founded 1840. WEEKLY. International Catholic journal with an ecumenical outlook. Authoritative articles on religion, politics, literature and the arts by writers of international standing.

## Take a Break

Academic House, 24–28 Oval Road, London NW1 7DT
📞 020 7241 8000
tab.features@bauer.co.uk
Owner *H. Bauer Publishing Ltd*
Editor *John Dale*
Circulation 900,016

Founded 1990. WEEKLY. True-life feature magazine. Approach with ideas in writing.
NEWS/FEATURES Always on the look-out for good, true-life stories. Max. 2,000 words.
FICTION Sharp, succinct stories which are well told and often with a twist at the end. All categories, provided it is relevant to the magazine's style and market. Max. 1,000 words.
£ Payment negotiable.

## Take It Easy
▷ The People under National Newspapers

## Take5
▷ Daily Star Sunday under National Newspapers

## Tate Etc

Millbank, London SW1P 4RG
📞 020 7887 8724    📠 020 7887 3940
tateetc@tate.org.uk
www.tate.org.uk
Owner *Tate*
Editorial Director *Bice Curiger*
Editor *Simon Grant*

Circulation 100,000

Relaunched May 2004. THREE ISSUES YEARLY. Visual arts magazine aimed at a broad readership with articles blending the historic and the contemporary. Please send material by post.

## Tatler

Vogue House, Hanover Square, London W1S 1JU
📞 020 7152 3600    📠 020 7409 0451
www.tatler.co.uk
Owner *Condé Nast Publications Ltd*
Editor *Catherine Ostler*
Features Editor *Ticky Hedley-Dent*
Editor's Assistant *Karen Deeks*
Features Associate *Richard Dennen*
Circulation 86,345

Up-market glossy from the Condé Nast stable. New writers should send in copies of either published work or unpublished material. The magazine works largely on a commission basis: they are unlikely to publish unsolicited features.

## The Teacher

Hamilton House, Mabledon Place, London WC1H 9BD
📞 020 7380 4708    📠 020 7387 8458
teacher@nut.org.uk
www.teachers.org.uk
Owner *National Union of Teachers*
Editor *Elyssa Campbell-Barr*
Circulation 320,000

Journal of the National Union of Teachers, published EIGHT TIMES YEARLY. News, advice, information and special features on educational matters relating to classroom teaching. 'Please send a brief outline of any feature ideas for consideration.' Approach by letter or email.

## Tesco Magazine

85 Strand, London WC2R 0DW
📞 020 7550 8900
tesco@cedarcom.co.uk
www.tesco.com/magazine
Owner *Cedar Communications*
Editor *Dawn Alford*
Circulation 2.1 million

Launched 2004. BI-MONTHLY glossy women's magazine aimed at 25–44-year-olds. Includes beauty, fashion, health, travel, real-life, entertainment features, practical ideas and tips. No unsolicited material.

## TGO (The Great Outdoors)

Newsquest (Herald & Times) Magazines Ltd, 200 Renfield Street, Glasgow G2 3QB
📞 0141 302 7700    📠 0141 302 7799
cameron.mcneish@tgomagazine.co.uk
emily.rodway@tgomagazine.co.uk
www.tgomagazine.co.uk

Owner *Newsquest*
Editor *Cameron McNeish*
Circulation 10,705

Founded 1978. MONTHLY. Deals with walking, backpacking and wild country topics. Unsolicited mss are welcome.
FEATURES Well-written and illustrated items on relevant topics. Max. 2,500 words. Quality high resolution digital colour images only, please.
NEWS All news items are written in-house.
£ Payment negotiable.

## That's Life!

3rd Floor, Academic House, 24–28 Oval Road, London NW1 7DT
☎ 020 7241 8000    ▣ 020 7241 8008
firstname.lastname@bauer.co.uk
www.thatslife.co.uk
Owner *H. Bauer Publishing Ltd*
Editor *Jo Checkley*
Circulation 433,921

Founded 1995. WEEKLY. True-life stories, puzzles, health, homes, parenting, cookery and fun.
FEATURES *Tiffany Sherlock* Max. 1,600 words.

## The Third Alternative
▷ Black Static

## This England

PO Box 52, Cheltenham GL50 1YQ
☎ 01242 537900    ▣ 01242 537901
Owner *This England Publishing Ltd*
Editor *Stephen Garnett*
Circulation 140,000

Founded 1968. QUARTERLY, with a strong overseas readership. Celebration of England and all things English: famous people, natural beauty, towns and villages, history, traditions, customs and legends, crafts, etc. Generally a rural basis, with the 'Forget-me-nots' section publishing readers' recollections and nostalgia. Up to a hundred unsolicited pieces received each week. Unsolicited mss/ideas welcome. Length 250–1,500 words.
£ £25 per 1,000 words.

## Time

Blue Fin Building, 110 Southwark Street, London SE1 0SU
☎ 020 3148 3000
edit_office@timemagazine.com
www.time.com
Owner *Time Warner*
Senior Editor *Catherine Mayer*
Circulation 130,264 (UK)

Founded 1923. WEEKLY current affairs and news magazine. There are few opportunities for freelancers on *Time* as almost all the magazine's content is written by staff members from various bureaux around the world. No unsolicited mss.

## Time Out

Universal House, 251 Tottenham Court Road, London W1T 7AB
☎ 020 7813 3000    ▣ 020 7813 6001
www.timeout.com
Publisher & Managing Director *Mark Elliott*
Editor *Mark Frith*
Deputy Editor *Rachel Halliburton*
Managing Editor *Claire Hojem*
Circulation 61,397

Founded 1968. WEEKLY magazine of news, arts, entertainment and lifestyle in London plus listings.
FEATURES *Simone Baird* 'Usually written by staff writers or commissioned, but it's always worth submitting an idea if particularly apt to the magazine.' Word length varies; up to 2,000 max.
CONSUMER SECTION Food and drink, shopping, services, travel, design, property, health and fitness.
£ Payment negotiable.

## The Times Educational Supplement

26 Red Lion Square, Holborn, London WC1R 4HQ
☎ 020 3194 3000
▣ 020 3194 3200 (editorial)/3277 (newsdesk)
www.tes.co.uk
Owner *Charterhouse*
Editor *Gerard Kelly*
Circulation 48,867

Founded 1910. WEEKLY. New contributors are welcome and should fax or email ideas on one sheet of A4 for news (newsdesk@tes.co.uk), features (features@tes.co.uk) or reviews. The main newspaper accepts contributions in the following sections:
COMMENT Weekly slot for a well-informed and cogently argued viewpoint. Max. 750 words (comment@tes.co.uk).
LEADERSHIP Practical issues for school governors and managers. Max. 500 words.
FE FOCUS Weekly pull-out section covering post-16 education and training in colleges, work and the wider community. Aimed at everyone from teachers/lecturers to leaders and opinion formers in lifelong learning. News, features, comment and opinion on all aspects of college life welcome. Length from 350 words (news) to 1,000 max. (features). Contact *Alan Thomson* (alan.thomson@tes.co.uk).
TES MAGAZINE Weekly magazine for teachers focusing on looking at their lives, inside and outside the classroom, investigating the key issues of the day and highlighting good practice. Max. 800 words (features@tes.co.uk).

## The Times Educational Supplement Scotland

Thistle House, 21–23 Thistle Street, Edinburgh
EH2 1DF
☎ 0131 624 8332    📠 0131 467 8019
scoted@tes.co.uk
www.tes.co.uk/scotland
Owner *TSL Education Ltd*
Editor *Neil Munro*
Circulation 5,987

Founded 1965. WEEKLY. Unsolicited mss welcome.
FEATURES Articles on education in Scotland.
Max. 1,000 words.
NEWS Items on education in Scotland. Max. 600 words.

## Times Higher Education

26 Red Lion Square, London WC1R 4HQ
☎ 020 3194 3000    📠 020 3194 3300
editor@tsleducation.com
www.timeshighereducation.co.uk
Owner *Charterhouse Private Equity*
Editor *Ann Mroz*
Circulation 21,843

Founded 1971. WEEKLY. Unsolicited items are
welcome but most articles and almost all book
reviews are commissioned. 'In most cases it is
better to write or email, but in the case of news
stories it is all right to phone.'
FEATURES *Ann Mroz* Most articles are
commissioned from academics in higher
education.
NEWS *John Gill* Freelance opportunities very
occasionally.
SCIENCE *Zoe Corbyn*
BOOKS *Karen Shook*
FOREIGN *Phil Baty*
💷 Payment by negotiation.

## The Times Literary Supplement

Times House, 1 Pennington Street, London
E98 1BS
☎ 020 7782 5000    📠 020 7782 4966
www.the-tls.co.uk
Owner *The Times Literary Supplement Ltd*
Editor *Peter Stothard*
Circulation 31,958
Founded 1902. WEEKLY review of literature,
history, philosophy, science and the arts.
Contributors should approach in writing and be
familiar with the general level of writing in the
*TLS*.
LITERARY DISCOVERIES AND POEMS *Alan Jenkins*
NEWS Reviews and general articles concerned
with literature, publishing and new intellectual
developments anywhere in the world. Length by
arrangement.
💷 Payment by arrangement.

## Today's Golfer

Media House, Lynchwood, Peterborough
Business Park, Peterborough PE2 6EA
☎ 01733 468000    📠 01733 468843
editorial@todaysgolfer.co.uk
www.todaysgolfer.co.uk
Owner *Bauer Media*
Editor-in-Chief *Andy Calton*
Circulation 72,007

Founded 1988. MONTHLY. Golf instruction,
features, player profiles and news. Most features
written in-house but unsolicited mss will be
considered. Approach in writing with ideas. Not
interested in instruction material from outside
contributors.
FEATURES/NEWS *Kit Alexander* Opinion, player
profiles and general golf-related features.

## Top of the Pops Magazine

Media Centre, 201 Wood Lane, London W12 7TQ
☎ 020 8433 3847
www.totpmag.com
Owner *BBC Worldwide Publishing Ltd*
Editor *Peter Hart*
Circulation 107,576

Founded 1995. FOUR-WEEKLY teenage celebrity
magazine with a lighthearted and humorous
approach. No unsolicited material. Freelance
contributions only after commissioning by the
editor.

## Total Coarse Fishing

DHP Ltd, 2 Stephenson Close, Drayton Fields,
Daventry NN11 8RF
☎ 01327 311999    📠 01327 311190
kevin.wilmot@dhpub.co.uk
www.tcfmagazine.com
Owner *David Hall Publishing Ltd*
Editor *Kevin Wilmot*
Deputy Editor *Steve Martin*
Circulation 35,000

Founded 2006. MONTHLY magazine for anglers
who like to keep their fishing varied and target
quality coarse fish from exceptional venues. No
unsolicited contributions.
FEATURES Overseas freshwater angling features
with 'exceptional' photographs. Max. 2,200
words. Approach by email. 💷 £200.

## Total Film

2 Balcombe Street, London NW1 6NW
☎ 020 7042 4000    📠 020 7042 4839
totalfilm@futurenet.com
Owner *Future Publishing*
Editor *Aubrey Day (aubrey.day@futurenet.com)*
Reviews *Matthew Leyland (matthew.leyland@
futurenet.com)*
Features *Jane Crowther (jane.crowther@futurenet.
com)*

News *Rosie Fletcher (rosie.fletcher@futurenet.com)*
Circulation 81,029

Founded 1997. MONTHLY reviews-based movie magazine. Interested in ideas for features, not necessarily tied in to specific releases, and humour items. No reviews or interviews with celebrities/directors. Approach by post or email.

## Total Flyfisher
DHP Ltd, 2 Stephenson Close, Drayton Fields, Daventry NN11 8RF
☎ 01327 311999   🖷 01327 311190
steve.cullen@dhpub.co.uk
www.totalflyfisher.com
Owner *David Hall Publishing Ltd*
Editor *Steve Cullen*
Circulation 17,000

Founded 2003. MONTHLY instructional fly fishing magazine. Welcomes contributions including fishery reports and fly tying articles. No foreign features or advanced skills. Approach by email.
FEATURES Stillwater trout tactics, salmon tactics, river trout/grayling tactics, UK saltwater tactics. Max. 2,500 words. 💷 Payment (including pictures) negotiable.
NEWS Stories relevant to UK fly fishers.
FICTION Funny fishing stories. Max. 400 words.
💷 Payment negotiable.

## Total TV Guide
H. Bauer Publishing, Academic House, 24–28 Oval Road, London NW1 7DT
☎ 020 7241 8000   🖷 020 7241 8042
barbara.miller@bauer.co.uk
www.bauer.co.uk
Owner *H. Bauer Publishing*
Editor *Jon Peake*
Circulation 110,748

Founded 2003. WEEKLY TV listings for those with access to multichannel television, covering everything from movies to sport, drama to children's programmes.
FEATURES *Ben Lawrence* Commissions features relating to current TV programmes; chiefly celebrity interviews and set visits. No unsolicited contributions.

## Total Vauxhall
Future Publishing, 30 Monmouth Street, Bath BA1 2BW
☎ 01225 442244   🖷 01225 446019
info@totalvauxhall.co.uk
www.totalvauxhall.co.uk
Owner *Future Publishing*
Editor *Dougie Rankine*
Circulation 11,196

Founded 2001. MONTHLY independent newsstand magazine aimed at the Vauxhall enthusiast. Covers new, modified, historical/classic, race and rally Vauxhalls of all kinds. Substantial amount of technical content. Also covers the more interesting parts of the GM family, particularly Opel and Holden. Uses a lot of freelance contributors; 'those who hit deadlines and fulfil the brief get regular work and lots of it.' Feature ideas, news items, historical pieces and potential feature cars welcome but must have a Vauxhall/GM tilt. Call the editor directly for an informal discussion on style, approach and angle.
💷 Payment 'surprisingly generous'.

## Trail
Media House, Lynchwood, Peterborough PE2 6EA
☎ 01733 468205
trail@bauermedia.co.uk
www.LFTO.com
Owner *Bauer Consumer Media*
Editor *Matthew Swaine*
Circulation 40,674

Founded 1990. MONTHLY. Gear reports, where to walk and practical advice for the hill walker and long distance walker. Inspirational reads on people and outdoor/walking issues. Health, fitness and injury prevention for high level walkers and outdoor lovers. 'We always welcome ideas from writers who understand our readers' needs. No travel pieces. We are interested in pieces that allow our readers to get out more, to enjoy their time outdoors and improve their skills. So don't tell us what *you've* done. Tell our readers what they could do, why they should do it and how best to do it. Inspiration and practical advice in equal measure.' Approach by phone or in writing in the first instance.
FEATURES *Simon Ingram* Very limited requirement for overseas articles, 'written to our style.' Ask for guidelines. Max. 2,000 words. Limited requirement for guided walks articles. Specialist writers only. Ask for guidelines. 750–2,000 words (depending on subject).
💷 Payment varies.

## Traveller
45–49 Brompton Road, London SW3 1DE
☎ 020 7589 0500
traveller@and-publishing.co.uk
www.traveller.org.uk
Owner *Wexas International Ltd*
Editor *Amy Sohanpaul*
Circulation 18,511

Founded 1970. QUARTERLY travel magazine.
FEATURES High quality, personal narratives of remarkable journeys. Articles should be off-beat, adventurous, authentic. For guidelines, see website. Articles may be accompanied by professional quality, original slides or high-res

digital photographs. Freelance articles considered. Max. 1,000 words. Initial contact by email.

(£) £200 per 1,000 words.

## Tribune

9 Arkwright Road, London NW3 6AN
(T) 020 7433 6410    (F) 020 7433 6410
mail@tribunemagazine.co.uk
www.tribunemagazine.co.uk
Owner *Tribune Publications 2009*
Editor *Chris McLaughlin*

Founded 1937. WEEKLY. Independent Labour publication covering parliament, politics, trade unions, public policy and social issues, international affairs and the arts. Welcomes freelance contributions; approach by telephone or email.
FEATURES *George Osgerby* Anything in line with the topics above. Max. 1,000 words.
NEWS *Keith Richmond* Breaking or forthcoming news events/stories. Max. 480 words.
OTHER PAGES Interviews, reviews, cartoons.

## Trout Fisherman

Bauer Media, Media House, Lynch Wood Business Park, Peterborough PE2 6EA
(T) 01733 395131
russell.hill@bauermedia.co.uk
www.gofishing.co.uk
Owner *Bauer Media*
Editor *Russell Hill*
Circulation 20,355

Founded 1977. THIRTEEN ISSUES YEARLY. Instructive magazine on trout fishing. Most of the articles are commissioned, but unsolicited mss and quality colour images welcome.
FEATURES Max. 2,500 words.
(£) Payment varies.

## TVTimes

IPC Media Ltd, Blue Fin Building, 110 Southwark Street, London SE1 0SU
(T) 020 3148 5615    (F) 020 3148 8115
Owner *IPC Media (A Time Warner Company)*
Editor *Ian Abbott*
Circulation 321,005

Founded 1955. WEEKLY magazine of listings and features serving the viewers of independent television, BBC, satellite and radio. Freelance contributions by commission only. No unsolicited contributions.

## TwelvePoint.com (previously ScriptWriter magazine)

2 Elliott Square, London NW3 3SU
(T) 020 7586 4853
jonquil@twelvepoint.com
www.twelvepoint.com
Owner *Scriptease Ltd*

Managing Editor *Jonquil Florentin*
Editor *Julian Friedmann*
Circulation 1,500

Launched 2001. Now an online website publishing new articles every week, together with many web-based extras (Forum, blog, Ask the Expert, member profiles) covering all aspects of the business and craft of writing for the small and large screen. Interested in serious, in-depth analysis; max. 1,500–3,500 words. Email with synopsis, sample material and c.v.

## 20x20 magazine

36 Osborne Road, London N4 3SD
info@20x20magazine.com
20x20magazine.com
Owner/Editors *Giovanna Paternò, Francesca Ricci*
Circulation 500

Launched 2008. BIANNUAL. 'A square platform for writings, visuals and cross-bred projects.' Includes three sections – Words: in the shape of fiction, essays, poetry; Visions: drawings and photography; The Blender: where words and visions cross path. Submissions: email only and in response to the meta-words set for the forthcoming issue (announced via the website or by joining the emailing list). Max. 1,500 words; submit up to three written pieces or poems. Visual artists are encouraged to submit images keeping the format of the page in mind; send high resolution images (300 dpi). 'If you would like to submit a visual project consisting of a selection of images, send low res first. We regret that at present we are unable to pay authors for their work, Also, due to the large amount of submissions received, we are usually able to notify successful contributors only.'

## Ulster Business

5b Edgewater Business Park, Belfast Harbour Estate, Belfast BT3 9JQ
(T) 028 9078 3200    (F) 028 9078 3210
davidelliott@greerpublications.com
www.ulsterbusiness.com
Owner *James Greer*
Editor *David Elliott*
Circulation 7,088

Founded 1987. MONTHLY. General business content with coverage of all sectors: ICT, retail, agriculture, commercial property, etc. 'Opportunities exist for well-written local orientated features, particularly those dealing with local firms or the issues they face; local business personality interviews also useful.' Contact the editor by email. Max. 650 words.

## Ulster Tatler

39 Boucher Road, Belfast BT12 6HR
(T) 028 9066 3311    (F) 028 9038 1915

edit@ulstertatler.com
www.ulstertatler.com
Owner/Editor *Chris Sherry*
Circulation 10,552

Founded 1965. MONTHLY. Articles of local interest
and social functions appealing to Northern
Ireland's ABC1 population. Welcomes unsolicited
material; approach by phone or in writing in the
first instance.
FEATURES *James Sherry* Max. 1,500 words.
FICTION *Richard Sherry* Max. 3,000 words

## Ultimate Advertiser

PO Box 7086, Clacton on Sea CO15 5WN
℡ 01255 428500    ℻ 0871 918 4756
mail@apexpublishing.co.uk
www.ultimateadvertiser.co.uk
Owner *Apex Publishing Ltd*
Editor *Chris Cowlin*
Circulation 5,000

Founded 2005. MONTHLY advertising magazine
including guides and information, features,
listings and reviews. Approach by post, only.

## Uncut

IPC Media Ltd, Blue Fin Building, 110 Southwark
Street, London SE1 0SU
℡ 020 3148 6985
www.uncut.co.uk
Owner *IPC Media (A Time Warner Company)*
Editor *Allan Jones*
Associate Editor *Michael Bonner*
Deputy Editor *John Mulvey*
Music Reviews Editor *John Robinson*
Art Editor *Marc Jones*
Online News Editor *Farah Ishaq*
Circulation 75,518

Launched 1997. MONTHLY music and film
magazine with free CD each issue. Welcomes
music and film reviews; approach by email.

## The Universe

Gabriel Communications Ltd, 4th Floor,
Landmark House, Station Road, Cheadle Hulme
SK8 7JH
℡ 0161 488 1700    ℻ 0161 488 1701
joseph.kelly@totalcatholic.com
www.totalcatholic.com
Owner *Gabriel Communications Ltd*
Chief Executive *Joseph Kelly*
Circulation 60,000

Occasional use of new writers, but a substantial
network of regular contributors already exists.
Interested in a very wide range of material:
all subjects which might bear on Catholic life.
Fiction not normally accepted.
£ Payment negotiable.

## The Vegan

Donald Watson House, 21 Hylton Street,
Birmingham B18 6HJ
℡ 0121 523 1730
editor@vegansociety.com
www.vegansociety.com
Owner *Vegan Society*
Editor *Rosamund Raha*
Circulation 7,000

Founded 1944. QUARTERLY. Deals with the
ecological, ethical and health aspects of veganism.
Unsolicited mss welcome. Max. 2,000 words.
£ Payment negotiable.

## Vogue

Vogue House, Hanover Square, London W1S 1JU
℡ 020 7499 9080    ℻ 020 7408 0559
www.vogue.com
Owner *Condé Nast Publications Ltd*
Editor *Alexandra Shulman*
Circulation 210,526

Launched 1916. Condé Nast Magazines tend
to use known writers and commission what's
needed, rather than using unsolicited mss.
Contacts are useful.

FEATURES *Harriet Quick* Upmarket general
interest rather than 'women's'. Good proportion of
highbrow art and literary articles, as well as travel,
gardens, food, home interest and reviews.

## The Voice

GV Media Group Ltd, Northern & Shell Tower,
6th Floor, 4 Selsdon Way, London E14 9GL
℡ 020 7510 0340    ℻ 020 7510 0341
yourviews@gvmedia.co.uk
www.voice-online.co.uk
Managing Director *George Ruddock*
Editor *Steve Pope*
Circulation 40,000

Founded 1982. Leading WEEKLY newspaper for
black Britons and other minority communities.
Includes news, features, arts, sport and a
comprehensive jobs section. Illustrations: colour
and b&w photos. Open to ideas for news and
features on sport, business, community events
and the arts.

## Volkswagen Golf+

Kelsey Publishing Ltd, PO Box 978, Peterborough
PE1 9FL
℡ 01733 347559
golf.ed@kelsey.co.uk
www.kelsey.co.uk
Owner *Kelsey Publishing*
Editor *Ian Cushway*
Circulation 25,000

Founded 1995. MONTHLY. Features Volkswagen
Golfs and other watercooled VAG cars. Articles

on tuning, styling, performance, technical and
potential feature cars. 'Not interested in anything
to do with the game of golf.' Submissions should
be made directly to the editor. Articles must show
excellent specialist knowledge of the subject
matter. Ⓔ Payment by negotiation.

## W.I. Life

104 New Kings Road, London SW6 4LY
Ⓣ 020 7731 5777　　Ⓕ 020 7736 4061
Owner *National Federation of Women's Institutes*
Editor *Neal Maidment*
Circulation 205,000

Women's Institute membership magazine. First
issue February 2007. EIGHT ISSUES YEARLY.
Contains articles on a wide range of subjects of
interest to women. Strong environmental country
slant with crafts and cookery plus gardening.
Contributions, photos and illustrations from WI
members welcome. Ⓔ Payment by arrangement.

## Walk Magazine

2nd Floor, Camelford House, 87–90 Albert
Embankment, London SE1 7TW
Ⓣ 020 7339 8500　　Ⓕ 020 7339 8501
walkmag@ramblers.org.uk
www.ramblers.org.uk
www.walkmag.co.uk
Owner *Ramblers' Association*
Editor *Dominic Bates*
Assistant Editor *Denise Noble*
Circulation 105,000

QUARTERLY. Official magazine of the Ramblers'
Association, available to members and on sale to
the public through selected outlets. Unsolicited
mss welcome. S.a.s.e. required for return.
FEATURES Freelance features are invited on any
aspect of walking in Britain. Length up to 1,500
words, preferably with good photographs. No
general travel articles.

## Wallpaper

IPC Media Ltd, Blue Fin Building, 110 Southwark
Street, London SE1 0SU
Ⓣ 020 3148 5000　　Ⓕ 020 3148 8119
contact@wallpaper.com
www.wallpaper.com
Owner *IPC Media (A Time Warner Company)*
Editor *Tony Chambers*
Circulation 105,028

Launched 1996. MONTHLY magazine with the
best of international design, architecture, fashion,
travel and lifestyle.
FEATURES *Nick Compton* Ideas welcome but
contributors must be familiar with the voice and
subject matter of the magazine. No celebrity
items. Approach by email.

## The War Cry

101 Newington Causeway, London SE1 6BN
Ⓣ 020 7367 4900　　Ⓕ 020 7367 4710
warcry@salvationarmy.org.uk
www.salvationarmy.org.uk/warcry
Owner *The Salvation Army*
Editor *Major Nigel Bovey*
Circulation 48,000

Founded 1879. WEEKLY full-colour magazine
containing Christian comment on current
issues. Unsolicited mss welcome if appropriate
to contents. No fiction or poetry. Approach by
phone or email with ideas.
NEWS relating to Christian Church or social
issues. Max. length 500 words.
FEATURES Human interest articles aimed at the
'person-in-the-street'. Max. 500 words.
Ⓔ Payment negotiable.

## Waterways World

151 Station Street, Burton on Trent DE14 1BG
Ⓣ 01283 742950　　Ⓕ 01283 742957
editorial@waterwaysworld.com
www.waterwaysworld.com
Owner *Waterways World Ltd*
Editor *Richard Fairhurst*
Circulation 15,246

Founded 1972. MONTHLY magazine for inland
waterway enthusiasts. Unsolicited mss welcome,
provided the writer has a good knowledge of the
subject. No fiction. See 'Notes for Contributors'
on website.
FEATURES *Keith Goss* Articles (preferably
illustrated) are published on all aspects of
inland waterways in Britain and abroad but
predominantly recreational boating on rivers and
canals.
NEWS *Chris Daniels* Max. 500 words.
Ⓔ £70 per published page.

## Web User

IPC Media Ltd, Blue Fin Building, 110 Southwark
Street, London SE1 0SU
Ⓣ 020 3148 5000　　Ⓕ 020 3148 8122
webuser@ipcmedia.com
www.webuser.co.uk
Owner *IPC Media (A Time Warner Company)*
Editor *Claire Woffenden*
Circulation 31,619

Founded 2001. FORTNIGHTLY bestseller
internet magazine for all users of the internet.
No unsolicited material; send email or letter of
enquiry in the first instance.

## Wedding

IPC Media Ltd, Blue Fin Building, 110 Southwark
Street, London SE1 0SU
Ⓣ 020 3148 7800

wedding@ipcmedia.com
www.weddingmagazine.co.uk
Owner *IPC Media (A Time Warner Company)*
Editor *Catherine Westwood*
Circulation 46,602

Founded 1985. BI-MONTHLY offering ideas
and inspiration for women planning their
wedding. Most features are written in-house
or commissioned from known freelancers.
Unsolicited mss are not welcome, but approaches
may be made in writing.

## Wedding Ideas Magazine

Giraffe Media Ltd, 8 Hammet Street, Taunton
TA1 1RZ
℡ 01823 288344    🖷 01823 288239
info@weddingideasmagazine.co.uk
www.weddingideasmagazine.co.uk
Owner *Rachel Southwood*
Editor *Rachel Morgan*
Circulation 28,531

Launched 2004. FOUR-WEEKLY guide for brides
on a budget with over 30 pages of real-life
weddings. Welcome specific feature ideas; study
the magazine for current editorial style. Approach
by email.

## Weekly News

D.C. Thomson & Co. Ltd, Albert Square, Dundee
DD1 9QJ
℡ 01382 223131    🖷 01382 201390
weeklynews@dcthomson.co.uk
Owner *D.C. Thomson & Co. Ltd*
Editor *David Burness*
Circulation 53,619

Founded 1855. WEEKLY. Newsy, family-orientated
magazine designed to appeal to the busy
housewife. Regulars include true-life stories told
in the first person, showbiz, royals and television.
Two or three general interest fiction stories
each week. Usually commissions, but writers of
promise will be taken up. 🄴 Payment negotiable.

## Welsh Country Magazine

Aberbanc, Llandysul SA44 5NP
℡ 01559 372010    🖷 01559 371995
editor@welshcountry.co.uk
www.welshcountry.co.uk
Owner *Equine Marketing Limited*
Editor *Kath Rhodes*
Circulation 16,798

Founded 2004. BI-MONTHLY. Outside
contributions welcome. Contact the editor with
samples of work before submitting full article.
The magazine is written in English.
FEATURES Pesonal experiences or viewpoints,
historical or modern, but there needs to be a
strong Welsh connection in some form.

NEWS All Welsh news welcome; anything at all
commercial is not paid for.
🄴 Payment negotiable.

## What Car?

Teddington Studios, Teddington Lock, Broom
Road, Teddington TW11 9BE
℡ 020 8267 5688
www.whatcar.com
Owner *Haymarket Motoring Publications Ltd*
Editor *Steve Fowler*
Circulation 83,102

MONTHLY. The car buyer's bible, *What Car?*
concentrates on road test comparisons of new
cars, news and buying advice on used cars, as well
as a strong consumer section. Some scope for
freelancers. No unsolicited mss.
🄴 Payment negotiable.

## What Hi-Fi? Sound & Vision

Teddington Studios, Teddington Lock, Broom
Road, Teddington TW11 9BE
℡ 020 8943 5000    🖷 020 8267 5019
whathifi@haymarket.com
www.whathifi.com
Owner *Haymarket Consumer Media Ltd*
Editor *Dominic Dawes*
Circulation 47,444

Founded 1976. THIRTEEN ISSUES YEARLY.
Features on hi-fi and home cinema. No
unsolicited contributions. Prior consultation with
the editor essential.
FEATURES General or more specific items on
hi-fi and home cinema pertinent to the consumer
electronics market.
REVIEWS Specific product reviews. All material
is now generated by in-house staff. Freelance
writing no longer accepted.

## What Investment

Vitesse Media Plc, Octavia House, 50 Banner
Street, London EC1Y 8ST
℡ 020 7250 7044    🖷 020 7250 7011
joe.mcgrath@vitessemedia.co.uk
www.whatinvestment.co.uk
Owner *Vitesse Media plc*
Editor *Joe McGrath*
Deputy Editor *Jenny Lowe*
Circulation 16,320

Founded 1982. MONTHLY. Features articles on a
variety of savings and investment matters. Max.
1,200 words. Email ideas to the editor.

## What Satellite and Digital TV

2 Balcombe Street, London NW1 6NW
℡ 020 7042 4000
wotsat@futurenet.co.uk
www.wotsat.com
Owner *Future Publishing Ltd*
Editor *Alex Lane*

Circulation 40,000

Founded 1986. MONTHLY including news, technical information, reviews, equipment tests, programme background, listings. Contributions welcome – phone first.

FEATURES In-depth guides to popular/cult shows. Technical tutorials.

NEWS Industry and programming. Max. 250 words.

## Wild Times

The RSPB, UK Headquarters, The Lodge, Sandy SG19 2DL

☏ 01767 680551   ⓕ 01767 683262

derek.niemann@rspb.org.uk

Owner *Royal Society for the Protection of Birds*
Editor *Derek Niemann*

Founded 1965. BI-MONTHLY. Bird, wildlife and nature conservation for under-8-year-olds (RSPB Wildlife Explorers members). No unsolicited mss. No 'captive/animal welfare' articles.

FEATURES Unsolicited material rarely used.

NEWS News releases welcome. Approach in writing in the first instance.

## Wingbeat

The RSPB, UK Headquarters, The Lodge, Sandy SG19 2DL

☏ 01767 680551   ⓕ 01767 683262

derek.niemann@rspb.org.uk

Owner *Royal Society for the Protection of Birds*
Editor *Derek Niemann*

Founded 1965. QUARTERLY. Bird, wildlife and nature conservation for 13–18-year-olds (RSPB Wildlife Explorers teenage members). No unsolicited mss. No 'captive/animal welfare' articles.

FEATURES Unsolicited material rarely used.

NEWS News releases welcome. Approach in writing in the first instance.

## The Wisden Cricketer

123 Buckingham Palace Road, London SW1W 9SL

☏ 020 7705 4911   ⓕ 020 7921 9151

twc@wisdencricketer.com

www.thewisdencricketer.com

Owner *BSkyB Publications*
Editor *John Stern*
Deputy Editor *Edward Craig*
Assistant Editor *Daniel Brigham*
Editorial Assistant *Beni Moorhead*
Circulation 32,599

Founded 2003. MONTHLY. Result of a merger between *The Cricketer International* (1921) and *Wisden Cricket Monthly* (1979). Very few uncommissioned articles are used, but would-be contributors are not discouraged. Approach in writing. ⓔ Payment varies.

## Woman

IPC Media Ltd, Blue Fin Building, 110 Southwark Street, London SE1 0SU

☏ 020 3148 5000

woman@ipcmedia.com

www.ipcmedia.com

Owner *IPC Media (A Time Warner Company)*
Editor *Karen Livermore*
Deputy Editor *Jenny Vereker*
Circulation 316,216

Founded 1937. WEEKLY. Long-running, popular women's magazine which boasts a readership of over 2.5 million. No unsolicited mss. Most work commissioned. Approach with ideas in writing.

FEATURES *Anna Kingsley* Max. 1,250 words.

BOOKS *Claire Snewin*

## Woman and Home

IPC Media Ltd, Blue Fin Building, 110 Southwark Street, London SE1 0SU

☏ 020 3148 7836

www.womanandhome.com

Owner *IPC Media (A Time Warner Company)*
Editor *Sue James*
Circulation 368,388

Founded 1926. MONTHLY. No unsolicited mss. Prospective contributors are advised to email ideas to the features editor. All freelance work is specially commissioned.

## Woman's Own

IPC Media Ltd, Blue Fin Building, 110 Southwark Street, London SE1 0SU

☏ 020 3148 5000   ⓕ 020 3148 8112

Owner *IPC Media (A Time Warner Company)*
Editor *Karen Livermore*
Features Editor *Anna Wharton*
Circulation 298,472

Founded 1932. WEEKLY. Prospective contributors should contact the features editor *in writing* in the first instance before making a submission.

## Woman's Weekly

IPC Media Ltd, Blue Fin Building, 110 Southwark Street, London SE1 0SU

☏ 020 3148 5000

womansweeklypostbag@ipcmedia.com

Owner *IPC Media (A Time Warner Company)*
Editor *Diane Kenwood*
Deputy Editor *Geoffrey Palmer*
Features Editor *Sue Pilkington*
Fiction Editor *Gaynor Davies*
Circulation 344,553

Founded 1911. Mass-market WEEKLY for the mature woman.

FEATURES General features covering any subject of interest to women of forty-plus. Can be newsy and/or emotional but should be informative, engaging and entertaining. Could be campaigning, issue-based, first or third person. Strong real-life stories always of interest. Words: ranging from 800 to 2,000. Only experienced journalists. Synopses and ideas should be submitted by email to pat_miller@ipcmedia.com FICTION Short stories, 1,000–2,000 words; serials, 12,000 words. Guidelines: should be contemporary and engaging. Humour and wit also welcome. Short stories up to 8,000 words considered for *Woman's Weekly Fiction Special* (see entry). Emails to maureen_street@ipcmedia.com

## Woman's Weekly Fiction Special

IPC Media Ltd, Blue Fin Building, 110 Southwark Street, London SE1 0SU
☏ 020 3148 6600
womansweeklypostbag@ipcmedia.com
www.ipcmedia.com
Owner *IPC Media (A Time Warner Company)*
Editor *Gaynor Davies*

Launched 1998. EIGHT ISSUES A YEAR. Welcomes short stories of between 1,000 and 8,000 words. Guidelines are available from the Fiction Department or via the postbag (above). See also *Woman's Weekly*.

## The Woodworker

MyHobbyStore Publishing Ltd, Berwick House, 8–10 Knoll Rise, Orpington BR6 0EL
☏ 01689 899210
ralph.laughton@myhobbystore.com
www.getwoodworking.com
Owner *MyHobbyStore Publishing Ltd.*
Editor *Ralph Laughton*

Founded 1901. MONTHLY. Contributions welcome; approach with ideas in writing.
FEATURES Articles on woodworking with good photo support appreciated. Max. 2,000 words.
£ Payment negotiable.

## World Fishing

Mercator Media Ltd, The Old Mill, Lower Quay, Fareham PO16 0RA
☏ 01329 825335    ℻ 01329 825330
cwills@worldfishing.net
www.worldfishing.net
Owner *Mercator Media Ltd*
Editor *Carly Wills*
Circulation 3,300

Founded 1952. MONTHLY. Unsolicited mss welcome; approach by phone or in writing with an idea.
NEWS/FEATURES of a technical or commercial nature relating to the commercial fishing and fish processing industries worldwide (the magazine is read in over 100 different countries). Max. 1,500 words.
£ Payment by arrangement.

## The World of Fine Wine

226 City Road, London EC1V 2TT
☏ 020 7812 8673
www.finewinemag.com
Editor *Dr Neil Beckett*
Editorial Adviser *Hugh Johnson*
Publisher *Sara Morley*
Managing Director *Piers Spence*

QUARTERLY book-length journal covering fine and noteworthy wine, established in 2004 and owned by the Quarto Group. Prospective contributors should contact the editor in the first instance.

## The World of Interiors

Vogue House, Hanover Square, London W1S 1JU
☏ 020 7499 9080    ℻ 020 7493 4013
interiors@condenast.co.uk
www.worldofinteriors.co.uk
Owner *Condé Nast Publications Ltd*
Editor *Rupert Thomas*
Circulation 62,032

Founded 1981. MONTHLY. Best approach by fax or letter with an idea, preferably with reference snaps or guidebooks.
FEATURES *Rupert Thomas* Most feature material is commissioned. 'Subjects tend to be found by us, but we are delighted to receive suggestions of interiors, archives, little-known museums, collections, etc. unpublished elsewhere, and are keen to find new writers.'

## World Ski Guide

Mountain Leisure Limited, Parman House, 30–36 Fife Road, Kingston-upon-Thames KT1 1SY
☏ 020 8547 9822    ℻ 020 8546 0984
edit@goodholidayideas.com
www.goodskiguide.com
Owner *Mountain Leisure Ltd*
Editors *John Hill, Christian Berger*
Circulation 40,000

Launched 2002. ANNUAL. Translation of *Good Ski Guide* into German. See website for editorial guidelines.
FEATURES Destination features on ski resorts worldwide. Max. 900 words.

## World Soccer

IPC Media Ltd, Blue Fin Building, 110 Southwark Street, London SE1 0SU
☏ 020 3148 4817    ℻ 020 3148 8130
world_soccer@ipcmedia.com
www.worldsoccer.com
Owner *IPC Media (A Time Warner Company)*
Editor *Gavin Hamilton*
Circulation 40,003

Founded 1960. MONTHLY. News and features on world soccer. Unsolicited material welcome but initial approach by email or in writing.

## Writers' Forum

PO Box 6337, Bournemouth BH1 9EH
℡ 01202 586848
editorial@writers-forum.com
www.writers-forum.com
Owner *Select Publisher Services*
Editor *Carl Styants*
Publisher *Tim Harris*

Founded 1993. MONTHLY. Magazine covers all aspects of the craft of writing. Well written articles welcome. Write to the editor in the first instance.
*Writers Forum Short Story Competition* Prizes: £300 (1st), £150 (2nd), £100 (3rd). Entrance fee, non-subscribers: £10; subscribers: £7. Winners published in every issue.
*Writers' Forum Poetry Competition* First prize £100; runners up receive a dictionary. Entrance fee: £5 for one poem or £7 for two. Winners published in every issue. Annual subscription: £33 UK; £46 Worldwide. Send s.a.s.e. for free back issue.

## Writers' News/Writing Magazine

Warners Group Publications, Fifth Floor, 31–32 Park Row, Leeds LS1 5JD
℡ 0113 200 2929   0113 200 2928
www.writersnews.co.uk
Owner *Warners Group Publications plc*
Publisher *Janet Davison*
Editor *Jonathan Telfer*

Founded 1989. MONTHLY magazines containing news and advice for writers. *Writers' News* is available exclusively by subscription; *Writing Magazine*, a full-colour glossy publication, is available by subscription or on newsstands. No poetry or general items on 'how to become a writer'. Receive 1,000 mss each year. Approach in writing or by email.
NEWS Exclusive news stories of interest to writers. Max. 350 words.
FEATURES How-to articles of interest to professional writers. Max. 1,000 words.

## Yachting Monthly

IPC Media Ltd, Blue Fin Building, 110 Southwark Street, London SE1 0SU
℡ 020 3148 4872    020 3148 8128
paul_gelder@ipcmedia.com
www.yachtingmonthly.com
Owner *IPC Media (A Time Warner Company)*
Editor *Paul Gelder*
Circulation 30,044

Founded 1906. MONTHLY magazine for yachting and cruising enthusiasts – not racing. Unsolicited mss welcome, but many are received and not used. Prospective contributors should make initial contact in writing.

FEATURES A wide range of features concerned with maritime subjects and cruising under sail; well-researched and innovative material always welcome, especially if accompanied by high resolution digital images. Max. 1,800 words.
 £80–£110 per 1,000 words.

## Yachting World

IPC Media Ltd, Blue Fin Building, 110 Southwark Street, London SE1 0SU
℡ 020 3148 4846    020 3148 8127
yachting_world@ipcmedia.com
www.yachtingworld.com
Owner *IPC Media (A Time Warner Company)*
Editor *Andrew Bray*
Circulation 23,529

Founded 1894. MONTHLY with international coverage of yacht racing, cruising and yachting events. Will consider well researched and written sailing stories. Preliminary approaches should be by phone for news stories and in writing for features.
 Payment by arrangement.

## The Yellow Room Magazine

1 Blake Close, Bilton, Rugby CV22 7LJ
℡ 01788 334302
yellowjo@me.com
www.theyellowroom-magazine.co.uk
Owner/Editor *Jo Derrick*

Launched 2008. BI-ANNUAL magazine aiming to publish quality short stories only. 'The Yellow Room is a place where female writers can find support, encouragement and friendship.' Welcomes well written short stories which appeal to a predominantly female readership. These are stories which you wouldn't find in a weekly women's magazine, however. Submit one story with either s.a.s.e. for reply or email address. Max. 4,000 words.  Payment £10 per story. See website and guidelines for further details. No poetry, articles or novel extracts.

## Yorkshire Women's Life Magazine

PO Box 113, Leeds LS8 2WX
ywlmagenquiries@btinternet.com
www.yorkshirewomenslife.co.uk
Editor/Owner *Dawn Maria France*
Fashion *Sky Taylor*
Magazine PA *Anna Jenkins*
Diary *Giles Smith*
Circulation 15,000

Founded 2001. THREE ISSUES YEARLY. Features of interest to women along with regional, national, international news and lifestyle articles. Past issues covered health, stress management, living with domestic abuse, pampering breaks for city women, coping with a difficult boss, Windrush awards. 'It is important to study the style of the magazine before submitting material. Send A4 s.a.s.e. with 46p stamp for copy of submission guidelines. Unsolicited mss and new writers actively encouraged; approach in writing in the first instance with s.a.s.e.' Magazine available on subscription at the website address.

## You & Your Wedding

National Magazine Company, 72 Broadwick Street, London W1F 9EP
℡ 020 7439 5000 (editorial)    ℻ 020 7439 2985
cathy.howes@natmags.co.uk
www.youandyourwedding.co.uk
Owner *The National Magazine Company Ltd*
Editor *Colette Harris*
Circulation 50,142

Founded 1985. BI-MONTHLY. Anything relating to weddings, setting up home, and honeymoons. No unsolicited mss. Submit ideas by email only. No phone calls.

## You – The Mail on Sunday Magazine
▷ The Mail on Sunday under National Newspapers

## Young Writer

Fifth Floor, 31–32 Park Row, Leeds LS1 5JD
℡ 0113 200 2929    ℻ 0113 200 2928
mhill@writersnews.co.uk
www.youngwriter.org
Editor *Matthew Hill*

Describing itself as 'The Magazine for Children with Something to Say', *Young Writer* is issued four times a year, at the back-to-school times of January, March, June and September. A forum for young people's writing – fiction and non-fiction, prose and poetry – the magazine is an introduction to independent writing for young writers aged five to 18. £ Payment from £20 to £100 for freelance commissioned articles (these can be from adult writers).

## Your Cat Magazine

Roebuck House, 33 Broad Street, Stamford PE9 1RB
℡ 01780 766199    ℻ 01780 766416
yourcat@bournepublishinggroup.co.uk
www.yourcat.co.uk
Owner *BPG (Stamford) Ltd*
Editor *Sue Parslow*
Circulation 20,000

Founded 1994. MONTHLY magazine giving practical information on the care of cats and kittens, pedigree and non-pedigree, plus a wide range of general interest items on cats. Will consider 'true life' cat stories (max. 900 words) and quality fiction by published novelists. Send synopsis in the first instance by email or post. 'No articles written as though by a cat and no poetry.'

## Your Dog Magazine

Roebuck House, 33 Broad Street, Stamford PE9 1RB
℡ 01780 766199    ℻ 01780 754774
s.wright@bournepublishinggroup.co.uk
www.yourdog.co.uk
Owner *BPG (Stamford) Ltd*
Editor *Sarah Wright*
Circulation 26,789

Founded 1995. MONTHLY. Practical advice for pet dog owners. Will consider practical features and some personal experiences (no highly emotive pieces or fiction). Telephone in the first instance.
NEWS Max. 300–400 words.
FEATURES Max. 2,500 words; limited opportunities.
£ Payment negotiable.

## Your Hair

Origin Publishing, 9th Floor, Tower House, Fairfax Street, Bristol BS1 3BN
℡ 0117 927 9009    ℻ 0117 314 8310
michelletiernan@originpublishing.co.uk
www.yourhair.co.uk
Editor *Michelle Tiernan*
Commissioning Editor *Sophie Jordan*
Circulation 25,615

Launched 2001. MONTHLY magazine for women aged 21-plus who need some style inspiration, a whole new look or to keep a finger on the pulse of catwalk trends. 'We pride ourselves on wearable, achievable solutions.' No unsolicited contributions but welcome product information from PR agencies.

## Your Horse

Media House, Lynchwood, Peterborough PE2 6EA
℡ 01733 395052
www.yourhorse.co.uk
Owner *Bauer Media*
Editor *Julie Brown*
Circulation 29,802

'The magazine that aims to make owning and riding horses more rewarding.' Most writing produced in-house but well-targeted articles will always be considered.

## Yours Magazine

Media House, Peterborough Business Park, Peterborough PE2 6EA
℡ 01733 468000
yours@bauermedia.co.uk
www.yours.co.uk
Owner *Bauer Media*
Editor *Valery McConnell*
Circulation 284,560

Founded 1973. FORTNIGHTLY. Aimed at a readership aged 50 and over. Submission guidelines on request.

FEATURES Unsolicited mss welcome but must enclose s.a.s.e. Max. 1,400 words.
FICTION One or two short stories used in each issue. Max. 1,500 words.

£ Payment negotiable.

## Zest

National Magazine House, 72 Broadwick Street, London W1F 9EP
T 020 7439 5000    F 020 7312 3750
zest.mail@natmags.co.uk
www.zest.co.uk
Owner *National Magazine Company*
Editor *Mandie Gower*
Circulation 93,130

Founded 1994. MONTHLY. Health, beauty, fitness, nutrition and general well-being. No unsolicited mss. Prefers ideas in synopsis form; approach in writing.

## ZOO

Bauer Consumer Media, Mapin House, 4 Winsley Street, London W1W 8HF
T 020 7182 8355    F 020 7182 8300
info@zooweekly.co.uk
www.zootoday.com
Owner *Bauer Publishing*
Editor *Tom Etherington*
Circulation 102,043

Launched 2004. WEEKLY men's lifestyle magazine featuring showbiz, sport, humour, sex, fashion and news.

# National Newspapers

## Daily Express

The Northern & Shell Building, No. 10 Lower Thames Street, London EC3R 6EN
☎ 020 8612 7000
www.express.co.uk
Owner *Northern & Shell Media/Richard Desmond*
Editor *Peter Hill*
Circulation 677,750

Under owner Richard Desmond, publisher of *OK!* magazine, the paper features a large amount of celebrity coverage. The general rule of thumb is to approach in writing with an idea; all departments are prepared to look at an outline without commitment. Ideas welcome but already receives many which are 'too numerous to count'.
News Editor *Geoff Maynard*
Diary (Hickey Column) *Lizzie Catt*
Features Editor *Fergus Kelly*
City Editor *Peter Cunliffe*
Political Editor *Macer Hall*
Sports Editor *Bill Bradshaw*
Planning Editor (News Desk) should be circulated with copies of official reports, press releases, etc., to ensure news desk cover at all times.

*Saturday* magazine. Editor *Graham Bailey*
💷 Payment negotiable.

## Daily Mail

Northcliffe House, 2 Derry Street, London W8 5TT
☎ 020 7938 6000
www.dailymail.co.uk
Owner *Associated Newspapers/Lord Rothermere*
Editor *Paul Dacre*
Circulation 2.1 million

In-house feature writers and regular columnists provide much of the material. Photo-stories and crusading features often appear; it is essential to hit the right note to be a successful Mail writer.

Close scrutiny of the paper is strongly advised. Not a good bet for the unseasoned. Accepts news on savings, building societies, insurance, unit trusts, legal rights and tax.
News Editor *Keith Poole*
City Editor *Alex Brummer*
'Money Mail' Editor *Tony Hazell*
Political Editor *James Chapman*
Education Correspondent *Laura Clark*
Diary Editor *Richard Kay*
Features Editor *Leaf Kalfayan*
Literary Editor *Sandra Parsons*
Head of Sport *Lee Clayton*
Assistant Editor (*Femail*) *Andrew Morrod*

*Weekend* Saturday supplement. Editor *Lisa Collins*

## Daily Mirror

1 Canada Square, Canary Wharf, London E14 5AP
☎ 020 7293 3000      📠 020 7293 3409
www.mirror.co.uk
Owner *Trinity Mirror Plc*
Editor *Richard Wallace*
Circulation 1.2 million

No freelance opportunities for the inexperienced, but strong writers who understand what the tabloid market demands are always needed.
Deputy Editor *Conor Hanna*
Head of News *Anthony Harwood*
News Editor *Barry Rabbetts*
Features Editor *Carole Watson*
Political Editor *Bob Roberts*
Business Editor *Clinton Manning*
Showbusiness Editor *Chris Bucktin*
Sports Editor *Dean Morse*

## Daily Record

One Central Quay, Glasgow G3 8DA
☎ 0141 309 3000      📠 0141 309 3340
reporters@dailyrecord.co.uk
www.dailyrecord.co.uk
Owner *Trinity Mirror plc*
Editor-in-Chief *Bruce Waddell*
Circulation 314,753

Mass-market Scottish tabloid. Freelance material is generally welcome.
News Editor *Andy Lines*

Features Editor *Melanie Harvey*
Political Editor *Magnus Gardham*
Executive Sports Editor *James Traynor*
Magazine Editor *Liz Cowan*

*Scotland Means Business* Quarterly business magazine, launched 2002. Editor *Magnus Gardham*

## Daily Sport

19 Great Ancoats Street, Manchester M60 4BT
☎ 0161 236 4466    🖷 0161 236 4535
www.dailysport.net
Owner *Sport Media Group*
Editor-in-Chief *Murray Morse*
Editor *Pam McVitie*
Circulation 78,000

Tabloid catering for young male readership. Unsolicited material welcome; send to News Editor *Neil Goodman*
Sports Editor *Marc Smith*

*Lads Mag* Monthly glossy magazine. Editor *Mark Harris.*

## Daily Star

The Northern & Shell Building, No. 10 Lower Thames Street, London EC3R 6EN
☎ 0871 434 1010
Owner *Northern & Shell Media/Richard Desmond*
Editor *Dawn Neesom*
Circulation 784,958

In competition with *The Sun* for off-the-wall news and features. Freelance opportunities available.
Deputy Editor *Kieron Saunders*
News Editor *Jon Lockett*
Sports Editor *Howard Wheatcroft*

## Daily Star Sunday

The Northern & Shell Building, No. 10 Lower Thames Street, London EC3R 6EN
☎ 020 8612 7424    🖷 0871 434 2941
michael.booker@dailystar.co.uk
www.dailystarsunday.co.uk
Owner *Northern & Shell Media/Richard Desmond*
Editor *Gareth Morgan*
Circulation 353,249

Launched in September 2002 in direct competition with *News of the World* and *The People.*

*Take5* Lifestyle/showbiz magazine.

## The Daily Telegraph

111 Buckingham Palace Road, London SW1W 0DT
☎ 020 7931 2000
www.telegraph.co.uk
Owner *Telegraph Media Group*
Editor-in-Chief *William Lewis*
Editor *Tony Gallagher*

Circulation 703,249

Unsolicited mss not generally welcome – 'all are carefully read and considered, but very few published'. Contenders should approach the paper in writing, making clear their authority for writing on that subject. No fiction.
Executive Editor *Mark Skipworth*
Head of Business *Damian Reece*
Online City Editor *Richard Blackden*
Political Editor *Andrew Porter*
Diary Editor *Tim Walker* Always interested in diary pieces.
Arts Editor *Sarah Crompton*
Education *Graeme Paton*
Features Editor *Liz Hunt* Most material supplied by commission from established contributors. New writers are tried out by arrangement with the features editor. Approach in writing. Maximum 1,500 words.
Assistant Editor (Books) *Gaby Wood*
💷 Payment by arrangement.

*Daily Telegraph Weekend* Saturday supplement. Editor *Jon Stock*

*Telegraph Magazine.* Editor *Michele Lavery*

## Financial Times

1 Southwark Bridge, London SE1 9HL
☎ 020 7873 3000    🖷 020 7873 3076
www.ft.com
Owner *Pearson*
Editor *Lionel Barber*
Circulation 400,827

Founded 1888. UK and international coverage of business, finance, politics, technology, management, marketing and the arts. All feature ideas must be discussed with the department's editor in advance. Not snowed under with unsolicited contributions – they get less than any other national newspaper. Approach by email with ideas in the first instance.
UK Business & Employment Editor *Brian Groom*
Arts Editor *Jan Dalley*
City Editor *Andrew Hill*
UK News Editor *Sarah Neville*
FT.com News Editor *Andrew Slade*
Books and Arts Editor, *FT Magazine Rosie Blau*
Literary Editor *Jan Dalley*
Education *David Turner*
Environment Correspondent *Fiona Harvey*
Political Editor *George Parker*
Leisure Industries *Roger Blitz*
*Weekend FT* Editor *Andy Davis*

*How to Spend It* Monthly magazine. Editor *Gillian de Bono*

## The Guardian

Kings Place, 90 York Way, London N1 9GU
☎ 020 3353 2000

firstname.secondname@guardian.co.uk
www.guardian.co.uk
Owner *The Scott Trust*
Editor *Alan Rusbridger*
Circulation 300,540

Of all the nationals *The Guardian* probably offers the greatest opportunities for freelance writers, if only because it has the greatest number of specialized pages which use freelance work. Read specific sections before submitting mss.
Deputy Editor *Paul Johnson* No opportunities except in those regions where there is presently no local contact for news stories.
Literary Editor *Claire Armitstead*
City Editor *Julia Finch*
G2 Features Editors *Clare Margitson, Emily Wilson*
Diary Editor *Jon Henley*
Education Editor *Alice Woolley*
Environment *John Vidal*
Features Editor *Katharine Viner*
'Guardian Society' *Patrick Butler* Focuses on all public service activity.
Media Editor *Jane Martinson*
Political Editor *Patrick Wintour*
Sports Editor *Ian Prior*
Women's Page *Kira Cochrane* Runs every Friday.

*Guardian Weekend* Glossy Saturday issue. Editor *Merope Mills*

*The Guide* Editor *Malik Meer*

## The Herald (Glasgow)

200 Renfield Street, Glasgow G2 3QB
℡ 0141 302 7000     ℻ 0141 302 7007
www.heraldscotland.com
Owner *Gannett UK Ltd*
Editor *Donald Martin*
Circulation 55,811

One of the oldest national newspapers in the English-speaking world, *The Herald*, which dropped its 'Glasgow' prefix in February 1992, was bought by Scottish Television in 1996 and by Newsquest in 2003. Lively, quality, national Scottish daily broadsheet. Approach with ideas in writing or by phone in first instance.
News Editor *Calum Macdonald*
Arts Editor *Keith Bruce*
Business Editor *Ian McConnell*
Diary *Ken Smith*
Features Editor *Matt Roper*
Education *Andrew Denholm*
Sports Editor *Donald Cowey*

*Herald Magazine* Editor *Garry Scott*

## The Independent

2 Derry Street, London W8 5HF
℡ 020 7005 2000
www.independent.co.uk

Owner *Independent Print Ltd/Alexander Lebedev*
Editor-in-Chief *Simon Kelner*
Circulation 186,940

Founded 1986. Particularly strong on its arts/media coverage, with a high proportion of feature material. Theoretically, opportunities for freelancers are good. However, unsolicited mss are not welcome; most pieces originate in-house or from known and trusted outsiders. Ideas should be submitted in writing.
Executive News Editor *Dan Gledhill*
Features *Adam Leigh*
Arts Editor *David Lister*
Business Editor *David Prosser*
Education *Richard Garner*
Environment *Michael McCarthy*
Literary Editor *Boyd Tonkin*
Political Editor *Andrew Grice*
Sports Editor *Christian Broughton*
Travel Editor *Simon Calder*

*The Independent Magazine* Saturday colour supplement. Editor *Laurence Earle*

*The Independent Traveller* Saturday supplement.

## Independent on Sunday

5 Derry Street, London W8 5HF
℡ 020 7005 2000
www.independent.co.uk
Owner *Independent Print Ltd/Alexander Lebedev*
Editor *John Mullin*
Circulation 155,460

Founded 1986. Regular columnists contribute most material but feature opportunites exist. Approach with ideas in first instance.
Executive Editor (News) *Peter Victor*
Arts Editor *Mike Higgins*
Comment Editor *James Hanning*
Environment *Geoffrey Lean*
Political Editor *Jane Merrick*
Sports Editor *Marc Padgett*
Travel Editor *Kate Simon*

*The New Review* supplement. Editor *Bill Tuckey*

## International Herald Tribune

6 bis, rue des Graviers, 92521 Neuilly Cedex
℡ 00 33 1 4143 9322
℻ 00 33 1 4143 9429 (editorial)
iht@iht.com
www.iht.com
Editor, Global Editions *Martin Gottlieb*
Executive Editor *Alison Smale*
Assistant Managing Editor *Katherine Knorr*
Circulation 239,689 (2008)

Published in France, Monday to Saturday, and circulated in Europe, the Middle East, North Africa, the Far East and the USA. General news, business and financial, arts and leisure. Uses regular freelance contributors. Contributor policy can be found on the website at: www.ihtinfo.com/press/contributorpolicy.html

## The Mail on Sunday

Northcliffe House, 2 Derry Street, London W8 5TS
☎ 020 7938 6000   ℻ 020 7937 3829
Owner *Associated Newspapers/Lord Rothermere*
Editor *Peter Wright*
Circulation 2.0 million

Sunday paper with a high proportion of newsy features and articles. Experience and judgement required to break into its band of regular feature writers.
News Editor *David Dillon*
Financial Editor *Lisa Buckingham*
Diary Editor *Katie Nicholl*
Features Editor/Women's Page *Sian James*
Books *Marilyn Warnick*
Political Editor *Simon Walters*
Sports Editor *Malcolm Vallerius*
'Live Night & Day' Editor *Gerard Greaves*
'Review' Editor *George Thwaites*

*You – The Mail on Sunday Magazine* Colour supplement. Many feature articles, supplied entirely by freelance writers. Editor *Sue Peart*, Features Editor *Rosalind Lowe*

## Morning Star

William Rust House, 52 Beachy Road, London E3 2NS
☎ 020 8510 0815   ℻ 020 8986 5694
newsed@peoples-press.com
www.morningstaronline.co.uk
Owner *Peoples Press Printing Society*
Editor *Bill Benfield*
Circulation 13,000

Launched in 1930 as the *Daily Worker*, the paper was relaunched in 1966 as the *Morning Star*. Not to be confused with the *Daily Star*, the *Morning Star* is the farthest left national daily. Those with a penchant for a Marxist reading of events and ideas can try their luck, though feature space is as competitive here as in the other nationals.
News Editor *Adrian Roberts*
Features *Ros Sitwell*
Foreign Editor *Tom Mellen*
Arts Editor *Katie Lambert*
Sports Editor *Greg Leedham*

## News of the World

1 Virginia Street, London E98 1NW
☎ 020 7782 1000   ℻ 020 7583 9504
www.newsoftheworld.co.uk
Owner *News International plc/Rupert Murdoch*
Editor *Colin Myler*
Circulation 2.79 million

Highest circulation Sunday paper. Freelance contributions welcome. News and features editors welcome tips and ideas.

Deputy Editor *Victoria Newton*
Assistant Editor (News) *Ian Edmondson*
Features Editor *Jules Stenson*
Political Editor *Ian Kirby*
Sports Editor *Paul McArthy*

*Fabulous* Colour supplement. Editor *Sally Eyden* Glossy magazine, launched in February 2008. Celebrity, fashion and real-life features. Unsolicited mss and ideas welcome.

## The Observer

Kings Place, 90 York Way, London N1 9GU
☎ 020 3353 2000
firstname.surname@observer.co.uk
www.observer.co.uk
Owner *Guardian Newspapers Ltd*
Editor *John Mulholland*
Circulation 351,019

Founded 1791. Acquired by Guardian Newspapers from Lonrho in May 1993. Occupies the middle ground of Sunday newspaper politics. Unsolicited material is not generally welcome, 'except from distinguished, established writers'. Receives far too many unsolicited offerings already. No news, fiction or special page opportunities. The newspaper runs annual competitions which change from year to year. Details are advertised in the newspaper.
News Editor *Chris Boffey*
Home News Editor *Lucy Rock*
Foreign News Editor *Julian Coman*
Crime & Defence Correspondent *Mark Townsend*
Foreign Affairs Editor *Peter Beaumont*
Home Affairs Editor *Jamie Doward*
Health Correspondent *Denis Campbell*
Social Affairs *Amelia Hill*
Arts and Media Correspondent *Vanessa Thorpe*
Political Editor *Lisa Bachelor*
Whitehall Correspondent *Toby Helm*
Arts Editor *Sarah Donaldson*
'The New Review' Editor *Jane Ferguson*
Comment Editor *Ruaridh Nicoll*
Deputy Business Editor/City Editor *Richard Wachman*
Business Editor *Ruth Sunderland*
Personal Finance Editor *Jill Insley*
Science Editor *Robin McKie*
Environment Editor *Juliette Jowit*
Literary Editor *William Skidelsky*
Travel Editor *Joanne O'Connor*
Sports Editor *Brian Oliver*

*The Observer Magazine* Glossy lifestyle supplement. Editor *Tim Lewis*

*The Observer Food Monthly* Launched in 2001. Editor *Nicola Jeal*

## The People

1 Canada Square, Canary Wharf, London E14 5AP
☎ 020 7293 3000   ℻ 020 7293 3517

www.people.co.uk
Owner *Trinity Mirror plc*
Editor *Lloyd Embley*
Circulation 532,680

Popular tabloid. Keen on exposés and big-name gossip. Interested in ideas for investigative articles. Phone in the first instance.
News Editor *Lee Harpin*
Features Editor *Caroline Waterston*
Political Editor *Nigel Nelson*
Sports Editor *James Brown*

*Take It Easy* Magazine supplement. Editor *Hanna Tavner* Approach by phone with ideas in first instance.

## Scotland on Sunday

Barclay House, 108 Holyrood Road, Edinburgh
EH8 8AS
☎ 0131 620 8620    🖷 0131 620 8491
www.scotlandonsunday.com
Owner *Scotsman Publications Ltd*
Editor *Ian Stewart*
Circulation 54,949

Scotland's top-selling quality broadsheet. Welcomes ideas rather than finished articles.
Deputy Editor *Kenny Farquharson*
News Editor *Jeremy Watson*
Group Magazines Editor *Fiona Leith*
Group Arts & Entertainments Editor *Andrew Eaton*

## The Scotsman

Barclay House, 108 Holyrood Road, Edinburgh
EH8 8AS
☎ 0131 620 8620    🖷 0131 620 8617
enquiries@scotsman.com
www.scotsman.com
Owner *Johnston Press*
Editor *John McLellan*
Circulation 44,972

Scotland's national newspaper. Many unsolicited mss come in, and stand a good chance of being read, although a small army of regulars supply much of the feature material not written in-house. See website for contact details.
News Editor *Frank O'Donnell*
Business Editor *Terry Murden*
Education *Fiona MacLeod*
Group Magazine & Arts Editor *Fiona Leith*
Group Arts & Entertainment Editor *Andrew Eaton*
Book Reviews *David Robinson*
Sports Editor *Donald Walker*

## The Sun

1 Virginia Street, London E98 1SN
☎ 020 7782 4000    🖷 020 7782 4108
firstname.lastname@the-sun.co.uk
www.the-sun.co.uk

Owner *News International Ltd/Rupert Murdoch*
Editor *Dominic Mohan*
Circulation 2.9 million

Highest circulation daily. Populist outlook; very keen on gossip, pop stars, TV soap, scandals and exposés of all kinds. No room for non-professional feature writers; 'investigative journalism' of a certain hue is always in demand, however.
Head of News *Chris Pharo*
Head of Features *Victoria Newton*
Head of Sport *Mike Dunn*
Health Editor *Emma Morton*
Fashion Editor *Erica Davies*

## Sunday Express

The Northern & Shell Building, Number 10 Lower Thames Street, London EC3R 6EN
☎ 020 8612 7000    🖷 020 8612 7766
Owner *Northern & Shell Media/Richard Desmond*
Editor *Martin Townsend*
Circulation 590,596

The general rule of thumb is to approach in writing with an idea; all departments are prepared to look at an outline without commitment.
News Editor *Stephen Rigley*
Features Editor *Amy Packer*
Business Editor *Tracey Boles*
Political Editor *Kirsty Buchanan*
Sports Editor *Scott Wilson*

*S* Fashion and lifestyle magazine for women.
Editor *Louise Robinson* No unsolicited mss. All contributions are commissioned. Ideas in writing only.

💷 Payment negotiable.

## Sunday Herald

200 Renfield Street, Glasgow G2 3QB
☎ 0141 302 7800    🖷 0141 302 7963
richard.walker@sundayherald.com
www.heraldscotland.com
Owner *Newsquest*
Editor *Richard Walker*
Circulation 40,619

Launched February 1999. Scottish seven-section compact.
Head of News *Neil Mackay*
Scottish Political Editor *Tom Gordon*
Sports Editor *Jonathan Jobson*
Opinion Editor *Susan Flockhart*

## Sunday Mail

One Central Quay, Glasgow G3 8DA
☎ 0141 309 3000    🖷 0141 309 3587
www.sundaymail.co.uk
Owner *Trinity Mirror plc*
Editor-in-Chief *Bruce Waddell*

Circulation 386,920

Popular Scottish Sunday tabloid.
News Editor *Andy Lines*
Sports Editor *Austin Barrett*

*7Days* Weekly supplement. Head of Magazines *Liz Cowan*

## Sunday Mirror

1 Canada Square, Canary Wharf, London E14 5AP
☏ 020 7293 3000     📠 020 7293 3939 (news desk)
www.sundaymirror.co.uk
Owner *Trinity Mirror Plc*
Editor *Tina Weaver*
Circulation 1.1 million

In general terms contributions are welcome, though the paper patiently points out it has more time for those who have taken the trouble to study the market. Initial contact in writing preferred, except for live news situations. No fiction.
News Editor *James Saville* The news desk is very much in the market for tip-offs and inside information. Contributors would be expected to work with staff writers on news stories. Approach by telephone or fax in the first instance.
Finance *Melanie Wright*
Features Editor *Jill Main* 'Anyone who has obviously studied the market will be dealt with constructively and courteously.' Cherishes its record as a breeding ground for new talent.
Sports Editor *David Walker*

*Celebs on Sunday* Colour supplement. Editor *Mel Brodie*

## Sunday Post

2 Albert Square, Dundee DD1 9QJ
☏ 01382 223131     📠 01382 201064
mail@sundaypost.com
www.sundaypost.com
Owner *D.C. Thomson & Co. Ltd*
Editor *David Pollington*
Circulation 334,737

Contributions should be addressed to the editor.

*post plus* Monthly colour supplement. Editor *Jan Gooderham*

## Sunday Sport

19 Great Ancoats Street, Manchester M60 4BT
☏ 0161 236 4466     📠 0161 236 4535
www.sundaysport.com
Owner *Sport Media Group*
Editor *Nick Appleyard*
Circulation 75,000

Founded 1986. Sunday tabloid catering for a particular sector of the male 15–35 readership. As concerned with 'glamour' (for which, read: 'page 3') as with human interest, news, features and sport. Unsolicited mss are welcome; receives about 90 a week. Approach should be made by email to the news editor.
News Editor *Neil Goodwin* Off-beat news, human interest, preferably with photographs.
Showbiz Editor *Clare Chapman* Regular items.
Sports Editor *Marc Smith* Hard-hitting sports stories on major soccer clubs and their personalities, plus leading clubs/people in other sports. Strong quotations to back up the news angle essential.
💷 Payment negotiable and on publication.

## The Sunday Telegraph

111 Buckingham Palace Road, London SW1W 0DT
☏ 020 7931 2000     📠 020 7931 2936
stnews@telegraph.co.uk
www.telegraph.co.uk
Owner *Press Holdings Limited*
Editor *Ian MacGregor*
Circulation 525,088

Right-of-centre quality Sunday paper which, although traditionally formal, has pepped up its image to attract a younger readership. Unsolicited material from untried writers is rarely ever used. Contact with idea and details of track record.
Deputy Editor *Tim Jotischky*
News Editor *James Hall*
'Seven' Editor *Ross Jones*
City Editor *Damian Reece*
Political Editor *Patrick Hennessy*
Education Editor *Julie Henry*
Arts Editor *Roya Nikkah*
Environment Editor *David Harrison*
Literary Editor *Michael Prodger*
Diary Editor *Tim Walker*

*Stella Magazine* Colour supplement. Editor *Anna Murphy*

## The Sunday Times

1 Pennington Street, London E98 1ST
☏ 020 7782 5000     📠 020 7782 5658
www.sunday-times.co.uk
Owner *News International plc/Rupert Murdoch*
Editor *John Witherow*
Circulation 1.1 million

Founded 1820. Tendency to be anti-establishment, with a strong crusading investigative tradition. Approach the relevant editor with an idea in writing. Close scrutiny of the style of each section of the paper is strongly advised before sending mss. No fiction. All fees by negotiation.
News Editor *Nicholas Hellen* Opportunities are very rare.
News Review Editor *Susannah Herbert* Submissions are always welcome, but the paper commissions its own, uses staff writers or works with literary agents, by and large. The features sections where most opportunities exist are *Style* and *The Culture*.

'The Culture' Editor *Helen Hawkins*
Business Editor *John Waples*
City Editor *Jenny Davey*
Science and Environment Editor *Jonathan Leake*
Literary Editor *Andew Holgate*
Sports Editor *Alex Butler*
'Style' Editor *Tiffanie Darke*

*Sunday Times Magazine* Colour supplement. Editor *Sarah Baxter* No unsolicited material. Write with ideas in first instance.

## The Times

1 Pennington Street, London E98 1TT
Ⓣ 020 7782 5000/5971    Ⓕ 020 7488 3242
www.thetimes.co.uk
Owner *News International plc/Rupert Murdoch*
Editor *James Harding*

Circulation 521,535

Generally right (though features can range in tone from diehard to libertarian). *The Times* receives a great many unsolicited offerings. Writers with feature ideas should approach by letter in the first instance. No fiction.

Deputy Editor *Keith Blackmore*
Deputy Head of News *John Wellman*
Business & City Editor *David Wighton*
Arts Editor *Tim Teeman*
Literary Editor *Erica Wagner*
Political Editor *Roland Watson*
Sports Editor *Tim Hallissey*

*Weekend* Editor *Nicola Jeal*
*times2life* Editor *Emma Tucker*

# Regional Newspapers

## England

### Berkshire

#### Reading Post

8 Tessa Road, Reading RG1 8NS
☏ 0118 918 3000    🅕 0118 959 9363
editorial@reading-epost.co.uk
Owner *Trinity Mirror Plc*
Editor *Andy Murrill*
Circulation 12,446

Unsolicited mss welcome; one or two received every day. Fiction rarely used. Interested in local news features, human interest, well-researched investigations. Special sections include holidays & travel; style page; business; food page; gardening; reviews; rock music; motoring. Also magazines, *Food Monthly* and *Business Post* (monthly) *24 seven* entertainment supplement.

### Cambridgeshire

#### Cambridge News

Winship Road, Milton, Cambridge CB24 6PP
☏ 01223 434434    🅕 01223 434415
Owner *Cambridge Newspapers Ltd*
Editor *Paul Brackley*
Deputy Editor *John Deeks*
Circulation 24,970
News Editor *Paul Holland*
Business Editor *Jenny Chapman*
Sports Editor *Paul Stimpson*

#### Peterborough Evening Telegraph

57 Priestgate, Peterborough PE1 1JW
☏ 01733 555111    🅕 01733 313147 (editorial)
www.peterboroughtoday.co.uk
Owner *East Midlands Newspapers Ltd*
Editor *Mark Edwards*
Circulation 16,523
News Editor *Paul Grinnell*
Assistant Editor (Content) *Alex Gordon*
Business Editor *John Kralevich*

### Cheshire

#### Chester Chronicle

Chronicle House, Commonhall Street, Chester
CH1 2AA
☏ 01244 340151    🅕 01244 606498
eric.langton@cheshirenews.co.uk
www.chesterchronicle.co.uk
Owner *Trinity Mirror Plc*
Editor-in-Chief *Eric Langton*
Circulation 22,070

All unsolicited feature material will be considered. No payment.

### Cleveland

#### Hartlepool Mail

New Clarence House, Wesley Square, Hartlepool
TS24 8BX
☏ 01429 239333    🅕 01429 869024
mail.news@northeast-press.co.uk
www.hartlepoolmail.co.uk
www.peterleemail.co.uk
Owner *Johnston Press Plc*
Editor *Joy Yates*
Deputy Editor *Gavin Foster*
Circulation 17,044
News Editor *Ian Willis*
Sports Editor *Roy Kelly*

### Cumbria

#### News & Star

Newspaper House, Dalston Road, Carlisle
CA2 5UA
☏ 01228 612600    🅕 01228 612601
Owner *CN Group Ltd*
Editor *Neil Hodgkinson*
Circulation 20,624
Assistant Editor *Andy Nixon*
Associate Editor/Woman's Page *Anne Pickles*
Sports Editor *Phil Rostron*

#### North West Evening Mail

Abbey Road, Barrow in Furness LA14 5QS
☏ 01229 821835    🅕 01229 840164

news@nwemail.co.uk
www.nwemail.co.uk
Owner *Robin Burgess*
Editor *Jonathan Lee*
Deputy Editor *Phil Pearson*
Circulation 16,400

All editorial material should be addressed to the editor.
Assistant Editor (Features) *Peter Leach*
News Editor *Rob Johnson*

## Derbyshire

### Derby Telegraph Media Group Ltd

Northcliffe House, Meadow Road, Derby DE1 2BH
☎ 01332 291111
newsdesk@derbytelegraph.co.uk
Owner *Northcliffe Media Group Ltd*
Editor *Steve Hall*
Circulation 39,152

## Devon

### Express & Echo

Heron Road, Sowton, Exeter EX2 7NF
☎ 01392 442211    🖷 01392 442287 (editorial)
echonews@expressandecho.co.uk
www.thisisexeter.co.uk
Owner *South West Media Group*
Editor *Marc Astley*
Circulation 19,132

Weekly supplements: *Business Week*; *Property Echo*; *Motoring*; *Weekend Echo*; *East Devon Supplement*; *Exmouth Supplement*.
Head of Content *Lynne Turner*
Features Editor *Sue Kemp*
Sports Editor *Richard Davies* (🖷 *01392 442416*)

### The Herald

17 Brest Road, Derriford Business Park, Derriford, Plymouth PL6 5AA
☎ 01752 765500    🖷 01752 765527
news@theplymouthherald.co.uk
www.thisisplymouth.co.uk
Owner *South West Media Group Ltd*
Editor *Bill Martin*
Circulation 34,342

All editorial material to be addressed to the editor or the News Editor, *James Garnett*.

### Herald Express

Harmsworth House, Barton Hill Road, Torquay TQ2 8JN
☎ 01803 676000    🖷 01803 676228 (editorial)
newsdesk@heraldexpress.co.uk
www.thisissouthdevon.co.uk
Owner *Northcliffe Media Group Ltd*
Editor *Andy Phelan*

Circulation 22,980

Drive scene, property guide, *On The Town* (leisure and entertainment guide), Monday sports, special pages, rail trail, Saturday surgery, nature and conservation column, dance scene, shop scene, The Business. Supplements: *Gardening* (weekly) and *Visitors Guide.* Unsolicited mss generally not welcome. All editorial material should be addressed to the editor in writing.

### Western Morning News

17 Brest Road, Derriford Business Park, Derriford, Plymouth PL6 5AA
☎ 01752 765500    🖷 01752 765535
wmnnewsdesk@westernmorningnews.co.uk
www.thisisdevon.co.uk
www.thisiscornwall.co.uk
www.thisiswesternmorningnews.co.uk
Owner *South West Media Group Ltd*
Editor *Alan Qualtrough*
Circulation 37,819

Unsolicited mss welcome, but must be of topical and local interest and addressed to the Deputy Editor, *Philip Bowern.*
News Editor *Steve Grant*
Features Editor *Su Carroll*
Sports Editor *Mark Stevens*

## Dorset

### Daily Echo

Richmond Hill, Bournemouth BH2 6HH
☎ 01202 554601    🖷 01202 297543
neal.butterworth@bournemouthecho.co.uk
www.bournemouthecho.co.uk
Owner *Newsquest Media Group Ltd (a Gannett company)*
Editor *Neal Butterworth*
Circulation 30,754

Founded 1900. Has a strong news and features content and invites specialist articles, particularly on unusual subjects, either contemporary or historical, but only with a local angle or flavour and by prior arrangement. Special review sections each day, including sport, property, entertainment and culture, heritage, motoring, gardening, the environment and the coastline. Also *Saturday Magazine* and monthly *Society* glossy magazine. All editorial material should be addressed to the Editor, *Neal Butterworth.* 🖷 Payment on publication.

### Dorset Echo

Fleet House, Hampshire Road, Granby Industrial Estate, Weymouth DT4 9XD
☎ 01305 830930    🖷 01305 830956
www.dorsetecho.co.uk
Owner *Newsquest Media Group Ltd (a Gannett company)*

Editor *Toby Granville*
Circulation 18,026

By-gone days, films, arts, showbiz, motoring, property, weekend leisure and entertainment, schools, food, news, sport and weekend magazine.
News Editor *Diarmuid MacDonagh*
Sports Editor *Nigel Dean*

## Co. Durham

### The Northern Echo

Priestgate, Darlington DL1 1NF
☏ 01325 381313  📠 01325 380539
newsdesk@nne.co.uk
www.thenorthernecho.co.uk
Owner *Newsquest (Yorkshire and North East) Ltd (a Gannett company)*
Editor *Peter Barron*
Circulation 46,729

Founded 1870. Freelance pieces welcome but telephone first to discuss submission. Assistant Editor (News) *Nigel Burton* Interested in reports involving the North East or North Yorkshire. Preferably phoned in.
Features Production Editor *Jenny Neeham* Background pieces to topical news stories relevant to the area. Must be arranged with the features editor before submission of any material.
Business Editor *Owen McAteer*
Sports Editor *Nick Loughlin*
£ Payment and length by arrangement.

## Essex

### Colchester Gazette

Oriel House, 43–44 North Hill, Colchester CO1 1TZ
☏ 01206 506000  📠 01206 508274
gazette.newsdesk@nqe.com
www.gazette-news.co.uk
Owner *Newsquest (Essex) Ltd*
Editor *Martin McNeil*
Circulation 19,656

Monday to Friday daily newspaper servicing north and mid-Essex including Colchester, Harwich, Clacton, Braintree, Witham, Maldon and Chelmsford. Unsolicited mss not generally used. Relies heavily on regular contributors.
Features Editor *Claire Borley*

### Echo Newspapers

Newspaper House, Chester Hall Lane, Basildon SS14 3BL
☏ 01268 522792  📠 01268 469281
Owner *Newsquest Media Group (a Gannett company)*
Editor *Martin McNeill*

Circulation 33,854

Relies almost entirely on staff and regular outside contributors, but will very occasionally consider material sent on spec. Approach the editor in writing with ideas. Although the paper is Basildon-based, its largest circulation is in the Southend area.

## Gloucestershire

### The Citizen

6–8 The Oxebode, Gloucester GL1 1RZ
☏ 01452 420621  📠 01452 420664 (editorial)
citizen.news@glosmedia.co.uk
www.thisisgloucestershire.co.uk
Owner *Northcliffe Newspapers Group Ltd*
Editor *Ian Mean*
Circulation 22,593

News or features material should be addressed to Managing Editor (News & Pictures) *Jenny Eastwood*, Deputy News Editor *Sally Munro* or Assistant Deputy News Editor *Jenni Silver*.

### Gloucestershire Echo

1 Clarence Parade, Cheltenham GL50 3NY
☏ 01242 271900  📠 01242 271848
echo.editor@glosmedia.co.uk
www.thisisgloucestershire.co.uk
Owner *Northcliffe Media Group Ltd*
Editor *Kevan Blackadder*
Managing Editor (News & Pictures) *Jenny Eastwood*
Circulation 18,850

All material, other than news, should be addressed to the editor

## Hampshire

### The News

The News Centre, Hilsea, Portsmouth PO2 9SX
☏ 023 9266 4488  📠 023 9267 3363
newsdesk@thenews.co.uk
www.portsmouth.co.uk
Owner *Portsmouth Printing & Publishing Ltd*
Editor *Mark Waldron*
Circulation 48,191

Unsolicited mss not generally accepted. Approach by letter or email.
Assistant Editor *Graeme Patfield* General subjects of S.E. Hants interest. Maximum 600 words. No fiction.
Sports Editor *Howard Frost* Sports background features. Maximum 600 words.

### The Southern Daily Echo

Newspaper House, Test Lane, Redbridge, Southampton SO16 9JX
☏ 023 8042 4777  📠 023 8042 4545

newsdesk@dailyecho.co.uk
www.dailyecho.co.u
Owner *Newsquest Plc*
Editor *Ian Murray*
Circulation 36,092

Unsolicited mss 'tolerated'. Approach the editor in writing with strong ideas; staff supply almost all the material.

## Kent

### Kent Messenger
6 & 7 Middle Row, Maidstone ME14 1TG
℡ 01622 695666    ℻ 01622 664988
messengernews@thekmgroup.co.uk
www.kentonline.co.uk
www.kentmessenger.co.uk
Owner *Kent Messenger Group*
Editor *Bob Bounds*
Circulation 46,733

Very little freelance work is commissioned.

### Medway Messenger
Medway House, Ginsbury Close, Sir Thomas Longley Road, Medway City Estate, Strood, Rochester ME2 4DU
℡ 01634 227800    ℻ 01634 715256
medwaymessenger@thekmgroup.co.uk
www.kentonline.co.uk
www.medwaymessenger.co.uk
Owner *Kent Messenger Group*
Editor *Bob Bounds*
Circulation 11,875 (Mon)/20,000 (Fri)

Published Mondays and Fridays with the free *Medway Extra* on Wednesday.
Senior News Editor *Sarah Clarke*
Business Editor *Trevor Sturgess*
Sports Editor *Mike Rees*

## Lancashire

### The Bolton News
Newspaper House, Churchgate, Bolton BL1 1DE
℡ 01204 522345    ℻ 01204 537427
newsdesk@theboltonnews.co.uk
www.theboltonnews.co.uk
Owner *Newsquest Media Group Ltd (a Gannett company)*
Editor-in-Chief *Ian Savage*
Circulation 26,064

Business, children's page, travel, local services, motoring, fashion and cookery.
News Editor *James Higgins*
Features Editor/Women's Page *Andrew Mosley*

### The Gazette (Blackpool)
Avroe House, Avroe Crescent, Blackpool FY4 2DP
℡ 01253 400888    ℻ 01253 361870

jon.rhodes@blackpoolgazette.co.uk
www.blackpoolgazette.co.uk
Owner *Johnston Press plc*
Managing Director *Darren Russell*
Editor *David Helliwell*
Circulation 26,019

Unsolicited mss welcome in theory. Approach in writing with an idea. Supplements: *Eve* (women, Tues); *All Stars* (football, youth sport, Wed); *Homes* (Thurs); *The Weekend* (entertainment, Fri); *Wheels* (motoring, Fri); *Life!* magazine (entertainment & leisure, Sat); *Jobs Plus* (jobs and careers, Sat).

### Lancashire Evening Post
Olivers Place, Eastway, Fulwood, Preston PR2 9ZA
℡ 01772 254841    ℻ 01772 880173
www.lep.co.uk
Owner *Johnston Press plc*
Editor *Simon Reynolds*
Circulation 28,275

Unsolicited mss are not generally welcome; many are received and not used. All ideas in writing to the editor.

### Lancashire Telegraph
Newspaper House, High Street, Blackburn BB1 1HT
℡ 01254 678678    ℻ 01254 680429
lt_editorial@lancashire.newsquest.co.uk
www.lancashiretelegraph.co.uk
Owner *Newsquest Media Group Ltd (a Gannett company)*
Editor *Kevin Young*
Circulation 27,481

News stories and feature material with an East Lancashire flavour (a local angle, or written by local people) welcome. Approach in writing with an idea in the first instance. No fiction.
News Editor *Ian Singleton*
Features Editor *John Anson*
Sports Editor *Paul Plunkett*
Pictures Editor *Neil Johnson*

### Oldham Evening Chronicle
PO Box 47, Union Street, Oldham OL1 1EQ
℡ 0161 633 2121    ℻ 0161 652 2111
news@oldham-chronicle.co.uk
www.oldham-chronicle.co.uk
Owner *Hirst Kidd & Rennie Ltd*
Editor *Jim Williams*
Circulation 18,062

Motoring, food and wine, lifestyle supplement, business page.
News Editor *Mike Attenborough*

## Leicestershire

### Leicester Mercury
St George Street, Leicester LE1 9FQ
☎ 0116 251 2512   📠 0116 253 0645
newsdesk@leicestermercury.co.uk
www.thisisleicestershire.co.uk
Owner *Northcliffe Media Group Ltd*
Editor *Keith Perch*
Circulation 64,919
News Editor *Mark Charlton*
Features Editor *Alex Dawson*

## Lincolnshire

### Grimsby Telegraph
80 Cleethorpe Road, Grimsby DN31 3EH
☎ 01472 360360   📠 01472 372257
newsdesk@grimsbytelegraph.co.uk
Owner *Northcliffe Media Group Ltd*
Editor *Michelle Lalor*
Circulation 31,194

Sister paper of the *Scunthorpe Telegraph*.
Unsolicited mss generally welcome. Approach in
writing. No fiction. Weekly supplements: *Business
Telegraph*; *Property Telegraph*; *Motoring*. All
material to be addressed to the News Editor, *Lucy
Wood*. Particularly welcomes hard news stories;
approach in haste by telephone.

### Lincolnshire Echo
Ground Floor, Witham Wharf, Brayford Wharf
East, Lincoln LN5 7AY
☎ 01522 820000   📠 01522 804493
news@lincolnshireecho.co.uk
Owner *Northcliffe Media Group Ltd*
Editor *Jon Grubb*
Circulation 19,720

News and sport, best buys, holidays, motoring,
dial-a-service, restaurants, leisure, home
improvement, record reviews, gardening corner,
stars.

### Scunthorpe Telegraph
4–5 Park Square, Scunthorpe DN15 6JH
☎ 01724 273273   📠 01724 273101
Owner *Northcliffe Media Group Ltd*
Editor *Mel Cook*
Circulation 18,304

All correspondence should go to the Deputy
Editor, *Dave Atkin*.

## London

### Evening Standard
Northcliffe House, 2 Derry Street, London W8 5EE
☎ 020 7938 6000   📠 020 7937 2648
www.standard.co.uk

Owner *Alexander Lebedev*
Editor *Geordie Greig*
Deputy Editor *Sarah Sands*
Circulation 608,533

Long-established evening paper, serving
Londoners with both news and feature material
(free of charge since October 2009). Genuine
opportunities for London-based features.
Produces a weekly colour supplement, *ES
Magazine* (Fri).
Managing Editor *Doug Wills*
Features Editor *Charlotte Ross*
News Editor *Hugh Dougherty*
Arts Editor *Fiona Hughes*
Sports Editor *Steve McKenlay*
ES Editor *Sasha Slater*

## Manchester

### Manchester Evening News
Number 1 Scott Place, Manchester M3 3RN
☎ 0161 832 7200   📠 0161 211 2030
www.manchestereveningnews.co.uk
Owner *Trinity Mirror Plc*
Acting Editor *Maria McGeoghan*
Circulation 153,724

One of the country's major regional dailies.
Initial approach in writing preferred. No fiction.
Sections: Personal Finance; Health; Holidays;
Homes & Property; Small Business; Lifestyle.
News Editor *Sarah Lester*
Features Editor *Deanna Delamotta* Regional news
features, personality pieces and showbiz profiles
considered. Maximum 1,200 words.
Sports Editor *Peter Spencer*
Women's Page *Deanna Delamotta*

💷 Payment based on house agreement rates.

## Merseyside

### Liverpool Daily Post
PO Box 48, Old Hall Street, Liverpool L69 3EB
☎ 0151 227 2000   📠 0151 472 2506
www.liverpooldailypost.co.uk
Owner *Trinity Mirror Plc*
Editor *Mark Thomas*
Circulation 11,648

Unsolicited mss welcome. Receives about six a day.
Approach in writing with an idea. No fiction. Local,
national/international news, current affairs, profiles
(with pictures). Maximum 800–1,000 words.
Head of Content (Advanced) *Jane Haase*
Assistant Head of Content (Advanced) *Emma
Johnson*
News Editor *Andy Kelly*
Business Editor *Bill Gleeson*
Sports Editor *John Thompson*

## Liverpool Echo

PO Box 48, Old Hall Street, Liverpool L69 3EB
📞 0151 227 2000   📠 0151 472 2474
letters@liverpoolecho.co.uk
www.liverpoolecho.co.uk
Owner *Trinity Mirror Merseyside*
Editor *Alastair Machray*
Circulation 97,779

One of the country's major regional dailies.
Unsolicited mss welcome; initial approach with
ideas in writing preferred.
News Editor *Maria Breslin*
Features Editor *Jane Haase*
Sports Editor *John Thompson*
Women's Editor *Susan Lee*

## Norfolk

### Eastern Daily Press

Prospect House, Rouen Road, Norwich NR1 1RE
📞 01603 628311   📠 01603 623872
EDP@archant.co.uk
www.EDP24.co.uk
Owner *Archant Regional*
Editor *Peter Waters*
Circulation 61,143

Most pieces by commission only. Supplements:
*EDP2* (daily); business (twice-weekly); rental,
motoring, property pages, agriculture,
employment, puzzles (all weekly); *Event*
full-colour magazine (Fri); *Saturday* full colour
56-page magazine; *Sunday* 32-page supplement;
*Pink 'Un* 16-page sport supplement (Sat).
Senior Content Editors Peter Hannam, David
Powles
Head of Features *Sarah Hardy*
Head of Sport *Richard Willner*
Assistant Editor (*Saturday*) *Steve Snelling*
Business Editor *Paul Hill*

### Evening News

Prospect House, Rouen Road, Norwich NR1 1RE
📞 01603 628311   📠 01603 623872
tim.williams@archant.co.uk
www.eveningnews24.co.uk
Owner *Archant Ltd*
Editor *Tim Williams*
Deputy Editor *Terry Redhead (terry.redhead@
archant.co.uk)*
Circulation 19,903

Includes special pages on local property,
motoring, pop, fashion, arts, entertainments
and TV, gardening, local music scene, home and
family.
Features Editor *Derek James (derek.james@
archant.co.uk)*

## Northamptonshire

### Chronicle and Echo

Upper Mounts, Northampton NN1 3HR
📞 01604 467000   📠 01604 467200
Owner *Northamptonshire Newspapers*
Editor *David Summers*
Circulation 19,016

Unsolicited mss are 'not necessarily unwelcome
but opportunities to use them are rare'. Approach
in writing with an idea. No fiction. Supplements:
*Sport* (Mon); *Property Today* (Wed); *Jobs Today*
(Thurs); *Motors Today* and *The Guide* (Friday);
*Weekend Life* (Sat).
News Editor *Richard Edmondson*
Features Editor/Women's Page *Lily Canter*
Sports Editor *Jeremy Casey*

### Evening Telegraph

Newspaper House, Ise Park, Rothwell Road,
Kettering NN16 8GA
📞 01536 506100   📠 01536 506195 (editorial)
et.newsdesk@northantsnews.co.uk
www.northantset.co.uk
Owner *Johnston Press Plc*
Editor *Jeremy Clifford*
Deputy Editor *Neil Pickford*
Circulation 21,648

*Northamptonshire Business Guide* (weekly);
*Entertainment Guide* and *Jobs* supplements
(Thurs); films and eating out (Fri); and *Home &
Garden* – monthly lifestyle supplement including
gardening, etc.
News Editor *David Brennan*
Sports Editor *Jim Lyon*

## Nottinghamshire

### Evening Post Nottingham

Castle Wharf House, Nottingham NG1 7EU
📞 0115 948 2000   📠 0115 964 4032
Owner *Northcliffe Media Group Ltd*
Editor *Malcolm Pheby*
Deputy Editor *Martin Done*
Circulation 51,526

Unsolicited mss occasionally used. Good local
interest only. Maximum 800 words. No fiction.
Send ideas in writing.
News Editor *Steven Fletcher*
Sports Editor *Dave Parkinson*

## Oxfordshire

### Oxford Mail

Osney Mead, Oxford OX2 0EJ
📞 01865 425262   📠 01865 425554

news@oxfordmail.co.uk
www.oxfordmail.co.uk
Owner *Newsquest (Oxfordshire & Wiltshire) Ltd*
Editor *Simon O'Neill*
Circulation 22,830

Unsolicited mss are considered but a great many unsuitable offerings are received. Approach in writing with an idea, rather than by phone. No fiction.

£ All fees negotiable.

## Shropshire

### Shropshire Star

Ketley, Telford TF1 5HU
☎ 01952 242424   ℻ 01952 254605
Owner *Shropshire Newspapers Ltd*
Editor *Sarah-Jane Smith*
Circulation 67,726

No unsolicited mss; approach the editor with ideas in writing in the first instance. No news or fiction.
Head of Features *Carl Jones* Limited opportunities; uses mostly in-house or syndicated material. Maximum 1,200 words.
Sports Editor *Dave Ballinger*

## Somerset

### The Bath Chronicle

Westpoint, James Street West, Bath BA1 2DA
☎ 01225 322322   ℻ 01225 322291
www.thisisbath.co.uk
Owner *Northcliffe Media Group Ltd*
Editor *Sam Holliday (s.holliday@bathchron.co.uk)*
Deputy Editor *Paul Wiltshire*
Circulation 18,415

Local news and features especially welcomed (news@bathchron.co.uk).
Features Editor *Christopher Hansford*
Sports Editor *Julie Riegal*

### Evening Post

Temple Way, Bristol BS99 7HD
☎ 0117 934 3000   ℻ 0117 934 3575
epnews@bepp.co.uk
www.thisisbristol.co.uk
Owner *Bristol News & Media Ltd (part of Northcliffe Media Group Ltd)*
Editor-in-Chief *Mike Norton*
Circulation 43,997
News Editor *Steve Mellen*
Features Editor *Tim Davey*
Sports Editor *Robin Perkins*

### Western Daily Press

Temple Way, Bristol BS99 7HD
☎ 0117 934 3000   ℻ 0117 934 3574
WDEditor or WDNews or
features@westpress.co.uk
www.westerndailypress.co.uk
Owner *Bristol News & Media Ltd (part of Northcliffe Media Group Ltd)*
Editor *Andy Wright*
Circulation 38,211
Sports Editor *Rob Perkins*
Women's Page *Natalie Hale*

## Staffordshire

### Burton Mail

65–68 High Street, Burton upon Trent DE14 1LE
☎ 01283 512345   ℻ 01283 515351
editorial@burtonmail.co.uk
Owner *Staffordshire Newspapers Ltd*
Editor *Kevin Booth*
Circulation 13,221

Fashion, health, wildlife, environment, nostalgia, financial/money; consumer; women's world, rock; property; motoring, farming; what's on, leisure *Weekend Supplement* (Sat).
Sports Editor *Rex Page*

### The Sentinel

Sentinel House, Etruria, Stoke on Trent ST1 5SS
☎ 01782 602525   ℻ 01782 201167
www.thisisstaffordshire.co.uk
Owner *Staffordshire Sentinel News & Media Ltd*
Editor-in-Chief *Mike Sassi*
Circulation 56,609

Weekly sports final supplement. All material should be sent to the Assistant Editor, *Martin Tideswell*.

## Suffolk

### East Anglian Daily Times

Press House, 30 Lower Brook Street, Ipswich IP4 1AN
☎ 01473 230023   ℻ 01473 324776
news@eadt.co.uk
www.eadt.co.uk
Owner *Archant Ltd*
Editor *Terry Hunt*
Circulation 32,208

Founded 1874. Unsolicited mss generally not welcome; three or four received a week and almost none are used. Approach in writing in the first instance. No fiction. Supplements: sport (Mon); business (Tues); jobs (Wed); property (Thurs); motoring (Fri); magazine (Sat).
News Editor *Brad Jones* Hard news stories involving East Anglia (Suffolk, Essex particularly) or individuals resident in the area are always of interest.

Features *Dominic Castle* Mostly in-house, but will occasionally buy in when the subject is of strong Suffolk/East Anglian interest. Photo features preferred (extra payment). Special advertisement features are run regularly. Some opportunities here. Maximum 1,000 words.
Deputy Editor, Sport *Peter Blythe*
Women's Page *Sheline Clarke*

## Evening Star

Press House, 30 Lower Brook Street, Ipswich IP4 1AN
☎ 01473 230023   🖷 01473 324850
Owner *Archant Ltd*
Editor *Nigel Pickover*
Circulation 17,548
News Editor *Jess Gallagher*

# Sussex

## The Argus

Argus House, Crowhurst Road, Hollingbury, Brighton BN1 8AR
☎ 01273 544544   🖷 01273 505703
editor@theargus.co.uk
www.theargus.co.uk
Owner *Newsquest (Sussex) Ltd*
Editor *Michael Beard*
Circulation 29,006
News Editor *Lee Gibbs*
Sports Editor *Chris Giles*

# Teesside

## Evening Gazette

Borough Road, Middlesbrough TS1 3AZ
☎ 01642 234242
www.gazettelive.co.uk
Owner *Trinity Mirror plc*
Editor *Darren Thwaites*
Circulation 46,692

Special pages: health, education, family, business, property, entertainment, leisure, motoring, recruitment.
News *Elaine Blackburne, Julie Gibson*
Business Features/Commercial Features *Sue Scott*
Sports Editor *Philip Tallentire*
Councils *Sandy McKenzie*
Crime *Simon Walton*

# Tyne & Wear

## Evening Chronicle

Groat Market, Newcastle upon Tyne NE1 1ED
☎ 0191 232 7500   🖷 0191 232 2256
ec.news@ncjmedia..co.uk
www.chroniclelive.co.uk
Owner *Trinity Mirror Plc*

Editor *Paul Robertson*
Circulation 63,972

Receives a lot of unsolicited material, much of which is not used. Family issues, gardening, pop, fashion, cooking, consumer, films and entertainment guide, home improvements, motoring, property, angling, sport and holidays. Approach in writing with ideas. Limited opportunities for features due to full-time feature staff.
News Editor *James Marley*
Sports Editor *Paul New*
Women's Interests *Angela Upex*

## The Journal

Groat Market, Newcastle upon Tyne NE1 1ED
☎ 0191 201 6006   🖷 0191 221 0172
jnl.newsdesk@ncjmedia.co.uk
www.journallive.co.uk
Owner *Trinity Mirror Plc*
Editor *Brian Aitken*
Deputy Editor *Peter Montellier*
Circulation 31,817

Daily platforms include farming and business.
Sports Editor *Kevin Dinsdale*
Arts & Entertainment Editor *David Whetstone*
Environment Editor *Tony Henderson*
Business Editor *Iain Laing*

## Shields Gazette

Chapter Row, South Shields NE33 1BL
☎ 0191 427 4800   🖷 0191 456 8270
www.shieldsgazette.com
Owner *Northeast Press Ltd*
Editor *John Szymanski*
Circulation 17,110
News & Features Editor *Helen Charlton*

## Sunday Sun

ncjMedia, Groat Market, Newcastle upon Tyne NE1 1ED
☎ 0191 201 6299   🖷 0191 201 6180
colin.patterson@ncjmedia.co.uk
www.sundaysun.co.uk flnm.co.uk (football website)
Owner *Trinity Mirror Plc*
Editor *Colin Patterson*
Circulation 61,869

All material should be addressed to the appropriate editor (phone to check), or to the editor.

Sports Editor *Neil Farrington*

## Sunderland Echo

Echo House, Pennywell, Sunderland SR4 9ER
☎ 0191 501 5800   🖷 0191 534 5975
rob.lawson@northeast-press.co.uk
Owner *Johnston Press Plc*
Editor *Rob Lawson*

Circulation 39,159

All editorial material to be addressed to the News Editor, *John Corney* (echo.news@northeast-press.co.uk).

## Warwickshire

### Coventry Telegraph

Corporation Street, Coventry CV1 1FP
℡ 024 7663 3633    ℻ 024 7655 0869
news@coventrytelegraph.net
www.coventrytelegraph.net
Owner *Trinity Mirror Plc*
Editor *Darren Parkin*
Circulation 41,152

Unsolicited mss are read, but few are published. Approach in writing with an idea. No fiction. All unsolicited material should be addressed to the editor. Maximum 600 words for features.
News Managers *Steve Chilton, Steve Williams*
Features Manager/Women's Page *Bryan Jones*
Sports Editor *Richard Pagett*
£ Payment negotiable.

### Leamington Spa Courier

32 Hamilton Terrace, Leamington Spa CV32 4LY
℡ 01926 457737    ℻ 01926 339960
news@leamingtoncourier.co.uk
www.leamingtoncourier.co.uk
Owner *Johnston Press Publishing*
Editor *Martin Lawson*
Deputy Editor *Simon Steele*
Circulation 12,419

One of the Leamington Spa Courier series which also includes the *Warwick Courier* and *Kenilworth Weekly News*. Unsolicited feature articles considered, particularly matter with a local angle. Email with idea in the first instance.

## West Midlands

### Birmingham Mail

Bpm Media (Midlands) Ltd, Floor 6, Fort Dunlop, Fort Parkway, Birmingham B24 9FF
℡ 0121 236 3366    ℻ 0121 233 0271 (editorial)
steve.dyson@birminghammail.net
www.birminghammail.net
Owner *Trinity Mirror Plc*
Editor *David Brookes*
Executive Editor *Stacey Barnfield*
Circulation 56,495

Freelance contributions are welcome, particularly topics of interest to the West Midlands and feature pieces offering original and lively comment on family issues, health or education. Also news tips and community picture contributions.

Content Editor *Andy Richards*
Regional Sports Editor *Ken Montgomery*
Regional Images Editor *Steve Murphy*
Regional Business Editor *Alun Thorne*

### Birmingham Post

Bpm Media (Midlands) Ltd, Floor 6, Fort Dunlop, Fort Parkway, Birmingham B24 9FF
℡ 0121 236 3366    ℻ 0121 234 5667
www.birminghampost.net
Owner *Trinity Mirror Plc*
Editor *Marc Reeves*
Executive Editor *Mike Hughes*
Circulation 12,791

One of the country's leading regional newspapers. Relaunched as a weekly publication in November 2009. Freelance contributions are welcome. Topics of interest to the West Midlands and pieces offering lively, original comment are particularly welcome.

Content Editor *Sarah Probert*

### Express & Star

Queen Street, Wolverhampton WV1 1ES
℡ 01902 313131    ℻ 01902 319721
Owner *Midlands News Association*
Editor *Adrian Faber*
Deputy Editor *Keith Harrison*
Circulation 129,000
Assistant Editor *Mark Drew*
Features Editor *Emma Farmer*
Business Editor *Jim Walsh*
Sports Editor *Tony Reynolds*
Internet News Editor *Stuart Austen*
Women's Editor *Maria Cusine*

### Sunday Mercury (Birmingham)

Bpm Media (Midlands) Ltd, Floor 6, Fort Dunlop, Fort Parkway, Birmingham B24 9FF
℡ 0121 234 5567    ℻ 0121 234 5877
sunday.mercury@sundaymercury.net
www.sundaymercury.net
Owner *Trinity Mirror Plc*
Executive Editor *Paul Cole*
Circulation 54,375
Content Editor *Tony Larner*

## Wiltshire

### Swindon Advertiser

100 Victoria Road, Swindon SN1 3BE
℡ 01793 501806    ℻ 01793 501888
newsdesk@swindonadvertiser.co.uk
www.swindonadvertiser.co.uk
Owner *Newsquest (Wiltshire) Ltd*
Editor *Dave King*
Deputy Editor *Pauline Leighton*

Circulation 20,845

Copy and ideas invited. 'All material must be strongly related or relevant to the town of Swindon or the county of Wiltshire.' Little scope for freelance work. Fees vary depending on material.
News Editor *George Hamilton*
Sports Editor *Steve Butt*

## Worcestershire

### Worcester News

Berrow's House, Hylton Road, Worcester WR2 5JX
☏ 01905 748200    🖷 01905 742277
Owner *Newsquest (Midlands South) Ltd*
Editor *Kevin Ward*
Circulation 15,975

Community (Tues); jobs (Wed); property (Thurs); entertainment and motoring (Fri).
News Editor *Stephanie Preece*
Sports Editor *Paul Ricketts*

## Yorkshire

### The Doncaster Star

Sunny Bar, Doncaster DN1 1NB
☏ 01302 348500    🖷 01302 348528
doncaster@thestar.co.uk
www.thestar.co.uk/doncaster
Owner *Sheffield Newspapers Ltd*
Editor *Alan Powell*
Circulation 3,263

All editorial material to be addressed to the News Editor, *David Kessen*.
Sports Writer *Steve Hossack*

### Huddersfield Daily Examiner

Queen Street South, Huddersfield HD1 3DU
☏ 01484 430000    🖷 01484 437789
editorial@examiner.co.uk
www.examiner.co.uk
Owner *Trinity Mirror Plc*
Editor *Roy Wright*
Deputy Editor *Michael O'Connell*
Circulation 22,601

Home improvement, home heating, weddings, dining out, motoring, fashion, services to trade and industry.
News Editor *Neil Atkinson*
Features Editor *Val Javin*
Sports Editor *Mel Booth*
Women's Page *Hilarie Stelfox*

### Hull Daily Mail

Blundell's Corner, Beverley Road, Hull HU3 1XS
☏ 01482 315350    🖷 01482 315353
news@mailnewsmedia.co.uk
www.thisishullandeastriding.co.uk
Owner *Northcliffe Media Group Ltd*

Editor *John Meehan*
Circulation 51,886
News Editor *Rick Lyon*
Features Coordinator *Jeanette Minns*

### The Press

PO Box 29, 76–86 Walmgate, York YO1 9YN
☏ 01904 653051    🖷 01904 612853
newsdesk@thepress.co.uk
www.yorkpress.co.uk
Owner *Newsquest Media Group (a Gannett company)*
Managing Editor *Steve Hughes*
Deputy Editor *Fran Clee*
Circulation 30,722

Unsolicited mss not generally welcome unless submitted by journalists of proven ability. *Business Press Pages* (daily); *Property Press* (Thurs); *Twenty4Seven* – entertainment (Fri).
Picture Editor *Martin Oates*
Sports Editor *Stuart Martel*

💷 Payment negotiable.

### Scarborough Evening News

17–23 Aberdeen Walk, Scarborough YO11 1BB
☏ 01723 363636    🖷 01723 379033
editorial@scarboroheveningnews.co.uk
www.scarboroughtoday.co.uk
Owner *Yorkshire Regional Newspapers Ltd*
Editor *Ed Asquith*
Deputy Editor *Sue Wilkinson*
Circulation 12,062

Special pages include property (Mon); jobs (Thurs); motors (Fri).
News Editor *John Ritchie*
Sports Editor *Charles Place*

All other material should be addressed to the editor.

### The Star

York Street, Sheffield S1 1PU
☏ 0114 276 7676    🖷 0114 272 5978
starnews@sheffieldnewspapers.co.uk
www.thestar.co.uk
Owner *Johnston Press*
Editor *Alan Powell*
Circulation 47,216

Unsolicited mss not welcome, unless topical and local.
News Editor *Charles Smith* Contributions only accepted from freelance news reporters if they relate to the area.
Features Editor *Martin Smith* Very rarely requires outside features, unless on specialized local subjects.
Sports Editor *Bob Westerdale*
Women's Page *Jo Davison*

💷 Payment negotiable.

## Telegraph & Argus (Bradford)

Hall Ings, Bradford BD1 1JR
☎ 01274 729511    🖷 01274 723634
www.thetelegraphandargus.co.uk
Owner *Newsquest Media Group Ltd (a Gannett company)*
Editor *Perry Austin-Clarke*
Circulation 31,963

No unsolicited mss – approach in writing with samples of work. No fiction.
Head of News *Martin Heminway*
Features Editor *Emma Clayton* Local features and general interest. Showbiz pieces. 600–1,000 words (maximum 1,500).

## Yorkshire Evening Post

Wellington Street, Leeds LS1 1RF
☎ 0113 238 8984    🖷 0113 238 8536
eped@ypn.co.uk
www.yorkshireeveningpost.co.uk
Owner *Johnston Press*
Editor *Paul Napier*
Circulation 49,064
Evening sister of the *Yorkshire Post*.
News Editor *Gillian Haworth*
Features Editor *Jayne Dawson*
Sports Editor *Mark Absolom*
Women's Editor *Jayne Dawson*

## Yorkshire Post

Wellington Street, Leeds LS1 1RF
☎ 0113 243 2701    🖷 0113 238 8537
yp.editor@ypn.co.uk
www.yorkshirepost.co.uk
Owner *Johnston Press*
Editor *Peter Charlton*
Circulation 44,690

A serious-minded, quality regional daily with a generally conservative outlook. Three or four unsolicited mss arrive each day; all will be considered but initial approach in writing preferred. All submissions should be addressed to the editor. No fiction, poetry or family histories.
Features Editor *Sarah Freeman* Open to suggestions in all fields (though ordinarily commissioned from specialist writers).
Sports Editor *Matt Reeder*

# Northern Ireland

## Belfast News Letter

Johnston Press, Northern Ireland Division, 2 Esky Drive, Portadown BT63 5YY
☎ 028 3839 3939    🖷 028 3839 3940
Owner *Johnston Press*
Editor *Darwin Templeton*

Circulation 25,250
Supplements: *Farming Life; Business News Letter; Lifestyle; PM; Sports Ulster.*
News Editors *Karen Grunason, Ben Lowry*
Business Editor *Adrienne McGill*
Agricultural Editor *Ruth Rodgers*
Sports Editor *Richard Mulligan*

## Belfast Telegraph

Royal Avenue, Belfast BT1 1EB
☎ 028 9026 4000    🖷 028 9055 4506
newseditor@belfasttelegraph.co.uk
www.belfasttelegraph.co.uk
Owner *Independent News & Media (UK)*
Editor *Mike Gilson*
Managing Editor *Paul Connolly*
Circulation 71,074 (Mon–Fri)

Weekly business, property and recruitment supplements.
Deputy Editors *Ed McCann (News & Production), Gail Walker (Features)*
News Editor *Ronan Henry*
Sports Editor *Steven Beacom*

## The Irish News

113/117 Donegall Street, Belfast BT1 2GE
☎ 028 9032 2226    🖷 028 9033 7505
newsdesk@irishnews.com
www.irishnews.com
Owner *Irish News Ltd*
Editor *Noel Doran*
Circulation 46,800

All material to appropriate editor (phone to check), or to the news desk.
Assistant Editor *Fiona McGarry*
Features Editor *Will Scholes*
Sports Editor *Thomas Hawkins*

## Sunday Life

124–144 Royal Avenue, Belfast BT1 1EB
☎ 028 9026 4300    🖷 028 9055 4507
sgordon@belfasttelegraph.co.uk
www.sundaylife.co.uk
Owner *Independent News & Media (Northern Ireland)*
Editor *Martin Breen*
Associate Editor *David O'Dornan*
Circulation 65,872
News Editor *Steven Gordon*
Sports Editor *Jim Gracey*

# Scotland

## The Courier and Advertiser

80 Kingsway East, Dundee DD4 8SL
☎ 01382 223131    🖷 01382 454590
courier@dcthomson.co.uk
www.thecourier.co.uk
Owner *D.C. Thomson & Co. Ltd*

Editor *Bill Hutcheon*
Circulation 69,414

Circulates in East Central Scotland. Features occasionally accepted on a wide range of subjects, particularly local/Scottish interest, including finance, insurance, agriculture, motoring, modern homes, lifestyle and fitness. Maximum length: 500 words.
News Editor *Michael Alexander*
Features Editor/Women's Page *Catriona McInnes*
Sports Editor *Graeme Dey*

## Daily Record (Glasgow)
▷ National Newspapers

## Evening Express (Aberdeen)
PO Box 43, Lang Stracht, Mastrick, Aberdeen AB15 6DF
☎ 01224 690222  🖷 01224 341820
ee.news@ajl.co.uk
Owner *D.C. Thomson & Co. Ltd*
Editor *Damian Bates*
Circulation 52,029

Circulates in Aberdeen and the Grampian region. Local, national and international news and pictures, sport. Family platforms include *What's On*; *Counter* (consumer news); *Eating Out Guide*; *Family Days Out*. Unsolicited mss welcome 'if on a controlled basis'.
Deputy Editor *Richard Prest* Freelance news contributors welcome.
💷 Payment negotiable.

## Evening News
108 Holyrood Road, Edinburgh EH8 8AS
☎ 0131 620 8620  🖷 0131 620 8696
www.edinburghnews.com
Owner *Johnston Press*
Editor *Tom Little*
Circulation 46,222

Founded 1873. Circulates in Edinburgh, Fife, Central and Lothian. Coverage includes: entertainment, gardening, motoring, shopping, fashion, health and lifestyle, showbusiness. Occasional platform pieces, features of topical and/or local interest. Unsolicited feature material welcome. Approach appropriate editor in writing.
News Editor *Alan Young*
Sports Editor *Graham Lindsay*
💷 NUJ/house rates.

## Evening Telegraph
80 Kingsway East, Dundee DD4 8SL
☎ 01382 223131  🖷 01382 454590
general@eveningtelegraph.co.uk
www.eveningtelegraph.co.uk
Owner *D.C. Thomson & Co. Ltd*
Editor *Gordon Wishart*

Circulation 23,048

Circulates in Tayside, Dundee and Fife. All material should be addressed to the editor.

## Evening Times
200 Renfield Street, Glasgow G2 3QB
☎ 0141 302 7000  🖷 0141 302 6677
news@eveningtimes.co.uk
www.eveningtimes.co.uk
Owner *Newsquest*
Editor-in-Chief *Donald Martin*
Group Content Editor *Tony Carlin*
Circulation 72,662 (Mon.–Fri.)/44,036 (Sat.)

Circulates in Glasgow and the west of Scotland. Supplements: *Job Search*; *City Living*; *Drive Times*; *Extra Times*; *Times Out*; *Life & Times*.
Head of News *Hugh Boag*
Sports Editor *Graham Shields*

## Greenock Telegraph
2 Crawfurd Street, Greenock PA15 1LH
☎ 01475 726511  🖷 01475 783734
Owner *Clyde & Forth Press Ltd*
Editor *Wendy Metcalfe*
Circulation 15,965

Circulates in Greenock, Port Glasgow, Gourock, Kilmacolm, Langbank, Inverkip, Wemyss Bay, Skelmorlie, Largs. Unsolicited mss considered 'if they relate to the newspaper's general interests'. No fiction. All material to be addressed to the editor.

## The Herald/Sunday Herald (Glasgow)
▷ National Newspapers

## Paisley Daily Express
14 New Street, Paisley PA1 1YA
☎ 0141 887 7911  🖷 0141 887 6254
pde@s-un.co.uk
www.paisleydailyexpress.co.uk
Owner *Scottish & Universal Newspapers Ltd*
Editor *Anne Dalrymple*
Circulation 8,764

Circulates in Paisley, Linwood, Renfrew, Johnstone, Elderslie, Neilston and Barrhead. Unsolicited mss welcome only if of genuine local (Renfrewshire) interest. The paper does not commission work and will consider submitted material. Maximum 1,000–1,500 words. All submissions to the news editor.
News Editor *Gavin Penny*
Sports Reporter *Paul Behan*

## The Press and Journal
PO Box 43, Lang Stracht, Mastrick, Aberdeen AB15 6DF
☎ 01224 690222  🖷 01224 663575
Owner *D.C. Thomson & Co. Ltd*
Editor *Derek Tucker*

Circulation 78,121

A well-established daily that circulates in Aberdeen, Grampians, Highlands, Tayside, Orkney, Shetland and the Western Isles. Most material is commissioned but will consider ideas. News Editor *Andrew Kellock* Wide variety of hard or off-beat news and features relating especially, but not exclusively, to the north of Scotland. Sports Editor *Alex Martin*
Your Life *Sonja Cox* Saturday lifestyle tabloid pullout. Features food, fashion, travel, books, arts and lifestyle.

£ Payment by arrangement.

### Scotland on Sunday (Edinburgh)
▷ National Newspapers

### The Scotsman (Edinburgh)
▷ National Newspapers

### Sunday Mail (Glasgow)
▷ National Newspapers

### Sunday Post (Dundee)
▷ National Newspapers

# Wales

### Daily Post

PO Box 202, Vale Road, Llandudno Junction, Conwy LL31 9ZD
☎ 01492 574455    📠 01492 574433
welshnews@dailypost.co.uk
www.dailypost.co.uk
Owner *Trinity Mirror Plc*
Editor, Daily Post, Wales/Publishing Director, Trinity Mirror, North Wales *Rob Irvine*
Executive Editor *Andy Gilpin*
Circulation 34,601

Very rarely commissions freelance writers

### The Leader

Mold Business Park, Wrexham Road, Mold CH7 1XY
☎ 01352 707707    📠 01352 752180
news@eveningleader.co.uk
Owner *NWN Media*
Editor-in-Chief *Barrie Jones*
Circulation 20,073

Circulates in Wrexham, Flintshire, Deeside and Chester. Special pages/features: motoring, travel, arts, women's, children's, local housing, information and news for the disabled, music and entertainment.
Assistant Editor *Martin Wright*
Sports Editor *Nick Harrison*

### South Wales Argus

Cardiff Road, Maesglas, Newport NP20 3QN
☎ 01633 810000    📠 01633 777202

www.southwalesargus.co.uk
Owner *Newsquest*
Editor *Gerry Keighley*
Circulation 27,101

Circulates in Newport, Gwent and surrounding areas.
News Editor *Maria Williams*
Sports Editor *Phil Webb*

### South Wales Echo

Media Wales Ltd, Six Park Street, Cardiff CF10 1XR
☎ 029 2022 3333    📠 029 2058 3624
echo.newsdesk@mediawales.co.uk
www.walesonline.co.uk
Owner *Media Wales (Trinity Mirror Plc)*
Editor *Mike Hill*
Circulation 39,361

Circulates in South and Mid Glamorgan.
Head of Content *Nick Machin*
Head of Features *Margaret O'Reilly*
Head of Sport *Paul Abbandonato*

### South Wales Evening Post

Adelaide Street, Swansea SA1 1QT
☎ 01792 510000    📠 01792 514697
postnews@swwmedia.co.uk
www.thisisswansea.co.uk
Owner *Northcliffe Media Group Ltd*
Editor *Spencer Feeney*
Circulation 46,069

Circulates throughout south west Wales.
News Editor *Chris Davies*
Features Editor *Peter Slee*
Sports Editor *Liam Sullivan*

### Wales on Sunday

Media Wales Ltd, Six Park Street, Cardiff CF10 1XR
☎ 029 2022 3333    📠 029 2058 3725
www.walesonline.co.uk
Owner *Media Wales (Trinity Mirror Plc)*
Editor *Tim Gordon*
Circulation 34,696

Launched 1989. Tabloid with sports supplement. Does not welcome unsolicited mss.
Head of Content *Nick Machin*
Sports Editor *Paul Abbandonato*

### Western Mail

Media Wales Ltd, Six Park Street, Cardiff CF10 1XR
☎ 029 2022 3333    📠 029 2024 3640 (editorial)
www.walesonline.co.uk
Owner *Media Wales (Trinity Mirror Plc)*
Editor *Alan Edmunds*
Executive Editor *Ceri Gould*

Circulation 33,693

Circulates in Cardiff, Merthyr Tydfil, Newport, Swansea and towns and villages throughout Wales. Mss welcome if of a topical nature, of Welsh interest. No short stories or travel. Approach in writing to the features editor. 'Usual subjects already well covered, e.g. motoring, travel, books, gardening. We look for the unusual.' Maximum 1,000 words. Opportunities also on women's page. Supplements: *Saturday Magazine*; *Education*; *WM*; *Box Office*; *Welsh Homes*; *Business*; *Sport*; *Motoring*; *Country & Farming*.
Head of Content *Nick Machin*
Head of Features *Margaret O'Reilly*
Head of Sport *Philip Blanche*

## Channel Islands

### Guernsey Press & Star
Braye Road, Vale, Guernsey GY1 3BW
☏ 01481 240240   🄵 01481 240235
newsroom@guernsey-press.com
www.guernsey-press.com

Owner *Guiton Group*
Editor *Richard Digard*
Circulation 15,586

Special pages include children's and women's interest, gardening and fashion.
Sports Editor *Rob Batiste*
Features Editor *Diane Digard*

### Jersey Evening Post
PO Box 582, Jersey JE4 8XQ
☏ 01534 611611   🄵 01534 611622
editorial@jerseyeveningpost.com
www.thisisjersey.com
Owner *Claverley Company*
Editor *Chris Bright*
Circulation 20,070

Special pages: gardening, motoring, property, boating, technology, food and drink, personal finance, health, business.
News Editor *Carl Walker*
Features Editor *Anna Plunkett-Cole*
Sports Editor *Ron Felton*

# News Agencies

## Associated Press Limited

▷ entry under US Media Contacts in the UK

## Dow Jones Newswires

10 Fleet Place, London EC4M 7QN

T 020 7842 9900   F 020 7842 9361 (admin)

A real-time financial and business newswire operated by Dow Jones & Co., publishers in the USA of *The Wall Street Journal*. No unsolicited material.

## National News Press and Photo Agency

4–5 Academy Buildings, Fanshaw Street, London N1 6LQ

T 020 7684 3000   F 020 7684 3030

news@nationalnews.co.uk

All press releases are welcome. Most work is ordered or commissioned. Coverage includes courts, tribunals, conferences, general news, etc. – words and pictures – as well as PR.

## The Press Association Ltd

292 Vauxhall Bridge Road, London SW1V 1AE

T 020 7963 7000   F 020 7963 7192 (news desk)

copy@pressassociation.com

www.pressassociation.com

No unsolicited material. Most items are produced in-house though occasional outsiders may be used. A phone call to discuss specific material may lead somewhere 'but this is rare'.

## Reuters

30 South Colonnade, Canary Wharf, London E14 5EP

T 020 7250 1122

www.thomsonreuters.com

No unsolicited material.

## Solo Syndication Ltd

17–18 Hayward's Place, London EC1R 0EQ

T 020 7566 0360   F 020 7566 0388

wgardiner@solosyndication.com

www.solosyndication.com

Exclusive syndication agents for the *Daily Mail*, *Mail on Sunday*, *London Evening Standard*, *Universal Press Syndicate* (Middle East and Africa) and the *CS Monitor* (Middle East and Africa).

## South Yorkshire Sport

6 Sharman Walk, Apperknowle, Sheffield S18 4BJ

T 01246 414767/07970 284848   F 01246 414767

nickjohnson@uwclub.net

Contact *Nick Johnson*

Provides written/broadcast coverage of sport in the South Yorkshire area.

## Space Press News and Pictures

Bridge House, Blackden Lane, Goostrey CW4 8PZ

T 01477 533403

Scoop2001@aol.com

Editor *John Williams (07970 213528)*

Pictures *Emma Williams (evwphoto@aol.com 07976 795494)*

Founded 1972. Press and picture agency covering Cheshire and the North West, North Midlands, including Knutsford, Macclesfield, Congleton, Crewe and Nantwich, Wilmslow, Alderley Edge, serving national, regional and local press, TV, radio, and digital picture transmission. Copy and pictures produced for in-house publications and PR. Property, countryside and travel writing. A member of the National Association of Press Agencies (NAPA).

# US Media Contacts in the UK

## ABC News Intercontinental Inc.
3 Queen Caroline Street, London W6 9PE
℡ 020 8222 5000    ℻ 020 8222 5020
Bureau Chief/VP, International Digital *Marcus Wilford*

## Associated Press
12 Norwich Street, London EC4A 1BP
℡ 020 7353 1515    ℻ 020 7353 8118
London Bureau Chief *Paisley Dodds*

## Bloomberg News
City Gate House, 39–45 Finsbury Square, London EC2A 1PQ
℡ 020 7330 7500/020 7392 6666 (news)
London Bureau Chief (Print) *Heather Harris*

## Business Week
20 Canada Square, Canary Wharf, London E14 5LH
℡ 020 7176 6060    ℻ 020 7176 6070
Bureau Chief *Stanley Reed*

## CBS News
Building 10, Chiswick Park, 566 Chiswick High Road, London W4 5XS
℡ 020 7887 3000    ℻ 020 7887 3092
VP/Bureau Chief *Jennifer Siebens*

## CNBC
10 Fleet Place, Limeburner Lane, London EC4M 7QS
℡ 020 7653 9451    ℻ 020 7653 9393
Assignment Editor *Jennifer Callegher*

## CNN (Cable News Network)
▷ entry under Television and Radio

## Dow Jones Newswires
10 Fleet Place, Limeburner Lane, London EC4M 7QN
℡ 020 7842 9900
News Editor *Gren Manuel*

## Fairchild Publications of New York
20 Shorts Gardens, London WC2H 9AU
℡ 020 7240 0420
Bureau Chief *Samantha Conti*

## Forbes Magazine
Malta House, 36/38 Picadilly, London W1J 0DP
℡ 020 7534 3900
Managing Director & Publisher (Europe, Middle East & Africa) *Bob Crozier*

## Fox News Channel
6 Centaurs Business Park, Grant Way, Isleworth TW7 5QD
℡ 020 7805 7143    ℻ 020 7805 1111
Correspondent *Amy Kellogg*

## The Globe and Mail
dsaunders@globeandmail.com
European Bureau Chief *Doug Saunders*

## International Herald Tribune
▷ entry under National Newspapers

## Los Angeles Times
5 Lancaster Drive, London NW3 4EY
Bureau Chief *Henry Chu*

## Market News International
50 Cannon Street, London EC4N 6JJ
℡ 020 7634 1655    ℻ 020 7236 7122
ukeditorial@marketnews.com
London Bureau Chief *David Thomas*

## National Public Radio
Room G-10 East Wing, Bush House, Strand, London WC2B 4PH
℡ 020 7557 1087    ℻ 020 7379 6486
Bureau Chief *Robert Gifford*

## NBC News
200 Gray's Inn Road, London WC1X 8XZ
℡ 020 7843 8777
Bureau Coordinator *Lucy Muccini*

## The New York Times
66 Buckingham Gate, London SW1E 6AU
℡ 020 7799 5050    ℻ 020 7799 2962
Bureau Chief *John Burns*

## Newsweek
Academy House, 36 Poland Street, London W1F 7LU
Bureau Chief *Stryker McGuire*

Includes FREE online access to **www.thewritershandbook.com**

## People Magazine

Blue Fin Building, 110 Southwark Street, London
SE1 0SU
🐵 020 3148 3000
Chief Foreign Correspondent *Simon Perry*

## Reader's Digest Association Ltd

▷ See entries under UK Publishers and Magazines.

## Time

▷ entry under Magazines

## Voice of America

London News Centre, 167 Fleet Street, London
EC4A 2EA
🐵 020 7410 0960
Correspondent *Sonja Pace*

## Wall Street Journal

10 Fleet Place, Limeburner Lane, London
EC4M 7QN
🐵 020 7842 9200
Bureau Chief *Bruce Orwall*

## Washington Post

1 Southwark Bridge, London SE1 9HL
🐵 020 7873 5940

# Television and Radio

## BBC TV and Radio

### BBC Vision, BBC Audio & Music
www.bbc.co.uk
Director-General *Mark Thompson*
Deputy Director-General *Mark Byford*

The BBC consists of three main divisions: BBC Vision, Audio and Music, and Journalism. BBC Vision runs production, commissioning and services and includes the terrestrial and digital television channels; also, Drama, Entertainment and Children's; and Factual and Learning. The Audio & Music Division is responsible for TV Music Entertainment, In-house Factual, Specialist Factual and Drama Audio production. The Journalism division covers News, Sport, Global News, Nations and Regions.

BBC Television Centre, Wood Lane, London W12 7RJ
℡ 020 8743 8000
Broadcasting House, Portland Place, London W1A 1AA
℡ 020 7580 4468

TV & RADIO COMMISSIONING:
www.bbc.co.uk/commissioning
Director, BBC Vision *Jana Bennett*
Director of Audio & Music Division *Tim Davie*
Controller, BBC One *Jay Hunt*
Controller, BBC Two *Janice Hadlow*
Controller, Daytime *Liam Keelan*
Controller, BBC Three *Danny Cohen*
Controller, BBC Four *Richard Klein*
Controller, BBC Radio 1, 1Xtra, Asian Network and Switch *Andy Parfitt*

Radio 1 documentary submission enquiries: *Joe Harland*, Executive Producer (joe.harland@bbc. co.uk; 020 7765 3552)
1Xtra commissioning enquiries: *Debbie Ramsay*, Acting Commissioning Editor, Documentaries (debbie.ramsay@bbc.co.uk)
1Xtra commissions weekly documentaries from independent producers and an in-house production team. Welcomes proposals that may be suitable as a co-commission with another network such as Radio 1 or Asian Network.
Controller, BBC Radio 2 & 6 Music *Bob Shennan*
Radio 2 commissioning enquiries: *Robert Gallacher*, Editor, Planning & Station Sound (robert.gallacher@bbc.co.uk; 020 7765 4373)
Controller, BBC Radio 3 *Roger Wright*
Radio 3 commissioning enquiries: *David Ireland*, Commissions & Schedules Manager (david. ireland@bbc.co.uk; 020 7765 4943)
Controller, BBC Radio 4 & BBC 7 *Mark Damazer*
Radio 4 proposals must be submitted through an in-house department or a registered independent production company (lists of departments and companies are available at www.bbc.co.uk/ commissioning/radio/network/radio4.shtml).
Controller, BBC Radio Five Live and Five Live Sports Extra *Adrian van Klaveren*
Radio Five Live commissioning enquiries: *Jonathan Wall*, Commissioning Editor (jonathan. wall.01@bbc.co.uk; 020 8624 8948)
Head of Programmes, BBC Radio 6 Music *Lewis Carnie*
Radio 6 Music commissioning enquiries: *Paul Rodgers*, Editor (paul.rodgers@bbc.co.uk; 020 7765 4763)

### BBC Drama, Comedy and Children's

BBC *writersroom* champions new writers across all BBC platforms for drama, comedy and children's programmes, running targeted schemes and workshops linked directly to production. It accepts and assesses unsolicited scripts for all departments: film, TV drama, radio drama, TV narrative comedy and radio narrative comedy. The *Writersroom* website (www.bbc. co.uk/writersroom) offers a diary of events/ opportunities, competitions, interviews with established writers, submission guidelines and free formatting software. BBC *Writersroom* also has a Manchester base which focuses on new writing in the north of England. Writers should address hard copies of original, completed scripts to: BBC *Writersroom*, Grafton House, 379–381 Euston Road, London NW1 3AU. Before sending in scripts, please go to the website for the latest guidelines on submitting unsolicited work. See also www.bbc.co.uk/commissioning/structure/public.shtml

WRITERSROOM
Creative Director, New Writing *Kate Rowland*
Development Manager *Paul Ashton*
FILMS
Commissioning Editor and Executive Producer *Christine Langan*
Executive Producers *Jamie Laurenson, Joe Oppenheimer*
TV DRAMA
Controller, Drama Commissioning *Ben Stephenson*
Controller, Drama Production & New Talent *John Yorke*
Controller, Series and Serials *Kate Harwood*
Creative Director, Drama Development *Victoria Vea*
Executive Producer, EastEnders *Brian Kirkwood*
RADIO DRAMA
Head of Radio Drama *Alison Hindell*
Executive Producer (Birmingham) & Editor, The Archers *Vanessa Whitburn*
Executive Producer (World Service Drama) *Marion Nancarrow*
Executive Producer (Manchester Radio Drama) *Sue Roberts*
COMEDY
Head of Comedy *Mark Freeland*
Head of Radio Entertainment *Paul Schlesinger*
Creative Director, Mainstream Comedy *Micheal Jacob*
CHILDREN'S
Controller, BBC Children's *Joe Godwin*
Creative Director, CBBC *Damian Kavanagh*
Controller, CBeebies *Michael Carrington*
Head of Children's Drama, Animation & Acquisitions *Steven Andrew*
Head of CBeebies Production, Animation & Acquisitions *Kay Benbow*

## BBC News
www.bbc.co.uk/news
news.bbc.co.uk

BBC News is the world's largest news-gathering organization with over 2,500 journalists, 45 bureaux worldwide and 15 networks and services across TV, radio and new media.
Director, BBC News *Helen Boaden*
Deputy Director of BBC News & Head of

Multimedia Programmes *Stephen Mitchell*
Head of Newsgathering *Francesca Unsworth*
Director of News Content, BBC World *Richard Porter*
Controller, BBC News Channel & BBC News at 6 o'clock & 10 o'clock *Kevin Bakhurst*
TELEVISION EDITORS
Editor, 6 o'clock & 10 o'clock News *James Stephenson*
Editor, Newsnight *Peter Rippon*
RADIO EDITORS
Editor, Today *Ceri Thomas*
Editor, PM and Broadcasting House *Joanna Carr*
Editor, The World at One/ The World This Weekend *Jamie Angus*
Editor, The World Tonight *Alistair Burnett*
CEEFAX
Room 7540, BBC Television Centre, Wood Lane, London W12 7RJ
SUBTITLING
Red Bee Media, Broadcast Centre, 201 Wood Lane, London W12 7TP www.redbeemedia.com
Available via Ceefax page 888.

## BBC Sport
www.bbc.co.uk/sport
Director, Sport *Barbara Slater*
Head BBC TV Sport *Philip Bernie*

Sports news and commentaries across television, online and Radios 1, 4 and 5 Live.

## BBC Religion
Room 5039, New Broadcasting House, Oxford Road, Manchester M60 1SJ
☎ 0161 200 2020  🖷 0161 244 3183
Head of Religion and Ethics *Aaqil Ahmed*

Regular programmes for television include *Songs of Praise*. Radio output includes *Good Morning Sunday*; *Sunday Half Hour*; *Thought for the Day*; *The Daily Service*.

## BBC World Service
PO Box 76, Bush House, Strand, London WC2B 4PH
☎ 020 7240 3456  🖷 020 7557 1258
www.bbc.co.uk/worldservice
Director, BBC Global News *Peter Horrocks*
Director, English Networks & News, BBCWS *Gwyneth Williams*

BBC World Service broadcasts in English and 31 other languages. With over 188 million weekly viewers and listeners, excluding countries where research is not possible, it reaches a bigger audience than its five closest international broadcasting competitors combined. The English radio service is round-the-clock, with news and current affairs as the main component. It attracts 40 million listeners each week. Like the other 31 language services, its radio services are increasingly available throughout the world on local FM stations, via satellite and online as well as through short-wave frequencies. Coverage includes world business, politics, people/events/ opinions, development issues, the international scene, developments in science and technology, sport, religion, music and the arts. BBC World Service broadcasting is financed by a grant-in-aid voted by Parliament amounting to £272 million for 2009/2010.

## BBC writersroom
▷ BBC Drama, Comedy and Children's

# BBC Nations

## BBC Northern Ireland
Broadcasting House, Ormeau Avenue, Belfast
BT2 8HQ
☎ 028 9033 8000
bbc.co.uk/northernireland
Director, NI *Peter Johnston*
Head of Programmes *Ailsa Orr*
Chief Operating Officer *Mark Taylor*
Head of News & Current Affairs *Andrew Colman*
Head of Drama *Patrick Spence*
Head of Entertainment & Events *Mike Edgar*
Head of Radio *Susan Lovell*
Head of Factual *Paul McGuigan*
Head of Multi Media Commissioning *Fergus Keeling*
Managing Editor Learning, Languages and Social Action *Jane Cassidy*
Senior Producer Irish Language *Antaine O'Donnaile*
Editor, TV News *Angelina Fusco*
Editor, Radio News *Kathleen Carragher*
Editor, News Gathering *Michael Cairns*
Editor TV Current Affairs *Jeremy Adams*
Editor News Online *Seamus Boyd*
Editor – Foyle *Paul McCauley*
Editor, Sport *Shane Glynn*

Regular television programmes include *BBC Newsline 6.30* and a wide range of documentary, popular factual and entertainment programmes. Radio stations: BBC Radio Foyle and BBC Radio Ulster (see entries). For further details on television and film scripts contact: *Susan Carson*, Programme Development Executive Television, BBC Northern Ireland Drama, Room 3.07, Blackstaff House, Great Victoria Street, Belfast, BT2 7BB (☎ 028 9033 8498; tvdrama.ni@bbc.co.uk). Radio scripts contact: *Anne Simpson*, Manager, Radio Drama at the same address.

## BBC Scotland
40 Pacific Quay, Glasgow G51 1DA
☎ 0141 422 6000
bbc.co.uk/scotland

Director, Scotland *Ken MacQuarrie*
Chief Operating Officer *Bruce Malcolm*
Head of Programmes and Services *Donalda MacKinnon*
Head of Drama, Television *Anne Mensah*
Head of Drama, Radio *Patrick Rayner*
Head of Factual Programmes *Andrea Miller*
Head of Service, BBC ALBA *Margaret Mary Murray*
Managing Editor, Gaelic *Marion MacKinnon*
Head of Radio, Scotland *Jeff Zycinski*
Head of News and Current Affairs *Atholl Duncan*
Head of Entertainment and Events *Eileen Herlihy*
Commissioning Editor, Television & Head of Sport *Ewan Angus*
Head of Children's *Simon Parsons*
Head of Learning, Scotland *Nick Simons*
Executive Editor, Cross Media *Matthew Lee*

BBC Scotland provides television and radio programmes for Scotland and the UK networks, online and interactive content and the BBC Scottish Symphony Orchestra. Based in the new digital headquarters building since 2007, there are also centres throughout Scotland, details below:

**Aberdeen**
Broadcasting House, Beechgrove Terrace, Aberdeen AB15 5ZT
☎ 01224 625233
**Dumbarton**
Strathleven Bottling Plant, Dumbarton G82 2AP
☎ 01389 736666
**Dumfries**
Elmbank, Lover's Walk, Dumfries DG1 1NZ
☎ 01387 268008
**Dundee**
4th Floor, Nethergate Centre, 4th Floor, 66 Nethergate, Dundee DD1 4ER
☎ 01382 202481
**Edinburgh**
The Tun, 4 Jackson's Entry, 111 Holyrood Road, Edinburgh EH8 8PJ
☎ 0131 557 5888
**Glasgow City Halls (BBC Scottish Symphony Orchestra)**
87 Albion Street, Glasgow G1 1NQ
☎ 0141 552 0909
**Inverness**
7 Culduthel Road, Inverness IV2 4AD
☎ 01463 720720
**Orkney**
Castle Street, Kirkwall, Orkney KW15 1DF
☎ 01856 873939
**Portree**
Clydesdale Bank Buildings, Somerled Square, Portree, Isle of Skye IV51 9BT
☎ 01478 612005
**Selkirk**
Unit 1, Ettrick Riverside, Dunsdale Road, Selkirk TD7 5EB

☎ 01750 724567
**Shetland**
Pitt Lane, Lerwick, Shetland ZE1 0DW
☎ 01595 694747
**Stornoway**
Radio Nan Gaidheal, Rosebank, 52 Church Street,
Stornoway, Isle of Lewis HS1 2LS
☎ 01851 705000

## BBC Wales

Broadcasting House, Llandaff CF5 2YQ
☎ 029 2032 2000 📠 029 2055 2973
www.bbc.co.uk/wales
Director, Wales *Menna Richards*
Head of Programmes (Welsh Language)
*Keith Jones*
Head of Programmes (English Language)
*Clare Hudson*
Head of News & Current Affairs
*Mark O'Callaghan*
Head of Drama *Piers Wenger*
Producer, Pobol y Cwm *Ynyr Williams*
Editor, New Media *Iain Tweedale*

Headquarters of BBC Wales, with regional centres
in Bangor, Aberystwyth, Carmarthen, Wrexham
and Swansea. BBC Wales television produces up
to 12 hours of English language programmes a
week, 12 hours in Welsh for transmission on **S4C**
and an increasing number of hours on network
services, including *Doctor Who*, *Last Chance to
See* and *Torchwood*. Regular programmes include
*Wales Today*; *Newyddion* (Welsh-language daily
news); *Week In Week Out*, *X-Ray* and *Pobol y
Cwm* (Welsh-language soap) on television and
*Good Morning Wales*; *Good Evening Wales*; *Post
Cyntaf* and *Post Prynhawn* on Radio Wales and
Radio Cymru. Ideas for programmes should
be submitted either by post to *Phillip Moss*,
Commissioning Editor, Room 3021 at the address
above or by email to commissioning@bbc.co.uk

**Bangor**
Broadcasting House, Meirion Road,
Bangor LL57 2BY
☎ 01248 370880 📠 01248 351443
Head of Centre *Marian Wyn Jones*

# BBC Regions

## BBC Asian Network

The Mailbox, Birmingham B1 1RF
☎ 0121 567 6000/0116 201 6772 (newsdesk)
www.bbc.co.uk/asiannetwork
Managing Editor *Vijay Sharma*
Network Manager *Mike Curtis*

Commenced broadcasting in November 1996.
Broadcasts nationwide. Programmes in English,
Bengali, Gujerati, Hindi, Punjabi and Urdu.

## BBC Birmingham

The Mailbox, Birmingham B1 1RF
☎ 0121 567 6767 📠 0121 567 6875
birmingham@bbc.co.uk
www.bbc.co.uk/birmingham
Head of Regional and Local Programmes
*Cath Hearne*
Regional News Editor *Mark Hayman*

BBC Birmingham's Documentaries &
Contemporary Factual Department produces a
broad range of programming for BBC One and
BBC Two, including *Gardener's World*; *Coast*;
*Desi DNA*; *To Buy or Not to Buy*. Network
radio includes: Radio 2, Radio 4 and the Asian
Network. Regional & Local Programmes: TV:
*Midlands Today*; *Inside Out*; *The Politics Show*;
Radio: BBC WM (see entry). BBC Birmingham is
the headquarters for BBC English Regions. There
is a television Drama Village at the University of
Birmingham.

**BBC Birmingham Drama**
Archibald House, 1059 Bristol Road, Selly Oak,
Birmingham B29 6LT
☎ 0121 567 7350
Programmes produced have included *Doctors*;
*The Afternoon Play* and *Brief Encounters* (which
champions new writers).

**BBC East Midlands (Nottingham)**
East Midlands Broadcasting Centre, London
Road, Nottingham NG2 4UU
☎ 0115 955 0500 📠 0115 902 1983
www.bbc.co.uk/eastmidlandstoday
Head of Regional Programming *Stuart Thomas*
Output Editor *Emma Agnew*

**BBC East (Norwich)**
The Forum, Millennium Plain, Norwich NR2 1BH
☎ 01603 619331 📠 01603 667865
www.bbc.co.uk/lookeast
Head of Regional Programming *Tim Bishop*
Output Editor *Nikki O'Donnell*

## BBC Bristol

Broadcasting House, Whiteladies Road, Bristol
BS8 2LR
☎ 0117 973 2211
www.bbc.co.uk/bristol
Head of Regional and Local Programmes
*Lucio Mesquita*
Managing Editor, BBC Bristol *Tim Pemberton*
Head of Factual *Ben Gale*
Head of Natural History Unit *Andrew Jackson*

BBC Bristol is the home of the BBC's Natural
History Unit, producing programmes such as
*Blue Planet*; *Life of Mammals*; *British Isles, A
Natural History*; *Abyss* and *Planet Earth* for
BBC One and BBC Two. It also produces natural
history programmes for Radio 4 and Radio 5
Live. Regional TV includes *BBC Points West*; *The

*Politics Show* and *Inside Out West.* The Factual department produces a wide range of television programmes, including *DIY SOS; Antiques Roadshow; Soul Deep* and *Ray Mears* in addition to radio programmes specializing in history, travel, literature and human interest features for Radio 4 including the long-running *Poetry Please.*

## BBC South East (Tunbridge Wells)

The Great Hall, Mount Pleasant Road, Tunbridge Wells TN1 1QQ
☎ 01892 670000
www.bbc.co.uk/kent
www.bbc.co.uk/surrey
www.bbc.co.uk/sussex
Head of Regional and Local Programmes *Mick Rawsthorne* (Responsible for BBC South East [TV], BBC Kent, BBC Sussex and BBC Surrey)
Editor, South East Today *Quentin Smith*
Managing Editor, BBC Sussex and BBC Surrey *Nicci Holiday*
Managing Editor, BBC Kent *Paul Leaper*
Executive Producer, Inside Out *Linda Bell*

## BBC West/BBC South/BBC South West

The three regional television stations, BBC West, BBC South and BBC South West produce breakfast, lunchtime and evening news programmes, as well as *Inside Out*, regular 30-minute local current affairs programmes and regional parliamentary programmes, *The Politics Show*. Each of the regions operates a comprehensive local radio and online service as well as a range of correspondents specializing in subjects such as health, education, business, home affairs and the environment.

### BBC West (Bristol)
Broadcasting House, Whiteladies Road, Bristol BS8 2LR
☎ 0117 973 2211
Head of Regional and Local Programmes *Lucio Mesquita* (Responsible for the BBC West region. TV: *BBC Points West* (regional news programme), *Inside Out West* (regional current affairs series) and *The Politics Show West*. Local Radio: BBC Bristol, BBC Somerset, BBC Gloucestershire, BBC Wiltshire and the BBC local websites)
Editor, TV News *Stephanie Marshall*
TV Current Affairs & Features Editor *Roger Farrant*
Political Editor *Paul Barltrop*

### BBC South (Southampton)
Broadcasting House, Havelock Road, Southampton SO14 7PU
☎ 023 8022 6201
Head of BBC South *Chris Carnegy* (Responsible for BBC South TV, BBC Berkshire, BBC Oxford, BBC Solent and online content for the region)

TV News Editor, BBC South Today *Lee Desty*
Series Producer, Inside Out *Jane French*

### BBC South West (Plymouth)
Broadcasting House, Seymour Road, Mannamead, Plymouth PL3 5BD
☎ 01752 229201
Head of BBC South West *Jane McCloskey* (Responsible for the BBC South West region. TV: *BBC Spotlight* (regional news programme), *Spotlight Channel Islands* (regional news programme), *Inside Out* (regional current affairs series) and *The Politics Show South West*. Local Radio: BBC Devon, BBC Cornwall, BBC Guernsey, BBC Jersey and the local BBC websites)
Output Editor *Simon Read*
Series Producer, Inside Out *Simon Willis*
Political Editor *Martyn Oates*

## BBC Yorkshire/BBC North West/BBC North East & Cumbria

The regional centres at Leeds, Manchester and Newcastle make their own programmes on a bi-media approach, each centre having its own head of regional and local programmes.

### BBC Yorkshire
Broadcasting Centre, 2 St Peter's Square, Leeds LS9 8AH
☎ 0113 244 1188
Head of Regional and Local Programmes *Helen Thomas*
Editor, News *Tim Smith*
Assistant News Editor, Look North *Denise Wallace*
Political Editor, North of Westminster *Len Tingle*
Weeklies Editor, Close Up North *Ian Cundall*
Producers, Close Up North *Richard Taylor, Paul Greenan*

### BBC North West (Manchester)
New Broadcasting House, Oxford Road, Manchester M60 1SJ
☎ 0161 200 2020
Head of Regional and Local Programmes *Aziz Rashid*
Editor, Newsgathering *Jim Clarke*
Output Editor *Cerys Griffiths*
Producer, Inside Out *Deborah van Bishop*
Producer, The Politics Show *Michelle Mayman*

### BBC North East & Cumbria (Newcastle upon Tyne)
Broadcasting Centre, Barrack Road, Newcastle Upon Tyne NE99 2NE
☎ 0191 232 1313
Head of Regional and Local Programmes *Phil Roberts*
News Editor *Andy Cooper*

# BBC Local Radio

## BBC Local Radio

bbc.co.uk/england

BBC local radio stations have their own newsroom which supplies local bulletins and national news service. Many have specialist producers. A comprehensive list of programmes for each is unavailable and would soon be out of date. For general information on programming visit the BBC website and select the local station you want or contact the relevant station direct.

## BBC Berkshire

PO Box 104.4, Reading RG4 8FH

T 0118 946 4200 (newsroom)    F 0118 946 4555

bbc.co.uk/berkshire

Editor *Marianne Bell*

## BBC Bristol

PO Box 194, Bristol BS99 7QT

T 0117 974 1111    F 0117 923 8323

radio.bristol@bbc.co.uk

bbc.co.uk/bristol

Managing Editor  *Tim Pemberton*

Wide range of feature material used.

## BBC Cambridgeshire

104 Hills Road, Cambridge CB2 1LQ

T 01223 259696

cambridgeshire@bbc.co.uk

bbc.co.uk/cambridgeshire/local_radio

Managing Editor *Jason Horton*

Commenced broadcasting in May 1982.

## BBC Cornwall

Phoenix Wharf, Truro TR1 1UA

T 01872 275421

radio.cornwall@bbc.co.uk

bbc.co.uk/cornwall/local_radio

Managing Editor *Pauline Causey*

On air from 1983 serving Cornwall and the Isles of Scilly. Broadcasts 110 hours of local programmes weekly including news, phone-ins and debate.

## BBC Coventry & Warwickshire

Priory Place, Coventry CV1 5SQ

T 024 7655 1000

coventry@bbc.co.uk

warwickshire@bbc.co.uk

bbc.co.uk/coventrywarwickshire

Station Manager *David Clargo*

Commenced broadcasting in January 1990 as CWR. News, current affairs, public service information and community involvement, relevant to its broadcast area. Details of any special events are available on the website.

## BBC Cumbria

Annetwell Street, Carlisle CA3 8BB

T 01228 592444

radio.cumbria@bbc.co.uk

bbc.co.uk/cumbria

Editor *Nigel Dyson*

Occasional opportunities for plays and short stories are advertised on-air.

## BBC Radio Cymru

Broadcasting House, Llandaf, Caerdydd CF5 2YQ

T 029 20 322018    F 029 20 322473

radio.cymru@bbc.co.uk

bbc.co.uk/radiocymru

Editor *Siân Gwynedd*

Welsh language station serving Welsh-speaking communities. Rich mix of news, music, chat, sport, features, current affairs, drama, comedy and religion. Writing opportunities: DRAMA During 2010–11, up to 18 new radio plays/monologues; COMEDY Occasional opportunities during the year (for details of independent commissions see www.bbc.co.uk/wales/commissioning/).

## BBC Derby

PO Box 104.5, Derby DE1 3HL

T 01332 361111

radio.derby@bbc.co.uk

bbc.co.uk/derby

Editor *Simon Cornes*

News, sport, information and entertainment.

## BBC Devon

Broadcasting House, Seymour Road, Mannamead, Plymouth PL3 5BD

T 01752 260323

radio.devon@bbc.co.uk

bbc.co.uk/devon

*Also at:* Walnut Gardens, St David's Hill, Exeter EX4 4DH

Managing Editor *Mark Grinnell*

On air since 1983.

## BBC Essex

PO Box 765, Chelmsford CM2 9XB

T 01245 616000

essex@bbc.co.uk

bbc.co.uk/essex

Managing Editor *Gerald Main*

Programmes are a mix of news, interviews, expert contributors, phone-ins, sport and special interest topics.

## BBC Radio Foyle

8 Northland Road, Londonderry BT48 7GD

T 028 7137 8600    F 028 7137 8638 (news)

radio.foyle@bbc.co.uk

bbc.co.uk/northernireland

Managing Editor *Paul McCauley*
News Producer *Eimear O'Callaghan*

Radio Foyle broadcasts about seven hours of original material a day, seven days a week to the north west of Northern Ireland. Other programmes are transmitted simultaneously with Radio Ulster. The output ranges from news, sport, and current affairs to live music recordings and arts reviews.

## BBC Gloucestershire

London Road, Gloucester GL1 1SW
☏ 01452 308585
radio.gloucestershire@bbc.co.uk
bbc.co.uk/radiogloucestershire
Managing Editor *Mark Hurrell*
Assistant Editor *Mark Jones*

News and information covering the large variety of interests and concerns in Gloucestershire. Leisure, sport and music, plus African Caribbean and Asian interests. Regular book reviews and interviews with local authors in the *Steve Kitchen Show*.

## BBC Guernsey

Broadcasting House, Bulwer Avenue, St Sampsons, Guernsey GY2 4LA
☏ 01481 200600    ☒ 01481 200361
bbcguernsey@bbc.co.uk
bbc.co.uk/guernsey
Managing Editor *Robert Wallace*

Opened with its sister station, BBC Radio Jersey, in March 1982. Broadcasts 80 hours of local programming a week.

## BBC Hereford & Worcester

Hylton Road, Worcester WR2 5WW
☏ 01905 748485
bbchw@bbc.co.uk
worcester@bbc.co.uk
bbc.co.uk/herefordworcester
*Also at:* 43 Broad Street, Hereford HR4 9HH
Managing Editor *James Coghill*

Has an interest in writers/writing with local connections.

## BBC Humberside

Queen's Court, Queen's Gardens, Hull HU1 3RH
☏ 01482 323232
humberside.news@bbc.co.uk (news)
bbc.co.uk/humber
Managing Editor *Simon Pattern*

On air since 1971. BBC Radio Humberside has broadcast short pieces by local writers.

## BBC Jersey

18–21 Parade Road, St Helier, Jersey JE2 3PL
☏ 01534 837200    ☒ 01534 732569
radiojersey@bbc.co.uk
bbc.co.uk/jersey
Managing Editor *Denzil Dudley*
Assistant Editor *Matthew Price*

Local news, current affairs and community items.

## BBC Kent

The Great Hall, Mount Pleasant Road, Tunbridge Wells TN1 1QQ
☏ 01892 670000    ☒ 01892 518279
radio.kent@bbc.co.uk
bbc.co.uk/kent/local_radio
Managing Editor *Paul Leaper*

Occasional commissions are made for local interest documentaries and other one-off programmes.

## BBC Lancashire

20–26 Darwen Street, Blackburn BB2 2EA
☏ 01254 262411    ☒ 01254 680821
radio.lancashire@bbc.co.uk
bbc.co.uk/lancashire/local_radio
Managing Editor *John Clayton*

Journalism-based radio station, interested in interviews with local writers. Contact *Alison Brown*, Daily Programmes Producer, Monday to Friday (alison.brown.01@bbc.co.uk).

## BBC Leeds

2 St Peter's Square, Leeds LS9 8AH
☏ 0113 244 2131    ☒ 0113 224 7316
radio.leeds@bbc.co.uk
bbc.co.uk/leeds/local_radio
Managing Editor *Rozina Breen*

BBC Radio Leeds is the station for West Yorkshire, serving 1.3 million potential listeners since 1968. Broadcasts a mix of news, sport, talk, music and entertainment, 24 hours a day.

## BBC Leicester

9 St Nicholas Place, Leicester LE1 5LB
☏ 0116 251 6688
leicesternews@bbc.co.uk (news)
bbc.co.uk/leicester
Managing Editor *Kate Squire*

The first local station in Britain. Occasional interviews with local authors.

## BBC Lincolnshire

PO Box 219, Newport, Lincoln LN1 3XY
☏ 01522 511411
radio.lincolnshire@bbc.co.uk
bbc.co.uk/lincolnshire
Managing Editor *Charlie Partridge*

Unsolicited material considered only if locally relevant. Maximum 1,000 words: straight narrative preferred, ideally with a topical content.

Includes FREE online access to **www.thewritershandbook.com**

## BBC London

2nd Floor, Egton Wing, Broadcasting House,
Portland Place, London W1A 1AA
Ⓣ 020 7224 2424    Ⓕ 020 7208 9661 (news)
londonnews@bbc.co.uk (news)
bbc.co.uk/london
Managing Editor *David Robey*

Formerly Greater London Radio (GLR), launched
in 1988, BBC London 94.9 broadcasts news,
information, travel bulletins, sport and music to
Greater London and the Home Counties.

## BBC Manchester

New Broadcasting House, Oxford Road,
Manchester M60 1SJ
Ⓣ 0161 200 2000
radiomanchester@bbc.co.uk
bbc.co.uk/manchester/local_radio
Managing Editor *John Ryan*
News Editor *Mark Elliot*

One of the largest of the BBC local radio stations,
broadcasting news, current affairs, phone-ins,
help, advice and sport.

## BBC Merseyside

PO Box 95.8, Liverpool L69 1ZJ
Ⓣ 0151 708 5500    Ⓕ 0151 794 0988
radio.merseyside@bbc.co.uk
bbc.co.uk/liverpool
Managing Editor *Mick Ord*

Launched in 1967. BBC Radio Merseyside's
editorial area covers the whole of Merseyside
(Liverpool, Bootle, Birkenhead, Southport, St
Helens, etc.) much of North Cheshire (Chester,
Warrington, Runcorn, Widnes, Ellesmere Port)
and West Lancashire (Skelmersdale, Orrell,
Burscough, Ormskirk). *Saturday Book Club* is
broadcast at 8.05 a.m. in Andy Ball's programme.

## BBC Newcastle

Broadcasting Centre, Barrack Road, Newcastle
upon Tyne NE99 1RN
Ⓣ 0191 232 4141
bbcnewcastle.news@bbc.co.uk
bbc.co.uk/tyne/local_radio bbc.co.uk/wear/
local_radio
Managing Editor *Andrew Robson*
Senior Producer (Programmes) *Sarah Miller*

Commenced broadcasting in January 1971. One of
the BBC's big city radio stations in England, Radio
Newcastle reaches an audience of 233,000.

## BBC Norfolk

The Forum, Millennium Plain, Norwich NR2 1BH
Ⓣ 01603 619331    Ⓕ 01603 284488
radionorfolk@bbc.co.uk
bbc.co.uk/radionorfolk
Managing Editor *David Clayton*

Good local ideas and material welcome for
features *if* directly related to Norfolk.

## BBC Northampton

Broadcasting House, Abington Street,
Northampton NN1 2BH
Ⓣ 01604 239100    Ⓕ 01604 230709
northampton@bbc.co.uk
bbc.co.uk/northamptonshire
Managing Editor *Louise Dow*

Books of local interest are regularly featured and
authors and poets are interviewed on merit.

## BBC Nottingham

London Road, Nottingham NG2 4UU
Ⓣ 0115 955 0500    Ⓕ 0115 902 1984
radio.nottingham@bbc.co.uk
bbc.co.uk/nottingham/local_radio
Managing Editor *Mike Bettison*

Rarely broadcasts scripted pieces of any kind but
interviews with authors form a regular part of the
station's output.

## BBC Radio Orkney

Castle Street, Kirkwall KW15 1DF
Ⓣ 01856 873939
radio.orkney@bbc.co.uk
Senior Producer *Dave Gray*

Regular programmes include *Around Orkney*
(news magazine programme), *Bruck* (swapshop
and general features) and *Orky-Ology*
(archaeology magazine). As a BBC Community
station, all Radio Orkney's programme material is
generated from within the Orkney Islands.

## BBC Oxford

269 Banbury Road, Oxford OX2 7DW
Ⓣ 08459 311444    Ⓕ 08459 311555
oxford@bbc.co.uk
bbc.co.uk/oxford/local_radio
Executive Editor *Steve Taschini*

Restored to its original name in 2000 having
been merged with BBC Radio Berkshire in
1995 to create BBC Thames Valley. The station
occasionally carries interviews with local authors
in Jo Thoenes' afternoon show.

## BBC Radio Scotland (Dumfries)

Elmbank, Lovers Walk, Dumfries DG1 1NZ
Ⓣ 01387 268008
dumfries@bbc.co.uk
Senior Producer *Willie Johnston*

Previously Radio Solway. The station mainly
outputs news bulletins (four daily) although it
has become more of a production centre with
programmes being made for Radio Scotland as
well as BBC Radio 2 and 5 Live. Freelancers of
a high standard, familiar with Radio Scotland,
should contact the producer.

## BBC Radio Scotland (Selkirk)

Unit 1, Ettrick Riverside, Dunsdale Road, Selkirk
TD7 5EB

℡ 01750 724567  ℻ 01750 724555

selkirk.news@bbc.co.uk

Senior Broadcaster *Cameron Buttle*

Formerly BBC Radio Tweed. Local news bulletins only.

## BBC Sheffield

54 Shoreham Street, Sheffield S1 4RS

℡ 0114 273 1177  ℻ 0114 267 5454

radio.sheffield@bbc.co.uk

bbc.co.uk/southyorkshire

Managing Editor *Gary Keown*

Assistant Editor *Mike Woodcock*

Writer interviews, writing-related topics and readings on the Rony Robinson show at 11.00 am.

## BBC Radio Shetland

Pitt Lane, Lerwick ZE1 0DW

℡ 01595 694747  ℻ 01595 694307

radio.shetland@bbc.co.uk

Senior Producer *John Johnson*

Regular programmes include *Good Evening Shetland*. An occasional books programme highlights the activities of local writers and writers' groups.

## BBC Shropshire

2–4 Boscobel Drive, Shrewsbury SY1 3TT

℡ 01743 248484

radio.shropshire@bbc.co.uk

bbc.co.uk/shropshire

Managing Editor *Tim Beech*

On air since 1985. Unsolicited literary material rarely used, and then only if locally relevant.

## BBC Solent

Broadcasting House, Havelock Road,
Southampton SO14 7PU

℡ 023 8063 1311

radiosolentnews@bbc.co.uk

bbc.co.uk/radiosolent

Managing Editor *Chris Carnegy*

Broadcasting since 1970.

## BBC Somerset

Broadcasting House, Park Street, Taunton
TA1 4DA

℡ 01823 323956

somerset@bbc.co.uk

bbc.co.uk/somerset

Managing Editor *Jess Rudkin*

Informal, speech-based programming with strong news and current affairs output and regular local interest features, including local writing. Poetry and short stories on the breakfast-time *Adam Thomas Programme* and Emma Britton's mid-morning programme; also arts features on Elise Rayner's *Somerset's Drive* at 5.00 pm.

## BBC Stoke

Cheapside, Hanley, Stoke on Trent ST1 1JJ

℡ 01782 208080

radio.stoke@bbc.co.uk

bbc.co.uk/stoke

Managing Editor *Sue Owen*

On air since 1968, one of the first eight 'experimental' BBC stations. Emphasis on news, current affairs and local topics. Music represents one fifth of total output. Unsolicited material of local interest is welcome – send to News Editor, *Joel Moors*.

## BBC Suffolk

Broadcasting House, St Matthew's Street, Ipswich
IP1 3EP

℡ 01473 250000

radiosuffolk@bbc.co.uk

bbc.co.uk/suffolk

Managing Editor *Peter Cook*

Strongly locally speech-based, dealing with news, current affairs, community issues, the arts, agriculture, commerce, travel, sport and leisure. Programmes sometimes carry interviews with writers.

## BBC Surrey

Broadcasting Centre, Guildford GU2 7AP

℡ 01483 306306/734312 (news)

surrey@bbc.co.uk

bbc.co.uk/surrey

Managing Editor *Nicci Holliday*

Formerly part of BBC Southern Counties Radio.

## BBC Sussex

Broadcasting House, 40–42 Queens Road,
Brighton BN1 3XB

℡ 01273 320400

sussexnews@bbc.co.uk

bbc.co.uk/sussex

Managing Editor *Nicci Holliday*

Formerly part of BBC Southern Counties Radio.

## BBC Tees

Broadcasting House, Newport Road,
Middlesbrough TS1 5DG

℡ 01642 225211  ℻ 01642 211356

tees.news@bbc.co.uk

bbc.co.uk/tees/local_radio

Managing Editor *Matthew Barraclough*

Assistant Editor *Ben Thomas*

Material used is local to Teesside, Co. Durham and North Yorkshire. News, current affairs and features about life in the north east.

## BBC Three Counties Radio

1 Hastings Street, Luton LU1 5XL
☎ 01582 637400 ☏ 01582 401467
3CR@bbc.co.uk
bbc.co.uk/threecounties/local_radio
Managing Editor *Mark Norman*

Covers Bedfordshire, Hertfordshire and
Buckinghamshire. Encourages freelance
contributions from the community across a wide
range of radio output, including interview and
feature material. Interested in local history topics
(five minutes maximum).

## BBC Radio Ulster

Broadcasting House, Ormeau Avenue, Belfast
BT2 8HQ
☎ 028 9033 8000
Head of Radio Ulster *Susan Lovell*

Programmes broadcast from 6.30 am to 1.00
am weekdays and from 6.55 am to 1.00 am at
weekends. Programmes include: *Good Morning
Ulster*; *The Stephen Nolan Show*; *Gerry Anderson*;
*Talk Back*; *Evening Extra*; *On Your Behalf*; *Your
Place and Mine*; *Sunday Sequence* and *Saturday
Magazine*. Comedy, documentary, specialist
music and community programming are also
included.

## BBC Radio Wales

Broadcasting House, Llandaff, Cardiff CF5 2YQ
☎ 029 2032 2000 ☏ 029 2032 2674
radio.wales@bbc.co.uk
bbc.co.uk/radiowales
Editor *Sally Collins*
Editor, Radio Wales News *Matthew Evans*

Broadcasts news on the hour and half hour
throughout weekday daytime programmes;
hourly bulletins in the evenings and at weekends.
Programmes include *Good Morning Wales* and
*The Richard Evans Programme*.

## BBC Wiltshire

BBC Broadcasting House, 56–58 Prospect Place,
Swindon SN1 3RW
☎ 01793 513626
bbcwiltshire@bbc.co.uk
bbc.co.uk/wiltshire
Managing Editor *Rose Aston*

Regular programmes include Mark O'Donnell's
mid-morning show (reviews and author
interviews).

## BBC WM

The Mailbox, Birmingham B1 1AY
☎ 0121 567 6000
bbcwm@bbc.co.uk
bbc.co.uk/radiowm
Managing Editor *Keith Beech*

Commenced broadcasting in November 1970 as
BBC Birmingham and has won four gold Sony
awards in recent years. Speech-based station
broadcasting to the West Midlands, South
Staffordshire and North Worcestershire.

## BBC York

20 Bootham Row, York YO30 7BR
☎ 01904 641351
northyorkshire.news@bbc.co.u (news)
bbc.co.uk/northyorkshire
Editor *Sarah Drummond*

Books of local interest are sometimes previewed.

# Independent Television

## Channel 4

124 Horseferry Road, London SW1P 2TX
☎ 020 7396 4444
www.channel4.com
Chief Executive *David Abraham*
Director of Television and Content *Kevin Lygo*
Head of Channel 4 Commissioning
*Julian Bellamy*
Head of More4 & Head of Documentaries
*Hamish Mykura*
Head of E4 *Angela Jain*
Controller of Film and Drama *Tessa Ross*

Channel 4 launched in 1982 to broadcast
programmes that catered to audiences not
served by the BBC or ITV. Over the years, its
purpose has evolved and grown and today
Channel 4 provides online content as well as TV
programming while continuing to expand into
new areas. There is also a range of 4 channels
reaching a variety of audiences including E4,
More4, Film4, 4Music, and Channel 4 +1. See
channel4.com/nexton4 for more details.

## Channel Television

Television Centre, La Pouquelaye, St Helier, Jersey
JE1 3ZD
☎ 01534 816816 ☏ 01534 816817
www.channelonline.tv
*Also at:* Television House, Bulwer Avenue,
St Sampsons, Guernsey GY2 4LA ☎ 01481 241888
☏ 01481 241878
Managing Director, Broadcast *Karen Rankine*
Managing Director, Commercial *Mike Elsey*
Director of Resources & Transmission
*Kevin Banner*

Channel Television is the independent television
broadcaster to the Channel Islands, serving
150,000 residents, most of whom live on the main
islands, Jersey, Guernsey, Alderney and Sark. The
station has a weekly reach of more than 60% with
local programmes (in the region of five and a half
hours each week) at the heart of the ITV service
to the islands.

## Five

22 Long Acre, London WC2E 9LY
☎ 020 7550 5555
www.five.tv
CEO *Dawn Airey*
Channel Controller *Richard Woolfe*
Controller of Children's Programmes *Nick Wilson*
Senior Programme Controller (News & Current Affairs) *Chris Shaw*

Channel 5 Broadcasting Ltd won the franchise for Britain's third commercial terrestrial television station in 1995 and came on air at the end of March 1997. Regular programmes include *The Wright Stuff* (weekday morning chat show), *Live From Studio Five* (talk show), *CSI: NY* (drama).

## GMTV

*Head Office:* The London Television Centre, Upper Ground, London SE1 9TT
☎ 020 7827 7000   ℻ 020 7827 7001
www.gm.tv
Chairman *Clive Jones*
Chief Operating Officer *Clive Crouch*
Managing Editor *Neil Thompson*
Deputy Editor *Malcolm Douglas*
Operations Director *Di Holmes*
Head of Entertainment *Corinne Bishop*
Deputy Head of Entertainment *Amy Vosburgh*

Fully owned by ITV plc following a buyout of Disney's stake in November 2009. GMTV took over from TV-am on 1 January 1993, with live programming from 6.00 am to 9.25 am. Regular news bulletins, current affairs, topical features, showbiz and lifestyle. Also competitions, travel and weather reports. The week's news and political issues are reviewed on Sundays, followed by children's programming. Launched its digital service, GMTV2, in January 1999, with daily broadcasts from 6.00 am to 9.25 am. GMTV2 moved from ITV2 to ITV4 in March 2008 and is now known as GMTV Digital. Children's programming with some simulcast with GMTV1.

## ITN (Independent Television News Ltd)

200 Gray's Inn Road, London WC1X 8XZ
☎ 020 7833 3000   ℻ 020 7430 4868
contact@itn.co.uk
www.itn.co.uk
Editor-in-Chief, ITV News *David Mannion*
Editor, Channel 4 News *Jim Gray*

Provider of the main national and international news for ITV and Channel 4 and radio news for IRN. Programmes on ITV: *Lunchtime News*; *London Today*; *Evening News*; *London Tonight*; *News at Ten*, plus regular news summaries and three programmes a day at weekends. Programmes on Channel 4 include the in-depth news analysis programmes *Channel 4 News at Noon, Channel 4 News* and *More 4 News*.

## ITV plc

200 Gray's Inn Road, London WC1X 8HF
☎ 020 7843 8000
www.itvplc.com
www.itv.com
Executive Chairman *Archie Norman*
Director of Television *Peter Fincham*

See individual listings below for ITV Anglia, ITV Border, ITV Central, ITV Granada, ITV London, ITV Meridian, ITV Tyne Tees, ITV Wales, ITV West, ITV Westcountry and ITV Yorkshire.

## ITV Network

ITV1 comprises 15 regional networks with ITV plc holding the English and Welsh regional ITV licences as well as the digital channels, ITV2, ITV3, ITV4, CiTV and ITVPlay. The remaining ITV licences belong to: SMG plc (Grampian and Scottish), Ulster Television plc (UTV) and Channel Television. Granada is the production arm of ITV plc.
ITV COMMISSIONING
Commissioning & Production:
www.itv.com/commissioning
Drama: www.itv.com/dramacommissioning

## ITV Anglia Television

Anglia House, Norwich NR1 3JG
☎ 0844 881 6900
www.itvlocal.com/anglia
Regional Director *Neil Thompson*
Head of News *David Jennings*

Broadcasting to the east of England, Anglia Television is a major producer of programmes for the ITV network. Its network factual department currently boasts the largest portfolio of North American documentary commissions for a UK-based production company.

## ITV Border Television

Television Centre, Durranhill, Carlisle CA1 3NT
☎ 0844 881 5850   ℻ 01228 541384
www.itvlocal.com/border
Regional Director *Neil Thompson*
Head of News *Catherine Houlihan*

Border's region covers three different cultures – English, Scottish and Manx. Programming concentrates on documentaries rather than drama. Most scripts are supplied in-house but occasionally there are commissions. Apart from notes, writers should not submit written work until their ideas have been fully discussed.

## ITV Central

Gas Street, Birmingham B1 2JT
☎ 0844 881 4000/0808 100 7888 (newsdesk)
www.itvlocal.com/central
Regional Director *Mike Blair*
Head of News *Alan Rook*

Formed in 2004 by the merger of Carlton and Granada, ITV Central is the largest of the UK commercial television networks outside of London.

## ITV Granada

Quay Street, Manchester M60 9EA
℡ 0161 832 7211    🖷 0161 827 2180
granada@itvlocal.com
www.itvlocal.com/central
Regional Director *Mike Blair*
Head of News *Richard Frediani*
Controller of Drama *Kieran Roberts*
Director of Production *Claire Poyser*

The longest continuous ITC licence holder, broadcasting for over 50 years. Programmes include *Coronation Street* and *Disappearing World*. See also **ITV Studios** under *Film, TV and Radio Producers*.

## ITV London

London Television Centre, Upper Ground, London SE1 9LT    🖷 020 7261 8163
newsdesk@itvlondon.com
www.itvlocal.com/london
Regional Director *Mark Southgate*

ITV London, formed in February 2004 by the merger of Carlton and LWT, broadcasts to the Greater London region, extending into the counties on its border.

## ITV Meridian

Solent Business Park, Whiteley PO15 7PA
℡ 0844 881 2000
www.itvlocal.com/meridian
Regional Director *Mark Southgate*
Head of News *Robin Britton*

Serves viewers across the South and South East region.

## ITV Thames Valley

Solent Business Park, Whiteley PO15 7PA
℡ 0844 881 2000
Managing Director *Mark Southgate*
Head of News *Robin Britton*

Formed in December 2004 from the two former ITV sub-regions, Central South and Meridian West, ITV Thames Valley is ITV's newest regional news service. It covers an area stretching from Banbury in the north to Winchester in the south, from Swindon in the west to Bracknell in the east.

## ITV Tyne Tees

Television House, The Watermark, Gateshead NE11 9SZ
℡ 0844 881 5000    🖷 0844 881 5010
Regional Director *Neil Thompson*
Head of News *Catherine Houlihan*
Managing Director, ITV Signpost *Malcolm Wright*

Programming covers politics, news and current affairs and sport. Regular programmes include *North East Tonight with Pam Royle and Ian Payne* and *Around the House* (politics).

## ITV Wales

The Television Centre, Culverhouse Cross, Cardiff CF5 6XJ
℡ 0844 881 0100    🖷 029 2059 7183
wales@itvlocal.com
www.itvlocal.com/wales
Regional Director *Mike Blair*
Head of News & Programmes *Phil Henfrey*
Associate Editor *Jonathan Hill*

Fomerly HTV Wales. Programmes include the consumer affairs magazine, *The Ferret* and current affairs series, *Wales this Week*. The company also makes Welsh language programming for S4C, including the series voted the most popular with viewers on the Welsh channel – *Cefn Gwlad* and the current affairs series, *Y Byd ar Bedwar*.

## ITV West

470 Bath Road, Bristol BS4 3HG
℡ 0844 881 2345/2308 (newsdesk)
🖷 0844 881 2346
itvwestnews@itv.com (newsdesk)
www.itvlocal.com/west
Regional Director *Mark Southgate*
Head of News *Liz Hannam*

Formerly HTV. Broadcasts to over two million viewers in the West of England.

## ITV West Country

Langage Science Park, Western Wood Way, Plymouth PL7 5BQ
℡ 0844 881 4900/4800 (newsdesk)
🖷 0844 881 4901
news@westcountry.co.uk (newsdesk)
www.itvlocal.com/westcountry
Regional Director *Mark Southgate*
Head of News *Liz Hannam*

Broadcasts to Cornwall, Devon, Somerset and west Dorset. Produces regional news, current affairs and features programmes.

## ITV Yorkshire

The Television Centre, Leeds LS3 1JS
℡ 0113 222 7000
www.itvlocal.com/yorkshire
Regional Director *Neil Thompson*
Head of News *Will Venters*
Controller of Drama, Leeds *Keith Richardson*
Controller of Comedy Drama and Drama Features *David Reynolds*

Drama series, comedy drama, single drama, adaptations and long-running series like *Emmerdale* and *Heartbeat*. Always looking for strong writing in these areas, but prefers to find it through an agent. Documentary/current affairs material tends to be supplied by producers; opportunities in these areas are rare but adaptations of published work as a documentary subject are considered. In theory, opportunity exists within series, episode material but the best approach is through a good agent.

## S4C

Parc Ty Glas, Llanishen, Cardiff CF14 5DU
☎ 029 2074 7444    🖷 029 2075 4444
s4c@s4c.co.uk
www.s4c.co.uk
Chief Executive *Iona Jones*
Director of Commissioning *Rhian Gibson*

The Welsh 4th Channel, established by the Broadcasting Act 1980, is responsible for a schedule of Welsh and English programmes on the Fourth Channel in Wales. Known as S4C, the analogue service is made up of about 34 hours per week of Welsh language programmes and more than 85 hours of English language output from Channel 4. S4C digidol broadcasts in Welsh exclusively for 80 hours per week. Ten hours a week of the Welsh programmes are provided by the BBC; the remainder are commissioned from ITV1 Wales and independent producers. Drama, comedy, sport, music and documentary are all part of S4C's programming. Commissioning guidelines can be viewed on www.s4c.co.uk/production

## stv

Pacific Quay, Glasgow G51 1PQ
☎ 0141 300 3000    🖷 0141 300 3030
www.stv.tv
*Also at:* Television Centre,
Craigshaw Business Park, West Tullos,
Aberdeen AB12 3QH
☎ 01224 848848 🖷 01224 848800
Managing Director, Broadcasting, stv *Bobby Hain*
Head of News & Current Affairs (Scotland) *Gordon Macmillan*

stv is the ITV (Channel 3) licence holder for the north and central regions of Scotland. The company produces a range of television programmes, covering current affairs, sport, entertainment, documentary and drama, as well as flagship news programmes *Scotland Today* and *North Tonight.*

## Teletext Ltd

10 Chiswick Park, 566 Chiswick High Road, London W4 5TS
☎ 020 8323 3000    🖷 020 8323 5204
www.teletext.co.uk
Managing Director *Dr Mike Stewart*
Editor-in-Chief *John Sage*

Teletext operates the digital TV text service on Freeview (behind channels ITV1, 2, 3 and 4, Channel 4, E4 and More4), on Sky (behind all the ITV channels and Channel 4) and on the analogue TV service behind ITV1, Channel 4 and Five. In addition, Teletext's editorial content and holiday offers can be accessed online or via mobile services.

## UTV

Ormeau Road, Belfast BT7 1EB
☎ 028 9032 8122    🖷 028 9024 6695
info@u.tv
www.u.tv
Managing Director, Television *Michael Wilson*
Head of News, Current Affairs and Sport *Rob Morrison*

Regular programmes on news and current affairs, sport and entertainment.

# Cable and Satellite Television

## British Sky Broadcasting Ltd (BSkyB)

Grant Way, Isleworth TW7 5QD
☎ 020 7705 3000    🖷 020 7705 3030
www.sky.com
Chief Executive/Executive Director *Jeremy Darroch*
Managing Director, Entertainment & News *Sophie Turner Laing*
Head of Sky News *John Ryley*
Managing Director, Sky Sports *Barney Francis*

Launched 1989. British Sky Broadcasting gives around 23 million viewers (in 9.5 million households) access to movies, news, entertainment and sports channels, and interactive services on Sky digital. In addition, all customers can receive more than 330 free-to-air television and radio channels and services.

SKY CHANNELS
**Sky News**
Launched in 1989 as Europe's first rolling 24-hours news service. Accessible on TV, through the web, on mobile phones, on big screens, at railway stations and on radio.
**Sky 1/Sky 1 HD/Sky 2/Sky 3**
The best in entertainment from blockbuster acquisitions such as *24* and *Lost* to breakthrough commissions such as *Ross Kemp on Gangs* and *Gladiators.*
**Sky Sports 1/Sky Sports 2/Sky Sports 3/Sky Sports News/Sky Sports Extra/Sky Sports HD1, HD2 & HD3**
Around 40,000 hours of sport are broadcast every year across five Sky Sports channels and three HD channels. Sky Sports 1, 2 and 3 are devoted to live events, support programmes and in-depth sports coverage seven days a week.

Sky Sports News provides sports news and the latest results and information 24 hours a day. Sky Sports Extra carries additional sports programming including the award-winning live interactive coverage.

**Sky Movies**

A choice of movies from the latest Hollywood blockbusters to cult classics.

**Sky Arts 1/Sky Arts 2/Sky Arts 1 HD/Sky Arts 2 HD**

The only channels in the UK and Ireland dedicated to the arts.

**Sky Real Lives**

Extraordinary stories of ordinary people, 24 hours a day.

JOINT VENTURES

National Geographic; Nickelodeon; Nick Jr; The History Channel; Paramount Comedy Channel; MUTV; Adventure One; The Biography Channel; Attheraces; Chelsea Digital Media.

## CNBC Europe

10 Fleet Place, London EC4M 7QS
☎ 020 7653 9300    📠 020 7653 5956
www.cnbc.com
CEO/President *Mick Buckley*

A service of NBC Universal. 24-hour business and financial news service. Programmes include *Capital Connection; Worldwide Exchange; Strictly Money; European Closing Bell; Squawk Box Europe*.

## CNN (Cable News Network)

Turner House, 16 Great Marlborough Street, London W1P 1DF
☎ 020 7693 1000    📠 020 7693 1552
edition.cnn.com
Managing Editor *Deborah Rayner*

CNN, the leading global 24-hour news network, is available to one billion people worldwide via the 26 CNN branded TV, internet, radio and mobile services produced by CNN News Group, a Time Warner company. CNN has major production centres in Atlanta, New York, Los Angeles, London, Hong Kong and Mexico City. The London bureau, the largest outside the USA, is CNN's European headquarters and produces over 50 hours of programming per week. Live business and news programmes, including *Business International; World News* and *World Business Today*.

## MTV Networks Europe

180 Oxford Street, London W1D 1DS
☎ 020 7478 6000
www.mtvne.com

Established 1987. Europe's 24-hour music and youth entertainment channel, available on cable,

via satellite and digitally. Transmitted from London in English across Europe.

## Travel Channel International

64 Newman Street, London W1T 3EF
☎ 020 7636 5401    📠 020 7636 6424
www.travelchannel.co.uk

Launched in 1994, Travel Channel broadcasts in 18 languages to 117 countries across Europe, the Middle East and Africa. Programmes cover every corner of the globe with backpacking adventures, culinary quests, eco-friendly holidays and exclusive, luxury getaways. The channel's core programming includes *Globe Trekker; Journey into Wine; Great Scenic Railway Journeys* and *Essential* guides.

# National Commercial Radio

## Absolute Radio

1 Golden Square, London W1F 9DJ
☎ 020 7434 1215    📠 020 7434 1197
www.absoluteradio.co.uk
Group Programme Director *Clive Dickens*

Previously known as Virgin Radio. A music and entertainment station launched in 1973.

## Classic FM

30 Leicester Square, London WC2H 7LA
☎ 020 7343 9000    📠 020 7766 6100
www.classicfm.com
Managing Director *Darren Henley*

Classic FM, Britain's largest national commercial radio station, started broadcasting in September 1992. Plays accessible classical music 24 hours a day and broadcasts news, weather, travel, business information, political/celebrity/general interest talks, features and interviews. Winner of the 'Station of the Year' Sony Award in 1993, 2000 and 2007.

## Digital One

33–4 Alfred Place, London WC1E 7DP
☎ 020 7299 8670    📠 020 7299 8671
info@digitalone.co.uk
www.ukdigitalradio.com

The UK's national commercial digital radio multiplex operator. Owned by Arqiva, Digital One began broadcasting on 15 November 1999.

## talkSPORT

18 Hatfields, London SE1 8DJ
☎ 020 7959 7800
www.talksport.net
Programme Director *Moz Dee*

Commenced broadcasting in February 1995 as Talk Radio UK. Re-launched January 2000 as TalkSport, the UK's first sports radio station. Acquired by Ulster Television in 2005. Broadcasts

24 hours a day. News items can be emailed via the website.

## Virgin Radio
▷ Absolute Radio

# Independent Local Radio

## 95.8 Capital FM
30 Leicester Square, London WC2H 7LA
📞 020 7766 6000  📠 020 7766 6012
www.capitalradio.com
Programme Director *Paul Jackson*

Commenced broadcasting in October 1973 as the country's second commercial radio station (the first being LBC, launched a week earlier). Europe's largest commercial radio station.

## Central FM
201–203 High Street, Falkirk FK1 1DU
📞 01324 611164
email@centralfm.co.uk
www.centralfm.co.uk
Programme Controller *Gary Muircroft*

Broadcasts music, sport and local news to central Scotland, 24 hours a day.

## Clyde 1/Clyde 2
Radio Clyde, 3 South Avenue, Clydebank Business Park, Clydebank G81 2RX
📞 0141 565 2200
www.clyde1.com
www.clyde2.com
Owner *Bauer Radio*
Station Director *Tracey McNellan*

Programmes usually originate in-house or by commission. All documentary material is made in-house. Good local news items always considered.

## Forth One/Forth 2
Forth House, Forth Street, Edinburgh EH1 3LE
📞 0131 556 9255  📠 0131 557 4645 (news)
www.forthone.com
www.forth2.com
Programme Director *Duncan Campbell*
Programme Producer, Forth 2 *Moira Millar*
News Editor *Paul Robertson*
Book Reviewer *Lesley Fraser-Taylor*

News stories welcome from freelancers. Music-based programming.

## Isle of Wight Radio
Dodnor Park, Newport PO30 5XE
📞 01983 822557/821777 (news)  📠 01983 822109
news@iwradio.co.uk
www.iwradio.co.uk
Station Manager *Claire Willis*
Programme Controller *Paul Topping*

Independently-owned Isle of Wight Radio is the island's only radio station broadcasting local news, music and general entertainment including phone-ins and interview based shows. Music, television, film, popular culture are the main areas of interest. *John Hannam Meets* – long-running interview programme – on Sundays.

## kmfm
*Head Office:* Express House, 34–36 North Street, Ashford TN24 8JR
📞 01233 895825
www.kmfm.co.uk
KM Group Head of Radio *Steve Fountain*

A wide range of music programming plus news, views and local interest. Part of the KM Group.

## LBC Radio Ltd
30 Leicester Square, London WC2H 7LA
📞 0845 6066 0973
www.lbc.co.uk
Programme Director & Editorial Director *Jonathan Richards*

LBC 97.3 FM is a talk-based station broadcasting 24 hours a day, providing entertainment, interviews, celebrity guests, music, chat shows, local interest, news and sport. LBC News 1152 AM, the sister station of LBC 97.3 FM, provides rolling news (7.00 am to 7.15 pm Mon. to Fri.; 7.00 am to 5.00 pm Sat. and Sun.).

## NorthSound Radio
Abbotswell Road, Aberdeen AB12 3AJ
📞 01224 337000
www.northsound1.com
www.northsound2.com
Managing Director *Luke McCullough*
Programme Director *Chris Thomson*

Features and music programmes 24 hours a day.

## Premier Radio
22 Chapter Street, London SW1P 4NP
📞 020 7316 1300
response@premier.org.uk
www.premier.org.uk
Chief Executive Officer *Peter Kerridge*

Broadcasts programmes that reflect the beliefs and values of the Christian faith, 24 hours a day on 1305, 1413, 1332 MW, Sky digital 0123, Virgin Media 968.

## Radio XL 1296 AM
KMS House, Bradford Street, Birmingham B12 0JD
📞 0121 753 5353  📠 0121 753 3111
Managing Director *Arun Bajaj*

Asian broadcasting for the West Midlands, 24 hours a day. Broadcasts a mixture of music, information, debate and entertainment. Writers should send material for the attention of the admin office.

## Sabras Radio

Radio House, 63 Melton Road, Leicester LE4 6PN
Ⓣ 0116 261 0666    Ⓕ 0116 266 7776
www.sabrasradio.com
Programme Controller *Don Kotak*

Programmes for the Asian community, broadcasting 24 hours a day.

## Spectrum Radio

4 Ingate Place, London SW8 3NS
Ⓣ 020 7627 4433    Ⓕ 020 7622 1953
jogden@spectrumradio.net
www.spectrumradio.net
Station Director *John Ogden*

Programmes for a broad spectrum of ethnic groups in London.

## Sunrise Radio (Yorkshire)

55 Leeds Road, Bradford BD1 5AF
Ⓣ 01274 735043    Ⓕ 01274 728534
www.sunriseradio.fm
Programme Controller, Chief Executive & Chairman *Usha Parmar*

Programmes for the Asian community in West Yorkshire.

## Swansea Sound 1170 MW/96.4 FM The Wave

Victoria Road, Gowerton, Swansea SA4 3AB
Ⓣ 01792 511170 (MW)/ 511964 (FM)
Ⓕ 01792 511171 (MW)/511965 (FM)
www.swanseasound.co.uk
Station Director *Helen Bowden*
Programme Director *Steve Barnes*
News Editor *Emma Grant*

Music-based programming on FM while Swansea Sound is interested in a wide variety of material, though news items must be of local relevance. An explanatory letter, in the first instance, is advisable.

## 96.4 FM The Wave

▷ Swansea Sound 1170 MW

# European TV Companies

## Austria

**ORF(Österreichisher Rundfunk)1/ORF2**
Würzburggasse 30, A–1136 Vienna
☎ 00 43 1 87 878-0
www.orf.at

## Belgium

**RTBF (Radio-Télévision Belge de la Communauté Française) (French language TV & radio)**
Boulevard Auguste Reyers 52, B–1044 Brussels
☎ 00 32 2 737 2111
www.rtbf.be

**Vlaamse Televisie Maatschappij (VTM) (cable)**
Medialaan 1, B–1800 Vilvoorde
☎ 00 32 2 255 3211
www.vtm.be

**VRT (Vlaamse Radio- en Televisieomroep) (Dutch language TV)**
Auguste Reyerslaan 52, B–1043 Brussels
☎ 00 32 2 741 3111
www.vrt.be

## Bosnia and Herzegovina

**BHRT (Radiotelevizija Bosne i Hercegovine)**
Bulevar Mese Selimovica 12, 71000 Sarajevo
www.bht.ba

## Bulgaria

**BNT**
29 San Stefano str., BG–1504 Sofia
☎ 00 359 2 944 4999
www.bnt.bg

## Croatia

**HRT (Hrvatska Televizija)**
Prisavlje 3, CR–10000 Zagreb
☎ 00 385 1 634 3683
www.hrt.hr

## Czech Republic

**CT (Ceska Televize)**
Kavci Hory, CZ–140 70 Prague 4
☎ 00 420 2 6113 7000
www.czech-tv.cz

## Denmark

**DR (Danmarks Radio)**
Emil Holms Kanal 20, DK–0999 Copenhagen C
☎ 00 45 35 20 3040
www.dr.dk

**SBS TV A/S (cable & satellite)**
Mileparken 20A, DK–2740 Skovlunde
☎ 00 45 70 10 1010
www.sbstv.dk

**TV–2 Danmark**
Rugaardsvej 25, DK–5100 Odense C
☎ 00 45 65 91 9191
www.tv2.dk

**TV3 (Viasat) (cable & satellite)**
Wildersgade 8, DK–1408 Copenhagen K
☎ 00 45 77 30 5500
www.viasat.dk

## Estonia

**ETV (Eesti Televisioon)**
Faehlmanni 12, EE–15029 Tallinn
☎ 00 372 628 4100
www.etv.ee

# Finland

### MTV3 Finland
Ilmalantori 2, SF–00033 Helsinki
☎ 00 358 10 300300
www.mtv3.fi

### YLE FST5 (Swedish language TV)
Yleisradio Oy, Radiokatu 5, SF–00024 Helsinki
☎ 00 358 9 14801
www.yle.fi/fst

### YLE TV1/YLE TV2
Yleisradio Oy, Radiokatu 5, SF–00024 Helsinki
☎ 00 358 9 14801
www.yle.fi/tv1
www.yle.fi/tv2

# France

### Arte France (cable & satellite)
8 rue Marceau, F–92785 Issy-les-Moulineaux
Cedex 9
☎ 00 33 1 55 00 77 77
www.artefrance.fr

### Canal+ (pay TV)
25 rue Leblanc, F–75015 Paris
☎ 00 33 1 40 60 39 01
www.canalplus.fr

### France 2/France 3
7 Esplanade Henri de France, F–75907 Paris
Cedex 15
☎ 00 33 1 56 22 42 42 (F2)/30 30 (F3)
www.france2.fr
www.france3.fr

### France 5
10 rue Horace Vernet, F–92136 Issy-les-
Moulineaux
☎ 00 33 1 56 22 91 91
www.france5.fr

### Planète (documentary channel)
48 quai du Pont du Jour, F–92659 Boulogne
☎ 00 33 1 71 10 10 20
www.planete.com

### Voyage (cable & satellite)
241 Boulevard Péreire, F–75838 Paris
☎ 00 33 1 058 05 58 05
www.voyage.fr

# Germany

### ARD – Das Erste
Arnulfstrasse 42, D–80335 Munich
☎ 00 49 89 5900 3344
www.daserste.de

### Arte Germany (cable & satellite)
Postfach 10 02 13, D–76483 Baden-Baden
☎ 00 49 7221 93690
www.arte.de

### DW-TV (Deutsche Welle) (international TV service)
Voltastrasse 6, D–13355 Berlin
☎ 00 49 30 4646 0
www.dw-world.de

### ZDF (Zweites Deutsches Fernsehen)
ZDF–Strasse 1, D–55127 Mainz
☎ 00 49 61 31/701
www.zdf.de

# Greece

### ERT S.A. (Elliniki Radiophonia Tileorassi)
4 Riglis st, GR–10674 Athens
☎ 00 30 2 10 740 7070
www.ert.gr

# Hungary

### DUNA TV (Duna Televízió)
Mészáros utca 48, H–1016 Budapest
☎ 00 36 1 489 1200
www.dunatv.hu

### MTV 1/MTV 2 (Magyar Televízió)
Szabadság tér 17, H–1810 Budapest
☎ 00 36 1 353 3200
www.mtv.hu

# Republic of Ireland

### Radio Telefís Éireann (RTÉ One – RTÉ Two)
Donnybrook, Dublin 4
☎ 00 353 1 208 3111
www.rte.ie

### TG4 (Teilefís na Gaelige) (Irish language TV)
Baile na hAbhann, Co. na Gaillimhe
☎ 00 353 91 505050
www.tg4.ie

# Italy

### RAI (RadioTelevisione Italiana)
Viale Mazzini 14, I–00195 Rome
☎ 00 39 06 3878-1
www.rai.it

### RTI SpA (Reti Televisive Italiane)
Viale Europa 44, I–20093 Cologno Monzese
Milano
☎ 00 39 02 2514-1
www.rete4.com

# Latvia

### LTV1 & LTV7
Zakusalas Embankment 3, LV–1509 Riga
☎ 00 37 6 720 0315
www.ltv.lv

# Lithuania

### LRT (Lietuvos radijas ir televizija)
S. Konarskio 49, LT–03123 Vilnius
☎ 00 370 5 236 3209
www.lrt.lt

# The Netherlands

### AVRO (Algemene Omroep Vereniging)
Postbus 2, NL–1200 JA Hilversum
☎ 00 31 35 671 79 11
www.avro.nl

### EO (Evangelische Omroep) (Evangelical TV)
Postbus 21000, NL–1202 BA Hilversum
☎ 00 31 35 647 4747
www.eo.nl

### IKON (Interkerkelijke Omroep Nederland) (Protestant TV)
Postbus 10009, NL–1201 DA Hilversum
☎ 00 31 35 672 72 72
www.omroep.nl/ikon

### KRO (Katholieke Radio Omroep) (Catholic TV)
PO Box 23000, NL–1202 EA Hilversum
☎ 00 31 35 671 3911
www.kro.nl

### NCRV (Nederlandse Christelijke Radio Vereniging)
Postbus 25000, NL–1202 HB Hilversum
☎ 00 31 35 671 99 11
info.omroep.nl/ncrv

### NOS (Nederlandse Omroep Stichting) – Nederland 1, 2 & 3
Postbus 26600, NL–1202 JT Hilversum
☎ 00 31 35 677 92 22
www.nos.nl

### NPS (Nederlandse Programma Stichting)
Postbus 29000, NL–1202 MA Hilversum
☎ 00 31 35 677 9333
www.nps.nl

### VPRO
Postbus 11, NL–1200 JC Hilversum
☎ 00 31 35 671 2911
www.vpro.nl

# Norway

### NRK (Norsk Rikskringkasting)
Bjørnstjerne Bjørnsonsplass 1, N–0340 Oslo
☎ 00 47 23 04 7000
www.nrk.no

### TV 2 Norge
Postboks 7222, Nøstegatan 72, N–5020 Bergen
☎ 00 47 55 90 8070
www.tv2.no

# Poland

### TVP (Telewizja Polska S.A.)
ul. Jana Pawla Woronicza 17, PL–00 999 Warsaw
☎ 00 48 22 547 8000
www.tvp.com.pl

# Portugal

### RTP 1 & RTP 2 (Radiotelevisão Portuguesa)
Avenida Marechal, Gomes da Costa 37, P–1849–030 Lisbon
☎ 00 351 21 794 7000
www.rtp.pt

### SIC (Sociedade Independente de Comunicacão) (cable TV)
Estrada da Outurela 119, P–2799-526 Carnaxide
☎ 00 351 21 417 9406
www.six.pt

### TVI (Televisão Independente S.A.)
Rua Mário Castelhano 40, Queluz de Baixo, P–2734–502 Barcarena
☎ 00 351 21 434 7500
www.tvi.pt

# Romania

### TVR (The Romanian Television Corporation)
Calea Dorobantilor 191, Sector 1 Bucharest
☎ 00 40 21 319 9112
www.tvr.ro

# Serbia & Montenegro

### RTS
Takovska 10, 11000 Beograda
☎ 00 381 11 321 1000
www.rts.co.yu

# Slovenia

### TV Slovenija
Kolodvorska 2, SL–1550 Ljubljana
☎ 00 386 1 475 2153
www.rtvslo.si

# Spain

## Canal Sur Televisión (Andalucian TV)
Ctra. San Juan de Aznalfarache-Tomares,
E–41920 San Juan de Aznalfarache
☎ 00 34 95 505 4600
www.canalsur.es

## Canal+ Espana (pay TV)
Avenida Artesanos 6, E–28760 Tres Cantos
(Madrid)
☎ 00 34 91 736 7000
www.plus.es

## EiTB-1/EiTB-2 (Basque TV)
Calle Capuchinos de Basurto 2, E–48013 Bilbao
☎ 00 34 94 603 1000
www.eitb.com

## RTVE (Televisión Española y Radio Nacional de España)
Prado del Rey, E–28023 Madrid
☎ 00 349 1 346 4000
www.rtve.es

## TVC (Televisió de Catalunya S.A.) (Catalan TV)
Carrer de la TV 3, E–08970 Saint Joan Despi
☎ 00 34 93 499 9333
www.tvcatalunya.com

## TVV (TVV-Televisio Valenciana) (Valencian TV)
Poligon Accés Ademús s/n, E–46100 Burjassot
(Valencia)
☎ 00 34 96 318 3000
www.rtvv.es

# Sweden

## SVT (Sveriges Television)
Oxenstierngatan 26–34, S–105 10 Stockholm
☎ 00 46 8 784 0000
www.svt.se

## TV4
Tegeluddsvägen 3–5, S–115 79 Stockholm
☎ 00 46 8 459 4000
www.tv4.se

# Switzerland

## RSI (Radiotelevisione svizzera) (Italian language TV)
Casella postale, CH–6903 Lugano
☎ 00 41 91 803 51 11
www.rsi.ch

## RTR (Radio e Televisiun Rumantscha) (Romansch language TV)
Plazza dal teater, Via da Masans 2, CH–7002
Cuira
☎ 00 41 81 255 7575
www.rtr.ch

## Schweizer Fernsehen (German language TV)
Fernsehstrasse 1–4, CH–8052 Zurich
☎ 00 41 1 305 66 11
www.sfdrs.tv

## SRG SSR idée suisse (Swiss Broadcasting Corp.)
Belpstrasse 48, CH-3000 Berne 14
☎ 00 41 31 350 91 11
www.srg-ssr.ch

## TSR (Télévision Suisse Romande) (French language TV)
Quai Ernest Ansermet 20, CH–1211 Geneva 8
☎ 00 41 22 708 20 20
www.tsr.ch

# Film, TV and Radio Producers

## Aardman

Gas Ferry Road, Bristol BS1 6UN
☎ 0117 984 8485   🖷 0117 984 8486
www.aardman.com

Founded 1976. Award-winning animation studio producing feature films and television programmes. Won Oscars for *Creature Comforts* and *Wallace and Gromit*. Other OUTPUT includes *Angry Kid*; *Chicken Run*; *Flushed Away*; *Shaun The Sheep*; *Chop Socky Chooks*; *Timmy Time*. No unsolicited submissions.

## Above The Title

Level 2, 10/11 St Georges Mews, London NW1 8XE
☎ 020 7916 1984   🖷 020 7722 5706
mail@abovethetitle.com
www.abovethetitle.com
CEO *Simon Clegg*
Contact *Vicky Bennett*

Producer of radio drama, comedy, music, arts, documentary and factual programmes. OUTPUT includes *Clive Anderson's Chat Room* and *Till the End of the Day – The Kinks Story* (R2); *Unreliable Evidence*; *The Glittering Prizes*; *The Hitchhiker's Guide to the Galaxy*; *Fame & Fortune* and *Dirk Gently's Holistic Dectective Agency* (R4); *Features Like Mine* and *Iain Burnside* (R3). 'We encourage and support new writing in every way we can.' Ideas welcome from writers who have representation from an agent or those with extensive previous writing credits. Send short synopsis only by email in the first instance (see website for details).

## Absolutely Productions Ltd

Unit 19, 77 Beak Street, London W1F 9DB
☎ 020 7644 5575
www.absolutely.biz

Founded in 1988 by a group of writer/performer/producers including Morwenna Banks, Jack Docherty, Moray Hunter, Pete Baikie, John Sparkes and Gordon Kennedy. Long-established producer for TV, has recently moved into feature films. OUTPUT *Absolutely*; *Welcome to Strathmuir* (comedy pilot for BBC Scotland) and *Baggage* (Radio 4).

## Abstract Images

117 Willoughby House, Barbican, London EC2Y 8BL
☎ 020 7638 5123
productions@abstract-images.co.uk
Contact *Howard Ross*

Television documentary and drama programming. Also theatre productions. OUTPUT includes *Balm in Gilead*; *Road* and *Bent* (all dramas); *God: For & Against* (documentary); *This Is a Man* (drama/doc). New writers should send synopsis in the first instance.

## Acacia Productions Ltd

80 Weston Park, London N8 9TB
☎ 020 8341 9392
acacia@dial.pipex.com
www.acaciaproductions.co.uk
Contact *J. Edward Milner*

Producer of award-winning television and video documentaries; also news reports, video for websites, corporates and programmes for educational charities. Undertakes basic video training and has a video manual in preparation. OUTPUT includes documentary series entitled *Last Plant Standing*; *A Farm in Uganda*; *Montserrat: Under the Volcano*; *Spirit of Trees* (8 programmes); *Vietnam: After the Fire*; *Macroeconomics – the Decision-makers*; *Greening of Thailand*; *A Future for Forests*. No unsolicited scripts.

## Acrobat Television

107 Wellington Road North, Stockport SK4 2LP
☎ 0161 477 9090   🖷 0161 477 9191
info@acrobat-tv.co.uk
www.acrobat-tv.co.uk
Contact *David Hill*

All script genres for broadcast and corporate television, including training and promotional scripts, comedy and drama-based material. OUTPUT includes *Make a Stand* (Jack Dee, Gina Bellman and John Thompson for Video Arts); *The Customer View* (Roy Barraclough for Air Products); *Serious About Waves* series (Peter Hart for the Royal Yachting Association); *Fat Face Night* series (Extreme). No unsolicited mss.

## Actaeon Films Ltd

50 Gracefield Gardens, London SW16 2ST
☏ 020 8769 3339   🖷 0870 134 7980
info@actaeonfilms.com
www.actaeonfilms.com
Producer *Daniel Cormack*
Head of Development *Becky Connell*

Founded 2004. Specializes in short and feature length theatrical motion pictures. OUTPUT includes *Amelia and Michael* (short drama starring Anthony Head). In development: *Golden Apples* (feature-length drama); *The Dead Letters* and *My Brother's Keeper* (both feature-length psychological thrillers). Actaeon Films is a member of the **New Producers Alliance**, **The Script Factory** and the London Filming Partnership. 'We actively encourage new writers and new writing. However, we can only give feedback on scripts with strong potential for developments (all mss are read by a professional reader within four weeks of submission). Scripts returned only if accompanied by s.a.s.e.'

## All Out Productions

50 Copperas Street, Manchester M4 1HS
☏ 0161 834 9955   🖷 0161 834 6978
mail@allout.co.uk
www.allout.co.uk
Contacts *Richard McIlroy, Jo Meek*

Founded 1994. Producer of documentaries, features and current affairs programmes across the radio network including Radio 1, Radio 2, Radio 3, Radio 4, Five Live and BBC World Service. Ideas welcome but not mss. Approach by email.

## Alomo Productions

▷ **FremantleMedia Group Ltd**

## Anglo/Fortunato Films Ltd

170 Popes Lane, London W5 4NJ
☏ 020 8932 7676   🖷 020 8932 7491
anglofortunato@aol.com
Contact *Luciano Celentino*

Film, television and video producer/director of action comedy and psych-thriller drama. No unsolicited mss.

## Arc

Dalintober Hall, 40 Dalintober Street, Glasgow G5 8NW
☏ 0141 420 0909   🖷 0141 429 3723
jr@arccreative.co.uk
www.arccreative.co.uk
Head of Production *John Rocchiccioli*

Produces screen-based content for education, training, sales and marketing, HR and induction. Operates in a variety of areas including health, education, social policy, etc. 'Aims to inform through entertainment.' Unsolicited mss welcome; 'keen to encourage new writing.'

## Arlington Productions Limited

Cippenham Court, Cippenham Lane, Cippenham, Nr Slough SL1 5AU
☏ 01753 516767   🖷 01753 691785

Television producer. Specializes in popular international drama, with *occasional* forays into other areas. 'We have an enviable reputation for encouraging new writers but only accept unsolicited submissions via agents.'

## Art & Training Films Ltd

33 Shakespeare Street, Stratford-upon-Avon CV37 6RN
☏ 01789 294910
andrew.haynes@atf.org.uk
www.atf.org.uk
Contact *Andrew Haynes*

Producer of documentaries, drama, commecials and corporate films and video. Submissions via agents considered.

## The Ashford Entertainment Corporation Ltd

20 The Chase, Coulsdon CR5 2EG
☏ 0844 357 6393   🖷 087 0116 4142
info@ashford-entertainment.co.uk
www.ashford-entertainment.co.uk
Managing Director *Frazer Ashford*

Founded in 1996 by award-winning producer Frazer Ashford to produce character-led documentaries for TV. Examples of work include *Serial Killers*, *Great Little Trains* (winner of RTS Best Regional Documentary) and *Streetlife* (winner of bronze award at the Flagstaff Film Festival). Ashford Entertainment also incorporates The Reel Thing Ltd, suppliers of corporate event management, DVD and video production; also production facilities. Happy to look at any submissions falling within the remit of character-led documentaries. Send hard copy with an s.a.s.e.

## Avalon Television

4a Exmoor Street, London W10 6BD
☏ 020 7598 8000
www.avalonuk.com/tv

Founded in 1993, Avalon Television is a BAFTA-winning production company supplying TV programming to all the British terrestrial channels and most leading satellite channels. Best known for light entertainment programmes such as documentary *Frank Skinner on Frank Skinner* (ITV1) or comedy *Fantasy Football* (BBC2). Now broadening output to include drama, factual and formatted shows. Script, treatment or ideas are welcomed via email.

## Baby Cow Productions
33 Foley Street, London W1W 7TL
☎ 020 7612 3370    🖷 020 7612 3352
www.babycow.co.uk

Founded in 1989 by Steve Coogan and Henry Normal. Specializes in cutting-edge comedy for television. Has divisions for animation, radio and films. OUTPUT *Marion and Geoff*; *Human Remains*; *The Sketch Show*; *24 Hour Party People*; *I'm Alan Patridge*; *Nighty Night*; *The Mighty Boosh*; *Dating Ray Fenwick*. No unsolicited material.

## BBC Films
Room 6023, Television Centre, Wood Lane, London W12 7RJ
www.bbc.co.uk/bbcfilms
Contact *Nish Panchal*

The feature film-making arm of the BBC. OUTPUT includes *Fish Tank*; *The Meerkats*; *The Damned United*; *Revolutionary Road*; *Creation*; *The Duchess*. No unsolicited mss. Awards development deals to National Film & Television School students and has BBC *writersroom* and BBC Film Network for young, aspiring writers.

## Beckmann Visual Publishing
Milntown Lodge, Lezayre Road, Ramsey, Isle of Man IM8 2TG
☎ 01624 816585    🖷 01624 816589
sales@beckmanngroup.co.uk
www.beckmanngroup.co.uk
Contacts *Jo White, Stuart Semark*

Isle of Man-based company. Video and television documentary distributor. OUTPUT *Practical Guide to Europe* (travel series); *Music Maestro* (12-part series on classical composers); *Christian the Lion* (wildlife collection); *Aerial Britain* (British heritage/travel).

## Big Heart Media
32 Clerkenwell Green, London EC1R 0DU
☎ 020 7608 0352    🖷 020 7250 1138
info@bigheartmedia.com
www.bigheartmedia.com
Contact *Colin Izod*

Producer of drama and documentaries for television and video. Ideas/outlines welcome by email, but not unsolicited mss.

## Bigbox Network Ltd
Ard-Daraich Studio B, Ardgour, Nr Fort William PH33 7AB
☎ 01855 841384    🖷 01855 841384
norriemaclaren@bigboxnetwork.com
www.bigboxnetwork.com
Contacts *Norrie Maclaren, Alistair Murray*

Film and television: documentary and drama programming such as *Dig* (gardening series for Ch4); various 'Dispatches' for Ch4 and 'Omnibus' for BBC. Keen to encourage and promote new writing.

## Bona Broadcasting Limited
2nd Floor, 9 Gayfield Square, Edinburgh EH1 3NT
☎ 0131 558 1696    🖷 0131 558 1694
enquiries@bonabroadcasting.com
www.bonabroadcasting.com
Contact *Turan Ali*

Producer of award winning-drama and documentary programmes for BBC radio, TV and film projects. No unsolicited mss but send a one-paragraph summary by email in the first instance. Runs radio and TV drama training courses in the UK and internationally.

## Brighter Pictures
▷ Endemol UK Productions

## Buccaneer Films
5 Rainbow Court, Oxhey WD19 4RP
☎ 01923 254000
Contact *Michael Gosling*

Corporate video production and still photography specialists in education and sport. No unsolicited mss.

## Cactus TV
373 Kennington Road, London SE11 4PS
☎ 020 7091 4900    🖷 020 7091 4901
cactustv.co.uk

Founded 1994. Specializes in broad-based entertainment, features and chat shows. Produces around 300 hours of broadcast material per year. OUTPUT includes *Saturday Kitchen*; *The Hairy Bikers Food Tour of Britain* (BBC); *The TV Book Club* (C4 & More4); *The Crime Thriller Awards* (ITV3) and *The British Book Awards*.

## Calon TV
3 Mount Stuart Square, Butetown, Cardiff CF10 5EE
☎ 029 2048 8400    🖷 029 2048 5962
enquiries@calon.tv
www.calon.tv
Head of Development *Andrew Offiler*

Founded 2005. Animated series, mainly for children. OUTPUT includes *Meeow*; *Hilltop Hospital*; *Hana's Helpline*; *The Hurricanes*; *Tales of the Toothfairies*; *Billy the Cat*; *The Blobs*; as well as the feature films, *Under Milk Wood* and *The Princess and the Goblin*. Write with ideas and sample script in the first instance.

## Carnival Film & Television Limited
Oxford House, 76 Oxford Street, London W1D 1BS
☎ 020 7307 6600    🖷 020 7307 6666
info@carnivalfilms.co.uk
www.carnivalfilms.co.uk

Managing Director *Gareth Neame*

TV and film drama producer. OUTPUT, TV: *Any Human Heart*; *Traffik*; *Porterhouse Blue*; *As If 1–3*; *Rosemary & Thyme 1–3* (all for Ch4); *Downton Abbey*; *Whitechapel 1–2*; *Harley Street*; *Midnight Man*; *The Whistleblowers*; *The Old Curiosity Shop*; *Poirot*; *Christmas at the Riviera* (ITV); *Material Girl*; *Enid*; *Hotel Babylon 1–4*; *Empathy*; *Lifeline*; *Sea of Souls*; *Bugs 1–4* (BBC); *The Philanthropist* (NBC); *The Grid* (TNT/Fox/BBC); *The Infiltrator* (HBO); *The Tenth Kingdom* (Sky/NBC); *Jeeves & Wooster* (Granada). FILM: *Shadowlands* (Savoy/Spelling); *Mill on the Floss* (BBC); *Firelight* (Hollywood Pictures/Wind Dancer); *Under Suspicion* (Columbia/Rank/LWT); *Up on the Roof* (Rank/Granada); *Wilt* (Rank/LWT). No unsolicited mss

## Cartwn Cymru

12 Queen's Road, Mumbles, Swansea SA3 4AN
☏ 07771 640400
production@cartwn-cymru.com
Producer *Naomi Jones*

Animation production company. OUTPUT *Toucan 'Tecs* (YTV/S4C); *Funnybones* and *Turandot: Operavox* (both for S4C/BBC); *Testament: The Bible in Animation* (BBC2/S4C); *The Miracle Maker* (S4C/BBC/British Screen/Icon Entertainment International); *Faeries* (HIT Entertainment plc for CITV); *Otherworld* (animated feature film for S4C Films, British Screen, Arts Council of Wales).

## Celtic Films Entertainment Ltd

3–4 Portland Mews, London W1F 8JF
☏ 020 7494 6886      ℻ 020 7494 9134
info@celticfilms.co.uk
www.celticfilms.co.uk
Contact *Steven Russell*

Founded in 1986 producing high-end TV drama and mainstream feature films. OUTPUT includes sixteen *Sharpe* television films, *Girl from Rio* and *Stefan Kiszko: A Life for a Life*. No unsolicited submissions.

## Chameleon Television Ltd

Church House, 14 Town Street, Horsforth, Leeds LS18 4RJ
☏ 0113 205 0040      ℻ 0113 281 9454
allen@chameleontv.com
Contacts *Allen Jewhurst, Alice Taylor*

Film and television drama and documentary producer. OUTPUT includes *Edge of the City*; 'Dispatches' – *Channel 4 News*; *The Family Who Vanished*; *Killing for Honour*; *College Girls*; *Ken Dodd in the Dock*; *Unholy War* (all for Ch4); *Diary of a Mother on the Edge*; *Divorces From Hell*; *Shipman*; *Love to Shop*; *The Marchioness* (all for ITV); *Ted & Sylvia – Love, Loss*; *Hamas*

*Bombers* (BBC); *Liverpool Poets* (Ch5). Scripts not welcome unless via agents but new writing is encouraged.

## Channel Television Ltd

Television Centre, St Helier, Jersey JE1 3ZD
☏ 01534 816873      ℻ 01534 816889
david@channeltv.co.uk
www.channeltv.co.uk
Senior Producer *David Evans*

Producer of TV commercials and corporate material: information, promotional, sales, training and events coverage. CD and DVD production; promotional videos for all types of businesses throughout Europe. No unsolicited mss; new writing/scripts commissioned as required. Interested in hearing from local writers resident in the Channel Islands.

## Channel X

4 Candover Street, London W1W 7DJ
☏ 0845 900 2940      ℻ 020 7580 8016
info@channelx.co.uk
www.channelx.co.uk
Contact *Gary Matsell*

Producer of television comedy such as *Lunch Monkeys*; *Chelsey: OMG!*; *Snuff Box*; *Jason Manford Live at the Manchester Apollo*. Unsolicited mss welcome; email in the first instance.

## Charles Dunstan Communications Ltd

42 Wolseley Gardens, London W4 3LS
☏ 020 8994 2328      ℻ 020 8994 2328
Contact *Charles Dunstan*

Producer of film, video and TV for documentary and corporate material. OUTPUT *Renewable Energy* for broadcast worldwide in 'Inside Britain' series; *The Far Reaches* travel series; *The Electric Environment*. No unsolicited scripts.

## Cheetah Television
▷ Endemol UK Productions

## The Children's Film & Television Foundation Ltd

c/o Simon George, Head of Finance, Ealing Studios, Ealing Green, London W5 5EP
☏ 07887 573479
info@cftf.org
www.cftf.org.uk
Chief Executive *Anna Home*

Involved in the development and co-production of films for children and the family, both for the theatric market and for TV.

## Cleveland Productions

5 Rainbow Court, Oxhey, Near Watford WD19 4RP
☏ 01923 254000

Contact *Michael Gosling*

Communications in sound and vision A/V production and still photography specialists in education and sport. No unsolicited mss.

### Cloud Nine Films

39 Long Acre, London WC2E 9LG

☎ 020 7845 6990

www.celador.co.uk

Head of Creative Affairs *Ivana MacKinnon*

Producer of feature films, launched in 2009 by Christian Colson, whose producing credits include *The Descent*; *Separate Lies*; *Eden Lake* and *Slumdog Millionaire*. Submissions from established literary and film agents only or talent already known to the company, or from third party producers seeking possible co-production.

### COI

Hercules House, Hercules Road, London SE1 7DU

☎ 020 7928 2345

www.coi.gov.uk

Government advertising and marketing communications, and public information films.

### Collingwood O'Hare Productions Ltd

10–14 Crown Street, London W3 8SB

☎ 020 8993 3666    🖷 020 8993 9595

info@crownstreet.co.uk

www.collingwoodohare.com

Director *Tony Collingwood*

Head of Development *Helen Stroud*

Film and TV; specializes in children's animation. OUTPUT *The Secret Show* (BBC/BBC Worldwide); *RARG* (award-winning animated film); *Yoko! Jakamoko! Toto!*; *Daisy-Head Mayzie* (Dr Seuss animated series for Turner Network and Hanna-Barbera); *Gordon the Garden Gnome* (CBBC); *Eddy and the Bear* and *The King's Beard* (both for CITV). Unsolicited mss not welcome 'as a general rule as we do not have the capacity to process the sheer weight of submissions this creates. We therefore tend to review material from individuals recommended to us through personal contact with agents or other industry professionals. We like to encourage new writing and have worked with new writers but our ability to do so is limited by our capacity for development. We can usually only consider taking on one project each year, as development/finance takes several years to put in place.'

### The Comedy Unit

6th Floor, 53 Bothwell Street, Glasgow G2 6TS

☎ 0141 220 6400

scripts@comedyunit.co.uk

www.comedyunit.co.uk

Managing Director *April Chamberlain*

Creative Director *Colin Gilbert*

Producers of scripted comedy and entertainment shows for TV and radio. OUTPUT includes *Rab C. Nesbitt*; *Fags, Mags & Bags*; *The Incredible Will & Greg*; *Still Game*; *The Karen Dunbar Show*; *Chewin' the Fat*; *Only an Excuse?*; *Watson's Wind Up*. Unsolicited sketches welcome. Sitcoms should be submitted as a one-page outline with a couple of sample scenes.

### Company Pictures

Suffolk House, 1–8 Whitfield Place, London W1T 5JU

☎ 020 7380 3900    🖷 020 7380 1166

enquiries@companypictures.co.uk

www.companypictures.co.uk

Managing Directors *George Faber, Charles Pattinson*

One of the UK's largest independent film and television drama production companies. Established in 1998. Won Best Independent Production Company at the 2008 and 2005 Broadcast Awards and the European Producers of the Year Award at the 2004 Monte Carlo Awards. OUTPUT includes *Shameless* (Ch4; BAFTA and RTS Awards); *Life and Death of Peter Sellers* (HBO, two Golden Globe Awards, 15 Emmies); *Elizabeth I*, starring Helen Mirren and Jeremy Irons (Ch4/HBO, three Golden Globe Awards, 9 Emmies); *Skins* (E4 Best Drama Series Broadcast Awards); *The Devil's Whore* (Ch4, three RTS Awards); *Generation Kill* (HBO, Ch4, three Emmys, 11 nominations). Proposals accepted via agents only.

### CPL Productions

38 Long Acre, London WC2E 9JT

☎ 020 7240 8101

info@cplproductions.co.uk

www.celadorproductions.com

Director of Production *Heather Hampson*

Head of Radio *Liz Anstee*

Producer of a range of TV and radio programmes, from situation comedy to quizzes, game shows, popular factual entertainment and international co-productions. OUTPUT *You Are What You Eat*; *Perfect Strangers*; *Popcorn*; *It's Been a Bad Week*; *Commercial Breakdown*; *How to Dump Your Mates*; *Three Off the Tee*; *All Star Mr & Mrs*; *Sexy Ads*; *Celebrating the Carpenters*. 'We are interested in radio scripts but do not accept unsolicited proposals for comedy or entertainment formats. As a relatively small company our script-reading resources are limited.'

### The Creative Partnership

13 Bateman Street, London W1D 3AF

☎ 020 7439 7762    🖷 020 7437 1467

sarah.fforde@thecreativepartnership.co.uk

www.thecreativepartnership.co.uk

Contact *Sarah Fforde*

'Europe's largest "one-stop shop" for advertising and marketing campaigns for the film and television industries.' Clients include most major and independent film companies. No unsolicited material

## Cricket Ltd

Medius House, 63–69 New Oxford Street, London WC1A 1EA
☏ 020 7845 0300    📠 020 7845 0303
team@cricket-ltd.com
www.cricket-ltd.com
Head of Recruitment *Mary McDonnell*

Film and video, live events and conferences, print and design. 'Communications solutions for business clients wishing to influence targeted external and internal audiences.'

## CSA Word

6A Archway Mews, 241A Putney Bridge Road, London SW15 2PE
☏ 020 8871 0220    📠 020 8877 0712
victoria@csaword.co.uk
www.csaword.co.uk
Contacts *Victoria Williams, Clive Stanhope*

Producer of drama, documentaries and readings for radio. OUTPUT *Alfie Elkins & His Little Life* (drama for BBC World Service); *The Hungry Years* (reading for BBC R4); John O'Farrell's *An Utterly Impartial History of Britain*; *A Stable Relationship* (features for BBC R4); *Bob Dylan's Chronicles 1*; *Eric Clapton's Autobiography* (readings for BBC R2); *It's a Girl*; *Chat Snaps & Videotape – Two* (documentaries for BBC World Service); *The Glenn Miller Story*; *Berlin...Soundz Decadent* and *Original Soundtrack Recordings* (documentaries for BBC R2), Unsolicited drama ideas welcome by post or email. 'We encourage new drama ideas.'

## Cutting Edge Productions Ltd

27 Erpingham Road, London SW15 1BE
☏ 020 8780 1476
juliannorridge@btconnect.com
Contact *Julian Norridge*

Corporate and documentary video and television. OUTPUT includes US series on evangelicalism, 'Dispatches' on US tobacco and government videos. No unsolicited mss; 'we commission all our writing to order, but are open to ideas.'

## Diverse Production Limited

1 Ariel Way, London W12 7SL
☏ 020 7603 4567    📠 020 3189 3200
www.diverse.tv
www.zodiakentertainment.com

Independent production company specializing in popular prime-time formats, strong documentaries, specialist factual, historical, cultural, religious, arts, music and factual

entertainment. Joined Zodiak Entertainment in 2006. Recent OUTPUT includes *Bear Grylls' Man vs. Wild*; *The Frankincense Trail*; *Human Zoo: Science's Dirty Secret*; *Beat the Star*; *The Prisoner*; *Headshrinkers of the Amazon*; *Gonzo: The Life of Hunter S. Thompson*; *Rocket Science*; *Think Tank*; *Hope for the Last Chance Kids* (follow-up to *Last Chance Kids* part of Channel 4's 'Lost For Words' season of campaigning programmes to get kids reading); *The Seven Sins of England*; *Mind Your F*cking Language*; *Not Cricket: The Captain & The Bookmaker*; *Britain's Rich List: Give It Away* and *Codex*.

## DoubleBand Films

3 Crescent Gardens, Belfast BT7 1NS
☏ 028 9024 3331    📠 028 9023 6980
www.doublebandfilms.com
Contacts *Michael Hewitt, Dermot Lavery*

Established 1988. Specializes in documentaries and some drama. Recent productions include *High Flyers*; *Caravans – The British Love Affair* (both for BBC4); *Best: His Mother's Son* (BBC2); *Seven Days that Shook the World* and *War in Mind* (both for Ch4); *Christine's Children* (BBC Northern Ireland; nominated for both the RTS and Celtic Film Festival); *D-Day: Triumph and Tragedy* (BBC NI). No unsolicited scripts.

## Drake A-V Video Ltd

89 St Fagans Road, Fairwater, Cardiff CF5 3AE
☏ 029 2056 0333
info@drakeav.com
www.drakeav.com

Specialists in corporate and educational films. No unsolicited mss.

## Endemol UK Productions

Shepherds Building Central, Charecroft Way, London W14 0EE
☏ 0870 333 1700    📠 0870 333 1800
info@endemoluk.com
www.endemoluk.com

Major independent producer of TV and digital media content, responsible for over 5,000 hours of programming per year. Incorporates several production brands including Brighter Pictures (factual entertainment shows, reality programming, live events and popular documentaries), Cheetah Television (factual), Initial (prime time entertainment, children's and teens, events and arts shows) and Zeppotron (comedy). OUTPUT includes *Big Brother*; *Deal or No Deal*; *Total Wipeout*; *Gok's Fashion Fix*; *Supersize v Superskinny*; *Golden Balls*.

## Farnham Film Company Ltd

34 Burnt Hill Road, Lower Bourne, Farnham GU10 3LZ
☏ 01252 710313    📠 01252 725855

info@farnfilm.com
www.farnfilm.com
Contact *Ian Lewis*

Television and film: intelligent full-length film. No unsolicited mss; check website for current requirements.

## Festival Film and Television Ltd

Festival House, Tranquil Passage, Blackheath, London SE3 0BJ
℡ 020 8297 9999   ℻ 020 8297 1155
info@festivalfilm.com
www.festivalfilm.com
Managing Director *Ray Marshall*
Producer *Matt Marshall*
Development *Johanna Devereaux*

Television drama and feature films. Best known for its Catherine Cookson Dramas, which became one of ITV's long-running brands. Now developing both TV drama (singles, adaptations and series) and features. Only accepts material submitted through an agent. Will not consider unsolicited material.

## Fiction Factory

14 Greenwich Church Street, London SE10 9BJ
℡ 020 8853 5100   ℻ 020 8293 3001
radio@fictionfactory.co.uk
www.fictionfactory.co.uk
Creative Director *John Taylor*

Production company specializing in intelligent entertainments. Recent OUTPUT for BBC Radio includes *Chronicles of Ait* Michael Batt; *Rage on the Road*; *Witches of Eastwick*. Ideas considered if sent by email. Mss only from agents or writers with a professional track record in the broadcast media.

## Film and General Productions Ltd

4 Bradbrook House, Studio Place, London SW1X 8EL
℡ 020 7235 4495
cparsons@filmgen.co.uk
Contacts *Clive Parsons, Davina Belling*

Film and television drama. Feature films include *True Blue*; *Tea with Mussolini* and *I Am David*. Also *Seesaw* (ITV drama), *The Greatest Store in the World* (family drama, BBC) and *The Queen's Nose* (children's series, BBC). Interested in considering new writing but subject to prior telephone conversation.

## Firehouse Productions

25 Beak Street, London W1F 9RT
℡ 020 7439 2220
hello@firehouse.co.uk
www.firehouse.co.uk
Contacts *Peter Granger, Mike MacLeod*

Corporate films and live production events. OUTPUT includes work for Shell; Vodafone; government and various agencies.

## First Writes Theatre Company

Lime Kiln Cottage, High Starlings, Banham, Norwich NR16 2BS
℡ 01953 888525   ℻ 01953 888974
ellen@first-writes.co.uk
www.first-writes.co.uk
Director *Ellen Dryden*

Producer of numerous afternoon, Friday and Saturday plays, classic serials and comedy narrative series for BBC Radio 3 and Radio 4. No unsolicited material.

## Flannel

21 Berwick Street, London W1F 0PZ
℡ 020 7287 9277   ℻ 020 7287 7785
mail@flannel.net
Contact *Kate Haldane*

Producer of drama, documentaries and comedy for television and radio. OUTPUT includes *Woman's Hour* (BBC R4); *The Hendersons' Christmas Party* (five-part Christmas drama, BBC R4). No unsolicited mss. 'Keen to encourage new writing, but must come via an agent. Particularly interested in 45–50 minute dramas for radio. Not in a position to produce plays for stage, but very happy to consider adaptations. Welcomes comedy with some track record.'

## Flashback Television Limited

58 Farringdon Road, London EC1R 3BP
℡ 020 7253 8768   ℻ 020 7253 8765
mailbox@flashbacktv.co.uk
www.flashbacktelevision.com
Managing Director *Taylor Downing*
Creative Director *David Edgar*

Founded in 1982, award-winning producer of drama, factual entertainment and documentaries for television. OUTPUT includes *Beau Brummell*. No unsolicited submissions.

## Focus Films Ltd

The Rotunda Studios, Rear of 116–118 Finchley Road, London NW3 5HT
℡ 020 7435 9004   ℻ 020 7431 3562
focus@focusfilms.co.uk
www.focusfilms.co.uk
Managing Director *David Pupkewitz*
Contact *Raimund Berens*

Founded 1982. Independent feature film development and production company. OUTPUT *The Book of Eve* (Canadian drama); *The Bone Snatcher* (Horror, UK/Can/SA); *Julia's Ghost* (German co-production); *The 51st State* (feature film); *Secret Society* (comedy drama feature film); *Crimetime* (feature thriller); *Diary of a Sane Man*;

*Othello*; *Chemical Wedding*; *Master Harold and the Boy*; *Surviving Evil*. Projects in development include *Heaven and Earth*. No unsolicited scripts.

## FremantleMedia Group Ltd

1 Stephen Street, London W1T 1AL
☎ 020 7691 6000   ℻ 020 7691 6100
www.fremantlemedia.com
CEO *Tony Cohen*
CEO, talkbackTHAMES *Lorraine Heggessey*

FremantleMedia, formerly known as Pearson Television, is the production arm of the RTL Group, Europe's largest TV and radio company. The company acquired Thames Television in 1993 (producer of *The Bill*) and Grundy Worldwide (*Neighbours*) in 1995. Further acquisitions were Witzend Productions (*Lovejoy*) and Alomo Productions in 1996 and TalkBack Productions in 2000. FremantleMedia is a global leader in content production with 60 global formats and more than 320 titles in production at any one time. *No unsolicited submissions, please.*

## Fulcrum TV Limited

14 Bowden Street, London SE11 4DS
☎ 020 3326 6996
info@fulcrumtv.com
www.fulcrumtv.com
Contact *Richard Belfield*

Producer of film, video and television: drama, documentary and corporate output. Interested in new writing; send short email (richard@fulcrumtv.com) in the first instance.

## Gaia Communications

20 Pevensey Road, Eastbourne BN21 3HP
☎ 01323 734809/727183
production@gaiacommunications.co.uk
www.gaiacommunications.co.uk
Producer *Robert Armstrong*

Established 1987. Video and TV, corporate and documentary. OUTPUT *Discovering* (south east regional tourist and local knowledge series); *Holistic* (therapies and general information); local interest audiobooks.

## Ginger Productions
▷ STV Productions & Ginger Productions

## Goldcrest Films International Ltd

65–66 Dean Street, London W1D 4PL
☎ 020 7437 8696   ℻ 020 7437 4448
mail@goldcrestfilms.com
www.goldcrestfilms.com
Chairman *John Quested*
Contact *Laura Wu*

Since it was established in 1977, Goldcrest Films has become a leading independent film production company winning many prizes at international festivals including 19 Academy Awards and 28 Baftas. Finances, produces and distributes films and television programmes. OUTPUT includes *Chariots of Fire*; *Gandhi*; *The Killing Fields*; *A Room With a View*; *Local Hero*; *The Mission*; *To End All Wars*. Scripts via agents only.

## The Good Film Company Ltd

The Studio, 5–6 Eton Garages, Lambolle Place, London NW3 4PE
☎ 020 7794 6222   ℻ 020 7794 4651
yanina@goodfilms.co.uk
www.goodfilms.co.uk
Contact *Yanina Barry*

Film production of all kinds, including commercials, pop videos, documentaries, TV, etc., specializing in clients from outside the UK. *No* unsolicited mss.

## Granite Productions

Easter Davoch, Tarland, Aboyne AB34 4US
☎ 01339 880175
Contact *Simon Welfare*

Producer of television documentary programmes such as *Nicholas & Alexandra*; *Victoria & Albert* and *Arthur C. Clarke's Mysterious Universe*.

## Greenpark Productions Ltd
▷ See entry under Library Services

## Greenwich Village Productions

14 Greenwich Church Street, London SE10 9BJ
☎ 020 8853 5100   ℻ 020 8293 3001
video@fictionfactory.co.uk
www.fictionfactory.co.uk
Contact *John Taylor*

Features, arts, corporate and educational projects, web-movies and new-media productions. Recent OUTPUT includes *Bags*. Mss via agents only or from writers with a professional track record in the chosen medium.

## H2 Business Communication

Shepperton Studios, Shepperton TW17 0QD
☎ 01932 593717   ℻ 01932 593718
mail@h2bc.co.uk
www.h2bc.co.uk

Conferences, videos, awards presentations and speaker training.

## Hammerwood Film Productions

info@filmangels.co.uk
www.filmangel.co.uk

Film, video and TV drama. OUTPUT *Iceni* (film; co-production with Pan-European Film Productions and Boudicca Film Productions Ltd); *Boudicca – A Celtic Tragedy* (TV series). In pre-production: *The Black Egg* (witchcraft in 17th century England); *The Ghosthunter*; *Iceni* (documentary of the rebellion of AD61);

*No Case to Answer* (legal series). 'Authors are recommended to access www.filmangel.co.uk' (see **FilmAngels** under *Useful Websites*).

## Hartswood Films Ltd

Twickenham Studios, The Barons, St Margaret's TW1 2AW

☎ 020 8607 8736    🖷 020 8607 8744

films.tv@hartswoodfilms.co.uk

Producer *Elaine Cameron*

Film and TV production for drama, comedy and documentary. OUTPUT *The Cup*; *Jekyll*; *Coupling*; *Men Behaving Badly*; *Border Cafe*; *Carrie & Barry*; *Fear, Stress and Anger* (all for BBC); *After Thomas* (ITV drama).

## Hat Trick Productions Ltd

33 Oval Road, London NW1 7EA

☎ 020 7184 7777    🖷 020 7184 7778

info@hattrick.com

www.hattrick.com

Managing Director *Jimmy Mulville*

Founded in 1986. Award winning programme maker whose productions include *Have I Got News For You*; *Room 101*; *Father Ted*; *Armstrong & Miller*; *Outnumbered*; *Whose Line Is It Anyway?*.

## Healthcare Productions Limited

The Great Barn, Godmersham Park, Canterbury CT4 7DT

☎ 01227 738279

contact@healthcareproductions.co.uk

www.healthcareproductions.co.uk

Contact *Andrew Wilson*

Established in 1989, producing health-based dramas and documentaries for TV and the DVD market. Has won numerous awards from the BMA. Happy to look at new material, which should be submitted by email.

## Holmes Associates

Lower Ground Floor, 37 Redington Road, London NW3 7QY

☎ 020 7813 4333

holmesassociates@blueyonder.co.uk

Contact *Andrew Holmes*

Prolific originator, producer and packager of documentary, drama and music television and films. OUTPUT has included *Ashes and Sand* (Film 4); *Chunky Monkey* (J&V Films); *Prometheus* (Ch4 'Film 4'); *The Shadow of Hiroshima* (Ch4 'Witness'); *The House of Bernarda Alba* (Ch4/WNET/Amaya); *Piece of Cake* (LWT); *The Cormorant* (BBC/Screen 2); *John Gielgud Looks Back*; *Rock Steady*; *Well Being*; *Signals*; *Ideal Home?* (all Ch4); *Timeline* (with MPT, TVE Spain & TRT Turkey); *Seven Canticles of St Francis* (BBC2). Submissions only accepted by email in synopsis form.

## Hourglass Productions

27 Prince's Road, London SW19 8RA

☎ 020 8540 8786

productions@hourglass.co.uk

www.hourglass.co.uk

Partners *Martin Chilcott, Jacqueline Chilcott*

Film and video; documentary and drama. OUTPUT BAFTA nominated scientific television documentaries and educational programming. Also current affairs, health and social issues.

## Icon Films

1–2 Fitzroy Terrace, Bristol BS6 6TF

info@iconfilms.co.uk

www.iconfilms.co.uk

Contact *Harry Marshall*

Film and TV documentaries. OUTPUT *Nick Baker's Weird Creatures* (five/Animal Planet/Granada International); *Tom Harrisson – The Barefoot Anthropologist* (BBC). Specializes in factual documentaries. Open-minded to new documentary proposals.

## The Illuminated Film Company

2 Glenthorne Mews, London W6 0LJ

☎ 020 8748 3030    🖷 020 8748 3725

info@illuminatedfilms.com

www.illuminatedfilms.com

Contact *Iain Harvey*

Producers of animation: film, TV, video, drama and commercials. Recent OUTPUT includes four series of *Little Princess* for Ch5; animated special of *War Game* for Ch4 and *Christmas Carol – The Movie*. Prefers to see an outline of the project/story in the first instance; send by email. 'Our requirement is story, story, story!'

## Imari Entertainment Ltd

PO Box 158, Beaconsfield HP9 1AY

☎ 01494 677147    🖷 01494 677147

info@imarientertainment.com

Contact *Jonathan Fowke*

TV and video producer, covering all areas of drama, documentary and corporate productions.

## Initial

▷ Endemol UK Productions

## Isis Productions

Goldcrest Post, 1 Lexington Street, London W1F 9AF

☎ 020 7220 2929    🖷 020 7990 8229

hello@isis-productions.com

www.isis-productions.com

Directors *Nick de Grunwald, George Scott, Martin Smith*

Formed in 1991, Isis Productions specializes in the production of music and documentary programmes and are developing entertainment and drama formats. Isis' main production credit is Classic Albums, an ongoing documentary series for TV and DVD. Five series to date have included films on albums by Elvis Presley, Stevie Wonder, Elton John, Paul Simon, The Who, Jimi Hendrix, U2, Bob Marley & The Wailers, Lou Reed, Sex Pistols, Steely Dan, Pink Floyd, Nirvana, Queen, Cream, Frank Zappa, Duran Duran, John Lennon, The Doors and Jay-Z. Other major profile films include Elizabeth Taylor, John Simpson and Carla Bruni. Isis produced the first ever televised UK Urban Asian Music Awards, a major opera film, *Dido & Aeneas* and the classical music series, *The Score*. No unsolicited mss. Send email detailing general subject of proposal before detailed correspondence.

### Isolde Films

28 Twyford Avenue, London W3 9QB
☎ 020 8896 2860
isolde@btinternet.com
www.tonypalmer.org

Film and TV documentaries. OUTPUT *Wagner*; *Margot*; *Menuhin*; *Maria Callas*; *Testimony*; *In From the Cold*; *Pushkin*; *England, My England* (by John Osborne). Unsolicited material is read, but please send a written outline first.

### ITV Studios

The London Television Centre, Upper Ground, South Bank, London SE1 9LT
☎ 020 7620 1620
www.itv.com
Creative Director, ITV Studios UK *John Whiston*

ITV Studios UK is the in-house production arm of ITV which makes around half of all ITV's programming plus many hours for other broadcasters. It produces many of this country's top programme titles from production centres in Leeds, Manchester, Norwich and London. Many of its original formats are now produced internationally for different territories and its programme archive is sold and licensed around the world. OUTPUT includes: *Coronation Street*; *Identity*; *Headcases*; *Emmerdale*;*The Street*; *Lewis*; *Poirot*; *The Prisoner*; *Miss Marple*.

### JAM Pictures and Jane Walmsley Productions

8 Hanover Street, London W1S 1YE
☎ 020 7290 2676    ☑ 020 7256 6818
producers@jampix.com
Contacts *Jane Walmsley, Michael Braham*

JAM Pictures was founded in 1996 to produce drama for film, TV and stage. Projects include: *Rain Man* (UK tour, 2009, starring Neil Morrissey) and West End production, 2008); *Hillary's Choice* (TV film, A&E Network); *Son of Pocahontas* (TV film, ABC); *Rudy: the Rudy Giuliani Story* (TV film, USA Network); *One*

*More Kiss* (feature, directed by Vadim Jean); *Bad Blood* (UK theatre tour). Jane Walmsley Productions, formed in 1985 by TV producer, writer and broadcaster, Jane Walmsley, has completed award-winning documentaries and features such as *Hot House People* (Ch4). No unsolicited mss. 'Letters can be sent to us, asking if we wish to see mss; we are very interested in quality material, from published or produced writers only, please'.

### Jane Marshall Productions

The Coach-House, Westhill Road, Blackdown, Leamington Spa CV32 6RA
☎ 01926 831680
jane@janemarshallproductions.co.uk
Contact *Jane Marshall*

Producer of readings of published work both fiction and non-fiction for BBC Radio. Published work only.

### Jane Walmsley Productions
▷ JAM Pictures

### Juniper Communications Ltd

Unit 3, 52 Lant Street, London SE1 1RB
☎ 020 7407 9292    ☑ 020 7407 3940
juniper@junipertv.co.uk
www.junipertv.co.uk
Contacts *Howard Coop, Belkis Bhegani*

Producer of television and radio documentaries. OUTPUT includes *The Qur'an*; *The Blair Years*; *Invitation to a Hanging*; *The Great Plague*. No unsolicited mss; approach by phone in the first instance.

### Justice Entertainment Ltd

PO Box 6933, London W1A 6UT
☎ 020 7935 7902    ☑ 020 7935 7902
aj@timwestwood.com
youtube.com/timwestwoodtv
PA/Booking Manager *A.J. Foronda*

TV and radio production company. No unsolicited material.

### Keo Films.com Ltd

101 St John Street, London EC1M 4AS
☎ 020 7490 3580    ☑ 020 7490 8419
keo@keofilms.com
www.keofilms.com
Head of Production *Maddy Allen*

Television documentaries and factual entertainment. OUTPUT includes programmes for all the main broadcasters including *River Cottage Spring*; *Meet the Natives*; *Hugh's Chicken Run* (all for Ch4); *Return to Lullingstone Castle*; *Blizzard: Race to the Pole* (BBC2). No unsolicited mss.

## Kudos Film and Televison Ltd

12–14 Amwell Street, London EC1R 1UQ
☎ 020 7812 3270  📠 020 7812 3271
info@kudosfilmandtv.com
www.kudosfilmandtv.com
Creative Director *Jane Featherstone*
Joint Managing Directors *Daniel Isaacs,
Simon Crawford Collins*

Television dramas include *Life on Mars*; *Spooks*; *Hustle*; *Tsunami – The Aftermath*; *The Amazing Mrs Pritchard*; *Wide Sargasso Sea*; *Scars*; *MI High*; *Child of Mine*; *Comfortably Numb*; *Pleasureland*; *The Magician's House*; *Psychos*; *Ashes to Ashes*; *The Fixer*; *Moving Wallpaper*; *Echo Beach*; *Plus One*; *Occupation*; *Nearly Famous*; *Spooks: Code 9* and *Law & Order (UK)*. Feature films such as *Among Giants*; *Pure*; *Miss Pettigrew Lives for a Day*; *The Crimson Wing* and *Eastern Promises*. No unsolicited mss.

## Ladbroke Productions (Radio) Ltd

17 Leicester Road, East Croydon CR0 6EB
info@ladbrokeradio.com
www.ladbrokeradio.com
Contact *Neil Gardner*

Producer of radio drama, documentary, corporate, music, music documentaries and readings. OUTPUT includes *The Woman in White*; *The Darling Buds of May*; *The True History of British Pop*; *The World on a String*; *The Colour of Music* (all for BBC R2); *Don Carlos*; *In the Company of Men* (BBC R3 drama); *Sitting in Limbo* (BBC World Service drama); *Your Vote Counts* (Electoral Commision audio CDs). Unsolicited mss and ideas welcome; send letter in the first instance but 'we are unable to accept mss for BBC Radio Four afternoon plays'.

## Landseer Productions Ltd

27 Arkwright Road, London NW3 6B
☎ 020 7794 2523
www.landseerfilms.com
Directors *Derek Bailey, Ken Howard*

Film and video production: documentary, drama, music and arts. OUTPUT *Swinger* (BBC2/Arts Council); *Auld Lang Syne* and *Retying the Knot – The Incredible String Band* (both for BBC Scotland); *Benjamin Zander* ('The Works', BBC2); *Zeffirelli, Johnnie Ray, Petula Clark, Bing Crosby, Maxim Vengerov* (all for 'South Bank Show', LWT); *Death of a Legend – Frank Sinatra* ('South Bank Show' special); *Routes of Rock* (Carlton); *See You in Court* (BBC); *Nureyev Unzipped, Gounod's Faust, The Judas Tree, Ballet Boyz, 4Dance* and *Bourne to Dance* (all for Ch4), *Proms in the Park* (Belfast).

## Lilyville Screen Entertainment Ltd

7 Lilyville Road, London SW6 5DP
☎ 020 7471 8989
tony.cash@btclick.com
Contact *Tony Cash*

Drama and documentaries for TV. OUTPUT *Poetry in Motion* (series for Ch4); 'South Bank Show': *Ben Elton* and *Vanessa Redgrave*; *Musique Enquête* (drama-based French language series, Ch4); *Sex and Religion* (ITV); *Landscape and Memory* (arts documentary series for the BBC); Jonathan Miller's production of the *St Matthew Passion* for the BBC; major documentary on the BeeGees for the 'South Bank Show'. Scripts with an obvious application to TV may be considered. Interested in new writing for documentary programmes.

## Lime Pictures Limited

Campus Manor, Childwall Abbey Road, Childwall, Liverpool L16 0JP
☎ 0151 722 9122  📠 0151 722 6839
www.limepictures.com
Creative Director *Tony Wood*

Originally known as Mersey Television until All3Media's acquisition in 2005 and a change of name to Lime Pictures in 2006. Creators of television dramas: *Brookside*; *Hollyoaks*; *Runners* (Ch4); *Grange Hill*; *Apparitions*; *The Well* (BBC); *Living On the Edge* (MTV).

## Loftus Audio Ltd

2a Aldine Street, London W12 8AN
☎ 020 8740 4666
ask@loftusaudio.co.uk
www.loftusaudio.co.uk
Contact *Matt Thompson (radio drama
☎ 01620 893876)*
Head of Radio *Jo Coombs*

Producer of factual radio programmes, documentaries, readings (mostly non-fiction) and drama, as well as audio books and audio guides for museums and galleries. OUTPUT includes titles for Radio 4's 'Book of the Week'. 'We are happy to look at brief emailed synopses of radio features and documentaries.'

## Lucida Productions

5 Alleyn Crescent, London SE21 8BN
☎ 020 8761 6766
pj.lucida@tiscali.co.uk
www.lucidaproductions.com
Contact *Paul Joyce*

Television and cinema: arts, adventure, current affairs, documentary, drama and music. OUTPUT has included *Motion and Emotion: The Films of Wim Wenders*; *Dirk Bogarde – By Myself*; *Sam Peckinpah – Man of Iron*; *Kris Kristofferson – Pilgrim*; *Wild One: Marlon Brando*; *Stanley*

Kubrick: 'The Invisible Man'; 2001: the Making of a Myth (Ch4); Mantrap – Straw Dogs, the final cut (with Dustin Hoffman). Restoration of the Director's Cut of The Devils plus the documentary Hell on Earth with Ken Russell and Vanessa Redgrave, and Hockney on Photography (Sky Arts, 2009).

## Mark Forstater Productions Ltd

11 Keslake Road, London NW6 6DJ
📞 020 8933 5475
mforstater@msn.com
Contact *Mark Forstater*

Producer of film drama: comedy, horror, thriller and film noir. No unsolicited mss.

## Maverick Television

Progress Works, Heath Mill Lane, Birmingham B9 4AL
📞 0121 771 1812    📠 0121 771 1550
mail@mavericktv.co.uk
www.mavericktv.co.uk
*Also at:* 40 Church Way, London NW1 1LW
📞 020 7383 2727
Head of Production *Alison Todd*

Established in 1994, Maverick has a strong reputation for popular factual programming. It is now one of network television's most prolific independent suppliers. OUTPUT includes *How to Look Good Naked; 10 Years Younger; Who'll Age Worst?; Bollywood Star; Fat Chance; Born Too Soon; VeeTV; Trade Secrets; Embarrassing Illnesses; 10 Things You Didn't Know About …; How To Live Longer; The Property Chain; Male, 33, Seeks Puberty; Extreme Engineering; Picture This: Accidental Hero; Up Your Street; Motherless Daughters; Highland Bollywood: Black Bag; Health Alert: My Teenage Menopause; Long Haul; Learning to Love the Grey.*

## Maya Vision International Ltd

6 Kinghorn Street, London EC1A 7HW
📞 020 7796 4842    📠 020 7796 4580
www.mayavisionint.com
Contact *Tamsin Ranger*

Film and TV: drama and documentary. OUTPUT *Saddam's Killing Fields* (for 'Viewpoint', Central TV); *3 Steps to Heaven* and *A Bit of Scarlet* (feature films for BFI/Ch4); *A Place in the Sun* and *North of Vortex* (dramas for Ch4/Arts Council); *The Real History Show* (Ch4); *In Search of Myths and Heroes; In Search of Shakespeare; In the Footsteps of Alexander the Great; Conquistadors* (BBC documentaries); *Hitler's Search for the Holy Grail; Once Upon a Time in Iran* (Ch4 documentaries); *The Story of India* (BBC/PBS). Absolutely no unsolicited material; commissions only.

## MBP TV

Saucelands Barn, Coolham, Horsham RH13 8QG
📞 01403 741620    📠 01403 741647
info@mbptv.com
www.mbptv.com
Contact *Phil Jennings*

Maker of film and video specializing in programmes covering equestrianism and the countryside. No unsolicited scripts, but always looking for new writers who are fully acquainted with the subject.

## Melendez Films

Julia House, 44 Newman Street, London W1T 1QD
📞 020 7323 5273    📠 020 7323 5373
Contact *Steven Melendez*

Independent production company specializing in 2D animation. Also involved in production and film design for clients in England, Spain, Sweden, India and the US, plus website design and 3D animation on the web. Clients include book publishers, TV companies and advertisers. Winner of international awards for films, particularly of classic books, stories and comic characters. 'We will look at unsolicited projects in outline or synopsis form only. Enclose s.a.s.e.'

## Mendoza Films

52–53 Dean Street, London W1D 5BL
📞 020 7209 9617
office@mendozafilms.com
www.mendozafilms.com

Commercials, title sequences (e.g. Alan Bleasdale's *G.B.H.*); party political broadcasts. Currently in pre-production on a feature-length comedy film. Involved with the Screenwriters' Workshop. Unsolicited mss welcome but 'comedies only, please'. Material will not be returned without s.a.s.e.

## Mersey Television

▷ Lime Pictures Limited

## Midsummer Films Productions Ltd

33 St Lawrence Terrace, London W10 5SR
📞 020 8932 8870    📠 020 8932 8871
terhi@midsummerfilms.com
Contact *Terhi Kylliainen*

Feature film producer. OUTPUT includes *An American Haunting; Dreaming of Joseph Lees; Relative Values.* Send treatment and/or synopsis by email in the first instance.

## The Mob Film Company

10–11 Great Russell Street, London WC1B 3NH
📞 020 7580 8142    📠 020 7255 1721
mail@mobfilm.com
www.mobfilm.com
Head of Film *Andrew Boswell*

Head of TV *Rod Brown*

Feature films and television drama, documentaries and commercials. OUTPUT includes *Stone of Destiny* (feature film); *Terry Prachett's Hogfather* (TV mini-series); *In the Land of the Free* (feature documentary). Welcomes unsolicited mss; send synopsis by email in the first instance.

## Moonstone Films Ltd

☎ 020 8144 9940
info@moonstonefilms.co.uk
www.moonstonefilms.co.uk
Executive Producer *Tony Stark*

Television: current affairs, science and history documentaries. OUTPUT *Arafat's Authority* and *Arafat Investigated*, both for BBC 'Correspondent'. Plus various Ch4 News commissions. Unsolicited mss welcome.

## Neon

Studio Two, 19 Marine Crescent, Kinning Park, Glasgow G51 1HD
☎ 0141 429 6366/07802 403033 ⚫ 0141 429 6377
stephy@go2neon.com
www.go2neon.com
Contact *Stephanie Pordage*

Television and radio: drama and documentary producers. OUTPUT includes *Brand New Country*; *Asian Overground*; *Peeking Past the Gates of Skibo*. Supports and encourages new writing 'at every opportunity'. Welcomes unsolicited material but telephone in the first instance.

## Noel Gay Television

Shepperton Studios, Studios Road, Shepperton TW17 0QD
☎ 01932 592569 ⚫ 01932 592172
www.charlesarmitage.com
CEO *Charles Armitage*

NGTV is willing to accept unsolicited material from writers but 1–2-page treatments only. No scripts, please.

## Number 9 Films

Linton House, 24 Wells Street, London W1T 3PH
☎ 020 7323 4060 ⚫ 020 7323 0456
info@number9films.co.uk
Contacts *Stephen Woolley, Elizabeth Karlsen*

Leading feature film producer. OUTPUT includes *How to Lose Friends and Alienate People*; *And When Did You Last See Your Father?*; *Sounds Like Teen Spirit*. Forthcoming productions: *Great Expectations*; *The Lonely Doll*. No unsolicited material.

## Odyssey Productions Ltd

72 Tay Street, Newport-on-Tay DD6 8AP
☎ 01382 542070 ⚫ 01382 542070

billykay@sol.co.uk
www.billykay.co.uk
Contact *Billy Kay*

Producer of radio documentaries. OUTPUT includes *Scotland's Black History*; *Gentle Shepherds* (oral history); *Street Kids* (Scottish missionaries working with street kids in Brazil). Ideas for radio documentaries welcome; send letter with one page outlining the idea and programme content.

## Omnivision

Pinewood Studios, Iver Heath SL0 0NH
☎ 01753 656329 ⚫ 01753 631145
info@omnivision.co.uk
www.omnivision.co.uk
Contacts *Christopher Morris, Alex Mead*

TV and video producers of documentary, corporate, news and sport programming. Also equipment and facilities hire. Interested in ideas; approach by letter or email.

## oooi

36 Atholl Crescent Lane, Edinburgh EH3 8ET
☎ 07911 542982
www.oooi.co.uk
Contact *Ryan Blackwood*

Established in 2008, oooi focuses on advertising, short and features. CLIENTS include: Holland & Sherry, Sofas-in-stock, Meridian Productivity. OUTPUT includes: *Entwined* (Skena UP Film Festival, 2010); *Lweena* (Britspotting Film Festival, 2009); *Seaside* (Scottish Students on Screen 'Best Fiction Film' nominee, 2009); *Drifters* (Super Shorts Film Festival, 2008). 'We accept unsolicited material.'

## Orlando Television Productions Ltd

Up-the-Steps, Little Tew, Chipping Norton OX7 4JB
☎ 01608 683218 ⚫ 01608 683364
info@orlandomedia.co.uk
www.orlandomedia.co.uk
Contact *Mike Tomlinson*

Producer of TV documentaries and digital multimedia content, with science, health and information technology subjects as a specialization. Approaches by established writers/ journalists to discuss proposals for collaboration are welcome.

## Ovation

Upstairs at the Gatehouse, Highgate, London N6 4BD
☎ 020 8340 4256 ⚫ 020 8340 3466
www.ovation.eu
Contact *John Plews*

Corporate video and conference scripts. Unsolicited mss not welcome. 'We talk to new writers from time to time.' Ovation also runs the fringe theatre, 'Upstairs at the Gatehouse'.

## Paper Moon Productions

Wychwood House, Burchetts Green Lane, Littlewick Green, Nr. Maidenhead SL6 3QW
℡ 01628 829819 ℻ 01628 829819
insight@paper-moon.co.uk
Producer/Director *David Haggas*

Broadcast documentaries and corporate communications. OUTPUT includes *Bilbo & Beyond*, an affectionate glimpse into the life and work of the dedicated philologist and fantasy writer J.R.R. Tolkien.

## Passion Pictures

3rd Floor, 33–34 Rathbone Place, London W1T 1JN
℡ 020 7323 9933 ℻ 020 7323 9030
info@passion-pictures.com
www.passion-pictures.com
Managing Director *Andrew Ruhemann*
Executive Producer *John Battsek*

Documentary and drama includes: *One Day in September* (Academy Award-winner for Best Feature Documentary, 2000); also commercials and music videos. Unsolicited mss welcome.

## Pathé Pictures

6 Ramillies Street, London W1F 7TY
℡ 020 7323 5151 ℻ 020 7631 3568
Head of Creative Affairs *Colleen Woodcock*

Produces 4–6 theatrical feature films each year. 'We are pleased to consider all material that has representation from an agent or production company.'

## Pearson Television
▷ FremantleMedia Group Ltd

## Pelicula Films

59 Holland Street, Glasgow G2 4NJ
℡ 0141 287 9522
Contact *Mike Alexander*

Television producer. Maker of drama documentaries and music programmes for the BBC and Ch4. OUTPUT *As an Eilean (From the Island)*; *The Trans-Atlantic Sessions 1&2, 3&4*; *Nanci Griffith, Other Voices 2*; *Follow the Moonstone*; *Haston*.

## Photoplay Productions Ltd

21 Princess Road, London NW1 8JR
℡ 020 7722 2500 ℻ 020 7722 6662
info@photoplay.co.uk
www.photoplay.co.uk
Contact *Patrick Stanbury*

Documentaries for film, television and video plus restoration of silent films and their theatrical presentation. OUTPUT includes *The Cat and the Canary*; *Orphans of the Storm*; *Cecil B. DeMille:*

*American Epic* and the 'Channel 4 Silents' series of silent film restoration, including *The Wedding March* and *The Iron Mask*. Recently completed *Garbo* and *I'm King Kong!* No unsolicited mss; 'we tend to create and write all our own programmes'.

## Picture Palace Films Ltd

13 Egbert Street, London NW1 8LJ
℡ 020 7586 8763 ℻ 020 7586 9048
info@picturepalace.com
www.picturepalace.com
Producer & Chief Executive *Malcolm Craddock*

Leading independent producer of film and TV drama. OUTPUT *Rebel Heart* (BBC1); *Sharpe's Peril*; *Sharpe's Challenge*; *Extremely Dangerous*; *A Life for A Life* and *Frances Tuesday* (all for ITV); *Sharpe's Rifles* (14 films for Carlton TV); *Little Napoleons* (comedy drama, Ch4); *The Orchid House* (drama serial, Ch4); *Tandoori Nights*; *4 Minutes*; *When Love Dies* (all for Ch4); *Ping Pong* (feature film); *Acid House* (Picture Palace North). Material will only be considered if submitted through an agent.

## Portobello Pictures

12 Addison Avenue, London W11 4QR
℡ 020 7605 1396 ℻ 020 7605 1391
mail@portobellopictures.com
www.portobellopictures.com
Producer *Eric Abraham*
Development Assistant *Kristin Irving*

Oscar-winning film, television and theatre production company. Not accepting unsolicited submissions.

## Pozzitive Television

Paramount House, 162–170 Wardour Street, London W1F 8AB
℡ 020 7734 3258 ℻ 020 7437 3130
david@pozzitive.co.uk
www.pozzitive.co.uk
Contact *David Tyler*

Producer of comedy and entertainment for television and radio. OUTPUT *Dinner Ladies*; *Coogan's Run*; *The 99p Challenge*; *The Comic Side of 7 Days*; *Armando Iannucci's Charm Offensive*; *Cabin Pressure*; *For One Night Only*. Unsolicited mss of TV and radio comedy welcome. 'No screenplays or stage plays or novels, please. Send hard copy of full sample script. We read everything submitted this way. Sorry, we don't return scripts unless you send an s.a.s.e.'

## Promenade Enterprises Limited

6 Russell Grove, London SW9 6HS
℡ 020 7582 9354
info@promenadeproductions.com
www.promenadeproductions.com
Producer *Nicholas Newton*

Producer of drama for radio and theatre predominantly. Supports new writing but only accepts unsolicited mss via agents or producers.

## Ragdoll Productions

9 Timothy's Bridge Road, Stratford upon Avon
CV37 9NQ
☎ 01789 404100   📠 01789 404178
reception@ragdoll.co.uk
www.ragdoll.co.uk
Contact *Mary Hagger*

Producer of children's television programmes.
OUTPUT includes *In the Night Garden*;
*Teletubbies*; *Tronji*; *Brum*; *Rosie & Jim*; *Boohbah*
and *Open a Door*. 'We do not accept unsolicited
work as all our ideas are generated in-house.
However, we do look for writers who would be
prepared to work with us collaboratively as part
of a team.'

## Redweather

Easton Business Centre, Felix Road, Bristol
BS5 0HE
☎ 0117 941 5854   📠 0117 941 5851
production@redweather.co.uk
www.redweather.co.uk

Broadcast documentaries on arts and disability,
corporate video and CD-ROM, Water Aid, British
Oxygen, etc.

## The Reel Thing Ltd
▷ The Ashford Entertainment Corporation Ltd

## Richmond Films & Television Ltd

PO Box 33154, London NW3 4AZ
☎ 020 7722 6464
mail@richmondfilms.com
Contact *Development Executive*

Film and TV: drama and comedy. OUTPUT *Press
Gang*; *The Lodge*; *The Office*; *Wavelength*; *Privates*;
*in2minds*. No unsolicited scripts.

## RS Productions

191 Trewhitt Road, Newcastle upon Tyne NE6 5DY
☎ 07710 064632
info@rsproductions.co.uk
www.rsproductions.co.uk
Contact *Mark Lavender*

Feature films and television: drama series/serials
and singles. TV documentaries and series.
Working with established and new talent.

## Sands Films

119 Rotherhithe Street, London SE16 4NF
☎ 020 7231 2209   📠 020 7231 2119
info@sandsfilms.co.uk
www.sandsfilms.co.uk
Contacts *Christine Edzard, Olivier Stockman*

Film and TV drama. OUTPUT *Little Dorrit*; *The
Fool*; *As You Like It*; *A Dangerous Man*; *The Long

*Day Closes*; *A Passage to India*; *Topsy Turvy*;
*Nicholas Nickleby*; *The Gangs of New York*;
*The Children's Midsummer Night's Dream*. No
unsolicited scripts.

## Scala Productions Ltd

2nd Floor, 37 Foley Street, London W1W 7TN
☎ 020 7637 5720   📠 020 7637 5734
scalaprods@aol.com
Contacts *Ian Prior, Nik Powell*

Production company set up by ex-Palace
Productions Nik Powell and Stephen Woolley,
who have an impressive list of credits including
*Company of Wolves*; *Absolute Beginners*; *Mona
Lisa*; *Scandal*; *The Crying Game*; *Backbeat*; *Neon
Bible*; *24:7*; *Little Voice*; *Divorcing Jack*; *The Last
September*; *Wild About Harry*; *Last Orders*; *A
Christmas Carol – The Movie*; *Black and White*;
*Leo*; *The Night We Called It a Day*; *Ladies in
Lavender*; *Stoned*. In development: *Kickback*; *Two
Way Split*; *Johnny Bollywood*; *The Rough*; *Black
Cockatoo*; *Nightcab*; *Romantic Friction*; *White
Turks*; *The Next Gael*.

## Scope Productions Ltd

180 West Regent Street, Glasgow G2 4RW
☎ 0141 221 4312
laurakingwell@scopeproductions.co.uk
www.scopeproductions.co.uk
Corporate *Laura Kingwell*

Established 1984. Corporate video and multimedia
communications for clients across all sectors.

## Screenhouse Productions Ltd

Danebury Lodge, 9 The Drive, Roundhay, Leeds
LS8 1JF
☎ 0113 266 8881   📠 0113 266 8882
paul.bader@screenhouse.co.uk
www.screenhouse.co.uk
Contact *Paul Bader*

Specializes in science TV, documentary, stunts
and events, including outside broadcasts.
OUTPUT includes *Cosmos: A Beginner's Guide*
(BBC2 astronomy series); *Timewatch: The Hidden
Children* (BBC2); *Britain's Tallest Teenagers*
(ITV1); TV skills training courses, e.g. Location
Directing, Factual TV Research, and Media Skills
for Scientists. 'More likely to consider written up
proposals.'

## September Films Ltd

Glen House, 22 Glenthorne Road, London
W6 0NG
☎ 020 8563 9393   📠 020 8741 7214
september@septemberfilms.com
www.septemberfilms.com
Director of Production *Elaine Day*

Factual entertainment and documentary
specialists. Feature film OUTPUT includes

*Breathtaking*; *House of America*; *Solomon &
Gaenor*. No unsolicited submissions, please.

## Shell Like

81 Whitfield Street, London W1T 4HG
☏ 020 7255 5224   🖷 020 7255 5255
enquiries@shelllike.com
www.shelllike.com
Contact *Mike Blunt*

Produces radio commercials.

## Sianco Cyf

Pen-y-graig, Llanfaglan, Caernarfon LL54 5RF
☏ 01286 676100/07831 726111
sian@sianco.tv
Contact *Siân Teifi*

Children's, youth and education programmes,
children's drama, people-based documentaries
for adults. 'Please note, *we do not accept any
unsolicited scripts*.'

## Silent Sound Films Ltd

Cambridge Court, Cambridge Road, Frinton on
Sea CO13 9HN
☏ 01255 676381   🖷 01255 676381
thj@silentsoundfilms.co.uk
www.silentsoundfilms.co.uk
www.londonfoodfilmfiesta.co.uk
Contact *Timothy Foster*

Active in European film co-production with
mainstream connections in the USA. Special
interest in developing stage and film musicals,
art house fiction and documentaries on the arts.
Initial enquiry via email.

## Skyline Productions

10 Scotland Street, Edinburgh EH3 6PS
☏ 0131 556 2026
leslie@skyline.uk.com
www.skyline.uk.com
Producer/Writer *Leslie Hills*

Produces film and television drama. No
unsolicited material.

## Somethin Else

20-26 Brunswick Place, London N1 6DZ
☏ 020 7250 5500   🖷 020 7250 0937
info@somethinelse.com
www.somethinelse.com
Contact *Jez Nelson*

Producer of television, video and radio
documentaries, DVD and interactive content.
Ideas for TV shows welcome; send letter in the
first instance.

## Specific Films

25 Rathbone Street, London W1T 1NQ
☏ 020 7580 7476   🖷 020 7636 6866
info@specificfilms.com
Contact *Michael Hamlyn*

Founded 1991. OUTPUT includes *Mr Reliable*
(feature film co-produced by PolyGram and
the AFFC); *The Adventures of Priscilla, Queen
of the Desert*, co-produced with Latent Image
(Australia) and financed by PolyGram and AFFC;
*U2 Rattle and Hum*, full-length feature – part
concert film/part cinema verité documentary;
*Paws* (executive producer); *The Last Seduction* 2
(Polygram); *Priscilla, Queen of the Desert – The
Musical* (West End) and numerous pop promos
for major international artists.

## Spice Factory (UK) Ltd

14 Regent Hill, Brighton BN1 3ED
☏ 01273 739182   🖷 01273 749122
info@spicefactory.co.uk
www.spicefactory.co.uk

Founded 1995. Film producers. OUTPUT *Plots
With a View* (Christopher Walken, Brenda
Blethyn, Alfred Molina, Lee Evans); *Bollywood
Queen* (Preeya Kallidas, James McAvoy, Ian
McShane); *The Bridge of San Luis Rey* (Robert
De Niro, Kathy Bates, Harvey Keitel); *A Different
Loyalty* (Sharon Stone, Rupert Everett); *Head in
the Clouds* (Charlize Theron, Penelope Cruz);
*The Merchant of Venice* (Al Pacino, Jeremy Irons,
Joseph Fiennes, Lynn Collins). No unsolicited
material accepted.

## 'Spoken' Image Ltd

Studio 6, Riverside Mews, 4 Commercial Street,
Manchester M15 4RQ
☏ 0161 236 7522   🖷 0161 832 3708
info@spoken-image.com
www.spoken-image.com
Contact *Geoff Allman*

Film, video and TV production for documentary
and corporate material. Specializes in high-quality
brochures and reports, CD-ROMs, exhibitions,
conferences, film and video production for
broadcast, industry and commerce.

## Square Dog Radio LLP

Studio 6, Riverside Mews, 4 Commercial Street,
Manchester M15 4RQY
☏ 0161 236 7522   🖷 0161 832 3708
mike@squaredogradio.co.uk
www.squaredogradio.co.uk
Producers *Mike Hally, Mark Whitaker*

Producer of radio documentaries and features.
'No current contract for book or short story
readings and unlikely to do any in the future.' No
unsolicited material.

## Stirling Film & TV Productions Limited

137 University Street, Belfast BT7 1HP
☏ 028 9033 3848   🖷 028 9043 8644
anne@stirlingtelevision.co.uk
www.stirlingtelevision.co.uk
Contact *Anne Stirling*

Producer of broadcast and corporate programming – documentary, sport, entertainment and lifestyle programmes.

## STV Productions & Ginger Productions

STV: Pacific Quay, Glasgow G51 1PQ
☏ 0141 300 3000
www.stv.tv
Ginger Productions: 3 Waterhouse Square, 138–142 Holborn, London EC1N 2NY
☏ 020 7882 1020 www.ginger.tv
Director of Content *Alan Clements*

STV Productions is a major player in the UK and international television production industry, producing drama, children's factual and entertainment. STV Productions, together with sister company Ginger Productions, has produced a wide range of programmes for a variety of networks, including drama such as *Taggart* and *Rebus* for ITV1, high profile factual programming for ITV1 and ITV2 including *Jack Osbourne – Adrenaline Junkie*, and *DNA Stories* with Lorraine Kelly for Sky Real Lives.

## Sunset + Vine Productions Ltd

Elsinore House, 77 Fulham Palace Road, London W6 8JA
☏ 020 7478 7300   ☒ 020 7478 7403
www.sunsetvine.co.uk

Sports, children's and music programmes for television. No unsolicited mss. 'We hire freelancers only upon receipt of a commission.'

## Table Top Productions

1 The Orchard, Chiswick, London W4 1JZ
☏ 020 8742 0507   ☒ 020 8742 0507
alvin@tabletopproductions.com
Contact *Alvin Rakoff*

TV and film. OUTPUT *Paradise Postponed*; *The Adventures of Don Quixote*; *A Voyage Round My Father*; *The First Olympics 1896*; *Dirty Tricks*; *A Dance to the Music of Time*; *Too Marvellous for Words*. Also Dancetime Ltd. No unsolicited mss.

## talkbackTHAMES

20–21 Newman Street, London W1T 1PG
☏ 020 7861 8000   ☒ 020 7861 8001
www.talkbackthames.tv
Chief Executive Officer *Lorraine Heggessey*
Chief Operating Officer *Sara Geater*

talkackTHAMES is a **FremantleMedia** company. OUTPUT includes: comedy, comedy drama, drama, entertainment and factual programmes. OUTPUT *The Apprentice*; *Green Wing*; *The Bill*; *Britain's Got Talent*; *The X Factor*; *Never Mind the Buzzcocks*; *Grand Designs*; *Property Ladder*; *QI*.

## Talkingheads Production Ltd

Argyll House, All Saint's Passage, London SW18 1EP
☏ 020 7292 7575   ☒ 020 8875 8301
johnsachs@talkingheadsproductions.com
www.talkingheadsproductions.com
Contact *John Sachs*

Feature films. OUTPUT includes *The Merchant of Venice* starring Al Pacino. Will consider scripts; contact by email. 'Somewhere out there is the new Tom Stoppard.'

## Tandem Creative

Charleston House, 13 High Street, Hemel Hempstead HP1 3AA
☏ 01442 261576   ☒ 01442 219250
info@tandem.tv
www.tandem.tv
Production Director *Barbara Page*
Creative Director *Terry Page*
Production Controller *Jevan Green*

Internal and external communications, documentaries, drama-documentaries, public relations, sales, marketing and training programmes for, among others, the construction, civil engineering, transport, local government and charitable sectors. Welcomes unsolicited mss.

## Tern Television Productions Ltd

73 Crown Street, Aberdeen AB11 6EX
☏ 01224 211123   ☒ 01224 211199
aberdeen@terntv.com
www.terntv.com
*Also at:* 4th Floor, 114 Union Street, Glasgow G1 3QQ ☏ 0141 204 1717 glasgow@terntv.com
1st Floor, Cotton Court, 38–42 Waring Street, Belfast BT1 2ED ☏ 028 9032 6061 belfast@terntv.com
2nd Floor, 4 Great Newport Street, London WC2H 7HY ☏ 020 7836 9500 london@terntv.com
Contacts *David Strachan, Gwyneth Hardy (Aberdeen), Harry Bell (Glasgow)*

Award-winning factual content creator. The company output ranges from returning features brands, specialist factual series, comedy animation and archive, traditional observational documentary, constructed reality and drama documentary. Also, digital content department creating Multi Platform.

## Testimony Films

12 Great George Street, Bristol BS1 5RH
☏ 0117 925 8589   ☒ 0117 925 7608
steve.humphries@testimonyfilms.com
Contact *Steve Humphries*

TV documentary producer. Specializes in social history exploring Britain's past using living memory. OUTPUT includes *Hooked: History of Addictions*; *Married Love* (both Ch4 series); *A*

*Secret World of Sex* (BBC series); *The 50s & 60s in Living Colour*; *Some Liked It Hot* (both ITV series). Welcomes ideas from those working on life stories and oral history.

## Tiger Aspect Productions

5 Soho Square, London W1D 3QA
📞 020 7434 6700   📠 020 7434 1798
general@tigeraspect.co.uk
www.tigeraspect.co.uk
Head of Drama *Greg Brennan*
Head of Entertainment *Clive Tulloh*
Head of Comedy *Sophie Clarke-Jervoise*

Part of IMG Entertainment. Television producer for children's animation, comedy, drama, factual and entertainment. OUTPUT *Ross Kemp on Gangs*; *Robin Hood*; *Secret Diary of a Call Girl*; *The Catherine Tate Show*; *Horne and Corden*; *Harry and Paul*; *The Vicar of Dibley*. Sister company: Tiger Aspect Pictures (feature films; OUTPUT includes *Boys Are Back*; *Billy Elliot*; *The League of Gentlemen's Apocalypse*). Only considers material submitted via an agent or from writers with a known track record.

## Tigress Productions Ltd

2 St Paul's Road, Bristol BS8 1LT
📞 0117 933 5600   📠 0117 933 5666
general@tigressproductions.co.uk
www.tigressproductions.co.uk
Creative Director *Dick Colthurst*

Specializes in wildlife, science adventure and documentary projects. Output includes *Jaguar Adventure with Nigel Marven*; *Monty Halls' Great Escape*; *Everest: Beyond the Limit*.

## Tony Staveacre Productions

Channel View, Blagdon BS40 7TP
📞 01761 462161
newstaving@btinternet.com
Contact *Tony Staveacre*

Producer of dramas and documentaries as well as music, arts and comedy progammes. Recent OUTPUT *The Wodehouse Notebooks*; *The Liberation of Daphne*, *Speaking from the Belly*; *Standing Up for Liverpool*; *Kington's Last Tapes*; *Wodehouse in Hollywood* (BBC R4); *The Very Thought of You* and *Jigsy* (theatre); *Tango Maestro* (BBC4); *The Bands that Mattered* (BBC R2). No unsolicited mss.

## Touch Productions Ltd

18 Queen Square, Bath BA1 2HN
📞 01225 484666   📠 01225 483620
info@touchproductions.co.uk
www.touchproductions.co.uk
Contacts *Erica Wolfe-Murray, Malcolm Brinkworth*

Over the last 20 years, Touch has made a wide range of programmes including award-winning

investigations, popular documentaries, medical and science films, revelatory history productions as well as observational, social, religious and arts programmes. Current commissions include *The Human Footprint*, a Ch4 documentary special and various series and documentaries for the BBC, National Geographic, TLC and Animal Planet. Other projects include *Thanks for the Memories*; *The Boy Who Couldn't Stop Running*; *Parish in the Sun*; *Revival* and *Angela's Dying Wish*.

## Transatlantic Films Production and Distribution Company

Cabalva Studios, Whitney-on-Wye HR3 6EX
📞 01497 831800   📠 01497 831808
revel@transatlanticfilms.com
www.transatlanticfilms.com
Executive Producer *Revel Guest*

Producer of TV documentaries. OUTPUT *Belzoni* (Ch4 Schools); *Science of Sleep and Dreams*; *Science of Love* and *Extreme Body Parts* (all for Discovery Health); *Legends of the Living Dead* (Discovery Travel/S4C International); *2025* (Discovery Digital); *How Animals Tell the Time* (Discovery); *Trailblazers* (Travel Channel). No unsolicited scripts. Interested in new writers to write 'the book of the series', e.g. for *Greek Fire* and *History's Turning Points*, but not usually drama script writers.

## TV Choice Ltd

PO Box 597, Bromley BR2 0YB
📞 020 8464 7402   📠 020 8464 7845
tvchoiceuk@aol.com
www.tvchoice.uk.com
Contact *Norman Thomas*

Produces a range of educational videos for schools and colleges on subjects such as history, geography, business studies and economics. No unsolicited mss; send proposals only.

## Twentieth Century Fox Film Limited

Twentieth Century House, 31–32 Soho Square, London W1D 3AP
📞 020 7437 7766
www.fox.co.uk

London office of the American giant. Does not accept unsolicited material.

## Twenty Twenty Television

20 Kentish Town Road, London NW1 9NX
📞 020 7284 2020   📠 020 7284 1810
execs@twentytwenty.tv
www.twentytwenty.tv
Head of Development *Ana De Moraes*
Head of Production *Dominic Barlow*
Head of Factual *Sam Whittaker*

Founded 1982. Large independent television production company. Concentrates on documentaries, lifestyle programmes, popular drama and living history series. OUTPUT includes *The Choir* (BBC2); *Bad Lads Army* (ITV1); *How to Divorce Without Screwing Up Your Children* (Ch4). Both *Brat Camp* and *That'll Teach Em* have now aired a third season.

## Twofour Broadcast

5th Floor, 6–7 St Cross Street, London EC1N 8UA
☎ 020 7438 1800    🖷 020 7438 1850
enq@twofour.co.uk
www.twofour.co.uk
Managing Director *Melanie Leach*

Television production company founded in 1988. Specializes in factual and factual entertainment.

## Tyburn Film Productions Limited

Cippenham Court, Cippenham Lane, Cippenham, Nr Slough SL1 5AU
☎ 01753 516767    🖷 01753 691785

Feature films. Subsidiary of **Arlington Productions Limited**. No unsolicited submissions.

## Vera Productions

66–68 Margaret Street, London W1W 8SR
☎ 020 7436 6116    🖷 020 7436 6117
ideas@tvera.co.uk
Contact *Phoebe Wallace*

Produces television comedy such as *Bremner, Bird and Fortune*.

## Video Enterprises

12 Barbers Wood Road, High Wycombe HP12 4EP
☎ 01494 534144/07831 875216    🖷 01494 534144
videoenterprises@ntlworld.com
www.videoenterprises.co.uk
Proprietor *Maurice R. Fleisher*

Video and TV, mainly corporate: business and industrial training, promotional material and conferences. No unsolicited material 'but always ready to try out good new writers'.

## VIP Broadcasting

8 Bunbury Way, Epsom KT17 4JP
☎ 01372 721196
mail@vipbroadcasting.co.uk
Contact *Chris Vezey*

Produces a wide range of radio programmes, particularly interviews, documentaries, music programmes and live concerts. Awards at Sony's and New York Festivals. Approach with idea by email in the first instance; no unsolicited mss.

## Wall to Wall

8–9 Spring Place, London NW5 3ER
☎ 020 7485 7424    🖷 020 7267 5292
mail@walltowall.co.uk
www.walltowall.co.uk

Chief Executive *Alex Graham*

Factual and drama programming. OUTPUT includes *Who Do You Think You Are?*; *New Tricks*; *A Rather English Marriage*; *Glasgow Kiss*; *Sex, Chips & Rock 'n' Roll*; *The 1940s House*; *Body Story*; *Neanderthal*; *The Mafia*; *Not Forgotten*; *H.G. Wells*; *Dispatches*.

## Walsh Bros. Limited

29 Trafalgar Grove, London SE10 9TB
☎ 020 8858 6870
john@walshbros.co.uk
www.walshbros.co.uk
Producer/Director *John Walsh*
Producer/Head of Finance *David Walsh, ACA*
Producer/Head of Development *Maura Walsh*

BAFTA-nominated producers of television, film drama and documentaries. OUTPUT *Karate Kids* and *Sofa Surfers* (BBC); *Prime Minister's Global Fellowship* (Ch4); *Monarch* (feature film); BAFTA-winning *Don't Make Me Angry* (Ch4); Grierson-nominated *Headhunting the Homeless*; *Trex* (factual series on teenagers at work in China, Mexico, Vancouver and Alaska); *Trex2* (follow-up series covering Romania, India, Iceland and Louisiana); *Boyz & Girlz*, (Derbyshire dairy farm documentary series); *Cowboyz & Cowgirlz* (US sequel to hit series of Brit teens working on a ranch in Montana). Also arts documentaries: *The Comedy Store*; *Ray Harryhausen* (the work of Hollywood special effects legend). Drama: *The Sleeper*; *The Sceptic and the Psychic*; *A State of Mind*. Initial enquiries to: development@walshbros.co.uk

## Whistledown Productions Ltd

8a Ayres Street, London SE1 1ES
☎ 020 7407 8001
info@whistledown.net
www.whistledown.net
Managing Director *David Prest*

Producer of Sony Award-winning Landmark Series for BBC Radio 4. Features and documentaries on a wide range of social and historical subjects, contemporary issues and popular culture. OUTPUT includes *Questions Questions* and *The Reunion* (BBC R4); music-based documentaries for BBC R2. 'We welcome contributions and ideas, but phone or email first.'

## Wise Buddah Creative Ltd

74 Great Titchfield Street, London W1W 7QP
☎ 020 7307 1600    🖷 020 7307 1601
info@wisebuddah.com
www.wisebuddah.com
Contact *Chris North*

Radio production company: documentaries and commercials. Interactive department: video production, content prodution and podcasts.

Jingles music and imaging department. Also studio facilities, sound-to-picture/sound design, talent management. No unsolicited material.

## Witzend Productions
▷ FremantleMedia Group Ltd

## Working Title Films Ltd
2 Aybrook Street, London W1U 4AN
Ⓣ 020 7307 3000    Ⓕ 020 7307 3001/2/3
Co-Chairmen *Tim Bevan, Eric Fellner*
Head of Film *Debra Hayward*
Development Executive *Amelia Granger*

Feature films OUTPUT (since 2004) *Shaun of the Dead*; *The Calcium Kid*; *Thunderbirds*; *Wimbledon*; *Bridget Jones: The Edge of Reason*; *Inside I'm Dancing*; *The Interpreter*; *Mickybo & Me*; *Nanny McPhee*; *Pride and Prejudice*; *Catch a Fire*; *United 93*; *Sixty Six*; *Smokin' Aces*; *Gone*; *Mr Bean's Holiday*; *Atonement*; *The Golden Age*; *Definitely, Maybe*; *Hot Fuzz*; *Wild Child*; *Frost/Nixon*; *The Soloist*; *Burn After Reading*; *State of Play*; *Hippie Hippie Shake*; *The Boat that Rocked*; *Green Zone*; *A Serious Man*. No unsolicited mss at present.

## Zeppotron
▷ Endemol UK Productions

## Zodiak Entertainment
▷ Diverse Production Limited

# Theatre Producers

## Key advice

For latest minimum rates for writers of plays for subsidized repertory theatres (not Scotland) access the following websites: Theatrical Management Association at www.tmauk.org; and the Writers' Guild at www.writersguild.org.uk

## Actors Touring Company
> ATC

## Almeida Theatre Company

Almeida Theatre, Almeida Street, London N1 1TA
020 7288 4900    020 7288 4901
www.almeida.co.uk
Artistic Director *Michael Attenborough*

Founded 1980. The Almeida is a full-time producing theatre, presenting a year-round theatre programme in which new light is shed on an eclectic mix of new plays, classic revivals, adaptations and new versions of international work. Previous productions: *Festen*; *The Goat, or Who is Sylvia?*; *The Lady From the Sea*; *The Late Henry Moss*; *Enemies*. No unsolicited mss.

## ATC

The Tab Centre, 3 Godfrey Place, London E2 7NT
020 7033 7360    020 7739 8298
atc@atctheatre.com
www.atctheatre.com
Artistic Director *Bijan Sheibani*

Collaborates with writers on new work, adaptations and/or translations.

## Bill Kenwright Ltd

BKL House, 1 Venice Walk, London W2 1RR
020 7446 6200    020 7446 6222

Presents both revivals and new shows for West End and touring theatres. Although new work tends to be by established playwrights, this does not preclude or prejudice new plays from new writers. The company has an in-house dramaturg. Scripts should be addressed to Bill Kenwright with a covering letter and s.a.s.e. 'We have enormous amounts of scripts sent to us although we very rarely produce unsolicited work. Scripts are read systematically. Please do not phone; the return of your script or contact with you will take place in time.'

## Birmingham Repertory Theatre

Centenary Square, Broad Street, Birmingham B1 2EP
0121 245 2000    0121 245 2100
writing@birmingham-rep.co.uk
www.birmingham-rep.co.uk
Contact *Arts Team*

The Birmingham Repertory Theatre aims to provide a platform for the best work from new writers from both within and beyond the West Midlands region. The Rep is committed to the production of new work which reflects both the diversity of contemporary experience and of approaches to writing for the stage. The commissioning of new plays takes place across the full range of the theatre's activities: in the Main House, The Door (which is a dedicated new writing space) and on tour to community venues in the region. The theatre runs a programme of writers' attachments every year in addition to its commissioning policy and maintains close links with *Script* (the regional writers' training agency) and the MPhil in Playwriting Studies at the University of Birmingham. For more information contact the Arts Team.

## Black Theatre Co-op
> Nitro

## Bootleg Theatre Company

23 Burgess Green, Bishopdown, Salisbury SP1 3EL
01722 421476
www.boottheatre.co.uk
Contact *Colin Burden*

Founded 1984. Tries to encompass as wide an audience as possible and has a tendency towards plays with socially relevant themes. A good bet for new writing since unsolicited mss are very welcome. 'Our policy is to produce new and/or rarely seen plays and anything received is given the most serious consideration.' Actively seeks to obtain grants to commission new writers for the company. New productions include: *Girls Allowed* by Trevor Suthers; *A Rainy Night in Soho* Stephen

Includes FREE online access to **www.thewritershandbook.com**

Giles; *Counting the Days* Roger Goldsmith plus new plays by Alec Taylor and Mark Bromley.

## Bristol Old Vic

King Street, Bristol BS1 4ED
☎ 0117 949 3993
sharon@bristol-old-vic.co.uk
www.bristololdvic.org.uk
Artistic Director *Tom Morris*
Literary Contact *Sharon Clark*

Bristol Old Vic's mission is to place writing for performance at the heart of its organization, providing a strong base for creative excellence, curiosity, adventure and passion for writers, practitioners and audiences. Please contact us before sending unsolicited scripts. There is a database for writers working and living in the South West – please contact by email to be included.

## The Bush Theatre

Shepherds Bush Green, London W12 8QD
☎ 020 8743 3584
info@bushtheatre.co.uk
www.bushtheatre.co.uk
www.bushgreen.org
Artistic Director *Josie Rourke*

One of the country's leading new writing venues. In the 38 years since its founding in 1972, the Bush Theatre has gained an impressive reputation for its ability to spot and nurture new playwrights. It has launched the careers of such highly respected talents as Jack Thorne, Lucy Kirkwood, Simon Stephens, Sharman Macdonald, Tony Kushner, Catherine Johnson, Stephen Poliakoff, Victoria Wood, Billy Roche, Jonathan Harvey, Conor McPherson and David Eldridge. In addition to productions at their home base, many Bush plays go on national tours, transfer to London's West End, Ireland, the USA, and development for TV and film. 'The Bush Theatre's aim is to discover, produce and promote the most innovative and exciting new playwrights of the future.' With bushfutures, the theatre offers an ambitious professional development and community outreach programme. In 2009, the Bush Theatre launched bushgreen a social networking website for people in theatre to connect, collaborate and publish plays in innovative ways. The mission of bushgreen is to connect playwrights with theatre practitioners, plays with producers and to lead and inspire best practice in the putting on of new plays.

## Citizens Theatre

Gorbals, Glasgow G5 9DS
☎ 0141 429 5561    📠 0141 429 7374
info@citz.co.uk
www.citz.co.uk
Artistic Directors *Jeremy Raison, Guy Hollands*

Administrative Director *Anna Stapleton*

No formal new play policy. The theatre has a play reader but opportunities to do new work are limited.

## Clwyd Theatr Cymru

Mold, Flintshire CH7 1YA
☎ 01352 756331    📠 01352 701558
drama@celtic.co.uk
www.clwyd-theatr-cymru.co.uk
Literary Manager *William James (william.james@ clwyd-theatr-cymru.co.uk)*

Clwyd Theatr Cymru produces plays performed by a core ensemble in Mold and tours them throughout Wales (in English and Welsh). Productions are a mix of classics, revivals and contemporary drama. Recent new writing includes: *A History of Falling Things* James Graham; *Drowned Out/Porth y Byddar* Manon Eames; *Two Princes* Meredydd Barker; *Memory* Jonathan Lichtenstein; *Stone City Blue* Ed Thomas; *The Rabbit* Meredydd Barker; *The Journey of Mary Kelly* Siân Evans; *Yesterday*; *Tales from Europe*; *Tales from Small Nations*; *The Way It Was*; *Flights of Fancy*; *The Ballad of Megan Morgan*; *Flora's War/Rhyfel Flora* all by Tim Baker. Plays by Welsh writers or on Welsh themes will be considered.

## Colchester Mercury Theatre Limited

Balkerne Gate, Colchester CO1 1PT
☎ 01206 577006    📠 01206 769607
info@mercurytheatre.co.uk
www.mercurytheatre.co.uk
Chief Executive & Artistic Director *Dee Evans*

Producing theatre with a wide-ranging audience. New writing encouraged. The theatre has a free playwright's group for adults with a serious commitment to writing plays.

## The Coliseum, Oldham

Fairbottom Street, Oldham OL1 3SW
☎ 0161 624 1731    📠 0161 624 5318
mail@coliseum.org.uk
www.coliseum.org.uk
Chief Executive/Artistic Director *Kevin Shaw*

The artistic policy of the theatre is to present a high quality and diverse theatre programme with the ambition to commission a new play each year. The Coliseum employs a reader to read all submitted scripts; a written report will be sent to the writer within approximately four months. An s.a.s.e. should be enclosed for return of the script along with an s.a.s.e. for acknowledgement of receipt, if required. All submitted scripts must be clearly typed, using one side of A4 for each page, and must contain a page of casting requirements.

## Concordance

Finborough Theatre, 118 Finborough Road, London SW10 9ED

☎ 020 7244 7439    📠 020 7835 1853
admin@concordance.org.uk
www.concordance.org.uk
Artistic Director *Neil McPherson*

Founded in 1981, Concordance is the resident company based at the **Finborough Theatre** (see entry). Presents world premières of new writing and revivals of neglected work with a special commitment to music theatre as well as integrating music into its work and productions featuring the writing of non-theatrical artists – poets, artists, novelists, etc. – presenting their work in a theatrical setting. We accept unsolicited scripts through the Finborough Theatre's submission department (details at www.finboroughtheatre.co.uk).

## Contact Theatre Company

Oxford Road, Manchester M15 6JA

☎ 0161 274 0623
www.contact-theatre.org

An award-winning company, Contact focuses on theatre for young adults. There is a wide range of activities, from outreach projects to intensive artistic development programmes such as Young Actors and Young Writers groups. Contact is currently reviewing its submission process; see the website for updates.

## Crucible Theatre

Sheffield Theatres, 55 Norfolk Street, Sheffield S1 1DA

☎ 0114 249 5999    📠 0114 249 6003
www.sheffieldtheatres.co.uk
Executive Producer *Daniel Bates*
Artistic Director *Daniel Evans*

Produces a rich variety of plays (classical, 20th Century and new) and musicals in the three spaces (a thrust stage, a studio space and a proscenium arch theatre).

## Druid Theatre Company

Flood Street, Galway, Republic of Ireland

☎ 00 353 91 568660    📠 00 353 91 563109
info@druid.ie
www.druid.ie
Artistic Director *Ms Garry Hynes*
New Writing Manager *Thomas Conway*
General Manager *Tim Smith*

Founded in 1975 and based in Galway, 'Druid has always worked to reinvigorate perceptions of classic dramatic works and to engage with new dramatic works of a challenging, innovative and daring kind'. Druid consistently commissions, develops, and produces new plays by a wide range of emerging and established writers both from Ireland and abroad.

## The Dukes

Moor Lane, Lancaster LA1 1QE

☎ 01524 598500    📠 01524 598519
info@dukes-lancaster.org
www.dukes-lancaster.org
Chief Executive/Artistic Director *Joe Sumsion*

Founded 1971. The only producing house in Lancashire. Wide target market for cinema and theatre. Plays in a 313-seater end-on auditorium and in a 198-seater in-the-round studio (The Dukes' newly refurbished space). Host for **Litfest** – Lancaster's annual festival of literature. In the summer months open-air promenade performances are held in Williamson Park. DT3 – The Education Centre is dedicated to work by and for young people.

## Dundee Repertory Theatre

Tay Square, Dundee DD1 1PB

☎ 01382 227684    📠 01382 228609
hwatson@dundeereptheatre.co.uk
www.dundeereptheatre.co.uk
Artistic Director *James Brining*

Founded 1939. Plays to a varied audience. Translations and adaptations of classics, and new local plays. Most new work is commissioned. Interested in contemporary plays in translation and in new Scottish writing. No scripts except by prior arrangement.

## Eastern Angles Theatre Company

Sir John Mills Theatre, Gatacre Road, Ipswich IP1 2LQ

☎ 01473 218202    📠 01473 384999
admin@easternangles.co.uk
www.easternangles.co.uk
Artistic Director *Ivan Cutting*
General Manager *Jill Streatfeild*

Founded 1982. Plays to a rural audience for the most part. New work only: some commissioned, some devised by the company, some researched documentaries. Unsolicited mss welcome but scripts or writers need to have some connection with East Anglia. 'We are always keen to develop and produce new writing, especially that which is germane to a rural area.'

## Edinburgh Royal Lyceum Theatre
▷ Royal Lyceum Theatre Company

## English Touring Theatre

25 Short Street, London SE1 8LJ

☎ 020 7450 1990
admin@ett.org.uk
www.ett.org.uk
Director *Rachel Tackley*

Founded 1993. National touring company visiting middle and large-scale receiving houses and arts centres throughout the UK. Produces a mixed programme of plays. *Unable to accept or process unsolicited scripts.*

## Exeter Northcott

Stocker Road, Exeter EX4 4QB
☎ 01392 223999    🅕 01392 223996
www.exeternorthcott.co.uk

Exeter Northcott offers a programme of in-house and touring theatre productions, contemporary dance, music and comedy. Currently accepting unsolicited scripts from writers in the South West region only. Please check the news section of the website for updates on open calls for submissions.

## Finborough Theatre

118 Finborough Road, London SW10 9ED
☎ 020 7244 7439    🅕 020 7835 1853
admin@finboroughtheatre.co.uk
www.finboroughtheatre.co.uk
Artistic Director *Neil McPherson*

Founded 1980. 'One of London's leading new writing venues' (*Time Out*). Presents new British writing, revivals of neglected work from the 19th and 20th centuries, music theatre and UK premières of foreign work, particularly from Ireland, the United States and Canada. The theatre is available for hire and the fee is sometimes negotiable to encourage interesting work. Premièred work by Chris Lee, Anthony Neilson, Naomi Wallace, Tony Marchant, Mark Ravenhill, Laura Wade, James Graham. Five times winner of the Pearson Award. Unsolicited scripts are accepted, but please read carefully the details on submissions policy on the website before sending any scripts.

## Gate Theatre Company Ltd

11 Pembridge Road, London W11 3HQ
☎ 020 7229 5387    🅕 020 7221 6055
gate@gatetheatre.co.uk
www.gatetheatre.co.uk
Artistic Directors *Natalie Abrahami, Carrie Cracknell*

Founded 1979. Plays to a mixed, London-wide audience, depending on production. Aims to produce British premières of plays which originate from abroad and translations of neglected classics. Productions include *The Sexual Neuroses of Our Parents* by Lukas Bärfuss, transl. by Neil Blackadder; *The Car Cemetery* by Fernando Arrabal, transl. by Barbara Wright; *I am Falling*, devised in collaboration with choreographer Anna Williams and dramturg Jenny Worton; *Press*, co-production created and performed by Pierre Rigal and his company, dernière minute; *The Internationalist*, by US

writer Anne Washburn. Positively encourages writers from abroad to send in scripts or translations. Most unsolicited scripts are read but plays by writers from the UK will not be accepted. Please address submissions to the Artistic Directors. Always enclose s.a.s.e. if play needs returning.

## Graeae Theatre Company

Bradbury Studios, 138 Kingsland Road, London E2 8DY
☎ 020 7613 6900    🅕 020 7613 6919
info@graeae.org
www.graeae.org
Minicom 020 7700 8184
Joint CEO/Artistic Director *Jenny Sealey*
Joint CEO/Executive Director *Judith Kilvington*

Europe's premier theatre company of disabled people, the company tours nationally and internationally with innovative theatre productions highlighting both historical and contemporary disabled experience. Graeae also runs educational programmes available to schools, youth clubs, students and disabled adults nationally and provides vocational training in theatre arts (including playwriting). Unsolicited scripts, from disabled writers, welcome. New work is commissioned.

## Hampstead Theatre

Eton Avenue, Swiss Cottage, London NW3 3EU
☎ 020 7449 4200    🅕 020 7449 4201
literary@hampsteadtheatre.com
www.hampsteadtheatre.com
Contact *Katy Silverton*

Hampstead Theatre has been associated with new writing since its foundation in 1959. Based in a state-of-the art building, the theatre is an intimate space with a flexible stage and an auditorium capable of expanding to seat 325. The artistic policy continues to be the production of British and international new plays and the development of important young writers. 'We are looking for writers who recognize the power of theatre and who have a story to tell. All plays which meet our submission requirements are read and discussed. We give feedback to all writers with potential.' Writers produced in the last five years at Hampstead include Gregory Burke, Dennis Kelly, Nell Leyshon, Sharman Macdonald, Tamsin Oglesby, Debbie Tucker Green and Roy Williams. In earlier years, breakthrough plays by Harold Pinter, David Hare, Mike Leigh, Pam Gems and Stephen Jeffreys were produced.

## Harrogate Theatre

Oxford Street, Harrogate HG1 1QF
☎ 01423 502710    🅕 01423 563205
www.harrogatetheatre.co.uk

'Due to funding cuts, Harrogate Theatre cannot accept any new writing at present. It is hoped that this situation will change in the future.'

## Headlong Theatre

34–35 Berwick Street, London W1F 8RP
☎ 020 7478 0270    🖷 020 7434 1749
info@headlongtheatre.co.uk
www.headlongtheatre.co.uk
Artistic Director *Rupert Goold*

A middle-scale touring company producing established and new plays. On average, one new play or new adaptation a year. Due to forthcoming projects the company is not considering unsolicited scripts at present.

## Hull Truck Theatre Company

Ferensway, Hull HU2 8LB
☎ 01482 224800    🖷 01482 581182
admin@hulltruck.co.uk
www.hulltruck.co.uk
Operations Manager *Paul Marshall*
Artistic Directors *John Godber, Gareth Tudor-Price*
Associate Director *Nick Lane*

John Godber, of *Teechers, Bouncers* and *Up 'n' Under* fame (the artistic director of this high-profile Northern company since 1984), has very much dominated the scene in the past with his own successful plays. The emphasis is still on new writing but Godber's works continue to be toured extensively. Recent premières/commissions include *On a Shout*; *Blue Cross Xmas*; *Sold*; *Wilde Boyz*; *Upon on Roof*; *Christmas Crackers*; *Crown Prince*; *My Favourite Summer*; *Ladies Down Under*; *Sully*. Most new plays are commissioned. 'New scripts should be addressed to *Nick Lane*, Associate Director, who will attempt as quick a response as possible. Bear in mind the artistic policy of Hull Truck, which is accessibility and popularity.' In general, not interested in musicals or in plays with casts of more than seven.

## Komedia

44–47 Gardner Street, North Laine, Brighton BN1 1UN
☎ 01273 647101    🖷 01273 647102
admin@komedia.co.uk
www.komedia.co.uk
Contact *Marina Cobler*

Founded in 1994, Komedia promotes, produces and presents new work. Mss of new plays welcome.

## Library Theatre Company

St Peter's Square, Manchester M2 5PD
☎ 0161 234 1913    🖷 0161 274 7055
ltcadmin@manchester.gov.uk
www.librarytheatre.com
Artistic Director *Chris Honer*

Produces new and contemporary work, as well as the classics. No unsolicited mss. Send outline of the nature of the script in the first instance. Encourages new writing through the commissioning of new plays and through a programme of rehearsed readings to help writers' development.

## Live Theatre Company

Broad Chare, Quayside, Newcastle upon Tyne NE1 3DQ
☎ 0191 261 2694    🖷 0191 232 2224
www.live.org.uk
Artistic Director *Max Roberts*
Chief Executive *Jim Beirne*

Live Theatre offers support and advice to writers of all abilities, from first-time writers to successful playwrights. Runs a series of events, courses and workshops and offers a free script reading service (priority given to those from and/or living in the North East region of England). See the website for submission details. New Writing Studio Sessions, held on the first Thursday of every month, feature some of the country's leading playwrights talking about their work while Short Cuts, Live Theatre's open access night presents rehearsed readings of new plays. Also runs an 'Introduction to Playwriting' course. Productions include *The Pitmen Painters, Cooking With Elvis* and *NE1* by Lee Hall; *A Nightingale Sang* by C.P. Taylor; *Laughter When We're Dead* by Sean O'Brien; *Falling Together* by Tom Hadaway; *Bones* by Peter Straughan.

## London Bubble Theatre Company

5 Elephant Lane, London SE16 4JD
☎ 020 7237 4434    🖷 020 7231 2366
admin@londonbubble.org.uk
www.londonbubble.org.uk
Creative Director *Jonathan Petherbridge*

London Bubble has existed within its community of southeast London for 36 years offering everyone access to enjoying theatre whether as an audience member or theatre-maker. Works with individuals and communities that may otherwise not have a chance to experience theatre, and provides access into the arts for young professionals. London Bubble has an extensive participatory programme, offering the chance to work collaboratively with theatre professionals to create and stage a piece of theatre which is then toured to parks and unusual open spaces around S.E. London. Runs innovative arts education projects in schools with a focus on Greenwich, Lewisham and Southwark.

## Lyric Hammersmith

Lyric Square, King Street, London W6 0QL
☎ 0871 221 1722    🖷 020 8741 5965

enquiries@lyric.co.uk
www.lyric.co.uk
Executive Director *Jessica Hepburn*

The main theatre stages an eclectic programme of new and revived classics. Interested in developing projects with writers, translators and adaptors. The Lyric does not accept unsolicited scripts. Its 110-seat studio focuses on work for children, young people and families.

## mac

Cannon Hill Park, Birmingham B12 9QH
www.macarts.co.uk

Closed in 2008 for a £14.8 million rebuild and refurbishment. Due to re-open in 2010. See website for details.

## Manchester Library Theatre
▷ Library Theatre Company

## Michael Codron Plays Ltd

Aldwych Theatre Offices, Aldwych, London
WC2B 4DF
📞 020 7240 8291    📠 020 7240 8467

Michael Codron Plays Ltd manages the Aldwych Theatre in London's West End. The plays it produces do not necessarily go into the Aldwych but always tend to be big-time West End fare. Previous productions: *Entertaining Angels*; *Democracy*; *Blue Orange*; *Copenhagen*; *The Invention of Love*; *Hapgood*; *Uncle Vanya*; *Rise and Fall of Little Voice*; *Arcadia*; *Dead Funny*; *My Brilliant Divorce*; *Dinner*; *Democracy*; *Glorious!*. No particular rule of thumb on subject matter or treatment. The acid test is whether 'something appeals to Michael'. Straight plays rather than musicals.

## New Vic Theatre

Etruria Road, Newcastle under Lyme ST5 0JG
📞 01782 717954    📠 01782 712885
admin@newvictheatre.org.uk
www.newvictheatre.org.uk
Artistic Director *Theresa Heskins*

The New Vic is a purpose-built theatre-in-the-round. Produces ten in-house plays each year and is active within the education sector and community. New plays produced are the result of specific commissions. Send a one-page synopsis, the first ten pages of the script and a c.v., *not* unsolicited scripts. 'We cannot guarantee that unsolicited scripts will be read; they will be returned on receipt of an s.a.s.e.'

## Nimax Theatres Limited

1 Lumley Court, off 402 Strand, London
WC2R 0NB
📞 020 7395 0780
general@nimaxtheatres.com
www.nimaxtheatres.com

CEO *Nica Burns*

West End theatre managers of the Apollo, Duchess, Garrick, Lyric and Vaudeville theatres. Unsolicited scripts are not considered.

## Nitro

6 Brewery Road, London N7 9NH
📞 020 7609 1331    📠 020 7609 1221
info@nitro.co.uk
www.nitro.co.uk
Artistic Director *Felix Cross*

Founded 1979. Formerly Black Theatre Co-op. A music theatre company, producing dynamic new work that explores stories by black people in Britain and elsewhere. 'Nitro productions present popular and epic themes, embracing the vibrant spectrum of black music and dance, appealing to diverse audiences of all ages.'

## Northumberland Theatre Company

The Playhouse, Bondgate Without, Alnwick
NE66 1PQ
📞 01665 602586    📠 01665 605837
admin@northumberlandtheatre.co.uk
www.northumberlandtheatre.co.uk
Artistic Director *Gillian Hambleton*
Tours Manager *Hilary Burns*

Founded 1978. An Arts Council England Regularly Funded Organization. Predominantly rural, small-scale touring company, playing to village halls and community centres throughout the northern region, the Scottish Borders and countrywide. Productions range from established classics to new work but must be appropriate to their audience. Unsolicited scripts welcome but are unlikely to be produced. All scripts are read and returned with constructive criticism within six months. Writers whose style is of interest may then be commissioned. The company encourages new writing and commissions when possible. Financial constraints restrict casting to a *maximum* of five/six and minimal set.

## Norwich Puppet Theatre

St James, Whitefriars, Norwich NR3 1TN
📞 01603 615564    📠 01603 617578
info@puppettheatre.co.uk
www.puppettheatre.co.uk
General Manager *Ian Woods*

Young/family-centred audience (aged 3–12) but developing shows for adult audiences interested in puppetry. All year round programme plus tours to schools and arts venues. Most productions are based on traditional stories but unsolicited mss welcome if relevant.

## Nottingham Playhouse

Nottingham Playhouse Trust, Wellington Circus, Nottingham NG1 5AF
☎ 0115 947 4361   🖷 0115 947 5759
www.nottinghamplayhouse.co.uk
Artistic Director *Giles Croft*

Aims to make innovation popular, and present the best of world theatre, working closely with the communities of Nottingham and Nottinghamshire. Unsolicited mss will be read. It normally takes about six months, however, and 'we have never yet produced an unsolicited script. All our plays have to achieve a minimum of 60 per cent audiences in a 732-seat theatre. We have no studio.'

## Nottingham Playhouse Roundabout Theatre in Education

Wellington Circus, Nottingham NG1 5AF
☎ 0115 947 4361
andrewb@nottinghamplayhouse.co.uk
www.nottinghamplayhouse.co.uk
Artistic Director *Andrew Breakwell*

Founded 1973. Theatre-in-Education company of the **Nottingham Playhouse**. Plays to a young audience aged 5–18 years of age. 'We are committed to the encouragement of new writing and commission at least one new play for young people each year. With other major producers in the East Midlands we share the resources of the *Theatre Writing Partnership* which is based at the Playhouse.' Please make contact before submitting scripts.

## Nuffield Theatre

University Road, Southampton SO17 1TR
☎ 023 8031 5500   🖷 023 8031 5511
alison.thurley@nuffieldtheatre.co.uk
www.nuffieldtheatre.co.uk
Artistic Director *Patrick Sandford*
Script Executive *John Burgess*

Well known as a good bet for new playwrights, the Nuffield gets an awful lot of scripts. Produces two new main stage plays every season. Previous productions: *Exchange* by Yuri Trifonov (transl. Michael Frayn) which transferred to the Vaudeville Theatre; *The Floating Light Bulb* Woody Allen (British première); *Nelson*, a new play by Pam Gems; *The Dramatic Attitudes of Miss Fanny Kemble* Claire Luckham; *The Winter Wife* Claire Tomalin and *Tchaikovsky and the Queen of Spades* John Clifford. Open-minded about subject and style, producing musicals as well as straight plays. Also opportunities for some small-scale fringe work. Scripts preferred to synopses in the case of writers new to theatre. All will, eventually, be read 'but please be patient. We do not have a large team of paid readers. We read everything ourselves.'

## Octagon Theatre Bolton

Howell Croft South, Bolton BL1 1SB
☎ 01204 529407   🖷 01204 556502
info@octagonbolton.co.uk
www.octagonbolton.co.uk
Executive Director *John Blackmore*

Founded in 1967, the award-winning Octagon Theatre Bolton stages at least eight main auditorium home-produced shows a year, and hosts UK touring companies such as **Out of Joint, Nitro**, and **Live Theatre Company** as well as the work of partner companies as part of its commitment to creative partnerships. The Theatre also programmes BoltON, a comprehensive and innovative series of special events to supplement theatregoers' experience. The Theatre boasts a thriving and constantly developing participatory department, activ8, which operates a highly successful youth theatre as well as initiating and facilitating education and outreach programmes. The Theatre is keen to encourage new writing through its relationships with **North West Playwrights** and the University of Bolton. Contact *Elizabeth Newman* on 01204 529407 for details of new writing opportunities with the Octagon Theatre.

## Orange Tree Theatre

1 Clarence Street, Richmond TW9 2SA
☎ 020 8940 0141   🖷 020 8332 0369
admin@orangetreetheatre.co.uk
www.orangetreetheatre.co.uk
Artistic Director *Sam Walters*
Executive Director *Gillian Thorpe*

The Orange Tree, a theatre-in-the-round venue, presents a broad cross section of work. Past productions have included plays by Rodney Ackland, John Galsworthy and new translations of Havel, Lorca and Michael Vinaver as well as new plays by Oliver Ford Davies, David Lewis, Ben Brown and Kenneth Jupp. The theatre no longer considers unsolicited mss; writers who may wish to approach the theatre are asked to write first. Submissions deemed to have potential can now have their work developed involving the Orange Tree's network of actors and directors in their rehearsal space. Play readings are also accommodated.

## Out of Joint

7 Thane Works, Thane Villas, London N7 7NU
☎ 020 7609 0207   🖷 020 7609 0203
ojo@outofjoint.co.uk
www.outofjoint.co.uk
Director *Max Stafford-Clark*
Producer *Graham Cowley*
Literary Associate *Alex Yates*

Founded 1993. Award-winning theatre company with new writing central to its policy. Produces new plays which reflect society and its concerns, placing an emphasis on education activity to attract young audiences. Welcomes unsolicited mss. Productions include: *Blue Heart* Caryl Churchill; *Our Lady of Sligo*, *The Steward of Christendom* and *Hinterland* Sebastian Barry; *Shopping and Fucking* and *Some Explicit Polaroids* Mark Ravenhill; *A State Affair* and *Talking to Terrorists* Robin Soans; *Sliding with Suzanne* Judy Upton; *The Positive Hour* and *A Laughing Matter* April De Angelis; *Duck* and *O go my Man* Stella Feehily; *The Permanent Way* David Hare; *The Overwhelming* J.T. Rogers; *Feelgood* and *King of Hearts* Alistair Beaton; *Testing the Echo* David Edgar.

## Paines Plough

4th Floor, 43 Aldwych, London WC2B 4DN
℡ 020 7240 4533    🖷 020 7240 4534
office@painesplough.com
www.painesplough.com
Artistic Directors *James Grieve, George Perrin*
Executive Producer *Anneliese Davidsen*
Literary Director *Tessa Walker*

Founded 1974. Award-winning company commissioning and producing new plays by British and international playwrights. Tours 2–4 plays a year, nationally and worldwide, for small and middle-scale venues. Also runs a range of projects focused on identifying and launching emerging playwrights (see website for details). 'We welcome unsolicited scripts and will respond to all submissions. We are looking for original plays that engage with the contemporary world and are written in a distinctive voice.' Recent productions: *Orphans* Dennis Kelly; *Roaring Trade* Steve Thompson; *Dallas Sweetman* Sebastian Barry; *Shoot/Get Treasure/Repeat* and *Product* Mark Ravenhill; *Long Time Dead* Rona Munro; *After the End* Dennis Kelly; *If Destroyed True* Douglas Maxwell; *Pyrenees* David Greig; *Mercury Fur* Philip Ridley; *The Small Things* Enda Walsh.

## Perth Theatre

Horsecross Arts Ltd, 185 High Street, Perth PH1 5UW
℡ 01738 472700    🖷 01738 624576
info@horsecross.co.uk
www.horsecross.co.uk
Creative Director – Theatre *Ian Grieve*

Founded 1935. Combination of one to four-weekly repertoire of plays, incoming tours and studio productions. Unsolicited mss are read when time permits, but the timetable for return of scripts is lengthy. New plays staged by the company are usually commissioned under the SAC scheme.

## Plymouth Theatre Royal
▷ Theatre Royal

## Polka Theatre
240 The Broadway, Wimbledon SW19 1SB
℡ 020 8543 4888 (box office)    🖷 020 8545 8365
admin@polkatheatre.com
www.polkatheatre.com
Artistic Director *Jonathan Lloyd*
Executive Director *Stephen Midlane*

Founded in 1968 and moved into its Wimbledon base in 1979. Leading children's theatre committed to commissioning and producing plays for children age 0 to 11. Programmes are planned two years ahead and at least three original new plays are commissioned each year. 'Our scripts are commissioned from established writers, though we are keen to develop work from writers new to children's theatre. We are unable to read unsolicited scripts from writers without professional productions to their credit.'

## Queen's Theatre, Hornchurch
Billet Lane, Hornchurch RM11 1QT
℡ 01708 462362    🖷 01708 462363
info@queens-theatre.co.uk
www.queens-theatre.co.uk
Artistic Director *Bob Carlton*

The Queen's Theatre is a 503-seat theatre producing eight main house productions per year, including pantomime. Aims for distinctive and accessible performances in an identifiable house style focused upon actor/musician shows but, in addition, embraces straight plays, classics and comedies. Also runs an occasional writers' group for adults (contact the Education Manager for information). The group has close links with the Queen's Community Company (amateur actors) as well as the main house company, which workshops and showcases the group's work.

## The Really Useful Group Ltd
22 Tower Street, London WC2H 9TW
℡ 020 7240 0880    🖷 020 7240 1204
www.reallyuseful.com

Commercial/West End, national and international theatre producer/co-producer/licensor whose output has included *Joseph and the Amazing Technicolor Dreamcoat*; *Jesus Christ Superstar*; *Cats*; *Evita*; *Song & Dance*; *Daisy Pulls It Off*; *Lend Me a Tenor*; *Starlight Express*; *The Phantom of the Opera*; *Aspects of Love*; *Sunset Boulevard*; *By Jeeves*; *Whistle Down the Wind*; *The Beautiful Game*; *Tell Me On a Sunday*; *Bombay Dreams*; *The Woman in White*; *The Sound of Music*; *Love Never Dies*.

## Red Ladder Theatre Company

3 St Peter's Buildings, York Street, Leeds LS9 8AJ
☎ 0113 245 5311 📠 0113 245 5351
rod@redladder.co.uk
www.redladder.co.uk
Artistic Director *Rod Dixon*

Founded 1968. National company touring
1–2 shows a year with a strong commitment
to new work and new writers. Aimed at an
audience of young people, performances are
held in community and theatre venues. Recent
productions: *Kaahini* Maya Chowdhry; *Door:
This Life Was Given to Me* Madani Younis;
*Where's Vietnam?* Alice Nutter; *Forgotten Things*
Emma Adams. The company is particularly keen
to enter into a dialogue with writers with regard
to creating new work for young people. Email
the artistic director for more information at the
address above.

## Red Shift Theatre Company

PO Box 60151, London SW19 2TB
☎ 020 8540 1271 📠 020 8540 1271
jane@redshifttheatreco.co.uk
www.redshifttheatreco.co.uk
Artistic Director *Jonathan Holloway*

Founded in 1982, Red Shift is a middle-scale
touring company staging 'commercial'
productions. Interested in new plays on subjects
and themes with broad audience appeal. Willing
to read treatments but full scripts should be sent
by invitation only.

## Ridiculusmus

c/o BAC, Lavender Hill, London SW11 5TN
☎ 020 7223 9959
enquiries@ridiculusmus.com
www.ridiculusmus.com
Artistic Directors *Jon Haynes, David Woods*

Founded 1992. Touring company which plays
to a wide range of audiences. Productions have
included adaptations of *The Importance of
Being Earnest; Three Men In a Boat; The Third
Policeman; At Swim Two Birds* and original work:
*The Exhibitionists; Yes, Yes, Yes; Say Nothing* and
*Ideas Men.* Most recent work: *Tough time, nice
time.* Unsolicited scripts not welcome.

## Royal Court Theatre

Sloane Square, London SW1W 8AS
☎ 020 7565 5050
literary@royalcourttheatre.com
www.royalcourttheatre.com
Literary Manager *Ruth Little*

The Royal Court is a leading international theatre
producing up to 17 new plays each year in its
400-seat proscenium theatre and 80-seat studio.
In 1956 its first director George Devine set out to
find 'hard-hitting, uncompromising writers whose
plays are stimulating, provocative and exciting'.
This artistic policy helped transform post-war
British theatre, with new plays by writers such
as John Osborne, Arnold Wesker, John Arden,
Samuel Beckett, Edward Bond and David Storey,
through to Caryl Churchill, Jim Cartwright, Kevin
Elyot and Timberlake Wertenbaker. Since 1994
it has produced a new generation of playwrights
such as Joe Penhall, Rebecca Prichard, Sarah
Kane, Jez Butterworth, Martin McDonagh, Mark
Ravenhill, Ayub Khan-Din, Conor McPherson,
Roy Williams and many other first-time writers.
The Royal Court has programmes for young
writers and international writers, and it is always
searching for new plays and new playwrights.

## Royal Exchange Theatre Company

St Ann's Square, Manchester M2 7DH
☎ 0161 615 6709 📠 0161 832 0881
jo.combes@royalexchange.co.uk
www.royalexchange.co.uk
Artistic Directors *Braham Murray, Greg Hersov,
Sarah Frankcom*
Associate Director (Literary) *Jo Combes*

Founded 1976. The Royal Exchange has developed
a new writing policy which it finds is attracting a
younger audience to the theatre. Over the past 34
years, Manchester's Royal Exchange has produced
over 100 new pieces of work by playwrights
ranging from Ronald Harwood, Iain Heggie and
Brad Fraser to Debbie Horsfield, Chloe Moss,
Simon Stephens and Bridget O'Connor. 'We
accept unsolicited work from writers based in
the UK, and as we will often include a full script
report, we limit our reading to one script per
writer per year. If you would like your script
returned at the end of the process, please include
the appropriate postage. Please note that we do
not accept musicals or electronic copies of scripts
and our current turnaround time is four months.
Send scripts to *Jo Combes*.'

## Royal Lyceum Theatre Company

Grindlay Street, Edinburgh EH3 9AX
☎ 0131 248 4800 📠 0131 228 3955
www.lyceum.org.uk
Artistic Director *Mark Thomson*
Administration Manager *Ruth Butterworth*

Founded 1965. Repertory theatre which plays to a
mixed urban Scottish audience. Produces classic,
contemporary and new plays. Would like to stage
more new plays, especially Scottish. No full-time
literary staff to provide reports on submitted
scripts.

## Royal National Theatre

South Bank, London SE1 9PX
☎ 020 7452 3333 📠 020 7452 3344
info@nationaltheatre.org.uk
www.nationaltheatre.org.uk

Associate Director (Literary) *Sebastian Born*
Deputy Literary Manager *Chris Campbell*
Literary Administrator *Sarah Clarke*
Assistant Literary Manager *Clare Slater*

The majority of the National's new plays come about as a result of direct commission or from existing contacts with playwrights. There is no quota for new work, though many of the plays presented have been the work of living playwrights especially in the Cottesloe Theatre. Writers new to the theatre would need to be of exceptional talent to be successful with a script here, however the Royal National Theatre Studio helps a limited number of playwrights, through readings, workshops and discussions. Unsolicited scripts considered from the UK but not by email (send s.a.s.e).

## Royal Shakespeare Company

1 Earlham Street, London WC2H 9LL
☏ 020 7845 0515
Daniel.Usztan@rsc.org.uk
www.rsc.org.uk
Company Dramaturg *Jeanie O'Hare*
Literary Manager *Pippa Hill*

The RSC is a classical theatre company based in Stratford-upon-Avon. The Company produces a core repertoire of Shakespeare alongside modern classics, new plays and the work of Shakespeare's contemporaries. The RSC works proactively with contemporary writers, encouraging them to write epic plays, seeking out the writers they wish to commission. They are unable to read unsolicited work but do monitor the work of emerging playwrights in production nationally and internationally. The Company is currently undergoing a four-year rebuilding programme transforming the Royal Shakespeare Theatre and the entire Waterside complex of studios, rehearsal rooms, actors' cottages and workshops. The theatre is being re-modelled to create a thrust stage within a 'one room' theatre, while Waterside will house state-of-the-art facilities for artists.

## Shared Experience

13 Riverside House, 27/29 Vauxhall Grove, London SW8 1SY
☏ 020 7587 1596    ℻ 020 7735 0374
admin@sharedexperience.org.uk
www.sharedexperience.org.uk
Joint Artistic Directors *Nancy Meckler, Polly Teale*

Founded 1975. Varied audience depending on venue, since this is a touring company. Recent productions have included: *Anna Karenina*, *Mill on the Floss*, and *War and Peace* (1996; all adapt. by Helen Edmundson); *Jane Eyre* (adapt. Polly Teale); Euripides' *Orestes* and *War and Peace* (2008; both adapt. by Helen Edmundson); *Kindertransport* Diana Samuels; *A Doll's House*

(transl. Michael Meyer); *The Magic Toyshop* (adapt. by Bryony Lavery); *The Clearing* and *Gone to Earth* Helen Edmundson; *A Passage to India* (adapt. by Martin Sherman); *After Mrs Rochester* and *Brontë* Polly Teale. No unsolicited mss. Primarily not a new writing company. New plays always commissioned.

## Sherman Cymru

Senghennydd Road, Cardiff CF24 4YE
☏ 029 2064 6901    ℻ 029 2064 6902
www.shermancymru.co.uk
Director *Chris Ricketts*

Originally founded in 1973, April 2007 saw a merger between The Sherman Theatre Company and Sgript Cymru to form a new company, Sherman Cymru, with the aim of becoming a powerhouse for drama, contemporary theatre and new writing in Wales and beyond. With a strong commitment to producing new work by emerging and experienced writers in both Welsh and English, Sherman Cymru now has a literary team which provides a reading service, advice and support, mentorship and script development for writers at every level. The company offers bursaries, full commissions and various one-off events and schemes to promote and nurture excellence in new writing for Welsh and Wales-based writers.

## Show of Strength

74 Chessel Street, Bedminster, Bristol BS3 3DN
☏ 0117 902 0235
info@showofstrength.org.uk
www.showofstrength.org.uk
Creative Producer *Sheila Hannon*

Founded 1986. Plays to an informal, younger than average audience. Aims to stage at least one new play each season with a preference for work from Bristol and the South West. Unable to read unsolicited scripts. Working increasingly in non-theatre venues as in 2008 performing *Trade It?* – a commissioned work by ten writers – performed on the streets of Bristol. OUTPUT: *The Wills Girls* Amanda Whittington; *Lags* Ron Hutchinson; *So Long Life* and *Nicholodeon* Peter Nichols. Also, rehearsed readings of new work.

## Soho Theatre Company

21 Dean Street, London W1D 3NE
☏ 020 7287 5060    ℻ 020 7287 5061
writers@sohotheatre.com
www.sohotheatre.com
Artistic Director *Lisa Goldman*

The company has an extensive development programme for writers with all levels of experience and various disciplines consisting of a free script-reading service, 'Open Access Workshops', Masterclasses, talks and events.

The company produces and co-produces year round. Recent productions include: *Iya-Ile* by Oladipo Agboluaje (co-production with Tiata Fahodzi); *Everything Must Go!* (eight short pieces written in response to the economic crisis); *This Isn't Romance* by In-Sook Chappell (2007 Verity Bargate Award winner); *The Diver* by Hideki Noda and Colin Teevan (co-production with the Setagaya Public Theatre, Tokyo); *Piranha Heights* by Philip Ridley; *A Couple of Poor Polish-Speaking Romanians* by Dorota Maslowska; *Joe Guy* by Roy Williams (co-production with Tiata Fahodzi) and *Pure Gold* by Michael Bhim (co-production with Talawa). Runs the **Verity Bargate Award**, a biennial competition (see entry under *Prizes*).

## Sphinx Theatre Company

13 Riverside House, 27/29 Vauxhall Grove, London SW8 1SY
☎ 020 7587 1596
info@sphinxtheatre.co.uk
www.sphinxtheatre.co.uk
Artistic Director *Sue Parrish*

'Sphinx is a feminist ensemble company that places women at the centre of its artistic endeavour.' Since 1973 the company has toured ground-breaking productions to small and mid-scale venues. Synopses and ideas are welcome by email or post with s.a.s.e.

## Stephen Joseph Theatre

Westborough, Scarborough YO11 1JW
☎ 01723 370540   🖷 01723 360506
enquiries@sjt.uk.com
www.sjt.uk.com
Artistic Director *Chris Monks*

A two-auditoria complex housing a 165-seat end stage theatre/cinema (the McCarthy) and a 400-seat theatre-in-the-round (the Round). Positive policy on new work. Alan Ayckbourn's work features quite strongly but a new writing programme ensures plays from other sources are actively encouraged. Also runs a lunchtime season of one-act plays each summer. Writers are advised however that the SJT is very unlikely to produce an unsolicited script – synopses are preferred. Recent commissions and past productions include: *Jack Lear* Ben Benison; *Howard and Mimi* Caroline Gold; *Awaking Beauty* and *My Wonderful Day* Alan Ayckbourn. 'Writers are welcome to send details of rehearsed readings and productions as an alternative means of introducing the theatre to their work.' Submit to the Literary Department enclosing an s.a.s.e. for return of mss.

## Talawa Theatre Company Ltd

53–55 East Road, London N1 6AH
☎ 020 7251 6644   🖷 020 7251 5969
hq@talawa.com
www.talawa.com

Founded 1985. 'Aims to provide high quality productions that reflect the significant creative role that black theatre plays within the national and international arena and also to enlarge theatre audiences from the black community.' Previous productions include *Pure Gold*; *High Heel Parrotfish*; *Blues for Mister Charlie*; *The Key Game*. Seeks to provide a platform for new work from up and coming black British writers. Send a copy of the script by post or email. Runs a black script development project and black writers' group. Talawa is funded by Arts Council England, London.

## Theatre Absolute

Institute for Creative Enterprise, Technology Park, Puma Way, Coventry CV1 2TT
☎ 024 7615 8340
www.theatreabsolute.co.uk
Artistic Director/Writer *Chris O'Connell*
Producer *Julia Negus*

Founded 1992. An independent theatre company which commissions, produces and tours new plays based on a strong narrative text and aimed at audiences aged 15 and upwards. Productions include: *Breathe*; *Car*; *Raw, Kid* (Street Trilogy); *cloud:burst*; *Hang Lenny Pope*; *Zero*.

## Theatre Royal Plymouth & Drum Theatre Plymouth

Royal Parade, Plymouth PL1 2TR
☎ 01752 230347   🖷 01752 230499
david.prescott@theatreroyal.com
www.theatreroyal.com
Artistic Director *Simon Stokes*
Artistic Associate *David Prescott*

Stages small, middle and large-scale drama and music theatre. Commissions and produces new plays. The theatre no longer accepts unsolicited playscripts but will consider plays through known channels, i.e. theatre practitioners, regional and national scriptwriters' groups and agents.

## Theatre Royal Stratford East

Gerry Raffles Square, London E15 1BN
☎ 020 8534 7374   🖷 020 8534 8381
theatreroyal@stratfordeast.com
www.stratfordeast.com
Artistic Director *Kerry Michael*
Contact *Karen Fisher (Associate Producer)*

Situated in the heart of London's East End, the theatre caters for a very mixed audience, in terms of both culture and age range. Produces plays and musicals, youth theatre and local community plays/events, all of which is new work. Special interest in plays and musicals reflecting the culturally diverse communities of London and the

UK. New initiatives in developing contemporary British musicals. No longer accepts unsolicited scripts but instead asks for a) synopsis of script, b) sample ten pages of writing, c) brief writer's biography. 'From this information we will decide whether or not to ask for a full-length script. We only respond to projects of interest. If no reply within three months, assume that the proposal is not right for TRSE.'

## Theatre Royal Windsor

Windsor SL4 1PS
☎ 01753 863444    ℻ 01753 831673
info@theatreroyalwindsor.co.uk
www.theatreroyalwindsor.co.uk
Executive Producer *Bill Kenwright*
Theatre Director *Simon Pearce*

Plays to a middle-class, West End-type audience. Produces thirteen plays a year and 'would be disappointed to do fewer than two new plays in a year; always hope to do half a dozen'. Modern classics, thrillers, comedy and farce. Only interested in scripts along these lines.

## Theatre Workshop Edinburgh

34 Hamilton Place, Edinburgh EH3 5AX
☎ 0131 225 7942    ℻ 0131 220 0112
www.theatre-workshop.com
Artistic Director *Robert Rae*

First ever professional producing theatre to fully include disabled actors in all its productions. Plays to a young, broad-based audience with many pieces targeted towards particular groups or communities. Output has included *D.A.R.E.*; *Threepenny Opera*; *Black Sun Over Genoa* and Beckett's *Endgame*. Particularly interested in issues-based work for young people and minority groups. Frequently engages writers for collaborative work and devised projects. Commissions a significant amount of new writing for a wide range of contexts, from large-scale community plays to small-scale professional productions. Favours writers based in Scotland, producing material relevant to a contemporary Scottish audience.

## The Torch Theatre

St Peter's Road, Milford Haven SA73 2BU
☎ 01646 694192    ℻ 01646 698919
info@torchtheatre.co.uk
www.torchtheatre.co.uk
Artistic Director *Peter Doran*

Founded 1977. Stages a mixed programme of in-house and middle-scale touring work. Unsolicited scripts will be read and guidance offered but production unlikely due to restricted funding. Please include s.a.s.e. for return of script and notes; mark clearly, 'FAO Peter Doran'. Torch Theatre Company productions include: *Dancing at Lughnasa*; *Neville's Island*; *The Woman in Black*; *Abigail's Party*; *Taking Steps*; *Blue Remembered Hills*; *A Prayer for Wings*; *Little Shop of Horrors*; *The Caretaker*; *One Flew Over the Cuckoo's Nest*; *One for the Road*; *Noises Off*; *Dead Funny* plus annual Christmas musicals.

## Traverse Theatre

Cambridge Street, Edinburgh EH1 2ED
☎ 0131 228 3223    ℻ 0131 229 8443
literary@traverse.co.uk
www.traverse.co.uk
Artistic Director *Dominic Hill*
Literary Manager *Katherine Mendelsohn*
Literary Officer *Louise Stephens*

The Traverse is Scotland's only new writing theatre, with a particular commitment to producing new Scottish plays. However, it also has a strong international programme of work in translation and visiting companies. Previous productions include *The People Next Door* Henry Adam; *Dark Earth* David Harrower; *15 Seconds* François Archambarlt, version by Isabel Wright; *Iron* Rona Munro; *Damascus* David Greig; *Gagarin Way* Gregory Burke; *Midsummer* David Greig and Gordon McIntyre. Please address hard copies of unsolicited scripts to *Louise Stephens*, Literary Officer.

## Trestle Theatre Company

Trestle Arts Base, Russet Drive, St Albans AL4 0JQ
☎ 01727 850950    ℻ 01727 855558
admin@trestle.org.uk
www.trestle.org.uk
Artistic Director *Emily Gray*

Founded 1981. Some devised work, but increasing collaboration with writers to create new writing for touring visual/physical performance work nationally and internationally and also local community projects. No unsolicited scripts.

## Tricycle Theatre

269 Kilburn High Road, London NW6 7JR
☎ 020 7372 6611    ℻ 020 7328 0795
info@tricycle.co.uk
www.tricycle.co.uk
Artistic Director *Nicolas Kent*

Founded 1980. Presenting an eclectic, culturally diverse programme that reflects the local area of Kilburn – particularly Black, Irish, Jewish, Asian and South African work. Known for Tribunal Plays, including *Guantanamo, Bloody Sunday, The Stephen Lawrence Inquiry* and *The Colour of Justice*, adapted from the enquiry transcripts by Richard Norton-Taylor and seen by more than 25 million people worldwide. Previous productions include: *Radio Golf*; *Joe Turner's Come and Gone*; *The Piano Lesson* and *Two Trains Runnin'* all by August Wilson; and *Let There Be Love* by Kwame

Kwei-Armah. New writing welcome. Looks for a strong narrative drive with popular appeal. Fee: £15 per script. Supplies written reader's report. Can only return scripts if s.a.s.e. is enclosed with original submission.

## Unicorn Theatre

147 Tooley Street, London SE1 2HZ
☎ 020 7645 0500    ℻ 020 7645 0550
admin@unicorntheatre.com
www.unicorntheatre.com
Artistic Director *Tony Graham*
Associate Artistic Director *Rosamunde Hutt*
Associate Director (Literary) *Carl Miller*

Founded 1947, resident at the Arts Theatre from 1967 to 1999 and now based at the new Unicorn Theatre with two performance spaces. Produces full-length professionally performed plays for children and adults. Recent work includes *DUCK!* by Philip Osment, *Twin Stars* by Mike Kenny, *The Flying Machine* by Phil Porter, *Red Fortress* by Carl Miller and a new version of *Sleeping Beauty* by Rosy Fordham. Does not produce unsolicited scripts but works with commissioned writers. Writers interested in working with the company should send an email to artistic@unicorntheatre.com requesting further information.

## Upstairs at the Gatehouse

▷ Ovation Productions under Film, TV and Radio Producers

## Warehouse Theatre

Dingwall Road (adjacent to East Croydon Station), Croydon CR0 2NF
☎ 020 8681 1257    ℻ 020 8688 6699
info@warehousetheatre.co.uk
www.warehousetheatre.co.uk
Artistic Director *Ted Craig*

South London's new writing theatre seats 90–100 and produces up to six new plays a year. Also co-produces with, and hosts, selected touring companies who share the theatre's commitment to new work. The theatre continues to build upon a tradition of discovering and nurturing new writers through the **International Playwriting Festival** (see entry under *Festivals*). Also runs a vigorous writers' workshop and hosts youth theatre workshops and Saturday morning children's theatre. Scripts will get a guaranteed response if submitted through the International Playwriting Festival.

## Watford Palace Theatre

Clarendon Road, Watford WD17 1JZ
☎ 01923 235455    ℻ 01923 819664
enquiries@watfordpalacetheatre.co.uk
www.watfordpalacetheatre.co.uk
Contact *Assistant Administrator*

An important part of artistic policy is developing new work suitable for this 600-seat proscenium arch theatre, which generally means the writer has some professional production experience. Runs an open writers' workshop for local playwrights. Unable to offer a script reading and reporting service for unsolicited scripts and are unable to return unsolicited scripts.

## West Yorkshire Playhouse

Playhouse Square, Leeds LS2 7UP
☎ 0113 213 7800    ℻ 0113 213 7250
alex.chisholm@wyp.org.uk
www.wyp.org.uk
Associate Director *Alex Chisholm*

Committed to working with new writing originating from the Yorkshire and Humberside region. New writing from outside the region is programmed usually where writer or company is already known to the theatre. For more information, call 0113 213 7286 or email to address above. Scripts should be submitted with s.a.s.e. for return. 'Not all submitted scripts will be read. You are strongly advised to check guidelines on the website *before* submitting.'

## Whirligig Theatre

14 Belvedere Drive, Wimbledon, London SW19 7BY
☎ 020 8947 1732    ℻ 020 8879 7648
whirligig-theatre@virgin.net
Contact *David Wood*

Occasional productions and tours to major theatre venues, usually a musical for primary school audiences and weekend family groups. Interested in scripts which exploit the theatrical nature of children's tastes. Previous productions: *The See-Saw Tree*; *The Selfish Shellfish*; *The Gingerbread Man*; *The Old Man of Lochnagar*; *The Ideal Gnome Expedition*; *Save the Human*; *Dreams of Anne Frank*; *Babe, the Sheep-Pig*.

## White Bear Theatre Club

138 Kennington Park Road, London SE11 4DJ
www.whitebeartheatre.co.uk
*Administration:* 3 Dante Road, Kennington, London SE11 4RB ☎ 020 7793 9193
Artistic Director *Michael Kingsbury*

Founded 1988. Output primarily new work for an audience aged 20–35. Unsolicited scripts welcome, particularly new work with a keen eye on contemporary issues, though not agitprop. *Absolution* by Robert Sherwood was nominated by the Writers' Guild for 'Best Fringe Play' and *Spin* by the same author was the *Time Out* Critics' Choice in 2000. The theatre received the *Time Out* award for Best Fringe Venue in 2001 and a Peter Brook award for best up-and-coming venue. In 2004 *Round the Horne ... Revisited* transferred to The Venue, Leicester Square and ran for 14 months.

## Windsor Theatre Royal
> Theatre Royal Windsor

## The Young Vic
66 The Cut, London SE1 8LZ
☎ 020 7922 2822   🅕 020 7922 2807
info@youngvic.org
www.youngvic.org

Artistic Director *David Lan*
Executive Director *Gregory Nash*

Founded in 1970, The Young Vic is a theatre for everyone. Produces revivals of classics – old and new – as well as music theatre, international collaborations, new plays and annual events that embrace both young people and adults.

# UK and Irish Writers' Courses

## England

### Berkshire

#### The Write Coach
2 Rowan Close, Wokingham RG41 4BH
☎ 0118 978 4904
enquiries@thewritecoach.co.uk
www.thewritecoach.co.uk
Contact *Bekki Hill*

Workshops and one-to-one coaching to assist both professional and aspiring writers to become more successful, break through their blocks, build confidence, increase motivation and expand creativity, define direction and find time and space to write. Free initial consultation available. See website for details.

### Buckinghamshire

#### Missenden Abbey
Evreham Adult Learning Centre, Swallow Street, Iver SL0 0HS
☎ 01296 383582
dcevreham@buckscc.gov.uk
www.arca.uk.net/missendenabbey

Residential and non-residential weekend and summer school workshops include: *Writing from Your Own Experience*; *Unleash Your Writing Power*; *Writing for Children*; *Poetry Writing*.

#### National Film and Television School
Beaconsfield Studios, Station Road, Beaconsfield HP9 1LG
☎ 01494 731425   📠 01494 674042
info@nfts.co.uk
www.nfts.co.uk

Skill-set accredited, two-year programme covering all aspects of screenwriting from the development of ideas through to final production. Studying alongside other filmmaking students allows writers to have their work tested in workshops and productions. Unlike a Screenwriting MA set in an academic institution, this course is set in a working studio and emphasizes practical filmmaking and contact with industry personnel through seminars and pitching sessions. The course requires and encourages a high level of dedication and a prolific output. Year One deals with the basic principles of storytelling, the craft of screenwriting for film and television and the collaborative nature of production, and includes several short writing assignments. Year Two focuses on longer writing assignments, with students writing a feature script and TV project as well as their MA dissertation. Opportunities may be available to write short fiction or animation scripts for production. In partnership with **The Script Factory**, The NFTS also offers an 18-month, part-time Diploma in *Script Development*.

### Cambridgeshire

#### National Extension College
Michael Young Centre, Purbeck Road, Cambridge CB2 8HN
☎ 0800 389 2839   📠 01223 400321
info@nec.ac.uk
www.nec.ac.uk
Contact *Customer Relations Adviser*

The National Extension College (NEC) was set up as a not for profit organization to help people of all ages fit learning into their lives. Supports more than 10,000 learners each year on over 100 distance learning courses. 'If you want to discover or develop writing skills our creative courses help to develop your practical skills or begin to make money for you.' Home study courses include *Essential Editing*; *Creative Writing* and *Writing Short Stories*; *Writing for Money*; *Writing Humour* and *Songwriting*. 'You can enrol at any time, study at your own pace, in your own time with the support of a specialist tutor for up to two years.' The NEC is fully accredited by the Open Distance Learning Quality Council. Contact the NEC for a free copy of the *Guide to Courses* or visit the website.

## University of Cambridge Institute of Continuing Education

Madingley Hall, Madingley, Cambridge CB23 8AQ
☎ 01223 746262　🖷 01223 746200
registration@cont-ed.cam.ac.uk
www.cont-ed.cam.ac.uk

A wide range of weekend and five-day creative writing courses for adults is offered by the University at the Institute of Continuing Education's residential headquarters at Madingley Hall. Evening courses are also available in Cambridgeshire and surrounding areas. Details of all courses can be found on the website or phone for a brochure.

## Cheshire

## Burton Manor

The Village, Burton, Neston CH64 5SJ
☎ 0151 336 5172　🖷 0151 336 6586
enquiry@burtonmanor.com
www.burtonmanor.com

Wide variety of short courses, residential and non-residential, on writing and literature, including *Kick-start Your Creative Writing* and *Poetry Writing Workshop*. Full details in brochure.

## University of Chester

Parkgate Road, Chester CH1 4BJ
☎ 01244 392715　🖷 01244 511330
a.chantler@chester.ac.uk
www.chester.ac.uk
Contact *Dr Ashley Chantler*

The Department of English offers two programmes: *Creative Writing* (full-time Combined Honours degree over three years) and *Creative Writing* (full-time MA over one year). See the website for details.

## Cornwall

## University College Falmouth

Woodlane, Falmouth TR11 4RH
☎ 01326 211077　🖷 01326 213880
admissions@falmouth.ac.uk
www.falmouth.ac.uk
Contact *Admissions*

MA *Professional Writing*. An intensive vocational writing programme developing skills in fiction, magazine journalism/features, screenwriting. Students work on an extended writing project and form links with other postgraduate programmes such as television production and broadcast journalism.

## Cumbria

## Higham Hall College

Bassenthwaite Lake, Cockermouth CA13 9SH
☎ 01768 776276　🖷 01768 776013
admin@highamhall.com
www.highamhall.com

Winter and summer residential courses. Programme includes *Creative Writing*; *Writing for Radio and TV*. Brochure available.

## Derbyshire

## Real Writers

PO Box 170, Chesterfield S40 1FE
☎ 01246 520834　🖷 01246 520834
info@real-writers.com
www.real-writers.com

Appraisal service with personal tuition by post or email from working writers. Send s.a.s.e. or see website for details.

## Swanwick – The Writers' Summer School

The Hayes Conference Centre, Nr Alfreton
☎ 01292 442786
fionamcfadzean@hotmail.com
www.wss.org.uk
Secretary *Fiona McFadzean*

Offers six days of speakers, courses, workshops, informal talks, discussion groups, and evening entertainment. Competitions for free places. All levels welcome. Held mid-August from Saturday to Friday morning. Cost (2010) £348–£498 per person, all inclusive. Full details on website or s.a.s.e. to the Secretary at 40 Pemberton Valley, Ayr KA7 4UB (telephone and email contact as above).

## University of Derby

Admissions Office, University of Derby, Kedleston Road, Derby DE22 1GB
☎ 01332 622236　🖷 01332 622754
J.Bains@derby.ac.uk
www.derby.ac.uk
Subject Leader *Professor Carl Tighe*

With upwards of 300 students, this is one of the oldest *Creative Writing* operations at degree level in the UK. The subject offers a BA Hons *Creative Writing* with particular strengths in short stories and storytelling. Teaching is done mainly through practical workshops led by established and experienced writers. In the final year students produce an independent research project on a subject of their choice. The subject also contributes to the BA Joint Honours where Creative Writing may be combined with another subject.

## Devon

### Arvon Foundation (Devon)
▷ entry under Greater London

### Exeter Phoenix
Bradninch Place, Gandy Street, Exeter EX4 3LS
☎ 01392 667080    🖷 01392 667599
www.exeterphoenix.org.uk

Regular literature events, focusing on performances by living poets and writers, often linked to wider programmes. Tutors in a wide range of writing skills run classes and workshops. 'Uncut Poets' group holds monthly meetings.

### Fire in the Head Courses & Writers' Services
PO Box 17, Yelverton PL20 6YF
www.fire-in-the-head.co.uk
Contact *Roselle Angwin*

Comprehensive year-round writing programme in poetry, fiction (novel-writing), reflective writing and creative development. Also available: online and snailmail correspondence courses and mentoring.

### University College Falmouth – Dartington Campus
Dartington Hall, Totnes TQ9 6EJ
☎ 01803 861600    🖷 01803 861666
registry@dartington.ac.uk
www.falmouth.ac.uk
Writing Admissions Tutor *Jerome Fletcher*
(*j.fletcher@dartington.ac.uk*)

BA (Hons) *Writing* or *Writing (Contemporary Practices)* or *Writing (Scripted Media)*; *Textual Practices* minor award; MA *Performance Writing*: exploratory approaches to writing as it relates to performance, visual arts, sound arts, poetics, narrative and contemporary culture. Encourages the interdisciplinary, with minor awards and electives at BA level in choreography, music, theatre, digital arts, visual performance, community arts. Contact the Writing Admissions Tutor. (As of July 2010 the College will relocate to Falmouth; see website for details.)

### University of Exeter
Department of English, University of Exeter, Queen's Building, Queen's Drive, Exeter EX4 4QH
☎ 01392 725305    🖷 01392 264361
soe.pgoffice@ex.ac.uk
www.sall.ex.ac.uk/english

Offers BA (Hons) in *English* with 2nd and 3rd-year options in creative writing, poetry, short fiction, screenwriting and creative non-fiction. MA *Creative Writing*: poetry, fiction. Also, PhD *Creative Writing*.

## Dorset

### Bournemouth University
The Media School, Weymouth House, Talbot Campus, Fern Barrow, Poole BH12 5BB
☎ 01202 965553    🖷 01202 965099
Programme Administrator *Janice Jeffrey*

Three-year, full-time BA(Hons) course in *Scriptwriting for Film and Television*.

## Essex

### National Council for the Training of Journalists
The New Granary, Station Road, Newport, Saffron Walden CB11 3PL
☎ 01799 544014    🖷 01799 544015
info@nctj.com
www.nctj.com

Journalism courses, short courses, full-time and via distance learning. For full details, please visit the website.

## Gloucestershire

### Chrysalis – The Poet In You
5 Oxford Terrace, Uplands, Stroud GL5 1TW
☎ 01453 759436/020 7794 8880
jay@ramsay3892.fsnet.co.uk
www.lotusfoundation.org.uk
Contact *Jay Ramsay, BAHons (Oxon), PGDip, UKCP member*

Two-part correspondence course combining poetry and personal development. Offers postal courses, day and weekend workshops (including `The Sacred Space of the Word'), an on-going poetry group based in London and Gloucestershire and individual therapy related to the participant's creative process. The Chrysalis course itself consists of Part 1, `for those who feel drawn to reading more poetry as well as wanting to start to write their own', and Part 2, `a more in-depth course designed for those who are already writing and who want to go more deeply into its process and technique, its background and cultural history'. Both courses combine course notes with in-depth individual correspondence and feedback. Editing and information about publication also provided. Brochure and workshop dates available from the address above. Students can if they wish buy Part 1 of the course as a book: *The Poet in You* (available from O Books, 2009, www.o-books.net or direct from Orca on 01202 665432, £11.95).

## Wye Valley Arts Centre

Hephzibah Gallery, Llandogo NP25 4TW
☎ 01594 530214/01291 689463
info@wyearts.co.uk
www.wyearts.co.uk

Courses ('held at our beautiful gallery in Llandogo in the Wye Valley') include *Writing and Poetry* and *Writing Workshop*. All styles and abilities.

## *Hampshire*

## Highbury College, Portsmouth

Dovercourt Road, Cosham, Portsmouth PO6 2SA
☎ 023 9238 3131
journalism@highbury.ac.uk
www.highbury.ac.uk
Contact *Course Administrator*

The 20-week fast-track postgraduate courses include: *Pre-entry Magazine Journalism*; *Pre-entry Newspaper Journalism*; and Diploma in *Broadcast Journalism*, run under the auspices of the Periodicals Training Council, the National Council for Training of Journalists and the Broadcast Journalism Training Council respectively.

## University of Portsmouth

School of Creative Arts, Film and Media, University of Portsmouth, St George's Building, 141 High Street, Portsmouth PO1 2HY
☎ 023 9284 5138    ☏ 023 9284 5372
creative@port.ac.uk
www.port.ac.uk/departments/academic/scafm
www.hackwriters.com (The International Writer's Magazine)
BA Course Leader *Alison Habens*
MA Course Leader *Sam North*
PhD Queries *Steven O'Brien*

Offers creative writing and combined creative writing degrees across a wide range of undergraduate programmes and in all genres. BA, MA and PhD study is available. Staff include a number of national and/or international award-winning writers, short story writers/novelists, playwrights, screenwriters, new-media writers and poets. The MA in Creative Writing is practice led and the majority of students write novels complemented by visiting authors, editors and agents.

## University of Winchester

Winchester SO22 4NR
☎ 01962 841515    ☏ 01962 842280
course.enquiries@winchester.ac.uk
www.winchester.ac.uk
Contact *Enquiries (01962 827234)*

The University of Winchester offers a range of courses for budding writers or English enthusiasts both at undergraduate and postgraduate level. Undergraduate Programmes: BA *American Literature*; BA *Creative Writing*; BA *English*; BA *English Literature and Language*; BA *English with American Literature*; BA *Journalism*. Postgraduate: MA *Creative and Critical Writing*; MA *English: Contemporary Literature*; MA *Journalism*; MA *Writing for Children*.

## Winchester Writers' Conference, Festival, Bookfair

Research & Knowledge Exchange Centre, University of Winchester, West Hill, Winchester SO22 4NR
☎ 01962 827238
barbara.large@winchester.ac.uk
www.writersconference.co.uk

The 30th Winchester Writers' Conference, Festival and Bookfair will be held in 2010 over the weekend of 25th–27th July, preceded by Week-long Writing Courses (28th June–2nd July). This international conference offers all writers the opportunity to harness their creative ideas and to develop technical skills under the guidance and instruction of professional writers, literary agents, commissioning editors and book industry specialists during mini courses, workshops, seminars, lectures and 500 one-to-one appointments. Enter the writing competitions attached to the conference, even if you can't attend. Book online or download the application form and post. For further information, contact *Barbara Large, MBE*, Conference Founder-Director at the University of Winchester address above.

## *Hertfordshire*

## Liberato Breakaway Writing Courses

16 Middle King, Braintree CM7 3XY
☎ 01376 551379
liberato@talktalk.net
www.liberato.co.uk
Contact *Maureen Blundell*

Weekend residential fiction writing courses from October to March. Manuscript critiques by email or post for novels and short stories. Mentoring service with in-depth attention and advice for bringing novels up to publishable standard.

## *Isle of Wight*

## Writing Courses & Workshops

F * F Productions, 39 Ranelagh Road, Sandown PO36 8NT
☎ 01983 407772    ☏ 01983 407772
felicity@writeplot.co.uk
www.learnwriting.co.uk

Contact *Felicity Fair Thompson*

Regular residential weeks and weekends on creative writing for beginners and experienced writers – individual advice and workshops. Also offers postal manuscript critiques and one-to-one advice on fiction and film scripts

## Kent

### North West Kent College
Oakfield Lane, Dartford DA1 2JT
℡ 01322 629436/0800 074 1447 (Freephone helpline)
℻ 01322 629468
www.nwkcollege.ac.uk
Contact *Neil Nixon, Head of School, Media & Communications*

Two-year, full-time course that explores writing from a number of angles, teaching essential skills, market and academic aspects of the subject. Successful students progress to work or the University of Greenwich, the latter option allowing them to gain a BA(Hons) in *Humanities*, BA (Hons) in *Creative Writing* or BA (Hons) in *Media Writing* from a further year of study. Staff include scriptwriters, novelists and a book publisher. Students become published in the first year and compile a portfolio in the final year.

### University of Kent
The Registry, University of Kent, Canterbury CT2 7NZ
℡ 01227 827272
information@kent.ac.uk
www.kent.ac.uk/studying
www.kent.ac.uk/english
Contact *Dr Emma Bainbridge (01227 823402)*

Diploma and degree level course in English and American Literature with Creative Writing in a vibrant department. Undergraduate certificate courses in *Creative Writing*. Also Combined Studies courses in *English and Creative Writing*.

## Lancashire

### Alston Hall College
Alston Lane, Longridge, Preston PR3 3BP
℡ 01772 784661   ℻ 01772 785835
alstonhall.general@lancashire.gov.uk
www.alstonhall.com

Holds regular day and residential *Literature* and *Creative Writing* workshops. Full colour brochure available.

### Edge Hill University
St Helen's Road, Ormskirk L39 4QP
℡ 01695 575171
shepparr@edgehill.ac.uk
www.edgehill.ac.uk

Contact *Professor Robert Sheppard*

Offers a two-year, part-time and one-year, full time MA in *Creative Writing*. Combines advanced-level writers' workshops with closely related courses in the poetics of writing and contemporary writing in English. There is also provision for MPhil and PhD-level research in writing and poetics. A full range of creative writing courses is available at undergraduate level, in poetry, fiction and scriptwriting which may be taken as part of a modular BA (contact: *Daniele Pantano*, daniele.pantano@edgehill.ac.uk).

### Lancaster University
English & Creative Writing, Lancaster University, County College, Bailrigg, Lancaster LA1 4YD
℡ 01524 594169   ℻ 01524 594247
l.kellett@lancaster.ac.uk
www.lancs.ac.uk/fass/english/
Contact *Lyn Kellett*

Offers practical graduate and undergraduate courses in writing fiction, poetry and scripts. All based on group workshops – students' work-in-progress is circulated and discussed. Distance learning MA now available. Graduates include Andrew Miller, Justin Hill, Monique Roffey, Alison MacLeod, Jacob Polley.

## Leicestershire

### Writing School Leicester
c/o Leicester Adult Education College, 2 Wellington Street, Leicester LE1 6HL
℡ 0116 233 4343   ℻ 0116 233 4344
val.moore@writingschoolleicester.co.uk
www.writingschoolleicester.co.uk
Contact *Valerie Moore*

Offers a wide range of quality part-time creative writing and journalism courses throughout the year in Leicester and elsewhere. The writing school is staffed by professional writers and offers a mix of critical workshops, one-day and term-length courses and short craft modules. Supports writers through to publication and has strong links with local media.

## Lincolnshire

### Creuse Writers' Workshop & Retreat
45 Browning Drive, Lincoln LN2 4HF
℡ 01522 880565   ℻ 01522 880565
sandra.doran50@yahoo.co.uk
www.cresuewritersworkshopandretreat.com

A selection of home study is available with personal tuition from professional published writers. One-to-one tutoring. Visit the website for further information.

## Greater London

### Arvon Foundation

*National Administration:* Free Word, 60 Farringdon Road, London EC1R 3GA
☎ 020 7324 2554  📠 0800 756 1349
www.arvonfoundation.org

**Devon:** Totleigh Barton, Sheepwash, Beaworthy EX21 5NS ☎ 01409 231338 📠 01409 231144 totleighbarton@arvonfoundation.org

**Yorkshire:** Lumb Bank, The Ted Hughes Arvon Centre, Heptonstall, Hebden Bridge HX7 6DF ☎ 01422 843714 📠 01422 843714 lumbbank@arvonfoundation.org

**Inverness-shire:** Moniack Mhor, Teavarran, Kiltarlity, Beauly IV4 7HT ☎ 01463 741675 moniackmhor@arvonfoundation.org

**Shropshire:** The John Osborne Centre, The Hurst, Clunton, Craven Arms SY7 0JA ☎ 01588 640658 📠 01588 640509 thehurst@arvonfoundation.org
Joint Presidents *Terry Hands, Sir Robin Chichester-Clark*
Chairman *Nigel Pantling*

National Director *Ruth Borthwick*
Founded 1968. Offers people of any age (over 16) and any background the opportunity to live and work with professional writers. Four-and-a-half-day residential courses are held throughout the year at Arvon's four centres, covering poetry, fiction, drama, writing for children, songwriting and the performing arts. Bursaries towards the cost of course fees are available for those on low incomes, the unemployed, students and pensioners. Runs a biennial international poetry competition (see entry under *Prizes*).

### Birkbeck College, University of London

School of English and Humanities, Malet Street, London WC1E 7HX
☎ 020 7631 6000
a.taylor@english.bbk.ac.uk
www.bbk.ac.uk/eh/eng
Contact *Anne Marie Taylor*

Taught by published writers, Birkbeck offers an MA course in *Fiction Writing*. The course will extend and cultivate existing writing skills, help develop writing to a professional level and is supported by masterclasses and readings from visiting professionals. All classes are held in the evening. Study part-time over two years or full-time over one year. Applications must be supported by a portfolio of creative writing. Download an application form from the website.

### The Central School of Speech and Drama

Embassy Theatre, Eton Avenue, London NW3 3HY
☎ 020 7722 8183  📠 020 7722 4132

enquiries@cssd.ac.uk
Contact *Academic Registry*

MA in *Advanced Theatre Practice*. One-year, full-time course aimed at providing a grounding in principal areas of professional theatre practice – *Writing for Performance, Dramaturgy, Directing, Performing, Puppetry* and *Object Theatre*, with an emphasis on collaboration between the various strands. *Writing for Performance* students have the opportunity of working with companies to create new and innovative work for the theatre. Also MA in *Writing for Stage & Broadcast Media*. Focuses on texts for the theatre, television, cinema and radio. Prospectus available.

### City Lit

Keeley Street, Covent Garden, London WC2B 4BA
☎ 020 7492 2652  📠 020 7492 2735 (Humanities)
humanities@citylit.ac.uk
www.citylit.ac.uk

The writing school offers a wide range of courses, with workshops catering for every level of student – from beginners to previously published authors – in a variety of disciplines. Classes include *Ways into Creative Writing; Writing for Children; Playwriting (stages 1 & 2); Writing Short Stories; Screenwriting (Stages 1 & 2); Comedy Writing; Writing Novels; Advanced Critical Workshop.* Various lengths of course are available, and the department offers information and advice all year round, with the exception of August.

### City University London

Northampton Square, London EC1V 0HB
☎ 020 7040 8268  📠 020 7040 8256
ell@city.ac.uk
www.city.ac.uk/evening
www.city.ac.uk/journalism

The Short Courses (Writing) programme includes: *Short Story Writing; Certificate in Novel Writing* (three-term course); *Writing for Children; Writing Poetry; Narrative Non-Fiction.* Contact *Alison Burns,* Course Coordinator, Short Courses (Writing), a.burns@city.ac.uk (☎ 020 7040 8843) The University's Journalism Department offers an MA in *Creative Writing (Non-fiction)* which focuses on writing and research skills including how to identify a subject, how to use archives, exploring existing genres and structuring material to complete a book. The course offers a taught component, one-to-one tutorials and opportunities to hear guest speakers. The course advisory board, made up of leading non-fiction writers, provides students with expert advice and direct contact with the publishing industry. Contact Programme Director *Julie Wheelwright,* Postgraduate Admissions, Journalism, julie.wheelwright.1@city.ac.uk (☎ 020 7040 8221).

## The Complete Creative Writing Course at the Groucho Club

☎ 020 7503 6285  📠 020 7503 6285
maggie@writingcourses.org.uk
www.writingcourses.org.uk
Contact *Maggie Hamand*

Courses of six three-hour sessions held at the Groucho Club in London's Soho and the New Cavendish Club at Marble Arch, starting in January, April and September, on Mondays, Tuesdays or Saturdays. There are courses for beginners, intermediate and advanced students as well as screenwriting and writing for children. All tutors are experienced writers and teachers, and courses include stimulating exercises, discussion and weekly homework. Fee: £275–£295 for the whole course including VAT.

## Euroscript Ltd

▷ entry under Writers' Circles and Workshops

## Fiction Writing Workshops and Tutorials

5 Queen Elizabeth Close, London N16 0HL
☎ 020 8809 4725
henrietta@writtenwords.net
www.WrittenWords.net
Contact *Henrietta Soames*

Workshops for fiction writers led by author Henrietta Soames. 'Lively discussions, stimulating exercises, valuable feedback.' Fiction tutorials/mentoring for writers who require concentrated attention and ongoing, committed support. Short stories/novels, full-length mss welcomed. 'Very reasonable fees.'

## London College of Communication

Elephant & Castle, London SE1 6SB
☎ 020 7514 2105
shortcourses@lcc.arts.ac.uk
www.lcc.arts.ac.uk/training

Intensive courses in writing, editing and journalism. Courses include: *News Writing*; *Feature Writing*; *Brush Up Your Grammar*; *Proofreading*; *Introduction to Copy-editing*; *Effective Copywriting*; *Speechwriting*; *Writing Direct Mail*; *Sub-editing*. Prospectus available; telephone or access the website.

## London School of Journalism

126 Shirland Road, London W9 2BT
☎ 020 7289 7777  📠 020 7432 8141
info@lsjournalism.com
www.home-study.com
www.lsj.org
Contact *Student Administration Office*

Distance learning and attendance courses with an individual and personal approach. Distance learning options include *Novel Writing*; *Short Story Writing*; *Writing for Children*; *Poetry*;

*Freelance Journalism* (with *Sport, Travel or Music* as add-ons); *Media Law*; *Improve Your English*; *Cartooning*; *Writing a Personal History*; *English Literature and English History*. Fees range from £220 to £425, dependent upon course. Postgraduate Courses are taught in London (three months, full-time; six months, part-time; nine months, evening classes) as are short courses and evening classes (*Creative Writing*; *News and Features*; *Journalism*; *Womens' Magazines*; *Media Law*; *Freelance Writing* and *Travel Writing*, *Playwriting*, *Teeline Shorthand*). An Online Postgraduate Course (four modules, 24–48 months' duration) is also available.

## Middlesex University

Trent Park, Bramley Road, London N14 4YZ
☎ 020 8411 5000  📠 020 8411 6652
admissions@mdx.ac.uk
www.mdx.ac.uk

The UK's longest established writing degree offers a Single or Combined Honours programme in *Creative and Media Writing* (full or part-time). This modular programme gives an opportunity to explore creative non-fiction, poetry, prose fiction and dramatic writing for a wide range of genres and audiences. Option for work experience in the media and publishing industries. Contact *Admissions* or *James Martin Charlton* (j.charlton@mdx.ac.uk).

MA *Creative Writing* (full or part-time, taught at Hendon Campus). Concentrates on fiction writing. Includes writing workshops; critical seminars; lectures and workshops from established writers; introduction to agents and publishers. Contact *David Rain* (d.rain@mdx.ac.uk).

Also offers research degrees M.Phil/PhD in *Creative Writing*. Contact *Maggie Butt* (m.butt@mdx.ac.uk). The University has a thriving Writing Centre running an annual literary festival, weekly talks, community projects and writers in residence.

## National Academy of Writing

☎ 0121 331 5675  📠 0121 331 6692
Rena.Brannan@thenationalacademyofwriting.org.uk
www.thenationalacademyofwriting.org.uk
Contact *Rena Brannan*

From 2011, the National Academy of Writing will run a course solely designed to prepare full-length works of prose fiction and non-fiction for publication. Twelve writers will be selected each year as members of the National Academy. They will receive close attention to their work with a mixture of tuition, editorial feedback, one-to-one mentoring and workshops. This advanced, intensive course will take place in

London over an eight-month period between April and November, and will be delivered by the Academy's Director and NAW patrons. See the website for information on how to apply.

## PMA Media Training

The PMA Centre for Media Excellence, 7a Bayham Street, London NW1 0EY
☏ 01480 300653/020 7278 0606
🄵 01480 496022
training@pma-group.com
www.pma-group.com
www.becomeajournalist.co.uk
Contacts *Vicky Chandler, Melanie Gilbert*

Fast-track nine-week Postgraduate Diploma and more than 150 one and two-day editorial, PR, design and publishing courses held in central London. Intensive workshops run by top journalists, print and online, designers and PR professionals. All short courses are designed to lead to the PMA Gold Standard qualification. Apple, Adobe, Quark and the Periodical Publishers' Association accreditation. NCTJ, NUJ and Communicators in Business recommended. Edexcel Exam Centre. Special discount for freelancers. See website for dates and fees.

## The Poetry School
▷ entry under Organizations of Interest to Poets

## Roehampton University

Roehampton Lane, London SW15 5PU
☏ 020 8392 3232     🄵 020 8392 3470
enquiries@roehampton.ac.uk
www.roehampton.ac.uk/postgraduate-courses/
creative-and-professional-writing/index.html
Marketing Officer *Katie Meehan*

The MA/MRes *Creative and Professional Writing* includes modules in: Creative reading and viewing: research and theory for writers; Research and theory for writers; Fiction: how to grow stories; Writing for a child audience; Poetry: form and innovation; Creative non-fiction and journalism; Knowing and subverting the rules: screenwriting for independent film; Screenwriting for independent film.

## Royal Court Theatre – Young Writers Programme and Critical Mass

Royal Court Theatre, Sloane Square, London SW1W 8AS
☏ 020 7565 5050
studio@royalcourttheatre.com
www.royalcourttheatre.com
Studio Administrator *Clare McQuillan*

Young Writers Programme: an 11-week *Introduction to Playwriting* group led by the Royal Court Theatre's resident playwriting tutor, Leo Butler. The group is open to anyone aged 18–25 and costs £100 or £50 if you are a concession. To apply: submit a play to and the most promising writers will be selected to join the group. There are three intakes a year, with the following script deadlines for applicants: 1st December for the January/March group, 1st March for the April/June group and 1st September for the October/December group.

Critical Mass: 'The Royal Court is committed to generating a critical mass of black, Asian and minority ethnic playwrights.' Each autumn they run an 11-week *Introduction to Playwriting* group for ethnic minority writers aged 18+ (no upper age limit). The next Critical Mass playwriting group takes place in autumn 2010.

## The Script Factory

The Square, 61 Frith Street, London W1D 3JL
☏ 020 7851 4890
general@scriptfactory.co.uk
www.scriptfactory.co.uk

Established in 1996, The Script Factory is a screenwriter and script development organization set up to bridge the gap between writers and the industry, and to promote excellence in screenwriting. Specializing in training, screenings, masterclasses and various services, The Script Factory operates throughout the UK and internationally. For the best source of information on upcoming activities, courses and latest news, and to join the free mailing list, check out the website

## Soho Theatre Company
▷ entry under Theatre Producers

## Travellers' Tales

92 Hillfield Road, London NW6 1QA
info@travellerstales.org
www.travellerstales.org

'Travellers' Tales is Britain's foremost provider of travel writing and travel photography training for non-professionals. Offers practical courses with top professionals and organizes the annual Travellers' Tales Festival. Tutors include award-winning authors William Dalrymple and Colin Thubron, travel editors Simon Calder (*The Independent*), Lyn Hughes (*Wanderlust*) and Jonathan Lorie (*Traveller*). Courses include beginners' weekends, UK masterclasses and creative holidays overseas.

## University of Westminster

School of Media, Arts and Design, Harrow Campus, Watford Road, Harrow HA1 3TP
☏ 020 7911 5903     🄵 020 7911 5955
harrow-admissions@wmin.ac.uk
www.wmin.ac.uk

Courses include part-time MAs available in *Film and Television Studies*; *Theory, Culture & Industry*; *Screenwriting and Producing*; *Public*

*Relations; Communication and Communication Policy; Screenwriting and Production.* Full-time MAs in: *Journalism (Broadcast); Journalism (Print & Online)* and *Global Media.*

## Greater Manchester

### Manchester Metropolitan University – The Manchester Writing School

Department of English, Geoffrey Manton Building, Rosamond Street West, off Oxford Road, Manchester M15 6LL
☎ 0161 247 1787   🖷 0161 247 6345
j.draper@mmu.ac.uk
www.mmu.ac.uk/writingschool
Course Convenor *James Draper*

The Manchester Writing School offers three 'routes' for students to follow at Masters level: *Novel, Poetry* and *Writing for Children.* Key features include: regular author readings; lectures, workshops and masterclasses with writers, publishers, producers, booksellers, librarians and agents; continuing professional development units in teaching creative writing; an optional unit in literary translation; and opportunities to collaborate with musicians. Tutors include Simon Armitage, Sherry Ashworth, Heather Beck, Andrew Biswell, Carol Ann Duffy, Paul Magrs, Jenny Mayhew, Livi Michael, Michael Symmons Roberts, Jackie Roy, Nicholas Royle and Jean Sprackland. The programme is also available to study by online distance learning.

### University of Manchester

Centre for New Writing, School of Arts, Histories and Cultures, Mansfield Cooper Building, Oxford Road, Manchester M13 9PL
☎ 0161 306 1259   🖷 0161 275 5987
englishpg@manchester.ac.uk
www.arts.manchester.ac.uk/newwriting
Course Directors *John McAuliffe, Ian McGuire*

Offers a one-year MA in *Creative Writing* (the novel, short story and poetry) and a PhD.

### University of Salford

Postgraduate Admissions, School of Media, Music & Performance, University of Salford, Adelphi Building, Peru Street, Salford M3 6EQ
☎ 0161 295 6026   🖷 0161 295 6023
r.humphrey@salford.ac.uk
www.smmp.salford.ac.uk

MA in *Television and Radio Scriptwriting.* Two-year, part-time course taught by professional writers and producers. Also offers masterclasses with leading figures in the radio and television industry.

### The Writers Bureau

Sevendale House, 7 Dale Street, Manchester M1 1JB
☎ 0161 228 2362   🖷 0161 236 9440
studentservices@writersbureau.com
www.writersbureau.com

Comprehensive home-study writing course with personal tuition service from professional writers. Fiction, non-fiction, articles, short stories, novels, TV, radio and drama all covered in detail. Trial period, guarantee and no time limits. Writing for children, using the internet to sell your writing and biographies, memoirs and family history courses also available. ODLQC accredited. Quote Ref. EH10. Free enquiry line: 0800 389 7360.

### The Writers Bureau College of Journalism

Address etc. as The Writers Bureau above

Home-study course covering all aspects of journalism. Real-life assignments assessed by qualified tutors with the emphasis on getting into print. Comprises 28 modules and three supplements. Ref: EHJ10. Free enquiry line: 0800 389 7360.

### The Writers College

Address etc. as The Writers Bureau above

*The Art of Writing Poetry Course* from The Writers Bureau sister college. A home-study course with a more `recreational' emphasis. The 60,000-word course has 17 modules and lets you complete 12 written assignments for tutorial evaluation. Quote Ref. EHP10. Free enquiry line: 0800 389 7360.

## Merseyside

### University of Liverpool

Continuing Education, University of Liverpool, 126 Mount Pleasant, Liverpool L69 3GR
☎ 0151 794 6900/6952 (24 hours)
🖷 0151 794 2544
conted@liverpool.ac.uk
www.liv.ac.uk/conted

Courses include: *Exploring Writing; Scriptwriting* (for radio and TV, and for film and TV); *Comedy Writing; Writing for Women by Women; Writing for the Stage; Writing Science Fiction; Writing for Children; Writing Poetry.* Most courses take place once weekly (evening or daytime) and there is also a series of Saturday courses on aspects of writing. CE courses offer University credits which can be accumulated for personal development purposes, or towards an award such as the Certificate in Higher Education (*Creative Writing*) – 120 credits. For most courses no previous knowledge is required. Fee concessions available if you receive certain benefits or are retired. Full programme on the website or free printed prospectus on request.

## Norfolk

### University of East Anglia

School of Literature and Creative Writing, University of East Anglia, Norwich NR4 7TJ
℡ 01603 592154   ℻ 01603 250599
pgt.hum.admiss@uea.ac.uk
www.uea.ac.uk/creativewriting
Contact *Postgraduate Admissions*

UEA has a history of concern with contemporary literary culture. Among its programmes is the MA in *Creative Writing* (founded by Angus Wilson and Malcolm Bradbury in 1970/71). The course has three parallel entry points: prose fiction, poetry and scriptwriting. A series of weekly workshops provide intensive examination of students' own work, and also draw on aspects of teaching in nineteenth and twentieth century literature, literary theory and film and cultural studies.

## Northamptonshire

### Knuston Hall Residential College for Adult Education

Irchester, Wellingborough NN29 7EU
℡ 01933 312104   ℻ 01933 357596
enquiries@knustonhall.org.uk
www.knustonhall.org.uk

Writing courses have included: *Articles for Magazines*; *My Life in Poems and Stories*; *Writer's Workshops* and *Creative Writing*.

## Northumberland

### Mavis Maureen Raper – Community Creative Writing

'Sea Winds', 2 St Helens Terrace, Berwick-upon-Tweed TD15 1RJ
℡ 01289 305213
mavismaureen@btinternet.com
Author/Tutor/Moderator *Maureen Raper, MBE*

The courses, which are held at Berwick Library, Walkergate, Berwick-upon-Tweed as well as by distance learning for those unable to travel into class, include: *Creative Writing for Beginners*; *Creative Writing for Intermediates*; *Writing for Radio and Television*. Also, classes on *Writing Crime and Mystery*; *Writing Life Stories* and *Writing Romantic Fiction*.

## Nottinghamshire

### The Nottingham Trent University

College of Arts, Humanities and Education, Nottingham Trent University, Clifton Campus, Nottingham NG11 8NS
℡ 0115 848 8977
acc.postgrad@ntu.ac.uk
www.ntu.ac.uk/postgrad/

MA in *Creative Writing*. Hands-on and workshop-based, the course concentrates primarily on the practice and production of writing. A choice of options from *Fiction*, *Poetry*, *Children's and Young Adult Fiction* and *Scriptwriting*. Assignments and a dissertation, but no formal exams. Visiting professors: David Almond and Michael Eaton, MBE. There is a full programme of visiting speakers. Study either full-time or part-time.

## Oxfordshire

### University of Oxford Department for Continuing Education

Rewley House, 1 Wellington Square, Oxford OX1 2JA
℡ 01865 280973   ℻ 01865 270309
pp@conted.ox.ac.uk
www.conted.ox.ac.uk
MSt Administrator *Claire Kelly*

Creative writing classes held during the autumn and spring terms. There are also one-week summer school courses, a two-year part-time Diploma and a two-year part-time MSt in *Creative Writing*. Early booking is advised as courses are very popular.

## Shropshire

### Arvon Foundation (Shropshire)
▷ entry under Greater London

## Somerset

### Ammerdown Conference and Retreat Centre

Ammerdown Park, Radstock, Bath BA3 5SW
℡ 01761 433709   ℻ 01761 433094
centre@ammerdown.org
www.ammerdown.org
www.ammerdown-conference.co.uk

Courses have included *Creative Journaling* with Ann Beazer, *Writing for Well Being* with Sue Ashby and *Writing the Spirit* with Judy Clinton. En suite residential facilities. Brochure available or full details on the website.

## Bath Spa University

Newton Park, Newton St. Loe, Bath BA2 9BN
T 01225 875875   F 01225 875444
r.kerridge@bathspa.ac.uk
www.bathspa.ac.uk

*MA in Creative Writing.* A full time course designed to help writers bring their work to a more publishable, performable and broadcastable level. Teaching is by published writers in the novel, poetry and short stories and includes visits by literary agents and publishers. MA in *Writing for Young People.* A full or part time course for writers for children of all ages, from the picture-book age through to adolescent and 'crossover' writing, which aims at markets among adults as well as young people. An experienced teaching team helps and encourages students to create a significant body of writing, with practical plans for its place in the real world of publishing. *MA in Scriptwriting* taught in 17 weekends of intensive 11-hour workshops. A professional training course for working writers across several media, it aims to turn out writers who understand the structure and craft of drama, have a finished script they can use as a calling card, know the industry in all its variety and can pitch and sell their work. In recent years several students from the courses have won the Waterstone's Children's Book of the Year Award, the Glen Dimplex New Writers Award and the Desmond Elliot Prize; have been listed for the Man Booker, Orange and Costa book prizes, and have had work produced on BBC Radio.

## Dillington House

Ilminster TA19 9DT
T 01460 258648   F 01460 258613
dillington@somerset.gov.uk
www.dillington.com

Dillington House is one of the finest historic houses in Somerset. Provides short residential courses across a wide range of subjects, including writing and literary appreciation. Full details of the programme are available in the free brochure and on the website.

## Institute of Copywriting

Overbrook Business Centre, Poolbridge Road, Blackford, Wedmore BS28 4PA
T 0800 781 1715   F 01934 713492
copy@inst.org
www.inst.org/copy

Comprehensive copywriting home-study course, including advice on becoming a self-employed copywriter. Each student has a personal tutor who is an experienced copywriter and who provides detailed feedback on the student's assignments. Other courses include: Diploma in *Creative Writing* and Diploma in *Screenwriting*.

## Staffordshire

## Keele University

The Centre for Continuing and Professional Education, Keele University, Newcastle under Lyme ST5 5BG
T 01782 733436

Daytime, evening and weekend courses on literature and creative writing. The 2008/09 programme included courses on novel writing, writers workshops and sessions on poetry writing.

## Surrey

## The Guildford Institute

Guildford Institute, Ward Street, Guildford GU1 4LH
T 01483 562142   F 01483 451034
info@guildford-institute.org.uk
www.guildford-institute.org.uk

The Guildford Institute is the venue for various creative writing courses. It also hosts Guildford Writers, a writers' circle meeting on alternate Tuesday evenings from 7.30 pm to 9.30 pm. Members of the group are writing short stories, poetry or novels, and bring their work to read aloud to the group for other members' advice, constructive criticism and general comments. New members are always welcome.

## Royal Holloway University of London

Department of Drama and Theatre, Royal Holloway University of London, Egham Hill, Egham TW20 0EX
T 01784 443922   F 01784 431018
drama@rhul.ac.uk
www.rhul.ac.uk/Drama
www.rhul.ac.uk/english
Contact *Dan Rebellato*

Three-year BA courses in *Drama and Creative Writing* or *English and Creative Writing*, during which students progressively specialize in playwriting, poetry or fiction. Playwriting can be studied as part of the BA *Drama and Creative Writing* degree and poetic practice is an option in the *English* programme. Writing for Performance can be studied at postgraduate level in the MA *Theatre (Playwriting)* and there are MAs in *Poetic Practice* and *Screenwriting*. Students can also study for a practice-based PhD in all forms of creative writing.

## Sussex

### Earnley Concourse

Earnley, Chichester PO20 7JN
☏ 01243 670392    🖶 01243 670832
info@earnley.co.uk
www.earnley.co.uk

Offers a range of residential and non-residential courses throughout the year. Previous programme has included *Creative Writing: Getting Started*. Brochure available.

### The University of Chichester

Bishop Otter Campus, University of Chichester, College Lane, Chichester PO19 6PE
☏ 01243 816000    🖶 01243 816080
www.chi.ac.uk

MA in *Creative Writing*, both full and part-time. Contact: *Stephanie Norgate*, MA Programme Coordinator, Creative Writing (s.norgate@chi.ac.uk; 01243 816296). Also MPhil/PhD in *Creative Writing, Creative/Critical*. Contact: *Sue Breakspear*, Postgraduate Research Coordinator (s.breakspear@chi.ac.uk, 01243 812137).

### University of Sussex

Centre for Continuing Education, The Sussex Institute, Essex House, Brighton BN1 9QQ
☏ 01273 606755
cce@sussex.ac.uk
www.sussex.ac.uk/cce
Lecturer in Creative Studies *Dr Sue Roe*

MA in *Creative Writing and Authorship*: a unique opportunity for graduate writers to develop writing practice in the context of the study of cultural and aesthetic issues of authorship, past and present, in workshops, masterclasses and seminars. One year, full-time, two years, part-time. Convenor: *Sue Roe. Creative Writing Programme*: short fiction, novel and poetry for imaginative writers. Two-years, part-time. Convenor: *Mark Slater*. Both courses include CCE Agents' and Publishers' Day. Short Courses in *Creative Writing*. Convenor: *Chris Sykes*.

### West Dean College

West Dean, Nr Chichester PO18 0QZ
☏ 01243 811301
diplomas@westdean.org.uk
www.westdean.org.uk

PGDIP/MA *Creative Writing* (University of Sussex Award, subject to validation). Postgraduate study in Creative Writing is available full-time or part-time. It can be pursued to Postgraduate Diploma level only or, with an additional component of individual study and creative writing, can be awarded a full MA postgraduate degree. The full-time course must be attended on two days per week for a full academic year of 36 weeks. The additional MA component can then be completed over the summer vacation, with a submission deadline of early September. The part-time course takes two academic years and attendance would be one day per week. The additional MA component is then completed over the summer vacation of year two, with a submission deadline of early September

## Tyne & Wear

### University of Sunderland

North East Centre for Lifelong Learning, University of Sunderland, 2nd Floor, Bedson Building, Kings Road, Newcastle upon Tyne NE1 7RU
☏ 0191 515 2800    🖶 0191 515 2890
lifelong.learning@sunderland.ac.uk
cll.explore.ac.uk

Courses, held in Newcastle and across the North East, include: *Creative Writing* and *Feature Writing*. Contact the North East Centre for Lifelong Learning.

## Warwickshire

### University of Warwick

Dept. of English and Comparative Literary Studies, Humanities Building, University of Warwick, Coventry CV4 7AL
☏ 024 7652 4928
english@warwick.ac.uk
www2.warwick.ac.uk/fac/arts/english/writingprog/

The Warwick Writing Programme, which offers optional writing courses at both undergraduate and MA levels, 'encourages good reading as well as writing, develops sound skills in academic and professional writing, and bridges "academic" and "creative" approaches to literature in a fully integrated range of activities.' Genres include biography, 'creative non-fiction', fantasy/science fiction, journalism, the novel, poetry, reviewing, the short story, translation, writing for the young. The Programme has three resident writers, Maureen Freely, Michael Hulse and David Morley.

## West Midlands

### Birmingham City University

School of English, City North Campus, Perry Barr, Birmingham B42 2SU
☏ 0121 331 5675    🖶 0121 331 6692
samantha.malkin@bcu.ac.uk
www.bcu.ac.uk

MA in *Writing* in association with the **National Academy of Writing**. This course has a strong professional focus and aims to produce working writers. Modules are available in a number of disciplines, including fiction, creative non-fiction, screenwriting and writing for the theatre. Teaching includes taught modules and supervision by Birmingham City University staff, all of whom are established writers. There are also regular masterclasses and visits by staff and patrons of the National Academy of Writing. Applications are welcome from all, regardless of previous academic experience. Admission is based on talent and commitment to a career in writing.

## University of Birmingham

Department of Drama and Theatre Arts, University of Birmingham, The Old Library, 998 Bristol Road, Selly Oak, Birmingham B29 6LQ
☎ 0121 414 5998
drama@contacts.bham.ac.uk
www.drama.bham.ac.uk
Course Convenor *Steve Waters (s.waters@bham. ac.uk)*

The MPhil in *Playwriting Studies*, established by playwright David Edgar in 1989, was the UK's first postgraduate course in playwriting. An intensive course which encourages students to think critically about dramatic writing, assisting them to put these insights into practice in their own plays.

## Wiltshire

## Marlborough College Summer School

Marlborough SN8 1PA
☎ 01672 892388    🅕 01672 892476
admin@mcsummerschool.org.uk
www.mcsummerschool.org.uk

Summer School with literature and creative writing included in its programme. Caters for residential and day students. Brochure available giving full details and prices.

## Urchfont Manor College

Urchfont, Devizes SN10 4RG
☎ 01380 840495    🅕 01380 840005
urchfontmanor@wiltshire.gov.uk
www.urchfontmanor.co.uk

Short courses – days, residential weeks and weekends – offered in a varied programme which includes literature and creative writing. 'Beautiful location; historic environment; delicious home cooking.' Send for a brochure.

## Yorkshire

## Arvon Foundation (Yorkshire)
▷ entry under Greater London

## Leeds Metropolitan University

School of Film, Television & Performing Arts, Leeds Metropolitan University, Electric Press Building, Leeds LS2 3AD
☎ 0113 812 3860
screenwriting@leedsmet.ac.uk
www.leedsmet.ac.uk
Administrator *Chris Pugh*

Offers a Diploma/MA in *Screenwriting (Fiction)*.

## Open College of the Arts

Unit 1B, Redbrook Business Park, Wilthorpe Road, Barnsley S75 1JN
☎ 0800 731 2116    🅕 01226 730838
enquiries@oca-uk.com
www.oca-uk.com

The OCA correspondence course, *Starting to Write*, offers help and stimulus from experienced writers/tutors. Emphasis is on personal development rather than commercial genre. Subsequent levels available include specialist poetry, fiction, autobiographical and children's writing courses. OCA writing courses are accredited by the Buckinghamshire New University. Prospectus and Guide to Courses available on request.

## Sheffield Hallam University

Faculty of Development & Society, City Campus, Sheffield Hallam University, Sheffield S1 1WB
☎ 0114 225 5555    🅕 0114 225 2167
dsmarketing@shu.ac.uk
www.shu.ac.uk

Offers MA in *Writing* (one year to complete seminar courses, two years to full MA, full-time; also part-time) and MA *English Studies* (one year full-time, also part-time).

## University of Hull

School of Arts and New Media, University of Hull, Scarborough Campus, Filey Road, Scarborough YO11 3AZ
☎ 01723 362392
a.head@hull.ac.uk
www.hull.ac.uk

BA Single Honours in *Theatre and Performance Studies* incorporates opportunities in writing and other media across each level of the programme. The campus hosts the annual National Student Drama Festival which includes the International Student Playscript Competition (see www.nsdf. org.uk).

## University of Leeds – Lifelong Learning Centre

Level 11, Marjorie and Arnold Ziff Building, University of Leeds, Leeds LS2 8JT
☎ 0113 343 3212
part-time@leeds.ac.uk
www.leeds.ac.uk/lifelonglearningcentre

The Lifelong Learning Centre provides accredited part-time courses in creative writing on the Leeds campus during the day and evening. These change from year to year but can include introductory workshop courses and more specialized courses in poetry and fiction writing. The courses form part of the Lifelong Learning Centre's Open Study programme, which is designed to provide adults with opportunities for continuing education and to create routes into higher education.

## University of Leeds

School of Performance and Cultural Industries, University of Leeds, Leeds LS2 9JT
☎ 0113 343 8735     📠 0113 343 8711
enquiries-pci@leeds.ac.uk
www.leeds.ac.uk/paci
Contact *Garry Lyons or Jane Richardson (Admissions Secretary)*

MA in *Writing for Performance and Publication*. This postgraduate programme is particularly relevant to aspiring writers with professional ambitions in the areas of theatre, film, television and radio drama, while also allowing opportunities to work in prose fiction and other culturally significant genres. Studies are offered over one-year full-time and two-years part-time and the tutors are established authors in their chosen fields. The course director is award winning playwright and screenwriter Garry Lyons (*The Worst Witch, Leah's Trials, Britain's First Suicide Bombers*). The MA is one of a cutting edge portfolio of post-graduate degrees to be offered by Leeds University's School of Performance and Cultural Industries and is housed in a new theatre complex – stage@leeds – at the heart of the University's city centre campus. Students are not only given the opportunity to work on their own writing projects but are also encouraged to collaborate with colleagues on the other MAs in Performance, Choreography and Cultural Entrepreneurship, with a view to seeing their work produced and staged.

## University of Sheffield

Institute for Lifelong Learning, School of Education, University of Sheffield, 196–198 West Street, Sheffield S1 4ET
☎ 0114 222 7000     📠 0114 222 7001
till@sheffield.ac.uk
www.shef.ac.uk/till

Certificate in *Creative Writing* (Degree Level 1) is part-time and open to all. Courses in poetry, journalism, scriptwriting, short story writing, travel writing. Brochures and information available from the address above.

# Ireland

## Dingle Writing Courses

Ballintlea, Ventry, Co. Kerry
☎ 00 353 66 915 9815
info@dinglewritingcourses.ie
www.dinglewritingcourses.ie
Directors *Abigail Joffe, Nicholas McLachlan*

An autumn programme of weekend courses for beginners and experienced writers alike. All courses are residential and take place over three days, with all meals and accommodation included in the course fee. Tutored by professional writers, the courses include poetry, fiction, starting to write, writing for children and writing for theatre as well as special themed courses. Recent tutors have included Anne Enright, Maurice Riordan, Claire Kilroy, Carlo Gébler, Paula Meehan, Kate Thompson, Andrew O'Hagan, Mary O'Malley, Marina Carr and Niall Williams. Also organizes tailor-made courses for schools, writers' groups or students on a creative writing programme.

## INKwell Writers' Workshops

The Old Post Office, Kilmacanogue, Co. Wicklow
☎ 00 353 1 276 5921/00 353 87 2835382
info@inkwellwriters.ie
www.inkwellwriters.ie
Contact *Vanessa O'Loughlin*

A series of one-day (Saturday) intensive fiction writing workshops with bestselling Irish and UK writers, held at the Fitzpatrick's Castle Hotel, Killiney, Co. Dublin. 'Designed to inspire and guide new and accomplished writers, giving them invaluable access to authors at the top of their field.' Workshops include: *Start Writing*: Julie Parsons and Sarah Webb; *Short Stories*: Eilis ni Dhuibhne and Martina Devlin; *Memoir: Tell Your Own Story*: Ferdia MacAnna and Martina Devlin; *Web Wise: Writing for the Web*: Beth Morrissey; *Writing for Children*: Sarah Webb and Oisin McGann; *Getting Published*: Paula Campbell of Poolbeg Ireland, Ciara Considine of Hachette Ireland, Sheila Crowley of Curtis Brown, London and Patricia O'Reilly. Full details on the website.

Inkwell Online/Distance Learning Workshops: *Write that Book*; *Writing Historical Fiction*; *Screenwriting for Beginners*; *Writing Romance for Online Markets*; *Getting Connected – Social Networking for Writers*.

Inkwell for Kids: Write How? Write Now! Summer Camps for 7–9s and 9–12s (July 5th–9th, 2010) at Rathdown School, Glenageary, Co.

Dublin. Free Inkwell enewsletter featuring latest industry news, competitions and publishing opportunities.

### Queen's University of Belfast

School of Education, Queen's University, Belfast BT7 1LN

☎ 028 9097 3323   📠 028 9097 1084

openlearning.education@qub.ac.uk
www.qub.ac.uk/edu/ol

Courses have included *Creative Writing* and *Writing for Profit and Pleasure*.

### University of Dublin (Trinity College)

Creative Writing Office, Arts Building, Trinity College, Dublin 2

☎ 00 353 1 896 1166

www.tcd.ie/graduate_studies

Course Administrator *Lilian Foley*

Offers an MPhil *Creative Writing* course. A one-year, full-time course intended for students who are seriously committed to writing or prospective authors. (Applications must be made online through www.pac.ie/tcd)

### The Writer's Academy

Carrig-on-Bannow, Co. Wexford

☎ 00 353 51 561789

thewritersacademy@eircom.net
www.thewritersacademy.net

Distance learning freelance article writing and short story writing courses, critique service and marketing advice. Personal tuition plus expert information and support for writers of all abilities.

## Scotland

### Arvon Foundation (Inverness-shire)
▷ entry under Greater London

### University of Dundee

Continuing Education, University of Dundee, Nethergate, Dundee DD1 4HN

☎ 01382 384809

www.dundee.ac.uk/conted/

A number of daytime and evening short courses are available through the University's Continuing Education Department. Telephone for further details or check the website.

### University of Edinburgh

Office of Lifelong Learning/IALS, University of Edinburgh, 11 Buccleuch Place, Edinburgh EH8 9LW

☎ 0131 650 4400   📠 0131 667 6097

openstudies@ed.ac.uk
www.ed.ac.uk/openstudies

Creative writing courses run throughout the year. Full details available on the website.

### University of Glasgow

Department of Adult and Continuing Education, University of Glasgow, 11 Eldon Street, Glasgow G3 6NH

☎ 0141 330 1835   📠 0141 330 1821

dace-query@educ.gla.ac.uk
www.glasgow.ac.uk/dace

Runs writers' workshops and courses at all levels; all friendly and informal. Daytime and evening meetings. Tutors are all experienced published writers. Call 0141 330 1829 for course brochure.

### University of St Andrews

School of English, The University, St Andrews KY16 9AR

☎ 01334 462666   📠 01334 462655

english@st-andrews.ac.uk
www.st-andrews.ac.uk

Offers postgraduate study in *Creative Writing*. Modules in either Writing Prose or Writing Poetry. Students submit a dissertation of original writing – prose fiction of 15,000 words or a collection of around 30 short poems. Taught by John Burnside, Robert Crawford, Kathleen Jamie, Meaghan Delahunt and Don Paterson. Also offers PhD in *Creative Writing*.

## Wales

### Aberystwyth University

Department of English and Creative Writing, Aberystwyth University, Hugh Owen Building, Aberystwyth SY23 3DY

☎ 01970 622534/5   📠 01970 622530

www.aber.ac.uk/english

Secretary *Mrs June Baxter*
Administrator *Ms Julie Roberts*

BA in *English and Creative Writing*, a three-year course whose creative modules are taught by practising writers: novelist and poet Professor Jem Poster; poet Dr Tiffany Atkinson; poet and novelist Dr Matthew Francis; poet and life-writer Dr Kelly Grovier; poet Dr Damian Walford Davies; and poet Dr Richard Marggraf Turley. Also offers a taught MA in *Creative Writing* with modules in writing poetry, writing fiction, research for writers, and writing and publication; and a PhD in *Creative Writing*.

### Tŷ Newydd Writers' Centre

Llanystumdwy, Criccieth LL52 0LW

☎ 01766 522811   📠 01766 523095

post@tynewydd.org
www.tynewydd.org

Residential writers' centre set up by the Taliesin Trust with the support of the **Arts Council of Wales** to encourage and promote writing in both English and Welsh. Most courses run from

Monday evening to Saturday morning. Each course has two tutors and a maximum of 16 participants. A wide range of courses for all levels of experience. Early booking essential. Fees start at £230 for weekends and £465 for week-long courses, all inclusive. People on low incomes may be eligible for a grant or bursary. Course leaflet available. (See also *Organizations of Interest to Poets*.)

## University of Glamorgan

Division of English, Department of Humanities and Languages, University of Glamorgan, Pontypridd CF37 1DL
☎ 0800 716925
www.glam.ac.uk

MPhil in *Writing*: a two-year part-time Masters degree for writers of fiction and poets. Established 1993. Contact *Professor Philip Gross* in the Division of English (pgross@glam.ac.uk).

BA in *Creative and Professional Writing*: a three-year course (5–6 years part-time) for undergraduates.

MA in *Scriptwriting (Theatre, Film, TV or Radio)*: a two-year part-time Masters degree for scriptwriters (held at the Cardiff School for Creative and Cultural Industries). Contact: *Wyn Mason* (wmason@glam.ac.uk).

## University of Wales, Bangor

www.bangor.ac.uk

The School of English (College Road, Bangor LL57 2DG ☎ 01248 382102 elso29@bangor.ac.uk postgrad-english@bangor.ac.uk) offers MPhil/PhD *Creative and Critical Writing*; Diploma/MA *Creative Studies (Creative Writing)*; BA

(Hons) *English with Creative Writing*; BA (Hons) *English with Songwriting*; BA (Hons) *English with Journalism*.

The School of Creative Studies and Media (Creative Studies and Creative Industries Building, John Phillips Hall, College Road, Bangor LL57 2DG ☎ 01248 383215) offers PhD *Creative and Critical Writing*; MA *Creative Writing*; Diploma *Creative Writing*; BA *Creative Studies*; BA *Creative Writing and Communications*; BA *Music and Creative Writing*; also, specialist creative writing short-courses, national and international conferences, workshops and discussions in all genre, including: prose fiction (short stories and novels), poetry, screenwriting, playwriting, writing for new media (including hypertext writing), creative non-fiction.

## Writers' Holiday at Caerleon

School Bungalow, Church Road, Pontnewydd, Cwmbran NP44 1AT
☎ 01633 489438
writersholiday@lineone.net
www.writersholiday.net
Contact *Anne Hobbs*

Annual six-day comprehensive conference including 14 courses for writers of all standards held in the summer at the University of Wales' Caerleon Campus. Courses, lectures, concert and excursion all included in the fee. Private, single and en-suite, full board accommodation. Courses have included *Writing for Publication*; *Writing Poetry*; *Writing Romantic Fiction* and *Writing for the Radio*.

# US Writers' Courses

## Alaska

### University of Alaska, Fairbanks
Dept. of English, University of Alaska, Fairbanks, PO Box 755720, Fairbanks AK 99775–5720
☏ 001 907 474 7193
faengl@uaf.edu
www.alaska.edu/english/
Contact *Professor Gerri Brightwell (gabrightwell@ alaska.edu)*

MFA in *Creative Writing* (fiction, non-fiction, poetry and dramatic writing). Offers a three-year program with teaching assistantships available that cover tuition and provide a stipend. A combined degree is also available: MFA/MA. The creative writing program hosts the Midnight Sun Visiting Writer Series that brings in writers from across North America to read and offer masterclasses to students. Deadline for applications: 1st February. For application information, see the department website.

## Arizona

### University of Arizona
MFA Program in Creative Writing, Dept. of English, Modern Languages 445, University of Arizona, PO Box 210067, Tucson AZ 85721–0067
☏ 001 520 621 3880  🖷 001 520 621 7397
mcooksey@email.arizona.edu
cwp.web.arizona.edu
Program Assistant *Marlene Cooksey*

Established in 1974, the MFA in *Creative Writing* is a two-year program with concentrations in Poetry, Fiction and Creative Non-Fiction. 'It boasts a faculty of 14 distinguished writers who share an uncommon commitment to teaching and supporting their students in the completion of publishable manuscripts.'

## Arkansas

### University of Arkansas
Programs in Creative Writing & Translation, Dept. of English, University of Arkansas, 333 Kimpel Hall, Fayetteville AR 72701
☏ 001 479 575 4301  🖷 001 479 575 5919
mfa@uark.edu
www.uark.edu/depts/english/PCWT.html

MFA in *Creative Writing*, offering small, intensive workshops and innovative classes in fiction and poetry.

## California

### American Film Institute
AFI Conservatory, 2021 N. Western Avenue, Los Angeles CA 90027
☏ 001 323 856 7628 (admissions)
admissions@afi.com
www.afi.com
Contact *Admissions Manager*

Screenwriting at AFI focuses on narrative storytelling. In their First Year, Screenwriting Fellows are immersed in the production process in order to learn how screenplays are visualized. Initially writing short screenplays – one of which will be the basis for a first year production – writers collaborate with Producing and Directing Fellows to see their work move from page to screen. The remainder of the first year is devoted to the completion of a feature-length screenplay. Second Year Fellows may write a second feature-length screenplay, or develop materials for television, including biopics, television movies and spec scripts for sitcoms and one-hour dramas. They also have the opportunity to work closely with other disciplines by writing a Second Year thesis script.

### California Institute of the Arts (CalArts)
Office of Admissions, MFA Writing Program, CalArts, 24700 McBean Parkway, Valencia CA 91355
☏ 001 661 255 1050

admissions@calarts.edu
http://calarts.edu/

The two-year MFA *Writing* program is a flexible program giving writers the opportunity to work in the areas of cultural commentary, fiction, criticism, writing for performance, poetry, playwriting and writing for interactive media. Although the MFA Writing Program covers traditional forms of literary and scholarly writing in depth, students are encouraged to work across genres and often work in various multi-media forms in addition to their literary production.

## Chapman University

Graduate Admission Office, Chapman University, One University Drive, Orange CA 92866
☎ 001 714 997 6711
www.chapman.edu
Contact *Graduate Admission*

The programme is a two-year Master of Fine Arts (MFA) degree in *Creative Writing* intended for graduate students who wish to specialize in writing literary fiction. There is an emphasis on fiction but secondary studies include screenwriting, playwriting and poetry.

## Saint Mary's College of California

MFA Program in Creative Writing, Saint Mary's College of California, PO Box 4686, Moraga CA 94575–4686
☎ 001 925 631 4762
writers@stmarys-ca.edu
www.stmarys-ca.edu
Program Coordinator *Thomas Cooney*

Two-year MFA course in *Creative Writing* with Fiction, Non-Fiction or Poetry. The core of the program is the Writing Workshop which provides the opportunity for students to work with established writers.

## San Francisco State University

Creative Writing Department, College of Humanities, San Francisco State University, 1600 Holloway Avenue, San Francisco CA 94132
☎ 001 415 338 1891
cwriting@sfsu.edu
www.sfsu.edu/~cwriting

Offers BA and MA in *English:* with a concentration on *Creative Writing*; and MFA in *Creative Writing*. Undergraduate topics include Personal Narrative, American Poetics, Style in Fiction, The Short Story, Plays: Reading and Viewing, while graduate classes include Characterization, Contemporary Non-Fiction, Centering on Language, Elements of Playwriting, Experimental Fiction, Feminist Poetics, The Novella and The Prose Poem.

## University of Southern California (USC)

School of Cinematic Arts, Attn: Writing for Screen & Television Program, USC, University Park, 900 W. 34th Street, Los Angeles CA 90089
☎ 001 213 740 3303
admissions@cinema.usc.edu
www-cntv.usc.edu

Offers a Bachelor of Fine Arts or Master of Fine Arts degree in *Writing for Screen and Television*. There is a strong international presence, having accepted students from over 44 countries.

# Colorado

## Colorado State University

Creative Writing Program, 1773 Campus Delivery, Colorado State University, 359 Eddy Hall, Fort Collins CO 80523–1773
☎ 001 970 491 2403
english@lamar.colostate.edu
www.colostate.edu/Depts/English/programs/mfa.htm
Contact *Assistant to the Director of Creative Writing or Director of Creative Writing*

Three-year Master of Fine Arts (MFA) program in *Creative Writing* with concentrations in Fiction or Poetry. The program offers a balance of intimate and intensive writing and translation. Course work culminates in a thesis – a collection of poetry or short stories or a novel. Students have the opportunity to teach introductory creative-writing courses, intern with literary journals including the *Colorado Review* and be a part of a thriving community.

# District of Columbia

## American University

Dept. of Literature, American University, 237 Battelle-Tompkins, 4400 Massachusetts Avenue, N.W., Washington DC 20016–8047
☎ 001 202 885 2971
www.american.edu/cas/lit/mfa-lit.cfm
Contact *Catherine Johnson (Graduate Programs Assistant)*

Established in 1980, the MFA in *Creative Writing* is a 48-semester-hour program allowing students to concentrate in Poetry, Fiction or Creative Non-Fiction. The program is allied with the University's MA in *Literature* which offers numerous courses and master's-level seminars enabling creative writing students to explore both classic texts and contemporary literature.

# Florida

## Florida International University

English Dept., Florida International University, Biscayne Bay Campus, 3000 NE 151st Street, North Miami FL 33181
℡ 001 305 919 5857
crwriting@fiu.edu
www.fiu.edu/crwriting
Director *Les Standiford*

The MFA Program in *Creative Writing* includes writing workshop, literature, form and theory and thesis (most students complete the course in about three years; completion of study within eight years is required). There is no language requirement. Graduate workshops include short fiction, the novel, popular fiction, screenwriting, creative non-fiction and poetry. Great emphasis is placed upon preparation and completion of a book-length thesis. Admission is based primarily on the strength of the applicant's submitted writing sample. Limited number of teaching assistantships are available. Application deadline: 15th January.

## University of Florida

MFA@FLA Program in Creative Writing, University of Florida, 4008 Turlington Hall, PO Box 117310, Gainesville FL 32611–7310
℡ 001 352 392 6650   📠 001 352 392 0860
crw@english.ufl.edu
www.english.ufl.edu/crw
Co-Directors, Program in Creative Writing *David Leavitt, Padgett Powell*

MFA@FLA, the graduate program in *Creative Writing* at the University of Florida, is one of the oldest writing programs in the United States, begun in 1948. 'We require an equal interest in writing and in reading literature. We don't believe in any particular school of writing; we have no wish to foster or found one. All the members of our faculty – four fiction writers and three poets – live in Gainesville and are tenured. We put out *Subtropics*, a new and important literary magazine.' In the autumn of 2008 the program changed from a two-year to a three-year format; all students will receive full tuition waivers and teaching assistantships.

# Idaho

## Boise State University

MFA in Creative Writing, Boise State University, 1910 University Drive, Boise ID 83725
℡ 001 208 426 2413
www.boisestate.edu
Program Director *Martin Corless-Smith*

The MFA Program in *Creative Writing* offers degree tracks in fiction and poetry, emphasizing the art and craft of literary writing and concentrating on the student's written work. Close work with faculty and visiting writers is encouraged through seminars, conferences and classroom interaction during the three-year course, the third year of which is normally devoted to thesis preparation. Students who intend to pursue a career in teaching literature and writing at college level have the opportunity to study the pedagogy of creative writing. Also offered are classes in the craft of literary publishing, with coursework in both the production of a literary annual (*The Idaho Review*) and of books for a small press (*Ahsahta Press*); internships and graduate assistantships are also available with these publishers. The program also publishes *Free Poetry* featuring essays and poetry from today's leading poets.

# Illinois

## Chicago State University

MFA Program in Creative Writing, Division of Graduate Studies, Chicago State University, 9501 South King Drive, LIB338, Chicago IL 60628
℡ 001 773 995 2189
www.csu.edu/GraduateSchool/

Offers an MFA with courses in *Creative Writing*; writing workshops in fiction, poetry, creative non-fiction, playwriting and scriptwriting. Students undertake coursework in African American Literature and non-African American Literature. Students will show competency in the genre of their choice, choosing from fiction, creative non-fiction and poetry. Students may take course work in playwriting and scriptwriting for film and television as electives.

# Indiana

## Purdue University

MFA Program in Creative Writing, Dept. of English, Purdue University, West Lafayette IN 47907
℡ 001 765 494 3740
www.cla.purdue.edu/mfacw
Program Director *Porter Shreve*

MFA in *Creative Writing* offering 'full funding, generous stipends and editorial experience with the award-winning literary journal *Sycamore Review*'. The three-year program in either fiction or poetry is small and intensive and includes workshops, literature and craft courses, and thesis tutorials toward a book-length manuscript.

### Taylor University
Dept. of Professional Writing, Taylor University, 236 West Reade Avenue, Upland IN 46989–1001
☎ 001 765 998 5590
DNHensley@Taylor.edu
Contact *Dr Dennis E. Hensley*

Summer Honours Program, one-week seminars in *Professional Writing*; evening courses in *Freelance Writing* and *Fiction Writing*; one year college certificate in *Professional Writing*. Dr Hensley is a professor of English and director of the professional writing department.

### University of Notre Dame
Creative Writing Program, Dept. of English, University of Notre Dame, Notre Dame IN 46556
☎ 001 574 631 7526     🖷 001 574 631 4795
creativewriting@nd.edu
www.nd.edu/~alcwp

The MFA in *Creative Writing* is a two-year degree program centred around workshops in poetry and fiction, offering literature courses, translation, a literary publishing course and twelve credits of thesis preparation with an individual faculty adviser.

## Louisiana

### Louisiana State University
English Department, Louisiana State University, 260 Allen Hall, Baton Rouge LA 70803–5001
☎ 001 225 578 5922
www.lsu.edu/creativewriting/
Director of Creative Writing *James Wilcox*

Master of Fine Arts in *Creative Writing* course with a focus in Poetry, Fiction, Drama, Screenwriting. Includes opportunity to edit literary journals.

## Maryland

### Goucher College
Welch Center for Graduate Studies, Goucher College, 1021 Dulaney Valley Road, Baltimore MD 21204–2794
☎ 001 410 337 6200
www.goucher.edu/mfa
Program Director *Patsy Sims*

The two-year MFA Program in *Creative Nonfiction* is a limited-residency program that allows students to complete most of the requirements off campus while developing their skill as non-fiction writers under the close supervision of a faculty mentor. Provides instruction in the following areas: narrative non-fiction, literary journalism, memoir, the personal essay, travel/nature/science writing and biography/profiles.

## Massachusetts

### Boston University
Creative Writing Program, Boston University, 236 Bay State Road, Boston MA 02215
☎ 001 617 353 2510     🖷 001 617 353 3653
crwr@bu.edu
www.bu.edu/writing
Director *Leslie Epstein*
Contact *Program Coordinator*

Intensive Master of Fine Arts in *Creative Writing*: Fiction, Poetry or Drama, can be completed in one or two years, depending on a student's needs. The Boston University Creative Writing Program is one of the oldest in the country. Students participate in workshops or seminars concentrating on their chosen specialization and are expected to balance that work with an equal number of graduate literature courses. Internships at the literary journal, *AGNI*, assistantships at the Boston Arts Academy, as well as teaching fellowships are available on a limited basis. 'We are pleased to announce new travel fellowships for up to half of our students, allowing them to go to any country and do there what they wish for, typically, three months (visit http://blogs.bu.edu/world).'

### Emerson College
120 Boylston Street, Boston MA 01226–4624
☎ 001 617 824 8610
gradapp@emerson.edu
www.emerson.edu/graduate_admission
Graduate Program Director *Douglas Whynott*

Offers MFA in *Creative Writing* with a focus on writing fiction, non-fiction and poetry. Includes writing workshops, internship opportunities, a student reading series and student teaching opportunities. Publishes literary magazines, *Ploughshares* and *Redivider*. Contact the Graduate Admission Office for more information.

### Lesley University
MFA Program in Creative Writing, Lesley University, Cambridge MA 02138
☎ 001 617 349 8369
jvanderv@lesley.edu
www.lesley.edu/gsass/creative_writing/index.html
Contact *Jana Van der Veer, Assistant Director*
Director *Steve Cramer*

The MFA Program in *Creative Writing* is a low-residency program that allows students, with the oversight of a faculty mentor, to design their own concentrations in fiction, poetry, creative non-fiction, writing for stage and screen, or writing for young people. A nine-day residency begins each semester with a program of seminars, workshops, readings, and the chance to design the semester's programme of study. Students work independently during the semester, under the guidance of their faculty mentor. Students also design individual, three-credit interdisciplinary projects for the first three of their four distance-learning semesters. The options for these projects are multiple – in some cases, students take courses; others develop independent studies or work as publishing interns, teachers, or teaching assistants. In their final semester, students work on their creative thesis as well as prepare a craft seminar to be presented to continuing students during the following residency. The program consists of four full residencies preceding four semesters of work, plus partial attendance at a fifth residency, for a total of 48 credits. Most students complete the program in two years of full-time study.

# Michigan

## The University of Michigan

Dept. of English, University of Michigan, 435 South State Street, 3187 Angell Hall, Ann Arbor MI 48109–1003
☎ 001 734 763 2267
janburg@umich.edu
www.lsa.umich.edu/english/grad/mfa
Program Director *Eileen Pollack*
Contact *Jan Burgess (001 734 763 2267)*

The MFA in *Creative Writing* is a two-year program with concentrations in fiction and poetry. At the heart of the program are the writing workshops, where students assemble as a community of writers to read and comment on one another's work in progress. In addition to workshops, students take four additional courses, which include English literature, craft, and cognate courses (graduate level courses outside of the English Department). The program culminates with a thesis in the student's chosen genre. Students are fully funded for two years, including funding for conference and research travel. Through the Visiting Writers Series, students are given the opportunity, through readings, residencies, and dinners, to meet established writers in their genres. The program also offers professionalization roundtables with agents and editors. Upon graduation, students have the opportunity to apply for year-long post-graduate fellowships, such as the Meijer and Zell.

## Western Michigan University

Dept. of English, Western Michigan University, 1903 W. Michigan Avenue, Kalamazoo MI 49008-5331
☎ 001 269 387 2584
english-graduate@wmich.edu
www.wmich.edu/english
Director of Graduate Studies *Jana Schulman*

BA English Major/Minor with *Creative Writing Emphasis* (poetry, fiction, playwriting). The English Major aims at giving students intensive practise in writing and criticism in various genres in a workshop format; for general writing careers or for prospective candidates for the MFA in Creative Writing.

MFA in *Creative Writing* (poetry, fiction, playwriting, non-fiction) For students who wish to become professional writers of poetry, fiction, drama or non-fiction. Qualifies them to teach the craft at college or university level.

PhD in *English, with an Emphasis on Creative Writing* (poetry, fiction, playwriting, non-fiction). Qualifies students to teach creative writing and literature at university level.

# Minnesota

## Minnesota State University, Mankato

Dept. of English, Minnesota State University, Mankato, 230 Armstrong Hall, Mankato MN 56001
☎ 001 507 389 2117   ☎ 001 507 389 5362
richard.robbins@mnsu.edu
www.english.mnsu.edu/cw
Program Director *Richard Robbins*

The MFA Program in *Creative Writing* meets the needs of students who want to strike a balance between the development of individual creative talent and the close study of literature and language. The program gives appropriate training for careers in freelancing, college-level teaching, editing and publishing and arts administration.

## Minnesota State University, Moorhead

Graduate Studies Office, MSUM, 1104 7th Avenue South, Moorhead MN 56563
☎ 001 218 477 2344
graduate@mnstate.edu
www.mnstate.edu/finearts
Contact *MFA Coordinator*

MFA in *Creative Writing*. Students specialize in poetry, fiction, scriptwriting or creative non-fiction and are encouraged to experiment in more than one genre. The program offers the opportunity to take workshops, seminars and tutorials in chosen areas and to work with New Rivers Press (www.newriverspress.com) or with *Red Weather*, the campus literary magazine.

# Missouri

## University of Missouri

MFA Program, Dept. of English, University
of Missouri, One University Drive, St Louis
MO 63121
☎ 00 314 516 6845
www.umsl.edu/~mfa
Contact *Mary Troy*

The MFA in *Creative Writing* provides
opportunities for growth in the writing of fiction
and poetry (with some work in non-fiction) as
well as practical training in literary editing. While
normally a studio/academic program mixing the
study of literature and criticism with workshops
and independent study and editing, the plan
of study is flexible and individual. Students
ordinarily specialize in one genre, either fiction
or poetry, and regular workshops in these forms
are at the heart of the program. Five workshops,
at least four in the student's chosen genre, are
required for the degree though more may be
taken as electives. Students also take from five
to nine courses on offer, choosing from graduate
courses in literary journal editing; in poetry and
fiction form, theory and technique; in literature
and literary criticism; in composition theory;
and in linguistics. A creative thesis of six hours
completes the 39-hour program.

## University of Missouri-Columbia

Creative Writing Program, Dept. of English,
University of Missouri-Columbia, 202 Tate Hall,
Columbia MO 65211–1500
☎ 011 573 884 7773
creativewriting@missouri.edu
creativewriting.missouri.edu
Contact *Sharon Fisher*

*Creative Writing* MA and PhD in English
programs in fiction, poetry and non-fiction.
Access the website for further information.

# New York

## Adelphi University

MFA Program in Creative Writing, Dept. of
English, Adelphi University, PO Box 701, Garden
City NY 11530
☎ 001 516 877 4044
mfa@adelphi.edu
www.adelphi.edu/mfacw
Director of Creative Writing *Judith Baumel*

The MFA in *Creative Writing* offers students the
opportunity to specialize in three major genres:
fiction, poetry and dramatic writing. Its unique
Professional Development Practicum introduces
students to the professional and practical life

of writers across many disciplines and prepares
them for careers in writing, teaching and/or more
advanced graduate studies through training in
creative writing, language and literary studies,
research and teaching. Students are required
to complete 37 credits in a plan of study that
includes writing workshops and literature classes.
A student thesis is a degree requirement.

## Brooklyn College of the City University of New York

MFA Program in Creative Writing, Dept. of
English, Brooklyn College, 2900 Bedford Avenue,
Brooklyn NY 11210–2889
☎ 001 718 951 5197
JCDavis@brooklyn.cuny.edu
HPhillips@brooklyn.cuny.edu
http://depthome.brooklyn.cuny.edu/english/
graduate/mfa/geninfo.htm
Deputy Chairperson for Graduate Studies *James Davis*
MFA Assistant *Helen Phillips*

The MFA in *Creative Writing* is a small, highly
personal two-year program which confers degrees
in fiction, poetry and playwriting. Admission is
highly competitive. Students must complete 12
courses (four workshops, four tutorials and four
literature courses), submit a book-length work
in their chosen genre near the end of the second
year and pass a comprehensive examination that
tests their knowledge and expertise in literature.
The workshops, seminars and one-on-one
tutorials with established writers particularly
emphasize relationships between instructors and
students.

## Hunter College of the City University of New York

MFA Program in Creative Writing, Dept. of
English, Hunter College, 695 Park Avenue, New
York NY 10065
☎ 001 212 772 5164
mfa@hunter.cuny.edu
www.hunter.cuny.edu/creativewriting
Associate Director *Gabriel Packard*

Located in the heart of New York City, Hunter
offers two-year MFA programs in fiction, memoir
and poetry. All students take a workshop and
a craft class every semester, plus two literature
classes in their first year and one-on-one
supervisions with a member of faculty throughout
their second year. Hunter's Distinguished Writers
Series brings world-class writers to teach and
give readings. Their Hertog Fellowship in fiction
and memoir pays students to work as researchers
for distinguished writers (Toni Morrison,
Jonathan Safran Foer, Jonathan Franzen, E.L.
Doctorow) and their Norma Lubetsky Friedman
Scholarships pay students to work for an

established poet (Philip Levine, Marie Howe, Jean Valentine). Students may also apply to teach the undergraduate course, *Introduction to Creative Writing*, in their second year.

## New York University

NYU Creative Writing Program, Lillian Vernon Creative Writers House, 58 West 10th Street, New York NY 10011

☏ 001 212 998 8816
creative.writing@nyu.edu
www.cwp.fas.nyu.edu
Program Coordinator *Jessica Flynn*

Offers MFA in *English with a Concentration in Creative Writing* and MA in *Creative Writing* in the genres of either poetry or fiction. Includes writing workshops and craft courses, literary outreach programs, a public reading series, student readings, special literary seminars and student teaching opportunities. Publishes a literary journal, *Washington Square*. Contact the department via email.

## North Carolina

## University of North Carolina at Greensboro

MFA Writing Program, 3302 HHRA Building, UNC Greensboro, Greensboro NC 27402

☏ 001 336 334 5459    ☐ 001 336 256 1470
terry_kennedy@uncg.edu
http://mfagreensboro.org
Contact *Terry Kennedy, Associate Director*

MFA in *Creative Writing (Poetry, Fiction)*. One of the oldest of its kind in the country, the MFA Writing Program at Greensboro is a two-year residency with an emphasis on providing students with studio time in which to study the writing of poetry or fiction. Gives a flexibility that permits students to develop their particular talents through small classes in writing, literature and the arts.

## University of North Carolina Wilmington

Dept. of Creative Writing, University of North Carolina Wilmington, 601 S. College Road, Wilmington NC 28403

☏ 001 910 962 3070
mfa@uncw.edu
www.uncw.edu/writers
Contact *MFA or BFA Coordinator*

An intensive studio-academic program in the writing of fiction, poetry and creative non-fiction leading to either the Master of Fine Arts or Bachelor of Fine Arts degree in *Creative Writing*. Courses include workshops in the three genres, special topics and forms courses as well as a range of courses in literature. Course work in publishing and editing is also offered in the Publishing

Laboratory, a university press imprint which supports local, regional and national publishing projects. The MFA program is home to the literary journal, *Ecotone: reimagining place*.

# Ohio

## Bowling Green State University

Creative Writing Program, English Dept., Bowling Green State University, Bowling Green OH 43403

☏ 001 419 372 6864    ☐ 001 419 372 0333
www.bgsu.edu/departments/creative-writing/

Undergraduate BFA program: a four-year program which offers comprehensive and rigorous training in the art of writing and develops students' skills in preparation for numerous post-graduate careers. Graduate MFA program: a two-year studio/academic composite, mostly work in writing itself in either poetry or fiction. Also offers ten new teaching assistantships each autumn. See the website for details on both programs.

# Oregon

## University of Oregon

Creative Writing Program, 104 Alder Building, 5243 University of Oregon, Eugene OR 97403–5243

☏ 001 541 346 3944    ☐ 001 541 346 0537
crwrweb@uoregon.edu
www.uoregon.edu

Offers MFA in *Creative Writing* in poetry or fiction. Includes writing workshops, craft courses, individualized tutorials with faculty, public reading series, student readings and student teaching opportunities. Admission is highly competitive.

# Pennsylvania

## Chatham University

Master of Fine Arts in Creative Writing Program, Chatham University, Woodland Road, Pittsburgh PA 15232

☏ 001 412 661 1809
www.chatham.edu/mfa
Director of MFA Program *Sheryl St. Germain*

Offers two MFA programs in *Creative Writing* (full-residency/low-residency). The heart of the program focuses on nature, environmental and travel writing but expands the interpretation to include any place-based writing and all genres shaped by human relationship with place. In addition to creative writing workshops and craft courses in poetry, fiction, creative non-fiction, screenwriting and children's writing, the program includes instruction in nature and environmental writing, travel writing and field seminars focused on the literature of wild, urban and rural landscapes. Annual creative writing field seminars offer students the opportunity to travel. Past seminars have included trips to Costa Rica, New Zealand, Greece, India, New Orleans and western Pennsylvania.

## Seton Hill University

1 Seton Hill Drive, Greensburg PA 15601
☎ 001 724 838 4283
bartos@setonhill.edu
www.setonhill.edu
Contact *Director of Graduate Studies*

The MFA in *Writing Popular Fiction* program at Seton Hill University allows students to earn a graduate degree by writing fiction that people *actually* read. Students attend week-long residencies in January and June and complete writing projects off-campus, working with a faculty mentor who is a published author in the chosen genre. Students can choose to specialize in science fiction, fantasy, horror, children's literature, romance or mystery.

## Wilkes University

Graduate Creative Writing Programs, Wilkes University, 84 West South Street, Wilkes-Barre PA 18766
☎ 001 570 408 4534
cwriting@wilkes.edu
www.wilkes.edu/creativewriting
Contact *Assistant Director, Graduate Creative Writing Programs*

The low-residency MA/MFA *Graduate Creative Writing* program of Wilkes University offers five areas of study: poetry, fiction, creative nonfiction, screenwriting and playwriting. Throughout the program, students study the personal life of a writer, the craft, technique and analysis of creative writing and the delivery method for one's work. Low-residency means that students are on the campus of Wilkes University for two eight-day residencies per year, in January and June. The regular coursework will be delivered through a password-protected website. Students can complete their MA in a year-and-a-half and their MFA in two-and-a-half years. The faculty includes writers such as Toi Derricotte, William Kennedy, Larry Heinemann, H.L. Hix, William Heyen and Lawrence Schiller. For more information, please visit the website or send an email.

# Rhode Island

## Brown University

Literary Arts, Box 1923, Brown University, Providence RI 02912
☎ 001 401 863 3260      ℻ 001 401 863 1535
writing@brown.edu
www.brown.edu/cw
Contact *Director of Literary Arts*

Two-year MFA in *Literary Arts*. Students attend two courses per semester: a writing workshop and an elective. In their final semester, graduate students work independently with a faculty advisor on thesis preparation. The thesis is a substantial work of electronic arts, fiction, or poetry – or may cross genre boundaries. Admission is highly selective and is based primarily on the quality of the applicant's writing. Application deadline: 15th December.

# Tennessee

## University of Memphis

MFA Creative Writing Program, University of Memphis, English Dept., 471A Patterson Hall, Memphis TN 38152
☎ 001 901 678 4692      ℻ 001 901 678-2226
creativewriting@memphis.edu
www.mfainmemphis.com
www.memphis.edu
MFA Administrator *Jan Colman*

The MFA *Creative Writing* program offers students a wide choice in the courses they can take. It has four options: Studio Option: only creative writing courses are required, for those who want to focus entirely on their creative work; Literary Studies Option: combines the study of literature with creative writing, for those who want to build a background in literature while they write; Creative/Professional Writing Option: for those who want to gain an excellent preparation for a flexible career in writing; and Cross Disciplinary Option: allows a wide range of options for taking courses in other departments, so that a student can use other interests to inform their writing and study foreign languages, history, art, theatre and other disciplines.

# Texas

## University of Houston

Creative Writing Program, Dept. of English, University of Houston, 229 Roy Cullen Building, Houston TX 77204–5008
☎ 001 713 743 2255

cwp@uh.edu
www.uh.edu
Program Coordinator *Shatera Anderson*

Offers MFA and PhD in *Creative Writing*. Admission is highly competitive; students can apply for admission in the genres of poetry, fiction and non-fiction. Normally admits 20 students a year. The MFA is a three-year degree program; the PhD normally takes five years. Fellowships and teaching assistantships are available. In collaboration with Inprint, it hosts the Inprint/Brown Reading series and the Inprint Studio series. The program offers opportunities to collaborate with artists in other disciplines. Publishes a literary journal, *Gulf Coast.*

## University of Texas at Austin

The James A. Michener Center for Writers, University of Texas at Austin, 702 East Dean Keeton Street, Austin TX 78705
☏ 001 512 471 1601    ✉ 001 512 471 9997
mcw@www.utexas.edu
www.utexas.edu/academic/mcw
Contact *Graduate Coordinator*

A three-year interdisciplinary MFA program in fiction, poetry, playwriting and screenwriting in which students work in two fields of writing, with thesis emphasis in a primary field. Courses include workshops and readings/studies courses. All students must be enrolled full-time, in residence, and are supported by annual fellowships of $25,000, plus remission of tuition and fees. Admission is extremely competitive and applicants must submit all materials by an annual 15th December deadline for the following autumn.

## University of Texas Pan American

MFA in Creative Writing Program, Department of English, University of Texas Pan American, 1201 West University Drive, Edinburg TX 78539
☏ 001 956 381 3421
www.utpa.edu/dept/english
Program Director *José Skinner*

Located in the Rio Grande Valley near the Mexican border, the university is uniquely situated to take advantage of multicultural opportunities and a local population with a high degree of bilingualism. In addition to the traditional genres of fiction, poetry, drama, creative non-fiction and screenwriting, UTPA also offers literary translation as a track in the MFA, culminating in an English to Spanish or Spanish to English literary translation as the thesis. Most workshops and seminars are taught in English but some creative workshops are also available in Spanish, and students proficient in Spanish are encouraged to use bilingualism or code-switching in their thesis work if it suits their artistic goals. The objective of the program is to prepare students for careers as writers, teachers at college or university level, editors, literary translators and other related professional careers. See the website for further information.

# Vermont

## Vermont College of Fine Arts

36 College Street, Montpelier VT 05602
☏ 001 802 828 8839
rachel.muehlmann@vermontcollege.edu
www.vermontcollege.edu /mfa
Program Directors *Louise H. Crowley, Melissa Fisher*
Contact *Rachel Muehlmann, Assistant Director*

The MFA in *Writing* program includes two options: MFA in *Writing* and MFA in *Writing for Children & Young Adults* and is a low-residency two-year program. Genres cover poetry, fiction, creative non-fiction, writing for children and young adults, and a concentration in translation for students earning a degree in poetry or prose. The literary journal, *Hunger Mountain, The Vermont College Journal of Arts & Letters* showcases work by some of the finest writers and artists and also provides students and alumni with opportunities to gain experience working with a literary magazine. Students come from all parts of the USA and many foreign countries and range in age from the twenties to the seventies.

# Virginia

## George Mason University

Graduate Creative Writing Program, George Mason University, 4400 University Drive, 3E4, Fairfax VA 22030
☏ 001 703 993 1180
writing@gmu.edu
creativewriting.gmu.edu
Contact *Graduate Coordinator*

The MFA in *Creative Writing* offers three concentrations – fiction, non-fiction, and poetry, each of which require 48 semester hours – and takes at least three years to complete. Course work blends writing workshops with craft seminars and the study of literature. Each concentration also requires completion of a thesis (a book-length manuscript). In addition, the poetry concentration requires completion of a comprehensive exam. The fiction and non-fiction concentrations offer several options, including an oral exam, completion of a reading list coupled with an annotated bibliography, and designing an individualized project to meet a student's long-term goals.

## Hollins University

MFA in Creative Writing, Hollins University
Graduate Center, PO Box 9603, Roanoke
VA 24020–1603
T 001 540 362 6575   F 001 540 362 6288
hugrad@hollins.edu
www.hollins.edu/grad
Manager, Graduate Studies *Cathy Koon*
Director, MFA & Undergrad Creative Writing
Program *Cathryn Hankla*

MFA in *Creative Writing*: a small, highly
selective two-year, full-time program which
allows students to specialize in poetry, fiction,
non-fiction or multi-genres, while also offering
opportunities to study and write screenplays.
'The program is characterized by an individual
approach, a lively community of writers and
a stimulating combination of challenge and
support.' Also three summer-term MA/MFA
programs in *Children's Literature*; *Playwriting*;
and *Screenwriting & Film Studies*. Students
typically study for three to five six-week terms.

## Old Dominion University

Batten Arts and Letters 5000, Old Dominion
University, Norfolk VA 23529
T 001 757 683 4770
sreynolds@odu.edu
www.al.odu.edu/english/mfacw/
Graduate Program Director *Luisa A. Igloria*

The MFA Program in *Creative Writing* registers
25–35 students each semester in fiction, poetry
and non-fiction. In the three-year, 54-hour
programme, students graduate not only as writers
but also as individuals competent in literary
studies. The Writers-in-Community Program
gives MFA students the opportunity to facilitate
writing workshops in schools, hospitals, homeless
shelters, housing projects and retirement homes
in the area. Students can also participate in the
reading series at Prince Books in downtown
Norfolk and assist with ODU's annual literary
festival. ODU's Writer-in-Residency Program
(offered in both the autumn and spring semesters)
allows students to work one-on-one with
respected poets and writers from around the
country.

## Virginia Commonwealth University

Dept. of English, Virginia Commonwealth
University, PO Box 842005, Richmond
VA 23284–2005
T 001 804 828 1329
tndidato@vcu.edu
www.has.vcu.edu/eng
Graduate Programs Coordinator *Thom Didato*

Three-year MFA in *Creative Writing* with tracks
in fiction and poetry, and workshops in short
fiction, the novel, poetry, drama, screenwriting
and creative non-fiction. Opportunities to work
with *Blackbird*, an online journal of literature
and the arts (www.blackbird.vcu.edu), *Stand
Magazine*, the Levis Reading Prize and the VCU
Cabell First Novelist Award.

# Washington

## Eastern Washington University

Creative Writing, Inland Northwest Center for
Writers, Eastern Washington University, 501 N.
Riverpoint Blvd., Suite 425, Spokane WA 99202
T 001 509 359 4956   F 001 509 455 7252
prussell@ewu.edu
www.ewumfa.com
Contact *Gregory Spatz, Program Director*

The MFA *Creative Writing* program is an
intensive two-year, pre-professional course
of study with an emphasis on the practice of
literature as a fine art. Includes course work in
the study of literature from the vantage point
of its composition and history, but the student's
principal work is done in advanced workshops
and in the writing of a book-length thesis of
publishable quality in fiction, literary non-fiction
or poetry. Students have the opportunity to
work as interns on several projects: the literary
magazine, *Willow Springs*; the community
outreach program, Writers in the Community;
and Literary Editing & Design. The MFA is a
terminal degree program.

## Seattle Pacific University

MFA Program in Creative Writing, Seattle Pacific
University, 3307 Third Avenue West, Seattle
WA 98119–1997
T 001 206 281 2727   F 001 206 281 2335
mfa@spu.edu
www.spu.edu/mfa
Contact *Gregory Wolfe, Program Director*

The low-residency MFA at SPU is a two-year
program for apprentice writers who not only want
to pursue excellence in the craft of writing but
also place their work within the larger context of
the Judeo-Christian tradition of faith. Students
work closely with a faculty mentor in their chosen
genre (poetry, fiction or creative non-fiction)
toward the completion of a creative thesis. They
also write a critical paper each quarter and
read a minimum of 62 books over the course
of the program. Students attend two intensive
ten-day residencies per year, one on Whidbey
Island in Washington State and one in Santa Fe,
New Mexico. Graduating students attend a final
residency in which they deliver a public reading
and a lecture.

# Wyoming

## University of Wyoming

MFA in Creative Writing, Dept. of English –
3353, University of Wyoming, 1000 E. University
Avenue, Laramie WY 82071–2000
☎ 001 307 766 2867
cw@uwyo.edu
www.uwyo.edu/creativewriting
Contact *Rachel Ferrell (Program Assistant)*

The two-year Master of Fine Arts in *Creative
Writing* is an intensive two-year studio degree
in Poetry, Fiction or Creative Non-Fiction.
Special features include opportunities for
interdisciplinary study, supported by a wide
range of university departments, and a required
professional internship on campus, in the
community or further afield, ensuring the
acquisition or polishing of 'real-life' writing skills.

# Writers' Circles and Workshops

## Directory of Writers' Circles, Courses and Workshops
39 Lincoln Way, Harlington LU5 6NG
☎ 01525 873197
diana@writers-circles.com
www.writers-circles.com
Editor *Diana Hayden*

The Directory contains contact details for the UK's 1,000 groups and courses.

## Bernard Kops and Tom Fry Drama Writing Workshops
41B Canfield Gardens, London NW6 3JL
☎ 07976 352837/020 7624 2940/020 8533 2472
artwrite@dsl.pipex.com
bernardkops@blueyonder.co.uk
Tutors *Bernard Kops, Tom Fry*

Three 11-week terms per year. Small group drama writing workshops for stage, screen and radio with actors in attendance. Tutorials also available. Call or email for details.

## Carmarthen Writers' Circle
Brynymor, Pembury Road, Kidwelly SA17 4TF
☎ 01554 891404
Contact *John Birch*

Founded 1989. The Circle meets monthly on the second Monday of the month at the Indoor Bowls Centre, Carmarthen. All levels and genres welcome. Holds occasional workshops.

## Chiltern Writers' Group
Wendover Library, High Street, Wendover
HP22 6DU
info@chilternwriters.org
www.chilternwriters.org

Guest speakers, workshops, manuscript critiques and monthly meetings, held at Wendover Library every second Thursday of the month at 8.00 pm. Regular newsletter. Subscription: £17 p.a.; £12 (concessions); £3 (non-members, meetings).

## Coleg Harlech WEA
Harlech, LL46 2PU
☎ 01766 781900   📠 01766 817621
www.harlech.ac.uk

Coleg Harlech WEA was formed in 2001 with the merger of Coleg Harlech and the Workers' Educational Association North Wales.

## Creuse Writers' Workshop & Retreat
45 Browning Drive, Lincoln LN2 4HF
☎ 07724 191077   📠 01522 880565
info@creusewritersworkshopandretreat
paul@creusewritersworkshopandretreat
www.creusewritersworkshopandretreat
Director of Studies *Sandra Doran*
Bookings & Information *Paul Bett*

One week writers' workshops held twice yearly in Creuse, central France. Run by a team of professional writers to encourage creativity in writers of all grades. The week includes guest speakers, fun practical discussions and social events. Opportunities to write exercises, discuss work as a group and one-to-one. Check the website for special offers and details.

## Cumbrian Literary Group
'Calgarth', The Brow, Flimby, Maryport CA15 8TD
☎ 01900 813444
President *Glyn Matthews*
Secretary *Joyce E. Fisher*

Founded 1946 to provide a meeting place for readers and writers in Cumbria. The Group meets once a month (April to November) in Keswick. Invites speakers to meetings and holds annual competitions for poetry and prose. Publishes *Bookshelf* magazine. Further details from the Secretary at the address above.

## East Anglian Writers
chair@eastanglianwriters.org.uk
www.eastanglianwriters.org.uk
Chairman *Victoria Manthorpe*

An informal association for professional writers living in Norfolk, Suffolk, Essex, Cambridgeshire and Bedfordshire. Affiliated to the **Society of Authors**. Largely social events with occasional speakers' evenings, literary day tours and promotional readings. Authors' profiles on the website. Events and publications on the blog.

## éQuinoxe TBC

8–10 rue de Normandie, 75003 Paris, France
📞 00 33 1 47 55 11 89 📠 00 33 1 48 04 52 58
www.equinoxetbc.fr

Week-long intensive screenwriting workshops, held twice a year in English and French, open to experienced European screenwriters. Participants are selected by an international jury: candidates must have written a script for a feature film, which is sufficiently developed and ready to go into production. Entry form available on the website.

## Euroscript Ltd

64 Hemingford Road, London N1 1DB
📞 07958 244656
ask@euroscript.co.uk
www.euroscript.co.uk

Established in 1995 by the Screenwriters' Workshop with funding by the EU's Media II programme. Script development organization offering creative and editorial input to writers and production companies. Develops screenplays through an intensive consultancy programme, including residential script workshops; also offers a broad programme of short courses, provides script reports and runs an annual Screen Story Competition. Open to writers of any nationality. Email, telephone or access the website for further information.

## Fiction Writing Workshops and Tutorials
▷ entry under UK and Irish Writers' Courses

## Foyle Street Writers Group
▷ Sunderland City Library and Arts Centre under Library Services

## Guildford Writers
▷ The Guildford Institute under UK and Irish Writers' Courses

## Historical Novel Folio

187 Baker Street, Alvaston, Derby DE24 8SG
Contact *Mrs Heidi Sullivan*

An independent postal workshop – single folio dealing with any period before World War II. Send s.a.s.e. for details.

## The Indian King Poets
▷ entry under Organizations of Interest to Poets

## Leicester Writers' Club

www.leicesterwriters.org.uk

Founded 1958. With over 50 members, the club meets weekly in central Leicester. Offers manuscript sessions, workshops, speaker evenings and competitions.

## N. Quentin Woolf – Writers' Workshops

workshops@nquentinwoolf.co.uk
www.blog.nquentinwoolf.com

*Beginners' 8-Week Creative Writing Course*, Brick Lane, London, Mondays 7.00 pm: for new writers looking to develop a portfolio of writing skills. The workshop aims to strengthen writing skills, practice techniques and to improve artistic range. By the end of the course, students will have learnt a core base of creative writing skills and planned, written and edited a short story.

*Writers' Critique Group*, Brick Lane, Tuesdays and Wednesdays 7.00 pm to 9.00 pm: weekly workshops for experienced writers with a demonstrable track-record of literary creation. Participants present work they are currently developing outside of the group in order to receive detailed criticism from the other members.

*Keep Fit for Writers*, South Bank, London, Thursdays 7.00 pm: sessions designed to develop skills encountered on beginners' courses. Each week exercises are covered to keep writers in shape and provide a springboard to help them produce new work regularly. Email for further information.

## New Writing South

9 Jew Street, Brighton BN1 1UT
📞 01273 735353
www.newwritingsouth.com
Direector *Chris Taylor*

Works throughout the South East to create an environment in which creative writers and their works can flourish. Offers resources and development and also builds partnerships between writers and those able to produce their work. Membership open to professional, emerging and aspiring creative writers in the region and to companies with a professional interest in new writing.

## North West Playwrights (NWP)

18 Express Networks, 1 George Leigh Street, Manchester M4 5DL
📞 0161 237 1978
newplaysnw@hotmail.com
www.newplaysnw.co.uk

NWP is the regional development agency for new theatre writing in the north west of England. Operates a script-reading service, provides script development and other support to writers at all stages of their careers, and presents a range of classes. Also collaborates with theatres and other bodies in a variety of writing-related projects. Bi-monthly subscription newsletter, *The Lowdown*. Services available to writers in the region only.

## The Original Writers Group

info@theoriginalwriter.com
www.originalwriters.ning.com
Group Moderator *Rupert Davies-Cooke*

Membership the Original Writers Group is made up of poets, novelists, playwrights, scriptwriters, historians, philosophers. Meets on the first and third Wednesdays of each month from 7.00 pm to 9.00 pm at the Battersea Arts Centre on Lavender Hill. A £3 attendance fee goes towards the room hire (NB the meeting room changes according to availability so please check with the box office on arrival). 'Even if you are just starting out and are just interested in writing but don't have anything to read, come along and experience the evening. Sometimes the best motivation to writing is to hear how others got there before you.'

## Player–Playwrights

▷ entry under Professional Associations

## Script Yorkshire

45 Market Street, Thornton, Bradford BD13 3EN
members@scriptyorkshire.co.uk
www.scriptyorkshire.co.uk
Membership *Joanna Loveday*

'Script Yorkshire is the region's only support and advocacy organization for scriptwriters across performance and broadcast media. Run democratically by writers for writers, it puts the needs of members at the very core of its activity, developing both their practice and their industry connections. Script Yorkshire aims to empower Yorkshire's writers for performance and broadcast media by giving them the skills and insights they need to develop their own careers.' Send an email or visit the website for information on becoming a member.

## 'Sean Dorman' Manuscript Society

3 High Road, Britford, Salisbury SP5 4DS
Director *Jan Smith*

Founded 1957. Provides mutual help among aspiring writers in England, Wales and Scotland. By means of circulating manuscript parcels, members receive constructive comment on their own work and read and comment on the work of others. Full details and application forms available on receipt of s.a.s.e.

## Short Story Writers' Folio/Children's Novel Writers' Folio

5 Park Road, Brading, Sandown, Isle of Wight
PO36 0HU
☏ 01983 407697
dawnwnott@talktalk.net
Contact *Mrs Dawn Wortley-Nott*

Postal workshops; members receive constructive criticism of their work and read and offer advice on fellow members' contributions. Send an s.a.s.e. for further details.

## Society of Sussex Authors

Dolphin House, 51 St Nicholas Lane, Lewes BN7 2JZ
☏ 01273 470100   🖷 01273 470100
sussexbooks@aol.com
www.room155.lewesonline.com
Contact *David Arscott*

Founded 1968. Six meetings per year, held in Lewes, plus social events. Membership restricted to Sussex-based writers only.

## Southwest Scriptwriters

www.southwest-scriptwriters.co.uk
Artistic Director *Tim Massey*

Founded 1994 to offer support to regional drama writers for all media. The group meets at the Bristol Old Vic. Annual subscription: £6.

## Speakeasy – Milton Keynes Writers' Group

PO Box 5948, Stoke Hammond, Milton Keynes MK10 1GX
☏ 01908 663860
speakeasy@writerbrock.co.uk
www.mkweb.co.uk/speakeasy
Chairman *Martin Brocklebank*

Monthly meetings on the first Friday of each month at 7.30 pm. Full and varied programme includes Local Writer Nights where work can be read and peformed and Guest Nights where writers, poets and journalists are invited to speak. Mini-workshops, critique nights and information nights are also in the programme. Invites entries to Open Creative Writing Competitions closing 31st October each year; see website for entry form and rules. Phone, email or send s.a.s.e. for details to address above.

## Spread the Word

77 Lambeth Walk, London SE11 6DX
☏ 020 7735 3111   🖷 020 7735 2666
annette@spreadtheword.org.uk
www.spreadtheword.org.uk
Programme Coordinator *Annette Brook*

Writer development agency. 'A catalyst for developing writers with a strong reputation for providing bold, playful and accessible support for writers of all genres, at all levels and from diverse backgrounds.' From networking events to publisher and agent talks, advice surgeries and online shared stories, community programmes and mentoring. 'We take a collaborative approach to developing and showcasing writers, ensuring that both they and the communities in which they live and work benefit from what we do.'

### Susan Grossman Writing Workshops and Media Coach

☎ 020 7794 0288
susangrossman@tiscali.co.uk
www.susangrossman.co.uk
Contact *Susan Grossman*

Media coach and trainer. Established travel writer, freelance journalist, editor and BBC broadcaster, now a lecturer at two London universities. Runs workshops for journalists and authors looking to focus ideas, write compelling synopses and approach editors and publishers with confidence. Pitching to Editors and Travel Writing workshops held regularly in central London. One-to-one coaching and mentoring sessions are also available, online, by email or telephone.

### Sussex Playwrights' Club

2 Brunswick Mews, Hove BN3 1HD
☎ 01273 730106
www.sussexplaywrights.com
Secretary *Dennis Evans*

Founded 1935. Monthly readings of members' work by experienced actors. Membership open to all. Meetings held at New Venture Theatre, Bedford Place, Brighton.

### Ver Poets

☎ 01727 864898
daphneschiller@ntlworld.com
www.verpoets.org.uk
Secretary *Daphne Schiller*

Founded 1966. Postal and local members. Meets in St Albans, runs evening meetings and daytime workshops and organizes competitions.

### Walton Wordsmiths

27 Braycourt Avenue, Walton on Thames KT12 2AZ
☎ 01932 702874
wendy@stickler.org.uk
www.waltonwordsmiths.co.uk
Contact *Wendy Hughes*

Founded in 1999 by Wendy Hughes. Supports writers of all grades and capabilities by offering constructive criticism and advice.

### Workers' Educational Association

*National Office*: 3rd Floor, 70 Clifton Street, London EC2A 4HB
☎ 020 7426 3450
national@wea.org.uk
www.wea.org.uk

Founded in 1903, the WEA is a charitable provider of adult education with students drawn from all walks of life. It runs writing courses and workshops in many parts of the country which are open to everyone. Contact your local WEA office for details of courses in your region.

**Eastern** Cintra House, 12 Hills Road, Cambridge, B2 1JP eastern@wea.org.uk

**East Midlands** 39 Mapperley Road, Mapperley Park, Nottingham NG3 5AQ ☎ 0115 962 8400 eastmidlands@wea.org.uk

**London** 4 Luke Street, London EC2A 4XW
☎ 020 7426 1950 london@wea.org.uk

**North East** 21 Portland Terrace, Newcastle upon Tyne NE2 1QQ ☎ 0191 212 6100 northeast@wea.org.uk

**North West** Suite 405–409, The Cotton Exchange, Old Hall Street (Bixteth Street Entrance), Liverpool L3 9JR ☎ 0151 243 5340 northwest@wea.org.uk

**Southern** Unit 57 Riverside 2, Sir Thomas Longley Road, Rochester ME2 4DP
☎ 01634 298600 southern@wea.org.uk

**South West** Bradninch Court, Castle Street, Exeter EX4 3PL ☎ 01392 490970 southwest@wea.org.uk

**West Midlands** 4th Floor, Lancaster House, 67 Newhall Street, Birmingham, B3 1NQ westmidlands@wea.org.uk

**Yorkshire and Humber** 6 Woodhouse Square, Leeds LS3 1AD ☎ 0113 245 3304 yorkshumber@wea.org.uk

### Workers' Educational Association North Wales
▷ **Coleg Harlech WEA**

### Workers' Educational Association Northern Ireland

1–3 Fitzwilliam Street, Belfast BT9 6AW
☎ 028 9032 9718   🖷 028 9023 0306
info@wea-ni.com
www.wea-ni.com

Autonomous from the WEA National Association, WEA Northern Ireland was founded in 1910.

### Workers' Educational Association Scotland

Riddles Court, 322 Lawnmarket, Edinburgh EH1 2PG
☎ 0131 226 3456
hq@weascotland.org.uk
www.weascotland.org.uk
Scottish Secretary *Joyce Connon*

WEA Scotland (registered charity no: 2806910) was founded in 1905 and is part of the national **Workers' Educational Association**.

### Workers' Educational Association South Wales

7 Coopers Yard, Curran Road, Cardiff CF10 5NB
☎ 029 2023 5277   🖷 029 2023 3986
weasw@swales.wea.org.uk
www.swales.wea.org.uk

Autonomous from the WEA National Association, WEA South Wales has been active in community education for over 100 years.

## Writers in Oxford
wio_membership@yahoo.co.uk
www.writersinoxford.org

Membership Secretary *Cherry Mosteshar*

Founded 1992. Open to published authors, playwrights, poets and journalists. Literary seminars and social functions. Publishes *The Oxford Writer* newsletter. Subscription: £25 p.a.

# Bursaries, Fellowships and Grants

## The Airey Neave Trust

PO Box 36800, 40 Bernard Street, London
WC1N 1WJ
☎ 020 7833 4440
hanthoc@aol.com
www.theaireyneavetrust.org.uk
Contact *Hannah Scott*

Initiated 1980. Annual research fellowships for
up to three years – towards a book or paper – for
serious research connected with national and
international law, and human freedom. Preferably
attached to a particular university in Britain.

## Alfred Bradley Bursary Award

c/o BBC Radio Drama, Room 2130, New
Broadcasting House, Oxford Road, Manchester
M60 1SJ
☎ 0161 244 4253   ☒ 0161 244 4248
Contact *Coordinator*

Established 1992. Biennial award in
commemoration of the life and work of the
distinguished radio producer Alfred Bradley.
Aims to encourage and develop new radio writing
talent in the BBC North region. Entrants must
have been born, currently resident or previously
lived in the north of England for at least five
years. The award is given to help writers to pursue
a career in writing for radio. Next award will be
launched late 2010. (Further information can
be found on bbc.co.uk/writersroom) Previous
winners: Chris Wilson, Anthony Cropper, Peter
Straughan, Lee Hall.
£ Award of up to £5,000 over two years;
potential BBC Radio Drama commissions and
opportunities for mentoring with producers.

## Arts Council of Wales ACW's Creative Wales Awards
▷ entry under Arts Councils and Regional Offices

## The Authors' Contingency Fund

The Society of Authors, 84 Drayton Gardens,
London SW10 9SB
☎ 020 7373 6642   ☒ 020 7373 5768
info@societyofauthors.org
www.societyofauthors.org

This fund makes modest grants to published
authors who find themselves in sudden financial
difficulties. Contact *Sarah Baxter* at the **Society
of Authors** for an information sheet and
application form.

## The Authors' Foundation

The Society of Authors, 84 Drayton Gardens,
London SW10 9SB
☎ 020 7373 6642   ☒ 020 7373 5768
info@societyofauthors.org
www.societyofauthors.org

The Foundation provides grants to writers to
assist them while writing books. Application is
by letter and there are two rounds a year. The
deadlines are 30th April and 30th September.
Contact *Paula Johnson* at the **Society of Authors**
for guidelines.

## The Betty Trask Awards

The Society of Authors, 84 Drayton Gardens,
London SW10 9SB
☎ 020 7373 6642   ☒ 020 7373 5768
info@societyofauthors.org
www.societyofauthors.org

These annual awards are for authors who are
under 35 and Commonwealth citizens, awarded
on the strength of a first novel (published or
unpublished) of a traditional or romantic nature.
The awards must be used for a period or periods
of foreign travel. Final entry date: 31st January.
Contact the **Society of Authors** for an entry form.
2009 winners: Samantha Harvey *The Wilderness*
(Jonathan Cape); Eleanor Catton *The Rehearsal*
(Granta).
£ Awards £20,000 (total).

## British Academy Small Research Grants

10 Carlton House Terrace, London SW1Y 5AH
☎ 020 7969 5217   ☒ 020 7969 5414
grants@britac.ac.uk
www.britac.ac.uk
Contact *Research Grants Department*

Award to further original academic research at
postdoctoral level in the humanities and social
sciences. Entrants must be resident in the UK.

See website for final entry dates and application details.

£ Award £7,500 (maximum).

## The Charles Pick Fellowship

School of Literature and Creative Writing, University of East Anglia, Norwich NR4 7TJ

☎ 01603 592286    ℻ 01603 507728

charlespickfellowship@uea.ac.uk

www1.uea.ac.uk/cm/home/schools/hum/lit/awards/pick

Contact *The Fellowship Administrator*

Founded in 2001 by the Charles Pick Consultancy in memory of the publisher and literary agent who died in 2000, to support a new unpublished writer of fictional or non-fictional prose. This residential Fellowship is awarded annually and lasts six months. There is no interview; candidates will be judged on the quality and promise of their writing, the project they describe and the strength of their referee's report. 2009 winners: Birgit Larsson and Simidele Awosika. Deadline for applications: 31st January each year.

£ Award up to £10,000 plus accommodation on campus.

## Cholmondeley Awards

The Society of Authors, 84 Drayton Gardens, London SW10 9SB

☎ 020 7373 6642    ℻ 020 7373 5768

info@societyofauthors.org

www.societyofauthors.org

Founded in 1965 by the late Dowager Marchioness of Cholmondeley. Annual honorary awards to recognize the achievement and distinction of individual poets. 2009 winners: Bernard O'Donoghue, Alice Oswald, Fiona Sampson, Pauline Stainer.

£ Awards £6,000 (total).

## The David T.K. Wong Fellowship

School of Literature and Creative Writing, University of East Anglia, Norwich NR4 7TJ

☎ 01603 592286    ℻ 01603 507728

davidtkwongfellowship@uea.ac.uk

www1.uea.ac.uk/cm/home/schools/hum/lit/awards/wong

Contact *The Fellowship Administrator*

Named for its sponsor, David Wong (retired Hong Kong businessman, teacher, journalist, senior civil servant and writer of short stories); this unique and generous award was launched in 1997 to enable a fiction writer who wants to write in English about the Far East. This residential Fellowship is awarded annually and lasts one academic year. Unpublished or established published writers are welcome to apply. There is no interview; candidates will be judged on the quality and promise of their writing and the

project they describe. 2009 winner: Hanh Hoang. Deadline for applications: 31st January each year.

£ Award £26,000.

## The Economist/Richard Casement Internship

The Economist, 25 St James's Street, London SW1A 1HG

☎ 020 7830 7000

casement2010@economist.com

www.economist.com

Contact *Geoffrey Carr, Science Editor (re. Casement Internship)*

For an aspiring journalist to spend three months in the summer writing for *The Economist* about science and technology. Applicants should write a letter of introduction along with an article of approximately 600 words suitable for inclusion in the Science and Technology Section. 'Our aim is more to discover writing talent in a scientist or science student than scientific aptitude in a budding journalist.' Competition details normally announced in the magazine late January or early February and 4–5 weeks allowed for application.

## The Elizabeth Longford Grants

The Society of Authors, 84 Drayton Gardens, London SW10 9SB

☎ 020 7373 6642    ℻ 020 7373 5768

info@societyofauthors.org

www.societyofauthors.org

Biannual grant, sponsored by Flora Fraser and Peter Soros, to a historical biographer for essential research, travel or other necessary expenditure. Contact *Paula Johnson* at the **Society of Authors** for full entry details. Final entry dates: 30th April and 30th September.

£ Grant £2,500.

## Eric Gregory Trust Fund

The Society of Authors, 84 Drayton Gardens, London SW10 9SB

☎ 020 7373 6642    ℻ 020 7373 5768

info@societyofauthors.org

www.societyofauthors.org

Annual awards of varying amounts are made for the encouragement of poets under the age of 30 on the basis of a submitted collection. Open only to British-born subjects resident in the UK. Final entry date: 31st October. Contact *Paula Johnson* at the **Society of Authors** for full entry details. 2009 winners: Liz Berry, James Brookes, Swithun Cooper, Alex McRae, Sam Riviere.

£ Awards £20,000 (total).

## European Jewish Publication Society

PO Box 19948, London N3 3ZL

☎ 020 8346 1668

www.ejps.org.uk

www.ejps.org.uk

Contact *Professor Colin Shindler*

Established in 1995 to help fund the publication of books of European Jewish interest which would otherwise remain unpublished. Helps with the marketing, distribution and promotion of such books. Publishers who may be interested in publishing works of Jewish interest should approach the Society with a proposal and manuscript in the first instance. Books which have been supported include: *The History of Zionism* Walter Laqueur; *Photographing the Holocaust* Janina Struk; *The Arab-Israeli Cookbook* Robin Soans; *Daughters of Sarah: An Anthology of Jewish Women Writing in French* eds. Eva Martin Sartori and Madeleine Cottenet-Hage; *Whistleblowers and the Bomb: Vanunu, Israel and Nuclear Secrecy* Yoel Cohen. Also supports the publication of poetry, and translations from and into other European languages.

Ⓔ Grant £3,000 (maximum).

## Francis Head Bequest

The Society of Authors, 84 Drayton Gardens, London SW10 9SB

Ⓣ 020 7373 6642    Ⓕ 020 7373 5768
info@societyofauthors.org
www.societyofauthors.org

Provides grants to published British authors over the age of 35 who need financial help during a period of illness, disablement or temporary financial crisis. Contact *Sarah Baxter* at the **Society of Authors** for an information sheet and application form.

## Fulbright Awards

The Fulbright Commission, Fulbright House, 62 Doughty Street, London WC1N 2JZ

Ⓣ 020 7404 6880    Ⓕ 020 7404 6834
programmes@fulbright.co.uk
www.fulbright.co.uk
Contact *Awards Assistant*

The Fulbright Commission offers a number of scholarships given at postgraduate level and above, open to any field of study/research to be undertaken in the USA. Length of award is typically an academic year. The application deadline for awards is 31st May of the preceding year of study. Full details and application forms are available on the Commission's website.

## The Guardian Research Fellowship

Nuffield College, Oxford OX1 1NF
Ⓣ 01865 278542    Ⓕ 01865 278666
Contact *The Academic Administrator*

One-year fellowship endowed by the Scott Trust, owner of *The Guardian*, to give someone working in the media the chance to put their experience into a new perspective, publish the outcome and give a *Guardian* lecture. Applications welcomed from journalists and management members,

in newspapers, periodicals or broadcasting. Research or study proposals should be directly related to experience of working in the media. Accommodation and meals in college will be provided and a stipend. Advertised biennially in November.

## Hawthornden Literary Institute

Hawthornden International Retreat for Writers, Lasswade EH18 1EG

Ⓣ 0131 440 2180    Ⓕ 0131 440 1989
office@hawthornden.com
Contact *The Director*

Established 1982 to provide a peaceful setting where published writers can work in silence. The Castle houses up to six writers at a time, who are known as Hawthornden Fellows. Writers from any part of the world may apply for the fellowships. No monetary assistance is given, nor any contribution to travelling expenses, but once arrived at Hawthornden, the writer is the guest of the Retreat. Applications on forms provided must be made by the end of June for the following calendar year. Previous winners include: Les Murray, Alasdair Gray, Helen Vendler, David Profumo, Hilary Spurling.

## Hosking Houses Trust

33 The Square, Clifford Chambers, Stratford-upon-Avon CV37 8HT
sarahhosking@btinternet.com
www.hoskinghouses.co.uk
Contact *The Secretary*

Established in 2001 to offer women writers over the age of 40 a period of financially-protected, domestic peace. Residence of Church Cottage, two miles from Stratford-upon-Avon; furnished and equipped; offered for between two months and one year, all bills paid (except BT) also a personal bursary. No duties except to complete personal work, report to the Trustees every three months, to acknowledge the Trust in all work produced during tenure or published as a result of it. Advertised as available: appointed on merit and need to an established, published woman writer on any subject, and who has the legal right to be in the UK.

## Jerwood Awards
▷ The Royal Society of Literature/Jerwood Awards

## The John Masefield Memorial Trust

The Society of Authors, 84 Drayton Gardens, London SW10 9SB

Ⓣ 020 7373 6642    Ⓕ 020 7373 5768
info@societyofauthors.org
www.societyofauthors.org

This trust makes occasional grants to professional poets (or their immediate dependants) who are faced with sudden financial problems. Contact

*Sarah Baxter* at the **Society of Authors** for an information sheet and application form.

## Journalists' Charity

Dickens House, 35 Wathen Road, Dorking RH4 1JY

📞 01306 887511    📠 01306 888212

enquiries@journalistscharity.org.uk
www.journalistscharity.org.uk

Director/Secretary *David Ilott*

Aims to relieve distress among journalists and their dependants. Continuous and/or occasional financial grants; also retirement homes for eligible beneficiaries. Further information available from the Director.

## The K. Blundell Trust

The Society of Authors, 84 Drayton Gardens, London SW10 9SB

📞 020 7373 6642    📠 020 7373 5768

info@societyofauthors.org
www.societyofauthors.org

Grants to writers to fund travel, important research or other expenditure. Author must be under 40, has to submit a copy of his/her previous book and the work must 'contribute to the greater understanding of existing social and economic organization'. Application by letter. Contact the **Society of Authors** for full entry details. Final entry dates: 30th April and 30th September.

## Lannan Literary Awards and Fellowships

Lannan Foundation, 313 Read Street, Santa Fe, New Mexico 87501–2628, USA

📞 001 505 986 8160    📠 001 505 986 8195

www.lannan.org

Established 1989. Annual awards given to writers of exceptional poetry, fiction and non-fiction who have made a significant contribution to English-language literature, as well as emerging writers of distinctive literary merit who have demonstrated potential for outstanding future work. On occasion, the Foundation recognizes a writer for lifetime achievement. Candidates for the awards and fellowships are recommended to the Foundation by a network of writers, literary scholars, publishers and editors. Applications or unsolicited nominations for the awards and fellowships are not accepted.

## Laurence Stern Fellowship

Department of Journalism, City University, Northampton Square, London EC1V 0HB

📞 020 7040 4036

a.r.mckane@city.ac.uk
www.city.ac.uk/journalism

Contact *Anna McKane*

Founded 1980. Awarded each year to a young journalist experienced enough to work on national stories. It gives a chance to work on the national desk of the *Washington Post*. A senior executive from the newspaper selects from a shortlist drawn up in February/March. 2009 winner: Alexi Mostrous of *The Times*. Full details about how to apply available on the City University website in January.

## The Luke Bitmead Bursary for New Writers

2 London Wall Buildings, London EC2M 5UU

📞 020 7448 5137

bitmeadbursary@legendpress.co.uk
www.legendpress.co.uk

Contacts *Tom Chalmers, Lucy Boguslawski*

Founded in 2006 by the family of Luke Bitmead and Legend Press following his death in 2006 aged 34. Annual award to support and encourage the work of fledgling unpublished authors. Writers are invited to submit the first three or four chapters of an unpublished novel for consideration together with details of personal background and why the bursary will be of benefit. Previous winners: Andrew Blackman *On the Holloway Road*; Ruth Dugdall *The Woman Before Me*.

💷 Award of £2,500 from the Luke Bitmead Memorial Fund and publication by Legend Press.

## Northern Writers' Awards
▷ New Writing North under Professional Associations and Societies

## Olive Cook Prize
▷ Tom-Gallon Trust Award

## PAWS (Public Awareness of Science) Drama Fund

The PAWS Office, OMNI Communications, First Floor, 155 Regents Park Road, London NW1 8BB

pawsomni@btconnect.com
www.pawsdrama.com

Contacts *Andrew Millington, Andrée Molyneux*

Established 1994. PAWS encourages and supports television drama drawing on science, engineering and technology. It offers a range of activities to bring science and technology to television writers and producers in an easily accessible way. Events: cover developments in science and technology and enable writers to meet scientists and engineers; Contacts and Information Service: provides specialist advice for writers to help them develop their television programme ideas; Periodic Funding Support: intended to help writers with some successful television drama experience bring their work closer to production; Annual Awards: for the best European programmes with scientific content in drama, documentary and all AV media.

## Pearson Playwrights' Scheme

c/o Pearson Plc, 80 Strand, London WC2R 0RL
playwrightscheme@googlemail.com
www.pearson.com
Administrator *Jack Andrews, MBE*

Awards up to five bursaries to playwrights annually, each worth £7,000. Applicants must be sponsored by a theatre which then submits the play for consideration by a panel. Each award allows the playwright a 12-month attachment. Applications invited via theatres in September each year. A further award of £10,000, for Best Play, is on offer each year to previous bursary winners. Email for up-to-date information.

## Peggy Ramsay Foundation

Hanover House, 14 Hanover Square, London
W1S 1HP
T 020 7667 5000    F 020 7667 5100
laurence.harbottle@harbottle.com
www.peggyramsayfoundation.org
Contact *G. Laurence Harbottle*

Founded in 1992 in accordance with the will of the late Peggy Ramsay, the well-known agent. Grants are made to writers for the stage who have some experience and who need time and resources to make writing possible. Grants are also made for writing projects by organizations connected with the theatre. The Foundation does not support production costs or any project that does not have a direct benefit to playwriting. Writers applying must have had one full-length play professionally produced for adults.
£ Grants total £150,000 to £200,000 per year.

## Ralph Lewis Award

University of Sussex Library, Brighton BN1 9QL
T 01273 678158    F 01273 873413
a.timoney@sussex.ac.uk
www.sussex.ac.uk/library/

Established 1985. Occasional award set up by Ralph Lewis, a Brighton author and art collector who left money to fund awards for promising manuscripts which would not otherwise be published. The award is given in the form of a grant to a UK-based publisher in respect of a publication of literary works by new authors. No direct applications from writers. Previous winners: **Peterloo Poets**; **Serpent's Tail**.

## The Royal Literary Fund

3 Johnson's Court, off Fleet Street, London
EC4A 3EA
T 020 7353 7159    F 020 7353 1350
egunnrlf@globalnet.co.uk
www.rlf.org.uk
Secretary *Eileen Gunn*

Grants and pensions are awarded to published authors of several works in financial need, or to their dependants. Examples of author's works are needed for assessment by Committee. Contact the Secretary for further details and application form.

## The Royal Society of Literature/Jerwood Awards for Non-Fiction

The Royal Society of Literature, Somerset House, Strand, London WC2R 1LA
T 020 7845 4676    F 020 7845 4679
paulaj@rslit.org
www.rslit.org
Submissions *Paula Johnson*

The awards offer financial assistance to authors engaged in writing their first major commissioned works of non-fiction. Three awards: one of £10,000 and two of £5,000 will be offered annually to writers working on substantial non-fiction projects. Open to UK and Irish writers and writers who have been resident in the UK for at least three years. Applications for the awards should be submitted by mid-October. Further details available on the website.

## Scottish Arts Council Writers' Bursaries

Scottish Arts Council, 12 Manor Place, Edinburgh
EH3 7DD
T 0131 226 6051    F 0131 225 9833
gavin.wallace@scottisharts.org.uk
www.scottisharts.org.uk
Head of Literature *Dr Gavin Wallace*

Bursaries to enable published writers of literary work, playwrights and storytellers to devote more time to their writing. Around 20 bursaries of up to £15,000 awarded annually; deadline for applications in July and January. Application open to writers based in Scotland.

## Scottish Book Trust Mentoring Scheme

Scottish Book Trust, Sandeman House, Trunk's Close, 55 High Street, Edinburgh EH1 1SR
T 0131 524 0160    F 0131 524 0161
www.scottishbooktrust.com
Contact *Caitrin Armstrong (Writer Development Manager)*

The mentoring scheme aims to support published writers living and working in Scotland who have a specific project on which they'd like some focused, dedicated support and advice. Successful applicants will be paired with another writer or industry professional who has appropriate experience and will support them as they work together over an intensive period of nine months. Applicants should see the website for full details.

## Scottish Book Trust New Writers' Awards

Scottish Book Trust, Sandeman House, Trunk's Close, 55 High Street, Edinburgh EH1 1SR
T 0131 524 0160    F 0131 524 0161
www.scottishbooktrust.com

Contact *Caitrin Armstrong (Writer Development Manager)*

(Formerly the Scottish Arts Council New Writers' Bursaries.) Eight awards available for writers living and working in Scotland who have little or no previous publication or production track record. Successful applicants will receive a cash award of £2,000 plus a bespoke professional development package which may include nine months working with a mentor, plus networking opportunities with publishers/agents. The award is open to writers of fiction, literary non-fiction and poets who can demonstrate their commitment to writing. A c.v., sample of work and referee will be required. Applicants must see website for full application details.

## The Somerset Maugham Awards

The Society of Authors, 84 Drayton Gardens, London sw10 9sb
☎ 020 7373 6642   ℻ 020 7373 5768
info@societyofauthors.org
www.societyofauthors.org

Annual awards designed to encourage writers under the age of 35 to travel. Given on the basis of a published work of fiction, non-fiction or poetry. Open only to British-born subjects resident in the UK. Final entry date: 30th November. 2009 winners: Adam Foulds *The Broken Word* (Jonathan Cape); Alice Albinia *Empires of the Indus* (John Murray); Rodge Glass *Alasdair Gray: A Secretary's Life* (Bloomsbury); Henry Hitchings *The Secret Life of Words: How English Became English* (John Murray); Thomas Leveritt *The Exchange Rate Between Love and Money* (Harvill Secker); Helen Wash *Once Upon a Time in England* (Canongate).
£ Awards £10,000 (total).

## Tom-Gallon Trust Award and the Olive Cook Prize

The Society of Authors, 84 Drayton Gardens, London sw10 9sb
☎ 020 7373 6642   ℻ 020 7373 5768
info@societyofauthors.org
www.societyofauthors.org

An award of £1,000 is made on the basis of a submitted story to fiction writers of limited means who have had at least one short story accepted for publication. Both awards are biennial and are awarded in alternate years. Contact the **Society of Authors** for an entry form. Final entry date: 31st October.

## The Travelling Scholarships

The Society of Authors, 84 Drayton Gardens, London sw10 9sb
☎ 020 7373 6642   ℻ 020 7373 5768
info@societyofauthors.org
www.societyofauthors.org

Annual honorary grants to established British writers. No submissions. 2009 winners: Paul Farley, Eva Hoffman.
£ Award £2,000 each.

# Prizes

## Academi Cardiff International Poetry Competition

Academi, PO Box 438, Cardiff CF10 5YA
☎ 029 2047 2266    ℻ 029 2049 2930
post@academi.org
www.academi.org
Contact *Peter Finch (Chief Executive)*

Established 1986. An annual competition supported by Cardiff Council for unpublished poems in English of up to 50 lines. Closing date in January each year.
£ Prizes total £6,000; £5,000 (1st); £500 (2nd); £250 (3rd); five runners-up receive £50 each.

## Adrienne Benham Award

c/o Theatre Centre, Shoreditch Town Hall, 380 Old Street, London EC1V 9LT
☎ 020 7729 3066
admin@theatre-centre.co.uk
www.theatre-centre.co.uk
Contact *Awards*

The Adrienne Benham Award is presented to an emerging playwright who displays an interest in writing for young audiences. See website for full details and application form. 2009 winner: Steve Bloomer.
£ £2,000 Seed Commission.

## Aldeburgh First Collection Prize

The Poetry Trust, The Cut, 9 New Cut, Halesworth IP19 8BY
☎ 01986 835950    ℻ 01986 874524
info@thepoetrytrust.org
www.thepoetrytrust.org
Director *Naomi Jaffa*

The prize is awarded to the author of what in the opinion of the judges is the best first full collection of poetry published in book form in the UK and the Republic of Ireland in the preceding twelve months. The winner receives £3,000 plus an invitation to read (fee paid) at the **Aldeburgh Poetry Festival** the following year. Please refer to the website for details on how to enter. 2009/10 winner: J.O. Morgan *Natural Mechanical* (C.B. Editions).

## The Alexander Cordell Literature Competition

Cordell Country, c/o Blaenavon Tourist Information Centre, Church Road, Blaenavon NP4 9AS
☎ 01495 742333
blaenavontic@torfaen.gov.uk
www.cordellcountry.org
Contact *Bogda Smreczak*

Founded 2003. Annual award in memory of Alexander Cordell, one of Wales' most popular writers whose *Rape of the Fair Country* became an international bestseller. The competition encourages a new generation of writers to look more closely at South Wales. Entries must be an original, unpublished work, maximum 1,500 words, in English and be typed in double-spacing on one side of A4 paper only. No covering letter or other material should be enclosed with the entry; the entrant's name must not appear on the ms which should be attached to the entry form. Entries accepted only if accompanied by form and £5 entry fee (cheques and postal orders should be made payable to Torfaen County Borough Council). Receipt of entry will be acknowledged only if s.a.s.e. marked 'ACKNOWLEDGEMENT' is enclosed. See application form for details of theme (available online or via contacts given above). 2010 winners: Patricia Davies (1st); Diana Adams (2nd); Frances Green (3rd).
£ £500 (1st); £300 (2nd); £150 (3rd).

## Alexander Prize

Royal Historical Society, University College London, Gower Street, London WC1E 6BT
☎ 020 7387 7532    ℻ 020 7387 7532
royalhistsoc@ucl.ac.uk
www.royalhistoricalsociety.org/grants.htm
Contact *Administrative Secretary*

Established 1897. Awarded for a published journal article or essay based upon original research. The article/essay must have been published during the period 1st January 2009 to 31st December 2009. Competitors may choose their own subject for the essay. Closing date: 31st December.
£ Prize of £250.

## The Alfred Fagon Award

Talawa Theatre Company, 53–55 East Road,
London N1 6AH
www.alfredfagonaward.co.uk

First presented in 1997. An annual award, in
memory of playwright Alfred Fagon, which is
open to any playwright of Caribbean descent,
resident in the UK, for the best new stage play
in English, which need not have been produced
(television and radio plays and film scripts not
eligible). Two copies of the script (plus s.a.s.e.
for return), together with a brief history of the
play and c.v. (including author's Caribbean
connection), should be sent to the address above
by 31st August. Previous winner: Paula B. Stanic.
£ Award of £5,000.

## The And/or Book Awards

Colman Getty, 28 Windmill Street, London
W1T 2JJ
☎ 020 7631 2666    ℻ 020 7631 2699
chris@colmangetty.co.uk
truda@colmangetty.co.uk
www.andorbookawards.org
Awards Administrator *Sarah Jackson (awards@
kraszna-krausz.co.uk)*

Awards for books published in the fields of
photography and the moving image (including
film, television and new media). Established in
1985 by Andor Kraszna-Krausz, the Hungarian
founder of Focal Press, the awards celebrate
excellence in photography and moving image
publishing. Submissions by publishers only.
Please refer to the website for entry letter, rules
& conditions, entry form and other information.
Initial books and entry form to be sent to
Chris Baker/Truda Spruyt at Colman Getty.
Sarah Jackson, the Awards Administrator, can
be reached on the email address above. 2009
winners: Best Photography Book: *Susan Meiselas:
In History* ed. Kristen Lubben (Steidl); Best
Moving Image Book: *Photography and Cinema*
David Campany (Reaktion Books).

## André Simon Memorial Fund Book Awards

1 Westbourne Gardens, Glasgow G12 9XE
☎ 07801 310973
katie@andresimon.co.uk
www.andresimon.co.uk
Contact *Katie Lander*

Established 1978. Awards given annually for
the best book on drink, best on food and
special commendation in either. 2009 winners:
Carla Capalbo *Collio* (Pallas Athene); Darina
Allen *Forgotten Skills of Cooking* (Kyle Cathie);
Commendation: Richard Wrangham *Catching
Fire* (Profile).
£ Awards of £2,000 (best book); £1,000 (special
commendation); £200 to shortlisted books.

## Angus Book Award

Angus Council, Educational Resources Service,
Bruce House, Arbroath DD11 3TL
☎ 01241 435045    ℻ 01241 435038
cularbers@angus.gov.uk
www.angus.gov.uk/bookaward
Contact *Moyra Hood, Educational Resources
Librarian*

Established 1995. Designed to try to help
teenagers develop an interest in and enthusiasm
for reading. Shortlisted books are read and voted
on by third-year schoolchildren in all eight Angus
secondary schools. Authors visit schools and
take part in the award ceremony in May of each
year. 2009 winner: Anne Cassidy *Forget Me Not*
(Scholastic).
£ Prize of £500 cheque, plus trophy in the form
of a replica Pictish stone.

## Annual Theatre Book Prize

▷ The Society for Theatre Research Annual Theatre Book
Prize

## The Anthony Hecht Poetry Prize

The Waywiser Press, Bench House, 82 London
Road, Chipping Norton OX7 5FN
☎ 01608 644755
waywiserpress@aol.com
waywiser-press.com/hechtprize.html
Contact *Philip Hoy, Managing Editor*

Established 2005. Annual prize awarded for
an outstanding, unpublished, full collection of
poems. Entrants must be at least 18 years of
age and may not have published more than one
previous collection. Detailed guidelines and entry
forms available from August on the website or by
sending an A4-sized s.a.s.e.
£ Prize of £1,500 and publication by Waywiser in
the UK and USA.

## Arthur C. Clarke Award for Science Fiction

246e Bethnal Green Road, London E2 0AA
clarkeaward@gmail.com
www.clarkeaward.com
Administrator *Tom Hunter*

Established 1986. The Arthur C. Clarke Award is
given annually to the best science fiction novel
with first UK publication in the previous calendar
year. Both hardcover and paperback books
qualify. Made possible by a generous donation by
the late Arthur C. Clarke, this award is selected
by a rotating panel of judges nominated by
the **British Science Fiction Association**, the
**Science Fiction Foundation** and the website
SFCrowsnest.com. 2009 winner: Ian R. MacLeod
*Song of Time* (PS Publishing).
£ Award of £2,010 (award increases by £1 per
year), plus trophy.

## Arts Council Children's Award
> Brian Way Award

## Arvon Foundation International Poetry Competition

Free Word Centre, 60 Farringdon Road, London
EC1R 3GA

☎ 020 7324 2554    ᴇ 0800 756 1349
poetry@arvonfoundation.org
www.arvonfoundation.org
Contact *The National Office*

Established 1980. Biennial competition (next
in 2010) for poems written in English and not
previously broadcast or published. There are no
restrictions on the number of lines, themes, age of
entrants or nationality. No limit to the number of
entries. Entry fee: £7 per poem; August deadline.
Previous winners: Paul Farley *Laws of Gravity*;
Don Paterson *A Private Bottling*.
ᴇ Prize of £5,000 (1st), and £5,000 worth of other
prizes.

## Asham Award for Women

Asham Literary Endowment Trust, c/o Town
Hall, High Street, Lewes BN7 2QS
☎ 01273 483159
carole.buchan@btinternet.com
www.ashamaward.com
Contact *Carole Buchan*

Founded in 1996, the award (named after the
house in Sussex where Virginia Woolf lived) is
a biennial short story competition for women
and is administered by the Asham Literary
Endowment Trust which encourages new
writing through competitions, mentoring,
training and publication. Candidates must be
female, over 18, resident in the UK and must
not have had a novel or complete anthology
of work published previously. The 2010 award
is sponsored by The Booker Prize and Much
Ado Books of Alfriston, East Sussex and will be
launched at the end of June. All details can be
found on the website.
ᴇ Prize money up to £3,600; 12 winning writers
are published by Bloomsbury alongside specially
commissioned stories by leading professional
writers, including Margaret Atwood and Esther
Freud.

## The Astrid Lindgren Memorial Award (ALMA)

Swedish Arts Council, PO Box 27215, SE-102 53
Stockholm, Sweden
☎ 00 46 8 519 264 00    ᴇ 00 46 8 519 264 99
literatureaward@alma.se
www.alma.se/en
Director *Erik Titusson*

Established 2002 by the Swedish government in
memory of the children's author Astrid Lindgren.
Administered by the Swedish Arts Council, it is
an international award for children's and young
people's literature given annually to one or more
recipients, irrespective of language or nationality.
Writing, illustrating and storytelling, as well as
reading promotion activities may be awarded.
Selected organizations worldwide are invited to
submit nominations once a year; jury members
may also contribute nominations. 2009 winner:
Tamer Institute.
ᴇ Award of SEK 5 million (approx. £400,000)

## Authors' Club First Novel Award

Authors' Club, 40 Dover Street, London W1S 4NP
☎ 020 7499 8581
stella@theartsclub.co.uk
Contact *Stella Kane*

Established 1954. This award is made for the most
promising work published in Britain by a British
author, and is presented at a dinner held at The
Arts Club. Entries for the award are accepted
from publishers during September of the year in
question and must be full-length (short stories are
not eligible). For further details/application form,
contact *Stella Kane* at the email address above or
access the website. 2010 winner: Anthony Quinn,
*The Rescue Man* (Cape).

## The BBC National Short Story Award

Booktrust, Book House, 45 East Hill, London
SW18 2QZ
☎ 020 8516 2973    ᴇ 020 8516 2977
www.bbc.co.uk/radio4/nssa
www.theshortstory.org.uk
Literature Promotions Coordinator *Rosa
Anderson*

Launched at the 2005 Edinburgh International
Book Festival, this annual short story award
is funded by the BBC and is administered
by **Booktrust**. It was formerly known as the
National Short Story Prize. Open to authors
with a previous record of publication who are
UK nationals or residents, aged 18 years or over.
Entries can be stories published during 2009 or
previously unpublished and must not be more
than 8,000 words. Entrants must have a prior
record of publication and submit original work
that does not infringe the copyright or any other
rights of any third party. The stories must be
written in English and only one will be accepted
per author. Full entry details from the address
above or via the website. 2009 winner: Kate
Clanchy *The Not-Dead and The Saved*.
ᴇ Prizes of £15,000 (winner); £3,000 (second
prize) £500 (3 runners-up).

## The BBC Samuel Johnson Prize for Non-Fiction

Colman Getty, 28 Windmill Street, London
W1T 2JJ
☎ 020 7631 2666    ᴇ 020 7631 2699

lois@colmangetty.co.uk
Contact *Lois Tucker*

Established 1998. Annual prize sponsored by BBC. Eligible categories include the arts, autobiography, biography, business, commerce, current affairs, history, natural history, popular science, religion, sport and travel. Entries submitted by publishers only. 2009 winner: Philip Hoare *Leviathan or, The Whale* (Fourth Estate). Ⓔ Prize of £30,000.

## BBC Wildlife Magazine Nature Writer of the Year Award

BBC Wildlife Magazine, 14th Floor, Tower House, Fairfax Street, Bristol BS1 3BN
Ⓣ 0117 314 8359    Ⓕ 0117 934 9008
wildlifemagazine@bbcmagazinesbristol.com
www.bbcwildlifemagazine.com
Editor *Sophie Stafford*

This competition aims to revive the art of nature writing. Entrants should submit an original and unpublished 800-word essay on any aspect of the natural world that interests them, aiming for originality, story-telling and a sense of connection. The overall winner will have their story published in *BBC Wildlife Magazine* and receive an additional prize (tbc). Last year's top prize was a place on an Earthwatch expedition. See the March issue and BBC Wildlife website for the call for entries and entry form.

## BBC Wildlife Magazine Travel Writing Award

BBC Wildlife Magazine, 14th Floor, Tower House, Fairfax Street, Bristol BS1 3BN
Ⓣ 0117 927 9009    Ⓕ 0117 934 9008
wildlifemagazine@bbcmagazinesbristol.com
www.bbcwildlifemagazine.com
Editor *Sophie Stafford*

This competition looks for travel writing that transports the reader to another place. Entrants should submit an 800-word essay that includes a true account of intimate encounter with wildlife, either local or exotic, and conveys a sense of travel, discovery and something of the essence of the natural world. The top prize is a wildlife 'holiday of a lifetime' and publication in the magazine. See the January issue and BBC Wildlife website for the call for entries and entry form.

## The Bernard Levin Award

Society of Indexers, Woodbourn Business Centre, 10 Jessell Street, Sheffield S9 3HY
Ⓣ 0114 244 9561    Ⓕ 0114 244 9563
admin@indexers.org.uk
www.indexers.org.uk
Secretary *John Silvester*

Established in 2000 to celebrate the late Bernard Levin, a journalist and author whose writings showed untiring and eloquent support for indexers and indexing. An occasional award for outstanding services to the Society of Indexers. Last awarded in 2007 to Jill Halliday.

## Bernard Shaw Translation Prize
▷ The Translation Prizes

## Besterman/McColvin Awards

555 Silbury Boulevard, Milton Keynes MK9 3HL
Ⓣ 01908 254078
ISGRefAwards@cilip.org.uk
www.cilip.org.uk
Awards Administrator *Helen Bowlt*

ISG (CILIP) Reference Awards. The Besterman McColvin awards are given biennially for an outstanding reference work available and relevant to the library and information sector in the UK. The publication should be a first or a new edition published in the UK during the preceding two years. Consists of two categories: printed and electronic. Works eligible for consideration include: encyclopaedias, general and special dictionaries; annuals, yearbooks and directories; handbooks and compendia of data; atlases. Nominations are invited from members of **CILIP**, publishers and others. 2007 winner: *Biographical Dictionary of British Quakers in Commerce and Industry 1775–1920* by Ted Milligan.
Ⓔ Medal and certificate for each category.

## The Betty Trask Awards
▷ entry under Bursaries, Fellowships and Grants

## Bill Naughton Short Story Competition

Box No. [year of competition], Aghamore, Ballyhaunis, Co. Mayo, Republic of Ireland
Ⓣ 00 353 9 4936 7019
paulwdrogers@eircom.net
www.aghamoreireland.com/kennynaughton/shortstory.htm
Contact *Paul Rogers*

Established 1994. Annual award organized by the Kenny/Naughton Autumn School. Stories can be on any topic and no more than 2,500 words in length. All work must be unpublished; typed mss only with no name or address appearing on the work. Entry fee: £5, €7 or $8 per story (three stories may be submitted for the price of two). Closing date: first Friday in September.
Ⓔ Awards of €200 (1st); €130 (2nd); €65 (3rd).

## Biographers' Club Prize
▷ The Tony Lothian Biographers' Club Prize

## Birdwatch Bird Book of the Year

c/o Birdwatch Magazine, B403A The Chocolate Factory, 5 Clarendon Road, London N22 6XJ
Ⓣ 020 8881 0550
editorial@birdwatch.co.uk
www.birdwatch.co.uk
Contact *Dominic Mitchell*

Established in 1992 to acknowledge excellence in ornithological publishing – an increasingly large market with a high turnover. Annual award. Entries, from publishers, must offer an original and comprehensive treatment of their particular ornithological subject matter and must have a broad appeal to British-based readers. 2008 winner: D.W. Yalden and U. Albarella *The History of British Birds* (Oxford University Press).

## Bisto Book of the Year Awards
▷ CBI Bisto Book of the Year Awards

## Blue Peter Children's Book Awards
Booktrust, Book House, 45 East Hill, London SW18 2QZ
☎ 020 8516 2973
www.bbc.co.uk/bluepeter
www.booktrust.org.uk
Contact *Claire Shanahan*

Judged by a panel of adults and children. The three categories are: Book I Couldn't Put Down, Most Fun Story with Pictures and Best Book with Facts. Books must be published in paperback between 1st November and 31st October the following year. Winners are announced in the spring in a special broadcast on Blue Peter. 2010 winners: Book I Couldn't Put Down and winner of Blue Peter Book of the Year 2010: *Frozen in Time* by Ali Sparkes (OUP); Most Fun Story with Pictures: *Dinkin Dins and the Frightening Things* by Guy Bass, illus. Pete Williamson (Stripes); Best Book with Facts: *Why Eating Bogeys is Good for You* by Mitchell Symons (Red Fox).

## Boardman Tasker Award
Pound House, Llangennith, Swansea SA3 1JQ
☎ 01792 386215    ☏ 01792 386215
margaretbody@lineone.net
www.boardmantasker.com
Contact *Maggie Body, Honorary Secretary*

Established 1983, this award is given for a work of fiction, non-fiction or poetry, whose central theme is concerned with the mountain environment and which can be said to have made an outstanding contribution to mountain literature. Authors of any nationality are eligible, but the book must have been published or distributed in the UK for the first time between 1st November the preceding year and 31st October of the award year. Entries from publishers only. 2008 winner: Andy Kirkpatrick *Psychovertical* (Hutchinson).
£ Prize of £3,000 (at Trustees' discretion).

## Bollinger Everyman Wodehouse Prize
Colman Getty, 28 Windmill Street, London W1T 2JJ
☎ 020 7631 2666
lois@colmangetty.co.uk
Contact *Lois Tucker*

Annual prize, launched in 2000 on the 25th anniversary of the death of P.G. Wodehouse. Presented at the **Guardian Hay Festival**, awarded to the best comic novel published in the preceding twelve months. 2009 winner: Geoff Dyer *Jeff in Venice, Death in Varanasi* (Canongate).

## The Booker Prize for Fiction
▷ The Man Booker Prize for Fiction

## Booktrust Early Years Awards
Booktrust, Book House, 45 East Hill, London SW18 2QZ
☎ 020 8516 2972    ☏ 020 8516 2978
prizes@booktrust.org.uk
www.booktrust.org.uk
Contact *Prize Administrator*

Formerly the Sainsbury's Baby Book Award, established in 1999. Annual awards with three categories: the Baby Book Award, the Best Book for Pre-School Children (up to the age of 5), and an award for the Best Emerging Illustrator. Books to be submitted by publishers only; authors and illustrators must be of British nationality or other nationals who have been resident in the UK for at least five years. 2009 winners, Baby Book: Ed Vere *Chick* (Puffin); Pre-School: Mara Bergman, illus. Nick Maland *Oliver Who Travelled Far and Wide* (Hodder Children's Books); Best Emerging Illustrator: Katie Cleminson *Box of Tricks* (Jonathan Cape).
£ Prizes of £2,000 for each category; in addition the Best Emerging Illustrator receives a specially commissioned piece of artwork.

## Booktrust Teenage Prize
Book House, 45 East Hill, London SW18 2QZ
☎ 020 8516 2973    ☏ 020 8516 2978
katherine@booktrust.org.uk
www.booktrust.org.uk
Contact *Megan Farr*

Established 2003. Annual prize that recognizes and celebrates the best in writing for teenagers. Funded and administered by Booktrust. Open to writers of contemporary fiction for teenagers in the UK, the books to be published between 1st July and 30th June. Final entry date in March. 2009 winner: Neil Gaiman *The Graveyard Book* (Bloomsbury).
£ Prize of £2,500.

## Bram Stoker Awards for Superior Achievement
Horror Writers Association, 244 Fifth Ave., Suite 2767, New York, NY 10001, USA
hwa@horror.org
www.horror.org

Founded in 1988 and named in honour of Bram Stoker, author of *Dracula*. Presented annually by the **Horror Writers Association** (HWA) for works of horror first published in the English language. Works are eligible during their first year of publication. HWA members recommend works for consideration in eight categories: Novel, First Novel, Short Fiction, Long Fiction, Fiction Collection, Poetry Collection, Anthology and Non-fiction. In addition, Lifetime Achievement Stokers are occasionally presented to individuals whose entire body of work has substantially influenced horror.

## The Branford Boase Award

8 Bolderwood Close, Bishopstoke, Eastleigh SO50 8PG
☎ 01962 826658   🖷 01962 856615
anne.marley@tiscali.co.uk
www.branfordboaseaward.org.uk
Administrator *Anne Marley*

Established in 2000 in memory of children's novelist, Henrietta Branford and editor and publisher, Wendy Boase. To be awarded annually to encourage and celebrate the most promising novel by a new writer of children's books, while at the same time highlighting the importance of the editor in nurturing new talent. 2009 winner: Bridget Collins *The Traitor Game* (author); Emma Matthewson, **Bloomsbury Publishing** (editor).
£ Award of a specially commissioned box, carved and inlaid in silver with the Branford Boase Award logo; winning author receives £1,000.

## Brian Way Award

c/o Theatre Centre, Shoreditch Town Hall, 380 Old Street, London EC1V 9LT
☎ 020 7729 3066
admin@theatre-centre.co.uk
www.theatre-centre.co.uk
Contact *Awards*

Formerly the Arts Council Children's Award, founded 2000. An annual award for playwrights who write for children and young people. The plays, which must have been produced professionally, must be at least 45 minutes long. The playwright must be resident in the UK. See website for full details and application form. 2009 winner: Douglas Maxwell *The Mothership*.
£ Award of £6,000.

## The Bridport Prize

PO Box 6910, Bridport DT6 9BQ
☎ 01308 428333
frances@bridportprize.org.uk
www.bridportprize.org.uk
Contact *Frances Everitt, Administrator*

Annual competition for poetry and short story writing. Unpublished work only, written in English. Winning stories are read by a literary agent, the winning poems are put forward to the **Forward Prize**, and an anthology of winning entries is published. Final entry date: 30th June. Send s.a.s.e. for entry forms.
£ Prizes of £5,000, £1,000 & £500 in each category, plus 10 supplementary prizes of £50 each. Winning stories are submitted to the National Short Story Prize.

## Bristol Festival of Ideas Book Prize

Bristol Cultural Development Partnership (BCDP), Leigh Court, Abbots Leigh, Bristol BS8 3RA
☎ 07778 932778
ideas@gwebusinesswest.co.uk
www.ideasfestival.co.uk
Contact *Andrew Kelly, Director*

This annual award was established to mark the fifth anniversary of the Bristol Festival of Ideas in 2009 and was founded by Andrew Kelly, Director of the Festival, in association with **Arts & Business**. The award is given to the book which is considered to present new, important and challenging ideas, which is rigorously argued and which is engaging and accessible. To be eligible, books must be first published in the year prior to each award (i.e. 2010 for the 2011 award). The submission deadline for the 2011 award is 30 November 2010, although proofs and manuscripts will be accepted for books published in December 2010. Submissions from publishers only. Inaugural winner: Nick Davies *Flat Earth News* (Vintage Books).
£ Prize of £10,000 in association with **Arts & Business**.

## British Association for Applied Linguistics (BAAL) Book Prize

BAAL Publications Secretary, School of Education, Communication and Language Sciences, Newcastle University,, Newcastle upon Tyne NE1 7RU
☎ 0191 222 5094   🖷 0191 222 8170
steve.walsh@ncl.ac.uk
www.baal.org.uk
Contact *Steve Walsh*

Annual award made by the British Association for Applied Linguistics to an outstanding book in the field of applied linguistics. Final entry in mid-December. Nominations from publishers only. 2009 winner: L. Wei and M.G. Moyer (eds.) *The Blackwell Guide to Research Methods in Bilingualism and Multilingualism* (Blackwell).

## British Book Awards

Publishing News Ltd, 1 Wellington Business Park, Dukes Ride, Crowthorne RG45 6LS
☎ 01252 861861   🖷 01344 750066

mailbox@publishingnews.co.uk

Established 1988. Viewed by the book trade as the one to win, 'The Nibbies' are presented annually. The awards are made in various categories. Each winner receives the prestigious Nibbie and the awards are presented to those who have made the most impact in the book trade during the previous year. 2009 winners included: Michael Palin, Aravind Adiga, Stieg Larsson, Barack Obama, Kate Atkinson.

## British Czech & Slovak Association Prize

The BCSA Prize Administrator, 24 Ferndale, Tunbridge Wells TN2 3NS
☎ 01892 543206
prize@bcsa.co.uk
www.bcsa.co.uk
Contact *Prize Administrator*

The British Czech & Slovak Association offers an annual prize for the best piece of original writing, in English, on the links between Britain and the Czech and Slovak Republics, or describing society in transition in those Republics since the Velvet Revolution in 1989. Entries can be fiction or factual, should not have been previously published and should be up to 2,000 words in length. Submissions are invited from individuals of any age, nationality or educational background. Closing date: 30th June. Entry details for 2010 should be checked with the Prize Administrator. 2009 winners: Zuzana Demcakova *The Laugh Is On Me* (1st); Pearl Harris *Driver Beware!* (2nd). 
£ Prize of £300, presented at the BCSA's annual dinner in London; 2nd prize, £100. The winning entry is published in the *British Czech & Slovak Review*.

## British Fantasy Awards

56 Leyton Road, Birmingham B21 9EE
bfsawards@britishfantasysociety.org
www.britishfantasysociety.org
Awards Administrator *Guy Adams*

Awarded by the **British Fantasy Society** by members at its annual conference for Best Novel and Best Short Story categories, among others. Not an open competition. Previous winners include: Ramsey Campbell, Dan Simmons, Michael Marshall Smith, Thomas Ligotti.

## British Press Awards

WLR Media and Entertainment, 3rd Floor, 6–14 Underwood Street, London N1 7JQ
☎ 020 7549 8671   📠 020 7549 8668
acrane@wilmington.co.uk
www.pressgazette.co.uk

Celebrates the best of national newspaper journalism and publishing. Open to all British national newspapers and news agencies. March event. Run by WLR Media & Entertainment.

## British Science Fiction Association Awards

www.bsfa.co.uk

Established 1970. Categories for novel, short fiction, non-fiction and artwork. The awards are announced and presented at the British National Science Fiction Convention. 2010 winners: China Miéville *The City and the City* (novel); Ian Watson and Robert Quaglia *The Beloved Time of their Lives* (short fiction); Nick Lowe *Mutant Popcorn* (non-fiction); Stephan Martiniere – for cover of *Desolation Road* (artwork).

## British Society for the History of Mathematics: BSHM Peter G. Neumann Prize, 2011

Warwick University, Department of Computer Science, Coventry CV4 7AL
☎ 024 7652 3193   📠 024 7657 3024
M.Campbell-Kelly@warwick.ac.uk
www.bshm.ac.uk
Contact *Professor Martin Campbell-Kelly*

Established 2009. Annual prize awarded for a book in English (including books in translation) dealing with the history of mathematics and aimed at a broad audience. There is no further restriction on subject matter, nor on the nationality of the author or the country of publication. Send three copies of books for consideration to Professor Martin Campbell-Kelly, chair of the judging panel, to whom all enquiries should be directed. Final entry date: Spring 2011. Inaugural winners: Reviel Netz and William Noel *The Archimedes Codex: Revealing the Blueprint of Modern Science* (Phoenix). 
£ Prize of £300 and a Notice in the *BSHM Bulletin*.

## British Sports Book Awards

National Sporting Club, Brompton House, 97–99 Kew Road, Richmond TW9 2PN
☎ 020 7036 9890
www.nationalsportingclub.co.uk

Established 2003. Annual awards presented at a Café Royal lunch in London in March. Seven categories: Best Autobiography, Best Biography, Best New Writer, Best Illustrated Title, Best Cricket Book, Best Football Book, Best Rugby Book. Nominations received in November each year by email (address above). Sponsors include Skysports, *The Times*, Llanllyr Spring Water, Littlehampton Book Services and Ladbrokes.

## BSHM Peter G. Neumann Prize

▷ British Society for the History of Mathematics: BSHM Peter G. Neumann Prize, 2011

## Buxton Poetry Competition

c/o Buxton Festival, 3 The Square, Buxton
SK17 6AZ
☎ 01298 70395    📠 01298 72289
claire@buxtonfestival.co.uk
www.derby.ac.uk/buxtonpoetrycompetition
Contact *Claire Rhodes*

Founded in 2007 by Buxton Festival and the
University of Derby, the competition aims to
develop interest in creative writing across the
country and particularly in Derbyshire. Also
aims to encourage children and young people
to get involved with poetry by giving free entry
to everyone under 19 years old. Schools in
Derbyshire are offered poetry workshops to
help the children get their poems ready for the
competition. Final entry date: 1st April. All poems
must be submitted by post with a completed
entry form and cheque (where applicable); entry
details and form available on the website. Open
Poetry: for poets over 18 years old, entry fee: £5
per poem; Young People's Poetry: for poets aged
between 12 and 18, free entry; Children's Poetry:
for poets aged under 12 years old, free entry. 2009
Open Poetry winner: Mandy Coe with her poem
entitled *Quarry*.
💷 Prizes – Open Poetry: cash prizes of £300,
£200 and £100; Young People's Poetry and
Children's Poetry: Waterstone's book tokens.
All winning poets have the chance to have
their poems displayed in an exhibition at the
Devonshire Dome in Buxton during the Festival.

## C.B. Oldman Prize

r.turbet@abdn.ac.uk
Contact *Richard Turbet, Special Collections
Cataloguer and Music Subject Specialist*

Established 1989 by the International Association
of Music Libraries, UK & Ireland Branch. Annual
award for best book of music bibliography,
librarianship or reference published the year
before last (i.e. books published in 2008
considered for the 2010 prize). Previous winners:
Pamela Thompson, Malcolm Lewis, Andrew
Ashbee, Michael Talbot, Donald Clarke, John
Parkinson, John Wagstaff, Stanley Sadie, William
Waterhouse, Richard Turbet, John Gillaspie,
David Fallows, Arthur Searle, Graham Johnson,
Michael Twyman, John Henderson, David Golby,
David Wyn Jones.
💷 Prize of £200.

## The Caine Prize for African Writing

51a Southwark Street, London SE1 1RU
☎ 020 7378 6234    📠 020 7378 6235
info@caineprize.com
www.caineprize.com
Administrator *Nick Elam*
Secretary *Jan Hart*

Annual award founded in 1999 in memory of Sir
Michael Caine, former chairman of Booker plc, to
recognize the worth of African writing in English.
Awarded for a short story (3,000–10,000 words) by
an African writer, published in English anywhere in
the world. An 'African writer' is someone who was
born in Africa, or who is a national of an African
country, or whose parents are African and whose
work has reflected African sensibilities. Final
entry date: 31st January; submissions by publishers
only. 2009 winner: E.C. Osondu *Waiting* from
Guernicamag.com, October 2008 issue.
💷 Prize of £10,000.

## The Calouste Gulbenkian Prize
▷ The Translation Prizes

## Calvin & Rose G. Hoffman Prize

King's School, 25 The Precincts, Canterbury
CT1 2ES
☎ 01227 595544
bursar@kings-school.co.uk
Contact *The Bursar*

Annual award for distinguished publication on
Christopher Marlowe, established by the late
Calvin Hoffman, author of *The Man Who was
Shakespeare* (1955) as a memorial to himself
and his wife. For unpublished works of at least
5,000 words written in English for their scholarly
contribution to the study of Christopher Marlowe
and his relationship to William Shakespeare. Final
entry date: 1st September each year.

## Canadian Poetry Association (Annual Poetry Contest)

Canadian Poetry Association, 331 Elmwood Dr.,
Suite 4-212, Moncton, New Brunswick E1A 1X6,
Canada
www.canadian-poetry-association.com

Annual contest now open to non-members of the
Association. See the website for entry details. All
winning poems are published on the CPA website
and in *Poemata*.
💷 Three cash prizes.

## Cardiff International Poetry Competition
▷ Academi Cardiff International Poetry Competition

## Carey Award

Society of Indexers, Woodbourn Business Centre,
10 Jessell Street, Sheffield S9 3HY
☎ 0114 244 9561    📠 0114 244 9563
admin@indexers.org.uk
www.indexers.org.uk
Secretary *John Silvester*

A private award made by the Society to a member
who has given outstanding services to indexing.
The recipient is selected by the Executive Board
with no recommendations considered from
outside the Society.

## Carnegie Medal
▷ CILIP Carnegie Medal

## CBI Bisto Book of the Year Awards
17 North Great Georges Street, Dublin 1, Republic of Ireland
☎ 00 353 1 872 7475   📠 00 353 1 872 7476
info@childrensbooksireland.ie
www.childrensbooksireland.ie
Contact *Jenny Murray*

Founded 1990 as Bisto Book of the Decade Awards. This led to the establishment of an annual award made by the Irish Children's Book Trust, later to become Children's Books Ireland. Open to any author or illustrator of children's books born or resident in Ireland; open to English or Irish language. 2008/09 winners: Bisto Book of the Year: Siobhan Dowd *Bog Child* (David Fickling Books); Eilís Dillon Award: Mary Finn *Anila's Journey* (Walker).
💷 Prizes of €10,000 (Bisto Book of the Year); €3,000 (Eilís Dillon Award); three honour awards of €1,000 each.

## The Centaur Essay Prize
centaurfound@gmail.com

A new prize which marks the first stage in founding a charitable educational trust, the Centaur Foundation, supporting teachers and parents wanting innovative humane standards encouraged from primary school onwards. The submitted essay of 2,000–3,000 words should be on a theme of non-violent solutions and methods for the 21st century and beyond. Entry deadline: 30th November 2010. Winners will be contacted by 28th February 2011. The essay may be academic, even partly anecdotal, with a religious or secular slant. All essayists *must* in the first instance email the address above to request a copy of the entry guidelines (essay submissions are not accepted by email).
💷 Prize of £500; plus two runner-up prizes of £250.

## Chapter One Promotions Children's Story Competition
Canterbury Court, 1–3 Brixton Road, London SW9 6DE
☎ 0845 456 5364
kidskorner2@chapteronepromotions.com
www.chapteronepromotions.com
Contact *Johanna Bertie*

Established in 2007, the competition is open to all young writers between the ages of seven and 18 to submit unpublished short stories. Five titles or themes are given to choose from. Maximum length 1,000 words. Contact details on a separate sheet. Submissions and payment accepted online or by post. Closing date: 31st October.

💷 Prizes of £150, £75, £50 in each of the four age categories, plus publication of the winners and runners-up in the Children's Story Book. Best school entry receives £200 worth of books plus a trophy.

## Chapter One Promotions International Short Story Competition
Canterbury Court, 1–3 Brixton Road, London SW9 6DE
☎ 0845 456 5364
info@chapteronepromotions.com
www.chapteronepromotions.com
Contact *Johanna Bertie*

Established in 2005, the competition is open to new and established writers to submit unpublished short stories of fewer than 2,500 words. Author details on separate sheet. Submissions and payment accepted online. Closing date: 14th January. 2009 winner: Robert Grossmith.
💷 Prizes of £2,500, £1,000, £500, plus publication of winners, runners-up and highly recommended in anthology.

## Chapter One Promotions Novel Competition
Canterbury Court, 1–3 Brixton Road, London SW9 6DE
☎ 0845 456 5364
info@chapteronepromotions.com
www.chapteronepromotions.com
Contact *Johanna Bertie*

Established in 2008, the Novel Competition is for writers looking to get their manuscript published. Judged by a leading literary agency, the winning submission will have the chance to work with an editor to bring the manuscript up to a publishable standard, where it will then be reviewed by the agency who will work on the writer's behalf to get a publishing deal. Submit the first three chapters and a one-page synopsis. Contact details on a separate sheet only. Submissions and payment accepted online or by post. Closing date: midnight on 29th February 2012.

## Chapter One Promotions Poetry Competition
Canterbury Court, 1–3 Brixton Road, London SW0 6DE
☎ 0845 456 5364
info@chapteronepromotions.com
www.chapteronepromotions.com
Contact *Johanna Bertie*

Established in 2005. Unpublished poems of no more than 30 lines requested. Judge decides on the top 20 poems which are displayed on the website and the public votes for their favourite poem. The three poems with the most votes win. Closing date: 1st June. Online voting period: 1st–15th July. Winning poems appear on the

website in August. Submissions and payment accepted online. 2009 winner: Siân Thomas.
£ Prizes £1,000, £500, £250.

## The Children's Laureate

Booktrust, Book House, 45 East Hill, London SW18 2QZ
T 020 8875 4580  F 020 8516 2978
childrenslaureate@booktrust.org.uk
www.childrenslaureate.org.uk

Established 1998. The Laureate is awarded biennially to an eminent British writer or illustrator of children's books both in celebration of a lifetime's achievement and to highlight the role of children's book creators in inspiring, informing and entertaining young readers. 2009–11 winner: Anthony Browne.
£ Award winner receives a medal and a bursary.

## Children's Poetry Bookshelf Poetry Competition
▷ Old Possum's Children's Poetry Competition

## CILIP Carnegie Medal

7 Ridgmount Street, London WC1E 7AE
T 020 7255 0650  F 020 7255 0651
ckg@cilip.org.uk
www.ckg.org.uk

Established 1936. Presented for an outstanding book for children written in English and first published in the UK during the preceding year. Fiction, non-fiction and poetry are all eligible. 2009 winner: Siobhan Dowd *Bog Child* (David Fickling Books).
£ Award winner receives a medal.

## CILIP Kate Greenaway Medal

7 Ridgmount Street, London WC1E 7AE
T 020 7255 0650  F 020 7255 0651
ckg@cilip.org.uk
www.ckg.org.uk

Established 1955. Presented annually for the most distinguished work in the illustration of children's books first published in the UK during the preceding year. 2009 winner: Catherine Rayner *Harris Finds His Feet* (Little Tiger Press).
£ Award winner receives a medal. The Colin Mears Award of £5,000 is given annually to the winner of the Kate Greenaway Medal.

## The Clarissa Luard Award
▷ David Cohen Prize for Literature

## CLPE Poetry Award

Centre for Literacy in Primary Education, Webber Street, London SE1 8QW
T 020 7902 2287  F 020 7928 4624
ann@clpe.co.uk
www.clpe.co.uk
Contact *Ann Lazim*

Founded in 2003, the CLPE Poetry Award honours excellence in poetry written for children. It is presented annually for a collection or anthology of poetry for children, published for the first time in the UK or the Republic of Ireland in the preceding year. Entry deadline: end of February. 2009 winner John Agard *The Young Inferno*, illus. Satoshi Kitamura (Frances Lincoln).

## Colin Mears Award
▷ CILIP Kate Greenaway Medal

## The Commonwealth Writers Prize

Commonwealth Foundation, Marlborough House, Pall Mall, London SW1Y 5HY
T 020 7747 6262  F 020 7839 8157
j.sobol@commonwealth.int
www.commonwealthfoundation.com/culturediversity/writersprize/

Established 1987. An annual award to reward and encourage the upsurge of new Commonwealth fiction. Any work of prose or fiction is eligible, i.e. a novel or collection of short stories. No drama or poetry. The work must be first written in English by a citizen of the Commonwealth and be first published in the year before its entry for the prize. Entries must be submitted by the publisher to the region of the writer's Commonwealth citizenship. The four regions are: Europe and South Asia, the Caribbean and Canada, South East Asia and South Pacific and Africa. 2009 winners: Best Book Award: Christos Tsiolkas (Australia) *The Slap* (Allen & Unwin); Best First Book Award: Mohammed Hanif (Pakistan) *A Case of Exploding Mangoes* (Vintage).
£ Prizes of £10,000 for Best Book; £5,000 for Best First Book; 8 prizes of £1,000 for each best and best first book in four regions.

## Cork City – Frank O'Connor International Short Story Award

The Munster Literature Centre, Frank O'Connor House, 84 Douglas Street, Cork, Republic of Ireland
T 00 353 21 431 2955
competitions@munsterlit.ie
www.munsterlit.ie
Contact *Patrick Cotter*

Founded in 2005 to reward and encourage writers and publishers of short stories. Annual competition for a collection of original short stories by a single living author published between September of the previous year and August of the award year. Final entry date: 31st March. Vanity press publications not accepted. 2009 winner: Simon van Booy *Love Begins in Winter* (Beautiful Books).
£ Award of €35,000.

## Costa Book Awards

The Booksellers Association, Minster House,
272 Vauxhall Bridge Road, London SW1V 1BA
☎ 020 7802 0802   🖷 020 7802 0803
naomi.gane@booksellers.org.uk
www.costabookawards.com
Contact *Naomi Gane*

Established 1971. Formerly the Whitbread Book
Awards. The awards celebrate and promote the
best contemporary British writing. They are
judged in two stages and offer a total of £50,000
prize money. The awards are open to novel, first
novel, biography, poetry and children's book, each
judged by a panel of three judges. The winner of
each award receives £5,000. The Costa Book of
the Year (£25,000) is chosen from the category
winners. Writers must have lived in Britain and
Ireland for three or more years. Submissions
received from publishers only. Closing date:
early July. Sponsored by Costa. 2009 winners:
Poetry and Overall Winner: Christopher Reid
*A Scattering* (Areté Books); Novel: Colm Tóibín
*Brooklyn* (Viking); First Novel: Raphael Selbourne
*Beauty* (Tindal Street Press); Biography: Graham
Farmelo *The Strangest Man: The Hidden Life of
Paul Dirac, Quantum Genius* (Faber); Children's:
Patrick Ness *The Ask and the Answer* (Walker).

## The Coventry Inspiration Book Awards

Coventry Schools Library & Resource Service,
Central Library, Smithford Way, Coventry
CV1 1FY
☎ 02476 832338   🖷 02476 832358
sls@coventry.gov.uk
www.myvotescoventry.org
Children & Young People's Categories and
Management of Awards *Joy Court*
Adult Categories *Mandy Hayward*

Established 2006/07. Annual award inspired
by the Family Reading Campaign uniquely
has an award for every age-group of children
and young people and by genre categories for
adults. The awards go live in October during
Children's Book Week and every shortlisted
title is open for reader's comments and votes.
From January each week the two books with
the least votes in each category get 'eliminated'.
Voting then starts again each week so fans have
to keep voting to keep their favourites in. The
final winners are announced in February half
term. An awards ceremony is held later in the
year to celebrate the winners in each category
and to present individual trophies. One winning
author is invited to contribute a story starter for
a Coventry Inspires Writing competition run
in conjunction with the *Coventry Telegraph*.
Winners are published and receive prizes and
certificates at the awards ceremony. Books in
all C&YP categories should not have won major

national awards before. The exception is the 'Our
Special Favourites' which by its very nature and
the nomination process could not make that
distinction.

## The Crashaw Prize for Poetry
▷ Salt Publishing Limited under UK Publishers

## Crime Writers' Association (Cartier Diamond Dagger)

PO Box 273, Boreham Wood WD6 2XA
secretary@thecwa.co.uk
www.thecwa.co.uk
Contact *The Secretary*

Established 1986. An annual award for a lifetime's
outstanding contribution to the genre. 2010
winner: Val McDermid.

## Crime Writers' Association (The Dagger in the Library)

PO Box 273, Boreham Wood WD6 2XA
secretary@thecwa.co.uk
www.thecwa.co.uk
Contact *The Secretary*

Reinstated 2002. Annual award (sponsored by
**Random House**) to the author whose work has
given most pleasure to readers. Nominated and
judged by librarians. 2009 winner: Colin Cotterill
💷 Award winner receives a Dagger, plus cheque.

## Crime Writers' Association (The Debut Dagger)

The Debut Dagger Competition, PO Box 273,
Boreham Wood WD6 2XA
debut.dagger@thecwa.co.uk
www.thecwa.co.uk
Contact *The Secretary*

Annual competition (sponsored by **Orion**) for
unpublished writers to submit the first 3,000
words and 500-word outline of a crime novel
(entry fee and form required). For full details see
website or send s.a.se. to address above. 2009
winner: Catherine O'Keefe, *The Pathologist*.
💷 Award winner receives a Dagger, plus cheque.

## Crime Writers' Association (The Ian Fleming Steel Dagger)

PO Box 273, Boreham Wood WD6 2XA
secretary@thecwa.co.uk
www.thecwa.co.uk
Contact *The Secretary*

Founded 2002. Annual award for the best
thriller, adventure or spy novel. Sponsored by Ian
Fleming Publications Ltd to celebrate the best of
contemporary thriller writing. 2009 winner: John
Hart *The Last Child* (John Murray).
💷 Award winner receives a Dagger, plus cheque.

## Crime Writers' Association (The CWA Ellis Peters Award)

PO Box 273, Boreham Wood WD6 2XA
secretary@thecwa.co.uk
www.thecwa.co.uk
Contact *The Secretary*

Sponsored by the Estate of Ellis Peters, Time Warner and **Headline**. Established 1999. Annual award for the best historical crime novel. Nominations from publishers only. 2009 winner: Philip Kerr *If the Dead Rise Not* (Quercus).
£ Award plus cheque.

## Crime Writers' Association (The CWA Gold Dagger for Non-Fiction)

PO Box 273, Boreham Wood WD6 2XA
secretary@thecwa.co.uk
www.thecwa.co.uk
Contact *The Secretary*

Biennial award for the best non-fiction crime book published during the year. Nominations from publishers only. 2008 winner: Kester Aspden *Nationality: Wog – The Hounding of David Oluwale* (Cape).
£ Award winner receives a Dagger, plus cheque (sum varies).

## Crime Writers' Association (The CWA Gold Dagger)

PO Box 273, Boreham Wood WD6 2XA
secretary@thecwa.co.uk
www.thecwa.co.uk
Contact *The Secretary*

An annual award for the best crime fiction published during the year. Nominations from publishers only. 2009 winner: William Brodrick *A Whispered Name* (Little, Brown).

## Crime Writers' Association John Creasey (New Blood) Dagger

PO Box 273, Boreham Wood WD6 2XA
secretary@thecwa.co.uk
www.thecwa.co.uk
Contact *The Secretary*

Established in 1973 following the death of crime writer John Creasey, founder of the **Crime Writers' Association**. This award is given annually for the best crime novel by an author who has not previously published a full-length work of fiction. Fomerly known as the John Creasey Dagger. Nominations from publishers only. 2009 winner: Johan Theorin *Echoes From the Dead* (Doubleday).
£ Award winner receives a Dagger, plus cheque.

## Crime Writers' Association Short Story Dagger

PO Box 273, Boreham Wood WD6 2XA
dagger.liaison@thecwa.co.uk
www.thecwa.co.uk
Contact *Dagger Liaison officer*

Founded 1982. Awarded annually by the Crime Writers' Association to the best short (crime) story published in Britain within the preceding year. Submissions from publishers only. 2009 winner: Sean Chercover 'One Serving of Bad Luck' from *Killer Year* (Mira).
£ Award winner receives a Dagger, plus cheque (sum varies).

## CrimeFest Last Laugh Award/CrimeFest Sounds of Crime Awards/CrimeFest Crime eBook Award

Basement Flat, 6 Rodney Place, Bristol BS8 4HY
info@crimefest.com
www.crimefest.com

The **CrimeFest Last Laugh Award** is for the best humorous crime novel first published in the British Isles in a calendar year. Publishers are asked to submit eligible titles for the longlist. Leading British crime fiction reviewers vote to establish the shortlist and winner. The winner is announced at the CrimeFest Gala Dinner and receives a piece of commemorative glass and a cash prize.

**CrimeFest Sounds of Crime Awards** are for the best abridged and unabridged crime audiobooks first published in the British Isles in a calendar year in both printed and digital formats. The award goes to both the author and the reader of the winning entry. Publishers are asked to submit eligible titles for the longlist. The winners are announced at the CrimeFest Gala Dinner and receive a piece of commemorative glass.

**CrimeFest Crime eBook Award** is for the best crime ebook first published in a calendar year in both printed and digital formats. Publishers are asked to submit eligible titles for the longlist. Leading British crime fiction reviewers vote to establish the shortlist and winner. The winners are announced at the CrimeFest Gala Dinner and receive a piece of commemorative glass and a cash prize. Email info@crimefest.com or visit the website for more information.

## David Berry Prize

Royal Historical Society, University College London, Gower Street, London WC1E 6BT
☎ 020 7387 7532    🖷 020 7387 7532
royalhistsoc@ucl.ac.uk
www.royalhistoricalsociety.org/grants.htm
Contact *Administrative Secretary*

Annual award for an essay of not more than 10,000 words on Scottish history. Closing date: 31st December.
£ Prize of £250.

## David C. Horn Prize
> The Yale Drama Series

## David Cohen Prize for Literature
Arts Council England, 14 Great Peter Street, London SW1P 3NQ
☏ 0845 300 6200
www.artscouncil.org.uk/davidcohenprize

Established in 1993 by the Arts Council and awarded biennially, the David Cohen Prize for Literature is one of the most distinguished literary prizes in Britain. It recognizes writers who use the English language and who are citizens of the UK and the Republic of Ireland, encompassing dramatists, novelists, poets and essayists. The prize is for a lifetime's achievement and is donated by the John S. Cohen Foundation.

The Clarissa Luard Award (a £12,500 award provided by Arts Council England) is awarded by the winner, to encourage new work, with the dual aim of encouraging young writers and readers. Submissions are not accepted for either award. Previous winners of the David Cohen Prize for Literature: William Trevor, Dame Muriel Spark, Harold Pinter, Doris Lessing, Beryl Bainbridge and Thom Gunn; Derek Mahon. 2009 winner: Seamus Heaney.
£ Award of £40,000, plus £12,500 towards new work.

## The David Watt Prize
Rio Tinto plc, 2 Eastbourne Terrace, London W2 6LG
☏ 01985 844613   ☒ 01985 844002
davidwattprize@riotinto.com
celiabeale@globalnet.co.uk
www.riotinto.com
Contact *The Administrator*

Initiated in 1988 to commemorate the life and work of David Watt. Annual award, open to writers currently engaged in writing for English language newspapers and journals, on international and national affairs. The winner is judged as having made 'an outstanding contribution towards the greater understanding of national, international or global issues'. Entries must have been published during the year preceding the award. 2008 winner: Jonathan Freedland for his *New York Review of Books* article 'Bush's Amazing Achievement'. The prize will not be awarded in 2010.

## The Desmond Elliott Prize
The Desmond Elliott Charitable Trust, 84 Godolphin Road, London W12 8JW
☏ 020 8222 6580
ema.manderson@googlemail.com
www.desmondelliottprize.com
Contact *Mrs Emma Manderson*

Established in 2008 in memory of literary agent and publisher, Desmond Elliott, this annual prize is for a first full-length novel, written in English and published in book form in the UK, by an author who is permanently resident in the UK. Final entry date: 30th November 2009. 2009 winner: Edward Hogan *Blackmoor* (Simon & Schuster).
£ Prize of £10,000.

## Dingle Prize
PO Box 3401, Norwich NR7 7JF
☏ 01603 516236
lucy.tetlow@bshs.org.uk
www.bshs.org.uk

Biennial award made by the British Society for the History of Science to the best book in the history of science (broadly construed) which is accessible to a wide audience of non-specialists. Next award: 2011. 2009 winner: Thomas Dixon *Science and Religion: A Very Short Introduction* (OUP).
£ Prize of £300.

## Dolman Best Travel Book Award
Authors' Club, 40 Dover Street, London W1S 4NP
☏ 020 7408 5092
stella@theartsclub.co.uk
Contact *Stella Kane*

William Dolman, a former chairman of the **Authors' Club**, instituted this prize for the most promising book of travel literature. Books that have had their first English language publication in Great Britain by British writers are eligible. Submissions are accepted in January of each year. 2009 winner: Alice Albinia *Empires of the Indus* (John Murray).

## Drama Association of Wales Playwriting Competition
The Old Library, Singleton Road, Splott, Cardiff CF24 2ET
☏ 029 2045 2200   ☒ 029 2045 2277
teresa@dramawales.org.uk
Contact *Teresa Hennessy*

Annual competition held to promote the writing of one-act plays in English and Welsh of between 20 and 50 minutes' playing time. Application forms from the address above. Closing date: 31st January.

## The Duff Cooper Prize

54 St Maur Road, London SW6 4DP
artemiscooper@btopenworld.com
Contact *Artemis Cooper*

An annual award for a literary work of biography, history, politics or poetry, published by a recognized publisher (member of the **Publishers Association**) during the previous 12 months. The book must be submitted by the publisher, not the author. Financed by the interest from a trust fund commemorating Duff Cooper, first Viscount Norwich (1890–1954). 2009 winner: Robert Service *Trotsky: A Biography* (Macmillan).
£ Prize of £5,000.

## The Duke of Westminster's Medal for Military Literature

Royal United Services Institute for Defence and Security Studies, Whitehall, London SW1A 2ET
☎ 020 7747 2637   📠 020 7321 0943
www.rusi.org/westminstermedal

Established in 1997, this annual award is sponsored by the Duke of Westminster and aims to mark a notable and original contribution to the study of international or national security, or the military profession. Work must be in English, by a living author, and have been published as a book, rather than an article, in the preceding or next six months of the closing date for entries. 2009 winner: Sir Lawrence Freeman *A Choice of Enemies, America Confronts the Middle East.*
£ Silver medal, plus £1,000. Winning author is invited to give lecture at RUSI where the Duke of Westminster will present the Medal.

## The Dylan Thomas Prize

The Dylan Thomas Centre, Somerset Place, Swansea SA1 1RR
☎ 01792 474051   📠 01792 463993
tim@dylanthomasprize.com
www.dylanthomasprize.com
Chief Executive *Tim J. Prosser*

Founded 2004. An annual award of £30,000 will be given to the winner of this prize, which was established to encourage, promote and reward exciting new writing in the English-speaking world and to celebrate the poetry and prose of Dylan Thomas. Entrants should be the author of a published book (in English), under the age of 30 (when the work was published), writing within one of the following categories: poetry, novel, collection of short stories by one author, play that has been professionally performed, a broadcast radio play, a professionally produced screenplay that has resulted in a feature-length film. Authors need to be nominated by their publishers, or producers in the case of performance art. Final entry date: 30th April. The award will be

presented at the **Dylan Thomas Festival** in Swansea in November 2010. Further entry details can be found on the website. 2008 winner: Nam Le *The Boat* (Canongate).

## Edge Hill Prize for the Short Story

Dept. of English, Edge Hill University, St Helen's Road, Ormskirk L39 4QP
☎ 01695 584121   📠 01695 579997
coxa@edgehill.ac.uk
www.edgehill.ac.uk
Contact *A. Cox*

Annual award founded in 2007 following a one-day conference on the short story. Its aim is to reward high achievement in writing the short story and to promote the genre. Awarded to the author of a published short story collection from the UK or Ireland. Entries are submitted through publishers and must be published in the previous year. 2009 winner: Chris Beckett *The Turing Test* (Elastic Press).
£ 1st prize of £5,000, plus specially commissioned artwork sponsored by Blackwells; Readers' Choice prize, £1,000. A £500 prize is also awarded for a short story written by a student from Edge Hill University's MA in Creative Writing.

## The Educational Writers' Award

The Society of Authors, 84 Drayton Gardens, London SW10 9SB
☎ 020 7373 6642
info@societyofauthors.org
www.societyofauthors.org
Awards Secretary *Paula Johnson*

Founded 2008. Supported by the **Authors' Licensing & Collecting Society** and the **Society of Authors**. Annual award given to an outstanding example of traditionally published non-fiction (with or without illustrations) that enhances teaching and learning. For work first published in the UK in the English language within the previous two calendar years. Final entry date: 1st June 2010. The age group alternates each year between the 5–11 age group and the 12–18 age group. 2009 winner (5–11 age range): *The Gooey, Chewy, Rumble, Plop Book* Steve Alton and Nick Sharratt (illus.) (Bodley Head).
£ Prize of £2,000.

## The Elizabeth Longford Prize for Historical Biography

The Society of Authors, 84 Drayton Gardens, London SW10 9SB
☎ 020 7373 6642   📠 020 7373 5768
info@societyofauthors.org
www.societyofauthors.org

Established in 2003 in memory of Elizabeth Longford and sponsored by Flora Fraser and Peter Soros. Awarded annually for an historical biography published in the year preceding the prize. No unsolicited submissions. 2009 winner: Mark Bostridge *Florence Nightingale: The Woman and Her Legend* (Viking).

£ Prize of £5,000.

## The Encore Awards

The Society of Authors, 84 Drayton Gardens, London SW10 9SB

T 020 7373 6642    F 020 7373 5768
info@societyofauthors.org
www.societyofauthors.org

Established 1990. Awarded biennially for the best second published novel or novels of the year. Final entry date: 30th November 2010. Details from the **Society of Authors**. 2009 winner: Julia Leigh *Disquiet* (Faber).

£ Prize of £10,000.

## Enid McLeod Prize

Franco-British Society, 2 Dovedale Studios, 465 Battersea Park Road, London SW11 4LR

T 020 7924 3511
www.francobritishsociety.org.uk
Executive Secretary *Kate Brayn*

Established 1982. Annual award to the author of the work of literature published in the UK which, in the opinion of the judges, has contributed most to Franco-British understanding. Any full-length work written in English by a citizen of the UK Commonwealth is eligible. No English translation of a book written originally in any other language will be considered. Nominations from publishers for books published between 1st January and 31st December of the year of the prize. Closing date: 31st December. Recent winner: Maria Fairweather *Madame de Stael* (Constable).

£ Prize winner receives a cheque.

## Envoi Poetry Competition

Ty Meirion, Glan yr afon, Tanygrisiau, Blaenau Ffestiniog LL41 3SU

T 01766 832112
Contact *Jan Fortune-Wood*

Run by *Envoi* poetry magazine three times a year (deadlines: 20th February, June, October) with prizes to the value of £300 given in poetry books, plus three annual subscriptions to *Envoi*. Winning poems along with full adjudication report are published. Send s.a.s.e. to Competition Secretary at the address above.

## The European Prize for Literature

ec.europa.eu/culture/our-programmes-and-actions/doc627_en.htm

A new literature prize (to run each year between 2009 and 2011) aimed at putting the spotlight on the creativity and diverse wealth of Europe's contemporary literature, to promote more circulation of literature within Europe and encourage greater interest in non-national literary works. The prize, which is financed through the Culture Programme of the European Union and organized by the European Booksellers Federation, the European Writers' Council and the Federation of European Publishers, is for emerging talents in the field of contemporary literature (fiction) from each of the participating countries (Belgium, Cyprus, Denmark, Estonia, Germany, Finland, Luxembourg, Macedonia, Romania, Slovenia and Spain in 2010). Each country selects their respective winner.

## Felicia Hemans Prize for Lyrical Poetry

University of Liverpool, Liverpool L69 7WZ

T 0151 794 2429    F 0151 794 3646
sarah.okeeffe@liv.ac.uk
www.liv.ac.uk
Contact *Sarah O'Keeffe, Faculties Support Office, Faculty of Humanities and Social Sciences*

Established 1899. Annual award for published or unpublished verse. Open to past or present members and students of the University of Liverpool. One poem per entrant only. Closing date: 1st May.

£ Prize of £30.

## Felix Dennis Prize for Best First Collection
▷ The Forward Prizes for Poetry

## Fish Fiction Prizes

Fish Publishing, Durrus, Bantry, Co. Cork, Republic of Ireland
info@fishpublishing.com
www.fishpublishing.com
Contact *Clem Cairns*

Offers several writing contests including the Short Story Prize, an annual award, founded in 1994, which aims to discover, encourage and publish new literary talent. The best ten stories are published in an anthology. Also, the Fish Poetry and One Page Prizes, with the winners also published. Details of these and other contests can be found on the website.

## The Forward Prizes for Poetry

*Administrator*: Colman Getty, 28 Windmill Street, London W1T 2JJ

T 020 7631 2666    F 020 7631 2699
kate@colmangetty.co.uk
www.forwardartsfoundation.co.uk
Contact *Kate Wright-Morris*

Established 1992. Three awards: the Forward Prize for Best Collection, the Felix Dennis Prize for Best First Collection and the Forward Prize for Best Single Poem in memory of Michael Donaghy, which is not already part of an anthology or collection. All entries must be published in the UK or Republic of Ireland and submitted by poetry publishers (collections) or newspaper and magazine editors (single poems). Individual entries of poets' own work are not accepted. 2009 winners: Best Collection: Don Paterson *Rain* (Faber); Best First Collection: Emma Jones *The Striped World* (Faber); Best Single Poem: Robin Robertson *At Roane Head*.

£ Prizes of £10,000 (best collection); £5,000 (best first collection); £1,000 (best single poem).

## The Foyle Young Poets of the Year Award

The Poetry Society, 22 Betterton Street, London WC2H 9BX

☎ 020 7420 9880    🖷 020 7240 4818

fyp@poetrysociety.org.uk
www.foyleyoungpoets.org
Contact *Lucy Wood*

This annual award was founded by the Poetry Society in 1997 in order to help young poets thrive as readers and writers. Since 2001 the Foyle Foundation has sponsored the award which encourages creativity, imagination and literacy amongst young people at school. Two top poets are selected as judges and they choose 100 of the best young poets, out of over 6,000 poets who enter each year. Poets must be aged 11–17 on the closing date of 31st July each year. Entry is free and can be via the web entry form, or by post. Poems can be submitted with no restrictions on length, content, style, form or theme.

£ Fifteen overall winners aged 11–17, including three international winners, have their poems printed in the winners' anthology; five 11–14-year-olds win a visit to their school by a leading poet; fifteen 15–17 year-old-winners attend a week-long residential course at one of the prestigious Arvon Centres. The three schools who inspire the most entries receive a special selection of books from **Faber & Faber**, **Bloodaxe Books**, **tall-lighthouse**, **Picador** and **Salt Publishing**.

## The Frances Lincoln Diverse Voices Children's Book Award

30 Lime Street, Ouseburn Valley, Newcastle upon Tyne NE1 2PQ

☎ 0845 271 0777    🖷 0191 261 1931

diversevoices@sevenstories.org.uk
www.sevenstories.org.uk
Diverse Voices Administrator *Helena McConnell*
Office Administrator *Liz Paton*

Founded in 2009 by **Frances Lincoln Ltd** and Seven Stories, the Centre for Children's Books, an innovative cultural centre for children's literature. The aim of the award, in memory of Frances Lincoln (1945–2001), is to take positive steps to increase the representation of people writing from or about different cultural perspectives; promote new writing for children, especially by or about people whose culture and voice are currently under-represented; and recognize that

as children's books shape earliest perceptions of the world and its cultures, promoting writing that represents diversity will contribute to social and cultural tolerance. Awarded to the best work of unpublished fiction for 8-to-12-year-olds by a writer who has not previously published a novel for children (although may have contributed to an anthology of prose or poetry). The work must be written in English and be a minimum of 15,000 words and a maximum of 35,000 words. Entrants must be aged 16 years or over by the closing date; only one submission per writer will be accepted. Application form is available on the website. (Seven Stories is a registered charity, number 1056812.)

£ Prize of £1,500, plus the option for Frances Lincoln Children's Books to publish the novel.

## Frank O'Connor International Short Story Award
▷ Cork City – Frank O'Connor International Short Story Award

## The Frogmore Poetry Prize

21 Mildmay Road, Lewes BN7 1PJ

☎ 07751 251689

www.frogmorepress.co.uk
Contact *Jeremy Page*

Established 1987. Awarded annually and sponsored by the Frogmore Foundation. The winning poem, runners-up and short-listed entries are all published in the magazine. Closing date 31st May. Previous winners: David Satherley, Bill Headdon, John Latham, Diane Brown, Tobias Hill, Mario Petrucci, Gina Wilson, Ross Cogan, Joan Benner, Ann Alexander, Gerald Watts, Katy Darby, David Angel, Howard Wright, Caroline Price, Julie-ann Rowell, Arlene Ang, Peter Marshall, Gill Andrews.

£ The winner receives 200 guineas and a two-year subscription to the biannual literary magazine, *The Frogmore Papers*.

## FT & Goldman Sachs Business Book of the Year Award

Financial Times, One Southwark Bridge, London SE1 9HL

☎ 020 7873 3000    🖷 020 7873 3072

bookaward@ft.com
www.ft.com/bookaward
Contact *Lizzie Talbot*

Established in 2005 to identify the book that provides the most compelling and enjoyable insight into modern business issues, including management, finance and economics. Titles must be published for the first time in the English language, or in English translation, between 31st October 2008 and 1st November 2009. Submissions by a publisher or *bona fide* imprint which holds English language rights in the book. 2009 winner: Liaquat Ahamed *Lords of Finance:*

*1929, The Great Depression and the Bankers Who Broke the World* (Heinemann).
£ Award of £30,000 (winner); shortlisted authors receive £5,000 each.

## Geoffrey Faber Memorial Prize

Faber & Faber Ltd, Bloomsbury House, 74–77 Great Russell Street, London WC1B 3DA
T 020 7927 3800 F 020 7927 3801
www.faber.co.uk

Established 1963 as a memorial to the founder and first chairman of **Faber & Faber**, this prize is awarded in alternate years for the volume of verse and the volume of prose fiction published in the UK in the preceding two years, which is judged to be of greatest literary merit. Authors must be under 40 at the time of publication and citizens of the UK, Commonwealth, Republic of Ireland or South Africa. 2009 winner: Nick Laird *On Purpose* (Faber).
£ Prize of £1,000.

## George Devine Award

9 Lower Mall, Hammersmith, London W6 9DJ
Administrator *Harriet Devine*

Annual award for a promising new playwright writing for the stage in memory of George Devine, artistic director of the **Royal Court Theatre**, who died in 1965. The play, which can be of any length, does not need to have been produced. Send two copies of the script, plus outline of work, to *Harriet Devine*. Closing date: 1st March. Send s.a.s.e. for the script to be returned, if required. Information leaflet available from January on receipt of s.a.s.e. 2009 winner: Nick Payne.
£ Prize of £10,000.

## GGBT Creative Writing Awards

> Graham Greene Birthplace Trust entry under Literary Societies

## The Gladstone History Book Prize

Royal Historical Society, University College London, Gower Street, London WC1E 6BT
T 020 7387 7532 F 020 7387 7532
royalhistsoc@ucl.ac.uk
www.royalhistoricalsociety.org/grants.htm
Contact *Administrative Secretary*

Established 1998. Annual award for the best new work on any historical subject which is not primarily related to British history, published in the UK in the preceding calendar year. The book must be the author's first (solely written) history book and be an original and scholarly work of historical research. Closing date: 31st December.
£ Prize of £1,000.

## Golden Hearts Awards

> Romance Writers of America under Professional Associations and Societies

## Golden PEN Award for Lifetime Distinguished Service to Literature

English PEN, Free Word Centre, 60 Farringdon Road, London EC1R 3GA
T 020 7324 2535
enquiries@englishpen.org
www.englishpen.org

Awarded to a senior writer, with a distinguished body of work written over many years, who has made a significant and constructive impact on fellow writers, the reading public and the literary world. Nominations by members of English PEN only. Previous winners include: Harold Pinter, Doris Lessing, Michael Frayn, Nina Bawden, J.G. Ballard, John Berger.

## Gourmand World Cookbook Awards

Pintor Rosales 50, 28008 Madrid, Spain
T 00 34 91 541 6768 F 00 34 91 541 6821
icr@virtualsw.es
www.cookbookfair.com
Contact *Edouard Cointreau*

Founded in 1995 by Edouard Cointreau to reward those who 'cook with words'. The only world competition for food and wine books in all languages. Annual event with the winners of the worldwide local or regional competitions competing for the 'Best Book in the World' award. The 2010 Awards were held at the Paris Cookbook Fair in Februay.

## Gregory O'Donoghue International Poetry Competition

The Munster Literature Centre, Frank O'Connor House, 84 Douglas Street, Cork, Republic of Ireland
T 00 353 21 431 2955
competitions@munsterlit.ie
www.munsterlit.ie
Contact *Patrick Cotter*

Founded 2009. Annual competition for original, unpublished poems in the English language of 40 lines or less. The poem can be on any subject, any style, by a writer of any nationality, living anywhere in the world. Final entry date: 15th December. Please visit the website for submission guidelines.
£ Awards of €1,000 (1st); €500 (2nd); €350 (3rd), plus publication for 10 runners-up.

## The Griffin Poetry Prize

The Griffin Trust for Excellence in Poetry, 363 Parkridge Crescent, Oakville, Ontario, Canada L6M 1A8
info@griffinpoetryprize.com
www.griffinpoetryprize.com
Contact *Ruth Smith, Manager*

Annual award established in 2000 by Toronto-based entrepreneur, Scott Griffin, for books of poetry written in or translated into English. Trustees include Margaret Atwood and Michael Ondaatje. Submissions from publishers only. See website for submission procedures and deadline. ⓔ Prizes total C$100,000, divided into two categories: International and Canadian.

## The Guardian Children's Fiction Award

The Guardian, Kings Place, 90 York Way, London N1 9GU
ⓣ 020 3353 2000
Children's Book Editor *Julia Eccleshare*

Established 1967. Annual award for an outstanding work of fiction for children aged eight and over by a British or Commonwealth author, first published in the UK in the year of the award, excluding picture books. No application form necessary. 2009 winner: Ma Peet *Exposure* (Walker Books). ⓔ Award of £1,500.

## The Guardian First Book Award

The Guardian, Kings Place, 90 York Way, London N1 9GU
ⓣ 020 3353 2000
Contact *Carla Dowling*

Established 1999. Annual award for first time authors published in English in the UK. All genres of writing eligible, apart from academic, guidebooks, children's, educational, manuals, reprints and TV, radio and film tie-ins. All books must be published between January and December 2010 and have an ISDN number or equivalent. Submissions are only received direct from publishers, not from individual authors. Self-published work is not eligible. 2009 winner: Petina Gappah *An Elegy for Easterly* (Faber). ⓔ Award of £10,000, plus *Guardian/Observer* advertising package.

## The Guild of Food Writers Awards

255 Kent House Road, Beckenham BR3 1JQ
ⓣ 020 8659 0422
awards@gfw.co.uk
www.gfw.co.uk
Contact *Jonathan Woods*

Established 1985. Annual awards in recognition of outstanding achievement in all areas in which food writers work and have influence. Entry is not restricted to members of the Guild. Entry form available from the address above. 2009 winners: Kate Whiteman Award for Work on Food and Travel: Fuchsia Dunlop *Shark's Fin and Sichuan Pepper*; Derek Cooper Award for Campaigning and Investigative Food Writing and Broadcasting: Channel Four's *Jamie's Ministry of Food* and Geoff Tansey and Tasmin Rajotte *The Future Control of Food: A Guide to International Negotiations and Rules on Intellectual Property, Biodiversity and Food Security*; Michael Smith Award for Work on British Food: Guy Watson and Jane Baxter *Riverford Farm Cook Book*; Jeremy Round Award for the Best First Book: Guy Watson and Jane Baxter *Riverford Farm Cook Book*; Food Book of the Year: Heston Blumenthal *The Big Fat Duck Cookbook*; Cookery Book of the Year: Mark Hix *British Seasonal Food*; Food Journalist of the Year: Bee Wilson for articles published in the *Sunday Telegraph's Stella* magazine; Evelyn Rose Award for Cookery Journalist of the Year: Diana Henry for articles published in the *Sunday Telegraph's Stella* magazine; Restaurant Reviewer of the Year: Emma Sturgess for reviews published in *Metro*; Miriam Polunin Award for Work on Healthy Eating: BBC Radio 4's *The Food Programme: Nutritionism*; New Media Award: Tim Hayward for his blog on *The Guardian* and the *Observer's* Word of Mouth (www.guardian.co.uk/lifeandstyle/wordofmouth); Food Broadcast Award: BBC Radio 4's *You and Yours: Lunch*; Lifetime Achievement Awards: Mary Berry and Egon Ronay.

## Gwobr Llyfr y Flwyddyn
▷ Wales Book of the Year Award

## H.H. Wingate Prize
▷ Jewish Quarterly Wingate Prize

## Hans Christian Andersen Awards

IBBY, Nonnenweg 12, Postfach, CH-4003 Basel, Switzerland
ⓣ 00 41 61 272 2917    ⓕ 00 41 61 272 2757
ibby@ibby.org
www.ibby.org
Deputy Director of Administration *Forest Zhang*
Executive Director *Liz Page*

The highest international prizes for children's literature: The Hans Christian Andersen Award for Writing, established 1956; The Hans Christian Andersen Award for Illustration, established 1966. Candidates are nominated by National Sections of IBBY (The International Board on Books for Young People). Biennial prizes are awarded, in even-numbered years, to an author and an illustrator whose body of work has made a lasting contribution to children's literature. 2010 winners: Award for Writing: David Almond (UK); Award for Illustration: Jutta Bauer (Germany). ⓔ Awards: gold medals.

## The Harri Webb Prize

10 Heol Don, Whitchurch, Cardiff CF14 2AU
☎ 029 2062 3359
Contact *Professor Meic Stephens*

Established 1995. Annual award to commemorate
the Welsh poet, Harri Webb (1920–94), for a
single poem in any of the categories in which
he wrote: ballad, satire, song, polemic or a first
collection of poems. The poems are chosen by
three adjudicators; no submissions. Previous
winner: Grahame Davies.
£ Prizes of £100/£200.

## The Harry Bowling Prize for New Writing

c/o 16 St Briac Way, Exmouth EX8 5RN
☎ 01395 279659
fiona@mbalit.co.uk
www.harrybowlingprize.net
Contacts *Margaret James, Marina Oliver
(StorytrackS Consultants)*

Established 2000 in honour of Harry Bowling,
'the king of Cockney sagas' (died 1999). Sponsored
by **Headline Book Publishing** and in partnership
with **the Romantic Novelists' Association**. Open
to anyone who has not been published previously,
judges are looking for great storytelling in an
urban setting, with a romantic theme. Entry
fee charged; forms from address above (enclose
s.a.s.e.) or via the website.
£ Prize winner receives £1,000; runners-up,
£100.

## Hellenic Foundation Prize
▷ The Translation Prizes

## The Hessell-Tiltman Prize for History

English PEN, Free Word Centre, 60 Farringdon
Road, London EC1R 3GA
☎ 020 7324 2535
enquiries@englishpen.org
www.englishpen.org

Founded 2002. Awarded for a history book
covering any period up to the end of the Second
World War, written in English (including
translations) and aimed at a wide audience.
Submissions cannot be made as books are
nominated by an independent judging panel.
2009 winner: Mark Thompson *The White War:
Life & Death on the Italian Front 1915–1919* (Faber)
£ Prize of £3,000.

## HISSAC Annual Short Story Award

20 Lochslin, Balintore, Easter Ross IV20 1UP
☎ 01862 832266    ℻ 01862 832266
info@hissac.co.uk
www.hissac.co.uk
Judge Chair *Clio Gray*

Established 2003. Annual short story award
organized by HISSAC, the Highlands & Islands
Short Story Association, an international network
of writers aiming to critique and encourage each
other's work and encourage all manner of creative
writing, including the short story form. Entry
fees: £4 per entry, £10 for three. Paypal facilities
on website. Maximum wordage: 2,500. No
connection to Scotland is needed. Double-spaced,
single-sided entries; no children's stories. Final
entry date: 31st July 2010. Winning entries are
published on the website.
£ Prizes of £400 (1st); £50 each (2nd & 3rd).

## The Hugo Awards

info@thehugoawards.org
www.thehugoawards.org
Contact *Kevin Standlee*

Founded in 1953, the Hugos are awards for
excellence in the field of science fiction and
fantasy. Awarded annually since 1953, and voted
on by members of the World Science Fiction
Convention (Worldcon), they cover written
fiction, art, media, editing and writing about
the field. The Hugos are awarded for works first
published or presented in the previous year. Work
may not be submitted for consideration as the
nomination process is carried out by about 500
members of the Worldcon.
£ Award of a trophy in the shape of a classic
rocket ship.

## Hunter Davies Prize
▷ Lakeland Book of the Year Awards

## The Imison Award

The Society of Authors, 84 Drayton Gardens
SW10 9SB
☎ 020 7373 6642    ℻ 020 7373 5768
sbaxter@societyofauthors.org
www.societyofauthors.org
Contact *Sarah Baxter*

Annual award established 'to perpetuate the
memory of Richard Imison, to acknowledge
the encouragement he gave to writers working
in the medium of radio, and in memory of the
support and friendship he invariably offered
writers in general, and radio writers in particular'.
Administered by the **Society of Authors** and
generally sponsored by the **Peggy Ramsay
Foundation**, the purpose is 'to encourage new
talent and high standards in writing for radio
by selecting the radio drama by a writer new
to radio which, in the opinion of the judges, is
the best of those submitted.' An adaptation for
radio of a piece originally written for the stage,
television or film is not eligible. Any radio drama
first transmitted in the UK between 1st January
and 31st December is eligible, provided the work
is an original piece for radio and it is the first
dramatic work by the writer(s) that has been

broadcast. Submission may be made by any party to the production in the form of three copies of both script and recording (non-returnable), accompanied by a nomination form and 250 word synopsis and c.v. 2009 winner: Lucy Caldwell *The Girl From Mars*.
£ Prize of £1,500.

## The Independent Foreign Fiction Prize

c/o Literature Department, Arts Council England, 14 Great Peter Street, London SW1P 3NQ
T 0845 300 6200
www.artscouncil.org.uk/iffp

The Independent Foreign Fiction Prize celebrates an exceptional work of fiction by a living author which has been translated into English from any other language and published in the United Kingdom in the last year. 2009 winner: *The Armies* by Evelio Rosero, translated by Anne McLean (MacLehose Press).
£ Prize of £10,000 shared equally between author and translator.

## The International Dagger (in association with Crime Writers' Association)

PO Box 273, Boreham Wood WD6 2XA
secretary@thecwa.co.uk
www.thecwa.co.uk
Contact *The Secretary*

Established 2006. An annual award for the best foreign crime fiction translated into English during the year. Nominations from publishers only. 2009 winner: Fred Vargas *The Chalk Circle Man*, translated by Sîan Reynolds.

## The International Dundee Book Prize

City of Discovery Campaign, 3 City Square, Dundee DD1 3BA
T 01382 434214    F 01382 434650
book.prize@dundeecity.gov.uk
www.dundeebookprize.com
Contact *Karin Johnston*

'The Dundee International Book Prize has established itself as the UK's premier prize for emerging novelists. Its £10,000 cash award together with publication by **Birlinn Ltd**, publishers of the Polygon imprint, make the Prize highly valued by tomorrow's great new writers seeking to break into the publishing world.' The award is for an unpublished novel on any theme and in any genre. Telephone or email *Karin Johnston* for further information. 2009 winner: Chris Longmuir *Dead Wood*.

## The International IMPAC Dublin Literary Award

Dublin City Library & Archive, 138–144 Pearse Street, Dublin 2, Republic of Ireland
T 00 353 1 674 4802/4801    F 00 353 1 674 4879
literaryaward@dublincity.ie
www.impacdublinaward.ie

Established 1995. Sponsored by Dublin City Council and US-based productivity improvement firm, IMPAC, this prize is awarded for a work of fiction written and published in the English language or written in a language other than English and published in English translation. Initial nominations are made by municipal public libraries in major and capital cities worldwide, each library putting forward up to three books to the international panel of judges in Dublin. 2009 winner: *Man Gone Down* by Michael Thomas (Grove Press).
£ Prize of €100,000 (if the winning book is in English translation, the prize is shared €75,000 to the author and €25,000 to the translator).

## International Prize for Arabic Fiction

IPAF Office, PO Box 280, Jounieh, Lebanon
T 00 961 9 935333
info@arabicfiction.org
www.arabicfiction.org/en/
Prize Administrator *Joumana Haddad*

Launched in Abu Dhabi, UAE, in April 2007 in association with the Booker Prize Foundation and with the support of the Emirates Foundation. Annual award which aims to reward excellence in contemporary Arabic creative writing and to encourage wider readership of quality Arabic literature internationally. Awarded specifically for prose fiction in Arabic (no short stories or poetry) with each of the six shortlisted authors receiving $10,000 with a further $50,000 going to the winner. Submissions from publishers only. Final entry date: 31st July. 2010 winner: Abdo Khal *Spewing Sparks as Big as Castles* (Al-Jamal Publications).

## International Student Playscript Competition
▷ **University of Hull under UK and Irish Writers' Courses**

## Irish Book Awards

Unique Media Ltd, 600A Cathedral Court, New Street, Dublin 8, Republic of Ireland
T 00 353 1 522 5200    F 00 353 1 522 5220
joanneleahy@uniquemedia.ie
www.irishbooksawards.ie

Established in 2005 to acknowledge the wealth of talent in Irish literature. Sponsored by Bord Gáis Energy. Annual awards comprising nine categories which are open to all Irish authors who had a book published in the previous twelve months. 2009 winners: Sebastian Barry *The Secret Scripture* (The Hughes & Hughes Irish Novel of the Year; The Tubridy Show Listeners' Choice Award ); Seamus Heaney and Dennis O'Driscoll *Stepping Stones: Interviews with Seamus Heaney* (The Argosy Irish Non-Fiction

Book of the Year); Marian Keyes *This Charming Man* (Eason Irish Popular Fiction Book of the Year); Benji Bennett *Before You Sleep* and Derek Landy *Skulduggery Pleasant: Playing with Fire* (The Dublin Airport Authority Irish Children's Book of the Year – junior and senior); Alice Taylor *The Parish* (Best Irish-Published Book of the Year); Alex Barclay *Blood Runs Cold* (The Ireland AM Irish Crime Fiction Book of the Year); Ronan O'Gara *Ronan O'Gara: My Autobiography* (Energise Sport Irish Sports Book of the Year); Ronan O'Brien *Confessions of a Fallen Angel* (International Education Services Irish Newcomer of the Year); *Edna O'Brien* (The 2009 Bob Hughes Lifetime Achievement Award).

## J.R. Ackerley Prize

English PEN, Free Word Centre, 60 Farringdon Road, London EC1R 3GA
℡ 020 7324 2535    🖷 020 7837 7838
enquiries@englishpen.org
www.englishpen.org

Commemorating the novelist/autobiographer J.R. Ackerley, this prize is awarded for a literary autobiography, written in English and published in the year preceding the award. Entry restricted to nominations from the Ackerley Trustees only ('please do not submit books'). 2009 winner: Julia Blackburn *The Three of Us* (Vintage).
💷 Prize of £1,000.

## The James Ackerman Award in the history of architecture (Premio James Ackerman per la storia dell'architettura)

Centro Internazionale di Studi di Architettura Andrea Palladio, Contra' Porti 11, Vicenza, Italy 36100
℡ 00 39 444 323014    🖷 00 39 444 322869
mostre@cisapalladio.org
www.cisapalladio.org
www.premioackerman.it
Contact *Ilaria Abbondandolo, Curatorial Assistant*

Established 2005. Annual award devoted to the publication of the first book by a young scholar in any field of architectural history. Administered by the International Centre for the Study of Architecture Andrea Palladio, and supported financially by James S. Ackerman who chairs the international jury awarding the prize. All periods are eligible within the field and there are no restrictions on the subject treated or on the nationality of the author. Those wishing to enter must have a PhD or have published an original study in architectural history in a qualified scientific journal. Texts may be submitted in Italian, English, German, French or Spanish. 2009 winner: Mantha Zarmakoupi *Designing for*

*Luxury on the Bay of Naples (c. 100 BCE–79 CE): Villas and Landscapes.*
💷 The winning book is published free of charge in a series dedicated to 'The James Ackerman Award in the history of architecture' and is distributed to libraries and bookstores.

## James Cameron Award

City University, Department of Journalism, Northampton Square, London EC1V 0HB
℡ 020 7040 8221    🖷 020 7040 8594
H.Stephenson@city.ac.uk
Contact *Hugh Stephenson*

Annual award for journalism to a reporter of any nationality, working for the British media, whose work is judged to have contributed most during the year to the continuance of the Cameron tradition. Previous winner: Peter Taylor.

## James Tait Black Memorial Prizes

University of Edinburgh, David Hume Tower, George Square, Edinburgh EH8 9JX
℡ 0131 650 3619    🖷 0131 650 6898
www.ed.ac.uk/about/2.419/tait-black/intro
Contact *Department of English Literature*

Established 1918 in memory of a partner of the publishing firm of **A.&C. Black Ltd**. Two prizes, one for biography and one for fiction. Closing date for submissions: 1st December. Each prize is awarded for a book first published in Britain in the previous calendar year (1st January–31st December). 2008 winners: Sebastian Barry *The Secret Scripture* (fiction); Michael Holroyd *A Strange Eventful History: The Dramatic Lives of Ellen Terry, Henry Irving and their Remarkable Families* (biography).
💷 Prizes of £10,000 each.

## Jewish Quarterly Wingate Prize

PO Box 37645, London NW7 1WB
℡ 020 8343 4675
www.jewishquarterly.org
Administrator *Pam Lewis*

Formerly the H.H. Wingate Prize. Annual award (for fiction or non-fiction) for a work that best stimulates an interest in and awareness of themes of Jewish interest. Books must have been published in the UK in the year of the award (written in English originally or in translation) by an author resident in Britain, the Commonwealth, Israel, Republic of Ireland or South Africa. 2009 winner: Fred Wander *The Seventh Well* (Granta).
💷 Prize of £4,000.

## The John Betjeman Young People's Poetry Competition

72 Vicars Hill, London SE13 7JL
www.johnbetjeman.com
Contact *Justin Gowers*

The John Betjeman Young People's Poetry Competition, sponsored by Shell, was inaugurated in 2006 to celebrate the centenary of one of the nation's best-loved poets. Annual competition. Entrants are limited to one poem each about their local surroundings or any aspect thereof, whether it be a house, a street, a garden, a park, a city or a wider landscape. The spirit behind the competition is to encourage young people to understand and appreciate the importance of place. The competition is open to 11–14 year olds in the British Isles and Republic of Ireland. Approximate final date for entries: 30th June 2010. Send one poem only about any aspect of your local surroundings, with a title. (Please note that for administrative reasons entries will not be acknowledged.)
ⓔ Prize of £1,000, (£500 to the winner and £500 to the English department of their school), is donated by **John Murray (Publishers) Ltd**.

## The John Dryden Competition

School of Literature and Creative Writing, University of East Anglia, Norwich NR4 7TJ
ⓣ 01603 593360    ⓕ 01603 250599
transcomp@uea.ac.uk
www.bcla.org
Competition Organizer *Professor Jean Boase-Beier*

Established 1983. Annual competition open to unpublished literary translations from all languages. Maximum submission: 25 pages. 2009 winner: John R.G. Turner *Seven Tricky Verlaines*.
ⓔ Prizes of £350 (1st); £200 (2nd); £100 (3rd); plus publication for all winning entries in the British Comparative Literature Association's journal. Other entries may receive commendations.

## The John Florio Prize
⊳ The Translation Prizes

## John Llewellyn Rhys Prize

Booktrust, Book House, 45 East Hill, London SW18 2QZ
ⓣ 020 8516 2972    ⓕ 020 8516 2978
prizes@booktrust.org.uk
www.booktrust.org.uk
Contact *Prize Administrator*

Established 1942. An annual literary award for a memorable work of fiction, non-fiction, drama or poetry. Entrants must be 35 or under at the time of publication; books must have been published in the UK during the calendar year of the award. The author must be a citizen of the UK or the Commonwealth, writing in English. 2009 winner: Evie Wild *After the Fire, a Still Small Voice* (Cape).
ⓔ Prize of £5,000 (1st); £500 for shortlisted entries.

## John Whiting Award
⊳ The Peter Wolff Theatre Trust supports the Whiting Award

## Kate Greenaway Medal
⊳ CILIP Kate Greenaway Medal

## Katharine Briggs Folklore Award

The Folklore Society, c/o The Warburg Institute, Woburn Square, London WC1H 0AB
ⓣ 020 7862 8564
enquiries@folklore-society.com
www.folklore-society.com
Contact *The Convenor*

Established 1982. An annual award in November for the book, published in Britain and Ireland between 1st June in the previous calendar year and 30th May, which has made the most distinguished non-fiction contribution to folklore studies. Intended to encourage serious research in the field which Katharine Briggs did so much to establish. The term folklore studies is interpreted broadly to include all aspects of traditional and popular culture, narrative, belief, custom and folk arts. 2009 winner: Kathryn Marsh *The Musical Playground: Global Tradition and Change in Children's Songs and Games* (OUP).
ⓔ Prize of £200, plus engraved goblet.

## Keats–Shelley Prize

Keats–Shelley Memorial Association, 117 Cheyne Walk, London SW10 0ES
ⓣ 020 7352 2180    ⓕ 020 7352 6705
hello@keats-shelley.co.uk
www.keats-shelley.co.uk
Contact *Harriet Cullen*

Established 1998. Annual award to promote the study and appreciation of Keats and Shelley, especially in the universities, and of creative writing inspired by the younger Romantic poets. Two categories: essay and poem; open to all ages and nationalities. 2009 winners: D.H. Maitreyabandhu (poem), Jillian Hess (essay).
ⓔ Prize of £3,000 distributed between the winners of the two categories.

## Kelpies Prize

Floris Books, 15 Harrison Gardens, Edinburgh EH11 1SH
ⓣ 0131 337 2372    ⓕ 0131 347 9919
floris@florisbooks.co.uk
www.florisbooks.co.uk/kelpiesprize
Contact *Prize Administrator*

Established 2004. Annual prize to encourage and reward new Scottish writing for children. The prize is for an unpublished novel (40–70,000 words) for children aged 9–12, set wholly or mainly in Scotland. The author does not need to be Scottish. Application form, guidelines, terms and conditions are available on the website. 2009 winner: Janis Mackay *Magnus Fin and the Ocean Quest*.

£ Prize of £2,000 and publication in **Floris Books**' Kelpies series.

## Kent & Sussex Poetry Society Open Competition

13 Ruscombe Close, Southborough, Tunbridge Wells TN4 0SG
☎ 01892 543862
www.kentandsussexpoetrysociety.org
Chairman *Clive R. Eastwood*

Annual competition. Entry fee: £4 per poem, maximum 40 lines. Full details available on the website.
£ Prizes total £1,550.

## L. Ron Hubbard's Writers of The Future Contest

PO Box 218, East Grinstead RH19 4GH
Contest Administrator *Andrea Grant-Webb*

Established in 1984 by L. Ron Hubbard to encourage new and amateur writers of science fiction and fantasy. Quarterly awards with an annual grand prize. Entrants must submit a short story of up to 10,000 words, or a novelette of fewer than 17,000 words, which must not have been published previously. The contest is open only to those who have not been published professionally. Send s.a.s.e. for entry form. Previous winners: Roge Gregory, Malcolm Twigg, Tom Brennan, Alan Smale, Janet Barron, C.L. Holland, Mike Wood.
£ Prizes of £600 (1st); £450 (2nd); £300 (3rd) each quarter; Annual Grand Prize: £3,000. All winners are awarded a trip to the annual L. Ron Hubbard Achievement Awards which include a series of professional writers' workshops, and are published in the *L. Ron Hubbard Presents Writers of the Future* anthology.

## Lakeland Book of the Year Awards

Cumbria Tourism, Windermere Road, Staveley LA8 9PL
☎ 01539 825052    📠 01539 825076
slindsay@cumbriatourism.org
www.golakes.co.uk
Contact *Sheila Lindsay*

Established in 1984 by local author and broadcaster Hunter Davies in conjunction with the Cumbria Tourist Board, the books entered can be about any aspect of life in the county of Cumbria, from local history books and walking guides to novels and poetry. The contest attracts entries from both new and established authors. Since the establishment of the awards they have grown in importance and are now attracting in the region of 60 entries annually, all competing for the Hunter Davies Prize for the Lakeland Book of the Year.

In addition to the Hunter Davies Prize there are currently five categories including Award for Guides, Walks and Places; Award for People and Social History; Award for Arts and Culture; Award for Heritage and Tradition; and the Best Illustrated Book. Closing date for entries is mid-March. The awards are presented at a charity luncheon in early July.
£ Awards of £100 for each category together with a framed certificate. Overall winner of the Hunter Davies Prize for the Lakeland Book of the Year also receives a cheque for £100 and a framed certificate.

## Lancashire County Library and Information Service Book Award

Lancashire County Library Headquarters, 'The Park Hotel, Eastcliff, Preston PR1 3EA
☎ 01772 534751    📠 01772 534880
jacob.hope@lancashire.gov.uk
Award Coordinator *Jake Hope*

Established 1986. Annual award, presented in June for a work of original fiction suitable for 12–14-year-olds. The winner is chosen by 13–14-year-old secondary school pupils in Lancashire. Books must have been published between 1st September and 31st August in the previous year of the award and authors must be UK and Republic of Ireland residents. Final entry date: 1st September each year. 2009 winner: Sophie McKenzie *Blood Ties*.
£ Prize of £1,000, plus engraved glass decanter.

## Le Prince Maurice Prize

Constance Hotels Experience, Poste de Flacq, Mauritius, Indian Ocean
☎ 00 230 402 29 30    📠 00 230 415 10 82
events@constancehotels.com
www.constancehotels.com
Events Coordinator *Jannick Chedumbrum*

Founded in 2003 and sponsored by one of Mauritius's five-star resorts, Constance Le Prince Maurice. Annual award designed to celebrate the literary love story. Administered in both the UK and France, it is awarded alternately to an English-speaking and French-speaking writer. The aim is to strengthen the cultural links between Mauritius and Europe. Submissions are made by publishers only. Previous winner: James Meek *We Are Now Beginning Our Descent*.
£ Prize winner receives a trophy, plus all-expenses-paid two-week 'writer's retreat' at Constance Le Prince Maurice.

## Legend Writing Awards

39 Emmanuel Road, Hastings TN34 3LB
www.legendwritingaward.com
Contact *Legend Coordinator*

Established 2001. Annual awards to encourage new fiction writers resident in the UK. The competition

is for short stories of 2,000 words maximum and flash fiction of 100 words *exactly* (excluding title). There is no set theme. Closing date: 31st August. Entry fees: short story, £7 (first entry); £5 (each subsequent entry); flash fiction, £3 per entry. Rules/entry form available from website or by sending s.a.s.e. to address above. The competition is organized and judged by Hastings Writers' Group.
£ Prizes of £500 (1st); £250 (2nd); £100 (3rd); plus three runners-up prizes of £50 for short stories; £50 (1st), £30 (2nd), £20 (3rd) plus three runners-up prizes of £10 for flash fiction. For 2010, the 10th anniversary year, there will be additional prizes of £50, £30, & £20 for young writers aged 15–17 years.

## Local Poem Competition

United Press, Admail 3735, London EC1B 1JB
☎ 0844 800 9177   ℻ 0844 800 8178
poems@unitedpress.co.uk
www.unitedpress.co.uk
Contact *Julie Embury*

Free to enter annual poetry competition. Up to three poems (20 lines and 160 words max. each) to be submitted by the annual closing date of 31st December. United Press select one winning poem which must have a local theme and be about a person, place or event.
£ Prize of £1,000 cash.

## Longman-History Today Book of the Year Award

c/o History Today, 20 Old Compton Street, London W1D 4TW
☎ 020 7534 8000
www.historytoday.com
Contact *Andy Patterson*

Established 1993. Annual award set up as a joint initiative between the magazine *History Today* and the publisher Longman (**Pearson**) to mark the past links between the two organizations, to encourage new writers, and to promote a wider public understanding of, and enthusiasm for, the study and publication of history. Award for author's first or second book.
£ Prize of £2,000.

## McColvin Medal
▷ Besterman/McColvin Medal

## McKitterick Prize

Society of Authors, 84 Drayton Gardens, London SW10 9SB
☎ 020 7373 6642   ℻ 020 7373 5768
info@societyofauthors.org
www.societyofauthors.org
Contact *Awards Secretary*

Annual award for a full-length novel in the English language, first published in the UK or unpublished. Open to writers over 40 who have not had any novel published other than the one submitted (excluding works for children). Closing date: 31st October. 2009 winner: Chris Hannan *Missy* (Chatto & Windus).
£ Prize of £4,000.

## The Macmillan Prize for Children's Picture Book Illustration

Macmillan Children's Books, 20 New Wharf Road, London N1 9RR
☎ 020 7014 6124   ℻ 020 7014 6142
lindsey.evans@macmillan.co.uk
www.panmacmillan.com
Contact *Lindsey Evans, Macmillan Children's Books*

Established by Macmillan Children's Books in order to stimulate new work from young illustrators in art schools, and to help them start their professional careers. Fiction or non-fiction. Macmillan have the option to publish any of the prize winners.
£ Prizes of £1,000 (1st); £500 (2nd); £250 (3rd).

## The Mail on Sunday Novel Competition

Postal box address may change annually (see below)
Contact *Paula Johnson*

Annual award established 1983. Judges look for a story/character that springs to life in the 'tantalizing opening 50–150 words of a novel'. Details of the competition, including the postal box address, are published in *The Mail on Sunday* in July/August. 2009 winner: Margaret Davies.
£ Awards of £400 book tokens and a writing course at the **Arvon Foundation** (1st); £300 tokens (2nd); £200 tokens (3rd); three further prizes of £150 tokens each.

## The Man Asian Literary Prize

201 Kai Tak Commercial Building, 66–72 Stanley Street, Central, Hong Kong
☎ 00 852 2877 9770
info@manasianliteraryprize.org
www.manasianliteraryprize.org
Chairman *David Heywood Parker*

Initiated by the Man Group plc and the Hong Kong Literary Festival in 2007. An annual award which aims to recognize the best of new Asian literature and to bring it to the attention of the world literary community. Open to a single work of Asian fiction, first published in English (translations into English of works originally in another language also eligible) of no less than 30,000 words by an Asian author residing in an Asian country or territory. Final entry date: end March. 2009 winner: Su Tong *The Boat to Redemption* (Doubleday).
£ Award of US$30,000; additional US$5,000 to the translator if winning entry is a translation.

## The Man Booker International Prize

Colman Getty, 28 Windmill Street, London
W1T 2JJ

ⓣ 020 7631 2666 ⓕ 020 7631 2699
info@colmangetty.co.uk
www.manbookerinternational.com

Established in 2004 to complement **The Man Booker Prize for Fiction** by recognizing one writer's achievement in literature and their significant influence on writers and readers worldwide. Sponsored by the Man Group, this biennial award is given to a living author who has published fiction either originally in English or whose work is generally available in translation into the English language. Submissions for the prize are not invited. Where the winning author's work has been translated into English an additional prize of £15,000 is awarded to the translator. The winning author chooses who the translator's prize should go to and whether it is to be awarded to one translator or divided between several. 2009 winner: Alice Munro.

## The Man Booker Prize for Fiction

Colman Getty, 28 Windmill Street, London
W1T 4JE

ⓣ 020 7631 2666 ⓕ 020 7631 2699
info@colmangetty.co.uk
www.themanbookerprize.com
Contact *Lois Tucker (Submissions)*

The Booker Prize for Fiction was originally set up by Booker plc in 1968 to reward merit, raise the stature of the author in the eyes of the public and encourage an interest in contemporary fiction. In April 2002 it was announced that the Man Group had been chosen by the Booker Prize Foundation as the new sponsor of the Booker Prize. The sponsorship is due to run until 2011. United Kingdom publishers may enter up to two full-length novels, with scheduled publication dates between 1st October and 30th September. 2009 winner: Hilary Mantel *Wolf Hall* (Fourth Estate).
ⓔ The winner receives £50,000. The six shortlisted authors each receive £2,500.

## Manchester Writing Competition

Writing School, Dept. of English, Manchester Metropolitan University, Geoffrey Manton Building, Rosamond Street West, off Oxford Road, Manchester M16 6LL

ⓣ 0161 247 1787
j.draper@mmu.ac.uk
www.manchesterwritingcompetition.co.uk
Project Manager *James Draper*

An annual competition, launched in 2008 by the Writing School at Manchester Metropolitan University, the Manchester Writing Competition alternates between poetry and fiction. The inaugural prize was for the best portfolio of poems with the winner receiving cash. In addition, a bursary was awarded to an entrant aged 18–25 for study at MMU as part of the Manchester Young Writer of the Year Award. Open to UK and international writers. See website for entry details. 2008 Manchester Poetry Prize joint winners: Lesley Saunders *Christina the Astonishing* and *The Uses of Greek* and Mandy Coe *Pinning the Tail on the Donkey*; Manchester Young Writer of the Year Award: Helen Mort.

## Manchester Young Writer of the Year Award
▷ Manchester Writing Competition

## Marsh Award for Children's Literature in Translation

The English-Speaking Union, Dartmouth House, 37 Charles Street, London W1J 5ED

ⓣ 020 7529 1550 ⓕ 020 7495 6108
education@esu.org
Contact *Katherine Plummer*

Established 1995 and sponsored by the Marsh Christian Trust, the award aims to encourage translation of foreign children's books into English. It is a biennial award (next award: 2011), open to translators of books for 4–16-year-olds, published in the UK by a British publisher. Any category will be considered with the exception of encyclopaedias and reference books. No electronic books. 2009 winner: Sarah Ardizzone for her translation of *Toby Alone* by Timothee de Fombelle (Walker Books).
ⓔ Prize of £2,000.

## Marsh Biography Award

The English-Speaking Union, Dartmouth House, 37 Charles Street, London W1J 5ED

ⓣ 020 7529 1564 ⓕ 020 7495 6108
gillian_parker@esu.org
www.esu.org
Contact *Gillian Parker*

A biennial award for the most significant biography published over a two-year period by a British publisher. Next award October 2011. 2009 winner: Rosemary Hill *God's Architect: Pugin and the Building of Romantic Britain* (Penguin).
ⓔ Award: Membership of the ESU and £5,000, plus a silver trophy presented at a gala dinner.

## Martha Gellhorn Trust Prize

Crosscombe, Town's Lane, Loddiswell TQ7 4QY
ⓣ 01548 550344 ⓕ 01548 550344
sandyandshirlee@phonecoop.coop

Annual prize for journalism in honour of one of the twentieth century's greatest reporters. Open for journalism published in English, giving 'the view from the ground – a human story that penetrates the established version of events and illuminates an urgent issue buried by prevailing fashions of what makes news'. The subject matter

can involve the UK or abroad. Six copies of each entry should be sent to the address above by 31st March. Previous winners include: Hala Jaber, Michael Tierney, Ghaith Abdul-Ahad, Patrick Cockburn, Robert Fisk, Jeremy Harding, Geoffrey Lean.
£ Prize of £5,000.

## Mary Vaughan Jones Award

Cyngor Llyfrau Cymru (Welsh Books Council), Castell Brychan, Aberystwyth SY23 2JB
☎ 01970 624151    ℻ 01970 625385
wbc.children@wbc.org.uk
www.wbc.org.uk
Contact *The Administrator*

Triennial award for distinguished services in the field of children's literature in Wales over a considerable period of time. 2009 winner: Angharad Tomos.
£ Award winner receives a silver trophy.

## Medical Book Awards

The Society of Authors, 84 Drayton Gardens, London SW10 9SB
☎ 020 7373 6642    ℻ 020 7373 5768
sbaxter@societyofauthors.org
www.societyofauthors.org
Contact *Sarah Baxter*

Awards honouring excellence in medical writing. Contact *Sarah Baxter* for information and entry details.

## Melissa Nathan Award for Comedy Romance

Melissa Nathan Foundation and Award, PO Box 56923, London N10 3YU
☎ 020 8671 8424    ℻ 020 8883 6694
information@melissanathan.co.uk
www.melissanathan.com
Chair *Andrew Saffron*
Committee *Maggie Phillips*

Established in 2007 and named after bestselling novelist, Melissa Nathan, who died of cancer aged 37, to honour her memory and writers of comedy romance novels. Awarded to the novel that best combines laugh-out-loud comedy with believable, heart-warming romance. To be eligible, novels must be published in the UK in the year before the award, including paperback version of hardbacks published the previous year. Final entry date: 31st December. 2009 winner: Farahad Zama *The Marriage Bureau for Rich People* (Abacus).
£ Prize of £5,000, plus trophy.

## The Mercedes-Benz Award for the Montagu of Beaulieu Trophy

Guild of Motoring Writers, 40 Baring Road, Bournemouth BH6 4DT
☎ 01202 422424
generalsec@gomw.co.uk
www.guildofmotoringwriters.co.uk

Contact *Patricia Lodge*

First presented by Lord Montagu on the occasion of the opening of the National Motor Museum at Beaulieu in 1972. Awarded annually to a member of the **Guild of Motoring Writers** who, in the opinion of the nominated jury, has made the greatest contribution to recording in the English language the history of motoring or motor cycling in a published book.
£ Cash prize sponsored by Mercedes-Benz UK.

## Mere Literary Festival Open Competition

'Lawrences', Old Hollow, Mere BA12 6EG
☎ 01747 860475
www.merelitfest.co.uk
Contact *Mrs Adrienne Howell (Events Organizer)*

Annual open competition which alternates between short stories and poetry. The winners are announced at the **Mere Literary Festival** during the second week of October. The 2011 competition is for poetry. For further details, including entry fees and form, access the website or contact the address above from 1st March with s.a.s.e.
£ Cash prizes.

## Meyer-Whitworth Award

Playwrights' Studio Scotland, CCA, 350 Sauchiehall Street, Glasgow G2 3JD
☎ 0141 332 4403
info@playwrightsstudio.co.uk
www.playwrightsstudio.co.uk
Contact *Emma McKee*

The Meyer-Whitworth Award is one of the largest annual monetary prizes for playwriting in the UK. Funded by the Royal National Theatre Foundation and managed by the Playwrights' Studio, Scotland in association with the UK Playwrights Network, it is intended to help further the careers of UK playwrights who are not yet established. Nominations by directors of professional theatre companies only. Plays must be original work, be in the English language and have been produced professionally in the UK for the first time in the year between 1st November and 30th November of the previous year. Translations are not eligible. Candidates will have had no more than two of their plays professionally produced, including the play submitted. No writer who has previously won the award may reapply and no play that has previously been submitted for the award is eligible. Writers must be resident in the British Isles or Republic of Ireland. Nomination forms available from the Playwrights' Studio, Scotland. 2009 winner: Ali Taylor *Cotton Wool*.
£ Award of £10,000.

## Michael Marks Awards for Poetry Pamphlets

Poetry Book Society, Fourth Floor, 2 Tavistock Place, London WC1H 9RA
☏ 020 7833 9247   🖷 020 7833 5990
pamphlets@poetrybooks.co.uk
www..poetrybookshoponline.com/pamphlets
Contacts *Hilary Davidson, Chris Holifield*

Established in 2009 and run by the British Library and the Poetry Book Society (funded by the Michael Marks Foundation), this competition consists of two awards, one for poetry pamphlets published in the UK, including self-published work and one for UK poetry pamphlet publishers. Final entry date: late March. Eligibility guidelines available from the Poetry Book Society or from the website. Inaugural winners: Pamphlet Award: Elizabeth Burns *The Shortest Days* (Galdragon Press); UK Poetry Pamphlet Publisher: Oystercatcher Press.
💷 Prize of £5,000 for the most outstanding poetry pamphlet of the year and £5,000 for the most outstanding poetry pamphlet publisher of the year.

## Michael Ramsey Prize for Theological Writing

SPCK, 36 Causton Street, London SW1P 4ST
☏ 020 7592 3900
prize@spck.org.uk
www.michaelramseyprize.org.uk
Coordinator *Primavera Quantrill*
Press Enquiries *Mark Casserley*

Founded in 2005 by Archbishop of Canterbury Dr Rowan Williams to encourage the most promising contemporary theological writing and to identify it for a wider Christian readership. This biennial prize commemorates Dr Ramsey (Archbishop of Canterbury, 1961–1974) and his commitment to increasing the breadth of theological understanding of people in general. Awarded to the author of a theological work that is judged to contribute most towards advancing theology and making a serious contribution to the faith and life of the Church. The book must be by a living author; must have been first published between March 2010 and December 2011; must be written in or translated into English and available in the UK (other terms and conditions on website). Nominations are made by an invited group, including Anglican Primates, by leaders of UK Christian church bodies, Church of England Bishops and by principals of theological colleges and by the sponsors of the prize, the Lambeth Fund Trustees. Closing date for nominations: 14th March 2012. 2009 winner: Richard Bauckham *Jesus and the Eyewitnesses*. The Michael Ramsey Prize is administered by SPCK on behalf of the Archbishop of Canterbury, and is sponsored by the Lambeth Partners.
💷 Prize of £15,000.

## Miles Franklin Literary Award

PO Box 229, Plymble BC, NSW 2073, Australia
☏ 00 61 2 9144 2415
trustawards@cauzgroup.com.au
www.trust.au/awards/miles-franklin
Contact *Petrea Salter*

Annual award, founded in 1957 by author Miles Franklin, for the novel of the year which is of the highest literary merit and which presents Australian life in any of its phases. All works must be in English and can be published in any country in the year preceding the award. More than one entry may be submitted by each author; a novel or play written by two or more authors in collaboration is eligible. Fee of A$75 for each entry; December deadline. See website for further information. 2009 winner: Tim Winton *Breath* (Penguin).

## MIND Book of the Year

Granta House, 15–19 Broadway, London E15 4BQ
☏ 020 8215 2301   🖷 020 8215 2269
j.bird@mind.org.uk
www.mind.org.uk

Established 1981. Annual award, in memory of Sir Allen Lane, for the author of a book published in the current year (fiction or non-fiction), which furthers public understanding of mental health problems. 2009 winner: Sathnam Sanghera *The Boy With the Topknot* (Penguin).

## Momaya Press Short Story Writing Competition

Momaya Press, Flat 1, 189 Balham High Road, Rear Building, London SW12 9BE
☏ 020 8673 9616
infouk@momayapress.com
www.momayapress.com
Contact *Monisha Saldanha*

Established 2004. Annual short story competition sponsored by Momaya Press. Open to writers in the English language worldwide. Entries, which should not have been published previously, can be in any style and format and on any subject; 2,500 words maximum. Entry fee: £7. Submissions by post or via entry form on the website.
💷 Prizes of £110 (1st); £60 (2nd); £30 (3rd); all winners are published in the *Momaya Annual Review*.

## nasen Awards 2011 – Celebrating Inclusive Practice

nasen House, 4–5 Amber Business Village, Amber Close, Amington, Tamworth B77 4RP
☏ 01827 311 500   🖷 01827 313 005

welcome@nasen.org.uk
www.nasen.org.uk

Awards presented in nine categories for 2011:
Book to Promote Professional Development;
ICT Accessibility (Software) Resource; ICT
Accessibility (Hardware) Resource; Book
to support teaching and learning – The
Pupil Book; Book to support teaching and
learning – The Educational Practitioner's Book;
Inclusive Resource for Primary Classrooms;
Inclusive Resource for Secondary Classrooms;
The Inclusive Children's Book; The Special
Educational Needs Academic Book. Deadline:
December 2010. Eligibility for 2011 Awards: books
and resources must have been published in the
UK between 1st May 2009 and December 2010.
Contact nasen for more information.
£ Prize of £500 for each category.

## National Association of Writers' Groups (NAWG)

Annual competitions with several categories. See
entry under *Professional Associations*.

## National Poetry Anthology

United Press, Admail 3735, London EC1B 1JB
☎ 0870 240 6190    📠 0870 240 6191
mail@unitedpress.co.uk
www.unitedpress.co.uk
Contact *Julie Embury*

Free-to-enter annual poetry competition.
Organizers United Press allow up to three poems
(20 lines and 160 words maximum each) by
annual closing date of 30th June. They select
around 250 regional winners in the UK. These
receive a free copy of the annual book, *The
National Poetry Anthology* and vote for the overall
winner who receives £1,000 and a trophy to keep
for life.

## National Poetry Competition

The Poetry Society, 22 Betterton Street, London
WC2H 9BX
☎ 020 7420 9895    📠 020 7240 4818
marketing@poetrysociety.org.uk
www.poetrysociety.org.uk
Contact *Competition Organizer (WH)*

One of Britain's major open poetry competitions.
Closing date: 31st October. Poems on any theme,
up to 40 lines. For rules and entry form send
s.a.s.e. to the competition organizer at the address
above or enter the competition via the website.
£ Prizes of £5,000 (1st); £1,000 (2nd); £500 (3rd);
plus 10 commendations of £50.

## 'The Nibbies'
▷ British Book Awards

## Nielsen Book Gold & Platinum Awards

Nielsen BookScan, 3rd Floor, Midas House, 62
Goldsworth Road, Woking GU21 6LQ
☎ 01483 712222    📠 01483 712220
awards.bookscan@nielsen.com
www.nielsenbookscan.co.uk
Head of Marketing *Mo Siewcharran*

Established in 2000 to award sales achieved for
actual consumer purchases of a title through
UK bookshops. Awarded to any title, in all its
editions, that sells more than 500,000 copies
(Gold) or one million copies (Platinum) over a
period of five consecutive years.
£ Award winner receives commemorative plaque
issued by the publisher to the author.

## Nobel Prize

The Nobel Foundation, PO Box 5232, 102 45
Stockholm, Sweden
☎ 00 46 8 663 0920    📠 00 46 8 660 3847
www.nobelprize.org
Contact *Information Section*

Awarded yearly for outstanding achievement
in physics, chemistry, physiology or medicine,
literature and peace. Founded by Alfred Nobel,
a chemist who proved his creative ability by
inventing dynamite. In general, individuals
cannot nominate someone for a Nobel Prize.
The rules vary from prize to prize but the
following are eligible to do so for Literature:
members of the Swedish Academy and of other
academies, institutions and societies similar
to it in constitution and purpose; professors of
literature and of linguistics at universities or
colleges; Nobel Laureates in Literature; presidents
of authors' organizations which are representative
of the literary production in their respective
countries. British winners of the literature prize,
first granted in 1901, include Rudyard Kipling,
John Galsworthy and Winston Churchill. Winners
since 2000: Gao Xingjian (France); V.S. Naipaul
(UK); Imre Kertész (Hungary); J.M. Coetzee
(South Africa); Elfriede Jelinek (Austria); Harold
Pinter (UK); Orhan Pamuk (Turkey); Doris
Lessing (UK); Jean-Marie Gustave Le Clézio
(France). Nobel Laureate in Literature 2009:
Herta Mueller (Germany).

## The Noma Award for Publishing Africa

PO Box 128, Witney OX8 5XU
☎ 01993 775235    📠 01993 709265
maryljay@aol.com
www.nomaaward.org
Contact *Mary Jay, Secretary to the Managing
Committee*

Established 1979. Annual award, founded by the
late Shoichi Noma, President of Kodansha Ltd,
Tokyo. The award is for an outstanding book,
published in Africa by an African writer, in three

categories: scholarly and academic; literature and creative writing; children's books. Entries, by publishers only, by 28th February for a title published in the previous year. Maximum number of three entries. Previous winners: Hamdi Sakkut *The Arabic Novel: Bibliography and Critical Introduction 1865–1995*; Elinor Sisulu *Walter and Albertina Sisulu. In Our Lifetime*; Werewere-Liking *La mémoire amputée*; Lebogang Mashile *In a Ribbon of Rhythm*; Shimmer Chinodya *Strife*; Zachariah Rapola *In a Ribbon of Rhythm*; Sefi Atta *Lawless and Other Stories*.
ⓔ Prize of US$10,000 and presentation plaque.

## Old Possum's Children's Poetry Competition

Fourth Floor, 2 Tavistock Place, London
WC1H 9RA
ⓣ 020 7833 9247    ⓕ 020 7833 5990
info@childrenspoetrybookshelf.co.uk
www.childrenspoetrybookshelf.co.uk
Contacts *Hilary Davidson, Chris Holifield*

Founded in 2005 under the name Children's Poetry Bookshelf Poetry Competition and sponsored by the Old Possum's Practical Trust. The judges were chaired by Michael Rosen in 2007 and 2008, and by Carol Ann Duffy in 2009. Annual poetry competition for the best individual poem written by children aged 7–8 and 9–11. The poem must be no more than 25 lines in length and on the subject of the theme of National Poetry Day. Individuals can enter their poems online, with their parents' consent, and teachers can send them by post.
ⓔ Cash prizes and books for the children and Children's Poetry Bookshelf memberships for their schools.

## Ondaatje Prize
▷ The Royal Society of Literature Ondaatje Prize

## Orange Prize for Fiction/Orange Award for New Writers

Booktrust, 45 East Hill, London SW18 2QZ
ⓣ 020 8516 2972    ⓕ 020 8516 2978
prizes@booktrust.org.uk
www.orangeprize.co.uk
Contact *Prizes Administrator*

Established 1996. Annual award founded by a group of senior women in publishing to 'create the opportunity for more women to be rewarded for their work and to be better known by the reading public'. Awarded for a full-length novel written in English by a woman of any nationality, and published in the UK between 1st April and 31st March of the following year. 2009 winner: Marilynne Robinson *Home* (Virago).
ⓔ Prize of £30,000 and a work of art (a limited edition bronze figurine known as 'The Bessie' in acknowledgement of anonymous prize endowment).
    Established 2005, the **Orange Award for**

**New Writers** is open to all first works of fiction, written by women of any age or nationality, published in the UK between 1 April and 31 March (short story collections and novellas also eligible). 2009 winner: Francesca Kay *An Equal Stillness* (Weidenfeld & Nicolson).

## The Orwell Prize

5/7 Vernon Yard, Portobello Road, London
W11 2DX
ⓣ 020 7229 5722
gavin.freeguard@mediastandardstrust.org
www.theorwellprize.co.uk
www.youtube.com/theorwellprize
http://orwelldiaries.wordpress.com
Administrator *Gavin Freeguard*

'Britain's most prestigious prize for political writing.' Each year, prizes are awarded to the works which come closest to George Orwell's ambition 'to make political writing into an art'. The Prize was established in its present form by the late Professor Sir Bernard Crick in 1993, via the Orwell Trust and *Political Quarterly*; the Media Standards Trust became a partner from the 2008 Prize, and Reuters, A.M. Heath and Richard Blair (Orwell's son) are sponsors. Submissions may be made by publishers, editors, agents or authors. 2009 winners (for work published in 2008): Andrew Brown *Fishing in Utopia* (Book); Patrick Cockburn *London Review of Books* and *The Independent* (Journalism); Jack Night (Special Prize for Blogs); Tony Judt (Special Prize for Lifetime Achievement). The Prize also organizes literary and political discussions around the country, and is publishing George Orwell's diary entries from 1938–42 as a blog (http://orwelldiaries.wordpress.com).
ⓔ Prize of £3,000 for each category.

## OWPG Awards for Excellence

Outdoor Writers & Photographers Guild,
1 Waterside, Garstang PR3 1HJ
ⓣ 01995 605340    ⓕ 0871 266 8621
secretary@owpg.org.uk
www.owpg.org.uk
Secretary *Jon Sparks*

Established 1980. Annual awards by the **Outdoor Writers & Photographers Guild** to raise the standard of outdoor writing, journalism, broadcasting and photography. Categories include guidebook, outdoor book, outdoor feature, travel feature, words and pictures, photography. Open to OWPG members only. Final entry date: April.

## The Oxford Weidenfeld Translation Prize

St Anne's College, Oxford OX2 6HS
ⓣ 01865 274820    ⓕ 01865 274895
sandra.madley@st-annes.ox.ac.uk
www.stannes.ox.ac.uk/about/translationprize.html
Contact *The Principals' Secretary*

Established in 1996 by publisher Lord Weidenfeld and now includes St. Anne's, Queen's and New colleges to encourage good translation into English. Annual award to the translator(s) of a work of fiction, poetry or drama written in any living European language. Translations must have been published in the previous calendar year. Submissions from publishers only.
£ Prize of £2,000.

## Paperbooks Publishing Writing Competition

Paperbooks Publishing Ltd, 2 London Wall Buildings, London Wall, London EC2M 5UU
info@legend-paperbooks.co.uk
www.paperbooks.co.uk

Paperbooks, the publishers of *A Tale of Two Halves*, are inviting writers to complete the book, the first half having been written by Gary Davison. There is no restriction on the style and format of the entries which should be a minimum of 10,000 words. Submissions can be completed inside the printed book or written in a Word document. Final entry date: 1st August 2010. Send submissions by post or by email with 'Writing Competition' in the subject line.

## Partners Annual Poetry Competition
▷ See entry under Poetry Magazines

## The Paul Foot Award

Private Eye, 6 Carlisle Street, London W1D 3BN
☎ 020 7437 4017    F 020 7437 0705
www.private-eye.co.uk
Contact *Sheila Molnar*

Annual award, established in 2005 by *The Guardian* and *Private Eye* for campaigning journalism in memory of former contributor Paul Foot. Submissions can be made by individual journalists, teams of journalists or publications for work appearing between October and August. Single pieces or entire campaigns are eligible. Final entry date: 1st September of the year in question. No broadcast material. 'Please do not send original material as no correspondence can be entered into.' 2009 winner: Ian Cobain (*The Guardian*).
£ Award of £5,000 with five runners-up receiving £1,000.

## PEN Awards

See **J.R. Ackerley Prize; Golden PEN Award for Lifetime Distinguished Service to Literature; The Hessell-Tiltman Prize for History; PEN/Pinter Prize**

## PEN/Pinter Prize

English PEN, Free Word Centre, 60 Farringdon Road, London EC1R 3GA
www.englishpen.org
Contact *Amy Oliver*

Established in 2009, in honour of the late playwright, Harold Pinter. Annual award presented to a British writer of outstanding literary merit whose work encapsulates the principles of freedom and truth that Pinter upheld throughout his writing career. Submissions are not accepted. Inaugural winner: Tony Harrison.

## Peter Wolff Theatre Trust supports the Whiting Award

Hampstead Theatre, Eton Avenue, London NW3 3EU
☎ 020 7749 4200
www.hampsteadtheatre.com
Contact *Neil Morris (General Manager)*

The original award was founded in 1965 by the Arts Council to commemorate the life and work of the playwright John Whiting (*The Devils, A Penny for a Song*). In 2008 it was taken over by a consortium of theatres, headed by Hampstead Theatre and funded by the Peter Wolff Theatre Trust. Annual, open award, for a play that is an original piece of writing (no adaptations or pieces of verbatim theatre), produced by a company in the subsidized sector and must have had its first performance in the award time period. The judges look for a play in which the writing is of special quality, of relevance and importance to contemporary life and of potential value to British theatre. Each year a different member of the consortium will host the award. 2009 winner: Alexi Kaye Campbell *The Pride*. The 2010 award is hosted by Nottingham Playhouse. For more information, contact *Bianca Winter*, General Manager, Theatre Writing Partnership (bianca@theatrewritingpartnership.org.uk).
£ Prize of £6,000.

## Poetry Business Competition

Bank Street Arts, 32–40 Bank Street, Sheffield S1 2DS
office@poetrybusiness.co.uk
www.poetrybusiness.co.uk
Contact *The Competition Administrator*

Established 1986. Annual award which aims to discover and publish new writers. Entrants should submit a short manuscript of poems. Winners will have their work published by the **Poetry Business** under the Smith/Doorstop imprint. Contact for conditions of entry. Previous winners include: Pauline Stainer, Michael Laskey, Mimi Khalvati, David Morley, Moniza Alvi, Selima Hill, Catherine Smith, Daljit Nagra.
£ Prizes: Publication of full collection; runners-up have pamphlets; 20 complimentary copies. Also cash prize (£1,000) to be shared equally between all winners.

## The Poetry Society's National Poetry Competition
▷ National Poetry Competition

## The Portico Prize for Literature
The Portico Library, 57 Mosley Street, Manchester M2 3HY
☎ 0161 236 6785   🖷 0161 236 6803
librarian@theportico.org.uk
www.theportico.org.uk
Contact *Miss Emma Marigliano*

Established 1985. Administered by the Portico Library in Manchester. Biennial awards for a work of fiction and of non-fiction published between the two closing dates. Set wholly or mainly in the North of England. Previous winners include: Anthony Burgess *Any Old Iron*; Jenny Uglow *Elizabeth Gaskell: A Habit of Stories*; Andrew Biswell *The Real Life of Anthony Burgess*.

## Practical Art Book of the Year
PO Box 3, Huntingdon PE28 0QX
☎ 01832 710201   🖷 01832 710488
award@acaward.com
www.acaward.com
www.artists-choice.co.uk
Contact *Henry Malt*

Jointly sponsored by **Artists' Choice** book club and *Leisure Painter* and *The Artist* magazines. A short list is drawn up by the editors and the winner voted by readers. Shortlist will be announced in June and the winner in November.

## The Premio Valle Inclán
▷ The Translation Prizes

## Prose & Poetry Prizes
The New Writer, PO Box 60, Cranbrook TN17 2ZR
☎ 01580 212626
admin@thenewwriter.com
www.thenewwriter.com
Contact *Merric Davidson*

Established 1997. Annual award. Open to all poets writing in the English language for an original, previously unpublished poem or collection of six to ten poems. Also open to writers of short stories and novellas/serials, features, articles, essays and interviews. Final entry date: 30th November. Full guidelines on the website. Previous winners: Clare Girvan, Leslie Forbes, Graham Clifford, Mario Petrucci, Ros Barber, Celia de Fréine, Andrew McGuinness, Katy Darby, David Grubb, Julia Widdows.
💷 Prizes total £2,000, plus publication in annual collection published by *The New Writer*.

## Pulitzer Prizes
The Pulitzer Prize Board, 709 Journalism Building, Columbia University, 2950 Broadway, New York, NY 10027, USA
☎ 001 212 854 3841   🖷 001 212 854 3342
pulitzer@pulitzer.org
www.pulitzer.org

Awards for journalism in US newspapers, and for published literature, drama and music by American nationals. 2009 winners include: Elizabeth Strout *Olive Kitteridge* (fiction); Jon Meacham *American Lion: Andrew Jackson in the White House* (biography); W.S. Merwin *The Shadow of Sirius* (poetry).

## The Red House Children's Book Award
The Federation of Children's Book Groups, 2 Bridge Wood View, Horsforth, Leeds LS18 5PE
☎ 0113 258 8910
info@fcbg.org.uk
www.redhousechildrensbookaward.co.uk
National Coordinator *Sinead Kromer*

Established 1980. Three categories. Awarded annually for best book of fiction suitable for children. Unique in that it is judged by the children themselves. 2009 overall winner: Sophie McKenzie *Blood Ties* (Simon & Schuster).
💷 Award winners receive silver bowls and trophy, portfolio of letters, drawings and comments from the children who take part in the judging.

## Reginald Taylor and Lord Fletcher Essay Prize
Institute of Archaelogy, c/o School of Art History, St Andrew's University, 79 North Street, St Andrew's KY16 9AR
Contact *Dr Julian Luxford*

A biennial prize, in memory of the late E. Reginald Taylor and of Lord Fletcher, for the best unpublished scholarly essay, not exceeding 7,500 words, on a subject of archaeological, art history or antiquarian interest within the period from the Roman era to AD 1830. The essay should show *original* research on its chosen subject, and the author will be invited to read the essay before the Association. The essay may be published in the journal of the Association if approved by the Editorial Committee. All enquiries by post, please. No phone calls. Send s.a.s.e. for details.
💷 Prize of £300 and a medal.

## Regional Press Awards
WLR Media and Entertainment, 3rd Floor, 6–14 Underwood Street, London N1 7QJ
☎ 020 7549 8671   🖷 020 7549 8668
acrane@wilmington.co.uk
www.pressgazette.co.uk

Open to all regional newspapers and regional journalists, whether freelance or staff. June event. Run by WLR Media & Entertainment.

## Renault UK Journalist of the Year Award

Guild of Motoring Writers, 40 Baring Road, Bournemouth BH6 4DT
☎ 01202 422424
generalsec@gomw.co.uk
www.guildofmotoringwriters.co.uk
Contact *Patricia Lodge*

Originally the Pierre Dreyfus Award, established in 1977. Awarded annually by Renault UK Ltd in honour of Pierre Dreyfus, president director general of Renault 1955–75, to the member of the **Guild of Motoring Writers** who is judged to have made the most outstanding journalistic effort in any medium during the year. Particular emphasis is placed on initiative and endeavour.

## RIBA International Book Awards

The RIBA International Book Awards have been suspended.

## RITA Awards

▷ Romance Writers of America under Professional Associations and Societies

## The Roald Dahl Funny Prize

Booktrust, Book House, 45 East Hill, London SW18 2QZ
☎ 020 8516 2972    📠 020 8516 2978
prizes@booktrust.org.uk
www.booktrust.org.uk
Contact *Prize Administrator*

Established in 2008 by Michael Rosen as part of his work as Children's Laureate, and administered by Booktrust. The prize is to honour those books that simply make us laugh and aims to reward those authors and artists who write and illustrate their books using humour in their stories, poetry and fiction. Books must have been first published in the UK or Ireland between 1st October and 31st September. Authors and illustrators must be British or Irish, or other nationals living in Britain or Ireland at the time of entering. The prize is awarded to the funniest book in two categories: for children aged six and under (picture book) and for children aged 7–14. 2009 winners: six and under: *Mr Pusskins Best in Show* Sam Lloyd (Orchard Books); 7–14: *Grubtown Tales: Stinking Rich and Just Plain Stinky* Philip Ardagh, illus. Jim Paillot (Faber).
💷 Prizes of £2,500 for each category.

## Romantic Novel of the Year

www.rna-uk.org
Honorary Administrator *Mrs P. Fenton*

Established 1960. Annual award for the best romantic novel of the year, open to non-members as well as members of the **Romantic Novelists' Association**. Novels must be first published in the UK between specified dates. Full details on the RNA website. Contact the organizer for entry form via the website address. To celebrate the 50th anniversary of the RNA, a new set of prizes were awarded in 2010: Romantic Novel of the Year: Lucy Dillon *Lost Dogs and Lonely Hearts* (Hodder); Love Story of the Year: Nell Dixon *Animal Instincts* (Little Black Dress); Romantic Film of the Year: *An Education*; Romantic Comedy: Jane Costello *The Nearly-Weds* (Simon & Schuster); The People's Choice: Louise Douglas *Missing You* (Pan); Harry Bowling Prize for New Writing: Debbie Johnson *Fear No Evil*; RNA Lifetime Achievement Awards: Maeve Binchy and Joanna Trollope.

## Rose Mary Crawshay Prize

The British Academy, 10 Carlton House Terrace, London SW1Y 5AH
☎ 020 7969 5200    📠 020 7969 5300
events@britac.ac.uk
www.britac.ac.uk
Contact *British Academy Chief Executive and Secretary*

Established 1888 by Rose Mary Crawshay, this prize is given for an historical or critical work to a woman of any nationality on English literature, with particular preference for a work on Keats, Byron or Shelley. The work must have been published in the preceding three years.

## Rossica Translation Prize

Academia Rossica, 4th Floor, 76 Brewer Street, London W1F 9TX
☎ 020 7287 2614    📠 020 7287 5712
rossica-prize@academia-rossica.org
www.academia-rossica.org
Contact *Svetlana Adjoubei*

Founded in 2005 by Academia Rossica (AR), a UK registered charity that supports cultural collaboration between Russia and the West. A biennial prize, which is awarded for the best new literary translation from Russian into English, published anywhere in the world. The aim is to promote the best of Russian literature and a better understanding of Russian culture in English-speaking countries. Four copies of the (published) English translation and of the Russian original should be submitted to Academia Rossica. 2009 winner: Amanda Love Darragh, translator of Maria Galina's novel, *Iramifications* (Glas).
💷 Prizes of £3,000 (translator); £1,000 (publisher).

## Royal Economic Society Prize

The Economic Journal, Department of Economics, London Business School, Regent's Park, London NW1 4SA
☎ 020 7000 8413    📠 020 7000 8401

econjournal@london.edu
www.res.org.uk/society.resprize.asp
Contact *Heather Daly*

Annual award for the best article published in
*The Economic Journal*. Open to members of the
Royal Economic Society only. Previous winners:
Professors Marcos Rangal, Tilman Börgers,
Christian Dustmann, Paul Cheshire, Stephen
Sheppard.
£ Prize of £3,000.

## Royal Mail Awards for Scottish Children's Books

Scottish Book Trust, Sandeman's House, Trunk's
Close, 55 High Street, Edinburgh EH1 1SR
☎ 0131 524 0160    📠 0131 524 0161
anna.gibbons@scottishbooktrust.com
www.scottishbooktrust.com
Children's Programme Manager *Anna Gibbons*

Awards are given to new and established authors
of published books in recognition of high
standards of writing for children in three age
group categories: younger children (0–7 years);
younger readers (8–11), older readers (12–16).
There is also a creative writing competition
in Gaelic and Scots for children. A shortlist is
drawn up by a panel of children's book experts
and the winner in each category is selected by
25,000 children and young people voting for their
favourites in schools and libraries across Scotland.
Authors should be Scottish or resident in
Scotland. Final entry date: 31st January; guidelines
available on request. 2009 winners: 0–7 years:
John Fardell *Manfred the Baddie* (Quercus); 8–11:
Lari Don *First Aid for Fairies and Other Fabled
Beasts* (Floris Books); 12–16: Keith Gray *Ostrich
Boys* (Random House).
£ Awards total £13,000.

## Royal Mail/Saltire Society Scottish First Book of the Year
▷ The Saltire Literary Awards

## Royal Society of Literature Awards

The Royal Society of Literature Ondaatje
Prize, V.S. Pritchett Memorial Prize and (under
*Bursaries, Fellowships and Grants*) The Royal
Society of Literature/Jerwood Awards

## The Royal Society of Literature Ondaatje Prize

The Royal Society of Literature, Somerset House,
Strand, London WC2R 1LA
☎ 020 7845 4676    📠 020 7845 4679
paulaj@rslit.org
www.rslit.org
Submissions *Paula Johnson*

Administered by the **Royal Society of Literature**
and endowed by Sir Christopher Ondaatje, the

prize is awarded annually to a book of literary
merit, fiction, non-fiction or poetry, best evoking
the spirit of a place. All entries must be published
within the calendar year 2009 and should be
submitted by mid-December. The writer must
be a citizen of the UK, Commonwealth or
Ireland. Further details available on the website.
2009 winner: Adam Nicolson *Sissinghurst: An
Unfinished History* (HarperPress).
£ Prize of £10,000.

## The Royal Society Prizes for Science Books

The Royal Society, 6–9 Carlton House Terrace,
London SW1Y 5AG
☎ 020 7451 2513    📠 020 7930 2170
sciencebooks@royalsociety.org
http://royalsociety.org
Head of Science Communication *Katherine Jarrett*

Annual prize established in 1988 to celebrate the
best in popular science writing. Awarded to books
that make science more accessible to readers of all
ages and backgrounds. Owned and managed by
the Royal Society and supported by the Beecroft
Trust. All entries must be written in English and
their first publication must have been between
1st January and 31st December; submission by
publishers only. Educational textbooks published
for professional or specialist audiences are not
eligible. 2009 winner: Richard Holmes *Age of
Wonder* (Harper Press).
£ Prizes total £15,000; £10,000 to the winner;
£1,000 to each of the five shortlisted authors.

## Runciman Award

The Anglo-Hellenic League, c/o The Hellenic
Centre, 16–18 Paddington Street, London
W1U 5AS
☎ 020 7486 9410
info@anglohellenicleague.org
www.anglohellenicleague.org
Contact *Dr Sophia Economides, Administrator*

Established 1985. Annual award, sponsored by
the National Bank of Greece. Founded by the
Anglo-Hellenic League to promote Anglo-Greek
understanding and friendship, for a work wholly
or mainly about some aspect of Greece or the
world of Hellenism, published in English in
any country of the world in its first edition
during 2009. It is a condition of the award
that shortlisted books should be available for
purchase to readers in the UK at the time of the
award ceremony. Named after the late Professor
Steven Runciman, former chairman of the
Anglo-Hellenic League. No category of writing
will be excluded from consideration. Works in
translation, with the exception of translations
from Greek literature, will not be considered.

Award presented in May/June. Additional information available on the website.
£ Award of £9,000.

## Saif Ghobash–Banipal Prize
▷ The Translation Prizes

## Sainsbury's Baby Book Award
▷ Booktrust Early Years Awards

## The Saltire Literary Awards

Saltire Society, 9 Fountain Close, 22 High Street, Edinburgh EH1 1TF
☎ 0131 556 1836    📠 0131 557 1675
saltire@saltiresociety.org.uk
www.saltiresociety.org.uk
Office Manager *Sarah Mason*

Established 1982 and 1988. Annual awards, one for Book of the Year and one for a First Book by an author publishing for the first time. Open to any author of Scottish descent or living in Scotland, or to anyone who has written a book which deals with either the work and life of a Scot or with a Scottish problem, event or situation. Nominations are invited from editors of leading newspapers, magazines and periodicals. 2009 winners: Saltire Society Scottish Book of the Year: Robert Crawford *The Bard: Robert Burns, A Biography* (Princeton University Press); Royal Mail/Saltire Society Scottish First Book of the Year: Eleanor Thom *The Tin-Kin* (Gerald Duckworth).
£ Prizes of £5,000 (Scottish Book); £1,500 (First Book).

## Samuel Johnson Prize for Non-Fiction
▷ The BBC Samuel Johnson Prize for Non-Fiction

## Samuel Pepys Award

Haremoor House, Faringdon SN7 8PN
Contact *Paul Gray*

Established 2003. Biennial award (closing date: 30th June 2011) given by the Samuel Pepys Award Trust for a book that makes the greatest contribution to the understanding of Samuel Pepys, his times or his contemporaries. 2009 winner: J.D. Davis *Pepys's Navy, Ships, Men and Warfare 1649–1689*.
£ Award of £2,000, plus commemorative medal.

## Schlegel-Tieck Prize
▷ The Translation Prizes

## Scott Moncrieff Prize
▷ The Translation Prizes

## The Scott Prize for Short Stories
▷ Salt Publishing Limited under UK Publishers

## Scottish Arts Council Book of the Year Awards
▷ Scottish Mortgage Investment Trust Book Awards

## Scottish History Book of the Year

The Saltire Society, 9 Fountain Close, 22 High Street, Edinburgh EH1 1TF
☎ 0131 556 1836    📠 0131 557 1675
saltire@saltiresociety.org.uk
www.saltiresociety.org.uk
Office Manager *Sarah Mason*

Established 1965. Annual award in memory of the late Dr Agnes Mure Mackenzie for a published work of distinguished Scottish historical research of scholarly importance (including intellectual history and the history of science). Editions of texts are not eligible. Nominations are invited and should be sent to the Office Manager. 2009 winner: Alexander Broadie *A History of Scottish Philosophy* (Edinburgh University Press).
£ Prize of £1,500.

## Scottish Mortgage Investment Trust Book Awards (In partnership with the Scottish Arts Council)

Scottish Arts Council, 12 Manor Place, Edinburgh EH3 7DD
☎ 0131 226 6051    📠 0131 225 9833
aly.barr@scottisharts.org.uk
www.scottisharts.org.uk
Literature Officer *Aly Barr*

Sponsored by Scottish Mortgage Investment Trust, these annual awards are the largest literary prizes of their kind in Scotland and the third biggest in the UK. Given in recognition of outstanding literary merit in fiction, poetry and literary non-fiction, there are four categories of awards: Fiction (including the short story); Poetry; Literary Non-Fiction; First Book. Authors should be Scottish or resident in Scotland, but books of Scottish interest by other authors are eligible for consideration. Submissions are made by publishers only, on behalf of their authors. Final closing date: 31st December for books published in that calendar year. Guidelines are available on request. The 2009 Book of the Year: *Kieron Smith, boy* by James Kelman (Hamish Hamilton).
£ £5,000 is given to the winning author of each category and the four winning authors are considered for the Scottish Mortgage Investment Trust Book of the Year, worth an additional £25,000.

## The Self-Publishing Awards

The David St John Thomas Charitable Trust, PO Box 6055, Nairn IV12 4YB
☎ 01667 453351    📠 01667 452365
dsjtcharitynairn@fsmail.net
Contact *Anne Hill*

Established in 1990 and run jointly by the David St John Thomas Charitable Trust and *Writers'*

*News*. Annual awards for books self-published in the previous year. Final entry date: 31st January. Full details available from the Trust at the address above. Previous winner: Meriel Buxton.

£ Awards total £2,250, plus two silver cups.

## SES Book Prizes

University of York, Department of Educational Studies, York YO10 5DD

☏ 01904 433452    🖷 01904 433459

id5@york.ac.uk

www.soc-for-ed-studies.org.uk

Acting Honorary Secretary *Professor Ian Davies*

Annual awards given by the Society for Educational Studies for the best books on education published in the UK during the preceding year. Nominations by publishers based in the UK.

£ Prizes of £2,000 (1st); £1,000 (2nd); £750 (highly commended).

## Seán Ó Faoláin Short Story Prize

The Munster Literature Centre, Frank O'Connor House, 84 Douglas Street, Cork, Republic of Ireland

☏ 00 353 21 431 2955

competitions@munsterlit.ie

www.munsterlit.ie

Contact *Patrick Cotter*

Founded 2003. Annual competition for an original, unpublished story of under 3,000 words. Final entry date: 31st July. Please visit the website for submission guidelines. 2009 winner: Alexa Beattie.

£ Awards of €1,500 (1st); €500 (2nd).

## The Sheridan Morley Prize

c/o Oberon Books, 521 Caledonian Road, London N7 9RH

☏ 020 7607 3637    🖷 020 7607 3629

info@oberonbooks.com

Contact *Berit*

Established in 2007/08 as a lasting memorial to Sheridan Morley, who during a long and illustrious career as critic, broadcaster and author, wrote more than 30 books. The prize celebrates the art of theatre biography and is awarded to the best biography on a theatrical subject, written in the English language and published in the given year. Final entry date: early March. 2009 winner: *A Strange Eventful History* Michael Holroyd (Vintage).

£ Prize of £2,000.

## Sid Chaplin Short Story Competition

Shildon Town Council, Civic Hall Square, Shildon DL4 1AH

☏ 01388 772563    🖷 01388 775227

Contact *Mrs J.M. Stafford*

Established 1986. Annual short story competition. Maximum 3,000 words; £2.50 entrance fee (juniors free). All stories must be unpublished and not broadcast and/or performed.

£ Prizes of £300 (1st); £150 (2nd); £75 (3rd); £30 (Junior).

## Sir Banister Fletcher Award

Authors' Club, 40 Dover Street, London W1S 4NP

☏ 020 7408 5092

stella@theartsclub.co.uk

Contact *Stella Kane*

Founded in 1954, this award was created by the late Sir Banister Fletcher, former President of the **Authors' Club** and the Royal Institute of British Architects, and is presented for 'the most deserving book on architecture or the arts'. The award is open to titles written by British authors or those resident in the UK and published under a British imprint. 2009 winner: Bruce Altshuler *Salon to Biennial – Exhibitions that Made Art History* (Phaidon Press).

## Sir Peter Ustinov Television Scriptwriting Award

Foundation of the International Academy of Television Arts & Sciences, 888 Seventh Avenue, 5th Floor, New York, NY 10019, USA

☏ 001 212 489 6969    🖷 001 212 489 6557

www.iemmys.tv

Contact *Award Administrator*

Established 1998. The late Sir Peter Ustinov gave his name to the Foundation's Television Scriptwriting Award. This annual competition is designed to motivate novice writers worldwide. The scriptwriter cannot be a United States citizen nor resident and must be below 30 years of age. Further details and entry form available on the website. 2009 winner: Claire Tonkin of Australia.

£ The award winner will receive US$2,500 and an invitation to the International Emmy Awards Gala.

## Sir William Lyons Award

The Guild of Motoring Writers, 40 Baring Road, Bournemouth BH6 4DT

☏ 01202 422424

generalsec@gomw.co.uk

www.guildofmotoringwriters.co.uk

Contact *Patricia Lodge*

An annual competitive award, sponsored by Jaguar, to encourage young people in automotive journalism and to foster interests into motoring and the motor industry. Entrance by two essays and interview with Awards Committee. Applicants must be British, aged 17–23 and resident in UK. Final entry date: 1st October.

## SJA British Sports Journalism Awards

> Sports Journalists' Association of Great Britain under Professional Associations and Societies

## The Society for Theatre Research Annual Theatre Book Prize

The Society for Theatre Research, PO Box 53971, London SW15 6UL
theatrebookprize@btinternet.com
www.str.org.uk

Established 1997. Annual award for books, in English, of original research into any aspect of the history and technique of British or British-related theatre. Not restricted to authors of British nationality nor books solely from British publishers. Books must be first published in English (no translations) during the calendar year. Play texts and those treating drama as literature are not eligible. Publishers submit books directly to the independent judges and should contact the Book Prize Administrator for further details. 2008 winner: *Theatre and Globalization: Irish Drama in the Celtic Tiger Era* Patrick Longeran (Palgrave Macmillan).
£ Award of £400.

## Sony Radio Academy Awards

Alan Zafer & Associates, 47–48 Chagford Street, London NW1 6EB
☎ 020 7723 0106    📠 020 7724 6163
secretariat@radioawards.org
www.radioawards.org
Contact *The Secretariat*

Established 1982 in association with the **Society of Authors** and Sony UK. Presented in association with the **Radio Academy**. Annual awards to recognize excellence in radio broadcasting. Entries must have been broadcast in the UK between 1st January and 31st December in the year preceding the award. The categories for the awards are reviewed each year and announced in November.

## Southport Writers' Circle Open Short Story Competition

16 Ormond Avenue, Westhead L40 6HT
☎ 01695 577938
southportwriterscircle@yahoo.co.uk
www.swconline.co.uk

Founded 2005. Open to fiction of any genre, up to 2,000 words, the story must be entered anonymously. No entry form required; include a cover sheet supplying title of story, name and contact details (send s.a.s.e. for results if no email). Fee: £3 per story (payable to Southport Writers' Circle). Final entry date: 31st October.
£ Prizes of £150 (1st); £75 (2nd); £25 (3rd).

## Southport Writers' Circle Poetry Competition

18 Ditchfield, Formby L37 4EQ
Contact *Steve Beattie*

For previously unpublished work. Entry fee: £3 per poem. Any subject, any form; maximum 40 lines. Closing date: end April. Poems must be entered anonymously, accompanied by a sealed envelope marked with the title of poem, containing s.a.s.e. Entries must be typed on A4 paper and be accompanied by the appropriate fee payable to Southport Writers' Circle. No application form is required. Envelopes should be marked 'Poetry Competition'. Postal enquiries only. No calls.
£ Prizes of £200 (1st); £100 (2nd); £50 (3rd); additional £25 Humour Prize.

## Strokestown International Poetry Competition

Strokestown International Poetry Festival Office, Strokestown, Co. Roscommon, Republic of Ireland
☎ 00 353 71 963 3759
www.strokestownpoetry.org
Administrator *Melissa Newman*

Annual poetry festival and competition. The festival takes place over the first weekend in May with readings and competitions. A centrepiece of the festival is the Strokestown International Poetry Competition for unpublished poems not exceeding 70 lines. Entry forms available on the website or call the Festival Office. Entry fee: €5.
£ Prizes of €4,000, €2,000 and €1,000 for a poem in English; €4,000, €2,000 and €1,000 for a poem in Irish or Scottish Gaelic.

## The Sunday Times EFG Private Bank Short Story Award

Booktrust, Book House, 45 East Hill, London SW18 2QZ
☎ 020 8516 2973
rosa@booktrust.org.uk
www.booktrust.org.uk/Prizes-and-awards/Sunday-Times-EFG-Award
Contact *Rosa Anderson*

Annual award for the best single short story by a UK or Irish author. Launched in 2009, this is the largest prize for an individual short story. Founded by Cathy Galvin, deputy editor of *The Sunday Times Magazine*, it is sponsored by *The Sunday Times* and EFG Private Bank and administered by **Booktrust**. Open to authors with a record of publication in fiction, drama, poetry or non-fiction in the UK and Ireland. Final entry date: 30th November; maximum wordage: 7,000; entries limited to one per author. Inaugural winner: C.K. Stead *Last Season's Man*.
£ £25,000 (1st); five runner-up prizes of £500 each.

## The Sunday Times Writer of the Year Award

The Sunday Times, 1 Pennington Street, London
E98 1ST

☎ 020 7782 5770    🖷 020 7782 5798
www.societyofauthors.org

Established 1987. Annual award to fiction and
non-fiction writers. The panel consists of *Sunday
Times* journalists and critics. Previous winners:
Anthony Burgess, Seamus Heaney, Stephen
Hawking, Ruth Rendell, Muriel Spark, William
Trevor, Martin Amis, Margaret Atwood, Ted
Hughes, Harold Pinter, Tom Wolfe, Robert
Hughes. No applications; prize at the discretion
of the Literary Editor.

## The Sunday Times Young Writer of the Year Award

The Society of Authors, 84 Drayton Gardens,
London SW10 9SB

☎ 020 7373 6642    🖷 020 7373 5768
info@societyofauthors.org
Contact *Awards Secretary*

Established 1991. Annual award given on the
strength of the promise shown by a full-length
published work of fiction, non-fiction, poetry
or drama. Entrants must be British citizens,
resident in Britain and under the age of 35. The
panel consists of *Sunday Times* journalists and
critics. Closing date: 31st October. The work must
be by one author, in the English language, and
published in Britain. Applications by publishers
via the **Society of Authors**. 2009 winner: Ross
Raisin *God's Own Country* (Penguin).

## T.S. Eliot Prize

Poetry Book Society, Dutch House, 307–308 High
Holborn, London WC1V 7LL

☎ 020 7833 9247    🖷 020 7833 5990
info@poetrybooks.co.uk
www.poetrybookshoponline.com
Contact *Chris Holifield*

Established in 1993 to mark the 40th anniversary
of the founding of the Poetry Book Society, the
T.S. Eliot prize is awarded by the PBS in January
of each year for the best new single author
collection of poetry published in the previous
year. Submissions must be made by the publisher
and are for a single-author poetry collection
published in the preceding calendar year in the
UK or Republic of Ireland. No self-published
work is eligible. Final entry date: beginning of
August. The prize-winners' cheques are given by
Mrs Valerie Eliot, who presents them at an awards
ceremony in central London. The T.S. Eliot Prize
Readings are held on the evening before the
ceremony. The Prize is distinctive in having a
Shadowing Scheme, run in association with the
English and Media Centre's *emagazine*. 2009

winner: Philip Gross *The Water Table* (Bloodaxe
Books).
💷 Prizes of £15,000 for the winner and £1,000 for
each of the ten shortlisted poets.

## T.S. Eliot Prize for Poetry

Truman State University Press, 100 East Normal
Avenue, Kirksville, MO 63501, USA
tsup@truman.edu
tsup.truman.edu
Contact *Nancy Rediger*

An annual award, established in 1997 in honour
of native Missourian, T.S. Eliot, to publish and
promote contemporary English-language poetry
regardless of a poet's nationality, reputation,
stage in career or publication history. Entry
requirements: 60–100 pages of original poetry
with $25 fee. Final entry date: 31st October. See
website for full entry details. 2009 winner: David
Moolten *Primitive Mood*.
💷 Prize of $2,000, plus publication.

## Theakstons Old Peculier Prize for the Crime Novel of the Year

Raglan House, Raglan Street, Harrogate HG1 1LE
☎ 01423 562303    🖷 01423 521264
crime@harrogatefestival.org.uk
www.harrogate-festival.org.uk/crime
Chief Executive *Sharon Canavar*
Manager *Erica Morris*

Established 2005. Sponsored by Theakstons Old
Peculier in association with Asda, the award is
open to full-length crime or mystery novels by
British authors, published in the UK. It is the only
crime fiction prize to be voted for by the general
public following announcement of a long list in
April. 2009 winner: (announced on the opening
night of the **Theakstons Old Peculier Crime
Festival, Harrogate**) Mark Billingham *Death
Message* (Sphere).
💷 Prize of £3,000 cash.

## The Times/Chicken House Children's Fiction Competition

The Chicken House, 2 Palmer Street, Frome
BA11 1DS
☎ 01373 454488    🖷 01373 454499
chickenhouse@doublecluck.com
www.doublecluck.com

Established 2007. Annual competition for
a finished novel, suitable for a worldwide
children's audience aged nine to 16, which, in the
opinion of the judges, demonstrates the greatest
entertainment value, quality and originality. Mss
of no more than 80,000 words should submitted
to the address above by the author (*no email
submissions*) who must be over 18. The novel must
be the original, unpublished work of the entrant.
No unsolicited mss. Full terms and conditions are

available on timesonline.co.uk/childrensauthor or write to The Times/Chicken House Children's Fiction Competition Ts&Cs request, Customer Liaison Department, 1 Pennington Street, London E98 1RL enclosing an s.a.s.e. 2009 winner: Sophia Bennett *Threads*.

£ The prize is the offer of a worldwide publishing contract with Chicken House, subject to negotiation and completion.

## The Tinniswood Award

The Writers' Guild of Great Britain, 40 Rosebery Avenue, London EC1R 4RX

T 020 7833 0777
anne@writersguild.org.uk
www.writersguild.org.uk
Contact *Anne Hogben*

Annual award established in 2004 by the **Society of Authors** and the **Writers' Guild of Great Britain** to perpetuate the memory of playwright Peter Tinniswood as well as to celebrate and encourage high standards in radio drama. Producers are invited to send in any radio drama first transmitted within the UK and Northern Ireland during the period 1st January to 31st December 2010. Submissions must come from the producers and are restricted to a maximum of *two entries only* per producer. The work must be an original piece for radio, and may also include the first episode from an original series or serial. If submitting 15-minute episodes from a series or serial, consecutive episodes are required (including the first episode) to make up at least 45 minutes. An adaptation for radio of a piece originally written for any other medium e.g. stage, television, film, novel, poem or a short story will not be eligible. 2009 winner: Peter Souter *Goldfish*.

£ Prize of £1,500, donated by the **ALCS**.

## The Tir Na N-Og Award

Cyngor Llyfrau Cymru (Welsh Books Council), Castell Brychan, Aberystwyth SY23 2JB

T 01970 624151   F 01970 625385
wbc.children@wbc.org.uk
www.wbc.org.uk

An annual award given to the best original book published for children in the year prior to the announcement. There are three categories: Welsh Language Book – Primary Sector; Welsh Language Book – Secondary Sector; Best English Book with an authentic Welsh background.

£ Awards of £1,000 (each category).

## The Tony Lothian Biographers' Club Prize

119A Fordwych Road, London NW2 3NJ
T 020 8452 4993
anna@annaswan.co.uk
www.biographersclub.co.uk
Contact *Anna Swan*

Recently renamed, the Tony Lothian Biographers' Club Prize aims to finance and encourage first-time writers researching a biography. Sponsored by the Duchess of Buccleuch, daughter of the late Tony Lothian. Open to previously unpublished writers, applicants should submit a proposal of no more than 20 pages, including outline, sample chapter, a note on the market for the book, sources consulted, competing books and c.v. Entry fee: £10 (cheques made payable to the Biographers' Club). 2009 winner: Harriet Tuckey *The Forgotten Hero of Everest*.

£ Prize of £2,000.

## The Translation Prizes

The Society of Authors, 84 Drayton Gardens, London SW10 9SB

T 020 7373 6642   F 020 7373 5768
pjohnson@societyofauthors.org
www.societyofauthors.org
Contact *Awards Secretary*

Various awards for published translations into English from Arabic (Saif Ghobash–Banipal Prize), Dutch and Flemish (The Vondel Translation Prize), French (Scott Moncrieff Prize), German (Schlegel-Tieck Prize), Modern Greek (Hellenic Foundation Prize), Italian (The John Florio Prize), Portuguese (The Calouste Gulbenkian Prize), Spanish (The Premio Valle Inclán) and Swedish (Bernard Shaw Translation Prize).

## Trevor Reese Memorial Prize

Institute of Commonwealth Studies, School of Advanced Study, University of London, 2nd Floor, South Block, Senate House, London WC1E 7HU

T 020 7862 8853   F 020 7862 8813
troy.rutt@sas.ac.uk
commonwealth.sas.ac.uk/reese.htm
Events & Publicity Officer *Troy Rutt*

Established in 1979 with the proceeds of contributions to a memorial fund to Dr Trevor Reese, Reader in Commonwealth Studies at the Institute and a distinguished scholar of Imperial History (d. 1976). Triennial award (next award, 2010 for works published 2007, 2008 and 2009) for a scholarly work, usually by a single author, which has made a wide-ranging, innovative and scholarly contribution in the broadly-defined field of Imperial and Commonwealth History. All correspondence relating to the prize should be marked 'Trevor Reese Memorial Prize'.

£ Prize of £1,000.

## V.S. Pritchett Memorial Prize

The Royal Society of Literature, Somerset House, Strand, London WC2R 1LA

T 020 7845 4676   F 020 7845 4679
info@rslit.org
www.rslit.org

Contact *Rachel Page*

Established 1999. Awarded for a previously unpublished short story of between 2,000 and 5,000 words. For entry forms, please contact *Rachel Page*. Open to UK and Irish writers. Winning entry to be published in *Prospect* magazine. 2009 winner: Kate Clanchy *The Not-Dead and the Saved*.
£ Prize of £1,000.

## Ver Poets Open Competition
181 Sandridge Road, St Albans AL1 4AH
☎ 01727 762601
gillknibbs@yahoo.co.uk
www.verpoets.org.uk
Contact *Gillian Knibbs*

Various competitions are organized by **Ver Poets**, the main one being the annual Open for unpublished poems of no more than 30 lines written in English. Entry fee: £3 per poem; £10 for four poems; £2 per poem thereafter. Entry form available from address above and/or website. Two copies of poems typed on A4 white paper. The anthology of winning and selected poems, and the adjudicators' report are normally available from mid-June. Final entry date: 30th April. Back numbers of the anthology are available for £4, post-free; one copy each free to those included.
£ Prizes of £500 (1st); £300 (2nd); £100 (3rd). Young Writers' Prize (16–21), £100.

## Verity Bargate Award
Soho Theatre, 21 Dean Street, London W1D 3NE
☎ 020 7287 5060    ℻ 020 7287 5061
www.sohotheatre.com

The award was set up to commemorate the late Verity Bargate, co-founder and director of the **Soho Theatre**. This national award is presented biennially for a new and unperformed play. Access details on the website. Previous winners include: In-Sook Chappell, Matt Charman, Shan Khan, Fraser Grace, Lyndon Morgans, Adrian Pagan, Diane Samuels, Judy Upton and Toby Whithouse.

## Vogue Talent Contest
Vogue, Vogue House, Hanover Square, London W1S 1JU
☎ 020 7152 3003    ℻ 020 7408 0559
Contact *Frances Bentley*

Established 1951. Annual award for young writers and journalists (under 25 on 1st January in the year of the contest). Final entry date is in April. Entrants must write three pieces of journalism on given subjects.
£ Prizes of £1,000, plus a month's paid work experience with *Vogue*; £500 (2nd).

## The Vondel Translation Prize
▷ The Translation Prizes

## W.J.M. Mackenzie Book Prize
Political Studies Association, Department of Politics, Newcastle University, Newcastle upon Tyne NE1 7RU
☎ 0191 222 8021    ℻ 0191 222 3499
psa@ncl.ac.uk
www.psa.ac.uk
Acting Company Secretary *Sandra McDonagh*

Established 1987. Annual award to best work of political science published in the UK during the previous year. Submissions from publishers only. Final entry date: 31st October. Prizes are judged in the year following publication and awarded the year after that. 2008 winner: Professor Philip Schofield *Utility & Democracy* (OUP).

## Wadsworth Prize for Business History
School of Management, University of Southampton, Building 2, Highfield, Southampton SO17 1BJ
R.A.Edwards@soton.ac.uk
Contact *Dr Roy Edwards*

Now in its 33rd year, the Wadsworth Prize is awarded annually by the Business Archives Council to an individual judged to have made an outstanding contribution to the study of British business history. Books are nominated by publishers. 2009 winners: Jock McCulloch and Geoffrey Tweedale *Defending the Indefensible: The Global Asbestos Industry and its Fight for Survival* (OUP).

## Wales Book of the Year Award
Academi, 3rd Floor, Mount Stuart House, Mount Stuart Square, Cardiff CF10 5FQ
☎ 029 2047 2266    ℻ 029 2047 0691
post@academi.org
www.academi.org
Contacts *Peter Finch (Chief Executive), Lleucu Siencyn (Deputy)*

Annual prize awarded for work of exceptional literary merit during the previous calendar year. There is one major prize in English, the Wales Book of the Year Award, and one major prize in Welsh, Gwobr Llyfr y Flwyddyn. 2009 winners: English language: Deborah Kay Davies *Grace, Tamar and Laszlo the Beautiful* (Parthian); Welsh language: William Owen Roberts *Petrograd* (Barddas).
£ Prizes of £10,000 for winners in both languages and four runner-up prizes of £1,000 each.

## Walford Award
555 Silbury Boulevard, Milton Keynes MK9 3HL
☎ 01908 254078

ISGRefAwards@cilip.org.uk
www.cilip.org.uk
Awards Administrator *Helen Bowlt*

Part of the ISG (CILIP) Reference Awards. Awarded to an individual who has made a sustained and continual contribution to the world of reference and information services over a period of years. The nominee need not be resident in the UK. The award is named after Dr A.J. Walford, a bibliographer of international repute. Previous winner: Dr Diana Dixon.
£ Cash prize and certificate.

## The Walter Scott Prize

c/o StonehillSalt PR, Haddington House, 28 Sidegate, Haddington EH41 1BJ
☎ 01620 829800    🖷 01620 829600
walterscottprize@bordersbookfestival.org
www.bordersbookfestival.org/walterscottprize
Contact *Rebecca Salt*

Founded in 2010, the Walter Scott Prize is open to historical novels first published in the UK in 2009. The prize, which is not restricted to Scottish books or authors, is funded by the Duke and Duchess of Buccleuch whose forebears were friends of Sir Walter Scott. Awarded at the 2010 **Borders Book Festival**, the £25,000 award puts the Walter Scott Prize in the top five biggest book awards in the UK.

## The Warwick Prize for Writing

The Writer's Room, Millburn House, University of Warwick, Warwick CV4 7AL
☎ 024 761 50708    🖷 024 765 24578
prizeforwriting@warwick.ac.uk
www2.warwick.ac.uk/fac/cross_fac/prizeforwriting
Contact *Helen May*

Founded in 2008, the award, which is run by the University of Warwick, is a biennial international award for a substantial piece of writing in the English language, in any genre or form, on a theme that will change with every award. Nominations are made by the current staff of the University, honorary graduates and honorary professors. Self nominations are ineligible. The theme for the 2010 award was Colour. Inaugural winner: Naomi Klein *The Shock Doctrine* (Penguin).
£ Award of £50,000 and the opportunity to take up a short placement at the University of Warwick.

## Waterstone's Children's Book Prize

Waterstone's, Capital Court, Capital Interchange Way, Brentford TW8 0EX
☎ 020 8742 3800    🖷 020 8742 0215
sarah.clarke@waterstones.com
waterstones.com
Contact *Sarah Clarke*

Founded 2005. An annual award which aims to support new children's authors and introduce them to a wide audience while introducing Waterstone's customers to some of the brightest new talent writing today. Books submitted should be fiction for children between the ages of seven and 13 with a focus on text rather than illustration. Open to new authors of any nationality with not more than two previously published fiction titles (adults' or children's). Publishers must declare any titles written under another name, including series fiction. All submitted titles must be available in paperback for a period of shortlist promotion and can be the first part of a series or trilogy but must also read as a stand alone novel. See website for submission deadline. 2010 winner: Katie Davies *The Great Hamster Massacre* (Simon & Schuster).
£ Prize of £5,000 and major promotion in Waterstone's.

## Waverton Good Read Award

8 Waverton Mill Quays, Waverton, Chester CH3 7PX
☎ 01244 336587/332053
gwen.goodhew@btinternet.com
petsmed@aol.com
www.wavertongoodread.org.uk
Contacts *Gwen Goodhew, Wendy Smedley*

Founded in 2003 and based on La Prix de la Cadière – an award made by the villagers of La Cadière d'Azur to the best French debut novel. WGRA is an annual award that aims to give readers of fiction opportunities to meet up and share their enthusiasms, to broaden their reading, to give ordinary readers a 'voice' in deciding what makes a 'good read', to provide a platform for new writers and to foster British talent. A team of up to 100 villagers reads and reviews submitted books. Over a period of six months the entries are whittled down to a short list of five. Copies of these are placed at meeting places around the village to encourage others to read them. Every reader is given a ballot paper and the winner is selected in July. Books must be debut novels by British (including those who have acquired British citizenship) and Irish writers and first published between 1st September of previous year and 31st August of current year. Final entry date: 31st August. Self-published books are not eligible. Previous winners: Mark Haddon *The Curious Incident of the Dog in the Night-Time*, Jonathan Trigell *Boy A*, Marina Lewycka *A Short History of Tractors in Ukrainian*, Nicola Monaghan *The Killing Jar*, Tom Rob Smith *Child 44*.
£ Sponsorship by Borders provides a prize of £1,000 to the winner, which is presented at a dinner during **Chester Literature Festival** in October/November.

## The Wellcome Trust Book Prize

Colman Getty, 28 Windmill Street, London
W1T 2JJ
☎ 020 7631 2666
bookprize@wellcome.ac.uk
www.wellcomebookprize.org
Contact *Lois Tucker*

Established in 2008, The Wellcome Trust
Book Prize is an annual award which aims to
stimulate interest, excitement and debate about
medicine and literature and will be awarded to
an outstanding work of fiction or non-fiction
that is centred around the theme of medicine.
The work must have been published in a given
12-month period. Submitted books, in English
or English translation, must be published by
a UK-based publisher during the prize year.
Reissues, anthologies and self-published books
are not eligible for the prize, and authors must
be living at the time of submission. Any book of
fiction or non-fiction will be considered, so long
as medicine or biomedical science is central to
its theme. This can include a collection of short
stories by the same author. Children's fiction may
also be eligible, as long as the book is published
by an adult imprint within the specified dates.
Inaugural winner: Andrea Gilles *Keeper: Living
with Nancy, A Journey into Alzheimer's* (Short
Books).
£ Prize of £25,000.

## The Wheatley Medal

Society of Indexers, Woodbourn Business Centre,
10 Jessell Street, Sheffield S9 3HY
☎ 0114 244 9561
admin@indexers.org.uk
www.indexers.org.uk
Chair *Ann Kingdom*

The Wheatley Medal is awarded for an
outstanding index and was established by the
**Society of Indexers** and the Library Association
(now **CILIP**) to highlight the importance of good
indexing and confer recognition on the most
highly regarded practitioners and their publishers.
First awarded in 1963, it has since been presented
for indexes to a wide range of publications, from
encyclopaedias to journals. Closing date for
entries 30th April.
£ Cash prize, plus medal and certificate.

## Whitbread Book Awards
▷ Costa Book Awards

## Whitfield Prize

Royal Historical Society, University College
London, Gower Street, London WC1E 6BT
☎ 020 7387 7532    📠 020 7387 7532

royalhistsoc@ucl.ac.uk
www.royalhistoricalsociety.org/grants.htm
Contact *Administrative Secretary*

Established 1977. An annual award for the best
new work within a field of British or Irish history,
published in the UK in the preceding calendar
year. The book must be the author's first (solely
written) history book and be an original and
scholarly work of historical research. Final entry
date: 31st December.
£ Prize of £1,000.

## Whiting Award
▷ The Peter Wolff Theatre Trust supports the Whiting
Award

## The Wilfred Owen Award for Poetry

21 Culverden Avenue, Tunbridge Wells TN4 9RE
☎ 01892 532712    📠 01892 532712
mmccrane@ukonline.co.uk
www.wilfredowen.org.uk
Contact *Meg Crane*

Established in 1998 by the **Wilfred Owen
Association**. Given to a poet whose poetry
reflects the spirit of Owen's work in its thinking,
expression and inspiration. Applications are not
sought; the decision is made by the Association's
committee. Previous winners: Seamus Heaney,
Michael Longley, Harold Pinter, Tony Harrison,
Dannie Abse, Professor Jon Stallworthy (2010
winner).
£ Award winner receives a work of art, suitably
decorated and engraved.

## William Hill Sports Book of the Year

Greenside House, Station Road, Wood Green,
London N22 7TP
☎ 020 8918 3731    📠 020 8918 3728
pressoffice@williamhill.co.uk
www.williamhillmedia.com
Contact *Graham Sharpe*

Established 1989. Annual award introduced by
Graham Sharpe of bookmakers William Hill.
Sponsored by William Hill and thus dubbed the
'Bookie' prize, it is the first, and only, Sports Book
of the Year award. Final entry date: September.
2009 winner: Duncan Hamilton *Harold Larwood*
(Quercus).
£ Prize (reviewed annually) of £25,000 package,
including £21,000 cash, hand-bound copy, £2,000
free bet. Runners-up prizes.

## Winchester Writers' Conference Prizes
▷ Winchester Writers' Conference, Festival, Bookfair
under UK and Irish Writers' Courses

## The Winston Graham Historical Prize

Royal Cornwall Museum, River Street, Truro
TR1 2SJ
☎ 01872 272205  📠 01872 240514
enquiries@royalcornwallmuseum.org.uk
www.royalcornwallmuseum.org.uk
Contact *Prize Administrator*

Established in 2008 the prize has been funded
by a generous legacy to the Royal Institution of
Cornwall from author Winston Graham. It will be
awarded in its inaugural year for an unpublished
work of historical fiction, preferably with a clear
connection to Cornwall. (NB The prize will not be
running in 2010.)

## Wisden (Cricket) Book of the Year

John Wisden & Co., 13 Old Aylesfield, Golden Pot,
Alton GU34 4BY
☎ 01420 83415
almanack@wisdengroup.com
www.wisden.com
Contact *Hugh Chevallier*

Founded 2002. Annual award. The Wisden Book
of the Year is chosen exclusively by the Wisden
Cricketers' Almanack book reviewer. In each
annual *Wisden Almanack*, a different person
is commissioned to review all cricket books
published in the previous calendar year, and that
reviewer selects the Wisden Book of the Year.
Any cricket book sent to John Wisden & Co.
for review in Wisden Cricketers' Almanack will
be considered. Final entry date: 1st November.
Previous winner: *Tom Cartwright: The Flame
Still Burns* by Stephen Chalke, selected by book
reviewer Patrick Collins.

## Wolfson History Prize

Wolfson Foundation, 8 Queen Anne Street,
London W1G 9LD
☎ 020 7323 5730  📠 020 7323 3241
www.wolfson.org.uk
Contact *Prize Administrator*

The Wolfson History Prize, established in 1972,
is awarded annually to promote and encourage
standards of excellence in the writing of history
for the general reading public. Submissions are
made through publishers. 2008 winners (awarded
in 2009): Mary Beard *Pompeii: The Life of a
Roman Town* (Profile Books) and Margaret M.
McGowan *Dance in the Renaissance: European
Fashion, French Obsession* (Yale University Press).

## The Writers Bureau Poetry and Short Story Competition

The Writers Bureau, Sevendale House, 7 Dale
Street, Manchester M1 1JB
☎ 0161 228 2362  📠 0161 228 3533
studentservices@writersbureau.com
www.wbcompetition.com

Competition Secretary *Vicky Bailey*

Established 1994. Annual award. Poems should be
no longer than 40 lines and short stories no more
than 2,000 words. £5 entry fee. Closing date: 30th
June 2010.
£ Prizes in each category: £1,000 (1st); £400
(2nd); £200 (3rd); £100 (4th); £50 (x 6).

## The Writers' Circle Antholoogy Trophy

The David St John Thomas Charitable Trust,
PO Box 6055, Nairn IV12 4YB
☎ 01667 453351  📠 01667 452365
dsjtcharitynairn@fsmail.net
Contact *Anne Hill*

Established in 1990 and run jointly by the David
St John Thomas Charitable Trust and *Writers'
News*. Annual prize for an anthology produced
by a writers' group. Final entry date: 31st January.
Full details available from the Trust at the
address above. Previous winner: Wingerworth
Wordsmiths.
£ Prize of £100 plus a silver cup.

## Writers' Forum Short Story & Poetry Competitions
▷ Writers' Forum under Magazines

## The Yale Drama Series/David C. Horn Prize

PO Box 209040, New Haven, CT 06520–9040,
USA
yalepress.yale.edu/yupbooks/drama.asp

Major annual playwriting competition intended
to support emerging playwrights. Submissions
must be original, unpublished full-length plays
written in English (translations, musicals and
children's plays are not accepted). Only one
manuscript may be submitted per year; plays that
have had professional productions are not eligible.
2011 submissions accepted between 1st June and
15th August 2010. Further information available
on the website. 2010 winner: Virginia Grise *blu*.
£ Prize winner receives the David C. Horn Prize
of $10,000, publication by Yale University Press
and a staged reading at Yale Rep.

## Yeovil Literary Prize

The Octagon Theatre, Hendford, Yeovil BA20 1UX
☎ 01935 422884 (messages only)
www.yeovilprize.co.uk
Administrator *Lizzie Pike*

Annual competition, established in 2004, to
stimulate interest in literary pursuits and to
provide funds for the Yeovil Community Arts
Association. 'Past finalists have achieved success
after winning the Prize.' Categories: novel (send
synopsis and opening pages; maximum 15,000
words); short story: (maximum 2,000 words);
poetry (maximum 40 lines). Entries, which can be
sent by post or online, must be in English and not

previously published; submissions from overseas welcome. Entrants must be over 18 years of age. Final entry date: end May.

£ Prizes: Novel: £1,000 (1st); £250 (2nd); £100 (3rd). Short Story/Poetry: £500 (1st); £200 (2nd); £100 (3rd).

## YoungMinds Book Award

Youngminds, 48–50 St John Street, London EC1M 4DG

T 020 7336 8445   F 020 7336 8446

hannah.smith@youngminds.org.uk
www.youngminds.org.uk/bookaward
Contact *Hannah Smith*

The award will be given to a work of fiction or biography for young people aged 12+, published between 1st June 2009 and 31st May 2010, which encourages their self esteem and helps them to cope with the stresses and challenges of growing up. 2009 winner: Chris Higgins, *A Perfect Ten* (Hodder).

£ Prize of £2,000.

# Professional Associations and Societies

## ABSW

Wellcome Wolfson Building, 165 Queen's Gate, London SW7 5HD
absw@absw.org.uk
www.absw.org.uk
Chair *Natasha Loder*
MEMBERSHIP £40 (full) p.a.; £36 (associate); £20 (student)

ABSW has worked to improve the standards of science journalism in the UK over the last 40 years. The Association seeks to improve standards by means of networking, lectures and organized visits to institutional laboratories and industrial research centres. A member of the European Union of Science Journalists' Associations, ABSW is able to offer heavily subsidized places on visits to research centres in most other European countries, and hosts reciprocal visits to Britain by European journalists. Membership open to those who are considered to be *bona fide* writers/editors, or their film/TV/radio equivalents, who work to promote public interest in and understanding of science. ABSW is hosting an annual conference in the summer of 2010.

## Academi (Yr Academi Gymreig)

3rd Floor, Mount Stuart House, Mount Stuart Square, Cardiff CF10 5FQ
☏ 029 2047 2266   ℻ 029 2049 2930
post@academi.org
www.academi.org
*Glyn Jones Centre:* Wales Millennium Centre, Bute Place, Cardiff Bay, Cardiff CF10 5AL
*North Wales office:* Tŷ Newydd, Llanystumdwy, Cricieth, Gwynedd LL52 0LW
☏ 01766 522811 ℻ 01766 523095
post@tynewydd.org
*West Wales office:* Dylan Thomas Centre, Somerset Place, Swansea SA1 1RR
☏ 01792 463980 ℻ 01792 463993
dylanthomas.centre@swansea.gov.uk
Chief Executive *Peter Finch*

Academi is the trading name of Yr Academi Gymreig, the Welsh National Literature Promotion Agency and Society for Writers. Yr Academi Gymreig was founded in 1959 as an association of Welsh language writers. An English language section was established in 1968. Membership, for those who have made a significant contribution to the literature of Wales, is by invitation. The Academi runs courses, competitions (including the **Cardiff International Poetry Competition**), conferences, tours by authors, festivals, and represents the interests of Welsh writers and Welsh writing both inside Wales and beyond. Its publications include *Taliesin*, a quarterly literary journal in the Welsh language, *The Oxford Companion to the Literature of Wales*, *The Welsh Academy English-Welsh Dictionary* and *The Welsh Academy Encyclopaedia of Wales*.

The Academi won the franchise from the Arts Council of Wales to run the Welsh National Literature Promotion Agency. In 2008 Academi won an ACW Beacon Company Award. The new, much enlarged organization now administers a variety of schemes including bursaries, the annual **Wales Book of the Year Award**, critical services, writers' mentoring, Writers on Tour, Writers Residencies and a number of literature development projects. It promotes an annual literary festival alternating between North and South Wales, runs its own programme of literary activity and publishes *A470*, a quarterly literature information magazine. Those with an interest in literature in Wales can become an Associate of the Academi (which carries a range of benefits). Rates are £15 p.a. (waged); £7.50 (unwaged).

## ALCS
▷ Authors' Licensing & Collecting Society Limited

## Alliance of Literary Societies

59 Bryony Road, Selly Oak, Birmingham B29 4BY
☏ 0121 475 1805
l.j.curry@bham.ac.uk
www.allianceofliterarysocieties.org.uk
Chair *Linda J. Curry*

Founded 1974. Aims to help and support its 100+ member societies and, when necessary, to act as a pressure group. Produces an annual journal (*ALSo*).

## Arts & Business (A&B)

Nutmeg House, 60 Gainsford Street, Butlers
Wharf, London SE1 2NY
☎ 020 7378 8143    ℻ 020 7407 7527
contactus@artsandbusiness.org.uk
www.artsandbusiness.org.uk
Director of Press & Public Affairs *Jonathan
Tuchner*
Marketing & Communications Director *Sebastian
Paul*

'We bring together business and the arts to create
a better future for both.' Arts & Business provides
consultancy, advice, research, training and events.
Offices across the UK; visit the website to find
your local office.

## Arvon Foundation
▷ entry under UK and Irish Writers' Courses

## ASLA
▷ The Association of Scottish Literary Agents

## The Association for Publishers and Providers of Consumer, Customer & Business Media in the UK
▷ PPA

## Association for Scottish Literary Studies

c/o Department of Scottish Literature, 7
University Gardens, University of Glasgow,
Glasgow G12 8QH
☎ 0141 330 5309    ℻ 0141 330 5309
office@asls.org.uk
www.asls.org.uk
General Manager *Duncan Jones*
SUBSCRIPTION £38 (individual); £67 (institutional)

Founded 1970. ASLS is an educational charity
promoting the languages and literature of
Scotland. Publishes works of Scottish literature;
essays, monographs and journals; and *Scotnotes*,
a series of comprehensive study guides to major
Scottish writers. Also produces *New Writing
Scotland*, an annual anthology of contemporary
poetry and prose in English, Gaelic and Scots (see
entry under *Magazines*).

## Association of American Correspondents in London

PO Box 645, Pinner HA5 9JJ
info@theaacl.co.uk
www.theaacl.co.uk
Secretary/Treasurer *Monique Jessen*
SUBSCRIPTION £110 (organizations); £50
(freelancers)

Founded in 1919 to serve the professional interests
of its member organizations, promote social
cooperation among them, and maintain the ethical
standards of the profession. (An extra £40 is
charged for each department of an organization
which requires separate listing on the Association's

website and a charge of £10 for each full-time
editorial staff listed, up to a maximum of £120
regardless of the number listed.)

## Association of American Publishers, Inc

71 Fifth Avenue, 2nd Floor, New York, NY 10003,
USA
☎ 001 212 255 0200    ℻ 001 212 255 7007
www.publishers.org
*Also at:* 50 F Street, NW, 4th Floor, Washington,
DC 20001 ☎ 001 202 347 3375 ℻ 001 202 347 3690

Founded 1970. For information, visit the
Association's website.

## Association of Authors' Agents (AAA)

c/o David Higham Associates Ltd, 5–8 Lower
John Street, Golden Square, London W1F 9HA
www.agentsassoc.co.uk
President *Anthony Goff*
MEMBERSHIP £150 p.a. (£75 for agencies with
fewer than 3 members of staff)

Founded 1974. Membership voluntary. The AAA
maintains a code of practice, provides a forum for
discussion and represents its members in issues
affecting the profession. For a full list of members
and a list of frequently asked questions visit the
AAA website. The AAA is a voluntary body and
unable to operate as an information service to the
public.

## Association of Authors' Representatives, Inc. (AAR)

www.aar-online.org
Contact *Administrative Secretary*

Founded in 1991 through the merger of the
Society of Authors' Representatives and the
Independent Literary Agents Association.
Membership of this US organization is restricted
to agents of at least two years' operation.
Provides information, education and support
for its members and works to protect their best
interests.

## Association of British Editors
▷ Society of Editors

## Association of British Science Writers
▷ ABSW

## Association of Canadian Publishers

174 Spadina Avenue, Suite 306, Toronto, Ontario
M5T 2C2, Canada
☎ 001 416 487 6116    ℻ 001 416 487 8815
admin@canbook.org
www.publishers.ca
Executive Director *Carolyn Wood*

Founded 1971. ACP represents 125 Canadian-
owned book publishers country-wide from the
literary, general trade, scholarly and education
sectors. Aims to encourage the writing,

publishing, distribution and promotion of Canadian books and to support the development of a 'strong, independent and vibrant Canadian-owned publishing industry'. The organization's website has information on getting published and links to many of their member publishers' websites. The ACP does not accept manuscripts and cannot provide assistance to authors who wish to find a Canadian publisher.

## Association of Christian Writers

23 Moorend Lane, Thame OX9 3BQ
☏ 01844 213673
admin@christianwriters.org.uk
www.christianwriters.org.uk
President *Adrian Plass*
Chairman *Lin Ball*
Administrator *Rev. Simon Baynes*
SUBSCRIPTION single: £22 (£18.50 Direct Debit); joint husband/wife: £25 (£21.50 DD); overseas: £30 (£25.50 DD on UK a/c)

Founded in 1971 'to inspire and equip men and women to use their talents and skills with integrity to devise, write and market excellent material which comes from a Christian worldview. In this way we seek to be an influence for good and for God in this generation'. Publishes a quarterly magazine. Runs three training events each year, biennial conference, competitions, postal workshops, area groups, prayer support and manuscript criticism. Charity No. 1069839.

## Association of Freelance Editors, Proofreaders & Indexers (Ireland)

11 Clonard Road, Sandyford, Dublin 16, Republic of Ireland
☏ 00 353 1 295 2194
brenda@ohanlonmedia.com
slq@ireland.com
www.afepi.ie
Co-chairs *Brenda O'Hanlon (at address above), Sine Quinn, 11 Summerville Place, Rathmines, Dublin 6; ☏ 00 353 1 497 7766)*
SUBSCRIPTION €50 p.a. (full); €35 (associate)

The organization was established in Ireland to protect the interests of its members and to provide information to publishers on freelancers working in the relevant fields. Full membership is restricted to freelancers with experience and/or references (but the association does not test or evaluate the skills of members). A new category of membership, Associate Member, is available for trainees in proofreading/editing who are taking the Publishing Training Centre correspondence courses in Proofreading and Copy-editing.

## Association of Freelance Writers

Sevendale House, 7 Dale Street, Manchester M1 1JB
☏ 0161 228 2362    ⒻⓉ 0161 228 3533

fmn@writersbureau.com
www.freelancemarketnews.com
Contact *Angela Cox*
SUBSCRIPTION £29 p.a.

Founded in 1995 to help and advise new and established freelance writers. Members receive a copy of *Freelance Market News* each month which gives news, views and the latest advice and guidelines about publications at home and abroad. Other benefits include one free appraisal of prose or poetry each year, reduced entry to **The Writers Bureau** writing competition, reduced fees for writing seminars and discounts on books for writers.

## Association of Golf Writers

1 Pilgrims Bungalow, Mulberry Hill, Chilham CT4 8AH
☏ 01227 732496    Ⓕ 01227 732496
enquiries@agwgolf.org
Honorary Secretary *Peter Dixon*
SUBSCRIPTION £50

Founded 1938. Aims to cooperate with golfing bodies to ensure best possible working conditions.

## Association of Illustrators

2nd Floor, Back Building, 150 Curtain Road, London EC2A 3AT
☏ 020 7613 4328    Ⓕ 020 7613 4417
info@theaoi.com
www.theaoi.com
Contact *Membership Coordinator*

The AOI is a non-profit-making trade association and members consist primarily of freelance illustrators as well as agents, clients, students and lecturers. As the only body to represent illustrators and campaign for their rights in the UK, the AOI has successfully increased the standing of illustration as a profession and improved the commercial and ethical conditions of employment. Organizes annual events programme and provides an advisory service for members. Publications include: triannual magazine, *Varoom – the journal of illustration and made images*; *The Illustrator's Guide to Law and Business Practice*; *The Illustrator's Guide to Success (not guaranteed)*, client directories (online) and *Images*, the only jury-selected annual of British contemporary illustration.

## Association of Independent Libraries

Leeds Library, 18 Commercial Street, Leeds LS1 6AL
☏ 0113 245 3071
enquiries@theleedslibrary.org.uk
www.independentlibraries.co.uk
Chairman *Geoffrey Forster*
Contact *Catherine Levy*

Established in 1989 to 'further the advancement, conservation and restoration of a little-known but important living portion of our cultural heritage'. Members include the **London Library, Devon & Exeter Institution, Linen Hall Library** and **Plymouth Proprietary Library.**

## Association of Learned and Professional Society Publishers

9 Stanbridge Road, London SW15 1DX

☎ 020 8789 2394    ℻ 0870 051 1921

nick.evansalpsp.org

www.alpsp.org

Chief Executive *Ian Russell*

Chief Operating Officer *Nick Evans*

Editor-in-Chief, Learned Publishing *Alan Singleton (editor@alpsp.org)*

The Association of Learned and Professional Society Publishers (ALPSP) is the international trade association for not-for-profit publishers and those who work with them. It currently has over 360 members in 36 countries. ALPSP provides representation of its sector, cooperative services such as the ALPSP Learned Journals Collection and ALPSP eBooks Collection, a professional development programme of training courses and seminars and a wealth of information and advice. ALPSP runs an annual international conference and awards programme. See the website for further information.

## The Association of Scottish Literary Agents (ASLA)

info@asla.org.uk

www.asla.org.uk

Chair *Jenny Brown (Jenny Brown Associates)*

Secretary/Treasurer *Lindsey Fraser (Fraser Ross Associates)*

Founded in 2009 as the national trade body for literary agents in Scotland. ASLA serves as the Scottish point of liaison with the UK-wide **Association of Authors' Agents.** Provides a forum to discuss matters of common professional interest. Membership is open to any agency or individual independent agent who has been actively engaged in representing authors for a period of at least one year.

## Association of Scottish Motoring Writers

32 Caledonia Crescent, Ardrossan KA22 8LW

☎ 07775 930686

spark@cfpress.co.uk

Secretary *Stephen Park*

SUBSCRIPTION £75

Founded 1961. Aims to co-ordinate the activities of, and provide shared facilities for, motoring writers resident in Scotland. Welcomes applications from active journalists.

## Audiobook Publishing Committee
▷ The Publishers Association

## Australian Copyright Council

PO Box 1986, Strawberry Hills, NSW 2012, Australia

☎ 00 612 8815 9777    ℻ 00 612 8815 9799

info@copyright.org.au

www.copyright.org.au

Contact *Customer Service*

Founded 1968. The Council's activities and services include producing a range of publications, organizing and speaking about copyright at seminars, and conducting research and consultancies. Aims include assistance for copyright owners to exercise their rights effectively, raising awareness about the importance of copyright and seeking changes to the law of copyright.

## Australian Publishers Association

60/89 Jones Street, Ultimo, NSW 2007, Australia

☎ 00 612 9281 9788    ℻ 00 612 9281 1073

www.publishers.asn.au

Chief Executive *Maree McCaskill*

SUBSCRIPTION Annual membership subscription open to Australian publishers

Founded 1948. The APA initiates programmes that contribute to the development of publishing in Australia, vigorously protects and furthers the interests of copyright owners, agents and licensees and actively represents members' interests to government and other organizations as appropriate. The Association encourages excellence in writing, editing, design, production, marketing and distribution of published works in Australia, protects freedom of expression and manages members' funds to further the interests of the industry.

## Australian Society of Authors

PO Box 1566, Strawberry Hills, NSW 2012, Australia

☎ 00 612 9318 0877    ℻ 00 612 9318 0530

asa@asauthors.org

www.asauthors.org

Executive Director *Dr Jeremy Fisher*

SUBSCRIPTION A$150 (full); A$110 (affiliate); A$350 (organization)

Founded in 1963, the Australian Society of Authors is the professional association for Australia's literary creators with 3,000 members in Australia and overseas.

## Authors' Club

40 Dover Street, London W1S 4NP

☎ 020 7499 8581

stella@theartsclub.co.uk

www.theartsclub.co.uk

Secretary *Stella Kane*

Founded in 1891 by Sir Walter Besant, the Authors' Club welcomes as members writers, agents, publishers, critics, journalists, academics and anyone involved with literature and the written word. Administers the **Authors' Club Best First Novel Award**, the **Dolman Best Travel Book Award** and **Sir Banister Fletcher Award**, and organizes regular talks and lunches with well-known guest speakers. Membership fee: apply to the Secretary.

## Authors' Licensing & Collecting Society Limited (ALCS)

The Writers' House, 13 Haydon Street, London EC3N 1DB

Ⓣ 020 7264 5700    Ⓕ 020 7264 5755

alcs@alcs.co.uk

www.alcs.co.uk

Chief Executive *Owen Atkinson*

SUBSCRIPTION One-off membership fee: £25 incl. VAT (UK; free to members of Society of Authors, Writers' Guild)

Founded 1977. The UK collective rights management society for writers and their successors, ALCS is a non-profit organization whose principal purpose is to ensure that hard-to-collect revenues due to writers are efficiently collected and speedily distributed. Established to give assistance to writers through the protection and exploitation of collective rights, ALCS has distributed over £205 million in secondary royalties to writers since its creation.

ALCS represents all types of writer, fiction and non-fiction, including educational, research and academic authors, scriptwriters, playwrights, poets, editors and freelance journalists across the print and broadcast media. On joining, members give ALCS a mandate to administer on their behalf those rights which the law determines must be received or which are best handled collectively. Chief among these are: photocopying, cable retransmission, rental and lending rights (but not British Public Lending Right), off-air recording, electronic rights, the performing right and public reception of broadcasts. The Society is a prime resource and a leading authority on copyright matters and writers' collective interests. It maintains a watching brief on all matters affecting copyright both in Britain and abroad, making representations to UK government authorities and the EU. Visit the ALCS website or contact the office for registration forms and further information.

## BAPLA (British Association of Picture Libraries and Agencies)

18 Vine Hill, London EC1R 5DZ

Ⓣ 020 7713 1780    Ⓕ 020 7713 1211

enquiries@bapla.org.uk

www.bapla.org.uk

Everything you need to know about finding, buying and selling pictures. Represents over 400 image supplying members.

## BASCA (British Academy of Composers and Songwriters)

2nd Floor, British Music House, 25–27 Berners Street, London W1T 3LR

Ⓣ 020 7636 2929    Ⓕ 020 7636 2212

info@basca.org.uk

www.basca.org.uk

Membership Manager *Graham Jackson*

BASCA exists to support and protect the artistic, professional, commercial and copyright interests of songwriters, lyricists and composers of all genres of music and to celebrate and encourage excellence in British music writing. Campaigns in the domestic, European and international political arenas on copyright and numerous other issues and aims to foster a sense of community amongst British songwriters, lyricists and composers. Provides a wide range of member support services through publications, websites and seminars. Administers the annual Ivor Novello Awards, the British Composer Awards and the Gold Badge Awards.

## BFI

21 Stephen Street, London W1T 1LN

Ⓣ 020 7255 1444    Ⓕ 020 7436 0439

www.bfi.org.uk

Chair *Greg Dyke*

Director *Amanda Nevill*

'There's more to discover about film and television through the BFI. Our world-renowned archive, cinemas, festivals, films, publications and learning resources are here to inspire you.' To find out more, visit the website.

## The Bibliographical Society

c/o Institute of English Studies, University of London, Senate House, Malet Street, London WC1E 7HU

Ⓣ 020 7862 8679    Ⓕ 020 7862 8720

admin@bibsoc.org.uk

www.bibsoc.org.uk

President *John Barnard*

Honorary Secretary *M.L. Ford*

SUBSCRIPTION £33 p.a.

Aims to promote and encourage the study and research of historical, analytical, descriptive and textual bibliography, and the history of printing, publishing, bookselling, bookbinding and collecting; to hold meetings at which papers are read and discussed; to print and publish works concerned with bibliography; to maintain a bibliographical library. Awards grants and bursaries for bibliographical research. Publishes a quarterly journal called *The Library*.

## Booksellers Association of the UK & Ireland Ltd

Minster House, 272 Vauxhall Bridge Road, London SW1V 1BA
☎ 020 7802 0802    📠 020 7802 0803
mail@booksellers.org.uk
www.booksellers.org.uk
Chief Executive *Tim Godfray*

Founded 1895. The BA helps 4,400 independent, chain and multiple retail outlets to sell more books, reduce costs and improve efficiency. It represents members' interests to the UK Government, European Commission, publishers, authors and others in the trade as well as offering marketing assistance, conferences, seminars and exhibitions. Together with **The Publishers Association**, coordinates World Book Day. Publishes directories, catalogues, surveys and various other publications connected with the book trade and administers the **Costa Book Awards**.

## Booktrust

Book House, 45 East Hill, London SW18 2QZ
☎ 020 8516 2977    📠 020 8516 2978
query@booktrust.org.uk
www.booktrust.org.uk
Director *Viv Bird*

Founded 1925. Booktrust is an independent national charity that encourages readers of all ages and cultures to discover and enjoy reading. Their website includes reviews of thousands of books, as well as information about Children's Book Week and the **Children's Laureate**, both of which are administered by Booktrust. Booktrust also administers literary prizes, including the **Orange Prize for Fiction**, the **John Llewellyn Rhys Prize**, the **Booktrust Teenage Prize** and **The Roald Dahl Funny Prize** (see entries under *Prizes*), and promotes reading through campaigns such as Get London Reading (www.getlondonreading.com), Story (www.theshortstory.org.uk) and a website about translated fiction (www.translatedfiction.org.uk). Booktrust runs Bookstart (www.bookstart.org.uk), the acclaimed national scheme that works through locally based organizations to give a free pack of books to babies, toddlers and pre-school children, along with guidance materials for parents and carers. They also run two other gifting schemes for older children (Booktime and Booked Up), as well as a scheme for looked after children (the Letterbox Club), and a programme to encourage writing in schools (Everybody Writes).

## British Academy of Composers and Songwriters
▷ BASCA

## British Association of Communicators in Business (CiB)

Suite GA2, Oak House, Woodlands Business Park, Linford Wood, Milton Keynes MK14 6EY
☎ 01908 313755    📠 01908 313661
enquiries@cib.uk.com
www.cib.uk.com
Chief Executive *Kathie Jones*

The professional body for in-house, freelance and agency staff involved in internal and corporate communications. Services to members include a business support helpline, an online forum, a monthly e-zine, the CiB Freelance Directory, an annual conference and gala dinner, automatic eligibility to FEIEA, a European network of 5,000 communicators.

## British Association of Journalists

89 Fleet Street, London EC4Y 1DH
☎ 020 7353 3003    📠 020 7353 2310
office@bajunion.org.uk
www.bajunion.org.uk
General Secretary *Steve Turner*
SUBSCRIPTION National newspaper staff, national broadcasting staff, national news agency staff: £17.50 per month. Other seniors, including magazine journalists and freelance writers and photographers who earn the majority of their income from journalism: £10 p.m. Journalists under 24: £7.50 p.m.

Founded 1992. Aims to protect and promote the industrial and professional interests of journalists.

## British Association of Picture Libraries and Agencies
▷ BAPLA

## British Centre for Literary Translation

University of East Anglia, Norwich NR4 7TJ
☎ 01603 592785    📠 01603 592737
bclt@uea.ac.uk
www.uea.ac.uk/bclt
Director *Amanda Hopkinson*

Founded 1989, BCLT is funded jointly by Arts Council England and the University of East Anglia. It aims to raise the profile of literary translation in the UK through events, publications, activities and research aimed at professional translators, students and the general reader. Member of the international RECIT literary translation network. Activities include the annual Sebald Lecture, Summer School and 'Translation Days' events. BCLT has a PhD programme in literary translation as well as an MA in Literary Translation. Joint sponsor with BCLA of the **John Dryden Translation Prize**.

Publishes a journal *In Other Words* and *New Books in German*. Free mailing list.

## British Copyright Council
Copyright House, 29–33 Berners Street, London W1T 3AB
☏ 01986 788122
secretary@britishcopyright.org
www.britishcopyright.org
Contact *Janet Ibbotson*

Works for the national and international acceptance of copyright and acts as a lobby/watchdog organization on behalf of creators, publishers and performers on copyright and associated matters.

## British Council
10 Spring Gardens, London SW1A 2BN
☏ 020 7389 3194  📠 020 7389 3199
arts@britishcouncil.org
www.britishcouncil.org/arts

The British Council uses cultural relations to address the need for intercultural dialogue, harnessing the power of education, the arts and creativity, sport, science, English and governance to build relationships of mutual benefit worldwide and to address global issues. 'Our vision says the world needs more cultural relations; our purpose is to respond.'

## British Equestrian Writers' Association
Priory House, Station Road, Swavesey, Cambridge CB24 4QJ
☏ 01954 232084  📠 01954 231362
gnewsumn@aol.com
Contact *Gillian Newsum*
SUBSCRIPTION £15

Founded 1973. Aims to further the interests of equestrian sport and improve, wherever possible, the working conditions of the equestrian press. Membership is by invitation of the committee. Candidates for membership must be nominated and seconded by full members and receive a majority vote of the committee.

## British Film Institute
▷ BFI

## British Guild of Beer Writers
Woodcote, 2 Jury Road, Dulverton TA22 9DU
☏ 01398 324314
tierneyjones@btinternet.com
www.beerwriters.co.uk
Secretary *Adrian Tierney-Jones*
SUBSCRIPTION £40 p.a.

Founded 1988. Aims to improve standards in beer writing and at the same time extend public knowledge of beers and brewing. Publishes an annual directory of members with details of their publications and their particular areas of expertise; this is then circulated to newspapers, magazines, trade press and broadcasting organizations. Also publishes a monthly newsletter, the *BGBW Newsletter*, edited by Adrian Tierney-Jones. As part of the plan to improve writing standards and to achieve a higher profile for beer, the Guild offers annual awards, including a Travel Bursary, to writers and broadcasters judged to have made the most valuable contribution towards this end in their work.

## British Guild of Travel Writers
26 Needwood House, Woodberry Down, London N4 2TN
☏ 020 8144 8713
secretariat@bgtw.org
www.bgtw.org
Chairman *Melissa Shales*
Secretariat *Robert Ellison*

The professional association of travel writers, broadcasters, photographers and editors which aims to serve its members' professional interests by acting as a forum for debate, discussion and 'networking'. The Guild publishes an annual Year Book and has a website giving full details of all its members and useful trade contacts, holds monthly meetings and has a monthly newsletter. Members are required to spend a significant proportion of their working time on travel.

## British Science Fiction Association
www.bsfa.co.uk

Founded originally in 1958 by a group of authors, readers, publishers and booksellers interested in science fiction. With a worldwide membership, the Association aims to promote the reading, writing and publishing of science fiction and to encourage SF fans to maintain contact with each other. Publishes *Matrix* e-zine, media review newsletter via the web with comment and opinions, news of conventions, etc. Contributions from members welcomed; *Vector* quarterly critical journal – reviews of books and magazines; *Focus* biannual magazine with articles, original fiction and letters column. Also offers postal and online writer's workshop. For further information, contact the Membership Secretary *Peter Wilkinson* via the website.

## British Society of Comedy Writers
61 Parry Road, Ashmore Park, Wolverhampton WV11 2PS
☏ 01902 722729  📠 01902 722729
comedy@bscw.co.uk
www.bscw.co.uk
Contact *Ken Rock, President*

Founded 1999. The Society aims to develop good practice and professionalism among comedy writers while bringing together the best creative professionals, and working to standards of excellence agreed with the light entertainment industry. Offers a network of industry contacts and a range of products, services and training initiatives including specialized workshops, an annual international conference, script assessment service and opportunities to visit international festivals.

## British Society of Magazine Editors (BSME)

137 Hale Lane, Edgware HA8 9QP
☎ 020 8906 4664    🖷 020 8959 2137
admin@bsme.com
www.bsme.com
Contact *Gill Branston*

Holds regular industry forums and events as well as an annual awards dinner.

## Broadcasting Press Guild

c/o Torin Douglas, Room 7540, BBC Television Centre, Wood Lane, London W12 7RJ
☎ 020 8624 9052
torin.douglas@bbc.co.uk
www.broadcastingpressguild.org
Membership Secretary *Tim Dams*
Lunch Secretary & Treasurer *Torin Douglas*
SUBSCRIPTION £15 p.a.

Founded 1974 to promote the professional interests of journalists specializing in writing or broadcasting about the media. Organizes monthly lunches addressed by leading industry figures, and annual TV and radio awards. Membership by invitation.

## BSME
▷ British Society of Magazine Editors

## Bureau of Freelance Photographers

Focus House, 497 Green Lanes, London N13 4BP
☎ 020 8882 3315    🖷 020 8886 3933
info@thebfp.com
www.thebfp.com
Membership Secretary *Angela Kidd*
SUBSCRIPTION £54 p.a. (UK); £70 p.a. (overseas)

Founded 1965. Assists members in selling their pictures through monthly *Market Newsletter*, and offers advisory, legal assistance and other services.

## Campaign for Press and Broadcasting Freedom

Second Floor, 23 Orford Road, London E17 9NL
☎ 020 8521 5932
freepress@cpbf.org.uk
www.cpbf.org.uk
SUBSCRIPTION £15 p.a. (concessions available); £25 p.a. (institutions/organizations)

Broadly based pressure group working for more accountable and accessible media in Britain. Members receive *Free Press* (bi-monthly), discounts on publications and news of campaign progress.

## Canadian Authors Association

*National Office:* PO Box 581, Station Main, Orillia, Ontario L3V 6K5, Canada
☎ 001 705 653 0323    🖷 001 705 653 0593
admin@canauthors.org
www.canauthors.org
Interim Executive Director *Anita Purcell*

Founded 1921. The CAA was founded to promote recognition of Canadian writers and their works, and to foster and develop a climate favourable to the creative arts. It is Canada's national association for writers of every kind; for those actively seeking to become writers and for those who want to support writers. The Association has branches across the country providing support to local members in the form of meetings, workshops, writers-in-residence, advice, local contests, publications and writers' circles. Publishes *National Newsline* (quarterly) and *The Canadian Writer's Guide*. 'The motto is Writers Helping Writers.'

## Canadian Federation of Poets

1248 E. Elmwood Avenue, Burbank, CA 91501, USA
info@federationofpoets.com
www.federationofpoets.com
President *Tracy Repchuk*

Founded 2003. Federations and members from around the world can post messages, 'meet' via forums and live chats and learn about themselves and their international colleagues with regard to poetry. Membership includes poetry workshops, anthologies, publishing opportunities, online forums, free book promotion, calendar of events and free subscription to *Poetry Canada* magazine (www.poetrycanada.com).

## Canadian Publishers' Council

250 Merton Street, Suite 203, Toronto, Ontario M4S 1B1, Canada
☎ 001 416 322 7011    🖷 001 416 322 6999
www.pubcouncil.ca
Executive Director, External Relations *Jacqueline Hushion*

Founded 1910. Trade association of English-language publishers which represents the domestic and international interests of member companies.

## Chartered Institute of Journalists

2 Dock Offices, Surrey Quays Road, London SE16 2XU
☎ 020 7252 1187    🖷 020 7232 2302
memberservices@cioj.co.uk
www.cioj.co.uk
General Secretary *Dominic Cooper*

SUBSCRIPTION £195 p.a.; £16 (monthly)

Founded 1884. The Institute is concerned with professional journalistic standards and with safeguarding the freedom of the media. It is open to writers, broadcasters and journalists (including self-employed) in all media. Affiliate membership (£133) is available to part-time or occasional practitioners and to overseas journalists who can join the Institute's International Division. Non-employing members also belong to the IOJ (TU), an independent trade union which protects, advises and represents them in their employment or freelance work; negotiates on their behalf and provides legal assistance and support. The IOJ (TU) is a certificated independent trade union which represents members' interests in the workplace, and the CIOJ is also a constituent member of the Media Society, the **British Copyright Council** and the Journalists Copyright Fund. Editor of the Institute's quarterly magazine is *Andrew Smith*.

## Chartered Institute of Linguists (IoL)

Saxon House, 48 Southwark Street, London SE1 1UN

☎ 020 7940 3100    🖷 020 7940 3101
info@iol.org.uk
www.iol.org.uk
Chief Executive *John Hammond*
SUBSCRIPTION Rates on application

Founded 1910. Professional association for translators, interpreters and trainers; examining body for languages at degree level and above for vocational purposes; the National Register of Public Service Interpreters is managed by NRPSI Limited, an IoL subsidiary. The Institute's limited company, Language Services Ltd, provides customized assessments of language-oriented requirements, skills, etc.

## Children's Books Ireland

17 North Great George's Street, Dublin 1, Republic of Ireland
☎ 00 353 1 872 7475    🖷 00 353 1 872 7476
info@childrensbooksireland.ie
www.childrensbooksireland.ie
Contact *Mags Walsh*
SUBSCRIPTION €30 p.a. (individual); €50 (institution); €20 (student); €45/US$40 (overseas)

Irish national children's books organization. Holds annual conferences as well as other occasional seminars and events. Quarterly magazine, *Inis*. Annual children's book festival in October; **CBI Bisto Book of the Year Award** (see entry under *Prizes*). Partners other organizations to promote and develop ways of bringing young people and books together.

## CiB
▷ British Association of Communicators in Business

## CILIP: The Chartered Institute of Library and Information Professionals

7 Ridgmount Street, London WC1E 7AE
☎ 020 7255 0500    🖷 020 7255 0501
info@cilip.org.uk
www.cilip.org.uk
Chief Executive *Bob McKee*

CILIP is the leading professional body for librarians, information specialists and knowledge managers. CILIP provides practical support for members throughout their entire careers, helping them with their academic education, professional qualifications, job hunting and continuing professional development. **Facet Publishing**, CILIP's publishing arm, produces books for library and information professionals worldwide.

## CILIP Wales

University of Wales Aberystwyth, Department of Information Studies, Llanbadarn Fawr, Aberystwyth SY23 3AS
☎ 01970 622174    🖷 01970 622190
mdp@aber.ac.uk
www.cilip.org.uk/walescymru
Development Officer *Mandy Powell (07837 032536)*

The Chartered Institute of Library and Information Professionals Wales was founded in 2002 as a branch of **CILIP**. Publishes a thrice-yearly journal, *Y Ddolen*.

## Circle of Wine Writers

5 Ingatestone Hall Cottages, Ingatestone CM4 9NS
julie.arkell@btinternet.com
www.winewriters.org
Administrator *Andrea Warren*
MEMBERSHIP £50 p.a.

Founded 1960. Membership is open to all those professionally engaged in communicating about wines and spirits, with the exception of people primarily doing so for promotional purposes. Aims to improve the standard of writing, broadcasting and lecturing about wines, spirits and beers; to contribute to the growing knowledge and interest in wine; to promote wines and spirits of quality and to comment adversely on faulty products or dubious practices; to establish and maintain good relations with the news media and the wine trade; to provide members with a strong voice with which to promote their views; to provide a programme of workshops, meetings, talks and tastings.

## CLÉ – Irish Book Publishers' Association

Guinness Enterprise Centre, Taylor's Lane,
Dublin 8, Republic of Ireland
℡ 00 353 1 415 1210
info@publishingireland.com
www.publishingireland.com
President *Alan Hayes*
Administrator *Karen Kenny*
Project Manager *Jolly Ronan*

Founded 1970 to promote Irish publishing,
protect members' interests and train the industry.

## Comhairle nan Leabhraichean/The Gaelic Books Council

22 Mansfield Street, Glasgow G11 5QP
℡ 0141 337 6211   ℻ 0141 341 0515
brath@gaelicbooks.net
www.gaelicbooks.org
www.ur-sgeul.com
Chairman *Professor Roibeard Ó Maolalaigh*
Director *Ian MacDonald*

Founded 1968 and now a charitable company
with its own bookshop. Encourages and
promotes Gaelic publishing by giving grants to
publishers and writers; providing editorial and
word-processing services; retailing Gaelic books;
producing a catalogue of all Gaelic books in print
and answering enquiries about them; mounting
occasional literary evenings and training courses.
Stock list on website.

## The Copyright Licensing Agency Ltd

Saffron House, 6–10 Kirby Street, London
EC1N 8TS
℡ 020 7400 3100/0800 085 6644 (new licences)
℻ 020 7400 3101
cla@cla.co.uk
licence@cla.co.uk (new licences)
www.cla.co.uk
Chief Executive *Kevin Fitzgerald*

CLA is the UK's reproduction rights
organization which looks after the interests
of authors, publishers and visual creators in
the photocopying and scanning of extracts
from books, journals, magazines and digital
publications. Founded in 1983 by the **Authors'
Licensing and Collecting Society Limited
(ALCS)** and the **Publishers Licensing Society
Ltd (PLS)** to promote and enforce intellectual
property rights of UK rightsholders both at
home and abroad. CLA works closely with
Reproduction Rights Organizations (RROs) from
other countries and has an agency agreement
with the **Design and Artists Copyright Society
(DACS)** which represents artists and illustrators.
A not-for-profit organization, CLA licenses the
business, education and government sectors and
distributes fees collected to authors, publishers
and visual creators. CLA has also developed
licences which enable digitization of existing
print material. Licences allow photocopying and
scanning from original print material as well as
copying from born digital publications. Scanning
and digital re-use of publications is available for
works from the UK, US, Australia, Spain, Ireland,
South Africa and Canada. Since its inception in
1983, CLA has distributed over £450 million.

## Council for British Archaeology

St Mary's House, 66 Bootham, York YO30 7BZ
℡ 01904 671417   ℻ 01904 671384
info@britarch.ac.uk
www.britarch.ac.uk
Head of Information & Communications *Dan
Hull*
Publications Officer *Catrina Appleby*

Founded in 1944 to represent and promote
archaeology at all levels. It is an educational
charity working throughout the UK to involve
people in archaeology and to promote the
appreciation and care of the historic environment
for the benefit of present and future generations.
The CBA publishes a wide range of academic,
educational and general works about archaeology,
ranging from monographs and handbooks to
online articles and the popular magazine, *British
Archaeology* (see **CBA Publishing** under *UK
Publishers*).

## Crime Writers' Association (CWA)

PO Box 273, Boreham Wood WD6 2XA
secretary@thecwa.co.uk
www.thecwa.co.uk
Hon. Secretary *Liz Evans*
Membership Secretary *Rebecca Tope (Crossways
Cottage, Walterstone HR2 0DX)*
MEMBERSHIP £50, £65 (associate); £50/$100
(overseas)

Full membership is limited to professional
crime writers, but publishers, literary agents,
booksellers, etc. who specialize in crime,
are eligible for Associate membership. The
Association has regional chapters throughout the
country, including Scotland. Meetings are held
regularly with informative talks frequently given
by police, scenes of crime officers, lawyers, etc.,
and a weekend conference is held annually in
different parts of the country. Produces a monthly
newsletter for members called *Red Herrings* and
presents various annual awards (see entries under
*Prizes*).

## The Critics' Circle

c/o Catherine Cooper, 69 Marylebone Lane,
London W1U 2PH
℡ 020 7224 1410
www.criticscircle.org.uk
President *Charles Spencer*

Honorary General Secretary *William Russell*
SUBSCRIPTION £25 p.a.

Membership by invitation only. Aims to uphold and promote the art of criticism (and the commercial rates of pay thereof) and preserve the interests of its members: professionals involved in criticism of film, drama, music, dance and art.

### Cyfnewidfa Lên Cymru
▷ Wales Literature Exchange

### Cyngor Llyfrau Cymru
▷ Welsh Books Council

### DACS
▷ Design and Artists Copyright Society

### Data Publishers Association
Queen's House, 28 Kingsway, London WC2B 6JR
☎ 020 7405 0836
info@dpa.org.uk
www.dpa.org.uk
Contact *Sarah Gooch*
SUBSCRIPTION £450 to £2450 p.a.

Since 1970, the DPA has existed to serve, promote and protect the interests of all companies operating in the directory, data and search publishing sector. Today, the industry contributes well over £1 billion to the UK economy. The DPA membership not only includes mainstream data providers but also a whole host of organizations providing services and products to the data provision and publishing sector, such as printers, software solution providers and contract directory publishers. The DPA's members are predominantly UK companies but it also offers membership to overseas companies with UK interests. It provides its members with opportunities to develop their businesses by providing a complete and diverse range of services, including communications, networking, training and seminars, statistics, representation to government, publicity and promotion, legal support and recognition.

### Department for Culture, Media and Sport
2–4 Cockspur Street, London SW1Y 5DH
☎ 020 7211 6200    🖷 020 7211 6270
michael.panayi@culture.gsi.gov.uk
www.culture.gov.uk
Director of Communications *Jane Cooper*

The Department for Culture, Media and Sport has responsibilities for Government policies relating to the arts, museums and galleries, sport, the 2012 Olympic & Paralympic Games, gambling, broadcasting, Press freedom and regulation, the built historic environment, the film and music industries, tourism and the National Lottery. It funds the Arts Council, national museums and galleries, the British Library, the **Public Lending**

**Right** and the Royal Commission on Historical Manuscripts. It is responsible within Government for the public library service in England, and for library and information matters generally, where they are not the responsibility of other departments.

### Design and Artists Copyright Society (DACS)
33 Great Sutton Street, London EC1V 0DX
☎ 020 7336 8811    🖷 020 7336 8822
info@dacs.org.uk
www.dacs.org.uk
Chief Executive *Gilane Tawadros*

DACS is the UK's copyright and collecting society for artists and visual creators. Represents 26,000 fine artists and their heirs as well as photographers, illustrators, craftspeople, cartoonists, architects, animators and designers.

### Drama Association of Wales
The Old Library, Singleton Road, Splott, Cardiff CF24 2ET
☎ 029 2045 2200    🖷 029 2045 2277
teresa@dramawales.org.uk
www.dramawales.org.uk
Contact *Teresa Hennessy, Member Services Officer*

Runs a large playscript lending library; holds an annual playwriting competition (see entry under *Prizes*); offers a script-reading service (£25 full mss; £17.50 one-act play) which usually takes one month from receipt of play to issue of reports. From plays submitted to the reading service, selected scripts are considered for publication of a short run (70–200 copies). Writers receive a percentage of the cover price on sales and a percentage of the performance fee.

### Edinburgh Bibliographical Society
Centre for Research Collections, Edinburgh University Library, George Square, Edinburgh EH8 9LJ
joseph.marshall@ed.ac.uk
http://mcs.qmuc.ac.uk/EBS/
Treasurer *Dr Joseph Marshall*
SUBSCRIPTION £15; £20 (institution); £10 (student)

Founded 1890. Organizes lectures on bibliographical topics and visits to libraries. Publishes an annual journal, which is free to members, and other occasional publications.

### Educational Publishers Council
▷ The Publishers Association

### Electronic Publishers' Forum
▷ The Publishers Association

## The English Association

University of Leicester, University Road, Leicester
LE1 7RH
☎ 0116 252 3982    📠 0116 252 2301
engassoc@leicester.ac.uk
www.le.ac.uk/engassoc
Chief Executive *Helen Lucas*
Administrator *Julia Hughes*

Founded 1906 to promote understanding and appreciation of the English language and its literatures. Activities include sponsoring a number of publications and organizing lectures and conferences for teachers, plus annual sixth-form conferences. Publications include *The Year's Work in Critical and Cultural Theory*; *English, The Use of English*; *English 4–11*; *Essays and Studies* and *The Year's Work in English Studies*. Also runs an annual poetry competition, The Fellows' Poetry Prize.

## English PEN

Free Word Centre, 60 Farringdon Road, London
EC1R 3GA
☎ 020 7324 2535    📠 020 7490 0566
enquiries@englishpen.org
www.englishpen.org
Director *Jonathan Heawood*
MEMBERSHIP Cheque/standing order: £45/£40 (London/overseas); £40/£35 (country); £15 (students)

English PEN is part of International PEN, a worldwide association of writers and other literary professionals which promotes literature, fights for freedom of expression and speaks out for writers who are imprisoned or harassed for having criticized their governments, or for publishing other unpopular views.

Founded in London in 1921, International PEN now consists of over 145 centres in almost 100 countries. PEN originally stood for poets, essayists and novelists, but membership is now open to all literary professionals. It is also possible to become a 'Friend of English PEN'. A programme of talks and discussions, and other activities such as social gatherings, is supplemented by mailings, website and annual congress at one of the centre countries.

## Federation of Entertainment Unions

c/o Equity, Guild House, Upper St Martin's Lane,
London WC2H 9EG
☎ 07973 714206
entertainment.unions@gmail.com
Secretary & Coordinator *Paul Evans*

Plenary meetings four times annually. Additionally, there are Training, Equalities & European Committees. Represents the following unions: British Actors' Equity Association; Broadcasting Entertainment Cinematograph and Theatre Union; Musicians' Union; Unite – the union; Professional Footballers' Association; **National Union of Journalists; Writers' Guild of Great Britain**.

## Film Agency for Wales

33–35 West Bute Street, Cardiff CF10 5LH
☎ 029 2046 7480    📠 029 2045 7491
adam@filmagencywales.com
www.filmagencywales.com
Contacts *Adam Partridge, Keith Potter*

Founded in July 2006, the Film Agency for Wales is the sole Agency for film in Wales, with a remit to ensure that the economic, cultural and educational aspects of film are effectively represented in Wales, the UK and the world. As a strategic agency, it aims to facilitate the emergence of a viable and sustainable Welsh film industry and to promote a vibrant and dynamic film culture.

## Foreign Press Association in London

c/o Royal Commonwealth Society, London
WC2N 5AP
☎ 020 7930 0445
briefings@foreign-press.org.uk
www.foreign-press.org.uk
Director *Christoph Wyld*

Founded 1888. Non-profit-making association providing accreditation, information, briefings and visits for foreign correspondents in London.

## The Gaelic Books Council

▷ **Comhairle nan Leabhraichean**

## The Garden Media Guild

Katepwa House, Ashfield Park Avenue, Ross-on-Wye HR9 5AX
☎ 01989 567393    📠 01989 567676
admin@gardenmediaguild.co.uk
www.gardenmediaguild.co.uk
Contact *Gill Hinton*
SUBSCRIPTION £60 (full); £85 (associate); £45 (probationary); £18 (retired); £25 (members of The Garden Photographers' Association)

Founded 1990. Aims to raise the quality of gardening communication, to help members operate efficiently and profitably, to improve liaison between garden communicators and the horticultural industry. Administers an annual awards scheme. Operates a mailing service and organizes press training days.

## Graphic Enterprise Scotland

48A Palmerston Place, Edinburgh EH12 5DE
☎ 0131 220 4353    📠 0131 220 4344
info@graphicenterprisescotland.org
www.graphicenterprisescotland.org
Director *B. Hodgson*

Founded 1910. Employers' organization and trade association for the Scottish printing industry. Represents the interests of the industry to government, public and other bodies and provides a range of services including industry promotion, industrial relations, education, training and commercial activities. Negotiates a national wages and conditions agreement with Unite GPM Sector. The Federation is a member of Intergraf, the international confederation for employers' associations in the printing industry, which represents its interests at European level.

## Guidebookwriters.com

27 Camwood, Clayton-le-Woods, Bamber Bridge PR5 8LA
☎ 01772 321243  📠 07053 491743
tm@wpu.org.uk
www.guidebookwriters.com
Contact *Terry Marsh*
SUBSCRIPTION £50 p.a.

Guidebookwriters.com exists to promote the work of its members, all of whom are professional guidebook writers, to the trade through direct contact with public relations companies, tourist boards, editors, publishers and consultants and, through its website, regular news bulletins and press releases.

## Guild of Agricultural Journalists

Isfield Cottage, Church Road, Crowborough TN6 1BN
☎ 01892 610628
don.gomery@btinternet.com
www.gaj.org.uk
Honorary General Secretary *Don Gomery*
SUBSCRIPTION £60 p.a.

Founded 1944 to promote a high professional standard among journalists who specialize in agriculture, horticulture and allied subjects. Represents members' interests with representative bodies in the industry; provides a forum through meetings and social activities for members to meet eminent people in the industry; maintains contact with associations of agricultural journalists overseas; promotes schemes for the education of members and for the provision of suitable entrants into agricultural journalism.

## Guild of Editors
▷ Society of Editors

## The Guild of Food Writers

255 Kent House Road, Beckenham BR3 1JQ
☎ 020 8659 0422
gfw@gfw.co.uk
www.gfw.co.uk
Administrator *Jonathan Woods*

SUBSCRIPTION £85

Founded 1985. The objects of the Guild are to bring together professional food writers including journalists, broadcasters and authors, to extend the range of members' knowledge and experience by arranging discussions, tastings and visits, and to encourage new writers through competitions and awards. The Guild aims to contribute to the growth of public interest in, and knowledge of, the subject of food and to campaign for improvements in the quality of food.

## Guild of Motoring Writers

40 Baring Road, Bournemouth BH6 4DT
☎ 01202 422424
generalsec@gomw.co.uk
www.guildofmotoringwriters.co.uk
General Secretary *Patricia Lodge*

Founded 1944. Represents members' interests and provides a forum for members to exchange information.

## Horror Writers Association

244 Fifth Avenue, Suite 2767, New York, NY 10001, USA
hwa@horror.org
www.horror.org
www.horror.org/UK (UK Chapter)

Founded 1987. World-wide organization of writers and publishers dedicated to promoting the interests of writers of horror and dark fantasy. Publishes a monthly newsletter, issues email bulletins, gives access to lists of horror agents, reviewers and bookstores; and keys to the 'Members Only' area of the HWA website. Presents the annual **Bram Stoker Awards** (see entry under *Prizes*).

## Independent Alliance

Founded in 2005, the Independent Alliance comprises ten UK publishers and their international partners, linked by a common vision of 'editorial excellence, original, diverse publishing, innovation in marketing and commercial success'. Founding publishers were Faber & Faber, Atlantic Books, Canongate, Icon Books, Profile Books and Short Books, later joined by Quercus Publishing, Serpent's Tail and Granta/Portobello. Faber & Faber provides sales and admin for the organization's publishers and The Book Service Ltd are the sole distributors.

## Independent Publishers Guild

PO Box 12, Llain, Whitland SA34 0WU
☎ 01437 563335  📠 01437 562071
info@ipg.uk.com
www.ipg.uk.com
Executive Director *Bridget Shine*

Founded 1962. Actively represents the interests of independent publishers. With more than 500 members, and steadily growing with combined revenues of £500 million, the IPG provides a networking base whereby members receive regular e-newsletters, training courses and seminars covering important areas for its members, from e-commerce to international trade. Runs a collective stand for members at leading international book fairs including Frankfurt and London. Membership is open to new and established independent publishers. Supplier membership is also available. Check the website for further information.

## Independent Theatre Council

12 The Leathermarket, Weston Street, London SE1 3ER

℡ 020 7403 1727   ℻ 020 7403 1745
admin@itc-arts.org
www.itc-arts.org
Chief Executive *Charlotte Jones*

The Independent Theatre Council (ITC) is the management association and political voice of around 700 performing arts professionals and organizations. ITC provides its members with legal and management advice, training and professional development, networking, regular newsletters and a comprehensive web resource. Additionally, ITC initiates and develops projects to enrich, enhance and raise the profile of the performing arts. Publications available from ITC include *The ITC Practical Guide for Writers and Companies*. ITC negotiates contracts and has established standard agreements for theatre professionals with Equity. The rights and fee structure agreement, reached with the **Writers' Guild** in 1991, was updated in December 2002 and rates of pay are reviewed annually. Contact the Writers' Guild or visit ITC's website for further details.

## Institute of Copywriting

Overbrook Business Centre, Poolbridge Road, Blackford, Wedmore BS28 4PA
℡ 0800 781 1715   ℻ 01934 713492
services@inst.org
www.inst.org/copy
Secretary *Lynn Hall*

Founded 1991 to promote copywriters and copywriting (writing publicity material). Maintains a code of practice. Membership is open to students as well as experienced practitioners. Runs training courses (see entry under *UK and Irish Writers' Courses*). Has a list of approved copywriters. Answers queries relating to copywriting. Contact the Institute for a free booklet.

## Institute of Translation and Interpreting (ITI)

Fortuna House, South Fifth Street, Milton Keynes MK9 2EU
℡ 01908 325250   ℻ 01908 325259
info@iti.org.uk
www.iti.org.uk

Founded 1986. The ITI is the only independent professional association of practising translators and interpreters in the UK. Its aim is to promote the highest standards in the profession. It has strong corporate membership and runs professional development courses and conferences. Membership is open to those with a genuine and proven involvement in translation and interpreting (including students). ITI's bi-monthly *Bulletin* and other publications are available from the Secretariat. ITI's *Directory of Members* offers a free referral service for those in need of a professional translator/interpreter and is available online. ITI is a full and active member of FIT (International Federation of Translators).

## International Association of Scientific, Technical & Medical Publishers

Prama House, 267 Banbury Road, Oxford OX2 7HT
℡ 01865 339321   ℻ 01865 339325
www.stm-assoc.org

STM is an international association of scientific, technical, medical and scholarly publishers which aims to create a platform for exchanging ideas and information, and to represent the interest of the STM publishing community in the fields of copyright, technology developments and end user/library relations. The membership consists of large and small companies, not-for-profit organizations, learned societies, traditional, primary, secondary publishers and new entrants to global publishing.

## IoL
▷ Chartered Institute of Linguists

## Irish Book Publishers' Association
▷ CLÉ

## Irish Copyright Licensing Agency Ltd

25 Denzille Lane, Dublin 2, Republic of Ireland
℡ 00 353 1 662 4211   ℻ 00 353 1 662 4213
info@icla.ie
www.icla.ie
Executive Director *Samantha Holman*

Founded in 1992 by writers and publishers in Ireland to provide a scheme through which rights holders can give permission, and users of copyright material can obtain permission to copy.

## Irish Playwrights and Screenwriters Guild

Art House, Curved Street, Temple Bar, Dublin 2, Republic of Ireland
℡ 00 353 1 670 9970
info@script.ie
www.script.ie
Contact *David Kavanagh*
SUBSCRIPTION Details available on request

Founded in 1969 to safeguard the rights of scriptwriters for radio, stage and screen.

## Irish Translators' & Interpreters' Association (ITIA)

Irish Writers' Centre, 19 Parnell Square, Dublin 1, Republic of Ireland
℡ 00 353 87 673 8386
itiasecretary@gmail.com
www.translatorsassociation.ie
Honorary Secretary *Mary Phelan*
Treasurer *Graziano Ciulli*
ANNUAL MEMBERSHIP €250 (corporate); €150 (institutional); €100 (professional); €50 (associate); €25 (student)

Founded 1986. The Association is open to all translators: technical, commercial, literary and cultural, and to all classes of interpreters. It is also for those with an interest in translation such as teachers and third-level students. Information on the profession of interpreting and translation and a register of members are available on the website. Publishes a newsletter, *Translation Ireland* and an eZine, the *ITIA Bulletin*. Organizes certification tests (for legal documents) for professional members and tests for new professional members.

## Irish Writers' Union

Irish Writers' Centre, 19 Parnell Square, Dublin 1, Republic of Ireland
℡ 00 353 1 872 1302
iwu@ireland-writers.com
www.ireland-writers.com
Chairperson *Helen Dwyer*
SUBSCRIPTION €50 p.a.; €30 (associate)

Founded in 1986 to promote the interests and protect the rights of writers in Ireland.

## ITC

▷ Independent Theatre Council

## ITI

▷ Institute of Translation and Interpreting

## IVCA (International Visual Communications Association)

19 Pepper Street, Glengall Bridge, London E14 9RP
℡ 020 7512 0571    🖷 020 7512 0591
info@ivca.org
www.ivca.org
CEO *Marco Forgione*

The IVCA is a professional association representing the interests of the users and suppliers of visual communications. In particular it pursues the interests of producers, commissioners and manufacturers involved in the non-broadcast and independent facilities industries and also business event companies. It represents all sizes of company and freelance individuals, offering information and advice services, publications, a professional network, special interest groups, a magazine and a variety of events including the UK's Film and Video Communications Festival.

## The Library Association

▷ CILIP: The Chartered Institute of Library and Information Professionals

## literaturetraining

PO Box 23595, Leith, Edinburgh EH6 7YX
℡ 0131 553 2210
info@literaturetraining.com
www.literaturetraining.com
Director *Philippa Johnston*

literaturetraining acts a first stop shop for writers and literature professionals across the UK looking for professional development information and advice. Services include an online directory of jobs and opportunities, training and events, organizations and resources; a fortnightly jobs and opportunities e-bulletin; professional development planning guidance; and specially commissioned features and other resources. It is a wing of the **National Association of Writers in Education** and is run in conjunction with its other partner literature organizations **Academi**, **Apples & Snakes**, Lapidus, **National Association for Literature Development**, renaissance one, **Scottish Book Trust** and **Survivors Poetry**.

## The MCPS–PRS Alliance

▷ PRS for Music

## Medical Journalists' Association

Fairfield, Cross in Hand, Heathfield TN21 0SH
℡ 01435 868786    🖷 01435 865714
secretary@mja-uk.org
www.mja-uk.org
Chairman *David Delvin*
Honorary Secretary *Philippa Pigache*
SUBSCRIPTION £40 p.a. (full); £30 (associate); £10 (junior)

Founded 1967. Aims to improve the quality and practice of medical and health journalism and to improve relationships and understanding between medical and health journalists and the health and medical professions. Over 420 members. Regular meetings with senior figures in medicine and medico politics; educational workshops on important subject areas or issues of the day;

debates; awards for medical journalists from commercial sponsors, plus the MJA's own annual awards for journalism and books. Publishes a newsletter five times a year.

## Medical Writers' Group

The Society of Authors, 84 Drayton Gardens, London SW10 9SB

☎ 020 7373 6642  📠 020 7373 5768
sbaxter@societyofauthors.org
www.societyofauthors.org
Contact *Sarah Baxter*

Founded 1980. A specialist group within the **Society of Authors** offering advice and help to authors of medical books.

## Mystery Writers of America, Inc.

1140 Broadway, Suite 1507, New York, NY 10001, USA

☎ 001 212 888 8171  📠 001 212 888 8107
mwa@mysterywriters.org
www.mysterywriters.org
Administrative Manager *Margery Flax*
SUBSCRIPTION $95 (US)

Founded 1945. Aims to promote and protect the interests of writers of the mystery genre in all media; to educate and inform its membership on matters relating to their profession; to uphold a standard of excellence and raise the profile of this literary form to the world at large. Holds an annual banquet at which the 'Edgars' are awarded (named after Edgar Allan Poe).

## National Association for Literature Development

PO Box 243, Ilkley LS29 7AT
☎ 01943 862107
director@nald.org
www.nald.org
Contact *Steve Dearden*

NALD is the largest membership organization for literature professionals. The only UK-wide body for all those involved in developing writers, readers and literature audiences. Offers networking advice, training and advocacy.

## National Association of Writers in Education

PO Box 1, Sheriff Hutton, York YO60 7YU
☎ 01653 618429
info@nawe.co.uk
www.nawe.co.uk
Contact *Paul Munden, Director*
SUBSCRIPTION £30 p.a. (professional); £20 (associate/student); £75 (institution); £30 (overseas)

Founded 1991. Aims to promote the contribution of living writers to education and to encourage both the practice and the critical appreciation of creative writing. Has over 1,000 members.

Organizes national conferences and training courses. A directory of writers who work in schools, colleges and the community is available online. Publishes a magazine, *Writing in Education*, issued free to members three times per year.

## National Association of Writers' Groups

PO Box 3266, Stoke-on-Trent ST10 9BD
nawg@live.co.uk
www.nawg.co.uk
Chairman *Nicolette Ward*
Secretary *Jo Sadler*

Founded 1995 with the object of furthering the interests of writers' groups and individuals throughout the UK. A registered charity (No. 1059047). Publishes *Link* magazine and books for writers (including *Running Good Writing Groups*); gives free entry to competitions for group anthologies, poetry, short stories, articles, novels and sketches; holds an annual open festival of writing with a variety of workshops, seminars, individual surgeries led by professional writers. Membership is open to all writers' groups and individual writers – there are no restrictions or qualifications required for joining.

## National Campaign for the Arts

1 Kingly Street, London W1B 5PA
☎ 020 7287 3777  📠 020 7287 4777
nca@artscampaign.org.uk
www.artscampaign.org.uk
Director *Louise de Winter*
Membership and Office Manager *Simon Cole*

The UK's only independent campaigning organization representing all the arts. The NCA seeks to safeguard, promote and develop the arts and win public and political recognition for their importance as a key element in the national culture. Its members shape its campaign work, provide its mandate and the core funding, through subscriptions that enable the NCA to act independently on their behalf; the organization receives no public subsidy. Members have access to the NCA's information, advice, seminars, conferences and publications. Membership is open to organizations and individuals working in or with an interest in the arts. Literature subscriptions available.

## National Centre for Research in Children's Literature (NCRCL)

Bede House, School of Arts, Digby Stuart College, Roehampton University, Roehampton Lane, London SW15 5PH
☎ 020 8392 3000
g.lathey@roehampton.ac.uk
l.sainsbury@roehampton.ac.uk
www.ncrcl.ac.uk

The National Centre for Research in Children's Literature facilitates and supports research exchange in the field of children's literature. Runs an internationally acclaimed MA Programme and annual conference in association with the International Board on Books for Young People (IBBY).

## National Literacy Trust

68 South Lambeth Road, London SW8 1RL
☏ 020 7587 1842   📠 020 7587 1411
contact@literacytrust.org.uk
www.literacytrust.org.uk
Director *Jonathan Douglas*

The National Literary Trust is an independent charity that changes lives through literacy. 'We have a vision of a society in which everyone has the reading, writing, speaking and listening skills that they need to fulfil their own and, ultimately, the nation's potential. We aim to empower learners, support professionals and influence policy and practice.' Supports those who work with learners through their innovative programmes, information and research, and brings together key organizations to lead literacy promotion in the UK. Programmes include National Young Readers' Programme – providing free books for children to choose and keep; Reading The Game – involving the professional football community; Reading Connects – developing a whole-school approach to reading for pleasure; Reading Champions – putting the boys in charge of their reading; and the Talk To Your Baby campaign which supports parents and professionals.

## National Union of Journalists

Headland House, 308 Gray's Inn Road, London WC1X 8DP
☏ 020 7278 7916   📠 020 7837 8143
info@nuj.org.uk
www.nuj.org.uk
General Secretary *Jeremy Dear*
SUBSCRIPTION £189 p.a. (freelance) or 1% of annual income if lower; or 0.5% if income less than £13,600 p.a.

Represents journalists in all sectors of publishing, print, broadcast and online media. Responsible for wages and conditions agreements which apply across the industry. Provides advice and representation for its members, as well as administering unemployment and other benefits. Publishes various guides and magazines: *On-Line Freelance Directory*; *On-Line Fees Guide*; *The Journalist* and *The Freelance*.

## NCRCL

▷ National Centre for Research in Children's Literature

## New Writing North

Holy Jesus Hospital, City Road, Newcastle upon Tyne NE1 2AS
☏ 0191 233 3850
office@newwritingnorth.com
www.newwritingnorth.com
Director *Claire Malcolm*
Administrator *Catherine Robson*

New Writing North is the literature development agency for the Arts Council England, North East region and offers many useful services to writers, organizing events, readings and workshops. NWN has a website and mailing list by which members of the public can receive news of literary events, courses and work opportunities. Administers the Northern Writers' Awards, which include tailored development packages (mentoring and financial help) for local writers at different stages of their careers. NWN works to develop theatre writing and creative radio writing projects within the region, as well as a strong education and community-based programme, working with schools, LEAs and both public and private sector organizations.

## The Newspaper Society

St Andrew's House, 18–20 St Andrew Street, London EC4A 3AY
☏ 020 7632 7425   📠 020 7632 7401
collette_yates@newspapersoc.org.uk
www.newspapersoc.org.uk
Director *David Newell*

The Newspaper Society is the voice of Britain's regional press which delivers local news to 40 million print readers a week and 24 million web users a month. It promotes the interests of the regional and local press, provides legal advice and lobbying services, holds a series of conferences and seminars, and runs the annual Local Newspaper Week.

## Outdoor Writers & Photographers Guild

1 Waterside Close, Garstang PR3 1HJ
☏ 01995 605340   📠 0871 266 8621
secretary@owg.org.uk
www.owpg.org.uk
Secretary *Jon Sparks*

Founded 1980 to promote, encourage and assist the development and maintenance of professional standards among creative professionals working in the outdoors. Members include writers, photographers, broadcasters, filmmakers, editors, publishers and illustrators. Publishes a quarterly journal, *Outdoor Focus*. Presents six Awards for Excellence plus other awards in recognition of achievement. Maintains a website of members' details; informal members' email 'chatline' to exchange news and information; sends out regular media news bulletins; organizes press trips.

## Pact (Producers Alliance for Cinema and Television)

3rd Floor, Fitzrovia House, 153–157 Cleveland Street, London W1T 6QW

℡ 020 7380 8230    🖷 020 7380 8248

info@pact.co.uk

www.pact.co.uk

Chief Executive *John McVay*

Chief Operations Officer *Andrew Chowns*

Pact is the UK trade association that represents and promotes the commercial interests of independent feature film, television, children's, animation and interactive media companies. Headquartered in London, Pact has regional representation throughout the UK, in order to support its members. A highly effective lobbying organization, it has regular and constructive dialogues with government, regulators, public agencies and opinion formers on all issues affecting its members and contributes to key public policy debates on the media industry, both in the UK and in Europe. Pact negotiates terms of trade with all public service broadcasters in the UK and supports members in their business dealings with cable and satellite channels. It also lobbies for a properly structured and funded UK film industry and maintains close contact with the **UK Film Council** and other relevant film organizations and government departments.

## PEN
▷ English PEN

## Performing Right Society
▷ PRS for Music

## The Personal Managers' Association Ltd

PO Box 6319, London N1P 1HL

℡ 0845 602 7191

info@thepma.com

www.thepma.com

Co-chairs *Fiona Williams, Nichola Young*

Secretary *Catherine O'Shea*

The Personal Managers' Association is the leading professional body for literary and theatrical agencies in the UK. Between them, members manage the careers of almost all the UK's dramatists and screenwriters, actors, directors and creatives in their work in film, television, theatre and radio, both in the UK and abroad.

## The Picture Research Association

Box 105 Hampstead House, 176 Finchley Road, London NW3

℡ 07771 982308

chair@picture-research.org.uk

www.picture-research.org.uk

Chair *Veneta Bullen*

Membership Secretary *Suzanne Williams*

SUBSCRIPTION Members: £45 (introductory); £55 (full); £50 (associate)

Founded in 1977 as the Society of Picture Researchers & Editors. The Picture Research Association is a professional body for picture researchers, managers, picture editors and all those involved in the research, management and supply of visual material to all forms of the media. The Association's main aims are to promote the interests and specific skills of its members internationally; to promote and maintain professional standards; to bring together those involved in the research and publication of visual material; to provide a forum for the exchange of information and to provide guidance to its members. Free advisory service for members, regular meetings, quarterly magazine, monthly newsletter and Freelance Register.

## Player-Playwrights

9 Hillfield Park, London N10 3QT

aoby37@dsl.pipex.com

www.playerplaywrights.co.uk

Presidents *Laurence Marks, Maurice Gran*

Contact *Peter Thompson (at the address above)*

SUBSCRIPTION Members: £10 (joining fee); £6 p.a. thereafter, plus £2 per attendance

Founded 1948. A society giving opportunity for writers new to stage, radio, television and film, as well as others finding difficulty in achieving results, to work with writers established in those media. At weekly meetings (7.30 pm–10.00 pm, Mondays, downstairs at The Porters Bar, The Green Man Pub, 383 Euston Road, London NW1), members' scripts are read or performed by actor members and afterwards assessed and dissected in general discussion. New writers and new acting members are always welcome.

## Playwrights' Studio, Scotland

350 Sauchiehall Street, CCA, Glasgow G2 3JD

℡ 0141 332 4403

info@playwrightsstudio.co.uk

www.playwrightsstudio.co.uk

The Playwrights' Studio, Scotland is a national initiative designed to engage the people of Scotland directly with new playwriting and raise the standard of plays for presentation to the public. Aims to increase awareness of the wealth of creative talent within Scotland, both at home and internationally, and work to develop the art form of theatre, examining and expanding the playwright's role within it. Delivers a wide range of activities which address the development needs of aspiring, emerging and established playwrights, encourage producers' interest in new plays and playwrights, and promote Scotland's plays and playwrights to the widest possible markets.

**PLR**
> Public Lending Right

**PLS**
> Publishers Licensing Society Ltd

**PMA**
> The Personal Managers' Association Ltd

**Poetry Book Society**
> under Organizations of Interest to Poets

**Poetry Ireland**
> under Organizations of Interest to Poets

**The Poetry Society**
> under Organizations of Interest to Poets

**PPA The Association for Publishers and Providers of Consumer, Customer & Business Media in the UK**
Queen's House, 55–56 Lincoln's Inn Fields, London WC2A 3LJ
☎ 020 7404 4166   🖷 020 7404 4167
info1@ppa.co.uk
www.ppa.co.uk

Founded 1913 to promote and protect the interests of its members in particular and the magazine industry as a whole.

**Press Complaints Commission**
Halton House, 20/23 Holborn, London EC1N 2JD
☎ 020 7831 0022   🖷 020 7831 0025
pcc@pcc.org.uk
www.pcc.org.uk
Chairman *Baroness Buscombe*
Information Officer *Tonia Milton*

Founded in 1991 to deal with complaints from members of the public about the editorial content of newspapers and magazines. Administers the editors' Code of Practice covering such areas as accuracy, privacy, harrassment and intrusion into grief. Publications available: *Code of Practice, How to Complain* and *Annual Review*.

**Private Libraries Association**
Ravelston, South View Road, Pinner HA5 3YD
dchambrs@aol.com
www.pla.org
Honorary Secretary *Dr Dean Sewell*
(d.a.sewell@hw.ac.uk)
SUBSCRIPTION £30 p.a.

Founded in 1956 for the advancement of the education of the public in the study, production and ownership of books. Publishes work concerned with the objects of the Association, particularly those works which are not commercially viable; holds meetings at which papers on cognate subjects are read and discussed;

arranges lectures and exhibitions which are also open to non-members.

**Producers Alliance for Cinema and Television**
> Pact

**PRS for Music**
Copyright House, 29–33 Berners Street, London W1T 3AB
☎ 020 7580 5544   🖷 020 7306 4455
www.prsformusic.com
JOINING FEE Refer to the website for full details

'One of the world's most efficient combined rights collecting operations.' PRS for Music is at the centre of the UK's music industry, helping businesses and community groups get access to the music they need, while making sure the people who created and published the music are rightfully rewarded for their creativity.

**Public Lending Right**
Richard House, Sorbonne Close, Stockton-on-Tees TS17 6DA
☎ 01642 604699   🖷 01642 615641
authorservices@plr.uk.com
www.plr.uk.com
Registrar *Jim Parker*

Founded 1979. Public Lending Right is funded by the Department for Culture, Media and Sport to recompense authors for books borrowed from public libraries. Authors and books can be registered for PLR only when application is made during the author's lifetime. To qualify, an author must be resident in a European Union member state, or in Iceland, Liechtenstein or Norway, and the book must be printed, bound and put on sale with an ISBN. In 2009, £6.63 million was distributed to over 23,500 authors (£6.66 million in the previous year) with 232 authors reaching the maximum payment threshold of £6,600. The Rate Per Loan was 5.98 pence in 2009. Consult the website for further information.

**The Publishers Association**
29b Montague Street, London WC1B 5BW
☎ 020 7691 9191   🖷 020 7691 9199
mail@publishers.org.uk
www.publishers.org.uk

The national UK trade association for books, learned journals, and electronic publications, with around 200 member companies in the industry. Very much a trade body representing the industry to Government and the European Commission, and providing services to publishers. Publishes the *Directory of Publishing* in association with **Continuum**. Also home of the Educational Publishers Council (school books), PA's International Division, the Academic and Professional Division, the Trade Publishers

Council, the Electronic Publishers' Forum and the Audiobook Publishing Committee.

## Publishers' Association of South Africa

PO Box 106, Green Point, Cape Town, South Africa 8051

☎ 00 27 21 425 2721   🖷 00 27 21 421 3270

pasa@publishsa.co.za
www.publishsa.co.za
Contact *Dudley Schroeder, Executive Director (dudley@publishsa.co.za)*

Founded in 1992 to represent publishing in South Africa, a key industry sector. With a membership of 170 companies, the Association includes commercial organizations, university presses, one-person privately-owned publishers as well as importers and distributors.

## Publishers Licensing Society Ltd

37–41 Gower Street, London WC1E 6HH

☎ 020 7299 7730   🖷 020 7299 7780

pls@pls.org.uk
www.pls.org.uk
Chief Executive *Alicia Wise*
Operations Manager *Tom West*

Founded in 1981, the PLS obtains mandates from publishers which grant PLS the authority to license photocopying of pages from published works. Some licences for digitization of printed works are available. PLS aims to maximize revenue from licences for mandating publishers and to expand the range and repertoire of mandated publishers available to licence holders. It supports the **Copyright Licensing Agency Ltd (CLA)** in its efforts to increase the number of legitimate users through the issuing of licences and vigorously pursues any infringements of copyright works belonging to rights' holders.

## Publishers' Publicity Circle

65 Airedale Avenue, London W4 2NN

☎ 020 8994 1881

ppc-@lineone.net
www.publisherspublicitycircle.co.uk
Contact *Heather White*

Enables book publicists from both publishing houses and freelance PR agencies to meet and share information regularly. Meetings, held monthly in central London, provide a forum for press journalists, television and radio researchers and producers to meet publicists collectively. A directory of the PPC membership is published each year and distributed to over 2,500 media contacts.

## Publishing Scotland

Scottish Book Centre, 137 Dundee Street, Edinburgh EH11 1BG

☎ 0131 228 6866   🖷 0131 228 3220

enquiries@publishingscotland.org
www.publishingscotland.org
Chief Executive *Marion Sinclair*
Membership Services Manager *Jane Walker*
BooksfromScotland.com Editor *Liam Davison*
Training and Information Manager *Joan Lyle*

Publishing Scotland, (formerly Scottish Publishers Association) is the trade, training and development body for publishing in Scotland. It represents a wide range of Scottish publishers, from multinationals to small presses, and has developed a Network membership from all sections of the book world, including the **Society of Authors in Scotland**. Publishing Scotland also acts as an information centre for the trade and general public. Publishes annual *Publishing in Scotland Yearbook*. Attends international book fairs; runs an extensive training programme; carries out industry research.

## The Radio Academy

5 Market Place, London W1W 8AE

☎ 020 7927 9920   🖷 020 7636 8924

info@radioacademy.org
www.radioacademy.org
Chief Executive *Trevor Dann*

The Radio Academy is a registered charity dedicated to the encouragement, recognition and promotion of excellence in UK broadcasting and audio production. Since 1983 the Academy has represented the radio industry to outside bodies including the government and offered neutral ground where everyone from the national networks to individual podcasters is encouraged to discuss the broadcasting, production, marketing and promotion of radio and audio.

## RadioCentre

4th Floor, 5 Golden Square, London W1F 9BS

☎ 020 3206 7800   🖷 020 3206 7801

www.radiocentre.org
Chief Executive *Andrew Harrison*

Founded 1986. RadioCentre represents commercial radio stations in the UK. Its role is to maintain and build a strong and successful commercial radio industry, both in terms of listening hours and revenues.

## The Reading Agency

Free Word Centre, 60 Farringdon Road, London EC1R 3GA

☎ 0871 750 1207

info@readingagency.org.uk
www.readingagency.org.uk

The Reading Agency was formed in 2002 by merging three smaller agencies – LaunchPad, The Reading Partnership and Well Worth Reading. Runs national partnership schemes, library reading promotions, projects that target specific readers, and their own research and publishing programme. The Agency assists people working directly with readers by providing training and advice as well as toolkits, websites and promotional materials, and uses connections to join up organizations interested in promoting reading, for example, helping libraries form partnerships with major organizations such as publishers and the BBC. Programmes include the Summer Reading Challenge for children and The Vital Link for adults trying to improve their literacy skills. Active in public policy and advocacy, working with government departments and think tanks, the Agency also works with the corporate sector, through sponsorship or Corporate Social Responsibility programmes.

## Romance Writers of America

14615 Benfer Road, Houston, TX 77069, USA
Ⓣ 001 832 717 5200    Ⓕ 001 832 717 5201
info@rwanational.org
www.rwanational.org
Professional Relations Coordinator *Carol Ritter*
MEMBERSHIP $75 p.a. plus $25 processing fee

Founded 1980. RWA, a non-profit association with more than 9,000 members worldwide, provides a service to authors at all stages of their careers as well as to the romance publishing industry and its readers. Anyone pursuing a career in romantic fiction may join RWA. Holds an annual conference, and provides contests for both published and unpublished writers through the RITA Awards and Golden Hearts Awards.

## The Romantic Novelists' Association

www.rna-uk.org
Chair *Katie Ford (katiefforde@katiefforde.com)*
Hon. Membership Secretary *Nicola Cornick (ncornick@madasfish.com)*
SUBSCRIPTION £40/£45 p.a. (full members/non-EU; £50/£55 (associate/non-EU); £93/£100.50 (New Writers' Scheme/non-EU)

Membership is open to published writers of romantic fiction (modern or historical), or those who have had one or more full-length serials published. Associate membership is open to publishers, editors, literary agents, booksellers, librarians and other professionals having a close connection with novel writing and publishing. Membership in the New Writers' Scheme is available to a limited number of writers who have not yet had a full-length novel published. New writers may submit a manuscript each year which will be appraised by experienced published members. Meetings are held in London and the regions with guest speakers and an annual conference is held each year in July. *Romance Matters* magazine is published quarterly and issued free to members. The Association presents awards, including the **Romantic Novel of the Year** and the Love Story of the Year at the annual awards luncheon in London.

## Royal Society of Literature

Somerset House, Strand, London WC2R 1LA
Ⓣ 020 7845 4676    Ⓕ 020 7845 4679
info@rslit.org
www.rslit.org
Chair *Anne Chisholm, FRSL*
Secretary *Maggie Fergusson*
SUBSCRIPTION £50 p.a.

Founded 1820. Membership by application to the Secretary. Fellowships are conferred by the Society on the proposal of two Fellows. Membership benefits include lectures, discussion meetings and poetry readings. Lecturers have included Julian Barnes, Michael Holroyd, Philip Pullman, Zadie Smith and Tom Stoppard. Presents the **Royal Society of Literature Ondaatje Prize**, the **Royal Society of Literature/Jerwood Awards** and the **V.S. Pritchett Memorial Prize.**

## Royal Television Society

Kildare House, 3 Dorset Rise, London EC4Y 8EN
Ⓣ 020 7822 2810    Ⓕ 020 7822 2811
info@rts.org.uk
www.rts.org.uk
Contact *Membership Manager (members@rts.org.uk)*
MEMBERSHIP from £65 p.a.

Founded 1927. Covers all disciplines involved in the television industry. Provides a forum for debate and conferences on technical, social and cultural aspects of the medium. Presents various awards including journalism, programmes, technology and design. Publishes *Television Magazine* ten times a year for members and subscribers.

## Science Fiction Foundation

75 Rosslyn Avenue, Harold Wood RM3 0RG
www.sf-foundation.org
Contact *Roger Robinson*

Founded 1970. The SFF is an international academic body for the furtherance of science fiction studies. Publishes a thrice-yearly magazine, *Foundation*, which features academic articles and reviews of new fiction. Publishes a series of books on critical studies of science fiction, and arranges conferences, lectures and seminars on related topics. It also has a reference library (see entry under *Library Services*), housed at Liverpool University.

## Scottish Book Trust

Sandeman House, Trunk's Close, 55 High Street, Edinburgh EH1 1SR
Ⓣ 0131 524 0160    Ⓕ 0131 524 0161

info@scottishbooktrust.com
www.scottishbooktrust.com
Contact *Marian Boubouze*

Scottish Book Trust is the leading agency
for the promotion of literature, reading and
writing in Scotland. Its innovative programmes
are designed to inspire readers, writers and
learning professionals all over the country.
Scottish Book Trust funds author events all
over Scotland; runs the biggest children's book
awards in the UK involving over 30,000 children;
supports the professional development of
Scottish writers through mentoring, awards and
training opportunities; supports teachers and
librarians to improve literacy; operates a touring
programme for children's authors; develops
public participation projects such as Days Like
This and promotes other readership development
opportunities.

### Scottish Daily Newspaper Society
▷ Scottish Newspaper Society

### Scottish Newspaper Publishers Association
▷ Scottish Newspaper Society

### Scottish Newspaper Society
21 Lansdowne Crescent, Edinburgh EH12 5EH
☎ 0131 535 1064    📠 0131 535 1063
Director *J.B. Raeburn*

Formed by the amalgamation of the Scottish
Daily Newspaper Society and the Scottish
Newspaper Publishers Association in January
2010. Represents publishers of Scottish daily and
Sunday newspapers including the major Scottish
editions of UK national newspapers.

### Scottish Print Employers Federation
▷ Graphic Enterprise Scotland

### Scottish Publishers Association
▷ Publishing Scotland

### Scottish Screen
2nd Floor, 249 West George Street, Glasgow
G2 4QE
☎ 0141 302 1700    📠 0141 302 1711
info@scottishscreen.com
www.scottishscreen.com
Contact *Christine McMillan*

Scottish Screen is the national development
agency for the screen industries in Scotland.
'We aim to inspire audiences, support new and
existing talent and businesses, educate young
people, and promote Scotland as a creative place
to make great films, award-winning television and
world renowned digital entertainment.'

### Society for Editors and Proofreaders (SfEP)
Enrico House, 93–99 Upper Richmond Road,
London SW15 2TG
☎ 020 8785 5617    📠 020 8785 5618
administration@sfep.org.uk
www.sfep.org.uk
Chair *Sarah Price*
Company Secretary *Justina Amenu*
SUBSCRIPTION £88 p.a. (individuals) plus £27.50
joining fee; corporate membership available.

Founded in 1988 in response to the growing
number of freelance editors and their increasing
importance to the publishing industry.
Aims to promote high editorial standards by
disseminating information through advice
and training, and to achieve recognition of
the professional status of its members. The
Society also supports moves towards recognized
standards of training and accreditation for
editors and proofreaders. It launched its own
Accreditation in Proofreading test in 2002.
Following discussions with the SfEP, City &
Guilds have developed the Licentiateship
in Editorial Skills. This was launched in
September 2008. The Licentiateship offers a
form of recognition based around Continuing
Professional Development.

### Society for Technical Communications (STC)
9401 Lee Highway, Suite 300, Fairfax, Virginia
22031, USA
☎ 001 703 522 4114    📠 001 703 522 2075
stc@stc.org
www.stc.org
www.stcuk.org
UK and Ireland Chapter Membership Manager
*Karen Lewis (membership@stcuk.org)*

Dedicated to advancing the arts and sciences
of technical communication, the STC has more
than 15,000 members worldwide, including
technical writers, editors, graphic designers,
videographers, multimedia artists, web and
intranet page information designers, translators
and others whose work involves making technical
information available to those who need it. The
UK and Ireland Chapter, established in 1985, hosts
talks and seminars, runs an annual technical
publications competition and publishes its own
newsletter quarterly.

### The Society of Authors
84 Drayton Gardens, London SW10 9SB
☎ 020 7373 6642    📠 020 7373 5768
info@societyofauthors.org
www.societyofauthors.org
General Secretary *Mark Le Fanu*
Deputy General Secretary *Kate Pool*

SUBSCRIPTION £90/85 p.a.

The Society of Authors is an independent trade union, chaired by Tom Holland, with some 9,000 members. As well as campaigning for the profession, it advises members on negotiations with publishers, agents, broadcasting organizations, theatre managers and film companies and on queries or concerns members may have about the business aspects of their writing. The Society assists with complaints and takes action for breach of contract, copyright infringement, etc. Among the Society's publications are *The Author* (quarterly) and the *Quick Guides* series to various aspects of writing (all free of charge to members). Other services include vetting of contracts, a searchable database of writers, emergency funds for writers, meetings and seminars, and various special discounts, including money off books, hotels and insurance. There are groups within the Society for broadcasters, children's writers and illustrators, educational writers, and translators, as well as regional groups.

Authors under 35 or over 65, not earning a significant income from their writing, may apply for lower subscription rates (after their first year, in the case of over-65s). Contact the Society for an application pack or visit the website.

## The Society of Authors in Scotland

17 Pittville Street Lane, Edinburgh EH15 2BU
☎ 0131 657 1391  🖷 0131 657 1391
caroline.dunford@googlemail.com
www.societyofauthors.org/
Honorary Secretary *Caroline Dunford*

The Scottish branch of the **Society of Authors**, which organizes business meetings, social and bookshop events throughout Scotland.

## Society of Children's Book Writers & Illustrators

ra@britishscbwi.org
www.britishscbwi.org
www.scbwi.org
British Isles Regional Adviser *Natascha Biebow*

Founded in 1971 by a group of Los Angeles-based writers, the SCBWI is an international organization for the exchange of information between writers, illustrators, editors, publishers, agents and others involved with literature for young people. With a membership of more than 22,000 worldwide, it holds two annual international conferences plus a number of regional events, publishes a bi-monthly journal and awards grants for works in progress. The British Isles region sponsors international conferences each year devoted entirely to writing and illustrating for children. It also facilitates local critique groups and publishes a quarterly regional

newsletter. For membership enquiries, visit the website or contact membership@britishscbwi.org

## Society of Civil and Public Service Writers

17 The Green, Corby Glen, Grantham NG33 4NP
joan@lewis5634.fsnet.co.uk
www.scpsw.co.uk
Membership Secretary *Mrs Joan Lewis*
SUBSCRIPTION £15 p.a.

Founded 1935. Welcomes serving and retired members of the Civil Service, armed forces, Post Office, BT, nursing profession and other public servants who are aspiring or published writers. Offers competitions for short story, article and poetry; postal folios for short story and article; AGM, occasional meetings and luncheon held in London; quarterly magazine, *Civil Service Author*, to which members may submit material; Poetry Workshop (extra £5) offers annual weekend outside London, anthology, newsletter, postal folio, competitions. S.a.s.e. to the Secretary for details.

## Society of Editors

University Centre, Granta Place, Mill Lane, Cambridge CB2 1RU
☎ 01223 304080  🖷 01223 304090
info@societyofeditors.org
www.societyofeditors.org
Executive Director *Bob Satchwell*

Formed by a merger of the Association of British Editors and the Guild of Editors, the Society of Editors has nearly 500 members in national, regional and local newspapers, magazines, broadcasting, new media, journalism education and media law. Campaigns for media freedom and self-regulation. For further information call *Bob Satchwell* (07860 562815).

## Society of Indexers

Woodbourn Business Centre, 10 Jessell Street, Sheffield S9 3HY
☎ 0114 244 9561  🖷 0114 244 9563
admin@indexers.org.uk
www.indexers.org.uk
Secretary *Judith Menes*
Administrator *Wendy Burrow*
SUBSCRIPTION £95.50 p.a.; institutions: from £191

Founded 1957. Publishes *The Indexer* (biannual, April and October), a quarterly newsletter, an online directory, *Indexers Available (IA)*, which lists members and their subject expertise. In addition, the Society runs an open-learning course entitled *Training in Indexing* and recommends rates of pay (currently £18.50–£30 per hour or £2–£5 per page). Awards annually **The Wheatley Medal** and **The Bernard Levin Award** (see entries under *Prizes*).

## The Society of Medical Writers
*Membership Secretary*: 30 Dollis Hill Lane, London NW2 6JE
richard_cutler_novelist@yahoo.co.uk
www.somw.org
Contact *Dr R.C. Cutler*
SUBSCRIPTION £50 p.a.; £45 (retired); £30 (nurses/ancilliary workers); £15 (students)

Now in its 22nd year, the Society is open to anyone in any branch of medicine or its allied professions who publishes or aspires to publish their written work of whatever nature – fact or fiction, prose or poetry. The Society aims to improve standards of writing; to encourage literacy; to provide meetings for members for the exchange of views, skills and ideas; to provide education on the preparation, presentation and submission of written material for publication; to consider questions of ethics relating to writing and publication; to further developments in the art of writing. Biannual conferences held in the spring and autumn at various locations; details published in *The Writer* or available from the secretary (copy for *The Writer* may be submitted to the editor throughout the year).

## Society of Women Writers & Journalists
14 Laburnum Walk, Rustington BN16 3QW
wendy@stickler.org
www.swwj.co.uk
Membership Secretary *Wendy Hughes*
SUBSCRIPTION £40 p.a. (full); £30 (associate); £25 (probationary/student); £30 (overseas). £15 joining fee; £15 (Friends of SWWJ)

Founded 1894. For women writers, the SWWJ upholds professional standards and literary achievements through regional group meetings; residential weekends; postal critique service; competitions; outings and major London meetings. For Full members, membership card doubles as a press card. Publishes *The Woman Writer*. Male writers are accepted as Associate Members. A new category has been introduced for Friends of the Society.

## Society of Young Publishers
29b Montague Street, London WC1B 5BW
www.thesyp.org.uk
Contact *Membership Secretary*
SUBSCRIPTION £30 p.a.; £24 (student/unwaged)

Provides facilities whereby members can increase their knowledge and widen their experience of all aspects of publishing, and holds regular social events. Run entirely by volunteers, it is open to those in related occupations, students, and associate membership is available for over-35s. Publishes a bi-monthly newsletter called *In*Print and holds monthly meetings featuring trade professionals. For other events, see the website.

Please enclose an s.a.s.e. when writing to the Society.

## Sports Journalists' Association of Great Britain
c/o Start2Finish Event Management, Unit 92, Capital Business Centre, 22 Carlton Road, Croydon CR2 0BS
☎ 020 8916 2234  📠 020 8916 2235
stevenwdownes@btinternet.com
www.sportsjournalists.co.uk
Secretary *Steven Downes (80 Southbridge Road, Croydon CR0 1AE; 07710 428562)*
SUBSCRIPTION £30 p.a. incl. VAT (London); £20 (regional)

Founded in 1948 as the Sports Writers' Association of Great Britain to promote and maintain a high professional standard among journalists who specialize in sport in all its branches – including photography, broadcasting and online – and to serve members' interests. Publishes a quarterly *Bulletin* and a Yearbook, and promotes the annual SJA British Sports Awards in Nov/Dec, and the SJA British Sports Journalism Awards and SJA British Sports Photography Awards every March.

## STM
▷ International Association of Scientific, Technical & Medical Publishers

## Tees Valley Arts
Melrose House, Melrose Street, Middlesbrough TS1 2HZ
☎ 01642 264651  📠 01642 264955
info@teesvalleyarts.org.uk
www.teesvalleyarts.org.uk
Director *Rowena Sommerville*
Education Officer *Jane Gray*
Communications Officer *Simon Smith*

TVA is an arts development agency based in Middlesbrough in the north east of England, championing participation in high quality arts and stimulating creative activities as a tool for improving the quality of life and learning for individuals and communities. TVA works in education: designing and delivering stimulating programmes for able students, for disengaged learners, for excluded or at risk pupils, and to deliver the science curriculum and environmental learning and awareness in unexpected and enjoyable ways. 'TVA works in diversity and inclusion: using the arts to support and empower people who face life's challenges.'

## Trade Publishers Council
▷ The Publishers Association

## The Translators Association

84 Drayton Gardens, London SW10 9SB
☎ 020 7373 6642　📠 020 7373 5768
jmccrum@societyofauthors.org
www.societyofauthors.org
Contact *Jo McCrum*

Founded 1958 as a subsidiary group within the **Society of Authors** to deal exclusively with the special problems of literary translators into the English language. Benefits to members include free legal and general advice and assistance on all business matters relating to translators' work, including the vetting of contracts and advice on rates of remuneration. Membership is normally confined to translators who have had their work published in volume or serial form or produced in this country for stage, television or radio. The Association administers translation prizes (see *Prizes*) and maintains an online database of members' details for the use of publishers who are seeking a translator for a particular work.

## UK Film Council

10 Little Portland Street, London W1W 7JG
☎ 020 7861 7861　📠 020 7861 7862
info@ukfilmcouncil.org.uk
www.ukfilmcouncil.org.uk

The UK's leading film body. Uses lottery money and Government Grant-in-Aid to encourage the development of new talent, skills and creative and technological innovation in UK film; help new and established filmmakers make distinctive British films; support the creation and growth of stable businesses in the film sector; provide access to finance and help the UK film industry to compete in the global marketplace. Also promotes enjoyment and understanding of the cinema and ensures that film's economic and creative and cultural interests are properly represented in public policy making.

## Voice of the Listener & Viewer Ltd (VLV)

PO Box 401, Gravesend DA12 9FY
☎ 01474 352835
info@vlv.org.uk
www.vlv.org.uk
Chairman *Richard Lindley*
Administrative Secretary *Sue Washbrook*

VLV represents the citizen and consumer interests in broadcasting. It is an independent, non-profit-making organization working to ensure independence, quality and diversity in broadcasting. VLV is the only consumer body speaking for listeners and viewers on the full range of broadcasting issues. It is funded by its members and is free from sectarian, commercial and political affiliations. Holds public lectures, seminars and conferences throughout the UK, and has frequent contact with MPs, civil servants, the BBC and independent broadcasters, regulators, academics and other consumer groups. Provides an independent forum where all with an interest in broadcasting can speak on equal terms. Produces a quarterly news bulletin and regular briefings on broadcasting issues. Holds its own archive and those of the former Broadcasting Research Unit (1980–90) and BACTV (British Action for Children's Television). Maintains a panel of speakers, the VLV Forum for Children's Broadcasting, the VLV Forum for Educational Broadcasting, and acts as secretariat for the European Alliance of Listeners' and Viewers' Associations (EURALVA). VLV has responded to all major public enquiries on broadcasting since 1984. VLV does not handle complaints.

## Wales Literature Exchange/Cyfnewidfa Lên Cymru

Mercator Institute for Media, Languages and Culture, Dept. of Theatre, Film & Television Studies, Aberystwyth University, Aberystwyth SY23 3AS
☎ 01970 622544　📠 01970 621524
post@walesliterature.org
post@cyfnewidfalen.org
www.walesliterature.org
www.cyfnewidfalen.org
Director *Sioned Puw Rowlands*

Founded 2000. Wales Literature Exchange was formerly known as Welsh Literature Abroad. Works to facilitate the translation of Wales's literature and to promote exchange between the culture and literatures of Wales and the wider world. Translation grants are available to publishers. Participates in international book fairs, and works with translators, publishers and festivals in Wales and abroad.

## WATCH
▷ Writers, Artists and their Copyright Holders

## Welsh Academy
▷ Academi

## Welsh Books Council (Cyngor Llyfrau Cymru)

Castell Brychan, Aberystwyth SY23 2JB
☎ 01970 624151　📠 01970 625385
castellbrychan@cllc.org.uk
www.wbc.org.uk
www.gwales.com
Director *Elwyn Jones*

Founded in 1961 to stimulate interest in books from Wales and to support publishing in Wales. The Council distributes publishing grants and promotes and fosters all aspects of book production in Wales in both Welsh and English. Its Editorial, Design, Marketing and Children's Books departments and wholesale distribution centre offer central services to publishers in Wales. Writers in Welsh and English are welcome to approach the Editorial Department for advice on how to get their manuscripts published or visit the Welsh Book Trade Info website (www.wbti.org.uk).

## Welsh Literature Abroad
▷ Wales Literature Exchange/Cyfnewidfa Lên Cymru

## Welsh National Literature Promotion Agency
▷ Academi

## West Country Writers' Association
Trevean, Yeolmbridge, Launceston PL15 8NH
☎ 01566 773516
secretarywcwa@fsmail.net
www.westcountrywriters.com
Secretary *Fiona McAughey*

Founded in 1951 in the interest of published authors living in or writing about the West Country. Annual congress and newsletters. Regional meetings to discuss news and views.

## Women in Publishing
info@wipub.org.uk
www.wipub.org.uk

Aims to promote the status of women working in publishing and related trades by helping them to develop their careers. Through WiP there are opportunities for members to learn more about every area of the industry, share information and expertise, give and receive support and take part in a practical forum for the exchange of information about different disciplines and industry sectors. Monthly meetings with speakers provide a forum for discussion. Meetings are held on the second Wednesday of each month. See website for further information.

## Writers and Photographers unLimited
27 Camwood, Clayton-le-Woods, Bamber Bridge PR5 8LA
☎ 01772 321243  📠 07053 491743
tm@wpu.org.uk
www.wpu.org.uk
Contact *Terry Marsh*
SUBSCRIPTION £52 p.a.

An online marketing service for English-language professional writers and photographers worldwide who specialize in all aspects of travel, tourism, the outdoors, adventure sports, food and drink. The USP of WPu is that all its members are full-time professionals. Promotes the work of its members by direct contact with public relations companies, tourist boards, editors, publishers and consultants and, through its website, regular news bulletins and press releases.

## Writers' Centre Norwich
14 Princes Street, Norwich NR3 1AE
☎ 01603 877177  📠 01603 625452
info@writerscentrenorwich.org.uk
www.writerscentrenorwich.org.uk
Contacts *Leila Telford, Richard White*

Founded in 2004, the Writers' Centre Norwich is a literature development agency catering for the East of England as well as nationally and internationally. Runs a host of activities for writers: regional and national writing competitions; writing workshops; networking sessions and events. Specializes in taking writers who are serious about progressing up to the next level and into the professional world. 'Many writers who have come through our various competitive programmes have gone on to get agents and publishers.' Professional writers are brought together in the Centre's annual Worlds Literary Festival which offers networking, a chance to share ideas and a space online to catch up with the latest news. As well as physical events, the Centre has information for writers online, including resources, the best new writing and training opportunities.

## Writers' Guild of Great Britain
40 Rosebery Avenue, London EC1R 4RX
☎ 020 7833 0777
admin@writersguild.org.uk
www.writersguild.org.uk
President *David Edgar*
Chair *Robert Taylor*
General Secretary *Bernie Corbett*
Deputy General Secretary *Anne Hogben*

SUBSCRIPTION Full member: 1.2% of earnings from professional writing (min. £180, max. £1,800); Candidate member: £100; Student member: £20; Affiliate member (agent or writers' group): £275

Founded in 1959 as the Screenwriters' Guild, the Writers' Guild is a trade union, affiliated to the TUC, representing professional writers in television, radio, theatre, film, books and new media. It negotiates Minimum Terms Agreements governing writers' contracts and covering minimum fees; advances; repeat fees; royalties and residuals; rights; credits; number of drafts; script alterations and the resolution of disputes. The most important MTAs cover BBC TV Drama; BBC Radio Drama; ITV; **Pact** (independent TV and film producers); TAC (Welsh language independent TV producers); Theatrical Management Association; **Independent Theatre Council**; and an agreement covering the **Royal National Theatre, Royal Shakespeare Company**, and **Royal Court Theatre**. These agreements are regularly renegotiated and in most cases the minimum fees are reviewed annually.

The Guild advises members on all aspects of their working lives, including contract vetting, legal advice, help with copyright problems and representation in disputes with producers, publishers or other writers.

The Guild organizes seminars, discussions and lectures on all aspects of writing. *UK Writer*, the Guild's biannual magazine, is free to members and contains features about professional writing, Guild news and details of work opportunities, training courses, literary competitions, etc. A detailed email of news, opportunities and other information is sent to members every Friday.

Full Membership is open to writers who have received payment for their work on terms not less than those negotiated by the Guild. Writers who do not qualify can join as Candidate Members and those on accredited writing courses or theatre attachments can become Student Members.

## Writers, Artists and their Copyright Holders (WATCH)

The Library, The University of Reading, PO Box 223, Whiteknights, Reading RG6 6AE
Ⓣ 0118 378 8783    Ⓕ 0118 378 6636
www.watch-file.com
Contact *Dr David Sutton*

Founded 1994. Provides an online database of information about the copyright holders of literary authors and artists. The database is available free of charge on the internet and the web. WATCH is the successor project to the Location Register of English Literary Manuscripts and Letters, and continues to deal with location register enquiries.

## Yachting Journalists' Association

36 Church Lane, Lymington SO41 3RB
Ⓣ 01590 673894
sec@yja.co.uk
www.yja.co.uk
Contact *Secretary*
SUBSCRIPTION £40 p.a.

To further the interest of yachting, sail and power, and to provide support and assistance to journalists in the field; current membership is just over 260 with 50 from overseas. A handbook, listing details of members and subscribing PR organizations, press facility recommendations, forthcoming events and other useful information, is published in alternate years, available from the Secretary at the address above. The YJA organizes the Yachtsman of the Year and Young Sailor of the Year Awards, presented annually at the beginning of January.

## Yr Academi Gymreig
▷ Academi

# Literary Societies

## The Adrian Bell Society
Apple Acre, Church Lane, Claxton NR14 7HY
☎ 01508 480665
Chairman *John Ford*
SUBSCRIPTION £5 p.a.

Founded in 1996, the Society aims to promote the study and enjoyment of the writing of Adrian Bell (1901–80); to publish journals twice a year and hold meetings; to help members obtain copies of out of print titles; to assemble a collection of photocopies of his 1500 *Countryman's Notebooks* that appeared in the *Eastern Daily Press* and to index them; to commemorate his career as *The Times* first crossword compiler; to share knowledge of the places and factual background mentioned in his books. The society has 180 members worldwide who keep in touch through the journal. *A Centenary Countryman's Notebook*, published 2001, comprised over 60 of his press essays selected by the Committee. 2009 saw re-publication of *Corduroy* and *Men and the Fields*.

## The Angela Thirkell Society
54 Belmont Park, London SE13 5BN
☎ 020 8244 9339
penny.aldred@ntlworld.com
www.angelathirkellsociety.com
www.angelathirkell.org (N. American branch)
Honorary Secretary *Mrs. P. Aldred*
SUBSCRIPTION £10 p.a.

Founded in 1980 to honour the memory of Angela Thirkell as a writer and to make her works available to new generations. Publishes an annual journal, holds an AGM in the autumn and a spring meeting which usually takes the form of a visit to a location associated with Thirkell. Has a flourishing North American branch which has frequent contact with the UK parent society.

## Anthony Powell Society
76 Ennismore Avenue, Greenford UB6 0JW
☎ 020 8864 4095    ☏ 020 8020 1483
secretary@anthonypowell.org
www.anthonypowell.org
Patron *John M.A. Powell*
President *Simon Russell Beale*
Honorary Secretary *Dr Keith C. Marshall*

SUBSCRIPTION £22 p.a. (individual); £28 (overseas); £33 (joint, UK); £39 (joint, overseas); £13 (student, UK); £19 (student, overseas); £100 (organization)

Founded in June 2000 to advance education and interest in the life and works of the English author Anthony Powell (1905–2000). The Society's major activity is a biennial Powell conference, the first five of which were at Eton College (2001), Balliol College, Oxford (2003), The Wallace Collection, London (2005), Bath (2007) and Washington, DC (2009). The Society organizes events for members, ranging from 'pub meets' to talks and visits to places of Powell interest as well as publishing *Secret Harmonies* (annual journal) and a quarterly *Newsletter*. Member of the **Alliance of Literary Societies**.

## Arnold Bennett Society
9 Penelope Close, Harbury CV33 9JD
arnoldbennettscty@btinternet.com
www.arnoldbennettsociety.org.uk
Secretary *Fiona Cole*
SUBSCRIPTION £12 p.a. (single); £14 (joint/family) plus £3 for membership outside the UK

Aims to promote interest in the life and works of 'Five Towns' author Arnold Bennett and other North Staffordshire writers. Annual dinner. Regular functions and talks in and around Burslem, plus annual seminar at Wedgwood College, Barlaston and annual conference at the Potteries Museum & Art Gallery, Stoke on Trent.

## The Arthur Conan Doyle Society
PO Box 1360, Ashcroft, British Columbia, Canada V0K 1A0
☎ 001 250 453 2045    ☏ 001 250 453 2075
sirhenry@telus.net
www.ash-tree.bc.ca/acdsocy.html
Joint Organizers *Christopher Roden, Barbara Roden*

Founded in 1989 to promote the study and discussion of the life and works of Sir Arthur Conan Doyle. Occasional meetings, functions and visits. Publishes an occasional journal together with reprints of Conan Doyle's writings.

## The Arthur Ransome Society Ltd

Abbot Hall Art Gallery & Museum, Kendal
LA9 5AL
tarsinfo@arthur-ransome.org
www.arthur-ransome.org
Trustee Chairman *Florence Galbraith*
Company Secretary *Peter Hyland*
SUBSCRIPTION UK: £5 (junior); £10 (student); £15
(adult); £10 (senior); £20 (family); £40 (corporate);
Overseas: £5 (junior); £10 (student); £20 (adult);
£15 (senior); £25 (family) Payable in local currency
in US, Canada, Australia, Japan and New Zealand

Founded in 1990 to celebrate the life and to
promote the works and ideas of Arthur Ransome,
author of *Swallows and Amazons* titles for
children, biographer of Oscar Wilde, works on
the Russian Revolution and extensive articles
on fishing. TARS seeks to encourage children
and adults to engage in adventurous pursuits, to
educate the public about Ransome and his works,
and to sponsor research into his literary works
and life.

## The Barbara Pym Society

St Hilda's College, Oxford OX4 1DY
☎ 01865 373753
eileen.roberts@st-hildas.ox.ac.uk
www.barbara-pym.org
Chairman *Clemence Schultz*
Archivist *Yvonne Cocking*
SUBSCRIPTION £18 p.a. (individual); £25 p.a.
(household); £10 p.a. (concession); £15 p.a.
(senior household)

Aims to promote interest in and scholarly
research into the works of Barbara Pym; to
bring together like-minded people to enjoy the
exploration of all aspects of her novels and to
continue to encourage publishers to keep her
novels in print. Annual weekend conference at St.
Hilda's College in September, focusing on a theme
or a particular novel. Annual North American
Conference in Boston. Annual one-day spring
meeting in London, with speaker. Publishes a
newsletter, *Green Leaves*, biannually. In addition
to reports on conferences and society activities,
the newsletter includes scholarly papers and
various unpublished items from the Pym archives
at the Bodleian Library (i.e. short stories).

## The Baskerville Hounds (The Dartmoor Sherlock Holmes Study Group)

6 Bramham Moor, Hill Head, Fareham PO14 3RU
☎ 01329 667325
samweller@90thentry.info
Chairman *Philip Weller*

Founded 1989. An international Sherlock Holmes
society specializing solely in studies of *The Hound
of the Baskervilles* and its Dartmoor associations.

Organizes many study functions, mostly on
Dartmoor. Currently involved in several large-
scale research projects. Membership is open only
to those who are willing to be actively involved, by
correspondence and/or attendance at functions,
in the group's study activities.

## The BB Society

*Secretary:* 8 Park Road, Solihull B91 3SU
☎ 0121 704 1002
enquiries@roseworldproductions.com
www.roseworldproductions.com
Chairman *Brian Mutlow*
Secretary *Bryan Holden*
SUBSCRIPTION £15 (individual); £25 (family); £50
(corporate); £5 (student); £20 (overseas)

Founded in 2000 to bring together devotees
of the writer/illustrator BB (Denys Watkins-
Pitchford). BB's oeuvre included books on
wildfowling, angling, children's fantasy and the
countryside. The Society holds two meetings a
year and publishes newsletters and an annual
journal, *Sky Gypsy*.

## The Beatrix Potter Society

The Lodge, Salisbury Avenue, Harpenden AL5 2PS
☎ 01582 769755
beatrixpottersociety@tiscali.co.uk
www.beatrixpottersociety.org.uk
Membership Secretary *Jenny Akester*
SUBSCRIPTION UK: £20 p.a. (individual); £25
(institution); Overseas: £25 (individual); £30
(institution)

Founded in 1980 to promote the study and
appreciation of the life and works of Beatrix
Potter (1866–1943). Potter was not only the
author of *The Tale of Peter Rabbit* and other
classics of children's literature; she was also a
landscape and natural history artist, diarist,
farmer and conservationist, and was responsible
for the preservation of large areas of the Lake
District through her gifts to the National Trust.
The Society upholds and protects the integrity of
the unique work of Potter, her aims and bequests;
holds regular talks and meetings in London and
elsewhere with visits to places connected with
Beatrix Potter. International Study Conferences
are held in the UK and the USA. The Society
has an active publishing programme. Members
receive a quarterly newsletter. (UK Registered
Charity No. 281198.)

## The Beckford Society

The Timber Cottage, Crockerton, Warminster
BA12 8AX
☎ 01985 213195
sidney.blackmore@btinternet.com
Secretary *Sidney Blackmore*

SUBSCRIPTION £20 (minimum) p.a.

Founded 1995 to promote an interest in the life and works of William Beckford (1760–1844) and his circle. Encourages Beckford studies and scholarship through exhibitions, lectures and publications, including an annual journal, *The Beckford Journal* and occasional newsletters.

## The Betjeman Society

386 Hurst Road, Bexley DA5 2JY
www.johnbetjeman.com
Honorary Secretary *Martin H. Revill*
SUBSCRIPTION £14 p.a.

Aims to promote the study and appreciation of the work and life of Sir John Betjeman. Annual programme includes poetry readings, lectures, discussions, visits to places associated with Betjeman, and various social events. Meetings are held in London and other centres. Regular newsletter and annual journal, *The Betjemanian*.

## The Bewick Society

c/o The Hancock Museum, Newcastle upon Tyne NE2 4PT
☎ 0191 232 6386
nhsn@ncl.ac.uk
www.bewicksociety.org.uk
Membership Secretary *June Holmes*
SUBSCRIPTION £10 p.a. (individual); £12 (family)

Founded in 1988 to promote an interest in the life and work of Thomas Bewick, wood-engraver and naturalist (1753–1828). Organizes related events and meetings, and is associated with the Bewick birthplace museum at Cherryburn, Northumberland.

## Birmingham Central Literary Association

23 Arden Grove, Ladywood, Birmingham B16 8HG
☎ 0121 454 9352
bruce@bakerbrum.co.uk
Contact *The Secretary*

Holds fortnightly meetings in central Birmingham to discuss the lives and work of authors and poets. Holds an annual dinner to celebrate Shakespeare's birthday.

## The Bram Stoker Memorial Association

Penthouse North, Suite 145, 29 Washington Square West, New York City, NY 10011–9180, USA
bramstoker145@yahoo.com
countdracula145@hotmail.com
President/Founder *Dr Jeanne Keyes Youngson*
European Vice President *Dr Mark Benecke*
Associate Executive Director *Jennifer O'Casey*

Founded in 1985 by Dr Jeanne Keyes Youngson who, in 1965, formed the Dracula Fan Club (now known as The Vampire Empire). News and announcements of Stoker events and relevant books appear in the club's quarterly newsletter. Society members are invited to join Dr Youngson's 'Bram Stoker Walks' in Dublin, starting at Trinity College and ending at St Michan's Church.

## British Fantasy Society

23 Mayne Street, Hanford, Stoke on Trent ST4 4RF
secretary@britishfantasysociety.org
www.britishfantasysociety.org
Honorary President *Ramsey Campbell*
Chairman *Guy Adams*
Secretary *Helen Hopley*
SUBSCRIPTION from £30 p.a.

Founded in 1971 for devotees of fantasy, horror and related fields in literature, art and the cinema. Publishes a regular newsletter with information and reviews of new books and films, plus related fiction and non-fiction magazines. Annual conference at which the **British Fantasy Awards** are presented. These awards are voted on by the membership and are not an open competition.

## The Brontë Society

Brontë Parsonage Museum, Haworth, Keighley BD22 8DR
☎ 01535 640195   📠 01535 647131
bronte@bronte.org.uk
www.bronte.info
Contact *Membership Officer*
SUBSCRIPTION Details from the Membership Officer or the website

Founded 1893. Aims and activities include the preservation of manuscripts and other objects related to or connected with the Brontë family, and the maintenance and development of the museum and library at Haworth. The Society holds regular meetings, lectures and exhibitions; publishes information relating to the family and a triannual *Gazette*. Freelance contributions for either publication should be sent to the Publications Secretary at the address above. Members can receive the journal, *Brontë Studies*, at a reduced subscription .

## The Browning Society

Department of English Literature, University of Glasgow, 5 University Gardens, Glasgow GL12 8QQ
r.williams@englit.arts.gla.ac.uk
Honorary Secretary *Dr Vicky Greenaway*
SUBSCRIPTION £15 p.a.

Founded in 1969 to promote an interest in the lives and poetry of Robert and Elizabeth Barrett Browning. Meetings are arranged in the London area, one of which occurs in December at Westminster Abbey to commemorate Robert Browning's death.

## The Captain W.E. Johns Appreciation Society

221 Church Road, Northolt UB5 5BE
☎ 07836 680048
brian@zhong-ding.com
Contact (Derby meeting) *Brian Woodruff (at address above)*
Contact (Twyford) *Joy Tilley (01785 240299)*

Society for the appreciation of W.E. Johns, creator of Biggles. Meets twice a year in Derby and Twyford, near Reading (see above for contacts). Biannual magazine, *Biggles Flies Again* (editor: *John Lester*, 01708 446297; johnstuartlester@hotmail.co.uk).

## The Charles Lamb Society

BM ELIA, London WC1N 3XX
Chairman *Nicholas Powell*
SUBSCRIPTION UK: £18 (single); £24 (double); £24 (corporate)

Founded 1935. Publishes studies of the life, works and times of Charles Lamb and his circle. Holds meetings in London five times a year. *The Charles Lamb Bulletin* is published four times a year (editor: *Professor Richard S. Tomlinson* british_romanticism@comcast.net).

## The Charles Williams Society

35 Broomfield, Stacey Bushes, Milton Keynes
MK12 6HA
☎ 01908 316779
charles_wms_soc@yahoo.co.uk
www.charleswilliamssociety.org.uk
Contact *Honorary Secretary*
SUBSCRIPTION £15 p.a.

Founded in 1975 to promote interest in, and provide a means for the exchange of views and information on the life and work of Charles Walter Stansby Williams (1886–1945).

## The Charlotte M. Yonge Fellowship

8 Anchorage Terrace, Durham DH1 3DL
www.dur.ac.uk/c.e.schultze/contact.html
Contact *Dr C.E. Schultze*

Founded in 1995 to provide the opportunity for all those who enjoy and admire the work of Charlotte M. Yonge (1823–1901), author of *The Heir of Radclyffe*, to learn more about her life and writings. Holds two meetings annually, one in November in London and the other, the spring AGM, at venues associated with Yonge's life, works or interests, or where buildings or gallery and museum collections relate to Victorian history, art and society. Publishes *Review* twice-yearly and an occasional *Journal*. A loan collection of Yonge's works enables members to borrow publications that are hard to obtain. The Fellowship's archive is held at St Hugh's College, Oxford and material is available to members and *bona fide* researchers.

## The Chesterton Society UK

11 Lawrence Leys, Bloxham, Near Banbury
OX15 4NU
☎ 01295 720869/07766 711984    🖷 01295 720869
roberthughes11@talktalk.net
Honorary Secretary *Rev. Deacon Robert Hughes, KCHS*
SUBSCRIPTION £12.50 p.a.

Founded in 1964 to promote the ideas and writings of G.K. Chesterton.

## The Children's Books History Society

26 St Bernards Close, Buckfast, Buckfastleigh
TQ11 0EP
☎ 01364 643568
cbhs@abcgarrett.demon.co.uk
Chair/Membership Secretary *Mrs Pat Garrett*
SUBSCRIPTION £10 p.a. (UK/Europe); write for overseas subscription details

Established 1969. Aims to promote an appreciation of children's books and to study their history, bibliography and literary content. Meetings held in London and the provinces, plus a summer meeting to a collection, or to a location with a children's book connection. Three substantial newsletters issued annually, sometimes with an occasional paper. Review copies should be sent to Newsletter co-editor *Mrs Pat Garrett* (address above). The Society constitutes the British branch of the Friends of the Osborne and Lillian H. Smith Collections in Toronto, Canada, and also liaises with **CILIP** (formerly The Library Association). In 1990, the Society established its biennial Harvey Darton Award for a book, published in English, which extends our knowledge of some aspect of British children's literature of the past. 2008 joint winners: Brian Alderson and Felix de Marez Oyens *Be Merry and Wise: Origins of Children's Book Publishing in England 1650–1850*; Tomoko Masaki *A History of Victorian Popular Picture Books*.

## The D.H. Lawrence Society

1 Church Street, Swepstone LE67 2SA
Secretary *Mrs B. Sumner*

Founded in 1974 to increase knowledge and the appreciation of the life and works of D.H. Lawrence. Monthly meetings, addressed by guest speakers, are held in the library at Eastwood (birthplace of DHL). Organizes visits to places of interest in the surrounding countryside, supports the activities of the D.H. Lawrence Centre at Nottingham University, and has close links with DHL Societies worldwide. Publishes two newsletters and one journal each year, free to members.

## The Dartmoor Sherlock Holmes Study Group
▷ The Baskerville Hounds

## The Dickens Fellowship
The Charles Dickens Museum, 48 Doughty Street, London WC1N 2LX

☎ 020 7405 2127
dickens.fellowship@btinternet.com
www.dickens.fellowship.org
Joint Honorary General Secretaries *Mrs Lee Ault, Mrs Joan Dicks*

Founded 1902. The Society's particular aims and objectives are: to bring together lovers of Charles Dickens; to spread the message of Dickens, his love of humanity ('the keynote of all his work'); to remedy social injustice for the poor and oppressed; to assist in the preservation of material and buildings associated with Dickens. Annual conference. Publishes journal called *The Dickensian* (founded 1905 and available at special rate to members) and organizes a full programme of lectures, discussions, visits and conducted walks throughout the year. Branches worldwide.

## The Dorothy L. Sayers Society
Rose Cottage, Malthouse Lane, Hurstpierpoint BN6 9JY
☎ 01273 833444  📠 01273 835988
www.sayers.org.uk
Contact *Christopher Dean*
SUBSCRIPTION £15 p.a. (UK); £17.50 (Europe); $36 (US)

Founded in 1976 to promote the study of the life, works and thoughts of Dorothy Sayers; to encourage the performance of her plays and publication of her books and books about her; to preserve original material and provide assistance to researchers. Acts as a forum and information centre, providing material for study purposes which would otherwise be unavailable. Annual conventions, usually in August and other meetings. Co-founder of the Dorothy L. Sayers Centre in Witham. Publishes bi-monthly bulletin, annual proceedings and other papers. Grants are currently made to support a young musician, an actor and a bell-ringer.

## The Dylan Thomas Society of Great Britain
4 Bishops Grove, Swansea SA2 8BE
☎ 01792 206363
Contact *John Rhys Thomas*
SUBSCRIPTION £10 (individual); £15 (two adults from same household)

Founded in 1977 to foster an understanding of the life and work of Dylan Thomas and to extend members' awareness of other 20th century writers, especially Welsh writers in English. Regular meetings are held, mainly in Swansea.

## Dymock Poets
▷ The Friends of the Dymock Poets

## E.F. Benson Society
The Old Coach House, High Street, Rye TN31 7JF
☎ 01797 223114
www.efbensonsociety.org
Secretary *Allan Downend*
SUBSCRIPTION £7.50 (UK/Europe); £12.50 (overseas)

Founded in 1985 to promote the life and work of E.F. Benson and the Benson family. Organizes social and literary events, exhibitions, talks and Benson interest walks in Rye. Publishes a quarterly newsletter and annual journal, *The Dodo*, postcards and reprints of E.F. Benson articles and short stories in a series called 'Bensoniana' plus other books of Benson interest. Holds an archive which includes the Seckersen Collection (transcriptions of the Benson collection at the Bodleian Library in Oxford). Also sells second-hand Benson books and organizes walks round Benson's Rye.

## Early English Text Society
Lady Margaret Hall, Oxford OX2 6QA
📠 01865 286581
www.eets.org.uk
Executive Secretary *Professor Vincent Gillespie (at address above)*
Editorial Secretary *Dr H.L. Spencer (Exeter College, Oxford OX1 3DP)*
Membership Secretary *Mrs J.M. Watkinson (12 North End, Durham DH1 4NJ)*
SUBSCRIPTION £20 p.a. (UK); $40 (US)

Founded 1864. Concerned with the publication of early English texts. Members receive annual publications (one or two a year) or may select titles from the backlist in lieu.

## The Edith Nesbit Society
26 Strongbow Road, Eltham, London SE9 1DT
www.edithnesbit.co.uk
Contact *Mrs Marion Kennett*
SUBSCRIPTION £7 p.a.; £9 (joint)

Founded in 1996 to celebrate the life and work of Edith Nesbit (1858–1924), best known as the author of *The Railway Children*. The Society's activities include a regular newsletter, booklets, talks and visits to relevant places.

## The Edward Thomas Fellowship
1 Carfax, Undercliff Drive, St Lawrence, Isle of Wight PO38 1XG
☎ 01983 853366
colingthornton@btopenworld.com
www.edward-thomas-fellowship.org.uk
Hon. Secretary *Colin G. Thornton*

SUBSCRIPTION £10 p.a. (single); £15 p.a. (joint)

Founded in 1980 to perpetuate and promote the memory of Edward Thomas and to encourage an appreciation of his life and work. The Fellowship holds a commemorative birthday walk on the Sunday nearest the poet's birthday (3rd March) and an autumn walk; issues newsletters and holds various events and seminars.

## Elinor Brent-Dyer
> Friends of the Chalet School and The New Chalet Club

## The Elsie Jeanette Oxenham Appreciation Society

32 Tadfield Road, Romsey SO51 5AJ
☎ 01794 517149
abbeybufo@gmail.com
www.bufobooks.demon.co.uk/abbeylnk.htm
Membership Secretary/Treasurer *Ms Ruth Allen*
Editor (The Abbey Chronicle) *Fiona Dyer*
*(fionadyer@btopenworld.com)*
SUBSCRIPTION £7.50 p.a.; enquire for overseas rates

Founded in 1989 to promote the works of Elsie J. Oxenham. Publishes a newsletter for members, *The Abbey Chronicle*, three times a year. 500+ members.

## The Followers of Rupert

31 Whiteley, Windsor SL4 5PJ
☎ 01753 865562
followersofrupert@hotmail.com
www.rupertthebear.org.uk
Membership Secretary *Ray Reeves*
SUBSCRIPTION UK: £25 (single); £27 (joint); Europe: £30 (single); £32 (joint); RoW: £35 (single); £37 (joint)

Founded 1983. The Society caters for the growing interest in the Rupert Bear stories, past, present and future. Publishes the *Nutwood Newsletter* quarterly which gives up-to-date news of Rupert and information on Society activities. A national get-together of members – the Followers Annual – is held during the summer.

## Folly (Fans of Light Literature for the Young)

21 Warwick Road, Pokesdown, Bournemouth BH7 6JW
☎ 01202 432562
folly@sims.abel.co.uk
www.follymagazine.co.uk
Contact *Mrs Sue Sims*
SUBSCRIPTION £9 p.a. (UK); £12 (Europe); £15 (RoW)

Founded in 1990 to promote interest in a wide variety of collectible children's authors, with a bias towards writers of girls' books and girls' school stories. Publishes three magazines a year. Please note that *Folly* does not review any fiction other than girls' school stories, nor any non-fiction save that which is relevant to the history of children's books. 'We do not publish any fiction of any kind.'

## The Ford Madox Ford Society

English Department, The Open University, Milton Keynes MK7 6AA
☎ 01908 653453   ☏ 01908 653750
s.j.haslam@open.ac.uk
www.open.ac.uk/Arts/fordmadoxford-society/
Chair *Dr Sara Haslam (at address above)*
Treasurer *Paul Skinner*
SUBSCRIPTION £15 (£12 by standing order) or £8 concession

Founded in 1996 to promote the works of Ford Madox Ford and to increase knowledge of his writing and impact on writing in the 20th century. Meets for academic conferences and more popular events annually. Distributes *International Ford Madox Ford Studies* free to members. Welcomes all with an interest in Ford and his works. Has members in the UK, USA, Italy, France and Germany and holds events in as many different places as possible.

## Francis Brett Young Society

92 Gower Road, Halesowen B62 9BT
☎ 0121 422 8969
www.fbysociety.co.uk
Honorary Secretary *Mrs Jean Hadley*
SUBSCRIPTION £7 p.a. (individual); £10 (couple sharing a journal); £5 (student); £7 (organization/overseas); £70 (life); £100 (joint, life)

Founded 1979. Aims to provide a forum for those interested in the life and works of English novelist Francis Brett Young and to collate research on him. Promotes lectures, exhibitions and readings; publishes a regular newsletter.

## The Franco-Midland Hardware Company

6 Bramham Moor, Hill Head, Fareham PO14 3RU
☎ 01329 667325
samweller@90thentry.info
Chairman *Philip Weller*

Founded 1989. A federation of societies in four continents. Two core groups, '221B' (Holmesian/Sherlockian) and 'The Arthur Conan Doyle Study Group' (Doylean), form the UK base. Membership is open only to those interested in being actively involved in the programme of correspondence studies and/or functions. Arranges certificated self-study projects and assists group and individual research programmes, plus scholarly gatherings.

## The Friars' Club

West Lodge, 47 The Terrace, Wokingham RG40 1BP
Contact *Dr Peter McCall,*

SUBSCRIPTION £12 (joining & first year subscription )

Founded in 1982 to promote the writings of Frank Richards (Charles Hamilton), author of the Greyfriars School stories in *The Magnet* and creator of Billy Bunter. The Club produces a quarterly magazine for members. Articles and comments are welcomed.

## The Friends of Arthur Machen

Apt 5, 26 Hervey Road, Blackheath, London SE3 8BS
www.machensoc.demon.co.uk
Contact *Jeremy Cantwell*
SUBSCRIPTION £17 p.a. (UK); £20 (Europe/RoW); $36 (US)

Brings together those who appreciate Machen's writings and generally promotes discussion and research. Publishes two journals and two newsletters annually, plus occasional publications for the membership only.

## The Friends of Coleridge

87 Richmond Road, Montpelier, Bristol BS6 5EP
☎ 0117 942 6366
gcdd@blueyonder.co.uk
www.friendsofcoleridge.com
Website Manager *Paul Cheshire (74 Wells Road, Bath BA2 3AR; website@friendsofcoleridge.com)*
Editor (Coleridge Bulletin) *Graham Davidson*

Founded in 1987 to advance knowledge about the life, work and times of Samuel Taylor Coleridge and his circle, and to support his Nether Stowey Cottage, with the National Trust, as a centre of Coleridge interest. Holds literary evenings, study weekends and a biennial international academic conference. Publishes *The Coleridge Bulletin* biannually. Short articles on Coleridge-related topics may be sent to the editor at the (Bristol) address above. For membership and subscription details, see website.

## The Friends of Shandy Hall (The Laurence Sterne Trust)

Shandy Hall, Coxwold, York YO61 4AD
☎ 01347 868465    ☎ 01347 868465
shandyhall@dsl.pipex.com
www.shandean.org
www.laurencesternetrust.org.uk
Curator *Patrick Wildgust*
SUBSCRIPTION £7 (annual); £70 (life)

Promotes interest in the works of Laurence Sterne and aims to preserve the house in which they were created (open to the public). Publishes annual journal, *The Shandean*. An Annual Memorial Lecture is delivered each summer. Regular exhibitions in gallery. Bookshop.

## Friends of the Chalet School

4 Rock Terrace, Coleford, Bath BA3 5NF
☎ 01373 812705
focs@rockterrace.org
www.chaletschool.org.uk
Contacts *Ann Mackie-Hunter, Clarissa Cridland*
SUBSCRIPTION Details on the website or on application

Founded in 1989 to promote the works of Elinor Brent-Dyer. The society has members worldwide; publishes four magazines a year and runs a lending library.

## The Friends of the Dymock Poets

16 Robinson's Meadow, Ledbury HR8 1SU
☎ 01531 635204
cateluck2003@yahoo.com
www.dymockpoets.co.uk
Chairman *Roy Palmer*
Hon. Secretary *Catharine Lumby*
Membership Secretary *Jeff Cooper*
SUBSCRIPTION £7 (individual); £12 (couple); £3 (student); £12 (society/family)

Founded 1993. Established to foster an interest in the work of the Dymock Poets – Edward Thomas, Robert Frost, Wilfrid Gibson, Rupert Brooke, John Drinkwater, Lascelles Abercrombie; help preserve places and things associated with them; keep members informed of literary and other matters relating to the poets; increase knowledge and appreciation of the landscape between May Hill (Gloucestershire) and the Malvern Hills. Members are offered lectures, poetry readings, social meetings, newsletters and annual journal, guided walks in the countryside around Dymock; links with other literary societies; and annual event to commemorate the first meeting between Edward Thomas and Robert Frost on 6th October 1913.

## The Gaskell Society

Far Yew Tree House, Chester Road, Tabley, Knutsford WA16 0HN
membership@gaskellsociety.co.uk
www.gaskellsociety.co.uk
Honorary Secretary *Joan Leach*

Founded in 1985 to promote and encourage the study and appreciation of the life and works of Elizabeth Cleghorn Gaskell. Meetings held in Knutsford, Manchester, Bath, London and York; annual journal and biannual newsletter. On alternate years holds either a residential weekend conference or visit.

## The George Borrow Society

60 Upper Marsh Road, Warminster BA12 9PN
www.clough5.fsnet.co.uk/gb.html
Chairman/Bulletin Editor *Dr Ann M. Ridler*
Membership Secretary *Michael Skillman*

Honorary Treasurer *David Pattinson*
SUBSCRIPTION £15 p.a.

Founded in 1991 to promote knowledge of the life and works of George Borrow (1803–81), traveller, linguist and writer. The Society holds one or two meetings a year (with published proceedings) all at places associated with Borrow. Publishes the *George Borrow Bulletin* twice yearly, containing scholarly articles, reviews of publications relating to Borrow, reports of past events and news of forthcoming events. Member of the **Alliance of Literary Societies**, the Norwich Society and the Friends of Brompton Cemetery, London, where Borrow's restored grave may be found.

## The George Eliot Fellowship

39 Lower Road, Barnacle, Coventry CV7 9LD
jkburton@tiscali.co.uk
www.george-eliot-fellowship.com
Chairman *John K. Burton*

Founded 1930. Exists to honour George Eliot and promote interest in her life and works. Readings, memorial lecture, birthday luncheon and functions. Issues newsletters and an annual journal. Awards an annual prize for a George Eliot essay. Contact the chairman for subscription details.

## The George MacDonald Society

38 South Vale, Upper Norwood, London SE19 3BA
☎ 020 8653 8768
r.lines878@btinternet.com
www.george-macdonald.com
Chairman *Richard Lines*
Membership Enquiries *Roger Bardet (r.bardet@ hotmail.co.uk)*
SUBSCRIPTION £10 p.a. (UK); £11 (overseas); US$18 (N. America)

Founded in 1980 to promote interest in the life and works of George MacDonald (1824–1905) and his friends and contemporaries. Organizes conferences, meetings, lectures and outings. Annual newsletter, *Orts* and an academic journal, *North Wind*, published by St Norbert College, De Pere, WA, USA, are distributed free to members. (Registered Charity No: 1024021.)

## The Ghost Story Society

PO Box 1360, Ashcroft, British Columbia, Canada V0K 1A0
☎ 001 250 453 2045   ℻ 001 250 453 2075
nebuly@telus.net
www.ash-tree.bc.ca/GSS.html
Joint Organizers *Barbara Roden, Christopher Roden*
SUBSCRIPTION UK: £25 (airmail); US: $40; Canada: Cdn$45

Founded 1988. Devoted mainly to supernatural fiction in the literary tradition of M.R. James,

Walter de la Mare, Algernon Blackwood, E.F. Benson, A.N.L. Munby, R.H. Malden, etc. Publishes a biannual journal, *All Hallows*, which includes new fiction in the genre and non-fiction of relevance to the genre.

## Graham Greene Birthplace Trust

17 North Road, Berkhamsted HP4 3DX
☎ 01442 866694
treasurer@grahamgreenebt.org
www.grahamgreenebt.org
Treasurer *Colin Garrett*
SUBSCRIPTION (Friends of the Trust) £10 (UK, £25 for 3 years); £14 (Europe, £33); £18 (RoW, £45)

Founded on 2nd October 1997, the 93rd anniversary of Graham Greene's birth, to promote the appreciation and study of his works. Publishes a quarterly newsletter, occasional papers, videos and compact discs. Organizes the annual four-day Graham Greene Festival during the weekend nearest to the writer's birthday and administrates the GGBT Creative Writing Awards.

## H.G. Wells Society

1 Nackinton Road, Canterbury CT1 3NU
www.hgwellsusa.50megs.com/
Contact *Paul Allen*
SUBSCRIPTION £18 (UK/EU); £21 (RoW); £22 (institutional); £12 (concessions)

Founded in 1960 to promote an interest in and appreciation of the life, work and thought of Herbert George Wells. Publishes *The Wellsian* (annual) and *The H.G. Wells Newsletter* (three issues yearly). Organizes meetings and conferences.

## The Hazlitt Society

c/o The Guardian, Kings Place, 90 York Way, London N1 9AG
correspondence@williamhazlitt.org
www.williamhazlitt.org
Founding President *Michael Foot*
Contact *Helen Hodgson (Secretary and Correspondent)*

Established to encourage appreciation of Hazlitt's work and to promote his values. The *Hazlitt Review*, launched in 2008, aims to foster and sustain the restitution of Hazlitt as a canonical author and to stimulate new scholarship in his work.

## The Henry Williamson Society

7 Monmouth Road, Dorchester DT1 2DE
☎ 01305 264092
zseagull@aol.com
www.henrywilliamson.co.uk
General Secretary *Mrs Sue Cumming*
Membership Secretary *Mrs Margaret Murphy (16 Doran Drive, Redhill RH1 6AX; 01737 763228 mm@misterman.freeserve.co.uk)*

SUBSCRIPTION £12 p.a.; £15 (family); £5 (student)

Founded in 1980 to encourage, by all appropriate means, a wider readership and deeper understanding of the literary heritage left by the 20th-century English writer Henry Williamson (1895–1977). Publishes annual journal.

## The Henty Society

205 Icknield Way, Letchworth SG6 4TT
davidwalmsley@hentysociety.org
www.hentysociety.org
Honorary Secretary *David Walmsley*
SUBSCRIPTION £16 p.a. (UK); £19 (overseas)

Founded in 1977 to study the life and work of George Alfred Henty, and to publish research, bibliographical data and lesser-known works, namely short stories. Organizes conferences and social gatherings in the UK and North America, and publishes bulletins to members. Published in 1996: *G.A. Henty (1832–1902) a Bibliographical Study* by Peter Newbolt (2nd edition with addenda and corrigenda 2005).

## Hilaire Belloc Society

1 Hillview, Elsted, Midhurst GU29 0JX
☎ 01730 825575
hilairebelloc1@aol.com
Contact *Dr Grahame Clough*
SUBSCRIPTION £12 p.a. (UK); $20 (US)

Founded in 1996 to promote the life and work of Hilaire Belloc through the publication of newsletters containing rare and previously unpublished material. Organizes walks, social events, a Belloc weekend and a three-day conference. Currently 70 members, of which 30% are from overseas.

## Housman Society

80 New Road, Bromsgrove B60 2LA
☎ 01527 874136
info@housman-society.co.uk
www.housman-society.co.uk
Contact *Jim Page*

Founded in 1973 to promote knowledge and appreciation of the lives and work of A.E. Housman and other members of his family, and to promote the cause of literature and poetry. Sponsors a lecture at **The Guardian Hay Festival** each year under the title of 'The Name and Nature of Poetry'. Publishes an annual journal and biannual newsletter.

## The Ivor Gurney Society

24 Claremont Road, Southport PR8 4DZ
☎ 01704 553903
rolf@rolfjordan.com
www.ivorgurney.org.uk
General Secretary *Rolf Jordan*

SUBSCRIPTION £14 (individual); £18 (joint); £12.50 (group); £10 (retired, individual); £12.50 (retired, joint); £5 (student)

Founded in 1995 to make Ivor Gurney's music and poetry available to a wider audience by way of performances, readings, conferences, recordings and publications; to enhance and promote informed scholarship in all aspects of his life and work through the publication of a regular newsletter and an annual journal, available free to Society members; and to assist the Gloucestershire Archives in maintaining the Ivor Gurney collection. Publishes occasional papers including unpublished letters, poems and compositions. Society events are held each spring and autumn when talks and recitals enable members to explore Gurney's verse and music.

## The J.B. Priestley Society

Eldwick Crag Farm, Otley Road, Bingley BD16 3BB
☎ 01274 563078
reavill@globalnet.co.uk
www.jbpriestley-society.com
President *Tom Priestley*
Chairman *Lee Hanson*
Honorary Secretary *R.E.Y. Slater*
Membership Secretary *Tony Reavill*
SUBSCRIPTION £12 (individual); £7 (concession); £18 (group/family)

Founded in 1997 to widen the knowledge and understanding of Priestley's works; promote the study of his life and his social, cultural and political influences; provide members of the Society with lectures, seminars, films, journals and stimulate education projects; promote public performances of his works and the distribution of material associated with him. For further information, contact the Membership Secretary at the address above.

## James Hanley Network

Old School House, George Green Road, George Green, Wexham SL3 6BJ
☎ 01753 578632
gostick@london.com
www.jameshanley.inspiron.co.uk/INDEX.HTM
Network Coordinator *Chris Gostick*

An informal international association founded in 1997 for all those interested in exploring and publicizing the works and contribution to literature of the novelist and dramatist James Hanley (1901–1985). Publishes an annual newsletter. Occasional conferences are planned for the future. All enquiries welcome.

## James Hilton Society

49 Beckingthorpe Drive, Bottesford, Nottingham NG13 0DN
www.jameshiltonsociety.co.uk

Honorary Secretary *J.R. Hammond*
SUBSCRIPTION £12 (UK/EU); £9 concession

Founded in 2000 to promote interest in the life and work of novelist and scriptwriter James Hilton (1900–54). Publishes *The Hiltonian* (annually) and *The James Hilton Newsletter* (quarterly) and organizes meetings and conferences.

## Jane Austen Society

9 Nicola Close, South Croydon CR2 6NA
www.janeaustensociety.org.uk
Secretary *Mrs Maureen Stiller*
Membership Secretary *Mrs Rosemary Culley*
SUBSCRIPTION UK: £10 (student for 3 years); £20 (standard annual); £25 (joint); £50 (corporate); £300 (life); Overseas: £80 (annual) all by credit card or UK banker's order only

Founded in 1940 to promote interest in and enjoyment of Jane Austen's novels and letters. The Society has the following Branches/Groups: Bath and Bristol, Cambridge, Northern, Midlands, London, Kent, Hampshire, Isle of Wight, Southern Circle, Wales and Scotland. There are independent Societies in North America, Australia and Japan.

## Jerome K. Jerome Society

c/o Fraser Wood, Mayo and Pinson, 15/16 Lichfield Street, Walsall WS1 1TS
☎ 01922 629000  ℻ 01922 721065
tonygray@jkj.demon.co.uk
www.jeromekjerome.com
Honorary Secretary *Tony Gray*
SUBSCRIPTION £12 p.a. (individual); £8 (student); £15 (family); £30 (corporate)

Founded in 1984 to stimulate interest in Jerome K. Jerome's life and works (1859–1927). Meetings, lectures, events and a newsletter, *Idle Thoughts*. Annual dinner in Walsall near Jerome's birth date (2nd May).

## The John Buchan Society

'The Toft', 37 Waterloo Road, Lanark ML11 7QH
☎ 01555 662103
glennismac2000@yahoo.co.uk
www.johnbuchansociety.co.uk
Secretary *Mrs Glennis McClemont*
Membership Secretary *Diana Durdon*
SUBSCRIPTION £15 (full/overseas); £120 (10 years)

To perpetuate the memory of John Buchan and to promote a wider understanding of his life and works. Holds regular meetings and social gatherings, publishes a journal, and liaises with the John Buchan Centre at Broughton in the Scottish borders.

## The John Clare Society

9 The Chase, Ely CB6 3DR
☎ 01353 668438
www.johnclare.org.uk human.ntu.ac.uk/clare/clare.html
Honorary Secretary *Miss Sue Holgate*
SUBSCRIPTION £12 (individual); £15 (joint); £10 (retired); £15 (group); £17 (library); £5 (student, full-time); £17 sterling draft/$34/€28 (eurocheque) (overseas)

Founded in 1981 to promote a wider appreciation of the life and works of the poet John Clare (1793–1864). Organizes an annual festival in Helpston in July; arranges exhibitions, poetry readings and conferences; and publishes an annual society journal and quarterly newsletter.

## The John Masefield Society

The Frith, Ledbury HR8 1LW
☎ 01531 633800
ies.sas.ac.uk/cmps/Projects/Masefield/Society/jmsws.htm
Chairman *Peter Carter*
SUBSCRIPTION £5 p.a. (individual); £8 (family, institution, library); £10 (overseas); £2.50 (junior, student)

Founded in 1992 to stimulate the appreciation of and interest in the life and works of John Masefield (Poet Laureate 1930–67). The Society is based in Ledbury, the Herefordshire market town of his birth and holds various public events in addition to publishing a journal and occasional papers.

## The John Meade Falkner Society

Greenmantle, Main Street, Kings Newton, Melbourne DE73 8BX
☎ 01332 865315
www.johnmeadefalknersociety.co.uk
Secretary *Kenneth Hillier*
SUBSCRIPTION £5 (UK); $10 (overseas)

Founded in 1999 to promote the appreciation and study of John Meade Falkner's life, times and works. Produces three newsletters a year and an annual journal.

## Johnson Society

Johnson Birthplace Museum, Breadmarket Street, Lichfield WS13 6LG
☎ 01543 264972
info@thejohnsonsociety.org.uk
www.thejohnsonsociety.org.uk
Chairman *Mary Baker*
SUBSCRIPTION £10 p.a.; £15 (joint)

Founded in 1910 to encourage the study of the life, works and times of Samuel Johnson (1709–84) and his contemporaries. The Society is committed to the support and the preservation of the Johnson Birthplace Museum and Johnson memorials. Journal, *Transactions*, published annually.

## Johnson Society of London

255 Baring Road, Grove Park, London SE12 0BQ
☎ 020 8851 0173
info@johnson.org.uk
www.johnsonsoc.org.uk
President *Lord Harmsworth*
Honorary Secretary *Mrs Z.E. O'Donnell*
SUBSCRIPTION £20 p.a.; £25 (joint); £15 (student)

Founded in 1928 to promote the knowledge and appreciation of Dr Samuel Johnson and his works. Publishes an annual journal, *New Rambler* and occasional newsletter, *The New Idler*. Regular meetings from October to April in the meeting room of Wesley's Chapel, City Road, London EC1 on the second Saturday of each month, and a commemoration ceremony around the anniversary of Johnson's death (December) held in Westminster Abbey.

## Joseph Conrad Society (UK)

c/o The Polish Social and Cultural Association (POSK), 238–246 King Street, London W6 0RF
theconradian@aol.com
www.josephconradsociety.org
Honorary Secretary *Hugh Epstein*
Treasurer/Editor (The Conradian) *Allan Simmons*
SUBSCRIPTION £25 p.a. (individual)

Founded in 1973 to promote the study of the works and life of Joseph Conrad (1857–1924). A scholarly society, supported by the Polish Library at the Polish Cultural Association where a substantial library of Conrad texts and criticism is held by the **Polish Library**. Publishes a journal of Conrad studies (*The Conradian*) biannually and holds an annual International Conference in the first week of July. Awards an essay prize and travel and study grants to scholars on application.

## The Just William Society

Easter Badbea, Dundonnell IV23 2QX
philandpaula@easter-badbea.co.uk
Membership Seretary *Paula Wild*
SUBSCRIPTION £10 p.a. (UK); £12 (overseas); £5 (juvenile/student); £15 (family)

Founded in 1994 to further knowledge of Richmal Crompton's *William* and *Jimmy* books. The *Just William Society Magazines* are published throughout the year and an annual 'William' meeting is held in April.

## The Keats–Shelley Memorial Association (Inc)

*Registered office:* Bedford House, 76A Bedford Street, Leamington Spa CV32 5DT
☎ 01926 427400   🖷 01926 335133
Contact *Honorary Secretary*

SUBSCRIPTION £15 p.a.

Founded in 1903 to promote appreciation of the works of Keats and Shelley, and their circle. One of the Society's main tasks is the preservation of 26 Piazza di Spagna in Rome as a memorial to the British Romantic poets in Italy, particularly Keats and Shelley. Publishes an annual review of Romantic Studies called the *Keats-Shelley Review*, arranges events and lectures for Friends and promotes bursaries and competitive writing on Romantic Studies (see **Keats-Shelley Prize** under *Prizes*). The *Review* is edited by *Professor Nicholas Roe*, c/o Dept. of English, St Andrews University KY16 9AL.

## Kenny/Naughton Society

Aghamore, Ballyhaunis,, Co. Mayo, Republic of Ireland
☎ 00 353 94 936 7019
paulwdrogers@eircom.net
www.aghamoreireland.com/kennynaughton/society.htm
Chairman *Paul Rogers*

Founded in 1993 for those with a special interest in the works of two famous writers with Aghamore links – P.D. Kenny and Bill Naughton (author of *Alfie*). Society members go on regular outings and meet monthly with the view to planning for the Annual Kenny/Naughton Autumn School, a literary gathering which honours the memory of Kenny and Naughton through lectures, talks, readings, workshops, writing competitions, drama, tours and entertainment.

## Kent & Sussex Poetry Society

39 Rockington Way, Crowborough TN6 2NJ
☎ 01892 662781
info@kentandsussexpoetrysociety.org
www.kentandsussexpoetrysociety.org
Publicity Secretary *John Arnold*
SUBSCRIPTION £15 p.a. (full); £10 (concessionary; country members living farther afield, senior citizen, under-16, unemployed)

Founded in 1946 to promote the enjoyment of poetry. Monthly meetings are held in Tunbridge Wells, including readings by major poets, a monthly workshop and an annual writing retreat week. Publishes an annual folio of members' work based on Members' Competition, adjudicated and commented upon by a major poet. Runs an annual **Open Poetry Competition** (see entry under *Prizes* and details on website) and Saturday workshops twice a year with leading poets.

## The Kilvert Society

30 Bromley Heath Avenue, Downend, Bristol BS16 6JP
☎ 0117 957 2030

www.communigate.co.uk/here/kilvertsociety
Honorary Secretary *Alan Brimson*
SUBSCRIPTION £12 p.a.; £15 (two persons at same address)

Founded in 1948 to foster an interest in the Diary kept by the Rev. Francis Kilvert from 1870 to 1879 and in the countryside he loved. The Society publishes three journals each year and during the spring and summer holds three weekends of walks, talks and commemoration services.

## The Kipling Society

6 Clifton Road, London W9 1SS
☎ 020 7286 0194    📠 020 7286 0194
jmkeskar@btinternet.com
www.kipling.org.uk
Honorary Secretary *Jane Keskar*
SUBSCRIPTION £24 p.a. (£22 p.a. by standing order)

Founded 1927. This is a literary society for all who enjoy the prose and verse of Rudyard Kipling (1865–1936) and are interested in his life and times. The Society's main activities are: maintaining a specialized library at the City University in Islington, London; answering enquiries from the public (schools, publishers, writers and the media); arranging a regular programme of lectures, especially in London and in Sussex, and an annual luncheon with guest speaker; maintaining a small museum and reference at The Grange, in Rottingdean near Brighton. The Society also publishes a quarterly journal. Please contact the Secretary by letter, telephone, fax or email for further information.

## The Kitley Trust

Woodstock, Litton Dale, Litton SK17 8QL
☎ 01298 871564
stottie2@waitrose.com
Contact *Rosie Ford*

Founded in 1990 by a teacher in Sheffield to promote the art of creative writing, in memory of her mother, Jessie Kitley. Activities include: biannual poetry competitions; a 'Get Poetry' day (distribution of children's poems in shopping malls); annual sponsorship of a writer for a school; campaigns; organizing conferences for writers and teachers of writing. Funds are provided by donations and profits (if any) from competitions.

## Lancashire Authors' Association

Heatherslade, 5 Quakerfields, Westhoughton, Bolton BL5 2BJ
☎ 01942 791390
eholt@cwctv.net
General Secretary *Eric Holt*

SUBSCRIPTION £10 p.a.; £13 (joint); £1 (junior)

Founded in 1909 for writers and lovers of Lancashire literature and history. Aims to foster and stimulate interest in Lancashire history and literature as well as in the preservation of the Lancashire dialect. Meets three times a year on Saturday at various locations. Publishes a quarterly journal called *The Record*, which is issued free to members, and holds annual competitions (open to members only) for both verse and prose. Comprehensive library with access for research to members.

## The Landor Society of Warwick

1 Avon Court, 51 Kenilworth Road, Leamington Spa CV32 6JH
☎ 01926 337874

Now a dormant society with occasional meetings only. Enquiries concerning W.S. Landor to *Jean Field* at the address above.

## The Laurence Sterne Trust
▷ The Friends of Shandy Hall

## The Leamington Literary Society

52 Newbold Terrace East, Leamington Spa CV32 4EZ
☎ 01926 425733
Honorary Secretary *Mrs Margaret Watkins*
SUBSCRIPTION £10 p.a.

Founded in 1912 to promote the study and appreciation of literature and the arts. Holds regular meetings every second Tuesday of the month (from September to June) at the Royal Pump Rooms, Leamington Spa. Also smaller groups which meet regularly to study poetry and the modern novel. The Society has published various books of local interest.

## Lewes Monday Literary Club

c/o 1c Prince Edward's Road, Lewes BN7 1BJ
☎ 01273 478512
chris.lutrario@btinternet.com
Contact *Christopher Lutrario*
SUBSCRIPTION £20 p.a.; £5 (guest, per meeting)

Founded in 1948 for the promotion and enjoyment of literature. Seven meetings are held during the winter generally on the last Monday of each month (from October to April) at the Pelham House Hotel in Lewes. The Club attracts speakers of the highest quality and a balance between all forms of literature is aimed for. Guests are welcome to attend meetings.

## Lewis Carroll Society

50 Lauderdale Mansions, Lauderdale Road, London W9 1NE
www.lewiscarrollsociety.org.uk

SUBSCRIPTION Details on the website

Founded in 1969 to bring together people with an interest in Charles Dodgson and promote research into his life and works. Publishes bi-annual journal, *The Carrollian*, featuring scholarly articles and reviews; a newsletter (*Bandersnatch*) which reports on Carrollian events and the Society's activities; and *The Lewis Carroll Review*, a book-reviewing journal. Regular meetings held in London with lectures, talks, outings, etc.

### Lewis Carroll Society (Daresbury)

Blue Grass Cottage, Clatterwick Lane, Little Leigh, Northwich CW8 4RJ
☎ 01606 891303 (day)/781731 (evening)
Honorary Secretary *Kenneth N. Oultram*
SUBSCRIPTION £7–10 p.a.

Founded 1970. To promote the life and work of Charles Dodgson, author of the world-famous *Alice's Adventures*. Holds meetings in the spring and autumn in Carroll's birthplace, Daresbury, Cheshire. Guest speakers, a newsletter and theatre visits.

### The Malcolm Saville Society

35 Chapel Road, Penketh, Warrington WA5 2NG
mystery@witchend.com
www.witchend.com
Secretary *George Jasieniecki*
SUBSCRIPTION £10 p.a. (UK); £12.50 (EU); £16 (RoW)

Founded in 1994 to remember and promote interest in the work of the popular children's author. Regular social activities, library, contact directory and four magazines per year.

### Margery Allingham Society

39 Morgan Street, London E3 5AA
www.margeryallingham.org.uk
Honorary Treasurer *Paula Cooze*
SUBSCRIPTION £14 p.a.

Founded in 1988 to promote interest in and study of the works of Margery Allingham. The Society publishes two issues of the journal, *The Bottle Street Gazette*, per year. Contributions welcome. Four social events a year. Open membership.

### The Marlowe Society

9 Middlefield Gardens, Hurst Green Road, Halesowen B62 9QH
☎ 0121 421 1482    ☎ 0121 421 1482
prwales@hotmail.com
www.marlowe-society.org
Membership Secretary *Frieda Barker*
Treasurer *Bruce Young*

SUBSCRIPTION £15 p.a. (individual); £10 p.a. (pensioner/student/unwaged); £20 p.a. (overseas); £20 (joint/couple); £200 (group)

Founded 1955. Holds meetings, lectures and discussions, stimulates research into Marlowe's life and works, encourages production of his plays and publishes a biannual newsletter. At a ceremony on 11th July 2002 the Society celebrated their success in establishing a memorial to the playwright and poet in Poets' Corner in Westminster Abbey.

### Mary Webb Society

8 The Knowe, Willaston, Neston CH64 1TA
☎ 07816 424712
suehigginbotham@yahoo.co.uk
www.marywebbsociety.co.uk
Secretary *Sue Higginbotham*
SUBSCRIPTION UK: £11 p.a. (individual); £14 p.a. (joint); Overseas: £14 p.a. (individual); £16 p.a. (joint)

Founded 1972. Attracts members from the UK and overseas who are devotees of the literature of Mary Webb and of the beautiful Shropshire countryside of her novels. Publishes biannual journal, organizes summer schools in various locations related to the authoress's life and works. Archives; lectures; tours arranged for individuals and groups.

### The Neil Munro Society

4 Randolph Road, Glasgow G11 7LG
☎ 0141 339 5425
ronnierenton@googlemail.com
www.neilmunro.co.uk
Secretary *Margaret Renton*

Founded in 1996 to encourage interest in the works of Neil Munro (1863–1930), the Scottish novelist, short story writer, poet and journalist. An annual programme of meetings is held in Glasgow and Munro's home-town of Inveraray. Publishes *ParaGraphs*, a twice-yearly magazine, sponsors reprints of Munro's work and is developing a Munro archive.

### The New Chalet Club

18 Nuns Moor Crescent, Newcastle upon Tyne NE4 9BE
www.newchaletclub.co.uk
Contact *Membership Secretary*
SUBSCRIPTION £6 p.a. (UK under 18); £10 (UK adult & Europe); £12 (RoW)

Founded in 1995 for all those with an interest in the books of Elinor Brent-Dyer. Publishes a quarterly journal and occasional supplements, including *Children's Series Fiction*. Regularly holds local and national meetings.

## The Oscar Wilde Society

22 Edric Road, London SE14 5EL
michael.seeney@btinternet.com
www.oscarwildesociety.co.uk
Chairman *Donald Mead*
Deputy Chairman *Michael Seeney*
SUBSCRIPTION £25 p.a. (UK); £30 p.a. (Europe);
£35 p.a. (RoW)

Founded in 1990 to promote knowledge,
appreciation and study of the life, personality
and works of the writer and wit Oscar Wilde.
Activities include meetings, lectures, readings
and exhibitions, and visits to locations associated
with Wilde. Members receive a journal of Oscar
Wilde studies, *The Wildean*, twice-yearly and a
newsletter journal, *Intentions* (six per year).

## The P.G. Wodehouse Society (UK)

26 Radcliffe Road, Croydon CR0 5QE
info@pgwodehousesociety.org.uk
www.pgwodehousesociety.org.uk
Membership Secretary *Christine Hewitt*
SUBSCRIPTION £15 p.a. (£20 Dec. to March, for
membership extended to May the following year)

Formed in 1997 to promote enjoyment of the
works of P.G. Wodehouse, author, lyricist,
playwright and journalist, creator of Lord
Emsworth, Jeeves & Wooster and many more.
Membership of 1,000+. Publications include the
quarterly journal, *Wooster Sauce*, and regular *By
The Way* papers. Meetings and events are held in
London and around the country. The Society is
allied to Wodehouse societies around the world.

## The Parson Woodforde Society

22 Gaynor Close, Wymondham NR18 0EA
01953 604124
editor@parsonwoodforde.org
www.parsonwoodforde.org.uk
Membership Secretary *Mrs Ann Elliott*
SUBSCRIPTION £16 (UK); £25 (overseas)

Founded 1968. Aims to extend and develop
knowledge of James Woodforde's life and the
society in which he lived and to provide the
opportunity for fellow enthusiasts to meet
together in places associated with the diarist.
Publishes a quarterly journal and newsletter. The
Society has produced a complete edition of the
diary of James Woodforde in 17 volumes covering
the period 1759–1802.

## The Philip Larkin Society

PO Box 44, Hornsea HU18 1WP
www.philiplarkin.com
General Secretary *Andrew Eastwood*
SUBSCRIPTION £20 (full rate); £14 (unwaged/
senior citizen); £6 (student); £50 (institutions)

Founded in 1995 to promote awareness of the
life and work of Philip Larkin (1922–85) and his
literary contemporaries; to bring together all
those who admire Larkin's work as a poet, writer
and librarian; to bring about publications on all
things Larkinesque. Organizes a programme of
events ranging from lectures to rambles exploring
the countryside of Larkin's schooldays and
publishes a biannual journal, *About Larkin*.

## The Powys Society

Flat D, 87 Ledbury Road, London W11 2AG
020 7243 0168
chris.d.thomas@hotmail.co.uk
www.powys-society.org
Honorary Secretary *Chris Thomas*
SUBSCRIPTION £18.50 (UK); £22 (overseas); £10
(student)

The Society (with a membership of 297) aims to
promote public education and recognition of the
writings, thought and contribution to the arts of
the Powys family; particularly of John Cowper,
Theodore and Llewelyn, but also of the other
members of the family and their close associates.
The Society holds a major collection of published
works, letters, manuscripts, photographs and
memoribilia devoted to the the Powys family
which is housed in the Dorset County Museum.
Publishes the *Powys Society Newsletter* in April,
June and November and *The Powys Journal*
in August. The Society organizes an annual
conference with lectures and walks to places with
Powysian associations, as well as other regional
meetings and events throughout the year.

## The Queen's English Society

1 Oban Gardens, Reading RG5 3RG
enquiries@queens-english-society.com
www.queens-english-society.com
President *Dr Bernard Lamb*
Honorary Secretary *Sidney Callis*
Honorary Media Consultant *Ken Thomson*
SUBSCRIPTION £20 p.a. (ordinary); £25 (joint); £12
(concession)

Founded in 1972 to promote and uphold the use
of good English and to encourage the enjoyment
of the language. Holds regular meetings and
conferences to which speakers are invited, an
annual luncheon and publishes a quarterly
journal, *Quest*, for which original articles are
welcome. (Registered Charity No. 272901.)

## The R.S. Surtees Society

Manor Farm House, Nunney, Near Frome
BA11 4NJ
01373 836937    01373 836574
rssurtees@fsmail.net
www.r.s.surteessociety.org
Contact *Orders and Membership Secretary*
(*01373 302155*)

Founded in 1979 to republish the works of R.S. Surtees and others.

## Randolph Caldecott Society

Blue Grass Cottage, Clatterwick Lane, Little Leigh, Northwich CW8 4RJ
☎ 01606 891303 (day)/781731 (evening)
Honorary Secretary *Kenneth N. Oultram*
SUBSCRIPTION £10–£15 p.a.

Founded in 1983 to promote the life and work of artist/book illustrator Randolph Caldecott. Meetings held in the spring and autumn in Caldecott's birthplace, Chester. Guest speakers, outings, newsletter, exchanges with the Society's American counterpart. (Caldecott died and was buried in St Augustine, Florida.) A medal in his memory is awarded annually in the US for children's book illustration.

## Rev. G. Bramwell Evens
▷ The Romany Society

## The Rhys Davies Trust

10 Heol Don, Whitchurch, Cardiff CF14 2AU
☎ 029 2062 3359
Contact *Professor Meic Stephens*

Founded in 1990 to perpetuate the literary reputation of the Welsh prose writer, Rhys Davies (1901–78), and to foster Welsh writing in English. Organizes competitions in association with other bodies such as the **Welsh Academy**, puts up plaques on buildings associated with Welsh writers, offers grant-aid for book production, etc.

## Richard Jefferies Society

Pear Tree Cottage, Longcot SN7 7SS
☎ 01793 783040
info@richardjefferiessociety.co.uk
richardjefferiessociety.co.uk
Honorary Secretary *Jean Saunders*
Membership Secretary *Mrs Margaret Evans*
SUBSCRIPTION £7 p.a. (individual); £8 (joint); life membership for those over 60

Founded in 1950 to promote understanding of the work of Richard Jefferies, nature/country writer, novelist and mystic (1848–87). Produces newsletters, reports, out-of-print works and an annual journal; organizes talks, discussions and readings. Library and archives. Assists in maintaining the museum in Jefferies' birthplace at Coate near Swindon. Membership applications should be sent to *Margaret Evans*, 23 Hardwell Close, Grove, Nr Wantage OX12 0BN.

## Rider Haggard Society

27 Deneholm, Whitley Bay NE25 9AU
☎ 0191 252 4516    ☏ 0191 252 4516
rb27allen@blueyonder.co.uk
www.riderhaggardsociety.org.uk
Contact *Roger Allen*

SUBSCRIPTION £10 p.a. (UK); £13 (overseas; £36 for 3 years)

Founded in 1985 to promote appreciation of the life and works of Sir Henry Rider Haggard, English novelist, 1856–1925. News/books exchange and annual meetings. Vast source for Haggard purchases, research and information.

## The Robert Burns World Federation Ltd

Dean Castle Country Park Dower House, Kilmarnock KA3 1XB
☎ 01563 572469    ☏ 01563 572469
admin@robertburnsfederation.com
www.worldburnsclub.com
Chief Executive *Sam Judge*
Office Administrator *Margaret Craig*
SUBSCRIPTION £25 p.a. (individual); £27 (family); £46 (club subscription )

Founded in 1885 to encourage interest in the life and work of Robert Burns and keep alive the old Scottish Tongue. The Society's interests go beyond Burns himself in its commitment to the development of Scottish literature, music and arts in general. Publishes the triannual *Burns Chronicle*.

## The Robert Graves Society

University of London Institute in Paris, 11 rue de Constantine, 75340 Paris Cedex 07, France
☎ 00 33 1 43 29 33 48
dunstanward@yahoo.com
www.robertgraves.org/
President *Professor Dunstan Ward*
Vice-President (Europe) *Fran Brearton* (*f.brearton@qub.ac.uk*)
Secretary/Treasurer *Patrick Villa* (*pjvilla@aol.com*)
SUBSCRIPTION £17, €30, $40 p.a.

The Society aims to promote interest in, and research on the life and works of Robert Graves (1895–1985), author of some 140 books of poetry, fiction, biography, criticism, anthropology, social history, mythology, biblical studies, translation and children's books. Organizes an international conference every two years. The 2010 conference will be held in Palma and Deià, Mallorca, 6th–10th July. Publishes annually the leading journal of Graves studies, *Gravesiana*, edited by Professor Dunstan Ward and Dr Fran Brearton. The Society's membership of more than 200 experts and enthusiasts forms the core of an international research community.

## The Robert Louis Stevenson Club

7 Stanley Road, Edinburgh EH6 4SE
Membership Secretary *Neil M. Ross*

SUBSCRIPTION £20 p.a. (individual); £26 p.a. (overseas); £150 (10-year)

Founded in 1920 to foster interest in Robert Louis Stevenson's life and works. The Club organizes an annual lunch and other events. Publishes *RLS Club News* twice a year.

## The Robert Southey Society

1 Lewis Terrace, Abergarwed, Neath SA11 4DL
☎ 01639 711480
Contact *Robert King*
SUBSCRIPTION £10 p.a.

Founded in 1990 to promote the work of Robert Southey. Publishes an annual newsletter and arranges talks on his life and work. Open membership.

## The Romany Society

62 Thornton Avenue, Macclesfield SK11 7XL
phil_shelley@btopenworld.com
www.romanysociety.org.uk
Contact *Membership Secretary (at address above)*
SUBSCRIPTION £5 (individual); £9 (family); £10 (institution); £3.50 (student/unwaged)

Promotes the life, work and conservation message of the Rev. G. Bramwell Evens – 'Romany' – the first broadcasting naturalist who influenced generations of listeners with his *Out With Romany* radio programmes on the BBC during the 1930s and 1940s and with his series of natural history books. The Romany Society was established originally following Evens' death at the age of 59 in 1943 and ran until 1965. Revived in 1996, the Society holds an annual members' weekend to areas of significance in the Romany story, erects memorial plaques where appropriate, encourages media coverage of Romany and supports young naturalists with the Romany Memorial Grant.

## The Rupert Brooke Society

40 Belmore Road, Thorpe, Norwich NR7 0PT
☎ 01603 438603
chairman@rupertbrooke.com
www.rupertbrooke.com
Chairman *Lorna Beckett*
Membership Secretary *Wes Smith (membership@rupertbrooke.com)*
SUBSCRIPTION £10 (UK); £15 (overseas)

Founded in 1999 to foster an interest in the life and work of Rupert Brooke and help preserve places associated with him. Members receive an annual journal with in-depth, academic articles relating to Brooke, and a newsletter with news, information about events, book reviews and activities. The Society stages three events a year, each held near a significant date associated with Rupert Brooke's life: 23rd April, the anniversary of Brooke's death in 1915 – an author of note is

usually invited to give a talk and sign books; 3rd August, Brooke's birthday in 1887 – the Society holds its AGM and accompanying event; 11th November, Armistice Day – an Armistice concert is held in Grantchester Church.

## The Ruskin Society of London

20 Parmoor Court, Summerfield, Oxford OX2 7XB
☎ 01865 310987
Honorary Secretary *Miss O.E. Forbes-Madden*
SUBSCRIPTION £15 p.a.

Founded in 1986 to promote interest in John Ruskin (1819–1900) and his contemporaries. All aspects of Ruskinia are introduced. Functions are held in London. Publishes the annual *Ruskin Gazette*, a journal concerned with Ruskin's influence. Affiliated to other literary societies.

## The Ruskin Society

49 Hallam Street, London W1W 6JP
☎ 020 7580 1894
c.gamble@zen.co.uk
www.theruskinsociety.co.uk
President *Professor Michael Wheeler*
Chairman *Dr Malcolm Hardman*
Vice-Chairman *Dr Cynthia J. Gamble*
SUBSCRIPTION £10 p.a. (payable on 1st Jan.)

Founded in 1997 to encourage a wider understanding of John Ruskin and his contemporaries. Organizes lectures and events which seek not only to explain to the public at large the nature of Ruskin's theories but also to place these in a modern context.

## The Shaw Society

Flat F, 10 Compton Road, London N1 2PA
☎ 020 7226 4266
shawsociety@blueyonder.co.uk
Honorary Secretary *Alan Knight*
SUBSCRIPTION £15 p.a. (individual); £22 (joint)

Founded in 1941 to promote interest in the life and works of G. Bernard Shaw. Meetings are held on the last Friday of every month (except July, August and December) at Conway Hall, Red Lion Square, London WC1 (6.30 pm for 7.00 pm) at which speakers are invited to talk on some aspect of Shaw's life or works. Monthly playreadings are held on the first Thursday of each month (except August and January). Over 40 years ago the Society began the annual open-air performance held at Shaw's Corner, Ayot St Lawrence in Hertfordshire, on the weekend nearest to Shaw's birthday (26th July). Publishes a biannual magazine, *The Shavian*, with a calendar of Shavian events. (No payment for contributors.)

## Sherlock Holmes

> The Baskerville Hounds and The Franco-Midland Hardware Company

## Sherlock Holmes Society (The Musgraves)

Hallas Lodge, Greenside Lane, Cullingworth, Bradford BD13 5AP
℡ 01535 273468
hallaslodge@btinternet.com
Contact *Anne Jordan*
SUBSCRIPTION £17 p.a.; £18 (joint)

Founded in 1987 to promote enjoyment and study of Sir Arthur Conan Doyle's Sherlock Holmes through publications and meetings. One of the largest Sherlock Holmes societies in Great Britain. Honorary members include Bert Coules, Edward Hardwicke and Clive Merrison. Past honorary members: Dame Jean Conan Doyle, Peter Cushing, Jeremy Brett, Michael Williams and Richard Lancelyn Green. Open membership. Lectures, presentations and consultation on matters relating to Holmes and Conan Doyle available.

## Sylvia Townsend Warner Society

2 Vicarage Lane, Fordington, Dorchester DT1 1LH
℡ 01305 266028
judithmbond@tiscali.co.uk
www.townsendwarner.com
Contact *Eileen Johnson*
SUBSCRIPTION £10 p.a.; $25 (overseas)

Founded in 2000 to promote a wider readership and better understanding of the writings of Sylvia Townsend Warner.

## The Tennyson Society

Central Library, Free School Lane, Lincoln LN2 1EZ
℡ 01522 552862
kathleen.jefferson@lincolnshire.gov.uk
www.tennysonsociety.org.uk
Honorary Secretary *Miss K. Jefferson*
SUBSCRIPTION £10 p.a. (individual); £12 (family); £15 (corporate); £160 (life)

Founded in 1960. An international society with membership worldwide. Exists to promote the study and understanding of the life and work of Alfred, Lord Tennyson. The Society is concerned with the work of the Tennyson Research Centre, 'probably the most significant collection of manuscripts, family papers and books in the world'. Publishes annually the *Tennyson Research Bulletin*, which contains articles and critical reviews; and organizes lectures, visits and seminars. Annual memorial service at Somersby in Lincolnshire.

## The Thomas Hardy Society

PO Box 1438, Dorchester DT1 1YH
℡ 01305 251501     📠 01305 251501
info@hardysociety.org
www.hardysociety.org
Honorary Secretary *Mike Nixon*
SUBSCRIPTION £18 (individual); £25 (corporate); £22.50 (individual, overseas); £30 (corporate, overseas)

Founded in 1967 to promote the reading and study of the works and life of Thomas Hardy. Thrice-yearly journal, events and a biennial conference. An international organization.

## Thomas Lovell Beddoes Society

9 Amber Court, Belper DE56 1HG
℡ 01773 828066     📠 01773 828066
john@beddoes.demon.co.uk
www.phantomwooer.org
Chairman *John Lovell Beddoes*

Formed to research the life, times and work of poet Thomas Lovell Beddoes (1803–49), encourage relevant publications, further the reading and appreciation of his works by a wider public and liaise with other groups and organizations. Publishes an annual journal.

## Thomas Paine Society

19 Charles Rowan House, Margery Street, London WC1X 0EH
℡ 020 7833 1305
postmaster@thomaspainesocietyuk.org.uk
www.thomaspainesocietyuk.org.uk
Chair *Chad Goodwin*
Secretary *Barbara Jacobson (at address above)*
Treasurer *Stuart Wright*
SUBSCRIPTION (minimum) £15 p.a. (UK); $35 (overseas); £5 (unwaged/pensioner/student)

Founded in 1963 to promote the life and work of Thomas Paine and continues to expound his ideals. Meetings, newsletters, lectures and research assistance. The Society has members worldwide and keeps in touch with American and French Thomas Paine associations. Publishes magazine, *The Journal of Radical History*, twice yearly (editor, *R.W. Morrell*, 43 Eugene Gardens, Nottingham NG2 3LF) and a newsletter. Holds occasional exhibitions and lectures, including the biannual Thomas Paine Memorial Lecture and the annual Eric Paine Memorial Lecture.

## The Tolkien Society

655 Rochdale Road, Walsden, Todmorden OL14 6SX
www.tolkiensociety.org
Membership Secretary *Marion Kershaw*
SUBSCRIPTION £21 p.a. (UK); £24 (EU); £27 (RoW)

An international organization which aims to encourage and further interest in the life and works of the late Professor J.R.R. Tolkien, CBE, author of *The Hobbit* and *Lord of the Rings*. Current membership stands at 1,100. Publishes *Mallorn* annually and *Amon Hen* bi-monthly.

## The Trollope Society

Maritime House, Old Town, Clapham, London SW4 0JW
☎ 020 7720 6789   📠 020 7627 2965
info@trollopesociety.org
www.trollopesociety.org

Founded in 1987 to study and promote Anthony Trollope's works. Publishes the complete works of Trollope's novels and travel books.

## Violet Needham Society

c/o 19 Ashburnham Place, London SE10 8TZ
☎ 020 8692 4562
richardcheffins@aol.com
violetneedhamsociety.org.uk
Honorary Secretary *R.H.A. Cheffins*
SUBSCRIPTION £7.50 p.a. (UK & Europe); £11 (RoW)

Founded in 1985 to celebrate the work of children's author Violet Needham and stimulate critical awareness of her work. Publishes thrice-yearly *Souvenir*, the Society journal, including a newsletter; organizes meetings and excursions to places associated with the author and her books. The journal includes articles about other children's writers of the 1940s and 1950s and on ruritanian fiction. Contributions welcome.

## The Virginia Woolf Society of Great Britain

Fairhaven, Charnleys Lane, Banks, Southport PR9 8HJ
stuart.n.clarke@btinternet.com
www.virginiawoolfsociety.co.uk
Membership Secretary *Stuart N. Clarke*
SUBSCRIPTION £16 p.a.; £21 (overseas)

Founded in 1998 to promote interest in the life and work of Virginia Woolf, author, essayist and diarist. The Society's activities include trips away, walks, reading groups and talks. Publishes a literary journal, *Virginia Woolf Bulletin*, three times a year.

## W.W. Jacobs Appreciation Society

3 Roman Road, Southwick BN42 4TP
☎ 01273 596217
Contact *A.R. James*

Founded in 1988 to encourage and promote the enjoyment of the works of W.W. Jacobs, and stimulate research into his life and works. No subscription charge. Material available for purchase includes *W.W. Jacobs*, a biography published in 1999, price £12, post paid, and *WWJ Book Hunter's Field Guide*, a narrative bibliography published in 2001, price £6, post

paid. An expanded biography is due to be published.

## Wainwright Society

Kendal Museum, Station Road, Kendal LA9 6BT
☎ 01539 721374
membership@wainwright.org.uk
www.wainwright.org.uk
Membership Secretary *Morag Clement*
SUBSCRIPTION £12 p.a. (per household, from 1st Jan. 2010)

Founded in 2002 to keep alive the things that Alfred Wainwright (1907–91) promoted through his guidebooks (*Pictorial Guides to the Lakeland Fells*), started 50 years ago, and the many other publications which were the 'labour of love' of a large portion of his life. Produces a newsletter three times per year, organizes walks, events, annual dinner and lecture.

## The Walmsley Society

April Cottage, No 1 Brand Road, Hampden Park, Eastbourne BN22 9PX
☎ 01323 506447
www.walmsleysoc.org
Honorary Secretary *Fred Lane*
SUBSCRIPTION £13 p.a.; £15 (family); £11 (student/senior citizen); £12 (overseas, £20, one year; £35, two years)

Founded in 1985 to promote interest in the writings of Leo Walmsley and to foster an appreciation of the work of his father, the artist Ulric Walmsley. Two annual meetings – one held in Robin Hood's Bay on the East Yorkshire coast, spiritual home of the author Leo Walmsley. Publishes a journal twice-yearly and newsletters, and is involved in other publications which benefit the aims of the Society. A biography of Leo Walmsley is now available.

## The Walter de la Mare Society

3 Hazelwood House, New River Crescent, London N13 5RE
☎ 020 8886 1771
fguthrie@talktalk.net
Honorary President *Professor John Bayley*, CBE
Honorary Secretary & Treasurer *Frances Guthrie*
SUBSCRIPTION £15 p.a.

Founded in 1997 to honour the memory of Walter de la Mare; to promote the study and deepen the appreciation of his works; to widen the readership of his works; to facilitate research by making available the widest range of contacts and information about de la Mare; and to encourage and facilitate new Walter de la Mare publications. Produces a regular newsletter and organizes events. Membership information available from the Secretary at the address above.

## The Wilfred Owen Association

29 Arthur Road, London SW19 7DN
☎ 020 8947 0476
www.wilfredowen.org.uk
Chairman *Mrs Meg Crane*
Secretary *Mrs Vanessa Davis (at address above)*
SUBSCRIPTION £10 adult (£8 concessions); £15
couple (£13 concessions); £15 overseas (adult); £20
overseas (couple)

Founded in 1989 to commemorate the life and
works of Wilfred Owen by promoting readings,
visits, talks and performances relating to Owen
and his work. The Association offers practical
support for students of literature and future
poets through links with education, support for
literary foundations and information on historical
and literary background material. Membership
is international with around 380 members.
Publishes regular newsletters. Speakers are
available.

## Wilkie Collins Society

4 Ernest Gardens, London W4 3QU
☎ 020 8747 0115
paul@wilkiecollins.org
www.wilkiecollins.org
Chairman *Andrew Gasson*
Membership Secretary *Paul Lewis (at address above)*
SUBSCRIPTION £10 (UK/Europe); £18 (RoW;
remittance must be made in UK sterling or by
PayPal to paul@paullewis.co.uk)

Founded in 1980 to provide information on and
promote interest in the life and works of Wilkie
Collins, one of the first English novelists to deal
with the detection of crime. *The Woman in White*
appeared in 1860 and *The Moonstone* in 1868.
Publishes newsletters, reprints of Collins' work
and an annual academic journal.

## William Cobbett Society

3 Park Terrace, Tillington, Petworth GU28 9AE
☎ 01798 342008
www.williamcobbett.org.uk
Chairman *Barbara Biddell*
*Also:* Boynell House, Outlands Lane, Curdridge,
Southampton SO30 2HD ☎ 01489 789893
Contact *David Chun*
SUBSCRIPTION £8 p.a.

Founded in 1976 to bring together those with an
interest in the life and works of William Cobbett
(1763–1835) and to extend the interest to a wider
public. Society activities include an annual
Memorial Lecture; publication of an annual
journal (*Cobbett's New Register*) containing
articles on various aspects of his life and times;
an annual expedition retracing routes taken by
Cobbett on his Rural Rides in the 1820s; visits to
his birthplace and his tomb in Farnham, Surrey.

In association with the Society, the Museum
of Farnham holds bound volumes of *Cobbett's
Political Register*, a large collection of Cobbett's
works, books about Cobbett and has various
Cobbett artefacts on display.

## William Morris Society

Kelmscott House, 26 Upper Mall, Hammersmith,
London W6 9TA
☎ 020 8741 3735    🖷 020 8748 5207
william.morris@care4free.net
www.morrissociety.org
Curator *Helen Elletson*
SUBSCRIPTION £18 p.a.; £8 (student); £30
(corporate)

Founded in 1955 to promote interest in the life,
work and ideas of William Morris (1834–96),
English poet and craftsman. Organizes events
and educational programmes. Members receive
newsletters and biannual journal. Archive and
library open by appointment. Museum open
Thursdays and Saturdays, 2.00 pm to 5.00 pm.

## The Wordsworth Trust

Dove Cottage, Grasmere LA22 9SH
☎ 01539 435544    🖷 01539 463538
enquiries@wordsworth.org.uk
www.wordsworth.org.uk
Literature Officer *Paul Kleia*

Founded in 1880, the Trust is a living memorial
to the life and poetry of William Wordsworth
and his contemporaries. As Centre for British
Romanticism, the Wordsworth Trust, with
its wealth of manuscripts, books, drawings
and pictures, provides 'the full context for
understanding and celebrating a major cultural
moment in history in which Britain played a
profound role'. Research facilities are available at
the Jerwood Centre; contact: *Jeff Cowton*, Curator.
The Wordsworth Trust is also a centre for
contemporary poetry in the North West. (See also
**The Wordsworth Trust** entry under *Festivals*).
(Registered Charity No. 1066184.)

## WW2 HMSO PPBKS Society

3 Roman Road, Southwick BN42 4TP
☎ 01273 596217
Contact *A.R. James*

Founded in 1994 to encourage collectors and to
promote research into HMSO's World War II
series of paperbacks. Most of them were written
by well-known authors, though in many cases
anonymously. No subscription charge. Available
for purchase: Collectors' Guide (£5); Handbook,
*Informing the People* (£10).

## The Yeats Society Sligo

Yeats Memorial Building, Hyde Bridge, Sligo,
Republic of Ireland
☎ 00 353 71 914 2693    🖷 00 353 71 914 2780

info@yeats-sligo.com
www.yeats-sligo.com
President *Joe Cox*
SUBSCRIPTION €25 (single); €40 (couple at same
address)

Founded in 1958 to promote the heritage of W.B.
Yeats and the Yeats family. Attractions include
a permanent Yeats Exhibition; films for public
viewing throughout the year; the annual Yeats
International Summer School in July/August; the
Winter School in January/February. The Yeats
Festival is also held in the summer, and lectures,
poetry readings, workshops, etc. are held in the
autumn and spring, together with a variety of
outings and other events for members. Visits are
made to schools, and contacts are made with
Yeats scholars and students worldwide. The 2010
Summer School (24th July to 6th August) is the
51st successive Summer School. Early booking is
advisable.

## Yorkshire Dialect Society

51 Stepney Avenue, Scarborough YO12 5BW
www.ydsociety.org.uk
Secretary *Michael Park*
SUBSCRIPTION £10 p.a.

Founded in 1897 to promote interest in and
preserve a record of the Yorkshire dialect.
Publishes dialect verse and prose writing.
Two journals to members annually. Details of
publications are available from YDS at address
above.

# Arts Councils and Regional Offices

## Arts Council England
14 Great Peter Street, London SW1P 3NQ
☎ 0845 300 6200 Textphone 020 7973 6564
enquiries@artscouncil.org.uk
www.artscouncil.org.uk
Chair *Dame Liz Forgan*
Chief Executive *Alan Davey*
Director of Literature Strategy *Antonia Byatt*

Founded 1946. Arts Council England is the national development agency for the arts in England, distributing public money from government and the National Lottery to artists and arts organizations. ACE works independently and at arm's length from government. Information about Arts Council England funding is available on the website, by email or by contacting the enquiry line on 0845 300 6200.

In 2009, Arts Council England announced plans to reduce staffing by grouping the nine regional offices in four areas: London, East and South East, Midlands and South West, and North. The London office will move to the national offices in Great Peter Street. At the time of writing, the nine regional offices are:

### Arts Council England, East
Eden House, 48–49 Bateman Street, Cambridge CB2 1LR
☎ 0845 300 6200 🖷 0870 242 1271
Textphone 01223 306893

### Arts Council England, East Midlands
St Nicholas Court, 25–27 Castle Gate, Nottingham NG1 7AR
☎ 0845 300 6200 🖷 0115 950 2467

### Arts Council England, London
2 Pear Tree Court, London EC1R 0DS
☎ 0845 300 6200 🖷 020 7608 4100
Textphone 020 7973 6564

### Arts Council England, North East
Central Square, Forth Street, Newcastle upon Tyne NE1 3PJ
☎ 0845 300 6200 🖷 0191 230 1020
Textphone 0191 255 8585

### Arts Council England, North West
Manchester House, 22 Bridge Street, Manchester M3 3AB
☎ 0845 300 6200 🖷 0161 834 6969
Textphone 0161 834 9131

### Arts Council England, South East
Sovereign House, Church Street, Brighton BN1 1RA
☎ 0845 300 6200 🖷 0870 242 1257
Textphone 01273 710659

### Arts Council England, South West
Senate Court, Southernhay Gardens, Exeter EX1 1UG
☎ 0845 300 6200 🖷 01392 498546
Textphone 01392 433503

### Arts Council England, West Midlands
82 Granville Street, Birmingham B1 2LH
☎ 0845 300 6200 🖷 0121 643 7239
Textphone 0121 643 2815

### Arts Council England, Yorkshire
21 Bond Street, Dewsbury WF13 1AX
☎ 0845 300 6200 🖷 01924 466522
Textphone 01924 438585

## The Arts Council/An Chomhairle Ealaíon
70 Merrion Square, Dublin 2, Republic of Ireland
☎ 00 353 1 618 0200 🖷 00 353 1 676 1302
artistsservices@artscouncil.ie
www.artscouncil.ie
Director *Mary Cloake*
Head of Literature *Sarah Bannon*

The Arts Council is the Irish government agency for developing the arts. It provides financial assistance to artists, arts organizations, local authorities and others for artistic purposes. It offers advice and information on the arts to Government and to a wide range of individuals and organizations. As an advocate for the arts and artists, the Arts Council undertakes projects and research, often in new and emerging areas of arts practice, and increasingly in cooperation with partner organizations. Its funding from government for 2010 was €69.15 million.

Of particular interest to individual writers is the Council's free booklet, *Arts Council Awards 2009*, which describes bursaries, awards and schemes on offer and how to apply for them. Applicants to these awards must have been born in, or be resident in, the Republic of Ireland.

## The Arts Council of Northern Ireland

MacNeice House, 77 Malone Road, Belfast
BT9 6AQ
Ⓣ 028 9038 5200    Ⓕ 028 9066 1715
dsmyth@artscouncil-ni.org
www.artscouncil-ni.org
Head of Arts *Damian Smyth*

Funds book production by established publishers, programmes of readings, literary festivals, mentoring services, writers-in-residence schemes, network service providers, literary magazines and periodicals. Annual awards and bursaries for writers are available. Holds information also on various groups associated with regional arts, workshops and courses.

## Scottish Arts Council

12 Manor Place, Edinburgh EH3 7DD
Ⓣ 0131 226 6051    Ⓕ 0131 225 9833
help.desk@scottisharts.org.uk
www.scottisharts.org.uk
Chief Executive *Jim Tough*
Head of Literature *Dr Gavin Wallace*
Literature Officers *Aly Barr, Emma Turnbull*
Literature Administrator *Catherine Allan*

Principal channel for government funding of the arts in Scotland. The Scottish Arts Council is funded by the Scottish Government and National Lottery. It aims to develop and improve the knowledge, understanding and practice of the arts, and to increase their accessibility throughout Scotland. It offers grants to artists and arts organizations concerned with the visual arts, crafts, dance and mime, drama, literature, music, festivals and traditional, ethnic and community arts. Scottish Arts Council's support for Scottish-based writers with a track record of publication includes bursaries, writing fellowships and book awards (see entries under *Bursaries, Fellowships and Grants* and *Prizes*). Information offered includes lists of literature awards, literary magazines, agents and publishers.

## The Arts Council of Wales/Cyngor Celfyddydau Cymru

9 Museum Place, Cardiff CF10 3NX
Ⓣ 029 2037 6500    Ⓕ 029 2022 1447
www.artswales.org.uk
www.celfcymru.org.uk

The Arts Council of Wales (ACW) is the development body for the arts in Wales. ACW's Creative Wales Awards offer substantial grants for established writers and playwrights. Funds the **Welsh Academy** and **Tŷ Newydd Writers' Centre** to provide services to individual writers, including bursaries, mentoring, the critical writers services and writers in residence/on tour. Responsibility for promoting the publishing and sales of books from Wales and funding literary magazines rests with the **Welsh Books Council** (see entry under *Professional Associations*).

# Library Services

## Aberdeen Central Library
Rosemount Viaduct, Aberdeen AB25 1GW
☎ 01224 652500　📠 01224 641985
centrallibrary@aberdeencity.gov.uk
www.aberdeencity.gov.uk
OPEN Central Library Lending and Information
Services: 9.00 am to 8.00 pm Monday and
Wednesday; 9.00 am to 5.00 pm Tuesday,
Thursday, Friday & Saturday.
OPEN ACCESS

General reference and loans. Books, pamphlets,
periodicals and newspapers; videos, CDs, DVDs;
arts equipment lending service; Internet and
Learning Centre for public access; photographs
of the Aberdeen area; census records, maps,
newspapers; online and remote access databases,
patents and standards. The library offers special
services to housebound readers.

## Armitt Collection, Museum & Library
Rydal Road, Ambleside LA22 9BL
☎ 015394 41212　📠 015294 31313
info@armitt.com
www.armitt.com
OPEN Museum: 10.00 am to 5.00 pm Monday to
Friday (last admission 4.30 pm); Library: times
vary but usually open most weekdays (phone in
advance).

A small but unique reference library of rare
books, manuscripts, pictures, antiquarian
prints, maps and museum items, mainly about
the Lake District. It includes early guidebooks
and topographical works, books and papers
relating to Ruskin, H. Martineau, Charlotte
Mason and others; fine art including work by
W. Green, J.B. Pyne, John Harden, K. Schwitters
and Victorian photographs by Herbert Bell; also
a major collection of Beatrix Potter's scientific
watercolour drawings and microscope studies.
Entry charge for museum (£2.50 adults, £2
concessions, £1 children).

## The Athenaeum, Liverpool
Church Alley, Liverpool L1 3DD
☎ 0151 709 7770　📠 0151 709 0418
reception@theathenaeum.org.uk
www.theathenaeum.org.uk

OPEN 9.00 am to 4.00 pm Monday and Tuesday;
9.00 am to 9.00 pm Wednesday to Friday
ACCESS To club members; researchers by
application only.

General collection, with books dating from the
15th century, now concentrated mainly on local
history with a long run of Liverpool directories
and guides. SPECIAL COLLECTIONS Liverpool
playbills; William Roscoe; Blanco White; Robert
Gladstone; 18th-century plays; 19th-century
economic pamphlets; Bibles; Yorkshire and other
genealogy. Some original drawings, portraits,
topographical material and local maps.

## Bank of England Information Centre
Threadneedle Street, London EC2R 8AH
☎ 020 7601 4715　📠 020 7601 4356
informationcentre@bankofengland.co.uk
www.bankofengland.co.uk
OPEN 9.00 am to 5.30 pm Monday to Friday
ACCESS For research workers only by prior
arrangement, when material is not readily
available elsewhere.

50,000 volumes of books and periodicals. 2,000
periodicals taken. UK and overseas coverage
of banking, finance and economics. SPECIAL
COLLECTIONS Central bank reports; UK
17th–19th-century economic tracts; government
reports in the field of banking.

## Barbican Library
Barbican Centre, London EC2Y 8DS
☎ 020 7638 0569/7638 0568 (24-hr renewals)
📠 020 7638 2249
barbicanlib@cityoflondon.gov.uk
www.cityoflondon.gov.uk/libraries
www.barbican.org
Librarian *John Lake, BA, MCLIP*
OPEN 9.30 am to 5.30 pm Monday and
Wednesday; 9.30 am to 7.30 pm Tuesday and
Thursday; 9.30 am to 2.00 pm Friday; 9.30 am to
4.00 pm Saturday
OPEN ACCESS

Situated on Level 2 of the Barbican Centre, this
is the City of London's largest lending library.
Study facilities are available plus free internet
access. In addition to a large general lending

department, the library seeks to reflect the Centre's emphasis on the arts and includes strong collections (including DVDs, MP3s, videos, spoken word CDs and CD-ROMs) on painting, sculpture, theatre, cinema and ballet, as well as a large music library with books, scores and CDs (sound recording loans available at a small charge). Also houses the City's main children's library and has special collections on basic skills, materials for young adults, finance, graphic novels and the history of London. Service available for housebound readers. A literature events programme is organized by the Library which supplements and provides cross-arts planning opportunities with the Barbican Centre artistic programme. Reading groups meet in the library on the first Thursday of every month and children's reading group and Questseeker group during term time.

## Barnsley Public Library

Central Library, Shambles Street, Barnsley S70 2JF
℡ 01226 773930    🖷 01226 773955
barnsleylibraryenquiries@barnsley.gov.uk
www.barnsley.gov.uk/Libraries
OPEN Lending & Reference: 9.30 am to 7.00 pm Monday and Wednesday; 9.30 am to 5.30 pm Tuesday, Thursday, Friday; 9.30 am to 4.00 pm Saturday. Telephone to check hours of other departments and other branch libraries
OPEN ACCESS

General library, lending and reference. Archives and Local Studies collection of family history and local material, local authors, Yorkshire and Barnsley collections; large junior library. (Specialist departments are closed on certain weekday evenings and Saturday afternoons.)

## BBC Written Archives Centre

Caversham Park, Reading RG4 8TZ
℡ 0118 948 6281    🖷 0118 946 1145
heritage@bbc.co.uk
www.bbc.co.uk/historyofthebbc/contacts/wac.shtml
Contact *Jacqueline Kavanagh*
OPEN 9.30 am to 5.30 pm Monday to Friday
ACCESS For reference, by appointment only, Wednesday to Friday.

Holds the written records of the BBC, including internal papers from 1922 to the 1990s and published material to date. 20th century biography, social history, popular culture and broadcasting. Charges for certain services.

## Beamish, The North of England Open Air Museum
▷ entry under Picture Libraries

## Bedford Central Library

Harpur Street, Bedford MK40 1PG
℡ 01234 718178    🖷 01234 342163
www.bedfordshire.gov.uk
OPEN 9.00 am to 6.00 pm Monday, Tuesday, Wednesday, Friday; 9.00 am to 1.00 pm Thursday; 9.00 am to 5.00 pm Saturday
OPEN ACCESS

Lending library with a wide range of stock, including books, music, audiobooks, DVDs and videos; reference and information library, children's library, local history library and internet facilities.

## Belfast Public Libraries: Central Library

Royal Avenue, Belfast BT1 1EA
℡ 028 9050 9150    🖷 028 9033 2819
info.belb@ni-libraries.net
www.belb.org.uk
www.ni-libraries.net
OPEN 9.00 am to 8.00 pm Monday to Thursday; 9.00 am to 5.30 pm Friday; 9.00 am to 4.30 pm Saturday
OPEN ACCESS is available to the Central Lending Library and to the stock of the specialist reference departments within the building.

Over two million volumes for lending and reference. SPECIAL COLLECTIONS United Nations depository; complete British Patent Collection; Northern Ireland Newspaper Library; British and Irish government publications. The Central Library provides the following Reference Departments: General Reference; Belfast, Ulster and Irish Studies; Music. A Learning Gateway includes public internet facilities. The Lending Library is one of 19 branches along with a range of outreach services to hospitals, care homes and housebound readers.

## Benjamin Britten Collection
▷ Suffolk County Council – Suffolk Libraries

## BFI National Library

21 Stephen Street, London W1T 1LN
℡ 020 7255 1444    🖷 020 7436 0165
library@bfi.org.uk
www.bfi.org.uk
OPEN 10.30 am to 5.30 pm Monday and Friday; 10.30 am to 8.00 pm Tuesday and Thursday; 1.00 pm to 8.00 pm Wednesday; Telephone Enquiry Service operates from 10.00 am to 5.00 pm (closed 12.00 noon to 3.00 pm)
ACCESS For reference only; annual, 5-day and limited day membership available.

The world's largest collection of information on film and television including periodicals, cuttings, scripts, related documentation, personal papers. Main library catalogue available via website. Free WiFi; Mediatheque; Screenonline.

## Birmingham and Midland Institute

9 Margaret Street, Birmingham B3 3BS
☎ 0121 236 3591  📠 0121 212 4577
admin@bmi.org.uk
www.bmi.org.uk
Administrator & General Secretary *Philip Fisher*
ACCESS Members only.

Established 1854. Later merged with the Birmingham Library which was founded in 1779. The Library specializes in the humanities, with approximately 100,000 volumes in stock. Founder member of the **Association of Independent Libraries**. Meeting-place of many affiliated societies devoted to poetry and literature.

## Birmingham Library Services

Central Library, Chamberlain Square, Birmingham B3 3HQ
☎ 0121 303 4511
central.library@birmingham.gov.uk
www.birmingham.gov.uk/libraries
OPEN 9.00 am to 8.00 pm Monday to Friday; 9.00 am to 5.00 pm Saturday

Over a million volumes. RESEARCH COLLECTIONS include the Shakespeare Library; War Poetry Collection; Parker Collection of Children's Books and Games; Johnson Collection; Milton Collection; Cervantes Collections; Early and Fine Printing Collection (including the William Ridler Collection of Fine Printing); Joseph Priestley Collection; Loudon Collection; Railway Collection; Wingate Bett Transport Ticket Collection; Labour, Trade Union and Co-operative Collections. PHOTOGRAPHIC ARCHIVES Sir John Benjamin Stone; Francis Bedford; Francis Frith; Warwickshire Photographic Survey; Boulton and Watt Archive. Also, Charles Parker Archive; Birmingham Repertory Theatre Archive and Sir Barry Jackson Library; Local Studies (Birmingham); Patents Collection; Song Sheets Collection; Oberammergau Festival Collection.

## Bournemouth Library

22 The Triangle, Bournemouth BH2 5RQ
☎ 01202 454848  📠 01202 454840
bournemouth@bournemouthlibraries.org.uk
www.bournemouth.gov.uk/libraries
OPEN 10.00 am to 7.00 pm Monday; 9.30 am to 7.00 pm Tuesday, Thursday, Friday; 9.30 am to 5.00 pm Wednesday; 10.00 am to 4.00 pm Saturday
OPEN ACCESS

Main library for Bournemouth with lending, reference and music departments, plus the Heritage Zone – local and family history.

## Bradford Central Library

Princes Way, Bradford BD1 1NN
☎ 01274 433600  📠 01274 395108
public.libraries@bradford.gov.uk
www.bradford.gov.uk/libraries
OPEN 9.00 am to 7.30 pm Monday to Friday; 9.00 am to 5.00 pm Saturday

Wide range of books and media loan services. Comprehensive reference and information services, including major local history collections and specialized business information service.

## Brighton Jubilee Library

Jubilee Street, Brighton BN1 1GE
☎ 01273 290800
libraries@brighton-hove.gov.uk
www.citylibraries.info/libraries/jubilee.asp
OPEN 10.00 am to 7.00 pm Monday and Tuesday; 10.00 am to 5.00 pm Wednesday, Friday and Saturday; 10.00 am to 8.00 pm Thursday; 11.00 am to 4.00 pm Sunday
ACCESS Stock on open access and in onsite store; material for reference use and lending

Specializations include art and antiques, history of Brighton, local illustrations, Hebrew and Oriental literature, natural history, children's books, World War Two and large bequests of rare and historical books.

## Bristol Central Library

College Green, Bristol BS1 5TL
☎ 0117 903 7200  📠 0117 922 1081
www.bristol.gov.uk/libraries
OPEN 9.30 am to 7.30 pm Monday, Tuesday and Thursday; 10.00 am to 5.00 pm Wednesday; 9.30 am to 5.00 pm Friday and Saturday; 1.00 pm to 5.00 pm Sunday
OPEN ACCESS

Lending, reference, art, music, business and local studies are particularly strong. DVD and CD collections on site. Facilities available: PCs (large screen with Jaws and Zoomtext), internet, printing; videophone on site; black & white and colour photocopiers.

## British Architectural Library

Royal Institute of British Architects, 66 Portland Place, London W1B 1AD
☎ 020 7580 5533  📠 020 7631 1802
info@inst.riba.org
www.architecture.com
OPEN 10.00 am to 5.00 pm Tuesday, Wednesday, Friday; 10.00 am to 1.30 pm Saturday; closed Sunday, Monday, Thursday and any Saturday preceding a Bank Holiday; full details on the website.
ACCESS Free to RIBA members and non-members on proof of identity and address; loans available to RIBA members only.

Collection of books, photographs and periodicals. All aspects of architecture, current and historical. Material both technical and aesthetic, covering related fields including: interior design, landscape architecture, topography, the construction industry and applied arts. Brochure available; queries by telephone, letter, email or in person. Charge for research (min. charge £40 + VAT).

**RIBA British Architectural Library Drawings & Archives Collections**

Drawings and manuscripts can be consulted at the RIBA Study Rooms, Henry Cole Wing, Victoria and Albert Museum: ☎ 020 7307 3708 drawings&archives@inst.riba.org
ACCESS Free to all.

## The British Cartoon Archive
▷ entry under Picture Libraries

## The British Library

Admission to St Pancras Reading Rooms – British Library Readers' Passes

Everyone is welcome to visit the British Library exhibition galleries or to tour the building. However, to use the reading rooms you will need to apply for a reader's pass, for which identification is required. The British Library issues passes to those who want to use its collections – researchers, innovators and entrepreneurs across all fields of study, and in academic, commerce or personal research – whether or not they are affiliated to a research institution. Two pieces of identification are required (original documents only). One proof of home address, e.g. a utility bill, bank statement or driving licence; and proof of signature, i.e. a bank or credit card, passport, driving licence or national identity card.

It would also be helpful to take with you anything to support your application (such as a student card, business card, professional membership card or details of the items you wish to see).

For further information, visit the website at www.bl.uk or contact Customer Services, The British Library, 96 Euston Road, London NW1 2DB ☎ 01937 546060 ⓕ 01937 546333; customer-services@bl.uk

## British Library Asia, Pacific and Africa Collections

96 Euston Road, London NW1 2DB
☎ 020 7412 7873    ⓕ 020 7412 7641
apac-enquiries@bl.uk
www.bl.uk
OPEN 10.00 am to 5.00 pm Monday; 9.30 am to 5.00 pm Tuesday to Saturday; closed for public holidays.
ACCESS By British Library reader's pass.

An extensive collection of printed volumes and manuscripts in the languages of Africa, the Near and Middle East and all of Asia, plus records of

the East India Company and British government in India until 1947. Also prints, drawings and paintings by British artists of India.

For information on British Library collections and services, visit the website.

## British Library Business & IP Centre

96 Euston Road, London NW1 2DB
☎ 020 7412 7454 (free enquiry service)
bipc@bl.uk
www.bl.uk/bipc
OPEN 10.00 am to 8.00 pm Monday; 9.30 am to 8.00 pm Tuesday to Thursday; 9.30 am to 5.00 pm Friday and Saturday; closed for public holidays.
ACCESS By British Library reader's pass.

The Business & IP Centre holds the most comprehensive collection of business and intellectual property information in the UK, including market research reports and journals, directories, company annual reports, trade and business journals, up-to-date literature on patents, trade marks, designs and copyright.

## British Library Early Printed Collections/ Rare Books and Music Reading Room

96 Euston Road, London NW1 2DB
☎ 020 7412 7564    ⓕ 020 7412 7691
rare-books@bl.uk
www.bl.uk
General enquiries about reader services & advance reservations: ☎ 020 7412 7676 ⓕ 01937 546321 reader-services-enquiries@bl.uk
OPEN 10.00 am to 8.00 pm Monday; 9.30 am to 8.00 pm Tuesday to Thursday; 9.30 am to 5.00 pm Friday and Saturday; closed for public holidays.
ACCESS By British Library reader's pass.

Early Printed Collections selects, acquires, researches and provides access to material printed in Britain and Western Europe from the 15th to early 20th centuries. The collections are available in the Rare Books and Music Reading Room at St Pancras which also functions as the focus for the British Library's extensive collection of humanities microforms.

Further information about Early Printed Collections can be found at the British Library website.

## British Library Humanities Reading Room

96 Euston Road, London NW1 2DB
☎ 020 7412 7676    ⓕ 020 7412 7794
humanities-enquiries@bl.uk
www.bl.uk/reshelp/inrooms/stp/rrbysubj/humrr/humrr.html
OPEN 10.00 am to 8.00 pm Monday; 9.30 am to 8.00 pm Tuesday, Wednesday, Thursday; 9.30 am to 5.00 pm Friday and Saturday; closed for public holidays.
ACCESS By British Library reader's pass.

This reading room is the focus for the Library's modern collections service in the humanities. It is on two levels, Humanities 1 and Humanities 2 and provides access to the Library's comprehensive collections of books and periodicals in all subjects in the humanities and social sciences and in all languages apart from Oriental. These collections are not available for browsing at the shelf. Material is held in closed access storage and needs to be identified and ordered from store using an online catalogue. A selective open access collection on most humanities subjects can be found in Humanities 1 whilst in Humanities 2 there are open access reference works relating to periodicals and theses, to recorded sound and to librarianship and information science.

To access British Library catalogues, go to the website at http://catalogue.bl.uk

## British Library Manuscript Collections

96 Euston Road, London NW1 2DB
T 020 7412 7513    F 020 7412 7745
mss@bl.uk
www.bl.uk
OPEN 10.00 am to 5.00 pm Monday; 9.30 am to 5.00 pm Tuesday to Saturday; closed for public holidays.
ACCESS Reading facilities only, by British Library reader's pass; a written letter of recommendation and advance notice is required for certain categories of material.

Two useful publications, *Index of Manuscripts in the British Library*, Cambridge 1984–6, 10 vols, and *The British Library: Guide to the Catalogues and Indexes of the Department of Manuscripts* by M.A.E. Nickson, help to guide the researcher through this vast collection of manuscripts dating from Ancient Greece to the present day. Approximately 300,000 mss, charters, papyri and seals are housed here.

For information on British Library collections and services and to access British Library catalogues, including the Manuscripts online catalogue, visit the website.

## British Library Map Collections

96 Euston Road, London NW1 2DB
T 020 7412 7702    F 020 7412 7780
maps@bl.uk
www.bl.uk
OPEN 10.00 am to 5.00 pm Monday; 9.30 am to 5.00 pm Tuesday to Saturday; closed for public holidays.
ACCESS By British Library reader's pass.

A collection of about 4.5 million maps, charts and globes, manuscript, printed and, increasingly, digital, with particular reference to the history of British cartography. Maps for all parts of the world in a wide range of scales, formats and dates,

including the most comprehensive collection of Ordnance Survey maps and plans. SPECIAL COLLECTIONS King George III Topographical Collection and Maritime Collection, the Crace Collection of maps and plans of London and the cartographic archive of the Ministry of Defence (i.e. GSGS).

To access main British Library catalogues, go to the website at http://catalogue.bl.uk

## British Library Music Collections

96 Euston Road, London NW1 2DB
T 020 7412 7772    F 020 7412 7751
music-collections@bl.uk
www.bl.uk
OPEN 10.00 am to 8.00 pm Monday; 9.30 am to 8.00 pm Tuesday to Thursday; 9.30 am to 5.00 pm Friday and Saturday; closed for public holidays.
ACCESS By British Library reader's pass.

SPECIAL COLLECTIONS The Royal Music Library (containing almost all Handel's surviving autograph scores), The Zweig Collection of Music & Literary Mss, The Royal Philharmonic Society Archive and the Paul Hirsch Music Library. Also a large collection (about one and a half million items) of printed music (UK via legal deposit) and about 100,000 items of manuscript music, both British and foreign.

The British Library website contains details of collections and services, and provides access to the catalogues.

## British Library Newspapers

Colindale Avenue, London NW9 5HE
T 020 7412 7353    F 020 7412 7379
newspaper@bl.uk
www.bl.uk
OPEN 10.00 am to 5.00 pm Monday to Saturday (last newspaper issued 4.15 pm); closed for public holidays.
ACCESS By British Library reader's pass or Newspaper Library pass (available from and valid only for Colindale). Proof of address and proof of signature required.

Major collections of English provincial, Scottish, Welsh, Irish, Commonwealth and selected overseas foreign newspapers from c.1700 are housed here. Some earlier holdings are also available. London newspapers from 1801 and many weekly and fortnightly periodicals are also in stock. (London newspapers pre-dating 1801 are housed at the new library building in St Pancras – 96 Euston Road, NW1 – though many are available at Colindale Avenue on microfilm.) Readers are advised to check availability of material in advance.

For information on British Library Newspapers collections and services, visit the website.

## British Library Science, Technology and Business Collections

96 Euston Road, London NW1 2DB
☏ 020 7412 7288 (general enquiries)
🖷 020 7412 7217
scitech@bl.uk
www.bl.uk
Business enquiries: ☏ 020 7412 7454 (Free enquiry service)
OPEN 10.00 am to 8.00 pm Monday; 9.30 am to 8.00 pm Tuesday to Thursday; 9.30 am to 5.00 pm Friday and Saturday; closed for public holidays.
ACCESS By British Library reader's pass.

Engineering, business information on companies, markets and products, physical science and technologies. See also **British Library Business & IP Centre**.

To access British Library catalogues go to the website at http://catalogue.bl.uk

## British Library Social Sciences & Official Publications

96 Euston Road, London NW1 2DB
☏ 020 7412 7676    🖷 020 7412 7794
social-sciences@bl.uk
www.bl.uk
OPEN Reading Room: 10.00 am to 8.00 pm Monday; 9.30 am to 8.00 pm Tuesday to Thursday; 9.30 am to 5.00 pm Friday and Saturday; closed for public holidays.
ACCESS By British Library reader's pass.

Provides an information service on the social sciences, law, public administration, and current and international affairs, and access to current and historical official publications from all countries and intergovernmental bodies. Material available in the reading room includes House of Commons sessional papers, UK legislation, UK electoral registers, up-to-date reference books on official publications and on the social sciences, a major collection of statistics and a browsing collection of recent social science books and periodicals; other collections are kept in closed stores and items have to be ordered for delivery to the reading room.

To access British Library catalogues, go to the website at http://catalogue.bl.uk

## British Library Sound Archive

96 Euston Road, London NW1 2DB
☏ 020 7412 7676    🖷 020 7412 7441
sound-archive@bl.uk
www.bl.uk/soundarchive
OPEN 10.00 am to 8.00 pm Monday; 9.30 am to 8.00 pm Tuesday to Thursday; 9.30 am to 5.00 pm Friday and Saturday; closed for public holidays.
**Listening service** (by appointment)
**Northern Listening Service**
British Library Document Supply Centre, Boston

Spa: 9.15 am to 4.30 pm Monday to Friday
OPEN ACCESS

An archive of over 1,000,000 discs and more than 200,000 hours of tape recordings, including all types of music, oral history, drama, literature, poetry, wildlife, selected BBC broadcasts and BBC Sound Archive material. Produces a twice-yearly newsletter, *Playback*.

For information on British Library Sound Archive collections and services, visit the website.

## British Psychological Society Library

c/o Psychology Library, Senate House Library, University of London, Senate House, Malet Street, London WC1E 7HU
☏ 020 7862 8451/8461    🖷 020 7862 8480
shl.enquiries@lon.ac.uk
www.shl.lon.ac.uk
OPEN Term-time: 9.00 am to 9.00 pm Monday to Thursday; 9.00 am to 6.30 pm Friday; 9.45 am to 5.30 pm Saturday (Holidays: 9.00 am to 6.00 pm Monday to Friday; 9.45 am to 5.30 pm Saturday)
ACCESS Members only; non-members £5 day ticket.

Reference library, containing the British Psychological Society collection of periodicals – over 140 current titles housed alongside the University of London's collection of books and journals. Largely for academic research. General queries referred to **Swiss Cottage Library** in London which has a good psychology collection.

## Bromley Central Library

High Street, Bromley BR1 1EX
☏ 020 8460 9955    🖷 020 8313 9975
central.library@bromley.gov.uk
www.bromley.gov.uk
OPEN 9.30 am to 6.00 pm Monday, Wednesday, Friday; 9.30 am to 8.00 pm Tuesday and Thursday; 9.30 am to 5.00 pm Saturday
OPEN ACCESS

A large selection of fiction and non-fiction books for loan, both adult and children's. Also DVDs, CDs and language courses. Other facilities include People's Network Internet, local studies library, 'Upfront' teenage section, large reference library with photocopying, fax, microfiche and film facilities, 'Bromley Knowledge' – online community information, reading groups and Local Links (access to council services). Library and Archives catalogues online. SPECIALIST COLLECTIONS include: H.G. Wells, Walter de la Mare, Crystal Palace, The Harlow Bequest, and the history and geography of Asia, America, Australasia and the Polar regions.

## Bromley House Library

Angel Row, Nottingham NG1 6HL
☏ 0115 947 3134

enquiries@bromleyhouse.org
www.bromleyhouse.org
Librarian *Carol Barstow*
OPEN 9.30 am to 5.00 pm Monday to Friday; 9.30 am to 12.30 pm Saturday
ACCESS For members only.

Founded 1816 as the Nottingham Subscription Library. Collection of 35,000 books including local history, topography, biography, travel and fiction.

## CAA Library and Information Centre

Aviation House, Gatwick Airport South, Gatwick RH6 0YR
☎ 01293 573725   🖷 01293 573181
infoservices@caa.co.uk
www.caa.co.uk
OPEN 9.00 am to 5.00 pm Monday to Friday
OPEN ACCESS

A collection of books, reports, directories, statistics and periodicals on most aspects of civil aviation and related subjects.

## Camomile Street Library

12–20 Camomile Street, London EC3A 7EX
☎ 020 7332 1855
camomile@cityoflondon.gov.uk
www.cityoflondon.gov.uk/camomilestlibrary
OPEN 9.30 am to 5.30 pm Monday, Tuesday, Thursday, Friday; 9.30 am to 6.30 pm Wednesday
OPEN ACCESS

City of London Corporation lending library. Wide range of fiction and non-fiction books and language courses on cassette and CD, foreign fiction, paperbacks, maps and guides for travel at home and abroad, children's books, a selection of large print, and collections of DVDs, videos and music CDs, books in Bengali. Spoken word recordings on cassette and CD. Free internet access.

## Cardiff Central Library

The Hayes, Cardiff Council, Cardiff CF10 1FL
☎ 029 2087 1588   🖷 029 2078 0989
nrichards@cardiff.gov.uk
www.cardiff.gov.uk/libraries
OPEN 9.00 am to 6.00 pm Monday, Tuesday, Wednesday, Friday; 9.00 am to 7.00 pm Thursday; 9.00 am to 5.30 pm Saturday; 11.00 am to 3.00 pm Sunday

Over 55,000 sq feet of floor space across five levels. Departments include: leisure, information, music, children's, local studies, Welsh, community languages, science and humanities. Meeting rooms and launch space for hire. Fax service available.

## Carmarthen Public Library

St Peter's Street, Carmarthen SA31 1LN
☎ 01267 224824   🖷 01267 221839
wtphillips@carmarthenshire.gov.uk
www.carmarthenshire.gov.uk
Regional Library Manager *William Phillips*
OPEN 9.30 am to 7.00 pm Monday, Tuesday, Wednesday, Friday; 9.30 am to 5.00 pm Thursday and Saturday
OPEN ACCESS

Comprehensive range of fiction, non-fiction, children's books and reference works in English and in Welsh. Large local history library. Free internet access and computer facilities. Large Print books, audiobooks, CDs, CD-ROMs and DVDs available for loan.

## Catholic National Library

St Michaels Abbey, Farnborough Road, Farnborough GU14 7NQ
☎ 01252 543818
library@catholic-library.org.uk
www.catholic-library.org.uk
OPEN ACCESS For reference (non-members must sign in; loans restricted to members).

Contains books, many not readily available elsewhere, on theology, religions worldwide, scripture and the history of churches of all denominations.

## Central Library Cambridge (Reference Library & Information Service)

7 Lion Yard, Cambridge CB2 3QD
☎ 0345 045 5225   🖷 01223 712011
your.library@cambridgeshire.gov.uk
www.cambridgeshire.gov.uk/leisure/libraries
OPEN seven days a week

Large stock of books, periodicals, newspapers, maps, plus comprehensive collection of directories and annuals covering UK, Europe and the world. Microfilm and fiche reading and printing services. Online access to news and business databases. News databases on CD-ROM; internet access; photocopiers.

## Chelmsford Library
▷ Essex County Council Libraries

## City Business Library

Guildhall, Aldermanbury, London EC2V 7HH
☎ 020 7332 1812/(3803 textphone)
cbl@cityoflondon.gov.uk
www.cityoflondon.gov.uk/citybusinesslibrary
OPEN 9.30 am to 5.00 pm Monday to Saturdays (except Saturdays before public holidays)
OPEN ACCESS

Local authority free public reference library run by the City of London Corporation. Subscription-based electronic resources available free of charge for the needs of the business community. Large directory collection for both the UK and overseas, plus companies information, market research reports, management, banking, insurance, investment and statistics. Free public internet access. No academic journals or textbooks.

## City of London Libraries
▷ Barbican Library; Camomile Street Library; City Business Library; Guildhall Library

## Colchester Library
▷ Essex County Council Libraries

## Commonwealth Secretariat Library
Marlborough House, Pall Mall, London SW1Y 5HX
☎ 020 7747 6164   🖷 020 7747 6168
library@commonwealth.int
www.thecommonwealth.org
Library Services Manager *Catherine Hume*
OPEN 10.00 am to 4.45 pm Monday to Friday
ACCESS For reference only, by appointment.

Extensive reference source concerned with economy, development, trade, production and industry of Commonwealth countries; also human resources including women, youth, health, management and education. Includes the archives of the Secretariat.

## Coventry Central Library
Smithford Way, Coventry CV1 1FY
☎ 024 7683 2314/2395 (text)   🖷 024 7683 2440
central.library@coventry.gov.uk
www.coventry.gov.uk
OPEN 9.00 am to 8.00 pm Monday to Friday; 9.00 am to 4.30 pm Saturday; 12.00 am to 4.00 pm Sunday
OPEN ACCESS

Located in the middle of the city's main shopping centre. Approximately 120,000 items (books, cassettes, CDs and DVDs) for loan, and a reference collection; free computer and internet access for first hour. SPECIAL COLLECTIONS Tom Mann Collection (trade union and labour studies); 'Peoplelink' community information database available. Local History and archive resources available in new purpose built History Centre next to the Museum and Art Gallery.

## Crace Collection
▷ British Library Map Collections

## Department for Environment, Food and Rural Affairs
Nobel House, 17 Smith Square, London SW1P 3JR
☎ 020 7238 3000
DEFRA Helpline 08459 335577 (local call rate): general contact point which can provide information on the work of DEFRA, either directly or by referring callers to appropriate

contacts. Available 9.00 am to 5.00 pm Monday to Friday (excluding Bank Holidays).
OPEN 9.30 am to 5.00 pm Monday to Friday
ACCESS For reference (but at least 24 hours notice must be given for intended visits).

## Derby Central Library
Wardwick, Derby DE1 1HS
☎ 01332 641702   🖷 01332 369570
www.derby.gov.uk/libraries
OPEN 9.30 am to 7.00 pm Monday, Tuesday, Thursday, Friday; 9.30 am to 1.00 pm Wednesday; 9.30 am to 4.00 pm Saturday
**Local Studies Library**
25B Irongate, Derby DE1 3GL
☎ 01332 255393 🖷 01332 255381
OPEN 9.30 am to 7.00 pm Monday and Tuesday; 9.30 am to 5.00 pm Wednesday, Thursday, Friday; 9.30 am to 4.00 pm Saturday
OPEN ACCESS

General library for lending, information and Children's Services. The Central Library also houses specialist private libraries: Derbyshire Archaeological Society; Derby Philatelic Society. The Local Studies Library houses the largest multimedia collection of resources in existence relating to Derby and Derbyshire. The collection includes mss deeds, family papers, business records including the Derby Canal Company, Derby Board of Guardians and the Derby China Factory. Both libraries offer free internet access.

## Devon & Exeter Institution Library
7 The Close, Exeter EX1 1EZ
☎ 01392 251017
J.P.Gardner@exeter.ac.uk
www.exeter.ac.uk/library
www.devonandexeterinstitution.org
OPEN 9.30 am to 5.00 pm Monday to Friday
ACCESS Members only (temporary membership available).

Founded 1813. Under the administration of Exeter University Library. Contains over 36,000 volumes, including long runs of 19th-century journals, theology, history, topography, early science, biography and literature. A large and growing collection of books, journals, newspapers, prints and maps relating to the South West.

## Doncaster Library and Information Services
Central Library, Waterdale, Doncaster DN1 3JE
☎ 01302 734305   🖷 01302 369749
reference.library@doncaster.gov.uk
library.doncaster.gov.uk
Reference Manager *Mrs Christine Zagraba*
OPEN 9.00 am to 6.00 pm Monday to Thursday; 10.00 am to 6.00 pm Friday; 9.00 am to 5.00 pm Saturday
OPEN ACCESS

Books, newspapers, periodicals, spoken word cassettes/CDs, DVDs, internet access. Reading aids unit for people with visual impairment; activities for children during school holidays and regular storytimes. Events programme, including visits by authors and talks, readers' groups, etc. Also reference library and Local Studies library.

## Dorchester Library (part of Dorset County Council)

Colliton Park, Dorchester DT1 1XJ
☎ 01305 224440 (lending)/224448 (reference)
🖷 01305 266120
dorchesterlendinglibrary@dorsetcc.gov.uk
dorchesterreferencelibrary@dorsetcc.gov.uk
www.dorsetforyou.com/libraries
OPEN 10.00 am to 5.30 pm Monday; 9.30 am to 7.00 pm Tuesday and Friday; 9.30 am to 1.00 pm Wednesday; 9.30 am to 5.30 pm Thursday; 9.00 am to 4.00 pm Saturday
OPEN ACCESS

General lending and reference library, including special collections on Thomas Hardy, the Powys Family and William Barnes. Periodicals, children's library, CD-ROMs, free internet access. Video lending service.

## Dr Williams's Library

14 Gordon Square, London WC1H 0AR
☎ 020 7387 3727
enquiries@dwlib.co.uk
www.dwlib.co.uk
OPEN 10.00 am to 5.00 pm Monday, Wednesday, Friday; 10.00 am to 6.30 pm Tuesday and Thursday.
ACCESS New users should contact the Library in advance of first visit.

Primarily a library of theology, religion and ecclesiastical history. Also philosophy, history (English and Byzantine). Particularly important for the study of English Nonconformity. The Dr Williams's Centre for Dissenting Studies is a collaboration with Queen Mary, University of London (see www.english.qmul.ac.uk/drwilliams).

## Dundee Central Library

The Wellgate, Dundee DD1 1DB
☎ 01382 431500     🖷 01382 431558
central.library@dundeecity.gov.uk
www.dundeecity.gov.uk
OPEN Departments of Leisure Reading, Children's, Connections Room, Reference & Information, and Local History Centre: 9.30 am to 6.00 pm Monday, Tuesday, Friday; 10.00 am to 6.00 pm Wednesday; 9.30 am to 8.00 pm Thursday; 9.30 am to 5.00 pm Saturday. Science & Business: 9.30 am to 9.00 pm Monday, Tuesday, Thursday, Friday; 10.00 am to 9.00 pm Wednesday; 9.30 am to 5.00 pm Saturday
ACCESS Reference services available to all; lending

services to those who live, work, study or were educated within Dundee City.

Adult lending, reference and children's services. Art, music, audio, video and DVD lending services. Internet access. Schools service (Agency). Housebound and mobile services.
SPECIAL COLLECTIONS The Wighton Collection of National Music; The Wighton Heritage Centre; The Wilson Photographic Collection; The Lamb Collection.

## Durning-Lawrence Library
▷ Senate House Library

## Edward Fitzgerald Collection
▷ Suffolk County Council – Suffolk Libraries

## Equality and Human Rights Commission

Arndale House, Arndale Centre, Manchester M4 3AQ
☎ 0161 829 8100     🖷 0161 829 8110
englandhelpline@equalityhumanrights.com
www.equalityhumanrights.com
The EHRC is open to the public via the Helpline (0845 604 6610) 9.00 am to 5.00 pm Monday, Tuesday, Thursday, Friday; 9.00 am to 7.45 pm Wednesday.

## Eric Frank Russell Archive
▷ Science Fiction Foundation Research Library

## Essex County Council Libraries

Goldlay Gardens, Chelmsford CM2 0EW
☎ 01245 284981     🖷 01245 492780
essexlib@essex.gov.uk
www.essex.gov.uk

Essex County Council Libraries has 73 static libraries throughout Essex as well as 12 mobile libraries. Services to the public include books, newspapers, periodicals, CDs, cassettes, videos, CD-ROMs and internet access. Specialist subjects and collections are listed below at the relevant library.

**Chelmsford Library**
PO Box 882, Market Road, Chelmsford CM1 1LH
☎ 01245 492758 🖷 01245 492536
chelmford.library@essex.gov.uk
OPEN 8.30 am to 7.00 pm Monday to Friday; 8.30 am to 5.30 pm Saturday; 12.30 pm to 4.30 pm Sunday

**Colchester Library**
Trinity Square, Colchester CO1 1JB
☎ 01206 245900 🖷 01206 245901
colchester.library@essex.gov.uk
OPEN 8.30 am to 7.30 pm Monday to Friday; 8.30 am to 5.00 pm Saturday; 12.30 pm to 4.30 pm Sunday
Local studies; Castle collection (18th-century subscription library); Cunnington collection; Margaret Lazell collection; Taylor collection.

## Harlow Library

The High, Harlow CM20 1HA
℡ 01279 413772 🖷 01279 424612
harlow.library@essex.gov.uk
OPEN 9.00 am to 7.00 pm Monday to Friday; 9.00 am to 5.00 pm Saturday; 1.00 pm to 4.00 pm Sunday
Sir John Newson Memorial collection; Maurice Hughes Memorial collection.

## Loughton Library

Traps Hill, Loughton IG10 1HD
℡ 020 8502 0181 🖷 020 8508 5041
loughton.library@essex.gov.uk
OPEN 9.00 am to 7.00 pm Monday to Friday; 9.30 am to 5.30 pm Saturday; 11.00 am to 3.00 pm Sunday
National Jazz Foundation Archive.

## Saffron Walden Library

2 King Street, Saffron Walden CB10 1ES
℡ 01799 523178 🖷 01799 513642
OPEN 9.00 am to 7.00 pm Monday to Friday; 9.00 am to 5.00 pm Saturday; 1.00 pm to 4.00 pm Sunday
Victorian studies collection.

## Witham Library

18 Newland Street, Witham CM8 2AQ
℡ 01376 519625 🖷 01376 501913
OPEN 9.00 am to 7.00 pm Monday to Friday; 9.00 am to 5.00 pm Saturday; 1.00 pm to 4.00 pm Sunday
Dorothy L. Sayers and Maskell collections.

## Family Records Centre

Part of the **National Archives** (see entry)

## The Fawcett Library
▷ The Women's Library

## Forestry Commission Library

Forest Research Station, Alice Holt Lodge, Wrecclesham, Farnham GU10 4LH
℡ 01420 22255 🖷 01420 23653
library@forestry.gsi.gov.uk
www.forestry.gov.uk
www.forestresearch.gov.uk
OPEN 9.00 am to 5.00 pm Monday to Thursday; 9.00 am to 4.30 pm Friday
ACCESS By appointment for personal visits.

Approximately 20,000 books on forestry and arboriculture, plus 400 current journals. CD-ROMS include Forest Science Database (1939 onwards). Offers a Research Advisory Service for advice and enquiries on forestry (℡ 01420 23000) with a charge for consultations and diagnosis of tree problems exceeding ten minutes.

## French Institute Library

Institut français du Royaume Uni, 17 Queensberry Place, London SW7 2DT
℡ 020 7073 1350 🖷 020 7073 1363
library@ambafrance.org.uk
www.institut-francais.org.uk
OPEN 12 noon to 7.00 pm Tuesday to Friday; 12 noon to 6.00 pm Saturday; Children's Library: 2.00 pm to 6.00 pm Tuesday to Friday; 12 noon to 6.00 pm Saturday; closed in August and for one week at Christmas
OPEN ACCESS For reference and consultation (loans restricted to members; leaflet available on demand).

A collection of over 55,000 volumes mainly centred on French cultural interests with special emphasis on French as a Foreign Language, literature and history. Books and periodicals, mainly in French, a few in English and some bilingual. Collection of DVDs, videos, CDs (French music), CD-ROMs, audiobooks; special collections: 'France Libre'. Denis Saurat MSSS, Paris. Recordings of lectures, press-cuttings on French current affairs; Campus France; internet access to members; Wifi. Group visits on request. Reading group; documentary screenings; talks. Children's library (8,000 documents). Bistro.

## The Froebel Archive for Childhood Studies
▷ entry under Useful Websites

## German Historical Institute Library

17 Bloomsbury Square, London WC1A 2NJ
℡ 020 7309 2050 🖷 020 7309 2055
library@ghil.ac.uk
www.ghil.ac.uk
OPEN 10.00 am to 5.00 pm Monday, Tuesday, Wednesday, Friday; 10.00 am to 8.00 pm Thursday
OPEN ACCESS Visitors must bring proof of address and recent photograph to be issued with a reader's ticket.

74,000 volumes; around 200 journals; databases. Devoted primarily to German history from the Middle Ages to the present day, with special emphasis on the 19th and 20th centuries, in particular Germany between 1933 and 1945, the development of the two German states after 1945/49 and German unification after 1989.

## Glasgow Women's Library

info@womenslibrary.org.uk
www.womenslibrary.org.uk

The collection includes reference, reports, education, politics, feminist theory, history, violence against women, abuse, pornography issues, prostitution, lesbian lives and cultures, myth, goddesses and religion; books, journals, videos and DVDs, photographs, posters, ephemera and newspaper cuttings. SPECIAL COLLECTIONS Lesbian Archive and Information Centre, Scottish Abortion Campaign archive; Edinburgh Women's Centre collection; Scottish Family Planning Association archive; Equal Opportunities Commission publications archive in Scotland; various collections relating to the Women's Liberation Movement in Scotland and women's history in general. (In temporary accommodation at present but due to move to the **Mitchell Library** in Glasgow.)

## Gloucestershire County Council Libraries & Information

Quayside House, Shire Hall, Gloucester GL1 2HY
☎ 08452 305420    🖷 01452 425042
libraryhelp@gloucestershire.gov.uk
www.gloucestershire.gov.uk/libraries
Head of Libraries & Information *David Paynter*
OPEN ACCESS

The service includes 39 local libraries and five mobile libraries. The website includes library opening hours, mobile library route schedules; the library catalogue; book renewal/reservations facility; and information about news and events.

## Goethe-Institut Library

50 Princes Gate, Exhibition Road, London SW7 2PH
☎ 020 7596 4040/4044    🖷 020 7594 0230
library@london.goethe.org
www.goethe.de/london
Head of Library Services *Elisabeth Pyroth*
OPEN 1.00 pm to 6.30 pm Monday to Thursday; 1.00 pm to 5.00 pm Saturday

Library specializing in German literature, especially contemporary fiction and drama, and German language teaching materials. Also books and audiovisual material on German culture and recent history. See website for membership details.

## Goldsmiths' Library
▷ Senate House Library

## Greater London Council History Library
▷ London Metropolitan Archives Library Services

## Greenpark Productions Ltd

PO Box 86, Launceston PL15 7WE
☎ 01566 782107    🖷 01566 782217
info@greenparkdirect.co.uk
www.greenparkimages.co.uk
Managing Director *Leonore Morphet*
Director of Business Affairs *Brian Norris*

Extensive film and video archive covering UK social and industrial history, including cinema and TV commercials from 1940s–1970s, the Middle East, particularly UAE from 1970–1996, plus Commonwealth. Also publishes DVDs of

vintage documentaries as well as special interest audiobooks based on diaries and other primary sources.

## Guildford Institute Library

Ward Street, Guildford GU1 4LH
☎ 01483 562142
library@guildford-institute.org.uk
www.guildford-institute.org.uk
OPEN 10.00 am to 3.00 pm Tuesday to Friday; 10.00 am to 1.00 pm each first Saturday of the month.
OPEN ACCESS to members only but open to enquirers for research purposes.

Founded 1834. Some 14,000 volumes of which 7,500 were printed before the First World War. The remaining stock consists of recently published works of fiction and non-fiction. Newspapers and periodicals also available. SPECIAL COLLECTIONS include an almost complete run of the *Illustrated London News* from 1843–1906, a collection of Victorian and early 20th century local history ephemera albums, and about 400 photos and other pictures relating to the Institute's history and the town of Guildford.

## Guildhall Library

Aldermanbury, London EC2V 7HH
☎ 020 7332 1868/1870    🖷 020 7600 3384
guildhall.library@cityoflondon.gov.uk
www.cityoflondon.gov.uk/guildhalllibrary
OPEN 9.30 am to 5.00 pm Monday to Saturday; NB closes on Saturdays preceding Bank Holidays; check for details.
ACCESS For reference (but much of the material is kept in storage areas and is supplied to readers on request; proof of identity is required for consultation of certain categories of stock). Free, *limited* enquiry service available. Also a fee-based service for in-depth research: ☎ 020 7332 1854 🖷 020 7600 3384 search.guildhall@cityoflondon.gov.uk www.cityoflondon.gov.uk/search_guildhall

Part of the City of London libraries. Seeks to provide a basic general reference service but its major strength, acknowledged worldwide, is in its historical collections. Strong on all aspects of London history, with wide holdings of English history, topography and genealogy, including local directories, poll books and parish register transcripts. Also good collections of English statutes, law reports, parliamentary debates and journals, and House of Commons papers. Home of several important collections deposited by London institutions: the Marine collection of the Corporation of Lloyd's, the Stock Exchange's historical files of reports and prospectuses, the Clockmakers' Company library and museum, the Gardeners' Company, Fletchers' Company, the

Institute of Masters of Wine, International Wine and Food Society and Gresham College.

## Guille–Alles Library

Market Street, St Peter Port, Guernsey GY1 1HB
☎ 01481 720392    🖷 01481 712425
ga@library.gg
www.library.gg
OPEN 9.00 am to 5.00 pm Monday, Thursday, Friday, Saturday; 10.00 am to 5.00 pm Tuesday; 9.00 am to 8.00 pm Wednesday
OPEN ACCESS For residents; payment of returnable deposit by visitors. Music CD collection: £10 for two-year subscription.

Lending, reference and information services. Public internet service.

## Harlow Library
▷ Essex County Council Libraries

## Herefordshire Libraries

Shirehall, Hereford HR1 2HX
☎ 01432 261644    🖷 01432 260744
libraries@herefordshire.gov.uk
www.herefordshire.gov.uk/leisure/
Principal Libraries Officer *Jon Chedgzoy*
Senior Stock Librarian *Carolyn Huckfield*
OPEN Opening hours vary in the libraries across the county.
ACCESS Information and reference services open to anyone; loans to members only (membership criteria: resident, being educated, visiting or working in Herefordshire).

Information service, reference and lending libraries. Non-fiction and fiction for all age groups, including normal and large print, spoken word CDs, pre-loaded mp3 players, cassettes, music CDs, DVDs, wii electronic games, playstation3. Hereford Library houses the largest reference section and the county local history collection, although some reference material and local history is available at all libraries. Internet access at all libraries. *Special collections* Cidermaking; Beekeeping; Alfred Watkins; John Masefield; Pilley.

## Highgate Literary and Scientific Institution Library

11 South Grove, London N6 6BS
☎ 020 8340 3343    🖷 020 8340 5632
librarian@hlsi.net
www.hlsi.net
Librarian *Margaret Mackay*
OPEN 10.00 am to 5.00 pm Tuesday to Friday; 10.00 am to 4.00 pm Saturday (closed Sunday and Monday)
ANNUAL MEMBERSHIP £60 (individual); £100 (household)

25,000 volumes of general fiction and non-fiction, with a children's section and extensive local

archives. SPECIAL COLLECTIONS on local history, London, and local poets Samuel Taylor Coleridge and John Betjeman.

## Highland Libraries, The Highland Council, Education, Culture and Sport Service

Library Support Unit, 31a Harbour Road, Inverness IV1 1UA
☎ 01463 235713    🖷 01463 236986
libraries@highland.gov.uk
www.highland.gov.uk
OPEN Library opening hours vary to suit local needs. Contact administration and support services for details (8.00 am to 5.00 pm Monday to Friday).
OPEN ACCESS

Comprehensive range of lending and reference stock: books, pamphlets, periodicals, newspapers, compact discs, audio and video cassettes, maps, census records, genealogical records, photographs, educational materials, etc. Free access to the internet in all libraries. Highland Libraries provides the public library service throughout the Highlands with a network of 43 static and 12 mobile libraries.

## Holborn Library

32–38 Theobalds Road, London WC1X 8PA
☎ 020 7974 4001    🖷 020 7974 6356
holbornlibrary@camden.gov.uk
www.camden.gov.uk/libraries
OPEN 10.00 am to 7.00 pm Monday and Thursday; 10.00 am to 6.00 pm Tuesday, Wednesday, Friday; 10.00 am to 5.00 pm Saturday
OPEN ACCESS

London Borough of Camden public library. Includes the London Borough of Camden Local Studies and Archive Centre.

## Holcenberg Jewish Collection
▷ City of Plymouth Library and Information Services

## House of Commons Research Library
▷ entry under Useful Websites

## Huntley Film Archive

22 Islington Green, London N1 8DU
☎ 020 7226 9260    🖷 020 7359 9337
films@huntleyarchives.com
www.huntleyarchives.com
Contact *Amanda Huntley*

Originally a private collection, the library is now a comprehensive archive of rare and vintage documentary film dating from 1895. 50,000 films on all subjects of a documentary nature. Online catalogue available.

## IHR Library
▷ The Institute of Historical Research

## Imperial College Central Library
▷ Science Museum Library

## Imperial War Museum
Department of Printed Books, Lambeth Road, London SE1 6HZ
☎ 020 7416 5342  🖷 020 7416 5246
collections@iwm.org.uk
www.iwm.org.uk
OPEN 10.00 am to 5.00 pm Monday to Saturday (restricted service Saturday; closed on Bank Holiday Saturdays and two weeks during the year for annual stock check).
ACCESS For reference (but at least 24 hours' notice must be given for intended visits).

A large collection of material on British and Commonwealth 20th-century life with detailed coverage of the two World Wars and other conflicts. This collection includes substantial holdings of European language material. Books, pamphlets and periodicals, including many produced for short periods in unlikely wartime settings; also maps, biographies and privately printed memoirs, and foreign language material. Additional research material available in the departments of Art, Documents, Exhibits and Firearms and in the Film, Photographs and Sound Archives. See www.iwmcollections.org.uk for their online catalogues.

## The Institute of Historical Research
Senate House, Malet Street, London WC1E 7HU
☎ 020 7862 8760
IHR.Library@sas.ac.uk
www.history.ac.uk/library
OPEN 9.00 am to 8.45 pm Monday to Friday; 9.30 am to 5.15 pm Saturday
ACCESS For reference only. Non-members are advised to telephone to check availability of items before visiting.

Reference collection of printed primary sources, bibliographies, guides to archives, periodicals and works covering the history of Western Europe from the fall of the Roman Empire. Holds a substantial microform collection of mainly British material from repositories outside London.

## Instituto Cervantes
102 Eaton Square, London SW1W 9AN
☎ 020 7201 0757  🖷 020 7235 0329
biblon@cervantes.es
londres.cervantes.es
OPEN 12 noon to 7.30 pm Monday to Thursday; 12 noon to 6.30 pm Friday; 9.30 am to 2.00 pm Saturday
OPEN ACCESS For reference and lending.

Spanish language and literature, history, art, philosophy. The library houses a collection of books, periodicals, videos, DVDs, slides, tapes, CDs, cassettes, CD-ROMs specializing entirely in Spain and Latin America.

## Italian Institute Library
39 Belgrave Square, London SW1X 8NX
☎ 020 7396 4425  🖷 020 7235 4618
www.icilondon.esteri.it
OPEN 10.00 am to 1.30 pm and 2.30 pm to 6.00 pm Monday to Friday
OPEN ACCESS For reference (telephone first to make an appointment).

A collection of over 30,000 volumes relating to all aspects of Italian culture, including DVDs and CDs relating to Italian cinema. Texts are mostly in Italian, with some in English.

## Jersey Library
Halkett Place, St Helier, Jersey JE2 4WH
☎ 01534 448700 (Jersey Library)/448701 (Reference)/448702 (Open Learning)/448733 (Branch Library)  🖷 01534 448730
je.library@.gov.je
www.gov.je/library
OPEN 9.30 am to 5.30 pm Monday, Wednesday, Thursday, Friday; 9.30 am to 7.30 pm Tuesday; 9.30 am to 4.00 pm Saturday
OPEN ACCESS

Books, periodicals, newspapers, CDs, DVDs and language packs, cassettes, videos, microfilm, specialized local studies collection, public internet access. Branch Library at Les Quennevais School, St Brelade. Mobile library and homes services. Open Learning Centre.

## John Frost Newspapers
22b Rosemary Avenue, Enfield EN2 0SS
☎ 020 8366 1392/0946  🖷 020 8366 1379
andrew@johnfrostnewspapers.com
www.johnfrostnewspapers.co.uk
Contacts Andrew Frost, John Frost

A collection of 80,000 original newspapers (1630 to the present day) and 200,000 press cuttings available, on loan, for research and rostrum/stills work (TV documentaries, book and magazine publishers and audiovisual presentations). Historic events, politics, sports, royalty, crime, wars, personalities, etc., plus many in-depth files.

## John Wyndham Archive
▷ Science Fiction Foundation Research Library

## Kent County Central Library
Kent Libraries and Archives, Springfield, Maidstone ME14 2LH
☎ 01622 696511  🖷 01622 696494
countycentrallibrary@kent.gov.uk
www.kent.gov.uk/libs
OPEN 9.00 am to 6.00 pm Monday to Friday; 9.00 am to 5.00 pm Saturday
OPEN ACCESS

50,000 volumes available on the floor of the library plus 250,000 volumes of non-fiction, mostly academic, available on request to staff. English literature, poetry, classical literature, drama (including play sets), music (including music sets). Strong, too, in sociology, art history, business information and government publications. Loans to all who live or work in Kent; those who do not may consult stock for reference or arrange loans via their own local library service.

## King George III Topographic Collection and Maritime Collection
▷ British Library Map Collections

## Leeds Central Library
Calverley Street, Leeds LS1 3AB
☎ 0113 247 6016     🅕 0113 247 8426
enquiry.express@leeds.gov.uk
www.leeds.gov.uk/libraries
OPEN 9.00 am to 8.00 pm Monday, Tuesday, Wednesday; 9.00 am to 5.00 pm Thursday and Friday; 10.00 am to 5.00 pm Saturday; 1.00 pm to 5.00 pm Sunday
ACCESS Level access and lift. Open access to lending libraries and some reference materials. Other reference materials on request. Free public internet access.

**Lending Library** ☎ 0113 247 8270
Adult and children's popular fiction and non-fiction for loan, including talking books, large print and books in world languages.

**Music Library** ☎ 0113 247 8273
Scores, books, DVD and audio.

**The Information Centre** ☎ 0113 247 8282 or 0113 395 1833
Company and product information, market research, statistics, directories, journals and computer-based information. Highlighted collections covering Law and Rights, Health, Jobs and Careers, Europe Direct, Education and Business. Extensive files of newspapers and periodicals plus all government publications since 1960. Special collections include military history, Judaic, early gardening books.

**Art Library** ☎ 0113 247 8247
Major collection of material on fine and applied arts.

**Local and Family History Library** ☎ 0113 247 8290
Extensive collection on Leeds and Yorkshire, including maps, books, pamphlets, local newspapers, illustrations and playbills. Census returns for the whole of Yorkshire also available. International Genealogical Index and parish registers.
    Leeds Library and Information Service has an extensive network of branch and mobile libraries.

## Leeds Library
18 Commercial Street, Leeds LS1 6AL
☎ 0113 245 3071     🅕 0113 245 1191
enquiries@theleedslibrary.org.uk
www.theleedslibrary.org.uk
OPEN 9.00 am to 5.00 pm Monday to Friday; 9.30 am to 1.00 pm on the first Saturday of each month.
ACCESS To members; research use upon application to the librarian.

Founded 1768. Contains over 140,000 books and periodicals from the 15th century to the present day. SPECIAL COLLECTIONS include Reformation pamphlets, Civil War tracts, Victorian and Edwardian children's books and fiction, European language material, spiritualism and psychical research, plus local material.

## Library of the Religious Society of Friends
Friends House, 173 Euston Road, London NW1 2BJ
☎ 020 7663 1135     🅕 020 7663 1001
library@quaker.org.uk
www.quaker.org.uk/library
OPEN 10.00 am to 5.00 pm Tuesday to Friday
OPEN ACCESS Reader registration required with proof of permanent address.

Quaker history, thought and activities from the 17th century onwards. Supporting collections on peace, anti-slavery and other subjects in which Quakers have maintained long-standing interest. Also archives and manuscripts relating to the Society of Friends.

## Lincoln Central Library
Free School Lane, Lincoln LN2 1EZ
☎ 01522 782010     🅕 01522 575011
lincoln_library@lincolnshire.gov.uk
www.lincolnshire.gov.uk
OPEN 9.00 am to 7.00 pm Monday to Friday; 9.00 am to 4.00 pm Saturday
OPEN ACCESS to the library; appointment required for the Tennyson Research Centre.

Lending and reference library. Special collections include Lincolnshire local history (printed and published material, photographs, maps, directories and census data) and the Tennyson Research Centre (contact *Grace Timmins*, grace.timmins@lincolnshire.gov.uk).

## Linen Hall Library
17 Donegall Square North, Belfast BT1 5GB
☎ 028 9032 1707     🅕 028 9043 8586
info@linenhall.com
www.linenhall.com
OPEN 9.30 am to 5.30 pm Monday to Friday; 9.30 am to 4.00 pm Saturday
OPEN ACCESS For reference (loans restricted to members).

Founded 1788. Contains about 250,000 books in the general lending library. Also houses a number of unique collections including the Irish and Local Studies Collection, which incorporates the world renowned Northern Ireland Political Collection relating to the 'Troubles' and ongoing peace process (c. 300,000 items).

## Literary & Philosophical Society of Newcastle upon Tyne

23 Westgate Road, Newcastle upon Tyne NE1 1SE
☎ 0191 232 0192   🖷 0191 261 4494
library@litandphil.org.uk
www.litandphil.org.uk
Librarian *Kay Easson*
OPEN 9.30 am to 7.00 pm Monday, Wednesday, Thursday; 9.30 am to 8.00 pm Tuesday; 9.30 am to 5.00 pm Friday; 9.30 am to 1.00 pm Saturday
ACCESS Members; research facilities for *bona fide* scholars on application to the Librarian.

200-year-old library of 140,000 volumes, periodicals (including 130 current titles), classical music on vinyl recordings and CD, plus a collection of scores. Free public lectures, events and recitals. Recent publications include: *The Reverend William Turner: Dissent and Reform in Georgian Newcastle upon Tyne* Stephen Harbottle; *History of the Literary and Philosophical Society of Newcastle upon Tyne, Vol. 2 (1896–1989)* Charles Parish; *Bicentenary Lectures 1993* ed. John Philipson.

## Liverpool Libraries and Information Services

William Brown Street, Liverpool L3 8EW
☎ 0151 233 5829   🖷 0151 233 5886
refbt.central.library@liverpool.gov.uk
www.liverpool.gov.uk/libraries
OPEN 9.00 am to 6.00 pm Monday to Friday; 9.00 am to 5.00 pm Saturday; 12 noon to 4.00 pm Sunday
OPEN ACCESS

**Learn Direct Centre – UK Online Centre** Free broadband internet access.

**Humanities Reference Library** A total stock in excess of 120,000 volumes and 24,000 maps, plus book plates, prints and autographed letters. SPECIAL COLLECTIONS Walter Crane and Edward Lear illustrations, Kelmscott Press, Audubon's *Birds of America*.

**Business and Technology Reference Library** Extensive stock dealing with all aspects of science, commerce and technology, including British and European standards and patents and trade directories.

**Audio Visual Library** Extensive stock relating to all aspects of music. Includes 128,000 volumes and music scores, over 4,000 CDs, 1,000 videos and 800 DVDs.

**Record Office and Local History Department** Material relating to Liverpool, Merseyside, Lancashire and Cheshire, together with archive material mainly on Liverpool. Proof of name and address required to obtain reader's ticket.

**Lending Library** Graphic novels, large print collection, children's collections, audio books, study support collections, Reader friendly displays of bestsellers and out of print titles.

## The London Library

14 St James's Square, London SW1Y 4LG
☎ 020 7930 7705   🖷 020 7766 4766
membership@londonlibrary.co.uk
www.londonlibrary.co.uk
Librarian *Miss Inez Lynn*
OPEN 9.30 am to 7.30 pm Monday, Tuesday, Wednesday; 9.30 am to 5.30 pm Thursday, Friday, Saturday
ACCESS For members only (2010 fees: £395 p.a.; £200 for 16–24-year-olds). Day and weekly reference tickets available for non-members (£10 and £30).

With over a million books and 8,000 members, The London Library 'is the most distinguished private library in the world; probably the largest, certainly the best loved'. Founded in 1841, it is a registered charity and wholly independent of public funding. Its permanent collection embraces most European languages as well as English. Its subject range is predominantly within the humanities, with emphasis on literature, history, fine and applied art, architecture, bibliography, philosophy, religion, and topography and travel. Some 8,000–9,000 titles are added yearly. Over 95 per cent of the stock is on open shelves to which members have free access. Members may take out up to 10 volumes at a time; 15 if they live more than 20 miles from the Library and may hold the books on loan until they are needed by another member. The comfortable Reading Room has an annexe for users of personal computers. There are photocopiers, CD-ROM workstations, free access to the internet, and the Library also offers a postal loans service.

Membership is open to all: prospective members are required to submit a refereed application form in advance of admission, but there is at present no waiting list for membership. The London Library Trust may offer reduced cost 'Carlyle Memberships' to those who are unable to afford the full annual fee; details on application.

## London Metropolitan Archives Library Services

*History Library:* 40 Northampton Road, London EC1R 0HB
☎ 020 7332 3820   🖷 020 7833 9136

ask.lma@cityoflondon.gov.uk
www.lma.gov.uk
Contact *The Enquiry Team*
OPEN 9.30 am to 4.45 pm Monday, Wednesday,
Friday; 9.30 am to 7.30 pm Tuesday and Thursday;
open some Saturdays – call for details.
ACCESS For reference only.

This 100,000 volume library covers all aspects
of the life and development of London, with
strong holdings on the history and organization
of London local government. As the former
Greater London Council History Library, the
collection covers all subjects of London life,
from architecture and biography to theatres
and transport. The collection includes London
directories from 1677 to the present, Acts of
Parliament, statistical returns, several hundred
periodical titles and public reports.

### London's Transport Museum Poster Archives
▷ London's Transport Museum Photographic Library
under Picture Libraries

### Lord Louis Library
Orchard Street, Newport, Isle of Wight PO30 1LL
☎ 01983 527655/823800 (Reference Library)
🖷 01983 825972
referencelibrary@iow.gov.uk
www.iwight.com/thelibrary
OPEN 9.00 am to 5.30 pm Monday, Tuesday,
Friday; 10.00 am to 8.00 pm Thursday; 9.00 am to
5.00 pm Saturday; 10.00 am to 1.00 pm Sunday;
closed Wednesday
OPEN ACCESS

General adult and junior fiction and non-fiction
collections; local history collection at Library HQ
(☎ 01983 203880 for details). Internet access in all
branches of the library on the Isle of Wight. Also
the county's main reference library.

### Loughton Library
▷ Essex County Council Libraries

### Manchester Central Library
libraries@manchester.gov.uk
www.manchester.gov.uk/libraries

The Central Library building will be closed during
2010–11 for major refurbishment to improve
access to the extensive collections. A temporary,
limited service will be available in Elliot House on
Deansgate during this time. Full details of services
on offer and contact details can be found on the
website.

### Marylebone Library (Westminster)
▷ Sherlock Holmes Collection

### Ministry of Defence Cartographic Archive
▷ British Library Map Collections

### The Mitchell Library
North Street, Glasgow G3 7DN
☎ 0141 287 2999    🖷 0141 287 2815
lil@cls.glasgow.gov.uk
www.glasgowlibraries.org
OPEN 9.00 am to 8.00 pm Monday to Thursday;
9.00 am to 5.00 pm Friday and Saturday
OPEN ACCESS

One of Europe's largest public reference libraries
with stock of over 1,200,000 volumes. It
subscribes to 48 newspapers and more than 1,200
periodicals. There are collections in microform,
records, tapes and videos, as well as CD-ROMs,
electronic databases, illustrations, photographs,
postcards, etc.

The library contains a number of special
collections, e.g. the Robert Burns Collection
(5,000 volumes), the Scottish Poetry Collection
(12,000 items) and the Scottish Drama Collection
(1,650 items).

### Morrab Library
Morrab House, Morrab Gardens, Penzance
TR18 4DA
☎ 01736 364474
Librarian *Annabelle Read*
OPEN 10.00 am to 4.00 pm Tuesday to Friday;
10.00 am to 1.00 pm Saturday
ACCESS Non-members may use the library for a
small daily fee but may not borrow books.

Formerly known as the Penzance Library. An
independent subscription lending library of over
40,000 volumes covering virtually all subjects
except modern science and technology, with large
collections on history, literature and religion.
There is a comprehensive Cornish collection of
books, newspapers and manuscripts including
the Borlase letters; a West Cornwall photographic
archive; many runs of 18th and 19th-century
periodicals; a collection of over 2,000 books
published before 1800.

### The National Archives
Kew, Richmond TW9 4DU
☎ 020 8876 3444
www.nationalarchives.gov.uk
www.nationalarchives.gov.uk/contact/form
OPEN 9.00 am to 5.00 pm Monday and Friday;
9.00 am to 7.00 pm Tuesday and Thursday; 10.00
am to 5.00 pm Wednesday; 9.30 am to 5.00 pm
Saturday (closed public holidays and for annual
stocktaking in early December). From April 2008
the Family Record Centre services have been
incorporated at the Kew site.
ACCESS Reader's ticket required for access to
original documents, available free of charge
on production of two forms of identification,
one proof of address and one proof of
identity, which must include a valid signature.

Further information is available online (www.nationalarchives.gov.uk/registration/) or by phone.

The National Archives is the official archive for England, Wales and the central UK government, containing 900 years of history from the Domesday Book to the present, with records ranging from parchment and paper scrolls through to recently created digital files and archived websites. Increasingly, these records are being put online, making them universally accessible.

The National Archives is a government department and an executive agency under the Secretary of State for Justice. Gives detailed guidance to government departments and the public sector on information management, in order to ensure the survival of records in whichever form they are created, be it paper or digital. Also advises custodians throughout the public and private sectors about the care of historical archives.

## National Art Library

Word & Image Department, Victoria and Albert Museum, Cromwell Road, London SW7 2RL
Ⓣ 020 7942 2400    Ⓕ 020 7942 2401
nal.enquiries@vam.ac.uk
www.vam.ac.uk
OPEN 10.00 am to 5.30 pm Tuesday, Wednesday, Thursday, Saturday; 10.00 am to 6.30 pm Saturday; closed Monday
ACCESS Registration necessary. Some special collections may only be viewed by appointment and require written recommendation, approved by the Deputy Keeper.

The NAL is both a major reference library and the V&A's curatorial department for the art, craft and design of the book. The subjects covered in its core material are central to the work of the Museum and are supplemented by literature from a broader subject field including, e.g., documentation about architecture and interior design. SPECIALIST COLLECTIONS include the art and craft of the book and 20th Century artists' books and book art; international exhibitions, including the Great Exhibition of 1851; letters, account books and other records relating to individual artists and the production and marketing of decorative and artistic objects. Databases available online; enquiry service, in person, by phone, email or post.

## National Library of Ireland

Kildare Street, Dublin 2, Republic of Ireland
Ⓣ 00 353 1 603 0200    Ⓕ 00 353 1 676 6690
info@nli.ie
www.nli.ie

Reading Rooms: OPEN 9.30 am to 9.00 pm Monday, Tuesday, Wednesday; 9.30 am to 5.00 pm Thursday and Friday; 9.30 am to 1.00 pm Saturday
ACCESS Passes are issued for genealogical research or to consult newspapers; Reader's Ticket required for access to other material.

The National Library of Ireland's holdings constitute the most comprehensive collection of Irish documentary material in the world and offer an invaluable representation of Ireland's history and heritage. Collections include printed books, manuscripts, newspapers, prints and drawings, photographs, ephemera.

## National Library of Scotland

George IV Bridge, Edinburgh EH1 1EW
Ⓣ 0131 623 3700    Ⓕ 0131 623 3701
enquiries@nls.uk
www.nls.uk
**Causewayside Building**: 33 Salisbury Place, Edinburgh EH9 1SL
**Scottish Screen Archive**: 39–41 Montrose Avenue, Hillington Park, Glasgow G52 4LA
Ⓣ 0845 366 4600 Ⓕ 0845 366 4601
ssaenquiries@nls.uk
http://ssa.nls.uk
OPEN See www.nls.uk/info/openinghours

With a collection of around 14 million printed items spanning centuries, the National Library of Scotland is the world's leading centre for the study of Scotland and the Scots. As well as printed items, the Library's collections include around two million maps, 30,000 films, manuscript archives of thousands of notable Scots and a growing digital collection. The Library has a year-round programme of free exhibitions, a newly-opened Visitor Centre with a café and shop, free Wi-Fi and an events room for regular cultural events, which can be booked online at www.nls.uk/events

## National Library of Wales

Penglais, Aberystwyth SY23 3BU
Ⓣ 01970 632800    Ⓕ 01970 615709
holi@llgc.org.uk
www.llgc.org.uk
OPEN 9.30 am to 6.00 pm Monday to Friday (reading rooms closed Bank Holidays and the week after Christmas).
ACCESS to reading rooms by reader's ticket, available on application. Open access to a wide-ranging exhibition programme.

Collection of over four million books and including large collections of periodicals, maps, manuscripts and audiovisual material. Particular emphasis on humanities in printed foreign material, and on Wales and other Celtic areas in all collections.

## National Meteorological Library and Archive

FitzRoy Road, Exeter EX1 3PB

☎ 01392 884841  📠 0870 900 5050

metlib@metoffice.gov.uk

www.metoffice.gov.uk

OPEN Library: 8.30 am to 5.00 pm Monday to
Thursday; 8.30 am to 4.30 pm Friday. Archive:
10.00 am to 6.00 pm

ACCESS Library: By Visitor's Pass available from
the reception desk; advance notice of a planned
visit is appreciated. Archive: Advance notice is
requested; bring ID and proof of address.

The major repository of most of the important
literature on the subjects of meteorology,
climatology and related sciences from the 16th
century to the present day. The Library houses
a collection of books, journals, articles and
scientific papers, plus published climatological
data from many parts of the world.

**The Archive** (National Meteorological Archive,
Great Moor House, Sowton Industrial Estate,
Bittern Road, Exeter EX2 7NL ☎ 01392 360987
📠 0870 900 5050; metlib@metoffice.gov.uk)
holds the document collection of meteorological
data and charts from England, Wales and British
overseas bases, including ships' weather logs.
Records from Scotland are stored in Edinburgh
and those from Northern Ireland in Belfast.

## National Monuments Record
▷ entry under Picture Libraries

## Natural England Library

Northminster House, Northminster Road,
Peterborough PE1 1UA

☎ 0300 060 0910

library@naturalengland.org.uk

www.naturalengland.org.uk

OPEN 8.30 am to 5.00 pm Monday to Thursday;
8.30 am to 4.30 pm Friday

ACCESS *Bona fide* students only (telephone the
library for appointment).

Information on nature conservation, nature
reserves, SSSIs, planning, legislation, countryside,
recreation, stewardship schemes and advice to
farmers.

## The Natural History Museum Library

Cromwell Road, London SW7 5BD

☎ 020 7942 5460  📠 020 7942 5559

library@nhm.ac.uk

www.nhm.ac.uk/library/index.html

OPEN 10.00 am to 4.30 pm Monday to Friday

ACCESS To *bona fide* researchers, by reader's
ticket on presentation of identification (telephone
first to make an appointment).

The library is in five sections: general; botany;
zoology; entomology; earth sciences. The
sub-department of ornithology is housed at
the Zoological Museum, Akeman Street, Tring
HP23 6AP (☎ 020 7942 6156). Resources available
include books, journals, maps, manuscripts,
drawings and photographs covering all aspects
of natural history, including palaeontology and
mineralogy, from the 15th century to the present
day. Also archives and historical collection on the
museum itself.

## Norfolk Library & Information Service

Norfolk and Norwich Millennium Library, The
Forum, Millennium Plain, Norwich NR2 1AW

☎ 01603 774774  📠 01603 774705

libraries@norfolk.gov.uk

www.library.norfolk.gov.uk

OPEN Lending Library, Reference and Information
Service and Norfolk Studies: 9.00 am to 8.00 pm
Monday to Friday; 9.00 am to 5.00 pm Saturday
EXPRESS: 9.00 am to 9.30 pm Monday to Friday;
9.00 am to 8.30 pm Saturday; 10.30 am to 4.30 pm
Sunday CHILDREN'S LIBRARY: 9.00 am to 8.00 pm
Monday to Friday; 9.00 am to 5.00 pm Saturday;
10.30 am to 4.30 pm Sunday

OPEN ACCESS

Reference lending library (stock merged together)
with wide range, including books, recorded
music, music scores, plays and videos. Houses the
2nd Air Division Memorial Library (www.2ndair.
org.uk) and has a strong Norfolk Heritage Library.
Extensive range of reference stock including
business information. Online databases. Public
fax and colour photocopying, free access to the
internet. EXPRESS (fiction, sound & vision library
within a library): selection of popular fiction,
videos, CDs and DVDs available, with extended
opening hours.

## Northamptonshire Libraries & Information
## Service

PO Box 216, John Dryden House, 8–10 The Lakes,
Northampton NN4 7DD

☎ 01604 237959  📠 01604 237937

nlis@northamptonshire.gov.uk

www.northamptonshire.gov.uk/Leisure/Libraries/
home.htm

'Northamptonshire Libraries and Information
Service provides its users with much more than
just book loans. There are clubs to join, events
to attend, skills to learn, local history resources
to use, CDs and DVDs to rent – to list but a
few services.' In addition, writers and readers
are supported by 'Literature Northants'. Run by
the Literature Development Officer, Literature
Northants (http://litnorthants.wordpress.com)
provides access to literature opportunities across
the county. The LDO is also the key contact for all
the latest literary news and events.

## Northumberland County Library

Beechfield, Gas House Lane, Morpeth NE61 1TA
☏ 01670 500390
www.northumberlandlibraries.com
OPEN 9.00 am to 7.30 pm Monday, Tuesday,
Wednesday, Friday; 9.00 am to 12.30 pm Saturday
(closed Thursday)
OPEN ACCESS

Books, periodicals, newspapers, story cassettes,
CD-ROMs, DVDs, CDs. Free internet access,
word processing facilities, prints, microforms,
vocal scores, playsets. SPECIAL COLLECTIONS
Northern Poetry Library: 15,000 volumes of
modern poetry (see entry under *Organizations
of Interest to Poets*); Cinema: comprehensive
collection of about 5,000 volumes covering all
aspects of the cinema; Family History.

## Nottingham Central Library

Angel Row, Nottingham NG1 6HP
☏ 0115 915 2828    🖷 0115 915 2840
enquiryline@nottinghamcity.gov.uk
www.nottinghamcity.gov.uk/libraries
OPEN 9.00 am to 7.00 pm Monday, Wednesday
and Thursday; 9.30 am to 7.00 pm Tuesday; 9.00
am to 4.00 pm Saturday; closed Friday
OPEN ACCESS

City Centre Library includes a wide range of
stock: Business collection; Sound and Vision
collection; comprehensive Nottinghamshire Local
Studies. Drama and music sets for loan to groups.
Free internet access for library members.

## Nottingham Subscription Library
▷ Bromley House Library

## Office for National Statistics, National Statistics Information and Library Service

Customer Contact Centre, Office for National
Statistics, Room 1.015 Government Buildings,
Cardiff Road, Newport NP10 8XG
☏ 0845 601 3034    🖷 01633 652747
info@statistics.gov.uk
www.statistics.gov.uk
OPEN 9.00 am to 5.00 pm; appointment required;
visitors must supply some form of identification.
*Also:* National Statistics Information and Library
Service, Government Buildings, Cardiff Road,
Newport NP10 8XG
OPEN as above.

Wide range of government statistical publications
and access to government internet-based data.
Census statistical data from 1801; population and
health data from 1837; government social survey
reports from 1941; recent international statistical
data (UN, Eurostat, etc.); monograph and
periodical collections of statistical methodology.
The library in south Wales holds a wide range
of government economic and statistical
publications.

## Olaf Stapledon
▷ Science Fiction Foundation Research Library

## Orkney Library and Archive

44 Junction Road, Kirkwall, Orkney KW15 1AG
☏ 01856 873166    🖷 01856 875260
general.enquiries@orkneylibrary.org.uk
archives@orkneylibrary.org.uk
www.orkneylibrary.org.uk
Principal Archivist *Alison Fraser*
Principal Librarian *Karen Walker*
OPEN 9.15 am to 7.00 pm Monday to Thursday;
9.15 am to 5.00 pm Friday and Saturday. Archives:
9.15 am to 5.00 pm Monday, Tuesday, Wednesday,
Friday; 9.15 am to 7.00 pm Thursday; 9.15 am to
5.00 pm Saturday

Local studies collection. Archive includes sound
and photographic departments.

## Oxford Central Library

Westgate, Oxford OX1 1DJ
☏ 01865 815549    🖷 01865 721694
oxfordcentral.library@oxfordshire.gov.uk
www.oxfordshire.gov.uk
OPEN 9.00 am to 7.00 pm Monday to Thursday;
9.00 am to 5.30 pm Friday and Saturday

General lending and reference library. Periodicals,
audio visual materials, music library and
children's library.

## Paul Hirsch Music Library
▷ British Library Music Collections

## Penzance Library
▷ Morrab Library

## City of Plymouth Library and Information Services

Central Library, Drake Circus, Plymouth PL4 8AL
☏ 01752 305923
library@plymouth.gov.uk
www.plymouth.gov.uk/libraries
OPEN ACCESS

CENTRAL LIBRARY LENDING DEPARTMENTS:
**Lending** ☏ 01752 305912
lendlib@plymouth.gov.uk

**Children's Department** ☏ 01752 305916
childrens.library@plymouth.gov.uk

**Music & Drama Department** ☏ 01752 305914
music@plymouth.gov.uk
OPEN 9.00 am to 7.00 pm Monday to Friday; 9.00
am to 5.00 pm Saturday

The Lending departments offer books on all
subjects; language courses on cassette and
foreign language books; the Holcenberg Jewish
Collection; books on music and musicians, drama
and theatre; music parts and sets of music parts;

play sets; DVDs; song index; CDs; public internet access.

CENTRAL LIBRARY REFERENCE DEPARTMENTS:
**Reference** ☎ 01752 305907/305908
ref@plymouth.gov.uk

**Local Studies & Naval Studies** ☎ 01752 305909
localstudies@plymouth.gov.uk
OPEN 9.00 am to 7.00 pm Monday to Friday; 9.00 am to 5.00 pm Saturday

The Reference departments include an extensive collection of Ordnance Survey maps and town guides; community and census information; marketing and statistical information; Patents; books on every aspect of Plymouth; naval history; GRO Index on microfiche; Baring Gould manuscript of 'Folk Songs of the West'; public internet access, including some electronic subscriptions.

## Plymouth Proprietary Library

Alton Terrace, 111 North Hill, Plymouth PL4 8JY
☎ 01752 660515
Librarian *Shirley Lamerton*
OPEN Monday to Saturday from 9.30 am (closing time varies).
ACCESS To members; visitors by appointment only.

Founded 1810. The library contains approximately 17,000 volumes of mainly 19th and 20th century work. Member of the **Association of Independent Libraries**.

## The Poetry Library
▷ entry under Organizations of Interest to Poets

## Polish Library POSK

238–246 King Street, London W6 0RF
☎ 020 8741 0474    📠 020 8741 7724
library@polishlibrary.co.uk
www.posk.org
OPEN 10.00 am to 8.00 pm Monday and Wednesday; 10.00 am to 5.00 pm Friday
ACCESS For reference to all interested in Polish affairs; limited loans to members and *bona fide* scholars only through inter-library loans.

Books, pamphlets, periodicals, maps, music, photographs on all aspects of Polish history and culture. SPECIAL COLLECTIONS Emigré publications; Joseph Conrad and related works; Polish underground publications; bookplates.

## Poole Central Library

Dolphin Centre, Poole BH15 1QE
☎ 01202 262424    📠 01202 262442
libraries@poole.gov.uk
www.boroughofpoole.com/libraries
Information Services Librarian *David Parish*
OPEN 9.00 am to 6.00 pm Monday to Friday; 9.00 am to 5.00 pm Saturday

OPEN ACCESS
Public library with books, audiobooks, CDs and DVDs for loan. Enquiry service and Healthpoint health information centre, business information, children's library, periodicals and newspapers. Free public computers and meeting room. Photocopier and fax service.

## The Portico Library

57 Mosley Street, Manchester M2 3HY
☎ 0161 236 6785    📠 0161 236 6803
librarian@theportico.org.uk
www.theportico.org.uk
OPEN 9.30 am to 4.30 pm Monday to Friday
ACCESS For members by subscription (researchers may consult freely upon application); free access to exhibitions; events from £5.

The strength of the collection lies in travel, biography, architecture and local history and typifies the mindset of nineteenth-century merchants, industrialists, clerics and the general cultural and intellectual debate. SPECIALIST COLLECTIONS Nineteenth century literature with some pre-nineteenth century works; North West fiction; local history. Member of the **Association of Independent Libraries**.

## Public Record Office
▷ The National Archives

## Racing Collection (Newmarket)
▷ Suffolk County Council – Suffolk Libraries

## Reading Central Library

Abbey Square, Reading RG1 3BQ
☎ 0118 901 5950    📠 0118 901 5954
www.readinglibraries.org.uk
OPEN 9.00 am to 5.30 pm Monday and Friday; 9.00 am to 7.00 pm Tuesday and Thursday; 9.00 am to 5.00 pm Wednesday; 9.30 am to 5.00 pm Saturday
OPEN ACCESS

Ground Floor: Fiction, audio-visual material, children's library; First Floor: Newspapers, non-fiction – biography, business, careers, cookery, DIY, engineering, gardening, health, languages, law, mind, body and spirit, pets, science, social studies, sport, transport, travel; Second Floor: LearnDirect, non-fiction – art, music, literature, plays, computing; Third floor: Non-fiction – world history, national history, local history. Magazines and periodicals and internet access on all floors.

## RIBA British Architectural Library Drawings & Archives Collection
▷ British Architectural Library

## Richmond Central Reference Library
Old Town Hall, Whittaker Avenue, Richmond
TW9 1TP
☎ 020 8940 5529  📠 020 8940 6899
reference.services@richmond.gov.uk
www.richmond.gov.uk
OPEN 9.30 am to 6.00 pm Monday, Tuesday,
Thursday, Friday; 9.30 am to 8.00 pm Wednesday;
9.30 am to 5.00 pm Saturday
OPEN ACCESS

General reference library serving the needs of
local residents and organizations. Internet access
and online databases for public use. Enquiries
received by visit, telephone and email.

## Robert Burns Collection
▷ The Mitchell Library

## Royal Geographical Society (with the Institute of British Geographers)
1 Kensington Gore, London SW7 2AR
☎ 020 7591 3044  📠 020 7591 3001
enquiries@rgs.org
www.rgs.org/collections
Foyle Reading Room: OPEN 10.00 am to 5.00 pm
Monday to Friday (except Bank Holidays and the
period from Christmas to New Year).

Around two million items on geography and
exploration, including monographs, periodicals,
images, maps and manuscripts. Coverage
primarily from 1830 onwards, with particular
strengths in Africa, Central Asia and Antarctica.
For information on commercial Picture Library
services, see entry under *Picture Libraries*.

## Royal Institute of Philosophy
▷ Senate House Library, University of London

## The Royal Philharmonic Society Archive
▷ British Library Music Collections

## Royal Society Library
6–9 Carlton House Terrace, London SW1Y 5AG
☎ 020 7451 2606  📠 020 7930 2170
library@royalsociety.org
royalsociety.org
OPEN 10.00 am to 5.00 pm Monday to Friday
(but closed until mid-2010 for refurbishment.
See http://royalsociety.org/page.asp?id=8712 for
updates)
ACCESS Open to all researchers with an interest
in the history of science, the Fellowship of the
Royal Society and science policy. Researchers are
advised to contact the Library in advance of their
first visit.

History of science, scientists' biographies, science
policy reports, and publications of international
scientific unions and national academies from all
over the world.

## RSA (Royal Society for the Encouragement of Arts, Manufactures & Commerce)
8 John Adam Street, London WC2N 6EZ
☎ 020 7451 6847  📠 020 7839 5805
library@rsa.org.uk
www.theRSA.org
Head of Archive and Library *Rob Baker*
Library and Information Management Assistant
*Chris Pugh*
OPEN Library: 8.00 am to 8.00 pm every weekday.
Archive material by appointment.
ACCESS to Fellows of RSA; Archive: by
appointment to all researchers (contact the
Archivist, 020 7451 6847; archive@rsa.org.uk).

Archives of the Society since 1754. A collection
of approximately 15,000 items including minutes
of the Society, correspondence, prints, original
drawings and international exhibition material
and an early library of over 700 volumes.

## Royal Society of Medicine Library
1 Wimpole Street, London W1G 0AE
☎ 020 7290 2940  📠 020 7290 2939
library@rsm.ac.uk
www.rsm.ac.uk
Contact *Wayne Sime, Director of Library Services*
OPEN 9.00 am to 7.00 pm Monday to Thursday;
9.00 am to 5.30 pm Friday; 10.00 am to 4.30 pm
Saturday
ACCESS For reference only, on introduction by
Fellow of the Society or temporary membership
is available to non-members; Half-day (after 1.00
pm or on Saturdays) £6; £10 per day; £20 per
week; £35 per month. Identification required.
Individual Library Membership rate: Medical
£378 p.a.; Non-medical £115 p.a.

Books, periodicals, databases on postgraduate
biomedical information. Extensive historical
collection dating from the fifteenth century and
medical portrait collection. Over 90% of titles are
on open access.

## Saffron Walden Library
▷ Essex County Council Libraries

## St Bride Library
Bride Lane, Fleet Street, London EC4Y 8EE
☎ 020 7353 4660
nigelroache@stbridefoundation.org
elizabethklaiber@stbridefoundation.org
www.stbridefoundation.org
OPEN 12 noon to 5.30 pm Tuesday and Thursday;
12 noon to 9.00 pm Wednesday
OPEN ACCESS

A public reference library maintained by the St
Bride Foundation. Appointments advisable for
consultation of special collections. Every aspect
of printing and related matters: publishing and
bookselling, newspapers and magazines, graphic

design, calligraphy and typography, papermaking and bookbinding. One of the world's largest specialist collections in its field, with over 50,000 volumes, over 3,000 periodicals (200 current titles), and extensive collection of drawings, manuscripts, prospectuses, patents and materials for printing and typefounding. Noted for its comprehensive holdings of historical and early technical literature.

## Science Fiction Foundation Collections

Special Collections & Archives, Sydney Jones Library, University of Liverpool, PO Box 123, Liverpool L69 3DA

☏ 0151 794 3142 ⓕ 0151 794 2681
a.p.sawyer@liverpool.ac.uk
www.sfhub.ac.uk
Contact *Andy Sawyer*
ACCESS For research, by appointment only (telephone first).

This is the largest collection outside the US of English-language science fiction and related material – including autobiographies and critical works. SPECIAL COLLECTIONS Runs of 'pulp' magazines dating back to the 1920s. Foreign-language material (including a large Russian collection), and the papers of the Flat Earth Society. The collection also features a growing range of archive and manuscript material, including material from Stephen Baxter, Ramsey Campbell and John Brunner. The University of Liverpool also holds the Olaf Stapledon, Eric Frank Russell and John Wyndham archives.

## Science Museum Library & Archives

Imperial College Road, London SW7 5NH
☏ 020 7942 4242 ⓕ 020 7942 4243
smlinfo@sciencemuseum.org.uk
www.sciencemuseum.org.uk/library
OPEN 9.30 am to 8.30 pm Monday to Friday; 10.00 am to 6.00 pm Saturday
OPEN ACCESS Reference only; no loans.

National reference library for the history and public understanding of science and technology, with a large collection of source material. Operates jointly with Imperial College Central Library.

## Scottish Drama Collection/Scottish Poetry Collection
▷ The Mitchell Library

## Scottish Poetry Library
▷ entry under Organizations of Interest to Poets

## Scottish Screen Archive
▷ National Library of Scotland

## Seckford Collection
▷ Suffolk County Council – Suffolk Libraries

## 2nd Air Division Memorial Library
▷ Norfolk Library & Information Service

## Senate House Library, University of London

Senate House, Malet Street, London WC1E 7HU
☏ 020 7862 8461/62 (Information Centre)
ⓕ 020 7862 8480
enquiries@shl.lon.ac.uk (Information Centre)
www.shl.lon.ac.uk
MEMBERSHIP DESK: ☏ 020 7862 8439/40
userservices@shl.lon.ac.uk
OPEN Term-time: 9.00 am to 9.00 pm Monday to Thursday; 9.00 am to 6.30 pm Friday; 9.45 am to 5.30 pm Saturday. Vacation: 9.00 am to 6.00 pm Monday to Friday; 9.45 am to 5.30 pm Saturday (closed on Sundays and at certain periods during Bank Holidays).

The Senate House Library is a major academic research library predominantly based across the Humanities and Social Sciences. Housed within its 16 floors are some two million titles including 5,500 current periodicals and a wide range of electronic resources. It contains a number of outstanding research collections which, as well as supporting the scholarly activities of the University, attract researchers from throughout the UK and internationally. These include: English (e.g. the Durning-Lawrence Library and Sterling Collection of first editions); Economic and Social History (the Goldsmiths' Library, containing 70,000 items ranging from 15th to early 19th century); Modern Languages (primarily Romance and Germanic); Palaeography (acclaimed as being the best open access collection in its field in Europe); History (complementary to the **Institute of Historical Research**); Music, Philosophy (acts as the Library of the Royal Institute of Philosophy); Psychology (includes the BPS library); Major area studies collections (Latin-American, including Caribbean; United States and Commonwealth Studies, British Government Publications and maps). The Library has a wide range of Special Collections holdings. Check the website for full details of collections and current access arrangements.

## Sheffield Libraries, Archives and Information

Central Library, Surrey Street, Sheffield S1 1XZ
☏ 0114 273 4712 ⓕ 0114 273 5009
libraries@sheffield.gov.uk
www.sheffield.gov.uk/libraries

**Central Lending Library**
☏ 0114 273 4727 (enquiries)/4729 (book renewals)
centrallending.library@sheffield.gov.uk
OPEN 10.00 am to 8.00 pm Monday; 9.30 am to 5.30 pm Tuesday, Thursday, Friday, Saturday; 9.30 am to 8.00 pm Wednesday

Books, talking books, large print, language courses, European fiction, books in cultural

languages, play sets. Free internet access.
**Writers' Resource Centre** (see details below).

## Sheffield Archives

52 Shoreham Street, Sheffield S1 4SP
☎ 0114 203 9395 🖷 0114 203 9398
archives@sheffield.gov.uk
OPEN 10.00 am to 5.30 pm Monday; 9.30 am to
5.30 pm Tuesday to Thursday; 9.00 am to 1.00
pm and 2.00 pm to 5.00 pm Saturday (documents
should be ordered by 5.00 pm Thursday for
Saturday); closed Friday
ACCESS By reader's card.

Holds documents relating to Sheffield and South
Yorkshire, dating from the 12th century to the
present day, including records of the City Council,
churches, businesses, landed estates, families
and individuals, institutions and societies. Free
internet access.

## Music and Film Library

☎ 0114 273 4733
musicandav.library@sheffield.gov.uk
OPEN As for Central Lending Library above.

Provides a comprehensive reference, information
and lending service with a wide range of music
and films. Free internet access.

## Local Studies Service

☎ 0114 273 4753
localstudies.library@sheffield.gov.uk
OPEN 10.00 am to 8.00 pm Monday; 9.30 am to
5.30 pm Tuesday to Saturday
ACCESS For reference.

Provides a wide range of printed material on
Sheffield and surrounding areas, past and present.
Free internet access. (Photograph collection
available on www.picturesheffield.com)

## Children's Library

☎ 0114 273 4734
kidsandteens.library@sheffield.gov.uk
OPEN 10.30 am to 5.30 pm Monday to Friday; 9.30
am to 5.30 pm Saturday

Books, spoken word cassettes, DVDs, videos;
under-five play area; teenage reference section;
readings and promotions; storytime and
Babytime sessions. Free internet access.

## Computer and Internet Centre

☎ 0114 273 4764
OPEN 10.00 am to 8.00 pm Monday; 9.30 am to
5.30 pm Tuesday, Thursday, Friday, Saturday; 9.30
am to 8.00 pm Wednesday

Free access to high speed internet, web-based
email, personal file storage and popular office
software packages for library members.

## Reception, Reference & Information

☎ 0114 273 4712 (reception)
☎ 273 4736/273 4737 (reference & information)
www.eventsheffield.com

www.sheffieldhelpyourself.org.uk
OPEN 10.00 am to 8.00 pm Monday; 9.30 am to
5.30 pm Tuesday, Thursday, Friday, Saturday; 9.30
am to 8.00 pm Wednesday

Provides a reference and information service
on aspects of arts, business, humanities, sports,
social sciences and science and technology. Also
provides the reception for the library service and
can give details of local clubs, organisations and
services.

## Writers Resource Centre

☎ 0114 273 4727
centrallending.library@sheffield.gov.uk
OPEN 5.30 pm to 7.30 pm Wednesday

Drop in base providing advice and information
for anyone interested in creative writing. Contact
Central Lending Library for more information

## Sheffield Community Information Service

☎ 0114 273 5634
scis@sheffield.gov.uk
www.sheffieldhelpyourself.org.uk
www.eventsheffield.com

Provides information on local events, and clubs,
organizations and services via the two websites.
Also, 28 community libraries; check the website
for further information or call 0114 273 4712.

## Sherlock Holmes Collection (Westminster)

Marylebone Library, Marylebone Road, London
NW1 5PS
☎ 020 7641 1206 🖷 020 7641 1019
ccooke@westminster.gov.uk
www.westminster.gov.uk/libraries/special/
sherlock.cfm
OPEN 9.30 am to 8.00 pm Monday, Tuesday,
Thursday, Friday; 10.00 am to 8.00 pm
Wednesday; closed Saturday and Sunday (unless
by prior arrangement).
ACCESS By appointment only.

Located in Westminster's Marylebone Library.
An extensive collection of material from all over
the world, covering Sherlock Holmes and Sir
Arthur Conan Doyle. Books, pamphlets, journals,
newspaper cuttings and photos, much of which
is otherwise unavailable in this country. Some
background material.

## Shetland Library

Lower Hillhead, Lerwick ZE1 0EL
☎ 01595 743868 🖷 01595 694430
shetlandlibrary@shetland.gov.uk
www.shetland-library.gov.uk
OPEN 9.30 am to 8.00 pm Monday and Thursday;
9.30 am to 5.00 pm Tuesday, Wednesday, Friday,
Saturday

General lending and reference library; extensive
local interest collection including complete set
of *The Shetland Times*, *The Shetland News* and

other local newspapers on microfilm; audio collection including talking books; talking newspaper service, children's area and children's events. Disabled access and Housebound Readers Service (delivery to reader's home). Mobile library services to rural areas. Learning packs and DVDs; photocopying; free public access computers and free Wi-Fi. Busy programme of community events and book promotions. Large collection of adult and junior books in a range of languages.

## Shoe Lane Library

Hill House, Little New Street, London EC4A 3JR
☎ 020 7583 7178
shoelane@cityoflondon.gov.uk
www.cityoflondon.gov.uk/shoelanelibrary
OPEN 9.00 am to 5.30 pm Monday, Wednesday, Thursday, Friday; 9.00 am to 6.30 pm Tuesday
OPEN ACCESS

Corporation of London general lending library, with a comprehensive stock of 50,000 volumes, most of which are on display. Free internet access.

## Shrewsbury Library and Reference & Information Service

Castlegates, Shrewsbury SY1 2AS
☎ 01743 255300    📠 01743 255309
shrewsbury.library@shropshire.gov.uk
www.shropshire.gov.uk/library.nsf
OPEN 9.30 am to 5.00 pm Monday, Wednesday, Friday; 9.30 am to 8.00 pm Tuesday and Thursday; 9.00 am to 5.00 pm Saturday; 1.00 pm to 4.00 pm Sunday
OPEN ACCESS

The largest public library in Shropshire. Books, cassettes, CDs, talking books, DVDs, videos, language courses. Open Learning and study centre with public use computers for word processing and internet access. Music, literature and art book collection for lending and reference. The West Midlands Literary Heritage Collection is housed at Shrewsbury Library.

## Spanish Institute Library
▷ Instituto Cervantes

## Sterling Collection
▷ Senate House Library

## Suffolk County Council – Suffolk Libraries

Endeavour House, Russell Road, Ipswich IP1 2BX
☎ 01473 584563    📠 01473 216847
help@suffolklibraries.co.uk
www.suffolk.gov.uk/LeisureAndCulture/Libraries/
OPEN See website for details of individual libraries or contact Endeavour House. Major libraries open seven days a week; all libraries open Sunday.
ACCESS A single library card gives access to the lending service of 44 libraries and resource centres across the county. Loans can be collected

and/or returned at any service point. Details on website.

Full range of lending and reference services, free public access to the internet. Wide range of online services for registered library users through the Suffolk Libraries Direct pages, e.g. self-service reservations, renewals, free access to subscription services. Suffolk InfoLink Plus database gives details of local organizations throughout the county. SPECIAL COLLECTIONS include Suffolk Archives and Local History Collection; Benjamin Britten Collection; Edward Fitzgerald Collection; Seckford Collection and Racing Collection (Newmarket). The Suffolk Infolink service gives details of local groups and societies and is available in libraries throughout the county and on the website.

## Sunderland City Library and Arts Centre

28–30 Fawcett Street, Sunderland SR1 1RE
☎ 0191 561 8407    📠 0191 565 5950
enquiry.desk@sunderland.gov.uk
www.sunderland.gov.uk/libraries
OPEN 9.30 am to 7.30 pm Monday and Wednesday; 9.30 am to 5.00 pm Tuesday, Thursday, Friday; 9.30 am to 4.00 pm Saturday

The city's main library for lending and reference services. Local studies and children's sections, plus Sound and Vision department (CDs, DVDs, talking books). Sunderland Public Libraries also maintain community libraries of varying size, offering a range of services, plus mobile libraries. Free internet access is available in all libraries across the city. A Books on Wheels service is available to housebound readers; the Schools Library Service serves teachers and schools. The Foyle Street Writers Group meets every Wednesday, 10.00 am to 12 noon, and Age Concern Writers Group meets every Monday, 1.30 pm to 3.30 pm.

## Swansea Central Library

Civic Centre, Oystermouth Road, Swansea SA1 3SN
☎ 01792 636464    📠 01792 637193
libraries.swansea@swansea.gov.uk
www.swansea.gov.uk/libraries
OPEN 8.30 am to 8.00 pm Tuesday to Friday; 10.00 am to 4.00 pm Saturday and Sunday. The Library has a lending service open the same hours, including books, DVDs, music CDs and talking books.
ONLINE REFERENCE & ENQUIRY SERVICE
OPEN 8.30 am to 6.00 pm Monday to Friday; 10.00 am to 4.00 pm Saturday Enquiries taken by phone, email, fax or letter (see contact details above).
ACCESS For reference only (many local studies

items now available on open access; other items must be requested on forms provided).

General reference material; also Welsh Assembly information, statutes, company information, maps, European Community information. Local studies: comprehensive collections on Wales; Swansea & Gower; Dylan Thomas. Local maps, periodicals, illustrations, local newspapers from 1804. B&w and colour photocopying and fax facilities, free computer and internet access available (library membership required and pre-booking recommended) and microfilm/microfiche copying facility.

## Swiss Cottage Central Library

88 Avenue Road, London NW3 3HA
☎ 020 7974 6522    ℉ 020 7974 6532
swisscottagelibrary@camden.gov.uk
OPEN 10.00 am to 8.00 pm Monday to Friday; 10.00 am to 5.00 pm Saturday; 11.00 am to 4.00 pm Sunday
OPEN ACCESS

Over 300,000 volumes in the lending and reference libraries. Home of the London Borough of Camden's Information and Reference Services.

## Tate Library and Archive

Hyman Kreitman Reading Rooms, Tate Britain, Millbank, London SW1P 4RG
☎ 020 7887 8838    ℉ 020 7887 3952
reading.rooms@tate.org.uk
www.tate.org.uk/research/researchservices/readingrooms/
OPEN Main Reading Room (Library Collection): 10.00 am to 5.45 pm Monday to Thursday; Archive and Special Collections Rooms: 11.00 am to 5.00 pm (appointments are required for this room)
FREE ACCESS but registration is required; please see: www.tate.org.uk/research/researchservices/readingrooms/registration.shtm

**Tate Library**: British art from 1500 and International art from 1900. The Library collection is unique with many of its items not found in any other art library in the world. The collection contains books, exhibition catalogues, Tate publications, journals, artists' books, auction catalogues, AV material, ephemera, press cuttings and electronic resources and extends to over 250,000 volumes. Holds a collection of about 4,500 artists' books published from the 1960s to the present. Includes books dating from the 1960s onwards and is international in scope with an emphasis on British artists. Browse the Library Catalogue and artists' books at: http://library.tate.org.uk

**Tate Archive**: Material on artists, art world figures and organizations in Britain from 1900. Amounting to over one million items in

more than 750 archive collections, significant series include letters, writings, sketchbooks & artworks, audio-visual material, photographs, printed ephemera and press cuttings. Notable catalogued collections include those on: Bacon, Bloomsbury, Burra, Colquhoun, Fluxshoe, Hepworth, the ICA, Nash, Nicholson, Paolozzi, Piper, Spencer, Sutherland and Vaughan. Also separate collections of audio-visual material (TAV collection), a large selection of artist-designed posters (TAP collection) and 100,000 images of artists and their studios in the Photographic Collection.

**Gallery Records**: Tate's institutional records documenting the museum's history are retained for public consultation under the Public Records Act. Dating from 1893 to the present, the records encompass the full range of Tate's activities since its inception. Key series include: exhibition files, acquisition files on works in the collection, Tate posters and photographs since 1911, Directors' correspondence, Board of Trustee minutes and building plans. Archive Catalogue and Gallery Records: http://archive.tate.org.uk

## Tennyson Research Centre
▷ Lincoln Central Library

## Thurrock Communities, Libraries & Cultural Services

Grays Library, Orsett Road, Grays RM17 5DX
☎ 01375 413973    ℉ 01375 370806
grays.library@thurrock.gov.uk
www.thurrock.gov.uk/libraries
OPEN 10.00 am to 7.00 pm Monday; 9.00 am to 7.00 pm Tuesday and Thursday; 9.00 am to 5.00 pm Wednesday, Friday, Saturday; branch library opening times vary.
OPEN ACCESS

General library lending and reference through ten libraries and a mobile library. Services include books, magazines, newspapers, CDs and language courses. Large collection of Thurrock materials. Free internet and Microsoft Office.

## Tom Mann Collection
▷ Coventry Central Library

## Truro Library

Union Place, Truro TR1 1EP
☎ 0300 123 4111 (lending)/01872 272702 (reference)/0845 607 6119 (24hr-renewals)
truro.library@cornwall.gov.uk
www.cornwall.gov.uk/library
OPEN 8.30 am to 6.00 pm Monday, Tuesday, Thursday, Friday; 9.30 am to 6.00 pm Wednesday; 9.00 am to 4.00 pm Saturday

Books, cassettes, CDs, videos, DVDs and PC games for loan through branch or mobile networks. Internet access and IT support and training. Reference collection. SPECIAL COLLECTIONS on local studies.

## University of the Arts London

London College of Communication Library, Elephant and Castle, London SE1 6SB
☎ 020 7514 6527
www.arts.ac.uk/library
ACCESS Appointment required.

Library and Learning Resources provides books, periodicals, slides, CD-ROMs, DVDs, videos and computer software on all aspects of the art of the book, printing, management, film/photography, graphic arts and retailing. SPECIAL COLLECTIONS History and the art of the western book.

## Vaughan Williams Memorial Library

English Folk Dance and Song Society, Cecil Sharp House, 2 Regent's Park Road, London NW1 7AY
☎ 020 7485 2206 exts. 33 or 34    ☏ 020 7284 0523
library@efdss.org
www.efdss.org
Library Director *Malcolm Taylor, OBE*
Assistant Librarians *Peta Webb, Elaine Bradtke*
OPEN 9.30 am to 5.30 pm Tuesday to Friday; 10.00 am to 4.00 pm 1st and 3rd Saturday (sometimes closed between 1.00 pm and 2.00 pm)
ACCESS For reference to the general public, on payment of a daily fee; EFDSS members may borrow books and use the library free of charge.

A multimedia collection: books, periodicals, manuscripts, tapes, records, CDs, films, videos, photographs. Mostly British traditional culture and how this has developed around the world. Some foreign language material, and some books in English about foreign cultures. Also, the history of the English Folk Dance and Song Society.

## Vin Mag Archive Ltd

84–90 Digby Road, London E9 6HX
☎ 020 8533 7588    ☏ 020 8533 7283
piclib@vinmag.com
www.vinmagarchive.com
Contact *Angela Maguire*

Wholly owned subsidiary of the Vintage Magazine Company. A collection of original printed material – magazines, newspapers, posters, adverts, books, ephemera and movie material. Also photographic collection; see entry under *Picture Libraries*.

## West Midlands Literary Heritage Collection

▷ Shrewsbury Library and Reference & Information Service

## Western Isles Libraries

Public Library, 19 Cromwell Street, Stornoway HS1 2DA
☎ 01851 708631    ☏ 01851 708676
www.cne-siar.gov.uk
OPEN 10.00 am to 5.00 pm Monday to Wednesday; 10.00 am to 6.00 pm Thursday and Friday; 10.00 am to 5.00 pm Saturday
OPEN ACCESS

General public library stock, plus local history and Gaelic collections including maps, videos, printed music, cassettes and CDs; census records and Council minutes; music collection (CDs and cassettes). Branch libraries on the isles of Barra, Benbecula, Harris and Lewis.

## City of Westminster Archives Centre

10 St Ann's Street, London SW1P 2DE
☎ 020 7641 5180/4879 (Minicom)
☏ 020 7641 5179
archives@westminster.gov.uk
www.westminster.gov.uk/archives
OPEN 10.00 am to 7.00 pm Tuesday, Wednesday, Thursday, 10.00 am to 5.00 pm Friday and Saturday (closed Monday)
ACCESS For reference.

Comprehensive coverage of the history of Westminster and selective coverage of general London history. 22,000 books, together with a large stock of maps, prints, photographs, local newspapers, theatre programmes and archives.

## Westminster Music Library

Victoria Library, 160 Buckingham Palace Road, London SW1W 9UD
☎ 020 7641 1300    ☏ 020 7641 4281
musiclibrary@westminster.gov.uk
www.westminster.gov.uk/libraries/special/music
OPEN 11.00 am to 7.00 pm Monday to Friday; 10 am to 5.00 pm Saturday
OPEN ACCESS

Located at Victoria Library, this is the largest public music library in the South of England, with extensive coverage of all aspects of music, including books, periodicals and printed scores. No recorded material, notated only. Lending library includes a small collection of CDs and DVDs.

## Westminster Reference Library

35 St Martin's Street, London WC2H 7HP
☎ 020 7641 1300    ☏ 020 7641 5247
referencelibrarywc2@westminster.gov.uk
www.westminster.gov.uk/libraries
OPEN: 10.00 am to 8.00 pm Monday to Friday. 10.00 am to 5.00 pm Saturday
ACCESS For reference only, but there are within the Art & Design and Performing Arts Collections several thousand older books available for loan to library members. Membership open to anyone with proof of a permanent address in Britain.

A general reference library with emphasis on the following: Art & Design – fine and decorative arts, architecture, graphics and design; Performing Arts – theatre, cinema, radio, television and dance; Official Publications – major collection of HMSO publications from 1947, plus parliamentary papers dating back to 1906 (TSO subscription ceased April 2006); Business –online and hard copy directories, market research, industry and company data; Periodicals – long files of many titles. One working day's notice is required for some government documents and some older periodicals. Westminster Reference Library is free and open to everyone 57 hours a week, with three study floors and strong collections in business information, performing arts and art & design.

## The Wiener Library

4 Devonshire Street, London W1W 5BH
Ⓣ 020 7636 7247    Ⓕ 020 7436 6428
info@wienerlibrary.co.uk
www.wienerlibrary.co.uk
Director *Ben Barkow*
Marketing and Fundraising Manager *Bridget McGing*
OPEN 10.00 am to 5.30 pm Monday to Friday
ACCESS The Library is open to the public. On your first visit please bring proof of address and some form of photo ID.

The Wiener Library is one of the world's leading and most extensive archives on the Holocaust and Nazi era. Founded in 1933 by Dr Alfred Wiener, the Library's unique collection of over one million items includes published and unpublished works, press cuttings, pamphlets, periodicals, photographs, audiovisual material and eyewitness testimony.

## The Wighton Collection of National Music
▷ Dundee Central Library

## Witham Library
▷ Essex County Council Libraries

## Wolverhampton Central Library
Snow Hill, Wolverhampton WV1 3AX
Ⓣ 01902 552025 (lending)/552026 (reference)
Ⓕ 01902 552024
libraries@wolverhampton.gov.uk
www.wolverhampton.gov.uk/libraries
OPEN 9.00 am to 7.00 pm Monday to Thursday; 9.00 am to 5.00 pm Friday and Saturday

General lending and reference stock, plus children's library and Learning Centre offering free internet access. Books, CDs, DVDs and music scores. Reference resources available online.

**Archives & Local Studies**

Molineux Hotel Building, Whitmore Hill, Wolverhampton WV1 1SF
Ⓣ 01902 552480
OPEN 10.00 am to 5.00 pm Tuesday, Thursday, Friday; 10.00 am to 7.00 pm Wednesday; closed Sunday and Monday

## The Women's Library
London Metropolitan University, Old Castle Street, London E1 7NT
Ⓣ 020 7320 2222    Ⓕ 020 7320 2333
moreinfo@thewomenslibrary.ac.uk
www.thewomenslibrary.ac.uk
OPEN Reading Room: 9.30 am to 5.00 pm Tuesday to Friday (8.00 pm Thursday); closed Saturday; Exhibition: 9.30 am to 5.30 pm Monday to Friday (8.00 pm Thursday); 10.00 am to 4.00 pm Saturday
OPEN ACCESS

The Women's Library, national research library for women's history, is the UK's oldest and most comprehensive research library on all aspects of women in society, with both historical and contemporary coverage. The Library includes materials on feminism, work, education, health, the family, law, arts, sciences, technology, language, sexuality, fashion and the home. The main emphasis is on Britain but many other countries are represented, especially the Commonwealth and the developing countries. Established in 1926 as the library of the London Society of Women's Service (formerly Suffrage), a non-militant organization led by Millicent Fawcett. In 1953 the Society was renamed after her and the library became the Fawcett Library.

Collections include: women's suffrage, work, education, women and the church, the law, sport, art, music, abortion, prostitution. Mostly British materials but some American, Commonwealth and European works. Books, journals, pamphlets, archives, photographs, posters, postcards, audiovisual materials, artefacts, scrapbooks, albums and press cuttings dating mainly from the 19th century although some materials date from the 17th century.

The Library's new building, which opened in 2002, includes a reading room, exhibition space, café, education areas and a conference room, and is the cultural and research centre for anyone interested in women's lives and achievements.

## Worcestershire Libraries and Learning
Libraries & Learning Services, Worcestershire County Council, County Hall, Spetchley Road, Worcester WR5 2NP
Ⓣ 01905 822722    Ⓕ 01905 766930
libraries@worcestershire.gov.uk
www.worcestershire.gov.uk/libraries
OPEN Opening hours vary in the 22 libraries, the History Centre and mobile libraries covering the

county; all full-time libraries open at least one evening a week until 8.00 pm, and on Saturday until 5.30 pm; part-time libraries vary.
ACCESS Information and reference services open to anyone; loans to members only (membership criteria: resident, being educated, working, or an elector in the county or neighbouring authorities; temporary membership to visitors. Proof of identity and address required). No charge for membership or for borrowing books.

Information service, and reference and lending libraries. Non-fiction and fiction for all age groups, including normal and large print, spoken word cassettes, sound recordings (CD, cassette), videos, maps, local history, CD-ROMs for reference at main libraries, free public internet access in all libraries. Joint Libraries Service/County Record Office History Centre with resources for local and family history. SPECIAL COLLECTIONS Carpets and Textiles; Needles & Needlemaking; Stuart Period; A.E. Housman.

## The Wordsworth Trust (Jerwood Centre)
▷ entry under Literary Societies

## Working Class Movement Library
51 The Crescent, Salford M5 4WX
☎ 0161 736 3601
enquiries@wcml.org.uk
www.wcml.org.uk
OPEN 10.00 am to 5.00 pm Tuesday to Friday
OPEN ACCESS By appointment.

The Working Class Movement Library records over 200 years of organizing and campaigning. Includes books, pamphlets, archives, photographs, plays, poetry, songs, banners, posters, badges, cartoons, journals, biographies and reports. Information on trade unions, politics and campaigns (from Chartism to the General Strike and more recent protests), international events such as the Spanish Civil War and aspects of Irish history. Oldest items date from the 1760s.

## York Library Learning Centre
Museum Street, York YO1 7DS
libraries@york.gov.uk
www.york.gov.uk/libraries

York Central Library has undergone major rerfurbishment to turn it into the city's third 'Explore Centre', the first two being at Acomb and New Earswick. The new Explore York Library Centre, opening in 2010, combines a modern city library with a hub for adult learning. The City Archives and Local Study Library have been merged and are based at the Explore Centre. Email or access the website for latest information and new opening hours.

## Zoological Society Library
Regent's Park, London NW1 4RY
☎ 020 7449 6293    ℻ 020 7586 5743
library@zsl.org
www.zsl.org
OPEN 9.30 am to 5.30 pm Monday to Friday
ACCESS To members and staff; non-members require proof of address and photographic ID.

160,000 volumes on zoology including 5,000 journals (1,300 current) and a wide range of books on animals and particular habitats. Historic image collection available and many historic zoological prints. Library catalogue available online (www.library.zsl.org).

## The Zweig Collection of Music & Literary Mss
▷ British Library Music Collections

# Picture Libraries

## actionplus sports images

2nd Floor, 29–31 Saffron Hill, London EC1N 8SW
☎ 020 7403 1558   🖷 020 7403 1558
info@actionplus.co.uk
www.actionplus.co.uk

Founded 1986. Specialist sports and action library
providing complete and creative international
coverage of over 300 varieties of professional and
amateur sport. Elite, junior, professional, amateur,
extreme, disabled, minority, offbeat, major and
minor events, adults, children, high and low, fast
and slow. Millions of images covering over 30
years of sport.

## The Advertising Archive Limited

45 Lyndale Avenue, London NW2 2QB
☎ 020 7435 6540   🖷 020 7794 6584
suzanne@advertisingarchives.co.uk
www.advertisingarchives.co.uk
Contacts *Suzanne Viner, Larry Viner*

With over one million images, the largest
collection of British and American press ads, TV
commercial stills and magazine cover illustrations
in Europe. Material spans the period from 1850 to
the present day. In-house research; rapid service,
competitive rates. On-line database and digital
delivery available.

## akg-images Ltd

5 Melbray Mews, 158 Hurlingham Road, London
SW6 3NS
☎ 020 7610 6103   🖷 020 7610 6125
enquiries@akg-images.co.uk
www.akg-images.co.uk

Specialists in fine art, history and photography.
Hundreds of thousands of images available online
with direct access to millions of images held by
the akg-images archive in Berlin.

## Alpine Garden Society

AGS Centre, Avon Bank, Pershore WR10 3JP
☎ 01386 554790   🖷 01386 554801
ags@alpinegardensociety.net
www.alpinegardensociety.net
Contact *Peter Sheasby*

Over 30,000 colour transparencies (35mm)
covering plants in the wild from many parts of
the world; particularly strong in plants from
mountain and sub-alpine regions, and from
Mediterranean climates; South Africa, Australia,
Patagonia, California and the Mediterranean.
Extensive coverage of show alpines in pots and in
gardens and of European orchids. Specific slide
lists available on request.

## Alvey & Towers Transport Picture Library

A37 The Springboard Centre, Mantle Lane,
Coalville LE67 3DW
☎ 01530 450011
www.alveyandtowers.com

Houses one of the country's most comprehensive
collections of transport images depicting not only
actual transport systems but their surrounding
industries as well. Also specialist modern railway
image collection.

## Andes Press Agency

26 Padbury Court, London E2 7EH
☎ 020 7613 5417   🖷 020 7739 3159
apa@andespressagency.com
www.andespressagency.com
Contacts *Val Baker, Carlos Reyes*

80,000 colour transparencies and 300,000
b&w, specializing in social documentary, world
religions, Latin America and Britain.

## Ann & Bury Peerless Picture Library

St David's, 22 King's Avenue, Minnis Bay,
Birchington-on-Sea CT7 9QL
☎ 01843 841428   🖷 01843 848321
www.peerlessimages.com
Contacts *Ann Peerless, Bury Peerless*

Specialist collection on world religions:
Hinduism, Buddhism, Confucianism,
Taoism, Jainism, Christianity, Islam, Sikhism,
Zoroastrianism (Parsees of India). Geographical
areas covered: India, Afghanistan (Bamiyan
Valley of the Buddhas), Pakistan, Bangladesh, Sri
Lanka, Cambodia (Angkor), Java (Borobudur),
Bali, Malaysia, Thailand, Taiwan, Russia, China,
Spain, Poland, Uzbekistan (Samarkand and
Bukhara), Vietnam. Basis of collection (35mm
colour transparencies), historical, cultural,
extensive coverage of art (sculpture and miniature
paintings), architecture including Pharaonic

Egypt, earlier coverage of significance on Iran, including oil industry of Iran, Kenya, Uganda and Zimbabwe, including the Victoria Falls, also Morocco, Malta, Libya and Tunisia.

## Architectural Association Photo Library

Architectural Association, 36 Bedford Square, London WC1B 3ES

☎ 020 7887 4066    ☏ 020 7414 0782

www.aaschool.ac.uk/photolib

Contacts *Valerie Bennett (valerie@aaschool. ac.uk), Sarah Franklin (sfranklin@aaschool.ac.uk), Henderson Downing (henderson@aaschool.ac.uk)*

150,000 35mm transparencies on architecture, historical and contemporary. Archive of large-format b&w negatives from the 1920s and 1930s.

## ArenaPAL

Thompson House, 42–44 Dolben Street, London SE1 0UQ

☎ 020 7403 8542    ☏ 020 7403 8561

enquiries@arenapal.com

www.arenapal.com

'Probably the best collection of entertainment images in the world.' Continually updated and covering classical music, opera, theatre, musicals, film, pop, rock, jazz, instruments, festivals, venues, circus, ballet and contemporary dance, props and personalities. Over two million images from the late 19th century onwards. Phone, fax or email requests.

## ArkReligion.com

57 Burdon Lane, Cheam SM2 7BY

☎ 020 8642 3593    ☏ 020 8395 7230

images@artdirectors.co.uk

www.arkreligion.com

Contacts *Helene Rogers, Bob Turner*

Extensive coverage of all religions.

## The Art Archive

2 The Quadrant, 135 Salusbury Road, London NW6 6RJ

☎ 020 7624 3500    ☏ 020 7624 3355

info@picture-desk.com

www.picture-desk.com

Picture library holding over 80,000 high-resolution images covering fine art, history and ancient civilizations.

## Art Directors & Trip Photo Library

57 Burdon Lane, Cheam SM2 7BY

☎ 020 8642 3593    ☏ 020 8395 7230

images@artdirectors.co.uk

www.artdirectors.co.uk

Contacts *Helene Rogers, Bob Turner*

Extensive coverage, with over 750,000 images, of all countries, lifestyles, peoples, etc. with detailed coverage of all religions. Backgrounds a speciality.

## aviation-images.com

Queens Road, London SW19 8LR

☎ 020 8944 5225    ☏ 020 8944 5335

pictures@aviation-images.com

www.aviation-images.com

Contacts *Mark Wagner, Denise Lalonde*

Two million 'stunning' aviation images, civil and military, archive and modern, from the world's best aviation photographers. Member of **BAPLA** and RAeS.

## Axel Poignant Archive

115 Bedford Court Mansions, Bedford Avenue, London WC1B 3AG

☎ 020 7636 2555    ☏ 020 7636 2555

Rpoignant@aol.com

Contact *Roslyn Poignant*

Anthropological and ethnographic subjects, especially Australia and the South Pacific. Also Scandinavia (early history and mythology), Sicily and England.

## B.M. Totterdell photography

Constable Cottage, Burlings Lane, Knockholt TN14 7PE

☎ 01959 532001    ☏ 01959 532001

info@thevolleyballlibrary.co.uk

www.the volleyballlibrary.co.uk

Contact *Barbara Totterdell, Owner*

Specialist volleyball library covering all aspects of the sport.

## Barnaby's Library
▷ Mary Evans Picture Library

## Barnardo's Photographic and Film Archive

Barnardo's House, Tanners Lane, Barkingside, Ilford IG6 1QG

☎ 020 8498 7345    ☏ 020 8550 0429

stephen.pover@barnardos.org.uk

www.barnardos.org.uk

Image Librarian *Stephen Pover*

The Barnardo's image archive is a unique resource of images of social history dating from 1874. Images can be sent either by post or email.

## Beamish, The North of England Open Air Museum Photographic Library

Regional Resource Centre, The North of England Open Air Museum, Beamish DH9 0RG

☎ 0191 370 4000    ☏ 0191 370 4001

museum@beamish.org.uk

www.beamish.org.uk

www.beamishcollections.com

An archive of over 300,000 images representing everyday life from 1860 to the present day. The images cover agricultural, industrial, topographical, advertising and shop scenes, and people at work and play. Major collections include: Durham Advertiser newspaper; Huwood Mining Machinery; ICI Teesside, Consett Iron Company, Farmer's Guardian and Agricultural Gazette. Digital images are available to view by appointment weekdays, and searches can be arranged. A sound archive and reference Library are also available at the Regional Resource Centre.

## BFI Stills, Posters & Designs Collection

BFI, 21 Stephen Street, London W1T 1LN
☏ 020 7957 4840
www.bfi.org.uk/collections/stills/index.html

'BFI Stills Sales is the world's most comprehensive collection of film and television images.' The collection captures on and off screen moments, portraits of the world's most famous stars – and those behind the camera who made them famous – as well as publicity posters, set designs, images of studios, cinemas, special events and early film and TV technologies.

## Boulton and Watt Archive

▷ Birmingham Library Services under Library Services

## The Bridgeman Art Library

17–19 Garway Road, London W2 4PH
☏ 020 7727 4065    🖷 020 7792 8509
research@bridgemanart.co.uk
www.bridgemanart.co.uk
Head of Picture Research *Jenny Page*

Fine art photo archive representing more than 1,000 museums, galleries and picture owners around the world. High res digital files and large-format colour transparencies of paintings, sculptures, prints, manuscripts, antiquities, photographs and the decorative arts. From prehistoric sculpture to contemporary paintings. Offices in London, Paris, New York and Berlin. Collections represented include the British Library, National Galleries of Scotland, Boston Museum of Fine Arts, the Wallace Collection, The Courtauld, Detroit Institute of Arts, The Barnes Foundation, the National Army Museum, the National Trust as well as the Giraudon archive.

## The British Cartoon Archive

The Templeman Library, University of Kent, Canterbury CT2 7NU
☏ 01227 823127    🖷 01227 823127
N.P.Hiley@kent.ac.uk
J.M.Newton@kent.ac.uk
www.cartoons.ac.uk
Contacts *Dr Nicholas Hiley, Jane Newton*

A national research archive of over 120,000 original cartoons and caricatures by 350 British cartoonists, supported by a library of books, papers, journals, catalogues, cuttings and assorted ephemera. A computer database of 120,000 cartoons provides quick and easy catalogued access via the web. A source for exhibitions and displays as well as a picture library service. Specializes in historical, political and social cartoons from British newspapers and magazines.

## British Library Images Online

British Library, 96 Euston Road, London NW1 2DB
**Images Online** ☏ 020 7412 7614 (research)/7755 (permissions)  🖷 020 7412 7771
imagesonline@bl.uk www.imagesonline.bl.uk

Images Online offers instant access to thousands of images from the British Library's historic worldwide collections which include manuscripts, rare books, music scores and maps spanning almost 3,000 years. A collection of 20,000 images including digital images of drawings, illustrations and photographs. Specialities: historical images.

## Brooklands Museum Photo Archive

Brooklands Museum, Brooklands Road, Weybridge KT13 0QN
☏ 01932 857381    🖷 01932 855465
info@brooklandsmuseum.com
www.brooklandsmuseum.com
Head of Collections *John Pulford*
General Manager – Operations *Julian Temple*

About 40,000 b&w and colour prints and slides. Subjects include: Brooklands Motor Racing 1907–39; British aviation and aerospace, 1908 to the present day, particularly BAC, Hawker, Sopwith and Vickers aircraft built at Brooklands.

## Bryan & Cherry Alexander Photography

Higher Cottage, Manston, Sturminster Newton DT10 1EZ
☏ 01258 473006    🖷 01258 473333
alexander@arcticphoto.co.uk
www.arcticphoto.com
Contact *Cherry Alexander*

Arctic and Antarctic specialists; indigenous peoples, wildlife and science in polar regions; Canada, Greenland, the Scandinavian Arctic, Siberia and Alaska.

## Capital Pictures

85 Randolph Avenue, London W9 1DL
☏ 020 7286 2212
sales@capitalpictures.com
www.capitalpictures.com

Up-to-date celebrity and entertainment photography from around the world and a film stills collection of all the great films and yet to be released blockbusters. The website is fully searchable.

## Cephas Picture Library

A1 Kingsway Business Park, Oldfield Road, Hampton TW12 2HD
☏ 020 8979 8647    🖷 020 8941 4001
pictures@cephas.com
www.cephas.com

Contact *Mick Rock*

The wine industry and vineyards of the world is the subject on which Cephas has made its reputation. '100,000 images make this the most comprehensive and up-to-date archive in Britain.' Almost all wine-producing countries and all aspects of the industry are covered in depth. Spirits, beer and cider also included. A major food and drink collection now also exists, through preparation and cooking, to eating and drinking.

## Charles Parker Archive
▷ Birmingham Library Services under Library Services

## Chris Bonington Picture Library
Badger Hill, Hesket Newmarket, Wigton CA7 8LA
☎ 01697 478286   📠 01697 478238
frances@bonington.com
www.bonington.com
Contact *Frances Daltrey*

Based on the personal collection of climber and author Chris Bonington and his extensive travels and mountaineering achievements; also work by Doug Scott and other climbers, including the Peter Boardman and Joe Tasker Collections. Full coverage of the world's mountains, from British hills to Everest, depicting expedition planning and management stages, the approach march showing inhabitants of the area, flora and fauna, local architecture and climbing action shots on some of the world's highest mountains.

## Chris Howes/Wild Places Photography
PO Box 100, Abergavenny NP7 9WY
☎ 01873 737707
photos@wildplaces.co.uk
Contacts *Chris Howes, Judith Calford*

Expanding collection of over 50,000 colour transparencies, b&w prints and digital photos covering travel, topography and natural history worldwide, plus action sports. Supply fully digital. Specialist areas include caves, caving and mines (with historical coverage using engravings and early photographs), wildlife, landscapes and the environment, including pollution and conservation. Europe (including the UK), Canada, USA, Africa and Australia are all well represented within the collection. Commissions undertaken.

## Christian Aid
▷ Specialist Stock

## Classical Collection Ltd
22 Avon, Hockley, Tamworth B77 5QA
☎ 01827 286086/07963 194921   📠 01827 286086
neil@classicalcollection.co.uk
Managing Director *Neil Arthur Williams, MA (Hum), MA (Mus)*

Archive specializing in classical music ephemera, particularly portraits of composers, musicians, conductors and opera singers, comprising old and very rare photographs, postcards, antique prints, cigarette cards, stamps, First Day Covers, concert programmes, Victorian newspapers, etc. Also modern photos of composer references such as museums, statues, busts, paintings, monuments, memorials and graves. Other subjects covered include ballet, musical instruments, concert halls, opera houses, 'music in art', manuscripts, opera scenes, music-caricatures, bands, orchestras and other music groups. Neil Arthur Williams is a qualified music historian, musicologist and freelance writer of music articles, concert programme notes and CD as well as being a commissionable composer.

## Colin Baxter Photography Limited
The Old Dairy, Grantown-on-Spey PH26 3NA
☎ 01479 873999   📠 01479 873888
sales@colinbaxter.co.uk
www.colinbaxter.co.uk
Contact *Mike Rensner (Editorial)*

Specializes in Scotland. Publishes guidebooks and maps, plus books, calendars and postcards on landscape, cityscape and natural history, and a range of books and stationery based on reproductons of the designs of Charles Rennie Mackintosh. Also publishers of the World*Life*Library series of natural history books.

## Collections
13 Woodberry Crescent, London N10 1PJ
☎ 020 8883 0083   📠 020 8883 9215
info@collectionspicturelibrary.com
www.collectionspicturelibrary.co.uk
www.collectionspicturelibrary.com
Contact *Simon Shuel*

Extensive coverage of the British Isles and Ireland, from the Shetlands to the Channel Islands and Connemara to East Anglia, including people, traditional customs, landscapes, buildings old and new – both well known and a bit obscure, 'plus a considerable collection of miscellaneous bits and pieces which defy logical filing'. Images supplied digitally or as original transparencies. Visitors welcome but please call first.

## Corbis Images
111 Salusbury Road, London NW6 6RG
☎ 020 7644 7644   📠 020 7644 7645
sales.uk@corbis.com
www.corbis.com

Corbis provides a comprehensive selection of photography and illustration to advertising, marketing and media professionals. Award-winning documentary, archival, fine art, current events and entertainment images.

## Country Images Picture Library

27 Camwood, Clayton-le-Woods, Bamber Bridge
PR5 8LA
☏ 01772 321243    🖷 0870 137 8888
tm@wpu.org.uk
www.countryimages.info
Contact *Terry Marsh*

35mm film and digital colour coverage of
landscapes and countryside features generally
throughout the UK (Cumbria, North Yorkshire,
Lancashire, southern Scotland, Isle of Skye, St
Kilda and Scottish islands, Wales, Cornwall),
France (Alps, Auvergne, Pyrenees, Provence,
Charente-Maritime, Hérault, Aube-en-
Champagne, Loire valley, Somme), Madeira, the
Azores and Australia. Commissions undertaken.

## Country Life Picture Library

Blue Fin Building, 110 Southwark Street, London
SE1 0SU
☏ 020 3148 4474
clpicturelibrary@ipcmedia.com
www.countrylifeimages.co.uk
Contact *Justin Hobson*

Over 150,000 b&w negatives, dating back
to 1897, and 100,000 colour transparencies.
Country houses, stately homes, churches and
town houses in Britain and abroad, interiors
of architectural interest (ceilings, fireplaces,
furniture, paintings, sculpture), and exteriors
showing many landscaped gardens, sporting and
social events, crafts, people and animals. Visitors
by appointment.

## Dale Concannon Collection
▷ Phil Sheldon Golf Picture Library

## Dalton–Watson Collection
▷ Ludvigsen Library

## David Hoffman Photo Library

c/o BAPLA, 18 Vine Hill, London EC1R 5DZ
☏ 020 8981 5041
info@hoffmanphotos.com
www.hoffmanphotos.com
Contact *David Hoffman*

David Hoffman has specialized in social issues
photography for more than 30 years. Primarily
shoots stock for his own photo library, rather
than working to commissions. Racial and social
conflict, policing, homelessness, drugs, poverty,
social exclusion and environmental protection are
documented, often through coverage of protest. A
founding member of Editorial Photographers UK
and Photo-Forum London, with a critical interest
in issues around copyright and intellectual
property.

## David King Collection

90 St Pauls Road, London N1 2QP
☏ 020 7226 0149
davidkingcollection@btopenworld.com
www.davidkingcollection.com
Contact *David King*

250,000 b&w original and copy photographs and
colour transparencies of historical and present-
day images. Russian history and the Soviet Union
from 1900 to the fall of Khrushchev; the lives
of Lenin, Trotsky and Stalin; the Tzars, Russo-
Japanese War, 1917 Revolution, World War I, Red
Army, the Great Purges, Great Patriotic War, etc.
Special collections on China, Eastern Europe,
the Weimar Republic, John Heartfield, American
labour struggles, Spanish Civil War. Open to
qualified researchers by appointment, Monday to
Friday, 10.00 am to 6.00 pm. Staff will undertake
research; negotiable fee for long projects. David
King's latest photographic books: *Red Star Over
Russia*, a visual history from 1917 to the death of
Stalin, *The Commissar Vanishes*, documenting
the falsification of photographs and art in Stalin's
Russia, and *Ordinary Citizens*, mugshots of
victims shot without trial from the archives
of Stalin's secret police. A display of Soviet
posters, 1917–60, from the David King Collection
continues at Tate Modern throughout 2010.

## David Williams Picture Library

Allt-na-Craobh, Old Shore Road, Connel
PA37 1PT
☏ 01631 710586    🖷 01631 710586
david@scotland-guide.co.uk

Specializes in travel photography with wide
coverage of Scotland, Iceland and Spain. Many
other European countries also included plus
smaller collections of Western USA and Canada.
The main subjects in each country are: cities,
towns, villages, 'tourist haunts', buildings of
architectural or historical interest, landscapes
and natural features. The Scotland and Iceland
collections include many pictures depicting
physical geography and geology. Photographic
commissions and illustrated travel articles
undertaken.

## The Defence Picture Library

14 Mary Seacole Road, Howeson Court, Plymouth
PL1 3JY
☏ 01752 312061    🖷 01752 312063
dpla@defencepictures.com
www.defencepictures.com
Contacts *David Reynolds, Jessica Kelly,
James Rowlands*

Specialist source of military photography
covering all areas of the UK Armed Forces,
supported by a research agency of facts and

figures. More than one million images. Visitors welcome by appointment.

## Dennis Hardley Photography, Scottish Photo Library

Rosslynn, Ceum Dunrigh, Benderloch, Nr Oban PA37 1ST

☎ 01631 720434   ℻ 01631 720434
info@dennishardley.com
www.ineedanimage.com
Contacts *Dennis Hardley, Tony Hardley*

Established 1974. About 30,000 images (6x7", 6x9" format colour transparencies) of Scotland: castles, historic, scenic landscapes, islands, transport, etc. Also English views – Liverpool, Chester, Bath, Weston Super Mare, Sussex, Somerset and Cambridge. All images available to buy and instantly download from the website. Includes Scotland, England and North Wales, covering lifestyle, landscapes, transport, cities, etc.

## E&E Heritage Image Library

Beggars Roost, Woolpack Hill, Smeeth TN25 6RR

☎ 01303 812608
info@eeimages.co.uk
www.eeimages.co.uk
Contact *Isobel Sinden*

Over 300,000 images of heritage, world religions, death, festivals, ancient, places, industrial archaeology, stained glass, manuscripts and illustrations. Access via the website or by email with a list of requirements. 'E&E will undertake research from your list until such time as all our images are available online.'

## Ecoscene

Empire Farm, Throop Road, Templecombe BA8 0HR

☎ 01963 371700
pix@ecoscene.com
www.ecoscene.com
Contact *Sally Morgan*

Specialist colour library of over 80,000 images covering all aspects of the environment: pollution, conservation, recycling, restoration, wildlife (especially underwater), habitats, education, industry, sustainable development and agriculture. All parts of the globe are covered with specialist collections covering Antarctica, Australia, North America. Sally Morgan, who runs the library, is a professional ecologist and expert source of information on all environmental topics. Photographic and writing commissions undertaken. Images delivered by email or available for download.

## Ed Geldard Picture Library

9 Sunderland Bridge Village, Durham DH6 5HB

☎ 0191 378 2592

camera-one@talktalk.net
Contact *Ed Geldard*

50,000 images of landscape and buildings covering the Lake District, Northumberland and Durham.

## Ed Lacey Collection

▷ Phil Sheldon Golf Picture Library

## Edifice

Cutterne Mill, Southwood, Evercreech BA4 6LY

☎ 01749 831400
info@edificephoto.com
www.edificephoto.com
Contacts *Philippa Lewis, Gillian Darley*

Colour material on exteriors of all possible types of building worldwide from grand landmarks to the small and simple. Specialists in domestic architecture, garden features, architectural detail, ornament, historical and period style, building materials and techniques. Website searchable online, but expert help and advice happily given over the phone or by email.

## Education Photos

8 Whitemore Road, Guildford GU1 1QT

☎ 01483 511666
johnwalmsley@educationphotos.co.uk
www.educationphotos.co.uk
Photographer/Owner *John Walmsley*

New website allowing searches and HiRes downloads. The library covers schools, colleges, universities and everyday life in the UK. Also, occupations, housing, portraits of ordinary people, in fact, much of the range required by educational publishers.

## English Heritage Photo Library

Kemble Drive, Swindon SN2 2GZ

☎ 01793 414903
images@english-heritage.org.uk
www.englishheritageimages.com
Contacts *Duncan Brown, Javis Gurr, Jonathan Butler*

Images of English castles, abbeys, houses, gardens, Roman remains, ancient monuments, battlefields, industrial and post-war buildings, interiors, paintings, artifacts, aerials, scenic views, landscapes. Special features for 2010: Aerofilms, Victorian England. See also **National Monuments Record**.

## Eric Whitehead Photography

7 Brow Close, Bowness on Windermere LA23 2HA

☎ 01539 448894
enquiry@ericwhiteheadphotography.com
www.ericwhiteheadphotography.com
Contact *Eric Whitehead*

The agency covers local news events, PR and commercial material, also leading library of snooker images (see **Snookerimages**).

## Erich Lessing Culture & Fine Arts Archives

c/o akg-images Ltd, 5 Melbray Mews, 158 Hurlingham Road, London SW6 3NS
☎ 020 7610 6103    🖷 020 7610 6125
enquiries@akg-images.co.uk
www.akg-images.co.uk

Top quality high resolution scans depicting the contents of many of the world's finest art galleries as well as ancient archaeological and biblical sites available in the UK via the akg-images website.

## Eye Ubiquitous

65 Brighton Road, Shoreham BN43 6RE
☎ 01273 440113    🖷 01273 440116
library@eyeubiquitous.com
www.eyeubiquitous.com

General stock specializing in social documentary worldwide, and an extensive travel related collection.

## Faces and Places

☎ 020 7602 1989/07930 622964
info@facesandplacespix.com
www.facesandplacespix.com

Extensive worldwide library built up over twenty years which covers 40,000 photographs focusing mainly on travel, tourism, unusual locations, indigenous peoples, scenery and underwater photography; Australia, Tibet, Mali, Yemen and Iran.

## Financial Times Pictures

One Southwark Bridge, London SE1 9HL
☎ 020 7873 3671    🖷 020 7873 4606
photosynd@ft.com
Contacts *Nicky Burr, Richard Pigden*

Photographs from around the world ranging from personalities in business, politics and the arts, people at work and other human interests and activities. Delivery via ftp or email.

## Fine Art Photographic Library Ltd

2A Milner Street, London SW3 2PU
☎ 020 7589 3127
mail@fineartphotolibrary.com
www.fineartphotolibrary.com

Over 30,000 large-format transparencies, with a specialist collection of 19th and 20th century paintings. CD-ROM available.

## Firepix International

68 Arkles Lane, Anfield, Liverpool L4 2SP
☎ 0151 260 0111/0777 5930419    🖷 0151 260 0111
info@firepix.com
www.firepix.com
Contact *Tony Myers*

The UK's only fire photo library. 23,000 images of fire-related subjects, firefighters, fire equipment manufacturers. Website contains 15 categories from industrial fire, domestic, digital images and abstract flame.

## Fogden Wildlife Photographs

16 Locheport, North Uist, Western Isles HS6 5EU
☎ 01876 580245
susan.fogden@fogdenphotos.co.uk
www.fogdenphotos.com
Contact *Susan Fogden*

Natural history collection, with special reference to rain forests and deserts. 'Emphasis on quality rather than quantity'; growing collection of around 25,000 images.

## Food Features

Beaconhurst, Chestnut Walk, Tangmere PO20 2HH
☎ 01243 532240    🖷 01243 532240
frontdesk@foodfeatures.net
www.foodfeatures.net
Contacts *Steve Moss, Alex Barker*

Specialized high-quality food and drink photography, features and tested recipes. Clients' specific requirements can be incorporated into regular shooting schedules. Commissions welcome.

## Forest Life Picture Library

231 Corstorphine Road, Edinburgh EH12 7AT
☎ 0131 314 6411
neill.campbell@forestry.gsi.gov.uk
www.forestry.gov.uk/pictures
Contact *Neill Campbell*

The official image bank of the Forestry Commission, the library provides a single source for all aspects of forest and woodland management. The comprehensive subject list includes tree species, scenic landscapes, employment, wildlife, flora and fauna, conservation, sport and leisure.

## Formula One Pictures

☎ 00 36 26 322 826/00 36 70776 9682 (mobile)
jt@f1pictures.com
www.f1pictures.com
Contact *John Townsend*

500,000 35mm colour slides, b&w and colour negatives and digital images of all aspects of Formula One Grand Prix racing from 1980 including driver profiles and portraits.

## Francis Bedford

▷ Birmingham Library Services under Library Services

## The Francis Frith Collection

Frith's Barn, Teffont, Salisbury SP3 5QP
☎ 01722 716376    🖷 01722 716881

www.francisfrith.com
Contact *Julia Skinner, Managing Editor*

Publishers of *Frith's Photographic Memories* series of illustrated local books, all featuring nostalgic photographs from the archive, founded by Frith in 1860. The archive contains over 360,000 images of 7,000 British towns.

## Frank Lane Picture Agency Ltd

Pages Green House, Wetheringsett, Stowmarket IP14 5QA
☎ 01728 860789  📠 01728 860222
pictures@flpa-images.co.uk
www.flpa.co.uk

Colour coverage of natural history, environment, pets and weather. Represents Sunset from France, Foto Natura from Holland, Minden Pictures from the US, Holt Studios and works closely with Eric and David Hosking, plus 270 freelance photographers. Website with 175,000 images.

## Fresh Food Images

2nd Floor, Waterside House, 9 Woodfield Road, London W9 2BA
☎ 020 7432 8200  📠 020 7432 8201
uksales@freshfoodimages.com
www.freshfoodimages.com

'Europe's premier source' of food and wine related images. From the farm and the vineyard to the plate and the bottle. Cooking and kitchens, top chefs and restaurants, country trades and markets, worldwide travel with an extensive Italian section.

## Galaxy Picture Library

34 Fennels Way, Flackwell Heath, High Wycombe HP10 9BY
☎ 01628 521338
robin@galaxypix.com
www.galaxypix.com
Contact *Robin Scagell*

Specializes in astronomy, space, telescopes, observatories, the sky, clouds and sunsets. Composites of foregrounds, stars, moon and planets prepared to commission. Editorial service available.

## Garden and Wildlife Matters Photo Library

'Marlham', Henley's Down, Battle TN33 9BN
☎ 01424 830566  📠 01424 830224
carol@gmpix
john@wildlifematters.com
www.gardenandwildlife.com/home.php
Contact *Dr John Feltwell*

Collection of 110,000 6x4" and 35mm images. General gardening techniques and design; cottage gardens and USA designer gardens. 10,000 species of garden plants and over 1,000 species of trees. Flowers, wild and house plants,

trees and crops. Environmental, ecological and conservation pictures, including sea, air, noise and freshwater pollution, SE Asian and Central and South American rainforests; Eastern Europe, Mediterranean. Recycling, agriculture, forestry, horticulture and oblique aerial habitat shots from Europe and USA. Digital images supplied worldwide.

## Garden Picture Library

2nd Floor, Waterside House, 9 Woodfield Road, London W9 2BA
☎ 020 7432 8200  📠 020 7432 8201
sales@gardenpicture.com
creative@gardenpicture.com
www.gardenpicture.com
Creative Director *Lee Wheatley*

'Our inspirational images of gardens, plants and gardening lifestyle offer plenty of scope for writers looking for original ideas to write about.' The collection covers all garden related subjects, everything from plant portraits to floral graphics and garden design details to whole landscapes. Brings together the work of over 100 professional photographers from across the gardening globe. Holds approximately 300,000 fully captioned images of which around 65,000 are available digitally. Free in-house picture research can be undertaken on request or searches can be made online via keyword or subject category to create lightboxes which can be emailed or downloaded. Visitors to the library are welcome by appointment and copies of promotional literature are available on request.

## Genesis Space Photo Library

tim@spaceport.co.uk
www.spaceport.co.uk
Contact *Tim Furniss*

Historical b&w spaceflight collection including rockets, spacecraft, spacemen, Earth, moon and planets. Catalogue available on the website.

## Geo Aerial Photography

4 Christian Fields, London SW16 3JZ
☎ 020 8764 6292/0115 981 9418  📠 0115 981 9418
geo.aerial@geo-group.co.uk
www.geo-group.co.uk
Contact *Kelly White*

Established 1990 and now a growing collection of aerial oblique photographs from the UK, Scandinavia, Asia and Africa – landscapes, buildings, industrial sites, etc. Commissions undertaken.

## GeoScience Features

6 Orchard Drive, Wye TN25 5AU
☎ 01233 812707  📠 01233 812707
gsf@geoscience.uk.com
www.geoscience.uk.com

Fully computerized and comprehensive library containing the world's principal source of volcanic phenomena. Extensive collections, providing scientific detail with technical quality, of rocks, minerals, fossils, microsections of botanical and animal tissues, animals, biology, birds, botany, chemistry, earth science, ecology, environment, geology, geography, habitats, landscapes, macro/microbiology, peoples, sky, weather, wildlife and zoology. Over 500,000 original colour transparencies in medium and 35mm-format. Subject lists and CD-ROM catalogue available. Incorporates the RIDA photolibrary and Landform Slides.

## Geoslides Photography

4 Christian Fields, London SW16 3JZ
℡ 0115 981 9418    ℻ 0115 981 9418
geoslides@geo-group.co.uk
www.geo-group.co.uk
Contact *John Douglas*

Established in 1968. Landscape and human interest subjects from the Arctic, Antarctica, Scandinavia, UK, Africa (south of Sahara), Middle East, Asia (south and southeast). Also specialist collections of images from British India (the Raj) and Boer War. Forty years of photography services.

## Getty Images

101 Bayham Street, London NW1 0AG
℡ 0800 376 7981/020 7428 6109
www.gettyimages.com
Contact *Sales Department*

'Getty Images is the world's leading imagery company, creating and providing the largest collection of still and moving images to communication professionals around the globe.' From sports and news photography to archival and contemporary imagery. For those who wish to commission photographers to fulfil specific needs, the company maintains a full-service department for custom-shot images.

## Giles Chapman Library

2 Bullfinch Close, Sevenoaks TN13 2BB
℡ 01732 452547
chapman.media@virgin.net
www.gileschapman.com
Contact *Giles Chapman*

Some 350,000 colour and b&w images of cars and motoring, from 1945 to the present day. No research fees.

## Greenpark Productions Ltd (Film Archive)
▷ entry under Library Services

## Hamish Brown MBE, D.Litt, FRSGS Scottish Photographic

3 Links Place, Burntisland KY3 9DY
℡ 01592 873546    ℻ 01592 873546
Contact *Hamish Brown*

Coverage of most topics and areas of Scotland (sites, historic, buildings, landscape, mountains), also travel, mountains, general (50,000 items) and Morocco. Commissions undertaken.

## Harry Price Library of Magical Literature
▷ Mary Evans Picture Library

## Heather Angel/Natural Visions

6 Vicarage Hill, Farnham GU9 8HG
℡ 01252 716700    ℻ 01252 727464
pictures@naturalvisions.co.uk
www.naturalvisions.co.uk
Contact *Heather Angel*

Constantly expanding worldwide natural history, wildlife and landscapes: polar regions, tropical rainforest flora and fauna, all species of plants and animals in natural habitats from Africa, Asia (huge Green China file), Australasia, South America and USA, urban wildlife, pollution, biodiversity, global warming. Also worldwide gardens and cultivated flowers. Can send a digital lightbox to authors with an email. 'We supply high resolution digital files direct to the publisher.'

## Hobbs Golf Collection

5 Winston Way, New Ridley, Stocksfield NE43 7RF
℡ 01661 842933/07941 445993    ℻ 01661 842933
info@hobbsgolfcollection.com
www.hobbsgolfcollection.com
Contact *Margaret Hobbs*

Specialist golf collection: players, courses, art, memorabilia and historical topics (1300–present). 40,000+ images, mainly 35mm colour transparencies and b&w prints. All images can be supplied in digital format. Commissions undertaken. Author of 30 golf books.

## Holt Studios
▷ Frank Lane Picture Agency Ltd

## Houghton's Horses/Kit Houghton Photography

Radlet Cottage, Spaxton, Bridgwater TA5 1DE
℡ 01278 671362
kit@kithoughton.com
www.houghtonshorses.com
Contacts *Kit Houghton, Kate Houghton*

Specialist equestrian library of over 300,000 transparencies and digital images on all aspects of the horse world, with images ranging from the romantic to the practical and competition pictures in all equestrian disciplines worldwide. Online picture delivery is now the norm. All transparencies in the process of being scanned. Commissions undertaken.

## Hutchison Picture Library

65 Brighton Road, Shoreham-by-Sea BN43 6RE
☎ 01273 440113   🖷 01273 440116
library@hutchisonpictures.co.uk
www.hutchisonpictures.co.uk

Worldwide contemporary images from the straight-forward to the esoteric and quirky. Over half a million documentary colour photographs on file, covering people, places, customs and faiths, agriculture, industry and transport. Special collections include the environment and climate, family life (including pregnancy and birth), ethnic minorities worldwide (including Disappearing World archive), conventional and alternative medicine, and music around the world. Search service available.

## Illustrated London News Picture Library

c/o Mary Evans Picture Library, 59 Tranquil Vale, Blackheath, London SE3 0BS
☎ 020 7805 5555   🖷 020 8852 7211
iln@maryevans.com
www.maryevans.com

Engravings, photographs and illustrations from 1842 to the present day, taken from magazines published by Illustrated Newspapers: *Illustrated London News*; *Graphic*; *Sphere*; *Tatler*; *Sketch*; *Illustrated Sporting and Dramatic News*; *Illustrated War News 1914–18*; *Bystander*; *Britannia & Eve*. Social history, London, Industrial Revolution, wars, travel. Now housed and managed by **Mary Evans Picture Library** (see entry). Research and scanning services available.

## Images of Africa Photobank

11 The Windings, Lichfield WS13 7EX
☎ 01543 262898
info@imagesofafrica.co.uk
www.imagesofafrica.co.uk
Contact *David Keith Jones, FRPS*

Images covering Botswana, Chad, Egypt, Ethiopia, Kenya, Lesotho, Madagasca, Malawi, Morocco, Namibia, Rwanda, South Africa, Swaziland, Tanzania, Uganda, Zaire, Zambia, Zanzibar and Zimbabwe. 'Probably the best collection of photographs of Kenya in Europe.' A wide range of topics are covered. Strong on African wildlife with over 80 species of mammals including many sequences showing action and behaviour. Popular animals like lions and elephants are covered in encyclopaedic detail. More than 100 species of birds and many reptiles are included. Other strengths include National Parks and reserves, natural beauty, tourism facilities, traditional and modern people. Most work is by David Keith

Jones, FRPS; several other photographers are represented. Many thousands of these images can now be accessed online. Go to the website and look for Online Images.

## Imperial War Museum Photograph Archive

All Saints Annexe, Austral Street, London SE11 4SJ
☎ 020 7416 5333/5338/5309   🖷 020 7416 5355
photos@iwm.org.uk
www.iwm.org.uk

A national archive of ten million photographs illustrating all aspects of 20th and 21st century conflict. Emphasis on the two World Wars but includes material from other conflicts involving Britain and the Commonwealth. Majority of material is b&w, although holdings of colour material increase with more recent conflicts. Visitors welcome by appointment, Monday, Wednesday and Friday, 10.00 am to 5.00 pm.

## Institution of Mechanical Engineers

Information & Library Service, Institution of Mechanical Engineers, 1 Birdcage Walk, London SW1H 9JJ
☎ 020 7973 1274   🖷 020 7222 8762
library@imeche.org
www.imeche.org

Historical images and archives on mechanical engineering. Open 9.15 am to 5.30 pm, Monday to Friday.

## International Photobank

PO Box 6554, Dorchester DT1 9BS
☎ 01305 854145   🖷 01305 853065
info@internationalphotobank.co.uk
www.internationalphotobank.co.uk
Contacts *Peter Baker, Gary Goodwin*

Over 450,000 images, in digital and transparency format. Colour coverage of travel subjects: places, people, folklore, events. Digital broadband service available for newspapers, magazines and other users. Pictures can also be selected from the comprehensive website.

## Jacqui Hurst

66 Richford Street, Hammersmith, London W6 7HP
☎ 020 8743 2315/07970 781336
jacquihurst@yahoo.co.uk
www.jacquihurstphotography.co.uk
Contact *Jacqui Hurst*

A specialist library of traditional and contemporary applied arts, regional food producers and markets. The photos form illustrated essays of how something is made and finish with a still life of the completed object. The collection is always being extended and a list is available on request. Commissions undertaken.

## Jayawardene Travel Photo Library

7A Napier Road, Wembley HA0 4UA
☎ 020 8795 3581   🖷 020 8795 4083
jaytravelphotos@aol.com
www.jaytravelphotos.com
Contact *Rohith Jayawardene (Partner)*

180,000 travel and travel-related images – high
end digital and original transparencies. Many
places photographed in depth. Over 20,000
images are available online.

## Jim Henderson Photography

Crooktree, Kincardine O'Neil, Aboyne AB34 4JD
☎ 01339 882149   🖷 01339 882149
JHende7868@aol.com
www.jimhendersonphotography.com
Contact *Jim Henderson*

Scenic and general activity coverage of north
east Scotland (Aberdeenshire) and Highlands
for tourist, holiday and activity illustration.
Specialist collection of over 300 Aurora Borealis
displays from 1989–2008 on Royal Deeside and
co-author of *The Aurora* (published 1997). Author
and photographer for *Aberdeen* and *Aberdeen &
Royal Deeside* titles for Aberdeen Journals. Recent
images of ancient Egyptian sites: Cairo through
to Abu-Simbel. **BAPLA** member. Contributor
to Alamy & jasonfriendimages. Commissions
undertaken.

## Joe Tasker Collection
▷ Chris Bonington Picture Library

## John Cleare/Mountain Camera

Hill Cottage, Fonthill Gifford, Salisbury SP3 6QW
☎ 01747 820320   🖷 01747 820320
cleare@btinternet.com
www.mountaincamera.com

A wide portfolio includes specialist coverage in
colour and b&w of mountains and wild places,
climbing, ski-touring, expeditions and travel,
landscapes, geographical features and people
from all continents, especially Britain, the Alps,
the Himalaya and the Americas. Topics range
from reindeer in Lapland and camels in Australia,
white-water rafting in Utah to skiing in China.
Recent projects have included photographing
books on rock climbing and canoeing for
youngsters and providing abstract and poetic
images for titles on Japanese haiku, Chinese
wisdom and Buddhist teaching. Commissions,
presentations and consultancy work undertaken.
Researchers welcome by appointment. Member
of **BAPLA** and the **OWPG**.

## John Frost Newspapers
▷ entry under Library Services

## John Heseltine Archive

Cotteswold House, The High Street, South
Woodchester GL5 5EL
☎ 01453 873792
john@heseltine.co.uk
www.heseltine.co.uk
Contact *John Heseltine*

A new website displays this distinctive collection
of mainly black and white photography dealing
with a range of subjects from Italy, France, the UK
and further afield.

## John Robert Young Collection

Paxvobiscum, 16 Greenacres Drive, Ringmer,
Lewes BN8 5LX
☎ 01273 814172
pax@freedom255.com
www.johnrobertyoung.com
Contacts *Jennifer Barrett, John Robert Young*

50,000 colour and b&w images. Military (French
Foreign Legion, Spanish Foreign Legion, Royal
Marines), religion (Christian), travel, China,
personalities of the 1960s and '70s, sixties'
fashion, also a growing portfolio of fine art
photography.

## The Kobal Collection

2 The Quadrant, 135 Salusbury Road, London
NW6 6RJ
☎ 020 7624 3500   🖷 020 7624 3355
info@picture-desk.com
www.picture-desk.com

Picture library holding colour and b&w coverage
of Hollywood films, portraits, stills, publicity
shots, posters and ephemera.

## Kos Picture Source Ltd

PO Box 104, Midhurst GU29 1AS
☎ 020 7801 0044   🖷 020 7801 0055
images@kospictures.com
www.kospictures.com

Specialists in water-related images including
international yacht racing and cruising, classic
boats and superyachts, and extensive range of
watersports. Also worldwide travel including
seascapes, beach scenes, underwater photography
and the weather.

## Landform Slides
▷ GeoScience Features

## Last Resort Picture Library

Manvers Studios, 12 Ollerton Road, Tuxford,
Newark NG22 0LF
☎ 01777 870166   🖷 01777 871739
dick@dmimaging.co.uk
www.dmimaging.co.uk
Contacts *Jo Makin, Dick Makin*

Subject areas include agriculture, architecture, IT, education, food, industry, landscapes, people at work, skiing, trees, flowers, mountains and winter landscapes. Images cover a wide variety of topics, ranging from the everyday to the obscure. 'Bespoke service available through Dick Makin Imaging, our linked photographic studio. Contact us for details.'

## LAT Photographic

Teddington Studios, Broom Road, Teddington TW11 9BE

☎ 020 8267 3010

lat.photo@haymarket.com

www.latphoto.co.uk

Motor sport collection of almost 12 million images dating from 1895 to the present day.

## Lee Miller Archives

Farley Farm House, Muddles Green, Chiddingly, Near Lewes BN8 6HW

☎ 01825 872691   🖷 01825 872733

archives@leemiller.co.uk

www.leemiller.co.uk

The work of Lee Miller (1907–77). As a photo-journalist she covered the war in Europe from early in 1941 to VE Day with further reporting from the Balkans. Collection includes photographic portraits of prominent Surrealist artists: Ernst, Eluard, Miró, Picasso, Penrose, Carrington, Tanning, and others. Surrealist and contemporary art, poets and writers, fashion, the Middle East, Egypt, the Balkans in the 1930s, London during the Blitz, war in Europe and the liberation of Dachau and Buchenwald.

## Lesley & Roy Adkins Picture Library

Ten Acre Wood, Whitestone, Exeter EX4 2HW

☎ 01392 811357

mail@adkinshistory.com

www.adkinshistory.com

Colour and some b&w coverage of archaeology, antiquity, heritage and related subjects in the UK, Europe, Egypt and Turkey. Subjects include towns, villages, housing, landscape and countryside, churches, temples, castles, monasteries, art and architecture, gravestones and tombs, inscriptions, maps, naval, nautical and antiquarian views. Images supplied in digital form. No service charge if pictures are used.

## Library of Art & Design
▷ V&A Images

## Lindley Library
▷ RHS Lindley Library

## Link Picture Library

41A The Downs, London SW20 8HG

☎ 020 8944 6933

library@linkpicturelibrary.com

www.linkpicturelibrary.com

Contact *Orde Eliason*

100,000 specialist images of South Africa, India and China. A more general collection of colour transparencies from 100 countries worldwide. Link Picture Library has an international network and can source material not in its file from China, Norway, India and South Africa. Original photographic commissions undertaken.

## London Aerial Photo Library

Unit 2B, Henley Business Park, Pirbright Road, Normandy GU3 2DX

☎ 01483 233395   🖷 01483 237081

info@londonaerial.co.uk

www.londonaerial.co.uk

www.flightimages.com

Contact *Amanda Campbell*

Extensive collections of aerial imagery. 450,000 images including continually updated photography of London plus excellent coverage of the UK in general. Covers landmarks, industrial/retail properties, sporting venues/football stadiums, conceptual shots. Library searches by expert staff, free of charge. Selection of digital images also available via website. Commissioned photography undertaken.

## London Metropolitan Archives

40 Northampton Road, London EC1R 0HB

☎ 020 7332 3820   🖷 020 7833 9136

ask.lma@cityoflondon.gov.uk

www.cityoflondon.gov.uk/lma

Contact *The Enquiry Team*

London Metropolitan Archives (LMA) is the largest local authority record office in the UK. Holds over 32 miles of archives – an enormous amount of information about the capital and its people. These include records of London government, hospitals, charities, businesses and parish churches. Types of record range from books and manuscript documents to photographs, maps and drawings. 'Nearly 900 years of London history can be brought to life at LMA.' There is also a 100,000 volume reference library specializing in London history. See entry under *Library Services*.

## London's Transport Museum Photographic Library

39 Wellington Street, London WC2E 7BB

☎ 020 7379 6344   🖷 020 7565 7252

www.ltmcollection.org/photos/index.html

www.ltmcollection.org/posters/index.html

Photographic collection reflecting London's public transport history from the 1860s to the present day. The core of the collection is made up of the old London Transport photographic

archive of over 150,000 b&w photographs. Also includes a smaller archive of colour material, as well as historic albums and prints from all periods. Over 16,000 images are now available to view and purchase on-line, with new photographs added regularly. Each image has a brief caption, stating location and date (where known) to assist retrieval. For more specialist enquiries, please make an appointment to view the photographic collection at Acton Depot (contact: dutycurator@ltmuseum.co.uk).
**London Transport Museum Poster Archives**: for 100 years, since Frank Pick commissioned the first graphic poster for London Underground, the company and its successors have kept copies of everything they produced. In the early 20th century, under Pick's guidance, London Transport commissioned work from the best artists and designers in the country. By the 1980s, when the collection was transferred to the Museum, it contained more than 5,000 printed posters and nearly 1,000 original artworks. The collection has grown steadily ever since and is now a uniquely comprehensive picture of a century of British graphic design.

## Ludvigsen Library

Scoles Gate, Hawkedon, Bury St Edmunds
IP29 4AU
☎ 01284 789246    ℻ 01234 252033
library@ludvigsen.com
www.ludvigsen.com
Contact *Karl Ludvigsen*

Extensive information research facilities for writers and publishers. Approximately 400,000 images (both b&w and many colour transparencies) of automobiles and motorsport, from 1890s through 1980s. Glass plate negatives from the early 1900s; Formula One, Le Mans, motor car shows, vintage, antique and classic cars from all countries. Includes the Dalton-Watson Collection and the work of noted photographers such as John Dugdale, Edward Eves, Peter Keen, Max le Grand, Karl Ludvigsen, Rodolfo Mailander, Ove Nielsen, Stanley Rosenthall and others.

## Magnum Photos Ltd

63 Gee Street, London EC1V 3RS
☎ 020 7490 1771    ℻ 020 7608 0020
magnum@magnumphotos.co.uk
www.magnumphotos.com

Founded in 1947 by Henri Cartier Bresson, George Rodger, Robert Capa and David 'Chim' Seymour. Represents over 70 of the world's leading photo-journalists and documentary photographers. Coverage of major world events from the Spanish Civil War to present day, daily life, social issues and reportage. Also a large collection of personalities and travel images.

## Martin and Dorothy Grace

Boxwood Cottage, The Row, Lyth, Kendal
LA8 8DD
☎ 015395 68569
martin.grace@tiscali.co.uk
Contact *Martin Grace (Partner)*

Colour coverage of natural history, topography and travel in Britain, the Galapagos Islands, France, southern Spain, Peru, Madagascar, South Georgia and the Antarctic. Specializes in trees, shrubs and wild flowers but also cover ferns, animals, birds and butterflies, habitats, landscapes, ecology. High quality digital stock supplied online.

## Mary Evans Picture Library

59 Tranquil Vale, Blackheath, London SE3 0BS
☎ 020 8318 0034    ℻ 020 8852 7211
pictures@maryevans.com
www.maryevans.com

Leading historical archive of artwork, photographs, prints and ephemera illustrating and documenting all aspects of the past, from ancient times to the later decades of the 20th century. Emphasis on everyday life plus extensive coverage of events, portraits, transport, and costume. 'Unrivalled material on folklore and paranormal phenomena.' Notable collections include the Weimar Archive documenting the Third Reich, and Barnaby's Library of social documentary photography from the 1930s to the 1970s. Own material complemented by important contributor collections, such as the **Illustrated London News Picture Library**; the archive of the National Magazine Company; the English Heritage Collection; the Harry Price Library of Magical Literature; Sigmund Freud Copyrights; the Women's Library and the Castle Howard Collection of fine art and prints. Arrangements with agencies in France (Rue des Archives); Italy (De Agostini); Germany (Interfoto); Austria (Imagno) and Spain (Aisa) have added thousands of additional photographs and fine art images to the website. Over 400,000 images searchable online, the majority available in high resolution for immediate download. Brochure on request. Founder member of **BAPLA**. Compilers of the *Picture Researcher's Handbook* published by Pira International.

## Michael Cole Camerawork

The Coach House, 27 The Avenue, Beckenham
BR3 5DP
☎ 020 8658 6120    ℻ 020 8658 6120
mikecole@dircon.co.uk
Contacts *Michael Cole, Derrick Bentley*

Probably the largest and most comprehensive collection of tennis pictures in the world. Over 50 years' coverage of the Wimbledon Championships. M.C.C. incorporates the tennis archives of Le Roye Productions, established in 1945. Picture requests and enquiries by email.

## Mick Sharp Photography

Eithinog, Waun, Penisarwaun, Caernarfon LL55 3PW

☎ 01286 872425   📠 01286 872425
mick.jean@virgin.net
www.windowonlosttime.co.uk
Contacts *Mick Sharp, Jean Williamson*

Archaeology, ancient monuments, buildings, churches, countryside, environment, history, inspirational, landscape, past cultures and topography. Emphasis on British Isles but material also from other countries. Features the photos of both Mick Sharp and Jean Williamson. Contacts with other specialist collections on related subjects. Commissions undertaken. Digital files available from medium format and 35mm colour transparencies and b&w prints from 5x4 negatives.

## Mirrorpix

22nd Floor, One Canada Square, Canary Wharf, London E14 5AP

☎ 020 7293 3700   📠 020 7293 0357
desk@mirrorpix.com
www.mirrorpix.com

Combines the photographic archives of the *Daily Mirror, Daily Herald, Sunday Mirror, The People, Daily Record* and *Sunday Mail*. 250,000 images online, 50 million negatives and prints offline. Ideal for illustrating autobiographical, entertainment, social and political history (domestic and foreign), the arts, culture, fashion, humour, industry, sport and royalty.

## Monitor Picture Library

The Forge, Roydon, Harlow CM19 5HH

☎ 01279 792700
sales@monitorpicturelibrary.com
www.monitorpicturelibrary.com

Colour and b&w coverage of 1960s, '70s and '80s personalities and celebrities from: music, entertainment, sport, politics, royals, judicial, commerce etc. Specialist files on Lotus cars. Syndication to international, national and local media.

## Motoring Picture Library

National Motor Museum, Beaulieu SO42 7ZN

☎ 01590 614656   📠 01590 612655
motoring.pictures@beaulieu.co.uk
www.motoringpicturelibrary.com
www.alamy.com/mpl
Contacts *Jonathan Day, Tim Woodcock*

Three-quarters of a million b&w images, plus over 140,000 colour images covering all forms of motoring history from the 1880s to the present day. Commissions undertaken. Own studio.

## Mountain Camera
▷ John Cleare

## Moving Image Communications

9 Faversham Reach, Faversham ME13 7LA

☎ 0845 257 2968   📠 01795 534306
mail@milibrary.com
nathalie@milibrary.com
mike@milibrary.com
www.milibrary.com
Contact *Nathalie Banaigs*

Over 16,000 hours of quality archive and contemporary footage, including: Channel X; TVAM Archive 1983–92; Drummer Films (travel classics, 1950–70); The Freud Archive (1930–39); Film Finders (early cinema); Adrian Brunel Films; Cuban Archives; Stockshots (timelapse, cityscapes, land and seascapes, chroma-key); Space Exploration (NASA); Wild Islands; Flying Pictures.

## Museum of Childhood
▷ V&A Images

## Museum of London Picture Library

150 London Wall, London EC2Y 5HN

☎ 020 7814 5604/5612   📠 020 7600 1058
picturelib@museumoflondon.org.uk
www.museumoflondonimages.org.uk
Library Manager *Sean Waterman*
Picture Researchers *Nikki Braunton, Sarah Williams, Jenna Collins*

The Museum of London picture library holds over 35,000 images illustrating the history of London and the life of its people from prehistoric times to the present day. The images are drawn from the Museum's extensive and unique collections of oil paintings, historic photographs, drawings, prints maps and artefacts including costume, jewellery and ceramics.

## The National Archives Image Library

Ruskin Avenue, Kew, Richmond TW9 4DU

☎ 020 8392 5225   📠 020 8487 1974
image-library@nationalarchives.gov.uk
www.nationalarchives.gov.uk/imagelibrary
Contacts *Paul Johnson, Hugh Alexander*

British and colonial history from the Domesday Book to the 1970s, shown in photography, maps, illuminations, posters, advertisements, textiles and original manuscripts. Digital scans supplied to order. Open 9.00 am to 5.00 pm, Monday to Friday.

## National Galleries of Scotland Picture Library
The Scottish National Gallery of Modern Art, 75 Belford Road, Edinburgh EH4 3DR
☎ 0131 624 6258/6260  📠 0131 623 7135
picture.library@nationalgalleries.org
www.nationalgalleries.org
Contact *Shona Corner*

Over 80,000 b&w and several thousand images in colour of works of art from the Renaissance to present day. Specialist subjects cover fine art (painting, sculpture, drawing), portraits, Scottish, historical, still life, photography and landscape.

## National Magazine Company
▷ Mary Evans Picture Library

## National Maritime Museum Picture Library
Greenwich, London SE10 9NF
☎ 020 8312 6631/6704  📠 020 8312 6533
pictures@nmmimages.com
www.nmmimages.com
Contact *Douglas McCarthy*

Over 400,000 images from the leading collection of maritime art and artefacts, including oil paintings from the 16th century to present day, prints and drawings, historic photographs, ships' plans, models, rare maps and charts, globes, manuscripts and navigation and scientific instruments. Images are available as digital files. Contact for further information or for picture research assistance.

## National Media Museum
▷ Science & Society Picture Library

## National Monuments Record
English Heritage, Kemble Drive, Swindon SN2 2GZ
☎ 01793 414600  📠 01793 414606
nmrinfo@english-heritage.org.uk
www.english-heritage.org.uk/nmr

The National Monuments Record (NMR), the public archive of English Heritage, holds over ten million photographs, plans, drawings, reports, records and publications covering England's archaeology, architecture, social and local history. The NMR is the largest publicly accessible archive in Britain and is the biggest dedicated to the historic environment.

## National Portrait Gallery Picture Library
St Martin's Place, London WC2H 0HE
☎ 020 7312 2474/5/6  📠 020 7312 2464
www.npg.org.uk
Contact *Head of Rights & Reproductions*

Pictures of brilliant, daring and influential characters who have made British history are available for publication. Images can be searched,

viewed and ordered on the website. Copyright clearance is arranged for all images supplied.

## National Railway Museum Search Engine – The NRM's Library & Archive Centre
Leeman Road, York YO26 4XJ
☎ 01904 686235  📠 01904 611112
search.engine@nrm.org.uk
www.nrm.org.uk

The National Railway Museum and the **Science and Society Picture Library** are both part of the National Museum of Science and Industry (NMSI) whose other constituents are the Science Museum in London and the National Media Museum in Bradford (formerly the National Museum of Photography, Film & Television). 1.5 million images, mainly b&w, covering every aspect of railways from the 1850s to the present day; over one million engineering drawings; the most comprehensive railway library in the UK; sound and oral history archives; a growing digital video record of today's railway industry; a large collection of British railway posters, graphic art and advertising material from the 1820s onwards. Search engine is open 10.00 am to 5.30 pm, seven days a week. Closed 24th–26th December.

## The National Trust Photo Library
Heelis, Kemble Drive, Swindon SN2 2NA
☎ 01793 817700
photo.library@nationaltrust.org.uk
www.ntpl.org.uk
Contact *Chris Lacey*

The National Trust Photo Library houses a unique collection of contemporary photography which vividly illustrates the rich diversity and historical range of properties in the National Trust's care, throughout England, Wales and Northern Ireland. The Library has an exceptional choice of images, created by commissioned specialist photographers, suitable for a broad range of creative applications. 40,000 images can be searched and ordered online.

## Natural History Museum
▷ NHM Image Resources

## Natural Visions
▷ Heather Angel

## Nature Photographers Ltd
West Wit, New Road, Little London, Tadley RG26 5EU
☎ 01256 850661  📠 01256 851157
paul@naturephotographers.co.uk
www.naturephotographers.co.uk
Contact *Dr Paul Sterry*

Over 150,000 images, digitized to order, on worldwide natural history and environmental subjects. The library is run by a trained biologist and experienced author on his subject.

## Nature Picture Library

5a Great George Street, Bristol BS1 5RR

📞 0117 911 4675 📠 0117 911 4699

info@naturepl.com

www.naturepl.com

Contacts *Tim Harris, Helen Gilks, Rachelle Macapagal*

A collection of 300,000 nature photos from around the world, including strong coverage of animal portraits and behaviour. Other subjects covered include plants, pets, landscapes and travel, environmental issues and wildlife film-makers at work. More than 160,000 images can be viewed online and downloaded direct for reproduction.

## NHM Image Resources

Natural History Museum, Cromwell Road, London SW7 5BD

📞 020 7942 5324/5401/5376 📠 020 7942 5443

nhmpl@nhm.ac.uk

www.nhm.ac.uk/piclib

Pictures from the Museum's collections, including dinosaurs, man's evolution, extinct species and fossil remains. Also pictures of gems, minerals, birds and animals, plants and insect specimens, plus historical artworks depicting the natural world.

## NHPA
▷ Photoshot

## Only Horses Picture Agency

27 Greenway Gardens, Greenford UB6 9TU

📞 020 8578 9047 📠 020 8575 7244

mike.roberts959@btinternet.com

www.onlyhorsespictures.com

Colour and b&w coverage of all aspects of the horse. Foaling, retirement, racing, show jumping, eventing, veterinary, polo, breeds, personalities.

## Oxford Picture Library

15 Curtis Yard, North Hinksey Lane, Oxford OX2 0LX

📞 01865 723404 📠 01865 725294

opl@cap-ox.com

www.cap-ox.co.uk

Contacts *Chris Andrews, Annabel Matthews*

Specialist collection on Oxford: the city, university and colleges, events, people, spires and shires. Also, the Cotswolds, architecture and landscape from Stratford-upon-Avon to Bath; the Thames and Chilterns, including Henley on Thames and Windsor; Channel Islands, especially Guernsey and Sark. Aerial views of all areas specified above. General collection includes

wildlife, trees, plants, clouds, sun, sky, water and teddy bears. Commissions undertaken.

## Oxford Scientific (OSF)

2nd Floor, Waterside House, 9 Woodfield Road, London W9 2BA

📞 020 7432 8200 📠 020 7432 8201

uksales@osf.co.uk

www.osf.co.uk

Collection of 350,000 colour transparencies and digital files of wildlife and natural science images supplied by over 300 photographers worldwide, covering all aspects of wildlife plus landscapes, weather, seasons, plants, pets, environment, anthropology, habitats, science and industry, space, creative textures and backgrounds and geology. Macro and micro photography. UK agents for Animals Animals, USA, Okapia, Germany and Dinodia, India. Research by experienced researchers for specialist and creative briefs. Visits welcome, by appointment.

## Panos Pictures

1 Honduras Street, London EC1Y 0TH

📞 020 7253 1424 📠 020 7253 2752

pics@panos.co.uk

www.panos.co.uk

Documentary colour and b&w library specializing in Third World and Eastern Europe, with emphasis on environment and development issues. Leaflet available. Fifty per cent of all profits from this library go to the Panos Institute to further its work in international sustainable development.

## Patrick Eagar Photography

📞 020 8940 9269

info@patrickeagar.com

www.patrickeagar.com

The cricket library consists of Patrick Eagar's work over the last 30 years with coverage of over 290 Test matches worldwide, unique coverage of all eight World Cups, countless one-day internationals and player action portraits of over 2,000 cricketers. The wine library consists of vineyards, grapes and festivals from Argentina to New Zealand. France and Australia are specialist areas. Photographs can be supplied by email, on CD or DVD.

## Paul Kaye B/W Archive
▷ Sylvia Cordaiy Photo Library

## Peter Boardman Collection
▷ Chris Bonington Picture Library

## Peter Newark's Picture Library

3 Barton Buildings, Queen Square, Bath BA1 2JR

📞 01225 334213 📠 01225 480554

Over one million images covering world history from ancient times to the present day. Includes an extensive military collection of photographs, paintings and illustrations. Also a special collection on American history covering Colonial times, exploration, social, political and the Wild West and Native-Americans in particular. Subject list available. Telephone, fax or write for further information.

## Peter Sanders Photography Ltd

24 Meades Lane, Chesham HP5 1ND
☎ 01494 773674/771372   🖷 01494 773674
photos@petersanders.com
www.petersanders.com

Specializes in the world of Islam in all its aspects from culture, arts, industry, lifestyles, celebrations, etc. Areas included are north, east and west Africa, the Middle East (including Saudi Arabia), China, Asia, Europe and USA. A continually expanding library.

## Phil Sheldon Golf Picture Library

40 Manor Road, Barnet EN5 2JQ
☎ 020 8440 1986   🖷 020 8440 9348
gill@philsheldongolfpics.co.uk
info@philsheldongolfpics.co.uk
www.philsheldongolfpics.co.uk

An expanding collection of over 600,000 quality images of the 'world of golf'. In-depth worldwide tournament coverage including every Major championship and Ryder Cup since 1976. Instruction, portraits, trophies and over 400 golf courses from around the world. Also the Dale Concannon collection covering the period 1870 to 1940, the classic 1960s collection by photographer Sidney Harris and the Ed Lacey Collection.

## Philip Craven Worldwide Photo-Library

Surrey Studios, 21 Nork Way, Nork, Banstead SM7 1PB
☎ 01737 814111
www.philipcraven.com
Contact *Philip Craven*

Extensive coverage of British scenes, cities, villages, English countryside, gardens, historic buildings and wildlife. Worldwide travel and wildlife subjects on medium and large-format transparencies.

## Photofusion

17A Electric Lane, London SW9 8LA
☎ 020 7733 3500   🖷 020 7738 5509
library@photofusion.org
www.photofusionpictures.org
Contact *Liz Somerville*

Colour and b&w coverage of contemporary social and environmental UK issues including babies and children, disability, education, the elderly, environment, family, health, housing, homelessness, people and work. Brochure available.

## The Photolibrary Wales

2 Bro-nant, Church Road, Pentyrch, Cardiff CF15 9QG
☎ 029 2089 0311   🖷 029 2089 0311
info@photolibrarywales.com
www.photolibrarywales.com
Contacts *Steve Benbow, Kate Benbow*

Over 100,000 digital images covering all areas and subjects of Wales. Represents the work of 260 photographers, living and working in Wales.

## Photoshot

29–31 Saffron Hill, London EC1N 8SW
☎ 020 7421 6000   🖷 020 7421 6006
www.photoshot.com
www.nhpa.co.uk
www.staystill.com
www.worldpictures.com
Contacts *Charles Taylor, Emma Hier, David Brenes*

Photoshot collections include U.P.P.A. (daily national and international business, political and establishment news); Stay Still (exclusive celebrity portraiture and TV publicity images); Starstock (live images of celebrity and entertainment personalities and events from around the world); World Illustrated (the world in pictures, art, culture, environment and heritage); NHPA (wildlife and nature); World Pictures (comprehensive archive of practical and stylish travel images); Woodfall Wild Images (wildlife, landscape and environmental).

## PictureBank Photo Library Ltd

Parman House, 30–36 Fife Road, Kingston upon Thames KT1 1SY
☎ 020 8547 2344   🖷 020 8974 5652
info@picturebank.co.uk
www.picturebank.co.uk

Over 400,000 images covering people (girls, couples, families, children), travel and scenic (UK and world), moods (sunsets, seascapes, deserts, etc.), industry and technology, environments and general. Commissions undertaken. Visitors welcome. Member of **BAPLA**. New material in digital format welcome.

## Pictures Colour Library

10 James Whatman Court, Turkey Mill, Ashford Road, Maidstone ME14 5SS
☎ 01622 609809   🖷 01622 609806
enquiries@picturescolourlibrary.co.uk
www.picturescolourlibrary.co.uk

Travel and travel-related images depicting lifestyles and cultures, people and places, attitudes and environments from around the world, including a comprehensive section on Great Britain.

## PPL (Photo Agency) Ltd

The Street, Bookers Yard, Walberton, Arundel BN18 0PF

📞 01243 555561   📠 01243 555562

www.pplmedia.com

Contacts *Barry Pickthall, Richard Johnson*

Two million pictures covering watersports, sub-aqua, business and commerce, travel and tourism; pictures of yesteryear and a fast growing archive on Sussex and the home counties. Pictures available in high resolution directly from the website.

## Premaphotos Wildlife

Amberstone, 1 Kirland Road, Bodmin PL30 5JQ

📞 01208 78258

enquiries@premaphotos.com

www.premaphotos.com

Contact *Jean Preston-Mafham, Library Manager*

Natural history worldwide. Subjects include flowering and non-flowering plants, fungi, slime moulds, fruits and seeds, galls, leaf mines, seashore life, mammals, birds, reptiles, amphibians, insects, spiders, habitats, scenery and cultivated cacti. Commissions undertaken. Searchable website.

## Professional Sport UK Ltd

18–19 Shaftesbury Quay, Hertford SG14 1SF

📞 01992 505000   📠 01992 505020

pictures@prosport.co.uk

www.professionalsport.com

Photographic coverage of tennis, soccer, athletics, golf, cricket, rugby, winter sports and many minor sports. Major international events including the Olympic Games, World Cup soccer and all Grand Slam tennis events. Also news and feature material supplied worldwide. Online photo archive; photo transmission services available for editorial and advertising.

## Public Record Office Image Library
▷ The National Archives

## Punch Cartoons

The Mansion, Bletchley Park, Bletchley MK3 6EB

📞 0845 658 2470

info@createonline.net

www.punchcartoons.com

www.punch.co.uk

Owner *Punch Limited*

Choose online from half a million images published from 1841 onwards. 'The world's largest cartoon archive' with a vast subject range including social history, politics, fashion, famous faces and much more, all drawn by some of the

worlds' greatest cartoonists such as Sir John Tenniel, E.H. Shepard and Norman Thelwell.

## The Raymond Mander & Joe Mitchenson Theatre Collection

Jerwood Library of the Performing Arts, Trinity College of Music, King Charles Court, Old Royal Naval College, London SE10 9JF

📞 020 8305 4426   📠 020 8305 9426

kdavis@tcm.ac.uk

www.mander-and-mitchenson.co.uk

Archive Officer/Cataloguer *Kristy Davis*

Enormous collection covering all aspects of the theatre: plays, actors, dramatists, music hall, theatres, singers, composers, etc. Visitors welcome by appointment.

## Redferns Music Picture Library

21–31 Woodfield Road, London W9 2BA

📞 020 3227 2720   📠 020 7266 2414

info@redferns.com

www.gettyimages.com/redferns

Picture library covering every aspect of music, from 18th century classical to present day pop. Over one million archived artists and other subjects including musical instruments, recording studios, crowd scenes, festivals, etc. Over 300,000 images available on the website. 'Contact us for free lightbox searches.'

## Report Digital

4 Clarence Road, Stratford-on-Avon CV37 9DL

📞 01789 262151

info@reportdigital.co.uk

www.reportdigital.co.uk

A collection of 70,000 images. Subjects available online include work issues and occupations, leisure, economy, health, education, politics, social issues, protest, trades union, environmental issues, culture. An increasing coverage of international issues of globalization, migration and climate change. Pictures are available online; search, browse and download high resolution images. Member of **BAPLA**.

## Retna Pictures Ltd

Units 1b, Farm Lane Trading Estate, 101 Farm Lane, London SW6 1QJ

📞 0845 034 0645   📠 0845 034 0646

info@retna.co.uk

sales@retna.co.uk

www.retna.co.uk

Established 1978, Retna Pictures Ltd is a leading picture agency with two libraries: celebrity/music and lifestyle. The former specializes in images of international and national celebrities, music from the 1960s to the current day, films and personalities. The lifestyle library specializes in people, family life, work, leisure and food.

Both libraries are constantly receiving new material from established and up and coming photographers.

## Rex Features Ltd

18 Vine Hill, London EC1R 5DZ

☎ 020 7278 7294    📠 020 7696 0974

enquiries@rexfeatures.com

www.rexfeatures.com

Contact *Sales Manager*

Extensive picture library established in the 1950s. Daily coverage of news, politics, personalities, showbusiness, glamour, humour, art, medicine, science, landscapes, royalty, etc.

## RHS Lindley Library

Lindley Hall, 80 Vincent Square, London
SW1P 2PE

☎ 020 7821 3050    📠 020 7821 3022

library.london@rhs.org.uk

www.rhs.org.uk/About-Us/RHS-Lindley-Library

Contact *Picture Librarian*

22,000 original drawings and approx. 8,000 books with hand-coloured plates of botanical illustrations. Appointments are necessary to visit the collection of unpublished material; all photography is carried out by their own photographer.

## RIDA Photolibrary

▷ **GeoScience Features**

## Robbie Jack Photography

45 Church Road, London W7 3BD

☎ 020 8567 9616    📠 020 8567 9616

robbie@robbiejack.com

www.robbiejack.com

Contact *Robbie Jack*

Built up over the last 25 years, the library contains over 500,000 colour images of the performing arts – theatre, dance, opera and music. Includes West End shows, the RSC and Royal National Theatre productions, English National Opera and Royal Opera. The dance section contains images of the Royal Ballet, English National Ballet, the Rambert Dance Company, plus many foreign companies. Also holds the largest selection of colour material from the Edinburgh International Festival. Researchers are welcome to visit by appointment. Colour transparencies as digital images are being added to on a daily basis.

## Robert Forsythe Picture Library

16 Lime Grove, Prudhoe NE42 6PR

☎ 01661 834511    📠 01661 834511

robert@forsythe.demon.co.uk

www.forsythe.demon.co.uk

Contacts *Robert Forsythe, Fiona Forsythe*

30,000 transparencies of industrial and transport heritage; plus a unique collection of 250,000

items of related publicity ephemera from 1945 (now largely located in the **National Railway Museum** in York). Image finding service available. Robert Forsythe is a transport/industrial heritage historian and consultant. Nationwide coverage, particularly strong on northern Britain. A bibliography of published material is available.

## Robert Harding World Imagery

Berkshire House, Queen Street, Maidenhead
SL6 1NF

☎ 020 7478 4000    📠 020 7478 4161

info@robertharding.com

www.robertharding.com

A leading source of stock photography with over two million colour images covering a wide range of subjects – worldwide travel and culture, geography and landscapes, people and lifestyle, architecture. Rights protected and royalty-free images. Visitors welcome; telephone or visit the website.

## The Roger Mann Collection

Wensley Court, 48 Barton Road, Torquay
TQ1 4DW

☎ 01803 323868    📠 01803 616448

rogermann48bart@aol.com

www.therogermanncollection.co.uk

Contact *R.F. Mann*

Comprehensive collection of cricket photographs covering the period 1750 to 1945. The photographs feature most of the first-class players, teams, Test match action and overseas tours of the period. This collection includes almost 2,000 original match scorecards, cartoons, images, prints, postcards, cigarette cards, letters and the personal memorabilia of many of the best-known players of the time. Also some coverage of the period 1946 to 1970.

## The Ronald Grant Archive

The Master's House, 2 Dugard Way, off Renfrew Road, London SE11 4TH

☎ 020 7840 2200    📠 020 7840 2299

pixdesk@rgapix.com

www.ronaldgrantarchive.com

Colour and b&w coverage (including stills) of the motion picture industry throughout its history. Smaller collections on theatre, variety, television and popular music.

## Royal Air Force Museum

Grahame Park Way, Hendon, London NW9 5LL

☎ 020 8205 2266    📠 020 8200 1751

photographic@rafmuseum.org

www.rafmuseum.org

www.rafmuseumphotos.com

About a quarter of a million images, mostly b&w, with around 1,500 colour in all formats, on the history of aviation. Particularly strong on the activities of the Royal Air Force from the 1870s to 1970s. Researchers are requested to enquire in writing only.

## The Royal Collection, Photographic Services

St. James's Palace, London SW1A 1JR

☎ 020 7839 1377   🖷 020 7024 5643

picturelibrary@royalcollection.org.uk

www.royalcollection.org.uk

www.royal.gov.uk

Contacts *Shruti Patel, Karen Lawson*

Photographic material of items in the Royal Collection, particularly oil paintings, drawings and watercolours, works of art, and interiors and exteriors of royal residences.

## Royal Geographical Society Picture Library

1 Kensington Gore, London SW7 2AR

☎ 020 7591 3060   🖷 020 7591 3001

www.rgs.org/images

A strong source of geographical images, both historical and modern, showing the world through the eyes of photographers and explorers dating from the 1830s to the present day. The RGS Contempory Collection provides up-to-date transparencies from around the world, highlighting aspects of cultural activity, environmental phenomena, anthropology, architectural design, travel, mountaineering and exploration. The RGS Picture Library offers a professional and comprehensive service for both commercial and academic use.

## Royal Photographic Society Collection

▷ Science & Society Picture Library

## RSPB Images

The Old Dairy, Broadfield Road, Sheffield S8 0XQ

☎ 0114 258 0001   🖷 0114 258 0101

info@rspb-images.com

www.rspb-images.com

RSPB Images represents some of the UK's leading wildlife photographers; holds thousands of images of insects, mammals, plants, people, landscapes and abstracts as well as environmental and conservation issues. 'Our site has excellent user facilities and high resolution downloads are available.' Experienced staff will assist with picture searches.

## RSPCA Photolibrary

RSPCA Trading Limited, Wilberforce Way, Southwater, Horsham RH13 9RS

☎ 0300 123 0150

bmurray@rspcaphotolibrary.com

www.rspcaphotolibrary.com

Picture Requests *Becky Murray*

With over 150,000 colour images, the RSPCA Photolibrary has a comprehensive collection of natural history images whose subjects include mammals, birds, domestic and farm animals, amphibians, insects and the environment, as well as a unique photographic record of the RSPCA's work. Also includes the Wild Images collection. Downloadable website and email lightboxes. No search fees.

## Russia and Eastern Images

'Sonning', Cheapside Lane, Denham, Uxbridge UB9 5AE

☎ 01895 833508

easteuropix@btinternet.com

www.easteuropix.com

Architecture, cities, landscapes, people and travel images of Russia and the former Soviet Union. Considerable background knowledge available and Russian language spoken.

## S&O Mathews Photography

Little Pitt Place, Brighstone, Isle of Wight PO30 4DZ

oliver@mathews-photography.com

www.mathews-photography.com

Specialist stock library of fine photographs of botanical and horticultural subjects, plants, plant portraits and plant associations, as well as gardens, including views, details and features.

## Sally and Richard Greenhill

357 Liverpool Road, London N1 1NL

☎ 020 7607 8549

sr.greenhill@virgin.net

www.srgreenhill.co.uk

Photo Librarian *Denise Lalonde*

Social documentary photography in colour and b&w of working lives: pregnancy and birth, child development, education, work, old people, medical, urban. Also Modern China, 1971 to the present; most London statues. Some material from Borneo, USA, India, Israel, Philippines and Sri Lanka.

## scenicireland.com

17 Clarence Street, Belfast BT2 8DY

☎ 028 9024 5038   🖷 028 9023 1942

sales@scenicireland.com

www.scenicireland.com

Contact *Christopher Hill*

A comprehensive collection of scenic landscapes of Ireland. Every aspect of Irish life is shown, concentrating on the positive. Also large miscellaneous section. The website contains over 15,000 images available to buy online.

## Science & Society Picture Library

Science Museum (North Entrance), Exhibition Road, London SW7 2DD

☎ 020 7942 4400   🖷 020 7942 4401

piclib@nmsi.ac.uk
www.scienceandsociety.co.uk

'The Science & Society Picture Library has one of the widest ranges of photographs, paintings, prints, posters and objects in the world.' The images come from the Science Museum, the **National Railway Museum** and the National Media Museum (formerly the National Museum of Photography, Film & Television) which now includes the Royal Photographic Society Collection. Images are available as high or low resolution files via email, FTP or direct from the website.

## Science Museum
▷ Science & Society Picture Library

## Science Photo Library (SPL)
327–329 Harrow Road, London W9 3RB
☎ 020 7432 1100    🖷 020 7286 8668
info@sciencephoto.com
www.sciencephoto.com

Subjects covered include the human body, health and medicine, research, genetics, technology and industry, space exploration and astronomy, earth science, satellite imagery, environment, flowers, plants and gardens, nature and wildlife and the history of science. The whole collection, more than 300,000 images, is available online.

## Sir John Benjamin Stone
▷ Birmingham Library Services under Library Services

## Skishoot
Hall Place, Upper Woodcott, Whitchurch RG28 7PY
☎ 01635 255527    🖷 01635 255528
pictures@skishoot.co.uk
www.skishoot.co.uk
Contact *Claire Randall*

Skishoot ski and snowboarding picture library has 500,000 images. Offshoot travel library specializes in France. Also produces the ski information website: www.welove2ski.com

## The Skyscan Photolibrary
Oak House, Toddington, Cheltenham GL54 5BY
☎ 01242 621357/07941 445993    🖷 01242 621343
info@skyscan.co.uk
www.skyscan.co.uk
Contact *Brenda Marks*

As well as the Skyscan Photolibrary collection of unique balloon's-eye views of Britain, the library includes the work of photographers from across the aviation spectrum; air to ground, aviation, aerial sports – 'in fact, anything aerial!' Links have been built with photographers across the world; photographs can be handled on an agency basis and held in house, or as a brokerage where the collection stays with the photographer; terms

50/50 for both. Commissioned photography arranged. Enquiries welcome. Historic aerial collections are now held in house and a search service provided to clients seeking evidence for land and legal enquiries is available.

## Snookerimages (Eric Whitehead Photography)
7 Brow Close, Bowness on Windermere LA23 2HA
☎ 01539 448894
snooker@snookerimages.co.uk
www.snookerimages.co.uk
www.snookerimages-pictures.com
Contact *Eric Whitehead*

Over 30,000 images of snooker from 1982 to the present day.

## Sotheby's Picture Library
34–35 New Bond Street, London W1A 2AA
☎ 020 7293 5383    🖷 020 7293 5062
piclib@sothebys.com
www.sothebys.com
Contact *Joanna Ling*

The library consists of over one million subjects sold at Sotheby's. Images from the 15th to the 20th century. Oils, drawings, watercolours, prints and decorative items as well as the unique archive of Cecil Beaton photographs. 'Happy to do searches or, alternatively, visitors are welcome by appointment.'

## South American Pictures
48 Station Road, Woodbridge IP12 4AT
☎ 01394 383963
info@southamericanpictures.com
www.southamericanpictures.com
Contact *Marion Morrison*

Colour and b&w images of South/Central America, Cuba, Mexico, New Mexico (USA), Dominican Republic and Haiti, including archaeology and the Amazon. There is an archival section, with pictures and documents from most countries. Now with 40 contributing photographers. All images are supplied digitally.

## Specialist Stock
1 Glen Cottages, Sandy Lane, Abbots Leigh, Bristol BS8 3SE
☎ 01275 375520    🖷 0705 061 3938
info@specialiststock.com
www.specialiststock.com
Contact *Julie-anne Wilce*

Founded 1970. High profile photo library specializing in environment, the Third World, social issues and nature. Represents 15 leading European and US agencies and over 400 photographers as well as the United Nations Environment Programme (UNEP) archive, the Christian Aid collection and a growing selection of images from Woodfall Wild Images. The website has nearly one million images online, ready for instant download.

## Starstock
> Photoshot

## Stay Still
> Photoshot

## Stockfile

5 High Street, Sunningdale SL5 0LX
℡ 01344 872249
info@stockfile.co.uk
www.stockfile.co.uk

Specialist cycling collection with emphasis on mountain biking. Expanding adventure sports section covering snow, land, air and water activities.

## Stockscotland.com

The Grange House, Lochside, Lairg IV27 4EG
℡ 01549 402295
info@stockscotland.com
www.stockscotland.com
Contact *Hugh Webster*

150,000 Scottish images with 10,000 currently available online.

## Sue Cunningham Photographic

56 Chatham Road, Kingston upon Thames
KT1 3AA
℡ 020 8541 3024
info@scphotographic.com
www.scphotographic.com
Contacts *Patrick Cunningham, Sue Cunningham*

Extensive coverage of many geographical areas: South America (especially Brazil), Eastern Europe from the Baltic to the Balkans, various African countries, Western Europe including the UK. Colour and b&w.

## Sylvia Cordaiy Photo Library

45 Rotherstone, Devizes SN10 2DD
℡ 01380 728328   ℻ 01380 728328
info@sylvia-cordaiy.com
www.sylvia-cordaiy.com

Over 180 countries on file from the obscure to main stock images – Africa, North, Central and South America, Asia, Atlantic, Indian and Pacific Ocean islands, Australasia, Europe, polar regions. Covers travel, architecture, ancient civilizations, world heritage sites, people worldwide, environment, wildlife, natural history, Antarctica, domestic pets, livestock, marine biology, veterinary treatment, equestrian, ornithology, flowers. UK files cover cities, towns villages, coastal and rural scenes, London collection. Transport, railways, shipping and aircraft (military and civilian). Aerial photography.

Backgrounds and abstracts. Also the Paul Kaye B/W archive. All images available by digital delivery.

## Tate Images

The Lodge, Tate Britain, Millbank, London SW1P 4RG
℡ 020 7887 8871/8890/4933   ℻ 020 7887 8805
tate.images@tate.org.uk
www.tate-images.com
Contact *Alison Fern*

Tate is one of the world's leading visual arts organizations and its collection encompasses the national collection of historic British art from 1500, including iconic masterpieces by Gainsborough, Constable, Turner, David Hockney and Henry Moore, and the national collection of international modern art that includes works by Dali, Picasso, Matisse, Rothko, Emin, Hirst, Warhol and Andreas Gursky. This vast collection of art imagery is available from the Tate Images website above. Clients can browse the entire Tate collection using a keyword-based search engine; select images and save their selections to a personal light box; download low-resolution imagery for layout purposes and buy images online, 24/7.

## Tessa Traeger Picture Library

7 Rossetti Studios, 72 Flood Street, London SW3 5TF
℡ 020 7352 3641   ℻ 020 7352 4846
info@tessatraeger.com
www.tessatraeger.com

Food, gardens, travel and artists.

## Theatre Museum
> V&A Images

## Thoroughbred Photography Ltd

The Hornbeams, 2 The Street, Worlington IP28 8RU
℡ 01638 713944
mail@thoroughbredphoto.com
www.thoroughbredphoto.com
Contacts *Gill Jones, Laura Green*

Photography by Trevor Jones. Extensive library of high-quality colour images depicting all aspects of thoroughbred horseracing dating from 1987. Major group races, English Classics, studs, stallions, mares and foals, early morning scenes, personalities, jockeys, trainers and prominent owners. Also international work: USA Breeders' Cup, Arc de Triomphe, French Classics, Irish Derby, Dubai World Cup Racing and scenes; and more unusual scenes such as racing on the sands at low tide, Ireland, and on the frozen lake at St Moritz. Visitors by appointment.

## TopFoto

PO Box 33, Edenbridge TN8 5PF
☎ 01732 863939    🖷 01732 860215
admin@topfoto.co.uk
www.topfoto.co.uk
Contact *Alan Smith*

International editorial distributor with over two million pictures available online to download and 12 million in hard copy, representing 40 leading suppliers.

## Travel Ink Photo Library

The Old Coach House, 14 High Street, Goring on Thames, Nr Reading RG8 9AR
☎ 01491 873011    🖷 01491 875558
info@travel-ink.co.uk
www.travel-ink.co.uk
Library Manager *Frances Honnor*
Sales and Administration *Felicity Bazell*

A collection of over 120,000 travel, tourism and lifestyle images, carefully edited and constantly updated, from countries worldwide. Specialist collections from the UK, Greece, France, Far East and Caribbean. The website offers a fully captioned and searchable selection of over 31,000 images.

## Tropix Photo Library

44 Woodbines Avenue, Kingston upon Thames KT1 2AY
☎ 020 8546 0823
veronica@tropix.co.uk
www.tropix.co.uk
Managing Director *Veronica Birley*

Specializes in images of developing nations: travel and editorial pictures emphasizing the attractive and progressive. Fully searchable, keyworded website. Evocative photos concerning the economies, environment, culture and society of 100+ countries across Africa, Central and South America, Caribbean, Eastern Europe, Middle East, Indian sub-continent, South East Asia, CIS and Far East. Worldwide travel collections also include UK, Europe, North America and Antarctica. Assignment photography available. All photos supplied with detailed captions. Established 1982. **BAPLA** member.

## True North

Louper Weir, Ghyll Head, Windermere LA23 3LN
☎ 07941 630420
www.northpix.co.uk
Contact *John Morrison*

The collection features the life and landscape of the north of England, photographed by John Morrison. No other photographer's work required.

## U.P.P.A.

▷ Photoshot

## Ulster Museum Picture Library

Ulster Folk & Transport Museum, Cultra, Holywood BT18 0EU
☎ 028 9042 8428
www.ulstermuseum.org.uk

Specialist subjects: art (fine and decorative, late 17th–20th century), particularly Irish art, archaeology, ethnography, treasures from the Armada shipwrecks, geology, botany, zoology, local history and industrial archaeology.

## United Nations Environment Programme (UNEP)

▷ Specialist Stock

## V&A Images

Victoria and Albert Museum, Cromwell Road, London SW7 2RL
☎ 020 7942 2487 (commercial)/2479 (academic)
🖷 020 7942 2482
vaimages@vam.ac.uk
www.vandaimages.com
www.vandaprints.com
Commercial Sales Manager *Elaine Lucas*
Academic Image Resource Manager *Roxanne Peters*

A vast collection of photographs from the world's largest museum of decorative and applied arts, reflecting culture and lifestyle spanning over 1,000 years of history to the present time. Digital delivery of contemporary and historical textiles, costumes and fashions, ceramics, furniture, metalwork, glass, sculpture, toys, and games, design and photographs from around the world. Unique photographs include 1960s fashion by John French, Harry Hammond's behind the scenes pop idols, Houston Rogers theatrical world of the 1930s to 1970s, images of royalty by Lafayette, Cecil Beaton, and the 19th century pioneer photographers. Images from the **National Art Library** (see entry under Library Services), the Library of Art & Design, the Theatre Museum and Museum of Childhood are readily available. For bespoke prints, a selection of over 5,000 images are available from the V&A's collections at www.vandaprints.com

## V.K. Guy Ltd

Browhead Cottage, Troutbeck, Windermere LA23 1PG
☎ 01539 433519    🖷 01539 432971
admin@vkguy.co.uk
www.vkguy.co.uk
Contacts *Vic Guy, Mike Guy*

25,000 5x4" UK and Ireland scenic transparencies from Shetland in the north to the Channel Isles in the south, suitable for tourism brochures, calendars, etc. Can be supplied digitally and browsed through via the online database.

## Vaughan Williams Memorial Library

English Folk Dance and Song Society, Cecil Sharp House, 2 Regent's Park Road, London NW1 7AY
☎ 020 7485 2206 ext. 29/33   🖷 020 7284 0523
library@efdss.org
www.efdss.org

Mainly b&w coverage of traditional/folk music, dance and customs worldwide, focusing on Britain and other English-speaking nations. Photographs date from the late 19th century to the present day.

## Vin Mag Archive Ltd

84–90 Digby Road, London E9 6HX
☎ 020 8533 7588   🖷 020 8533 7283
piclib@vinmagarchive.com
www.vinmagarchive.com
Contact *Angela Maguire*

Wholly owned subsidiary of the Vintage Magazine Company. 'A unique source of images covering just about any subject, period, theme and location.' Also a collection of half a million movie, TV, celebrity and sports images available online. Contact *Angela Maguire* for code and password. See also entry under *Library Services*.

## Warwickshire Photographic Survey
▷ Birmingham Library Services under Library Services

## Waterways Photo Library

39 Manor Court Road, Hanwell, London W7 3EJ
☎ 020 8840 1659   🖷 020 8567 0605
watphot39@aol.com
www.waterwaysphotolibrary.com
Contact *Derek Pratt*

A specialist photo library on all aspects of Britain's inland waterways. Top-quality 35mm and medium-format colour transparencies, plus a large collection of b&w and an increasing collection of digital photography. Rivers and canals, bridges, locks, aqueducts, tunnels and waterside buildings. Town and countryside scenes, canal art, waterway holidays, boating, fishing, windmills, watermills, watersports and wildlife.

## Weimar Archive
▷ Mary Evans Picture Library

## Wellcome Images

183 Euston Road, London NW1 2BE
☎ 020 7611 8348   🖷 020 7611 8577
images@wellcome.ac.uk
http://images.wellcome.ac.uk

Approximately 180,000 images on the history of medicine and human culture worldwide, including religion, astronomy, botany, genetics, landscape and cell biology.

## Werner Forman Archive Ltd

36 Camden Square, London NW1 9XA
☎ 020 7267 1034   🖷 020 7267 6026
wfa@btinternet.com
www.werner-forman-archive.com

Colour and b&w coverage of ancient civilizations, oriental and primitive societies around the world. A number of rare collections. Searchable website.

## West Cornwall Photographic Archive
▷ Morrab Library under Library Services

## Wild Images
▷ RSPCA Photolibrary

## Wild Places Photography
▷ Chris Howes

## The Wilson Photographic Collection
▷ Dundee Central Library under Library Services

## Woodfall Wild Images
▷ Photoshot

## World Illustrated
▷ Photoshot

## World Pictures
▷ Photoshot

## York Archaeological Trust Picture Library

47 Aldwark, York YO1 7BX
☎ 01904 663006   🖷 01904 663024
ckyriacou@yorkat.co.uk
www.yorkarchaeology.co.uk

Specialist library of rediscovered artifacts, historic buildings and excavations, presented by the creators of the highly acclaimed Jorvik Viking Centre. The main emphasis is on the Roman, Anglo-Saxon, Medieval and Viking periods.

# Press Cuttings Agencies

## Cision UK

Cision House, 16–22 Baltic Street West, London
EC1Y OUL
℡ 0870 736 0010
info.uk@cision.com
www.cision.com

Formerly Romeike, Cision monitors national
and international dailies and Sundays, provincial
papers, consumer magazines, trade and technical
journals, teletext services as well as national radio
and TV networks. Back research, advertising
checking and internet monitoring, plus analysis
and editorial summary service available.

## Durrants

Discovery House, 28–42 Banner Street, London
EC1Y 8QE
℡ 020 7674 0200    ℻ 020 7674 0222
sales@durrants.co.uk
www.durrants.co.uk

Wide coverage of all print media sectors including
foreign press plus internet, newswire and
broadcast monitoring. High speed, early morning
press cuttings from the national press by web
or email. Overnight delivery via courier to most
areas or first-class mail. Well presented, laser
printed, A4 cuttings. Rates on application.

## International Press-Cutting Bureau

224–236 Walworth Road, London SE17 1JE
℡ 020 7708 2113    ℻ 020 7701 4489
info@ipcb.co.uk
www.ipcb.co.uk
Contact *Robert Podro*

Covers national, provincial, trade, technical and
magazine press. Cuttings are normally sent twice
weekly by first-class post. Charges start at £80 per
month + £1.50 per cutting.

## We Find It (Press Clippings)

40 Galwally Avenue, Belfast BT8 7AJ
℡ 028 9064 6008    ℻ 028 9064 6008
Contact *Avril Forsythe*

Specializes in Northern Ireland press and
magazines, both national and provincial. Rates on
application.

# Festivals

## Aberdeen Arts Carnival

Aberdeen Arts Centre, 33 King Street, Aberdeen
AB24 5AA
☏ 01224 635208
www.aberdeenartscentre.org.uk
Venue Manager *Paula Gibson*

Performances by local amateurs and professional
theatre companies. Exhibitions, arts workshops in
drama, music, art, dance and creative writing take
place each summer during the school holidays.
Drama workshops also take place throughout the
year.

## The Aldeburgh Literary Festival

44 High Street, Aldeburgh IP15 5AB
☏ 01728 452389    🖷 01728 452389
johnandmary@aldeburghbookshop.co.uk
www.aldeburghbookshop.co.uk
Festival Organizers *John & Mary James*

Founded 2002. Held on the first weekend in
March featuring a programme of talks. Previous
speakers have included Beryl Bainbridge,
Alan Bennett, Craig Brown, Richard Dawkins,
Lady Antonia Fraser, Michael Frayn, Anthony
Horowitz, P.D. James, Doris Lessing, David
Lodge, Ian McEwan, Harold Pinter, Matt Ridley,
Alexander McCall Smith, Claire Tomalin, Salley
Vickers, A.N. Wilson.

## Aldeburgh Poetry Festival

The Poetry Trust, The Cut, 9 New Cut,
Halesworth IP19 8BY
☏ 01986 835950    🖷 01986 874524
info@thepoetrytrust.org
www.thepoetrytrust.org
Director *Naomi Jaffa*

Founded 1989. One of the UK's biggest
celebrations of international contemporary
poetry, held in the small coastal town of
Aldeburgh annually in early November (5th–7th
in 2010). Features readings, performances, craft
talks, public masterclass, workshops, discussions
and a family event. Includes a reading by the
winner of the annual **Aldeburgh First Collection
Prize** (see entry under *Prizes*).

## Arthur Miller Centre Literary Festival at UEA

University of East Anglia, School of American
Studies, Norwich NR4 7TJ
☏ 01603 592286
n.orr@uea.ac.uk
www.uea.ac.uk/litfest
Contact *Natalie Mitchell*

Founded 1991. Annual festival, held in the
autumn.

## Aspects Literature Festival

North Down Borough Council, Town Hall, The
Castle, Bangor BT20 4BT
☏ 028 9127 8032    🖷 028 9127 1370
gail.prentice@northdown.gov.uk
www.aspectsfestival.co.uk
Festival Director/Arts Officer *Gail Prentice*

Founded 1991. 'Ireland's Premier Literary Festival'
is held at the end of September and celebrates the
richness and diversity of living Irish writers with
occasional special features on past generations. It
draws upon all disciplines – fiction (of all types),
poetry, theatre, non-fiction, cinema, song-writing,
etc. It also includes a day of writing for young
readers and sends writers to visit local schools
during the festival. Highlights of recent festivals
include appearances by Bernard MacLaverty,
Marion Keyes, Alice Taylor, Frank Delaney,
Seamus Heaney, Brian Keenan and Fergal Keane.

## Aye Write! Bank of Scotland Book Festival

The Mitchell Library, North Street, Charing
Cross, Glasgow G3 7DN
☏ 0141 287 5114
www.ayewrite.com
Festival Director *Karen Cunningham*
Programme Director *Andrew Kelly*

Launched 2005. Annual festival, held in March,
which aims to increase the use of libraries,
encourage a love of books, reading and writing
and an awareness of Glasgow's writing heritage.
The 2010 festival guests included Germaine
Greer, Joanna Trollope, Michael Mansfield, Carol
Ann Duffy, David Dimbleby and William Boyd.

## Bath Festival of Children's Literature

PO Box 4123, Bath BA1 0FR
☎ 01225 462231 (tickets)   🖷 01225 445551
info@bathkidslitfest.co.uk
www.bathkidslitfest.co.uk
Festival Directors *John McLay, Gill McLay*

Founded 2007. Annual event celebrating
children's books and reading, held from 24th
September to 3rd October in 2010. Over 80
events for readers aged up to 16. Guest authors
have included Jacqueline Wilson, Eoin Colfer,
Anthony Horowitz, Michael Morpurgo, Michael
Rosen, Meg Cabot, Darren Shan, Lauren Child,
Louise Rennison, Francesca Simon, Michelle
Paver, Charlie Higson, Rick Riordan, Allan
Ahlberg and Harry Hill. Sponsored by *The Daily
Telegraph* and Waterstone's Booksellers.

## Bath Literature Festival

Bath Festivals, Abbey Chambers, Kingston
Buildings, Bath BA1 1NT
☎ 01225 462231   🖷 01225 445551
info@bathfestivals.org.uk
www.bathlitfest.org.uk
*Box Office:* Bath Festivals Box Office, 2 Church
Street, Abbey Green, Bath BA1 1NL
☎ 01225 463362
boxoffice@bathfestivals.org.uk
Artistic Director *James Runcie*

Founded 1995. This annual festival (held in
February and March) programmes over 100
different literary events from debates and
lectures to readers' groups and workshops in
venues throughout the city. In addition, there
are a number of events for children and young
people. Previous featured writers include Martin
Amis, Terry Pratchett, Margaret Drabble,
Joanna Trollope, A.C. Grayling, Steven Berkoff,
Hermione Lee.

## BayLit Festival

Academi, Mount Stuart House, Mount Stuart
Square, Cardiff CF10 5FQ
☎ 029 2047 2266   🖷 029 2049 2930
post@academi.org
www.academi.org
Chief Executive *Peter Finch*

Annual literature festival held in Cardiff Bay
(October in 2010), featuring writers from Wales
and beyond. Lectures, readings, performances,
workshops and book launches in English and
Welsh. Dates and further details available on
the Academi website. Writers who appeared in
previous festivals include Ian McMillan, Andrew
Motion, Gillian Clarke, Fay Weldon, Simon Singh,
Howard Marks and Will Self.

## Belfast Festival at Queen's
▷ The Ulster Bank Belfast Festival at Queen's

## Beverley Literature Festival

Wordquake, Libraries and Information, Council
Offices, Skirlaugh HU11 5HN
☎ 01482 392745   🖷 01482 392710
john@bevlit.org
www.bevlit.org
Festival Director *John Clarke*

Founded 2002. Annual October festival which has
acquired 'a national reputation for commissioning
and hosting quality poetry events'. Also covers
all other literary genres and mixes readings and
performances with author-led readers' groups and
creative writing workshops. Held in intimate and
historic venues across Beverley.

## Birmingham Book Festival

c/o Unit 116, The Custard Factory, Gibb Street,
Birmingham B9 4AA
☎ 0121 246 2770   🖷 0121 246 2771
jonathan@birminghambookfestival.org
www.birminghambookfestival.org
Contact *Jonathan Davidson*

Founded 1999. Annual literature festival held
during October at venues around Birmingham.
Also promotes events throughout the year.
Includes performances, lectures, discussion
events and workshops.

## Book Now – Richmond Literature Festival

The Arts Service, Orleans House Gallery,
Riverside, Twickenham TW1 3DJ
artsinfo@richmond.gov.uk
www.richmond.gov.uk/literature

Founded 1992. Annual three-to-four week
literature festival held in November, delivered by
the Arts Service of Richmond Borough. Plays host
to leading writers.

## Borders Book Festival

Harmony House, St Mary's Road, Melrose
TD6 9LJ
☎ 07929 435575
info@bordersbookfestival.org
www.bordersbookfestival.org
Festival Coordinator *Paula Ogilvie*

Founded in 2004. Annual literary festival held
over four days during the third weekend in
June (17th–20th in 2010). The event brings
internationally acclaimed and new authors to
the gardens of the National Trust for Scotland's
splendid Georgian property, Harmony House, and
other locations in Melrose, for discussions and
performances to audiences of all ages and tastes.
Past speakers include Ian Rankin, Iain Banks,
Michael Palin, Rory Bremner, Alexander McCall
Smith, Sarah Raven, Roy Hattersley, Claire
Messud, Sheila Hancock, William Dalrymple
and A.L. Kennedy. Festival bookshop is open
throughout the festival dates. Free mailing list.

## Bournemouth Literary Festival

☎ 01202 417535
info@bournemouthliteraryfestival.co.uk
www.bournemouthliteraryfestival.co.uk
Director & Founder *Lillian Avon*

Founded in 2004, the festival fuses literature with performing arts, celebrating the international world of words in venues all over Bournemouth.

## Brighton Festival

Brighton Dome & Festival Ltd, 12a Pavilion Buildings, Castle Square, Brighton BN1 1EE
☎ 01273 709709    🖷 01273 707540
info@brightonfestival.org
www.brightonfestival.org
Contact *Festival Office*

Founded 1966. For 24 days every May, Brighton hosts England's largest mixed arts festival. Music, dance, theatre, film, opera, literature, comedy and exhibitions. Literary enquiries will be passed to the literature officer. Deadline: October for following May.

## Bristol Poetry Festival

▷ Poetry Can under Organizations of Interest to Poets

## Broadstairs Dickens Festival

Dickens House, 2 Victoria Parade, Broadstairs CT10 1QD
☎ 01843 861232    🖷 01843 871133
cramptontower@hotmail.com
www.broadstairsdickensfestival.co.uk
Festival Coordinator *Peter Shaw*

Founded in 1937 to commemorate the 100th anniversary of Charles Dickens' first visit to Broadstairs in 1837, which he continued to visit until 1859. The festival lasts for five days in June and events include a parade, a performance of a Dickens play, melodramas, Dickens readings, a Victorian cricket match, Victorian bathing parties, talks, music hall, five-day Victorian country fair. Costumed Dickensian ladies in crinolines with top-hatted escorts promenade during the festival.

## Budleigh Salterton Literary Festival

11 East Terrace, Budleigh Salterton EX9 6PG
smward@mail.com
Contact *Artistic Director*

Annual festival held in September (24th–26th in 2010). Guests include Carol Ann Duffy, Anna Beer, Simon Brett, Ian Frost and Bill Studdiford, Susan Hill, Hilary Mantel, Sir Christopher Meyer, Ian Mortimer, Jeremy Musson, Mal Peet, John Sampson, Caroline Taggart, Alex Wade, Hugh Williams.

## Buxton Festival

3 The Square, Buxton SK17 6AZ
☎ 01298 70395    🖷 01298 72289
info@buxtonfestival.co.uk
www.buxtonfestival.co.uk
Chief Executive *Glyn Foley*
Press & Marketing Officer *Johanna Dollerson*
Literary Manager *Rosie Hughes*

Founded 1979. Held annually in July, this 19-day opera, literature and music festival includes a varied literary programme, attracting those with a broad interest in well-crafted writing, whether in biography, fiction, politics or personal memoir.

## Cambridge Wordfest

7 Downing Place, Cambridge CB3 2EL
☎ 01223 515335
admin@cambridgewordfest.co.uk
www.cambridgewordfest.co.uk
Festival Director *Cathy Moore*
Festival Patrons *Dame Gillian Beer, Rowan Pelling, Ali Smith*

An annual spring festival for writers as well as readers which takes place at various venues across Cambridge and the surrounding area. A weekend event packed with the best of contemporary writing, poetry, political debate, events for children and writing workshops. Also a one-day winter festival during the last weekend in November.

## Canterbury Festival

Christ Church Gate, The Precincts, Canterbury CT1 2EE
☎ 01227 452853    🖷 01227 781830
info@canterburyfestival.co.uk
www.canterburyfestival.co.uk
Festival Director *Rosie Turner*

Canterbury provides a unique location for the largest festival of arts and culture in the region (16th–30th October in 2010). Showcases local, national and international talent in classical and contemporary music, world theatre, opera, comedy, dance, walks and community events, plus an extended programme of literature and debate.

## Carnegie Sporting Words Festival

Harrogate International Festivals, Raglan House, Raglan Street, Harrogate HG1 1LE
☎ 01423 562303    🖷 01423 521264
erica@harrogate-festival.org.uk
www.sportingwords.co.uk
Contact *Erica Morris*

Founded in 2008, the Carnegie Sporting Words Festival (CSWF) takes place in Leeds and Harrogate over a weekend in the autumn. It is the UK's first festival celebrating the sporting word as expressed through books, newspapers, broadcast media and the web, and features sports celebrities, broadcasters, journalists and authors. Previous events have included a Sports Quiz Dinner hosted by BBC's Mark Pougatch, a seminar exploring the future of digital sports media and a Schools Activity Day at the Headingley Carnegie Stadium in partnership with BBC Sport.

## Centre for Creative & Performing Arts Spring Literary Festival at UEA

University of East Anglia, School of Literature and Creative Writing, Norwich NR4 7TJ
📞 01603 592286
n.orr@uea.ac.uk
www1.uea.ac.uk/cm/home/schools/hum/
booksandwriters/litfest
Contact *Natalie Mitchell*

Founded 1993. Annual event held in the spring.

## Charleston Festival

The Charleston Trust, Charleston, Nr Firle, Nr Lewes BN8 6LL
📞 01323 811626
info@charleston.org.uk
www.charleston.org.uk
Festival Programmer *Diana Reich*

Annual literary festival held in May over two weeks. Novelists, biographers, travel writers, broadcasters, poets, food writers, actors and artists gather at Charleston, the country home of the Bloomsbury Group. Past speakers include Margaret Atwood, Alan Bennett, Jeanette Winterson, Louis de Bernières, Patti Smith, Paula Rego, Clive James, Ali Smith, Roger McGough, Germaine Greer and Harold Pinter.

## Cheltenham Literature Festival
▷ The Times Cheltenham Literature Festival

## Chester Literature Festival

Chester Railway Station, 1st Floor West Wing Offices, Station Road, Chester CH1 3NT
📞 01244 405605    📠 01244 405601
info@chesterfestivals.co.uk
www.chesterfestivals.co.uk
Director *Tracy Lynn*

Founded 1989. Annual festival held in early October. Events include international and nationally known writers, as well as events by local literary groups. The performance section, the Chester Oyez, has performance poetry and storytelling, events for children and workshops for all ages. Free mailing list.

## Children's Books Ireland – Annual Festival
▷ Children's Books Ireland under Professional Associations

## CrimeFest

Basement Flat, 6 Rodney Place, Bristol BS8 4HY
info@crimefest.com
www.crimefest.com
Directors and co-hosts *Myles Allfrey, Adrian Muller*

Founded 2008. CrimeFest is an annual crime fiction convention which draws top crime novelists, readers (both casual and die-hard), editors, publishers and reviewers from around the world and gives delegates the opportunity to celebrate the genre in an informal atmosphere. The convention is made up of author panels and interviews; crime writing workshops and a pitch-an-agent strand; as well as a gala dinner that includes presentations of the **CrimeFest Last Laugh Award**, the **CrimeFest Sounds of Crime Awards** and the **CrimeFest Crime eBook Award** (see entries under *Prizes*).

## The Daphne du Maurier Festival of Arts & Literature

Cornwall Council, 39 Penwinnick Road, St Austell PL25 5DR
📞 01726 223398
www.dumaurierfestival.co.uk

Founded in 1997, this annual arts and literature festival is held in and around Fowey in Cornwall in May and includes a wide range of professional and community events.

## Dartington Literary Festival
▷ Ways With Words

## Derbyshire Literature Festival

c/o Arts Team, Derbyshire County Council, Cultural & Community Services Department, Alfreton Library, Severn Square, Alfreton DE55 7BQ
📞 01773 832497    📠 01773 831359
ann.wright@derbyshire.gov.uk
www.derbyshire.gov.uk
Festival Organizer *Ann Wright*

Founded 2000. The festival takes place in June every two years and covers the whole county. The festival programming includes all types of live literature events; performance poetry, theatre, talks, readers' groups, workshops, dramatized readings, signings, storytelling and literary trails, as well as a number of cross-art form events. The festival takes place in libraries and many other community venues including heritage centres, industrial buildings, stately homes, parks and moors, churches and schools and is specifically designed to reach as many different communities and geographical areas in the county as possible.

## Derwent Poetry Festival

Templar Poetry, PO Box 7082, Bakewell DE45 9AF
☎ 01629 582500
info@templarpoetry.co.uk
www.templarpoetry.co.uk

Poetry festival held in November. Guests at the third festival, held at Matlock Bath, included Maggie O'Dwyer, Jane Weir, David Morley, Paul Maddern, Pat Winslow and Katrina Naomi.

## Dorchester Festival

Dorchester Arts Centre, School Lane, The Grove, Dorchester DT1 1XR
☎ 01305 266926
enquiries@dorchesterarts.org.uk
www.dorchesterarts.org.uk
Artistic Director *Sharon Hayden*

Founded 1996. An annual large-scale mulitcultural festival held over late May Bank Holiday which includes performance, live music and visual arts with associated educational and community projects. Also three days of free events in various venues around the town, including literature and poetry. The theme for the 2010 festival is Africa and its wider cultural impact.

## Dumfries and Galloway Arts Festival

Gracefield Arts Centre, 28 Edinburgh Road, Dumfries DG1 1JQ
☎ 01387 260447
info@dgartsfestival.org.uk
www.dgartsfestival.org.uk
Chair *Dame Barbara Kelly*
Programme Planning & Development Officer *Peter Renwick*

Founded 1980. Annual week-long festival, held at the end of May, with a variety of events including classical, jazz and folk music, film, theatre, dance, literary events, exhibitions and children's events.

## Dundee Literary Festival

The Tower Building, Nethergate, Dundee DD1 4HN
☎ 01382 384413
literarydundee@gmail.com
www.literarydundee.co.uk
Festival Director *Anna Day*

Founded 2007. Annual literary festival held in June with a mix of internationally renowned writers, local authors, poetry and new writing. Speakers have included Philip Pullman, Jacqueline Wilson and Kirsty Gunn. Email to be added to the mailing list.

## Durham Book Festival

Durham City Arts, 2 The Cottages, Fowlers Yard, Back Silver Street, Durham DH1 3RA
☎ 0191 375 0763
enquiries@durhamcityarts.org.uk
www.durhamcityarts.org
Events Manager *Susie O'Hara-Sheader*

Founded 1989. Annual festival held in October at various locations in the city. Performances and readings plus workshops, cabaret, exhibitions and other events.

## Dylan Thomas Festival

Dylan Thomas Centre, Somerset Place, Swansea SA1 1RR
☎ 01792 463980    ⊞ 01792 463993
dylanthomas.lit@swansea.gov.uk
www.dylanthomas.com
Contacts *David Woolley, Jo Furber*

Two weeks of performances, talks, lectures, films, music, poetry, exhibitions and celebrity guests. The 2010 Festival (27th October–9th November) will include the award of the third **Dylan Thomas Prize** for young writers. The Dylan Thomas Centre also runs a year-round programme of literary events; please email for details.

## Edinburgh International Book Festival

5a Charlotte Square, Edinburgh EH2 4DR
☎ 0131 718 5666    ⊞ 0131 226 5335
admin@edbookfest.co.uk
www.edbookfest.co.uk
Director *Nick Barley*
Children and Education Programme Director *Sara Grady*
Programme Manager *Roland Gulliver*

Founded 1983. 'The world's largest and most dynamic annual book event' takes place across 17 days in Edinburgh each August. An extensive programme of over 750 events featuring 700 or more authors, showcasing the work of the world's top writers and thinkers to an audience of over 200,000 adults and children. Featuring workshops, readings, lectures and a high profile debate and discussion series, book signings, bookshops and cafés.

## Folkestone Book Festival

The Block, 65–69 Tontine Street, Folkestone CT20 1JR
☎ 01303 858500    ⊞ 01303 223761
www.folkestonebookfest.com
Festival Director *Roberta Spicer*

Annual festival held in November. Tours, discussions, talks, live performances and readings. Guests at the 2009 Festival included Jo Brand, Ion Trewin, Libby Purves, David Aaronovitch, Victoria Hislop, Andrew Roberts, Kate Mosse.

## Frome Festival

25 Market Place, Frome BA11 1AH
☎ 01373 453889

office@fromefestival.co.uk
www.fromefestival.co.uk

An annual festival held from the first Friday in July for ten days (9th–18th in 2010) celebrating all aspects of visual and performing arts and entertainment and with a strong literary element. Features music, film, dance, drama and visual arts plus readings, talks and workshops. Previous literary guests have included nationally renowned writers as well as local authors.

## Graham Greene Festival
> Graham Greene Birthplace Trust under Literary Societies

## The Guardian Hay Festival
25 Lion Street, Hay-on-Wye HR3 5AD
☎ 01497 822629  🖷 01497 821066
admin@hayfestival.com
www.hayfestival.com
Festival Director *Peter Florence*

Founded 1988. Annual May festival sponsored by *The Guardian*. Also has a children's programme which began in 2003. The event coincides with half term and caters for young children through to adult. Guests have included Paul McCartney, Bill Clinton, Salman Rushdie, Toni Morrison, Stephen Fry, Joseph Heller, Carlos Fuentes, Maya Angelou, Amos Oz, Arthur Miller, Jimmy Carter.

## Guildford Book Festival
c/o Tourist Information Centre, 14 Tunsgate, Guildford GU1 3QT
☎ 01483 444334
www.guildfordbookfestival.co.uk
Festival Director *Glenis Pycraft*

Held annually, during October in venues throughout the ancient town of Guildford. The festival includes events with established and new writers, workshops, poetry performances, children's events. This diverse festival aims to involve, instruct and entertain all who care about literature. Guest authors in 2009 included Val McDermid, Roger McGough, Peter Sallis, Marina Lewycka, William Nicholson, Kathy Lette.

## Harrogate Crime Writing Festival
> Theakstons Old Peculier Harrogate Crime Writing Festival

## Harrogate International Festival
Raglan House, Raglan Street, Harrogate HG1 1LE
☎ 01423 562303  🖷 01423 521264
info@harrogate-festival.org.uk
www.harrogate-festival.org.uk
Chief Executive *Sharon Canavar*

Founded 1966. Annual two-week festival at the end of July and beginning of August. Events include international symphony orchestras, chamber concerts, ballet, celebrity recitals, contemporary dance, opera, drama, world music, jazz and comedy.

## Harwich Festival of the Arts
2A Kings Head Street, Harwich CO12 3EG
☎ 01255 503571
anna@rendell-knights.freeserve.co.uk
Contact *Anna Rendell-Knights*

Founded 1980. Annual ten-day summer festival held in July. Events include concerts, recitals, drama, film, dance, art, exhibitions, historic town walks. The festival in 2010 (1st–11th) will commemorate the 70th Anniversary of the Evacuation of the Harwich Children, and Dunkirk in WWII.

## The Hay Festival
> The Guardian Hay Festival

## Hebden Bridge Arts Festival
New Oxford House, Albert Street, Hebden Bridge HX7 8AH
☎ 01422 842684
hbfestival@gmail.com
www.hebdenbridgeartsfestival.co.uk
www.hbaf.co.uk
Coordinator *Rebecca Yorke*
Programming *Stephen May*

Founded 1994. Annual arts festival with a strong strand of literature events. Previous guest writers include Roger McGough, Benjamin Zephaniah, Jacqueline Wilson, Quentin Blake, Ian McMillan, Carol Ann Duffy, AL Kennedy.

## Humber Mouth – Hull Literature Festival
City Arts Unit, Central Library, Albion Street, Kingston upon Hull HU1 3TF
☎ 01482 616961  🖷 01482 616827
humbermouth@gmail.com
www.humbermouth.org.uk
Contact *Maggie Hannan*

Founded 1992. Hull's largest festival, held in the summer, and one of the region's liveliest events. Features readings, talks, performances and workshops by writers and artists from around the world and from the city.

## Ilkley Literature Festival
The Manor House, 2 Castle Hill, Ilkley LS29 9DT
☎ 01943 601210  🖷 01943 817079
admin@ilkleyliteraturefestival.org.uk
www.ilkleyliteraturefestival.org.uk
Director *Rachel Feldberg*
Festival Manager *Gail Price*

Founded 1973. Major literature festival in the north held over 17 days every October. Includes children's literature weekend and festival fringe. Recent guests include Maya Angelou, Simon Armitage, Alan Bennett, Sarah Waters, Claire Tomalin, Benjamin Zephaniah, Richard Ford,

Carol Ann Duffy, P.D. James, Donna Tartt, Kate Adie, Germaine Greer, Michael Ondaatje, Chimamanda Adichie, Liz Lochhead, Tony Harrison, Tracy Chevalier, Ian Rankin. Telephone or email to join free mailing list.

## Imagine – Southbank Centre's Annual Children's Literature Festival

Belvedere Road, London SE1 8XX
☎ 0871 663 2500
www.southbankcentre.co.uk

A nine-day festival of children's literature held in February at London's Southbank Centre. The 2010 festival featured readings, storytelling, poetry, stand-up comedy, exhibitions and music. Guests included Michael Rosen, John Hegley and Jason Bradbury.

## The International Festival of Mountaineering Literature

130 Nicholson Road, Sheffield S8 9SX
☎ 0114 258 4063
niallgrimes@hotmail.com
Director *Niall Grimes*

Founded in 1987. Annual festival held over three days at Kendal Mountain Festivals, Kendal, Cumbria. Taking place in November, the festival celebrates recent books, commissions new writing, gives overviews of national literatures, holds debates of issues, book signings. Visit the website (www.mountainfest.co.uk) to join the free mailing list.

## International Playwriting Festival

Warehouse Theatre, Dingwall Road, Croydon CR0 2NF
☎ 020 8681 1257    🖷 020 8688 6699
info@warehousetheatre.co.uk
www.warehousetheatre.co.uk/ipf.html
Festival Administrator *Rose Marie Vernon*

The Warehouse Theatre Company's International Playwriting Festival will celebrate 25 successful years in 2010. It is held in two parts: the *first* is the competition with entries accepted between January and June from all over the world and judged by a panel of distinguished theatre practitioners. The *second* is the Festival itself which showcases the selected plays from the competition in the following February. The IPF also showcases the successful plays in Europe in association with its partners Extra Candoni in Italy and Theatro Ena in Cyprus. Rules and entry form on web page above. Previous winners produced at the theatre include: Guy Jenkin *Fighting for the Dunghill*; James Martin Charlton *Fat Souls*; Peter Moffat *Iona Rain*; Kevin Hood *Teach Yourself Italian*; Dino Mahoney *YoYo*; Dominic McHale *The Resurrectionists*; Neil Flynn *Clocked*; Roumen Shomov *The Dove*; Maggie

Nevill *The Shagaround*; Andrew Shakeshaft *Just Sitting*; Mark Norfolk *Knock Down Ginger*; Des Dillon *Six Black Candles*; Alex Evans *Mother Russia*.

## King's Lynn, The Fiction Festival

19 Tuesday Market Place, King's Lynn PE30 1JW
☎ 01553 691661 (office hours) or 761919
🖷 01553 691779
tony.ellis@hawkins-solicitors.com
www.lynnlitfests.com
Contact *Anthony Ellis*

Founded 1989. Annual weekend festival held in March. Over the weekend there are readings and discussions, attended by guest writers of which there are usually twelve. Previous guests have included Beryl Bainbridge, Louis de Bernière, Alan Sillitoe, J.P. Donleavy and D.J. Taylor.

## King's Lynn, The Poetry Festival

19 Tuesday Market Place, King's Lynn PE30 1JW
☎ 01553 691661 (office hours) or 761919
🖷 01553 691779
tony.ellis@hawkins-solicitors.com
www.lynnlitfests.com
Contact *Anthony Ellis*

Founded 1985. Annual weekend festival held at the end of September (24th–26th in 2010), with guest poets (usually eight). Previous guests have included Clive James, Les Murray, Peter Porter, D.M. Thomas, C.K. Williams and Kit Wright.

## King's Sutton Literary Festival

Glebe House, St Rumbold's Drive, King's Sutton, Banbury OX17 3PJ
☎ 01295 811473
info@kslitfest.co.uk
www.kslitfest.co.uk
Festival Organizer *Rona Rowe*

Founded 2004. Takes place over a weekend in early March. Celebrates reading in a village that has attracted many world-renowned writers to the festival, such as Margaret Drabble, Andrew Motion, Salley Vickers, Justin Cartwright, David Lodge, Rachel Billington. Takes place under one roof in the modern village centre and café with new and second-hand book sales, a programme involving the village primary school. Proceeds to the Parish Church Restoration Fund.

## Lambeth Readers & Writers Festival

London Borough of Lambeth, Brixton Library, Brixton Oval, London SW2 1JQ
☎ 020 7926 1105
to'dell@lambeth.gov.uk
www.lambethgov.uk/Libraries
Reader Development Officer *Tim O'Dell*

Founded in 2001. Held annually in May, the festival brings together internationally recognized, new and local authors for a month of talks, poetry and writing courses focusing on the reading experience. Previous guests have included Peter Ackroyd, Ben Okri, Armando Iannucci, Kate Mosse, Linton Kwesi Johnson, Doris Lessing, Buchi Emecheta, Jo Brand, Gillian Slovo and Lionel Shriver.

## Lancaster LitFest

The Storey, Meeting House Lane, Lancaster
LA1 1TH
☎ 01524 62166
andy.derby@litfest.org
www.litfest.org
Artistic Director *Andy Derby*
Admin Manager *Jonathan Bean*

Founded 1978. Regional Literature Development Agency, organizing workshops, readings, residencies, publications. Year-round programme of literature-based events, readings and annual festival in October featuring a wide range of writers from the UK and overseas. Publishes new North West prose and poetry writing under the Flax Books imprint.

## Ledbury Poetry Festival

Church Lane, Ledbury HR8 1DH
☎ 0845 458 1743
admin@poetry-festival.com
www.poetry-festival.com
Festival Director *Chloe Garner*

Founded 1997. Annual ten-day festival held in June/July. Includes readings, discussions, workshops, exhibitions, music and walks. There is also an extensive year-round community programme and a national poetry competition. Past guests have included Ruth Padel, Joan Bakewell, Iain Sinclair, Alice Oswald, Adam Foulds, Ben Okri, Alan Halsey, Michael Horovitz, Benjamin Zephaniah. Full programme available in May.

## Lennoxlove Book Festival

c/o Haddington House, 28 Sidegate, Haddington
EH41 4BJ
☎ 01620 829800
info@lennoxlovebookfestival.com
www.lennoxlovebookfestival.com
Festival Coordinators *Nicky Stonehill, Rebecca Salt*

New annual literary festival held in Lennoxlove House and grounds over three days in mid-November. The festival brings acclaimed authors to one of Scotland's great houses for an atmospheric winter festival, with events for adults and families. Past speakers include Kate Adie, Alexander McCall Smith, Rory Bremner, Michael Morpurgo, Simon King, Gervase Phinn and Hardeep Singh Kohli. The Festival site includes a bookshop and café, additional entertainments and exhibitions over the weekend. Free mailing list.

## Lewes Live Literature

PO Box 2766, Lewes BN7 2WF
☎ 07972 037612
info@leweslivelit.co.uk
www.leweslivelit.co.uk
Artistic Director *Mark Hewitt*

Lewes Live Lit is a project-based live literature production company working with writers to develop crafted performances and other outcomes as part of curated seasons or festivals of new work. Occasional/year-round.

## Lichfield Festival

7 The Close, Lichfield WS13 7LD
☎ 01543 306270    ℻ 01543 306274
info@lichfieldfestival.org
www.lichfieldfestival.org
Festival Director *Richard Hawley*

Annual July festival with events taking place in the 13th century Cathedral, the Lichfield Garrick Theatre and other venues. Mainly music but a growing programme of literary events such as poetry, plays and talks. New dedicated literature weekend each autumn; literary dinner plus two days of writer talks.

## Lincoln Book Festival

☎ 01522 873844
info@lincolnbookfestival.gov.uk
www.lincolnbookfestival.co.uk
Contact *Sara Bullimore & David Lambert (Joint Artistic Directors)*

Founded 2004. Annual festival held in May over five days. Guests across the years include The Hairy Bikers, Sir David Frost, Frank Gardner, Michael Morpurgo, Melvyn Bragg, Michael Connelly, John Boyne, Alistair Campbell, Jilly Cooper, Joanne Harris.

## City of London Festival

Fitz-Eylwin House, 25 Holborn Viaduct, London
EC1A 2BP
☎ 020 7583 3585    ℻ 020 7353 0455
admin@colf.org
www.colf.org
Director *Ian Ritchie*

Founded 1962. Annual three-week festival held in June and July. Features over fifty classical and popular music events alongside poetry and prose readings, street theatre and open-air extravaganzas, in some of the most outstanding performance spaces in the world.

## London Literature Festival

Southbank Centre, Belvedere Road, London
SE1 8XX
☎ 0871 663 2500 (box office)
www.southbankcentre.co.uk

Established in 2007, events take place at the Southbank Centre over two weeks in early July. Previous guests have included Tony Benn, Arundhati Roy, Buzz Aldrin, Benjamin Zephaniah, Jeanette Winterson, Peter Ackroyd, Lemn Sissay and Amartya Sen.

## Lowdham Book Festival

The Bookcase, 50 Main Street, Lowdham NG14 7BE
℡ 0115 966 4143
janestreeter@thebookcase.co.uk
www.lowdhambookfestival.co.uk
Contact *Jane Streeter*

Founded 1999. Annual nine-day festival held in June combining a village fête atmosphere with that of a major literature festival. Guests have included Carol Ann Duffy, Polly Toynbee, Ian McMillan, Alan Sillitoe, Jackie Kay, Gary Younge and Carole Blake. Talks on everything from St Kilda to the Jewish roots of rock 'n' roll. Also holds annual 'Winter Weekends' with guest speakers in November or January and, in the first weekend of March, 'Flicks in the Sticks', a mini film festival.

## Manchester Literature Festival

Beehive Mill, Jersey Street, Ancoats, Manchester M4 6JG
℡ 0161 236 5555
admin@manchesterliteraturefestival.co.uk
www.manchesterliteraturefestival.co.uk
Contact *Administrator*

Held in October, Manchester Literature Festival offers 'unique and imaginative literature experiences to its audiences' with a programme that features readings by some of the world's finest authors, freshly commissioned work, and a series of cutting-edge events exploring the crossover between new technology and literature.

## Mere Literary Festival

Lawrence's, Old Hollow, Mere BA12 6EG
℡ 01747 860475/861211 (Tourist Information)
www.merelitfest.co.uk
Contact *Adrienne Howell*

Founded 1997. Annual festival held in the second week of October in aid of registered charity, The Mere & District Linkscheme. Events include readings, quiz, workshop, writer's lunch and talks. Finale is adjudication of the festival's writing competition (see entry under *Prizes*) and presentation of awards.

## National Association of Writers' Groups – NAWG Open Festival of Writing

PO Box 3266, Stoke-on-Trent ST10 9BD
nawg@live.co.uk
www.nawg.co.uk

Founded 1995. A three-day festival which usually takes place during the first weekend in September (see the website for full details). Includes creative writing tuition covering poetry, short and long fiction, playwriting, journalism, TV sitcom and many other subjects, all led by professional writer-tutors. Workshops, one-to-one tutorials, opportunities to meet other writers and fringe events. Saturday gala dinner and awards ceremony. Full or part-residential weekend, or single workshops only. Open to all, no qualifications or NAWG membership required.

## National Eisteddfod of Wales

40 Parc Ty Glas, Llanishen, Cardiff CF14 5DU
℡ 029 2076 3777    ℻ 029 2076 3737
info@eisteddfod.org.uk
www.eisteddfod.org.uk

The National Eisteddfod, held in August, is the largest arts festival in Wales, attracting over 170,000 visitors during the week-long celebration of more than 800 years of tradition. Competitions, bardic ceremonies and concerts.

## National Student Drama Festival
▷ University of Hull under UK Writers' Courses

## Northern Children's Book Festival

Schools Library Service, Sandhill Centre, Grindon Lane, Sunderland SR3 4EN
℡ 0191 553 8866/7/8    ℻ 0191 553 8869
schools.library@sunderland.gov.uk
www.ncbf.co.uk
Secretary *Eleanor Dowley*

Founded 1984. Annual two-week festival during November. Events in schools and libraries for children in the North East region. One Saturday during the festival sees the staging of a large book event hosted by one of the local authorities involved.

## Off the Shelf Festival of Writing and Reading

Central Library, Surrey Street, Sheffield S1 1XZ
℡ 0114 273 4716/4400    ℻ 0114 273 5009
offtheshelf@sheffield.gov.uk
www.offtheshelf.org.uk
Festival Organizers *Maria de Souza, Susan Walker, Lesley Webster*

Founded 1992. Annual festival held over three weeks at the end of October into November. Lively and diverse mix of readings, workshops, children's events, storytelling and competitions. Previous guests have included Michael Frayn, Rabbi Lionel Blue, Germaine Greer, Ben Okri, Hilary Mantel, Michael Wood, Linton Kwesi Johnson, Sir Arnold Wesker, Jeremy Bowen, John Pilger, David Guterson, David Lodge, Margaret Drabble, Tariq Ali, Alan Sillitoe, Simon Armitage, Rose Tremain

## Oundle Festival of Literature

2 New Road, Oundle PE8 4LA

☎ 01832 273050

enquiries@oundlelitfest.org.uk

www.oundlelitfest.org.uk

Festival Deputy Chair *Liz Dillarstone*

Founded 2002. Annual festival running in the first two weeks of March for adults and children. Features talks by high profile authors and poets, writing workshops, poetry and prose showcases for local writers, children's writing competitions, school events, drama productions and community readings for young and old. All Year Round programme now offers opportunities outside of festival fortnight.

## Oxford Literary Festival
▷ The Sunday Times Oxford Literary Festival

## Redbridge Book and Media Festival

3rd Floor, Central Library & Museum, Clements Road, Ilford IG1 1EA

☎ 020 8708 2857    📠 020 8708 2431

shermain.philip@redbridge.org.uk

www.redbridge.gov.uk

Contact *Shermain Philip (Arts Development Officer)*

Founded in 2003, the festival celebrates media and literature in all forms, and takes place annually over three weeks in April/May. Features a programme of author readings, book signings, creative writing classes, workshops, community events, performances, poetry evenings, competitions, exhibitions and panel debates.

## Richmond Literature Festival
▷ Book Now

## Runnymede International Literature Festival

Royal Holloway, University of London, Egham TW20 0EX

☎ 01273 571700

info@rfest.org.uk

www.rfest.co.uk

www.rhul.ac.uk/english

Festival Director *Professor Robert Hampson*

Founded 2006. Annual literary festival held in Runnymede as a partnership beween Royal Holloway, University of London and Runnymede Borough Council. Held annually in the spring, the festival runs for a core weekend with associated community and education activities. Includes readings, performances, masterclasses and community workshops.

## Saffron Walden Literary Festivals

Harts, Hart House, Shire Hill, Saffron Walden CB11 3AQ

☎ 01799 523456 (box office)

events@harts1836.co.uk

www.hartsevents.co.uk

www.hartsbooks.co.uk

Festival Director *Jo Burch*

Founded 2006. Two themed literature festivals (spring and autumn) run by family-owned Harts Bookshop. The theme of the 2010 spring festival, held in March, was Words Around Britain and featured events and workshops for adults and children. Please see events website for full details.

## Southwold Festival
▷ Ways with Words

## StAnza: Scotland's International Poetry Festival

*Registered Office:* 57 Lade Braes, St Andrews KY16 9DA

☎ 01333 360491

info@stanzapoetry.org

www.stanzapoetry.org

Festival Director *Brian Johnstone (admin@ stanzapoetry.org)*

Artistic Director *Eleanor Livingstone (arts@stanzapoetry.org)*

Press & Media Manager *Annie Kelly (press@stanzapoetry.org)*

The only regular festival dedicated to poetry in Scotland, StAnza is international in outlook. Held annually in March in the ancient university town of St Andrews, the festival is an opportunity to hear world class poets reading in exciting and atmospheric venues. The 2010 festival themes were *Myth & Legend* and *The Director's Cut*. The festival features readings, discussions, conversations, performance poetry, poetry in exhibition, workshops and children's poetry. It has a strong showing of poets from across the UK and overseas as well as the festival's signature foreign language readings. Free programmes can be ordered from Fife Contemporary Art & Craft on 01334 474610 or by email to mail@fcac.co.uk

## Stena Line Wigtown Book Festival

Festival Office, County Buildings, Wigtown DG8 9JH

☎ 01988 403222

mail@wigtownbookfestival.com

www.wigtownbookfestival.com

Festival Director *Adrian Turpin*

Festival Admin *Anne Barclay*

Founded 1998. Annual Festival held over ten days (last week September/first week October). Author readings, poetry events, workshops, drama, music. Children's events and school outreach programme. Also Spring Festival (founded 2005) held over first May Bank Holiday weekend.

## Stratford-upon-Avon Poetry Festival

The Shakespeare Centre, Henley Street, Stratford-upon-Avon CV37 6QW

☎ 01789 204016/292176 (box office)

🖷 01789 296083

info@shakespeare.org.uk

www.shakespeare.org.uk

Festival Director *Paul Edmondson*

Sponsored by the Shakespeare Birthplace Trust, this Festival, now in its 57th year, takes place in July and celebrates poetry past and present with special reference to the work of William Shakespeare. Features weekly recitals given by established poets and by actors who present themed evenings of verse. The Festival includes opportunities for young people to write their own work and have it performed. There is a Poetry Mass and local poets' events. Full details are available from March and can be accessed on the Trust's website.

## Strokestown International Poetry Festival

▷ Strokestown International Poetry Competition under Prizes

## The Sunday Times Oxford Literary Festival

Christ Church, Oxford OX1 1DP

☎ 01865 286074

info@sundaytimes-oxfordliteraryfestival.co.uk

www.sundaytimes-oxfordliteraryfestival.co.uk

Festival Director *Sally Dunsmore*

Operations Manager *Alex Simmons*

Founded 1997. Annual week-long festival held in March/April. Authors speaking about their books, covering a wide variety of writing: fiction, poetry, biography, travel, food, art, gardening, children's books. There is also a range of events for schools. Previous guests have included William Boyd, Andrew Motion, Beryl Bainbridge, Sophie Grigson, Philip Pullman, Doris Lessing, Seamus Heaney, Richard Dawkins, Zandra Rhodes, Kazuro Ishiguro. Runs a programme of creative writing workshops during the festival; speakers at masterclasses during the 2010 festival included Helen Dunmore, Joanne Harris and Julie Myerson.

## Swindon Festival of Literature

Lower Shaw Farm, Shaw, Swindon SN5 5PJ

☎ 01793 771080

swindonlitfest@lowershawfarm.co.uk

www.swindonfestivalofliterature.co.uk

Festival Director *Matt Holland*

Founded 1994. Annual festival held in May, starting with 'Dawn Chorus' at sunrise. Includes a wide range of authors, speakers, discussions, performances and workshops, plus the Swindon Performance Poetry Slam competition. Guests at the 2009 Festival included David Attenborough, Margaret Drabble, Michael Morpurgo and Deborah Moggach.

## Theakstons Old Peculier Crime Writing Festival, Harrogate

Raglan House, Raglan Street, Harrogate HG1 1LE

☎ 01423 562303    🖷 01423 521264

crime@harrogate-festival.org.uk

www.harrogate-festival.org.uk

Chief Executive *Sharon Canavar*

Festival Manager *Erica Morris*

Launched 2003. Four days of events at the end of July each year featuring the best of British and international crime writers. The winner of the **Theakstons Old Peculier Prize for the Crime Novel of the Year** is announced at the Festival (see entry under *Prizes*). Events also include industry 'How to ...' sessions, social events and late night shows. Part of the **Harrogate International Festival**.

## The Times Cheltenham Literature Festival

109 Bath Road, Cheltenham GL53 7LS

☎ 01242 775861

clair.greenaway@cheltenhamfestivals.com

www.cheltenhamfestivals.com

Festival Manager *Clair Greenaway*

Founded 1949. Annual festival held in October (8th–17th in 2010). The first purely literary festival of its kind, it has over the past decade developed from an essentially local event into the largest and most popular in Europe. A wide range of events including talks and lectures, poetry readings, novelists in conversation, exhibitions, discussions and a large bookshop.

## Torbay Weekend Festival of Poetry

6 The Mount, Higher Furzeham, Brixham TQ5 8QY

☎ 01803 851098

pwoxley@aol.com

www.acumen-poetry.co.uk/events

Festival Organizer *Patricia Oxley*

Founded 2001. Held annually at the end of October, Thursday evening until Monday afternoon, with workshops, poetry readings, talks and debates. Encourages active participation through workshops, open mike events, etc. Guest poets include internationally known writers as well as local authors.

## Tŷ Newydd Festival

Tŷ Newydd, Llanystumdwy, Criceith LL52 0LW

☎ 01766 522811    🖷 01766 523095

post@tynewydd.org

www.tynewydd.org

Director *Sally Baker*

Biennial, bilingual literature festival held on alternate years. Run by the National Writers'

Centre for Wales, it is located at Tŷ Newydd Writers' Centre and other venues near Cricieth. Features writers and poets from Wales (working in both English and Welsh) and worldwide.

## The Ulster Bank Belfast Festival at Queen's

8 Fitzwilliam Street, Belfast BT9 6AW
☎ 028 9097 1034    🖷 028 9097 1336
g.farrow@qub.ac.uk
www.belfastfestival.com
Director *Graeme Farrow*

Founded 1963. Annual three-week festival held in October (15th–30th in 2010). Organized by Queen's University, the festival covers a wide variety of events, including literature. Programme available in September.

## University of Aberdeen Writers Festival
▷ Word

## Warwick Words

The Court House, Jury Street, Warwick CV34 4EW
☎ 01926 427056
info@warwickwords.co.uk
www.warwickwords.co.uk

Founded 2002. Annual festival featuring both living writers and those who have had connections with Warwick – Tolkien, Larkin and Landor in particular. Also workshops and an education programme. A festival of literature and spoken word for all the family. Events at the Bridge House Theatre, St Mary's Church, the Lord Leycester Hospital and other historic buildings around Warwick during October.

## Ways with Words

Droridge Farm, Dartington, Totnes TQ9 6JG
☎ 01803 867373    🖷 01803 863688
admin@wayswithwords.co.uk
www.wayswithwords.co.uk
Festival Director *Kay Dunbar*

Ways With Words runs three major annual literature festivals: at Dartington Hall, Devon for ten days in July, featuring over 150 writers giving talks, readings, interviews, discussions, performances and workshops; Words by the Water, Keswick, Cumbria in March, a ten-day festival, as Dartington; and a five-day festival at Southwold, Suffolk in November. Also organizes writing, reading and painting courses in Italy and France.

## Wellington Literary Festival

Civic Offices, Larkin Way, Tan Bank, Wellington, Telford TF1 1LX
☎ 01952 567697    🖷 01952 567690
welltowncl@aol.com
www.wellington-shropshire.gov.uk
Contact *Howard Perkins*

Founded 1997. Annual festival held throughout October. Events include storytelling, writers' forum, 'Pints and Poetry', children's poetry competition, story competition, theatre review and guest speakers.

## Wells Festival of Literature

The Old Vicarage, Henton Wells BA5 1PD
☎ 01749 675492
www.wlitf.co.uk

Founded 1992. Annual week-long festival held in the middle of October. Main venue is the historic, moated Bishop's Palace. A wide range of speakers caters for different tastes in reading. Recent guests include: John Le Carré, Douglas Hurd, Fay Weldon, Melvyn Bragg, Margaret Drabble, William Dalrymple, Timothy West. Short story and poetry competitions and writing workshops are run in conjunction with the festival.

## Wigtown Book Festival
▷ Stena Line Wigtown Book Festival

## Winchester Writers' Conference, Festival, Bookfair
▷ entry under UK and Irish Writers' Courses

## Word – University of Aberdeen Writers Festival

Office of External Affairs, University of Aberdeen, King's College, Aberdeen AB24 3FX
☎ 01224 273874
word@abdn.ac.uk
www.abdn.ac.uk/word
Artistic Director *Alan Spence*

Held annually in May, the Festival takes place over six days at the historic King's College Campus and at venues throughout the city. Attracts over 70 authors and over 10,000 visitors to a weekend of readings, lectures, debates, music, art exhibitions and film screenings. Also includes an extended schools' and children's programme and a Festival of Gaelic.

## Word Market

c/o Ulverston Library, Kings Road, Ulverston LA12 0BT
☎ 01229 580999
loretta@meetingvoice.co.uk
www.wordmarket.org.uk

Word Market celebrates the written word through a programme of exciting, live events and projects throughout the year. As well as an annual festival, regular workshops, master-classes and performances are held with well-known poets, best-selling authors, publishers, literary agents and theatre directors. Check the website for further details.

## Words by the Water
> Ways with Words

## The Wordsworth Trust
Dove Cottage, Grasmere LA22 9SH
☎ 01539 435544    ℻ 01539 463538
enquiries@wordsworth.org.uk
www.wordsworth.org.uk
Contact *Andrew Forster (Literature Officer)*

Ongoing contemporary poetry programme held over the summer and one-off events held around the country. Past readers have included Carol Ann Duffy, Adam O'Riordan, Simon Armitage, Seamus Heaney, Paul Muldoon, Sharon Olds, Philip Pullman and Don Paterson as well as a 'support' reader at most events.

The Trust also operates an annual poets-in-residence programme, giving time and space for new writers to develop their work in this setting: a cottage and stipend are supplied and the artist is given the space to develop their own work as well as opportunities to become involved in The Wordsworth Trust literary programme. Past artists-in-residence include Henry Shukman,

Owen Sheers, Paul Farley, Helen Farish, Jack Mapanje, Rebecca O'Connor, Jacob Polley, Neil Rollinson and John Hartley Williams.

## Worlds Literary Festival
> Writers' Centre Norwich under Professional Associations

## Writing on the Wall
First Floor, Mission Hall, 36 Windsor Street, Liverpool L8 1XF
☎ 0151 703 0020
info@writingonthewall.org.uk
www.writingonthewall.org.uk
Festival Coordinator *Madeline Heneghan*

Annual festival held in May in Liverpool that works alongside schools, young people, local communities and broader audiences to celebrate writing, diversity, tolerance, story-telling and humour through controversy, inquiry and debate. Performances, readings, workshops, screenings, high profile debates and discussions. Guest speakers have included Noam Chomsky, Irvine Welsh, Howard Marks, Roddy Doyle, Benjamin Zephaniah.

# International Book Fairs

## Abu Dhabi International Book Fair
P.O. Box 2380, Abu Dhabi, United Arab Emirates
① 00 971 2 657 6180    Ⓕ 00 971 2 444 5507
info@adbookfair.com
www.adbookfair.com

Week-long event held in March. The 19th fair, held in 2009, attracted over 500 exhibitors from 51 countries.

## American Library Association (ALA) Conferences
50 E. Huron, Chicago, IL 60611, USA
www.ala.org

The American Library Association holds two annual events. The ALA Annual Conference (Washington Convention Center, 24th–29th June in 2010) is attended by more than 25,000 librarians, publishers, writers and educators. The ALA Midwinter Meeting is held in January and includes some 450 exhibitors featuring the latest in books, videos, computers and other materials available to today's libraries and their users.

## Beijing International Book Fair
International Exhibition Centre, Beijing, China
www.bibf.net

Five-day book fair (30th August–3rd September in 2010), now in its 17th year. The 2009 Fair attracted more than 1,500 exhibitors from 56 countries.

## Bologna Children's Book Fair
Piazza Costituzione, 6, 40128 Bologna, Italy
① 00 39 051 282 242    Ⓕ 00 39 051 6374 011
bookfair@bolognafiere.it
www.bookfair.bolognafiere.it

Children's book fair held in March, now in its 47th year.

## BookExpo America (BEA)
Jacob K. Javits Convention Center of New York, 655 W 34th Street, New York, NY 1001, USA
inquiry@bookexpoamerica.com
www.bookexpoamerica.com

Annual event held in May in 2010. From 2009 to 2012, BEA will take place in New York.

## Cairo Book Fair
General Egyptian Book Organization, Corniche El-Nil, Ramlat Boulaq, Cairo, Egypt
info@egyptianbook.org.eg
www.egyptianbook.org.eg

Run by the General Egyptian Book Organization, the Cairo Book Fair, now in its 43rd year, takes place at the end of January.

## Cape Town Book Fair
PO Box 51498, Waterfront 8002, Cape Town, South Africa
① 00 27 21 418 5493    Ⓕ 00 27 21 418 5949
info@capetownbookfair.co.za
www.capetownbookfair.com

The Cape Town Book Fair is organized under a partnership of the Publishers Association of South Africa and the Frankfurt Book Fair. In 2010 the fair runs from 30th July to 2nd August.

## Frankfurt Book Fair
Ausstellungs- und Messe GmbH, Reineckstr. 3, 60313 Frankfurt am Main, Germany
① 00 49 69 2102-0    Ⓕ 00 49 69 2102-227/-277
info@book-fair.com
www.frankfurt-book-fair.com

Major annual book fair held in October. For the 2010 event (6th–10th), the 'Guest of Honour' is Argentina.

## Jerusalem Book Fair
PO Box 775, Jerusalem 91007, Israel
① 00 972 2 629 6415    Ⓕ 00 972 2 624 0663
stshira@jerusalem.muni.il
www.jerusalembookfair.com

Founded 1963. Biennial fair attracting publishers from more than 40 countries. Next fair takes place in 2011.

## Kolkata Book Fair
Guild House, 2B Jhama Pukur Lane, Kolkata 700 009, West Bengal, India
① 00 91 33 2360 4588    Ⓕ 00 91 33 2360 4566
guild@cal2.vsnl.net.in
www.kolkatabookfaironline.com

The 34th fair, held in January 2010, had Mexico as the 'Focal Theme' and 'Partner Country'.

## London Book Fair

Earls Court Exhibition Centre, Warwick Road, London SW5 9TA

Ⓣ 020 8271 2124　Ⓕ 020 8910 7930

www.londonbookfair.co.uk

Annual, three-day event held at Earls Court during April.

## Mumbai International Book Fair

Bandra Kurla Complex, Mumbai, India

bookfair@fairfest.com

www.internationalbookfair.com

Month-long book fair, held in January.

## New Delhi World Book Fair

National Book Trust of India, Phase–II, 5 Institutional Area, Vasant Kunj, New Delhi 110070, India

Ⓣ 00 91 11 2670 7700

nbtindia@ndb.vsnl.net.in

www.nbtindia.org.in

Biennial fair held in January/February. Organized by the National Book Trust of India, the theme of the 2010 fair was 'Reading for Common Wealth' and Moscow was 'Guest of Honour'.

## Paris Book Fair

⊵ Salon du Livre

## Prague International Book Fair

Association of Czech Booksellers & Publishers, Fügnerovo nám. 3, CZ–120 00 Prague 2, Czech Republic

Ⓣ 00 420 224 498 236　Ⓕ 00 420 224 498 754

info@svetknihy.cz

www.bookworld.cz/en

Held during May in the Prague Exhibition Grounds, the 17th fair, held in 2010, had Poland as 'Guest of Honour'.

## Salon du Livre

Syndicat national de l'edition (SNE), 115 Blvd. Saint Germain, 75006 Paris, France

Ⓣ 00 1 44 41 40 50

www.salondulivreparis.com

Established in 1981 and organized by the Syndicat national de l'édition (French Publishers Association). The 30th book fair was held at Porte de Versailles in March 2010, attracting around 1,200 publishers from 25 foreign countries.

## Tokyo International Book Fair

18F Shinjuku-Nomura Bldg., 1-26-2 Nishishinjuku, Shinjuku-ku, Tokyo 163-0570, Japan

Ⓣ 00 81 3 3349 8507　Ⓕ 00 81 3 3345 7929

tibf-eng@reedexpo.co.jp

http://web.reedexpo.co.jp/tibf/english/

Asia's leading book fair which attracted 776 exhibitors from 29 countries in 2009. The 17th fair takes place 8th–11th July 2010.

# Useful Websites

Many of these and other useful websites for writers can be found in *The Internet for Writers* by Nick Daws (ISBN 9781840253085), one of a series of books published by Internet Handbooks; *The Incredibly Indispensable Web Directory* by Clive and Bettina Zietman, published by Kogan Page (ISBN 9780749434250); *The Good Web Guide for Writers: The Simple Way to Explore the Internet* by Paul Chronnell, published by The Good Web Guide Limited (ISBN 9781903282380).

## AbeBooks
www.abebooks.co.uk

Online bookshop for new and second-hand publications including rare and out-of-print books. Includes search engine for textbooks and reference books.

## Academi (Welsh Academy/Yr Academi Gymreig)
www.academi.org

News of events, publications and funding for Welsh-based literary events. (See entry under *Professional Associations and Societies*.)

## Alibris
www.alibris.co.uk

Over 100 million used, new and hard-to-find books online. Bargain books, text and reference listings and rare books.

## Alliance of Literary Societies
www.allianceofliterarysocieties.org.uk

Details of societies and events. (See entry under *Professional Associations and Societies*.)

## Amazon
www.amazon.co.uk

The online shop.

## Ancestry
www.ancestry.co.uk

Family history information – databases, articles and other sources of genealogical data.

## Arts Council England
www.artscouncil.org.uk

Includes information on funding applications and publications. (See entry under *Arts Councils and Regional Offices*.)

## Arts Council of Northern Ireland
www.artscouncil-ni.org

Information on funding and awards, events and free E-Newsletter service. (See entry under *Arts Councils and Regional Offices*.)

## Arts Council of Wales/Cyngor Celfyddydau Cymru
www.acw-ccc.org.uk

Information on publications, grants, council meetings, the arts in Wales. (See entry under *Arts Councils and Regional Offices..*)

## Arvon Foundation
www.arvonfoundation.org

Residential creative writing courses. See entry under *UK and Irish Writers' Courses.*

## Ask About Writing
www.askaboutwriting.net

Resource site for writers.

## Association for Scottish Literary Studies
www.asls.org.uk

The educational charity promoting the languages and literature of Scotland. (See entry under *Professional Associations and Societies*.)

## Association of Authors' Agents (AAA)
www.agentsassoc.co.uk

UK literary agents' organization including list of current members. (See entry under *Professional Associations and Societies*.)

## Association of Authors' Representatives (AAR)
www.aar-online.org

US literary agents' organization including search engine for current members. (See entry under *Professional Associations and Societies*.)

## Author-Network
www.author-network.com

Writers' resource site.

## Authors' Licensing and Collecting Society Limited (ALCS)
www.alcs.co.uk

Details of membership, news and publications. (See entry under *Professional Associations and Societies*.)

## Bartleby.com
www.bartleby.com

An ever-expanding list of books published online for reference, free of charge.

## BBC
www.bbc.co.uk

Access to all BBC departments and services.

## Bibliomania
www.bibliomania.com

Over 2,000 classic texts, study guides and reference resources available online for free. Also selected books for sale in the Bibliomania shop.

## book2book/booktrade.info
www.booktrade.info/index.php

Established by a group of publishers, booksellers, website developers and trade journalists to provide up-to-date news, features and useful information for the book trade.

## BookBrunch
www.bookbrunch.co.uk

BookBrunch is a fully searchable website and daily news bulletin for the book industry, in Britain and abroad. Run by Liz Thomson and Nicholas Clee, writers and journalists who are both former editors of trade journals.

## BookBuzzr.com
www.bookbuzzr.com

Free, online book marketing technology for authors. Helps to market books on sites such as Facebook, Twitter, MySpace, blogs and author websites.

## Booktrust
www.booktrust.org.uk

Book information service, guide to prizes and awards, factsheets on getting published. (See entry under *Professional Associations and Societies*.)

## British Association of Picture Libraries and Agencies (BAPLA)
www.bapla.org.uk

Website includes the BAPLA Online Database search facility by category or name. (See entry under*Professional Associations and Societies*.)

## British Centre for Literary Translation
www.literarytranslation.com

A joint website with the British Council with workshops by leading translators, contacts and networks, and search engine for translation conferences, seminars and events. (See entry under *Professional Associations and Societies*.)

## The British Council
www.britishcouncil.org

Information on the Council's English Language services, education programmes, society and science links. (See entry under *Professional Associations and Societies*.)

## British Film Institute (bfi)
www.bfi.org.uk

See entry under *Professional Associations and Societies*.

## British Library
www.bl.uk

Reader service enquiries, access to main catalogues, information on collections, links to the various reading rooms and exhibitions. (See related entries under *Library Services*.)

## Chapter One Promotions
www.chapteronepromotions.com

Provides services to writers at various stages of their writing career: story and poetry critique services, proof reading, 'Kids Korner'.

## Chartered Institute of Linguists (IoL)
www.iol.org.uk

News on regional societies, job opportunities, 'Find a Linguist' service and *The Linguist* magazine. (See entry under *Professional Associations and Societies*.)

## CILIP
www.cilip.org.uk

The professional body for librarians and information professionals. (See entry under *Professional Associations and Societies*.)

## CILIP Wales
www.dil.aber.ac.uk/cilip_w/index.htm

See entry under *Professional Associations and Societies*.

## Complete Works of William Shakespeare
www-tech.mit.edu/Shakespeare/works.html

Access to the text of the complete works.

## Copyright Licensing Agency Ltd (CLA)

www.cla.co.uk

Copyright information, customer support and information on CLA services. (See entry under *Professional Associations and Societies*.)

## Crime Writers' Association (CWA)

www.thecwa.co.uk

Website of the professional crime writers' association. (See entry under *Professional Associations and Societies*.)

## Daily Express

www.express.co.uk

*Daily Express* and *Sunday Express* online.

## Daily Mail

www.dailymail.co.uk

*Daily Mail* online.

## Daily Mirror

www.mirror.co.uk

*Daily Mirror* and *Sunday Mirror* online.

## The Daily Telegraph

www.telegraph.co.uk

*Daily Telegraph* online.

## Dictionary of Slang

dictionaryofslang.co.uk

A guide to slang 'from a British perspective'. Research information and search facility.

## The Eclectic Writer

www.eclectics.com/writing/writing.html

US website offering a selection of articles giving advice for writers and an online discussion board.

## The English Association

www.le.ac.uk/engassoc

News, publications, conference and membership information. (See entry under *Professional Associations and Societies*.)

## FilmAngels

www.filmangel.co.uk

Established in conjunction with **Hammerwood Films** to create a shop window for writers and would-be film angels alike. Submitted synopses are displayed (for a fee) for a pre-determined period, while would-be angels are invited to finance a production

## Filmmaker Store

www.filmmakerstore.com

US site giving scriptwriting resources, listings and advice.

## Financial Times

www.ft.com

*Financial Times* online.

## Frankfurt Book Fair

www.frankfurt-book-fair.com/en

Provides latest news and market analysis of the book business plus information on the annual Book Fair. (See entry under *International Book Fairs*.)

## Froebel Archive for Childhood Studies

http://studentzone.roehampton.ac.uk/ library/specialcollectionsandarchives/ froebelarchiveforchildhoodstudies/

Collection of children's literature. Supports courses at Roehampton University and is available to *bona fide* researchers. Catalogue available online.

## The Good Web Guide Ltd

www.thegoodwebguide.co.uk

Guide to the best websites. Provides thousands of detailed and independent reviews of a wide range of websites.

## Google Images

www.images.google.com

'The most comprehensive image search on the web.'

## The Guardian/Observer

www.guardian.co.uk

Website of *The Guardian* and *The Observer* newspapers online.

## Guide to Grammar and Style

www.andromeda.rutgers.edu/~jlynch/Writing

A guide to grammar and style, organized alphabetically, plus articles and links to other grammatical reference sites.

## Hack Writers (The International Writer's Magazine)

www.hackwriters.com

Free online magazine. 'We do not pay any fees for work published, but aim to offer a forum of quality work by new or unpublished writers, or by established writers wishing to say something different or branch out. If you have something to say, we want to hear it.'

## Hansard (House of Commons Daily Debates)

www.parliament.the-stationery-office.co.uk/pa/ cm/cmhansrd.htm

The official record of debates and written answers in the House of Commons. The transcript of each day's business appears at noon on the following weekday.

## Help! I Need a Publisher! (And Maybe an Agent ...)

www.helpineedapublisher.blogspot.com/

Writer Nicola Morgan's blog for unpublished and newly published authors.

## The Herald

www.theherald.co.uk

Scottish daily broadsheet.

## House of Commons Research Library

www.parliament.uk/parliamentary_publications_ and_archives/research_papers.cfm

Gives access to the text of research reports prepared for MPs on a wide range of current issues.

## The Independent

www.independent.co.uk

*The Independent* newspaper online.

## Inpress Books

www.inpressbooks.co.uk

Funded by Arts Council England, Inpress Books provides sales, marketing and distribution for small independent presses.

## Institute of Translation and Interpreting (ITI)

www.iti.org.uk/indexMain.html

Website of the professional association of translators and interpreters, with the ITI directory of members, publications, training and membership information. (See entry under *Professional Associations and Societies*.)

## Internet Classics Archive

classics.mit.edu

Includes works of classical literature; mostly Greek and Roman with some Chinese and Persian. All are in English translation.

## Internet Movie Database (IMDb)

www.imdb.com

Essential resource for film buffs and researchers with search engine for cast lists, screenwriters, directors and producers; film and television news, awards, film preview information, video releases.

## The Irish Arts Council/An Chomhairle Ealaíon

www.artscouncil.ie

Monthly email newsletter available giving latest information on grants and awards, news and events, etc. (See entry under *Arts Councils and Regional Offices*.)

## Journalism UK

www.journalismuk.co.uk

A website for UK-based print journalists who write for text-based publications. Includes links to newspapers, magazines, e-zines, news sources plus information on training and organizations.

## Literature North East

www.literaturenortheast.co.uk

Monthly e-newsletter; listings of events and readings in the region; writing courses and local training opportunities.

## Literature North West

www.literaturenorthwest.co.uk

Promotional agency for the region's independent presses and literature organizations

## Literature South West

www.literaturesouthwest.co.uk

Information website, devised by **Poetry Can**, for writers in the South West region.

## literaturetraining

www.literaturetraining.com

Online directory of training and professional development opportunities for UK writers and literature professionals. Includes information on courses, workshops, jobs, residencies, submissions, competitions, organizations and funding for professional development. Also, information sheets and specially commissioned features.

## Location Register of 20th-Century English Literary Manuscripts and Letters

www.reading.ac.uk/library/about-us/projects/lib-location-register.asp

Reference source for the study of English literature. Information about the manuscript holdings of repositories of all sizes, from the British Library to small-town museums, of literary authors – from major poets to minor science fiction writers.

## The Mail on Sunday

www.mailonsunday.co.uk

*The Mail on Sunday* online.

## Mr William Shakespeare and the Internet

shakespeare.palomar.edu

Guide to scholarly Shakespeare resources on the internet.

## National Union of Journalists (NUJ)

www.nuj.org.uk

Represents those journalists who work in all sectors of publishing, print and broadcasting. (See entry under *Professional Associations and Societies*.)

## The Never Ending Story

www.TheNeverEndingStory.co.uk

A range of constantly evolving online publications which online users can read and contribute to. Includes stories and poems by well-known writers, personalities and site members (free membership).

### New Writing North

www.newwritingnorth.com

Essentially for writers based in the north of England but also a useful source of advice and guidelines. (See entry under *Professional Associations and Societies*.)

### New Writing Partnership
▷ Writers' Centre Norwich

### Newnovelist

www.newnovelist.com

Novel-writing software. Aids research, characterization, structure and plot, and breaks down the process of writing a novel into manageable chunks.

### Pact (Producers Alliance for Cinema and Television)

www.pact.co.uk

Publications, training, production companies, membership details. (See entry under *Professional Associations and Societies*.)

### PEN

www.englishpen.org

Website of the English Centre of International PEN. News of events, information on prizes, membership details. (See entry under *Professional Associations and Societies*.)

### Poets and Writers

www.pw.org

US site containing information and advice for writers. Founded in 1970.

### ProfWriting

www.thewritingcentre.com

University College Falmouth resource site for writers.

### The Publishers Association

www.publishers.org.uk

UK publishers' association. (See entry under *Professional Associations and Societies*.)

### Publishing North West
▷ Literature North West

### Publishing Scotland

www.publishingscotland.org

Scottish publishers' association. (See entry under *Professional Associations and Societies*.)

### RefDesk.com

refdesk.com

US website with links to free facts and statistics.

### Royal Society of Literature

www.rslit.org

Information on lectures, discussions and readings; membership details and prizes. (See entry under *Professional Associations and Societies*.)

### The Scotsman

www.scotsman.com

*The Scotsman, Scotland on Sunday* and *Edinburgh Evening News* newspapers online.

### Scottish Arts Council

www.sac.org.uk

Information on funding and events. (See entry under *Arts Councils and Regional Offices*.)

### Scottish Book Trust

www.scottishbooktrust.com

See entry under *Professional Associations and Societies*.

### Scottish Publishers Association
▷ Publishing Scotland

### Screenwriters Online

screenwriter.com/insider/news.html

US resources for 'professional screenwriters, students, serious beginners and lovers of movies' with screenplay analysis, expert articles and *The Insider Report*.

### The SF Hub

www.sfhub.ac.uk/

Science fiction research website created by the University of Liverpool. Includes links to the Science Fiction Foundation collection and the John Wyndham archives.

### Shots Magazine

www.shotsmag.co.uk

Electronic magazine of crime and mystery fiction.

### Slainte: Information & Libraries Scotland

www.slainte.org.uk

Links to various services and information on librarianship and information management in Scotland.

### Society for Editors and Proofreaders (SfEP)

www.sfep.org.uk

Information about the Society including online directory of members. (See entry under *Professional Associations and Societies*.)

## The Society of Authors
www.societyofauthors.org

Includes FAQs for new writers, diary of events, membership details, prizes and awards. (See entry under *Professional Associations and Societies*.)

## Society of Indexers
www.indexers.org.uk/

Indexing information for publishers and authors, 'Find an Indexer' pages and membership information. (See entry under *Professional Associations and Societies*.)

## South Bank Centre, London
www.sbc.org.uk

Links to the Royal Festival Hall, Purcell Room, Queen Elizabeth Hall, the Hayward Gallery and Poetry Library; news of literature events.

## Story Wizard
www.storywizard.co.uk

Software to enable children to write creative stories.

## The Sun
www.thesun.co.uk

*The Sun* newspaper online.

## The Times/The Sunday Times
www.thetimes.co.uk

*The Times* and *Sunday Times* online.

## trAce Online Writing Centre
tracearchive.ntu.ac.uk

Online archive of work published by the trAce Online Writing Centre between 1995–2005. Also holds articles and transcripts of discussions.

## TwelvePoint.com (previously ScriptWriter magazine)
▷ entry under Magazines

## UK Children's Books Directory
www.ukchildrensbooks.co.uk

Website created by Steve and Diana Kimpton to 'increase the profile of UK children's books on the internet'.

## The UK Public Libraries Page
dspace.dial.pipex.com/town/square/ac940/weblibs.html

Website links to public libraries throughout the UK, compiled by Sheila and Robert Harden.

## Webster Dictionary/Thesaurus
www.m-w.com/home.htm

Merriam-Webster Online. Includes a search facility for words in the Webster Dictionary or Webster Thesaurus; word games and daily podcast.

## Welsh Academy
▷ Academi

## Welsh Books Council (Cyngor Llyfrau Cymru)
www.cllc.org.uk

Information about books from Wales, editorial and design services, publishing grants. (See entry under *Professional Associations and Societies*.)

## The Word Pool
www.wordpool.co.uk

Children's book site with information on writing for children and a thriving discussion group for children's writers. Also Word Pool Design (www.wordpooldesign.co.uk): web design for writers, illustrators and publishers.

## WordCounter
www.wordcounter.com

Highlights the most frequently used words in a given text. Use as a guide to see what words are over-used. Also Political Vocabulary Analysis which measures indications of political leanings in given text.

## Write4kids.com
www.write4kids.com

US website for children's writers, whether published or beginners. Includes special reports, articles, advice, news on the latest bestsellers and links to related sites.

## Writers' Centre Norwich
www.writerscentrenorwich.org.uk

Formerly known as New Writing Partnership. Promotes and supports creative writing in the east of England. (See entry under *Professional Associations and Societies*.)

## Writers' Circles
www.writers-circles.com

Directory of writers' circles, courses and workshops.

## Writers' Guild of Great Britain
www.writersguild.org.uk

Information on contracts, copyright, news, writers' resources and industry regulations. (See entry under *Professional Associations and Societies*.)

## Writers, Artists and their Copyright Holders (WATCH)
www.watch-file.com

Database of copyright holders in the UK and North America. (See entry under *Professional Associations and Societies*.)

## WritersNet

www.writers.net

Writers' resources, news and discussion for writers, editors, publishers and literary agents.

## WritersServices

www.WritersServices.com

Established in March 2000 by Chris Holifield, former deputy managing director and publisher at Cassell. Offers factsheets, book reviews, advice, links and other resources for writers including editorial services, contract vetting and self-publishing.

## Writing-World.com

writing-world.com

US resource site for writers.

# Miscellany

## Apple Coaching (inc. CoachingWriters.co.uk)
8 Feering Road, Billericay CM11 2DR
☎ 01277 632085
eve@applecoaching.com
www.CoachingWriters.co.uk
Contact *Eve Menezes Cunningham*

Business coaching and NLP to help professional and aspiring writers write more, feel more creative, get organized, become more confident, handle setbacks and sell more. Writing and editorial services also available. Sign up for free e-newsletter on the website. Rates negotiable.

## Caroline Landeau
6 Querrin Street, London SW6 2SJ
☎ 07050 600420
Contact *Caroline Landeau*

Experienced research and production – films, multimedia, books, magazines, exhibitions, animation, general interest, art, music, crime, travel, food, film, theatre.

## ISBN Agency
3rd Floor, Midas House, 62 Goldsworth Road, Woking
☎ 0870 777 8712   🖷 0870 777 8714
isbn.agency@nielsen.com
www.isbn.nielsenbook.co.uk
Senior Manager, Registration Services ISBN, SAN & ISTC Agencies *Julian Sowa*
Manager, ISBN, SAN & ISTC Agencies *Diana Williams*

ISBNs are product numbers used by all sections of the book trade for ordering and listing purposes. The ISBN Agency is responsible for issuing ISBN prefixes for publishers based in the UK and Republic of Ireland. Nielsen Book also runs the **SAN** & **ISTC** agencies (see entries).

## ISTC Registration Agency
3rd Floor, Midas House, 62 Goldsworth Road, Woking
☎ 0870 777 8712   🖷 0870 777 8714
istc.agency@nielsen.com
www.istc.nielsenbook.co.uk
Senior Manager, Registration Services ISBN, SAN & ISTC Agencies *Julian Sowa*

Manager, ISBN, SAN & ISTC Agencies *Diana Williams*

ISTC (International Standard Text Code) is used by publishers, booksellers and libraries to identify textual works. Nielsen Book also runs the **ISBN** and **SAN** agencies (see entries).

## Jacqueline Edwards
104 Earlsdon Avenue South, Coventry CV5 6DQ
twigsbranches@yahoo.co.uk
Contact *Jacqueline Edwards, MA, LLB(Hons)*

Historical research: family, local and 19th and 20th century legal history. Covers Warwickshire, Gloucestershire, Northamptonshire, Worcestershire and the National Archives, Kew, London.

## Julia McCutchen, Writers' Coach, Mentor & Professional Publishing Consultant
PO Box 3703, Trowbridge
☎ 01380 871331   🖷 01380 871331
Julia@JuliaMcCutchen.com
www.JuliaMcCutchen.com
Contact *Julia McCutchen*

Julia McCutchen is the creator of the *How to Write the Ultimate Book Proposal Online Masterclass Course* and the author of *The Writer's Journey: From Inspiration to Publication*. She has 20 years experience of publishing (Element, Random House) and offers coaching, mentoring and masterclasses to support writers at each stage of the writing journey. Visit the website for free articles, audios and videos for writers.

## Leda Sammarco
☎ 07930 568516
leda.sammarco@btopenworld.com
www.ledasammarco.com
Contact *Leda Sammarco*

A coaching service for writers in the area of mind body spirit, self-help, popular psychology, health and wellbeing, business and inspirational memoir. Gives advice and guidance on the creative process, structure of your book, getting your book published, and provides an editing service.

## Literature Northants

Arts Development Team, Northamptonshire County Council, Music & Performing Arts Centre, 125–129 Kettering Road, Northampton NN1 4AZ

☎ 01604 824155   ⨊ 01604 824158
kwilkinson@northamptonshire.gov.uk
http://litnorthants.wordpress.com/
Literature Development Officer *Kate Wilkinson*

Writers and readers are supported by Literature Northants. Run by the Literature Development Officer, it provides access to literature opportunities across the country. The LDO is also the key contact for all the latest literary news and events.

## Nielsen BookData

3rd Floor, Midas House, 62 Goldsworth Road, Woking GU21 6LQ
☎ 01483 712200   ⨊ 01483 712201
sales@nielsen.com
www.nielsenbookdata.co.uk
Sales Director *Simon Skinner*
Publisher Subscription Manager *Vesna Nall*

BookData information services provide timely, accurate and enriched bibliographic data for English-language books to booksellers and libraries worldwide in a variety of formats including data feed, online and disk services.

## Nielsen BookNet

3rd Floor, Midas House, 62 Goldsworth Road, Woking GU21 6LQ
☎ 01483 712200   ⨊ 01483 712201
sales.booknet@nielsen.com
www.nielsenbooknet.co.uk
Head of BookNet Sales *Stephen Long*
Business Development Manager *Joanna de Courville*

BookNet's range of e-commerce services allows electronic trading between booksellers, publishers/distributors, libraries and other suppliers regardless of size or location.

## Nielsen BookScan

3rd Floor, Midas House, 62 Goldsworth Road, Woking GU21 6LQ
☎ 01483 712222   ⨊ 01483 712220
info.bookscan@nielsen.com
www.nielsenbookscan.co.uk
Publisher Account Manager *Reeta Windsor*
Research and Development Analyst *Andre Breedt*

The BookScan service is the largest continuous retail monitoring service for books in the UK, Ireland, US, South Africa, Spain, Italy, New Zealand and Denmark. Each week data is collected, coded and analyzed, producing complete market information for retailers, publishers, libraries, agents and the media. In the UK these reports are produced within 72 hours of the week ending Saturday.

## Ormrod Research Services

Weeping Birch, Burwash TN19 7HG
☎ 01435 882541
richardormrod@aol.com
www.writeonservices.co.uk
Contact *Richard Ormrod*

Established 1982. Comprehensive research service: literary, historical, academic, biographical, commercial. Verbal quotations available. Also editing, indexing, ghost-writing and critical reading.

## SAN Agency

3rd Floor, Midas House, 62 Goldsworth Road, Woking
☎ 0870 777 8712   ⨊ 0870 777 8714
san.agency@nielsen.com
www.san.nielsenbook.co.uk
Manager, ISBN, SAN & ISTC Agencies *Diana Williams*
Senior Manager, Registration Services ISBN, SAN & ISTC Agencies *Julian Sowa*

SANs (Standard Address Numbers) are unique identifiers for geographical locations and can be assigned to the addresses of organizations involved in the book industry. The SAN Agency is responsible for managing the scheme on behalf of Book Industry Communication in the UK and Republic of Ireland. Nielsen Book also runs the **ISBN** & **ISTC** agencies (see entries).

## The United Kingdom Copyright Bureau

110 Trafalgar Road, Portslade BN41 1GS
☎ 01273 277333
info@copyrightbureau.co.uk
www.copyrightbureau.co.uk
Contacts *Ralph de Straet von Kollman, Petra Ginman*

The UKCB provides a secure copyright service at reasonable cost, enabling multiple copyrights to be registered nominally when required. Prices are advertised on the website including the UKCB's solicitors, etc. Copyrights preferred on floppy disk or CD-ROM; manuscripts are not accepted due to storage space.

## Writers Direct

29 Great Southsea Street, Southsea PO5 3BY
☎ 023 9283 1289   ⨊ 023 9283 1289
info@writersdirect.net
www.writersdirect.net
Contacts *Mel Croucher (marketing), Ricky Foyle (design)*
Designs and manages author websites, and provides marketing and promotion to keep clients at the top of the search engine rankings.

# Index of Entries

# Subject Index

### Fiction: Literary Agents (US)

## Fiction: Publishers (European)

## Fiction: Publishers (International)

## Fiction: UK Packagers

## Fiction: Writers' Courses (UK and Ireland)

## Fiction: Writers' Courses (US)

## Film: Library Services

## Film: Literary Agents (Ireland)

## Film: Literary Agents (UK)

## History: Literary Agents (US)

## History: Magazines

## History: Miscellany

## History: Picture Libraries

## History: Poetry Presses

## History: Prizes

## History: Publishers (European)

Whispering Buffalo Literary
Agency Ltd, 258

**Music: Literary Agents (US)**
Doris S. Michaels Literary
Agency Inc, 269
Fine Print Literary Management,
270

**Music: Magazines**
*Acoustic*, 281
*Aesthetica Magazine*, 282
*BBC Music Magazine*, 286
*Church Music Quarterly*, 294
*Classic Rock*, 296
*Classical Guitar*, 296
*Classical Music*, 296
*Dazed & Confused*, 301
*Early Music Today*, 303
*Hi-Fi News*, 313
*i-D Magazine*, 315
*Kerrang!*, 318
*List, The*, 319
*Loaded*, 320
*Mixmag*, 322
*Mojo*, 323
*Music Week*, 324
*Musical Opinion*, 324
*New Musical Express*, 326
*Piano*, 331
*Q*, 334
*Record Collector*, 337
*Strad, The*, 345
*Top of the Pops Magazine*, 348
*Uncut*, 351

**Music: Miscellany**
Caroline Landeau, 642

**Music: PR Consultants**
Midas Public Relations, 264

**Music: Picture Libraries**
ArenaPAL, 596
Classical Collection Ltd, 598
Hutchison Picture Library, 604
Monitor Picture Library, 608
Raymond Mander & Joe
Mitchenson Theatre
Collection, The, 612
Redferns Music Picture Library,
612
Retna Pictures Ltd, 612
Robbie Jack Photography, 613
Ronald Grant Archive, The, 613
Vaughan Williams Memorial
Library, 618

**Music: Poetry Magazines**
*Eratica*, 194

**Music: Prizes**
C.B. Oldman Prize, 482
Pulitzer Prizes, 505

**Music: Professional
Associations and Societies**
BASCA (British Academy of
Composers and Songwriters),
522
PRS for Music, 536

**Music: Publishers (European)**
Adelphi Edizioni S.p.A., 150
Arnoldo Mondadori Editore
S.p.A., 150
Brepols Publishers NV, 144
Éditions Gallimard, 147
Éditions Larousse, 147
Editorial Caminho S.A., 153
Giunti Editoriale S.p.A., 151
Grupul Humanitas, 153
Gyldendal, 145
Hoffmann und Campe Verlag
GmbH, 148
Librairie Arthème Fayard, 148
Piper Verlag GmbH, 149
Verlag C.H. Beck (OHG), 149

**Music: Publishers
(International)**
Currency Press Pty Ltd, 176
Munshiram Manoharlal
Publishers Pvt Ltd, 180

**Music: Publishers (Ireland)**
An Gúm, 137
Merlin Publishing, 140
O'Brien Press Ltd, The, 141

**Music: Publishers (UK)**
A.&C. Black Publishers Ltd, 48
Anova Books, 50
Arc Publications Ltd, 52
Ashgate Publishing Ltd, 53
Aurum Press Ltd, 54
Austin & Macauley Publishers
Limited, 54
Calder Publications Ltd, 62
Carlton Publishing Group, 64
Continuum International
Publishing Group Ltd, The, 68
Elliott & Thompson, 73
Faber & Faber Ltd, 75

Giles de la Mare Publishers Ltd,
79
Harvard University Press, 84
Helter Skelter Publishing, 85
Kevin Mayhew Publishers, 91
Independent Music Press, 88
JR Books Ltd, 91
Kahn & Averill, 91
Luath Press Ltd, 94
Macmillan Publishers Ltd, 95
Omnibus Press, 104
Oxford University Press, 106
Quartet Books, 112
Random House Group Ltd, The,
114
Routledge, 118
Shetland Times Ltd, The, 122
Souvenir Press Ltd, 124
Stainer & Bell Ltd, 125
Usborne Publishing Ltd, 131
Wild Goose Publications, 134
Yale University Press (London),
135

**Music: Publishers (US)**
Indiana University Press, 162
Louisiana State University Press,
164
McFarland & Company, Inc.,
Publishers, 164
Perseus Books Group, 166
Scarecrow Press Inc., 168
University of Illinois Press, 171
University of Michigan Press,
The, 172
University of Nebraska Press, 172
University of North Texas Press,
172

**Music: Small Presses**
Allardyce, Barnett, Publishers,
208
Fand Music Press, 211
Parapress, 217
Witan Books, 220

**Mysteries: Book Clubs**
BCA (Book Club Associates), 280

**Mysteries: Literary Agents
(UK)**
Brie Burkeman & Serafina Clarke
Ltd, 235
Creative Authors Ltd Literary
Agency, 238
Feldstein Agency, The, 242
Henser Literary Agency, 244
J.M. Thurley Management, 245

## Poetry: Professional Associations and Societies

## Poetry: Publishers (European)

## Poetry: Publishers (International)

## Poetry: Publishers (Ireland)

## Poetry: Publishers (UK)

## Poetry: Publishers (US)

## Poetry: Small Presses

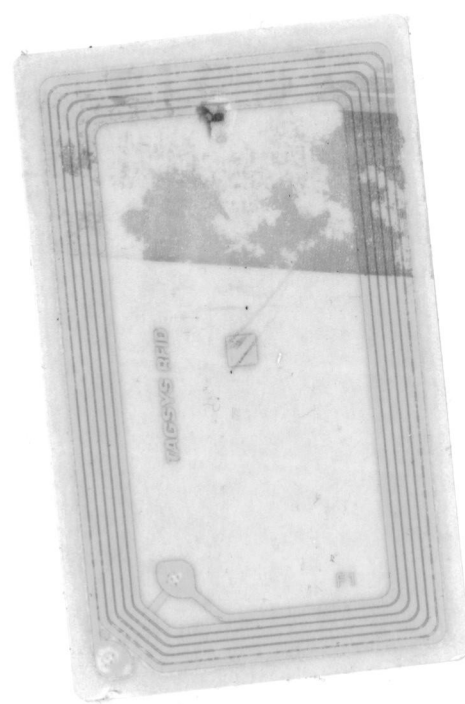